THE OXFORD HANDBOOK OF
EXPERTISE

THE OXFORD HANDBOOK OF

EXPERTISE

Edited by
PAUL WARD,
JAN MAARTEN SCHRAAGEN,
JULIE GORE, *and* EMILIE ROTH

Great Clarendon Street, Oxford, OX2 6DP,
United Kingdom

Oxford University Press is a department of the University of Oxford.
It furthers the University's objective of excellence in research, scholarship,
and education by publishing worldwide. Oxford is a registered trade mark of
Oxford University Press in the UK and in certain other countries

© Oxford University Press 2020

The moral rights of the authors have been asserted

First Edition published in 2020

Impression: 1

All rights reserved. No part of this publication may be reproduced, stored in
a retrieval system, or transmitted, in any form or by any means, without the
prior permission in writing of Oxford University Press, or as expressly permitted
by law, by licence or under terms agreed with the appropriate reprographics
rights organization. Enquiries concerning reproduction outside the scope of the
above should be sent to the Rights Department, Oxford University Press, at the
address above

You must not circulate this work in any other form
and you must impose this same condition on any acquirer

Published in the United States of America by Oxford University Press
198 Madison Avenue, New York, NY 10016, United States of America

British Library Cataloguing in Publication Data

Data available

Library of Congress Control Number: 2019950595

ISBN 978-0-19-879587-2

Printed and bound by
CPI Group (UK) Ltd, Croydon, CR0 4YY

Links to third party websites are provided by Oxford in good faith and
for information only. Oxford disclaims any responsibility for the materials
contained in any third party website referenced in this work.

The Volume Editor (Paul Ward's) affiliation with The MITRE Corporation is provided
for identification purposes only, and is not intended to convey or imply MITRE's concurrence
with, or support for, the positions, opinions or viewpoints expressed by the author.

Acknowledgments

The Oxford Handbook of Expertise had its origins in conversations within and between the self-organizing community of scholars engaged in the development of the field of Naturalistic Decision Making and primarily as a response to a shared observation that our wealth of research on expertise needed to be creatively collated in one place.

We were delighted that Gary Klein and Robert Hoffman having engaged an international community of scholars and practitioners and fostered a curiosity for examining and enhancing expert cognition, agreed to support the development of the handbook. We dedicate this Handbook to both of them. Paul, Jan Maarten, Julie and Emilie have been part of the NDM community in different guises, for several decades and supported by The MITRE Corporation, USA, Michigan Technological University, USA, Netherlands Organisation for Applied Scientific Research TNO, the University of Bath, UK, and Roth Cognitive Engineering, USA, were provided with the time and space to complete the task of putting the handbook together. Although much of this work was completed at the weekends and often during vacations via skype - our enthusiasm for the project was constantly ensured by Paul who kept us on track with his passion for including a multi-disciplinary range of work on expertise.

Our sincere thanks go to Martin Baum, Senior Commissioning Editor, and Charlotte Holloway, Senior Assistant Commissioning Editor for Psychology and Neuroscience at Oxford University Press for her continued support and insight. A particular thank you to Charles Lauder, Jr. Gayathri Manoharan, and the Oxford production team for their great attention to detail. Last, but not least our enormous thanks go to the authors of this volume whose work continues to inspire us alongside our families, colleagues and friends who are always happy to engage in great conversations about expertise.

Paul Ward, Jan Maarten Schraagen, Julie Gore, and Emilie Roth

Summary Table of Contents

Section One: Characterizing Expertise: Frameworks, Theories, and Models   31
SECTION EDITOR: JAN MAARTEN SCHRAAGEN

Section Two: Methods to Study, Test, Analyse, and Represent Expertise   287
SECTION EDITOR: EMILIE ROTH

Section Three: Domains and Applications   525
SECTION EDITOR: JULIE GORE

Section Four: Developing, Accelerating, and Preserving Expertise   899
SECTION EDITOR: PAUL WARD

Section Five: Current Issues and the Future of Expertise Research   1153
SECTION EDITOR: PAUL WARD

TABLE OF CONTENTS

List of Contributors — xv

1. An Introduction to the *Handbook*, Communities of Practice, and Definitions of Expertise — 1
   PAUL WARD, JAN MAARTEN SCHRAAGEN, JULIE GORE, AND EMILIE ROTH

SECTION ONE: CHARACTERIZING EXPERTISE: FRAMEWORKS, THEORIES, AND MODELS

Section Editor: Jan Maarten Schraagen

2. The Classic Expertise Approach and its Evolution — 35
   FERNAND GOBET

3. Domain-General Models of Expertise: The Role of Cognitive Ability — 56
   DAVID Z. HAMBRICK, ALEXANDER P. BURGOYNE, AND FREDERICK L. OSWALD

4. Studies of Expertise and Experience: A Sociological Perspective on Expertise — 85
   HARRY COLLINS AND ROBERT EVANS

5. Giftedness and Talent Development in Children and Youth — 103
   STEVEN I. PFEIFFER

6. Neural Mechanisms of Expertise — 128
   FREDRIK ULLÉN, ÖRJAN DE MANZANO, AND MIRIAM A. MOSING

7. Modeling Experts with Fast-and-Frugal Heuristics — 149
   ULRICH HOFFRAGE

8. Expertise: A Holistic, Experience-Centered Perspective — 173
   JOHN M. FLACH AND FRED A. VOORHORST

9. Macrocognitive Models of Expertise   190
    ROBERT J. B. HUTTON

10. Cognitive Systems Engineering: Expertise in Sociotechnical Systems   219
    NEELAM NAIKAR AND ASHLEIGH BRADY

11. Is Expertise All in the Mind? How Embodied, Embedded, Enacted, Extended, Situated, and Distributed Theories of Cognition Account for Expert Performance   243
    CHRIS BABER

12. Adaptive Expertise   262
    KATERINA BOHLE CARBONELL AND JEROEN J. G. VAN MERRIENBOER

SECTION TWO: METHODS TO STUDY, TEST, ANALYSE, AND REPRESENT EXPERTISE

Section Editor: Emilie Roth

13. Representative Test and Task Development and Simulated Task Environments   291
    KEVIN R. HARRIS, LINDSEY N. FOREMAN, AND DAVID W. ECCLES

14. Developing Mastery Models to Support the Acquisition and Assessment of Expertise   312
    KAROL G. ROSS AND JENNIFER K. PHILLIPS

15. Computational Models of Expertise   333
    ALEX KIRLIK AND MICHAEL D. BYRNE

16. Studying Expert Behavior in Sociotechnical Systems: Hierarchical Task Analysis   354
    PAUL M. SALMON, NEVILLE A. STANTON, GUY H. WALKER, AND GEMMA J. M. READ

17. A Historical Perspective on Introspection: Guidelines for Eliciting Verbal and Introspective-Type Reports   377
    PAUL WARD, KYLE WILSON, JOEL SUSS, WILLIAM DOUGLAS WOODY, AND ROBERT R. HOFFMAN

18. *Close to Practice* Qualitative Research Methods 408
SARAH YARDLEY, KAREN MATTICK, AND TIM DORNAN

19. Incident-based Methods for Studying Expertise 429
LAURA G. MILITELLO AND SHILO ANDERS

20. Cognitive Work Analysis: Models of Expertise 451
CATHERINE M. BURNS

21. Reflections on the Professional Practice of Knowledge Capture 468
BRIAN MOON

22. Stress, Skilled Performance, and Expertise: Overload and Beyond 490
GERALD MATTHEWS, RYAN W. WOHLEBER, AND JINCHAO LIN

SECTION THREE: DOMAINS AND APPLICATIONS

Section Editor: Julie Gore

23. Expertise in STEM Disciplines 529
DAVID. F. FELDON, SOOJEONG JEONG, AND JOANA FRANCO

24. A Cognitive Examination of Skill and Expertise in Word Games and Puzzles 551
SHANE T. MUELLER

25. Musical Expertise 574
JENNIFER MISHRA

26. Skilled Anticipation in Sport: Past, Present, and Future 594
A. MARK WILLIAMS, BRADLEY FAWVER, DAVID P. BROADBENT, COLM P. MURPHY, AND PAUL WARD

27. Diagnostic Reasoning and Expertise in Health care 618
VIMLA L. PATEL, DAVID R. KAUFMAN, AND THOMAS G. KANNAMPALLIL

28. Fire fighting and Emergency Responding 642
MARK WIGGINS, JAIME AUTON, AND MELANIE TAYLOR

29. Expertise in Aviation 662
CHRISTOPHER D. WICKENS AND FREDERIC DEHAIS

30. Uncovering Expertise for Safe and Efficient Performance in Railroad Operations — 690
   EMILIE M. ROTH, ANJUM NAWEED, AND JORDAN MULTER

31. The Cyber Domains: Understanding Expertise for Network Security — 718
   ROBERT THOMSON

32. Expertise in Intelligence Analysis — 740
   MICHAEL P. JENKINS AND JONATHAN D. PFAUTZ

33. Expertise in Law Enforcement — 765
   JOEL SUSS AND LAURA BOULTON

34. Military Expertise — 792
   J. D. FLETCHER AND DENNIS KOWAL

35. Expertise in Business: Evolving with a Changing World — 807
   LIA A. DIBELLO

36. Teamwork in Spaceflight Operations — 830
   UTE FISCHER AND KATHLEEN MOSIER

37. Developing Operator Expertise on Nuclear Power Production Facilities and Oil & Gas Installations — 850
   †MARGARET CRICHTON, SCOTT MOFFAT, AND LAUREN CRICHTON

38. Expertise in Weather Forecasting — 872
   DAPHNE S. LaDUE, PHAEDRA DAIPHA, REBECCA M. PLISKE, AND ROBERT R. HOFFMAN

SECTION FOUR: DEVELOPING, ACCELERATING, AND PRESERVING EXPERTISE

Section Editor: Paul Ward

39. Expertise for the Future: A New Challenge for Education — 903
   LAUREN B. RESNICK, JENNIFER LIN RUSSELL, AND FAITH SCHANTZ

40. Learning with Zeal: From Deliberate Practice to Deliberate Performance ... 927
   Peter J. Fadde and Mohammadreza Jalaeian

41. Cognitive Flexibility Theory and the Accelerated Development of Adaptive Readiness and Adaptive Response to Novelty ... 951
   Rand J. Spiro, Paul J. Feltovich, Aric Gaunt, Ying Hu, Hannah Klautke, Cui Cheng, Ian Clemente, Sean Leahy, and Paul Ward

42. Cognition and Expert-Level Proficiency in Intelligence Analysis ... 977
   David T. †Moore and Robert R. Hoffman

43. Team Reflection: A Catalyst of Team Development and the Attainment of Expertise ... 1001
   Kai-Philip Otte, Kristin Knipfer, and Michaéla Schippers

44. Learning at the Edge: The Role of Mentors, Coaches, and Their Surrogates in Developing Expertise ... 1021
   Erich Petushek, Güler Arsal, Paul Ward, Mark Upton, James Whyte IV, and Robert R. Hoffman

45. Acquiring and Maintaining Expertise in Aging Populations ... 1058
   Dan Morrow and Renato F. L. Azevedo

46. Skill decay: The Science and Practice of Mitigating Skill Loss and Enhancing Retention ... 1085
   Winfred Arthur, Jr. and Eric Anthony Day

47. Expertise and Resilience ... 1109
   Jop Havinga, Johan Bergström, Sidney Dekker, and Andrew Rae

48. Framing and Translating Expertise for Government ... 1132
   Gareth E. Conway and Julie Gore

SECTION FIVE: CURRENT ISSUES AND THE FUTURE OF EXPERTISE RESEARCH

Section Editor: Paul Ward

49. The "War" on Expertise: Five Communities that Seek to Discredit Experts ... 1157
    GARY KLEIN, BEN SHNEIDERMAN, ROBERT R. HOFFMAN, AND ROBERT L. †WEARS

50. Reflections on the Study of Expertise and Its Implications for Tomorrow's World ... 1193
    PAUL WARD, JAN MAARTEN SCHRAAGEN, JULIE GORE, EMILIE ROTH, ROBERT R. HOFFMAN, AND GARY KLEIN

Name Index ... 1215
Subject Index ... 1229

List of Contributors

Shilo Anders, Center for Research and Innovation in Systems Safety, Vanderbilt University Medical Center, USA

Güler Arsal, Envision Research Institute, Envision Inc., USA

Winfred Arthur, Jr., Department of Psychological and Brain Sciences, Texas A&M University, USA

Jaime Auton, Centre for Elite Performance, Expertise, and Training, Department of Psychology, Macquarie University, Australia

Renato F. L. Azevedo, Department of Educational Psychology, University of Illinois at Urbana-Champaign, USA

Chris Baber, School of Computer Science, University of Birmingham, UK

Johan Bergström, Lund University, Sweden

Laura Boulton, School of Forensic and Applied Sciences, University of Central Lancashire, UK

David P. Broadbent, Centre for Cognitive Neuroscience, Division of Sport, Health and Exercise Sciences, Brunel University London, UK

Katerina Bohle Carbonell, Department of Marketing and Supply Chain Management, School of Business and Economics, Maastricht University, The Netherlands

Ashleigh Brady, Centre for Cognitive Work and Safety Analysis, Joint and Operations Analysis Division, Defence Science and Technology Group, Australia

Alexander P. Burgoyne, Department of Psychology, Michigan State University, USA

Catherine M. Burns, Department of Systems Design Engineering, University of Waterloo, Canada

Michael D. Byrne, Departments of Psychological Science and Computer Science, Rice University, USA

Cui Cheng, Michigan State University, USA

Ian Clemente, Michigan State University, USA

Harry Collins, Centre for the Study of Knowledge, Expertise and Science, School of Social Sciences, Cardiff University, UK

Gareth E. Conway, Defence Science & Technology Laboratory; UK Ministry of Defence

Lauren Crichton, People Factor Consultants Ltd, UK

Margaret Crichton (†), People Factor Consultants Ltd, UK

Phaedra Daipha, Data, Discovery & Decision Science (D3) Department, Allstate, USA

Eric Anthony Day, Department of Psychology, University of Oklahoma, USA

Örjan de Manzano, Department of Neuroscience, Karolinska Institutet, Stockholm, Sweden

Frederic Dehais, ISAE-SUPAERO, Université de Toulouse, France

Sidney Dekker, Griffith University, Australia; and Delft University, the Netherlands

Lia A. DiBello, Workplace Technologies Research Inc., USA

Tim Dornan, School of Medicine, Dentistry and Biomedical Sciences, Queens University Belfast, UK

David W. Eccles, Department of Educational Psychology and Learning Systems, College of Education, Florida State University, USA

Robert Evans, Centre for the Study of Knowledge, Expertise and Science, School of Social Sciences, Cardiff University, UK

Bradley Fawver, Department of Health, Kinesiology, and Recreation, University of Utah, USA

Peter J. Fadde, Department of Curriculum and Instruction, Southern Illinois University, USA

David. F. Feldon, Utah State University, USA

Paul J. Feltovich, Institute for Human & Machine Cognition, USA

Ute Fischer, Georgia Institute of Technology, School of Literature, Media and Communication, USA

John M. Flach, Mile Two, LLC, USA

J. D. Fletcher, Science and Technology Division, Systems and Analysis Center, Institute for Defense Analyses, USA

Lindsey N. Foreman, Department of Psychological Science and Counseling, Austin Peay State University, USA

Joana Franco, Utah State University, USA

Aric Gaunt, Michigan State University, USA

Fernand Gobet, Centre for Philosophy of Natural and Social Science, London School of Economics and Political Science, UK

Julie Gore, University of Bath, UK

David Z. Hambrick, Department of Psychology, Michigan State University, USA

Kevin R. Harris, Department of Psychological Science and Counseling, Austin Peay State University, USA

Jop Havinga, Griffith University, Australia

Robert R. Hoffman, Institute for Human Machine Cognition, USA

Ulrich Hoffrage, Faculty of Business and Economics, University of Lausanne, Lausanne, Switzerland

Ying Hu, Michigan State University, USA

Robert J. B. Hutton, Trimetis Ltd, UK; Nottingham Trent University, UK.

Mohammadreza Jalaeian, Department of Curriculum and Instruction, Southern Illinois University, USA

Michael P. Jenkins, Charles River Analytics Inc., USA

Soojeong Jeong, Utah State University, USA

Thomas G. Kannampallil, Department of Anesthesiology & Institute for Informatics, School of Medicine, Washington University in St. Louis, USA

David R. Kaufman, DR Kaufman Consulting, USA

Alex Kirlik, Department of Computer Science, University of Illinois at Urbana-Champaign, USA

Hannah Klautke, Michigan State University, USA

Gary Klein, MacroCognition LLC, USA

Kristin Knipfer, Technical University of Munich, Germany

Dennis Kowal, Science and Technology Division, Systems and Analysis Center, Institute for Defense Analyses, USA

Daphne S. LaDue, University of Oklahoma, USA

Sean Leahy, Arizona State University, USA

Jinchao Lin, Institute for Simulation and Training, University of Central Florida, USA

Gerald Matthews, Institute for Simulation and Training, University of Central Florida, USA

Karen Mattick, Centre for Research in Professional Learning, University of Exeter, UK

Laura G. Militello, Applied Decision Science, LLC, USA

Jennifer Mishra, Kent State University, USA

Scott Moffat, People Factor Consultants Ltd, UK

Brian Moon, Perigean Technologies LLC, USA

David T. Moore (†), US Department of Defense, USA

Kathleen Mosier, TeamScape LLC, USA

Miriam A. Mosing, Department of Neuroscience and Department of Epidemiology and Biostatistics, Karolinska Institutet, Stockholm, Sweden

Dan Morrow, Department of Educational Psychology, University of Illinois at Urbana-Champaign, USA

Shane T. Mueller, Department of Cognitive and Learning Sciences, Michigan Technological University, USA

Jordan Multer, Volpe National Transportation Systems Center, USA

Colm P. Murphy, Faculty of Sport, Health and Applied Science, St Mary's University, UK

Neelam Naikar, Centre for Cognitive Work and Safety Analysis, Joint and Operations Analysis Division, Defence Science and Technology Group, Australia

Anjum Naweed, Appleton Institute for Behavioural Science, Central Queensland University, Australia

Frederick L. Oswald, Department of Psychological Sciences, Rice University, USA

Kai-Philip Otte, Kiel University, Germany

Vimla L. Patel, Center for Cognitive Studies in Medicine and Public Health, The New York Academy of Medicine, USA

Erich Petushek, Department of Cognitive and Learning Sciences, Michigan Technological University, USA

Jonathan D. Pfautz, Department of Defense, USA

Steven I. Pfeiffer, Florida State University, USA

Jennifer K. Phillips, Cognitive Performance Group, USA

Rebecca M. Pliske, Dominican University, USA

Andrew Rae, Griffith University, Australia

Gemma J. M. Read, Centre for Human Factors and Sociotechnical Systems, University of the Sunshine Coast, Australia

Lauren B. Resnick, University of Pittsburgh, USA

Karol G. Ross, Cognitive Performance Group, USA

Emilie M. Roth, Roth Cognitive Engineering, USA

Jennifer Lin Russell, University of Pittsburgh, USA

Paul M. Salmon, Centre for Human Factors and Sociotechnical Systems, University of the Sunshine Coast, Australia

Faith Schantz, professional writer/independent scholar

Michaéla Schippers, Rotterdam School of Management, Department of Technology & Operations Management, Erasmus University Rotterdam, The Netherlands

Jan Maarten Schraagen, TNO, The Netherlands and University of Twente, The Netherlands

Ben Shneiderman, University of Maryland, USA

Rand J. Spiro, Michigan State University, USA

Neville A. Stanton, Human Factors Engineering, Transportation Research Group, University of Southampton, UK

Joel Suss, Department of Psychology, Wichita State University, USA

Melanie Taylor, Centre for Elite Performance, Expertise, and Training, Department of Psychology Macquarie University, Australia

Robert Thomson, Army Cyber Institute, United States Military Academy, USA

Fredrik Ullén, Department of Neuroscience, Karolinska Institutet, Stockholm, Sweden

Mark Upton, myfastestmile c.i.c., UK

Jeroen J. G. van Merrienboer, Department of Educational Research and Development, Faculty of Health, Medicine, and Life Sciences, Maastricht University, The Netherlands

Fred A. Voorhorst, Oyster Lab by Alpiq, Switzerland

Guy H. Walker, Centre for Sustainable Road Freight, Heriot-Watt University, Edinburgh, UK.

Paul Ward, The MITRE Corporation, USA, and Michigan Technological University, USA

Robert L. Wears (†), University of Florida, USA

James Whyte IV, College of Nursing, The Florida State University, USA

Christopher D. Wickens, Department of Psychology, Colorado State University, USA

Mark Wiggins, Centre for Elite Performance, Expertise, and Training, Department of Psychology, Macquarie University, Australia

A. Mark Williams, University of Utah, USA

Kyle Wilson, University of Huddersfield, UK

Ryan W. Wohleber, Institute for Simulation and Training, University of Central Florida, USA

William Douglas Woody, University of Northern Colorado, USA

Sarah Yardley, Marie Curie Palliative Care Research Department, University College London & Central and North West London NHS Foundation Trust, London, UK

CHAPTER 1

AN INTRODUCTION TO THE *HANDBOOK*, COMMUNITIES OF PRACTICE, AND DEFINITIONS OF EXPERTISE

PAUL WARD, JAN MAARTEN SCHRAAGEN, JULIE GORE, AND EMILIE ROTH

Introduction

The study of expertise weaves its way through various communities of practice, across disciplines, and over millennia. Arguably, Aristotle's writings on the habitual acquisition of virtues were an important impetus for the field as one virtue was *excellence*. Ever since, the polarizing nature versus nurture debates have held center stage for researchers focused on identifying the catalyst for expertise. While this debate still rages on (e.g., *Outliers; The Talent Code; The Sports Gene*), and is frequently popularized by the media, over the past century the study of expertise has matured into a multidisciplinary field spanning scientific disciplines, such as psychology, engineering, computer science, and education.

To date, the study of expertise has been primarily concerned with how human beings perform at a superior level in complex environments and sociotechnical systems, and at the highest levels of proficiency. Early expertise researchers focused on relatively simple

---

Historical Development of the Study of Expertise and Communities of Practice; and an Outline of the Handbook by Paul Ward, Jan Maarten Schraagen, Julie Gore, Emilie Roth, Gary Klein, & Robert Hoffman

The author (Paul Ward's) affiliation with The MITRE Corporation is provided for identification purposes only, and is not intended to convey or imply MITRE's concurrence with, or support for, the positions, opinions or viewpoints expressed by the author.

tasks suited to college-level introductory courses in such fields as mechanics and mathematics, and on identifying underlying causal mechanisms—often with a special focus on the cognitive functions, processes, and requirements for operating at an expert level (e.g., Chi, Feltovich, & Glaser, 1981). Others have focused on the pathways leading to the attainment of expertise (e.g., Bloom, 1985; Ericsson, Krampe, & Tesch Römer, 1993). The more recent research has continued the search for better descriptions, and causal mechanisms that explain the complexities of expertise *in context*, with a view to translating this understanding into useful predictions and interventions capable of improving the performance of human systems as efficiently as possible.

Research on expertise is both topical and timely. As the nature of work in today's society becomes increasingly cognitive, and technological advances continue to accelerate the nature of work, the need for proficient and expert workers, and the ability to rapidly acquire skill, has increased in urgency. This is true not just for key professions that shape future society—including, industry, government, military, and healthcare—but for all walks of public and personal life where learning is central to daily activities.

The purpose of this *Handbook* is to provide a comprehensive representation of the development of this field of study. As such, we offer traditional and contemporary perspectives, and importantly, a multidiscipline-multimethod view of the state-of-the-science and -engineering research on expertise. Our aim is to present different perspectives, theories, and methods of conducting expertise research that have been influential in improving our current understanding of expert phenomena across a range of domains. Our second aim is to provide a particular focus on conveying how that understanding has been applied to address *practical* problems and societal challenges. In particular, the *Handbook* focuses on how this understanding has been translated into reliable predictions useful to society, and into effective interventions that can improve all facets of proficiency, performance, security, safety, health, and well-being.

Overview of this Chapter and Structure of the Handbook

In this chapter we provide a brief summary of the various communities of practice that have paved the way for current expertise researchers, and are formative of this *Handbook*. We then provide a synopsis of how expertise has been defined both historically and in present day. Our purpose in this chapter is threefold: To demonstrate the heterogeneity of approaches and conceptions of expertise, to contextualize current views of expertise presented in this *Handbook*, and to use these views as a springboard to examine how we should examine expertise in the future—which we address in the final chapter. Finally, we present an outline of the chapters that are presented in this *Handbook*. First, we provide a brief overview of the structure of the *Handbook*.

This *Handbook* is organized into five sections. Section I, edited by Jan Maarten Schraagen, presents frameworks, theories, and models that characterize current views of expertise. In line with the goals of the *Handbook*, multidisciplinary perspectives that range from early work in cognitive psychology to more recent work, for instance, in cognitive systems engineering are presented.

Section II, edited by Emilie Roth, presents a variety of methods, developed by researchers from different theoretical perspectives, that are vital to advancing our understanding of expert phenomena. These methods are sampled from the full range of methods used to study, test, analyze, and represent expertise and, employed collectively, would represent a truly mixed methods approach.

Section III, edited by Julie Gore, presents a diversity of application domains that offer several insights in to the nature of expertise when it matters: when the "rubber meets the road." Chapters that focus on advancing our knowledge in professions, traditional skill domains, decision making under uncertainty in naturalistic decision making (NDM) research areas, and emerging domains are presented.

Section IV, edited by Paul Ward, presents chapters with the central theme of developing, accelerating, or preserving expertise. Some of the chapters present a review of the current means by which expertise can be achieved and the constraints and barriers to maintaining expertise, whereas others present a challenge to the expertise community to begin to think differently about how we develop and support expertise. In Section V, we conclude with two chapters that examine topical issues of the value and future of expertise. We now turn to setting the scene by providing an overview of the communities of practice that have laid the groundwork for, or been most active in, pursuing studies of expertise.

Past and Current Communities of Practice and Definitions of Expertise

The history of expertise is pluralistic and multifaceted. Numerous researchers have documented our and related histories elsewhere (e.g., Amirault, & Branson, 2006; Feltovich, Prietula, & Ericsson, 2006; Hoffman & Deffenbacher, 1992; Hoffman & Militello, 2009; Klein, 2008; Chapter 17, "A Historical Perspective on Introspection," by Ward, Wilson, Suss, Woody, & Hoffman, this volume) and so we will not repeat those here. Instead, we present in this section an overview of various communities of practice that have had an impact on the development of this field of study and comment on past and current definitions and conceptions of expertise.

First, to demonstrate the impact of the study of expertise, we note that several books have already been published that cover some of the topics presented in this *Handbook*. For instance, some have reviewed expertise theories, methods, and applications, albeit with a specific or narrow focus (e.g., professional expertise—Ericsson, 2009; expertise in organizations—Wiggins & Loveday, 2015). Others have made recommendations about adopting a particular perspective on what constitutes a science of expertise (e.g., Ericsson & Smith, 1991). Another has provided a specific tribute to a particular individual (William Chase), representing the influence of this work on the field (e.g., expertise and skill acquisition—Staszewski, 2013). Yet others have presented research from a sole or specific theoretical perspective (e.g., naturalistic decision making—Hoffman, 2007; Mosier & Fischer, 2015; sociology—Mieg, 2001; philosophy—Selinger & Crease, 2006). Numerous edited texts that have examined expertise in specific application domains also exist (e.g., sport—Baker & Farrow, 2015; security—Berling & Bueger, 2015; technology—Hoc, Cacciabue, & Hollnagel, 1995; nursing—Benner, Tanner, & Chesla, 1996; science & policy—Collins & Evans, 2007). Only one other *Handbook* on expertise exists (see Ericsson, Hoffman, Kozbelt, & Williams, 2018). We think that our *Handbook* offers something different—it offers a range of perspectives and is more comprehensive in its scope, and we hope that it is more inclusive in its approach than previous publications. Moreover, in this *Handbook* we have an eye on both the past *and* the future. Many authors in our *Handbook* address questions about how future research and practice should and/or may be carried out in order to continue to make the scientific and engineering leaps needed to shape future intellectual discussions and societal contributions of this field.

Communities of Practice

When we started this *Handbook* project we asked chapter authors to send us some words describing the people who had influenced their careers, especially those that had helped position each of them as world-leading researchers in the study of expertise. What followed was a list of names constituting PhD and post-doctoral advisors, mentors, authors of scholarly works, and practitioners. The names provided did not always reflect an academic lineage—which was, in part, our original goal: to map this *Handbook* to the original pioneers of the study of expertise. When genealogical maps were provided, predecessors were not always readily identifiable as someone who one might think of as a pioneer of expertise research. Instead, predecessors and influential figures were pioneers of science, engineering, psychology, sociology, and philosophy—including theoreticians, philosophers, methodologists, practitioners, and stakeholders—that reflected multiple, often disparate, communities of practice in which we operate. Our history is much richer than just a handful of notable individuals.

Next, we describe some of the communities of practice that reflect or have emerged from these influences, and that represent the majority of perspectives, theories, methods, and applications presented in this *Handbook*. Many of these communities overlap with other communities (e.g., numerous NDM researchers use verbal reporting methods; perceptual-motor expertise researchers also conduct deliberate practice research), and some have been largely superseded by other communities or other areas of research (e.g., expert systems by AI). Although some of the early research occurred around the turn of the twentieth century, the 1970s and 1980s were a particularly busy time for the emergence of many of these communities.

Verbal Reports of Thinking Community of Practice

The verbal reports community is arguably one of the earliest communities of practice relevant to the study of expertise. The methods that unite this community have their roots in introspection—the history of which as it pertains to the study of expertise is presented in detail by Ward, Wilson, et al. (this volume). Hence, we refer the reader to that chapter for more information.

Despite Brentano's (1874/2009) observation that "outstanding people" (p. 26) are an important focus for introspective studies of psychological laws (see Ward, Wilson, et al., this volume), the history of these methods is far longer than their application to the study of expertise. De Groot (1946/1965) was arguably the first person to seriously study chess experts using methods employed by this community. However, and arguably, following Duncker (1935; who leveraged Selz's (e.g., 1924/1981) work before him), De Groot is also the source of a division in this community. On the one hand, there are those who employ verbal report methods almost exclusively as a process tracing methodology (e.g., Ericsson & Simon, 1980, 1993)—so-called think aloud reports. On the other, there are those who also employ knowledge elicitation methods to better understand expertise more broadly, as well as to represent expertise so that process-traced verbalizations can be interpreted in context (e.g., Crandall, Klein, & Hoffman, 2006; Hoffman & Militello, 2009). Generally speaking, this division also reflects those who have studied simpler tasks (e.g., chess) and those that have studied more complex ones (e.g., firefighting), respectively.

Regardless of the theoretical, methodological, and practical approach adopted, as a whole this community has given rise to several treatises that have paved the way for this community of practice to employ these methods to better understand expertise more systematically, and with greater rigor. Moreover, these methods have also played a large part in developing theories of expertise (e.g., long-term working memory theory—Ericsson & Kintsch, 1995; recognition-primed decision making—Klein, 1989). Various chapters in this *Handbook* describe these methods, their products, and/or their application to the study of expertise (e.g., Chapter 13, "Representative Test and Task Development and Simulated Task Environments," by Harris, Foreman, & Eccles; Chapter 9, "Macrocognitive Models of Expertise," by Hutton; Chapter 19, "Incident-Based Methods for Studying Expertise," by Militello & Anders; Ward, Wilson, et al.).

Skill Acquisition Community of Practice

This community of practice emerged from early research on learning (e.g., Ebbinghaus, 1885/1964) and transfer of training (e.g., Thorndike & Woodworth, 1901) where, amongst other things, researchers were interested in the *savings* from engaging in prior learning when relearning the same task, and the degree of transfer from having previously learned a similar task. One of the landmark studies with respect to expertise was conducted by Bryan and Harter (1897), who studied the acquisition of telegraphic language in field operators. Their research suggested that some skills were acquired more rapidly than others (i.e., sending rate was acquired faster than receiving rate), and habits were hypothesized to be acquired hierarchically—i.e., letters, then words, then clauses, then sentences, etc.—suggesting an automation of some aspects of the task before others are acquired. This phasing of skill acquisition is, perhaps, most representative of this community of practice, and exemplified in the framework proposed by Paul Fitts (e.g., 1964). Fitts argued that skill acquisition progresses through phases, from cognitively demanding to more direct (i.e., less verbally mediated) associations between stimulus and response that are typically accompanied with decreases in error rates and time to respond. The final phase, in contrast, is assumed to be more characteristic of automatic behavior and not to be consciously mediated. This is a point of departure for some communities of practice, who speculate that experts actively defer automating skill in order to maintain conscious control and/or access to underlying representational structures, especially those involving higher-order thinking (e.g., see Ericsson & Ward, 2007).

Many parallel lines of skill acquisition research emerged around the same time, including research focused on contextual interference (e.g., Battig, 1966), the power law of learning (e.g., Crossman, 1959; Newell & Rosenbloom, 1981; Snoddy, 1926), and reasoning more broadly (e.g., Bartlett, 1958). Some of this research has been instantiated subsequently as computational models of skill acquisition, with some emphasis on expertise and expert learning. Examples of approaches emanating (jointly) from this community of practice (and others) can be found in this *Handbook* (e.g., see Chapter 46, "Skill Decay," by Arthur & Day; Chapter 2, "The Classic Expertise Approach and its Evolution," by Gobet; Chapter 15, "Computational Models of Expertise," by Kirlik & Byrne).

Individual Differences Community of Practice

The individual differences community has its roots in early efforts (pre-WWI) to develop tests capable of measuring verbal, non-verbal, and performance-based tests of intelligence and mental ability, as well as physical and motor proficiency. The goal was often pragmatic—to be able to classify individual differences in various types of ability, to diagnose intellectual or performance disabilities, and to predict future accomplishment from current levels. Although many of these efforts would have been designed to differentiate "the best from the rest" on a range of ability measures, a landmark study in the area of individual differences as they pertain to the development of expertise was conducted by Ackerman (1988). Ackerman (1988) showed that different abilities (e.g., general intelligence, perceptual-motor speed) changed in their degree of contribution to skill development as individuals progressed through different phases of learning. As individuals become more skilled with training, ability measures tended to do less well at predicting *job* performance than those related to skill or knowledge.

More recently, however, there has been a resurgence in interest in the role of individual differences in expertise. Where previously discounted, researchers have recently asserted that some of the variance in skill is explained by domain-general factors, challenging the widely held conception that 10,000 hours of deliberate practice is sufficient to attain expertise. This community has been active in the past decade in their attempts to redress this balance (e.g., see Chapter 3, "Domain-General Models of Expertise," by Hambrick, Burgoyne, & Oswald; and Chapter 6, "Neural Mechanisms of Expertise," by Ullén, de Manzano, & Mosing, both this volume).

The Knowledge and Classic Expertise Approaches Communities of Practice

Two interrelated communities, often treated as one, emerged in the 1970s: one at the Learning Research and Development Centre (LRDC) at the University of Pittsburgh, and one at Carnegie Mellon University (CMU). Arguably, these communities of practice had the earliest concentration of effort on the experimental study of expertise, and provided specific methodologies that could be used by others to study this field. These lines of research clearly motivated those that followed to test the ideas put forward, but often at the cost of omitting other emerging frameworks that used different methods and/or had alternative theoretical orientations (e.g., Rasmussen, 1983, 1986).

Although equally known for his work in standardized testing, Robert Glaser, the first director of the LRDC, was instrumental in establishing a line of research focused on how people learn and, especially, how they progress within a particular field of expertise from novice to expert. This work led to a range of experimental and verbal report-based research (e.g., see Chi, Glaser, & Farr, 1988), including some citation classics (e.g., Chi, Feltovich, & Glaser, 1981) that examined the structure of *knowledge*, and the associated cognitive strategies and memory skills that support the transition to expertise.

Arguably, the expertise work at CMU was made possible by Herbert Simon and William Chase, and Allen Newell (e.g., Newell & Simon, 1972), which built on work by Adrian de Groot (1965; which in turn built on research by Karl Duncker and, indirectly, Otto Selz; see Ward, Wilson, et al., this volume). The *classic expertise approach* adopted by the CMU researchers is covered in some detail in the chapter by Gobet (this volume); hence, it will not be repeated here. Suffice to say, that this work, amongst many other concepts, led to the concept of *chunking*, a mechanism used by chess experts to circumvent the limits of short-term memory by grouping individual units of information (e.g., chess pieces) in to constellations of information (e.g., patterns of chess pieces). Skilled memory theory followed (e.g., see Ericsson & Staszewski, 1989), as did the concept of a retrieval structure to elaborate on the explanation of chunk storage in long-term memory, and subsequent derivations of this theory (e.g., Ericsson & Kintsch, 1995; Gobet & Simon, 1996). This community also explored differences between experts and novices in problem understanding, problem-solving strategies (e.g., forward-chaining versus backward-chaining), and differences in problem representations. As such, this work foreshadowed some of the later work on recognition-primed decision making (see Schraagen, 2018, for a discussion of the relations between these lines of work).

The work at LRDC and CMU has been extraordinarily influential, yet has tended to focus on providing computational and/or theoretical explanations of simpler domains of expertise, and of the study of single tasks. Various chapters in this *Handbook* have adopted one of the perspectives that emerged from the LRDC or CMU, or describe work that is consistent with it (e.g., Gobet; Harris et al.; Kirlik & Byrne; Chapter 39, "Expertise for the Future," by Resnick, Russell, & Schantz).

Dreyfus-ian Community of Practice

The Dreyfus-ian community emerged in the 1980s, in part, in response to the expert systems and computational model-based approaches to studying cognition, including those associated with the classic expertise approach. Rather than propose a model of expertise per se, Hubert Dreyfus extended traditional phase-based theories produced by the skill acquisition community by proposing particular stages of expertise development (see Dreyfus & Dreyfus, 1986). Learners were assumed to progress from performance that was verbal-, rule-, and problem-solving-based, to a form of thinking that was based on experience—where experts recognize important and relevant aspects of the situation and know intuitively what to do without any need to engage in effortful and deliberative thinking or problem solving. This approach has been most notably advocated in the domain of nursing (e.g., see Benner, 1984; Benner, Tanner, & Chesla, 1996) but its influence has been felt in other areas too (e.g., see Chapter 11, "Is Expertise All in the Mind?", by Baber, this volume; Chapter 40, "Learning with Zeal," by Fadde & Jalaeian, this volume; Chapter 14, "Developing Mastery Models to Support the Acquisition and Assessment of Expertise," by Ross & Phillips, this volume) (cf. Ericsson, Whyte, & Ward, 2007; Gobet & Chassy, 2008).

Social Studies of Science Community of Practice

Somewhat akin to the Dreyfus-ian community, the social studies of science community reacted to the expert systems and artificial intelligence communities' failure to consider the situated nature of human knowledge. Specifically, Harry Collins (e.g., 1990, 2018; Collins & Kusch, 1998) argued that these researchers had neglected the fact that humans and computers were socially embedded, which is needed for intelligence to emerge—because knowledge and understanding are societally determined (for a summary of this argument, see Hoffman & Militello, 2009). The strongest proponents of this view would argue that knowledge, including scientific facts, is socially constructed. For instance, rather than physical laws being universal, this community of practice might argue that they are a reflection of the received view of the world at a particular moment in time that is culturally and contextually specific, rather than a truly universal phenomenon. These views are captured in current definitions of expertise that have been forwarded by this community, such as contributory and interactional expertise—respectively, they reflect the practical competence within a domain and "the ability to master the language of a specialist domain in the absence of practical competence" (Collins & Evans, 2007, p. 14). Approaches that tap into the social nature of expertise are reviewed in multiple chapters within the *Handbook* (e.g., Chapter 4, "Studies of Expertise and Experience," by Collins & Evans; Chapter 18, "Close-to-Practice Qualitative Research Methods," by Yardley, Mattick, & Dornan).

The Deliberate Practice Community of Practice

The deliberate practice community emerged, initially, with the publication of a book by Bloom (1985) that contained several chapters examining the support mechanisms leveraged, and developmental pathways taken by current experts from a range of domains. In 1993, Ericsson et al. published a seminal paper on deliberate practice, which they defined as those solitary practice activities designed by a coach specifically to improve performance, that require substantial effort, and that are not inherently enjoyable. Ericsson et al. (1993) claimed that engagement in these activities is related monotonically to the level of expertise attained, and that the greatest improvements in performance are likely to be associated with the largest weekly amounts of deliberate practice. Therefore, individuals who have accumulated the largest number of practice hours throughout their career and consistently and deliberately engaged in high levels of practice for sustainable periods are more likely to attain expertise. The general rule of thumb to attain expert status reiterated in this research is 10 years or 10,000 hours of deliberate practice (see also Hayes, 1985; Simon & Gilmartin, 1973).

The concept of deliberate practice has permeated the expertise literature in the past quarter of a century and the influence and application of this community has been demonstrated in fields as diverse as soccer (Ward, Hodges, Starkes, & Williams, 2007), clinical psychology (Rosenberg, 2000), teacher education (Dunn & Schriner, 1999), and insurance agents (Sonnetag & Kleine, 2000). Several chapters in this *Handbook* discuss this research (e.g., Chapter 25, "Musical Expertise," by Mishra; Fadde & Jalaeian). Notably, the strong claims of this research, that deliberate practice is sufficient (rather than just necessary) to explain expertise have recently been challenged (see Hambrick et al., this volume; Ullén et al., this volume).

Perceptual–Motor Expertise Community of Practice

The perceptual–motor expertise community grew out of other communities of practice, such as: the individual differences community, and their early emphasis on general and specific motor abilities; the skill acquisition community, with its focus on a shift from deliberative to automatic behavior with increases in perceptual–motor skill; and the classic expertise approach, with its emphasis on intuition as a mechanism of perceptual automation. Arguably, the latter approach had most influence with early sports expertise researchers attempting to test chunking theory by building on Chase and Simon's recall studies using sport-specific stimuli. Poulton's (1957) work on perceptual anticipation was also influential in providing a pragmatic point of focus for researchers who realized that anticipation played a significant role in sporting expertise. Following the trend in motor control and sport psychology more broadly, a division in this community occurred with sports expertise being investigated largely from two dominant perspectives: ecological (e.g., Araujo & Kirlik, 2008) and cognitive psychology (e.g., Starkes & Allard, 1993). The work of this community led to numerous texts on expertise (e.g., Baker & Farrow, 2015; Starkes & Ericsson, 2003; Williams, Davids & Williams, 1999) and is well represented in this *Handbook* (see Fadde & Jalaeian; Harris et al.; Chapter 26, "Skilled Anticipation in Sport," by Williams, Fawyer, Broadbent, Murphy, and Ward).

Naturalistic Decision Marking and Macrocognition Community of Practice

The naturalistic decision making (NDM) and macrocognition community emerged in the mid-1980s, pioneered by Gary Klein and colleagues. The impetus for this community came from the military's desire to better understand high-stakes decisions under extreme time pressure and in dynamic, uncertain, and complex environments. The Army Research Institute for the Behavioral and Social Sciences supported several lines of NDM research during the mid-1980s, and incidents such as the 1988 *USS Vincennes* shoot-down led the US Navy to want to better understand naturalistic decisions. The first NDM conference was held in 1989 in Ohio and, over the years, the NDM community came to appreciate that its purview should not be on naturalistic decision making alone but on all aspects of cognitive activity in actual work environments (e.g., see Gore et al., 2018). Hence, NDM evolved to focus on macrocognition—or the way in which individuals cognitively adapt to complexity.

This community gave rise to the development of a host of cognitive field research-based methods (e.g., see Crandall et al., 2006), including the critical decision method (Klein, Calderwood, & MacGregor, 1989) and a range of models of expert thinking including the recognition-primed decision-making model (Klein, Calderwood, & Clinton-Cirocco, 1986; Klein, 1989), the data/frame model of sensemaking (Klein, Phillips, Rall, & Peluso, 2006), and the flexecution model of adaptive replanning (Klein, 2007a, b). Its impact has been far reaching (see Hutton, this volume) but like other communities of practice, it has not been without its critics (e.g., see Yates, 2001). These critics argue, first, that the explanations provided for expert performance by the NDM community are descriptive rather than predictive, and that these descriptions over-represent the importance of aspects of performance that are easy to verbalize; second, that the more mundane, daily decisions that people make are ignored because of its emphasis on expert performance in complex environments; third, that NDM relies on a case study approach that limits generalization to the population at large (Markman, 2018).

Cognitive Systems Engineering Community of Practice

The cognitive systems engineering (CSE) community emerged in the early 1980s, pioneered by Neville Moray, Thomas Sheridan, and Jens Rasmussen, among others. Based on some early verbal report work on electronics troubleshooting and plant diagnosis (e.g., Rasmussen & Jensen, 1974; Rasmussen, 1981), Rasmussen (1983) developed a model of cognitive control to identify the different ways in which humans at different levels of expertise exert cognitive control of performance (i.e., knowledge based, rule-based, and skill based control) and the resulting heuristic strategies and workarounds used by skilled performers. Following some of the pioneers in ecological psychology and dynamic systems theory (e.g., Bernstein, 1967; Brunswik, 1956), cognitive systems engineers who were most influenced by Rasmussen adopted an ecological and systems perspectives with subsequent models (e.g., abstraction-decomposition hierarchy) of the work environment. This community views expertise not as an individual phenomenon or a particular stage of information processing, but rather as a coupling between an expert with a problem ecology through a representation. After two seminal books on cognitive systems engineering (Rasmussen, 1986; Rasmussen, Pejtersen, & Goodstein, 1994), Kim Vicente synthesized this work into an accessible treatise detailing Rasmussen's collection of methods, known as cognitive work analysis (see Vicente, 1999). This and related perspectives, and the associated methods in particular, are detailed throughout this *Handbook* (e.g., Chapter 20, "Cognitive Work Analysis," by Burns; Chapter 8, "Expertise: A Holistic, Experience-Centered Perspective," by Flach & Voorhorst; Chapter 10, "Cognitive Systems Engineering," by Naikar & Brady). This work has led to numerous elaborations and offshoots of this perspective that are relevant to the study of expertise, including work on joint cognitive systems, applied cognitive work analysis, and resilience engineering (e.g., see Chapter 47, "Expertise and Resilience," by Havinga, Bergström, Dekker, and Rae).

The above list of communities is not exhaustive and others surely exist (e.g., ethnomethodology, expert systems, and resilience engineering communities of practice). In the next section, we examine the different ways in which expertise has been defined in this *Handbook* across communities of practice. We begin with a short discussion of the value of performance-based definitions.

Definitions of Expertise

We would need a book dedicated to this topic to record completely the range of definitions that have been used to capture what it means to be an expert. Rather than provide an exhaustive review of this literature without doing it justice, we focus on some of the common definitions of expertise and attempt to capture the variety of definitions presented in this *Handbook*. Perhaps the most widely cited definition of expertise is one of *reliably superior performance on representative tasks*—a key tenet of the expert performance approach (e.g., Ericsson & Smith, 1991). One could argue that this is one of the best approaches for studying expertise—where measurable differences in performance between those performing at different levels of proficiency are reproduced and scrutinized under standardized and controlled conditions (e.g., see Ericsson & Ward, 2007).

This definition of expertise is exemplified in Morrow and Azevedo's chapter (Chapter 45, "Acquiring and Maintaining Expertise in Aging Populations," this volume) where they describe expertise as "superior levels of performance on representative tasks" and as "performance improvements with increasing task-related experience...skills, knowledge, interests, and other changes that accompany improved performance" (see also, Harris et al., this volume; Williams et al, this volume). However, one could argue that this and related performance-based definitions, while popular, do not easily apply in the study of cognitive work in many complex domains where performance cannot be measured or simulated easily, or reduced to single tasks.

Winning a chess match, serving a tennis ace, typing without error at speed, and playing a piece of music flawlessly (or innovatively) are relatively easy tasks to simulate and standardize under controlled conditions—and this is likely one reason why much research has focused on these domains. Studying performance in these domains and on such tasks may retain the functional complexity of work without compromising its ecological validity, making what is captured during experimentation a good representation of actual performance in these domains. But this does not help us to understand expertise in the vast array of complex domains (such as many of those discussed in this *Handbook*) where performance measurement on a given task is difficult, is impossible, or does not reflect the totality of the domain expertise. This is not to say performance measurement should not be part of a definition of expertise. Perhaps, it always should be, at least, when it is possible. The real question is: What do you do (and how should one define expertise) when working in the majority of complex domains where performance measurement is particularly challenging or impractical? We address this issue in more detail in the final chapter.

How does one measure, for instance, expertise in those individuals or teams whose task is to contemplate a dilemma (i.e., where there is no right or wrong answer)? What about those areas of work that are purely knowledge based, or where the environment is highly uncertain, the problem or goals are ill-defined, or where there are high stakes? Other authors in this *Handbook* suggest a slightly revised definition, still relevant to performance, but where the emphasis is on the measurement of performance goals or the associated knowledge, skills, and/or attitudes; and the emphasis is on using a multi-measures approach as well. For instance, Matthews, Wohleber, and Lin (Chapter 22, "Stress, Skilled Performance, and Expertise: Overload and Beyond") suggest that expertise is about cognitive and/or psychomotor skill competence, where those skills are central to accomplishing performance goals. Burns (this volume) suggests the emphasis should be on the "expert's deep knowledge and experience that brings about significantly different and more effective behaviors than novices." Feldon, Jeong, and Franco (Chapter 23, "Expertise in STEM Disciplines") suggest that expertise in scientific fields is more about "mastery of the knowledge and skills capable of bringing about new knowledge that meets or exceeds current standards." Moreover, Crichton, Moffat, and Crichton (Chapter 37, "Developing Operator Expertise on Nuclear Power Production Facilities and Oil & Gas Installations") adds another dimension to a definition of expertise in the nuclear and oil & gas industries, suggesting that the ability to demonstrate and implement relevant knowledge and skills competently as well as *confidently* are what really matters.

Baber (this volume) and Ross and Phillips (this volume) both draw on Dreyfus' and Dreyfus' or Klein's conceptualizations suggesting that expertise is defined by the ability to see the signal in the noise—the critical situational elements—and intuitively generate the appropriate, or at least an effective, course of action. Ross and Phillips (this volume) go one step further suggesting that in times of uncertainty expertise is defined by the ability to mentally simulate those actions and *see* their success. Emphasizing similar elements but also highlighting the social nature of expertise, Yardley et al. (this volume) add:

> Experts need a fine-tuned moral compass and the ability *to navigate complex social situations where power is at play* as well as intellectual and psychomotor skills. They have to be tolerant of ambiguity and have a capacity to withhold action or act in the face of uncertainty, based on a fine balance of risks and benefits. [emphasis added]

Collins and Evans (this volume) operationalize expertise along these same lines, and include the role of *Individual accomplishment* of the type described by Dreyfus and Dreyfus (1986) and others. In addition, they add two social dimensions: *Exposure*—the ease with which tacit knowledge within the domain of expertise can be accessed in general and from others; and *esotericity*—the ease with which social aspects of expertise can be accessed. This is consistent with Otte, Knipfer, and Schippers' (Chapter 43, "Team Reflection: A Catalyst of Team Development and the Attainment of Expertise") definition of team expertise: "the ability to effectively leverage ... the knowledge and expertise of all team members."

In their definition of expertise, Ross and Phillips (this volume) also allude to the expert's ability to immediately recognize changes in the scenario and to flexibly apply knowledge and experience, even when the situation is novel. Baber and Flach and Voorhorst suggest that expertise is a matter of "sensitivity to environmental constraints and opportunities." These adaptive and context-sensitive components of expertise are not new. For instance, Hoffman (e.g., see Hoffman et al., 2014) has described expertise as context-dependent choice amongst alternatives and Bohle Carbonell and van Merrienboer (this volume) describe how routine expertise should be differentiated from what Hatano and Inagaki (1984, 1986) termed adaptive expertise. They describe the former as high-level performance on representative tasks, and the latter as the same but in unfamiliar situations, where adaptivity is attributed to a deeper conceptual understanding of the fit between a specific procedural skill and a specific situation (see also Ward et al., 2018).

Conversely, Wickens and Dehais (Chapter 29, "Expertise in Aviation") differentiate the type of (routine) *expert* performance of real experts from that which is more representative of competent journeymen—"the ability to successfully perform job-relevant duties and solve common problems quickly, reliably, and accurately." They suggest that, by definition, expertise conveys an "ability to successfully solve uncommon, unusually difficult, and/or strategic problems that others cannot." This perspective is consistent with Hoffman's (1998) delineation of proficiency across the skill continuum, which was amongst the first to present a proficiency scale—based on the craft guilds of the Middle Ages—that captured this adaptive feature within a definition of expertise.

Despite the limited evidence for positive skill transfer with expertise, the flexible and adaptive nature of expertise is emphasized by several authors in this *Handbook*. One view which exemplifies many is that of Resnick et al. (this volume), who suggest that experts "draw fluidly and flexibly on the information at hand and on the complex set of skills and attitudes (including the willingness to change one's mind) that comprise reasoning." We will revisit the concept of adaptive skill in the final chapter of the *Handbook*. In the next section of this chapter, we provide a brief summary of the chapters within each section of the Handbook.

Outline of the *Oxford Handbook* of *Expertise*

Characterizing Expertise: Frameworks, Theories, and Models

Section I provides an overview of frameworks, theories, and models used to characterize expertise. The chapters range from the classic approach to expertise, as exemplified by Chase and Simon's research on chess expertise in the early 1970s, to more recent approaches focusing on macrocognition and cognitive systems engineering. Although the study of expertise has been dominated by research in cognitive psychology, and most chapters build upon this tradition, this section also includes chapters from a sociological and neural point of view. A brief overview of each of the chapters in this section is presented next.

In Chapter 2, Gobet describes the classic expertise approach and its evolution. He starts off by briefly discussing early research on expertise that influenced the classic approach. He then describes in some detail Chase and Simon's classic papers and chunking theory. This leads the way to a presentation of some of the key experimental and theoretical research that was characterized by detailed analyses of the cognitive processes involved, use of verbal protocols, and a small number of participants. The chapter then discusses more recent theories that can be considered as outgrowths of the classic approach, providing a good opportunity to try to understand not only its key characteristic but also why it had such a large impact. The chapter concludes by a discussion of what this approach tells us about the means to address the challenges currently facing research on expertise.

In Chapter 3, Hambrick, Burgoyne, and Oswald review evidence concerning the contribution of cognitive ability to individual differences in expertise. Their review covers research in traditional domains for expertise research such as music, sports, and chess, as well as research from industrial-organizational psychology on job performance. The specific question that they seek to address is whether domain-general measures of cognitive ability (e.g., IQ, working memory capacity, executive functioning, processing speed) predict individual differences in domain-relevant performance, beyond beginning levels of skill. The authors note that evidence from the expertise literature relevant to this question is difficult to interpret, due to small sample sizes, restriction of range, and other methodological limitations. By contrast, there is a wealth of consistent evidence that cognitive ability is an important and statistically significant predictor of job performance, even after extensive job experience. The authors discuss ways that cognitive ability measures might be used in efforts to accelerate the acquisition of expertise.

Chapter 4 provides a sociological perspective on expertise. Collins and Evans build upon a research program they refer to as studies of expertise and experience (SEE), often referred to as the *third wave of science studies*, which treats expertise as real and as the property of social groups. This chapter explains the foundations of SEE and sets out the theoretical and methodological innovations created using this approach. These include the development of a new classification of expertise, which identifies a new kind of expertise called *interactional expertise*, and the creation of a new research method known as the imitation game designed to explore the content and distribution of interactional expertise. The authors conclude by showing how SEE illuminates a number of contemporary issues such as the challenges of interdisciplinary working and the role of experts in a *post-truth* society.

In Chapter 5, Pfeiffer discusses giftedness and talent development in children and youth, with a focus on talent development as a path toward expertise and eminence. The chapter briefly discusses a history of gifted education and then tackles some big picture issues and future possibilities. The chapter addresses a number of questions, including: Who is gifted? How are gifted individuals identified? Is giftedness domain-specific or domain-general? How malleable is giftedness? Does giftedness represent a qualitative or quantitative difference? How does the concept of expertise fit into gifted education? The author proposes a tripartite model of giftedness that offers three distinct lenses through which high-ability students can be viewed: (1) high intelligence; (2) outstanding accomplishments, and (3) potential to excel.

In chapter 6, Ullén, de Manzano, and Mosing provide an overview of some of the neuroanatomical and functional correlates of expertise, concluding that expertise is related to macroanatomical properties of domain-relevant brain regions and ultra-structural properties of both the gray and white matter. The consequence of these neural adaptations is a capacity for vastly more efficient performance of domain-specific tasks. In functional terms, this depends on multiple mechanisms that are situated at different levels of neural processing. These mechanisms include automation and alterations in functional connectivity, as well as specializations within memory systems and sensorimotor systems that optimize the processing of information which is relevant for the particular domain of expertise. The author concludes with a discussion of neural mechanisms of expertise from the perspective of new models that emphasize a multifactorial perspective and take into account both genetic and environmental influences on expertise and its acquisition.

In Chapter 7, Hoffrage provides an overview of the *fast-and-frugal heuristics* program of research on expertise. According to the program reviewed in this chapter, people—including experts—use fast-and-frugal heuristics. These heuristics are models of bounded rationality that function well under limited knowledge, memory, and computational capacities. These heuristics are ecologically rational: they are fitted to the structure of information in the environment. While studying experts in the context of this program amounts to modeling them with fast-and-frugal heuristics, studying the acquisition of expertise focuses on how laypeople learn such heuristics. Because fast-and-frugal heuristics do not require complex calculation and are typically easy to set up, this program offers a straightforward way to aid experts: After the heuristics' performance has been determined under various environmental conditions, experts can be educated about these results.

In Chapter 8, Flach and Voorhorst advance the claim that expertise is not a property of any particular stage of information processing, nor is it a property of an individual. Rather it is the property of a triadic semiotic system where the quality of performance depends on the coupling of an agent with a problem ecology through a representation. The dynamics of this coupling is akin to a self-organizing, adaptive control system. The authors argue that many of the debates about the nature of expertise arise from the different ways that people have parsed the triadic system into subcomponents (elements or dyads). Thus, the fundamental point of this chapter is that expertise is not something that can be isolated as a property of a mind, independent from a problem ecology, or vice versa.

In Chapter 9, Hutton provides an overview of so-called macrocognitive models of expertise, of which three are discussed in some detail: The recognition-primed decision model, the data-frame model of sensemaking, and the flexecution model of replanning and adaptation. Macrocognitive models are models of experienced, often expert performers and have been developed primarily from the study of decision making and cognitive work in naturalistic settings, as opposed to well-controlled laboratory experiments. They describe how people manage uncertainty and complexity in the world of work. The limitations and applications of these models are also illustrated in order to provide a future-oriented perspective on how the models might be improved and how they might be applied to support more effective cognitive work and more resilient work systems.

In Chapter 10, Naikar and Brady present a perspective of human expertise in sociotechnical systems based on the phenomenon of self-organization. Consistent with the ideals of the field of cognitive systems engineering, this perspective is based on empirical observations of how work is achieved in complex settings and incorporates an emphasis on design. The proposed perspective is motivated by the observation that workers in sociotechnical systems adapt not just their individual behaviors, but also their collective structures, in ways that are closely fitted to the evolving circumstances, such that these systems are necessarily self-organizing, a phenomenon that is essential for dealing with complexity in the task environment. Accordingly, the chapter explores in depth the theoretical and design implications of the phenomenon of self-organization for understanding and supporting human expertise in sociotechnical systems, and draws attention to the broader implications of this phenomenon for advancing a social basis for human cognition.

In Chapter 11, Baber reviews theories that explore the relationship between action and performance. These theories ask whether our cognitive activity depends on *internal representations* or whether it can be explained by our interaction with the world around us. In other words, rather than projecting a model of the world outwards in order to plan and guide our actions, these approaches see physical interaction with the world as a form of cognitive activity. These theories focus less on using mental representation and more on perception–action coupling between us and our world. Baber concludes that this points to an account of expertise which sees it as a matter of sensitivity to environmental constraints and opportunities, together with the ability to focus on optimal parameters in a given situation. From a practical point of view, he considers ways in which such sensitivity could be probed through field study and interview with experts.

This section concludes with Chapter 12 on adaptive expertise by Bohle Carbonell and van Merrienboer. The authors start by noting that the increasing number of changes at the workplace created through automation, political upheavals, and new technology frequently exposes individuals to unfamiliar situations. According to the chapter authors, mastering these situations requires individuals to possess adaptive expertise. By being an adaptive expert, individuals are able to deal with novel situations and remain performing at their original level. By drawing on recent literature, the goal of this chapter is to describe what adaptive expertise is. The authors contrast the concept of adaptive expertise with routine expertise to clarify when adaptive expertise produces superior performance to routine expertise. Subsequently they compare adaptive expertise to other expertise concepts. Following this, they describe how adaptive expertise can be developed and measured. The chapter ends with a number of recommendations of how individuals can be stimulated to develop adaptive expertise.

Methods to Study, Test, Analyze, and Represent Expertise

In Section II we present a range of methods that have been used to study expertise. These methods provide a sampling of techniques that draw from multiple theoretical perspectives and communities of practice. In combination they provide an excellent introduction to the variety of theoretical approaches and practical methods available for analyzing, representing, and testing the knowledge and skills associated with expertise. Several of the chapters provide practical *how to* guidance and describe pitfalls to avoid when conducting research on expertise.

In chapter 13 Harris, Foreman, and Eccles provide an introduction to designing representative tasks, tests, and simulated task environments for use in uncovering the basis of expertise. The authors describe how well-designed representative tasks can be used to discover the mechanisms that underlie superior performance, to stratify performers based on skill, and to discover the developmental steps involved in reaching superior levels of performance. The authors describe how representative tasks and simulated task environments can be used to understand the basis for expert performance as well as to develop training on the basis of expert performers. The authors present recent research illustrating the use of representative tasks and simulated task environments.

In Chapter 14, Ross and Phillips describe the origin, development, and application of a mastery model approach to the acquisition, representation, and assessment of expertise. The framework of the mastery model originates from the Dreyfus and Dreyfus general model of cognitive skill acquisition. Development of a mastery model is based on a semi-structured interview of experts and a qualitative analysis process. The model specifies the hallmarks of performance, characterizes the progression of skill, and provides performance indicators for each key area, categorized in five progressive levels. The chapter includes specific examples and discusses differences between the mastery model and competency modeling approaches.

In Chapter 15 Kirlik and Byrne provide a comprehensive introduction to computational models of expertise, reviewing both the foundational and contemporary body of research. The chapter discusses and provides examples of computational models built within the framework of a unified cognitive architecture as well as models that are more domain or task specific in their psychological assumptions. It highlights the requirements for effective computational modeling of expert behavior, including the need for extensive analysis, and possibly expert-level knowledge of both tasks and the environments in which expert behavior is manifest. The chapter ends with a discussion of promising future directions for research using computational modeling, as well as other emerging techniques such as neuroimaging, that can be combined to advance the scientific understanding of human expertise.

In Chapter 16 Salmon, Stanton, Walker, and Read discuss the use of hierarchical task analysis (HTA) to represent expert behavior and the factors influencing it. HTA is among the most widely used methods for conducting task analysis and a variety of ergonomics methods build on HTA outputs to provide in-depth analyses of behavior. The chapter describes HTA and its origins, discusses its strengths and weaknesses, and provides practical guidance on how to apply the method. Two rail level crossing cases studies are used to illustrate how HTA can be employed to describe and analyze both the behavior of individuals at the sharp end (i.e., at the rail level crossing itself) and the behavior of the overall sociotechnical system (i.e., the rail level crossing *system*).

In Chapter 17 Ward, Wilson, Suss, Woody, and Hoffman provide a historical examination of the philosophical roots and long-standing controversies associated with one of the most widely used methods for understanding expertise—elicitation and analysis of verbal reports. The chapter serves as a comprehensive introduction to the variety of introspective-type methods discussing their validity and utility. It begins with a historical review of the perspectives and contributions of the pioneers of introspective methods, highlighting key motivations, arguments, and disputes that have driven methodological development over the past 100+ years. It then turns to a review of current methods that rely on *thinking aloud* and other types of verbal reports to study expertise. It offers cautions and guidance on appropriate use of verbal reports, and ends with reflections on the future of introspection methodology and opportunities to improve the state-of-the-science and escape the legacies of behaviorism.

In Chapter 18 Yardley, Mattick, and Dornan provide an introduction to qualitative research methods, particularly as part of research grounded in practice. The authors argue that expertise is inherently linked to the context in which experts work. Qualitative methods enable researchers to examine the broader context, including social practices, within which expertise is manifest and to uncover and pursue unexpected findings. The chapter describes different qualitative methods and discusses their distinct contributions to understanding expertise development in professional work. The authors provide examples to illustrate how qualitative methods can be used to answer *how* and *why* questions, to disentangle the impact of different factors in complex and uncertain situations, and to explore the messiness and complexity of expertise more generally.

In Chapter 19 Militello and Anders provide a comprehensive introduction to incident-based interview methods for discovering and characterizing expertise. Incident-based methods are among the best known and easily applied methods for understanding expertise. The chapter presents four types of incident-based interview methods: critical decision method, knowledge audit, simulation interview, and cued-retrospective interviews. It describes strategies for analyzing the outputs of incident-based interviews, and provides examples of the types of products that are generated. Among practical applications described are uncovering requirements for training; identifying requirements for support tools; and developing testable hypotheses and descriptive models to inform basic research on expertise. The chapter ends with discussion of practical issues associated with the use of incident-based interview methods and experience-based advice in how to address real-world challenges.

In Chapter 20 Burns reviews the roots of the cognitive work analysis (CWA) framework and how it can be used to represent expertise. CWA methods are best known as tools for the design of novel displays that improve performance in complex domains such as process control, health, and military operations. Displays built on CWA principles have proven to be particularly effective in supporting performance in unanticipated complex situations where expert *knowledge-based* strategies are most useful. Less improvement in performance has been found in routine situations that are well supported by conventional displays and procedures. Burns argues that this may be because CWA focuses on the needs for expert performance. She points out that CWA was originally founded as part of attempts to understand human expertise and transfer the knowledge of human experts into a design so that those who are *less expert* could benefit. As a result, CWA methods are useful for understanding and transferring expertise. Burns reviews the steps of CWA and discusses their various contributions to the understanding and development of expertise. The chapter ends with a discussion of how CWA can be used to develop and transfer expertise through design.

In Chapter 21 Moon provides an invaluable perspective on how to understand and represent expertise through reflections on his own experiences in the professional practice of knowledge capture. The chapter describes the scope, uses, and origins of knowledge capture methods. Most particularly it covers protocols and methods for knowledge elicitation, focusing on the elements common to all capture methods—structure and probing questions. The issues of cognitive burden sharing across the capturer and holder, how purpose guides execution, and how constraints can shape practical developments in approaches to knowledge capture are also discussed. Throughout, the chapter offers insightful, first-hand stories of knowledge capture and provides invaluable advice in dealing with pragmatic complexities that inevitably arise when going out *into the field*. The chapter concludes by looking at future directions for the profession.

In Chapter 22 Matthews, Wohleber, and Lin examine the complex relationship that exists between stress, skilled performance and expertise. Stress generally impairs attention and working memory, increasing vulnerability to cognitive overload. The authors present a model characterizing the interrelation between stress and expertise called the standard capacity model (SCM) derived from theories of attention resources and cognitive skills acquisition. Matthews and colleagues argue that SCM, while having some empirical support, has serious limitations including neglecting contextual factors that can alter the pattern of findings. As an illustration, the authors describe the interplay between stress and expertise across four domains: test anxiety, sports performance, surgery, and vehicle driving. Consistent with the SCM, in some cases stress is associated with cognitive overload and expertise is shown to buffer the effects of stressors. However, the authors also provide evidence that expert performance is subject to domain-specific influences beyond cognitive capacity, including strategies for emotional regulation, choking under pressure, and aggressive behaviors that mediate the relationship between stress and expertise. They conclude that the relationships between stress and expertise must be examined contextually.

Domains and Applications

Section III focuses upon the wide variety of domains and applications exploring expertise. The chapters highlight great diversity of application and provide a range of insights from disparate domains. Two chapters focus on advancing professions of business and science, technology, engineering, and mathematics (STEM) education. Three chapters focus on advancing skill in domains in which expert performance has been studied traditionally: games, music, and sports. Seven chapters focus on advancing decision making under uncertainty in traditional naturalistic decision-making (NDM) areas of research: spaceflight, nuclear, oil & gas operations, healthcare, firefighting and emergency responding, weather forecasting, aviation, railroad transportation, law enforcement, and military. Two chapters focus on decision making under uncertainty in the emerging domains of cyber security and intelligence analysis. Notably, the authors who have contributed to this section are located in academia and practice across many disciplines around the world. A brief overview of each of the chapters in this section is outlined here.

In Chapter 23, Feldon, Jeong, and Franco provide a synthesis of literature which is most relevant for enhancing expertise in STEM. This synthesis focuses upon relevant findings from cognitive psychology and the psychology of science, sociology and anthropology, and educational research. The authors present the fundamental mechanisms of thinking and problem-solving practices in science and engineering that underlie expert performance within these disciplines. The chapter also examines issues pertaining to assessment and recognition of expertise in STEM fields, and the impact of training and education. The chapter ends by suggesting that further work is needed to explore and question the nature of expertise, the dynamic nature of STEM disciplines, and interdisciplinarity.

Mueller, in Chapter 24, examines word game expertise from a cognitive perspective and proposes a general taxonomic space of word games where the primary organizing axis distinguishes letter- versus meaning-centered games. His critical review of the area concludes with the hypothesis that word game expertise is supported by both practice *and* prior skills and ability, and suggests predisposition opportunity may be a fruitful framework for understanding skilled performance in this domain.

Chapter 25, by Mishra, suggests that music is a foundational domain in the development of the theory of expertise. Similar to Mueller's conclusions, Mishra reports that current thinking is that while deliberate practice is important, it is insufficient to entirely explain expertise. Mishra conjects that future research will aim to determine how genetics and practice interact in the development of experts.

In Chapter 26, Williams, Fawver, Broadbent, Murphy, and Ward provide a historical overview of research focusing on the topic of anticipation, with a particular emphasis on its importance in various high-performance domains, including sport. They review more than five decades of research which has highlighted some of the key perceptual–cognitive skills underpinning anticipation and how these interact and vary in importance from one situation to another. In the second half of their chapter, they highlight the need for methodological improvements and identify ways in which conceptual understanding may be enhanced.

Patel, Kaufman, and Kannampallil (Chapter 27) report on the study of diagnostic expertise which initially focused on characterizing the reasoning process and, later, on understanding the nature of expert knowledge and its impact on performance, including memory, comprehension, and reasoning. This chapter highlights that medical expertise is not a simple construct, and its development is characterized by non-linear growth in skills and knowledge. Facilitated by new technology, recent research has moved toward real-world studies (or a combination of both laboratory-based and naturalistic studies), with automated and often precise methods of data collection and analysis.

Chapter 28 by Wiggins, Auton, and Taylor examines the study of expertise in the context of firefighting and emergency responding. The outcomes of existing research initiatives are examined, emphasizing the importance of accurate and precise mental models acquired through active interaction within the operational environment. Future research directions are proposed that will ensure the development of a continued comprehensive understanding of the nature expertise in firefighting and emergency responding.

Wickens and Dehais (Chapter 29) explore proficiency and make the distinction between the experience of aviation professionals, often quantified in terms of hours of flight time, or flight qualifications, and expertise: proficiency at aviation tasks. They conclude from an extensive review of the literature in this area that experience of skills such as situation awareness, decision making, task management, and crew resource management may be only loosely coupled with proficiency and explore why this may be so.

Another well-established research area is covered by Roth, Naweed, and Multer in Chapter 30, who summarize methods used to uncover expertise in railroad research, followed by a review of the types of strategies that railroad workers exhibit. By providing a discussion of the impact of ongoing technological changes on the requirements for expertise the authors speculate on longer term changes in railroad technologies and the nature of expertise.

Thomson (Chapter 31) describes the historical and continuing evolution of the cyber domains, and how we can operationalize current research in cyber expertise. Research into cyber expertise is in its infancy; in fact, there is no clear definition of what constitutes cyber expertise or how it may be unique when compared to other technical fields. Thus, Thomson describes the work roles of cyber operators and reviews results from cognitive task analyses of their workplace. Finally, topics are presented for future research, including the use of realistic synthetic environments to study cyber operations with more ecological validity.

Jenkins and Pfautz (Chapter 32) focus upon intelligence analysis (IA) and highlight past research methods that have been applied to characterize the domain of IA from a high-level workflow perspective down to low-level models of analyst information processing. They argue that such characterizations of the domain provide opportunities for highlighting different characteristics of expert behaviors throughout the IA process. They conclude with implications for the design of training and propose new types of technologies to aid in IA and analogous domains.

In Chapter 33, Suss and Bolton examine expertise in law enforcement and provide guidance for those planning on conducting research in this field. Illustrative examples cover a broad range of methods and highlight the subtleties that researchers new to the domain should consider when designing and conducting expertise research.

Fletcher and Kowal (Chapter 34) review characteristics of expertise common to all domains as a context for the expertise needed by military personnel. Cognitive qualities needed for military expertise are discussed, including the emerging issue of cognitive readiness required for irregular as well as regular military operating environments. The chapter emphasizes that military expertise is similar to expertise elsewhere; however, the volatility, uncertainty, complexity, ambiguity, and lethality of the environment in which military decisions are made that affect large numbers of individuals may be unique.

Chapter 35 illustrates DiBello's work, which shows changes in society have influenced a greater need for expertise in business. DiBello's chapter provides an exceptional insight into her work with organizations over the past two decades, helping them adapt their expertise to an increasingly complex and interconnected world of business. DiBello's chapter suggests that the new *expert* in business may not be an individual at all, but rather a high performing and highly efficient team. The chapter ends with reflections about how business organizations may learn from expertise.

Fischer and Mosier's chapter (36) on teams in space captures the complexities associated with human space flights' multiteam effort requiring the coordination and collaboration not only of individuals within a team (mission control or space crew), but importantly also between teams. The chapter discusses the strategies and procedures these expert teams have established to ensure common task and team models, and to facilitate their communication and joint performance. The teamwork challenges of future long-duration space exploration are discussed, as are the continuing advancements and research needed to address them.

Chapter 37, by Crichton, Moffat, and Crichton, describes the current status of expertise development in nuclear power production and oil & gas facilities, for both routine operations and emergency response. They note simulator-based exercises increasingly being introduced. The chapter summarizes existing research into the content and format of the skills required by operators in these settings, highlighting many questions yet to be answered, including how do we measure this combination of task, duration of experience, and level of performance to determine expertise?

The final chapter of this section by LaDue, Daipham, Pliske, and Hoffman (Chapter 38) captures the latest insights on weather forecasting. The chapter summarizes four research programs ranging from organizational to individual analyses to provide unique, complementary insights about expertise in this highly technologically focused domain. Like many areas of expertise, the forecasters have extensive, complex knowledge about each type of weather process they forecast, a knowledge that may be lost if not captured and passed on to the next generation. LaDue et al. suggest that the empirical work on professional activity in context, has the potential to invigorate studies of expertise.

Developing, Accelerating, and Preserving Expertise

Section IV presents a collection of approaches that have been used, broadly speaking, to develop or maintain expertise. The chapters range in emphasis from envisioning new pathways for educating our children and teachers to leveraging our knowledge of how experts learn and adapt in complex environments. Chapters cover important topics from improving expert performance individually and in teams, to accelerating the development of expertise, maintaining or retaining expert skills, and avoiding breakdowns in system performance.

Resnick, Russell, and Schantz (Chapter 39) argue that current methods of classroom teaching may be unfit for purpose—they do not adequately prepare students to thrive and excel in a complex world. As an alternative, they discuss the role of argumentation in developing expert reasoning skills, and point to an emerging body of research which suggests that teaching these skills in the classroom can lead to the acquisition and retention of general knowledge, beyond the topics taught through discussion. They review the ways in which *dialogic* reasoning can be used as a form of teaching to support the development of argumentation skill, and discuss some of the barriers to extending these methods beyond those students already considered gifted. In addition, they examine the role of educational, organizational, and social systems in facilitating a transition toward greater adoption of these methods in the classroom and beyond.

In Chapter 40, Fadde and Jalaeian provide an overview of the concept of deliberate practice and review the associated research in teacher education, medicine/surgery, and sports. They examine the difference between domains that have a culture of *practice*—like sports and music—versus a culture of *study* or a culture of *experience*. They highlight that while strong correlations have been found between expertise level and domains that have a culture of practice, much weaker correlations have been found in those with a culture of study or experience, such as education or professions. Accordingly, they present an alternative to deliberate practice, termed deliberate performance, which captures the kinds of learning activities in which professionals might engage to deliberately improve their performance. Last, Fadde and Jalaeian compare three models of training based on related research and review their effectiveness and conclude that consciously incorporating deliberate practice during college-based professional education and deliberate performance during the career work of professionals (who typically have little time to *practice*) can accelerate the development of professionals to expert levels.

In Chapter 41, Spiro and colleagues examine expertise in complex and ill-structured domains from the perspective of *cognitive flexibility theory* (CFT). Their emphasis is on *adaptation* in modern situations that deviate from novelty in relatively ordinary yet unexpected ways. Spiro et al. build on and extend the adaptive skill framework proposed by Ward et al. (2018) by further specifying how one prepares for situational novelty via meta-features of an *adaptive mindset* that generalize across cases in ways that content does not. Spiro's view also specifies how these features support the novel rearrangement of previously encountered case features in ways that are adaptive to new situations. Computer-supported case-based learning environments are used as a means to apply CFT to *expertise acceleration*, and a theoretical rationale and empirical examples are provided for structuring these computer systems in terms of the principles of case and concept selection and sequencing. *New modes of deliberate practice* that foster *adaptive readiness* are proposed, including skill at situation-adaptive assembly of knowledge and experience for *adaptive performance* that require a rethinking of what constitutes deliberate practice. Spiro and colleagues conclude with a discussion of a wide range of practical *implications* to the accelerated fostering of adaptive response to novel situations.

In Chapter 42, Moore and Hoffman present a view of proficiency scaling in the domain of intelligence analysis. They highlight the role of what are termed essential competencies, and detail the many distinct analytical roles entailing a specialization of expertise in this domain. Moreover, they discuss models of analyst reasoning and knowledge as a function of proficiency level and consider the stability of individual differences in styles, and distinctiveness of approaches to critical thinking across proficiency levels. In an era when the intelligence community is calling for robust measures of performance, they review how analysts make sense of situations and events for which there is no single cause, and discuss the role of human agency and motivations in causal reasoning. Last, they present implications of this research for training future analysts.

Otte, Knipfer, and Schippers put forward the claim, in Chapter 43, that team reflection is a major driver for the development and attainment of expertise in teams. They define team reflection as the collective evaluation of prior team activities, review the associated research, elaborate the mechanisms that link team reflection to expertise in teams and discuss multiple catalysts of team reflection. In the final part of their chapter, they investigate the shortcomings of previous team reflection research. These include the level at which this research has been analyzed and the short- and long-term consequences of engaging in this activity. They make suggestions for future research in this area to deepen our understanding of the effects of team reflection on the development of expertise in teams.

In Chapter 44, Petushek, Arsal, Ward, Upton, Whyte, and Hoffman focus on the role of mentoring in the development of expertise and discuss the multifactorial nature of mentoring, coaching, and related learning enhancement methods. They pursue a specific goal of unpacking the complex interactional relationship between developmental functions and roles to more fully describe what it means to be an effective coach/mentor. They review the meta-analytic and empirical evidence supporting the effectiveness of developmental and mentoring-type roles on a range of outcomes, such as job performance and career progression, and provide some specific examples of studies that have documented mentoring activities in action. They conclude by summarizing some of the major issues yet to be resolved in this field, and make specific suggestions for how to advance this area of research.

In Chapter 45, Morrow and Azevedo focus on the relationships between expertise and aging. They begin their chapter by considering how experts excel on domain-relevant tasks despite cognitive limitations. Further they examine how these expertise-related advantages develop, providing possible ways that adults can offset age-related cognitive constraints to maintain performance in later years. Their chapter centers around two key issues related to expertise and aging: the extent to which superior levels of performance can be maintained by experts as they continue to age; and the extent to which knowledge and skill associated with experience can offset age-related declines in abilities and function.

Arthur and Day (Chapter 46) focus on the issue of skill and knowledge decay and retention and how this intersects with expertise. Their review highlights several important conclusions: First, decay is affected more by interference than forgetting of information. Second, decay is highly dependent on task and situational factors.

Third, there is less decay on complex tasks than is observed for simple tasks. Fourth, retention is a function of expertise—it is stronger with practice, elaborative rehearsal, and greater mastery of the task. Fifth, retention and transfer are distinct concepts. Sixth, there is a limited amount of empirical research on decay and expertise in complex real-world performance domains. They conclude by suggesting that much could be learned by closer integration of the literature on expertise and skill decay and make recommendations for how to proceed in this regard.

In Chapter 47, Havinga, Bergström, Dekker, and Rae present an argument for expertise from a resilience engineering perspective, suggesting that this has changed the value of expertise from meeting required standards to helping organizations adapt. They begin by introducing the concept of resilience and its application to safety in sociotechnical systems. Then they explore how to manage expertise in complex systems, considering both its costs and benefits to engineering resilient organizations. Their review considers the role of expertise at multiple levels of the system, including frontline workers, teams, and management, and on an organizational systems level.

In Chapter 48, Conway and Gore focus on the role of expertise in developing complex policy interventions for government. They begin by providing some context on the nature of government work and its challenges; and then examine the role of expertise across the system, including the overlap in expertise between different roles, how this has evolved over time, and how the evidence is derived, from whom, and how it is shaped by values. They highlight the difficulties of identifying and incorporating true expertise into policy interventions and consider whether expertise itself is under threat in this process. Last, they identify research gaps that if addressed would support the further professionalization of the policy function in government.

Current Issues and the Future of Expertise Research

Section V presents two chapters that address the current and future challenges of expertise research. In Chapter 49, Klein, Shneiderman, Hoffman, and Wears highlight the seeming irony that although expertise is increasingly sought out and needed in today's society, at the same time several communities have actively begun to disparage experts! They present a series of arguments that demonstrate why their criticisms are misguided and assert that the criticisms made can help the research community discover better methods for supporting experts and for developing expertise.

In the final chapter, Ward, Schraagen, Gore, Roth, Hoffman, and Klein discuss some of the future directions that the field of expertise studies might take in order to continue to allow people to thrive in a world whose complexities are ever increasing. In particular, they present a particular view of expertise focused on adaptive skill—a concept that has often been discussed but is an empirically neglected aspect of expertise research—as a potential remedy for advancing the field and better preparing individuals to cope with the uncertainties and complexities of tomorrow's society.

REFERENCES

Ackerman, P. L. (1988). Determinants of individual differences during skill acquisition: Cognitive abilities and information processing. *Journal of Experimental Psychology: General* 117(3), 288–318.

Amirault, R. J., & Branson, R. K. (2006). Educators and expertise: A brief history of theories and models. In K. A. Ericsson, N. Charness, P. J. Feltovich, & R. R. Hoffman (Eds), *The Cambridge handbook of expertise and expert performance* (pp. 69–86). Cambridge, UK: Cambridge University Press.

Araujo, D., & Kirlik, A. (2008). Towards an ecological approach to visual anticipation for expert performance in sport. *International Journal of Sport Psychology* 39(2), 157–165.

Baker, J., & Farrow, D. (2015). *Routledge handbook of sport expertise*. London: Routledge.

Bartlett, F. C. (1958). *Thinking: An experimental and social study*. London: Allen & Unwin.

Battig, W. F. (1966). Facilitation and interference. In E. A. Bilodeau (Ed.), Acquisition of skill (pp. 215–244). New York: Academic Press.

Benner, P. (1984). *From novice to expert: Excellence and power in clinical nursing practice*. Upper Saddle River, NJ: Prentice-Hall Health.

Benner, P., Tanner, C. A., & Chesla, C. A. (1996). *Expertise in nursing practice: Care, clinical judgment and ethics*. New York: Springer.

Berling, T. V. & Bueger, C. (2015). *Security expertise: Practice, power, responsibility*. London: Routledge.

Bernstein, N. (1967). *The coordination and regulation of movements*. New York: Pergamon.

Bloom, B. (1985). *Developing talent in young people*. New York: Balantine.

Brentano, F. (2009). *Psychology from an empirical standpoint*. London: Taylor & Francis. (Original work published 1874.)

Brunswik, E. (1956). *Perception and the representative design of psychological experiments* (2nd edn). Oakland, CA: University of California Press.

Bryan, W. L., & Harter, N., (1897). Studies in the physiology and psychology of the telegraphic language. *Psychological Review* 4, 27–53.

Chi, M. T. H., Feltovich, P. J., & Glaser, R. (1981). Categorization and representation of physics problems by experts and novices. *Cognitive Science* 5(2), 121–152.

Chi, M. T. H., Glaser, R., & Farr, M. J. (Eds.) (1988). *The nature of expertise*. Hillsdale, NJ: Lawrence Erlbaum.

Collins, H. (1990). *Artificial experts: Social knowledge and intelligent machines*. Cambridge, MA: MIT Press.

Collins, H. (2018). *Artifictional Intelligence: Against humanity's surrender to computers*. Cambridge, MA: Polity.

Collins, H., & Evans, R. (2007). *Rethinking expertise*. Chicago: University of Chicago Press.

Collins, H., & Kusch, M. (1998). *The shape of actions: What humans and machines can do*. Cambridge, MA: MIT Press.

Crandall, B., Klein, G., & Hoffman, R. R. (2006). *Working minds*. Cambridge, MA: MIT Press-Bradford.

Crossman, E. R. F. W. (1959). A theory of the acquisition of speed-skill. *Ergonomics* 2, 153–166.

De Groot, A. (1965). *Thought and choice in chess*. The Hague, Netherlands: Mouton. (Original work published 1946.)

Dreyfus, H. L. & Dreyfus, S. E. (1986). *Mind over machine: The power of human intuition and expertise in the era of the computer*. New York: The Free Press.

Duncker, K. (1935). On problem solving. *Psychological Monographs* 58(5), 1–113. Translated by Lees, L. S. (1945; *Zur Psychologie des produktiven Denkens*).

Dunn, T. G. & Schriner, C. (1999) Deliberate practice in teaching: What teachers do for self-improvement. *Teaching and Teacher Education* 15, 631–651.

Ebbinghaus, H. (1885). *On memory: A contribution to experimental psychology* (H. Ruger and C. Busenius, Trans.). New York: Teacher's College, Columbia University. (Reprinted 1964.)

Ericsson, K. A. (2009) (Ed.). Development of professional expertise. New York: Cambridge University Press.

Ericsson, K. A., Hoffman, R. R., Kozbelt, A., & Williams, A. M. (2018) (Eds). *The Cambridge handbook of expertise and expert performance* (2nd edn). Cambridge, UK: Cambridge University Press.

Ericsson, K. A., & Kintsch, W. (1995). Long term working memory theory. *Psychological Review* 102(2), 211–245.

Ericsson, K. A., Krampe, R. T. h., & Tesch-Römer, C. (1993) The role of deliberate practice in the acquisition of expert performance. *Psychological Review* 100, 363–406.

Ericsson, K. A., & Simon, H. (1980). Verbal reports as data. *Psychological Review* 87(3), 215–251.

Ericsson, K. A., & Simon, H. (1993). *Protocol analysis: Verbal reports as data* (rev. edn). Cambridge, MA: MIT Press.

Ericsson, K. A., & Smith, J. (1991). Prospects and limits of the empirical study of expertise: an introduction. In K. A. Ericsson & J. Smith (Eds), *Toward a general theory of expertise: Prospects and limits* (pp. 1–38). Cambridge, UK: Cambridge University Press.

Ericsson, K. A., & Staszewski, J. J. (1989). Skilled memory and expertise: Mechanisms of exceptional performance. In D. Klahr & K. Kotovsky (Eds), *Complex information processing: The impact of Herbert A. Simon* (pp. 235–267). Hillsdale, NJ: Lawrence Erlbaum.

Ericsson, K. A., & Ward, P. (2007). Capturing the naturally-occurring superior performance of experts in the laboratory: Toward a science of expert and exceptional performance. *Current Directions in Psychological Science* 16(6), 346–350.

Ericsson, K. A., Whyte, J., & Ward, P. (2007). Expert performance in nursing: Reviewing research on expertise within the framework of the expert-performance approach. *Advances in Nursing Science* 30(1), E58–E71.

Feltovich, P. J., Prietula, M. J., & Ericsson, K. A. (2006). Studies of expertise from psychological perspectives. In K. A. Ericsson, N. Charness, P. J. Feltovich, & R. R. Hoffman (Eds), *The Cambridge handbook of expertise and expert performance* (pp. 41–67). Cambridge, UK: Cambridge University Press.

Fitts, P. (1964). Perceptual-motor skill learning. In A. W. Melton (Ed.), *Categories of human learning* (pp. 243–285). New York: Academic Press.

Gobet, F., & Chassy, P. (2008). Towards an alternative to Benner's theory of expert intuition in nursing: A discussion paper. *International Journal of Nursing Studies* 45(1), 129–39.

Gobet, F., & Simon, H. A. (1996). Templates in chess memory. A mechanism for recalling several boards. *Cognitive Psychology* 31(1), 1–40.

Gore, J., Ward, P., Conway, G., Ormerod, T., Wong, W., & Stanton, N. (2018). Editorial. Naturalistic decision making: Navigating uncertainty in complex sociotechnical work—an introduction to the Special Issue. *Cognition, Technology, & Work*. Available from: http://static.springer.com/sgw/documents/1629808/application/pdf/CTW+SI+on+NDM+2018.pdf

Hatano, G., & Inagaki, K. (1984). Two courses of expertise. *Research and Clinical Center for Child Development Annual Report* 6, 27–36.

Hatano, G., & Inagaki, K. (1986). Two courses of expertise. In H. Stevenson, H. Azuma, & K. Hakuta (Eds), *Child development and education in Japan* (pp. 262-272). New York: W. H. Freeman.

Hayes, J. R. (1985). Three problems in teaching general skills. In S. F. Chipman, J. W. Segal, & R. Glaser (Eds), *Thinking and learning skills*, Vol. 2: *Research and open questions* (pp. 391-405). Hillsdale, NJ: Lawrence Erlbaum.

Hoc, J-M., Cacciabue, P. C., & Hollnagel, E. (1995). *Expertise and technology: Cognition & human–computer cooperation*. Hillsdale, NJ: Lawrence Erlbaum.

Hoffman, R. R. (1998). How can expertise be defined? Implications of research from cognitive psychology. In R. Williams, W. Faulkner, & J. Fleck (Eds), *Exploring expertise* (pp. 81-100). New York: Macmillan.

Hoffman, R. R. (Ed.). (2007) *Expertise out of context: Proceedings of the sixth international conference on naturalistic decision making*. New York: Lawrence Erlbaum.

Hoffman, R. R., & Deffenbacher, K. A. (1992). A brief history of applied cognitive psychology. *Applied Cognitive Psychology* 6(1) 1-48.

Hoffman, R. R., & Militello, L. G. (2009). *Perspectives on cognitive task analysis: Historical origins and modern communities of practice*. New York: Psychology Press.

Hoffman, R. R., Ward, P., Feltovich, P. J., DiBello, L., Fiore, S. M., & Andrews, D. (2014). *Accelerated expertise: Training for high proficiency in a complex world*. New York: Psychology Press.

Klein, G. (2007a). Flexecution, as a paradigm for replanning, part 1. *IEEE Intelligent Systems* 22, 79-83.

Klein, G. (2007b). Flexecution, part 2: Understanding and supporting flexible execution. *IEEE Intelligent Systems* 22, 108-112.

Klein, G. (2008). Natruralistic decision making. *Human Factors* 50(3), 456-460.

Klein, G., Phillips, J. K., Rall, E. L., & Peluso, D. A. (2006). A data/frame theory of sensemaking. In R. R. Hoffman (Ed.), *Expertise out of context: Proceedings of the 6th international conference on naturalistic decision making* (pp. 113-155). New York: Lawrence Erlbaum.

Klein, G. A. (1989). Recognition-primed decisions. In W. B. Rouse (Ed.), *Advances in man-machine systems research* (Vol. 5, pp. 47-92). Greenwich, CT: JAI Press.

Klein, G. A., Calderwood, R., & Clinton-Cirocco, A. (1986). Rapid decision making on the fire ground. In *Proceedings of the Human Factors Society 30th annual meeting* (Vol. 1, pp. 576-580). Santa Monica, CA: Human Factors Society.

Klein, G. A., Calderwood, R., & MacGregor, D. (1989). Critical decision method for eliciting knowledge. *IEEE Transactions on Systems, Man, and Cybernetics* 19(3), 462-472.

Markman, A. B. (2018). Combining the strengths of naturalistic and laboratory decision-making research to create integrative theories of choice. *Journal of Applied Research in Memory and Cognition* 7, 1-10.

Mieg, H. (2001). *The social psychology of expertise*. Mahwah, NJ: Lawrence Erlbaum.

Mosier, K. L., & Fischer U. M. (Eds). (2015) *Informed by knowledge: Expert performance in complex situations*. New York: Psychology Press.

Newell, A., & Rosenbloom, P. S. (1981). Mechanisms of skill acquisition and the law of practice. In J. R. Anderson (Ed.), *Cognitive skills and their acquisition* (pp. 1-55). Hillsdale, NJ: Lawrence Erlbaum.

Newell, A., & Simon, H. A. (1972). *Human problem solving*. Englewood Cliffs, NJ: Prentice-Hall.

Poulton, E. (1957). On prediction in skilled movements. *Psychological Bulletin* 54, 467-478.

Rasmussen, J. (1981). Models of mental strategies in process plant diagnosis. In J. Rasmussen & W. B. Rouse (Eds), *Human detection and diagnosis of system failures*. New York: Plenum Press.

Rasmussen, J. (1983). Skill, rules and knowledge: Signals, signs, and symbols, and other distinctions in human performance models. *IEEE Transactions on Systems, Man and Cybernetics SMC-13*(3), 257–266.

Rasmussen, J. (1986). *Information processing and human-machine interaction: An approach to cognitive engineering.* New York: North-Holland/Elsevier.

Rasmussen, J., & Jensen, A. (1974). Mental procedures in real-life tasks: A case study of electronic troubleshooting. *Ergonomics 17*, 293–297.

Rasmussen, R., Pejtersen, A. M., & Goodstein, L. P. (1994). *Cognitive systems engineering.* New York: Wiley & Sons.

Rosenberg, J. I. (2000). Reconstructing clinical training: Self-reflection, deliberate practice and critical thinking skills. *International Journal of Psychology 35*, 415.

Schraagen, J.M.C. (2018). Naturalistic decision making. In L. J. Ball & V. A. Thompson (Eds), *The Routledge international handbook of thinking and reasoning* (pp. 487–501). London and New York: Routledge.

Selinger, E., & Crease, R. P. (2006). *The philosophy of expertise.* New York: Columbia University Press.

Selz, O. (1981). *Die gesetze der produktiven und reproduktiven geistestätigkeit kurzgefasste darstellung.* Bonn. Reproduced in N. H. Frijda, & A. D. de Groot (Eds), *Otto Selz: His contribution to psychology* (pp. 21–75). The Hague: Mouton. (Original work published 1924.)

Simon, H. A., & Gilmartin, K. (1973). A simulation of memory for chess positions. *Cognitive Psychology 5*, 29–46.

Snoddy, G. S. (1926). Learning and stability: A psychophysiological analysis of a case of motor learning with clinical applications. *Journal of Applied Psychology 10*, 1–36.

Sonnetag, S., & Kleine, B. M. (2000) Deliberate practice at work: A study with insurance agents, *Journal of Occupational and Organizational Psychology 73*, 87–102.

Starkes, J., & Allard, F. (1993). *Cognitive issues in motor expertise.* Amsterdam: North-Holland.

Starkes, J. & Ericsson, K. A. (2003). *Expert performance in sport: Advances in research on sport expertise.* Champaign, IL: Human Kinetics.

Staszewski, J. J. (2013). *Expertise and skill acquisition: The impact of William G. Chase.* New York: Psychology Press.

Thorndike, E. L., & Woodworth, R. S. (1901). The influence of improvement on one mental function upon the efficiency of other functions, I: Extended training and experience are required to achieve high levels of proficiency. *Psychological Review 8*, 247–261.

Vicente, K. J. (1999). *Cognitive work analysis: Toward safe, productive and healthy computer-based work.* Mahwah, NJ: Lawrence Erlbaum.

Ward, P., Gore, J., Hutton, R., Conway, G., & Hoffman, R. (2018). Adaptive skill as the *conditio sine qua non* of expertise. *Journal of Applied Research in Memory and Cognition 7*(1), 35–50.

Ward, P., Hodges, N. J., Starkes, J. L., & Williams, A. M. (2007). The road to excellence: Deliberate practice and the development of expertise. *High Ability Studies 18*(2), 119–153.

Wiggins, M. W., & Loveday, T. (Eds). (2015). *Diagnostic Expertise in Organizational Environments.* London: Routledge.

Williams, A. M., Davids, K., & Williams, J. G. (1999). *Visual perception and action in sport.* London: E & FN Spon.

Yates, J. F. (2001). "Outsider:" Impressions of naturalistic decision making. In E. Salas & G. Klein (Eds), *Linking expertise and naturalistic decision making* (pp. 9–33). Mahwah, NJ: Lawrence Erlbaum.

SECTION I

CHARACTERIZING EXPERTISE

FRAMEWORKS, THEORIES, AND MODELS

SECTION EDITOR: JAN MAARTEN SCHRAAGEN

CHAPTER 2

THE CLASSIC EXPERTISE APPROACH AND ITS EVOLUTION

FERNAND GOBET

INTRODUCTION

CHASE and Simon's research on chess expertise, published in three papers (Chase & Simon, 1973a, b; Simon & Chase, 1973), was highly influential and defined research on expertise for the following decades. It was followed by a spate of research on expertise, particularly in Pittsburgh, which tended to share a number of characteristics: close links with cognitive psychology and adherence to the then-dominant information-processing paradigm, tests of Chase and Simon's chunking theory with elegant experiments, detailed analyses of the processes involved, use of verbal protocols, and small number of participants. I will call this approach the *classic expertise approach*.

The aim of this chapter is to describe this approach and to understand why it was (and still is) so influential. After distinguishing it from *non-classic* approaches, I briefly discuss early research on expertise that influenced—to various extents—the classic approach. I then describe in some detail Chase and Simon's classic papers and chunking theory. This leads the way to a presentation of some of the key experimental and theoretical research. Given the limited space available, this presentation will by necessity be very selective. The chapter then discusses more recent theories that can be considered as outgrowths of the classic approach, providing a good opportunity to try to understand not only its key characteristic but also why it had such a large impact. The chapter concludes by a discussion of what this approach tells us about the means to address the challenges currently facing research on expertise.

It might be helpful at this point to contrast the classic approach with other influential research paradigms that predated it. Alfred Binet is arguably the first psychologist to have studied expertise experimentally. He developed clever methods for studying great calculators (Binet, 1894b, 1966) and magicians (Binet, 1894a). In particular, he pioneered a number of chronometric techniques and instruments for understanding how skilled magicians create visual illusions in magic tricks. His research on calculators, including the experimental tasks he developed, would have a direct impact on skilled memory, one of the theories developed within the classic approach. By contrast, his research on chess players (Binet, 1894b), which relied on interviews and introspection, was not as influential, although it did emphasize the role of memory and knowledge in expertise.

In the tradition of research known as judgment and decision making, Meehl's (1954) review of the literature criticized expertise in psychotherapy and psychiatry. His central points were that experience is not a good predictor of expertise and that the diagnoses reached by simple mathematical models have a much higher reliability than those reached by human clinical experts. Research into individual differences, sometimes called differential psychology, used psychometrics to study expertise and creativity. In this approach, superior performance is considered as mostly due to innate talent (Djakow, Petrowski, & Rudik, 1927; Galton, 1869). Lehman (1953) was a prime example of the historiometric approach, which uses the statistical analysis of archival documents to study outstanding performance. He found interesting patterns in the way artistic and scientific creativity changes during one's career and how these patterns differ between different fields. Finally, skill[1] had been studied in the fields of human factors and engineering psychology, amongst other related fields, starting with the early work of Bryan and Harter (1899). Probably the best-known result in this field is Fitts's (1964) idea that the acquisition of perceptual and motor behavior consists of three stages: the *cognitive phase*, the *associative phase*, and the *autonomous phase*. In general, when one progresses through these phases, behavior moves from conscious effort (use of declarative rules and trial and error) to unconscious, automatic, and more efficient behavior. As behavior demands less attention in the last phase, it is then possible to perform several tasks in parallel.

While these approaches were almost certainly known to researchers of the classic approach to expertise, their influence was less than that of Dutch psychologist Adriaan de Groot, who pioneered some of the tools central to the classic approach, such as the recall task and the use of concurrent verbal protocols for studying problem solving, as discussed in the following section.

De Groot's Research on Chess

Just before World War II, De Groot collected data on chess players' thinking for his PhD thesis, which was later turned into a book (1946, 1965). Whereas the wisdom of

---

[1] Many authors use "skill" and "expertise" interchangeably, but others prefer using "skill" for the mastery of relatively simple, typically perceptual and motoric, behaviors and "expertise" for more complex, typically cognitive, behaviors. Following Chase and Simon's example who titled one of their papers "Skill in chess," this chapter will use both terms as synonyms.

the time was that there would be large differences between grandmasters and amateurs in the amount of search carried out (e.g., in measures such as the number of positions visited or the depth of search), he found that these variables hardly distinguished such players. Rather, the key differences were in perception and intuition (the speed with which grandmasters can identify the key features of a position) and in knowledge (e.g., typical positions and the playing methods to use in specific cases).

Numerous are De Groot's contributions to the classic approach to expertise.[2] The recall task he devised, originally aimed at understanding perception, became one of the central weapons in the experimental arsenal of the classic approach. So did his task of asking individuals of different skill levels to think aloud when trying to find a solution to a problem. This type of experiment was not totally new, of course. It was originally developed by Otto Selz (1922) to understand the processes underpinning productive thinking in (non-expert) individuals and was used by Bahle (1930) in a study of music composition. Selz's proposition that thinking consists of a linear chain of cognitive operations anticipated the idea of a production system (Newell & Simon, 1972) but was expressed only verbally and thus lacked precision.

De Groot's genius was to devise statistics to quantify the structure of problem solving, in particular using the mathematical idea of a tree. In doing so, not only was he able to study Selz's framework statistically but he also anticipated Newell and Simon's (1972) idea of problem-space states and many measures of search trees later used in computer science. With hindsight, this seems a rather obvious choice, as trees are a very natural way to describe how different variations and subvariations are explored when trying to solve a chess problem. However, applying this formalism to human thinking was ground-breaking.

De Groot's book is a gold mine of ideas and insights. While his experiments did not satisfy the canons of current methodology—and perhaps because of this!—he anticipated most of the key questions in the study of expertise: the roles of perception, memory and knowledge, the organization of expert problem solving, and the importance of both talent and practice.

Simon's Early Computer Models of Chess Cognition

Chase and Simon's three seminal papers on chess skill combined Chase's acumen in designing experiments with Simon's theoretical and methodological knowledge—a

---

[2] A good case could be made for including De Groot in the classic approach. After much deliberation, I decided against it, as this would destroy the classic Aristotelian unities in drama (action, time, and place) organizing this chapter. In addition, some of De Groot's views were clearly at variance with the classic approach, notably his defense of introspection as scientific method and his skepticism about the usefulness of computer simulations (for a discussion, see De Groot, Gobet, & Jongman, 1996).

very powerful mix! While Chase was a young associate professor when the papers were published, Simon had already received world fame for having created the fields of artificial intelligence and modern cognitive psychology.

Several previous lines of Simon's research are evident in these three papers, in particular when the discussion focuses on the mechanisms enabling expertise. The concept of bounded rationality (Simon, 1956), for which Simon would later earn the Nobel Prize in Economics, looms large in chunking theory, which strongly emphasizes the limits of short-term memory, the restricted span of attention, and the slow rate at which learning takes place. The idea of selective search is also used when Chase and Simon discuss problem solving.

Several computer programs developed by Simon and colleagues had considerable influence on chunking theory. Two programs implemented the idea of selective search and satisficing (the use of satisfactory rather than optimal solutions), a key concept in Simon's theory of bounded rationality: NSS (Newell, Shaw, & Simon, 1958a) and MATER (Baylor & Simon, 1966). A third program, PERCEIVER (Simon & Barenfeld, 1969), simulated a chess player's eye movements. The results showed that information-processing mechanisms similar to those used for problem solving can explain high-level perception.

Another important source of ideas was the EPAM model,[3] which was originally applied to verbal learning (Feigenbaum & Simon, 1962). Not only did EPAM implement several ideas linked to the notion of bounded rationality, such as limited short-term memory capacity and slow learning rates, but it also provided mechanisms crucial for chunking theory. It showed that it was possible to closely link perception, learning, and memory in a computational model; in addition, it provided mechanisms explaining how chunks can be incrementally learned as a function of the information provided by the environment. A chunk is essentially a node in an EPAM network. In later work, Simon (1990, p. 1) wrote that "human rational behavior is shaped by a scissors whose blades are the structure of task environments and the computational capabilities of the actor." EPAM, by incorporating both learning mechanisms exploiting the regularities of the environment and assumptions about cognitive limits, addresses the two blades of this analogy.

With respect to methodology, the classic expertise approach was clearly influenced by the views that Newell and Simon (1972) expounded in their opus magnum, *Human Problem Solving*. In particular, several researchers adopted their emphasis on processes, often focusing on the behavior of very few participants studied in great detail, rather than opting for a large number of participants and statistical power. There was also an emphasis on content, as is appropriate given that experts master a specific domain: being an expert in ornithology is not the same as being an expert in poker. Moreover, Newell and Simon stressed the importance of strategies and heuristics, even when they are specific to a single individual.

---

[3] EPAM stands for Elementary Perceiver and Memorizer.

A final methodological point made by Newell and Simon was of course about the best way to express theories. They advocated the use of computer programs and were ruthless in criticizing the inadequacies of informal, verbal theorizing (e.g., Newell, Shaw, & Simon, 1958b; Newell & Simon, 1961) (for a discussion, see Gobet & Lane, 2015). Computer modeling offers several important advantages: behavior can be decomposed into elementary processes, which makes it possible to provide mechanistic explanations; behavior can be studied as a process that evolves as a function of time, as opposed to using just one or at most a few snapshots, as is common in experimental psychology; and finally, models are *sufficient* in the sense that they offer explanations that can generate the behavior under study. For example, a computer model of chess memory can recall chess positions, ideally with the same errors as those committed by humans.

The methodology of verbal protocols, another important contribution of Simon, was not much used in the empirical work described in Chase and Simon's papers. However, it was important in other developments with the classic approach, most notably Chase and Ericsson's experiments on the digit span task (see the section "Digit span task").

Chase and Simon's Seminal Papers

"Perception in chess" (Chase & Simon, 1973b) reports an experiment with three chess players (one beginner, one class A player (good amateur), and one international master) and two tasks: De Groot's short-term recall task and a perception task, where players had to copy a position in plain view onto a second chessboard. In addition to the game positions used by De Groot (1946, 1965), Chase and Simon also used random positions. The originality of the paper consists in the converging methods developed for identifying chunks (configurations of pieces): glances at the stimulus position in the perception task, and, in both tasks, latencies between the placement of two pieces and pattern of chess relations between those pieces. According to the authors, the results strongly suggested that expertise in chess derives from the ability to encode positions as perceptual chunks.

"The mind's eye in chess" (Chase & Simon, 1973a) is a chapter in the *Proceedings of the Eighth Annual Carnegie Symposium on Cognition* (Chase, 1973), devoted to visual information processing. It starts by summarizing the data presented in the first paper, and then presents additional analyses and new experiments aiming at pinpointing the concept of a chunk. The impression left by the first part of the paper is that much can be learnt by clever experiments addressing well-specified hypotheses, even with a sample as small as three participants. The second part of the paper develops the narrow version of chunking theory, an information-processing theory of chess skill. The interest here is to explain skill differences in De Groot's recall task. Based on earlier computational models of chess perception and memory developed by Simon and colleagues (see previous section), the authors argue that chess skill is made possible by the acquisition of a large number of chunks. Attentional mechanisms detect salient pieces, to which eye movements are directed. Pointers to recognized chunks are then placed in short-term memory. The third part of the paper presents additional experiments, focusing on long-term memory of positions and games, and on immediate recall of moves. It also investigates the Knight's Tour, a task supposed to measure chess talent. The fourth and final part describes the full version of chunking theory, aimed not only at explaining behavior in memory tasks, but also how chess players select good moves. The theory combines the ideas of mental imagery, chunking, and production system (Newell & Simon, 1972).[4] With respect to mental imagery, the mind's eye not only stores visuo-spatial structures from external inputs and memory stores but can also manipulate them by mental operations. Its capacity is limited. Long-term memory (LTM) chunks, such as specific patterns of pieces on a chessboard, act as *conditions* to *actions*, which evoke possible moves or plans. Chunks are also linked to information allowing configurations of pieces to be manipulated in the mind's eye. Pattern recognition makes it possible for players to visualize the chessboard and to anticipate moves during look-ahead search. An important aspect of the theory is that pattern recognition occurs not only on the external board, but also with the boards imagined in the mind's eye. It is thus an important mechanism for explaining how players keep selecting reasonable moves during look-ahead search.

While the first two papers were mostly addressed to psychologists, the third paper, "Skill in chess" (Simon & Chase, 1973), wooed a broader audience and was published in a generalist journal, the *American Scientist*. The emphasis is on the computational models of search, perception, and memory developed by Simon and colleagues, and on the narrow and full versions of chunking theory. The summary of the empirical work described in the previous two papers is limited to the methods and data supporting the psychological reality of a chunk. A fair amount of discussion is devoted to the number of chunks (50,000) required to become a master, as estimated by Simon and Gilmartin (1973), and the time needed for becoming a class A player (from 1,000 to 5,000 hours) and a master (from 10,000 to 50,000 hours). In the final paragraphs of the article, Simon and Chase (p. 403) answer the question of how one becomes a master—"The answer is *practice*—thousands of hours of practice" and note that "clearly, practice also interacts with talent."

Key Empirical Work in the Classic Expertise Approach

Chase and Simon's paper spawned a flurry of experimental papers. Beyond replicating the skill effect in the memory recall task in several domains (sports, Allard & Starkes, 1980; bridge, Charness, 1979; electronics, Egan & Schwartz, 1979; computer programming,

---

[4] Productions are rules of the type (IF condition THEN action). For example: IF the traffic light is red, THEN stop.

Schneiderman, 1976), several papers aimed to test some of the theoretical assumptions of chunking theory. This section reviews some of the most influential papers, organizing them by domain of research (for a detailed discussion, see Ericsson, Charness, Feltovich, & Hoffman, 2006; Gobet, 2016; Gobet, de Voogt, & Retschitzki, 2004).

Games

Some studies tested the hypothesis that information is grouped as chunks in players' long-term memory. For example, Frey and Adesman (1976) incrementally displayed a chess position by groups of four pieces, each new group appearing every two seconds. Under a control condition, the position was displayed column by column. Players recalled the position better when it was presented by chunks than when it was presented by columns. An unexpected support for chunking theory was that recall was better under the chunk-by-chunk condition than when the position was shown in its totality for the same duration (12 seconds). It is possible that delineating chunks enables players to recognize them better than when they must extract them from a complete position.

Reitman (1976) replicated Chase and Simon's study with two Go players, one master and one beginner. Unlike the original experiment, the correspondence was poor between the chunks identified by glances in the perception task and those identified with latencies in the recall task. Reitman also performed a partitioning task, where participants were asked to draw the boundaries of the clusters of pieces they found meaningful. Contrary to Chase and Simon's hypothesis that chunks form a hierarchy, her master perceived the positions as overlapping groups, a result that was later replicated by Chi (1978) with chess.

Charness (1976) and Frey and Adesman (1976) tested the hypothesis that it takes a relatively long time to encode new information in LTM (8 seconds to create a new chunk). Together with the assumption that the capacity of short-term memory (STM) is limited to seven items, this hypothesis implies that recall should be much affected by interpolating a second task between the presentation of a chessboard and its recall. This is because the second task should wipe out the pointers to LTM chunks from STM, as is clearly the case in experiments using unfamiliar material (Kintsch, 1970). However, it turned out that this does not happen with familiar material such as chess, as only a small decrease in memory recall was observed (around 10 percent) (Charness, 1976; Frey & Adesman, 1976). Interference was minimal even when the interpolated task was of a similar nature as the main task, such as finding the best move in a chess position.

Several researchers argued that the chunks proposed by Chase and Simon are simply too small to be of any use, and that higher-level knowledge structures are employed rather. Some results speak in favor of this viewpoint. Showing participants the moves leading to a position produces better recall (Goldin, 1978). Similar results were obtained in an unexpected recall task by manipulating the level of semantic processing at which players study a position. Independently of skill level, recall was higher when the task was to evaluate a position and select a good move than to count the number of pieces placed on white and black squares (Lane & Robertson, 1979). However, the effect disappeared when participants were previously told that they would have to recall the position.

In an elegant experiment, Chi (1978) demonstrated that research on chess could have implications beyond the study of expertise—in her case, a challenge to leading theories in developmental psychology. A dominant theory in developmental psychology holds that cognitive development is in great part produced by neural changes affecting memory capacity, speed of information processing, and executive functions (e.g., Pascual-Leone, 1970). (Currently, this is probably the leading explanation in cognitive neuroscience.) Chi showed that this is a vast simplification, and that memory capacity interacts with knowledge, and chunking in particular. Children and adults were asked to recall digits and chess positions. While adults outperformed children in the digit span task, chess-playing children had a better recall in the chess task than adults who did not play chess. Thus, knowledge in a domain more than compensates (putative) differences due to neural development. Incidentally, age differences in the digit span can also, at least to some extent, be explained by the idea that, compared to children, adults have acquired more chunks about numbers, such as dates and prices (Gobet, 2016; Simon, 1974).

Physics

Several key papers were written on physics, often using protocol analysis to compare the way novices and experts solve problems. When trying to solve relatively simple problems, novices tend to begin from the goal, and move back to the givens of the problem (i.e., they *search backward*), while experts tend to do the opposite: they start from the current situation and move to the goal of the problem (i.e., they *search forward*) (Larkin, McDermott, Simon, & Simon, 1980a; Simon & Simon, 1978). However, experts revert to backward search with difficult problems. A similar pattern has later been identified in other, but by no means all domains of expertise.

Chi, Feltovich, and Glaser (1981) were interested in the types of representation used by experts in physics and the kind of knowledge that enables them. A sorting task, where participants had to categorize problems from an undergraduate textbook, yielded two main results: first, the categories used by the novices and experts did not have much overlap; and second, the kind of representations used by the two groups were vastly different: experts' problem representations employed fundamental principles of physics, such as the concept of force, whilst novices' representations were based on superficial aspects of the problems, such as the kind of device used (e.g., pulley vs inclined plane). These experimental results were supplemented by a theoretical analysis, where novice and expert protocols were coded as node–link structures (schemas) and production rules. Based on this very detailed analysis, the authors concluded that experts' schematic knowledge allows them to build efficient representations of a problem and is linked to likely solution methods.

The research on physics expertise had important implications for education, as it provided insight about the kind of representations and solving methods that students need to acquire in order to move from novices to experts. This does not mean that novices can become experts simply by searching forward and learning the fundamental principles of physics by rote. Experts' knowledge is encoded as productions, and acquiring them requires solving many problems—hence a considerable amount of practice. Learning occurs through doing. In this respect, Simon (1980) notes that textbooks in physics and other domains tend to underemphasize the condition part of conditions, and hence opportunities to acquire perceptual knowledge.

Writing

While the analysis of verbal protocols had historically been predominantly used to study problem solving, Flower and Hayes drew on this methodology for understanding the processes involved in expert writing (Flower & Hayes, 1981; Hayes & Flower, 1980). They used Newell and Simon's (1972) framework of task environment, long-term memory, and processes. The task environment includes many external aspects influencing writing, such as the topic to address, the targeted audience, and the writer's motivation. Long-term memory includes knowledge about the topic and the audience, as well as methods such as writing plans. Finally, Flower and Hayes assumed that there are three main writing processes: planning, translating, and reviewing, each of them being subdivided into several subprocesses. Contrasting with previous stage models of writing, their model assumes that the process of writing is not linear, but hierarchical in nature: processes form a complex hierarchy, where any process can be embedded within other processes. Writers follow two kind of goals: process goals, which are instructions writers give to themselves (e.g., "let's copy-edit this chapter"), and content goals, such as how a writer intends a section to affect the reader. Goals are dynamic: the initial goals are typically revised as writing progresses and new goals are created. Just like the study of expertise in physics, research into writing expertise ensured that the classic approach would have a high impact on the field of education. Flower and Hayes (1981) is a classic in the field, with more than 4,500 citations on Google Scholar.

Computer Programming

Although Chase and Simon's recall experiment had been replicated several times, several authors disputed their explanation, which is essentially based on the number of chunks in LTM. Rather, they argued that knowledge organization is the key factor. Using a statistical technique making it possible to infer tree structures from the order with which the information is recalled, McKeithen, Reitman, Rueter, and Hirtle (1981) found important skill differences in chunk organization. Compared to beginners and intermediates, expert programmers' organization relies less on simple mnemonics and common-language associations, and more on abstract and functional programming knowledge. Similar results were obtained by Adelson (1981), who found that novices' organization relied on syntax, while experts' organization was more hierarchical, semantic, and abstract in nature, and was related to the functions of the programs.

Music

Sloboda (1976b) applied Chase and Simon's recall experiment to music, comparing novices with experienced musicians in how they could memorize musical excerpts presented visually. After the brief presentation of notes, participants were required to reproduce them on an empty stave. To control for novices' lack of knowledge of musical notation, Sloboda used very simple stimuli (between one and six random notes). Presentation times ranged from 20 ms to 2 seconds.

Musicians were much better with the 2-second presentation, but not with the 20-ms presentation (both groups performed poorly in this case). However, when participants had to reproduce not the exact sequence of notes, but the contour of notes signaled by ups and downs, musicians outperformed novices even with 20 ms (Sloboda, 1978). Note that Sloboda found a skill effect with random sequences of notes, whilst there was no skilled difference with random positions in Chase and Simon's experiment.[5] Another interesting result, this time with an auditory presentation of the stimuli, was that recall was not affected when participants were asked to perform interfering tasks—even remembering a melody—when listening to musical stimuli. Encoding music is therefore an automated skill.

The hypothesis of automaticity, but also of the use of high-level schemas, was supported by a further experiment carried out by Sloboda (1976a) with competent pianists. Scores of classical music were modified by displacing one note by one scale step, creating dissonant sequences. The pianists sight read each piece twice. Whilst accuracy was very high (above 97 percent overall), about 40 percent of the misprints were incorrectly recalled, pianists substituting the incorrect note by the correct one. Interestingly, the number of mistakes increased the second time pianists played the piece, although they made fewer mistakes overall. Sloboda suggests that pianists built an overall representation of the structure of the piece, which led them to correct the misprints in the score.

---

[5] Later research showed that, in most domains of expertise, there is a skill effect even with randomized material (Sala & Gobet, 2017).

Digit Span Task

The research on extraordinary memory in the digit span (Chase & Ericsson, 1981; Ericsson, Chase, & Faloon, 1980) tested, and refuted, one of the key assumptions of chunking theory—that long-term memory encoding is slow. SF, a student with average STM capacity, was trained to improve his performance in the digit span task. Digits were dictated at the pace of one digit per second. After a fairly short but intensive practice (about 250 hours in 2 years), SF was able to recall eighty-four digits, which was equivalent to the performance of the best mnemonists at the time.[6] To do so, he developed a number of mnemonics, including recoding digits using his knowledge of race times and dates, and using a pre-learned hierarchical structure to store them. These results are hard to explain with chunking theory, which would have to make the implausible assumption that SF had learned very large chunks so that eighty-four digits could be encoded in STM. They led to the development of skilled memory theory, which will be discussed later.

KEY THEORETICAL WORK IN THE CLASSIC EXPERTISE APPROACH

The classic expertise approach also produced a large number of influential theoretical works. Some studies aimed to refine the idea of chunking, others to replace it with alternative mechanisms.

Computational Model of Chess Memory

MAPP (Memory-Aided Pattern Perceiver) (Simon & Gilmartin, 1973) is an application of EPAM (see earlier) to chess that implements a subset of chunking theory. It assumes that the human information-processing system is limited by constraints applying to all individuals, irrespective of their level of expertise. For example, STM is limited to seven items and it takes 8 seconds for creating a new chunk in LTM. Learning occurs by adding nodes (chunks) to a discrimination network, which can then be accessed through perceptual cues. MAPP simulates the recall task by directing its attention to salient pieces and trying to recognize chunks around them. If a chunk is recognized, a symbol pointing to it is placed in STM. During recall, MAPP uses the symbols in STM to access LTM chunks, and then unpacks the information contained in those chunks.

---

[6] The world record is currently held by Lance Tschirhart (USA) with 456 digits. Unlike SF but like most mnemonists, Tschirhart uses a mnemonic based on recoding numbers into words and creating vivid associations between these words.

In the computer simulations, MAPP could reproduce the performance of a good amateur, but not of a master. Extrapolating from the simulations, Simon and Gilmartin proposed that masters have learnt about 50,000 chunks.

Computational Models of Physics and Engineering Problem Solving

Larkin, McDermott, Simon, and Simon (1980b) built a production system (ABLE) explaining how novices become experts in physics; in particular, it accounts for change in search strategy—from backward search with novices to forward search with experts. ABLE can use declarative statements for solving problems and derive new results, and then reuse these results when solving new problems. Bhaskar and Simon (1977) modeled a more complex and semantically richer domain, thermodynamics as it is taught to students in engineering. A methodological contribution of this project was a computer program that codes verbal protocols semi-automatically.

Skilled Memory Theory

Skilled memory theory (Chase & Ericsson, 1981; Ericsson et al., 1980) was motivated by several anomalies in Chase and Simon's data (e.g., the fact that chunks were relatively small for their master) and also by the difficulty faced by chunking theory in explaining SF's extraordinary memory in the digit span task. Skilled memory theory consists of three principles. First, memory cues are used to link new information with previous long-term memory knowledge; second, retrieval structures allow rapid storage in LTM; and third, intensive practice leads to a decrease of LTM storage and retrieval times. Retrieval structures are assumed to be domain-specific: for example, a structure developed for memorizing briefly presented digits is of little help for memorizing colors. The main application of the theory was to explain mnemonists' memory and, later, mental calculation (Staszewski, 1988). An infelicity of the theory is that its principles belong to different levels of explanation. The first two are about memory mechanisms, while the third principle is a redescription of the empirical data.

High-Level Knowledge Theories

Chase and Simon's theory stressed the importance of acquiring a large number of chunks. Other theories emphasized the way knowledge is organized in LTM and qualitative differences between novices and experts. In addition, the knowledge structures postulated are more complex and abstract than chunks and often refer to schemas, which encode both static and variable information. A good example of this approach is the work of Chi and colleagues on physics (Chi et al., 1981), and some years later, the work of Patel and Groen (1986) on medical reasoning.

Outgrowths of the Classic Approach to Expertise

This section briefly discusses several more recent theories that were directly influenced by the classic approach to expertise.

Long-Term Working Memory

Long-term working memory is a theory of expertise and memory based on skilled memory (Ericsson & Kintsch, 1995). Schemas, retrieval structures, and associations with items and context enable rapid access to LTM, to the point that LTM can be used as an extension of working memory in domains where one has acquired expertise. More specifically, cognitive processing is considered as a succession of stable states representing end products, which can be encoded in LTM when individuals have acquired sufficient expertise in a domain. Beyond digit span and mental calculation, the theory has been applied to domains such as reading and problem solving in chess.

EPAM-IV

EPAM-IV (Richman, Staszewski, & Simon, 1995) is a computer model based on EPAM (see earlier). It simulated the behavior of DD, a mnemonists who was able to increase his memory for briefly presented digits up to 108 digits (Ericsson & Staszewski, 1989), using strategies similar to those employed by SF (see the earlier section "Digit Span Task"). Compared to earlier versions of EPAM, the model makes more detailed assumptions about the components of STM and LTM. However, the critical innovation is the assumption that DD uses retrieval structures to encode information in LTM quickly, in a matter of a few hundred milliseconds. These structures are fairly similar to those postulated by skilled memory theory; in particular, while using them is rapid, learning them takes a long time.

Being stated formally, EPAM-IV makes very precise predictions, which in general are in line with the observed data. Another important contribution of this work is that the concept of retrieval structure was specified formally and with great precision, which is in contrast with skilled and long-term working memory theories.

Template Theory and CHREST

A first aim of CHREST (Chunk Hierarchy and REtrieval STructures; Gobet & Simon, 2000), the computational implementation of template theory (Gobet & Simon, 1996), was to develop a theory seamlessly integrating perception, learning, and memory. A second aim was to show how low-level knowledge structures (i.e., chunks) could lead to the acquisition of high-level, schema-like structures, called templates. Templates make it possible to encode information rapidly into LTM. In this respect, they are comparable to the retrieval structures of skilled memory and EPAM-IV. However, a key difference is that, with CHREST, their learning is assumed to be unconscious and based on the statistical properties of the input rather than on deliberate goals as with mnemonists. CHREST associates every cognitive process to a time parameter, which enables precise predictions about the timing of behavior.

The first application was chess, where CHREST corrected several limitations of MAPP (Simon & Gilmartin, 1973): the new model simulated eye movements in detail, selected the chunks to learn itself (this was done by the programmer with MAPP), and could simulate grandmaster recall (MAPP was stuck at expert level). Later applications covered aspects of expertise in physics, board games such as Awele and Go, and computer programming (Gobet, Lloyd-Kelly, & Lane, 2017). In all these cases, the program learns chunks and templates by processing naturalistic input. The model has also been applied to the acquisition of language (both syntactic structures and vocabulary). An exciting implication is that acquiring one's first language can be considered as acquiring a type of expertise (Gobet et al., 2001; Jones, Gobet, & Pine, 2000). This view is certainly at odds with standard linguistics, which considers that important aspects of language, in particular those related to syntax, are innate.

It is worth noting that the key insights of skilled memory, long-term working memory, and template theory—that LTM storage is rapid with experts—has recently obtained support from brain imaging research (Guida, Gobet, Tardieu, & Nicolas, 2012). If there is sufficient practice in a domain, then cerebral functional reorganization occurs allowing LTM structures to be used as virtual working memory.

Naturalistic Decision Making

The naturalistic decision-making approach studies expert decision making in real-world and high-stake situations such as fire-fighting and hospital intensive care units (Klein, 1998; Zsambok & Klein, 1997). In most of these cases, decisions are made under severe time pressure. The recognition-primed decision model used in this approach emphasizes mechanisms similar to those used by chunking theory: pattern recognition, intuition, selective search, and satisficing in expert decision making. As argued by Gobet (2016), the model might be less applicable for explaining expert decision making in cases where there is sufficient time for carrying out search when solving difficult problems.

Deliberate Practice

As noted earlier, Simon and Chase argued that it took from 10,000 to 50,000 hours of practice to become an expert in chess, which is roughly equivalent to 10 years of study and practice (Simon, 1969). The framework of deliberate practice (Ericsson, Krampe, & Tesch-Römer, 1993) has focused on the characteristics of the practice necessary for becoming an expert. For example, it emphasizes that goal-directed, highly structured activities aimed at correcting errors and weaknesses are essential for developing expertise. In addition, it is crucial to receive rapid and veridical feedback, preferably by a coach or a teacher. The repetitive nature of training means that it is not enjoyable per se. A key prediction of the framework is that performance improvement is directly related to the amount of deliberate practice. Note that, contrary to most of the theories discussed in this chapter, the empirical evidence is mostly correlational and the focus is on the characteristics of practice and not on the cognitive mechanisms underpinning expertise.

Characteristics of the Classic Expertise Approach

The classic approach displays several characteristics that are worth discussing. A first striking feature is its strong unity. There is of course the clear geographical concentration, with very complementary research being carried out at Carnegie Mellon University and its close neighbor the University of Pittsburgh. But there is also a strong theoretical unity, which can be traced back to the works of Adriaan de Groot and Herbert Simon. De Groot provided not only empirical methods (recall task and problem solving with quantitative analysis of verbal protocols) but also key theoretical insights such as the central role of perception in expertise. Simon provided powerful theoretical ideas, such as bounded rationality and satisficing. Finally, the three Chase and Simon papers also contributed to this unity by providing a unifying theory and elegant methods for operationalizing the notion of a chunk.

A second important characteristic is that the approach was very successful in cross-fertilizing with other fields of research. With cognitive psychology, there was a fruitful exchange of methods and theoretical ideas. The classic approach was particularly influential because it cut across traditional boundaries of cognitive psychology (perception, memory, and problem solving). With artificial intelligence (AI) and particularly the then nascent subfield of expert systems, theoretical exchanges focused on the notion of knowledge. How do experts represent knowledge? Can we use what we know about human expert knowledge to build artificial expert systems? Conversely, representations were imported from AI, such as the concept of a schema. Other fields were also influenced by the classic approach. For example, Chi's (1978) study is well known in developmental psychology and, just like the research on physics, Flower and Hayes's (1981) work on reading impacted on education.

Third, the classic approach had a very distinctive theoretical flavor, inspired by Newell and Simon's (1972) framework of information-processing psychology. It put a strong emphasis on mechanisms and the relative invariants of cognition, such as the limited capacity of STM, originally assumed to be seven items but later downscaled to five items (Simon, 1974), 8 seconds to learn a chunk, and 50,000 chunks to become an expert. Computer modeling played an essential role in laying out the theoretical foundations, although it had relatively few followers even within the classic approach. While terms such as expertise, talent, and genius are often seen as ill defined and hard to study scientifically, computer modeling imposed rigorous constraints on theorizing and enabled clear-cut predictions.

Finally, the classic approach is characterized empirically by the detailed analyses of the processes in play rather than statistical elegance and power. Several key papers had very few participants indeed: Chase and Simon (1973a, b) had three participants; Reitman (1976) had two, and Ericsson, Chase, and Faloon (1980) had only one.

Looking at the Future

The aim of this section is to argue that some of the characteristics that made the classic approach so successful could be used with considerable benefits by current and future expertise researchers. While these characteristics are met every so often in contemporary expertise research, it is clear that they do not describe most mainstream approaches (e.g., deliberate practice). In my view, the main weakness of research on expertise today is that the theories that are developed tend to be unspecified and thus cannot make clear-cut predictions and be testable.

I believe that one important reason behind the success of the classic approach is that it was anchored in rigorously specified theories. Two of the Chase and Simon papers (Chase & Simon, 1973a; Simon & Chase, 1973) devoted substantial space to discuss the computer models developed by Simon and colleagues. Computational modeling is currently a rare citizen in expertise research, although there are some exceptions, as noted earlier. My first advice is then to develop more rigorous theories, implemented as computer programs. Recent developments in AI—e.g., deep learning (LeCun, Bengio, & Hinton, 2015)—open exciting prospects, both for theory and applied research.

Today, perhaps an inevitable price for specialization, researchers aim to understand expertise per se. By contrast, a second conspicuous feature of the classic approach is that it aimed to address general questions of cognition, using expertise as a means. For example, it was interested in the micro-structure of cognition and sought to find its relative invariants: the time to create a chunk in memory, the capacity of STM, and the time necessary to become an expert. Some expertise researchers have criticized this goal as misguided, as practice in some cases improves encoding and retrieval times

(Ericsson & Kintsch, 1995). It could also be argued that the classic approach did not investigate the possibility of individual differences due to talent (e.g., creating a new chunk might take 7.9 seconds for one individual and 8.1 seconds for another, a small difference that snowballs with the learning of 50,000 chunks; Chassy & Gobet, 2010). However, the possibility of using strategies for speeding memory processes and the presence of individual differences do not mean that such (approximate) constants do not exist. Identifying these parameters and setting their values, even approximatively, would bring considerable benefits not only for our understanding of expertise but also for education and coaching. Such an endeavor would also make it easier to develop rigorous theories of expertise. Thus, my second advice is to go back to fundamental questions, including setting the value of the parameters of cognition.

Another characteristic of the classic approach is that it was interested in the content of expertise, including the strategies used. Think, for example, of Chi et al.'s (1981) research on physics. Here, the interest was not only on the relationship between inputs and outputs but in the way inputs are processed, using domain-specific strategies and other kinds of knowledge, to produce outputs. This emphasis on knowledge and strategies has been lost recently,[7] possibly because publishing such research is not easy in the current culture that emphasizes sophisticated statistical analyses. My third advice—carrying out detailed analyses of experts' knowledge and behavior—dovetails naturally with the need to express theories as computer models.

My final piece of advice is without any doubt the most controversial. If we are to understand in great detail the mechanisms, knowledge, and strategies enabling expertise, it will be necessary, at least in some cases, to use single-subject designs such as those used by Ericsson, Chase, and Faloon (1980). Such designs do not mean that theory development is impossible. In fact, the research carried out by Ericsson and colleagues is a beautiful example of how detailed analyses can help answer theoretical questions. However, unless the research is purely descriptive, it is essential to have clear-cut hypotheses. Thus, my final advice is to carry more research with $n = 1$, ideally with a large number of experimental tasks and the parallel development of a computational model accounting for the data (Gobet, 2017; Gobet & Ritter, 2000).

Conclusion

This chapter has presented the classic expertise approach, which started with Chase and Simon's three seminal papers in 1973. The classic approach has broadly focused on chunking theory, often to refute it and extend it. It has had considerable impact, for both theory and applied research (e.g., education). Several characteristics of the classic approach—strong emphasis on theory, use of computer models, emphasis on

---

[7] An exception is the research carried out in the naturalistic decision-making tradition.

the micro-structure of cognition and multidisciplinary approach—contributed to its success. The chapter recommends that current and future students of expertise incorporate these characteristics in their research.

References

Adelson, B. (1981). Problem solving and the development of abstract categories in programming languages. *Memory and Cognition* 9, 422–433.

Allard, F., & Starkes, J. L. (1980). Perception in sport: volleyball. *Journal of Sport Psychology* 2, 22–33.

Bahle, J. (1930). *Zur Psychologie des musikalischen Gestaltens. Eine Untersuchung über das Komponieren auf experimenteller und historischer Grundlage.* Leipzig: Akademische Verlagsgesellschaft.

Baylor, G. W., & Simon, H. A. (1966). A chess mating combinations program. In *1966 Spring Joint Computer Conference* (Vol. 28, pp. 431–447). Boston.

Bhaskar, R., & Simon, H. A. (1977). Problem solving in semantically rich domains: An example from engineering thermodynamics. *Cognitive Science* 1, 193–215.

Binet, A. (1894a). La psychologie de la prestidigitation. *Revue des Deux Mondes* 25, 903–922.

Binet, A. (1894b). *Psychologie des grands calculateurs et joueurs d'échecs.* Paris: Hachette. (Re-edited by Slatkine Ressources, Paris, 1981.)

Binet, A. (1966). Mnemonic virtuosity: A study of chess players. *Genetic Psychology Monographs* 74, 127–162. (Translated from the *Revue des Deux Mondes* (1893), 1117, 1826–1859).

Bryan, W. L., & Harter, N. (1899). Studies on the telegraphic language. The acquisition of a hierarchy of habits. *Psychological Review* 6, 345–375.

Charness, N. (1976). Memory for chess positions: Resistance to interference. *Journal of Experimental Psychology: Human Learning and Memory* 2, 641–653.

Charness, N. (1979). Components of skill in bridge. *Canadian Journal of Psychology* 33, 1–16.

Chase, W. G. (1973). *Visual information processing.* New York: Academic Press.

Chase, W. G., & Ericsson, K. A. (1981). Skilled memory. In J. R. Anderson (Ed.), *Cognitive skills and their acquisition* (pp. 141–189). Hillsdale, NJ: Erlbaum.

Chase, W. G., & Simon, H. A. (1973a). The mind's eye in chess. In W. G. Chase (Ed.), *Visual information processing* (pp. 215–281). New York: Academic Press.

Chase, W. G., & Simon, H. A. (1973b). Perception in chess. *Cognitive Psychology* 4, 55–81.

Chassy, P., & Gobet, F. (2010). Speed of expertise acquisition depends upon inherited factors. *Talent Development and Excellence* 2, 17–27.

Chi, M. T. H. (1978). Knowledge structures and memory development. In R. S. Siegler (Ed.), *Children's thinking: What develops?* (pp. 73–96). Hillsdale, NJ: Erlbaum.

Chi, M. T. H., Feltovich, P. J., & Glaser, R. (1981). Categorization and representation of physics problems by experts and novices. *Cognitive Science* 5, 121–152.

De Groot, A. D. (1946). *Het denken van den schaker.* Amsterdam: Noord Hollandsche.

De Groot, A. D. (1965). *Thought and choice in chess* (first Dutch edition in 1946). The Hague: Mouton Publishers.

De Groot, A. D., Gobet, F., & Jongman, R. W. (1996). *Perception and memory in chess: Heuristics of the professional eye.* Assen: Van Gorcum.

Djakow, I. N., Petrowski, N. W., & Rudik, P. A. (1927). *Psychologie des Schachspiels.* Berlin: de Gruyter.

Egan, D. E., & Schwartz, E. J. (1979). Chunking in recall of symbolic drawings. *Memory & Cognition 7*, 149–158.

Ericsson, K. A., Charness, N., Feltovich, P. J., & Hoffman, R. R. (2006). *The Cambridge handbook of expertise and expert performance*. New York: Cambridge University Press.

Ericsson, K. A., Chase, W. G., & Faloon, S. (1980). Acquisition of a memory skill. *Science 208*, 1181–1182.

Ericsson, K. A., & Kintsch, W. (1995). Long-term working memory. *Psychological Review 102*, 211–245.

Ericsson, K. A., Krampe, R. T., & Tesch-Römer, C. (1993). The role of deliberate practice in the acquisition of expert performance. *Psychological Review 100*, 363–406.

Ericsson, K. A., & Staszewski, J. J. (1989). Skilled memory and expertise: Mechanisms of exceptional performance. In D. Klahr & K. Kotovski (Eds), *Complex information processing: The impact of Herbert A. Simon*. Hillsdale, NJ: Erlbaum.

Feigenbaum, E. A., & Simon, H. A. (1962). A theory of the serial position effect. *British Journal of Psychology 53*, 307–320.

Fitts, P. M. (1964). Perceptual-motor skill learning. In A. W. Melton (Ed.), *Categories of human learning* (pp. 243–285). New York: Academic Press.

Flower, L., & Hayes, J. R. (1981). A cognitive process theory of writing. *College Composition and Communication 32*, 365–387.

Frey, P. W., & Adesman, P. (1976). Recall memory for visually presented chess positions. *Memory & Cognition 4*, 541–547.

Galton, F. (1869). *Hereditary genius: An inquiry into its laws and consequences*. London: MacMillan.

Gobet, F. (2016). *Understanding expertise: A multidisciplinary approach*. London: Palgrave.

Gobet, F. (2017). Allen Newell's program of research: The video game test. *Topics in Cognitive Science 9*, 522–532.

Gobet, F., de Voogt, A. J., & Retschitzki, J. (2004). *Moves in mind: The psychology of board games*. Hove, UK: Psychology Press.

Gobet, F., & Lane, P. C. R. (2015). Human problem solving—Beyond Newell et al.'s (1958): Elements of a theory of human problem solving. In M. W. Eysenck & D. Groome (Eds), *Cognitive psychology: Revisiting the classic studies*. Thousand Oaks, CA: Sage.

Gobet, F., Lane, P. C. R., Croker, S., Cheng, P. C. H., Jones, G., Oliver, I., & Pine, J. M. (2001). Chunking mechanisms in human learning. *Trends in Cognitive Sciences 5*, 236–243.

Gobet, F., Lloyd-Kelly, M., & Lane, P. C. R. (2017). Computational models of expertise. In D. Z. Hambrick, G. Campitelli & B. N. Macnamara (Eds), *The science of expertise: behavioral, neural, and genetic approaches to complex skill* (pp. 347–364). New York: Psychology Press.

Gobet, F., & Ritter, F. E. (2000). Individual data analysis and Unified Theories of Cognition: A methodological proposal. In N. Taatgen & J. Aasman (Eds), *Proceedings of the Third International Conference on Cognitive Modelling* (pp. 150–157). Veenendaal, The Netherlands: Universal Press.

Gobet, F., & Simon, H. A. (1996). Templates in chess memory: A mechanism for recalling several boards. *Cognitive Psychology 31*, 1–40.

Gobet, F., & Simon, H. A. (2000). Five seconds or sixty? Presentation time in expert memory. *Cognitive Science 24*, 651–682.

Goldin, S. E. (1978). Effects of orienting tasks on recognition of chess positions. *American Journal of Psychology 91*, 659–671.

Guida, A., Gobet, F., Tardieu, H., & Nicolas, S. (2012). How chunks, long-term working memory and templates offer a cognitive explanation for neuroimaging data on expertise acquisition: A two-stage framework. *Brain and Cognition 79*, 221–244.

Hayes, J. R., & Flower, L. (1980). Identifying the organization of writing processes. In L. W. Gregg & E. R. Steinberg (Eds), *Cognitive processes in writing: An interdisciplinary approach* (pp. 3–30). Hillsdale, NJ: Lawrence Erlbaum.

Jones, G., Gobet, F., & Pine, J. M. (2000). A process model of children's early verb use. In L. R. Gleitman & A. K. Joshi (Eds), *Proceedings of the Twenty Second Annual Meeting of the Cognitive Science Society* (pp. 723–728). Mahwah, NJ: Erlbaum.

Kintsch, W. (1970). *Learning, memory, and conceptual processes.* New York: John Wiley.

Klein, G. A. (1998). *Sources of power: How people make decisions.* Cambridge, MA: MIT Press.

Lane, D. M., & Robertson, L. (1979). The generality of the levels of processing hypothesis: An application to memory for chess positions. *Memory & Cognition 7*, 253–256.

Larkin, J. H., McDermott, J., Simon, D. P., & Simon, H. A. (1980a). Expert and novice performance in solving physics problems. *Science 208*, 1335–1342.

Larkin, J. H., McDermott, J., Simon, D. P., & Simon, H. A. (1980b). Models of competence in solving physics problems. *Cognitive Science 4*, 317–345.

LeCun, Y., Bengio, Y., & Hinton, G. (2015). Deep learning. *Nature, 521*, 436–444.

Lehman, H. C. (1953). *Age and achievements.* Princeton, NJ: Princeton University Press.

McKeithen, K. B., Reitman, J. S., Rueter, H. H., & Hirtle, S. C. (1981). Knowledge organisation and skill differences in computer programmers. *Cognitive Psychology 13*, 307–325.

Meehl, P. E. (1954). *Clinical versus statistical prediction: A theoretical analysis and a review of the evidence.* Minneapolis: University of Minnesota Press.

Newell, A., Shaw, J. C., & Simon, H. A. (1958a). Chess-playing programs and the problem of complexity. *IBM Journal of Research and Development 2*, 320–335.

Newell, A., Shaw, J. C., & Simon, H. A. (1958b). Elements of a theory of human problem solving. *Psychological Review 65*, 151–166.

Newell, A., & Simon, H. A. (1961). Computer simulation of human thinking. *Science 134*, 2011–2017.

Newell, A., & Simon, H. A. (1972). *Human problem solving.* Englewood Cliffs, NJ: Prentice-Hall.

Pascual-Leone, J. A. (1970). A mathematical model for transition in Piaget's developmental stages. *Acta Psychologica 32*, 301–345.

Patel, V. L., & Groen, G. J. (1986). Knowledge based solution strategies in medical reasoning. *Cognitive Science 10*, 91–116.

Reitman, J. S. (1976). Skilled perception in go: Deducing memory structures from inter-response times. *Cognitive Psychology 8*, 336–356.

Richman, H. B., Staszewski, J. J., & Simon, H. A. (1995). Simulation of expert memory with EPAM IV. *Psychological Review 102*, 305–330.

Sala, G., & Gobet, F. (2017). Experts' memory superiority for domain-specific random material generalizes across fields of expertise: A meta-analysis. *Memory & Cognition 45*, 183–193.

Schneiderman, B. (1976). Exploratory experiments in programmer behavior. *International Journal of Computer and Information Sciences 5*, 123–143.

Selz, O. (1922). *Zur Psychologie des produktiven Denkens und des Irrtums.* Bonn: Friedrich Cohen.

Simon, D. P., & Simon, H. A. (1978). Individual differences in solving physics problems. In R. S. Siegler (Ed.), *Children's thinking: What develops?* (pp. 323–348). Hillsdale, NJ: Erlbaum.

Simon, H. A. (1956). Rational choice and the structure of the environment. *Psychological Review 63*, 129–138.
Simon, H. A. (1969). *The sciences of the artificial*. Cambridge, MA: MIT Press.
Simon, H. A. (1974). How big is a chunk? *Science 183*, 482–488.
Simon, H. A. (1980). Problem solving and education. In D. Tuma & F. Reif (Eds), *Problem solving and education* (pp. 81–96). Hillsdale, NJ: Erlbaum.
Simon, H. A. (1990). Invariants of human behavior. *Annual Review of Psychology 41*, 1–20.
Simon, H. A., & Barenfeld, M. (1969). Information processing analysis of perceptual processes in problem solving. *Psychological Review 7*, 473–483.
Simon, H. A., & Chase, W. G. (1973). Skill in chess. *American Scientist 61*, 393–403.
Simon, H. A., & Gilmartin, K. J. (1973). A simulation of memory for chess positions. *Cognitive Psychology 5*, 29–46.
Sloboda, J. A. (1976a). Effect of item position on likelihood of identification by inference in prose reading and music reading. *Canadian Journal of Psychology 30*, 228–237.
Sloboda, J. A. (1976b). Visual perception of musical notation: Registering pitch symbols in memory. *Quarterly Journal of Experimental Psychology 28*, 1–16.
Sloboda, J. A. (1978). Perception of contour in music reading. *Perception 7*, 323–331.
Staszewski, J. J. (1988). Skilled memory and expert mental calculation. In M. T. H. Chi, R. Glaser, & M. J. Farr (Eds), *The nature of expertise* (pp. 71–128). Hillsdale, NJ: Erlbaum.
Zsambok, C. E., & Klein, G. A. (Eds). (1997). *Naturalistic decision making*. Mahwah, NJ: Erlbaum.

CHAPTER 3

DOMAIN-GENERAL MODELS OF EXPERTISE
The Role of Cognitive Ability

DAVID Z. HAMBRICK,
ALEXANDER P. BURGOYNE,
AND FREDERICK L. OSWALD

Domain-General Models of Expertise: The Role of Cognitive Ability

WHY do some people reach higher levels of expertise in complex real-world tasks than other people? There is no doubt that domain-specific knowledge and skills contribute substantially to individual differences in expertise, whether it be in vocational or avocational pursuits (see Ward, Belling, Petushek, & Ehrlinger, 2017, for a review). Here, while not denying the major importance of domain-specific factors, we consider the contribution of *domain-general cognitive ability factors*, reflecting the efficiency and effectiveness of basic mental processes.

Scope and Organization

In everyday life, people often rely on reputation to identify individuals with expertise—physicians, carpenters, auto mechanics, and so on. However, reputation does not ensure a high level of expertise (Ericsson, 2006). As a scientific concept, expertise is better thought of as a person's objective level of performance in a domain, as quantified by domain-relevant tasks (Ericsson & Smith, 1991) or proxy measures (e.g., performance-based rankings). For some domains, a single type of task may be sufficient to measure expertise. For example, given that playing good chess obviously depends on making good chess moves, chess expertise can be measured with *move-choice* tasks (de Groot, 1965/1978). For other domains, no single type of task captures expertise. For example, musical expertise comprises playing music from memory, sight-reading, and improvising, among other activities. Some musicians may be strong in all these activities; others may be strong in some but weak in others. Similarly, some auto mechanics may specialize in repairing diesel engines, others in transmissions, and still others in body repair. In short, expertise may be multidimensional.

Here, we review evidence for the role of cognitive ability in acquiring expertise. Along with limited space, there are two major reasons for this restricted focus. First, much of the controversy in contemporary research on expertise revolves around the question of whether, and to what extent, cognitive ability plays an important role in acquiring expertise (see, e.g., Detterman, 2014). Second, as industrial–organizational psychologists have demonstrated, measures of cognitive ability (along with other measures) are useful in organizational settings for selecting qualified job applicants, because they are consistently and positively correlated with job performance (Schmidt & Hunter, 1998, 2004). Similarly, scores on standardized cognitive tests such as the Graduate Record Examination (GRE), the Law School Admission Test (LSAT), and the Graduate Management Admission Test (GMAT) are useful and valid predictors of success in advanced academic studies (Kuncel & Hezlett, 2007).

Table 3.1 lists the cognitive ability constructs that we consider, with a definition of each construct and examples of assessments. Though often treated as if they are empirically and conceptually distinct, measures of these constructs correlate positively, and sometimes nearly 1.0 after correcting for measurement error variance (e.g., Engelhardt et al., 2016; McCabe, Roediger, McDaniel, Balota, & Hambrick, 2010). This implies that common mechanisms underlie individual differences in these constructs, which could include acquired factors such as general problem-solving strategies, neural factors such as the functional connectivity of different brain regions, and genetic factors (see Haier, 2016). Any (or all) of these factors could contribute, directly or indirectly, to associations between cognitive ability factors and expertise. Cognitive ability constructs are also sometimes described as being *innate*, but heritability (i.e., estimated genetic contribution) of any human characteristic is always less than 100 percent (Turkheimer, 2000), leaving room for a contribution by environmental factors. At the population level, heritability is typically around 50 percent for measures of cognitive ability, indicating roughly equal contributions of genetic and environmental factors to individual differences (Knopik, Neiderhiser, DeFries, & Plomin, 2016).

Table 3.1 Domain-general cognitive ability factors, with representative definitions and examples of assessments

Construct	Definition/tests
Intelligence	Intelligence is a very general mental capability that, among other things, involves the ability to reason, plan, solve problems, think abstractly, comprehend complex ideas, learn quickly, and learn from experience. (Gottfredson, 1997, p. 13) - Wechsler Adult Intelligence Scale (full-scale IQ) - Raven's Progressive Matrices (fluid intelligence, or Gf) - Air Force Officer Qualifying Test (AFOT)
Executive functioning	Executive function can be thought of as the set of abilities required to effortfully guide behavior toward a goal, especially in nonroutine situations. (Banich, 2009, p. 89) - Wisconsin Card Sorting task - Tower of Hanoi - Trailmaking
Working memory capacity	[Working memory capacity refers to] the attentional processes that allow for goal-directed behavior by maintaining relevant information in an active, easily accessible state outside of conscious focus, or to retrieve that information from inactive memory, under conditions of interference, distraction, or conflict. (Kane et al., 2007, p. 23) - Operation span - n-back - Backward digit span
Attentional control	Attention control refers to the ability to protect items that are actively being maintained in working memory, to effectively select target representations for active maintenance, and to filter out irrelevant distractors and prevent them from gaining access to working memory. (Unsworth, Fukuda, Awh, & Vogel, 2015, p. 864) - Attention Network Task (ANT) - Stroop task - Flanker task
Speed of processing	Processing speed refers to the ability to quickly and efficiently carry out mental operations. (Tucker-Drob, 2011, p. 333) - Digit-symbol substitution - Letter/pattern comparison - Choice reaction time

Review of Evidence for Role of Cognitive Ability in Expertise

Classical theories of skill acquisition (e.g., Fitts & Posner, 1967) posit that domain-general processes impact performance early in training, after which procedural knowledge becomes the major determinant. Consistent with this assumption, there is ample evidence that cognitive ability predicts initial acquisition of knowledge/skill in complex domains. For example, measures of cognitive ability from test batteries such as the Armed Services Vocational Aptitude Battery (ASVAB) positively predict job training performance, with validity coefficients averaging around 0.50 (Schmidt & Hunter, 2004). It is less clear whether cognitive ability remains a valid predictor of performance differences after extensive training. This question is not only of theoretical interest to expertise researchers (e.g., the *circumvention-of-limits hypothesis;* Hambrick & Meinz, 2011), but one of applied interest: If a measure significantly predicts performance in a task, especially beyond the beginner level, then that measure might be used to help make decisions such as whom to select for a costly training program.

Next, we review evidence relevant to this question. We performed literature searches in Google Scholar and PsycINFO, using a wide range of search terms (e.g., "expertise" and "cognitive ability" with "sports," "chess," and "aviation"). We searched approximately 1,300 documents, identifying relevant studies in two primary literatures: the literature on expertise in domains such as music, chess, and sports; and the literature on job performance. Our review focuses on studies that tested cognitive ability–performance relations across different levels of expertise, or at least in non-beginners. (We excluded studies that measured specific aptitudes, such as music aptitude and mechanical aptitude.) The specific question we set out to address is whether expertise mitigates the effect of cognitive ability on domain-relevant performance. Throughout the chapter, we note correlations between domain-specific factors and domain-relevant performance for comparative purposes.

Games

There is evidence that cognitive ability predicts acquisition of chess skill at the beginner level (e.g., de Bruin, Kok, Leppink, & Camp, 2014), but it is unclear what role it plays at higher levels of expertise. Evidence is mixed. For example, in two studies, Unterrainer and colleagues (Unterrainer, Kaller, Halsband, & Rahm, 2006; Unterrainer, Kaller, Leonhart, & Rahm, 2011) found near-zero correlations between IQ measures and chess rating in small samples of chess players ($N = 25$ and 26, respectively) with intermediate-level average chess ratings, whereas Grabner, Stern, and Neubauer (2007) found a correlation of 0.35 between IQ and chess rating using a larger sample ($N = 90$) with a slightly higher average rating. Even the latter finding is tentative because the confidence interval around a correlation of 0.35 with a sample of 90 is quite wide, ranging from 0.15 to 0.52.

To try to make sense of the conflicting evidence, Burgoyne et al. (2016) performed a meta-analysis of the relationship between cognitive ability and chess expertise. Across 19 studies, four cognitive abilities were measured: fluid intelligence, crystallized intelligence, short-term/working memory, and processing speed. The meta-analytic average of the correlations was 0.22 ($p < 0.001$). (Correlations between chess rating and domain-specific factors are typically much larger (e.g., Pfau & Murphy, 1988, $r = 0.68$).)

Burgoyne et al. also found that the correlation between fluid intelligence and expertise was stronger for less skilled (unranked) chess players than for more skilled (ranked) players (0.33 vs 0.10; see Burgoyne et al., 2018, for a correction to the originally reported values). However, it is important to note that expertise was highly confounded both with age (i.e., nearly all ranked chess players were adults, nearly all unranked chess players were youths) and with type of skill measure (i.e., Elo ratings for ranked players, chess test scores for unranked players).

In another meta-analysis, Sala et al. (2017) found that chess players are, on average, significantly higher in measured cognitive ability than non-chess players. As most of the chess samples included relatively highly skilled players, this could be because people high in cognitive ability are more likely to enjoy success in chess than those lower in cognitive ability, and are thus less likely to quit the game (i.e., performance effects). Alternatively, it could be that playing chess enhances cognitive ability (i.e., training effects) or because higher-ability individuals are more likely to take up chess than lower-ability individuals (i.e., selection effects).

Summarized, evidence is inconclusive on whether the importance of cognitive ability declines with chess expertise. The same is true for other games. In neuroimaging studies, Lee et al. (2010, $N = 16$) and Jung et al. (2013, $N = 17$) reported IQ data on small samples of elite Baduk (Korean for *Go*) players. Full-scale IQ was lower for the Baduk players than for a control group by 8 points in Lee et al. ($M = 93.2$ vs $101.2, p = 0.052$) and 7.7 points in Jung et al. ($M = 93.1$ vs $100.8, p = 0.06$). The fact that the Baduk group in each study had a *lower* average IQ than the control group is somewhat puzzling and may partly reflect the fact that the Baduk group had less education on average than the control group (by 1.3 years in Lee et al., $p < 0.05$; and by 1.1 years in Jung et al., $p = 0.19$).

A much larger study of Go was carried out by Masunaga and Horn (2001). Participants ($N = 263$) representing wide ranges of Go expertise completed tests of both domain-general and domain-specific factors. The domain-general battery included standard tests of fluid reasoning, short-term memory, and perceptual speed; the domain-specific battery included *Go-embedded* tests designed to measure the same abilities but with Go-specific content. The Go reasoning test was modeled after move-choice tasks in chess (de Groot, 1965/1978), and can be considered a measure of Go skill. On average, the domain-general measures correlated 0.18 with Go move-choice. The highest correlations were for fluid intelligence (avg. $r = 0.27$); group average r values (obtained from Takagi, 1997) were as follows: beginner (avg. $r = 0.21, p = 0.001, n = 62$), intermediate (avg. $r = 0.33, p < 0.001, n = 89$), expert (avg. $r = 0.27, p < 0.001, n = 92$), and professional (avg. $r = 0.18, p = 0.14, n = 20$). These correlations are not significantly different from each other (z statistics < 1). The average correlations between the fluid intelligence measures and Go rank were non-significant: beginner (avg. $r = -0.03$), intermediate (avg. $r = -0.06$), expert (avg. $r = 0.03$), and professional (avg. $r = -0.26$). It is somewhat surprising that fluid intelligence correlated with move-choice performance but not with Go rank, given the high correlation between the latter measures ($r = 0.71$) and that move-choice must be critical for success in Go tournaments. It could be that the move-choice task was somewhat artificial in that it presented the player with novel positions, whereas in actual Go games a skilled player can steer a game toward familiar territory and thus encounter more familiar positions. The average correlation of the domain-specific measures with Go move-choice was 0.46 and with Go rank was 0.47.

Word games have also been used to investigate the relationship between cognitive ability and expertise. Tuffiash, Roring, and Ericsson (2007) compared groups of elite, average, and novice Scrabble players on tests of various domain-specific and domain-general cognitive abilities. There were significant group differences (favoring higher expertise) in the domain-relevant tasks (e.g., anagramming; medium-to-large effect sizes), but not in domain-general perceptual speed (i.e., digit-symbol substitution). However, the rated players (average and elite groups) outperformed the novices on tests of vocabulary and reading comprehension ($ds > 2$). More recently, Toma, Halpern, and Berger (2014) found that Scrabble and crossword puzzle experts tended to outperform the control subjects on two tests of working memory capacity (avg. $d = 1.23$). As in chess, these skill group differences could reflect performance effects, training effects, and/or selection effects.

There have been a few studies of poker expertise. In a study of undergraduate students described as being *familiar with* Texas Hold 'em poker, Leonard and Williams (2015) found that scores on several subtests from the Stanford–Binet Intelligence Scales correlated non-significantly with performance on a test of poker skills. However, in a sample of 155 undergraduates representing a wider range of Texas Hold 'em experience, Meinz et al. (2012) found that working memory capacity explained a significant amount of variance (avg. $R^2 = 0.071$) in measures of Hold 'em component skills (e.g., hand evaluation), above and beyond poker knowledge (avg. $R^2 = 0.358$). Moreover, there was no evidence for Poker Knowledge × Working Memory Capacity interactions, indicating that effects of working memory capacity on performance were similar across levels of poker knowledge.

Finally, Ceci and Liker (1986) found that groups of *nonexperts* ($n = 16$) and *experts* ($n = 14$) in horserace handicapping were not only nearly identical in average IQ, but both near the population mean of 100 ($Ms = 99.3$ and 100.8, respectively). However, in a re-analysis, Detterman and Spry (1988) found that the correlation between IQ and a key measure of success (correct top horse) was positive in the expert group ($r = 0.35$, or 0.59 after correction for unreliability) but negative in the novice group ($r = -0.25$, or -0.42 after correction for unreliability), casting some doubt on the argument that IQ is unrelated to success in horserace handicapping. That said, these sample sizes were very small, and the result would obviously need to be replicated in a larger sample.

Music

It is also unclear what role cognitive ability plays in music expertise beyond the beginner level. Ruthsatz, Detterman, Griscom, and Cirullo (2008) found that scores on a test of fluid intelligence (Raven's Progressive Matrices) correlated positively and significantly with musical achievement in high school band members ($r = 0.25$, $n = 178$), but not in university music majors ($r = 0.24$, $n = 19$) and music institute students ($r = 0.12$, $n = 64$)—although statistical power obviously differed across the samples. Moreover, the correlations did not differ between the lower- and higher-skill groups (tests of differences in rs, z statistics < 1). Correlations with estimated amount of *deliberate practice* (Ericsson, Krampe, & Tesch-Römer, 1993) in the high school, university, and music institute samples were 0.34, 0.54, and 0.31, respectively (all significant).

Meinz and Hambrick (2010) had 57 pianists provide estimates of deliberate practice and perform tests of both working memory capacity and sight-reading. Deliberate practice accounted for 45 percent of the variance in sight-reading performance; working memory capacity accounted for an additional 7.4 percent. (The correlation between deliberate practice and working memory capacity was near zero.) Moreover, the Deliberate Practice × Working Memory Capacity interaction was non-significant, indicating that the effect of working memory capacity on performance was similar across levels of deliberate practice. By contrast, perceptual speed did not contribute significantly to the prediction of sight-reading performance.

Using a sample of 52 pianists with a more uniform level of skill (piano majors at a music university), Kopiez and Lee (2008) found that although correlations between sight-reading performance and fluid reasoning ($r = 0.12$) and reaction time (avg. $r = -0.07$) were non-significant, there was a significant correlation for a measure of perceptual speed ($r = -0.44$; faster processing, higher performance). The correlation between working memory and sight-reading performance did not reach significance ($r = 0.26$, $p = 0.062$). Correlations between measures of domain-relevant motoric speed (*trilling*) and sight-reading performance averaged 0.50; the correlation between deliberate practice and sight-reading performance was 0.50.

Other studies have compared musicians of varying levels of skill on measures of cognitive ability, as well as musicians to non-musicians. Schellenberg and colleagues have found that musically trained individuals tend to score higher in full-scale IQ than non-musically trained individuals (see Schellenberg & Weiss, 2013, for a review). As with chess, this difference could reflect performance effects, training effects, and/or selection effects.

Sports

Evidence for the role of cognitive ability in sports expertise is inconsistent, as well. For example, Lyons, Hoffman, and Michel (2009) found that scores on the Wonderlic IQ test correlated near zero ($r = -0.04$) with future NFL performance in a large sample of elite college football players (total $N = 762$; see also Berri & Simmons, 2011), whereas Vestberg, Gustafson, Maurex, Ingvar, and Petrovic (2012) found that a measure of executive functioning (design fluency from the D-KEFs) significantly predicted goals scored in elite Swedish soccer players ($r = 0.54$, $N = 25$), albeit in a much smaller sample.

In a meta-analysis of 42 studies, Mann, Williams, Ward, and Janelle (2007) compared nonexpert and expert athletes on performance measures from sports-specific perceptual-cognitive tasks (e.g., occlusion paradigms). Across measures, there was a statistically significant advantage for experts ($ds = 0.23$ to 0.35). Given evidence for the importance of training in acquiring skill in sports (e.g., Ward, Hodges, Starkes, & Williams, 2007), these differences likely reflect domain-specific factors, but they could also reflect domain-general factors as well (Ward et al., 2017).

In a subsequent meta-analysis of 20 studies, Voss, Kramer, Basak, Prakash, and Roberts (2010) found a significant advantage for athletes over non-athletes on processing speed (Hedges' $g = 0.67$) and *varied* attention tasks (Hedges' $g = 0.53$) but not attentional cueing (Hedges' $g = 0.17$). (Hedges' g is similar to Cohen's d.) These results lend some support to the possibility that domain-general factors contribute to sports expertise (i.e., performance effects), but as before could also reflect selection effects and/or training effects (Ward et al., 2017).

Science

The relationship between cognitive ability and scientific expertise has also been of interest to psychologists. Early studies of this relationship yielded mixed evidence. Bayer and Folger (1966) reported a correlation of -0.05 between IQ and number of citations (a proxy for scientific expertise) in a sample of 224 biochemists, and Folger, Astin, and Bayer (1970) found correlations ranging from 0.04 to 0.10 between cognitive ability in high school and number of citations in a sample of 6,300 PhDs. However, Creager and Harmon (1966) found that scores on the GRE predicted citation counts 8–12 years later (median $r = 0.28$; cited in Clark & Centra, 1982) in NSF predoctoral fellowship applicants (see also Kaufman, 1972).

More convincing results come from a meta-analysis of 6,589 correlations from 1,753 independent samples by Kuncel, Hezlett, and Ones (2001). After applying psychometric corrections for statistical artifacts such as range restriction and measurement unreliability in the criterion measures, Kuncel et al. found that estimated validity coefficients (ρs) in the population for the General GRE test were positive and significant not only for first-year GPA (avg. $\rho = 0.36$; avg. $r = 0.24$) and overall GPA (avg. $\rho = 0.34$; avg. $r = 0.23$), but also for publication citation counts (avg. $\rho = 0.20$; avg. $r = 0.15$), and were positive for research productivity (avg. $\rho = 0.10$; avg. $r = 0.08$). Validities for the Subject GRE test (reflecting domain-specific knowledge) were higher for all outcomes, including publication citation counts ($\rho = 0.24$; $r = 0.20$) and research productivity ($\rho = 0.21$; $r = 0.17$).

This evidence corroborates the results of the Study of Mathematically Precocious Youth (SMPY). As part of a planned 50-year study, the Scholastic Aptitude Test (now just called the SAT) was administered to a large national sample of gifted youth by age 13, and those scoring in the top 1 percent were tracked into adulthood ($N > 2,300$). Analyses have since demonstrated that—even within this highly restricted range of ability—SAT scores are positively predictive of success in scientific fields. For example, Lubinski (2009) found that, compared with individuals in the 99.1 percentile, those in the 99.9 percentile were about 5 times more likely to have published in a STEM journal and about 3 times more likely to have been awarded a patent.

Thus, there is evidence that cognitive ability predicts general measures of scientific expertise. There is, however, some evidence that cognitive ability may become less important in specific scientific tasks. Hambrick et al. (2012) had a sample of 67 participants representing a wide range of knowledge and experience in geological fields perform a highly realistic *bedrock mapping* task in which the goal was to create a *field map* representing the geological structure of an area based on observable features (e.g., rock outcrops). There was a significant Geological Knowledge × Visuospatial Ability interaction, such that a composite measure of visuospatial ability positively predicted map accuracy, but only in those with lower levels of geological knowledge.

Surgery/Medicine

There is a growing literature on the role of cognitive ability in surgical expertise, but the results are no clearer than in other domains. In a study of 120 surgical residents (Schueneman, Pickleman, Hesslein, & Freeark, 1984), 4 of 5 measures of visuospatial ability correlated significantly with surgical performance (avg. $r = 0.28$), as evaluated by attending surgeons. Year of residency correlated 0.60 with surgical performance. Gibbons, Baker, and Skinner (1986) found that scores on a hidden figures test correlated significantly with surgical performance in small samples of surgical residents ($rs = 0.55$ and 0.60, $Ns = 42$ and 16), but Deary, Graham, and Maran (1992) found no significant positive correlations between expert ratings of surgical ability and intelligence test scores in trainee surgeons ($N = 22$).

Several studies have compared ability–performance correlations across different levels of surgical expertise. Wanzel et al. (2003) found that scores on two tests of "high-level" visuospatial ability (mental rotation and surface development) correlated significantly with expert ratings of surgical performance in dental students (*novices, n = 27*, avg. $r = 0.56$), but not in surgical residents (*intermediates, n = 12*) or staff surgeons (*experts, n = 8*). The correlations for the latter groups were not reported, but given the extremely small sample sizes here, they would not be significantly different from the novice correlation even if those correlations were assumed to be zero. Comparing groups of surgeons on a simulated videoscopic task, Keehner et al. (2004) found that a measure of visuospatial ability correlated significantly with mean skill rating in a low experience group ($r = 0.39$, $n = 48$), but not in a high experience group ($r = 0.02$, $n = 45$). But, again, the correlations are not significantly different ($z = 1.83$, $p = 0.067$).

Gallagher, Cowie, Crothers, Jordan-Black, and Satava (2003) found that scores on a test of visuospatial ability in which participants recover three-dimensional structures from two-dimensional images correlated significantly and similarly with performance on a laparoscopic laboratory cutting task in two samples of novices ($rs = 0.50$ and 0.50, $ns = 48$ and 32) and in experienced surgeons ($r = 0.54$, $n = 18$). These correlations also do not differ across skill level. Enochsson et al. (2006) compared 18 resident and 11 expert surgeons in a simulated gastroscopy task, and found that correlations between scores on a test of visuospatial ability (card rotation test) and various metrics of performance in these very small samples were generally non-significant for both groups (avg. $r = 0.06$).

Murdoch, Bainbridge, Fisher, and Webster (1994) found that both manual dexterity and visuospatial ability correlated significantly with medical students' performance on microsurgical tasks ($rs = -0.54$ and 0.36, respectively). And in a sample of surgeons ($N = 94$), Risucci, Geiss, Gellman, Pinard, and Rosser (2001) found that measures of visuospatial ability correlated moderately (and 10/12 significantly) with performance on four surgical tasks (avg. $r = -0.30$; higher ability, faster performance); a measure of domain-specific experience correlated significantly with two of the performance measures ($rs = 0.35$ and 0.29), as did a measure of domain-specific knowledge (post-test examination; $rs = 0.30$ and 0.39). Groenier, Schraagen, Miedema, and Broeders (2014) examined the validity of tests of cognitive ability for predicting performance in a laparoscopic training simulator in medical students ($N = 53$) over 2 months. In univariate analyses, visuospatial ability, spatial memory, perceptual speed, and reasoning ability significantly predicted one performance measure (motion efficiency), whereas visuospatial ability and reasoning ability predicted another performance measure (duration). By contrast, in multivariate analyses, which controlled for correlations among the predictor variables, only one of the preceding effects was significant. The finding that univariate effects became non-significant in the multivariate analyses suggests that variance common to the ability measures (a g factor) may have been predictive of surgical performance.

More recently, Louridas and colleagues performed a meta-analysis of 52 studies on the relationship between various measures of cognitive ability and performance in laparoscopic, open, and endoscopic surgery (Louridas, Szasz, de Montbrun, Harris, & Grantcharov, 2016). Only a few cognitive ability measures positively predicted surgical performance across multiple studies, among them the mental rotation test, a pictorial surface orientation test, and the grooved pegboard test. Louridas et al. concluded that "no single test has been reported to reliably predict technical performance across the range of techniques and skills required of surgical trainees" (p. 689).

One other study fits in this category. In a sample ($N = 428$) that included professionals in exercise science-related jobs (e.g., physicians, trainers) as well as participants from the general population, Petushek, Cokely, Ward, and Myer (2015) found that two measures of cognitive ability had non-significant effects on performance in a task designed to assess risk of injury to the anterior cruciate ligament (ACL). By contrast, domain-specific factors (i.e., ACL knowledge and use of particular visual cues) were positive and statistically significant predictors of performance ($r = 0.59$ for ACL knowledge; Petushek, 2014).

Aviation

Several studies have tested for cognitive ability–performance correlations in aviation. In a sample of 86 pilots representing a wide range of experience and skill, along with 96 non-pilots, Morrow, Menard, Stine-Morrow, Teller, and Bryant (2001) found that a cognitive ability composite (working memory, perceptual speed, and visuospatial ability) positively predicted aviation-related performance (i.e., a composite reflecting accuracy in recalling and understanding air traffic control, ATC, commands), accounting for 29 percent of the variance. An expertise composite (ATC knowledge and flight hours) accounted for an additional 37 percent of the variance, but the Expertise × Cognitive Ability interaction was non-significant for all performance measures, indicating that the effect of cognitive ability on performance was similar across levels of expertise.

In a similar study of pilots ($N = 91$), Morrow et al. (2003) found that a cognitive ability composite accounted for an average of 22 percent of the variance in ATC tasks; an expertise composite accounted for an additional 28 percent of the variance, on average. (Expertise × Cognitive Ability interactions are not reported for this study.) Consistent with these findings, in a study of 97 licensed pilots with a wide range of flight experience, Taylor, O'Hara, Mumenthaler, Rosen, and Yesavage (2005) found that performance in an aviation communication task correlated significantly with working memory ($r = 0.76$), processing speed ($r = 0.33$), and interference control ($r = 0.43$), but interactions of expertise (flight rating) with these factors were all non-significant.

Using a sample with 25 novice and 25 expert pilots, Sohn and Doane (2003) found that working memory capacity predicted success in an aviation situational awareness task, but only in pilots who scored low on an aviation-specific test measuring skilled access to long-term memory (i.e., long-term working memory; Ericsson & Kintsch, 1995), as evidenced by a significant Long-Term Working Memory × Working Memory Capacity interaction. In a similar study, Sohn and Doane (2004) found that two measures of working memory capacity (spatial span and verbal span) correlated more strongly with situational awareness in 25 novice pilots ($rs = 0.52, p < 0.01$, and 0.30, respectively) than in 27 expert pilots ($rs = 0.10$ and 0.10, respectively). However, these correlations are not significantly different from each other across skill groups (z statistics < 1.7). Sohn and Doane (2004) did not test the Long-Term Working Memory × Working Memory Capacity interaction using the full sample (as in their earlier study), but instead tested it separately in each skill group, finding significance only in the expert group.

Finally, in a small sample of private pilots ($N = 24$), Causse, Dehais, and Pastor (2011) examined the relationship of broad cognitive abilities (reasoning and processing speed) and executive functions (working memory updating, set-shifting, and inhibition) to performance during a 45-minute flight simulator task. Reasoning ability correlated significantly with flight-path deviations (rs = -0.63); the other correlations were non-significant.

Job Performance

Measures of general cognitive ability positively predict job performance (Schmidt & Hunter, 2004), but do they remain valid predictors after extended job experience? This question has long been of interest to industrial–organizational psychologists. Using laboratory perceptual–motor tasks, Fleishman and colleagues demonstrated that general ability factors become less important with training, whereas task-specific factors become more important (e.g., Fleishman & Rich, 1963; see Hulin, Henry, & Noon, 1990, for other examples). However, the general finding from large-scale studies of actual work performance (as opposed to laboratory tasks) is that cognitive ability remains a significant predictor of job performance even after extensive job experience.

McDaniel (1986) investigated the impact of job experience on the validity of general cognitive ability using the General Aptitude Test Battery (GATB) database.[1] Compiled by the U.S. Employment Service in the 1970s, this database includes information on a large sample of civilian workers, including measures of job performance (i.e., supervisor ratings), job experience, and cognitive ability. McDaniel computed correlations between an "intelligence" score from the GATB (based on visuospatial, vocabulary, and arithmetic reasoning scores) and job performance across different levels of job experience. As shown in Figure 3.1, the correlations decrease somewhat as a function of job experience, but are still significant at the maximum amount of job experience (10+ years, $r = 0.20$, corrected $r = 0.29$).

Again using the GATB database, Farrell and McDaniel (2001) extended Ackerman's (1988) model of skill acquisition to job performance. Briefly, Ackerman hypothesized that involvement of different cognitive abilities in skill acquisition is moderated by the consistency of the task: When the demands of the task are consistent, meaning that the stimuli, rules, and sequences of action remain constant, automaticity can develop and the influence of general cognitive ability (reflecting attentional resources) on performance decreases with training. Meanwhile, the influence of perceptual speed increases and later decreases (i.e., an inverted U function), and the influence of psychomotor speed increases. To test this model, Farrell and McDaniel classified jobs as consistent or inconsistent using two different definitions of consistency: the complexity of the job (low complexity = consistent, high complexity = inconsistent) and tolerance for repetition required to perform the job (high tolerance for repetition = consistent, low tolerance for repetition = inconsistent). They then computed correlations between two cognitive composites (along with psychomotor speed) from the GATB (intelligence and perceptual speed) and job performance for different levels of job experience.

---

[1] We thank Michael McDaniel for sending us a copy of this study.

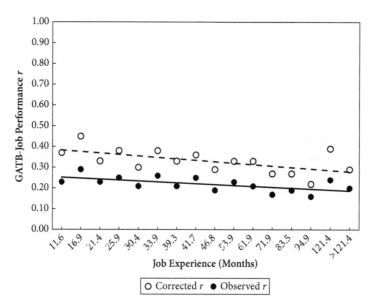

FIGURE 3.1 Correlations between GATB intelligence scores and job performance ratings as a function of job experience (total N = 16,058; across intervals, ns = 1,000 to 1,050, except for >121.4 months, n = 879). Solid circles represent observed (raw) correlations; open circles represent correlations after correction. Data from McDaniel, M. A., "The evaluation of a causal model of job performance: The interrelationships of general mental ability, job experience, and job performance," Tables 1 and 19, PhD thesis, George Washington University, Washington, D. C., 1986.

Support for Ackerman's model was mixed. For example, the intelligence correlations decreased as a function of job experience for low complexity jobs, but *increased* slightly for high-tolerance-for-repetition jobs. For the present discussion, the more important finding is simply that the cognitive ability factors significantly predicted job performance even at the maximum level of job experience: intelligence (avg. r = 0.25; avg. corrected r = 0.34) and perceptual speed (avg. r = 0.15; avg. corrected r = 0.20).

Studies of military personnel provide additional evidence that cognitive ability remains a significant predictor of job performance beyond initial training. Schmidt, Hunter, Outerbridge, and Goff (1988) tested for effects of cognitive ability and job experience on job performance in a sample of 1,474 soldiers in four jobs (armor repairman, armor crewman, supply specialist, and cook). Job performance was measured using work samples and supervisor ratings; cognitive ability was measured using the Armed Forces Qualification Test (AFQT) score from the ASVAB, which is based on Arithmetic Reasoning, Mathematics Knowledge, Paragraph Comprehension, and Word Knowledge subtests. (Job knowledge was also treated as a measure of job performance, though we think of it as a *predictor* of job performance.) Up to 5 years of job experience, correlations between AFQT scores and job performance were nearly constant. Across this span, correlations ranged from 0.38 to 0.42 for work samples and from 0.18 to 0.36 for supervisor ratings. Beyond 5 years of job experience (i.e., 61+ months), there was apparent convergence of ability groups for most measures, indicating a drop in validity beginning at 5 years. However, the average amount of job experience was actually much higher than 5 years in this group—from 9.5 years to 13 years, depending on the job. Moreover, only 1 of 12 AFQT × job experience interactions (work sample performance for armor crewman) was statistically significant, and it was not clearly interpretable as supporting convergence of the ability groups. Schmidt et al. concluded that "[a]t least out to 5 years, the validity of general mental-ability measures appears neither to decrease ... nor to increase .... Instead, the validity remains relatively constant" (p. 56).

Wigdor and Green (1991) reported results of the Joint-Service Job Performance Measurement/Enlistment (JPM) Standards Project, a large study initiated in 1980 by the U.S. Department of Defense to develop measures of military job performance. Wigdor and Green reported that, across 23 jobs ($N = 7,093$ military personnel), the median correlation between AFQT scores and *hands-on* job performance (HOJP) was 0.26 (0.38 after correction for range restriction). They also reported mean hands-on performance for four AFQT categories (representing different levels of cognitive ability) as a function of job experience. As shown in Figure 3.2, mean differences among AFQT categories were largest at 0–12 months (about 10 points, or 1 *SD*), but still sizeable thereafter (5–6 points, or 0.50–0.60 *SD*). Wigdor and Green concluded that "the level of performance is positively related to AFQT score category at each of the four levels of job experience" (p. 163) and noted that "the lowest aptitude group never reaches the initial performance level of the highest aptitude group" (p. 163).

To further investigate cognitive ability–job performance relations, we obtained the JPM dataset.[2] The final dataset included 31 jobs and a total sample size of 10,088 military personnel. We performed three new analyses. First, we computed the AFQT–HOJP correlation across the job experience intervals used by Wigdor and Green (1991). As shown in Figure 3.3A, the correlations are as follows: 0–12 months ($r = 0.34$, $p < 0.001$, $n = 747$), 13–24 months ($r = 0.21$, $p < 0.001$, $n = 5,234$), 25–36 months ($r = 0.19$, $p < 0.001$, $n = 2,338$), and 37+ months ($r = 0.22$, $p < 0.001$, $n = 1,769$).[3] There is a statistically significant drop in the correlation from the first year of service to the second ($z = 3.60$, $p < 0.001$), but AFQT is still a statistically significant predictor of individual differences in HOJP after the first year of service. Second, capitalizing on the large data set, we broke the 37+-month group into additional experience intervals,

---

[2] We are grateful to Dr. Jane M. Arabian (Assistant Director, Accession Policy Office of the Under Secretary of Defense, The Pentagon, Washington, DC) for granting us permission to use the data, and to Dr. Rodney McCloy (Principal Scientist, Human Resources Research Organization, Louisville, KY) for sending us the data, with helpful notes.

[3] Prior to conducting statistical analyses, we screened the variables for values greater than |3.5| *SD*s from the total sample mean (i.e., univariate outliers); 31 of the 30,264 values (0.1%) met this criterion, and we truncated these values to the |3.5| *SD* cutoff value. There was one participant with zero months of job experience; prior to log transforming the job experience variable, we set this value to 0.03 months (1 day).

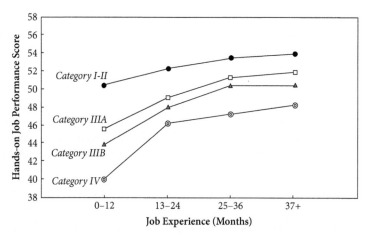

FIGURE 3.2 Mean Hands-on Job Performance Score by AFQT category (i.e., cognitive ability level). Percentile ranges for AFQT categories: I–II (65–99), IIIA (50–64), IIIB (31–49), and IV (10–30) (see Wigdor & Green, 1991, p. 53). Data from Wigdor, Alexandra K., and Green, Bert F., *Performance assessment for the workplace*, Volume 1, p. 53, Table 2.5, National Academy Press, 1991.

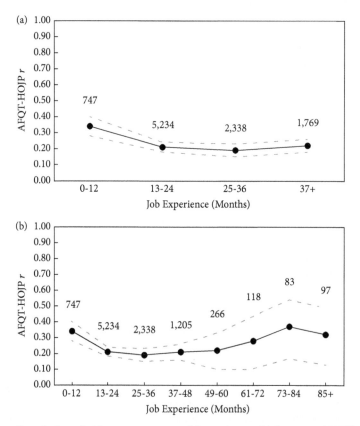

FIGURE 3.3 Correlations (with 95 percent confidence intervals) between AFQT scores and Hands-on Job Performance (HOJP) scores at 4 (A) and 8 (B) job experience intervals. Dashed lines are 95 percent confidence intervals; adjacent values are sample sizes. Data from Joint-Service Job Performance Measurement/Enlistment (JPM) Standards Project ($N = 10{,}088$).

to 85+ months (creating any more groups would result in small sample sizes, $ns < 50$). As shown in Figure 3.3B, the AFQT-HOJP correlation decreases from the first year to the second, stabilizes, and then increases—though the estimates become less precise as sample size decreases.

Finally, as the most statistically powerful analysis, we evaluated the Job Experience × AFQT interaction on HOJP using the entire data set via moderated multiple regression. (Prior to performing the regression analysis, we log-transformed job experience because it was non-normal, skewness = 2.40 and kurtosis = 9.56, and we mean-centered the predictors.) There were significant main effects of both AFQT ($\beta = 0.210$, $t = 21.92$, $p < 0.001$, part $r^2 = 0.044$) and log job experience ($\beta = 0.167$, $t = 17.37$, $p < 0.001$, part $r^2 = 0.028$) on HOJP. High levels of both AFQT and job experience were associated with higher HOJP. The AFQT × Log Job Experience interaction was also statistically significant and under-additive ($\beta = -0.023$, $t = -2.41$, $p = 0.016$, part $r^2 = 0.0005$), though the effect was virtually nil, indicating that AFQT was predictive of HOJP regardless of level of job experience (see Figure 3.4).

The overall picture to emerge from these large-scale employment studies is that cognitive ability remains a significant predictor of job performance, even after extensive job experience and even if validity drops initially. The question of how *far* beyond initial training cognitive ability predicts job performance is unanswered, but the results we have just reviewed indicate at least 5 years (Schmidt et al., 1986) to 10+ years (McDaniel, 1986). Reeve and Bonaccio (2011) reached a similar conclusion in their own review of the relationship between cognitive ability and job performance, noting

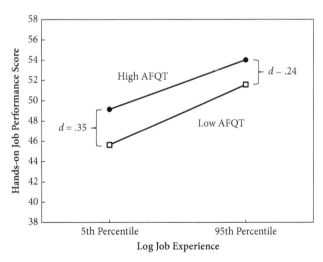

FIGURE 3.4 Predicted values for Hands-on Job Performance (HOJP) for low vs high AFQT (25th vs 75th percentile) at 5th vs 95th percentiles for Log Job Experience. Values generated using regression equation with AFQT score (mean-centered), Log Job Experience (mean-centered), and AFQT × Log Job Experience as predictors of HOJP score: HOJP score = 50.031 + 0.096(AFQT) + 8.119(Log Job Experience) + -0.052(AFQT × Log Job Experience). $N = 10{,}088$.

that "although validities might degrade somewhat over long intervals, we found no evidence to suggest that they degrade appreciably, thereby retaining practically useful levels of validity over very long intervals" (p. 269). Our analysis of the JPM data provide new support for this conclusion. Nevertheless, it remains possible that the validity of cognitive ability would drop to near zero over longer spans of time than have been examined in research (e.g., 20 years).

Discussion

What can be concluded about the role of cognitive ability in expertise? Table 3.2 summarizes findings from the expertise literature most directly relevant to the possibility of expertise-related mitigation of cognitive ability effects. These studies tested (or reported information to test) whether domain-specific factors mitigate effects of cognitive ability factors on domain-relevant performance, by either comparing ability-performance correlations across skill groups or testing interactions between domain-specific factors and cognitive ability factors. As shown, three studies provide evidence for expertise-related mitigation of cognitive ability effects and 10 studies do not; the results of two other studies are mixed or unclear. Based on sample size alone, the Burgoyne et al. (2016) chess meta-analysis might be seen as the best evidence for mitigation, but we reiterate that expertise level (i.e., ranked or unranked) was highly confounded with both age and type of skill measure. This evidence certainly does not warrant any strong conclusions about expertise-related mitigation of the effects of cognitive ability.

What can be made of these results? Unfortunately, not much, because studies in the expertise literature differ in methodological/design characteristics such as sample size, type of criterion task, tests used to measure cognitive ability factors, and use of single versus composite measures to index cognitive ability factors. Any (or all) of these differences could explain differences across studies in cognitive ability–performance relationships. With some of our own studies as examples (e.g., Meinz & Hambrick, 2010, $N = 57$), it is worth emphasizing that sample sizes in this literature are often very small for research on individual differences (see Table 3.2), leading not only to low statistical power but low precision. Consequently, it is not surprising when results do not replicate (it is, in fact, often more surprising when they do). Note also that correlations in the expertise literature are seldom corrected for measurement error and/or restriction of range, resulting in systematic underestimates of the true magnitude of underlying relationships (see McAbee & Oswald, 2017).

A more consistent picture emerges from large-scale studies of job performance. Though validity may drop somewhat initially, measures of cognitive ability significantly predict job performance well beyond initial training. Expertise research often focuses on a specific aspect or component of performance in a domain (e.g., flight path prediction, poker hand evaluation); job performance research more often uses global

Table 3.2 Summary of evidence for expertise-related mitigation of cognitive ability effects

Study	Domain	Sample Size		Cognitive Factor[b]	Evidence for Mitigation?	Empirical Test[c]
		N	Group n[a]			
Ceci & Liker (1986)	Handicapping	30	16/14	IQ	Unclear	Correlations
Masunaga and Horn (2001)	Go	263	62/89/92/23	Gf, Gc, PS	No	Correlations
Morrow et al. (2001)	Aviation	182	96/86	WMC, PS, VS	No	Interaction
Gallagher et al. (2003)	Surgery	98	48, 32/18	VS	No	Correlations
Sohn & Doane (2003)	Aviation	50	25/25	WMC	Yes	Interaction
Wanzel et al. (2003)	Surgery	47	27/12/8	VS	No	Correlations
Keehner et al. (2004)	Surgery	93	48/45	VS	No	Correlations
Sohn & Doane (2004)	Aviation	52	25/27	WMC	Mixed	Correlations
Taylor et al. (2005)	Aviation	97	25/53/19	WMC, PS, AC	No	Interaction
Enochsson et al. (2006)	Surgery	29	18/11	VS	No	Correlations
Ruthsatz et al. (2008)	Music	261	178/19/64	Gf	No	Correlations
Meinz & Hambrick (2010)	Music	57	NA	WMC	No	Interaction
Meinz et al. (2012)	Poker	155	NA	WMC	No	Interaction
Hambrick et al. (2012)	Geology	67	NA	VS	Yes	Interaction
Burgoyne et al. (2016)[d]	Chess	1,604	1,267/337	Gf	Yes	Correlations

Note. Studies are listed in chronological order.
[a] The skill group *n* values are listed in order of increasing expertise. NA for skill group *n* indicates that expertise was treated only as a continuous variable.
[b] Gf, fluid intelligence; Gc, crystallized intelligence; WMC, working memory capacity; PS, perceptual speed; VS, visuospatial ability; AC, attentional control.
[c] In the *interaction* test, mitigation is tested by evaluating the statistical interaction between a domain-specific factor and a cognitive ability factor. In the *correlations* test, mitigation is tested by testing for a difference in correlations between a cognitive ability factor and performance across groups representing different levels of a domain-specific factor.
[d] Meta-analysis.

measures of performance (e.g., overall supervisory ratings, total work sample scores). It could be that involvement of cognitive ability factors decreases as a function of skill in some components of a complex task or job but not in others (e.g., consistent but not variable components; Ackerman, 1992). This is one possible explanation for why correlations between cognitive ability and job performance may drop somewhat with job experience yet still remain statistically significant.

Before proceeding, we note that when cognitive ability and domain-specific factors are measured in the same study, the latter generally account for more variance in expertise than the former (see Ward et al., 2017, for examples). At the same time, cognitive ability and domain-specific knowledge cannot generally be assumed to be independent. For example, Schmidt, Hunter, and Outerbridge (1986) found a correlation of 0.46 between AFQT scores and job knowledge. One interpretation of this finding is that measures of cognitive ability (e.g., IQ, working memory capacity) capture basic mental processes involved in acquiring information in learning situations (Ackerman, 1996; Cattell, 1971; Jensen, 1998). Moreover, even if domain-specific factors explain far more of the variance in expertise than domain-general factors, this does not preclude the latter from being practically useful. We examine this issue next.

Potential Uses of Cognitive Ability Measures to Accelerate Acquisition of Expertise

There are two major ways that cognitive ability measures might be used in efforts to accelerate acquisition of expertise. The first is for personnel selection and classification. That is, cognitive ability measures might be used to make hiring decisions and to assign employees to jobs once hired. The second area of application is in the design of training programs. If a certain cognitive ability factor (e.g., attentional control) is found to be a significant predictor of performance in a domain, then designing training to augment or *bootstrap* that ability (e.g., prompts to direct attention to task-relevant information) might be particularly beneficial for individuals lower in the ability (though see Hoffman et al., 2014, for a cautionary note about removing *desirable difficulties* from training).

But how large must a validity coefficient for a cognitive ability test be to justify its use for these applications? What qualifies as a practically significant effect? Given real-world outcomes (vs outcomes that do not generalize easily to the real world), moderate correlations can prove to be very important. Moreover, although variance explained (r^2) may be of theoretical interest to researchers (e.g., Macnamara, Hambrick, & Oswald, 2014), it is r and not r^2 that is an index of the direct relationship or the utility of a measure in terms of prediction (see Schmidt, Hunter, McKenzie, & Muldrow, 1979). As Kuncel and Hezlett (2010) commented:

> Moderate relationships between predictors and criteria often are inappropriately discounted. For example, correlations of .30 have been dismissed as accounting for less than 10% of the variance in the criteria. However, this relationship is sufficiently large that hiring or admitting individuals who score better on the test can double the rate of successful performance.
> (p. 340)

This point was made nearly 80 years ago by Taylor and Russell (1939), who noted that interpreting the practical importance of correlation coefficients based on methods involving r^2

has led to some unwarranted pessimism on the part of many persons concerning the practical usefulness in an employment situation of validity coefficients in the range of those usually obtained. We believe that it may be of value to point out the very considerable improvement in selection efficiency which may be obtained with small correlation coefficients. (p. 571)

To that end, Taylor and Russell published a set of easy-to-use tables to determine the benefits of using selection tests of different validities in employment settings (see Law & Myors, 1993, for an automated approach). Three pieces of information are needed to use the tables: (3.1) the base rate of success in a job (i.e., the proportion of people who currently succeed in a job), (3.2) the selection ratio for the job (i.e., the ratio of applicants who are selected), and (3.3) the validity of the test. With these three pieces of information, one can consult a Taylor–Russell table and find the predicted improvement in using the test for selection versus not using it.

Table 3.3 gives an example where the base rate of success is 0.20. As shown, when the selection ratio is low, even a selection test with modest validity will lead to a substantial improvement in employee performance over the base rate. For example, if the selection ratio is 0.10, use of a test with validity of 0.20 to select applicants will lead to an 11 percent improvement over not using the test (or 17 percent for a test with validity of 0.30). However, if the selection ratio is high, even a test with high validity will yield little benefit. For example, if the selection ratio is 0.90, then even the use of a test with validity of 0.80 would lead to an improvement of only 2 percent. More sophisticated approaches to utility analysis have been developed since Taylor and Russell published their tables (see Hunter & Schmidt, 1996; Schmidt, Hunter, Outerbridge, & Trattner, 1986), but suffice it to say that use of a test with a moderate level of validity can be practically useful.

Table 3.3 Example of Taylor–Russell Utility Table

Base Rate of Success = 0.20

	Selection Ratio[a]								
Validity (r)	0.10	0.20	0.30	0.40	0.50	0.60	0.70	0.80	0.90
0.00	0.20	0.20	0.20	0.20	0.20	0.20	0.20	0.20	0.20
0.10	0.25	0.24	0.23	0.23	0.22	0.22	0.21	0.21	0.21
0.20	0.31	0.28	0.27	0.26	0.25	0.24	0.23	0.22	0.21
0.30	0.37	0.33	0.30	0.28	0.27	0.25	0.24	0.23	0.21
0.40	0.44	0.38	0.34	0.31	0.29	0.27	0.25	0.23	0.22
0.50	0.52	0.44	0.38	0.35	0.31	0.29	0.26	0.24	0.22
0.60	0.60	0.50	0.43	0.38	0.34	0.30	0.27	0.24	0.22
0.70	0.69	0.56	0.48	0.41	0.36	0.31	0.28	0.25	0.22
0.80	0.79	0.64	0.53	0.45	0.38	0.33	0.28	0.25	0.22
0.90	0.91	0.75	0.60	0.48	0.40	0.33	0.29	0.25	0.22

Note. Validity (r): correlation between predictor variable and criterion variable.
[a] Selection ratio: proportion of applicants who are hired. Values in the cells of the table indicate incremental validity, i.e., expected rate of success as a result of using the selection test.

| | Training Outcome | | |
Test Used to Select?	Fail	Pass	Total
No	60	40	100
Yes	40	60	100
Total	100	100	

FIGURE 3.5 Example of binomial effect size display (BESD) relevant to expertise research and application.

Rosenthal and Rubin's (1982) binomial effect size display (BESD) provides another way to indicate the practical significance of an effect size of a given magnitude (see Ward et al., 2007, for an example of how the BESD can be used in expertise research). Displaying the difference between two proportions (e.g., treatment vs no treatment; selection test vs no selection test), like the Taylor–Russell tables the BESD reveals that modest effect sizes can be practically important. For example, Rosenthal (2005) explained that "an r of 0.20 is said to account for 'only 4% of the variance', but the BESD shows that this proportion of variance accounted for is equivalent to increasing the success rate... from 40 to 60%." Figure 3.5 illustrates this point in terms of a hypothetical scenario where 100 individuals in an organization must be selected for a training program. In one case, a selection test with validity of 0.20 is used; in the other case, it is not used. As shown, using the selection test increases the chances that a trainee will pass the training program by 20 percent (i.e., 20 more people out of 100 pass), even though the scores on the test account for only 4 percent of the variance in the outcome.

Ethical Considerations

There are ethical issues associated with use of any psychological test to make decisions that affect people's lives (e.g., hiring decisions). Probably everyone would agree that it is unethical (not to mention legally unwise) to select individuals using a test with no demonstrated validity, but consider a situation where a test has modest validity—say, 0.30. One might argue that because the validity coefficient is far from perfect, it is unethical to use the test for selection because a considerable number of people with lower scores would be expected to succeed. However, one might also argue that *not* using the test for selection is unethical because lower-scoring individuals will be at a relatively high risk for failure, which may have adverse consequences for the individual (e.g., negative perceptions of other employees, lowered self-efficacy) and also the organization. Along with conducting a proper job analysis and validity study, any organization wishing to use a cognitive test for making personnel decisions must consider these sorts of ethical questions before putting the test into use (Landy & Conte, 2013).

Conclusions

Psychologists have long been interested in identifying traits that may help to explain individual differences in expertise (Hambrick, Campitelli, & Macamara, 2017). Here, we reviewed evidence for the contribution of cognitive ability. There is ample evidence that cognitive ability positively predicts individual differences in complex task performance early in training, but it is unclear whether it remains predictive after extensive practice or training. Evidence from research on traditional domains for expertise research (e.g., chess, music) is inconsistent. For some tasks, domain-specific factors may mitigate the effect of cognitive ability factors on performance (e.g., maintaining situational awareness in aviation), but for other tasks, this may not be the case (e.g., sight-reading music). Evidence from research on job performance is more consistent in indicating that measures of cognitive ability are predictive of job performance, well beyond initial training. In light of this evidence, we believe that at a broad level combining optimal procedures for training complex skills (Hoffman et al., 2014) with valid selection procedures holds tremendous promise for accelerating acquisition of expertise.

At a theoretical level, we believe that it is imperative for expertise researchers to develop and test formal models of expertise. Research on the involvement of cognitive ability factors in expertise has often proceeded somewhat haphazardly, with no general theory describing how mechanisms underlying performance differ across domains. There is no better illustration of this critical point than our own work. We have conducted a number of one-off studies—one in piano sight-reading (Meinz & Hambrick, 2010), another in geological bedrock mapping (Hambrick et al., 2012), another in Texas Hold 'em poker (Meinz et al., 2012)—with no theory to account for how results differ across these domains. Moving ahead, theories of expertise should draw on existing theoretical frameworks to identify potential predictors of expertise (e.g., Ackerman, 1996; Ericsson et al., 1993; Gagné, 2017). However, guided by both computational models (e.g., Altmann, Trafton, & Hambrick, 2014) and cognitive task analysis (Chipman, Schraagen, & Shalin, 2000), they must also specify the information processing mechanisms underlying performance in different types of tasks. Otherwise, there will continue to be no solid basis for comparing results across tasks, and evidence will remain fragmentary.

In the spirit of Hoffman et al.'s (2014) recommendations, we believe that it is also critical that expertise research expand beyond highly constrained activities such as chess, music, and sports, to messy real-world tasks in which the requirements of a job can change rapidly with technological developments and there is no well-circumscribed body of knowledge (as there is in, say, chess). We think that measures of cognitive ability factors hypothesized to underlie *adaptability* (e.g., attentional control, working memory capacity) may have particular promise for predicting performance in jobs such as these. These measures are also attractive because some research has suggested they may reduce group differences (e.g., by race/ethnicity) and resultant adverse impact in selection while still achieving high validity (Verive & McDaniel, 1996). More generally, we are optimistic that the scientific knowledge that will accumulate through programmatic research on individual differences in expertise has great potential to inform efforts to accelerate the acquisition of societally important skills.

Acknowledgements

We thank Paul Ward and Jan Maarten Schraagen for their comments on an earlier version of this chapter.

References

Ackerman, P. L. (1988). Determinants of individual differences during skill acquisition: Cognitive abilities and information processing. *Journal of Experimental Psychology: General 117*, 288–318.

Ackerman, P. L. (1992). Predicting individual differences in complex skill acquisition: Dynamics of ability determinants. *Journal of Applied Psychology 77*, 598–614.

Ackerman, P. L. (1996). A theory of adult intellectual development: Process, personality, interests, and knowledge. *Intelligence 22*, 227–257.

Altmann, E. M., Trafton, J. G., & Hambrick, D. Z. (2014). Momentary interruptions can derail the train of thought. *Journal of Experimental Psychology: General 143*, 215–226.

Banich, M. T. (2009). Executive function: The search for an integrated account. *Current Directions in Psychological Science 18*, 89–94.

Bayer, A. E., & Folger, J. (1966). Some correlates of a citation measure of productivity in science. *Sociology of Education 39*, 381–390.

Berri, D. J., & Simmons, R. (2011). Catching a draft: On the process of selecting quarterbacks in the National Football League amateur draft. *Journal of Productivity Analysis 35*, 37–49.

Burgoyne, A. P., Sala, G., Gobet, F., Macnamara, B. N., Campitelli, G., & Hambrick, D. Z. (2018). The relationship between cognitive ability and chess skill: A comprehensive meta-analysis. *Intelligence 59*, 72–83.

Burgoyne, A. P., Sala, G., Gobet, F., Macnamara, B. N., Campitelli, G., & Hambrick, D. Z. (2018). Corrigendum: The relationship between cognitive ability and chess skill: A comprehensive meta-analysis. *Intelligence*.

Cattell, R. B. (1971). *Abilities: Their structure, growth, and action*. Oxford, UK: Houghton Mifflin.

Causse, M., Dehais, F., & Pastor, J. (2011). Executive functions and pilot characteristics predict flight simulator performance in general aviation pilots. *International Journal of Aviation Psychology 21*, 217–234.

Ceci, S. J., & Liker, J. K. (1986). A day at the races: A study of IQ, expertise, and cognitive complexity. *Journal of Experimental Psychology: General 115*, 255–266.

Chipman, S. F., Schraagen, J. M., & Shalin, V. L. (2000). Introduction to cognitive task analysis. In J. M. Schraagen, S. F. Chipman, & V. L. Shalin (Eds), *Cognitive task analysis* (pp. 3–23). New York: Lawrence Erlbaum.

Clark, M. J., & Centra, J. A. (1982). *Conditions influencing the career accomplishments of Ph. D.s.* (GREB Research Report No. 76-2R). Princeton, NJ: Educational Testing Service.

Creager, J. A., & Harmon, L. R. (1966). *On-the-job validation of selection variables.* Office of Scientific Personnel, National Academy of Sciences-National Research Council.

Deary, I. J., Graham, K. S., & Maran, A. G. (1992). Relationships between surgical ability ratings and spatial abilities and personality. *Journal of the Royal College of Surgeons of Edinburgh 37*, 74–79.

de Bruin, A. B., Kok, E. M., Leppink, J., & Camp, G. (2014). Practice, intelligence, and enjoyment in novice chess players: A prospective study at the earliest stage of a chess career. *Intelligence 45*, 18–25.

de Groot, A.D. (1965/1978). *Thought and choice in chess.* Amsterdam: Amsterdam Academic Archive.

Detterman, D. K. (2014). Introduction to the intelligence special issue on the development of expertise: Is ability necessary? *Intelligence 45*, 1–5.

Detterman, D. K., & Spry, K. M. (1988). Is it smart to play the horses? Comment on "A day at the races: A study of IQ, expertise, and cognitive complexity" (Ceci & Liker, 1986). *Journal of Experimental Psychology: General 117*, 91–95.

Engelhardt, L. E., Mann, F. D., Briley, D. A., Church, J. A., Harden, K. P., & Tucker-Drob, E. M. (2016). Strong genetic overlap between executive functions and intelligence. *Journal of Experimental Psychology: General 145*, 1141–1159.

Enochsson, L., Westman, B., Ritter, E. M., Hedman, L., Kjellin, A., Wredmark, T., & Felländer-Tsai, L. (2006). Objective assessment of visuospatial and psychomotor ability and flow of residents and senior endoscopists in simulated gastroscopy. *Surgical Endoscopy and Other Interventional Techniques 20*, 895–899.

Ericsson, K. A. (2006). The influence of experience and deliberate practice on the development of superior expert performance. In K. A. Ericsson, N. Charness, P. J. Feltovich, & R. R. Hoffman (Eds.), *The Cambridge handbook of expertise and expert performance* (pp. 683–703). New York: Cambridge University Press.

Ericsson, K. A., & Kintsch, W. (1995). Long-term working memory. *Psychological Review 102*, 211–245.

Ericsson, K. A., Krampe, R. Th., & Tesch-Römer, C. (1993). The role of deliberate practice in the acquisition of expert performance. *Psychological Review 100*, 363–406.

Ericsson, K. A., & Smith, J. (1991). *Toward a general theory of expertise: Prospects and limits.* Cambridge, UK: Cambridge University Press.

Farrell, J. N., & McDaniel, M. A. (2001). The stability of validity coefficients over time: Ackerman's (1988) model and the General Aptitude Test Battery. *Journal of Applied Psychology 86*, 60–79.

Fitts, P. M., & Posner, M. I. (1967). *Human performance.* Oxford: Brooks/Cole.

Fleishman, E. A., & Rich, S. (1963). Role of kinesthetic and spatial-visual abilities in perceptual-motor learning. *Journal of Experimental Psychology 66*, 6–11.

Folger, J. K., Astin, H. S., & Bayer, A. E. (1970). *Human resources and higher education: Staff report on the commission on human resources and advanced education.* New York, NY: Russell Sage Foundation.

Gagné, F. (2017). Expertise development from an IMTD perspective. In D. Z. Hambrick, G. Campitelli, & B. N. Macnamara (Eds.), *The science of expertise: Behavioral, neural, and genetic approaches to complex skill* (pp. 307–327). New York: Routledge.

Gallagher, A. G., Cowie, R., Crothers, I., Jordan-Black, J. A., & Satava, R. M. (2003). PicSOr: an objective test of perceptual skill that predicts laparoscopic technical skill in three initial studies of laparoscopopic performance. *Surgical Endoscopy 17*, 1468–1471.

Gibbons, R. D., Baker, R. J., & Skinner, D. B. (1986). Field articulation testing: A predictor of technical skills in surgical residents. *Journal of Surgical Research 41*, 53–57.

Gottfredson, L. S. (1997). Mainstream science on intelligence: An editorial with 52 signatories, history, and bibliography. *Intelligence 24*, 13–23.

Grabner, R. H., Stern, E., & Neubauer, A. C. (2007). Individual differences in chess expertise: A psychometric investigation. *Acta Psychologica 124*, 398–420.

Groenier, M., Schraagen, J. M. C., Miedema, H. A., & Broeders, I. A. J. M. (2014). The role of cognitive abilities in laparoscopic simulator training. *Advances in Health Sciences Education 19*, 203–217.

Haier, R. J. (2016). *The neuroscience of intelligence*. Cambridge, UK: Cambridge University Press.

Hambrick, D. Z., & Campitelli, G., & Macamara, B. N. (2017). Introduction: A brief history of the science of expertise and overview of the book. In D. Z. Hambrick, G. Campitelli, & B. N. Macamara (Eds.), *The science of expertise: Behavioral, neural, and genetic approaches to complex skill* (pp. 1–10). New York: Routledge.

Hambrick, D. Z., Libarkin, J. C., Petcovic, H. L., Baker, K. M., Elkins, J., Callahan, C. N., . . . , & LaDue, N. D. (2012). A test of the circumvention-of-limits hypothesis in scientific problem solving: The case of geological bedrock mapping. *Journal of Experimental Psychology: General, 141*, 397–403.

Hambrick, D. Z., & Meinz, E. J. (2011). Limits on the predictive power of domain-specific experience and knowledge in skilled performance. *Current Directions in Psychological Science 20*, 275–279.

Hoffman, R. R., Ward, P., Feltovich, P. J., DiBello, L., Fiore, S. M., & Andrews, D. H. (2014). *Accelerated expertise: Training for high proficiency in a complex world*. New York: Psychology Press.

Hulin, C. L., Henry, R. A., & Noon, S. L. (1990). Adding a dimension: Time as a factor in the generalizability of predictive relationships. *Psychological Bulletin 107*, 328–340.

Hunter, J. E., & Schmidt, F. L. (1996). Intelligence and job performance: Economic and social implications. *Psychology, Public Policy, and Law 2*, 447–472.

Jensen, A. R. (1998). *The g factor: The science of mental ability*. Westport, CT: Greenwood Publishing Group.

Jung, W. H., Kim, S. N., Lee, T. Y., Jang, J. H., Choi, C. H., Kang, D. H., & Kwon, J. S. (2013). Exploring the brains of Baduk (Go) experts: Gray matter morphometry, resting-state functional connectivity, and graph theoretical analysis. *Frontiers in Human Neuroscience 7*, 633.

Kane, M. J., Conway, A. R., Hambrick, D. Z., & Engle, R. W. (2007). Variation in working memory capacity as variation in executive attention and control. In A. R. A. Conway, C. Jarrold, M. J. Kane, A. Miyake, J. N. Towse (Eds.), *Variation in working memory* (pp. 21–48). New York: Oxford University Press.

Kaufman, H. G. (1972). Relations of ability and interest to currency of professional knowledge among engineers. *Journal of Applied Psychology 56*, 495–499.

Keehner, M. M., Tendick, F., Meng, M. V., Anwar, H. P., Hegarty, M., Stoller, M. L., & Duh, Q. Y. (2004). Spatial ability, experience, and skill in laparoscopic surgery. *American Journal of Surgery 188*, 71–75.

Knopik, V. S., Neiderhiser, J. M., DeFries, J. C., & Plomin, R. (2016). *Behavioral genetics* (7th ed). New York: Macmillan Higher Education.

Kopiez, R., & In Lee, J. (2008). Towards a general model of skills involved in sight reading music. *Music Education Research 10*, 41–62.

Kuncel, N. R., & Hezlett, S. A. (2007). Standardized tests predict graduate students' success. *Science* 315(5815), 1080–1081.

Kuncel, N. R., & Hezlett, S. A. (2010). Fact and fiction in cognitive ability testing for admissions and hiring decisions. *Current Directions in Psychological Science 19*, 339–345.

Kuncel, N. R., Hezlett, S. A., & Ones, D. S. (2001). A comprehensive meta-analysis of the predictive validity of the graduate record examinations: implications for graduate student selection and performance. *Psychological Bulletin 127*, 162–181.

Landy, F. J., & Conte, J. M. (2013). *Work in the 21st century: An introduction to industrial and organizational psychology* (4th ed). London: Wiley.

Law, K. S., & Myors, B. (1993). Cutoff scores that maximize the total utility of a selection program: Comment on Martin and Raju's (1992) procedure. *Journal of Applied Psychology 78*, 736–740.

Lee, B., Park, J. Y., Jung, W. H., Kim, H. S., Oh, J. S., Choi, C. H., ..., & Kwon, J. S. (2010). White matter neuroplastic changes in long-term trained players of the game of "Baduk" (GO): A voxel-based diffusion-tensor imaging study. *Neuroimage 52*, 9–19.

Leonard, C. A., & Williams, R. J. (2015). Characteristics of good poker players. *Journal of Gambling Issues* 31, 45–68.

Louridas, M., Szasz, P., de Montbrun, S., Harris, K. A., & Grantcharov, T. P. (2016). Can we predict technical aptitude?: A systematic review. *Annals of Surgery 263*, 673–691.

Lubinski, D. (2009). Exceptional cognitive ability: The phenotype. *Behavior Genetics 39*, 350–358.

Lyons, B. D., Hoffman, B. J., & Michel, J. W. (2009). Not much more than g? An examination of the impact of intelligence on NFL performance. *Human Performance 22*, 225–245.

McAbee, S. T., & Oswald, F. L. (2017). Primer—Statistical methods in the study of expertise: Some key considerations. In D. Z. Hambrick, G. Campitelli, & B. N. Macnamara (Eds.), *The science of expertise: Behavioral, neural, and genetic approaches to complex skill* (pp. 13–30). New York: Routledge.

McCabe, D. P., Roediger III, H. L., McDaniel, M. A., Balota, D. A., & Hambrick, D. Z. (2010). The relationship between working memory capacity and executive functioning: evidence for a common executive attention construct. *Neuropsychology 24*, 222–243.

McDaniel, M. A. (1986). *The evaluation of a causal model of job performance: The interrelationships of general mental ability, job experience, and job performance* (Unpublished doctoral dissertation). George Washington University, Washington, DC.

Macnamara, B. N., Hambrick, D. Z., & Oswald, F. L. (2014). Deliberate practice and performance in music, games, sports, education, and professions: A meta-analysis. *Psychological Science 25*, 1608–1618.

Mann, D. T., Williams, A. M., Ward, P., & Janelle, C. M. (2007). Perceptual-cognitive expertise in sport: A meta-analysis. *Journal of Sport and Exercise Psychology 29*, 457–478.

Masunaga, H., & Horn, J. (2001). Expertise and age-related changes in components of intelligence. *Psychology and Aging 16*, 293–311.

Meinz, E. J., & Hambrick, D. Z. (2010). Deliberate practice is necessary but not sufficient to explain individual differences in piano sight-reading skill the role of working memory capacity. *Psychological Science 21*, 914–919.

Meinz, E. J., Hambrick, D. Z., Hawkins, C. B., Gillings, A. K., Meyer, B. E., & Schneider, J. L. (2012). Roles of domain knowledge and working memory capacity in components of skill in Texas Hold'Em poker. *Journal of Applied Research in Memory and Cognition 1*, 34–40.

Morrow, D. G., Menard, W. E., Ridolfo, H. E., Stine-Morrow, E. A. L., Teller, T., & Bryant, D. (2003). Expertise, cognitive ability, and age effects on pilot communication. *International Journal of Aviation Psychology 13*, 345–371.

Morrow, D. G., Menard, W. E., Stine-Morrow, E. A., Teller, T., & Bryant, D. (2001). The influence of expertise and task factors on age differences in pilot communication. *Psychology and Aging 16*, 31–46.

Murdoch, J. R., Bainbridge, L. C., Fisher, S. G., & Webster, M. H. (1994). Can a simple test of visual-motor skill predict the performance of microsurgeons? *Journal of the Royal College of Surgeons of Edinburgh 39*, 150–152.

Petushek, E. J. (2014). *Development and validation of the anterior cruciate ligament injury-risk-estimation quiz (ACL-IQ)*. Retrieved from ProQuest Dissertations & Theses Global. (Accession No. 3643834).

Petushek, E. J., Cokely, E. T., Ward, P., & Myer, G. D. (2015). Injury risk estimation expertise: cognitive-perceptual mechanisms of ACL-IQ. *Journal of Sport and Exercise Psychology 37*, 291–304.

Pfau, H. D., & Murphy, M. D. (1988). Role of verbal knowledge in chess skill. *American Journal of Psychology 101*, 73–86.

Reeve, C. L., & Bonaccio, S. (2011). On the myth and the reality of the temporal validity degradation of general mental ability test scores. *Intelligence 39*, 255–272.

Risucci, D., Geiss, A., Gellman, L., Pinard, B., & Rosser, J. (2001). Surgeon-specific factors in the acquisition of laparoscopic surgical skills. *American Journal of Surgery 181*, 289–293.

Rosenthal, R. (2005). Binomial effect size display. *Encyclopedia of statistics in behavioral science*. Wiley StatsRef: Statistics Reference Online.

Rosenthal, R., & Rubin, D. B. (1982). A simple, general purpose display of magnitude of experimental effect. *Journal of Educational Psychology 74*, 166–169.

Ruthsatz, J., Detterman, D., Griscom, W. S., & Cirullo, B. A. (2008). Becoming an expert in the musical domain: It takes more than just practice. *Intelligence 36*, 330–338.

Sala, G., Burgoyne, A. P., Macnamara, B. N., Hambrick, D. Z., Campitelli, G., & Gobet, F. (2017). Checking the "Academic Selection" argument. Chess players outperform non-chess-players in cognitive skills related to intelligence: A meta-analysis. *Intelligence 61*, 130–139.

Schellenberg, E. G., & Weiss, M. W. (2013). Music and cognitive abilities. In D. Deutsch (Ed.), *Psychology of music* (3rd ed, pp. 499–550). London: Academic Press.

Schmidt, F. L., Hunter, J. E., McKenzie, R. W., & Muldrow, T. W. (1979). Impact of valid selection procedures on work-force productivity. *Journal of Applied Psychology 64*, 609–626.

Schmidt, F. L., & Hunter, J. E. (1998). The validity and utility of selection methods in personnel psychology: Practical and theoretical implications of 85 years of research findings. *Psychological Bulletin 124*, 262–274.

Schmidt, F. L., & Hunter, J. (2004). General mental ability in the world of work: occupational attainment and job performance. *Journal of Personality and Social Psychology 86*, 162–173.

Schmidt, F. L., Hunter, J. E., & Outerbridge, A. N. (1986). Impact of job experience and ability on job knowledge, work sample performance, and supervisory ratings of job performance. *Journal of Applied Psychology 71*, 432–439.

Schmidt, F. L., Hunter, J. E., Outerbridge, A. N., & Goff, S. (1988). Joint relation of experience and ability with job performance: Test of three hypotheses. *Journal of Applied Psychology 73*, 46–57.

Schmidt, F. L., Hunter, J. E., Outerbridge, A. N., & Trattner, M. H. (1986). The economic impact of job selection methods on size, productivity, and payroll costs of the federal work force: An empirically based demonstration. *Personnel Psychology 39*, 1–29.

Schueneman, A. L., Pickleman, J., Hesslein, R., & Freeark, R. J. (1984). Neuropsychologic predictors of operative skill among general surgery residents. *Surgery 96*, 288–295.

Sohn, Y. W., & Doane, S. M. (2003). Roles of working memory capacity and long-term working memory skill in complex task performance. *Memory & Cognition 31*, 458–466.

Sohn, Y. W., & Doane, S. M. (2004). Memory processes of flight situation awareness: Interactive roles of working memory capacity, long-term working memory, and expertise. *Human Factors 46*, 461–475.

Takagi, H. (1997). *Cognitive aging: expertise and fluid intelligence* (Doctoral dissertation). Retrieved from ProQuest Dissertations and Theses database. (UMI No. 9733148).

Taylor, J. L., O'Hara, R., Mumenthaler, M. S., Rosen, A. C., & Yesavage, J. A. (2005). Cognitive ability, expertise, and age differences in following air-traffic control instructions. *Psychology and Aging 20*, 117–133.

Taylor, H. C., & Russell, J. T. (1939). The relationship of validity coefficients to the practical effectiveness of tests in selection: Discussion and tables. *Journal of Applied Psychology 23*, 565–578.

Toma, M., Halpern, D. F., & Berger, D. E. (2014). Cognitive abilities of elite nationally ranked SCRABBLE and crossword experts. *Applied Cognitive Psychology 28*, 727–737.

Tucker-Drob, E. M. (2011). Global and domain-specific changes in cognition throughout adulthood. *Developmental Psychology 47*, 331–343.

Tuffiash, M., Roring, R. W., & Ericsson, K. A. (2007). Expert performance in SCRABBLE: implications for the study of the structure and acquisition of complex skills. *Journal of Experimental Psychology: Applied 13*, 124–134.

Turkheimer, E. (2000). Three laws of behavior genetics and what they mean. *Current Directions in Psychological Science 9*, 160–164.

Unsworth, N., Fukuda, K., Awh, E., & Vogel, E. K. (2015). Working memory delay activity predicts individual differences in cognitive abilities. *Journal of Cognitive Neuroscience 27*, 853–865.

Unterrainer, J. M., Kaller, C. P., Halsband, U., & Rahm, B. (2006). Planning abilities and chess: A comparison of chess and non-chess players on the Tower of London task. *British Journal of Psychology 97*, 299–311.

Unterrainer, J. M., Kaller, C. P., Leonhart, R., & Rahm, B. (2011). Revising superior planning performance in chess players: The impact of time restriction and motivation aspects. *American Journal of Psychology 124*, 213–225.

Verive, J. M., & McDaniel, M. A. (1996). Short-term memory tests in personnel selection: Low adverse impact and high validity. *Intelligence 23*, 15–32.

Vestberg, T., Gustafson, R., Maurex, L., Ingvar, M., & Petrovic, P. (2012). Executive functions predict the success of top-soccer players. *PloS One 7*, e34731.

Voss, M. W., Kramer, A. F., Basak, C., Prakash, R. S., & Roberts, B. (2010). Are expert athletes "expert" in the cognitive laboratory? A meta-analytic review of cognition and sport expertise. *Applied Cognitive Psychology 24*, 812–826.

Wanzel, K. R., Hamstra, S. J., Caminiti, M. F., Anastakis, D. J., Grober, E. D., & Reznick, R. K. (2003). Visual-spatial ability correlates with efficiency of hand motion and successful surgical performance. *Surgery 134*, 750–757.

Ward, P., Belling, P., Petushek, E., & Ehrlinger, J. (2017). Does talent exist? A re-evaluation of the nature–nurture debate. In J. Baker, S. Cobley, J. Schorer, & N. Wattie (Eds), *Routledge handbook of talent identification and development in sport* (pp. 19–34). New York: Routledge.

Ward, P., Hodges, N. J., Starkes, J. L., & Williams, M. A. (2007). The road to excellence: Deliberate practice and the development of expertise. *High Ability Studies 18*, 119–153.

Wigdor, A. K., & Green, B. F. (1991). *Performance assessment for the workplace*, Volumes I & II. Washington, DC: National Academy Press.

CHAPTER 4

STUDIES OF EXPERTISE AND EXPERIENCE
A Sociological Perspective on Expertise

HARRY COLLINS AND ROBERT EVANS

INTRODUCTION

Is expertise a status that is attributed to someone or a capacity a person possesses? There is merit in both perspectives but here we describe an approach known as studies of expertise and experience (SEE) that focuses on the second and treats expertise as real rather than relational (Collins & Evans, 2002, 2007). This realist conception of expertise differs from many psychological and philosophical approaches, in that it turns on acquiring expertise through socialization and can be used whether this socialization is ubiquitous—as in native language speaking—or esoteric—as in a scientific specialty. The idea that expertise is acquired through socialization makes SEE entirely compatible with social constructivist theories of knowledge. For example, a natural language is a social construct but speaking it is a deep and difficult expertise that can be acquired only through socialization.

The origins of SEE are to be found in the field of science and technology studies (STS), where science is seen as a *hard case* for the study of knowledge and expertise more generally. We therefore begin our exposition of the approach with a summary of the debates within STS that led to its development. We then examine how the ideas that inform SEE relate to wider debates about embodiment and tacit knowledge. Next we set out some novel theoretical and methodological consequences that follow from this approach: a new classification of expertise, summarized in a *Periodic Table of Expertises*, and the creation of a new research method known as the *Imitation Game*. We conclude by setting out the future questions for research using SEE and illustrate its utility and relevance through the examples of interdisciplinary working and the role of experts in a *post-truth* society.

Three Waves of STS

STS is a diverse field whose history can be told in many ways (see Felt, Fouche, Miller, & Smith-Doerr, 2017 for an illustration of this diversity). Here we simplify the history by distinguishing between *three waves* of STS that illustrate the key transitions in its understanding of science and, over time, of expertise more generally (Collins & Evans, 2002). The first wave of science studies is characterized by the idea that science is, without question, an epistemically superior form of knowledge-making that is different to other kinds of expert knowledge. To the extent that the social analysis of science is possible, its questions concern the institutional structures that support and facilitate science. To try to explain the content of valid scientific knowledge by social analysis makes no sense under wave one, though scientific mistakes can be explained this way.

Wave one dominated the sociology of science from the early part of the twentieth century with Robert Merton being its most important theorist (e.g., Merton, 1973). In the early 1970s, however, the sociology of science, like many other fields in the social sciences, began to take an increasingly constructivist approach. Rather than seeing science as passively *holding up a mirror to nature*, sociologists began treating scientific knowledge as actively created through agreements made by scientists about how to interpret what they saw in the many different mirrors they had made (Collins, 1975, 1985/1992; Latour & Woolgar, 1979). Wave two thus rejected the asymmetry of wave one, in which scientific mistakes alone are open to sociological explanation, and advocated a *symmetrical* approach in which both scientific truth and scientific error are treated as the same kind of social agreement (Barnes, Bloor, & Henry, 1996; Bloor, 1973, 1991).

Under this second wave of STS, expertise becomes a relational phenomenon in which being an expert amounts to being granted expert status by others. For sociologists, this means expertise is an *actors' category* in that it reflects the success or otherwise of particular social groups in attaining positions of credibility and authority (Collins, 2008). It also makes scientific expertise look a lot less special, as the difference between science and other forms of expertise is no longer an epistemic difference but a sociological one: scientific expertise is different because it is treated differently, not because it is made differently. This approach has been enormously successful, producing rich descriptions of how knowledge is created and shared, as well as revealing the ways in which elite institutions can use science to marginalize the concerns raised by others (Carr, 2010 provides a summary and overview).

The third wave of STS endorses the symmetrical description of science provided by wave two but questions the policy conclusions drawn by many of its advocates (e.g., Funtowicz & Ravetz, 1993; Jasanoff, 2003, 2013; Wynne, 2003). According to the relational view of wave two, the symmetrical analysis of science shows that it can never be neutral and that it invariably reflects the dominant interests in that society.

Improving policy-making under wave two thus means challenging the institutions and assumptions that grant authority to science and deny legitimacy to others. The result is that the policy recommendations of wave two always push in the same direction: democratize expertise by opening up esoteric questions to ever wider groups of participants. This works well when the problem of expertise is that of technocracy in which it is mistakenly thought that those who possess an esoteric expertise either exhaust what can be known about a topic or have special abilities when it comes to the exercise of ubiquitous expertise. It is, however, much more difficult to justify this stance when the expertise required is genuinely esoteric and the policy failure is more accurately described as technological populism (Collins & Evans, 2002). Two examples illustrate this, newer, phenomena. In the UK in the late 1990s, a scare over the MMR (measles, mumps, and rubella) vaccine led to a fall in vaccination rates as parents' local experiences were misleadingly labelled as *lay expertise* and treated as equivalent to the consensus medical researchers had reached through the epidemiological analysis of population data. More recently, and more widely, the corrosive consequences of widespread scepticism about experts for the idea that science—and expertise in general—should have some special status when technological decisions are being made are being seen in public and political debate across the Western world. SEE responds to this challenge by proposing a realist model of expertise that is consistent with wave two's description of science but which also argues that scientific values—more so than scientific facts—are an important part of any modern society (Collins & Evans, 2017). As explained in more detail in the section 'Making Boundaries: Expertise as Real', this enables SEE to argue for more participation in some cases and less in others.

Philosophical Roots: Expertise as a Form of Life

SEE starts with early work in the sociology of scientific knowledge (e.g., Bloor, 1973, 1991; Collins, 1974, 1975, 1985/1992) that draws on Ludwig Wittgenstein's concept of a *form of life* (Winch, 1958; Wittgenstein, 1953) as the fundamental unit of analysis. The link with social constructivism comes from the way a form of life, which captures the set of ideas, actions, traditions, and values that define a social group, gives its members the tools to make sense of the world and the standards by which to hold each other accountable. The concept applies to all scales and topics, so a form of life can be very large and include many people spread over wide geographical areas (e.g., English speakers or soccer players) or it can be very local and/or highly specialized (e.g., Basque language speakers in the Pyrenees or Maori kapa haka performers).

The notion of form of life can be used to understand the continuity of knowledge and what happens when new ideas emerge. Both aspects are illustrated in the following passage from Peter Winch's (1958) analysis of the relationship between philosophy and sociology. Note that, although the example given relates to changes in scientific knowledge, the insight is intended to generalize to all forms of social and cultural innovation:

> Imagine a biochemist making certain observations and experiments as a result of which he discovers a new germ which is responsible for a certain disease. In one sense we might say that the name he gives this new germ expresses a new idea, but I prefer to say in this context that he has made a discovery within the existing framework of ideas. I am assuming that the germ theory of disease is already well established in the scientific language he speaks. Now compare with this discovery the impact made by the first formulation of that theory, the first introduction of the concept of germ into the language of medicine. This was a much more radically new departure, involving not merely a new factual discovery within an existing way of looking at things, but a completely new way of looking at the whole problem of the causation of diseases, the adoption of new diagnostic techniques, the asking of new kinds of questions about illnesses, and so on. In short it involved the adoption of new ways of doing things by people involved, in one way or another, in medical practice. (Winch, 1958, pp. 121–122)

As explained in the first part of the passage, the *germ theory of disease* is the basis of a form of life within which medical scientists and practitioners carry out their day-to-day work. This is very similar to Kuhn's concept of a paradigm (Kuhn, 1996, originally published in 1962) as well as many other sociological concepts—collectivity, culture, subculture, microculture, etc.—that also describe the shared values and practices that define a social group. The idea expressed in the second part of the passage is close to the idea of a Kuhnian revolution and gives STS its more radical, constructivist, edge. Rather than describing the world, here we see the germ theory of disease creating and constituting that world, giving meaning to entities, institutions, and practices that did not—and could not—exist without the idea of germs.

The second distinctive feature of STS, and of SEE in particular, is the importance attached to tacit knowledge. Although Wittgenstein does not use the term, which is usually attributed to Michael Polanyi (e.g. Polanyi, 1966), the importance of knowledge that is not explicitly articulated is central to Wittgenstein's writings about the nature of rules and the role of a form of life in determining how rules are to be interpreted and applied. Using the analogy of a signpost, Wittgenstein writes:

> A rule stands there like a sign-post.—Does the sign-post leave no doubt open about the way I have to go? Does it shew [sic] which direction I am to take when I have passed it; whether along the road or the footpath or cross-country? But where is it said which way I am to follow it; whether in the direction of its finger or (e.g.) in the opposite one?—And if there were, not a single sign-post, but a chain of adjacent ones or of chalk marks on the ground—is there only *one* way of interpreting them? (Wittgenstein, 1953, para. 85)

It is possible to distinguish between several different types of tacit knowledge (Collins, 2001, 2010), with two—collective tacit knowledge and somatic-limit tacit knowledge—being especially important for understanding expertise. Collective tacit knowledge is the sort of knowledge implied in Wittgenstein's example of the sign-post: how do we know that the *pointy end* of the sign indicates which way to go or that we should follow the path even as it bends around a corner and we are no longer travelling in the *same direction*? The answer is not that these things are written down somewhere as explicit knowledge. The answer is that these, and a myriad other social practices, are given to us by our form of life and we learn how to follow signs, and other social rules, by being socialized into the collectivity that shares that particular form of life.

The other kind of tacit knowledge that matters in debates about expertise is somatic-limit tacit knowledge (Collins, 2010). Here the focus is on the individual expert and the bodily experience of performing physical tasks such as riding a bike or catching a ball. Again, the knowledge needed to perform the task is unexplicated and, even if rules could be discovered that would enable machines or other humans to perform the task, they would not necessarily describe the experience of being an expert practitioner (Collins, 2007). Indeed, in many such practical tasks a lack of self-consciousness is said to be the mark of the expert as in the Dreyfus five-stage model of expertise (e.g., Dreyfus, 2004).

Making Boundaries: Expertise as Real

In contrast to tacit knowledge and form of life, where there is much in common between the second and third waves of STS, the idea of boundaries reveals—perhaps ironically—the differences between the two approaches. In the case of wave two, expert status, and the demarcation this creates between experts and non-experts, is typically treated as the outcome of *boundary work* carried out by competing groups of social actors (see, e.g., Gieryn, 1999). As a result, the research questions addressed are largely descriptive: how were claims to expertise negotiated and expert status recognized in this or that particular case?

In practice, however, this descriptive approach contains an implicit normative agenda. Just as accepting the new germ theory of disease requires that new ways of doing medicine are developed, so the new description of science as a social practice, and not a source of authoritative and unbiased knowledge, suggests that new ways of relating to science and expertise are needed. For wave two, recognizing that social judgements are essential parts of science has led to the idea that science's institutions and practices should become more representative of the wider society (e.g. Irwin, 1995; Ottinger, 2010; Shapin, 2007; Wynne, 1992). The idea is that a more inclusive approach will allow a wider range of social interests and positions to be represented in the making of scientific and technical knowledge and a wider range of questions to be asked about its applications (for examples of these kinds of arguments see Callon, Lascoumes, & Barthe, 2011; Douglas, 2009; Fisher et al., 2015; Harding, 2006; Longino, 1990).

The third wave recognizes these problems but distinguishes between questions of framing and those of knowledge production. Whilst there can be nothing wrong with calls for more democracy in the setting of public policy, to solve problems of knowledge-production we must look to the inclusion of a wider range of *expertises*. The third wave endorses a widening of expert debate but this must mean including more of the appropriate expert communities, not diluting the notion of technical expertise to include the general public. The solution is better understanding of expertise in order to recognize a wider range of esoteric experts that includes the previously excluded experience-based experts found outside the elite institutions of science (Collins & Evans, 2002, 2017; Collins, Weinel, & Evans, 2010). Under this approach expertise is an *analysts' category* as well as an actors' category and this requires a theory that can explain what expertise consists of, the kinds of decisions for which it is relevant, and a way of telling who is and who is not an expert.

In many of the case studies used to develop wave two the danger of too much participation does not arise. Instead the pressing problem is invariably an overly technocratic form of decision-making in which elite experts frame the question and determine what will be allowed to count as a relevant policy option. In this scenario, sociological analysis exposes the hidden assumptions and values of experts and justifies a more inclusive approach that recognizes the value of local, indigenous, or other experiential knowledge. But what happens when it is not the assumptions and values of the mainstream elite that are suspect but those of the groups that want to reject the consensus view? We have already mentioned the case of MMR vaccinations, where limited and partial data were overinterpreted by parents, whose fears were encouraged and exacerbated by press reporting, with the result that vaccination rates fell, herd immunity was lost, and the health of vulnerable people was put at unnecessary risk. A similar diagnosis can be made of Thabo Mbeki's decision, when he was President of South Africa, not to permit the use of the anti-retroviral drug AZT (azidothymidine) to reduce the risk of mother-to-child transmission of HIV in South Africa. The decision, which led to more children being infected than would otherwise have been the case, was based, at least in part, on Mbeki's belief that the mainstream medical theories about HIV and AIDS were wrong, though in this case the misinformation came not from the media but from Internet pages maintained by scientists whose work had long been dismissed within the scientific community (Weinel, 2010). In a similar way, Oreskes and Conway (2010) show that the political controversy about anthropogenic climate warming is being deliberately kept alive by trained and qualified scientists who are supported by the oil industry to use their knowledge to create the appearance of uncertainty even though mainstream scientific opinion is clear. Wave two of science studies has no conceptual apparatus to explain why these practices are wrong since everything is a matter of social construction. SEE's approach, with its emphasis on the norms and values of a scientific form of life, does distinguish this kind of activity from science proper and is, therefore, able to say who is and is not a legitimate contributor to an expert debate turning on technical expertise.

Not all technical debates turn on technical expertise, however, and the policy consequences of any technical consensus are a matter of politics not science. Thus, Mbeki, even if he had accepted the safety and efficacy of anti-retroviral drugs, could still have taken the political decision that South Africa was not going to allow itself to be presented as a disease-ridden society and robbed by the money-gouging pharmaceutical companies. Even the acceptance of anthropogenic climate warming still leaves open the question of what to do about it. This distinction between the *technical* and *political* is a central element of SEE that is radically different to the fact–value distinction that was refuted by wave two. For wave three, the *technical* and *political* are two distinct forms of life turning on different expertises and different sets of values—in other words, the untenable fact–value distinction is replaced with sociologically informed value–value distinction. For political decision-making, democratic norms set the standard and the rights of citizens to participate in such decisions are guaranteed by law; for technical decision-making, scientific values provide the moral framework, with the participants selected using the typology of expertises set out in the next section.

Theoretical and Methodological Innovations

The Periodic Table of Expertises

The classification of expertises developed by SEE is summarized in the Periodic Table of Expertises (see Table 4.1) first published in Collins and Evans (2007).

Table 4.1 The Periodic Table of Expertises

Ubiquitous Expertises					
Dispositions				Interactive ability / Reflective ability	
Specialist Expertises	Ubiquitous			Specialist	
	Beer-mat knowledge	Popular understanding	Primary source knowledge	Interactional expertise	Contributory expertise
				Polimorphic / Mimeomorphic	
	External		Internal		
Meta-expertises	Ubiquitous discrimination	Local discrimination	Technical connoisseurship	Downward discrimination	Referred expertise
Meta-criteria	Credentials		Experience		Track-record

Reproduced from Collins, Harry and Evans, Robert, *Rethinking Expertise*, p. 14, Table 1, © 2007, University of Chicago Press.

The principle behind the table is that different kinds of experiences give rise to different kinds of expertise. Working from the top, the first row acknowledges the society-wide *ubiquitous expertises* (e.g., cultural norms, speaking a natural language) derived from general socialization that are needed for a person to function within a given society. The second row identifies two individual *dispositions* that can facilitate the socialization needed to acquire the other kinds of expertise listed in the table but which are not, strictly speaking, necessary for socialization to take place.

The next row—*specialist expertises*—are the expertises associated with specific social groups such as scientists, car drivers, and plumbers. It is important to note that the difference between a specialist expertise and a ubiquitous expertise is sociological—access to the former is restricted in some way within a society, whereas access to the latter is not. Otherwise, they are both the product of successful socialization into a form of life.

Moving across the row of specialist expertises, there is a progression from very basic knowledge to complete mastery of the domain. The first three categories—beer-mat knowledge, popular understanding, and primary source knowledge—denote the kinds of understanding that can be achieved by using pre-packaged resources such as webpages, online videos, magazines, books, and academic journals. Acquisition of these expertises depends only on the deployment of ubiquitous expertises and, as they involve no direct interaction with specialist communities of practitioners, no socialization into the tacit knowledge of esoteric communities is possible.

In contrast, the last two categories of specialist expertise—interactional expertise and contributory expertise—can be attained only through sustained immersion in the relevant community. If this is successful, then expert performance can be attained as the learner will have acquired the specialist tacit knowledge needed to act independently but in ways that other members of the group would endorse as correct (for a recent assessment of contributory and interactional expertises see Collins & Evans, 2015; Collins, Evans, & Weinel, 2016).

Because high-level expertise is hard to acquire—for example, it is suggested that it takes thousands of hours of deliberate practice in the case of esoteric expertises (Ericsson, Krampe, & Tesch-Römer, 1993; Ericsson & Pool, 2016)—modern societies require citizens to make judgements about experts without having the relevant specialist expertise themselves. The meta-expertises row lists some of the ways in which this can be done. These include *external* social judgements based on discrimination and which make little or no reference to the content of what is being judged and *internal* judgements that require some familiarity with the specialist expertise in question. The final row of the table identifies criteria that might be used for identifying experts, with relevant experience being the best of the three options as it encompasses a wider range of possibilities and skills.

Using the Periodic Table of Expertises, it is possible to talk about experts and expertise in a more nuanced way. For example, the traditional idea of an expert is that of a skilled practitioner who has mastered the practice of a domain. Using Table 4.1, this usage corresponds to a contributory expert and the importance attached to acquiring tacit knowledge explains why achieving this type of expertise requires extensive immersion in the domain of practice. This is most obviously true for skills like speaking a language or driving a car, but the same is also true for other, more cognitive, domains of expertise including scientific specialties. Significantly, and despite the popularity of online learning, not even the most sophisticated descriptions or multimedia demonstrations will be enough to produce contributory expertise (or interactional expertise) in an esoteric domain, as participation in the community of practitioners is essential.

Unlike contributory expertise, which adds clarity to an old idea, the category of interactional expertise is novel and poses new questions about the relationship between language and practice. Interactional expertise is defined as fluency in the language that a community uses to describe its practices; hereafter its *practice language* (Collins, 2011). All contributory experts are also interactional experts but the novel claim is that it is possible to acquire interactional expertise to the level of that possessed by a contributory expert without mastering or even experiencing the physical practices that define the domain of expertise (Collins, 2004b, 2016b; Collins, Evans, & Weinel, 2017a).

The two new insights that interactional expertise provides are (1) that language itself is a social practice that cannot be reduced to written or recorded text and (2) that it is not necessary to perform a physical practice in order to acquire the practice language that describes it and which is used to make expert judgments about it. For these reasons, sustained immersion is always needed to achieve the necessary linguistic socialization but actually performing the practice, despite being an efficient way of entering a community and learning its practice language, is not necessary. As a result, although interactional expertise was initially seen as a relatively rare commodity—the property of the social scientist, specialist journalist, technically skilled manager, or peer reviewer, for example—we now see that it is far more widespread (Collins & Evans, 2015; Collins et al., 2017). Not only are contributory experts also interactional experts, interactional expertise may be shared wherever members of two different social groups interact on a regular basis and for a prolonged period of time.

The Imitation Game

The Imitation Game is a new research method that has been developed to explore the content and distribution of interactional expertise. It is a more rigorous version of the parlour game that inspired Alan Turing's famous *Turing Test* for the intelligence of computers (Turing, 1950). A basic Imitation Game consists of three players. One player, drawn from the *target group*, acts as the *Judge/Interrogator* and creates questions that are sent to the other two players. One of these, the *Non-Pretender*, is always drawn from the target group and answers naturally. The other, the *Pretender*, is recruited from a different group and asked to answer as if they were a member of the target group. The Judge/Interrogator then compares the answers and tries to work out which comes from the Pretender and which from the Non-Pretender. The hypothesis is that, where the Pretender has interactional expertise, the Judge/Interrogator will be unable to distinguish between the two sets of answers, no matter how demanding the questions they set. In many Imitation Games the players, who must be hidden from each other and may be in remote locations, interact via computers using specialist software but the game can also be played much more simply, over email via a *postman* who conceals the identities, or even with paper and pencil and some screens. (For more details of Imitation Game research see Collins et al., 2015, 2006; Collins & Evans, 2014; Evans & Collins, 2010.)

There are now several different variations of the Imitation Game method which have been used over a wide range of topics. The *classic*, three-player version is well suited to in-depth studies that probe the interactional expertise of individuals or small groups. This approach informed much of the early work using the method (Collins et al., 2006; Giles, 2006), where Collins demonstrated that his extended sociological fieldwork had enabled him to develop interactional expertise in gravitational wave physics. Other, more recent, work has used variants of this small-scale approach. Wehrens has used small-scale Imitation Games to prompt group discussions as part of a larger study investigating the extent to which therapists in an eating disorder clinic can take the perspective of their patients (Wehrens, 2014), whilst Evans and Crocker (2013) used the method to demonstrate that dieticians are able to take the perspective of patients with celiac disease and articulate the lived experience of the condition in ways that go beyond everyday understandings. Kubiak and Weinel (2016) have used Imitation Games to explore the change in East German identities following the reunification of Germany, showing that younger East Germans tend to see East Germany in geographical rather than cultural or institutional terms, and Collins (2016a) has used a variation on the method to explore how successful different kinds of Judges are at distinguishing between answers provided by experts and non-experts. This latter experiment used gravitational wave physics as the case study and showed that the expertise needed to make good judgements was restricted to contributory and interactional experts, with lay participants completely unable to distinguish between answers provided by experts and non-experts.

Larger-scale research has focused on more general topics such as religion, gender, and sexuality, with a significant amount of methodological work done to refine and develop protocols. Early reports of this research are published in Collins and Evans (2014), whilst Collins et al. (2015) include examples of content analysis and an analysis of the effect of running Imitation Games with the three roles played by small groups, rather than individuals. These early results show that Pretenders find it harder to succeed when questions are set by groups of Judges rather than individuals and that, even where groups are very diverse—as would be the case for gender—it is still possible for Judges to correctly identify the Pretender and Non-Pretender.

Three-Dimensional Model of Expertise

The sociological model of expertise put forward in this chapter can be summed up by saying that it operationalizes expertise along three dimensions (Collins, 2013b):

1. *Individual accomplishment*: this is similar to the model proposed by the philosopher Dreyfus (e.g., Dreyfus, 2004), as well as many analysts drawn from the psychological tradition, and captures the proficiency of the individual with respect to the domain of expertise in question.
2. *Esotericity*: this refers to ease with which the social collectivity that holds the expertise can be accessed, with ubiquitous expertises being at the most open end of this scale.
3. *Exposure to tacit knowledge*: this describes the way in which the learner has accessed the domain, ranging from published sources that add no specialist tacit knowledge through linguistic socialization in the practice language to full participation in the form of life.

Combining these gives the three-dimensional expertise space shown in Figure 4.1. As with the Periodic Table of Expertises, disaggregating expertise into different elements allows a much richer description of expertise and the ways in which it develops and spreads over time. For example, it is possible to see how an expertise such as natural language speaking can be ubiquitous, whilst other domains of expertise, such as the sciences, are more esoteric and, in addition, to see how this might change over time (e.g., car driving might once have been esoteric but it is now nearly ubiquitous). It also enables individual accomplishment and practice to be set within a broader scheme that suggests what kind of practice is needed—thus, if full mastery means reaching the right-hand edge of the back wall of the *expertise space diagram*, then practice must involve social interaction with the relevant community as, without this, the tacit knowledge of the domain can never be attained.

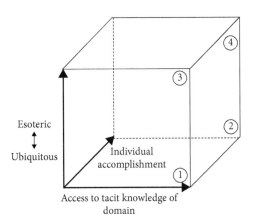

1: 5-year old language speaker   2: Accomplished adult language speaker
3: Novice sociologist   4: Poet (virtuoso)

FIGURE 4.1 Three dimensions of expertise.

Applications and Future Challenges

The SEE programme has an ambitious research agenda that includes a number of elements. The Periodic Table of Expertises has provided a useful starting point for research, with the category of interactional expertise being particularly important in terms of generating new insights. There is, however, still much work to do in understanding the nature of meta-expertises. For example, just in terms of the Periodic Table of Expertises, there is more work to be done on the potential overlap between the internal meta-expertises, which require some knowledge of the domain, and the higher level specialist expertises in which this knowledge is complete. Looking more broadly, we can also ask how meta-expertise categories like discrimination relate to psychological work on heuristics or decision-making under uncertainty. There is also much more that can be done with the Imitation Game, in terms of developing the protocols needed for comparative and longitudinal research and exploiting its potential for the in-depth analysis of particular domains of expertise. Finally, and perhaps most importantly given the concerns about policy-making that led to the development of SEE, there is a clear need for a more thorough engagement with political theory in order to better understand the relationship between expertise and democracy. As we cannot cover all these issues here, we use the remainder of this section to examine what SEE can contribute to the understanding of two contemporary issues in which expertise is central.

Interdisciplinary Working

Interdisciplinary working requires collaboration and co-operation between different groups of experts. A number of different models exist for understanding how this can happen, with trading zones (Galison, 1997) and boundary objects (Star & Griesemer, 1989) being two of the most common in the STS literature. The ideas associated with SEE can contribute to this literature in two different ways, with the idea of interactional expertise being particularly important. First, the typology of expertises provided by the Periodic Table of Expertises can be used to identify common themes and differences between different approaches and so provides an overarching framework within which different forms and modes of collaboration can be discussed and compared (Collins, Evans, & Gorman, 2007; Evans & Marvin, 2006; Fisher et al., 2015).

Second, the idea of interactional expertise can be used to design and facilitate interdisciplinary collaborations. Emphasizing socialization and the importance of tacit knowledge suggests that successful collaborations will require at least one group to learn new ways of seeing and doing in order to understand what the other is seeking to achieve and why (Galison, 1997; Gorman, 2002). As acquiring interactional expertise is not easy, time and space for this to occur will need to be created (Fisher et al., 2015).

Quite what form of interaction is most suitable for promoting and supporting collaboration needs much further research. Working within the SEE framework, methods for promoting interactional expertise are of most concern and build on the recognition that interactional expertise is more widespread than first thought and might, therefore, play a more significant role in the division of labour than previously imagined. This is particularly clear in Collins' (2004a, 2013a, 2017) studies of gravitational wave physics, where a collaboration of specialists was able to work because each was able to understand what their colleagues in other subfields were doing. The novel insight that comes from SEE is that the scientists did not need to be practitioners (i.e., contributory experts) in those other subfields in order to work together; what they needed to be was interactional experts who could understand the discourse of those subfields.

This finding has implications for professional development and training in many areas. In medical contexts, for example, the Imitation Game might be used to examine how well different kinds of medical practitioners are able to transcend the medical model and develop the interactional expertise needed to bridge the gap between biomedicine and the lived experience of patients (Evans & Crocker, 2013; Wehrens, 2014). This approach also has some resonances with the idea of *T-shaped* expertise (Conley, Foley, Gorman, Denham, & Coleman, 2017; Glushko, 2008) in which the vertical bar is the person's contributory expertise and the horizontal bar represents their growing interactional expertise in the domains where this is to be applied. It also suggests novel forms of teaching, not least that of students on multidisciplinary courses learning subjects like thermodynamics as a *second language* rather than *mathematical practice*, that is, as an interactional expertise rather than a contributory expertise (Berardy, Seager, & Selinger, 2011).

Role of Experts in a *Post-truth* Society

When the SEE program was first proposed, the problems affecting technological decision-making in the public domain were mostly associated with technocracy. Although there were some instances of what could be called excessive populism, such as the MMR controversy in the UK (Boyce, 2007), Thabo Mkeki's refusal to approve the use of AZT in South Africa (Weinel, 2010), and the deliberate attempts to create the impression of a controversy within climate change research (Oreskes & Conway, 2010) mentioned earlier, these were relatively rare. In 2002, there was no sense that the anti-establishment populism that characterizes contemporary political discourse in the USA and, to a lesser extent, Europe was only a decade or so away. For a more wide-ranging analysis of this *war on expertise* see Chapter 49, "The 'War' on Expertise", by Klein et al. (this volume).

The contribution of STS to these events is difficult to pin down. Despite the claims made by so-called *science warriors* (Gross & Levitt, 1998; Koertge, 2000) there is little evidence of any causal connection between STS and the rise of anti-science sentiments, not least because the vast majority of STS is very *scientific* in its approach (Labinger & Collins, 2001). That said, it is also clear that wave two STS struggles to respond to these events because the relational model of expertise it adopts provides no means to criticize the attributions of expertise made by others (for an example of the difficulties this creates see Sismondo, 2017; for a reply see Collins, Evans, & Weinel, 2017b).

In contrast, SEE provides a clear way to articulate concerns about the missapplication of expertise. The key ideas are the distinction between the technical and political elements of a decision (Collins & Evans, 2002; Collins et al., 2010), which determines when esoteric and ubiquitous expertises respectively are needed, and the normative claim that a scientific approach provides a morally superior way of answering the propositional questions that characterize the technical phase. The argument is set out in detail in *Why Democracies Need Science* (Collins & Evans, 2017), but the basic principle is that people who know what they are talking about should be preferred to those who do not in technical matters. More specifically, preferring someone who knows what they are talking about means preferring someone with substantial relevant experience who adheres to the norms of science such as being willing to admit they are wrong, listening to counter-claims and criticism irrespective of their source, striving to corroborate or falsify results, and doing their best to put aside vested interests. Crucially, the model turns on moral arguments about the kind of society that we want to live in as opposed to instrumental or utilitarian arguments about the probability of experts being correct. This matters because it is only in this way that SEE can provide a justification for preferring science even when it is sometimes obvious that science will be unable to provide certainty in the short term.

In summary, the studies of expertise and experience approach provides a theory of expertise that includes both a technical understanding of the nature of knowledge and a normative agenda that aims to shape how expertise is used in democratic societies. The core of the technical element is the idea of a form of life, with expertise understood as the outcome of successful socialization that can be demonstrated by social fluency. The political element relates to the use of expertise in democratic societies, where the novel contribution of SEE is the distinction between the technical and political elements of a decision, treating them as different domains of expertise. By making this distinction, SEE avoids both the reification of science that characterizes wave one and the collapse of expertise into nothing but attributions of status that paralyses wave two. Putting both the technical and political dimensions of SEE together provides an account of science that is faithful to its practices and which sees it as complementing—not usurping—the overarching role of democratic institutions.

References

Barnes, B., Bloor, D., & Henry, J. (1996). *Scientific knowledge: a sociological analysis*. Chicago: University of Chicago Press.

Berardy, A., Seager, T. P., & Selinger, E. (2011). Developing a pedagogy of interactional expertise for sustainability education. In *IEEE International Symposium on Sustainable*

Systems and Technology (ISSST) (pp. 1-4). Chicago: IEEE. https://doi.org/10.1109/ISSST.2011.5936891

Bloor, D. (1973). Wittgenstein and Mannheim on the sociology of mathematics. *Studies in History and Philosophy of Science Part A* 4(2), 173-191. https://doi.org/10.1016/0039-3681(73)90003-4

Bloor, D. (1991). *Knowledge and social imagery*, 2nd edn. Chicago: University of Chicago Press.

Boyce, T. (2007). *Health, risk and news: The MMR vaccine and the media*. New York: Peter Lang.

Callon, M., Lascoumes, P., & Barthe, Y. (2011). *Acting in an uncertain world: an essay on technical democracy*, trans. G. Burchell. Cambridge, MA: MIT Press.

Carr, E. S. (2010). Enactments of Expertise. *Annual Review of Anthropology* 39(1), 17-32. https://doi.org/10.1146/annurev.anthro.012809.104948

Collins, H. M. (1974). The TEA set: Tacit knowledge and scientific networks. *Science Studies* 4(2), 165-185.

Collins, H. M. (1975). The seven sexes: A study in the sociology of a phenomenon, or the peplication of experiments in physics. *Sociology* 9(2), 205-224. https://doi.org/10.1177/003803857500900202

Collins, H. M. (1985/1992). *Changing order: Replication and induction in scientific practice*. Chicago: University of Chicago Press.

Collins, H. M. (1996). In praise of futile gestures: How scientific is the sociology of scientific knowledge? *Social Studies of Science* 26(2), 229-244. https://doi.org/10.1177/030631296026002002

Collins, H. M. (2001). Tacit knowledge, trust and the Q of sapphire. *Social Studies of Science* 31(1), 71-85. https://doi.org/10.1177/030631201031001004

Collins, H. M. (2004a). *Gravity's shadow: The search for gravitational waves*. Chicago: University of Chicago Press.

Collins, H. M. (2004b). Interactional expertise as a third kind of knowledge. *Phenomenology and the Cognitive Sciences* 3(2), 125-143. https://doi.org/10.1023/B:PHEN.0000040824.89221.1a

Collins, H. M. (2007). Bicycling on the moon: Collective tacit knowledge and somatic-limit tacit knowledge. *Organization Studies* 28(2), 257-262. https://doi.org/10.1177/0170840606073759

Collins, H. M. (2008). Actors' and analysts' categories in the social analysis of science. In P. Meusburger, M. Welker, & E. Wunder (Eds), *Clashes of knowledge* (Vol. 1, pp. 101-110). Netherlands: Springer. Retrieved from http://dx.doi.org/10.1007/978-1-4020-5555-3_4

Collins, H. M. (2010). *Tacit and explicit knowledge*. Chicago; London: University of Chicago Press.

Collins, H. M. (2011). Language and practice. *Social Studies of Science* 41(2), 271-300. https://doi.org/10.1177/0306312711399665

Collins, H. M. (2013a). *Gravity's ghost and big dog: Scientific discovery and social analysis in the twenty-first century* (enlarged edn). Chicago: University of Chicago Press.

Collins, H. M. (2013b). Three dimensions of expertise. *Phenomenology and the Cognitive Sciences* 12(2), 253-273. https://doi.org/10.1007/s11097-011-9203-5

Collins, H. M. (2016a). An Imitation Game concerning gravitational wave physics. *arXiv:1607.07373 [Physics]*. Retrieved from http://arxiv.org/abs/1607.07373

Collins, H. M. (2016b). Interactional expertise and embodiment. In J. Sandberg, L. Rouleau, A. Langley, & H. Tsoukas (Eds), *Skillful performance: Enacting expertise, competence, and capabilities in organisations* (Vol. 7). Oxford: Oxford University Press.

Collins, H. M. (2017). *Gravity's kiss: The detection of gravitational waves.* Cambridge, MA: MIT Press.

Collins, H. M., & Evans, R. (2002). The Third Wave of science studies: Studies of expertise and experience. *Social Studies of Science* 32(2), 235–296. https://doi.org/10.1177/0306312702032002003

Collins, H. M., & Evans, R. (2007). *Rethinking expertise.* Chicago: University of Chicago Press.

Collins, H. M., & Evans, R. (2014). Quantifying the tacit: The Imitation Game and social fluency. *Sociology* 48(1), 3–19. https://doi.org/10.1177/0038038512455735

Collins, H. M., & Evans, R. (2015). Expertise revisited, Part I—Interactional expertise. *Studies in History and Philosophy of Science Part A* 54, 113–123. https://doi.org/10.1016/j.shpsa.2015.07.004

Collins, H. M., & Evans, R. (2017). *Why democracies need science.* Cambridge, UK; Malden, MA: Polity Press.

Collins, H. M., Evans, R., & Gorman, M. (2007). Trading zones and interactional expertise. *Studies in History and Philosophy of Science Part A* 38(4), 657–666. https://doi.org/10.1016/j.shpsa.2007.09.003

Collins, H. M., Evans, R., Ribeiro, R., & Hall, M. (2006). Experiments with interactional expertise. *Studies in History and Philosophy of Science Part A* 37(4), 656–674. https://doi.org/10.1016/j.shpsa.2006.09.005

Collins, H. M., Evans, R., & Weinel, M. (2016). Expertise revisited, Part II: Contributory expertise. *Studies in History and Philosophy of Science Part A* 56, 103–110. https://doi.org/10.1016/j.shpsa.2015.07.003

Collins, H. M., Evans, R., & Weinel, M. (2017a). Interactional expertise. In C. Miller, L. Smith-Doerr, U. Felt, & R. Fouche (Eds), *Handbook of Science and Technology Studies*, 4th edn. Cambridge, MA: MIT Press.

Collins, H. M., Evans, R., & Weinel, M. (2017b). STS as science or politics? *Social Studies of Science* 47(4), 580–586. https://doi.org/10.1177/0306312717710131

Collins, H. M., Evans, R., Weinel, M., Lyttleton-Smith, J., Bartlett, A., & Hall, M. (2015). The Imitation Game and the nature of mixed methods. *Journal of Mixed Methods Research* 11(4), 510–527. https://doi.org/10.1177/1558689815619824

Collins, H. M., Weinel, M., & Evans, R. (2010). The politics and policy of the Third Wave: new technologies and society. *Critical Policy Studies* 4(2), 185–201. https://doi.org/10.1080/19460171.2010.490642

Conley, S. N., Foley, R. W., Gorman, M. E., Denham, J., & Coleman, K. (2017). Acquisition of T-shaped expertise: an exploratory study. *Social Epistemology* 31(2), 165–183. https://doi.org/10.1080/02691728.2016.1249435

Douglas, H. E. (2009). *Science, policy, and the value-free ideal.* Pittsburgh, PA: University of Pittsburgh Press.

Dreyfus, S. E. (2004). The five-stage model of adult skill acquisition. *Bulletin of Science, Technology & Society* 24(3), 177–181. https://doi.org/10.1177/0270467604264992

Ericsson, K. A., Krampe, R. T., & Tesch-Römer, C. (1993). The role of deliberate practice in the acquisition of expert performance. *Psychological Review* 100(3), 363–406. https://doi.org/10.1037/0033-295X.100.3.363

Ericsson, K. A., & Pool, R. (2016). *Peak: secrets from the new science of expertise.* Boston: Houghton Mifflin Harcourt.

Evans, R. (2008). The sociology of expertise: The distribution of social fluency. *Sociology Compass* 2(1), 281–298. https://doi.org/10.1111/j.1751-9020.2007.00062.x

Evans, R., & Collins, H. (2010). Interactional expertise and the Imitation Game. In M. E. Gorman (Ed.), *Trading zones and interactional expertise creating new kinds of collaboration* (pp. 53–70). Cambridge, MA: MIT Press.

Evans, R., & Crocker, H. (2013). The Imitation Game as a method for exploring knowledge(s) of chronic illness. *Methodological Innovations Online* 8(1), 34–52. https://doi.org/10.4256/mio.2013.003

Evans, R., & Marvin, S. (2006). Researching the sustainable city: Three modes of interdisciplinarity. *Environment and Planning A* 38(6), 1009–1028. https://doi.org/10.1068/a37317

Felt, U., Fouche, R., Miller, C., & Smith-Doerr, L. (Eds). (2017). *The handbook of science and technology studies*, 4th edn. Cambridge, MA: The MIT Press.

Fisher, E., O'Rourke, M., Evans, R., Kennedy, E. B., Gorman, M. E., & Seager, T. P. (2015). Mapping the integrative field: Taking stock of socio-technical collaborations. *Journal of Responsible Innovation* 2(1), 39–61. https://doi.org/10.1080/23299460.2014.1001671

Funtowicz, S. O., & Ravetz, J. R. (1993). Science for the post-normal age. *Futures* 25(7), 739–755. https://doi.org/10.1016/0016-3287(93)90022-L

Galison, P. L. (1997). *Image and logic a material culture of microphysics*. Chicago: University of Chicago Press.

Gieryn, T. F. (1999). *Cultural boundaries of science: Credibility on the line*. Chicago: University of Chicago Press.

Giles, J. (2006). Sociologist fools physics judges. *Nature* 442(7098), 8. https://doi.org/10.1038/442008a

Glushko, R. J. (2008). Designing a service science discipline with discipline. *IBM Systems Journal* 47(1), 15–27. https://doi.org/10.1147/sj.471.0015

Gorman, M. E. (2002). Levels of expertise and trading zones: A framework for multidisciplinary collaboration. *Social Studies of Science* 32(5-6), 933–938. https://doi.org/10.1177/030631270203200511

Gross, P. R., & Levitt, N. (1998). *Higher superstition: The academic left and its quarrels with science*. Baltimore: Johns Hopkins University Press.

Harding, S. G. (2006). *Science and social inequality: Feminist and postcolonial issues*. Urbana: University of Illinois Press.

Irwin, A. (1995). *Citizen science: a study of people, expertise, and sustainable development*. London; New York: Routledge.

Jasanoff, S. (2003). Breaking the waves in science studies: Comment on H.M. Collins and Robert Evans, "The Third Wave of science studies." *Social Studies of Science* 33(3), 389–400. https://doi.org/10.1177/03063127030333004

Jasanoff, S. (2013). Fields and fallows: A political history of STS. In A. Barry & G. Born (Eds), *Interdisciplinarity: Reconfigurations of the natural and social sciences* (pp. 99–118). London: New York: Routledge.

Koertge, N. (Ed.). (2000). *A house built on sand: exposing postmodernist myths about science*. New York: Oxford University Press.

Kubiak, D., & Weinel, M. (2016). DDR-Generationen revisited—Gibt es einen Generationszusammenhang der "Wendekinder"? In A. Lettrari, C. Nestler, & N. Troi-Boeck (Eds), *Die Generation der Wendekinder* (pp. 107–129). Wiesbaden: Springer Fachmedien Wiesbaden. https://doi.org/10.1007/978-3-658-11480-0_8

Kuhn, T. S. (1996). *The structure of scientific revolutions*, 2nd edn, enlarged, Vol. 14. Chicago: University of Chicago Press.

Labinger, J. A., & Collins, H. M. (Eds). (2001). *The one culture?* Chicago: University of Chicago Press. Retrieved from http://www.press.uchicago.edu/ucp/books/book/chicago/O/bo3634845.html

Latour, B., & Woolgar, S. (1979). *Laboratory life: The social construction of scientific facts.* Beverly Hills: Sage Publications.

Longino, H. E. (1990). *Science as social knowledge: Values and objectivity in scientific inquiry.* Princeton, NJ: Princeton University Press.

Merton, R. K. (1973). *The sociology of science: Theoretical and empirical investigations.* Chicago: University of Chicago Press.

Oreskes, N., & Conway, E. M. (2010). *Merchants of doubt: How a handful of scientists obscured the truth on issues from tobacco smoke to global warming.* New York: Bloomsbury Press.

Ottinger, G. (2010). Buckets of resistance: Standards and the effectiveness of citizen science. *Science, Technology & Human Values* 35(2), 244–270. https://doi.org/10.1177/0162243909337121

Polanyi, M. (1966). *The tacit dimension.* Chicago; London: University of Chicago Press.

Shapin, S. (2007). *A social history of truth: Civility and science in seventeenth-century England.* Chicago: University of Chicago Press.

Sismondo, S. (2010). *An introduction to science and technology studies*, 2nd edn. Chichester, UK; Malden, MA: Wiley-Blackwell.

Sismondo, S. (2017). Post-truth? *Social Studies of Science* 47(1), 3–6. https://doi.org/10.1177/0306312717692076

Star, S. L., and Griesemer, J. R. (1989). Institutional ecology, 'translations' and boundary objects: amateurs and professionals in Berkeley's Museum of Vertebrate Zoology, 1907–39. *Social Studies of Science* 19(3), 387–420.

Turing, A. (1950). Computing machinery and intelligence. *Mind* 59(236), 433–460. https://doi.org/10.1093/mind/LIX.236.433

Wehrens, R. (2014). The potential of the Imitation Game method in exploring healthcare professionals' understanding of the lived experiences and practical challenges of chronically ill patients. *Health Care Analysis* 23(3), 253–271. https://doi.org/10.1007/s10728-014-0273-8

Weinel, M. (2010). *Technological decision-making under scientific uncertainty: Preventing mother-to-child transmission of HIV in South Africa*, PhD dissertation. Cardiff University. Retrieved from http://orca.cf.ac.uk/55502/

Winch, P. (1958). *The idea of a social science and its relation to philosophy.* London; New York: Routledge & Kegan Paul.

Wittgenstein, L. (1953). *Philosophical investigations*, trans. G. E. M. Anscombe. Oxford: Blackwell.

Wynne, B. (1992). Misunderstood misunderstanding: Social identities and public uptake of science. *Public Understanding of Science* 1(3), 281–304. https://doi.org/10.1088/0963-6625/1/3/004

Wynne, B. (2003). Seasick on the Third Wave? Subverting the hegemony of propositionalism: Response to Collins & Evans (2002). *Social Studies of Science* 33(3), 401–417. https://doi.org/10.1177/03063127030333005

CHAPTER 5

GIFTEDNESS AND TALENT DEVELOPMENT IN CHILDREN AND YOUTH

STEVEN I. PFEIFFER

INTRODUCTION

THIS chapter provides an overview on recent thinking about gifted and talented children and youth. The chapter briefly discusses a history of gifted education and then tackles some big picture issues and future possibilities. The chapter addresses a number of questions, including the following: Who are the gifted? How are gifted identified? Is giftedness domain-specific or domain-general? How malleable is giftedness? Does giftedness represent a qualitative or quantitative difference? How does the concept of expertise fit into gifted education? The chapter concludes with future directions and challenges facing the gifted field. Because of space limitations, the chapter does *not* tackle the topic of the social and emotional development of the gifted; the interested reader is directed to resources by Neihart, Pfeiffer, & Cross (2016), and Pfeiffer, Shaunessy-Dedrick, & Foley-Nicpon (2017).

WHO ARE THE GIFTED AND TALENTED?

To many, it may seem trivial to ask the question of exactly who are the gifted. If you work in schools, at least in the USA, then you know exactly who they are. The gifted are those students who meet the eligibility criteria that your school district or state stipulate to qualify as gifted students. And if you are a psychologist, then you also know who the gifted are. They are those youngsters who obtain an IQ test score that exceeds a certain threshold, according to what you learned demarcates giftedness—typically, two standard deviations above the mean in most places (Pfeiffer, 2015). These are the views of a great many practitioners, according to a recent national survey (McClain & Pfeiffer, 2012). This chapter departs from a traditional and, some might argue, antiquated view and provides a more recent and nuanced perspective on what we mean by the gifted and talented.[1] For example, recent conceptions of giftedness reflect an appreciation for domain-specific talent development, consideration of contextual and sociocultural factors, a stronger orientation toward processes that moderate and mediate the unfolding of gifts, and talent development over the lifespan. These are recent changes shared by leading researchers in the gifted and talent development field internationally (Stoeger, Balestrini, & Ziegler, 2017).

High IQ equals gifted dominated twentieth-century thinking. But we are now in a new millennium, and research in developmental psychology and the cognitive neurosciences has informed our thinking about high ability students—the gifted (Pfeiffer, 2002, 2003, 2013b, 2015; Ziegler, Stoeger, & Vialle, 2012). Shortly, the chapter will examine a few different models of giftedness that lead to different ways for conceptualizing and identifying giftedness. Most readers would agree that the young child who is reading at age three, excelling at competitive chess by age six, writing computer software programs at age eight, or performing piano in an orchestra by age ten is gifted. These four examples are indicative of children who are developmentally advanced, one hallmark of giftedness (Pfeiffer, 2001, 2002, 2009, 2012, 2013b).

Most gifted authorities, even today, agree that the gifted are those in the upper 3–10% compared to their same-age peers in general intellectual ability, distinguished performance in one or more academic domains, and/or evidence of creative work (Gagné, 2009; Pfeiffer, 2012, 2015; Wai, 2014). Not surprisingly, research confirms that there is a genetic influence in the expression of giftedness, at least at the high end of the IQ continuum (Plomin & Spinath, 2004). For example, the fields of music and mathematics are particularly rich with examples of child prodigies. Evidence also comes from the emergence of eminence among young children from impoverished environments (Nisbett, 2009). However, most authorities today also agree that the full unfolding of gifts and talents requires a nurturing and supportive environment, available resources, certain personality characteristics, and even good fortune (Foley-Nicpon & Pfeiffer, 2011; Pfeiffer, 2012, 2013b, 2015). This is a recent departure from traditional, twentieth-century views of giftedness seen as a fixed and immutable state of the young child.

The definition of gifted adopted for this chapter is based on the *tripartite model of giftedness*, which will be explained shortly. The definition of gifted is the following: "The gifted child demonstrates a greater likelihood, when compared to other students of the same age, experience and opportunity, to achieve extraordinary accomplishments in one or more culturally valued domains" (Pfeiffer, 2013b). Based on this definition, a child's gifts can be in any culturally valued domain, such as science,

---

[1] Whenever the chapter refers to "gifted," the meaning is "gifted and talented," terms often used interchangeably in the gifted literature.

mathematics, athletics, the performing arts, leadership, and even community volunteerism. The list of gifts is almost inexhaustible, limited only by what the culture values. As the child gets older, in most cultures and societies, there is increased opportunity for exposure to a growing number of different domains in which they can excel and gain expertise and eminence (Pfeiffer, 2015). For example, the young child who demonstrates precocious mathematical abilities at age seven has the potential to explore a wide variety of academic trajectories and become distinguished in a number of careers as an adult.

The gifted student's academic needs—in the USA and globally—are frequently *not* substantially met in the classroom or school, and quite often they require specialized programs or services not ordinarily provided in the regular classroom (Pfeiffer, 2015). This is the primary justification for gifted assessment in the schools—to determine whether a bright student has extraordinary intellectual abilities or evidence of potential for outstanding performance, frequently indicative of a need for special educational programs or resources not presently available in the regular classroom (Pfeiffer, 2018a).

Brief History of Gifted Education

There has been much written about the history of giftedness and gifted assessment (e.g., Mönks, Heller, & Passow, 2000; Pfeiffer, 2008, 2018a; Robinson & Clinkenbeard, 2008; Tannenbaum, 1983, 2000). As far back as Confucius in China and Plato in Greece, philosophers wrote about *heavenly* children. Their writings theorized on what high ability constituted, and also provided suggestions for how society should find and nurture these special young citizens (Mönks, Heller, & Passow, 2000).

In US and Western culture, we trace the early roots of gifted education to Lewis Terman's research at Stanford University. Terman conducted a large longitudinal study which followed a cohort of students tested with IQ scores at or above 140. Terman collected considerable data on these students over the course of fifty years. He stated that the "twofold purpose of the project was ... to find what traits characterize children of high IQ, and ... to follow them for as many years as possible to see what kind of adults they might become" (Terman, 1925, 223; Terman & Oden, 1951, 21). Terman concluded that children of high IQ (140 or higher) are healthier, better-adjusted, and higher achievers than unselected children (Robinson & Clinkenbeard, 2008).

There are other early Western studies and writings on the gifted, such as Galton's *Hereditary Genius* (1869) and Cattell's *A Statistical Study of American Men of Science* (1906–1910) (Whipple, 1924). However, nothing quite captured the imagination of the public as did Terman's *Genetic Studies of Genius* (Mönks, Heller, & Passow, 2000). Terman's work played a pivotal role in equating giftedness and high IQ in Western thinking. Almost one hundred years later, Terman's influence in Western culture remains pronounced. This traditional *gifted child focus* emphasizes general intelligence and assumes that the gifted reflect a clearly demarcated and fixed category of exceptional individuals who differ in quantitative *and* qualitative ways from their non-gifted peers. The *gifted child focus* dominated twentieth-century Western cultural thinking. It has been the major zeitgeist in gifted education through the twentieth century in Western societies.

Numerous publications offer insights into non-Western sociocultural conceptions of giftedness (cf., for an overview, Phillipson & McCann, 2007) and intelligence (cf., for an overview, Niu & Brass, 2011). Clearly, there have been, and still remain, large-scale East–West differences in giftedness conceptions. Perhaps the most extensive research on non-Western conceptions of giftedness has focused on East Asia (Stoeger, Balestrini, & Ziegler, 2017; Van Tassel-Baska, 2013). Typical of East-Asian conceptions of giftedness is less emphasis on entity theories of giftedness and greater focus on educability and Confucian holistic outlooks (Phillipson, 2013), as opposed to Western society, mind–body dualism. This translates in the East into more malleable views of learning and intelligence (and giftedness), and greater emphasis on effort and motivation (Li & Fischer, 2004; Pfeiffer, 2013a; Yeo & Pfeiffer, 2018). These Eastern views are consistent with the newly emerging talent development approach to the education of gifted students taking root in the West.

This new focus that is emerging and challenging the dominant *gifted child focus* in Western societies is a shift aligned more with Eastern philosophy and thinking. It is labeled a *talent development perspective* (Dai, 2010; Pfeiffer, 2013b, 2015; Worrell, Subotnik, & Olszewski-Kubilius, 2017). Major proponents of the talent development perspective advocate that this new zeitgeist in gifted education lends itself logically to *a path toward eminence* (Worrell, Subotnik, & Olszewski-Kubilius, 2017). The talent development approach doesn't necessarily define giftedness differently, but it clearly embraces a non-static, ecological, transactional, and developmental viewpoint in explaining the unfolding of gifts and talents over the individual's life. It elegantly links the education of the gifted, at an early age, with career planning and ultimate career eminence, at later ages. But before we get too far ahead of ourselves, let's return to some of the early figures in the history of gifted education.

Another early, influential figure was James Gallagher.[2] In 1960 Gallagher submitted to the Illinois legislature a report whose purpose was "to review and summarize all of the information now available relating to the education of gifted children" (Gallagher, 1960, 3). Gallagher's report, *Analysis of Research on the Education of Gifted Children*, concluded that "special programming for gifted children requires additional personnel and services" (131). Gallagher pointed out that only two cents out of every $100 spent on K–12 education in the USA supports the gifted, and that existing programs for the gifted do not reach nearly enough gifted students in America's schools (Pfeiffer, 2013b). He added that special programs for the gifted are a low priority, the federal role in services to the gifted is all but nonexistent, and that "gifted students have been relatively ignored in educational programs such as No Child Left Behind (and the more recent federal legislation, Reach for the Top)" (Gallagher, 2008, 7). In 2006, for

---

[2] One of my mentors in graduate school at the University of North Carolina—Chapel Hill.

example, the US Department of Education spent nearly $84 billion. The only program specifically funded to address the education of the gifted got $9.6 million, one-hundredth of 1% of federal education expenditures. Few countries, in fact, allocate anything but very nominal funds earmarked for the gifted (VanTassel-Baska, 2013).

A number of other individuals have influenced the history of gifted education. Leta Hollingsworth (1886–1939), for example, played an important early role with her case studies into high IQ students in the New York City schools. Hollingsworth was a psychologist who practiced in New York City about the same time as Terman was a professor at Stanford. Hollingsworth is author of the first textbook on gifted education, *Gifted Children: Their Nature and Nurture* (1926).

Most countries have been slow to respond to the educational needs of students of high ability (Pfeiffer, 2015). Many authorities believe that the ambivalence and disinclination of governments to address the unique needs of high ability students is the result of society's perception that they are already a privileged group, and will succeed without special funding or services (Stephens, 2008, 2011, 2018). There is also a sense that *equity* trumps *excellence* in driving educational policy. And yet the US National Science Board (2010) recognizes this dilemma. It stated in a report, "the opportunity for excellence is a fundamental American value and should be afforded to all" (5).

One final point bears mentioning. A recent survey indicates that changes in definitions and categories of giftedness are occurring, both in the USA and globally. For example, across the USA, states vary considerably in how they identify gifted students. A majority of states still adhere to Terman's view that giftedness equates to high IQ, although they don't use quite as high a threshold or cut-score for demarcating giftedness. States frequently endorse a multiple cutoff or averaging approach to gifted decision-making (McClain & Pfeiffer, 2012). Most states continue to embrace a *gifted child focus*, and have not considered an alternative *talent development perspective* that de-emphasizes measured IQ and emphasizes domain-specific definitions of gifted (Dai, 2010; Pfeiffer, 2015). A recent survey in Europe reported similar findings (Tourón & Freeman, 2017).

Alternative Ways of Conceptualizing Giftedness

There are many different ways to conceptualize giftedness. Sternberg & Davidson (2005) suggest at least twenty different ways to view giftedness. Many different authors have proposed models of giftedness (Pfeiffer, 2008, 2018a). Most of the widely cited models fit into one of four popular models that are described next. These different models imply different ways to define, identify, and nurture gifts.

The different models vary in their level of detail and in how easily they can be translated into assessment protocols and intervention programs. They also vary in terms of their relative emphasis on the role of individual differences, developmental antecedents, genetics, family, and the environment (Ackerman, 2013; Pfeiffer, 2015; Simonton, 2014; Wai, 2014). The four models discussed next are *traditional psychometric views, talent development models, expert performance perspectives,* and *multiple intelligences.* These models are not contradictory to the fifth model, the *tripartite model* (Pfeiffer, 2013b, 2015), which will also be introduced.

Julian Stanley's mathematically and verbally precocious talent search model, for example, reflects thinking that cuts across two models: traditional psychometric views and talent development models (Stanley, 1976, 1990, 2000). Francoys Gagné's developmental, differentiated model of giftedness and talent (Gagné, 2005), Joseph Renzulli's three-ring conception of giftedness (Renzulli, 1978, 2005, 2011), and Rena Subotnik's developmental model (Subotnik, 2003; Subotnik, Olszewski-Kubilius, & Worrell, 2011) all are *talent development models.* K. Anders Ericsson's work epitomizes a somewhat unique way of thinking about giftedness with a strong emphasis on the environment, from an *expert performance perspective* (Ericsson, 1996, 2014; Ericsson & Charness, 1995).

There is growing consensus among most gifted authorities that giftedness is best viewed as *specific,* that the expression of giftedness occurs within a particular domain (Mayer, 2005). I agree with this viewpoint, at least when we consider students of high ability beginning around the third or fourth grade (Pfeiffer, 2013b, 2015). In the earlier grades, one could make a compelling argument that giftedness—or rather the prediction of giftedness—is not necessarily yet specific to one particular domain but rather a reflection of general intellectual ability and potential to excel (Pfeiffer, 2015). For example, most would agree that a three year old who is reading at a second-grade level is gifted.

Traditional Psychometric Views

Some readers are probably familiar with the psychometric view of giftedness, which conceptualizes high IQ as the defining feature of the construct giftedness. The traditional psychometric model views high IQ and giftedness as synonymous (Pfeiffer, 2002, 2012, 2013b, 2015), although it is unclear just how high constitutes giftedness. Many of the earliest researchers investigated the scientific basis of giftedness from a domain-general perspective, using the terms gifted, genius, and talented interchangeably. Francis Galton's book *Hereditary Genius* (1869) introduced the notion of intellectual genius to the public. Galton carefully and scientifically analyzed the family lineage of distinguished men and concluded that genius is genetically inherited. His estimations of genius were subjective, not based on psychometric measures, but nonetheless set the stage for the scientific study of giftedness (Ackerman, 2013; Kaufman & Sternberg, 2008; Pfeiffer, 2015).

Galton's work was followed by Charles Spearman (1904), who used the newly developed statistical tool of factor analysis to demonstrate that there was a significant amount of shared variance across most cognitive tests. He called this ubiquitous shared ability *g*, or general intelligence (now called *psychometric g*). The analyses that he ran also uncovered specific abilities unique to one or two of the tests, labeled specific abilities *s*. At around the same time, Alfred Binet and Theodore Simon (1916) developed a mental scale to identify students struggling in the Paris schools who might need alternative education. Binet and Simon's scale was the first test to include assessment of higher-level cognitive skills.

Lewis Terman adapted Binet and Simon's scale and created the Stanford–Binet Intelligence Scale, one of the first tests to actually identify gifted students (Terman, 1916). Terman's scale yielded a global score that viewed giftedness from a domain-general perspective and intelligence as a single entity. He proposed a classification system in which a youngster who obtained an IQ score of 135 or above was labeled *moderately gifted* (Terman, 1925), above 150 as *exceptionally gifted*, and above 180 as *severely and/or profoundly gifted* (Kaufman & Sternberg, 2008; Webb, Meckstroth, & Tolan, 1982). This classification system is still popular today.

Psychometric *g* or IQ continues to be *the* leading index for identifying the gifted both in the USA and internationally (McClain & Pfeiffer, 2012; Pfeiffer, 2013a; Sternberg, Jarvin, & Grigorenko, 2011). The IQ enjoys wide popularity because it provides a seemingly precise, impartial, objective, and quantifiable integer representing human intelligence. There also is a large research literature supporting the validity of the IQ score. There is arguably more published research testing the validity of the IQ construct than any other psychological construct (Neisser et al., 1996; Nisbett, 2009). IQ does predict school performance moderately well, and also predicts many other important life outcomes, including attainment of expertise (Simonton, 2017). However, there are certainly other psychological constructs that have also shown great promise in predicting school performance and life success (Simonton, 2014; Sternberg, Jarvin, & Grigorenko, 2011).

There are many within the psychometric camp who don't endorse a domain-specific model of intelligence. Louis Thurstone (1938) was the first researcher who challenged the prevailing domain-general model and proposed the notion of specific abilities as an alternative way of conceptualizing intelligence. Thurstone used a different method of factor analysis and identified seven primary and independent mental abilities. A growing body of studies supports hierarchical factor models of intelligence. The hierarchical models have general ability at the very top of the apex, more general higher cognitive or intellectual abilities at the next level, and various more specific cognitive or intellectual skills lower in the hierarchy. The hierarchical model that has gained greatest acceptance in the psychometric community is Carroll's (1993) three-stratum theory. In Carroll's model, Stratum I consists of highly specialized cognitive skills, Stratum II somewhat less specific and more broad domains of intellectual abilities, and Stratum III, at the apex, only one ability, the *g* factor.

Recently, Carroll's model and a second hierarchical model, the Horn and Cattell (1966) model of fluid and crystallized intelligence, were synthesized into the Cattell–Horn–Carroll (CHC) theory (Flanagan & Harrison, 2012). Although the CHC model includes *g* at the apex, its main emphasis is on the measurement of those factors and cognitive abilities at the middle stratum (Kaufman & Sternberg, 2008). The CHC theory has influenced the development and revision of a number of IQ tests used in gifted identification, including the fifth edition of the Stanford–Binet (Roid, 2003), the second edition of the Kaufman Assessment Battery for Children (KABC-II; Kaufman & Kaufman, 2004), and the third edition of the Woodcock–Johnson Cognitive Abilities Assessment (WJ III; Mather, Wendling, & Woodcock, 2001).

Talent Development Models

The dominant and most familiar conceptualization of giftedness remains the traditional psychometric view. However, there has been growing interest in a group of new models, called *talent development models*. Two talent development models will be described to illustrate the richness and variety of talent developmental models within the gifted field, and how they depart from psychometric views. Because of space limitations, important talent development models will not be covered, most notably the *three-ring conception of giftedness*, proposed by Renzulli (Renzulli, 1984, 2005, 2009; Renzulli & Reis, 2017). Renzulli's writings have had a huge impact on the instructional pedagogy and the differentiated curriculum for gifted learners globally (Pfeiffer, 2013b; Reis & Renzulli, 2009; Renzulli & Reis, 2017).

The Differentiated Model of Giftedness and Talent

Professor Francoys Gagné conceptualizes giftedness as natural abilities which are transformed through learning and experience into high-level skills, in particular occupational fields. In this regard, he views gifts as residing within the child, the result of favorable genetics, prenatal environment, and neurobiological status. Gagné's conceptualization is particularly unique in distinguishing gifts from talents (Gagné, 2005, 2009, 2017). His model, which he calls the differentiated model of giftedness and talent (DMGT), proposes four broad aptitude domains: intellectual, creative, socioaffective, and sensorimotor. Each of these aptitude domains can be subdivided. Gagné acknowledges that many different and competing classification systems exist at this next level (e.g., the intellectual domain could be subdivided into verbal and nonverbal intelligence, fluid and crystallized intelligence, Caroll's (1993) three-level system, Sternberg's (1985; 2001) triarchic and expertise theories, and many other views). The same is true for the other three broad aptitude domains in the DMGT model.

Gagné proposes that talents progressively emerge from the *systematic transformation* of aptitudes—in the case of gifted, high aptitudes, into well-developed skills characteristic of a particular field or domain. This view reflects an appreciation for development over time, viewing "the talent development process consisting of transforming specific natural abilities into the skills that define competence or expertise in a given occupational field" (Gagné, 2005, 103). In this regard, his model is compatible with the ideas proposed by other theorists supportive of a developmental view, such as

Subotnik and her colleagues (Subotnik, 2003; Subotnik, Olszewski-Kubilius, & Worrell, 2011), and my own thinking (Pfeiffer, 2013b, 2015).

Gagné's DMGT conceptualization posits a five-level gifted classification system. He sets the first threshold at 10%, which he labels mildly gifted. This equates to an IQ of approximately 120 with, on average, one in ten students considered mildly gifted. Gagné sets the second threshold at 1%, which he labels moderately gifted: students with IQ scores of 135. The next three levels are for the highly gifted (145 IQ), exceptionally gifted (155 IQ), and extremely gifted (165 IQ) (Gagné, 2017). Gagné's elegant writings have influenced many, including my own thinking. It is one of the first developmental models formulated in response to the field's early emphasis on wholly genetic determinants of giftedness based on high IQ (Kaufman & Sternberg, 2008).

Subotnik's Developmental Transitions in Giftedness and Talent

Rena Subotnik's innovative ideas on talent development parallel the DMGT model proposed by Gagné—they both discuss factors that mediate the full unfolding of gifts. Subotnik's work has had a profound impact on my own thinking. During my tenure as director of Duke's gifted program (Duke TIP), I invited Dr. Subotnik to campus to share her provocative ideas on how a talent development model explains how general and specific abilities transform into competencies, then expertise and ultimately outstanding performance (Subotnik, 2009; Subotnik & Rickoff, 2010). What was particularly compelling was hearing Dr. Subotnik's first-hand experience observing gifted young performing artists, and how these experiences formulated her thinking on talent development and attaining expertise in dance. It parallels my own *learning on the sidelines* experience, observing and working with highly gifted young female soccer players, who reached the highest levels of expertise in their sport *on the pitch* (Pfeiffer, 2015).

Subotnik eloquently observes that giftedness is a dynamic construct and that eminence and real-world creativity develop over time. Her developmental model posits that gifted children transition first from broad educational experiences in the early years to more narrowly focused domains in college, institutes, and conservatories. And if these same highly able learners continue on a trajectory of talent development, they will engage in experiences and opportunities that afford them *the pursuit of scholarly productivity, innovation*, or *artistry*. In other words, her model views "talent development as the transformation of abilities into competencies, competencies into expertise, and expertise into outstanding performance or seminal ideas" (Subotnik, 2009, 155).

Subotnik's developmental model is similar to Gagné's conceptualization. One notable difference is that Subotnik's model extends the vision of gifted education and takes a long view in articulating the ultimate goal of our efforts with high ability kids. She emphasizes that the goal of gifted education should be recruiting and providing a large number of bright kids of high ability with a range of facilitative opportunities and experiences over childhood, adolescence, and even young adult life to maximize the likelihood that as many as possible ultimately reach the highest levels of expertise, creativity, or eminence in different fields. Subotnik envisions talent development as consisting of a series of transitions and stages, with environmental factors and psychosocial variables including motivation, persistence, drive, the will to overcome obstacles, and high interest in a field as all playing central roles in propelling the child along the talent development trajectory. At each stage of development, different factors come into play. For example, in Stage 1 which is a *transition from ability to competency*, high levels of intrinsic motivation, persistence, responsiveness to external rewards, and teachability are critical factors (Subotnik, 2009; Subotnik, Olszewski-Kubilius, & Worrell, 2017; Worrell, Subotnik, & Olszewski-Kubilius, 2017).

Subotnik recognizes that not every domain or field follows exactly the same path, and that future research may illuminate age and gender differences across various domain trajectories. For example, there are likely significant differences in the age on onset and relative influence of facilitative factors that promote expertise in the fields of ballet, architecture, psychotherapy, aerospace physics, and surgical medicine. Subotnik expects that future research may help develop algorithms that predict the relative role that family, school, mentors, psychosocial variables, personality, and community play in the unfolding of talents across different domains. She also believes that definitions of giftedness must change over the course of a child's development and path toward eminence (Subotnik, Olszewski-Kubilius, & Worrell, 2017). Similar to my own thinking, she contends that giftedness should be defined in terms of actual accomplishment (Subotnik, 2003; Worrell, Subotnik, & Olszewski-Kubilius, 2013). This remains a contentious idea in the gifted field.

Stanley's Talent Search Model

As mentioned earlier, Julian Stanley's model incorporates features of both the traditional psychometric view and a talent development model. During my tenure at Duke, I had the good fortune of visiting with Professor Stanley at Johns Hopkins University and became acquainted with the talent search model that he created (Stanley, 1976). Stanley's model is based on an *above-level* testing protocol that is both ingenious and elegant. Stanley was familiar with Hollingsworth's use of above-level testing, in which a student is given a test designed for older students—in other words, above-level (Stanley, 1990). Stanley piloted his model with math prodigies who were given the mathematics section of the Scholastic Aptitude Test in the seventh and eighth grades; he later expanded his talent search beyond math prodigies (Assouline & Lupkowski-Shoplik, 2012).

Stanley's talent search model is predicated on the principle of administering an above-level test (i.e., a test designed for older students) to already-identified bright students (in the top 3 to 5% on grade-level standardized tests); the above-level (also called *out-of-level*) test protocol provides a much higher ceiling to help further differentiate the range of abilities among extraordinarily bright youngsters. Using above-level testing, he was able to *cherry-pick* the very brightest from among an already-select group of high ability students (Park, Lubinski, & Benbow, 2008).

Stanley recognized that discovering exceptionally high ability among the very brightest was not enough. He provided these uniquely gifted youngsters with a different type of intensive (i.e., *high-powered*), highly challenging, and accelerated curriculum and educational experiences on the campus of Johns Hopkins University. At this writing, Stanley's talent search model has expanded exponentially with literally hundreds of summer programs—and weekend, home study, and online educational programs—offered on campuses around the globe for gifted students identified through regional talent searches.

The talent search model is one of the most well-researched and empirically supported models of talent development (Subotnik, Olszewski-Kubilius, & Worrell, 2011). Many talent search students complete one or more years of mathematics in a 3-week summer program (Brody & Benbow, 1987; Kolitch & Brody, 1992; Stanley, 2000). There is considerable empirical support for the predictive validity of this domain-specific gifted identification system used by talent search programs (Olszewski-Kubilius, 2004; Park, et al., 2008; Pfeiffer, 2015). Youths identified before age 13 as demonstrating profound mathematical or verbal reasoning abilities have been tracked longitudinally for nearly three decades. And their outcomes, as a group, have been impressive (Kell, Lubinski, & Benbow, 2013). The Florida Governor's School for Science and Space Technology pilot program, which I co-directed on the campus of Florida State University, in collaboration with NASA and Kennedy Space Center, was designed incorporating Stanley's talent search model. The pilot program considered general measures of intellectual ability in its admissions process, but the admissions review process also put considerable weight on evidence of each applicant's specific abilities and accomplishments in science and math. In selecting finalists for the pilot summer academy affiliated with Kennedy Space Center/NASA, the program also considered each applicant's level of motivation, persistence, and passion for learning—added psychosocial elements that the designers of the program believed sweetened the recipe predicting who would benefit most from our gifted summer academy (Pfeiffer, 2013b).

Expert Performance Perspective

Professor Anders Ericsson has enjoyed a highly successful career investigating the concept of expertise and expert performance and how it is accomplished. His research has focused on identifying and "specifying the mediating mechanisms that can be assessed by process-tracing and experimental studies" (Ericsson, Roring, & Nandagopal, 2007, 13). Ericsson does not believe that IQ tests or intellectual ability, for that matter, play a particularly useful role in predicting performance domains of expertise. He advocates for the power of environmental variables, including what he labels as the importance of *deliberate practice* in explaining extraordinary accomplishments. His writings downplay the relevance of innate ability, heritability, and individual differences in predicting gifted adults (Ericsson, 2014). Critics of his model label his position "the most extreme exemplar of the environmentalist viewpoint" (Ackerman, 2013, 1; 2017).

In a widely cited study of elite chess players, Nobel prizewinner Herbert Simon and William Chase (with whom Ericsson studied) proposed a *ten-year rule*, based on their observations that it took more or less a decade of intensive study and practice to reach the top ranks of chess (see Simon & Chase, 1973). Even Bobby Fisher was no exception (Colvin, 2008). Ericsson agrees with Simon and Chase and has conducted numerous studies that corroborate the idea that *deliberate practice* makes all the difference between expert performers and average adults across almost all domains. Deliberate practice is characterized by several components: it is activity designed specifically to improve performance, often under the watchful eye and close supervision of an instructor, mentor, or coach; it includes a good deal of specific and continuous feedback; it must be repeated a lot; it is highly demanding mentally (Colvin, 2008; Coyle, 2009). Deliberate practice is meant to stretch the individual beyond their comfort zone and beyond their current skills level; it requires that the learner and/or coach identify and isolate very specific elements of performance that need to be learned or improved upon to further development toward expertise (Ericsson, et al., 1993; Syed, 2010). Deliberate practice is effortful, it requires feedback to improve, and as Ericsson and his colleagues remind us, it is not inherently enjoyable (Ericsson, 1996, 2014).

Ericsson's work obviously deserves mention and other contributors to this handbook have prominently represented his important work on expertise. However, the reader may be wondering why a chapter on giftedness includes Ericsson's work on deliberate practice and the acquisition of expert performance. This is a fair question. Some authorities in the gifted field have embraced Ericsson's ideas as relevant to giftedness and talent development. This is despite his spirited opposition to the importance of natural abilities. For example, Ericsson et al. (2007) write, "With the exception of fixed genetic factors determining body size and height, we were unable to find evidence for innate constraints to the attainment of elite achievement for healthy individuals" (3). Some in the gifted field find this conclusion misguided and even heresy. Although I personally am not upset with Ericsson's de-emphasis of individual differences and his extreme environmental position, I don't agree with many of his foundational ideas. They run counter to over forty years of experience working firsthand with many young high ability children and youth. I think that an extreme environmental view ignores considerable research supporting the importance of natural ability, individual differences, early experiences, and critical periods (Ackerman, 2013, 2017; Ackerman & Lakin, 2018; Park, et al., 2008; Pfeiffer, 2015). Irrespective of one's thoughts and opinions about Ericsson's strong environmental position, his emphasis on the importance of deliberate practice, sustained effort in the face of frustration, and years of effortful practice to reach a high level of expertise, eminence, and creativity are important lessons for the gifted field. And as a colleague on the same campus at Florida State, I value his provocative ideas.

Multiple Intelligences Model

The *multiple intelligences model* proposed by Howard Gardner has enjoyed wide appeal by the lay public (Robertson, Pfeiffer, & Taylor, 2011). Gardner is a very recently retired professor at Harvard University and caught the attention of the public in 1983 with the pioneering publication of his highly popular and eminently readable book, *Frames of Mind* (Gardner, 1983), in which he proposed the idea of multiple intelligences. In his model, multiple intelligences are perceived as independent cognitive systems, *not* hierarchically nested under one general ability factor (Gardner, 1983, 1993; Pfeiffer, 2015). Gardner's theory of human intelligences was formulated by a selective analysis of the research literature, not psychometric techniques such as factor analysis. His review and synthesis of a wide-ranging literature in support of his multiple intelligences theory included studies of patients with brain damage, idiot savants, prodigies, evolutionary history, and research in psychometric and experimental psychology. Gardner concluded that there was compelling evidence for at least eight separate intelligences: linguistic, logical-mathematical, spatial, musical, bodily kinesthetic, interpersonal, intrapersonal, and naturalist. He has most recently added a ninth intelligence, existential intelligence, to his list (Dai, 2010; Kaufman & Sternberg, 2008).

Gardner's theory of multiple intelligences has had a significant impact on the educational field internationally, although much less so in the gifted field. His ideas have played a substantial role in greatly expanding educators' views on intelligence. In my international travel, his multiple intelligences theory is often among the first topics that educators want to discuss. Among the most significant aspects of Gardner's theory is the thesis that intelligence is not a single, unitary construct. This single idea, and his highly engaging writing style, helped make Gardner akin to a rock star for some in gifted education. Many intervention programs have been published by followers of the multiple intelligences model. However, there isn't a lot of hard empirical data supporting his theoretical model. The public and lay media have been infatuated with the model, particularly since many incorrectly interpret the multiple intelligences model as implying that everyone is gifted in something.

Hence, Gardner's theory is not without criticism. There is no published research that has tested the multiple intelligences theory. The multiple intelligences that Gardner proposes are based on highly selective reviews of the literature. He omits a considerable amount of the psychometric literature on intelligence, which arguably should be included in any unifying theory of intelligence (Sternberg, 2017). Finally, there exists only a handful of measures of his different intelligences, and most suffer from less-than-adequate psychometric rigor (Kaufman & Sternberg, 2008; Pfeiffer, 2013b).

Theory of Successful Intelligence: WICS

Robert Sternberg recently proposed an alternative way of looking at intelligence. His ideas aren't quite a well-developed model. But his *theory of successful intelligence* is provocative, as many of Sternberg's ideas are, and worth mentioning. Sternberg's theory of successful intelligence emphasizes how three components of intelligence work harmoniously. The components are creativity, intelligence (both academic and practical), and wisdom. Sternberg writes, "Successfully intelligent people balance adaptation to, shaping of, and selection of environments by capitalizing on strengths and compensating for or correcting weaknesses" (Sternberg, Jarvin, & Grigorenko, 2011, 43).

His theory of successful intelligence is referred to as WICS, representing the three components wisdom, intelligence, creativity, synthesized. He believes that giftedness involves both skills and attitudes; the skills are developing competencies and expertise, similar to Subotnik's and Ericsson's ideas. The attitudes are how the gifted individual employs the skills that they have developed. His proposal contends that gifted individuals do not necessarily excel at everything. He believes that gifted individuals are well aware of their strengths and limitations and make the most of their strengths and find ways to compensate for their weaknesses (Sternberg, 2017). This is a cheeky and seductive but yet untested hypothesis.

By creativity, Sternberg means the skills and attitudes needed to generate relatively novel, high quality, and appropriate ideas and products (Pfeiffer, 2015). Sternberg views intelligence as consisting of both those skills and attitudes that we think of when we consider conventional intelligence (or psychometric g), and practical intelligence as the skills and attitudes that individuals rely on to solve everyday problems. Sternberg contends that academic skills and attitudes are important for giftedness since gifted individuals need to be able to retrieve, remember, analyze, synthesize, and evaluate information. However, he also argues that practical intelligence is important; gifted individuals need to be able to adapt to their environment, change the environment to suit their needs, or seek a different, more facilitative environment. There is a flavor of Piaget in this notion of practical intelligence. Sternberg contends that ideally gifted individuals need to be high in practical as well as academic intelligence. "Their creativity may help them generate wonderful ideas, but it will not ensure that (gifted individuals) can implement the ideas or convince others to follow the ideas. Many creative individuals have ended up frustrated because they have been unable to convince others to follow their ideas" (Sternberg, Jarvin, & Grigorenko, 2011, 44).

Wisdom is the thorniest and most difficult component to define in his theory of successful intelligence. He believes that individuals are wise to the extent that they use their successful intelligence, creativity, and knowledge to pursue ethical values, balance their own and others' interests, and seek to reach common good. However, it is unclear how to operationalize or measure wisdom or the synthesis of intelligence, creativity, and wisdom. Without a doubt, wisdom is a highly valued character strength, but should not necessarily be viewed as a component of giftedness per se (Pfeiffer, 2013a, 2015). Many view wisdom as a highly valued character strength, virtue, or strength of the heart, but not a component of intelligence, that develops over time and experience (Pfeiffer, 2018b).

Parenthetically, in a seminar on intelligence that I first taught at Duke University, and now teach at Florida State University, as an *ice breaker* I routinely ask my students to rank order from a long list of adjectives those terms that they think most characterize an ideally intelligent, creative, and wise individual. I have used this classroom activity to illustrate the concept of implicit theories of intelligence, creativity, and wisdom, based on Sternberg's early work (Sternberg, 1985). When I have examined the ranking of adjectives over the years, fairly consistent findings exist, suggesting distinct implicit conceptions of how at least college students view the ideally intelligent, ideally creative, and ideally wise person! For example, seven characteristics are consistently selected most frequently as illustrative of an intelligent person: good problem-solving ability, inquisitive, reasons clearly, good at distinguishing between correct and incorrect answers, huge store of information, thinks quickly, and perceptive. The creative person, on the other hand, is characteristically described by college students as imaginative, unorthodox, takes chances, and is emotional, intuitive, and a free spirit. Finally, the wise person is viewed as reflecting four *signature* characteristics: a good listener, thoughtful, listens to all sides of an issue, and considers advice. Although falling short of carefully designed, rigorous research, this classroom activity lends anecdotal support to Sternberg's proposal for three components of successful intelligence.

TRIPARTITE MODEL OF GIFTEDNESS

There are many different ways to conceptualize giftedness. No one conceptualization is correct. They are all simply different ways to view children and youth who are in some way special. Each of the different models has implications for who are gifted, how to identify them, and what we should do to educate them and actualize their gifts and talents. As a result of many years' work with highly gifted youth during my tenure at Duke University, I proposed a *tripartite model of giftedness* (Pfeiffer, 2013b, 2015). The tripartite model provides three different ways of viewing students with high ability or extraordinary potential. The tripartite model offers three different, but complementary ways to conceptualize, identify, and program for gifted learners. The three distinct lenses through which high ability students can be viewed within this model are as follows:

- Giftedness through the lens of high intelligence;
- Giftedness through the lens of outstanding accomplishments; and
- Giftedness through the lens of potential to excel.

The first perspective, the *high intelligence* view, is by now familiar. Through this first lens, an IQ test or its proxy can be used to identify students functioning at a certain level considerably above average intellectually. Other tests can supplement the IQ test, but the criterion for high intelligence giftedness is based on compelling evidence that the child is advanced intellectually when compared to his or her same-age peers. This first gifted perspective can follow a general (g) or multidimensional view (e.g., C-H-C, structure of intellect, multiple intelligences) of intelligence. It can even be based on a neuroanatomical model of giftedness; recent work, for example, has postulated that more intelligent children demonstrate a more plastic cortex, with an initial acceleration and prolonged phase of cortical increase, followed by a period of vigorous cortical thinning by early adolescence (Shaw et al., 2006).

The rationale for gifted programs based on viewing giftedness through the lens of a high IQ is that students with superior intelligence need and/or are entitled to advanced, intellectually challenging, and/or more fast-paced academic material not typically found in the regular classroom (Pfeiffer, 2013b, 2015). Based on this perspective, gifted education consists of a highly accelerated and/or academically advanced and challenging curriculum. The Johns Hopkins and Duke TIP summer programs are two examples.

The second perspective, *outstanding accomplishments*, does not scoff at or disparage the importance of high intelligence. However, this second perspective emphasizes performance in the classroom and on real-world projects as the core defining characteristic for giftedness. According to this second perspective, evidence of real-world excellence is the *sine qua non* to qualify as a gifted student and to qualify for admittance into a gifted program, not high IQ (Pfeiffer, 2013b, 2015). Portfolio and rubric assessment of actual student products are used to identify high-performing students as gifted from this second perspective, not high IQ test scores. Gifted educators within this second perspective are looking for direct evidence of *authentic* academic excellence. Creativity is emphasized when viewing giftedness through this second lens (Pfeiffer, 2015). For example, teacher-rating scales list items such as displays an active imagination, generates many ideas to *what if* questions, and displays an active imagination (Pfeiffer & Jarosewich, 2003). Also, motivation, drive, persistence, and academic passion—clearly non-intellectual factors—are emphasized by many advocates of this second way of conceptualizing giftedness (Pfeiffer, 2012, 2013b, 2015).

The rationale for gifted programs based on an outstanding accomplishments perspective is that students who excel academically have earned and deserve special academic programs because of their outstanding effort and superior accomplishments. Gifted education, based on an outstanding accomplishments perspective, would be different from gifted education guided by a high intelligence perspective. Gifted programs would consist of highly enriched and academically challenging curricula, although not necessarily fast-paced or highly advanced (Pfeiffer, 2013b, 2015).

The third perspective is *potential to excel*. Some children and youth—for any number of reasons—have not been provided enough opportunity or the proper intellectual stimulation (or even *deliberate practice*) to develop what remains latent and as yet undeveloped or underdeveloped intellectual or academic gifts (Pfeiffer, 2013a, b, 2015). This third perspective is based on my experience working with a great many students of high potential, the experience of countless others, and an abundant body of research (Irving & Hudley, 2008; Nisbett, 2009; Pfeiffer, 2015).

Most would agree that not all children and youth start out on equal footing. Some children from poverty, those from families in which intellectual and educational activities are neither encouraged nor nurtured in the home, or in which their native language is not the language spoken in their schools or community, or children growing up in rural or overcrowded or dangerous communities where intellectual stimulation and educational opportunities are rare, are all at a distinct disadvantage to develop their gifts (Ford & Whiting, 2008; Nisbett, 2009; Pfeiffer, 2002, 2012, 2013b, 2015). This is the rationale for the third perspective within the tripartite model.

The third perspective implies a *prediction* that students of high potential will very likely excel when provided with special resources or programs. The assumption underlying this third perspective is that with time, an encouraging and highly stimulating environment, and the proper psycho-educational interventions, these students will actualize their yet unrealized high potential and distinguish themselves from among their peers as gifted. Gifted education, guided by a potential-to-excel perspective, consists of a highly motivating and enriched curriculum that may include compensatory interventions (Pfeiffer, 2015). This third category of gifted also carries with it a prediction. The prediction is that if the student is provided a well-conceived, comprehensive, intensive, evidence-based psycho-educational intervention—often requiring a home component—then he or she will ultimately appear indistinguishable, or at least very similar to, any student who is already identified as falling within one of the other two gifted categories, high intelligence or academically gifted learner. Unfortunately, there isn't yet empirical research supporting (or refuting, for that matter) the hypothesis that there exists a third type of gifted, the *diamonds in the rough*, who will flourish with well-designed and intensive psycho-educational interventions. It is clear that the interventions would need to be of *high dosage* to compensate for the early, missed familial and educational experiences and opportunities. And it is also apparent that the earlier educators identify young, potential-to-excel gifted students, the more likely they will respond favorably to the planned psycho-educational interventions. This is an area of exciting research opportunity, globally.

These three categories of the gifted constitute different types of children, with different levels and profiles of abilities, and in many instances, different skills and even personality characteristics (Pfeiffer, 2015). However, the groups are not necessarily mutually exclusive and there is considerable overlap. For example, there are many students with exceptionally high IQ scores who are academically gifted learners with a burning passion to learn. The tripartite model serves to eliminate much of the acrimony often found when schools adopt only one, typically narrowly defined high IQ view of giftedness. Details on how to operationalize the tripartite model are found in Pfeiffer (2015).

Concluding Comments

The chapter concludes by highlighting and further developing a few key points that reflect many of the gifted models and are shared by most authorities in the gifted field:

- Giftedness is a *social construct* and not something that is real. Giftedness is not something that students have or don't have (Borland, 2009). It is a social construction, a concept educators use to label a subset of students based on alternative criteria (such as high IQ or outstanding academic performance). However, there can never be a cut-score that precisely separates gifted from non-gifted students (Pfeiffer, 2015; Pfeiffer & Jarosewich, 2003, 2007). The decision on where to draw the line will always be based on human judgment—hopefully thoughtful, deliberate, and prudent judgment.

- There are different levels or degrees of giftedness. This view is grounded in my early training, experience over the years working with exceptionally bright students, and deep respect for individual differences in human abilities. Recall that Gagné proposes five levels of giftedness: the mildly gifted (top 10%; IQ approximately 120), moderately gifted (top 1%; IQ approximately 135), highly gifted (145 IQ), exceptionally gifted (155 IQ), and the extremely gifted (165 and above). Terman (1916, 79) provided a slightly different four-level classification system when he first published the Stanford–Binet: 110–20 (superior intelligence), 120–40 (very superior intelligence, and above 140 (near genius and genius). These are but two of many possible ways that one can reliably sort and distinguish the gifted by level or degree. There does not exist one correct classification system or most precise number of categories that best differentiates level or degree of giftedness, whatever the domain or field (Pfeiffer, 2015).
- General intellectual ability matters in school performance and in real-world success. There are various ways of defining and conceptualizing the construct of intelligence and no one way is correct. General ability is almost always important to measure when thinking about the gifted child, when conducting a gifted assessment, and when considering a student for gifted eligibility. Most, but not all models, acknowledge the importance of recognizing and assessing natural abilities in one or more culturally valued domains.
- In addition to general ability, specific abilities and skills, a constellation of attitudes, interests, and beliefs, opportunities provided and taken advantage of, and motivation, persistence, frustration tolerance, and passion contribute to the accomplishments and potential eminence that many but not all bright students reach in a given field. Factors beyond the school and gifted program contribute to the calculus that ultimately determines one's success in life (Pfeiffer, 2015; Wai, 2014). Opportunities, personal choices, personality type, unanticipated events, and good fortune all play a role at every stage in the talent development process. In my opinion, gifted assessment should go beyond measuring general ability on an IQ test and include collecting data and evidence on a number of cognitive, personality, and attitudinal factors that have been shown to play a role in the development of expertise, and even eminence in various fields (Pfeiffer, 2018b).
- There are many different ways of defining the gifted student; the tripartite model provides three different lenses through which we can conceptualize giftedness. Some in the gifted field argue that the number of students provided gifted services should be based on the actual need for services; in my opinion, it is very difficult, if not impossible, to operationalize educational need in a scientifically defensible way (Pfeiffer, 2015). Part of the reason is because the construct gifted is not

something real. Gifted is a concept that we humans have invented. All students, including students of uncommon or high ability, benefit from a challenging and differentiated curriculum (Borland, 2005). How many of these students should be provided a special gifted program, however, is ultimately a political, sociophilosophical, and practical decision guided by available resources and value judgment. Unfortunately, it is not a scientific question that can be answered by inserting data into a precise mathematical formula.

- There is growing support within the gifted field to focus on talent development as a path toward expertise and eminence (Worrell, Subotnik, & Olszewski-Kubilius, 2017). Critics of this position, however, argue that a shift in gifted education emphasizing developing expertise and eminence in culturally valued domains "results in sacrificing the needs of the many for those of the few" (Borland, 2012, 8). Borland (2012) also contends that talent development of the sort that leads to (expertise and) eminence in a number of fields involves activities, most of which are outside the mission of the schools (10, 12). This pushback to having expertise and eminence as a pre-eminent goal of gifted education has been widespread, although not universal (Pfeiffer, 2013b). Time will tell whether the talent development models assume primacy in gifted education in the USA and internationally (Dai, 2017). I suspect that there will remain a schism in gifted education between *advanced academics* and *high ability psychology* (McBee et al., 2012).

There are compelling economic, political, and cultural reasons for nations to provide significant financial and human resources for gifted students. And yet gifted education in the USA and globally has a history of being appropriated extraordinarily limited resources (Gallagher, 2008; personal communication, February 17, 2013). There are obvious long-range benefits to our society when governments and local school districts fund resources for gifted education. Ceci and Papierno (2005), for example, advocate for identifying the top 10% of the under-represented segments of society and providing them with the resources to develop their potential. This proposal is consistent with the tripartite model's third lens for viewing the gifted: those youngsters of unusually high potential to accomplish great things if provided the right opportunities at a high-enough dosage and over a long-enough period of time (Pfeiffer, 2015).

References

Ackerman, P. L. (2013). Nonsense, common sense, and science of expert performance: Talent and individual differences. *Intelligence* 22, 229–259.

Ackerman, P. L. (2017). Expertise: Individual differences, human abilities, and personality traits. In S. I. Pfeiffer, E. Shaunessy-Dedrick, & M. Foley-Nicpon (Eds), *APA handbook of giftedness and talent* (pp. 259–272). Washington, DC: APA Books.

Ackerman, P. L., & Lakin, J. M. (2018). Expertise and individual differences. In S. I. Pfeiffer (Ed.), *Handbook of giftedness in children: Psychoeducational theory, research, and best practice*, 2nd edn (pp. 65–80). New York: Springer.

Assouline, S. G., & Lupkowski-Shoplik, A. (2012). The talent search model of gifted identification. *Journal of Psychoeducational Assessment 30*, 45–59.

Binet, A., & Simon, T. (1916). *The development of intelligence in children (the Binet-Simon scale)*. Baltimore, MD: Williams & Wilkins.

Borland, J. H. (2005). Gifted education without gifted children: The case for no conception of giftedness. In R. J. Sternberg & J. E. Davidson (Eds), *Conceptions of Giftedness*, 2nd edn (pp. 1–19). New York: Cambridge University Press.

Borland, J. H. (2009) Myth 2: The gifted constitute 3 percent to 5 percent of the population. Moreover, giftedness equals high IQ, which is a stable measure of aptitude: Spinal tap psychometrics in gifted education. *Gifted Child Quarterly 53*, 236–238.

Borland, J. H. (2012). A landmark monograph in gifted education, and why I disagree with its major conclusion. Creativity Post. Retrieved from http://www.creativitypost.com/education/a_landmark_monograph_in_gifted_education_and_why_I_disagree_with_its_major

Brody, L. E., & Benbow, C. P. (1987). Accelerative strategies: How effective are they for the gifted? *Gifted Child Quarterly 31*, 105–109.

Carroll, J. B. (1993). *Human cognitive abilities: A survey of factor-analytic studies*. Cambridge, UK: Cambridge University Press.

Ceci, S. J., & Papierno, P. B. (2005). The rhetoric and reality of gap closing: When the "have-nots" gain but the "haves" gain even more. *American Psychologist 60*, 149–160.

Colvin, G. (2008). *Talent is overrated: What really separates world-class performers from everybody else*. New York: Portofolio.

Coyle, D. (2009). *The talent code: Greatness isn't born, it's grown*. New York: Bantam Books.

Dai, D. Y. (2010). *The nature and nurture of giftedness*. New York: Teachers College Press.

Dai. D. Y. (2017). A history of giftedness: A century of quest for identity. In S. I. Pfeiffer, E. Shaunessy-Dedrick, & M. Foley-Nicpon (Eds), *APA handbook of giftedness and talent* (pp. 3–24). Washington, DC: APA Books.

Ericsson, K. A. (Ed.) (1996). *The road to excellence: The acquisition of expert performance in the arts and sciences, sports, and games*. Mahwah, NJ: Erlbaum.

Ericsson, K. A. (2014). Why expert performance is special and cannot be extrapolated from studies of performance in the general population: A response to criticisms. *Intelligence 45*, 81–103.

Ericsson, K. A., & Charness, N. (1995). Abilities: Innate talent or characteristics acquired through engagement in relevant activities? *American Psychologist 50*, 803–804.

Ericsson, K. A., Krampe, R. T., & Tesch-Romer, C. (1993). The role of deliberate practice in the acquisition of expert performance. *Psychological Review 100*, 363–406.

Ericsson, K. A., Roring, R. W., & Nandagopal, K. (2007). Giftedness and evidence for reproducibly superior performance: An account based on the expert-performance framework. *High Ability Studies 18*, 3–56.

Feldhusen, J. F. (2005). Giftedness, talent, expertise, and creative achievement. In R. J. Sternberg & J. E. Davidson (Eds), *Conceptions of giftedness*, 2nd edn (pp. 64–79). New York: Cambridge University Press.

Flanagan, D. P., & Harrison, P. L. (2012). *Contemporary intellectual assessment: Theories, tests, and issues*, 3rd edn. New York: Guilford Press.

Foley-Nicpon, M., & Pfeiffer, S. I. (2011). High ability students: New ways to conceptualize giftedness and provide psychological services in the schools. *Journal of Applied School Psychology 27*, 293–305.

Ford, D. Y., & Whiting, G. W. (2008). Recurring and retaining underrepresented gifted students. In S. I. Pfeiffer (Ed.), *Handbook of giftedness* (pp. 293–308). New York: Springer.

Gagné, F. (2005). From gifts to talents: The DMGT as a developmental model. In R. J. Sternberg & J. E. Davidson (Eds), *Conceptions of giftedness*, 2nd edn (pp. 98–120). New York: Cambridge University Press.

Gagné, F. (2009). Debating giftedness: Pronat vs. antinat. In L. Shavinina (Ed.), *International handbook on giftedness* (pp. 155–198). New York: Springer.

Gagné, F. (2017). Academic talent development: Theory and best practices. In S. I. Pfeiffer, E. Shaunessy-Dedrick, & M. Foley-Nicpon (Eds), *APA handbook of giftedness and talent* (pp. 162–184). Washington, DC: APA Books.

Gallagher, J. J. (1960). *Analysis of research on the education of gifted children*. Springfield, IL: Office of the Superintendent of Public Instruction.

Gallagher, J. J. (2008). Psychology, psychologists, and gifted students. In S. I. Pfeiffer (Ed.), *Handbook of giftedness in children* (pp. 1–11). New York: Springer.

Galton, F. (1869). *Hereditary genius: An inquiry into its laws and consequences*. London: Macmillan.

Gardner, H. (1983). *Frames of mind*. New York: Basic Books.

Gardner, H. (1993). *Multiple intelligences: The theory in practice*. New York: Basic Books.

Hollingsworth, L. S. (1926). *Gifted children: Their nature and nurture*. New York: Macmillan.

Horn, J. L., & Cattell, R. B. (1966). Refinement and test of the theory of fluid and crystallized general intelligences. *Journal of Educational Psychology* 57(5), 253–270.

Irving, M. A., & Hudley, C. (2008). Cultural identification and academic achievement among African American males. *Journal of Advanced Academics* 19, 676–698.

Kaufman, A. S. (2013). Intelligent testing with Wechlser's Fourth Editions: Perspectives on the Weiss et al. studies and the eight commentaries. *Journal of Psychoeducational Assessment* 31, 224–234.

Kaufman, A. S., & Kaufman, N. L. (2004). *Kaufman assessment battery for children*, 2nd edn (KABC-II). Circle Pines, MN: American Guidance Service.

Kaufman, S., & Sternberg, R. J. (2008). Conceptions of giftedness. In S. I. Pfeiffer (Ed.), *Handbook of giftedness in children* (pp. 71–92). New York: Springer.

Kell, H. J., Lubinski, D., & Benbow, C. P. (2013). Who rises to the top? Early indicators. *Psychological Science* 24, 648–659.

Kolitch, E. R., & Brody, L. E. (1992). Mathematics acceleration of highly talented students: An evaluation. *Gifted Child Quarterly* 36, 78–86.

Li, J., & Fischer, K. W. (2004). Thought and affect in American and Chinese learners' beliefs about learning. In D. Y. Dai & R. J. Sternberg (Eds), *Motivation, emotion, and cognition: Integrative perspectives on intellectual functioning and development* (pp. 385–419). Mahwah, NJ: Lawrence Erlbaum.

McBee, M. T., McCoach, D. B., Peters, S. J., & Matthews, M. S. (2012). The case for a schism: A commentary on the Subotnik, Olszweski-Kubilius, & Worrell (2011). *Gifted Child Quarterly* 56, 210–214.

McClain, M. C., & Pfeiffer, S. I. (2012). Identification of gifted students in the U.S. today: A look at state definitions, policies, and practices. *Journal of Applied School Psychology* 28, 59–88.

Mather, N., & Wendling, B. J. (2014). *Examiner's Manual. Woodcock–Johnson IV Tests of Cognitive Abilities*. Rolling Meadows, IL: Riverside.

Mather, N., Wendling, B. J., & Woodcock, R. W. (2001). *Essentials of WJ III tests of achievement assessment*. New York: Wiley.

Mayer, R. E. (2005). The scientific study of giftedness. In R. J. Sternberg & J. E. Davidson (Eds), *Conceptions of giftedness*, 2nd ed (pp. 437–447). New York, NY: Cambridge University Press.

Mönks, F. J., Heller, K. A., & Passow, H. (2000). The study of giftedness: Reflections on where we are and where we are going. In K. A. Heller, F. J. Mönks, R. J. Sternberg, & R. F. Subotnik (Eds), *International Handbook of giftedness and talent*, 2nd edn (pp. 839–863). Oxford, UK: Elsevier Science.

National Science Board (2010). Preparing the next generation of STEM innovators: Identifying and developing our nations' human capital. Retrieved from http://www.nsf.gov/nsb/publications/2010/nsb1033.pdf

Neihart, M., Pfeiffer, S. I., & Cross, T. L. (Eds) (2016). *The social and emotional development of gifted children. What do we know?*, 2nd edn. Waco, TX: Prufrock Press.

Neisser, U., Boodoo, G., Bouchard, T. J., Boykin, A. W., Brody, N., Ceci, S. J., . . . Urbina, S. (1996). Intelligence: Knowns and unknowns. *American Psychologist 51*, 77–101.

Nisbett, R. E. (2009). *Intelligence and how to get it*. New York: Norton.

Niu, W., & Brass, J. (2011). Intelligence in worldwide perspective. In R. J. Sternberg & S. B. Kaufman (Eds), *The Cambridge handbook of intelligence* (pp. 623–646). Cambridge, UK: Cambridge University Press.

Olszewski-Kubilius, P. (2004). Talent searches and accelerated programming for gifted students. In N. Colangelo, S. G. Assouline, & M. U. M. Gross (Eds), *A nation deceived: How schools hold back America's brightest students* (Vol. 2, pp. 69–76). Iowa City, IA: Connie Belin & Jacqueline N. Blank International Center for Gifted Education and Talent Development.

Park, G., Lubinski, D., & Benbow, C. P. (2008). Ability differences among people who have commensurate dgrees matter for scientific creativity. *Psychological Science 19*, 957–961.

Pfeiffer, S. I. (2001). Professional psychology and the gifted: Emerging practice opportunities. *Professional Psychology: Research and Practice 32*, 175–180.

Pfeiffer, S. I. (2002). Identifying gifted and talented students: Recurring issues and promising solutions. *Journal of Applied School Psychology 19*, 31–50.

Pfeiffer, S. I. (2003). Challenges and opportunities for students who are gifted: What the experts say. *Gifted Child Quarterly 47*, 161–169.

Pfeiffer, S. I. (2008). *Handbook of giftedness in children*. New York: Springer.

Pfeiffer, S. I. (2009). The gifted: Clinical challenges for child psychiatry. *Journal of the American Academy of Child and Adolescent Psychiatry 48*, 787–790.

Pfeiffer, S. I. (2012). Current perspectives on the identification and assessment of gifted students. *Journal of Psychoeducational Assessment 30*, 3–9.

Pfeiffer, S. I. (2013a). Lessons learned from working with high ability students. *Gifted Education International 29*, 86–97.

Pfeiffer, S. I. (2013b). *Serving the gifted: Evidence-based clinical and psychoeducational practice*. New York: Routledge.

Pfeiffer, S. I. (2015). *Essentials of gifted assessment*. Hoboken, NJ: Wiley.

Pfeiffer, S. I. (2017). Success in the classroom and in life: Focusing on strengths of the head *and* strengths of the heart. *Gifted Education International 33*, 95–101.

Pfeiffer, S. I. (Ed.) (2018a). *Handbook of giftedness in children: Psychoeducational theory, research, and best practice*, 2nd edn. New York: Springer.

Pfeiffer, S. I. (2018b). Understanding success and psychological well-being of gifted kids and adolescents: Focusing on strengths of the heart. *Estudios de Psicologia 35* (3), 229–233.

Pfeiffer, S. I., & Jarosewich, T. (2003). *The Gifted Rating Scales*. San Antonio, TX: Pearson Assessment.

Pfeiffer, S. I., & Jarosewich, T. (2007). The Gifted Rating Scales-School Form: An analysis of the standardization sample based on age, gender, race, and diagnostic efficiency. *Gifted Child Quarterly 51*, 39–50.

Pfeiffer, S. I., Shaunessy-Dedrick, E., & Foley-Nicpon, M. (Eds) (2017). *APA handbook of giftedness and talent*. Washington, DC: APA Books.

Phillipson, S. N. (2013). Confucianism, learning self-concept and the development of exceptionality. In S. N. Phillipson, H. Stoeger, & A. Ziegler (Eds), *Exceptionality in East Asia: Explorations in the actiotope model of giftedness* (pp. 40–64). Abingdon, UK: Routledge.

Phillipson, S. N., & McCann, M. (2007). *Conceptions of giftedness: Socio-cultural perspectives*. Mahwah, NJ: Lawrence Erlbaum Associates.

Plomin, R., & Spinath, F. M. (2004). Intelligence: Genetics, genes, and gnomics. *Journal of Personality and Social Psychology 86*, 112–129.

Reis, S. M. (2006). Comprehensive program design. In J. H. Purcell & R. D. Ecket (Eds), *Designing services and programs for high-ability learners* (pp. 73–86). Thousand Oaks, CA: Corwin Press.

Reis, S. M., & Renzulli, J. S. (2009). Myth 1: The gifted and talented constitute one single homogeneous group and giftedness is a way of being that stays in the person over time and experiences. *Gifted Child Quarterly 53*(4), 233–235.

Renzulli, J. S. (1978). What makes giftedness? Reexamining a definition. *Phi Delta Kappan 60*, 180–184.

Renzulli, J. S. (1983). Rating the behavioral characteristics of superior students. *G/C/T 29*, 30–35.

Renzulli, J. S. (1984). The triad/revolving door system: A research based approach to identification and programming for the gifted and talented. *Gifted Child Quarterly 28*, 163–171.

Renzulli, J. S. (2005). The three-ring conception of giftedness: A developmental model for promoting creative productivity. In R. J. Sternberg & J. E. Davidson (Eds), *Conceptions of giftedness*, 2nd edn (pp. 246–279). New York: Cambridge University Press.

Renzulli, J. S. (2009). The multiple menu model for developing differentiated curriculum. In J. S. Renzulli, E. J. Gubbins, K. S. McMillen, R. D. Eckert, & C. A. Little (Eds), *Systems and models for developing the gifted and talented*, 2nd edn (pp. 353–381). Mansfield Center, CT: Creative Learning Press.

Renzulli, J. S. (2011). Theories, actions, and change: An academic journey in search of finding and developing high potential in young people. *Gifted Child Quarterly 55*, 305–308.

Renzulli, J. S., & Reis, S. M. (2017). The three-ring conception of giftedness: A developmental approach for promoting creative productivity in young people. In S. I. Pfeiffer, E. -Shaunessy-Dedrick, & M. Foley-Nicpon (Eds), *APA handbook of giftedness and talent* (pp. 185–200). Washington, DC: APA Books.

Robertson, S. G., Pfeiffer, S. I., & Taylor, N. (2011). Serving the gifted: A national survey of school psychologists. *Psychology in the Schools 48*, 786–799.

Robinson, A., & Clinkenbeard, P. R. (2008). History of giftedness: Perspectives from the past presage modern scholarship. In S. I. Pfeiffer (Ed.). *Handbook of giftedness in children* (pp. 13–31). New York: Springer.

Roid, G. H. (2003). *Stanford-Binet Intelligence Scales, 5th Edition*. Itasca, IL: Riverside.

Shaw, P., Greenstein, D., Lerch, J., Clasen, L., Lenroot, R., Gogtay, N., ... Giedd, J. (2006). Intellectual ability and cortical development in children and adolescents. *Nature 440*, 676–679.

Simon, H., & Chase, W. (1973). Skill in chess: Experiments with chess-playing tasks and computer simulation of skilled performance throw light on some human perceptual and memory processes. *American Scientist 61*, 394–403.

Simonton, D. K. (2008). Scientific talent, training, and performance: Intellect, personality, and genetic endowment. *Review of General Psychology 12*, 28–46.

Simonton, D. K. (2014). Creative performance, expertise acquisition, individual differences, and developmental antecedents: An integrative research agenda. *Intelligence 45*, 66–73.

Simonton, D. K. (2017). From giftedness to eminence: Developmental landmarks across the lifespan. In S. I. Pfeiffer, E. Shaunessy-Dedrick, & M. Foley-Nicpon (Eds), *APA handbook of giftedness and talent* (pp. 273–285). Washington, DC: APA Books.

Spearman, C. (1904). "General intelligence," objectively determined and measured. *The American Journal of Psychology 15*, 201–292.

Spearman, C. (1927). *The abilities of man*. London: Macmillan.

Stanley, J. C. (1976). The case for extreme educational acceleration of intellectually brilliant youths. *Gifted Child Quarterly 20*, 66–75.

Stanley, J. C. (1990). Leta Hollingworth's contributions to above-level testing of the gifted. *Roeper Review 12*, 166–171.

Stanley, J. C. (2000). Helping students learn only what they don't already know. *Psychology, Public Policy, and Law 6*, 216–222.

Stephens, K. R. (2008). Applicable federal and state policy, law, and legal considerations in gifted education. In S. I. Pfeiffer (Ed.), *Handbook of giftedness in children* (pp. 387–408). New York: Springer.

Stephens, K. R. (2011). Federal and state response to the gifted and talented. *Journal of Applied School Psychology 27*, 306–318.

Stephens, K. R. (2018). Update on federal and state policy, law, and legal considerations in gifted education. In S. I. Pfeiffer (Ed.), *Handbook of giftedness in children: Psychoeducational theory, research, and best practice*, 2nd edn (pp. 163–182). New York: Springer.

Sternberg, R. J. (1985). Implicit theories of intelligence, creativity, and wisdom. *Journal of Personality and Social Psychology 49*, 607–627.

Sternberg, R. J. (2001). Giftedness as developing expertise: A theory of the interface between high abilities and achieved excellence. *High Ability Studies 12*, 159–179.

Sternberg, R. J. (2017). Theories of intelligence. In S. I. Pfeiffer, E. Shaunessy-Dedrick, & M. Foley-Nicpon (Eds), *APA handbook of giftedness and talent* (pp. 145–161). Washington, DC: APA Books.

Sternberg, R. J., & Davidson, J. E. (Eds). (2005). *Conceptions of giftedness*. New York: Cambridge University Press.

Sternberg, R. J., Jarvin, L., & Grigorenko E. L. (2011). *Explorations in giftedness*. New York: Cambridge University Press.

Stoeger, H., Balestrini, D. P., & Ziegler, A. (2017). International perspectives and trends on giftedness and talent development. In S. I. Pfeiffer, E. Shaunessy-Kendrick, & M. Foley-Nicpon (Eds), *APA handbook of giftedness and talent* (pp. 25–38). Washington, DC: APA Books.

Subotnik, R. F. (2003). A developmental view of giftedness: From being to doing. *Roeper Review 26*, 14–15.

Subotnik, R. F. (2009). Developmental transitions in giftedness and talent: Adolescence into adulthood. In F. D. Horowitz, R. F. Subotnik, & D. J. Matthews (Eds), *The development of giftedness and talent across the lifespan* (pp. 155–170). Washington, DC: American Psychological Association.

Subotnik, R. F., Olszewski-Kubilius, & Worrell, F. C. (2011). Rethinking giftedness and gifted education: A proposed direction forward based on psychological science. *Psychological Science in the Public Interest 12*, 3–54.

Subotnik, R. F., Olszewski-Kubilius, & Worrell, F. C. (2017). Talent development as the most promising focus of giftedness and gifted education. In S. I. Pfeiffer, E. Shaunessy-Dedrick, & M. Foley-Nicpon (Eds), *APA handbook of giftedness and talent* (pp. 231–245). Washington, DC: APA Books.

Subotnik, R. F., & Rickoff, R. (2010). Should eminence based on outstanding innovation be the goal of gifted education and talent development? Implications for policy and research. *Learning and Individual Differences 20*, 358–364.

Syed, M. (2010). *Bounce: Mozart, Federer, Picasso, Beckham, and the science of success*. New York: Harper Collins.

Tannenbaum, A. J. (1983). *Gifted children: Psychological and educational perspectives*. New York: Macmillan.

Tannenbaum, A. J. (2000). A history of giftedness in school and society. In K. A. Heller, F. J. Mönks, R. J. Sternberg, & R. F. Subotnik (Eds), *International handbook of giftedness and talent*, 2nd edn (pp. 23–53). Oxford, UK: Elsevier Science.

Terman, L. M. (1916). *The measurement of intelligence*. Boston, MA: Houghton-Mifflin.

Terman, L. M. (1925). *Genetic studies of genius. Mental and physical characteristics of a thousand gifted children* (Vol. I). Stanford, CA: Stanford University Press.

Terman, L. M., & Oden, M. H. (1951). The Stanford studies of the gifted. In P. Witty (Ed.), *The gifted child*. Boston, MA: D. C. Heath.

Thurstone, L. (1938). *Primary mental abilities* (Psychometric monograph; no. 1). Chicago: University of Chicago Press.

Tourón, J., & Freeman, J. (2017). Gifted education in Europe: Implications for policymakers and educators. In S. I. Pfeiffer, E. Shaunessy-Dedrick, & M. Foley-Nicpon (Eds), *APA handbook of giftedness and talent* (pp. 55–70). Washington, DC: APA Books.

VanTassel-Baska, J. (2013). International perspectives on gifted education and talent development, part I [Special issue]. *Journal for the Education of the Gifted 36*(1).

Wai, J. (2014). Experts are born, then made: Combining prospective and retrospective data shows that cognitive ability matters. *Intelligence 45*, 74–80.

Webb, J. T., Meckstroth, E. A., & Tolan, S. S. (1982). *Guiding the gifted child*. Columbus, OH: Psychology Publishing.

Whipple, G. M. (Ed.) (1924). *The education of gifted children* (23rd Yearbook, Part I). National Society for the Study of Education. Bloomington, IL: Public School Publishing Company.

Worrell, F. C., Subotnik, R. F., & Olszewski-Kubilius, P. (2013). Giftedness and gifted education: Reconceptualizing the role of professional psychology. *The Register Report 39*, 14–22.

Worrell, F. C., Subotnik, R. F., & Olszewski-Kubilius, P. (2017). Talent development: A path toward eminence. In S. I. Pfeiffer, E. Shaunessy-Dedrick, & M. Foley-Nicpon (Eds), *APA handbook of giftedness and talent* (pp. 247–258). Washington, DC: APA Books.

Yeo, L. S., & Pfeiffer, S. I. (2018). Counseling gifted children in Singapore: Implications for evidence-based treatment with a multicultural population. *Gifted Education International 34*, 64–75.

Ziegler, A., Stoeger, H., & Vialle, W. J. (2012). Giftedness and gifted education: The need for a paradigm shift. *Gifted Child Quarterly 56*, 194–197.

CHAPTER 6

NEURAL MECHANISMS OF EXPERTISE

FREDRIK ULLÉN, ÖRJAN DE MANZANO,
AND MIRIAM A. MOSING

Introduction

INTEREST in the neurobiological basis of human skills and abilities has a long history, but dramatic progress in neurobiological research on expertise began only a few decades ago with the development of neuroimaging techniques, which allow us to study the structure and function of the living human brain. All forms of expertise depend critically on neural circuits in the brain and spinal cord that are optimized to enable performance at a very high level within a particular domain, such as music or sports. Some of the neural features that underpin expert performance are reflected in macroanatomical properties of the nervous system, and we will begin this review with summarizing some key findings from the neuroanatomical expertise literature. Thereafter, we will discuss the functional neural mechanisms that underlie expert performance, focusing on four general phenomena that appear to be important for many forms of expertise: automation, domain-specific modifications of memory systems, adaptations in sensorimotor processing, and changes in network properties, i.e., functional connectivity. The final section discusses our current knowledge of the neural basis of expertise from the perspective of new models that emphasize a multifactorial perspective and take into account both genetic and environmental influences on expertise and its acquisition.

Research on expertise and its brain basis is in a dynamic phase, and an increasing number of relevant papers are published each year. A comprehensive review of the earlier literature on neural mechanisms of expertise can be found in Hill and Schneider (2006). Here we will deal mainly with the rapid developments during the past decade. Needless to say, we will make no attempt to cover the vast literature on memory and learning during shorter periods of training, but focus on the brain mechanisms of professional experts with long periods of dedicated training. Expertise research in general has benefitted from studies on many different model behaviors, but neural mechanisms of expertise have so far arguably been most well studied for music. Findings on the neural basis of musical expertise will therefore usually be discussed in somewhat more detail, but we will also refer to results from other domains where applicable.

Neuroanatomical Correlates of Expertise

Gray Matter Anatomy in Experts

The gray matter of the brain is located in the cortex that lines the surface of the brain and the cerebellum, as well as in deeper structures such as the basal ganglia and the thalamus, and in the spinal cord. It contains cell bodies of nerve cells and their processes and synapses, but also non-neural tissue, including supporting cells (glia) and blood vessels. Studies of the gray matter anatomy of experts have typically used structural magnetic resonance imaging (MRI) of the brain and standard morphometric techniques based either on computerized image processing or on manual morphometry. Commonly used outcome measures include regional gray matter volume, gray matter density, and cortical thickness. As will be discussed later these measures may reflect differences in the ultrastructure of both nerve cells and non-neural tissue. The overwhelming majority of these studies are cross-sectional, and typically involve group comparisons between experts and non-experts. However, a few studies have also used longitudinal designs, as will be noted.

A general finding in these studies is that experts and non-experts differ in local gray matter structure in regions implicated in the studied form of expert performance. One of the most studied expertise domains in this context is, as mentioned, music (Zatorre, Fields, & Johansen-Berg, 2013). Regardless of musical instrument and genre, essential components of musical expertise include the learning and performance of complex sequential motor skills, timing, coordination, and auditory–motor integration. These functions rely on a distributed network of brain regions that comprise sensorimotor and premotor areas of the frontal and parietal lobes, auditory regions in the temporal lobe, and subcortical regions of importance for skill learning and performance, i.e., the basal ganglia and the cerebellum (Zatorre, Chen, & Penhune, 2007). Neuroanatomical differences between musically trained and non-trained participants have been demonstrated repeatedly in these parts of the brain. One key structure in the motor system is the precentral gyrus. It is located immediately rostral to the central sulcus, which forms the border between the external surfaces of the frontal and parietal lobes. The posterior wall of the precentral gyrus contains the primary motor cortex, whereas its anterior portions contain the lateral premotor cortex (Nieuwenhuys, Voogd, & Huijzen, 2008).

In a pioneering study, Amunts and co-workers used manual morphometry to measure the length of the posterior wall of the precentral gyrus, as a proxy for the size of the primary motor cortex, and found that this was larger in keyboard players than in non-musicians (Amunts et al., 1997). This finding has since been replicated using different morphometric methods. Bangert and Schlaug (2006) studied the size of the so-called *omega sign*, an anatomical landmark of the hand representation in primary motor cortex, and found a more pronounced omega sign in musicians as compared to non-musicians. Other studies comparing the brains of musicians and controls have found musicians to have a higher gray matter density and regional volume of both the primary motor and premotor cortical areas (Bermudez, Lerch, Evans, & Zatorre, 2009; Gaser & Schlaug, 2003). One frontal area that repeatedly has been found to be more well developed in musicians is Broca's area in the inferior frontal cortex, a region which appears to be essential for the control of sequential structures in both language and music (Abdul-Kareem, Stancak, Parkes, & Sluming, 2011; Gaser & Schlaug, 2003). Auditory regions of the temporal lobe have been shown to be larger in size in individuals with more musical training in a number of studies (see, e.g., Gaser & Schlaug, 2003; Palomar-Garcia, Zatorre, Ventura-Campos, Bueicheku, & Avila, 2017; Schneider et al., 2002 with references). At the subcortical level, musical expertise is associated with a larger volume of the cerebellum, a structure that plays essential roles for motor control and coordination (de Manzano & Ullén, 2018; Hutchinson, Lee, Gaab, & Schlaug, 2003).

In one of the still relatively few longitudinal studies of neural correlates of expertise acquisition, Hyde and co-workers (2009) compared two groups of children, one of which had chosen to start taking weekly keyboard lessons. Analyses with deformation-based morphometry demonstrated that after 15 months of musical training, children in the music group displayed significant increases in the regional volume of both the hand motor area of the precentral gyrus and auditory cortex, as compared to the control group, which did not receive instrumental training during the period. This finding supports that plastic changes in the macroanatomy of the gray matter can develop over time during long-term training. Correlational support was also found for that the observed gray matter effects were relevant for expertise: Performance on a manual motor sequence task and an auditory musical discrimination task correlated positively with anatomical changes in primary motor cortex ($r = 0.45$) and auditory cortex ($r = 0.40$), respectively.

Associations between expertise and the neuroanatomy of relevant gray matter regions have been observed also for other domains. To give a few examples, Hänggi and co-workers (2014) found a lower regional volume and cortical thickness of the occipitotemporal junction—a region implied in the visual processing of chess positions (Bilalić, Langner, Ulrich, & Grodd, 2011)—in expert chess players. Bernardi and co-workers have reported higher gray matter density in professional racecar drivers in several gray matter regions, including visual and motor cortical regions and the basal ganglia (Bernardi et al., 2014). In professional perfumers, Delon-Martin and colleagues found a larger gray matter volume of the orbitofrontal regions surrounding the olfactory cortex in the basal part of the frontal lobe (Delon-Martin, Plailly, Fonlupt,

Veyrac, & Royet, 2013). In one well-known early study, Maguire et al. (2000) observed a larger regional volume of the posterior hippocampus—a brain region important for spatial navigation which is located in the medial wall of the temporal lobe—in taxi drivers, as well as a positive relation ($r = 0.6$) between posterior hippocampal volume and time spent taxi driving. In a follow-up longitudinal study, Woollett and Maguire (2011) scanned taxi driver trainees before and after four years of training that included learning the spatial organization of the streets of London, demonstrating that the increase in size of the posterior hippocampus developed over time only in the group of trainees that passed the final test to become a licensed taxi driver. Longitudinal studies of expertise acquisition and gray matter anatomy have also been performed by Draganski and co-workers. In one study of medical students studying for an exam, extensive bilateral increases of gray matter in brain regions involved in long-term memory were found after three months of studies (Draganski et al., 2006). A similar longitudinal study of juggling training found focal increases in the regional volume of visual areas in the occipital and temporal lobes after three months of training to juggle three balls (Draganski et al., 2004).

A few general observations can be made regarding this literature. First, for many forms of domain-specific expertise, the number of studies is still small and individual findings in particular gray matter regions are often not well replicated across studies. One reason for this is presumably the use of relatively small sample sizes in many neuroimaging studies (Button et al., 2013). Other reasons could be method variance and sample heterogeneity. For musical expertise, it, e.g., appears clear that the relations between training and brain anatomy may be moderated by various variables, which include age, sex, and type of training (instrument, genre, etc.; Merrete, Peretz, & Wilson, 2013). However, in several cases—for example the involvement of auditory–motor areas in musical expertise, or certain visual areas in juggling expertise—there appears to be convincing convergent evidence from several studies with different methodological approaches that expert performance is indeed related to the regional anatomy of specific gray matter regions.

Secondly, many studies show a bias towards interpreting correlations between anatomy and training as being due to causal effects of the latter on the former (Ullén, Hambrick, & Mosing, 2016). However, as will be discussed later, this conclusion is simply not warranted based on the cross-sectional and observational longitudinal studies that dominate the literature. Although it is indisputable that causal effects of training are essential for expertise (de Manzano & Ullén, 2018), group differences between experts and non-experts are also very likely to reflect, at least to a large part, differences in genetic liability, including gene–environment interactions and gene–environment covariation (Ullén et al., 2016; Ullén, Mosing, & Hambrick, 2017).

Finally, associations between training, expertise, and performance, on the one hand, and measures of gray matter (e.g., volume, density, and thickness), on the other hand, are typically positive. This appears relatively straightforward to explain and in line with many general observations on neural structure and performance. Animal studies show that skill acquisition is accompanied by the generation of more neural processes and synapses in involved brain areas (Kleim et al., 2004), as well as hypertrophy of glial cells

(Zatorre et al., 2013). Training involving motor activity and exercise has also been shown to be accompanied by an increase in the amount of capillary blood vessels (Black, Isaacs, Anderson, Alcantara, & Greenough, 1990). It thus appears likely that the differences in gray matter observed at a macroanatomical level in human imaging studies reflect effects in both neural and non-neural tissue. As will be discussed further later, the regional pattern of effects also makes it likely that many of the anatomical findings reflect functional adaptations that optimize expert performance. Notably, some studies have found negative relations between regional gray matter and expertise. Hänggi and co-workers' (2014) observation of a smaller size of the cortex of the occipitotemporal junction in chess experts has already been mentioned. Other studies have reported regional decreases of gray matter in groups such as musicians (Vaquero et al., 2016) and dancers (Hänggi, Koeneke, Bezzola, & Jäncke, 2010). These negative relations may appear puzzling, and for many expertise domains, more studies and replications would certainly be useful. One possibility could be, in line with the previous point, that some of the findings do not reflect causal effects of training, but rather constitutional group differences in relevant traits (aptitudes, interests, personality, etc.) between experts in different domains and non-experts. However, interestingly, recent longitudinal studies using repeated scanning on multiple occasions have demonstrated that experience-dependent changes in neuroanatomy need not progress monotonically. Wenger and co-workers scanned participants learning to write with the non-dominant hand on up to 18 occasions during a seven-week period of training (Wenger et al., 2017). Gray matter volume in the primary motor cortices increased during the first four weeks but then renormalized during the following weeks, despite continued practice and skill improvement. These results are a strong indication that expertise acquisition may involve both increases and decreases in gray matter volume, perhaps corresponding to processes of synapse generation and selective pruning observed during brain development (Edelman, 1987; Wenger et al., 2017).

White Matter Anatomy in Experts

The white matter of the brain mainly consists of bundles of nerve fibers that connect different brain regions (Zatorre et al., 2013). Numerous imaging studies have been performed to study the organization of white matter connections in experts. Some of these studies rely on manual morphometry, but a common approach today is to use diffusion MRI, an imaging technique where information about the ultrastructure of white matter is gained by measuring water diffusion (Le Bihan, 2003; Zatorre et al., 2013). A common outcome in these studies is fractional anisotropy (FA), an index of the local directional dependence of water diffusion. FA ranges from 0, meaning that the diffusion is equally restricted in all directions, to a theoretical maximum of 1, implying that the diffusion occurs along a single axis (Le Bihan, 2003). As will be discussed later, FA may be influenced by a number of properties of both neural and non-neural tissue.

In musicians, one of the more replicated findings is a more well-developed corpus callosum (de Manzano & Ullén, 2018; Schlaug, Jäncke, Huang, Staiger, & Steinmetz, 1995; Steele, Bailey, Zatorre, & Penhune, 2013). This large fiber bundle comprises the main connection for coordination and information transfer between the left and right hemispheres, and plays an essential role, e.g., for bimanual tasks (Gooijers & Swinnen, 2014). Studies using manual morphometry have found the corpus callosum to have a larger cross-sectional area in professional musicians than in non-musicians (Schlaug et al., 1995). Hyde and co-workers (2009), in their previously discussed longitudinal study of music training in children, found an increase in size of the midbody of the corpus callosum after training in the music group, as compared to the control group. The size of this effect was correlated ($r = 0.45$) with performance on a manual sequential motor task. Bengtsson and co-workers (2005) used diffusion imaging to study white matter organization in pianists and non-musician controls, and found a positive association between musical training in childhood and adolescence, and FA of the corpus callosum, within the pianist group. In a more recent study, Steele and co-workers (2013) compared two groups of musicians who had started their training before or after age 7, but were matched for total hours of training and music education. Interestingly, the early starting musicians had a higher FA in a central portion of the corpus callosum that was shown, using fiber tracking, to interconnect the sensorimotor cortices of the two hemispheres. Furthermore, FA in the same callosal region correlated with performance on a timed sensorimotor synchronization task.

Musical expertise is also related to the ultrastructure of other white matter tracts. FA in the corticospinal tract has been found to be higher in pianists than in non-musicians (de Manzano & Ullén, 2018; Bengtsson et al., 2005; Rüber, Lindenberg, & Schlaug, 2015). In addition, Bengtsson and co-workers (Bengtsson et al., 2005) observed a correlation between corticospinal tract FA and childhood practicing within the pianist group (Bengtsson et al., 2005). This tract, which originates in primary motor, premotor, and primary somatosensory cortical areas and descends to motor neurons and interneurons in the spinal cord, is essential for the cortical control of finger movements (Armand, Olivier, Edgley, & Lemon, 1996). Recent studies support that its ultrastructure may be relevant for expert performance in musicians. Rüber and co-workers (Rüber et al., 2015) thus found correlations between ultrastructural properties of corticospinal fibers originating in the primary motor cortex and maximal tapping rate of the contralateral index finger, while Engel and colleagues found that a higher corticospinal tract FA was associated with faster learning of piano melodies in non-musicians (Engel et al., 2014). Another fiber bundle that has been related to musical expertise is the arcuate fasciculus. This large arc-shaped tract includes axons connecting auditory regions in the superior temporal lobe with areas in the frontal lobe, i.e., the inferior frontal cortex and the adjacent premotor cortex, and plays an important role for auditory–motor integration (Schlaug, Marchina, & Norton, 2009). Halwani and co-workers found that both singers and instrumental musicians have a higher regional volume and a higher FA in the arcuate fasciculus than non-musicians (Halwani, Loui, Rüber, & Schlaug, 2011). Recently, Moore and co-workers (Moore, Schaefer, Bastin, Roberts, & Overy, 2017) used a longitudinal design to study neuroanatomical effects of four weeks of training of a sequential finger movement task, which was presented with and without musical cues in two groups. The authors report a significant increase of FA in the arcuate fasciculus only in the group that received musical cues during training, providing further support for the notion that expertise-related differences in the ultrastructure of this pathway may reflect training effects related to auditory–motor integration.

Associations between expertise and local white matter structure of relevant fiber tracts have been observed for other domains than music, although this literature is still relatively small. In a longitudinal study of juggling training, Scholz and co-workers (Scholz, Klein, Behrens, & Johansen-Berg, 2009) found increased FA in the white matter underlying the right intraparietal sulcus in the training group but not in the control group, after six weeks of training. The intraparietal sulcus is one of the major sulci on the lateral surface of the parietal lobe, and earlier studies have found support for that cortex in this region is important for visuomotor coordination during juggling (Draganski et al., 2004). Debowska and colleagues (Debowska et al., 2016) reported increased FA in white matter pathways underlying primary and secondary somatosensory areas after three weeks of training to read Braille with touch, in a single group of participants. Hänggi and co-workers (Hänggi et al., 2015) observed higher FA in the right corticospinal tract of female professional handball players than in a control group, and concluded that this may be related to more efficient catching and throwing skills in this expert group.

The earlier cautionary comments on the literature about gray matter anatomy and expertise apply also to studies of white matter. The total number of studies in this field is still relatively small and one can note many inconsistencies in reported findings, which presumably reflect the use of small samples, method variance, and heterogeneity of the studied expert groups (Merrete et al., 2013). For music, there appears to be relatively good support for the notion that expert performance is associated with anatomical differences in white matter tracts that are involved in motor control, auditory–motor integration, and information exchange between the two hemispheres, i.e., the corticospinal tracts, the arcuate fasciculus, and the corpus callosum. As for studies on gray matter, it should be noted that many studies do not allow conclusions about the causal underpinnings of the observed effects, i.e., whether they reflect influences of training on the brain, reverse causality (i.e., brain differences affecting training), or common influences on training and brain anatomy from some third factor, such as genetic constitution , however see de Manzano & Ullén (2018).

A typical finding is that relations between expertise and white matter measures such as regional volume and FA are positive. The FA value of a given white matter voxel can be influenced by different local tissue properties, such as the diameter and densities of the axons, their spatial organization, and the thickness of the axonal myelin sheath (Le Bihan, 2003; Zatorre et al., 2013). The interpretation of training-related differences in diffusion parameters such as FA is therefore difficult, and numerous putative underlying mechanisms that involve both neural and non-neural elements have been suggested. Among these are angiogenesis, growth of glial cells, modulation of thickness of the myelin sheath, axonal sprouting, changes in axonal diameters, elimination of inactive axons, and even re-routing of long-range connectivity (for discussions, see Ullén, 2009; Zatorre et al., 2013). It should also be noted that it is not uncommon to find regions where experts display lower regional FA than non-experts, further underscoring that the relations between plastic processes underlying expertise at the ultrastructural level, and imaging measures such as FA may be complex (see e.g. Hänggi et al., 2010; Roberts, Bain, Day, & Husain, 2013).

A final note is that several studies on gray and white matter in experts suggest that the anatomical correlates of expertise are to some extent specific to the particular training of the participants. In their study of hand motor representations in the primary cortex of musicians, Bangert and Schlaug (2006) found that these were more well developed on the left side in pianists and on the right side in string players and, in line with this, Rüber and co-workers (Rüber et al., 2015) found FA values of right-hemispheric corticospinal tracts to be higher than in controls in both string players and keyboard players, whereas only the keyboard players had a higher FA in the left corticospinal tract. These findings are likely to reflect differences between string and keyboard instruments, where playing on the former requires fractionated finger movements mainly with the left hand. Another example of highly specific neural correlates of expertise was found by Halwani and co-workers (Halwani et al., 2011), who demonstrated that singers had a lower FA than instrumental musicians specifically in the left dorsal arcuate fasciculus, a finding which they suggest may reflect specific adaptations related to vocal control.

FUNCTIONAL NEURAL CORRELATES OF EXPERTISE

The previous section documents that neuroanatomical correlates of expertise are widely distributed across the nervous system and may involve systems for sensory processing, motor control, and memory and cognition. As discussed, many of these anatomical findings presumably reflect functionally relevant differences between experts and non-experts in the organization of neural circuitry and, as one would expect, adaptations in expertise indeed also appear to involve optimized information processing at different levels in the nervous system, all the way from the processing of incoming sensory stimuli to the generation of skilled performance. The next section is focused on general neural expertise mechanisms for which there is substantial empirical support, i.e., automation of behavior, domain-specific boosting of working memory through interactions with long-term memory, and optimized processing in sensorimotor systems, including so-called action–observation (*mirror*) systems. In a final subsection, we briefly discuss interesting new findings from studies that explicitly address expertise from a distributed network perspective, and demonstrate altered functional interactions between brain regions in experts.

Automation

One of the most pervasive consequences of training is automation. In general terms, automation implies a transition from controlled processing, which is effortful and flexible and relies on attention and working memory, to automatic processing, which in contrast is low effort, is difficult to modify, and does not require cognitive control (Hill & Schneider, 2006; Schneider & Shiffrin, 1977). Automation occurs effectively when tasks are consistent and do not involve processing of new information or variable task conditions. In line with this, Chein and Schneider (2005), in a meta-analysis of the early functional MRI and positron emission tomography (PET) literature on practice of explicit, consistent tasks, found that practice leads to reduced activity in prefrontal and parietal cortical regions involved in domain-general cognitive control. A comprehensive discussion of the importance of automation for expertise and dual-task performance, as well as putative neural mechanisms, can be found in Hill and Schneider (2006), and here we will add some observations from more recent studies.

One thing to note is that automation of simple, consistent tasks occurs after training intervals that are extremely short from an expertise perspective. For instance, the inclusion criterion in Chein and Schneider (2005) was that a study had employed at least 10 minutes of practice. Presumably, a consequence of the much more extensive training of experts is that progressively more and more complex aspects of the trained behavior can be automated. An interesting possibility is that the expansion of sensory and motor regions observed in neuroanatomical studies of musicians (see the previous section) reflects such processes, i.e., that regions at lower levels of the nervous system adapt and become capable of performing complex domain-specific functions that require frontoparietal control systems in novices.

Recent studies support that the load on neural systems for cognitive control in experts is highly specific to the task and particular training of the participant. Proverbio and Orlandi (2016) measured event-related potentials from musicians (clarinetists and violinists) while they judged the number of notes played in video clips of music performances on the clarinet or the violin. Anterior negativity responses over the prefrontal cortex were larger when the musicians watched performances with the unfamiliar instrument rather than their own instrument. De Manzano and Ullén (2012) used functional MRI to compare the neural correlates of improvisation of brief musical melodies and the free generation of pseudo-random key sequences of the same length in professional pianists. Frontoparietal regions involved in cognitive control had a significantly higher level of activity in the unfamiliar pseudo-random condition. Interestingly, the lower activity during improvisation indicates that experts may have lower load on executive control systems even during domain-specific creative tasks, i.e., tasks that involve the free generation of meaningful responses. More recent studies from our laboratory provide further support for this notion. Pinho and co-workers (Pinho, de Manzano, Fransson, Eriksson, & Ullén, 2014) measured brain activity with functional MRI while pianists performed brief improvisations that either had a particular emotional character (happy/fearful) or were limited to certain keys (pitch sets). The participants had a varying background in either jazz or classical piano playing, and improvisation practice. Activity in the frontoparietal network showed a significant negative relation to the total hours of improvisation experience, when statistically controlling for age and hours of classical training, with the most experienced improvisers showing lower activity in these areas during improvisation than during baseline (rest). In a follow-up analysis, we found that within participants, the frontoparietal network was less active under the more ecological emotional conditions than under the structural (pitch-set) conditions (Pinho, Ullén, Castelo-Branco, Fransson, & de Manzano, 2016). Deactivations of prefrontal cortex in jazz musicians during improvisation have also been reported by Limb and Braun (2008). All in all, these findings suggest that automation in expertise may also facilitate highly complex aspects of expert performance that involve creativity and free generation. Notably, professional improvisers have acquired extensive long-term stores of musical patterns and cognitive strategies that can be used during extemporization (Pressing, 1988). Improvisation thus does not only involve free choices and the processing of novel materials, but also relies on skills that presumably are more easily automated, such as the modification, recombination, and expressive rendering of previously learned musical structures.

One important consequence of automation—e.g., in forms of expertise that rely on complex sensorimotor skills, such as dance, music, and sports—could be reallocation of attentional resources. If, after extensive training, basic components of expert performance need little explicit control, attentional resources thus freed could instead be invested in the control of higher order cognitive processes, such as decision-making and strategic planning. As emphasized by Ericsson and co-workers (see, e.g., Ericsson, Krampe, & Tesch-Römer, 1993), a maintained, explicit effort to improve skill may also be required for long-term improvements of performance.

Interactions between long-term memory and working memory

A large psychological literature suggests that a key mechanism underlying the superior domain-specific information processing and problem-solving capacities of experts is an efficient interaction between working memory and stores of relevant information in long-term memory (Gobet, 2016; Ullén et al., 2016). One direct consequence of this is that experts show superior memory for complex ecological stimuli, something which has been documented for a wide range of expertise domains, including games such as chess (Charness, 1976; Chase & Simon, 1973b; Simon & Chase, 1973), bridge (Engle & Bukstel, 1978), Go (Reitman, 1976), Othello (Wolff, Mitchell, & Frey, 1984), computer programming (McKeithen, Reitman, Rueter, & Hirtle, 1981), and music (Halpern & Bower, 1982; Sloboda, 1976, 1978). A pioneering model of expert problem solving is

Chase and Simon's *chunking theory* (Chase & Simon, 1973a, 1973b), which was originally formulated to account for expert performance in chess. A key idea of this theory is that experts can represent large, complex structures consisting of several elements as integrated units, called *chunks*. The organization of chunks reflects functional properties and relations between the elements. An interesting further development of these ideas is Gobet and Simon's *template theory*, which has been implemented in the computer program CHREST (Gobet, 2005, 2016; Gobet & Simon, 1996). Template theory assumes that expert problem solving relies on a large number of chunks that are associated with other information in long-term memory, encoded as schemata and productions. For a recent comprehensive discussion of template theory and its relations to other models, such as the long-term working memory theory of Ericsson and Kintsch (Ericsson & Kintsch, 1995), see Gobet (2016) and this Handbook (Gobet, Chapter 2, "The Classic Expertise Approach and Its Evolution").

Imaging studies support that long-term memory systems are implied in expert performance. Guida and co-workers (Guida, Gobet, Tardieu, & Nicolas, 2012), in a review of the literature, found evidence for activation of long-term memory systems in the temporal lobe during performance of working memory tasks in experts, including mental calculators, mnemonists, and chess players. They hypothesize that two processes, with different time-courses, may be involved in the acquisition of expert working memory. The earlier phase would be characterized by binding of elements into chunks within working memory, and a concomitant decrease of activity in frontoparietal regions. During the second phase, a functional reorganization with increased involvement of long-term memory regions is seen, as stable representations of domain-specific patterns are formed in long-term stores (Guida, Gobet, & Nicolas, 2013; Guida et al., 2012).

There is thus both psychological and neurobiological evidence for the idea that long-term memory systems are involved in expert performance. However, the analysis of the actual neural mechanisms underlying the representation of domain-specific information in long-term stores, and interactions between working memory and long-term memory is still in its infancy. This seems like an important area for further research, given the central role of these mechanisms for the superior performance of experts. It appears likely that the neural mechanisms underlying these functions may differ between domains and tasks (e.g., depending on the sensory modality of processed stimuli). At present, the most studied examples can be found in domains of expertise that involve the processing of complex visual stimuli. The fusiform face area in the temporal lobe, which has been shown to be involved in the processing of faces in classical studies (Kanwisher, McDermott, & Chun, 1997; Kanwisher & Yovel, 2006), is also implicated in processing of domain-specific visual stimuli in experts. This has, e.g., been demonstrated in experts of chess, cars, and birds (Bilalić et al., 2011; Gauthier, Skudlarski, Gore, & Anderson, 2000). Bilalić and co-workers have also shown that in chess experts, responses to random and ecological chess positions differ in this brain area, supporting the notion that it is indeed involved in task-relevant processing of stimuli (Bilalić et al., 2011).

Sensorimotor Adaptations

The processes of automation and reorganization of memory systems discussed in the previous sections are likely to be crucial for most forms of expertise. In particular for domains of expertise that involve physical action, adaptations of the brain's sensorimotor regions also appear to be important. Specific adaptations in the sensory cortices have been demonstrated, e.g., by Elbert and co-workers (Elbert, Pantev, Wienbruch, Rockstroh, & Taub, 1995), who found an enlarged representation of the left hand in the somatosensory cortex of string musicians, and by Pantev and colleagues, in numerous studies of auditory cortical representations of sounds in musicians (for a review see Pantev & Herholz, 2011). These findings indicate that even relatively early sensory processing in experts may be adapted to process relevant information optimally.

The action observation system is a network of regions that include the motor and premotor areas, the inferior frontal cortex, the superior and inferior parietal cortices, and the cerebellum (Calvo-Merino et al., 2006; Caspers, Zilles, Laird, & Eickhoff, 2010). Neural circuits in these regions are activated when we observe others performing actions, and influential theories suggest that this represents an internal simulation of the act that may be important for functions such as action interpretation, prediction, imitation learning, and communication (Calvo-Merino, Glaser, Grezes, Passingham, & Haggard, 2005; Jeannerod, 2001; Rizzolatti, Fogassi, & Gallese, 2001). This naturally suggests that expertise involving sensorimotor skills—such as music, dance, and sports—could involve modifications of the action observation system, to optimize processing of information relevant for expert performance.

Indeed, several neuroimaging studies have found that patterns of activity in this system differ between experts and non-experts specifically for observation of domain-specific actions that the participant is familiar with (Liew, Sheng, Margetis, & Aziz-Zadeh, 2013; Zatorre et al., 2007). For instance, experts in tennis and volleyball display higher activity in superior parietal cortex, the supplementary motor area, and the cerebellum when they observe serves in their trained sport as compared to serves in the non-trained sport (Balser et al., 2014); skilled soccer players activate action observation regions more during visual discrimination of soccer moves than do unskilled participants (Wright, Bishop, Jackson, & Abernethy, 2013). Similar expertise-specific responses in the action observation system have been observed, e.g., for dancers (Calvo-Merino et al., 2005) and pianists (Haslinger et al., 2005). These findings clearly support the notion that the action observation is involved in advanced skill learning, and that neural circuits in these regions can represent and discriminate between spatiotemporally complex action sequences, even though their component movements are quite similar (Calvo-Merino et al., 2005).

Interestingly, not only the observation of acts but even the mere perception of stimuli that are associated with actions can activate sensorimotor areas in experts. For instance, sport-related environmental sounds were found to induce higher activity in sensorimotor areas in athletes than in novices (Woods, Hernandez, Wagner, &

Beilock, 2014), and Behmer and Jantzen (2011) found indications in an EEG study that visual observation of sheet music induced activation of the motor system in musicians. Lyons et al. (2010) report that listening to verbal stimuli referring to ice-hockey induced activity in motor areas in ice hockey players.

Expertise and Functional Connectivity

It is evident from the previous discussions that expert performance relies on information processing within networks of interconnected cortical and subcortical brain areas. While many studies have focused on regional effects, several recent studies have also started to analyze relations between expertise and the functional connectivity between brain regions that are active during the performance of relevant tasks. This literature is promising but still relatively small, and here we will provide some examples of findings from the musical domain, which underscore that expertise is reflected not just in activity patterns within individual brain regions, but also in the communication between them.

As discussed earlier, auditory–motor integration is essential for music performance, and musical expertise is related to anatomical and functional properties of auditory and motor regions, and the fiber tracts connecting them. Chen and co-workers (Chen, Penhune, & Zatorre, 2008) studied tapping in synchrony with auditory rhythms in musicians and non-musicians. Auditory regions in the planum temporale and dorsal premotor cortex were functionally connected in both musicians and non-musicians, but only the musicians showed significant bilateral functional connectivity between the right planum temporale and both left and right dorsal premotor cortex. Grahn and Rowe (2009) studied perception of auditory rhythms in musicians and non-musicians and similarly found that the functional connectivity between auditory and premotor areas under different conditions was influenced by musicianship. Pinho et al. (2014), in the previously mentioned study on improvisation in classical and jazz pianists, found that improvisation training was specifically related to higher functional connectivity between prefrontal, premotor, and motor regions during improvisation, and speculated that this may reflect a more efficient integration of representations of musical structures at different levels of abstraction. Finally, in a recent study, Alluri et al. (2017) investigated brain activity as musically trained and untrained participants listened to musical pieces in different genres, using a graph-theoretical analysis of whole-brain functional connectivity patterns. A main finding was that the primary hubs of musicians comprised sensorimotor regions, whereas the dominant hubs in non-musicians were related to the default mode network—a network of regions involved in introspective thought processes, daydreaming, imagery, etc., typically during rest (Mittner, Hawkins, Boekel, & Forstmann, 2016). The authors interpret the findings as an indication that musicians automatically employ action-based neural networks during music listening.

Neurobiological Models of Expertise— A Multifactorial Perspective

To summarize the main points of the previous sections, expertise is correlated with macroanatomical properties of domain-relevant brain regions and ultrastructural properties of both gray and the white matter. The specific regions involved depend on the domain of expertise. The consequence of these neural adaptations is a capacity for vastly more efficient performance of domain-specific tasks. This in turn depends on multiple functional mechanisms situated at different levels of neural processing. The mechanisms can include automation and alterations in functional connectivity, as well as specializations within memory systems and sensorimotor systems that optimize the processing of information relevant for the particular domain of expertise.

As we have emphasized earlier, there has been a strong tendency in the neurobiological expertise literature to interpret differences between experts and non-experts and correlations between practice and various outcomes as being essentially due to causal effects of practice (see, e.g., Ericsson et al, 1993; Ericsson, 2014). However, recent developments in expertise studies make it increasingly clear that this view is too simplistic, and that expertise is a function of many factors apart from causal influences of long-term practice.

First, a number of meta-analyses of the literature on deliberate practice and expert performance include estimates of how much variance in performance is explained by practice (Hambrick et al., 2014; Macnamara, Hambrick, & Oswald, 2014; Platz, Kopiez, Lehmann, & Wolf, 2014). These estimates are relatively modest in size, ranging from 1 percent for professional expertise in Macnamara et al. (2014), to 36 percent for music in Platz et al. (2014). In other words, a large proportion of the variance in expert performance is apparently unrelated to practice. As we have discussed elsewhere (Ullén et al., 2016), practice-independent influences on expertise are likely to be related to individual differences in both general traits (e.g., intelligence and personality) and more specific traits of relevance for the particular domain. This is relevant for a discussion of the neural mechanisms of expertise, since it suggests that neural correlates of implicated traits may influence expert performance independently of practice effects. For example, reported neural correlates of intelligence include total brain volume (Deary, Penke, & Johnson, 2010; Rushton & Ankney, 2009), cortical thickness (Karama et al., 2009), and topological properties of white matter connections (Li et al., 2009). Conceivably, then, such global brain properties could facilitate expert performance, over and above practice effects, in domains where expertise correlates with intelligence (Gobet, 2016).

Secondly, recent studies using twin modeling have found substantial genetic effects on long-term practice (Mosing, Madison, Pedersen, Kuja-Halkola, & Ullén, 2014), musical expert performance (Drayna, Manichaikul, de Lange, Snieder, & Spector, 2001; Mosing et al., 2014), and self-rated expertise in various domains (Vinkhuyzen, van der

Sluis, Posthuma, & Boomsma, 2009). Furthermore, studies from our laboratory have demonstrated that the association between music practice and musical expert performance, operationalized as auditory discrimination of musical stimuli, is mainly driven by genetic pleiotropy, i.e., common genetic influences on the two variables (Mosing et al., 2014). This is a dramatic empirical demonstration of the notion that correlations between practice and performance do not necessarily constitute evidence for causal effects of practice. Importantly, these findings on the behavioral level, of course, also strongly suggest, by implication, that genetic influences are important for differences between experts and non-experts at the neural level. In line with this, twin modeling studies have demonstrated substantial genetic influences on gray and white matter structure (Eyler et al., 2012; Shen et al., 2016), as well as on functional differences such as working memory-related brain activation (Blokland et al., 2011; Sinclair et al., 2015).

Based on these recent developments in expertise research, we have proposed a generic model of expert performance called the Multifactorial Gene environment Interaction Model (MGIM) (Ullén et al., 2016, 2017). An important feature of the MGIM is that the neural mechanisms of expert performance are influenced by practice but also by other variables, such as traits, in a domain-specific manner. Secondly, both practice-dependent and practice-independent shaping of the neural circuitry underlying expert performance is assumed to depend on genetic and non-genetic factors, and their interactions. This framework can easily accommodate existing findings on expertise and its neural basis, and serve as a guide for future research.

Research in recent years has provided exciting new discoveries about expertise and its neural basis and also suggests many interesting avenues for further studies. In general, we think a multifactorial approach to expertise studies will be essential for further progress in the field. One question that clearly needs more empirical work is, for example, how the acquisition of expertise and the shaping of its neural machinery depend on interactions between practice and other variables, traits as well as environmental factors. More studies on genetically informative samples, in order to characterize gene–environment interactions and correlations in expertise would also be important. Recent demonstrations that practice effects on the brain may involve both increases and decreases in regional brain volumes (Wenger et al., 2017), as well as theoretical accounts which propose that expertise acquisition occurs in different stages (Guida et al., 2013), underscore the importance of longitudinal designs, and studies in genetically informative samples (de Manzano & Ullén, 2018). Finally, many interesting questions remain unanswered regarding the actual neural mechanisms underlying expert performance, such as interactions between working memory and long-term memory.

References

Abdul-Kareem, I. A., Stancak, A., Parkes, L. M., & Sluming, V. (2011). Increased gray matter volume of left pars opercularis in male orchestral musicians correlate positively with years of musical performance. *Journal of Magnetic Resonance Imaging* 33(1), 24–32.

Alluri, V., Toiviainen, P., Burunat, I., Kliuchko, M., Vuust, P., & Brattico, E. (2017). Connectivity patterns during music listening: evidence for action-based processing in musicians. *Human Brain Mapping 38*(6), 2955–2970.

Amunts, K., Schlaug, G., Jäncke, L., Steinmetz, H., Schleicher, A., Dabringhaus, A., & Zilles, K. (1997). Motor cortex and hand motor skills: structural compliance in the brain. *Human Brain Mapping 5*(3), 206–215.

Armand, J., Olivier, E., Edgley, S. A., & Lemon, R. N. (1996). The structure and function of the developing corticospinal tract: some key issues. In A. M. Wing, P. Haggard, & J. R. Flanagan (Eds), *Hand and Brain* (pp. 125–146). San Diego: Academic Press.

Balser, N., Lorey, B., Pilgramm, S., Naumann, T., Kindermann, S., Stark, R., ... Munzert, J. (2014). The influence of expertise on brain activation of the action observation network during anticipation of tennis and volleyball serves. *Frontiers in Human Neuroscience 8*. doi:10.3389/fnhum.2014.00568

Bangert, M., & Schlaug, G. (2006). Specialization of the specialized in features of external human brain morphology. *European Journal of Neuroscience 24*(6), 1832–1834.

Behmer, L. P., & Jantzen, K. J. (2011). Reading sheet music facilitates sensorimotor mu-desynchronization in musicians. *Clinical Neurophysiology 122*(7), 1342–1347.

Bengtsson, S. L., Nagy, Z., Forsman, L., Forssberg, H., Skare, S., & Ullén, F. (2005). Extensive piano practicing has regionally specific effects on white matter development. *Nature Neuroscience 8*(9), 1148–1150.

Bermudez, P., Lerch, J. P., Evans, A. C., & Zatorre, R. J. (2009). Neuroanatomical correlates of musicianship as revealed by cortical thickness and voxel-based morphometry. *Cerebral Cortex 19*(7), 1583–1596.

Bernardi, G., Cecchetti, L., Handjaras, G., Sani, L., Gaglianese, A., Ceccarelli, R., ... Pietrini, P. (2014). It's not all in your car: functional and structural correlates of exceptional driving skills in professional racers. *Frontiers in Human Neuroscience 8*. doi:10.3389/fnhum.2014.00888

Bilalić, M., Langner, R., Ulrich, R., & Grodd, W. (2011). Many faces of expertise: fusiform face area in chess experts and novices. *Journal of Neuroscience 31*(28), 10206–10214.

Black, J. E., Isaacs, K. R., Anderson, B. J., Alcantara, A. A., & Greenough, W. T. (1990). Learning causes synaptogenesis, whereas motor activity causes angiogenesis, in cerebellar cortex of adult rats. *PNAS Proceedings of the National Academy of Sciences of the United States of America 87*, 5568–5572.

Blokland, G. A. M., McMahon, K. L., Thompson, P. M., Martin, N. G., de Zubicaray, G. I., & Wright, M. J. (2011). Heritability of working memory brain activation. *Journal of Neuroscience 31*(30), 10882–10890.

Button, K. S., Ioannidis, J. P. A., Mokrysz, C., Nosek, B. A., Flint, J., Robinson, E. S. J., & Munafo, M. R. (2013). Power failure: why small sample size undermines the reliability of neuroscience. *Nature Reviews Neuroscience 14*(5), 365–376. doi:10.1038/nrn3475

Calvo-Merino, B., Glaser, D. E., Grezes, J., Passingham, R. E., & Haggard, P. (2005). Action observation and acquired motor skills: An fMRI study with expert dancers. *Cerebral Cortex 15*(8), 1243–1249.

Calvo-Merino, B., Grèzes, J., Glaser, D. E., Passingham, R. E., Haggard, P., & Gre, J. (2006). Seeing or doing? Influence of visual and motor familiarity in action observation. *Curren Biology 16*, 1905–1910.

Caspers, S., Zilles, K., Laird, A. R., & Eickhoff, S. B. (2010). ALE meta-analysis of action observation and imitation in the human brain. *NeuroImage 50*, 1148–1167.

Charness, N. (1976). Memory for chess positions: Resistance to interference. *Journal of Experimental Psychology: Human Learning and Memory 2*, 641–653.

Chase, W. G., & Simon, H. A. (1973a). The mind's eye in chess. In W. G. Chase (Ed.), *Visual Information Processing* (pp. 215–281). New York: Academic Press.

Chase, W. G., & Simon, H. A. (1973b). Perception in chess. *Cognitive Psychology 4*, 55–81.

Chein, J. M., & Schneider, W. (2005). Neuroimaging studies of practice-related change: fMRI and meta-analytic evidence of a domain-general control network for learning. *Cognitive Brain Research 25*(3), 607–623.

Chen, J. L., Penhune, V. B., & Zatorre, R. J. (2008). Moving on time: brain network for auditory-motor synchronization is modulated by rhythm complexity and musical training. *Journal of Cognitive Neuroscience 20*(2), 226–239.

Deary, I. J., Penke, L., & Johnson, W. (2010). The neuroscience of human intelligence differences. *Nature Reviews Neuroscience 11*, 201–211.

Debowska, W., Wolak, T., Nowicka, A., Kozak, A., Szwed, M., & Kossut, M. (2016). Functional and structural neuroplasticity induced by short-term tactile training based on Braille reading. *Frontiers in Neuroscience, 10*. doi:10.3389/fnins.2016.00460

Delon-Martin, C., Plailly, J., Fonlupt, P., Veyrac, A., & Royet, J. P. (2013). Perfumers' expertise induces structural reorganization in olfactory brain regions. *NeuroImage 68*, 55–62.

de Manzano, Ö., & Ullén, F. (2012). Goal-independent mechanisms for free response generation: creative and pseudo-random performance share neural substrates. *NeuroImage 59*(1), 772–780.

de Manzano, Ö., & Ullén, F. (2018). Same Genes, Different Brains: Neuroanatomical Differences Between Monozygotic Twins Discordant for Musical Training. *Cerebral Cortex 28* (1), 387–394.

Draganski, B., Gaser, C., Busch, V., Schuierer, G., Bogdahn, U., & May, A. (2004). Neuroplasticity: changes in grey matter induced by training. *Nature 427*(6972), 311–312.

Draganski, B., Gaser, C., Kempermann, G., Kuhn, H. G., Winkler, J., Büchel, C., & May, A. (2006). Temporal and spatial dynamics of brain structure changes during extensive learning. *Journal of Neuroscience 26*(23), 6314–6317.

Drayna, D., Manichaikul, A., de Lange, M., Snieder, H., & Spector, T. (2001). Genetic correlates of musical pitch recognition in humans. *Science 291*(5510), 1969–1972.

Edelman, G. M. (1987). *Neural Darwinism: The theory of neuronal group selection*. New York: Basic Books.

Elbert, T., Pantev, C., Wienbruch, C., Rockstroh, B., & Taub, E. (1995). Increased cortical representation of the fingers of the left hand in string players. *Science 270*, 305–307.

Engel, A., Hijmans, B. S., Cerliani, L., Bangert, M., Nanetti, L., Keller, P. E., & Keysers, C. (2014). Inter-individual differences in audio-motor learning of piano melodies and white matter fiber tract architecture. *Human Brain Mapping 35*(5), 2483–2497.

Engle, R. W., & Bukstel, L. (1978). Memory processes among bridge players of differing expertise. *American Journal of Psychology 91*(4), 673–689.

Ericsson, K. A. (2014). Why expert performance is special and cannot be extrapolated from studies of performance in the general population: A response to criticisms. *Intelligence 45*, 81–103.

Ericsson, K. A., & Kintsch, W. (1995). Long-term working memory. *Psychological Review 102*(2), 211–245.

Ericsson, K. A., Krampe, R. T., and Tesch-Römer, C. (1993). The role of deliberate practice in the acquisition of expert performance. *Psychological Review 100*(3), 363–406.

Eyler, L. T., Chen, C. H., Panizzon, M. S., Fennema-Notestine, C., Neale, M. C., Jak, A., ... Kremen, W. S. (2012). A comparison of heritability maps of cortical surface area and thickness and the influence of adjustment for whole brain measures: a magnetic resonance imaging twin study. *Twin Research and Human Genetics 15*(3), 304–314.

Gaser, C., & Schlaug, G. (2003). Brain structures differ between musicians and non-musicians. *Journal of Neuroscience 23*(27), 9240–9245.

Gauthier, I., Skudlarski, P., Gore, J. C., & Anderson, A. W. (2000). Expertise for cars and birds recruits brain areas involved in face recognition. *Nature Neuroscience 3*(2), 191–197.

Gobet, F. (2005). Chunking models of expertise: Implications for education. *Applied Cognitive Psychology 19*(2), 183–204.

Gobet, F. (2016). *Understanding Expertise: A multi-disciplinary approach.* London: Palgrave.

Gobet, F., & Simon, H. A. (1996). Templates in chess memory: A mechanism for recalling several boards. *Cognitive Psychology 31*(1), 1–40.

Gooijers, J., & Swinnen, S. P. (2014). Interactions between brain structure and behavior: The corpus callosum and bimanual coordination. *Neuroscience and Biobehavioral Reviews 43*, 1–19.

Grahn, J. A., & Rowe, J. B. (2009). Feeling the beat: Premotor and striatal interactions in musicians and nonmusicians during beat perception. *Journal of Neuroscience 29*(23), 7540–7548.

Guida, A., Gobet, F., & Nicolas, S. (2013). Functional cerebral reorganization: a signature of expertise? Reexamining Guida, Gobet, Tardieu, and Nicolas' (2012) two-stage framework. *Frontiers in Human Neuroscience 7*, 590. doi: 10.3389/fnhum.2013.00590

Guida, A., Gobet, F., Tardieu, H., & Nicolas, S. (2012). How chunks, long-term working memory and templates offer a cognitive explanation for neuroimaging data on expertise acquisition: A two-stage framework. *Brain and Cognition 79*(3), 221–244.

Halpern, A. R., & Bower, G. H. (1982). Musical expertise and melodic structure in memory for musical notation. *American Journal of Psychology 95*(1), 31–50.

Halwani, G. F., Loui, P., Rüber, T., & Schlaug, G. (2011). Effects of practice and experience on the arcuate fasciculus: comparing singers, instrumentalists, and non-musicians. *Frontiers in Psychology 2*, 156. doi: 10.3389/fpsyg.2011.00156

Hambrick, D. Z., Oswald, F. L., Altmann, E. M., Meinz, E. J., Gobet, F., & Campitelli, G. (2014). Deliberate practice: Is that all it takes to become an expert? *Intelligence 45*, 34–45.

Hänggi, J., Brutsch, K., Siegel, A. M., & Jancke, L. (2014). The architecture of the chess player's brain. *Neuropsychologia 62*, 152–162.

Hänggi, J., Koeneke, S., Bezzola, L., & Jäncke, L. (2010). Structural neuroplasticity in the sensorimotor network of professional female ballet dancers. *Human Brain Mapping 31*(8), 1196–1206.

Hänggi, J., Langer, N., Lutz, K., Birrer, K., Merillat, S., & Jancke, L. (2015). Structural brain correlates associated with professional handball playing. *PLoS One 10*(4). doi:10.1371/journal.pone.0124222

Haslinger, B., Erhard, P., Altenmuller, E., Schroeder, U., Boecker, H., & Ceballos-Baumann, A. O. (2005). Transmodal sensorimotor networks during action observation in professional pianists. *Journal of Cognitive Neuroscience 17*(2), 282–293.

Hill, N. M., & Schneider, W. (2006). Brain changes in the development of expertise: Neuroanatomical and neurophysiological evidence about skill-based adaptations. In K. A. Ericsson, N. Charness, P. J. Feltovich, & R. R. Hoffman (Eds), *The Cambridge handbook of expertise and expert performance* (pp. 653–682). Cambridge, UK: Cambridge University Press.

Hutchinson, S., Lee, L. H., Gaab, N., & Schlaug, G. (2003). Cerebellar volume of musicians. *Cerebral Cortex 13*(9), 943–949.

Hyde, K. L., Lerch, J., Norton, A., Forgeard, M., Winner, E., Evans, A. C., & Schlaug, G. (2009). Musical training shapes structural brain development. *Journal of Neuroscience 29*(10), 3019–3025.

Jeannerod, M. (2001). Neural simulation of action: a unifying mechanism for motor cognition. *NeuroImage 14*(1 Pt 2), S103–109.
Kanwisher, N., McDermott, J., & Chun, M. M. (1997). The fusiform face area: A module in human extrastriate cortex specialized for face perception. *Journal of Neuroscience 17*(11), 4302–4311.
Kanwisher, N., & Yovel, G. (2006). The fusiform face area: a cortical region specialized for the perception of faces. *Philosophical Transactions of the Royal Society of London Series B, Biological Sciences 361*, 2109–2128.
Karama, S., Ad-Dab'bagh, Y., Haier, R. J., Deary, I. J., Lyttelton, O. C., Lepage, C., . . . Brain Development Cooperative Group. (2009). Positive association between cognitive ability and cortical thickness in a representative US sample of healthy 6 to 18 year-olds. *Intelligence 37*(2), 145–155.
Kleim, J. A., Hogg, T. M., Van den Berg, P. M., Cooper, N. R., Bruneau, R., & Remple, M. (2004). Cortical synaptogenesis and motor map reorganization occur during late, but not early, phase of motor skill learning. *Journal of Neuroscience 24*(3), 628–633.
Le Bihan, D. (2003). Looking into the functional architecture of the brain with diffusion MRI. *Nature Reviews Neuroscience 4*, 469–480.
Li, Y. H., Liu, Y., Li, J., Qin, W., Li, K. C., Yu, C. S., & Jiang, T. Z. (2009). Brain anatomical network and intelligence. *PLoS Computational Biology 5*(5), e1000395. doi: 10.1371/journal.pcbi.1000395
Liew, S. L., Sheng, T., Margetis, J. L., & Aziz-Zadeh, L. (2013). Both novelty and expertise increase action observation network activity. *Frontiers in Human Neuroscience 7*, 541. doi: 10.3389/fnhum.2013.00541
Limb, C. J., & Braun, A. R. (2008). Neural substrates of spontaneous musical performance: an fMRI study of jazz improvisation. *PLoS One 3* (2), e1679. doi: 10.1371/journal.pone.0001679
Lyons, I. M., Mattarella-Micke, A., Cieslak, M., Nusbaum, H. C., Small, S. L., & Beilock, S. L. (2010). The role of personal experience in the neural processing of action-related language. *Brain and Language 112*(3), 214–222.
McKeithen, K. B., Reitman, J. S., Rueter, H. H., & Hirtle, S. C. (1981). Knowledge organization and skill differences in computer programmers. *Cognitive Psychology 13*(3), 307–325.
Macnamara, B. N., Hambrick, D. Z., & Oswald, F. L. (2014). Deliberate practice and performance in music, games, sports, education, and professions: A meta-analysis. *Psychological Science 25*(8), 1608–1618.
Maguire, E. A., Gadian, D. G., Johnsrude, I. S., Good, C. D., Ashburner, J., Frackowiak, R. S., & Frith, C. D. (2000). Navigation-related structural change in the hippocampi of taxi drivers. *PNAS Proceedings of the National Academy of Sciences of the United States of America 97*(8), 4398–4403.
Merrete, D. L., Peretz, I., & Wilson, S. J. (2013). Moderating variables of music training-induced neuroplasticity: a review and discussion. *Frontiers in Psychology 4*. doi:10.3389/fpsyg.2013.00606
Mittner, M., Hawkins, G. E., Boekel, W., & Forstmann, B. U. (2016). A neural model of mind wandering. *Trends in Cognitive Sciences 20*(8), 570–578.
Moore, E., Schaefer, R. S., Bastin, M. E., Roberts, N., & Overy, K. (2017). Diffusion tensor MRI tractography reveals increased fractional anisotropy (FA) in arcuate fasciculus following music-cued motor training. *Brain and Cognition 116*, 40–46.

Mosing, M. A., Madison, G., Pedersen, N. L., Kuja-Halkola, R., & Ullén, F. (2014). Practice does not make perfect: No causal effect of music practice on music ability. *Psychological Science* 25(9), 1795–1803.

Nieuwenhuys, R., Voogd, J., & van Huijzen, C. (2008). *The human nervous system*, fourth edn. Berlin: Springer-Verlag.

Palomar-Garcia, M. A., Zatorre, R. J., Ventura-Campos, N., Bueicheku, E., & Avila, C. (2017). Modulation of functional connectivity in auditory-motor networks in musicians compared with nonmusicians. *Cerebral Cortex* 27(5), 2768–2778.

Pantev, C., & Herholz, S. C. (2011). Plasticity of the human auditory cortex related to musical training. *Neuroscience & Biobehavioral Reviews* 35(10), 2140–2145.

Park, I. S., Lee, K. J., Han, J. W., Lee, N. J., Lee, W. T., Park, K. A., & Rhyu, I. J. (2011). Basketball training increases striatum volume. *Human Movement Science* 30(1), 56–62.

Pinho, A. L., de Manzano, Ö., Fransson, P., Eriksson, H., & Ullén, F. (2014). Connecting to create—expertise in musical improvisation is associated with increased functional connectivity between premotor and prefrontal areas. *Journal of Neuroscience* 34(18), 6156–6163.

Pinho, A. L., Ullén, F., Castelo-Branco, M., Fransson, P., & de Manzano, Ö. (2016). Addressing a paradox: Dual strategies for creative performance in introspective and extrospective networks. *Cerebral Cortex* 26(7), 3052–3063.

Platz, F., Kopiez, R., Lehmann, A. C., & Wolf, A. (2014). The influence of deliberate practice on musical achievement: a meta-analysis. *Frontiers in Psychology* 5, 646. doi: 10.3389/fpsyg.2014.00646

Pressing, J. (1988). Improvisation: methods and models. In J. A. Sloboda (Ed.), *Generative Processes in Music* (pp. 129–178). New York: Oxford University Press.

Proverbio, A. M., & Orlandi, A. (2016). Instrument-specific effects of musical expertise on audiovisual processing (clarinet vs violin). *Music Perception* 33(4), 446–456.

Reitman, J. S. (1976). Skilled perception in Go—deducing memory structures from inter-response times. *Cognitive Psychology* 8(3), 336–356.

Rizzolatti, G., Fogassi, L., & Gallese, V. (2001). Neurophysiological mechanisms underlying the understanding and imitation of action. *Nature Reviews Neuroscience* 2, 661–670.

Roberts, R. E., Bain, P. G., Day, B. L., & Husain, M. (2013). Individual differences in expert motor coordination associated with white matter microstructure in the cerebellum. *Cerebral Cortex* 23(10), 2282–2292.

Rüber, T., Lindenberg, R., & Schlaug, G. (2015). Differential adaptation of descending motor tracts in musicians. *Cerebral Cortex* 25(6), 1490–1498.

Rushton, J. P., & Ankney, C. D. (2009). Whole brain size and general mental ability: a review. *International Journal of Neuroscience* 119, 691–731.

Schlaug, G., Jäncke, L., Huang, Y., Staiger, J. F., & Steinmetz, H. (1995). Increased corpus callosum size in musicians. *Neuropsychologia* 33(8), 1047–1055.

Schlaug, G., Marchina, S., & Norton, A. (2009). Evidence for plasticity in white-matter tracts of patients with chronic Broca's aphasia undergoing intense intonation-based speech therapy. *Annals of the New York Academy of Sciences* 1169(1), 385–394.

Schneider, P., Scherg, M., Dosch, H. G., Specht, H. J., Gutschalk, A., & Rupp, A. (2002). Morphology of Heschl's gyrus reflects enhanced activation in the auditory cortex of musicians. *Nature Neuroscience* 5(7), 688–694.

Schneider, W., & Shiffrin, R. M. (1977). Controlled and automatic human information-processing. 1. Detection, search, and attention. *Psychological Review* 84(1), 1–66.

Scholz, J., Klein, M. C., Behrens, T. E., & Johansen-Berg, H. (2009). Training induces changes in white-matter architecture. *Nature Neuroscience* 12(11), 1370–1371.

Shen, K. K., Dore, V., Rose, S., Fripp, J., McMahon, K. L., de Zubicaray, G. I., . . . Salvado, O. (2016). Heritability and genetic correlation between the cerebral cortex and associated white matter connections. *Human Brain Mapping* 37(6), 2331–2347.

Simon, H. A., & Chase, W. G. (1973). Skill in chess. *American Scientist* 61, 394–403.

Sinclair, B., Hansell, N. K., Blokland, G. A. M., Martin, N. G., Thompson, P. M., Breakspear, M., . . . McMahon, K. L. (2015). Heritability of the network architecture of intrinsic brain functional connectivity. *NeuroImage* 121, 243–252.

Sloboda, J. A. (1976). Visual perception of musical notation: registering pitch symbols in memory. *Quarterly Journal of Experimental Psychology* 28, 1–16.

Sloboda, J. A. (1978). Perception of contour in music reading. *Perception* 7(3), 323–331.

Steele, C. J., Bailey, J. A., Zatorre, R. J., & Penhune, V. B. (2013). Early musical training and white-matter plasticity in the corpus callosum: evidence for a sensitive period. *Journal of Neuroscience* 33(3), 1282–1290.

Ullén, F. (2009). Is activity regulation of late myelination a plastic mechanism in the human nervous system? *Neuron Glia Biology* 5(1–2), 29–34.

Ullén, F., Hambrick, D. Z., & Mosing, M. A. (2016). Rethinking expertise: a multi-factorial gene-environment interaction model of expert performance. *Psychological Bulletin* 142(4), 427–446.

Ullén, F., Mosing, M. A., & Hambrick, D. Z. (2017). The multifactorial gene–environment interaction model (MGIM) of expert performance. In D. Z. Hambrick, G. Campitelli, & B. Macnamara (Eds), *The science of expertise: Behavioral, neural, and genetic approaches to complex skill* (pp. 365–375). London: Routledge.

Vaquero, L., Hartmann, K., Ripolles, P., Rojo, N., Sierpowska, J., Francois, C., . . . Altenmuller, E. (2016). Structural neuroplasticity in expert pianists depends on the age of musical training onset. *NeuroImage* 126, 106–119.

Vinkhuyzen, A. A., van der Sluis, S., Posthuma, D., & Boomsma, D. I. (2009). The heritability of aptitude and exceptional talent across different domains in adolescents and young adults. *Behavior Genetics* 39, 380–392.

Wenger, E., Kuhn, S., Verrel, J., Martensson, J., Bodammer, N. C., Lindenberger, U., & Lovden, M. (2017). Repeated structural imaging reveals nonlinear progression of experience-dependent volume changes in human motor cortex. *Cerebral Cortex* 27(5), 2911–2925.

Wolff, A. S., Mitchell, D. H., & Frey, P. W. (1984). Perceptual skill in the game of Othello. *Journal of Psychology* 118, 7–16.

Woods, E. A., Hernandez, A. E., Wagner, V. E., & Beilock, S. L. (2014). Expert athletes activate somatosensory and motor planning regions of the brain when passively listening to familiar sports sounds. *Brain and Cognition* 87, 122–133.

Woollett, K., & Maguire, E. A. (2011). Acquiring "the Knowledge" of London's layout drives structural brain changes. *Current Biology* 21(24), 2109–2114.

Wright, M. J., Bishop, D. T., Jackson, R. C., & Abernethy, B. (2013). Brain regions concerned with the identification of deceptive soccer moves by higher-skilled and lower-skilled players. *Frontiers in Human Neuroscience* 7. doi:10.3389/fnhum.2013.00851

Zatorre, R. J., Chen, J. L., & Penhune, V. B. (2007). When the brain plays music: auditory-motor interactions in music perception and production. *Nature Reviews Neuroscience* 8(7), 547–558.

Zatorre, R. J., Fields, R. D., & Johansen-Berg, H. (2013). Plasticity in gray and white: neuroimaging changes in brain structure during learning. *Nature Neuroscience* 15(4), 528–536.

CHAPTER 7

MODELING EXPERTS WITH FAST-AND-FRUGAL HEURISTICS

ULRICH HOFFRAGE

Introduction

BEING an expert can be defined as "having, involving, or displaying special skill or knowledge derived from training or experience" (Merriam-Webster). For some types of experts, such as athletes or musicians, their expertise consists of perceptual–motor skills or established routines that enable them to perform better than laypeople. For other types, for instance physicians, financial analysts, or firefighters—and these may be those that come to mind first when we think of experts—outperforming laypeople basically means making better judgments and decisions. It hence does not come as a surprise that expertise is a fascinating topic for many judgment and decision-making researchers. As the field of expertise is by no means homogeneous but rather divided into many schools, its heterogeneity is also reflected in the multiplicity of approaches to expertise (see the variety of contributions to the first part of this book).

The present chapter focusses on one of these approaches. Its main goal is to characterize the perspective of the fast-and-frugal heuristics program on expertise: How to understand expertise when seen through the lenses of this program, how to study experts and expertise, and how to aid experts? The answers to these questions are as follows: Experts use fast-and-frugal heuristics. These heuristics are models of bounded rationality. Experts differ from novices because they are able—through experience, training, education, and feedback—to construct accurate and useful models of the environment. Studying experts in the context of this program amounts to modeling them with fast-and-frugal heuristics, and studying the acquisition of expertise amounts to studying how laypeople learn to use such heuristics. Experts can be aided by educating them about the success of fast-and-frugal heuristics when used prescriptively.

In the first part, I will introduce the fast-and-frugal heuristics program. Because some context and contrast will help in this endeavor, I will enrich the presentation of the program by clarifying the relationships to two other influential research programs: the heuristics-and-biases program, and the naturalistic decision-making program. In the second part, I will characterize the fast-and-frugal heuristics program's approach to expertise, on the basis of some selected examples.

The Conceptual Framework of the Fast-and-Frugal Heuristics Program

Heuristics-and-Biases as a Starting Point

In the 1970s, Daniel Kahneman and Amos Tversky initiated what can be considered one of the most influential programs in the social sciences in the twentieth century: the *heuristics-and-biases program* (Kahneman, Slovic, & Tversky, 1982). This program and the historical context in which it developed set the stage for the discovery of fast-and-frugal heuristics. To better understand the heuristics-and-biases program, it is useful to look at its roots—so let us go back in history by another 30 years.

In the 1940s, von Neumann and Morgenstern (1944) built on *subjective expected utility* (SEU) *theory* and formulated axioms for rational choice, such as comparability, transitivity, cancelation, and continuity. Soon afterwards, however, deviations from this theoretical approach were identified. For instance, in the *Allais paradox* (Allais, 1953) people reversed their preferences between two lotteries if the same number of balls with identical outcomes were added to each of the two urns representing these lotteries. Mathematically, these additional balls canceled out when comparing the two lotteries (hence it is called the "cancelation axiom"), but psychologically, they made a difference. It therefore does not come as a surprise that psychologists became interested in the question to what extent SEU theory is a good descriptive model of behavior (Edwards, 1954). Note that such descriptive approaches that referred to the question how people actually *do* make decisions need to be separated from normative considerations that asked how they *should* decide and what constitutes rational decisions. In fact, von Neumann and Morgenstern did not conceive their axioms of SEU theory as normative, and, likewise, Allais did not think that a violation of the cancelation axiom would be irrational. Others (including Savage, 1972, or Tversky & Kahneman, 1974), in contrast, conceived SEU theory to be *the* normative theory for rational decision making: a decision consistent with its axioms was considered to be rational, and conversely, a violation of (some of) its axioms was considered to be irrational.

The heuristics-and-biases program emerged in exactly this context and with such a normative stance. It can be seen as an attempt to document and explain deviations from SEU theory and other "normative" principles of rationality. Such deviations, so the argument went, can be explained by heuristics, for instance, by the *representativeness heuristic*, the *availability heuristic*, and the *anchoring-and-adjustment heuristic* (Kahneman et al., 1982). The "rhetoric" (Lopes, 1991) was that heuristics are often useful and yield good approximations, but that they also lead to systematic biases in some situations. Subsequent research focused more and more on those situations in which heuristics lead people astray and the overall conclusion of this program painted a pessimistic picture of humans' ability to make sound and rational judgments and decisions. In contrast to other traditions in which the term heuristic has been used before (e.g., Gestalt psychology), it acquired a new connotation in the heuristics-and-biases program: from neutral or positive, to negative (Hertwig & Pachur, 2015). This negative connotation can also be found when researchers affiliated with the heuristics-and-biases program studied experts. For instance, physicians' difficulties in estimating the positive predictive value of a medical test has been accounted for by the use of the representativeness heuristic in a situation in which it led to deviations from the statistical norm (Eddy, 1982).

From Heuristics-and-Biases to Fast-and-Frugal Heuristics

While the heuristic-and-biases program attracted a lot of attention, it has also been challenged. In particular Gerd Gigerenzer, whom Kahneman (2011) referred to as "our most persistent critic" (p. 449), and his colleagues scrutinized some of the flagship phenomena that have been frequently cited as landmarks of human irrationality. Here are two examples (for overviews that also include others, see Gigerenzer, 1994; Gigerenzer, Hertwig, Hoffrage, & Sedlmeier, 2008).

First, in the early 1990s, Gigerenzer, Hoffrage, and Kleinbölting (1991) became interested in overconfidence, and proposed, inspired by Egon Brunswik's probabilistic functionalism, the theory of probabilistic mental models. This theory specifies a mechanism that generates decisions in two-alternative-forced-choice tasks (e.g., Which of two cities has more inhabitants?) and the associated confidence judgments. If people are asked to compare two objects with respect to a quantitative criterion and do not have any knowledge about the criterion values for these two objects, then they construct a probabilistic mental model. A reference class that contains these two objects is generated and subsequently probabilistic cues are checked that allow one to infer the correct response. The cues are accessed hierarchically, in an order established by their predictive validity (i.e., the percentage of correct inferences if the cue discriminates). Once a cue is found that discriminates, search is stopped, the object to which the cue points is chosen, and confidence is given as the cue's validity (in later publications, a slightly modified version of this mechanism that places object recognition before the cue hierarchy and that differs with respect to the question when a cue discriminates between two objects has been labeled "take-the-best"; Gigerenzer & Goldstein, 1996). Gigerenzer et al. (1991) attributed the empirical finding that confidence judgments tend to be too high when compared to accuracy to the item selection procedure, and hence to the experimenter. When the objects in a pair comparison task were randomly drawn from a reference class with which the participants were familiar (i.e., where they knew the statistical structure), overconfidence disappeared.

Second, Gigerenzer and Hoffrage (1995) challenged Kahneman and Tversky's (1972) conclusion that in "his evaluation of evidence man is apparently not ... Bayesian at all" (p. 450). Gigerenzer and Hoffrage argued, and experimentally demonstrated, that performance in tasks that can be solved by Bayesian reasoning (i.e., by using Bayes' rule) can be boosted (from 16 percent correct inferences to 46 percent, without training) when information is not presented in terms of probabilities, but in terms of natural frequencies. Natural frequencies can be seen as a result of naturally sampled information and therefore correspond to the information format to which the human mind is adapted—which is not the case for probabilities, a format that emerged only quite recently in the history of science.

What were, initially, contributions to a research agenda that could have been labeled "How to make cognitive illusions disappear" (Gigerenzer, 1991) slowly turned into a full-blown alternative research program on heuristics that differed from the heuristics-and-biases program in two important aspects. First, when Kahneman and Tversky (1996) defended their work in an article entitled "On the Reality of Cognitive Illusions," Gigerenzer (1996) criticized, in a rejoinder, how vaguely their heuristics had been defined. The critique of vague heuristics prepared the ground for fast-and-frugal heuristics as precisely defined algorithms, specified in terms of their *building blocks* (explained later in "Modularity, Models, and Methodologies"). Second, Gigerenzer (1996) criticized Kahneman and Tversky's conception of context-independent, logical norms as the only relevant normative benchmark. This criticism of content-blind norms later morphed into an agenda for studying the ecological (= content-dependent) rationality of those fast-and-frugal heuristics (explained in the next section, "Bounded, Ecological, and Social Rationality").

Bounded, Ecological, and Social Rationality

Fast-and-frugal heuristics are models of bounded rationality (Gigerenzer & Selten, 2001; Gigerenzer, Todd, & the ABC Research Group, 1999). The notion of bounded rationality has been developed by Herbert Simon (1957, 1982). In contrast to models of unbounded rationality—which are widespread in neo-classical economics and aim at finding the optimal solution to a problem at hand—models of bounded rationality take into account that humans often have only limited information, time, and computational capacities when making judgments or decisions. Given these constraints, it is often impossible to find the optimal solution—and for many interesting problems the optimal solution cannot be known even if such constraints were not an issue.

Models of bounded rationality specify the (cognitive) processes that allow one to find a *satisficing* solution to a given problem. Simon's classic term satisficing is a blending of *satis*fying and suf*ficing*, and indeed, one can often be happy with a solution that is good enough. In the middle ground between these models of bounded and unbounded rationality are models that optimize (be it accuracy of inferences or expected outcomes of decisions) under constraints (be it constraints that come with the psychological limits of human nature, or those reflecting economic boundaries such as information acquisition costs). Note that fast-and-frugal heuristics have nothing in their DNA that resembles optimization, including optimization under constraints. Fast-and-frugal heuristics do *not* optimize, period.

Fast-and-frugal heuristics are not evaluated against a content-blind norm (be it logic, probability theory, or statistical principles) and consequently fast-and-frugal heuristics are not designed to produce, first and foremost, coherent and consistent inferences. Rather, the performance of a heuristic is evaluated against a criterion in the environment. For instance, *take-the-best* (Gigerenzer & Goldstein, 1996) chooses one among two or more alternatives and is evaluated by determining the proportion of correct choices. Similarly, the *QuickEst* heuristic (Hertwig, Hoffrage, & Martignon, 1999) makes inferences about the numerical values of objects, and is evaluated by comparing estimated and true values. For overviews of various heuristics and how they perform in various environments, see Gigerenzer and Gaissmaier (2011), Gigerenzer, Hertwig, and Pachur (2011), and Hafenbrädl, Waeger, Marewski, and Gigerenzer (2016).

A heuristic is ecologically rational to the extent that it is adapted to the structure of information in the environment. If such a match between heuristics and informational structures exists, heuristics do not need to trade-off accuracy for speed and frugality (for research on the effort–accuracy trade-off, see Payne, Bettman, & Johnson, 1993). The importance of considering the environment when studying the human mind is best illustrated in Simon's (1990) analogy of a pair of scissors, with the mind and environment as the two blades: "Human rational behavior (. . .) is shaped by a scissors whose two blades are the structure of task environments and the computational capabilities of the actor" (p. 7). When considering only one blade, one cannot fully understand how the human mind works, just as one cannot understand how scissors with one single blade would function. The simultaneous focus on the mind and its environment, past and present, puts research on decision making into an ecological framework, a framework that is missing in most alternative theories of reasoning, both descriptive and normative. This focus on the fit of a heuristic to the environment in which it is evaluated is an important aspect of the fast-and-frugal heuristics research program (Todd, Gigerenzer, & the ABC Research Group, 2012; see also Hogarth & Karelaia, 2007). Note that this fit of heuristics to environments benefits the experts who use heuristics in the environments in which they operate.

While some experts operate in the natural or technological world (e.g., meteorologists, car mechanics), others operate in the social world (e.g., diplomats, bankers). Correspondingly, a special form of ecological rationality is social rationality, which captures the fact that social species need to make decisions in environments that are typically also shaped by the actions of others. Models of social rationality describe the structure of social environments and their match with boundedly rational strategies people—including experts—use (Hertwig, Hoffrage, & the ABC Research Group, 2013). Conceptually, the social world we are embedded in constitutes a particular aspect of the world surrounding us. There is a variety of goals and heuristics unique to social environments. That is, in addition to the goals that define ecological rationality—to make fast, frugal, and fairly accurate decisions—social rationality is concerned with goals such as choosing an option that one can defend with argument or moral justification, or that can create a consensus. To a much higher degree than for the purely cognitive focus of most research on bounded rationality, socially adaptive heuristics include emotions and social norms that can act as heuristic principles for decision making.

Modularity, Models, and Methodologies

One major insight that results from taking the environment into account is that different environments call for different decision strategies. In contrast to SEU theory, fast-and-frugal heuristics are not a general-purpose decision-making algorithm. Rather, they are designed to solve a particular task (e.g., choice, numerical estimation, categorization) in a particular environment and they cannot solve tasks for which they are not designed—just like a hammer, which is designed to hammer nails but is useless for sawing a board. The collection of such task-specific and fit heuristics that, arguably, are shaped by phylogenetic (evolution), social (culture), and individual learning, and can be used by the human mind, is called the *adaptive toolbox* (Gigerenzer, Hoffrage, & Goldstein, 2008; Gigerenzer & Selten, 2001). The question of which heuristic to choose from the adaptive toolbox for a given task relates to the problem of strategy selection (see Marewski, Bröder, & Glöckner, 2018).

Although there are various fast-and-frugal heuristics that are adapted to different tasks and environments, they share the same guiding construction principles. In particular, they are composed of the same building blocks. The building blocks specify how information is searched for, be it in memory or in the environment (*search rule*), when information search is stopped (*stopping rule*), and how a decision is made based on the information acquired (*decision rule*). Thus, in contrast to models that assume all information is already known to the decision maker, fast-and-frugal heuristics specify the cognitive processes, including those involved in information acquisition (for related programs that explicitly include information search, see Busemeyer & Townsend, 1993, and Payne, Bettman, & Johnson, 1993; and for a discussion of process models, in contrast to outcome models, see Berg & Gigerenzer, 2010).

Fast-and-frugal heuristics are fast for two reasons. First, they are fast as a consequence of being frugal. That is, they may stop searching for further information early in the process of information acquisition. Second, they do not integrate the acquired information in a complex and time-consuming way. In this respect, many heuristics of the adaptive toolbox are as simple as possible because they do not combine pieces of information at all; instead, the decision is based on just one single reason, which is often also referred to as *one-reason decision making*. Recall that fast-and-frugal heuristics do not try to find the optimal solution. Likewise, their building blocks do not optimize either. Martignon and Hoffrage (2002) have even shown that, when making predictions out of sample (i.e., in cross-validation with the split-half method), simple search rules are more robust and outperform the cue order that is optimal when fitted within the training set.

Above we introduced two important elements of fast-and-frugal heuristics: their building blocks, and their ecological rationality. By specifying the building blocks, researchers essentially develop a process model of how decisions are made. Often such theories are formalized into computational models, or even integrated into cognitive architectures such as *ACT-R* (Anderson & Lebiere, 1998; Marewski & Schooler, 2011). Building process models, instead of merely trying to model the outcomes of decisions, allows researchers to predict other characteristics of the decision process, such as response times or confidence assessments (e.g., Hertwig, Fischbacher, & Bruhin, 2013). Moreover, conceptualizing decisions as a series of simple steps makes fast-and-frugal heuristics useful as decision aids.

Studies on fast-and-frugal heuristics address either descriptive or prescriptive questions (or both). Research on heuristics devoted to descriptive questions addresses the question how good heuristics are as behavioral models. Specifically, experimental and observational studies seek to explore whether and when people actually use these heuristics (e.g., Rieskamp & Hoffrage, 2008; Garcia-Retamero & Hoffrage, 2006). In contrast, prescriptive studies explore the performance of heuristics in a given environment, be it in real-world or artificially created environments (where information structures are systematically varied). The methods of choice for such studies are computer simulations (e.g., Czerlinski, Gigerenzer, & Goldstein, 1999) and the use of mathematical or analytical analyses (and sometimes a combination; e.g., Martignon & Hoffrage, 2002). Applications of the fast-and-frugal heuristics to expertise reflect this methodological heterogeneity: While descriptive approaches aim at modeling experts, prescriptive approaches aim at studying the performance of heuristics, which could, eventually, be used to inform and to aid experts (for various examples, see the second part of this chapter).

Fast-and-Frugal Heuristics, Naturalistic Decision Making, and Intuition

The fast-and-frugal heuristics program can be seen as a response to some of the blind spots of the heuristics-and-biases program. Very much the same can be said for the naturalistic decision-making approach. This program was initiated in 1989 in a conference that took place in Dayton, Ohio, where more than 30 professionals met to discuss their ideas for studying the decision making of experts in their natural environments. Given that this approach is presented in a chapter on its own (see Chapter 9, "Macrocognitive Models of Expertise," by Hutton, this volume), its description here can be reduced to a minimum (see also Klein, Orasanu, Calderwood, & Zsambok, 1993; Schraagen, 2018; and http://www.macrocognition.com).

Instead of testing undergraduates with statistical or logical toy problems in the lab, researchers of the naturalistic decision-making community study experts in their domain, for instance pilots in the cockpit, nurses at the patient's bedside, or firefighters in burning houses. The methods most often used are observation and interviews, rather than experiments. The central models of expert learning are cognitive flexibility theory and cognitive transformation theory (see Hoffman et al., 2014; Ward, Gore, Hutton, Conway, & Hoffman, 2018), and the central models of expertise are, arguably, sensemaking and flexecution (see Ward et al., 2018), and, much earlier, the *recognition-primed decision model* (Klein, 2003). According to the latter model, experts are able to recognize patterns. Their expertise consists of knowing which cues are important and which ones are less so when assessing a situation and when predicting outcomes. The experts learned how these cues intercorrelate, and they learned to distinguish certain configurations of cues—patterns that are meaningful to experts but that novices may not recognize as such (Simon, 1987). Most importantly, these experts also learned what the best course of action is for which pattern. Such learning does not necessarily lead to explicit knowledge that can be verbalized, but can also result in implicit, tacit knowledge. That is, people with a rich experience base may not necessarily be able to provide reasons for why they judge and decide as they do—sometimes they have a gut feeling and know intuitively how to act (for a discussion of this distinction between explicit and implicit knowledge, see Chassot, Klöckner, & Wüstenhagen, 2015).

The major commonality between the fast-and-frugal heuristics program and the naturalistic decision-making approach—which, at the same time, set both apart from the heuristics-and-biases program—is their focus on the environment in which cognition has developed and to which it is adapted. Accordingly, performance is assessed in terms of fit between environmental structure and an actor's goals. In contrast, in the context of the heuristics-and-biases program, performance is typically assessed against laws of logic, probability theory, or rational choice theory—not against the person–ecology fit. A major difference between the fast-and-frugal heuristics program and the naturalistic decision-making approach is that the former focuses on rule-based information processing whereas according to the latter, (experts') cognition is mainly exemplar-based (for other differences, see also Shan & Yang, 2017).

I will conclude the first part with a discussion of intuition. Experts have typically developed intuitions, and understanding how the three schools presented earlier conceptualize intuitions provides a nice opportunity to contrast them with each other (see also Klein, 2015, who offers seven suggestions for theory construction and research practice resulting from his synopsis of these three different approaches). The literature that emerged from the heuristics-and-biases program typically refers to intuition in the context of so-called dual-process theories (Evans & Frankish, 2009, Kahneman, 2003, Sloman, 1996, Stanovich & West, 2000). System 1 is often referred to as intuitive, fast, experiential, implicit, automatic, effortless, emotional, affective, and associative, whereas System 2 is analytic, slow, symbolic, explicit, controlled, effortful, deliberate, and rule-based. Reponses consistent with what is seen as rational are seen as originating from System 2, and responses that amount to norm deviations are seen to originate in System 1. Conversely, not every System 1 response is biased and produces errors. To the contrary, in "kind" environments (Hogarth, 2001) in which there are (many) valid cues and in which experts receive immediate and accurate feedback, these experts develop good intuitions (Kahneman & Klein, 2009). Whereas Kahneman and Klein, as representatives of the heuristics-and-biases and of the naturalistic decision-making communities, respectively, agree on this, they differ with respect to their research emphasis: Kahneman focuses on experts going astray with their intuitions, and Klein focuses on the successes of experts' intuitions.

According to the fast-and-frugal heuristics account, intuition is often a subjective experience that accompanies the use of fast-and-frugal heuristics (Hoffrage & Marewski, 2015). Because these heuristics are fast and can be executed automatically, people (which explicitly includes experts) may not necessarily be aware of the fact that they use a heuristic to find a solution to a problem at hand. It is hence straightforward that Gigerenzer's (2007) book *Gut Feelings* reads like an introduction to the fast-and-frugal heuristics framework. Both heuristics and intuitions are seen in a positive light in this program and the examples assembled in *Gut Feelings* are typically success stories—in contrast to the general picture that emerges from the heuristic-and-biases program.

The naturalistic decision-making community sees intuition as an expression of experience. Klein defines "intuition as the way we translate our experience into action" (2003, p. iv). This view of intuition is compatible with Simon (1992), who conceptualizes intuition as "nothing more and nothing less than recognition" (p. 155), and with Chase and Simon (1973), who posit that experts need to acquire thousands of patterns (see also Shanteau, 2015). These patterns are not generic tools, but rather "specific accumulations of direct and vicarious experiences" (Klein, 2015, p. 164).

The Fast-and-Frugal Heuristics Account of Expertise

In the following part of this chapter, some selected contributions from the fast-and-frugal heuristics program that might be relevant for researchers interested in expertise are presented. To reiterate, according to this program, experts use—and hence can be modeled with—fast-and-frugal heuristics. The same can be said for novices, but the models that experts develop for their domain are superior, and the heuristics they use are smarter. But before going into details, two general remarks are in order. The first remark is on the scope of the fast-and-frugal heuristics approach to expertise and the second is on modeling and methods.

Reconsider that fast-and-frugal heuristics can arise from individual, social, and phylogenetic learning (Gigerenzer, Hoffrage, & Goldstein, 2008). Which of these sources are relevant for the study of expertise? To reiterate Merriam-Webster's definition, being an expert means "having, involving, or displaying special skill or knowledge derived from training or experience." Note that this definition does *not* include phylogenetic learning. Granted, it may be tempting to view animals as experts. In fact, every species has developed special skills that enable it to survive and thrive under environmental conditions under which most other species would be doomed to death (just imagine a polar fox in an ocean, or a shark in the dessert). But according to Merriam-Webster's definition, the giraffe with its ability to eat leaves from high trees is still not an expert for living in the savanna—after all, it did not receive any special training or experience and its long neck is simply inherited. Consequently, not only adaptations, such as the giraffe's long neck, but also phylogenetically acquired heuristics do not constitute examples of expertise. Accepting the limits of Merriam-Webster's definition, the contribution of the fast-and-frugal heuristics program to expertise hence needs to be restricted to those heuristics that result from individual learning and from social learning during our lifetime (for an evolutionary account to cognitive psychology in general and to fast-and-frugal heuristics in particular, see Todd, Hertwig, & Hoffrage, 2015, and for the use of simple heuristics in the animal kingdom, see Hutchinson & Gigerenzer, 2005). Any evolved capacities that constitute integral elements of heuristics—such as the ability for recognition—are simply taken for granted and how these came about in the first place are not in the focus of attention when applying the program to research on experts and expertise.

The examples selected and presented in the following reflect, perhaps not surprisingly, the variety of methodological approaches pursued within the program. The structure of their presentation echoes this variety. I will start with three examples in which heuristics turned out to be good descriptions of experts who have been observed in their field—the first reporting anecdotal evidence, the second experimental evidence, and the third the results of modeling in an observational study. The subsequent section shifts the focus from experts to expertise—more precisely to the question of how either people (who are not yet experts) or strategies (that start from scratch, without any parameters carried over from other data sets) can develop expertise and thereby become experts. Thereafter I will present examples of how some of the heuristics that have been studied in this program from a prescriptive point of view, typically by means of computer simulations, could be used to aid experts.

Observing and Modeling Experts in their Fields

On January 15, 2009, a flock of Canadian wild geese hit the turbofans of US Airways Flight 1549 and the resulting engine failure turned the Airbus A320 into a glider. The pilots Chesley B. Sullenberger and Jeffrey Skiles had to make a decision. Where to land the plane with its 155 passengers? The options they considered first were to try to return to LaGuardia, their starting point, or try to reach Teterborough, another airport close by. The worst-case outcome was to hit a crowded spot on the ground, killing many people on top of those on board. The actual outcome made the front news: Sullenberger conducted a spectacular landing in the Hudson River, without a single fatality.

How did these pilots—Sullenberger was shortly before retirement and had decades of experience—decide in favor of the Hudson? When and why did they eliminate the airports from their choice set? As co-pilot Skiles later explained on *The Charlie Rose Show*: "It's not so much a mathematical calculation as visual, in that when you are flying in an airplane, things that—a point that you can't reach will actually rise in your windshield. A point that you are going to overfly will descend in your windshield."[1] What Skiles formulated based on his experience can easily be used as a recipe to eliminate options: *Fix your gaze on a potential landing spot. If this spot rises in your windshield, then you will not be able to reach it.* This elimination rule is a consequence of a simple rule of thumb known as the *gaze heuristic*, which, for the task of catching balls that come from high in the air, reads as follows: *Fix your gaze on the ball, start running, and adjust your running speed so that the angle of gaze remains constant* (Gigerenzer, 2007). This simple rule of thumb does not require any information other than the angle. Fielders (who can adjust their running speed) would not need to consider any other information such as sidewind or spin of the ball to determine their running speed, and the pilots of US Airways 1549 (who could no longer accelerate their plane) would not need to look at their instruments to determine whether they could reach the airport. The gaze heuristic is used by dogs catching frisbees, bats, birds, and dragonflies catching prey, professional baseball players catching balls, but also by sailors who, in contrast to the aforementioned examples, want to avoid a collision.

From Skiles' response to Charlie Rose one may infer that the pilots had used one piece of information, namely whether a potential landing point was rising in their windshield, to classify it as either reachable or not. I now turn from aviation experts to medical experts, from anecdotal evidence to experimental evidence, and from an example in which only one piece of information was used to a situation in which many were at the experts' disposal.

Doctors routinely must decide whether to send a patient with chest pain to the coronary care unit (CCU) or to a regular nursing bed. To be on the safe side (and also to protect themselves against potential lawsuits), the doctors of a particular Michigan Hospital sent about 90 percent of the patients to the CCU, even though only about 25 percent of them actually had a medical indication that warranted being sent there (Green & Mehr, 1997). In other words, they elected to use a conservative response strategy that resulted in few "misses," rather than prioritize discriminability of the

---

[1] https://charlierose.com/videos/14176, downloaded on Feb 15, 2018. Even though Skiles said this during an interview in which Charly Rose focused on the question of what crossed his mind during these critical seconds, one may still argue that his statement could still be a post-hoc explanation—but still, it is probably the best window that is available to get a glance into the black box of these pilots' minds.

signal. So, the CCU was constantly overcrowded, the quality of care suffered, and medical costs increased. To alleviate these problems, the Heart Disease Predictive Instrument (HDPI) was developed. This is a chart with about 50 probabilities that were generated by a logistic regression and that captures how likely it is that a particular patient with a particular combination of symptoms and other pieces of information has acute ischemic heart disease. To test for the instrument's effectiveness, the chart was given to and withdrawn from the doctors in alternating weeks. To the researchers' surprise, the availability of the HDPI did *not* affect the doctors' CCU utilization. However, CCU use had already markedly declined before the evaluation trial started—in fact, right after the HDPI had been introduced at a departmental clinical conference. Green and Mehr (1997) concluded that "simply in learning about the instrument, residents achieved sensitivity and specificity equal to the instrument's optimum, whether or not they actually used it" and they speculated that this "may have resulted from improved focus on relevant clinical factors identified by the tool" (Green & Mehr, 1997, p. 219).

After they had learned about fast-and-frugal heuristics, Green and Mehr developed a fast-and-frugal decision tree (they referred to it as "probabilistic mental model") to model how the doctors had decided after they had a chance to extract, from the HDPI's probability chart, which factors were most useful to predict death of heart failure. Fast-and-frugal trees, formally introduced and defined by Martignon, Vitouch, Takezawa, and Forster (2003), can be described in terms of their building blocks, much like other fast-and-frugal heuristics. Specifically, they, first, have a search rule: They inspect cues in a specific order. In the present medical example, the tree asks three questions: "Is there a certain anomaly (elevated ST segment) in the electrocardiogram?", "Is chest pain the chief complaint?", and "Are there are any other of four specific factors present such as myocardial infarction or nitroglycerin use for chest pain relief?" Second, they have a stopping rule: Each cue has one value that leads to an exit node and hence to a classification, and another value that leads to consulting the next cue in the cue hierarchy (the exception being the last cue in the hierarchy, which has two exit nodes). In the present example, if the first cue is in its high state (ST segment is elevated), then the patient is sent to the CCU; otherwise, the tree asks whether chest pain is the chief complaint. Finally, they have a decision rule, specifically, a classification rule. In the present example, they always classify the patient as having the disease if the value of the consulted cue is in its high state (for instance, if a medical test is positive, or if a continuous variable falls in a specific range) and as not having the disease if this value is in its low state (e.g., test negative). Green and Mehr's tree led to fewer misses and a better false-alarm rate than both the HDPI and physicians' decisions. For more details on tree construction principles, see Woike, Hoffrage, and Martignon (2017) and this chapter's section "Aiding Experts and Aiding the Development of Expertise," and for the link between risk representation and fast-and-frugal trees, see Hoffrage, Krauss, Martignon, and Gigerenzer (2015).

This example offers some interesting insights into the link between observed behavior of experts and modeling attempts using heuristics. At the outset there was a problem: doctors sent too many patients to the CCU. As a solution, some statisticians designed a therapy for the doctors and introduced it at a departmental clinical conference: the HDPI. This introduction can be seen as an intervention. Doctors received training, essentially a table with the probabilities of a heart attack conditioned on multiple factors and their combinations. These probabilities were calculated with a logistic regression. The training turned out to be effective: doctors clearly changed their admission policy. However, Green and Mehr reported that the doctors could not really understand the underlying statistical model, nor could they remember the probabilities. On top of that, in the evaluation period after that conference they used their new policy even under the condition without the probability charts at their disposal. So, the effect of the intervention was not due to the availability of the probability charts when classifying patients, as the researchers had intended, but it was more likely due to the training session in which the HDPI was presented.

We do not know how the doctors made their decisions before the training, but this was not what Green and Mehr were interested in. Rather, the question that interested them was: How did the doctors make their decisions after the training? The answer "with the HDPI" did not convince them. First, the doctors did not really understand it, and second, their behavior could still be correctly described by the HDPI even when they could no longer use the relevant probability chart. Green and Mehr then developed, post-hoc, a fast-and-frugal tree as a model that the experts might have constructed to represent the essence of the logistic regression, namely the answer to the question: what are the most predictive cues? This model was subsequently confirmed when discussing it with the experts (which is, admittedly, weak evidence; see also Nisbett & Wilson, 1977, on people's limited capacities to formulate insights from observing their own reasoning), and when using it as a descriptive model to account for their decisions (which can be considered as stronger evidence). Altogether, this example shows that a fast-and-frugal tree was useful as a descriptive model, even if the behavioral change was prompted by another statistical model and even though this tree was developed post-hoc.

Fast-and-frugal trees have also been used successfully as behavioral models of experts' decision making in other areas and for other tasks. For instance, they outperformed a weighted linear model when modeling the bailing decisions of practicing magistrates (with an average of 13 years of experience; ranging from 6 months to 35 years) from 44 different courts (Dhami, 2003; Dhami & Ayton, 2001). Moreover, they performed as well as a regression model when modeling the likelihood judgments of 36 general practitioners prescribing lipid-lowering drugs for 130 patients, described on 12 cues (Dhami & Harries, 2001).

I conclude this section by asking: Do laypeople and experts differ with respect to the heuristics they use, and if so, how? One should expect that experts know better what to look for, that is, what the important cues are. Whereas laypeople may need to explore predictor variables, consider more of them, and are still in the process of learning which are the most important ones, experts may already have acquired sufficient knowledge so that they can exploit the cues, thereby focusing on those they identified to be important. This was exactly the result in a study conducted by Garcia-Retamero and Dhami (2009). Laypeople (graduate students) and experts (professional burglars who had committed an average of 57.2 burglaries and police officers with an average of 19.4 years of experience) were asked to judge which of two residential properties was more likely to be burgled. Most students (38 of 40) were classified as users of a compensatory *weighted additive model*. Such a model weighs each predictor variable according to its importance (here, individual students' cue weight estimations), then adds up the weighted predictors for each option, and finally chooses the option with the higher sum score. In contrast, most experts (34 of 40 burglars, and 31 of 40 police officers) were classified as users of the noncompensatory *take-the-best heuristic* (Gigerenzer & Goldstein 1996). This heuristic selects one of two properties as being more likely to be burgled based on only one cue, specifically, the most important cue among all those that discriminate.

In another study, Pachur and Marinello (2013) obtained basically the same results. Both experienced airport customs officers and a novice control group were asked to decide, based on several cues, which passengers they would submit to a search. While the laypeople's decisions could best be modeled with a compensatory strategy, the experts could best be modeled with a noncompensatory heuristic. Moreover, the authors "also found that the experts' subjective cue validity estimates showed a higher dispersion across the cues and that differences in cue dispersion partially mediated differences in strategy use between experts and novices" (p. 97).

What do all these examples discussed in this section have in common? They all focus on real experts—specifically, pilots, physicians, bail officers, burglars, police officers, and airport customs officers—who have been studied in their domain of expertise. In each of these examples the experts knew where to focus their attention. Hence, it may not come as a surprise that these experts could be successfully modeled by a fast-and-frugal heuristic. After all, such heuristics prioritize more valid cues and start their search with such cues. In other words: Experts learned, through training and experience, the statistical structure of cues and outcome variables in their environment. That is why they perform well in the environment (Hogarth, 2001; Kahneman & Klein, 2009). And that is why fast-and-frugal heuristics—with their ability to adapt to statistical structures—perform well when it comes to modeling experts.

Studying and Simulating the Development of Expertise

I will now turn from modeling experts to the development of expertise. A number of studies conducted within the framework of fast-and-frugal heuristics have addressed the questions: How do people acquire expertise, or, equivalently, how do laypeople become experts? According to the fast-and-frugal heuristics program's account of expertise, these questions amount to asking: How do people learn which heuristics to use?

Rieskamp and Otto (2006) put naïve participants—most of them undergraduates—in the shoes of a bank consultant who had to select, in a series of pair comparisons, which of two companies applying for a loan was more creditworthy. In the first block, consisting of 24 items, during which no feedback was given, participants' choices were consistent with take-the-best in about 30 percent of the items. For the remaining 144 trials, feedback was given. How this feedback was determined varied between-subjects. For half of the participants, the "correct" company was determined by a noncompensatory strategy, specifically, by take-the-best (which cue was the most important one, second most important one, and so on, had been counterbalanced), and given to the participants in 92 percent of the time (the other 8 percent of the trials were strategy execution errors). In this group, the percentage of choices consistent with take-the-best rose to about 70 percent. For the other half of the participants, a compensatory strategy, specifically, a weighted linear rule was used to determine which feedback to give (again, with 8 percent execution errors). In this group, the percentage of choices consistent with take-the-best dropped to about 10 percent. Granted, this study did not use experts but students. However, these students gained expertise in an environment that was controlled by the experimenter, and which heuristics they learned to use over time hinged on the incentive scheme in their environment. Note that the participants were not informed that the feedback they received was determined by a specific strategy.

While Rieskamp and Otto (2006) determined the success of a particular heuristic as a behavioral model (of people, over time, and under various environmental conditions), Woike, Hoffrage, and Petty (2015) determined the success of various strategies, including heuristics, as a prescriptive model. Specifically, Woike et al. (2015) simulated the performance of hypothetical venture capitalists (VCs) who had to select business plans for a portfolio. The business plans were described by binary cues. The VCs (i.e., the simulated agents) saw the plans one after the other and had to decide whether to invest in a given plan (without being able to go back to a plan they had rejected). After they made a decision, they received feedback on the success of the plan. Woike et al. simulated various VCs, who differed with respect to the strategy they used to make their investment decisions. They used a simple heuristic, namely tallying (which is, in their setup, equivalent to equal weighting), a fast-and-frugal tree, or a more complex strategy, namely logistic regression or a classification and regression tree (CART). The parameters of each strategy[2] were updated continuously based on the feedback provided. To evaluate the ecological rationality of these strategies, the authors also analyzed how the strategies' performance was affected by (statistical) properties of the decision-making environment. These environmental properties included the selectivity

---

[2] For tallying, these were the cues' directionalities; for logistic regression, these were the beta-weights; and both for the fast-and-frugal tree and for CART, these were the architecture of the tree and the decisions at its exit nodes.

of feedback (information about the success or failure of a plan was given either for all plans or only for those in which a VC invested), the relative importance of cues, that is, how well the cues could predict the success of the plans and how they were distributed with respect to this predictive power, and the number of cues that characterized the plans.

A major result of this simulation study was that tallying turned out to be competitive with the more complex benchmark strategies and that, notably, its competitiveness was robust across environments. The finding that tallying was able to match—and in some conditions even to outperform—the two more complex strategies can largely be attributed to the robustness of the strategies' parameter estimates (tallying only had to estimate cue directionality). Complex strategies suffer from overfitting: when fitting their parameters to known data, they captured—more than simple heuristics did—a considerable amount of noise that did not transfer when making predictions for new data (see also Gigerenzer & Brighton, 2009, Meehl, 1954). The setup of this simulation implemented exactly this realistic feature, namely that strategies learned from the past but then had to make investment decisions for new business plans. Two other important results of this simulation study were that the performance of the decision strategies depended critically on the environment, specifically, on how cues differed with respect to their potential to predict success of a plan; and that learning only from those plans in which the VC invested drastically reduced their potential to learn from experience.

How do the studies of Rieskamp and Otto (2006) and of Woike et al. (2015) relate to each other and to the present topic, namely, expertise? Rieskamp and Otto showed that laypeople changed their decisions to be more "expert-like" toward the end of their exposure to feedback. The development of their expertise was modeled with various heuristics and, under the condition in which cues had a noncompensatory structure (that is, in which feedback given to the participants was determined by a noncompensatory strategy for which any cue outweighed any other cues further down in the cue hierarchy), the participants could best be modeled with the take-the-best heuristic that processed cues in a noncompensatory way. In contrast, in the study by Woike et al. the modeling was not done descriptively, but prescriptively. Rather than training (and modeling) participants, Woike et al. trained heuristics and other more complex strategies by giving those heuristics, in extensive computer simulations, outcome feedback that resulted from their decisions. To the extent that laypeople can be turned into experts through training via outcome feedback (recall Merriam-Websters definition), and to the extent that strategies can be trained via outcome feedback, one could transfer the development of expertise in artificial systems to human experts who operate in the same environment. This could be done by educating the experts about the success of various strategies under various conditions and assumptions (for the distinction between these two kinds of training, see Klayman, 1988, who distinguished outcome feedback in trial-by-trial learning and cognitive feedback about strategies' parameters and achievement in an educational setting). This possibility leads directly to the next section.

Aiding Experts and Aiding the Development of Expertise

Many examples of expertise and intuition reported in the context of the naturalistic decision-making program are such that the experts' knowledge is implicit. Arguably, these cases may constitute a minority (see Hoffman & Militello, 2009; Crandall, Klein, & Hoffman, 2006), but still, sometimes experts cannot verbalize and communicate why they decided as they did (not much different from laypeople, see Nisbett & Wilson, 1977). Likewise, there are studies conducted within the framework of the fast-and-frugal heuristics program (e.g., Dhami & Ayton, 2001) in which experts could be modeled with a heuristic—notwithstanding the fact that these experts were not able to formulate the heuristic, or that they reported a strategy that predicted their own decisions not as well as the model the researchers constructed.[3]

The fact that fast-and-frugal heuristics are precise algorithms that can be used in computer simulations offers interesting opportunities when it comes to helping experts to perform even better. With the help of researchers who model experts, implicit and explicit knowledge of how to process information can be mapped onto explicitly formulated heuristics, which may, in turn, increase experts' reliability and internal consistency. With the help of computer simulations, researchers are then able to inform experts how heuristics perform under various environmental conditions. Experts and modelers can thus collaborate to build an adaptive toolbox at the disposal of the experts. The work reviewed in this section is partly grounded in a descriptive approach (describing and modeling how experts function), or it is at least inspired by the question how experts might function. But at some point in the process a prescriptive approach kicked in and so the commonality of the studies reported in the following is that they all give an answer to the question: How can experts be aided with a fast-and-frugal heuristics perspective? Because of space constraints, I will focus on fast-and-frugal trees (see also Marewski & Gigerenzer, 2012).

Fast-and-frugal trees are transparent, easy to communicate, easy to execute, and occasionally even easy to set up. The doctors' decision making at the coronary care unit in Michigan presented in the section "Observing and Modeling Experts in their Fields" illustrates how the descriptive and the prescriptive approaches can go hand in hand. After Green and Mehr (1997) identified the fast-and-frugal tree that captured the doctors' lesson-learned from the introduction of the probability chart, and had tested it as a valid descriptive model, they could develop this and similar trees using a prescriptive model to aid and educate the doctors. More generally, such trees can easily be developed for new patient populations with other disease base rates or for other diseases with different tests.

Fast-and-frugal trees have also been proposed as suitable for social workers seeking to stabilize critical situations until fuller services are available (Taylor, 2016; see also Kirkman

---

[3] Note, however, that the opposite has been reported as well: Reisen, Hoffrage, and Mast (2008) found that the idiosyncratic strategies that their participants reported to have used in a consumer choice task outperformed, by far, any of the heuristics that competed when it came to correctly predicting their choices.

& Melrose, 2014). They have been tested as predictors of depression (Jenny, Pachur, Williams, Becker, & Margraf, 2013). They have been praised as an ideal data structure for information technology implementation, particularly in pharmacogenomics, where complex data need to be used at point of care (Van Rooij et al., 2015). They have been used by the Bank of England to assess bank vulnerability (Aikman et al., 2014; see also Haldane, 2015). They allow swift decisions to be made in potentially dangerous situations where time is of the essence—their performance on past data indicates that their use might reduce the number of civilian casualties at military checkpoints (Keller & Katsikopoulos, 2016). They have been discussed as smartly designed strategies and as tools to boost decision makers' performance and autonomy (Grüne-Yanoff & Hertwig, 2016).

Woike, Hoffrage, and Martignon (2017) proposed how trees can be constructed and they studied (in 11 medical datasets) their ecological rationality. Specifically, they compared the performance of the tree construction principles to a Bayesian classifier that is normative when fitting known data. They also compared it to a naïve Bayesian classifier that can be conceived as a heuristic approach because it simply treats cues as independent of each other. One of the major results was that the trees competed well against, and often outperformed, the benchmark strategies. Equally important, in particular for practitioners, Woike et al. could identify the environmental properties that gave a competitive advantage for each of the tree construction principles. Another project that has great potential to aid researchers and practitioners is currently being undertaken by Phillips, Neth, Woike, and Gaissmaier (2017), who are developing a software package (in R) for fast-and-frugal trees. Similar to chess grandmasters who nowadays use chess programs to improve their skills, medical and other experts could use this tool (or similar ones) to improve their diagnoses.

While fast-and-frugal trees are appealing to many practitioners in various domains, some of these practitioners may face an interesting dilemma. Wegwarth, Gaissmaier, and Gigerenzer (2009), for instance, report that some doctors felt pressured to hide the simple way they make decisions. Anticipating that their patients would not trust the fast-and-frugal approach, they preferred to pretend that their decisions stemmed from relatively complex and sophisticated strategies. Analyses such as those reported in Woike et al. (2017) can help to legitimize the use of fast-and-frugal trees and other simple heuristics in understanding and developing expertise.

Conclusion

According to the program presented in this chapter, people, including experts, are boundedly rational and they use fast-and-frugal heuristics when making decisions in an uncertain world. Consequently, they can also be modeled with fast-and-frugal heuristics. While there is ample evidence for the latter claim, some of which has been reviewed in this chapter, the proposition that people actually use such heuristics is an inference that an old-fashioned behaviorist would not want to make. In this regard, the fast-and-frugal heuristic program is not much different from the neo-Brunswikian approach to model people's (including experts') judgments and decisions with multiple regression, or from the heuristics-and-biases program with its judgmental heuristics, or from the naturalistic decision-making community when it accounts for experts' decisions with the recognition-primed decision model. Each of these programs have their examples and illustrations of successful outcome modeling, occasionally also supported by some process tracing, but whether the strategies and heuristics are merely in the researchers' toolbox or eventually also in people's—including experts'—adaptive toolboxes is a question that can possibly not be answered on an empirical basis (see also footnote 1).

The strengths of the fast-and-frugal heuristic program are the following. First, the heuristics it posits are psychologically plausible: they are designed to be executable by people, including experts, with limited memory, knowledge, and computational capacities. Second, the adaptive toolbox consists of heuristics that are ecologically rational so that, depending on the structure of information in the environment, most likely one can be found that fits. For instance, heuristics with sequential search rules that stop after relevant information has been retrieved are adaptive in environments in which some cues are more important than others. Conversely, in environments in which no cue stands out, simple tallying heuristics that weigh cues equally are robust and not as vulnerable as complex strategies that try to optimize weighing schemes (Dawes, 1979; Dawes & Corrigan, 1974; Meehl, 1954). In environments in which there are hardly any valid cues, for instance, it will be hard to find a useful heuristic, but very much the same could be said for any other strategy. Third, in contrast to models that are hard to describe verbally (for instance, neural networks with their weighing structure), fast-and-frugal heuristics are transparent and easy to explain to experts, thereby offering a huge potential to boost their performance after some explanations and education and without equipping them with software or other devices.

The ultimate judges of whether this program is useful to describe and to aid the decisions of experts should be those experts themselves.

Author Note

Parts of this chapter are based on text within Hoffrage, U., Hafenbrädl, S., & Marewski, J. N. (2018), The fast-and-frugal heuristics program, in L. J. Ball & V. A. Thompson (Eds), *The Routledge international handbook of thinking & reasoning*, pp. 325–345, London: Routledge.

Acknowledgments

I would like to thank Jan Maarten Schraagen and Paul Ward for their extremely helpful comments on a previous version of this chapter, and the Swiss National Science Foundation (Grant SNF 100014-140503/1) for its financial support.

References

Aikman, D., Galesic, M., Gigerenzer, G., Kapadia, S., Katsikopoulos, K., Kothiyal, A., ... & Neumann, T. (2014). *Taking uncertainty seriously: Simplicity versus complexity in financial regulation.* Bank of England Financial Stability Paper 28. London: Bank of England.

Allais, M. (1953). Le comportement de l'homme rationnel devant le risqué: Critique des postulats et axiomes de l'Ecole Americaine. *Econometrica* 21(4), 503–546.

Anderson, J., & Lebiere, C. (1998). The atomic components of thought. Mahwah: NJ: Lawrence Erlbaum.

Berg, N., & Gigerenzer, G. (2010). As-if behavioral economics: Neoclassical economics in disguise? *History of Economic Ideas* 18(1), 133–165.

Busemeyer, J., & Townsend, J. T. (1993). Decision field theory: A dynamic cognition approach to decision making. *Psychological Review* 100, 432–459.

Chase, W. G., & Simon, H. A. (1973). The mind's eye in chess. In W. G. Chase (Ed.), *Visual information processing* (pp. 215–272). New York: Academic Press.

Chassot, S., Klöckner, C. A., & Wüstenhagen, R. (2015). Can implicit cognition predict the behaviour of professional energy investors? An explorative application of the Implicit Association Test (IAT). *Journal of Applied Research in Memory and Cognition* 4(3), 285–293.

Crandall, B., Klein, G., & Hoffman R. R. (2006). *Working minds: A practitioner's guide to cognitive task analysis.* Cambridge, MA: MIT Press.

Czerlinski, J., Gigerenzer, G., & Goldstein, D. G. (1999). How good are simple heuristics? In G. Gigerenzer, P. M. Todd, & the ABC Research Group, *Simple heuristics that make us smart* (pp. 97–118). New York: Oxford University Press.

Dawes, R. M. (1979). The robust beauty of improper linear models in decision making. *American Psychologist* 34, 571–582.

Dawes, R. M., & Corrigan, B. (1974). Linear models in decision making. *Psychological Bulletin* 81(2), 95–106.

Dhami, M. K. (2003). Psychological models of professional decision making. *Psychological Science* 14, 175–180.

Dhami, M. K., & Ayton, P. (2001). Bailing and jailing the fast and frugal way. *Journal of Behavioral Decision Making* 14(2), 141–168.

Dhami, M. K., & Harries, C. (2001). Fast and frugal versus regression models of human judgement. *Thinking & Reasoning* 7(1), 5–27.

Eddy, D. M. (1982). Probabilistic reasoning in clinical medicine: Problems and opportunities. In D. Kahneman, P. Slovic, & A. Tversky (Eds), *Judgment under uncertainty: Heuristics and biases* (pp. 249–267). Cambridge, UK: Cambridge University Press.

Edwards, W. (1954). The theory of decision making. *Psychological Bulletin* 51(4), 380–417.

Evans, J. St. B. T., & Frankish, K. (Eds). (2009). *In two minds: Dual processes and beyond.* Oxford: Oxford University Press.

Garcia-Retamero, R., & Dhami, M. K. (2009). Take-the-best in expert-novice decision strategies for residential burglary. *Psychonomic Bulletin & Review* 16, 163–169.

Garcia-Retamero, R., & Hoffrage, U. (2006). How causal knowledge simplifies decision-making. *Minds and Machines* 16, 365–380.

Gigerenzer, G. (1991). How to make cognitive illusions disappear: Beyond "heuristics and biases." *European Review of Social Psychology* 2(1), 83–115.

Gigerenzer, G. (1994). Why the distinction between single-event probabilities and frequencies is relevant for psychology (and vice versa). In G. Wright & P. Ayton (Eds), *Subjective probability* (pp. 129–161). New York: Wiley.

Gigerenzer, G. (1996). On narrow norms and vague heuristics: A reply to Kahneman and Tversky. *Psychological Review 103*(3), 592–596.

Gigerenzer, G. (2007). *Gut feelings: The intelligence of the unconscious.* New York: Viking Press.

Gigerenzer, G., & Brighton, H. (2009). Homo heuristicus: Why biased minds make better inferences. *Topics in Cognitive Science 1*(1), 107–143.

Gigerenzer, G., & Gaissmaier, W. (2011). Heuristic decision making. *Annual Review of Psychology 62*, 451–482.

Gigerenzer, G., & Goldstein, D. G. (1996). Reasoning the fast and frugal way: Models of bounded rationality. *Psychological Review 103*(4), 650–669.

Gigerenzer, G., Hertwig, R., Hoffrage, U., & Sedlmeier, P. (2008). Cognitive illusions reconsidered. In C. R. Plott & V. L. Smith (Eds), *Handbook of experimental economics results* (vol. 1, pp. 1018–1034). Amsterdam: North Holland/Elsevier Press.

Gigerenzer, G., Hertwig, R., & Pachur, T. (Eds). (2011). *Heuristics: The foundations of adaptive behavior.* New York: Oxford University Press.

Gigerenzer, G., & Hoffrage, U. (1995). How to improve Bayesian reasoning without instruction: Frequency formats. *Psychological Review 102*, 684–704.

Gigerenzer, G., Hoffrage, U., & Goldstein, D. G. (2008). Fast and frugal heuristics are plausible models of cognition: Reply to Dougherty, Franco-Watkins, and Thomas (2008). *Psychological Review 115*(1), 230–239.

Gigerenzer, G., Hoffrage, U., & Kleinbölting, H. (1991). Probabilistic mental models: A Brunswikian theory of confidence. *Psychological Review 98*, 506–528.

Gigerenzer, G., & Selten, R. (2001). *Bounded rationality: The adaptive toolbox.* Cambridge, MA: MIT Press.

Gigerenzer, G., Todd, P. M., & the ABC Group (1999). *Simple heuristics that make us smart.* New York: Oxford University Press.

Green, L., & Mehr, D. R. (1997). What alters physicians' decisions to admit to the coronary care unit? *Journal of Family Practice 45*(3), 219–226.

Grüne-Yanoff, T., & Hertwig, R. (2016). Nudge versus boost: How coherent are policy and theory? *Minds and Machines 26*(1–2), 149–183.

Hafenbrädl, S., Waeger, D., Marewski, J. N., & Gigerenzer, G. (2016). Applied decision making with fast-and-frugal heuristics. *Journal of Applied Research in Memory and Cognition 5*(2), 215–231.

Haldane, A. G. (2015). Multi-polar regulation. *International Journal of Central Banking 11*(3), 385–401.

Hertwig, R., Fischbacher, U., & Bruhin, A. (2013). Simple heuristics in a social game. In R. Hertwig, U. Hoffrage, & the ABC Research Group, *Simple heuristics in a social world* (pp. 39–65). Oxford: Oxford University Press.

Hertwig, R., Hoffrage, U., & Martignon, L. (1999). Quick estimation: Letting the environment do the work. In G. Gigerenzer, P. M. Todd, and the ABC Research Group, *Simple heuristics that make us smart* (pp. 209–234). New York: Oxford University Press.

Hertwig, R., & Hoffrage, U., & the ABC Research Group (2013). *Simple heuristics in a social world.* Oxford: Oxford University Press.

Hertwig, R., & Pachur, T. (2015). Heuristics, history of. In J. D. Wright (Ed.), *International encyclopedia of the social & behavioral sciences*, 2nd edn (pp. 829–835). Oxford: Elsevier.

Hoffman, R. R., & Militello, L. G. (2009). *Perspectives on cognitive task analysis*. London: Taylor & Francis.

Hoffman, R. R., Ward, P., Feltovich, P. J., DiBello, L., Fiore, S. M., & Andrews, D. (2014). *Accelerated expertise: Training for high proficiency in a complex world*. New York: Psychology Press.

Hoffrage, U., Hafenbrädl, S., & Marewski, J. N. (2018). The fast-and-frugal heuristics program. In L. J. Ball & V. A. Thompson (Eds), *The Routledge international handbook of thinking & reasoning* (pp. 325–345). Abingdon, Oxon: Routledge.

Hoffrage, U., Krauss, S., Martignon, L., & Gigerenzer, G. (2015). Natural frequencies improve Bayesian reasoning in simple and complex inference tasks. *Frontiers in Psychology* 6, 1473.

Hoffrage, U., & Marewski, J. N. (2015). Unveiling the Lady in Black: Modeling and aiding intuition. *Journal of Applied Research in Memory and Cognition* 4(3), 145–163.

Hogarth, R. M. (2001). *Educating intuition*. Chicago: University of Chicago Press.

Hogarth, R. M., & Karelaia, N. (2007). Heuristic and linear models of judgment: Matching rules and environments. *Psychological Review* 114(3), 733–758.

Hutchinson, J. M. C., & Gigerenzer, G. (2005). Simple heuristics and rules of thumb: Where psychologists and behavioural biologists might meet. *Behavioural Processes* 69, 97–124.

Jenny, M. A., Pachur, T., Williams, S. L., Becker, E., & Margraf, J. (2013). Simple rules for detecting depression. *Journal of Applied Research in Memory and Cognition* 2(3), 149–157.

Kahneman, D. (2003). A perspective on judgment and choices: Mapping bounded rationality. *American Psychologist* 58(9), 697–720.

Kahneman, D. (2011). *Thinking, fast and slow*. New York: Farrar, Straus and Giroux.

Kahneman, D., & Klein, G. (2009). Conditions for intuitive expertise: A failure to disagree. *American Psychologist* 64, 515–526.

Kahneman, D., Slovic, P., & Tversky, A. (Eds). (1982). *Judgment under uncertainty: Heuristics and biases*. New York: Cambridge University Press.

Kahneman, D., & Tversky, A. (1972). Subjective probability: A judgment of representativeness. *Cognitive Psychology* 3, 430–454.

Kahneman, D., & Tversky, A. (1996). On the reality of cognitive illusions. *Psychological Review* 103(3), 582–591.

Keller, N., & Katsikopoulos, K. V. (2016). On the role of psychological heuristics in operational research; and a demonstration in military stability operations. *European Journal of Operational Research* 249(3), 1063–1073.

Kirkman, E., & Melrose, K. (2014). *Clinical judgement and decision-making in children's social work: An analysis of the "front door" system*. Research Report of the Department for Education, Behavioral Insight Team. Retrieved from https://www.gov.uk/government/publications/clinical-judgement-and-decision-making-in-childrens-social-work.

Klayman, J. (1988). On the how and why (not) of learning from outcomes. In B. Brehmer & C. R. B. Joyce (Eds), *Human judgment: The SJT view* (Advances in psychology, Vol. 54) (pp. 115–162). Amsterdam: Elsevier.

Klein, G. (2003). *The power of intuition: How to use your gut feelings to make better decisions at work*. New York: Currency.

Klein, G. (2015). A naturalistic decision making perspective on studying intuitive decision making. *Journal of Applied Research in Memory and Cognition* 4(3), 164–168.

Klein, G. A., Orasanu, J., Calderwood, R., & Zsambok, C. E. (Eds). (1993). *Decision making in action: Models and methods.* Norwood, NJ: Ablex.

Lopes, L. L. (1991). The rhetoric of irrationality. *Theory & Psychology* 1(1), 65–82.

Marewski, J. N., Bröder, A., & Glöckner, A. (2018). Some metatheoretical reflections on adaptive decision making and the strategy selection problem. *Journal of Behavioral Decision Making* 31(2), 181–198.

Marewski, J. N., & Gigerenzer, G. (2012). Heuristic decision making in medicine. *Dialogues in Clinical Neuroscience* 14(1), 77–89.

Marewski, J. N., & Schooler, L. J. (2011). Cognitive niches: An ecological model of strategy selection. *Psychological Review* 118(3), 393–437.

Martignon, L., & Hoffrage, U. (2002). Fast, frugal and fit: Simple heuristics for paired comparison. *Theory and Decision* 52, 29–71.

Martignon, L., Vitouch, O., Takezawa, M., & Forster, M. (2003). Naïve and yet enlightened: From natural frequencies to fast and frugal decision trees. In D. Hardman & L. Macchi (Eds), *Thinking: Psychological perspectives on reasoning, judgment, and decision making* (pp. 189–211). Chichester, UK: Wiley.

Meehl, P. E. (1954). *Clinical versus statistical prediction: A theoretical analysis and a review of the evidence.* Minneapolis: University of Minnesota Press.

Nisbett, R. E., & Wilson, T. D. (1977). Telling more than we can know: Verbal reports on mental processes. *Psychological Review* 84(3), 231.

Pachur, T., & Marinello, G. (2013). Expert intuitions: How to model the decision strategies of airport customs officers? *Acta Psychologica* 144, 97–103.

Payne, J. W., Bettman, J. R., & Johnson, E. J. (1993). *The adaptive decision maker.* New York: Cambridge University Press.

Phillips, N. D., Neth, H., Woike, J. K., & Gaissmaier, W. (2017). FFTrees: A toolbox to create, visualize, and evaluate fast-and-frugal decision trees. *Judgment and Decision Making* 12(4), 344–368.

Reisen, N., Hoffrage, U., & Mast, F. (2008). Identifying decision strategies in a consumer choice situation. *Judgment and Decision Making* 3(8), 641–658.

Rieskamp, J., & Hoffrage, U. (2008). Inferences under time pressure: How opportunity costs affect strategy selection. *Acta Psychologica* 127(2), 258–276.

Rieskamp, J., & Otto, P. E. (2006). SSL: A theory of how people learn to select strategies. *Journal of Experimental Psychology: General* 135, 207–236.

Savage, L. J. (1972). *The foundations of statistics.* North Chelmsford, MA: Courier Corporation.

Schraagen, J. M. C. (2018). Naturalistic decision making. In L. J. Ball & V. A. Thompson (Eds), *The Routledge international handbook of thinking and reasoning* (pp. 487–501). Abingdon, Oxon: Routledge.

Shan, Y., & Yang, L. (2017). Fast and frugal heuristics and naturalistic decision making: A review of their commonalities and differences. *Thinking & Reasoning* 23(1), 10–32.

Shanteau, J. (2015). Why task domains (still) matter for understanding expertise. *Journal of Applied Research in Memory and Cognition* 4(3), 169–175.

Simon, H. A. (1957). *Models of man.* New York: Wiley.

Simon, H. A. (1982). *Models of bounded rationality.* Cambridge, MA: MIT Press.

Simon, H. A. (1987). Making management decisions: The role of intuition and emotion. *Academy of Management Executive (1987-1989)* 1(1), 57–64.

Simon, H. A. (1990). Invariants of human behavior. *Annual Review of Psychology* 41, 1–20.

Simon, H. A. (1992). What is an explanation of behavior? *Psychological Science* 3, 150–161.

Sloman, S. A. (1996). The empirical case for two systems of reasoning. *Psychological Bulletin* 119, 3–22.

Stanovich, K. E., & West, R. F. (2000). Advancing the rationality debate. *Behavioral and Brain Sciences* 23(5), 701–717.

Taylor, B. J. (2016). Heuristics in professional judgement: A psycho-social rationality model. *British Journal of Social Work* 47(4), 1043–1060.

Todd, P. M., Gigerenzer, G., & the ABC Research Group (2012). *Ecological rationality: Intelligence in the world*. New York: Oxford University Press.

Todd, P. M., Hertwig, R., & Hoffrage, U. (2015). Evolutionary cognitive psychology. In D. M. Buss (Ed.), *The handbook of evolutionary psychology*, vol. 2: *Integrations*, 2nd edn (pp. 885–903). Hoboken, NJ: Wiley.

Tversky, A., & Kahneman, D. E. (1974). Judgments under uncertainty: Heuristics and biases. *Science* 185, 1124–1131.

Van Rooij, T., Roederer, M., Wareham, T., Van Rooij, I., McLeod, H. L., & Marsh, S. (2015). Fast and frugal trees: Translating population-based pharmacogenomics to medication prioritization. *Personalized Medicine* 12(2), 117–128.

Von Neumann, J., & Morgenstern, O. (1944). *Theory of games and economic behavior*. Princeton, NJ: Princeton University Press.

Ward, P., Gore, J., Hutton, R., Conway, G., & Hoffman, R. (2018). Adaptive skill as the *conditio sine qua non* of expertise. *Journal of Applied Research in Memory and Cognition* 7(1), 35–50.

Wegwarth, O., Gaissmaier, W., & Gigerenzer, G. (2009). Smart strategies for doctors and doctors-in-training: Heuristics in medicine. *Medical Education* 43(8), 721–728.

Woike, J. K., Hoffrage, U., & Martignon, L. (2017). Integrating and testing natural frequencies, Naive Bayes, and fast-and-frugal trees. *Decision* 4, 234–260.

Woike, J. K., Hoffrage, U., & Petty, J. S. (2015). Picking profitable investments: The success of equal weighting in simulated venture capitalist decision making. *Journal of Business Research* 68(8), 1705–1716.

CHAPTER 8

EXPERTISE

A Holistic, Experience-Centered Perspective

JOHN M. FLACH AND FRED A. VOORHORST

Introduction

In 1998, an article on an "Ecological Theory of Expertise Effects in Memory Recall" by Vicente and Wang sparked a debate about whether an "ecological" perspective offered anything new or innovative with respect to understanding expert performance (Ericsson, Patel, & Kintsch, 2000; Simon & Gobet, 2000; Vicente, 2000; Vicente & Wang, 1998). The goal of this chapter is to put that debate into a larger historical context that reflects two different ways of framing questions of human cognition in general and to explore the implications for understanding expert performance in particular. This historical debate has its roots in the beginnings of psychology in America with the distinction between structuralism (Titchener, 1899) and functionalism (Angell, 1907) (see Benjafield, 2010). However, this debate is still alive today and it is reflected in contemporary discussions of situation awareness (e.g., Flach, 2015), embodied cognition (e.g., Lakoff & Johnson, 1999), and expertise (e.g., Ericsson & Towne, 2010).

Structuralism versus Functionalism

The early *structuralists* framed psychology in terms of "mental chemistry" or in terms of the inner workings or computations of mind. Thus, it was assumed that the phenomena began with the encoding of a *stimulus* and ended with a *thought*, *intention*, or *plan*. In doing this, the phenomenon of psychology could be parsed out as something that existed and could be studied independently of particular situational or action constraints. In other words, the mind was seen as a computational mechanism that could be studied as an entity that existed independently from any particular ecological context or domain. The goal was to discover the particular laws or constraints (e.g., in more contemporary terms—the information processing or computational limitations) of this mechanism. In semiotic terms, the structuralist approach framed cognition in terms of a dyadic semiotic system that focused on the interpretation of *symbols* (stimuli) independently from the ecological context, as illustrated in Figure 8.1 (Eco, 1976). From the mid-fifties onwards (with the advent of computers as symbol-processing machines), this approach tended to be framed in terms of computer metaphors of mind that treated cognition as a purely logical symbol-processing problem—independent from the constraints of a body or an action context.

Today the structuralist framing of psychology is reflected in contemporary computational approaches to cognition that emphasize conformity with general normative mathematical models of rationality (e.g., Anderson, 1990). In this framing, rationality is defined in terms of conformity with context independent logical systems that can be applied generally across a wide range of contexts (e.g., Kahneman, 2011). The focus is on coherence with the logical norms, rather than on correspondence with the pragmatics of particular domains or situations. Much of the research inspired by this approach focuses on illusions (Gregory, 1974) or biases (Kahneman, 2011), because these are thought to be symptoms revealing the internal constraints of the

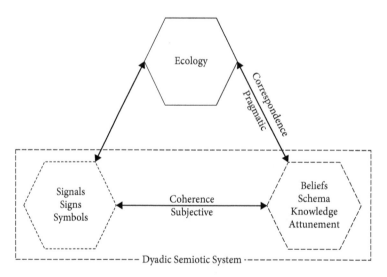

FIGURE 8.1 The full triadic semiotic system proposed by Peirce (1991). Computational and information processing approaches to cognition tend to study the dyadic component as an independent symbol processing mechanism and the constructivist ecological debate tends to reflect a different emphasis on the coherence versus correspondence dyads within the larger triadic system.

Data from Hoopes, James (Ed.), *Peirce on signs: Writings on semiotic by Charles Sanders Peirce*, Chapel Hill: University of North Carolina Press, 1991.

computational mechanisms (e.g., the capacity of working memory). Today, structuralism is seen in efforts to simulate the cognitive mechanism using production systems, such as Anderson's ACT-R (Anderson, 1990; see http://act-r.psy.cmu.edu).

In contrast to structuralism, functionalists such as James, Dewey, and Peirce saw cognition as a capacity for adapting to the pragmatic demands of everyday life (e.g., see Benjafield, 2010). In this framing, emphasis shifted to the pragmatic *functions* of mind in relation to meeting the specific demands of situations or ecologies. Functionalists were less infatuated with *illusions* and *biases*. They were more interested in understanding abilities and skills, such as people's capacity to control locomotion (Gibson, 1979) or people's ability to navigate (Hutchins, 1995).

The emphasis in a functionalist approach shifted from *coherence* (e.g., conformity with abstract mathematical norms) to *correspondence* (e.g., conformity with the pragmatic demands of specific situations). Thus, functionalists were less interested in whether thinking conformed with the principles of abstract, mathematical logic, than with whether the thinking led to successful action in the world. For the modern functionalist, it seemed obvious that cognition was *situated* (e.g., Suchman, 1987).

Functionalism has led to views of mind as a dynamically adaptive system (e.g., Flach & Voorhorst, 2016; Flach, Feufel, Reynolds, Henrikson, & Kellogg, 2017). This implies that the internal structures of the cognitive system (e.g., representations) change as a function of the situation. In control theoretic terms, this means that there is no stable independent transfer function that can be specified independently from the situation (Rasmussen, 1986). In semiotic terms, the implication of a functionalist framework is that the triadic semiotic system must be treated as a whole (see Figure 8.1). It means that the underlying dynamic of cognition has emergent properties (e.g., expertise) that cannot be understood as a sum of the parts (e.g., mind + matter or situations + awareness) or even as the sum of the dyads (coherence + correspondence) within the larger triadic system.

MISLEADING DICHOTOMIES AND CONFUSION

Over the years, the structuralist and functionalist approaches to psychology have continued to shape the debates about how to frame a science of psychology. However, the debate has often been cast in terms of misleading dichotomies that often lead to unproductive arguments in which each side creates a strawman of the other side. For example, the debate between ecological and constructivist approaches to perception tended to be cast in terms of *coherence* (symbol processing) versus correspondence (attunement to affordances).

On the constructivist side, Gregory (1974) framed perception as the "brain game" with the focus on understanding the "symbolic rules" (p. xxvii). Gregory (1974) argued that "perceptual phenomena are essentially illusory" (p. xxvi). Thus, the emphasis was on how internal knowledge is combined with the impoverished information available to the senses to construct hypotheses about the current and future states of the world. In this framework, the ability to correctly interpret the ambiguous *symbols* associated with sensation depends on associations with stored memories or knowledge that is used to construct *internal models* of the world that fill in the gaps and ambiguities that result from impoverished sensory representations (e.g., a two-dimensional retina) and that lead to guesses about the ecology (that are often wrong as reflected in the catalog of various illusions and biases).

In Piagetian terms, Gregory's constructivist approach tends to focus exclusively on the dynamic of *assimilation* in which all the action is associated with how existing internal schemata (i.e., representations) shape the interpretation of the symbols. Thus, constructivist theories tend to focus exclusively on the coherence associated with the internal logic of the mind and pay little attention to structure or constraints in the ecology. It is not that constructivists deny the existence of ecological constraints or the pragmatic constraints of specific problem domains (e.g., deep structure) or even that a person adapts to these constraints. It is simply that they consider these constraints to be secondary to the constraints inside the head—at least with respect to a science of psychology. For example, they are interested in the *deep structure* as a property of an associative network, but have little to say about the ecological basis for this structure.

In contrast to the constructivist approach, the ecological approach to perception (E. Gibson, 1969; J. Gibson, 1979) started with the assumption that the information couplings with the ecology (i.e., the signals) are richly structured in ways that specify many of the functionally significant constraints relative to skilled interaction with the world (i.e., affordances). For example, work inspired by this perspective has focused on the relation between structure in optical flow fields and pragmatic aspects of controlling locomotion, such as creating or avoiding collisions (e.g., Smith, Flach, Dittman, & Stanard, 2001; Flach, Smith, Stanard, & Dittman, 2004; Stanard, Flach, Smith, & Warren, 2012).

For the Gibsons skilled perception did not involve construction (e.g., adding knowledge to fill in gaps), but rather it involved selection or tuning to the relevant structures (e.g., optical invariants) for specific functions. For example, E. Gibson (2003) wrote:

> The process is one of differentiation, a narrowing down from a manifold of information to the minimal, optimal information for the affordance of an event or object. The process is one of selection, not addition. (p. 286)

J. Gibson (1979) made a similar claim in more provocative terms:

> Evidently the theory of information pickup does not need memory. It does not have to have as a basic postulate the effect of past experience on present experience by way of memory. It needs to explain learning, that is, the improvement of perceiving with practice and the education of attention, but not by an appeal to the catch-all of past experience or to the muddle of memory. (p. 254)

Note that in challenging the "muddle of memory," Gibson is not denying any role for internal constraints in shaping perception, any more than the computational camp is denying adaptation. Rather he is claiming that the information available to a moving observer (e.g., optical flow) does not need to be supplemented by additional knowledge in order for people to act skillfully (e.g., to avoid collisions). The function of the internal constraints is to shape selective processes (i.e., attention) so that relevant information (e.g., optical invariants) is picked up. Thus, in the case of controlling locomotion, Gibson suggested that solving the *correspondence* problem does not require an intervening internal model of *space*. Rather, he suggests that there is a direct correspondence between structure in optical flow fields (e.g., angles, and angular rates) and functional constraints on locomotion (e.g., imminence of collision) that allows people to skillfully move through the environment (e.g., Flach et al., 2004).

In Piagetian terms, the Gibsons (particularly E. Gibson) tended to emphasize the dynamic of accommodation. That is, they focused on perceptual learning or attunement in which the internal constraints were shaped by rich feedback resulting from interactions with the ecology. In contrast to Gregory, for the Gibsons the interesting action was in the ecology, not in the head. Though it is important to note that they did not deny the existence of internal constraints, any more than Gregory denied the existence of situational constraints. Rather, they conceptualized these constraints as *filters* that were selectively tuned to pick up information available to active perceptual systems, rather than as stored knowledge that needed to be *added* to the impoverished sensory information in order to compensate for ambiguities.

In approaching the problem of expertise, the constructivist framework tended to emphasize the difference between novices and experts in terms of knowledge or memories in the head. For example, in explaining differential performance of expert and novice chess players, Simon and colleagues have suggested that expertise in chess is due to the fact that experts have stored a large number of chess patterns in memory (> 50,000) (Chase & Simon, 1973; Simon & Gilmartin, 1973).

In contrast, the ecological framework would tend to emphasize differential attunement of attention. For example, Reynolds (1982) found differences between novices and experts in terms of how they directed their attention while playing chess. Novices tended to direct their attention to pieces on the board, but experts tended to attend to spaces on the board that were affected by many pieces. Reynolds wrote that "while the beginning tournament player is captivated by configurations of black and white pieces of wood, the masters and grandmasters center their attention on those squares effected by many pieces" (p. 391).

Reingold and Sheridan (2011) report support for Reynolds' observations from studies of eye movements. For example, they cite research by Reingold and Charness (2005) that showed that experts produced a greater number of fixations on empty squares than novices and they also found that relative to intermediate level players, experts produced substantially fewer fixations and a much greater proportion of long fixations over the viewing period. One interpretation of these data is that the figure–ground relations for experts are different than for less skilled players. That is, the novices tend to treat the pieces as the *objects* or *figures*, whereas the experts tend to treat larger patterns or *chunks* as the *figures* (Chase & Simon, 1973).

Adding to the noise associated with debates between constructivist and ecological approaches to expertise is the fact that terms like *chunk* can have different connotations, depending on the dominant perspective. From the constructivist perspective, *chunking* reflects a mental action or mnemonic device in which structure is imposed on the input, based on prior knowledge (e.g., stored patterns from past games). This perspective is bolstered by early experimental work that used stimuli with minimal structure (e.g., strings of binary digits) to demonstrate how "recoding" of the information into chunks using prior knowledge could improve memory recall (Anderson, 1990; Miller, 1956).

From the ecological perspective, *chunking* reflects selective attention to higher order functional relations inherent to the problem ecology (e.g., seeing the pieces in the context of potential actions and consequences). This perspective is bolstered by the fact that differences between experts and novices are significantly reduced or eliminated when the functional structure is disrupted as in the random control condition used by Chase and Simon (1973; see also Reynolds, 1982).

Which perspective (constructivism or ecological psychology) is right, with respect to cognition in general, and with respect to expertise in particular? Is expertise a function of *stored knowledge* or is it a function of the *attunement of attention*? As happens so often in science, after many years of contentious debate between two opposing and apparently contradictory views of cognition (and expertise), the answer turns out to be that they are both right—but they are both incomplete.

A Holistic Experience-Centered Alternative

In the past, the debate has been miscast in terms of particular dyads (coherence/inside the head or correspondence/outside the head; or subjective vs pragmatic). However, this misses the larger structure versus function issue with respect to semiotics. This is the question of whether the mind can be understood as an independent, general mechanism (e.g., a general-purpose computer or servomechanism) as opposed to a system that is intimately linked with its ecology (adaptive control system). Thus, the critical issue is whether to frame cognition in terms of a dyadic or triadic semiotic system. We contend that constructs such as information, schema, representation, pattern, affordance, and meaning can best be understood as holistic properties of the triadic semiotic system, rather than properties that can be associated with any of the components or dyads within the triadic semiotic system (Flach, Stappers, & Voorhorst, 2017; Flach & Voorhorst, 2016).

In our view, the essence of a functionalist framework is reflected in William James' (1976) ontological approach of radical empiricism. In contrast with the dualistic ontological frame that has dominated Western thought, the central tenet of radical empiricism is that the basic unit of analysis for psychology is *experience*, which is a joint function of situations (matter) and awareness (mind). Thus, an experience-centered approach is one that treats the triadic semiotic system holistically, rather than as a collection of parts or dyads. Experience involves both of the dynamics that Piaget described—it involves assimilation in which current events are interpreted in light of existing internal constraints (e.g., schemata) and it involves accommodation in which the internal constraints themselves are shaped through the rich feedback resulting from direct interaction with the ecology. Additionally, an experience-centered approach recognizes that the quality of processing (i.e., expertise) must be a joint function of internal (e.g., intentions, assumptions, expectations, preferences, information processing limits) and external constraints (e.g., physical dynamics such as the laws of motion, pragmatics associated with actual consequences, windows of opportunity, and constraints of the information media such as optics).

Simon (1969, 1990) explicitly recognized that human rationality was a joint function of the fit between two blades of a scissors—the structure of task environments and the computational abilities of the actor. His analogy of the ant on the beach (Simon, 1969) also recognized the contribution of the beach (i.e., the situation) in shaping behavior. Neisser (1976) also explicitly recognizes the contributions and limitations of both sides of the constructivist/ecological debate. His *perceptual cycle* has much in common with the triadic semiotic framework suggested by Peirce (1991). More recently, we have made our own attempt to articulate a holistic approach to human experience as an emergent property of a triadic semiotic system that recognizes the roles of both internal and external constraints in jointly determining *What Matters* (Flach & Voorhorst, 2016).

An Adaptive Control Model

> [A] thinking human being is an adaptive system; ... to the extent that [humans] ... are effectively adaptive, possessing the relevant knowledge and skills, their behavior will reflect characteristics largely of the outer environment (in the light of their goals) and will reveal only a few limiting properties of their inner environments—of the physiological machinery that enables them to think.
>
> Simon (1969; cited in Simon & Gobet, 2000, p. 599)

As the opening quote from Simon illustrates, those interested in expertise have long recognized the adaptive nature of cognition. Figure 8.2 illustrates cognition as an adaptive control system. This system has two interconnected loops. In the inner loop, perception and action are linked through a particular schema or representation. This loop functions like a servomechanism, where the focus is on actions in relation to some

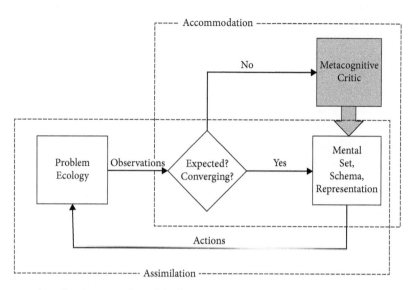

FIGURE 8.2 An adaptive control model of cognition.

Adapted from *Applied Ergonomics* 63, John M. Flach, Markus A. Feufel, Peter L. Reynolds, Sarah Henrickson Parker, and Kathryn M. Kellogg, Decision making in practice: The dynamics of muddling through, pp. 133-141, doi.org/10.1016/j.apergo.2017.03.017, © 2017 Elsevier Ltd., with permission from Elsevier.

goal or function. In this context, observations are typically considered to be *error* signals and actions are intended to reduce the errors—to achieve the goal or intention. The mental set or representation is essentially the *control law* that specifies the logic linking the observations with actions for reducing errors—for instance, how do I keep my car within the safe field of travel. In Piagetian terms, this loop reflects the assimilation dynamic where observations are seen and actions are selected in terms of what has been successful in the past.

For experts, the representation is typically heuristic in nature, but in the most positive sense of that term (e.g., see Todd, Gigerenzer, & the ABC Group, 2012). In this positive sense, a heuristic reflects an expert's ability to leverage constraints associated with their domains of expertise in ways that tend to reduce the demands (e.g., the computational complexity) with minimal sacrifices to accuracy. Thus, experts tend to be both faster and more accurate than novices with respect to problems in their specific domain. As Ericsson and Charness (1994) concluded, "acquired skill can allow experts to circumvent basic capacity limits of short-term memory and of the speed of basic reactions, making potential limits irrelevant" (p. 731).

The outer loop functions as a monitor or critic of the inner loop. That is, this loop is monitoring for surprises or deviations from expectations relative to the current schema. When surprises are encountered this outer loop functions to adjust the *control law* or *schema* governing behavior in the inner loop—with the goal of reducing the surprise or deviations from expectations. For example, in driving a rental car, you may discover that the braking dynamics are different or that the layout of the controls (e.g., turn signals) is different than in your own car, requiring a change in your schema. In Piagetian terms, the outer loop reflects the accommodation dynamic in which schemas are modified (i.e., tuned) with the hope of achieving a better fit to the demands of new situations. A shift in the schema governing inner loop control is akin to a figure–ground shift in which new aspects of the feedback information become salient for control (e.g., new patterns such as invariants in an optical flow field or relations among pieces on a chess board).

Consistent with the ecological construct of *attunement*, both of the loops are *recognition-primed* (Klein, 1989). And consistent with constructivist approaches, internal representations or schemas play a significant role in the dynamic. In the inner loop, actions (e.g., coordinative structures) or expectations (e.g., hypotheses) are primed by the recognition of deviations from an intention (error signals). This typically reflects generalizations based on prior experiences. In the outer loop, alternative hypotheses or schemas are primed by the recognition of deviations from expectations (surprises). For example, adjustments are made when the schemas/representations being employed in the inner loop are not yielding satisfying results, as intended or expected. The function of the outer loop is to switch schemas or to tune current schemas in ways that lead to more satisfying results in the inner loop.

In the context of this model, expertise reflects the ability to recognize whether a particular schema is appropriate to a situation and the ability to execute the schema. In complex domains, many different schemas may be required to meet the demands associated with a wide range of situations. For example, the expert golfer must select the best coordinate structure to fit the various situations associated with winning golf (driving, chipping, or putting) and must know how to control each coordinate structure. Or the physician in the Emergency Department may utilize multiple heuristics to guide the sensemaking processes associated with managing patient care (Flach et al., 2017). Thus, the general expectation is that experts have a collection of schemas (coordinative structures and/or heuristics) that are more compatible and more comprehensive with respect to the demands of a particular domain, than less experienced people. However, even experts can occasionally encounter situations for which they have no schema, and in these cases they must actively explore using more generic problem solving heuristics in order to make sense out of the situation.

What Makes a Good Representation?

> It turns out that the form in which information is represented can have substantial effects on later recall of the material.
>
> Anderson (1995, p. 210)

> Perceptual learning is not properly described as association or as an addition of any kind, as a response to a stimulus, or as a 'representation' of any thing.
>
> E .J. Gibson & Pick (2000, p. 149)

A central component in our adaptive control model of cognition is the construct of *schema* or *representation*. In this model, the quality of performance (e.g., expertise) will largely be a function of the quality of the representation. However, as indicated by the quotes at the beginning of this section—the construct of *representation* has had different significance for those who take a constructivist perspective and those who take an ecological perspective.

Representation has been a central theme for constructivist theories that tend to treat representation as a construct that has great significance for the internal cognitive structures. For example, in discussing chunking, Anderson (1995) emphasized the implications for memory structures—suggesting that a chunk is the result of storing information together as a single representation or unit in memory. In contrast, those adopting an ecological perspective have been quite skeptical about the construct of *representation*, as an unnecessary mentalistic mediator between perception and action. As noted earlier, from the ecological perspective a chunk tends to reflect functional properties of a domain (e.g., higher order relations or deep structure) that can be picked up or attended to, without the need for association with specific memories (e.g., specific stored patterns from past games). Note that as with the construct of schema in perception—the schema guiding recognition may be an abstraction from prior examples (e.g., prototype) that has never actually been seen before (e.g., Posner, 1969; Posner & Keele, 1968). Both the computational and ecological approaches recognize that experience results in *learning*. The debate is whether to characterize such learning as *remembering* or as *attuning*.

Again, we contend that this debate has been miscast and that the construct of representation can take on a broader connotation within the full triadic semiotic dynamic than has been suggested by the previous debates. On the one hand we see evidence in the literature on expertise that invites an explanation in terms of internal constraints (mental representations). For example, representations may be in the form of knowledge that leads to mental abstractions such as an analogy (Gentner, 1983), a metaphor (Lakoff & Johnson, 1980), or a categorical classification (Chi, Feltovich, & Glaser, 1981).

On the other hand, we can see evidence in skilled performance that suggests that the *representation* is more intimately related to constraints in the physical medium of perception and action. For example, optical flow fields resulting from self-motion have internal structure (invariants) that can provide a direct mapping to functional aspects of locomotion (Gibson, Olum, & Rosenblatt, 1955). Also, Bernstein's (1967) construct of coordinative structure, in which selectively locking out degrees of freedom can lead to *soft* structures that improve motor control, can be thought of as a non-mentalistic form of representation. Alternative scanning strategies, as suggested by Reynolds (1982), would be another change that might best be conceived of as a change in the perception–action coupling, rather than an exclusively *mental* constraint. The scanning strategies might reflect a shift in figure–ground relations that is analogous to a coordinative transformation in mathematics. This shift does not *add* new information or data, but it reorganizes (or filters) existing data in ways that make some relations more salient than others. The representation might also be characterized as a heuristic (Todd et al., 2012) or smart mechanism (Runeson, 1977). Finally, the representation may be artificially constructed interfaces or artifacts that have structure that map directly to important functional properties of a domain (Bennett & Flach, 2011; Hutchins, 1995).

In all of these cases, the representations are analogous to coordinate transformations. They often do not *add* meaning, but they can both amplify and filter properties of the information that is being observed. As Woods (1991) has noted, "there are no neutral representations" (p. 75). A representation is going to make some facets of the information more salient and it will make other facets of the information less salient. Thus, consistent with E. J. Gibson's view of perceptual learning, the role of a representation (as we are using it) is typically *selective*, not *additive*. We contend that this is true of both representations framed as mental abstractions (e.g., metaphors) and representations framed in terms of properties of the medium of perception–action (e.g., flow fields, coordinative structures, heuristics). Certainly, there are instances where new mental associations might lead to new ways of framing a problem (e.g., an Aha! moment)—but we think it is best to think of the new framing as a *filter* impacting the pick-up of information in real time, rather than as a supplement to the information.

The key point is that the available information is not a collection of independent *stimuli* (cues or symbols) that requires the addition of stored knowledge in order to become meaningful, but rather the information is structured or constrained relative to the pragmatics of the triadic semiotic system such that some aspects of the information (e.g., patterns) are functionally relevant (i.e., signals) and some aspects are functionally irrelevant (i.e., noise). Thus, the domain information has a property that Bowdle and Gentner (2005) refer to as *systematicity*. With respect to analogies and metaphor, they suggest that "alignments that form deeply interconnected structures, in which higher order relations constrain lower order relations, are preferred over less systematic sets of commonalities" (Bowdle & Gentner, 2005, p. 196). We contend that a nested structure of constraints is typical of all functional problem domains. Thus, a good representation is one in which the higher and lower order relations in the representation map well to higher (e.g., goals and values) and lower (alternative actions) functional constraints.

In the context of the constructivist approach, *meaning* typically is seen as a product of an agent's interpretation (i.e., a product of information processing or mental activities). In the context of an ecological approach, *meaning* typically is seen as a property of the ecology associated with the consequences of an agent's action (i.e., what the ecology affords). However, in a truly functionalist approach (i.e., an experience-centered approach), *meaning* is seen as an emergent property of the triadic semiotic dynamic, such that it reflects the nesting of functional constraints within the problem domain. Such constraints are *meaning-full* as properties of the domain, which reflect both subjective (e.g., goals and values) and situational (e.g., pragmatic consequences and the potential for action) facets of the functional dynamic.

VICENTE AND WANG (1998) REVISITED

An explicit goal of the Vicente and Wang paper was to develop a more "accurate functional account of the phenomenon" (of expertise) (p. 35). This is a goal that we share. However, in pursuing this goal, they may have made a tactical error by creating another misleading dichotomy in the distinction between *process* and *product* theories of expertise. They define a *process theory* as a theory that is

> directed at the question of *how* (i.e., by what psychological mechanism and knowledge structures) expertise in memory recall comes about. (p. 35)

In contrast, they define a *product theory* as one that focuses on "input–output" without committing to "any particular psychological mechanism" (p. 35). The product theory focuses on properties of the ecology or situation in order to better "predict and understand *what* expertise effects in memory are observed under various conditions" (p. 35).

Although Vicente and Wang suggest that these two theoretical perspectives are complementary descriptions of expertise, they suggest that a product perspective (understanding the what—i.e., situations) should take precedence over a process perspective:

> the question of what logically precedes that of how. That is, to develop an adequate theory of how expertise effects in memory recall are generated it may be very useful, and perhaps even necessary, to first have an accurate account of exactly what those effects are because the latter can put strong constraints on the former. (p. 35)

Thus, Vicente and Wang set up the debate along lines similar to those of the classical debate between constructivist and ecological approaches. Their *process theory* focused on the coherence dyad in the semiotic system; and their *product theory* focused on the correspondence dyad in the semiotic system. This parsing of the problem implies that these dyads can be explored independently and then combined to gain a more complete understanding of expertise. As we noted earlier, this contradicts a truly functionalist perspective that argues for a holistic approach to the triadic semiotic system. In particular, this contradiction is reflected in the first issue that they list which needs to be addressed by a viable product theory of expertise:

> How should one represent the constraints that the environment (i.e., the problem domain) places on expertise? (p. 35)

It is our contention that the "problem domain" is not a property of the *environment* that can be meaningfully described independently from an *agent*. A problem domain is a property of the full semiotic system reflecting the joint constraints of mind and matter on *satisfying* (situated consequences in relation to intentions and values), *specifying* (available information in relation to the tuning of attentional filters), and *affording* (physical properties in relation to effector capabilities) (Flach & Voorhorst, 2016). Each of these constructs—satisfying, specifying, and affording—are emergent properties of the triadic semiotic system that cannot be reduced to any of the components or dyads and that cannot be independently associated with either an agent or an ecology.

Despite the fact that we disagree with how Vicente and Wang framed their argument, we wholeheartedly agree with them with respect to the value of Rasmussen's (1985; 1986) *Abstraction Hierarchy* for gaining insights into the dynamic of the triadic semiotic system. The Abstraction Hierarchy decomposes problems in terms of a nested set of means-ends relations that in turn can be related to different ways to parse the problem into coherent parts or chunks.

In our view, the Abstraction Hierarchy provides an important framework for mapping Bowdle and Gentner's (2005) construct of *systematicity* to the specifics of particular problem domains. In essence, the Abstraction Hierarchy is a positive hypothesis about how functional constraints combine into a nested structure of distinctions that are *meaningful* with respect to sensemaking in a problem domain. In other words, the Abstraction Hierarchy provides a concrete framework for generating hypotheses about what is typically referred to as the *deep structure* of a problem (in contrast to the surface structure). As such, the Abstraction Hierarchy provides a framework for constructing a priori hypotheses about what relations within a problem domain will be critical to success; and thus for predicting both how experts *will* parse or chunk the problem (e.g., in order to predict memory recall—as Vicente and Wang did) and how experts *should* chunk the problem (e.g., as a basis for designing training systems and for creating interfaces that enhance perspicacity—e.g., see Bennett & Flach, 2011).

Thus, we totally concur with Vicente and Wang with respect to the need for there to be an a priori framework for parsing problems in order to distinguish signal (e.g., the systematicity, deep structure, high-order relations, meaningful patterns) from noise (e.g., surface or literal features, illusory patterns, irrelevant variables, distractions) relative to sensemaking. Further, we agree that parsing the problem to reflect means–ends relations as in the Abstraction Hierarchy is a promising direction. We encourage those interested in expertise to consult other sources in addition to the Rasmussen papers cited earlier to learn more about the Abstraction Hierarchy and Work Analysis and the implications for supporting productive thinking (e.g., Bennett & Flach, 2011; Flach, Mulder, & van Paassen, 2004; Flach, Schwartz, Bennett, Russell, & Hughes, 2008; Naikar, 2013; Vicente, 1999). However, we think it is misleading to treat *means* and *ends* as either uniquely mental/subjective (in peoples' heads) or uniquely physical/objective (in the environment). We suggest that it is best to think of them as properties of a functional triadic semiotic system. We suggest that it is best to think of them as properties of experience.

An Experience-Centered Approach

We were invited to write this chapter as an *ecological* perspective on expertise. However, though our approach has been strongly influenced by the ecological approach of both E. J. and J. J. Gibson, we have concluded that the ecological versus non-ecological (e.g., constructivist or computational) debate has been ill-posed. As we previously noted, both positions are right in some respect, but neither perspective provides a completely satisfying framework for understanding expertise. Expertise is about the quality of adaptation between an agent and an environment. Thus, the *environment* must be considered. As Simon and Gobet (2000) colorfully put it: "To predict the shape of Jell-O, look at the mold in which it is jelling" (p. 5). Thus, in the case of expertise—context matters, such that describing the ecological constraints on consequences, information, and action is essential to a full understanding.

However, the ecology is but one side of the "fit" equation. We also agree with Neisser (1976) that

> There must be definite kinds of structure in every perceiving organism to enable it to notice certain aspects of the environment rather than others, or indeed to notice anything at all. (p. 9)

We use *representation* as a broad term for that structure and suggest that it can take many forms—analogy, metaphor, categorization, coordinative structure, heuristic, or search strategy. In general, we suggest that, consistent with E. J. Gibson's (1969) ideas of perceptual learning, these structures primarily function as *filters* that determine what aspects of a problem are figure (e.g., salient) and what aspects of a problem are ground (i.e., what are the meaningful aspects of a domain). Thus, rather than *adding* information, these structures function like coordinative transforms in mathematics (e.g., a log transform). These transforms do not add information, but different transforms tend to alter the relative salience of different relations in the data (e.g., additive versus multiplicative relations). Thus, expertise can be seen as finding a coordinative transform (e.g., a way of looking) that makes the functionally significant patterns salient. This position is consistent with de Groot's (1965) vivid descriptions of chess expertise:

> The swift insight of the chessmaster into the possibilities of a newly shown position, his immediate *"seeing" of structural and dynamic essentials*, of possible combinatorial gimmicks, and so forth are only understandable, indeed, if we realize that as a result of his experience *he quite literally "sees" the position in a totally different (and much more adequate) way than a weaker player*. The vast difference between the two in efficiency, particularly in the time required to find out what the core problem is ("what's cooking really") and to discover highly specific, adequate means of thought and board action, need not and must not be primarily ascribed to large differences in "natural" power for (means) abstraction. *The difference is mainly due to differences in perception.* (p. 306, emphasis added)

We suggest the term *experience-centered* as a way to get past the conventional mind–matter dichotomy and the ecological versus constructivist debates. We see the term *experience* as a joint function of mind and matter that reflects both constraints of situations and constraints on awareness. We see experience as an emerging property of the triadic semiotic dynamic—that is more than the sum of the elements or of the component dyads.

We agree with Vicente and Wang (1998) that it is important to have a framework for generating hypotheses about what the "structural and dynamic essentials" (see de Groot quote above) of a problem are. We further agree that the Abstraction Hierarchy can be extremely useful in this regard. For example, Ericsson and Towne (2010) have argued for an *expert-performance* approach to expertise. The key idea is to use performance on representative tasks from a domain to judge levels of expertise (as opposed to general context free metrics such as years of experience or hours of practice). The Abstraction Hierarchy framework might prove to be a useful guide for parsing a domain into elemental tasks that preserve meaningful constraints. We suggest that preserving the domain constraints may be critical in order to generalize from success on a part task to overall domain success (e.g., see Flach, Lintern, & Larish, 1990).

In conclusion, the central point that we want to leave you with is that expertise is not something that can be isolated as a property of a mind, independent from a problem ecology, or vice versa. On the one hand, it is impossible to understand expertise in chess, without understanding the game of chess; or expertise in medicine without understanding the healthcare domain. On the other hand, it is essential to understand the nature of internal structural changes (representations) that result from learning—and that influence how experts see, think, and act when solving problems. To fully understand expertise, it is important to escape from the dichotomy of mind and matter, and to not get hung up on debates framed about the relative importance of elements or dyads within the triadic semiotic system. Our main point is that when it comes to expertise, the whole of experience is more than the sum of the pieces (i.e., nodes or dyads in Figure 8.1). Expertise can only be fully understood as an emergent property of the full triadic semiotic dynamic.

REFERENCES

Anderson, J. R. (1990). *The adaptive character of thought*. Hillsdale, NJ: Erlbaum.
Anderson, J. R. (1995). *Learning and memory*. New York: Wiley.
Angell, J. R. (1907). The province of functional psychology. *Psychological Review* 16, 152–169.
Benjafield, J. G. (2010). *History of psychology*, 3rd edn. Oxford: Oxford University Press.
Bennett, K. B., & Flach, J. M. (2011). *Display and interface design: Subtle science, exact art*. London: Taylor & Francis.
Berstein, N. (1967). *The coordination and regulation of movements*. New York: Pergamon.
Bowdle, B. F., & Gentner, D. (2005). The career of metaphor. *Psychological Review* 112(1), 193–216.
Chase, W. G., & Simon, H. A. (1973). Perception in chess. *Cognitive Psychology* 4, 55–81.
Chi, M. T. H., Fletovich, P. J., & Glaser, R. (1981). Categorization and representation of physics problems by experts and novices. *Cognitive Science* 5(2), 121–152.
De Groot, A. D. (1965). *Thought and choice in chess*. The Hague: Mouton Press.
Eco, U. (1976). *A theory of semiotics*. Bloomington: Indiana University Press.
Ericsson, K. A., & Charness, N. (1994). Expert performance: Its structure and acquisition. *American Psychologist* 49(8), 725–747.

Ericsson, K. A., Patel, V., & Kintsch, W. (2000). How experts' adaptations to representative task demands account for expertise effect in memory recall: Comment on Vicente and Wang (1998). *Psychological Review* 107(3), 578–592.

Ericsson, K. A., & Towne, T. J. (2010). Expertise. *WIREs Cognitive Science* 1(3), 404–416.

Flach, J. M. (2015). Situation awareness: Context matters! *Journal of Cognitive Engineering and Decision Making* 9(1), 59–72.

Flach, J. M., Feufel, M. A., Reynolds, P. L., Parker, S. H., & Kellogg, K. M. (2017). Decision making in practice: The dynamics of muddling through. *Applied Ergonomics* 63, 133–141.

Flach, J. M., Lintern, G., & Larish, J. F. (1990). Perceptual motor skill: A theoretical framework. In R. Warren & A. H. Wertheim (Eds), *Perception & control of self-motion* (pp. 327–355). Hillsdale, NJ: Erlbaum.

Flach, J., Mulder, M., & van Paassen, M. M. (2004). The concept of the "situation" in psychology. In S. Banbury & S. Tremblay (Eds), *A cognitive approach to situation awareness: Theory, measurement, and application* (pp. 42–60). Aldershot, UK: Ashgate.

Flach, J. M., Schwartz, D., Bennett, A., Russell, S., & Hughes, T. (2008). Integrated constraint evaluation: A framework for continuous work analysis. In A. M. Bisantz & C. M. Burns (Eds), *Applications of cognitive work analysis* (pp. 273–297). London: Taylor & Francis.

Flach, J. M., Smith, M. R. H., Stanard, T., & Dittman, S. M. (2004). Collision: Getting them under control. In H. Hecht & G. J. P. Savelsbergh (Eds), *Theories of time to contact*, Advances in Psychology Series (pp. 67–91). Elsevier, North-Holland.

Flach, J. M., Stappers, P. J., & Voorhorst, F. A. (2017). Beyond affordances: Closing the generalization gap between design and cognitive science. *Design Issues* 33(1), 76–89.

Flach, J. M., & Voorhorst, F. A. (2016). *What matters?* Dayton, OH: Wright State University Library.

Gentner, D. (1983). Structure-mapping: A theoretical framework for analogy. *Cognitive Science* 7, 155–170.

Gibson, E. J. (1969). *Principles of perceptual learning and development*. New York: Appleton-Century-Crofts.

Gibson, E. J. (2003). The world is so full of a number of things: On specification and perceptual learning. *Ecological Psychology* 15(4), 283–287.

Gibson, E. J., & Pick, A. D. (2000). *An ecological approach to perceptual learning and development*. New York: Oxford University Press.

Gibson, J. J. (1979). *The ecological approach to visual perception*. Boston: Houghton Mifflin.

Gibson, J. J., Olum, P., & Rosenblatt, F. (1955). Parallax and perspective during aircraft landings. *American Journal of Psychology* 68(3), 372–385.

Gregory, R. L. (1974). *Concepts and mechanisms of perception*. New York: Charles Scribner's Sons.

Hutchins, E. (1995). *Cognition in the wild*. Cambridge, MA: MIT Press.

James, W. (1976). *Essays in radical empiricism*. Cambridge, MA: Harvard University Press.

Kahneman, D. (2011). *Thinking, fast and slow*. New York: Farrar, Straus and Giroux.

Klein, G. A. (1989). Recognition-primed decisions. In W. B. Rouse (Ed.), *Advances in man-machine system research*, 5 (pp. 47–92). Greenwich, CT: JAI Press.

Lakoff, G., & Johnson, M. (1980). *Metaphors we live by*. Chicago: University of Chicago Press.

Lakoff, G., & Johnson, M. (1999), *Philosophy in the Flesh: The Embodied Mind and Its Challenge to Western Thought* (New York: Basic Books)

Miller, G. A. (1956). The magical number seven plus or minus two: Some limits on our capacity to process information. *Psychological Review* 63, 81–97.

Naikar, N. (2013). *Work domain analysis*. Boca Raton, FL: CRC Press.
Neisser, U. (1976). *Cognition and reality*. San Francisco: W. H. Freeman.
Peirce, C. S. (1991). *Peirce on signs*, ed. J. Hoopes. Chapel Hill: University of North Carolina Press.
Posner, M. (1969). Abstraction and the process of recognition. In G. H. Bower & J. T. Spence (Eds), *The psychology of learning and motivation*, Vol. 3. New York: Academic Press.
Poser, M. I., & Keele, S. W. (1968). On the genesis of abstract ideas. *Journal of Experimental Psychology 77*, 353–363.
Rasmussen, J. (1985). The role of hierarchical knowledge representation in decisionmaking and system management. *IEEE Transactions on Systems, Man, and Cybernetics SMC-15*, 234–243.
Rasmussen, J. (1986). *Information processing and human-machine interaction: An approach to cognitive engineering*. Amsterdam: North-Holland.
Reingold, E. M., & Charness, N. (2005). Perception in chess: Evidence from eye movements. In G. Underwood (Ed.), *Cognitive processes in eye guidance* (pp. 325–354). Oxford: Oxford University Press.
Reingold, E. M., & Sheridan, H. (2011). Eye movements and visual expertise in chess and medicine. In S. P. Liversedge, I. D. Gilchrist, & S. Everling (Eds), *Oxford handbook of eye movements* (pp. 528–550). Oxford: Oxford University Press.
Reynolds, R. I. (1982). Search heuristics of chess players of different calibers. *American Journal of Psychology 95*(3), 383–392.
Runeson, S. (1977). On the possibility of "smart" perceptual mechanisms. *Scandinavian Journal of Psychology 18*(1), 172–179.
Simon, H. A. (1969). *The sciences of the artificial*. Cambridge, MA: MIT Press.
Simon, H. A. (1990). Invariants of human behavior. *Annual Review of Psychology 41*, 1–19.
Simon, H. A., & Gilmartin, K. (1973). A simulation of memory for chess positions. *Cognitive Psychology 5*, 29–46.
Simon, H. A., & Gobet, F. (2000). Expertise effects in memory recall: Comment on Vicente and Wang (1998). *Psychological Review 107*(3), 593–600.
Smith, M. R. H., Flach, J. M., Dittman, S. M., & Stanard, T. W. (2001). Monocular optical constraints on collision control. *Journal of Experimental Psychology: Human Perception & Performance 27*(2), 395–410.
Stanard, T., Flach, J. M., Smith, M. R. H., & Warren, R. (2012). Learning to avoid collisions: A functional state space approach. *Ecological Psychology 24*(4), 328–360.
Suchman, L. (1987). *Plans and situated actions*. Cambridge, UK: Cambridge University Press.
Titchener, E. (1899). Structural and functional psychology. *Philosophical Review 8*, 290–299.
Todd, P. M., Gigerenzer, G., & the ABC Research Group (2012). *Ecological rationality*. Oxford: Oxford University Press.
Vicente, K. J. (1999). *Cognitive work analysis: Toward safe, productive, and healthy computer-based work*. Mahwah, NJ: Erlbaum.
Vicente, K. J. (2000). Revising the constraint attunement Hypothesis: Reply to Ericsson, Patel, and Kintsch (2000) and Simon and Gobet (2000). *Psychological Review 107*(3), 601–608.
Vicente, K. J., & Wang, J. H. (1998). An ecological theory of expertise effects in memory recall. *Psychological Review 105*, 33–57.
Woods, D. D. (1991). The cognitive engineering of problem representations. In G. R. S. Weir & J. L. Alty (Eds), *Human–computer interaction and complex systems* (pp. 169–188). London: Academic Press.

CHAPTER 9

MACROCOGNITIVE MODELS OF EXPERTISE

ROBERT J. B. HUTTON

INTRODUCTION

MACROCOGNITION is the adaptation of cognition to the complexities of real-world work. Macrocognitive models were developed in order to explore the boundaries of microcognitive models, addressing the cognitive phenomena of dynamic interactions of people, work, and the environment in which work takes place. Of particular relevance to the focus of this *Handbook*, macrocognitive models tend to be models of expert performance, given that they are based on empirical evidence from skillful professionals in their work contexts.

This chapter is intended to provide a brief understanding of macrocognition, some of its theoretical underpinnings, and some examples of macrocognitive models. It is not intended to be a comprehensive resource representing all the macrocognitive models that have been developed. The interested reader is encouraged to read Schraagen, Klein, and Hoffman (2008), who provide an overview of the concept as an introduction to a broader set of relevant chapters, and Hoffman and McNeese (2009), who provide an excellent historical perspective.

This chapter addresses the following:

- What do we mean by macrocognition?
- Theoretical foundations and influences
- Methods used to study and develop these models
- A few exemplar models
- Future challenges and forward thinking

What Is Macrocognition?

Macrocognitive models have emerged in the most part due to efforts to understand and remedy applied problems. Models from cognitive psychology have not provided sufficient support for understanding and solving the challenges of real-world work, and thus macrocognitive models have emerged to fill that gap. Models developed from the controlled contexts of experimental psychology have been termed *microcognitive* models in order to distinguish the levels of analysis represented by the macro- and micro- terminology. Cacciabue and Hollnagel (1995) summarized the distinction in the following way:

> Macro-cognition refers to the study of the role of cognition in realistic tasks, that is in interacting with the environment. Macro-cognition only rarely looks at phenomena that take place exclusively within the human mind or without overt interaction. It is thus more concerned with human performance under actual working conditions than with controlled experiments. (pp.57–58)

The emerging definition of macrocognition acknowledges two key elements (Schraagen, Klein, & Hoffman, 2008):

1. Cognitive work can only be understood through study at a number of levels or perspectives
2. Information processing models (exemplified by experimental cognitive psychological models, e.g., Wickens, Hollands, Banbury, & Parasuraman, 2015) provide an incomplete and distorted understanding of cognitive work.

Klein, Klein, and Klein (2000) suggested that the term macrocognition be used to "designate the more complex cognitive functions. These functions would include decision-making, situation awareness, planning, problem detection, option generation, mental simulation, attention management, uncertainty management, expertise and so forth" (p.173).

Klein et al.'s (2003) offer a contrast between microcognition and macrocognition in order to emphasize the complementarity of the two levels of description:

> These types of functions—detecting problems, managing uncertainty, and so forth—are not usually studied in laboratory settings. To some extent, they are emergent phenomena. In addition to describing these types of phenomena on a macrocognitive level, we can also describe them on a microcognitive level. The two types of description are complementary. Each serves its own purpose, and together they might provide a broader and more comprehensive view than either by itself. We do not suggest that the investigation of macrocognitive phenomena will supercede or diminish the importance of microcognition work—just that we need research to better understand macrocognitive functions in order to improve cognitive engineering. (p.81)

These authors characterized the key challenges associated with cognitive work which occurs in naturalistic settings but not laboratory studies as:

- Decisions are typically complex, often involving data overload;
- Decision are often made under time pressure and involve high stakes and high risk;
- Research participants are domain practitioners rather than college students;
- Goals are sometimes ill-defined, and multiple goals often conflict; and
- Decisions must be made under conditions in which few things can be controlled or manipulated; indeed, many key variables and their interactions are not even fully understood.

They identified the limitations of a purely microcognitive perspective with respect to the potential distortions that could result from studying cognitive processes in isolation from one another and in isolation from the contexts in which the practitioners apply those processes to achieve work or performance objectives.

The term macrocognition has also been used specifically to describe cognition at the level of team performance, where the *macro-* refers to cognition amongst multiple actors (Fiore et al., 2010; Letsky, Warner, Fiore, & Smith, 2008; Warner, Letsky, & Cowen, 2005). This use of this term is certainly relevant in the context of cognition in naturalistic environments where multiple players often characterize the work context. Understanding the challenge of cognitive performance in multiple-actor contexts (including intelligent/computational agents) is beyond the scope of this chapter; however, it should be noted that macrocognition is not merely the purview of team performance. The models described in this chapter focus primarily on individual decision makers.

Macrocognitive models provide attempts to describe and understand the challenging aspects of purposeful cognitive performance in the dynamic flow of complex and uncertain situations. The models acknowledge the antecedents of problem solving and decision making, the meaning and implications of situational factors, and they recognize the value and impact of those solutions and decisions on subsequent performance. They are arguably ecological models in the Gibsonian/Brunswikian sense (Brunswik, 1956; Gibson, 1979/2014) in that they often represent cycles of behavior, rather than mere input–output relationships. They are intended to provide a view of human performance that acknowledges the messiness of many of the work contexts that have provided data for these models. Hoffman, Norman, and Vagners (2009) defined macrocognition as a process of "adapting cognition to complexity" (p. 87), which describes macrocognition as a dynamic application of thinking to evolving events.

With respect to how we represent macrocognition as a model, this presents a number of challenges. Indeed there are disagreements within the community as to the key characteristics of macrocognitive models and how they are represented. For the purposes of this chapter we recognize that there are both *weak* and *strong* models of macrocognition. These may represent the evolution of macrocognitive models over the past 30 years, where some of the earlier (arguably *weaker*) models represented retrospective, descriptive, causal chain, input–output models that help tell stories, but not make predictions. Some also believe that they oversimplify cause–effect relationships in the cognitive dynamic. Later *strong* models, which are represented as closed-loop models, recognize the reciprocity of actor and environment, much like Gibson and Neisser's views of perception and action (Gibson, 1979/2014; Neisser, 1976) and the dynamics of cognition. They represent sets of processes that are continuous, parallel, highly interacting that represent the dynamics of cognition as a *stream of events*.

Finally, and most importantly for the purposes of the *Handbook*, we must address the role of experience and expertise in the people who are planning, making decisions, making sense of situations, re-planning, and so forth in their work contexts. In many cases, models of macrocognition are, by definition, models of expert cognition. Expertise and skilled performance are viewed as the gold standard for cognitive performance and thus many of the models represent expert cognition. The reason that this is important is that experts make sense, decisions, and plans in the context of work-based situations. To do so successfully, they must have developed a refined sense of the cues and factors that contribute to their thinking, and developed sophisticated mental models of how their world works to allow them to understand and predict situations effectively. Macrocognitive models reflect this actor–environment relationship in terms of how experts bring their experience, mental models, and knowledge to bear on complex problems to effect satisfactory outcomes.

Macrocognitive models describe how knowledge and experience contribute to effective adaptation to the complexity of scenarios which impact performance outcomes. This is in contrast to many *traditional* cognitive models which fail to account for these adaptation and complexity challenges because they have been generated based on relative novices working on tasks that are new to them (e.g., tower of Hanoi problem solving, or gamble or choice decision tasks).

Thus far we have only hinted at the focus of macrocognitive models. It should be noted that the variety of macrocognitive models has continued to evolve and grow over the past 30 years, and thus the framework presented in Figure 9.1 for understanding macrocognition represents only a snapshot. Figure 9.1 represents one version of the macrocognitive functions and processes as described in Klein et al., (2003), providing the reader with a sense of the language, scope, and focus of macrocognitive models.

The functions represented in Figure 9.1 are not a comprehensive list, nor is it complete. Iterations of current models are continuously being revised and updated based on new research. Crandall, Klein, and Hoffman (2006) provide brief descriptions of each of these functions and processes (pp. 137–142). Three example models will be presented in more depth later in the chapter.

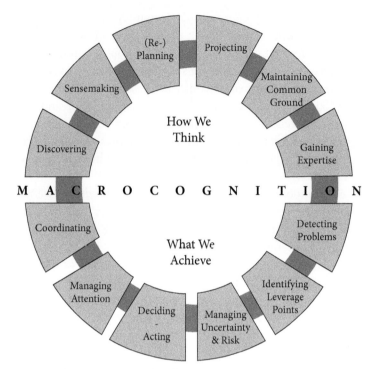

FIGURE 9.1 The macrocognitive "wheel" which effectively provides a list of the various macrocognitive processes and functions. The processes describe how we think, whereas the functions describe what we achieve.

© 2003 IEEE. Adapted, with permission, from G. Klein, K. G. Ross, B. M. Moon, D. E. Klein, R. R. Hoffman, and E. Hollnagel, "Macrocognition," *IEEE Intelligent System*, 18 (3), p. 83, Figure 1, doi: 10.1109/MIS.2003.1200735.

Theoretical Foundations

Macrocognition has broadened its scope from an initial focus on decision making, to addressing the span of cognitive performance (for individuals and teams). This means that it is comparing itself with a variety of *traditional* models including decision making, problem solving, memory, attention, levels of processing, learning and transfer. This chapter will focus on three macrocognitive models (recognition-primed decision making; sensemaking; and flexecution) which can be contrasted with *traditional* cognitive psychology models, especially of decision making (classical judgment and decision theorists), comprehension, and problem solving. However, it is difficult to pinpoint specific targets for comparison as macrocognitive models draw different boundaries around the activities and typically incorporate several components of what would traditionally be the realm of a single microcognitive model. The comparisons are therefore not like-for-like as a single macrocognitive model could be reliant on the *building blocks* of multiple microcognitive models. In some cases, because the unit of analysis for macrocognitive models tends to be at the level of *cognitive work* there are rarely close direct corollaries within the cognitive psychology models of the 1970s and 1980s. Klein et al. (2003) described the situation in the following way using the example of the recognition-primed decision model which will be described in more detail next:

> After considerable research on recognition-primed decision making, we realized that the model was basically a combination of three decision heuristics that had already been well-studied from the microcognition perspective: availability and representativeness to identify the typical course of action, and the simulation heuristic to evaluate the course of action. Therefore, in this case it was possible to trace the macrocognitive phenomenon back to hypothetical microcognitive components. However, several decades of research on the availability, representativeness, and simulation heuristics had not led to a discovery of recognitional decision making. That is why we see the macrocognitive functions as emergent. We discover them by investigating cognition in field settings rather than by continually pursuing explanations of lab findings. (p. 82)

On the whole, traditional cognitive models of cognition are fundamentally lacking in accounts of experienced people (but see other chapters in this section, including Chapter 2, "The Classic Expertise Approach and Its Evolution," by Gobet) who use their domain knowledge and experience to solve real work problems characterized by time pressure, uncertainty, complexity (emergent problems), and a variety of external constraints which impact cognitive performance. The macrocognitive models are derived from studying exactly these kinds of problems, and provide insights into how people achieve effective cognitive performance under *messy* conditions.

Methodological Foundations

In addition to the theoretical underpinnings, it is also useful to understand the methodological approaches that have provided the predominant source of data for the development of macrocognitive models. Fundamentally, the methods used are cognitive field research methods. In order to understand the influences of experience and expertise and of environmental factors on human cognition in real-world contexts, researchers have had to develop and refine methods for studying cognition outside the laboratory. These are broadly described as cognitive task analysis (CTA) methods (see Hoffman & Militello, 2009 for an excellent treatment of the methodological underpinnings of CTA).

Data gathering to understand the nature of participant experience, knowledge, and expertise, and to understand the environmental opportunities and constraints typically involve a triangulation of methods including:

1. Examination of documentation of the work, including training manuals, procedure guides, doctrine and policies, and so forth;
2. Observation of workers in action, working with tools, artifacts, and other people, and struggling with the realities of complex work;
3. *Think aloud* protocols from workers as they are doing their work (see Chapter 17, "A Historical Perspective on Introspection," by Ward et al., this volume);
4. Semi-structured interviews with workers about the challenges and complexities of their work contexts, with a particular emphasis on *tough cases* (see Chapter 19, Incident-Based Methods for Studying Expertise," by Militello and Anders, this volume); and
5. Modeling of work practice, with review and feedback by a variety of workers and subject matter experts (see Chapter 16, "Studying Expert Behavior in Sociotechnical Systems: Hierarchical Task Analysis," by Salmon et al.; Chapter 20, "Cognitive Work Analysis," by Burns; and Chapter 21, "Reflections on the Professional Practice of Knowledge Capture," by Moon, this volume).

The macrocognitive model development process has often been instigated by a practical, applied question that researchers have been unable to answer based on existing cognitive models, concepts, or theories. These applied questions have stimulated enquiry using a variety of the methods identified above (e.g., see Crandall, Klein, & Hoffman, 2006 for one useful guide to conducting CTA; for an overview of the full breadth of other cognitive engineering methods and their applications, see also Lee & Kirlik, 2013). A review of the large variety of methods that have contributed evidence to the development of these models is beyond the scope of this chapter, but the reader is encouraged to look at Schraagen, Chipman, and Shalin (2000) and Hoffman and Militello (2009) for reviews of the various perspectives, both theoretical and methodological, that have made valuable contributions. The next section provides three examples of macrocognitive models in order to provide illustrations of the character, content, and focus of these models.

Three Macrocognitive Models

Three models will be described here for purposes of illustration. These models represent examples of how the concept of macrocognition has evolved over time. The first model is Klein's recognition-primed decision (RPD) model (Klein, Calderwood, and Clinton-Cirocco, 1986; Klein, 1997; 1999). The second is a model that complements one aspect of the RPD model which represents a *diagnosis* loop that has since been elaborated into the data–frame model of *sensemaking* (Klein, Phillips, Rall, & Peluso, 2007; Sieck, Klein, Peluso, Smith, & Harris-Thompson, 2007). The final model is the *flexecution* model of adaptive replanning (Klein, 2007a, 2007b).

Recognition-Primed Decision Model

The RPD model is probably the oldest and best known model that has emerged from the naturalistic decision making community, a community of practice who have pioneered the macrocognitive perspective. First described by Klein et al. (1986), the purpose of the study that led to the development of the model was to examine the ways decisions are made by highly proficient personnel, under conditions of extreme time pressure, and where the consequences of the decisions could affect lives and property. The study was conducted on behalf of the US Army who wanted a better explanation of how military commanders could make effective decisions given the characteristics of operations including time pressure, information uncertainty, and high stakes. The firefighting domain was chosen as a surrogate for military decision making given the difficulty of observing military commanders in action. Observations and interviews were conducted with experienced fire ground commanders (FGCs) who are responsible for allocating personnel and resources at the scene of a fire. The interviews focused on the work challenges, and particularly tough cases where the FGC's expertise was challenged.

The original intent of Klein's study was to understand how FGCs identified options and selected courses of action from amongst those options. However, based on an analysis of 156 decision points, the researchers found that only 12 percent of the decisions discussed included any sort of option comparison, but that 80 percent of the examples were resolved based on matching current experienced situations to similar situations from past experience. The course of action (COA) that worked before was implemented again (Klein, Calderwood, & Clinton-Cirocco, 1988). This study provided the initial evidence for the development of a recognition-primed strategy for effective decision making that contradicted many of the classical normative rational decision models that had been proposed to date (e.g., see Hastie & Dawes, 2001/2010).

Evidence for the RPD Model

There is a growing amount of empirical evidence that supports RPD's descriptive account of the way that experienced people make decisions. Although challenging to test empirically, there are some assertions upon which RPD relies and for which empirical support has been found:

1. People can use experience to generate a plausible option as the first one they consider (Klein, Wolf, Militello, & Zsambok, 1995).
2. Time pressure need not cripple experienced decision makers (Calderwood, Klein & Crandall, 1988).
3. Experienced decision makers can adopt a course of action without comparing and contrasting possible courses of action (Driskell, Salas, & Hall, 1994; Kaempf, Klein, Thordsen, & Wolf, 1996; Mosier, 1991; Pascual & Henderson, 1997; Randel, Pugh, Reed, Schuler, & Wyman, 1996).

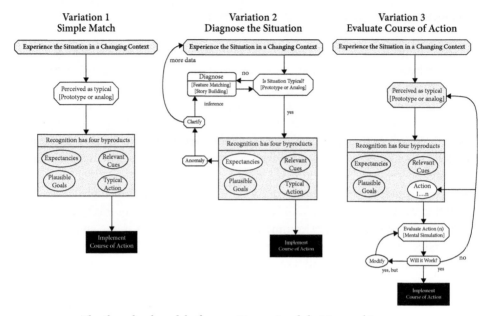

FIGURE 9.2 The three-level model of recognition-primed decision making.

Reproduced from Gary Klein, "The recognition-primed decision (RPD) model: Looking back, looking forward," in Caroline E. Zsambok and Gary Klein, Eds, *Naturalistic decision making*, pp. 285–292 © 1996, Taylor and Francis Group.

The RPD Model

The RPD model is illustrated in Figure 9.2. It was developed to describe and explain how experienced workers made effective decisions (i.e., the outcome was satisfactory in the context) using their knowledge and experience. Figure 9.2 presents three variations of the RPD model.

SIMPLE MATCH

In the left-hand panel of Figure 9.2, the *simple match* version of RPD is presented, which is akin to a simple pattern matching strategy or heuristic. The middle panel represents the *diagnosis* variation which is triggered by the detection of an anomaly in the situation compared to previous experiences, often a violation of expected cues, patterns, or trajectory of a situation.

DIAGNOSE SITUATION

This *diagnosis* variation of the RPD model describes a situation where a simple match is not identified or is called into question because the situation assessment is not clear or an *anomaly* has been detected with respect to violated expectations about the situation. This variation of the RPD model has since been further developed in the form of the data–frame sensemaking model which is described later in this chapter.

EVALUATE COURSE OF ACTION (COA).

The final panel represents the *COA evaluation* variation. This is an instance where a situation is recognized as typical, eliciting an initial candidate course of action. However, on initial consideration of the preferred course of action a potential anomaly occurs requiring the COA to be evaluated further. This analysis of the COA is conducted by a mental simulation whereby the decision maker conducts a mental walkthrough of how the candidate COA might play out given the current situation and the decision maker's mental models of situational factors and dynamics from experience. If the COA is considered workable, it might be implemented as is. However, if a deviation or anomaly is detected during the mental simulation, the COA might be tweaked or the decision maker's understanding of the situation may be sufficiently altered to warrant a different assessment and thus a different candidate COA. This process is described more fully in Klein and Crandall's model of mental simulation (Klein & Crandall, 1995).

The first variation of the RPD model has more recently been described in the context of *intuitive* decision making because it describes how people with expertise leverage their tacit knowledge and perceptual skills to make effective decisions without apparent conscious deliberation, hence appearing to use their *gut* or intuition (Klein, 1999; 2004). Kahneman and Klein (2009) provide a compelling description of the conditions required for developing expertise and the strengths and limitations of intuition in different decision-making contexts based on their different perspectives on skilled decision making.

The RPD model is often mistaken to be merely a pattern-matching process for decision making using implicit production rules: if situation X, then course of action Y. Although this does describe the simple match version of RPD there are also nuances in the other variations of the model that describe more conscious analytical processes for understanding anomalies and violated expectancies. RPD is not solely an automatic pattern matching process. The full model (including the variants described earlier) incorporate analytical resources and deliberation. However, the key distinction from deliberation in the rational models, and one of the key insights of the RPD model, is that the analyses are of the situations, not alternative courses of action. In addition, those deliberations occur serially rather than in parallel, against a criterion of satisficing as opposed to optimizing (cf. Kahneman & Klein, 2009; Simon, 1972).

The RPD model presents an empirically grounded description of rapid decision making when deliberate comparison of COAs is impractical and/or inappropriate. Hammond (1988) posited the cognitive continuum theory to address a spectrum of judgment and decision making from intuitive to analytical. More recently this has been characterized by Stanovich and West (2000), and popularized by Kahneman (2011) as System 1 and System 2, where System 1 refers to models like the first variant of the RPD model (simple match; intuitive), and System 2 refers to the more deliberate analytical models (e.g., Kahneman, 2011). The RPD model has characteristics of both System 1

(RPD Variation 1) and System 2 thinking (RPD Variations 2 and 3) (Kahneman & Klein, 2009).

Limitations

The RPD model is not without its critics, and there are doubtless aspects of the model that could be expanded upon, amended, or expanded. The second variation *diagnose situation* loop (Figure 9.2) was added to the RPD model subsequent to its original description, and this chapter will make the further assertion that the data–frame model of sensemaking takes this elaboration a step further (although this link has not previously been made explicit). Klein also suggests that processes for option identification or option generation might be aspects that are still missing from the current model (Klein & Wolf, 1998).

It should also be noted that this is a model of expert decision making, requiring knowledge and experience and a repertoire of situation models in order to apply what has also been referred to as the recognitional heuristic approach to decision making. In truly novel situations RPD potentially breaks down; however, research on experts has illustrated that they are still able to generate workable solutions even in the face of novelty (for a review, see Ward, Gore, Hutton, Conway, & Hoffman, 2018), suggesting that there is more to expert decision making than is described in the RPD model (this challenge rests on the definition of novelty and whether the situation falls within the domain of expertise; the notion of adaptive expertise is also explored further, see Chapter 10, "Cognitive Systems Engineering," by Naikar and Brady; Chapter 12, "Adaptive Expertise," by Bohle Carbonell & van Merrienboer; and Chapter 50, "The Future of Expertise in a Digital Society," by Woods & Cook, this volume).

Furthermore, it could be argued that a more comprehensive model of cognition could have been developed based on an integration of a number of macrocognitive models with the RPD as the core framework on which the additional processes hang. This is also arguably a limitation of the RPD model and the macrocognitive models more broadly, that there is a gap with respect to a more coherent overarching theoretical treatment of macrocognitive phenomena. We will revisit this issue in the final section of this chapter when we examine a more recent effort to consolidate across the macrocognitive landscape into a more coherent and integrated perspective which is intended to provide a macrocognitive model of cognitive work more broadly.

Applications

The recognition that RPD is actually broader than a model of decision making only, and more broadly as a model of *cognitive work*, has been evident in its application in a variety of what are referred to as *decision-centered* approaches to supporting improved decision/cognitive performance (Klein, 1993). This is evidently true based on the implicit conception in the naturalistic decision making (NDM) community that decision making is part of a cyclical perception–decision–action cycle rather than merely a choice point. However, human factors applications of RPD theory have inevitably strayed into decision support and cognitive work support through a broad variety of applications including visualization, support to situation assessment and situation awareness development, maintenance and recovery, training in situational dynamics, and building better mental models.

For example, the US military (US Army and US Marine Corps) revised their military decision making, planning, and command and control doctrines to recognize the role of rapid, *intuitive* decision-making processes, particularly in crisis or time-pressured situations. Various fire services (e.g., US, UK, and The Netherlands) have adopted rapid decision making as part of their incident command doctrine and processes, making it an explicit part of training curricula. A number of decision-centered training programs has been developed and tested utilizing implications from the RPD model to focus training on decision-making performance (see Phillips, Klein, & Sieck, 2004; Klein, 2004). With regard to designing engineered systems, decision-centered design (DCD; Hutton, Miller, & Thordsen, 2003) adopts methods and models for requirements capture and early concept design that are grounded in an RPD understanding of decision making and how it should be supported by technological applications. Fundamentally, engineered designs should rely on a clear understanding of the cognitive work that needs to be supported. Inevitably RPD, along with other macrocognitive models, provides a way of describing and understanding the key decision requirements and more broadly the cognitive work requirements.

Data–Frame Model of Sensemaking

The data–frame (DF) model of sensemaking represents an evolution in thinking from the RPD model as described earlier. It was designed partly to unpack the *diagnosis* loop of the RPD model which provided only *story-building* and *pattern matching* as processes and strategies for resolving anomalies in understanding evolving situations. The DF model also emerged as a natural progression from exploring commitments to a course of action to elaborating the role of situation understanding. It also satisfied the discovery that many of the *decisions* that were being unearthed by NDM researchers were actually related to assessments rather than courses of action.

Much had been written about the role of situation awareness in human–system performance and the role of assessment of ongoing situations (Endsley, 1995). The DF model of sensemaking was developed to address the cognitive activity that resulted from the emergence of a surprise or anomaly in the decision maker's understanding of the current situation. It was not originally intended to describe how people make sense of ongoing situations that are evolving as expected, or in normal circumstances. The critical driver for the DF model was the response to a sudden realization that what the decision maker thought was happening in the world was actually not the case. No psychological models previously addressed this process of making sense of a situation following a surprise. It is easy to see the progression from the RPD model where an anomaly triggers the *diagnosis* loop of the model. However, in order to provide the theoretical challenge with appropriate attention and focus, the DF model was developed independently of the RPD model.

Other theoretical treatments of sensemaking exist, most notably Karl Weick's work on organizational sensemaking (1993; 1995). Weick's work focuses on a larger scale of sensemaking in organizations and is mostly used in the context of post-hoc explanations of significant events (e.g., the Mann Gulch disaster, Weick, 1993). The focus of the DF model is on individual sensemaking, although further elaborations have been suggested with respect to the coordination of sensemaking across a team (Klein, Wiggins, & Dominguez, 2010; Hutton, Attfield, Wiggins, McGuinness, & Wong, 2012). The next section describes the DF model of sensemaking.

The Data–Frame Model

For a detailed description of the DF model of sensemaking, the reader is pointed to Sieck et al. (2007) and Klein, Phillips, Rall, and Peluso (2007). Figure 9.3 presents the DF model which depicts four core aspects, the data–frame relationship, a *questioning* the frame process, an *elaboration* cycle, and a *re-framing* cycle. Sensemaking is defined as the deliberate effort to understand events and is typically triggered by unexpected changes or surprises that make us doubt our prior understanding. Sensemaking as a process serves several cognitive functions. It supports our ability to detect problems and to focus attention on problematic features of a situation. It also supports making new discoveries and generating insights. It provides us with a way to form explanations about how the current situation came about, but also to anticipate how the situation will evolve in the future. It supports the identification of levers for action by helping to identify the critical causal and influential factors in a situation. Sensemaking helps identify critical relationships between cues and factors that support our explanations and expectancies. Finally, sensemaking enables problem identification (i.e., diagnosis) that supports our understanding of the critical cues and factors that might suggest a solution strategy.

Figure 9.3 illustrates the key aspects of how the sensemaking process works. The sensemaking process describes a data–frame relationship which provides an initial account that people generate to explain events based on the current data (bottom-up) in conjunction with some organizing frame (top-down). It then supports the elaboration of that account in terms of adding detail, accounting for more information available about the situation, and suggesting additional aspects of the situation based on the current frame. When faced with inconsistent data, the sensemaking process requires a questioning process that allows a person to challenge the current assessment. However, there is a tendency to fixate on the initial account for which the model of the sensemaking process must take account. The process supports the discovery of inadequacies in the initial account which must then be addressed with respect to comparisons of alternative accounts, and/or re-framing of the initial account, replacing it with an alternative that either is recognized from previous experience or must be deliberately constructed. This is a description of the sensemaking process at a high level based on evidence from a variety of examples from real-world decision contexts including

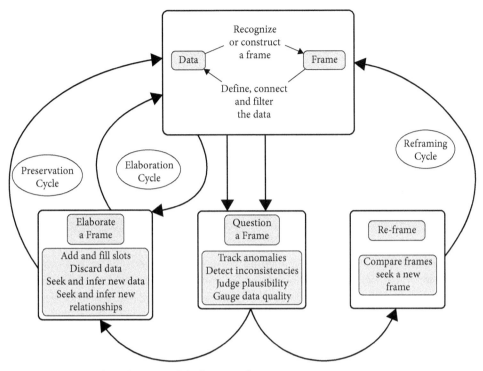

FIGURE 9.3 The data–frame model of sensemaking.
Adapted from Klein, G., Phillips, J. K., Rall, E. L., & Peluso, D. A., "A data-frame theory of sensemaking," in R. R. Hoffman, Ed., *Expertise out of context. Proceedings of the sixth international conference on naturalistic decision making*, pp. 113–155, © 2007, Taylor and Francis Group.

military operations, medical scenarios, and business and firefighting examples (Sieck et al., 2007).

The DF model postulates that elements of a situation, based on available data, are explained when they are fitted into a structure that links them to other elements. The concept of a *frame* is used to denote an explanatory structure that defines entities by describing their relationships to other entities. A frame is a structure for accounting for the data as well as guiding the search for more data. The frame could be based on a person's compiled experiences or it could be represented by a narrative or a physical structure, such as a map, that is used to piece together the existing data and provide the connections and relationships between the available data, as well as filling in the gap (inferences) and revealing gaps (future data collection requirements).

Evidence for the Data–Frame Model of Sensemaking

The empirical data which were used to develop this model came primarily from studies in a defense context. Information operations (IO) officers provide one source of data based on the completion of challenging assessment scenarios which were then used as a basis for CTA activities. Another source of data came from people making in situ navigational assessments, particularly in a situation where they were trying to recover from getting lost (Sieck et al., 2007). Finally, a retrospective analysis of a corpus of critical decision data was reviewed in order to identify interviews which focused on challenging decisions relating to assessments of situations and sensemaking, rather than on developing and evaluating course of action option.

The study with IO officers looked at expert/novice differences (Sieck et al., 2007). Officers were presented with a number of situation reports, and supporting materials, relating to complex military scenarios. They were interviewed with respect to their assessments of an evolving situation, about judgments and inferences relating to understanding the current situation, speculations and explanations for events, and the knowledge and previous experiences which supported those inferences, speculations, and explanations.

Independent coding of the interview protocols revealed a number of processes and strategies that were used to make sense of the situations, including active exploration of connections between reports (including reports which were included as *noise*). Experts generated more connections between the reports based on richer mental models developed from their experience. Both groups used similar sensemaking strategies, including inferring causes, effects, and their relationships, as well as inferring causes from effects, having an awareness of multiple causes, and identification of instances where the cause resulted in an unexpected effect. Experts were able to identify more associations between multiple causes and effects than novices, and more implications with respect to anticipating events and taking actions to develop their understanding.

Further interviews were conducted with IO officers using interview methods relying on the elicitation of retrospective incidents and generated a number of sensemaking incidents which were analyzed in depth (Sieck et al., 2007), contributing to an elaboration of the sensemaking model. Data were also collected relating to the corruption and recovery of sensemaking during real-life navigational experiences of getting lost and *getting found* again, which supported the initial versions of the data–frame model. Finally, Sieck et al. reviewed archival data from a number of projects to generate further evidence of sensemaking based on CTA interviews of past critical incidents. These data included incident examples from firefighting, neo-natal intensive care nurses, naval operations room (combat information centre) teams, and a number of small unit army commanders.

Key Features

The DF model is based on some key assertions relating to the processes of sensemaking.

RECIPROCAL DATA–FRAME RELATIONSHIP.

Firstly, sensemaking is the process of fitting data into a frame whilst also fitting a frame around the data. It is both a top-down and bottom-up process. This is critical with respect to how the model deviates from other models where there is often an assumption of one-way processing. The data are the interpreted signals of events. Frames are the explanatory structures that account for the data. Sensemaking balances these two entities; neither has primacy.

IT'S NOT JUST CONNECTING THE DOTS.

The notion of the *cognitive hierarchy/pyramid* or waterfall models of sensemaking suggests that more data generate information which is transformed into knowledge, and finally leads to understanding (for example, Army Field Manual 6-0 Mission Command: Command and Control of Army Forces; Appendix B: *Information*). This sort of thinking promotes the idea of sensemaking in a military intelligence context as *connecting the dots*; however, the DF model highlights the idea that a dot is defined in part by the frame with which the analyst is making sense of the situation. Thus, *connecting the dots* only makes sense in hindsight, because what counts as a dot is often hard to discern in the noise, uncertainty, and complexity of real-world scenarios (Klein, 2011, Ch. 12). Data elements are not perfect representations of the world but are constructed. Different people viewing the same events can perceive and recall different things, depending on their goals and experiences. The identification of what counts as data depends on the background experience and on the repertoire of frames.

DATA AND FRAMES ARE INFERRED THROUGH ABDUCTIVE REASONING.

Likewise, frames are influenced by the information that is available. Seeing the data–frame relationship as reciprocal presents a number of challenges for the sensemaker, but also explains some of the complexities of interpretation and assessment of data. Data are not seen as primitives in the DF model, rather as inferred based on the current frame. Likewise, the frame is inferred from a few key anchors in the situation. The sensemaking model relies on inference as a key mechanism for understanding; however, it relies on inference to the best explanation by abductive reasoning rather than the formal logics of inductive and deductive reasoning (Bennett & Flach, 2011, pp. 33–34; Klein, Phillips et al., 2007).

DATA–FRAME CONGRUENCE STOP SENSEMAKING.

One of the assertions from the model is that sensemaking will cease once the data and frame are brought into congruence; that is, the data fit the frame and the frame fits the data, without any anomalies or inconsistencies. It is not an endless effort to generate more and more inferences.

EXPERTS AND NOVICES DIFFER BY CONTENT NOT PROCESS.

With respect to expert/novice differences in sensemaking, Klein et al. (2007) and Sieck et al. (2007) found evidence to suggest that experts and novices reason about situations in the same way, but the experts have a richer repertoire of frames that support the sensemaking process, allowing them to perform at a higher level.

SENSEMAKING IS PRAGMATIC RATHER THAN RATIONAL.

Sensemaking is used to achieve a functional understanding and is therefore evaluated against effectiveness in supporting action and correspondence with respect to matching external events, rather than being evaluated against abstract standards of internal consistency and coherence of formal logical reasoning. Sensemakers want to know what can be accomplished and how capabilities can be expanded, which requires the application of understanding of available resources and action capabilities.

PEOPLE PRIMARILY RELY ON JUST-IN-TIME MENTAL MODELS.

The evidence from research by Sieck et al. (2007) suggested that the frames that people use can be external artifacts or representations as well as internal mental models that reflect causal understanding. Frames can take the form of mental models, stories, scripts, maps, and so forth; anything that provides a structure. These authors found that their use of internal representations, or mental models, was opportunistic and pragmatic with respect to their reliance on fragmentary mental models that were generated as needed, and which supported the immediate need for inference, rather than requiring complete and accurate models of the world.

SENSEMAKING TAKES DIFFERENT FORMS, EACH WITH THEIR OWN DYNAMICS.

The final assertion is that sensemaking (represented by the different bubbles in Figure 9.3) takes several different forms, each of which has its own dynamics, and thus which might require different forms of support if we were to propose interventions or tools to improve sensemaking. The key dynamics relate to *connecting data and frame, questioning a frame, elaborating a frame, preserving a frame, re-framing,* and *constructing or finding a new frame.*

Limitations

The sensemaking model was first generated in the early 2000s based on evidence generated from several tailored research studies (Sieck et al., 2007) as well as review of past CTA research data, which was used to identify examples of sensemaking from projects that were not necessarily designed to explore the sensemaking activities. There have been limited validations of this model since the original work, and the assertions made about how people make sense of situations largely draw on related models from other areas of psychological research (e.g., in the areas of reasoning, schema, and so forth). The model therefore represents an attempt to integrate the cognitive field research evidence with existing models of microcognition related to sensemaking into a coherent account of the macrocognitive challenges of sensemaking in complex scenarios.

Applications

The sensemaking models have been used to explore applications that support a variety of sensemaking activities including military UAV command and control (Klein et al.,

2004) and military signals intelligence analysis (Attfield et al., 2015; Blackford et al., 2015), and to provide general design guidance and principles for human computer interfaces (Hutton, Klein, & Wiggins, 2008). In addition, the DF model has been used to provide a framework to support the development of a technique for evaluating technologies intended to support collaborative analysis tasks, or team sensemaking tasks (Hutton et al., 2012). The sensemaking model supports one approach to a broader set of approaches which fall under the banner of cognitive systems engineering (Blackford, Bessell, Hutton, & Harmer, 2017; Hutton et al., 2003; Militello & Klein, 2013) and posits an explicit approach to understanding and designing to support sensemaking challenges. In addition, the DF model has been used to develop decision skills training with an emphasis on situation assessment and understanding components of a task (Phillips, Baxter, & Harris, 2003).

A Flexecution Model of Replanning & Adaptation

The final model used to illustrate macrocognition is the flexecution model of replanning (Klein 2007a; 2007b). The flexecution model was developed as a model of replanning and adaptation in the face of complex problems. It is derived from the term *flexible execution* where execution refers to activities conducted to achieve objectives based on some sort of plan. Complex problems are characterized by emergent and unpredictable challenges which render plans inappropriate with respect to the methods or courses of action being employed to achieve an objective, or with respect to the objectives themselves. In 1978, Klein and Weitzenfeld identified the challenges of problem solving in ill-structured problem scenarios. They identified a number of drivers and variations on how goals must change in response to emergent features of a problem situation. Subsequently, research into adaptive teams (e.g., Klein & Pierce, 2001) and in planning and replanning (e.g. Klein, 1996; Klein, Wiggins, & Schmitt, 1999) identified the limitations of a *management by objectives* approach to problem solving and planning. Management by objectives assumed static, specifiable objectives at the beginning of the planning process, and implied that replanning was limited to changes in methods or courses of action in pursuit of the same goals. However, the flexecution model evolved as a solution to a *management by discovery* approach (Klein, 2011) where both courses of action and goals need to change in order to meet operational demands of obsolete, conflicting, and/or emergent goals.

Flexecution Model

The flexecution model was derived from observations and systematic analysis of planning teams, primarily in military and emergency response domains. It was proposed as a means for describing how planners adapt to unforeseen circumstances as well as redefining goals during the operational phase of executing a plan, based on what is being learned as a plan is being executed. The need to replan and to clarify goals during an operation is rarely addressed in psychological descriptions of problem solving and planning (e.g., Hayes-Roth & Hayes-Roth, 1979). The work required to simultaneously achieve goals as well as define and redefine those goals is the focus of the flexecution model.

The flexecution model views goals as holding multiple simultaneous characteristics and serving multiple functions. Some goals are seen as foreground and providing the initial stimulus and objective for action. However, other goals, including individual and organizational values which are a source of often tacit objectives, remain in the background until the situation forces them into the foreground. The plan is formulated based on combinations of actions and resources to achieve the foreground goals often based on leverage points which provide additional value as the action provides a disproportionate positive influence on the outcome. However, emergent goals require the juggling of goals (between foreground and background), assessment and management of goal conflicts, and management of the inevitable trade-offs created by those conflicts.

The flexecution model recognizes that goals change as actions unfold, and that plans must be flexible with respect to their ways and means (i.e., methods and resources for achieving an objective) and their ends (i.e., the objectives themselves). Flexecution is a model of adaptive planning or re-planning, or *planning in-stride* (i.e., after a plan has been developed and communicated to those responsible for executing the plan). In a similar fashion to the processes associated with sensemaking, replanning is a continuous, closed-loop activity where the actions and objectives identified to meet the operational requirement are continuously evaluated. Inconsistencies or anomalies must be detected, validated, and understood with respect to the impact on meeting the overall intent of the actions. Plans (including both methods and goals) can be adjusted and elaborated in order to improve performance. However, sometimes goals must be *reframed* in terms of the level of aspiration for success, changing priorities, adding new goal properties or deleting/refining existing goal properties, or even identifying new goals in order to achieve the higher-level intent. This is a continuous process that must be recognized and supported in order to allow decision makers to *muddle through* complex problem spaces and achieve acceptable levels of success in the face of emergent challenges which prohibit *optimization* or specification of goals at the outset. In complex environments, rigid plans with pre-specified goals and no support or leeway to adjust performance mid-stream lead to brittle plans and ultimately failure to achieve the higher-level intent and operational requirement.

With reference to Figure 9.4, Klein remarks of the flexecution process:

> Because the goals are dynamic or ill defined, people need to act to learn more, taking their best understanding and pressing forward from there. Figure 9.1 [Figure 9.4] isn't a flowchart in any traditional sense, because the macrocognitive events it depicts are largely parallel or simultaneous. Instead, it illustrates the different kinds of pathways for elaborating or reframing goals. (Klein, 2007b)

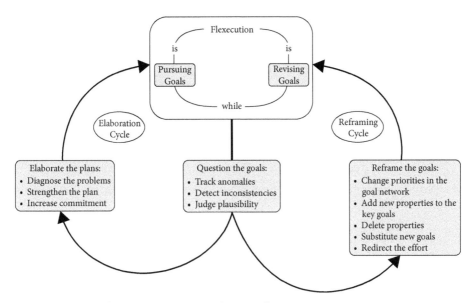

FIGURE 9.4 The flexecution process: replanning during execution.

Copyright © 2007, IEEE. Reprinted, with permission, from Gary Klein, Flexecution, Part 2: Understanding and supporting flexible execution, *IEEE Intelligent Systems Magazine* 22(6), 108–112, 2007.

Evidence for the Flexecution Model

The evidence for this model has, for the most part, come from a small group of researchers working primarily in military planning. The model emerged from work underscoring the challenges of wicked problems (Brehmer, 2005; Dorner, 1996; Klein, 1996; Klein & Weitzenfeld, 1978; Rittel & Webber, 1984) and field research and analysis based on exercises using observations and interviews with military planning teams (Klein & Miller, 1999; Klein et al., 1996; Klein et al., 2000; Ross, Klein, Thunholm, Schmitt, & Baxter, 2003; Ross, Thunholm, et al., 2003b; Schmitt & Klein, 1999). The pre-cursor to the flexecution model, the recognitional planning model, also fueled work by others to look at rapid replanning in complex operational environments (Cheah et al., 2005; Thunholm, 2005).

The biggest challenge to evaluating and validating aspects of the flexecution model lie in the need for decision-making environments that are complex enough to generate the cognitive demands on both the planners and the operators who execute those plans. Simulation and experimental paradigms often fail to provide either the levels of complexity and drivers of adaptation, and/or the requirement to execute a plan for long enough to generate feedback on progress and the demand signals for adaptation. This has hindered the generation of empirical evidence and evaluation of models of replanning, such as the flexecution model.

Limitations

The key limitation associated with this model is its lack of external empirical validation. It is primarily a descriptive model that captures characteristics of real-world operations in complex working environments. It has been used to sensitize planners to the challenges of planning and conducting operations in complex environments through education and training interventions in military command and staff colleges. In addition, there have been suggestions made with respect to implications for organizational planning processes and to the design and engineering of planning tools (e.g., Cheah et al., 2005; Klein & Miller,1999; Thunholm, 2005). However, there are few explicit evaluations of the flexecution model or testing of hypotheses that might be generated from the model.

Applications

In terms of implications and applications in real-world settings, as described earlier, the flexecution model has been used for purposes of education and training in a military context. In addition, there are implications and recommendations for doctrine, processes, and operating procedures which have been suggested (e.g. Hoffman & Shattuck, 2006; Ross, Klein, Thunholm, Schmitt, & Baxter, 2004; Schmitt & Klein, 1999; Thunholm, 2005). Likewise, there are implications for designing software tools to support planning and execution in complex operational environments, but there are no fielded examples nor rigorous evaluation studies.

FUTURE DIRECTIONS

The variety of macrocognitive models has evolved over time and continues to evolve and mature. Early models of cognition in context were more like information-processing models; however, more recent efforts have attempted to capture the continuous nature of thought and purposeful cognition in closed-loop macrocognitive models. What counts as a macrocognitive model is as much about the context in which cognition occurs as it is about what is going on inside the head (see Flach & Warren, 1995). Critically, macrocognitive models represent purposeful activity, based on experience, in context, sometimes distributed, often studied using cognitive field research methods.

Until recently, models of macrocognition have tended to develop as a series of related but not integrated models. As identified previously there are implied connections between the various macrocognitive models such as the diagnosis loop variation of the RPD model and the data–frame sensemaking model, as well as other examples of more elaborated subcomponents of models such as the mental simulation model (Klein & Crandall, 1995), which represents the COA evaluation loop of the RPD model. A case could be made that the RPD model actually represents a framework from which many of the macrocognitive models might hang, given the level of interrelatedness and complementarity.

More recently Hoffman (2013; Hoffman, & Hancock, 2017) has taken the macrocognitive modeling challenge a step further by proposing an integrated model of macrocognitive work in the context of sociotechnical system performance and particularly the issue of trust. Hoffman took the data–frame model of sensemaking and developed a revised version of the flexecution model of replanning (Klein, 2007b) in order to provide an isomorphic representation of the two models before integrating them into an integrated model of macrocognitive work. Hoffman describes the marriage thus:

> The closed loop at the top (*of the flexecution model*) is the counterpart to the topmost closed loop in the D/F model. Likewise, the other loops in the Flexecution model are counterparts to those in the D/F model. The two conceptual models are cut from the same cloth, one describing how people make sense of complex situations, and the other describing how people act on the basis of their understanding. (Hoffman, 2013, p. 24)

Figure 9.5 provides a representation of this integrated model of sensemaking and flexecution to describe adaptive cognitive performance.

Hoffman's integrated model is an attempt to reconcile two separate models originally developed in isolation. Hoffman suggested applications of this sort of modeling effort in the context of computational modeling and empirical validation of evidence for the model. He also suggested that this model could inform research efforts focused on cognitive-process tracing, such as verbal and introspective-type reporting techniques, and other *cognitive field research* methods, by providing potential coding themes against which to assess and understand elicited protocols (Hoffman, 2013; see Chapter 17, "A Historical Perspective on Introspection," by Ward et al., this volume). Developing unifying models of macrocognition is a valuable effort to tie together the conceptual framework for understanding cognitive work; however, there remains a challenge with respect to the evidence base available to validate these models, and the continued challenges of testing these models in order to generate that evidence base. Despite being in the open literature for over 10 years for example, the flexecution model has received limited feedback and challenge by the scientific community.

The diversity of models is arguably a result of the requisite variety of complex problems and applied challenges in the field, and so, it is no surprise that this has resulted in a diverse set of models. However, in researching this chapter it also became clear that there is not a recognized set of criteria for what counts as a macrocognitive model. In this chapter, I presented one perspective in an attempt to provide a coherent story about the extant set of macrocognitive models and to provide the reader with a way to make sense of this diversity. The remaining challenge is to extend these discussions by clarifying what constitutes a macrocognitive model, determining how to develop valid and useful macrocognitive models, and specifying how they differ in meaningful ways from other models that represent complex human–machine and human–context interactions (such as models of distributed and situated cognition).

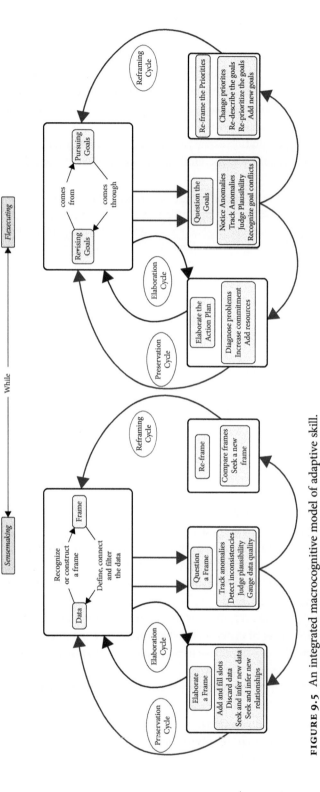

FIGURE 9.5 An integrated macrocognitive model of adaptive skill.

Adapted from Robert R. Hoffman, An integrated model of macrocognitive work and trust in automation, *AAAI Spring Symposium: Trust and Autonomous Systems*, Figure 3, p. 6, https://www.aaai.org/ocs/index.php/SSS/SSS13/paper/viewFile/5724/6002 (c) Robert R. Hoffman, 2013.

Conclusion

The need to understand, describe, explain, and predict the performance of sociotechnical systems accurately and reliably is a key challenge for the field of macrocognition, and increases the motivation for better models of macrocognitive work. As technologists build more and more complex machines with increasing capabilities (arguably *intelligence*), the need for us to understand how experienced people, sometimes experts in their fields, and their technologies will interact in the accomplishment of safe, productive, and healthy work only increases. Macrocognitive models are intended to provide windows onto the *true* nature of the cognitive work that must be supported by the variety of solutions at our disposal, be they work redesign, technology solutions (including autonomous and intelligent systems), ways of working and work processes, training solutions, team and organizational design, and so forth. In addition, macrocognitive models tend to describe expert levels of performance, given that they are developed based on evidence from professionals overcoming the challenges of their work contexts. These models describe the cognitive work required to work at the edge of performance envelopes driven by the contextual drivers of adaptation namely uncertainty, time pressure, dynamic, and unexpected situation trajectories. Macrocognition as the study of cognitive adaptations to complexity (Schraagen et al., 2008) provides a powerful way for us to begin to understand expertise at the edge of these performance envelopes and to begin to develop applied solutions to these performance challenges.

Acknowledgments

The author would like to thank Robert Hoffman and Jan Maarten Schraagen for their contributions to improving the arguments, content, and flow of this chapter. In addition, the production of this summary has been possible in large part due to the continuing interactions with and mentoring of Gary Klein, a valued friend and colleague.

References

Attfield, S., Fields, B., Wheat, A., Hutton, R. J. B., Nixon, J., Leggatt, A., & Blackford, H. (2015). Distributed sensemaking: a case study of military analysis. *Proceedings of the 12th international conference on naturalistic decision making* (pp. 9–12). McLean, VA: MITRE Corp.

Bennett, K. B., & Flach, J. M. (2011). *Display and interface design: Subtle science, exact art.* Boca Raton, FL: CRC Press.

Blackford, H., Bessell, K., Hutton, R. J. B., & Harmer, S. (2017) *Reducing the cognitive burden of system users via cognitive systems engineering.* DHCSTC Report O-DHCSTC_I2_H_T2_087_2/003. Yeovil, UK: BAE Systems.

Blackford, H., Nixon, J., Malin, A., Attfield, S., Hutton, R., & Leggatt, A. (2015). *Multi-source information assimilation to support decision making: final technical report*. DHCSTC, Yeovil: BAE Systems.

Brehmer, B. (2005). The dynamic OODA loop: Amalgamating Boyd's OODA loop and the cybernetic approach to command and control. *Proceedings to 10th International Command and Control Research and Technology Symposium: The Future of Command and Control, June 13–16, McLean, VA. Command and Control Research Program (CCRP), Washington, DC.*

Brunswik, E. (1956). *Perception and the representative design of psychological experiments*. Los Angeles: *University* of California Press.

Cacciabue, P. C., & Hollnagel, E. (1995). Simulation of cognition: Applications. In J-M. Hoc, P. C. Cacciabue, & E. Hollnagel (Eds), *Expertise and technology: Cognition and human-computer cooperation* (pp. 55–73). Hillsdale, NJ: Lawrence Earlbaum.

Calderwood, R., Klein, G. A., & Crandall, B. W. (1988). Time pressure, skill, and move quality in chess. *American Journal of Psychology* 101(4), 481–493.

Cheah, M., Thunholm, P., Chew, L. P., Wikberg, P., Andersson, J., & Danielsson, T. (2005). C2 Team collaboration experiment—A joint research by Sweden and Singapore on teams in a CPoF environment. *Proceedings to 10th International Command and Control Research and Technology Symposium: The Future of Command and Control* (pp. 13–16).

Crandall, B., Klein, G. A., & Hoffman, R. R. (2006). *Working minds: A practitioner's guide to cognitive task analysis*. Cambridge, MA; MIT Press.

Dörner, D. (1996). *The logic of failure: Why things go wrong and what we can do to make them right*. New York: Metropolitan Books.

Driskell, J. E., Salas, E., & Hall, J. K. (1994). The effect of vigilant and hypervigilant decision training on performance. *Annual Meeting of the Society of Industrial and Organizational Psychology, Nashville, TN.*

Endsley, M. R. (1995). Toward a theory of situation awareness in dynamic systems. *Human Factors* 37(1), 32–64.

Fiore, S. M., Rosen, M. A., Smith-Jentsch, K. A., Salas, E., Letsky, M., & Warner, N. (2010). Toward an understanding of macrocognition in teams: predicting processes in complex collaborative contexts. *Human Factors* 52(2), 203–224.

Flach, J. M., & Warren, R. (1995). Active psychophysics: The relation between mind and what matters. In J. Flach, J. Caird, & P. Hancock (Eds), *The ecology of man-machine systems* (pp. 189–209). Hillsdale, NJ: Lawrence Earlbaum.

Gibson, J. J. (1979/2014). *The ecological approach to visual perception* (classic edition). New York: Psychology Press.

Hammond, K. R. (1988). *Judgement and decision making in dynamic tasks*. ARI Research Note 88–81. US Army Research Institute for the Behavioral and Social Sciences.

Hastie, R., & Dawes, R. M. (2001/2010). *Rational choice in an uncertain world: The psychology of judgment and decision making*. Thousand Oaks, CA: Sage.

Hayes-Roth, B., & Hayes-Roth, F. (1979). A cognitive model of planning. *Cognitive Science* 3(4), 275–310.

Hoffman, R. (2013). An integrated model of macrocognitive work and trust in automation. *AAAI Spring Symposium: Trust and autonomous systems*. Available at https://www.aaai.org/ocs/index.php/SSS/SSS13/paper/viewFile/5724/6002. Retrieved 24 May 2018.

Hoffman, R. R., & Hancock, P. A. (2017). Measuring resilience. *Human Factors* 59(4), 564–581.

Hoffman, R. R., & McNeese, M. D. (2009). A history for macrocognition. *Journal of Cognitive Engineering and Decision Making* 3(2), 97–110.

Hoffman, R. R., & Militello, L. G. (2009). *Perspectives on cognitive task analysis: Historical origins and modern communities of practice*. London: Taylor & Francis, Psychology Press.

Hoffman, R. R., Norman, D. O., & Vagners, J. (2009). Complex sociotechnical joint cognitive work systems. *IEEE Intelligent Systems* 24(3), 82–89.

Hoffman, R. R., & Shattuck, L. G. (2006). Should we rethink how we do opords? *Military Review* 86(2), 100.

Hutton, R., Attfield, S., Wiggins, S., McGuinness, B., & Wong, W. (2012). Team sensemaking assessment method (TSAM): An inspection method for evaluating collaborative analysis technologies. In M. Anderson (Ed), *Contemporary ergonomics and human factors* (pp. 418–420). Boca Raton, FL: CRC Press.

Hutton, R., Klein, G., & Wiggins, S. (2008, April). Designing for sensemaking: A macrocognitive approach. *Sensemaking Workshop*, CHI (Vol. 8). Available at https://sites.google.com/site/dmrussell2/Hutton-sensemaking-final.pdf. Retrieved 25 May 2018.

Hutton, R. J., Miller, T. E., & Thordsen, M. L. (2003). Decision-centered design: Leveraging cognitive task analysis in design. In E. Hollnagel (Ed.), *Handbook of cognitive task design* (pp. 383–416). Mahwah, NJ: Lawrence Earlbaum.

Kaempf, G. L., Klein, G., Thordsen, M. L., & Wolf, S. (1996). Decision making in complex naval command-and-control environments. *Human Factors* 38(2), 220–231.

Kahneman, D. (2011). *Thinking, fast and slow*. London: Macmillan.

Kahneman, D., & Klein, G. (2009). Conditions for intuitive expertise: A failure to disagree. *American Psychologist* 64(6), 515.

Klein, D. E., Klein, H. A., & Klein, G. (2000). Macrocognition: Linking cognitive psychology and cognitive ergonomics. *Proceedings of the 5th international conference on human interactions with complex systems, University of Illinois at Urbana-Champaign, Urbana-Champaign* (pp. 173–177). Accessible at http://www.macrocognition.com/documents/K3.pdf. Retrieved April 20, 2017).

Klein, G. (1993). *Naturalistic decision making: Implications for design* (No. CSERIAC93-01). WPAFB, OH: University of Dayton Research Institute.

Klein, G. (1996). Nonlinear aspects of problem solving. *Information & Systems Engineering* 2(3/4), 195–204.

Klein, G. (1997). The recognition-primed decision (RPD) model: Looking back, looking forward. In C. E. Zsambok & G. A. Klein (Eds), *Naturalistic decision making* (pp. 285–292). New York: Psychology Press.

Klein, G. (2007a). Flexecution as a paradigm for replanning, part 1. *IEEE Intelligent Systems* 22(5), 79–83.

Klein, G. (2007b). Flexecution, part 2: Understanding and supporting flexible execution. *IEEE Intelligent Systems* 22(6), 108–112.

Klein, G., & Crandall, B. (1995). The role of mental simulation in problem solving. In P. Hancock, J. Flach, J. Caird, & K. Vicente (Eds), *Local applications of the ecological approach to man-machine systems* (Vol 2). Hillsdale, NJ: Lawrence Earlbaum.

Klein, G., Long, W. G., Hutton, R. J. B., & Shafer, J. (2004). *Battlesense: An innovative sensemaking-centered design approach for combat systems* (Final report prepared under Contract N00178-04-C-3017 for Naval Surface Warfare Center, Dahlgren, VA). Fairborn, OH: Klein Associates.

Klein, G., & Miller, T. E. (1999). Distributed planning teams. *International Journal of Cognitive Ergonomics* 3(3), 203–222.

Klein, G., Phillips, J. K., Rall, E. L., & Peluso, D. A. (2007). A data-frame theory of sensemaking. In R. R. Hoffman (Ed.), *Expertise out of context. Proceedings of the sixth international conference on naturalistic decision making,* (pp. 113–155). New York: Lawrence Erlbaum.

Klein, G., Ross, K. G., Moon, B. M., Klein, D. E., Hoffman, R. R., & Hollnagel, E. (2003). Macrocognition. *IEEE Intelligent Systems* 18(3), 81–85.

Klein, G., Schmitt, J., McCloskey, M., Heaton, J., Klinger, D., & Wolf, S. (1996). *A decision-centered study of the regimental command post.* Final Contract USC PO 681584 for the Naval Command, Control and Ocean Surveillance Center, San Diego, CA. Fairborn, OH: Klein Associates.

Klein, G., Wiggins, S., & Dominguez, C. O. (2010). Team sensemaking. *Theoretical Issues in Ergonomics Science* 11(4), 304–320.

Klein, G., Wiggins, S. L., & Schmitt, J. (1999). *Cognitive aspects of replanning at Army division-level command posts.* Final Report Contract DAAH01-99-C-R060 for US Army Aviation and Missile Command. Fairborn, OH: Klein Associates.

Klein, G., & Wolf, S. (1998). The role of leverage points in option generation. *IEEE Transactions on Systems, Man, and Cybernetics, Part C (Applications and Reviews)* 28(1), 157–160.

Klein, G., Wolf, S., Militello, L., & Zsambok, C. (1995). Characteristics of skilled option generation in chess. *Organizational Behavior and Human Decision Processes* 62(1), 63–69.

Klein, G. A (1999). *Sources of power: How people make decisions.* Cambridge, MA: MIT Press.

Klein, G. A (2004). *The power of intuition: How to use your gut feelings to make better decisions at work.* New York: Doubleday.

Klein, G. A (2011). *Streetlights & shadows: Searching for the keys to adaptive decision making.* Cambridge, MA: MIT Press.

Klein, G. A., Calderwood, R., & Clinton-Cirocco, A. (1986). Rapid decision making on the fire ground. In *Proceedings of the Human Factors and Ergonomics Society annual meeting* (Vol. 30, No. 6, pp. 576–580). Thousand Oaks: SAGE.

Klein, G. A., Calderwood, R., & Clinton-Cirocco, A. (1988). Rapid decision making on the fire ground. Army Research Institute Technical Report 796. Aberdeen, MA: Army Research Institute.

Klein, G. A., & Pierce, L. (2001). Adaptive teams. International Command and Control Research & Technology Symposium. Available at http://www.dodccrp.org/events/6th_ICCRTS/Tracks/Papers/Track4/132_tr4.pdf.

Klein, G. A., & Weitzenfeld, J. (1978). Improvement of skills for solving ill-defined problems. *Educational Psychologist* 13, 31–41.

Lee, J. D., & Kirlik, A. (2013). *The Oxford handbook of cognitive engineering.* New York: Oxford University Press.

Letsky, M., Warner, N. W., Fiore, S. M., & Smith, C. A. P. (2008). *Macrocognition in teams: Theories and methodologies.* New York: CRC Press.

Militello, L. G., & Klein, G. (2013). Decision-centered design. In J. D. Lee, & A. Kirlik. (Eds), *The Oxford handbook of cognitive engineering* (pp. 261–271). New York: Oxford University Press.

Mosier, K. L. (1991). *Expert decision-making strategies.* NASA Technical Report. Mountain View, CA: NASA Ames.

Neisser, U. (1976). *Cognition and reality: Principles and implications of cognitive psychology.* New York: Freeman.

Pascual, R., & Henderson, S. (1997). Evidence of naturalistic decision making in military command and control. In C. E. Zsambok & G. A. Klein (Eds), *Naturalistic decision making* (pp. 285–292). New York: Psychology Press.

Phillips, J. K., Baxter, H. C., & Harris, D. S. (2003). *Evaluating a scenario-based training approach for enhancing situation awareness skills.* Technical Report prepared for ISX Corporation under contract DASW01-C-0036. Fairborn, OH: Klein Associates.

Phillips, J. K., Klein, G., & Sieck, W. R. (2004). Expertise in judgment and decision making: A case for training intuitive decision skills. In D. J. Koehler & N. Harvey (Eds), *Blackwell handbook of judgment and decision making* (pp. 297–315). Malden, MA: Blackwell.

Randel, J. M., Pugh, H. L., Reed, S. K., Schuler, J. W., & Wyman, B. (1996). Methods for analyzing cognitive skills for a technical task. *International Journal Human Computer Studies 45*, 579–597.

Rittel, H. W. J., & Webber, M. M. (1984). Planning problems are wicked problems. In N. Cross (Ed.), *Developments in design methodology* (pp. 135–144). New York: Wiley.

Ross, K. G., Klein, G., Thunholm, P., Schmitt, J. F. & Baxter, H. (2003a). The recognitional planning model: Application for the objective force unit of action (UA). *Proceedings to 2003 CTA Symposium, Boulder, CO.* Fairborn, OH: Klein Associates.

Ross, K. G., Klein, G. A., Thunholm, P., Schmitt, J. F., & Baxter, H. C. (2004). The recognition-primed decision model. *Military Review 84(4)*, 6–10.

Ross, K. G., Thunholm, P., Uehara, M. A., McHugh, A., Crandall, B., Battaglia, D. A., Klein, G., & Harder, R. (2003b). *Unit of Action Battle Command: Decision-making process, staff organizations, and collaborations.* Report prepared through collaborative participation in the Advanced Decision Architectures Consortium sponsored by the US Army Research Laboratory under the Collaborative Technology Alliance Program, Cooperative Agreement DAAD19-01-2-0009. Fairborn, OH: Klein Associates.

Schmitt, J., & Klein, G. (1999). A recognitional planning model. A paper presented at the Command and control research and technology symposium. Fairborn, OH: Klein Associates. Available from http://www.dtic.mil/get-tr-doc/pdf?AD=ADA461179. Retrieved May 25, 2018.

Schraagen, J. M., Chipman, S. F., & Shalin, V. L. (Eds). (2000). *Cognitive task analysis.* Mahwah, NJ: Lawrence Earlbaum.

Schraagen, J. M., Klein, G., & Hoffman, R. R. (2008). The macrocognition framework of naturalistic decision making. In J. M. Schraagen, L. G. Militello, T. Ormerod, & R. Lipshitz (Eds), *Naturalistic decision making and macrocognition* (pp. 3–26). Aldershot, UK: Ashgate.

Sieck, W. R., Klein, G., Peluso, D. A., Smith, J. L., & Harris-Thompson, D. (2007). *FOCUS: A model of sensemaking.* Technical Report. Fairborn, OH: Klein Associates.

Simon, H. A. (1972). Theories of bounded rationality. *Decision and Organization 1(1)*, 161–176.

Stanovich, K E., & West, R F. (2000). "Individual difference in reasoning: implications for the rationality debate?" *Behavioral and Brain Sciences 23*, 645–726.

Thunholm, P. (2005). Planning under time-pressure: An attempt toward a prescriptive model of military tactical decision making. In H. Montgomery, R. Lipshitz, & B. Brehmer (Eds), *How professionals make decisions* (pp. 43–56). Hillsdale, NJ: Lawrence Erlbaum.

Ward, P., Gore, J., Hutton, R., Conway, G., & Hoffman, R. (2018). Adaptive skill as the *conditio sine qua non* of expertise. *Journal of Applied Research in Memory and Cognition 7(1)*, 35–50.

Warner, N., Letsky, M., & Cowen, M. (2005). Cognitive model of team collaboration: Macrocognitive focus. In *Proceedings of the Human Factors and Ergonomics Society annual meeting* (Vol. 49, No. 3, pp. 269–273). Los Angeles: SAGE.

Weick, K. E. (1993). The collapse of sensemaking in organizations: The Mann Gulch disaster. *Administrative Science Quarterly* 38(4), 628–652.

Weick, K. E. (1995). *Sensemaking in organizations* (Vol. 3). Thousand Oaks, CA: Sage.

Wickens, C. D., Hollands, J. G., Banbury, S., & Parasuraman, R. (2015). *Engineering psychology & human performance*. London: Psychology Press.

CHAPTER 10

COGNITIVE SYSTEMS ENGINEERING

Expertise in Sociotechnical Systems

NEELAM NAIKAR AND ASHLEIGH BRADY

Introduction

COGNITIVE systems engineering is a field that is concerned with managing the complexity of cognitive work through design, based on empirical enquiry of how work is achieved in context or in situ (Endsley, Hoffman, Kaber, & Roth, 2007; Hollnagel & Woods, 1983; Klein, Wiggins, & Deal, 2008; Militello, Dominguez, Lintern, & Klein, 2010; Woods & Roth, 1988). In the name given to this field, *cognitive* reflects the focus on cognitive work, rather than physical or manual work; *engineering* signifies the concern with design; and *systems* denotes the adoption of a systems perspective, whereby humans and technologies and the natural and engineered environments, or contexts, in which they function are studied as a single unit of analysis, which may be considered a joint cognitive system (Hollnagel & Woods, 2005), a distributed cognitive system (Hutchins, 1995), or a sociotechnical system (Vicente, 1999, 2006), depending on the emphasis required.

This conceptualization of the three terms reflects the view that the field is concerned with the design or engineering of cognitive systems, not with providing a cognitive approach for systems engineering (Hollnagel, 2016). Therefore, the name of the field should be read as *cognitive-systems engineering*, not as *cognitive systems-engineering*. According to Klein et al. (2008), the origins of the field can be traced to the Three Mile Island accident and the work of Jens Rasmussen, who drew attention to the fact that the designs of nuclear power plant control rooms presented information to workers in a way that sometimes interfered with their ability to understand the situation and adapt to the circumstances.

This special orientation to the study of complex systems has motivated some of the most influential perspectives of human expertise in the past fifty years. Widely accepted is Gary Klein's model of how experts make effective decisions under stressful and demanding conditions, including time pressure (Klein, 1989), which portrays expertise as involving recognition-primed decision making. Also well known is Jens Rasmussen's perspective of how experts are able to deal successfully with ongoing and significant instability, uncertainty, and unpredictability in their work (Rasmussen, 1986; Rasmussen, Pejtersen, & Goodstein, 1994), which depicts expertise as involving adaptive problem-solving. These, and other, perspectives from cognitive systems engineering, though distinct, provide complementary theoretical insights into the nature of expertise, which have shaped the design of complex systems.

This chapter seeks to conceptualize expertise from a cognitive systems engineering viewpoint. That is, it is concerned with characterizing expertise as it manifests in sociotechnical systems, based on empirical studies of actual work settings. Moreover, it is concerned with understanding expertise in a way that is useful for system design. The particular view of expertise considered, and expanded, in this chapter is Jens Rasmussen's representation of expertise as involving adaptive problem-solving, a perspective that is becoming increasingly widely appreciated and influential across a variety of research communities. The proposed expansions are motivated by the recognition that expertise must be understood in the context of systems with dynamic work organization, or self-organizing systems.

First, the chapter describes some essential characteristics of sociotechnical systems, focusing on those which make adaptation necessary for their viability. Then, based on recent and classic empirical studies of a range of work contexts, the chapter demonstrates that workers in sociotechnical systems function within groups or teams and that they adapt not only their individual behaviors but also their group structure or organization in line with the evolving situation. Following that, the chapter presents the concept of self-organization, as it manifests in social systems particularly, which suggests that in sociotechnical systems new spatial, temporal, and functional structures can emerge from the spontaneous actions of multiple, interacting actors—without external intervention, a priori planning, or centralized coordination—and that this phenomenon is essential for dealing with a dynamic and ambiguous work environment. The chapter then considers the theoretical and design implications of the self-organization phenomenon for understanding and promoting expertise in sociotechnical systems, particularly in relation to the observations that workers may have to carry out tasks beyond those they are trained or expected to perform normally and that workers may have to execute the same tasks under very different conditions, some of which may be completely new or unfamiliar to them. Finally, the chapter concludes by summarizing the key features of the self-organization perspective of expertise and outlining some of the future research challenges.

At the outset, it is worth noting explicitly that although this chapter is concerned with human expertise, the term *actor* or *actors* is used much of the time, especially in latter sections, to accommodate both human workers and automated agents in recognition of the increasing sophistication of automation being introduced into the workplace. Nevertheless, it is important to bear in mind that this chapter focuses on human expertise in sociotechnical systems. It does not consider how the notion of expertise might be conceptualized in the case of automated agents or how this conceptualization might differ from that of human expertise.

Sociotechnical Systems

Sociotechnical systems are systems with psychological, social, and technical dimensions (Vicente, 1999). In some systems, such as petrochemical refineries, the technical dimension may be more prominent, whereas in other systems, such as stock market trading, the psychological and social dimensions may be more apparent. Nevertheless, in all cases, all three dimensions play an important role in effective performance, so that each must be accounted for in system design. Designs that focus on technical functions while neglecting psychological and social factors, or vice versa, may compromise performance, sometimes catastrophically, as many industrial accidents highlight. Some high-profile cases are the Three Mile Island nuclear meltdown (President's Commission on the Accident at Three Mile Island, 1979), the Deepwater Horizon oil spill (Deepwater Horizon Study Group, 2011), and the Air France flight 447 crash on route from Brazil to France (Bureau d'Enquêtes et d'Analyses, 2012).

Many sociotechnical systems are complex in nature. That is, they tend to have, to varying degrees, some or all of the following factors (Perrow, 1984; Vicente, 1999; Woods, 1988). First, they tend to have *large problem spaces* composed of many different concepts and elements. In medicine, for example, the number of identified illnesses, and thus the number of potential diagnoses, is on the order of 500,000 (Vicente, 1999). Unsurprisingly then sociotechnical systems are composed of *multiple actors*, in some cases numbering in the hundreds or thousands, which creates the need for effective coordination in the workplace. In the health care sector, many different professional workers, with a variety of specializations, must integrate their activities, over both short and long timeframes, to ensure that patients are treated successfully.

It follows therefore that actors in sociotechnical systems usually bring to the problem space *heterogeneous perspectives* connected with different technical backgrounds or disciplines, each with particular objectives and values, which may be conflicting at times. In large-scale engineering design, for instance, the value structures of engineers, architects, accountants, managers, clients, and other stakeholders must often be reconciled during decision-making, such that a social negotiation process is necessary (Vicente, 1999). Moreover, the actors may be *distributed* across a variety of locations, sometimes globally so that they bring different cultural backgrounds to the work. Commercial aviation, for example, requires effective collaboration between workers from different cultures. Furthermore, modern systems are likely to be highly *automated*, so that the actors are composed of both human workers and sophisticated automated agents, which presents different kinds of collaborative challenges. For instance, military forces of several countries are already operating highly automated aerial vehicles, some armed with weapons, that are supervised by teams of workers on the ground.

Actors in sociotechnical systems must also deal with *dynamic* conditions, which means that the nature of the problems, demands, and pressures they are faced with may constantly change or evolve. For example, the specific challenges posed by a wildfire to emergency management workers may fluctuate widely as the fire spreads throughout a region, depending on such factors as the current and anticipated weather conditions or population sizes and infrastructure in the affected areas. Moreover, the actors face a high degree of potential *hazard* in that inappropriate actions can have catastrophic consequences, not just locally but societally or globally. An accidental missile firing of a commercial airliner, for instance, may have significant implications for global politics.

Actors in complex sociotechnical systems usually also have to contend with considerable *uncertainty*, especially in relation to the accuracy or completeness of the information available to them. In the naval domain, imperfections in acoustics sensors for detecting objects submerged underwater may mean that the true situation can never be established with complete certainty. Furthermore, actors may have to contend with significant *unpredictability*, or events that cannot be forecasted or fully specified a priori, such as an unforeseen chain of financial collapses in the aftermath of a natural disaster.

The fact that sociotechnical systems are composed of many highly *coupled*, or interacting, subsystems or components also adds to their complexity. As a result, it can be very difficult for actors to ascertain all of the effects of an unanticipated event or predict all of the effects of an action, so that reasoning about the problem space can be highly challenging and burdensome. For example, the consequences of a novel surgical intervention or implications of an unanticipated reaction of a patient to an anesthetic is challenging to establish fully because of the complicated network of anatomical and physiological subsystems that constitute the human body.

Finally, the complexity of actors' jobs is also compounded by the fact that they cannot observe or explore the problem space directly, at least not in its entirety. Rather, their interaction with the problem space is, to a large extent, *mediated* through computers. A military commander, for example, cannot observe the entire battlefield directly or even the activities of his co-workers in another location. Thus, the highly evolved perceptual, motor, cognitive, and social abilities that humans depend on for their usual or everyday activities may not always be sufficient. Instead, more complicated, laborious reasoning processes may be necessary.

In general, then, many sociotechnical systems have very large and complex problem spaces, characterized by a high degree of instability, uncertainty, and unpredictability. Given these problem characteristics, actors cannot rely solely on dealing with their work demands by executing well-established procedures or trained task sequences or by recognizing workable solutions from their prior experiences, even if they have considerable know-how (Rasmussen, 1986; Rasmussen et al., 1994; Vicente, 1999). Instead, actors must be capable of adapting their behavior, continuously and reliably, to the evolving demands of the problem, sometimes in highly original or creative ways.

Workers' abilities to adapt their work practices *on the job* are critical for responding to unforeseen events, which are widely regarded as posing the greatest threats to system safety (e.g., Leveson, 1995; Perrow, 1984; Rasmussen, 1969; Reason, 1990; Vicente, 1999). Furthermore, worker adaptation is essential for sustaining productivity in highly computerized or automated workplaces, especially when innovative solutions are required to deal with emerging problems for which rules or algorithms have not been, and perhaps cannot be, written. Finally, it is also well established that workers who have greater autonomy to improvise or adapt in doing their jobs, and to follow their individual preferences when it is appropriate to do so, tend to have better health, as indicated by such factors as longevity and the absence of stress or disease (Eason, 2014; Karasek & Theorell, 1990; Vicente, 1999).

Adaptation in the Workplace

Empirical studies of sociotechnical systems provide valuable insight into the nature of the adaptations that occur, and are necessary, in actual work settings. Of particular importance to the present discussion is the observation that workers in sociotechnical systems adapt not just their individual behaviors but also their collective structures, or relationships, in accordance with the demands of the unfolding situation (Naikar & Elix, 2016b). In this section, studies from three different contexts, emergency management, healthcare, and military organizations, are described to demonstrate the nature of behavioral and structural adaptation in the workplace.

Bigley and Roberts (2001) provide a compelling account of worker adaptation in an emergency management context in reporting a field study of a large fire department employing the incident command system, a common approach for emergency management in the USA. Some of the improvisations they observed related to workers adjusting their behaviors, or revising their tasks, plans, goals, actions, strategies, or priorities, sometimes completely in line with the emerging situation, which Bigley and Roberts described as involving tools, rules, and routines. When a truck arrives at the scene of a fire, for instance, workers must manage the situation with the available tools, which is limited by the load capacity of the truck. Consequently, workers may employ tools in novel ways, perhaps even using them to perform functions for which they were not originally intended. The execution of standard routines, such as those for *hose laying* and *ladder throwing*, and the coordination of standard routines also may be adjusted to accommodate unexpected occurrences. For example, in the case of a peculiar building construction, the routines for fighting a fire from inside a building while ventilating the structure from the outside may need to be coordinated differently by the respective teams. Bigley and Roberts also found that workers may even find it necessary to adopt behaviors that directly breach standard operating procedures. For example, a firefighter described an incident in which "opposing hose streams" were employed as the primary tactic for a rescue, despite the fact that this approach is prohibited due to the risk that one group can push the fire into the other (p. 1289). According to Bigley and Roberts, such improvisations are regarded as legitimate within the organization provided they are consistent with organizational goals and do not cause harm or injury.

During their field study, Bigley and Roberts (2001) also observed workers adjusting their structure or organization, including their roles and relationships, in step with the demands of the emerging situation. The incident command system is, in fact, highly formalized with a hierarchical organization, highly specialized jobs and training, and an extensive set of rules, procedures, policies, and instructions. However, Bigley and Roberts found that the fire department consistently employs a number of mechanisms for rapidly transforming this rigid organizational structure into highly flexible configurations, which they described as involving structure elaborating, role switching, authority migration, and system resetting.

Structure elaborating refers to the process of constructing an organization at the scene of an incident. The first captain arriving becomes the incident commander, at least temporarily, and begins to build an organization by assigning roles and tasks to incoming resources. Pre-existing roles or positions within the system are only populated with personnel to the extent required, and functions are not assigned to specialized positions unless it is necessary, so that some personnel may be responsible for multiple functions. Role switching occurs throughout an incident, with positions continually activated and deactivated and personnel either shifted into different positions or discharged. Authority migration also occurs, whereby informal decision-making responsibility is shifted temporarily to personnel possessing the most relevant expertise. Finally, system resetting involves disengaging or regrouping, whereby a team may be withdrawn completely from a situation because of unexpected occurrences and reconfigured or redirected as needed. Bigley and Roberts (2001) observed that "Within the most reliable systems, objectives and corresponding structural elements and relationships are adjusted swiftly in accordance with changing environmental contingencies" (p. 1287).

In the context of this discussion, it is worth highlighting another study of emergency management, conducted by Lundberg and Rankin (2014), who reported on role improvisations by Swedish emergency response workers to such an extent that they were performing jobs beyond their areas of specialization or "working outside of professional roles" (p. 145). Some examples of such role improvisations included workers without the necessary skills or experience carrying out logistics activities, such as arranging transport or food supplies; conducting crisis support talks; and cremating bodies. Although it was acknowledged that these workers may be less effective or efficient than specialists performing the same jobs, sometimes considerably less so, in the circumstances the arrangements were regarded as being better than having nobody performing the work or having workers sitting idle. In fact, some informants thought that having *team players* was more important than having specialists and related that "in contrast to regular work at home, in crisis work, it is normal to step into the territories of others' [*sic*] field of competences [*sic*]" (p. 151).

Similar findings in relation to adaptation in the workplace, both behavioral and structural, have been documented in the health care sector. Bogdanovic, Perry, Guggenheim, and Manser (2015) reported a study in which they described the kinds of adaptation that occur in surgical teams. For instance, when incidental findings occur during surgery, such as the presence of inflammation, the surgical plan may be modified or even halted altogether, so that the team can consider the next steps. Furthermore, the priorities may change during surgery to such an extent that the procedure is discontinued. In the case of deteriorating patient condition, for example, patient stabilization will become the priority over continuing with the surgical procedure. Team members' communication strategies may also change (see also Barth, Schraagen, & Schmettow, 2015; Schraagen, 2011). While team members generally keep up a high level of communication during surgery, exchanging information about the procedure, instruments, and patient condition, when complications occur or during difficult surgical steps, team members may keep the exchange of information to a minimum, with the intent of preventing distraction and enabling complete engagement in the task at hand. As Bogdanovic et al. stated, "situational variability requires the team to behave adaptively" (p. 1).

Bogdanovic et al. (2015) also described how the task distribution alters as a function of situational requirements during surgery. Specifically, only the general task distribution is established prior to surgery, with some tasks delegated on the basis of team members' professions, such as whether one is an anesthetist, nurse, or surgeon. In contrast, tasks that can be fulfilled by any person are not assigned beforehand but are delegated dynamically throughout the surgery, depending on the circumstances. For instance, the anesthetist might assist the scrub nurse if the circulating nurse is occupied with other important tasks. While some options for the task distribution in view of the anticipated challenges may be considered before surgery, in the event of unforeseen complications, new arrangements are conceived and instituted at the time. For example, if problems emerge for which a team member does not possess the necessary skills, a senior physician may take over a step of the procedure initially assigned to someone else. Alternatively, additional resources may be mobilized for the task, with a more experienced clinician being called into the group. Bogdanovic et al. observe that such open-ended fine tuning of the task distribution is necessary for providing the flexibility required for dealing with situational variability, minimizing the pressure on the team, and enabling a smoothly running procedure.

Finally, the need for adaptation in the workplace is also evident in military organizations. Rochlin, La Porte, and Roberts (1987) reported a field study of large, nuclear-powered naval aircraft carriers, noting the significant adaptability and flexibility involved in the organization's day-to-day performance. Routines and operating procedures are altered as necessary, sometimes dramatically. In one case, the work-up to a military exercise was shortened by two weeks. As a result, the ship was forced to complete its training during the middle of a difficult and demanding exercise and to adjust its evaluation procedures, which placed considerable strain on personnel and increased risk to operations.

Rochlin et al. (1987) also observed considerable flexibility in the organization's structure. While the formal organization of this system is rigid, hierarchical, and centralized, with clear chains of command and means to enforce authority, during complex operations, a radically different type of structure is exhibited. This informal structure is flat and distributed. As an example, lower-ranked personnel can make critical decisions without the approval of officials with higher rankings, as events on the flight deck can occur too quickly to allow for appeals through a chain of command. The informal organization is also flexible, in that there is no a priori plan for when it will be executed, and the particular structure that is adopted on any one occasion is emergent, so there is no single, fixed mapping between people and roles. Instead, the mapping varies as a function of the circumstances. According to Rochlin et al., such adaptability on the ship is essential for balancing the need for safety with the push for productivity in light of local contingencies (for a recent review on adaptive skill as the central component of expertise, especially in defense contexts, see Ward, Gore, Hutton, Conway, & Hoffman, 2018).

Self-Organization

The preceding case studies demonstrate that adaptations in sociotechnical systems can be observed at the levels of the behaviors of individual actors and structures of multiple actors. In this context, the concept of self-organization is important because it can provide an explanation of how such adaptations can be achieved spontaneously, continuously, and relatively seamlessly, without centralized coordination, a priori planning, or external intervention. Consequently, on the assumption that sociotechnical systems are, and even must be, self-organizing at least some of the time to withstand the challenges of a changeable work environment (Naikar, Elix, Dâgge, & Caldwell, 2017), this concept has implications for how expertise must be understood in a sociotechnical context, and how it must be developed and supported through system design.

A fundamental aspect of the self-organization concept that is key to the present discussion is the unambiguous acknowledgment of the tight interconnection between structure and behavior in complex systems (e.g., Haken, 1988; Heylighen, 2001). That is, it is recognized that structural properties, which are observed at the system level, vary with the behaviors of individual, interacting elements, and vice versa. Figure 10.1 expresses this relationship more precisely in the case of self-organizing *social* systems. The figure depicts that structures constrain and enable actors' behaviors or actions as well as emerge from their actions (e.g., Fuchs, 2004; Hofkirchner, 1998). Here it is proposed that this interplay between the structures of multiple actors and the behaviors

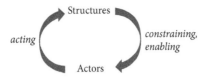

FIGURE 10.1 Social self-organization.

Reproduced from Hofkirchner, W., "Emergence and the logic of explanation. An argument for the unity of science," *Acta Polytechnica Scandinavica, Mathematics, Computing and Management in Engineering Series*, Volume 91, pp. 23–30 © TAF, Technology Academy Finland.

of individual actors is integral to the process of a sociotechnical system adapting to its environment.

Specifically, the empirical studies in the preceding section demonstrate that new structures may be observed in sociotechnical systems that are not planned a priori, centrally coordinated, or imposed by external agents, but rather seem to be a spontaneous response of the system itself to changes in its environment. As discussed by Naikar (2018), the concept of self-organization suggests that a system's structure may inhibit its response in ways that are unsuitable or ineffective in particular circumstances (Figure 10.1). However, in context of the local conditions, individual, interacting actors may engage in spontaneous behaviors from which new structures emerge that enable the system to respond appropriately to the circumstances. For example, actors may change the order in which jobs are performed (Bigley & Roberts, 2001), take over steps of a procedure not initially assigned to them (Bogdanovic et al., 2015), or assume greater decision-making responsibilities (Rochlin et al., 1987). A new structure may be suitable for a time, in that it constrains and enables behaviors in ways that are appropriate to the situation, until the circumstances change and the flexible actions of individual, interacting actors result in further structural changes.

Both actors' structures and behaviors then may evolve continuously in response to the challenges posed by the environment. Moreover, although this self-organizing process may not be flawless, it may be fairly seamless, particularly in well-established systems, and in any case, it is necessary. Conditions are too dynamic for a priori planning, centralized coordination, or external intervention to be feasible always. Instead, flexibility and adaptation within the system are required to achieve a proper balance between the system's safety and productivity goals in the circumstances (Rochlin et al., 1987).

The phenomenon of self-organization, therefore, has implications for how expertise must be understood in sociotechnical systems and how it must be developed and supported through design. Conceptualizations of expertise focused at the level of either behavior *or* structure will not be sufficient. Designs based primarily on the behaviors of individual actors may have undesirable consequences for the structural relationships between these actors, and vice versa (Naikar, 2018). Rather, expertise must be understood in the context of both structure and behavior—in an integrated manner—and thus in the context of self-organizing systems. The next two sections of this chapter are concerned explicitly with the theoretical and design implications of the self-organization phenomenon for understanding and promoting expertise in sociotechnical systems.

Theoretical Implications

The recognition that actors in sociotechnical systems may adapt both their behaviors and structures in a self-organizing manner—that is without a priori planning, centralized coordination, or external intervention, but instead in a spontaneous or opportunistic fashion in response to the demands of local contingencies—leads to particular insights about the nature of expertise in sociotechnical systems. These insights are substantiated by the field observations of sociotechnical systems reported earlier.

The phenomenon of self-organization suggests that expertise in sociotechnical systems may best be conceptualized as the ability of an actor to engage in flexible behaviors, in view of the structural possibilities of multiple actors in the system, rather than simply as the ability of an actor to undertake typical jobs or trained task sequences with a high level of proficiency. First, actors in sociotechnical systems may have to carry out tasks or assume responsibilities they are not trained or expected to perform normally. For example, in their field study of naval aircraft carriers, Rochlin et al. (1987) found that most officers as well as a number of senior enlisted personnel are familiar with several tasks other than those they usually perform and therefore can perform atypical tasks in emergencies. In a healthcare context, Bogdanovic et al. (2015) reported that members of a surgical team may be required to step outside of their professional roles temporarily, as tasks that can be fulfilled by any team member are not delegated upfront but assigned on an ad hoc basis in response to the emerging situation. Similarly, in the emergency management domain, Lundberg and Rankin (2014) uncovered a number of cases of workers performing jobs outside their areas of specialization, and found that personnel working in improvised roles as part of their regular work had less difficulty in adapting to new roles during crisis work.

Second, actors in sociotechnical systems may carry out their usual tasks in changed circumstances, some of which may be novel or unforeseen, so that they may perform those tasks in radically different ways or execute significantly different courses of action or behaviors from those they are expected, or trained, to adopt under routine conditions or from those they have adopted in prior situations. For instance, Bigley and Roberts (2001) found that firefighters may use the tools available on a fire truck in novel ways and adjust how standard routines are executed and coordinated because of unexpected occurrences, such as a peculiar building construction. Moreover, *most of* the informants in their study provided examples of cases when it was necessary to adopt tactics that violated standard operating procedures.

Third, because actors may undertake unusual jobs, or conduct their usual jobs under unusual conditions, their performance may not necessarily be regarded as skillful or flawless, in view of a normative or idealized benchmark, yet it may be considered as reasonable or sufficient, and likely even critical or masterful, in view of the circumstances. In other words, while actors' performance may be judged as falling short if it is assessed in terms of their ability to execute an idealized set of tasks or task sequences, their performance may be considered effective or even masterful if it is assessed in terms of their ability to engage in flexible action in the circumstances, which may be unfamiliar or unforeseen. In fact, if a normative perspective is adopted, the performance of actors in sociotechnical systems would perhaps often fall short of the mark. Yet many sociotechnical systems are regarded as high-reliability organizations (La Porte, 1996; Rochlin, 1993; Weick & Sutcliffe, 2001) because of their capacity to sustain high levels of safety and productivity in the face of considerable instability, uncertainty, and unpredictability in their work environments.

The perspective of expertise presented here does recognize that actors may often base their actions on standard or pre-defined plans, protocols, or procedures, or on solutions that have worked on prior occasions, and that many of the situations that actors experience may be routine or familiar to them. Furthermore, other situations, though unusual, may be anticipated by actors so that options for dealing with those events may be formulated beforehand. Nevertheless, the proposed perspective also recognizes that the most significant threats to system safety and productivity are posed by novel or unforeseen events (e.g., Leveson, 1995; Perrow, 1984; Rasmussen, 1969; Reason, 1990; Vicente, 1999), or events that cannot be predicted a priori in their entirety, and that actors must have the ability to shift flexibly between known behaviors that are suitable for dealing with routine, familiar, or anticipated events and novel behaviors that are effective in unforeseen circumstances. In addition, actors in sociotechnical systems may be required to carry out atypical jobs—for which they do not necessarily have prior training or experience—as evidenced by the case studies described earlier.

Furthermore, the proposed perspective of expertise recognizes that, aside from dealing with novel events, adaptations are necessary in *everyday* (Rankin, Lundberg, Woltjer, Rollenhagen, & Hollnagel, 2014) or commonly occurring situations (Gerson & Star, 1986; Naikar, 2013; Rasmussen, 1986; Rasmussen et al., 1994; Simon, 1969; Suchman, 1987; Vicente, 1999; Weick, 1993). Even minor variations in context can require adaptation (Vicente, 1999) and it is neither safe (Dekker, 2003) nor possible (Hoffman & Woods, 2011) to pre-specify an algorithm, plan, or procedure for every single occurrence. Accordingly, ongoing local adjustments or improvisations are necessary to accommodate the inevitable flux that arises in everyday work (Bigley & Roberts, 2001; Bogdanovic et al., 2015; Militello, Sushereba, Branlat, Bean, & Finomore, 2015; Rankin et al., 2014).

Finally, it is not the intention here to suggest that all adaptations by workers are necessarily ad hoc occurrences in the sense that they have never been considered, tried, or practiced before. Certainly, in many cases, workers may routinely prepare for deviations from standard behaviors, which is an important aspect of effective practice. For example, Bogdanovic et al. (2015) observe that, if possible, deviations from standards are communicated among members of a surgical team prior to entering the operating room. In general, however, it appears that although workers may be aware of the need for improvisation, and the general strategies that may be suitable, they do not always know in advance the specific adjustments that will be needed, as these may depend on local details that cannot be specified fully beforehand. Furthermore, plans, priorities, routines, and strategies may need to be altered significantly, and even revised completely, because of unanticipated complications or novel or unusual circumstances, such as a peculiar building construction in a fire fighting emergency (Bigley & Roberts, 2001). It is also clear that the adaptations may involve tasks that workers don't perform normally, tasks that workers haven't planned a priori, or tasks that workers don't have high levels of skill or experience for undertaking. Accordingly, in some cases, the adaptations are reported as being demanding for workers or as producing results that are less than ideal, at least when considered from a normative standpoint. For example, Rochlin et al. (1987) observed that shortening the work-up to a military exercise by two weeks necessitated adaptations to the ship's training procedures, requiring the training to be completed in the middle of a demanding exercise, which placed considerable strain on personnel and increased risk to operations. In addition, Lundberg and Rankin (2014) reported that emergency management workers without the necessary skills or experience for carrying out certain activities were less effective or efficient than specialists carrying out the same jobs. In such cases, it seems unlikely that the workers involved were prepared for or accustomed to such adaptations. Lastly, many case studies draw attention to the idiosyncrasies or uniqueness of the situations regularly experienced by workers, suggesting that, if not significant shifts or improvisations in behavior, ongoing, open-ended fine-tuning of behaviors is necessary in everyday work to meet the demands of a constantly evolving task environment. For example, Lundberg and Rankin state that each crisis brought "its unique problems, where specific previous solutions and plans cannot be certain to work" (p. 154). Similarly, Bigley and Roberts (2001) reported a fire captain's remark that "Every incident is dynamic, and so every incident may require a different approach" (p. 1289). Altogether, these observations of the nature of work practice in sociotechnical systems signify the importance of flexible action—that is, action that is closely fitted to evolving circumstances, which cannot be specified fully a priori. Such behavioral spontaneity is conducive to the emergence of new or different temporal, spatial, or functional structures, which is essential for a system to cope effectively with a complex work environment, one that is characterized by significant instability, uncertainty, and unpredictability.

Given the proposed view of expertise, then, as the ability of actors to engage in flexible action, a specific worker's level of expertise may be conceptualized as their current space of possibilities for behavior, or behavioral repertoire, which continues to expand as they gain skills and experience. Therefore, in this view, the development of expertise involves expanding an actor's current space of possibilities for behavior, and an experienced actor has a greater behavioral repertoire than a novice actor. An actor with a greater behavioral repertoire will not necessarily generate more options during problem-solving. Rather, by having a greater variety of possibilities for behavior, an actor can address a greater variety of problems, or tasks, with different cognitive strategies, so that courses of action fitted, or adapted, to the challenges or pressures of an ever-changing work environment can be enacted. Finally, an expert is considered a notional concept, signifying an actor whose behavioral repertoire matches the action possibilities afforded by the work context, which may be rarely realized in reality.

This conceptualization of expertise is important in the context of self-organizing systems. Specifically, the greater a worker's behavioral repertoire, the greater the system's structural possibilities. Consequently, a perspective of expertise that is concerned with supporting and expanding a worker's behavioral repertoire to match the action possibilities of the work environment, even if this is never realized in full, will facilitate the emergence of novel organizational forms or structures from the flexible behaviors of individual actors, irrespective of their current levels of expertise, the task or tasks in question, and the specific situations encountered, thereby enhancing a system's capacity to deal with instability, uncertainty, and unpredictability.

The action possibilities afforded by the work context may be considered at the level of the behavioral opportunities of individual actors and the structural possibilities of multiple actors (Naikar et al., 2017). In sociotechnical systems, the behavioral opportunities afforded to individual actors vary with the structural possibilities of multiple

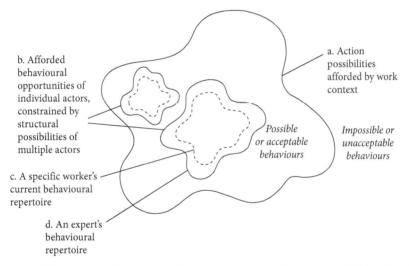

FIGURE 10.2 Illustration of the relationship between: (a) set of action possibilities afforded by the work context, irrespective of specific actors; (b) space of behavioral opportunities afforded to individual actors, which is constrained by the structural possibilities of multiple actors; (c) a particular worker's current space of possibilities for behavior, or behavioral repertoire, which continues to expand with training and experience; and (d) an expert's capacity or space of possibilities for behavior, which matches the action possibilities afforded to them by their work context, although this may never be realized in reality.

actors, as illustrated in Figure 10.1. Therefore, the full space of behavioral opportunities afforded to an individual actor is defined by their role in each of the possible structures or organizational forms the group *can* assume (rather than *do*, *should*, or *will* assume) to deal with the work demands of the system. As there is usually a very large number of structural possibilities (Naikar & Elix, 2016b), it seems unlikely that a worker will ever be able to exploit fully their afforded behavioral opportunities or, in other words, execute the set of behavioral opportunities at a high level of skill in all cases. Nevertheless, their ability to do so may expand with training and experience and, as noted earlier, their performance may be considered critical and even masterful given the circumstances. Consequently, in sociotechnical systems, which are necessarily self-organizing, expertise is best understood as the ability of individual actors to engage in flexible behaviors, in view of the structural possibilities of multiple actors. Figure 10.2 illustrates the relationships between some of the key concepts in this self-organization perspective of expertise.

Design Implications

The self-organization perspective of expertise recognizes that effective performance in sociotechnical systems is achieved through the spontaneous behaviors of individual actors, and the emergence of new structural forms or organizations from these behaviors, which make it possible for a system to adapt continuously to the changing demands of the work environment. This perspective has important design implications for promoting expertise in sociotechnical systems.

First, designs for sociotechnical systems must facilitate the acquisition of expertise by supporting workers in developing their abilities to engage in flexible behaviors. In particular, designs must make available to workers the full space of action possibilities afforded to each by their work context, given the structural forms that can be assumed by actors in the system, and the support to expand their behavioral repertoires, through training and experience, to match the action possibilities afforded to them as an ideal. This design objective recognizes that, rather than simply carrying out a narrow range of preconceived tasks or task sequences, workers in sociotechnical systems may be required to perform unusual jobs or their usual jobs under unusual conditions. Moreover, their levels of expertise for engaging in flexible action may continue to grow with training, both on and off the job, and with experience.

Second, designs must support workers in utilizing or exploiting their expertise in the workplace by systematically supporting them in engaging in flexible behaviors. This design objective recognizes that, regardless of their current levels of development, or expertise, workers' performance levels may be compromised by designs intended to support workers in carrying out fixed sets of tasks or task sequences. Instead, designs must deliberately support workers in shifting seamlessly between jobs, in attending to the demands of multiple jobs, and in adjusting or changing how jobs are carried out to fit the circumstances (some examples are provided later in this section).

Ultimately, these design objectives are consistent with the phenomenon of self-organization observed in sociotechnical systems. Given the necessity of self-organization in the workplace, designs aimed at developing and utilizing effectively each worker's abilities to capitalize on their afforded behavioral opportunities may promote greater collective structural possibilities. Consequently, resulting designs should be able to foster the emergence of novel group structures from individual actors' spontaneous behaviors, thereby promoting, rather than compromising, a system's capacity for dealing with instability, uncertainty, and unpredictability.

A suitable starting point for achieving the preceding design objectives is provided by the diagram of work organization possibilities, or WOP diagram (Figure 10.3), a modeling tool that was recently proposed as an extension to cognitive work analysis (CWA; Naikar & Elix, 2016b; the standard CWA method is described in Chapter 20, "Cognitive Work Analysis: Models of Expertise," by Burns, this volume). This diagram establishes the limits on how the action possibilities afforded by the work context can be distributed across actors in a system, independently of specific events or situations (Figure 10.2b). Thus, these limits, or organizational constraints, simultaneously reveal the action possibilities of the work context at the levels of the behavioral opportunities of individual actors and the structural possibilities of multiple actors within a single, integrated model, in a way that is consistent with many of the characteristics of

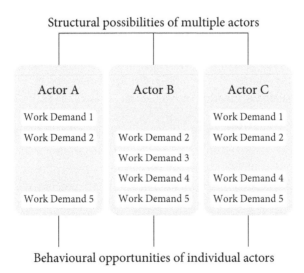

FIGURE 10.3 The diagram of work organization possibilities, or WOP diagram.

Reproduced from Naikar N and Elix B, Integrated system design: Promoting the capacity of sociotechnical systems for adaptation through extensions of cognitive work analysis, *Frontiers in Psychology*, 7, p. 962. doi: 10.3389/fpsyg.2016.00962. Copyright © 2016 Naikar and Elix. This is an Open Access article distributed under the terms of the Creative Commons Attribution License (http://creativecommons.org/licenses/by/4.0/), which permits unrestricted reuse, distribution, and reproduction in any medium, provided the original work is properly cited.

self-organizing systems (cf. Figure 10.1; Naikar et al., 2017). Therefore, it is proposed that designs based on the WOP diagram may promote the development and utilization of expertise in the workplace in a way that is consistent with the phenomenon of self-organization.

Specifically, designs based on the WOP diagram (Figure 10.3) have the potential to preserve the spaces of behavioral opportunities afforded to individual actors, in view of the structural possibilities of the group of actors (Figure 10.2b), and to support each worker in continuing to expand their existing behavioral repertoires, through training and experience, to match the action possibilities afforded to them by their work context as an ideal (Figure 10.2c). Furthermore, such designs have the potential to support workers in utilizing or exploiting their expertise in the workplace, regardless of their current levels of development or behavioral repertoires. That is, by taking account of each actor's afforded space of behavioral opportunities, designs may systematically support workers in shifting between jobs, attending to multiple jobs, or changing or adjusting how jobs are performed, thereby enabling them to exploit their behavioral opportunities, as much as possible, at their current levels of expertise.

Within the WOP diagram, the behavioral opportunities of individual actors are represented as a set of work demands inherent to the work context, or environmental constraints, that collectively define a bounded space for each actor, within which there are many degrees of freedom for behavior (Figure 10.2b; Naikar et al., 2017). These constraints distinguish behaviors that are possible or acceptable from those that are impossible or unacceptable, thereby defining boundaries on effective performance, which are applicable regardless of the specific circumstances or events a system may confront, including those that are novel or unforeseen. Therefore, designs based on these bounded spaces may safely and productively support spontaneous actions from individual, interacting actors in a variety of situations, such that new or different structural forms may emerge from the collective behaviors of multiple actors, in response to the challenges posed by a changeable, uncertain, and unpredictable work environment.

As a simplified example, Figure 10.3 shows that the afforded spaces of behavioral opportunities of both actors A and C, though not B, encompass work demand 1. Thus, some of the structural possibilities that may emerge from this set of actors are as follows: actor A is engaged in behaviors relating to work demand 1 while actors C and B are occupied in other behaviors; actor C is involved in behaviors relating to work demand 1 while actors A and B are engaged in other behaviors; or actors A and C are both occupied in behaviors relating to work demand 1 while actor B is involved in other behaviors.

In the case of a future maritime surveillance aircraft (Naikar & Elix, 2016b), for instance, both of the flight deck actors can be engaged in behaviors associated with the work demands or constraints of flying or navigating the aircraft. Therefore, how these behaviors are distributed across these actors may change opportunistically as a function of the circumstances, so that any one or both of these actors may be involved in these behaviors at any point in time. Moreover, the behavioral opportunities of actors at six workstations in the aircraft's cabin also encompass navigation, so that any one or more of these actors may be occupied in such behaviors as well.

Another example of the work organization possibilities on the maritime surveillance aircraft is that only actors on the flight deck and at the observer stations can be engaged in behaviors concerned with sighting or observing targets out of a window. However, at any point in time, any one or more of the two flight deck actors and two observer station actors, who are normally located at these stations, may be occupied in these behaviors. Moreover, if there is an electrical failure, such that none of the sensor systems available to the actors at the six workstations can be used for detecting, tracking, or identifying targets, any one or more of these actors might relocate to the stations with a window to increase the chances of sighting the target.

Within a system's organizational constraints, then, many different temporal, spatial, or functional structures can emerge from the behavioral opportunities available to individual, interacting actors. An approximate counting for the maritime surveillance aircraft showed the number of structural possibilities to be on the order of 10^{27} (Naikar & Elix, 2016a, b). Therefore, by coordinating the designs of such system features as the interfaces, teams, training, and human–automation interaction around a system's organizational constraints (Naikar, 2018; Naikar & Elix, 2016a, b; Naikar et al., 2017), solutions may be created for developing and utilizing expertise in the workplace in a way that is consistent with the self-organization phenomenon.

Consider, for instance, the case of training systems. Rather than focusing solely on typical tasks or task sequences, the designs of training systems may consider the learning requirements associated with the full spaces of behavioral opportunities afforded to individual actors, with the intent of expanding workers' behavioral repertoires over time, through on and off the job training and experience. Although workers' behavioral repertoires may never match the action possibilities afforded by the work context completely, field studies show that workers in sociotechnical systems do take responsibilities for tasks for which they have little or no prior training, because though their performance of these tasks may be far from ideal, it is necessary in the circumstances (e.g., Lundberg & Rankin, 2014; Rochlin et al., 1987). Therefore, approaches for developing expertise that take into account the range of work demands for which workers can be responsible have the potential to foster the development of more robust systems.

As an example, a specific goal of training may be to expand workers' behavioral repertoires systematically so that they develop overlapping capacities, at least partially. This could be achieved in a number of ways including through designing suitable pathways for progression. In the case of the future maritime surveillance system mentioned earlier, all of the sensor systems on the aircraft can be controlled from any of the six workstations in the cabin. Therefore, if the progression pathway is designed so that a radar sensor operator may progress to the role of an acoustics sensor operator on the aircraft, the acoustics sensor operator would retain some capacity for operating the radar and could carry out this responsibility if necessary, which creates flexibility within the system. A WOP diagram highlights the progression pathways that may be feasible by capturing the sets of work demands for which actors in a system can be responsible.

Similarly, interface designs may support the utilization of expertise in the workplace by accommodating the information requirements associated with the spaces of behavioral opportunities afforded to individual actors. Consequently, workers won't be restricted unnecessarily to a narrow range of tasks, simply because they haven't been provided with the information relevant for exploiting their abilities to engage in flexible action. One potential concern with this approach is that workers may become overloaded with information. However, as field studies show, workers may need to, and do, take responsibilities for tasks that are not considered to be part of their normal jobs (Bigley & Roberts, 2001; Bogdanovic et al., 2015; Lundberg & Rankin, 2014; Rochlin et al., 1987). Therefore, rather than expecting them to do so without adequate support, it makes sense to consider design strategies that can provide them with the necessary information in a way that takes account of the inherent cognitive capabilities and limitations of humans.

For example, the ecological interface design approach (Bennett & Flach, 2011; Burns & Hajdukiewicz, 2004; Rasmussen & Vicente, 1989; Vicente & Rasmussen, 1990, 1992) recognizes that workers may engage in flexible action, and that they may contribute to multiple system ends or purposes by exploiting multiple means or resources. Accordingly, an ecological interface organizes information according to means–ends relations, and several controlled experimental investigations have demonstrated that ecological interfaces may lead to better performance by workers when compared with conventional displays (Vicente, 2002). As Naikar and Elix (2016b) suggest, this approach may be extended to account for the full set of work demands for which workers can be responsible, based on a WOP diagram of the actors in the system. The resulting interface would have the potential to support workers in monitoring or considering the effects of their actions on multiple system functions, in an integrated way, so that the information-processing burden on workers may be reduced. If all of the work demands relevant to particular actors cannot be accommodated effectively, the WOP diagram provides a basis for systematically creating partially overlapping information spaces for workers, consistent with the training approach discussed earlier, to foster greater flexibility within the system. For instance, a radar sensor operator and an acoustics sensor operator may be provided with partially, if not fully, overlapping information spaces. The overall intent is to accommodate as many of the work organization possibilities encompassed in a WOP diagram as is feasible, thereby maximizing the behavioral opportunities of individual actors and the structural possibilities of multiple actors, such that the system has greater capacity for self-organization.

Finally, it is important to acknowledge the potential negative consequences of providing completely open designs to workers. Not giving workers specific procedures or instructions about which jobs to perform, and how to perform them, may have the effect of increasing the demands of their jobs, which could lead to errors or delays in the system. However, it is conceivably just as dangerous to have workers adapt their work practices without adequate support, when the guidance they are provided with is found to be unsuitable in the circumstances, as may be the case in unusual or unforeseen situations. Clearly, a balance must be achieved between the two extremes. Here, it is suggested that a constraint-based approach that systematically demarcates the boundaries that must be respected by workers for safe and productive performance, while giving them the flexibility to capitalize on their behavioral opportunities and structural possibilities within those limits, offers a means for striking this balance, specifically by providing workers with constrained flexibility in the workplace.

DISCUSSION

This chapter has presented a perspective of human expertise in sociotechnical systems based on the phenomenon of self-organization. Consistent with the ideals of the field of cognitive systems engineering, this perspective is based on empirical observations of how work is achieved in context or in situ. Furthermore, this perspective incorporates an emphasis on design.

Early in this chapter, some of the essential characteristics of sociotechnical systems were described, highlighting the need for adaptation in the workplace. In addition, empirical studies of a variety of systems that demonstrate that workers in sociotechnical systems adapt not only their individual behaviors, but also their group structure or organization, were presented. The concept of self-organization is important in this context because it recognizes that adaptations in behavior and structure are closely intertwined, and it explains how these adaptations may occur spontaneously, continuously, and relatively seamlessly, at least in well-established systems, without external intervention, a priori planning, or centralized coordination.

Given the necessity of self-organization in sociotechnical systems, it was argued that expertise in the workplace is best understood as the ability of an actor to engage in flexible behaviors, in view of the structural possibilities of multiple actors, rather than simply as the ability of an actor to undertake a set of typical tasks or task sequences with a high level of proficiency. This viewpoint is consistent with the observations that workers in sociotechnical systems may carry out unusual tasks, for which they have little prior training or experience; that they may undertake their usual tasks under changed or novel conditions, such that new courses of action or behaviors are necessary; and that although their execution may not be regarded as skillful or flawless if assessed against a normative benchmark, their performance may be considered critical or even masterful in the circumstances.

From this viewpoint, then, a specific worker's level of expertise is regarded as their current space of possibilities for behavior, or behavioral repertoire, and an experienced worker is seen as having a greater behavioral repertoire than an inexperienced worker. By having a greater variety of possibilities for behavior, an actor can attend to a greater variety of problems, or tasks, with different cognitive strategies, so that courses of action that are fitted to the circumstances may be enacted. Furthermore, as discussed earlier in this chapter, an expert is regarded as a notional concept representing someone whose behavioral repertoire matches their afforded behavioral opportunities or, in other words, the action possibilities afforded to them by their work context. Although this concept of an expert may never be realized in practice, workers' behavioral repertoires may expand over time with training and experience.

The self-organization perspective of expertise is important for guiding the design of sociotechnical systems so that they have greater capacity for dealing with unstable, uncertain, and unpredictable work environments. This perspective recognizes that the greater each worker's behavioral repertoire, the greater the system's structural possibilities. Therefore, designs that support workers in developing their behavioral repertoires to match their afforded behavioral opportunities, even if this is never realized in full, and in utilizing their expertise by systematically upholding them in undertaking flexible action, when it is necessary, have the potential to foster the emergence of novel group structures from individual actors' spontaneous behaviors.

In this chapter it was proposed that the WOP diagram provides a starting point for achieving these design objectives. This diagram models the action possibilities afforded by the work context at the level of the behavioral opportunities of individual actors and the structural possibilities of multiple actors. Therefore, by coordinating the designs of such features as the interfaces, teams, training, and human–automation interaction around the organizational constraints specified in a WOP diagram, it may be possible to support the development and utilization of expertise in the workplace in a way that is consistent with a system's inherent capacity for self-organization.

Future research challenges for the self-organization perspective of expertise include the need to demonstrate and validate the ideas presented in this chapter. One way in which this objective may be approached is to develop designs, for training systems and interfaces for instance, that take account of the work organization possibilities encompassed in a WOP diagram, and subsequently examine whether the resulting solutions have the capacity to support the development and utilization of expertise in the workplace in a way that is compatible with the self-organization phenomenon. In particular, it is necessary to examine whether such designs can support individual actors in engaging more effectively in flexible behaviors and whether, as a result, it is possible for new or different structural forms or organizations to emerge from the collective behaviors of multiple actors, which are conducive to a system dealing with a dynamic and ambiguous work environment.

Finally, the self-organization perspective has broader implications for the study of expertise. Research investigating the nature of expertise should not be limited to individuals' behaviors or cognitive structures but should extend to the social structures within which they enact action. In sociotechnical systems, tasks or responsibilities are necessarily distributed across many individuals, so that effective, orchestrated action at the system level is socially constructed out of the spontaneous interactions of multiple cognitive agents. Flexible action and structural emergence are especially critical in dynamic, ambiguous work contexts, when there is limited opportunity for centralized coordination, a priori planning, or external intervention. Accordingly, studies of expertise must be concerned with such questions as what individual cognitive capabilities can yield effective organization, or collective action, in such complex settings, and how social structures can bound individuals' cognitive acts in ways that are fitted to circumstances, which are constantly changing or evolving. Such studies would advance a social basis for human cognition and expertise that—critically—accounts for the phenomenon of self-organization.

ACKNOWLEDGMENTS

We thank Jan Maarten Schraagen and Paul Ward, who in their roles as co-editors of this *Handbook* provided insightful comments on our chapter, and Dr Seng Boey from the Defence Science and Technology Group for his careful reading of our chapter and detailed feedback as well. We are also grateful to Claire Dâgge for her assistance in preparing this manuscript for publication.

REFERENCES

Barth, S., Schraagen, J. M., & Schmettow, M. (2015). Network measures for characterising team adaptation processes. *Ergonomics* 58(8), 1287–1302.

Bennett, K. B., & Flach, J. M. (2011). *Display and interface design: Subtle science, exact art.* Boca Raton, FL: CRC Press.

Bigley, G. A., & Roberts, K. H. (2001). The incident command system: high-reliability organizing for complex and volatile task environments. *Academy of Management Journal* 44(6), 1281–1299.

Bogdanovic, J., Perry, J., Guggenheim, M., & Manser, T. (2015). Adaptive coordination in surgical teams: an interview study. *BMC Health Services Research* 15(1), 128.

Bureau d'Enquêtes et d'Analyses. (2012). Final report on the accident on 1st June 2009 to the Airbus A330-203 registered F-GZCP operated by Air France flight AF 447 Rio de Janeiro—Paris (English edition). Retrieved from https://www.bea.aero/docspa/2009/f-cp090601.en/pdf/f-cp090601.en.pdf

Burns, C.M., & Hajdukiewicz, J. R. (2004). *Ecological interface design.* Boca Raton, FL: CRC Press.

Deepwater Horizon Study Group. (2011). *Final report on the investigation of the Macondo Well blowout.* Berkeley, CA: University of California, Berkeley.

Dekker, S. (2003). Failure to adapt or adaptations that fail: Contrasting models on procedures and safety. *Applied Ergonomics* 34(3), 233–238.

Eason, K. (2014). Afterword: The past, present and future of sociotechnical systems theory. *Applied Ergonomics* 45(2), 213–220.

Endsley, M. R., Hoffman, R., Kaber, D., & Roth, E. (2007). Cognitive engineering and decision making: An overview and future course. *Journal of Cognitive Engineering and Decision Making* 1(1), 1–21.

Fuchs, C. (2004). Knowledge management in self-organizing social systems. *Journal of Knowledge Management Practice* 5. Available at http://www.tlainc.com/articl61.htm. Accessed December 9, 2017.

Gerson, E. M., & Star, S. L. (1986). Analyzing due process in the workplace. *ACM Transactions on Office Information Systems* 4(3), 257–270.

Haken, H. (1988). *Information and self-organization: A macroscopic approach to complex systems.* Berlin: Springer.

Heylighen, F. (2001). The science of self-organization and adaptivity. *The Encyclopedia of Life Support Systems* 5(3), 253–280.

Hoffman, R. R., & Woods, D. D. (2011). Beyond Simon's slice: Five fundamental trades-offs that bound the performance of macrocognitive work systems. *IEEE Intelligent Systems* 26(6), 67–71.

Hofkirchner, W. (1998). Emergence and the logic of explanation. An argument for the unity of science. *Acta Polytechnica Scandinavica, Mathematics, Computing and Management in Engineering Series* 91, 23–30.

Hollnagel, E. (2016). CSE—Cognitive systems engineering. Retrieved from http://erikhollnagel.com/ideas/cognitive-systems-engineering.html.

Hollnagel, E., & Woods, D. D. (1983). Cognitive systems engineering: New wine in new bottles. *International Journal of Man-Machine Studies* 18, 583–600.

Hollnagel, E., & Woods, D. D. (2005). *Joint cognitive systems: Foundations of cognitive systems engineering.* Boca Raton, FL: CRC Press.

Hutchins, E. (1995). *Cognition in the Wild.* Cambridge, MA: MIT Press.

Karasek, R., & Theorell, T. (1990). *Healthy work: Stress, productivity, and the reconstruction of working life.* New York: Basic Books.

Klein, G. A. (1989). Recognition-primed decisions. In W. B. Rouse (Ed.), *Advances in man-machine system research* (Vol. 5, pp. 47–92). Greenwich, CT: JAI Press.

Klein, G., Wiggins, S., & Deal, S. (2008). Cognitive systems engineering: The hype and the hope. *IT Systems Perspective*, March, 95–97.

La Porte, T. R. (1996). High reliability organizations: Unlikely, demanding and at risk. *Journal of Contingencies and Crisis Management* 4(2), 60–71.

Leveson, N. G. (1995). *Safeware: System safety and computers.* Reading, MA: Addison-Wesley.

Lundberg, J., & Rankin, A. (2014). Resilience and vulnerability of small flexible crisis response teams: Implications for training and preparation. *Cognition, Technology and Work* 16, 143–155.

Militello, L. G., Dominguez, C. O., Lintern, G., & Klein, G. (2010). The role of cognitive systems engineering in the systems engineering design process. *Systems Engineering* 13(3), 261–273.

Militello, L. G., Sushereba, C. E., Branlat, M., Bean, R., & Finomore, V. (2015). Designing for military pararescue: Naturalistic decision-making perspective, methods, and frameworks. *Journal of Occupational and Organizational Psychology* 88(2), 251–272.

Naikar, N. (2013). *Work domain analysis: Concepts, guidelines, and cases.* Boca Raton, FL: CRC Press.

Naikar, N. (2018). Human-automation interaction in self-organizing sociotechnical systems. *Journal of Cognitive Engineering and Decision Making* 12(1), 62–66.

Naikar, N., & Elix, B. (2016a). A consideration of design approaches based on cognitive work analysis: System design and integrated system design. *Proceedings of the 34th European conference on cognitive ergonomics* (pp. 1–7). New York: Association of Computing Machinery.

Naikar, N., & Elix, B. (2016b). Integrated system design: Promoting the capacity of socio-technical systems for adaptation through extensions of cognitive work analysis. *Frontiers in Psychology* 7, 962.

Naikar, N., Elix, B., Dâgge, C., & Caldwell, T. (2017). Designing for self-organisation with the diagram of work organisation possibilities. In J. Gore and P. Ward (Eds), *Proceedings of the 13th International Conference on Naturalistic Decision Making* (pp. 159–166). Bath, UK: University of Bath.

Perrow, C. (1984). *Normal accidents: Living with high risk technologies.* New York: Basic Books.

President's Commission on The Accident at Three Mile Island. (1979). *The need for change: The legacy of TMI.* Washington, DC: U.S. Government Printing Office.

Rankin, A., Lundberg, J., Woltjer, R., Rollenhagen, C., & Hollnagel, E. (2014). Resilience in everyday operations: a framework for analyzing adaptations in high-risk work. *Journal of Cognitive Engineering and Decision Making* 8(1), 78–97.

Rasmussen, J. (1969). *Man-machine communication in the light of accident records* (Report No. S-1-69). Roskilde, Denmark: Danish Atomic Energy Commission, Research Establishment Risø.

Rasmussen, J. (1986). *Information processing and human-machine interaction: An approach to cognitive engineering.* New York: North-Holland.

Rasmussen, J., Pejtersen, A. M., & Goodstein, L. P. (1994). *Cognitive systems engineering.* New York: Wiley.

Rasmussen, J., & Vicente, K. J. (1989). Coping with human errors through system design: Implications for ecological interface design. *International Journal of Man-Machine Studies* 31(5), 517–534.

Reason, J. (1990). *Human error.* Cambridge, UK: Cambridge University Press.

Rochlin, G. (1993). Defining "high reliability" organisations in practice: A taxonomic prologue. In K. Roberts (Ed.), *New challenges to understanding organisations.* New York: Macmillan.

Rochlin, G. I., La Porte, T. R., & Roberts, K. H. (1987). The self-designing high-reliability organization: aircraft carrier flight operations at sea. *Naval War College Review* 40(4), 76–90.

Schraagen, J. M. (2011). Dealing with unforeseen complexity in the OR: The role of heedful interrelating in medical teams, *Theoretical Issues in Ergonomics Science* 12(3), 256–272.

Simon, H. (1969). *The sciences of the artificial.* Cambridge, MA: MIT Press.

Suchman, L.A. (1987). *Plans and situated actions: The problem of human-machine communication.* Cambridge UK: Cambridge University Press.

Vicente, K. J. (1999). *Cognitive work analysis: Toward safe, productive, and healthy computer-based work.* Mahwah, NJ: Lawrence Erlbaum.

Vicente, K. J. (2002). Ecological interface design: Progress and challenges. *Human Factors* 44(1), 62–78.

Vicente, K. J. (2006). Cognitive engineering: A theoretical framework and three case studies. *International Journal of Industrial and Systems Engineering* 1(1/2), 168–181.

Vicente, K. J., & Rasmussen, J. (1990). The ecology of human-machine systems II: Mediating "direct perception" in complex work domains. *Ecological Psychology* 2(3), 207–249.

Vicente, K. J., & Rasmussen, J. (1992). Ecological interface design: Theoretical foundations. *IEEE Transactions on Systems, Man and Cybernetics* 22(4), 589–606.

Ward, P., Gore, J., Hutton, R., Conway, G., & Hoffman, R. R. (2018). Adaptive skill as the *conditio sine qua non* of expertise. *Journal of Applied Research in Memory and Cognition* 7(1), 35–50.

Weick, K., & Sutcliffe, K. M. (2001). *Managing the unexpected: Assuring high performance in an age of complexity.* Jossey Bass: San Francisco.

Weick, K. E. (1993). The collapse of sensemaking in organizations: The Mann Gulch disaster. *Administrative Science Quarterly 38*(4), 628–652.

Woods, D. D. (1988). Coping with complexity: The psychology of human behaviour in complex systems. In L. P. Goodstein, H. B. Andersen, & S. E. Olsen (Eds), *Tasks, errors, and mental models: A festschrift to celebrate the 60th birthday of Professor Jens Rasmussen* (pp. 128–148). London: Taylor & Francis.

Woods, D. D., & Roth, E. M. (1988). Cognitive engineering: Human problem solving with tools. *Human Factors 30*(4), 415–430.

CHAPTER 11

IS EXPERTISE ALL IN THE MIND?

How Embodied, Embedded, Enacted, Extended, Situated, and Distributed Theories of Cognition Account for Expert Performance

CHRIS BABER

Introduction

In a telling anecdote, Gleick (1992) records a conversation between historian Charles Weiner and physicist Richard Feynman. Weiner was looking at a collection of Feynman's notes, and commented that is was good to have a record of work. Feynman tetchily replied, "No, it's not a record, not really. It's working. You have to work on paper, and this is the paper. Okay?" (Gleick, 1992, p. 409). In a nutshell, we have the crux of the debate in this chapter—was writing on the paper a record of the thinking done in Feynman's head, or was the writing on the paper (in order to solve equations) an act of thinking?

In this chapter, *conventional*, or *good old-fashioned* cognitive psychology is compared with a collection of alternative approaches. The approaches considered challenge the idea that cognition involves thinking done in the head using a *mental representation* (such as a schema or mental model). Against this characterization, advocates of the alternative approaches argue that cognition is constructed through the dynamic interaction of person and environment without the need for mental representation. Cognition thus involves events which arise from the experience of having *a* body, with a set of sensorimotor capacities and capabilities, in *an* environment. In other words, the foundations of cognition shift from mental representations to perception–action couplings; this is what Shapiro (2011) has termed the *replacement hypothesis*. From this perspective, expertise is more a matter of *doing* than it is of *knowing*, and the expert is someone whose actions on their environment indicate mastery of this interaction (in a way that makes the less expert person seem clumsy, hesitant, or awkward).

In order to appreciate why these approaches claim to be presenting ideas which cannot be accommodated by *conventional* cognitive psychology, one needs to recall that Descartes (1637/2005) distinguished between *res cogitans* (things that think) and *res extensa* (things that extend into the world, and which, by definition, do not think). This means that *res cogitans* need to be able to represent their environment in order to think about it, and this requires a form of mental representation, where data from the senses are used to build a model of the world against which actions can be planned, and consequences of the actions imagined. This would be the stuff of schemata or mental models in cognitive psychology. This separates perception *of* the environment from action *on* the environment in ways that the approaches considered in this chapter profoundly disagree. In place of this separation, the approaches claim that the dynamic interplay between perception and action is such that there is no need for mental representation; one simply *sees* the environment in terms of opportunities to act. The initial impetus for much of the work covered in this chapter developed from discussions in philosophy, particularly in the field of phenomenology.

Philosophical Foundations

Merleau-Ponty (1945/2014) claimed that we are embodied perceivers who act upon the world (and have the world act upon us). Merleau-Ponty (1945/2014) developed the idea of *intentionality*, which concerns the manner in which one is conscious of those features of an object that influence a given use rather than the whole collection of its features. This leads to the concept of an *intentional object*, which anticipates the notion of *affordance*, in that the way in which one sees an object is affected by the way in which one will interact with it. To develop this notion further, imagine that you are asked to use blocks to make a model, copying a picture. In such a task, Ballard et al. (1995) show that participants looked at the picture at strategic moments; e.g., an initial glance might identify the colour of the block, and a subsequent glance might determine its precise position in the pattern. Such a *minimal memory strategy* shows how people respond to task constraints in their visual sampling of the environment as they progress through different stages in task performance. In other words, rather than referring to a detailed mental model to guide action, participants visually sampled the environment for specific information as they needed it. Eye movements are influenced by the type of task being performed, e.g., making a sandwich (Ballard et al., 1995) or engaging in sword-fighting (Hagemann et al., 2010). Key to this proposal is the idea that perception is continuous over time. This means that there is an ongoing interaction between the person and the environment, without the need to construct *percepts* (or other mental reconstruction of features from the environment).

For Gibson (1977, 1979) the purpose of the sensory system was to manage active movement through a sensation-rich environment. In broad terms, affordances are possibilities for action by organisms with certain abilities (Greeno, 1994). This suggests that objects in the environment can be regarded as resources for action, providing the individual is capable of using that resource (Shaw & Turvey, 1981). In this respect, environmental constraints (in terms of properties of objects) are responded to in terms of bodily constraints (in terms of effectivities). This means that a given individual responds to an *ecological niche* (Gibson, 1979).

Following Merleau-Ponty, Dreyfus (2002) emphasized the expert's "vast repertoire of situational discriminations" (Dreyfus, 2002, p. 372). This involves classifying high-level features which can enable the "intuitive situational response that is characteristic of expertise" (Dreyfus, 2002, p. 372). The *novice* is directed to specific features of the environment (by an expert mentor) and learns to associate these features with specific actions. For example, each piece on a chessboard has a particular way of moving and can be captured by other pieces. In this case, a set of rules that pair feature with action are learned. This can be a demanding process and limits the ability of the novice to adapt to changes in context. The Advanced Beginner begins to recognize collections of features, which can be associated with collections of rules. So, on a chessboard a *weakened king's side* could be recognized and a set of moves to exploit this could be learned. Here, the *features* become chunked into a larger pattern, and the associated actions chunked into sequences. The Competent Person has learned several of these patterns, and the challenge is to prioritize these in terms of plausibility. Over the course of this development, one can imagine the assimilation of feature–action pairings into larger patterns, and the ability to discriminate between these patterns. The transition from Competence to Proficiency begins to see a shift from recall to something that feels more like intuition. The Proficient Person's response to the chessboard is in terms of salient patterns of features *standing out*, which can allow a goal to be *seen*. In this case, rather than interpreting the state of the board and recalling similar states, perception becomes much more like Gibson's notion of *affordance*. The argument could be that repeated exposure to situations allows the development of experience which is different in kind to the previous forms of response. What the Proficient Person might lack is the ability to *see* the most appropriate strategy for achieving a goal. The Expert sees what needs to be done, in response to salient aspects of the situation, and also how best to achieve this.

So far, the argument has been about perception–action coupling as a way of responding to affordances in the environment. This implies that there is no need for any mental representation. However, there remains an impasse in such accounts in that they do not clearly articulate *how* the expert is able to remember which actions to perform (although, as we shall see in later sections, the question of *memory* could be side-stepped through an appeal to the idea that the body, through its adaptation to the environment, acts without recourse to mental representation of that environment). Appealing to a notion of mental representation (albeit one which does not require propositions), Barsalou (1999, 2008) proposed that successful actions on the environment can be stored in the form of perceptual symbols. These are multimodal traces of neural activity arising from affordances and motor information present during sensorimotor experience. When a given perceptual symbol is recalled then the traces are reactivated. By analogy, work on *mirror neurons* in the motor cortex suggests that watching an activity in which one has proficiency results in activation of the same cortical regions required to perform the movement. Calvo-Merino et al. (2005) had professional ballet dancers, experts in capoeira (a Brazilian martial art that combines dance with acrobatics), and non-expert control participants view a video of people performing ballet and capoeira while in a functional magnetic resonance imaging (fMRI) machine. The analysis showed increased activation in the premotor cortex when people watched the dance style with which they were familiar (and no such activation for either the controls or when an unfamiliar style was viewed).

Expert Performance

> An elite cricketer, for example, with less than half a second to execute an ambitious cover drive to a hard ball honing directly in at 140 km/h, draws not only on smoothly-practised stroke play, but somehow also on experience of playing this fast bowler in these conditions, and on dynamically-updated awareness of the current state of the match and of the opposition's deployments, to thread an elegant shot with extraordinary precision through a slim gap in the field. It's fast enough to be a reflex, yet it is perfectly context-sensitive. (Sutton et al., 2011, p. 80)

This example emphasizes how individuals coordinate movements in a well-controlled manner that is sufficiently flexible to allow adaptation to changing environmental demands. In one sense, expertise could simply be defined in terms of coordinating responses to environmental demands as efficiently as possible. Coordination (Bernstein, 1967) is a matter of managing the degrees of freedom of the body in order to create a controllable system. Thus, *coordination* is a process of constraining the degrees of freedom, *control* is a process of defining parameters for the coordinated system, and *expertise* is the ability to select optimal parameters for control (Kugler et al., 1982; Newell, 1985). For (some) cognitive psychologists, the ability to act in a coordinated manner suggests that the *expert* has developed a set of motor programs which are activated in response to situational demands, e.g., using motor schemata (Schmidt, 1975). For dynamic systems researchers, the action occurs as part of the dextrous, dynamic interplay between person and environment (Jagacinski & Flach, 2003). In the former, action is performed as a result of the feed-forward control from perception of the environment to mental representation to motor program to action. Here, the brain serves as a computing machine which samples the environment in order to represent it and then determines the most effective response to this representation. In the latter, the brain serves as a controller which (largely but not entirely) is continually responding through feedback from the movement of the body in relation to the environment.

How the Body and the Environment Influence and Constrain Cognition

Newell (1986) pointed out that activity in the environment emerges in response to specific constraints. Interpretation of the environment is influenced by the proportions of our body (see the next section, "Exploring Affordances"). The challenge presented by such bodily constraints relates to the selection of an appropriate action, given the range of options available (e.g., in terms of degrees of freedom) and the likelihood of successful completion of the action (while also minimizing risk, error, etc.). The environment constrains activity in terms of the opportunities it presents, for instance, in terms of the location or appearance of objects available to the individual, or the interaction between these objects and the bodily constraints of the individual.

While the bodily and environmental constraints can limit or encourage specific actions, the choice of action will also depend on task constraints. These could include a purpose for completing the action as well as some criteria that might define *good* or *acceptable* performance. Newell (1985, 1986) saw the development of skill in terms of a hierarchy in which variables can be usefully combined into a behavioral unit; some of these variables will be essential to performance, while others will be non-essential, but can still influence the performance. From this perspective, selection of variables to include in the behavioral unit becomes a defining feature of coordinated activity. Having defined the variables, the next level of the hierarchy involves specifying parameters for these variables to control the behavioral unit. Once the set of control parameters are applied to the most relevant set of variables, one can begin to speak of expert performance. Thus, people who exhibit higher levels of ability in jewelry-making (compared to less able participants) tend to apply less force and a greater movement amplitude when they use a piercing saw to cut metal (Baber et al., 2015) or people with greater dexterity in flint-knapping tend to manage kinetic energy more efficiently than novice performers (Bril et al., 2010) while also exhibiting a greater range of joint angles (Biryukova & Bril, 2008) and better ability to concurrently modify several parameters (Vernooij et al., 2012).

EXPLORING AFFORDANCES

Relating affordances to individual capabilities raises the question of how these capabilities are perceived. We can judge whether we can walk straight through the gap or whether we need to turn sideways, or avoid the gap entirely (Stefanucci & Geuss, 2009; Warren & Wang, 1987). Indeed, the ability to rapidly assess environmental features in order to judge whether a particular action could be performed can be applied to all manner of situations, for instance, step-onto-ability (Warren, 1984); walking-up-ability (Kinsella-Shaw et al., 1992); sit-on-ability (Mark, 1987); step-across-ability

(Cornus et al., 1999); and pass-underability (Stefanucci & Geuss, 2010). Related to this class of body-scaled affordances, people can also judge whether an action is possible, i.e., action-scaled affordances, such as reachability, where people with longer arms estimate objects to be close to them (Proffit & Linkenauger, 2013) or people holding tools estimate objects to be closer (Witt & Proffit, 2008). In terms of task-constraint, Rosenbaum et al. (2012) review the ways in which people adopt uncomfortable postures because they are seeking comfort in an end-state (wine glasses) or in order to exert maximal torque (faucets). These various examples relate to Shapiro's (2011) *conceptualization hypothesis*, which sees cognition grounded in, and constrained by, perception–action systems.

Another way to consider the *x-ability* (where x is any of the prefixes used in the previous paragraph) is to assume two distinct but complementary routes from object to action (Humphreys, 2001): (i) perception of specific features of an object (structural description) and then associating these with knowledge of how to use that object (action description); (ii) affordance, as a direct link between the structural description of an object and the action description of how to handle that object. This latter route is reminiscent of Merleau-Ponty's (1945/2014) notion of intentional object, as well as Gibson's (1979) notion of affordance which it quotes. Evidence in favor of this *direct link* comes from studies on the *orientation effect* (Tucker & Ellis, 1998). When the orientation of a handle (in an image of an object) points to the side of the screen that matches the hand used to make a response in a decision task, reaction times are faster. This implies a priming of the orientation of the handle with the movement of the hand. Thus, action selection and action specification interact with each other (rather than one following the other as hierarchical theories of motor control assume) and this requires some form of *motor attention* to manage this interaction.

Not only can we respond to the affordances of physical objects in anticipation of acting on them, we can also respond to the dynamics of moving objects. The best known example of this involves the ability to catch a flying object, i.e., the outfielder problem (Chapman, 1968; Fink et al., 2009). Imagine playing a sport in which a key challenge is to catch a ball that has been hit by another player. As the ball is hit, it moves toward you and you are going to catch it. One explanation of how this might be achieved involves the prediction of the ball's movement (and its most likely future location) through the application of the laws of physics, particularly through some estimation of ballistic flight. As Saxberg (1987a,b) explains, the catcher can estimate initial parameters (such as the size and mass of the ball and its starting point) to calculate direction, velocity, angle in order to simulate the ball's movement and then predict its final location. For a reasonably smart high school pupil, this is not a challenging mathematical problem and, given time and mental resources, it is a calculation that many people can perform. The question, though, is not whether people *could* do this calculation but whether, during the heat of the moment in a sports game, this is what people *actually* do. Two plausible accounts suggest that the catcher uses visual information about the ball's movement to influence their choice of location to move to in order to catch it. Optical acceleration cancellation (OAC) involves the catcher running along the ball's flight path at a speed which is enough to make the ball look as if it is moving at a constant velocity (Chapman, 1968). Linear optical trajectory (LOT) involves the catcher running sideways to make the ball appear to fly in a straight line (McBeath et al., 1995). What each strategy offers is a simple (non-mathematical, as far as the catcher is concerned) adaptation to the visually perceived movement of the ball. In other words, responding to the relationship between catcher and ball is sufficient to remove the need for internal representation, simulation, or calculation. In terms of the development of expertise, the notion of affordance can be used to explain how the skilled sports player develops increased sensitivity to perceptual information, such as a cricket batsman's ability to interpret how a ball will be delivered from the body movement, arm position, and head adjustments of a bowler prior to releasing the ball (Sutton et al., 2011).

Embodied Cognition

There are several schools of thought in embodied cognition: perhaps three (Shapiro, 2011), perhaps as many as six (Wilson, 2002). What these schools of thought have in common is a broad view that "the brain is not the sole cognitive resource we have available to us to solve problems. Our bodies and their perceptually guided motions through the world do much of the work required to achieve our goals, replacing the need for complex internal mental representations" (Wilson & Golonka, 2013, p. 1). We have already considered Barsalou's perceptual symbol system and Shapiro's conceptualization hypothesis (see the sections "Philosophical Foundations" and "Exploring Affordances", respectively). Lakoff and Johnson (1999) describe how common metaphors are typically grounded in the nature of our bodies and experiences in the world (the future is *forward*, power is *up*, relationships are *a journey*).

In the developing area of radical embodied cognitive science (RECS), Chemero (2009) elucidates the relationship between human activity and the environment in which it is performed. This approach moves forward from principles of Gibson (e.g., that perception is direct, for action, and of affordances) in order to explain cognition without recourse to the "mental gymnastics" (Chemero, 2009, p. 18) surrounding internal representation. In order to achieve this, RECS emphasizes the contextual interplay between human and environment, and builds explanations of how thinking can be captured through nonlinear dynamics models of human activity (e.g., Guastello & Gregson, 2016; Haken, Kelso, & Bunz, 1985; Kelso, 1997). Central to this view is that the interactions between people and their environments can be thought of as self-organizing, loosely coupled systems. This means that the *information* needed to act is available in the environment and that there is no need for the additional *representation* of this information through perceptual reconstruction. In this account, *information* is perceived in the environment in terms of its relation to specific states of affairs (for specific individuals). This takes the Gibsonian notion of *information* and extends it to higher order cognition. It also points to the various forms of *x-ability* that were discussed in the section "Exploring Affordances".

Embedded Cognition

If some aspect of cognition corresponds with some aspect of the environment, as a result of evolution and natural selection, then that aspect of the environment is part of the cognitive system (Rowlands, 1999). In this sense, *embedded* refers to the reciprocal interaction between the body and the physical world, which results in cognitive processes. Crucially, people make use of partial representations to support action in context (Gibbs, 2006; Rupert, 2009; Wheeler, 2005). Embedded cognition, therefore, relates to the manner in which we develop strategies that allow us to structure our environments in order to simplify cognitive demands and to support problem-solving tasks.

In terms of simplifying cognitive demands we can consider the ways in which people make rapid decisions, using fast-and-frugal heuristics (Gigerenzer & Todd, 1999). One such heuristic is termed *take-the-best* (TTB) (Gigerenzer & Todd, 1999) and assumes that people employ rules which define relevance of information (i.e., the probability that a specific piece of information will lead to the correct decision), together with rules for searching and criteria for stopping (i.e., completing the search with a correct decision). Thus, the decision maker will seek out information with high relevance first and then add further information (with decreasing relevance) until the stopping criterion is reached. This means that there is no need to conduct an exhaustive search of information. However, it does require some means of defining relevance. Perhaps the expert has prior experience with this type of decision and can define the information needed (say, in a *slot-and-frame* representation). More likely, the expert has learned a Bayesian weight for information values (Chen et al., 2017).

Enactive Cognition

In enactive cognition, the *situated cognitive agent* is *tuned* to opportunities in the environment around them (Varela et al., 1991). In order to be tuned, it is necessary for the objects, features, and opportunities to be relevant for the individual, and for this relevance to have been co-created (through interaction of person and the environment), which means that there is no need for explicit computation of *rules* to guide activity because the individual is tuned to their environment, and cognition can be described by the analogy of *laying down a path in walking* (Varela, 1987).

There is an implication, in Varela's view of enactivism, that the lack of representation (which we also see in Dreyfus' account of the "expert") could imply a lack of thought, and that expertise becomes a form of *embodied intelligence*. Gallagher (2017) discusses expert performance across a range of disciplines and notes the importance of *performative awareness*. In this, experts are able to attend to and reflect upon the movement and behavior of their body. For example, a tennis player might be aware that different muscle groups are active during the playing of different shots. In this way, the expert is able to develop a body schema which can be monitored and managed—but which is treated in an objective manner.

The Extended Mind

Extended cognition proposes that people offload not just the role of representation but also the role as *cognitive agent* to artifacts; for instance, electronic diaries do not just store dates but also actively remind us of them, and smartphones have *contacts* that absolve us from the need to remember phone numbers. In this respect, the *mind* (brain or however one wishes to characterize the *thing that thinks*) is extended into the environment, or the objects contained in the environment. In other words, cognition involves a hybrid of the biological and non-biological (Clark, 1997) and activity involving the use of artifacts *is* cognition (Clark & Chalmers, 1998). This is not simply saying that artifacts are repositories of information that we consult, but rather that they *participate* in cognitive activity. By way of a defining characteristics, Clark and Chalmers (1998) offered a *parity principle* which states that "If, as we confront some task, a part of the world functions as a process which, *were it to go on in the head*, we would have no hesitation in recognizing as part of the cognitive process, then that part of the world *is* (so we claim) part of the cognitive process" (Clark & Chalmers, 1998, p. 8).

In terms of expertise, Clark (2008) makes use of Beach's (1988) account of bartenders in a busy bar taking a large drinks order, and lining up the glasses that will be used as a reminder of the order.

> The problem of remembering which drink to prepare next is thus transformed, as a result of learning within this prestructured niche, into the problem of perceiving the different shapes and associating each shape with a kind of drink. The bartender, by creating persisting spatially arrayed stand-ins for the drink orders, actively structures the local environment to press more utility from basic modes of visually cued action and recall. In this way, the exploitation of the physical situation allows relatively lightweight cognitive strategies to reap large rewards. (p. 62)

While this example provides grist to the extended mind mill, it relates to the numerous examples that have been gathered together under the research umbrella of *situated cognition*.

Situated Cognition

For the approaches reviewed so far, the common emphasis on the interaction with the environment could lead one to assume that these approaches are merely variations on situated cognition. Situated cognition explores how knowledge is constructed within and linked to the activity, context, and the culture in which it was learned (Brown et al., 1989). Classic studies, reported by Lave and Wenger (1990), focus on contexts as diverse as midwives in Mexico, tailors in Liberia, butchers in US supermarkets, US Navy quartermasters, and recovering alcoholics in the USA. In these studies, novices engage in a cognitive apprenticeship (see Collins et al., 1988), which allows key skills, values, and traditions to be acquired. Apprentice tailors (Lave & Wenger, 1990), for instance, begin by ironing finished garments (which tacitly teaches them a lot about cutting and sewing). From this, expertise is full participation in a community of practice.

In a study of learning algebra by US 8th grade school-children, Kellman et al. (2008) found that pupils could solve equations of the form $x+4 = 12$, with 80 percent accuracy but in an average time of 28 s. This suggests that, while the students might have possessed sufficient knowledge and procedures to solve the equation, they were not able to *see* the solution. This is what E. Gibson (1969) termed *perceptual learning*. Kellman et al. (2008) showed that pupils could be taught to learn to perceive key attributes in each stage of the solution process (and that they could then generalize these attributes to other processes). This is a good example of *practical thinking* (Scribner, 1984). In similar fashion, objects in the environment are used to work out offers in supermarkets (Lave, 1988) or estimate quantity of milk in a milk-processing plant (Scribner, 1984). In these examples, expertise is less about mathematical ability or ability to apply algorithms, so much as the capability to develop, apply, and share a repertoire of approaches to solving problems that minimize cognitive effort and respond to opportunities in the environment (Suchman, 1987). In this sense, the environment *is* the representation of the problem space (Clancey, 1993). What experts learn is not a way of (internally) representing the environment, but a way of seeing it in terms of possibilities for action.

Distributed Cognition

Hutchins (1995) noted that cognitive processes are not solely in the head but arise through our interactions with things or other people in our environment. The actions could be physical, e.g., in the sense that some objects might *afford* picking up, or could be cognitive, in the sense that some objects might support a particular form of information-processing; for example, price labels on foodstuffs in a supermarket might support arithmetical calculation. This relates to Larkin's (1989) notion of *display-based problem solving*, in which a display is used to represent the current state of a problem and in which "important parts of problem solving be done by perceptual rather than logical inference" (Larkin, 1989, p. 340). A central tenet of distributed cognition is that the representations that are manipulated and exchanged by the agents within a system make up the system's *mental state* (Hollan et al., 2002). This distribution of cognitive activity can reduce cognitive overhead costs (i.e., less effort), improve efficiency (i.e., faster, fewer errors), and enhance the effectiveness (i.e., coping with harder problems) of that activity (Kirsh, 2013). Artifacts that are used in a work environment become changed by their use, and these changes provide cues for subsequent use (Nemeth, 2003).

In their study of people playing the computer game Tetris, Kirsh and Maglio (1994) showed that much of the activity involved *epistemic actions* in order to simplify the problem-solving task, rather than to implement a plan (pragmatic actions). In a similar manner, allowing people to tackle a problem with physical objects (as opposed to a paper-based version of the task) can influence the strategies that are applied. For example, in the *17-animal problem* (where participants have to place 17 animals into pens with an even number of animals in each pen), the ability to make a physical model improved the likelihood of solving the problem (Steffensen et al., 2016; Vallee-Tourangeau et al., 2016).

Baber et al. (2006) describe crime scene examination as a distributed cognition process in which the environment and the objects it contains become resources for action for experienced crime scene examiners, affording interpretations (such as cueing what evidence to recover) that are not available to the novice or untrained observer. The search strategies of experienced crime scene examiners differ from those of novices less in terms of how they consider the way in which the crime was committed and more in terms of how they consider the ways in which items might become evidence, that is, provide sufficient material for other people in the investigation (such as forensic scientists in the laboratory) (Baber & Butler, 2012).

Concluding Remarks

In broad terms, each of the approaches considered in this chapter make the uncontroversial argument that the brain is situated in a body; the body is situated in an environment; activity arises through the interaction between body and environment. Where they become controversial is in terms of the relationship between the brain (and its activity) and the way that the body and environment interact. In part, this chapter has been concerned with the question of where does cognition occur? If it occurs solely in the brain, then expertise must be about improving the brain. If it arises in the interactions between people and their environment, then expertise must be about enhancing sensitivity to these interactions. Before considering these questions, it might be useful to think about ways in which these approaches could be studied.

Research Methods

The question for methodology is how should expertise be studied from the approaches considered in this chapter? After all, much of the debate that has been covered could feel overly philosophical. The primary concern of the approaches considered in this chapter, I believe, would be less to do with quantifying the activity of expert practitioners and more to do with understanding their lived experience, and with practices that involve the dynamic interaction between human and environment. To this end, the methods ought to focus on issues relating to the phenomenology of expert performance, and in broad terms, this means asking the expert what it feels like to be an expert. I will discuss this in the section "Phenomenological Interview" but begin with consideration of in-the-field recording (see "Capturing Data to Analyze Expertise in Context") and analysis methods that focus on expert–environment dynamics.

Capturing Data to Analyse Expertise in Context

In order to capture data concerning expert performance, we need to decide whether it is more important to capture the person's actions or their thoughts. Given the arguments in this chapter about the role of representation in expertise ('representations' (if such exist) which underpin expertise could be difficult to articulate), it makes sense to consider how data could be collected from expert actions. Key to this question is the assumption that the primary concern is the way in which people interact with their environment. To this end, cameras could be mounted on the person's head or body, and this could record what the person is doing, where the person is looking, etc. This is the basis of data collection in previous work on crime scene examination, for example, Baber and Butler (2012). Alternatively, one could put sensors on the person and the objects that are being used. For example, in a study of jewelry-making, accelerometers were fitted to the handles of saws (Baber et al., 2015). In both cases, the initial aim is to provide an egocentric perspective on the performance of tasks in the expert's environment. The choice of recording is less important than the subsequent analysis.

For video recording, one could either have the person provide commentary "live," as they perform their work (Baber & Butler, 2012) but this can be subject to the usual problems associated with verbal protocol (Ericsson & Simon, 1993; see Chapter 17, "A Historical Perspective on Introspection: Guidelines for Eliciting Verbal and Introspective-Type Reports," by Ward, Wilson, Suss, Woody, & Hoffman, this volume). Alternatively, one could replay the video to the expert after recording and ask for a commentary on what they are watching (Mollo & Falzon, 2004), using auto-confrontation (where they comment on their own activity) or allo-confrontation (where they comment on the activity of another person). This could be particularly beneficial if combined with the interview approach considered in the next section, "Phenomenological Interview."

For data from sensors, the analysis would be concerned with the manner in which dextrous performance was handled. This could be explored through the analysis of movement parameters (Bril et al., 2010) or in terms of nonlinear dynamics, as recommended by Chemero's (2009) radical embodied cognitive science. Thus, $1/f$ scaling provides a straightforward approach to distinguishing levels of expertise (Baber & Starke, 2015) and can be used to characterize "the coordinative, metastable basis of cognitive function" (Kello et al., 2007).

Phenomenological Interview

As Gallagher and Zahavi (2008) point out, the aim of a phenomenological interview is not to capture idiosyncratic, personalized experience but to "capture the invariant structures of experience" (p. 28).

In terms of conducting a phenomenological interview, the interviewer begins with an explication phase. This involves the interviewee providing general accounts of their experience, and then focusing in on specific instances (cf. Chapter 18, "'Close-to-Practice' Qualitative Research Methods," by Yardley, Mattick, and Dornan, this volume). The reader might see in this approach some similarity with critical decision method (CDM) interviews (Hoffman et al., 1998), in which the interviewer is asked to focus on a specific instance (see also Chapter 19, "Incident-Based Methods for Studying Expertise," by Militello & Anders, and Chapter 17, "A Historial Perspective on Introspection," by Ward et al., both this volume). In the CDM, the interview could involve probing questions to get a rich picture of the *why* and the *what* of the experience. In the phenomenological interview, the emphasis is to move beyond these concerns to the *how* of doing something. For example, the interviewer invites elaboration of a statement by asking "how do you x...?" Initially, the account of the expert could be vague, abstract, and poorly articulated. This could be because this understanding is tacit (Polanyi, 1966). In a study of skilled coping in concert musicians, Høffding (2014) interviewed members of the Danish String Quartet. The study was based on Dreyfus's notion of skill (see the section "Philosophical Foundations"), and this influenced the general direction of the interviews but the questions were developed in response to the statements of the musicians themselves. For example, one of the musicians (F) spoke about "disappearing" in the music.

[F]: The deeper you are in, the less you observe the world around you... and I had this especially powerful experience... where I completely disappeared. I remember that it was an incredibly pleasant feeling in the body. And it was incredibly strange to come back and at that point I spent a few seconds to realize where I had been. I had been completely gone and with no possibility of observing .... It was this intense euphoric joy.

[INTERVIEWER]: OK, but if you are certain of having played, you cannot have been completely gone, so you must have known that you were playing, or?

[F]: Weeell... in this case I cannot completely answer you. You can say that it was easy for me to figure out that I had played at the time I was finished.

[INTERVIEWER]: But how can it be that it was easy for you to figure out?

[F]: Well because there was still, you can hear a bit of resonance in the room and you kind of feel "Wow, now I have been playing."

[INTERVIEWER]: Neither can you remember what you were playing?

[F]: Yes I can. Because I know that I was practising Bach 5th suite., & that is quite long. There is the overture and the fugue that together take 6–7 minutes. So I have probably been starting it.

[INTERVIEWER]: You can remember starting it?

[F]: Yes. And the in the course of, I have been playing and playing, and then in the course of the 6–7 minutes, I have just disappeared somewhere.

Once the interview has been transcribed, any sections that are overly terse or which feel to be overly reliant on opinion (rather than evidenced by experience) can be removed, so that the rich and detailed sections can be coded. The methodology could, therefore, be familiar to many readers of this chapter, but the goal (of discovering and distilling the experience of being an expert) might be new.

Conclusions

One overarching question relates to the question of what an *internal representation* might contain. Some of the approaches, particularly distributed cognition or extended mind, take a weaker view of the problem; regarding *representation*s as necessary, but only in terms of small-scale storage of local information for immediate use. As an extreme form of anti-representation, one could claim that there is little need for the brain itself to model the environment, as it can respond to opportunities already present (Brooks, 1986). Certainly, Chemero's (2009) Radical Embodied Cognitive Science and Varela et al's. (1991) enactivism explicitly argue against the need for internal representation, preferring to describe cognition as emerging from the interplay between person and environment. Another way to think of this *intelligence without representation* (Brooks, 1991) is to consider the brain as managing neuronal connection through experience (Freeman, 1991). In Barsalou's (1999) perceptual symbol system, representations are neural traces. This raises fundamental questions about the extent to which internal representations (schemata, mental models, etc.) underpin cognition, or how cognition emerges from our interactions with the environment, or how experts differ from novices? Perhaps the point at issue is whether *schemata* are mental representations which model the environment, or whether they are neural structures (in terms of neurons which fire together) that resonate with features of the world, or whether they are simply convenient fictions that we use when we talk about cognition. Indeed, it is not obvious that cognitive psychology needs a commitment to propositional representation. We speak of schema, perhaps, because we lack a vocabulary to express the neural activity involved in thought or because the ways in which thinking is accessed (through self-report) is primarily through verbal description.

How do these approaches change the ways in which cognitive psychology ought to approach expertise? One suggestion, particularly from Dreyfus' (2002) notion of skill acquisition, is that there is a difference in kind between novice and expert. This is echoed by the discussion of situated cognition, with its suggestion that the community of (expert) practice *sees* opportunities for action in the environment. The implication is that expertise exists *in situ*, and that placing participants in laboratories could render everyone a novice because the tasks that they face are unfamiliar and decontextualized. In a study of people using the Macwrite word processor, Mayes et al. (1988) found that the ability of experienced users to recall menu items in a paper and pencil test was surprisingly low (around 50 percent) but that when they were asked to use these commands while seated at a computer, recall rose to well over 90 percent. Their conclusion is that "Rather than being based on 'knowledge' consisting of mental models that replicate substantial aspects of the external world..., human action may be organized around a flow of information picked up from the environment during execution" (Mayes et al., 1988, p. 288). In conclusion, expertise is not simply a matter of having well-grooved paths of action from habit or repetition, nor having completely specified schemata (internal representation) for all occasions, but it requires the flexible, adaptive, and dynamic response to relevant environmental constraints. From this perspective, expertise is a matter of sensitivity to environmental constraints and opportunities.

References

Baber, C., & Butler, M. (2012). Expertise in crime scene examination: comparing search strategies of expert and novice crime scene examiners in simulated crime scenes. *Human Factors 54*, 413–424.

Baber, C., Cengiz, T. G., Starke, S., & Parekh, M. (2015). Objective classification of performance in the use of a piercing saw in jewellery making. *Applied Ergonomics 51*, 211–221.

Baber, C., Smith, P., Cross, J., Hunter, J. E., & McMaster, R. (2006). Crime scene investigation as distributed cognition. *Pragmatics & Cognition 14*, 357–385.

Baber, C., & Starke, S. D. (2015). Using 1/f scaling to study variability and dexterity in simple tool using tasks. In *Proceedings of the Human Factors and Ergonomics Society Annual Meeting* (pp. 431–435). Los Angeles, CA: SAGE.

Ballard, D. H., Hayhoe, M. M., & Pelz, J. B. (1995). Memory representations in natural tasks. *Journal of Cognitive Neuroscience 7*, 66–80.

Barsalou, L. W. (1999). Perceptual symbol systems. *Behavioral & Brain Sciences 22*, 577–660.

Barsalou, L. W. (2008). Grounded cognition. *Annual Review of Psychology 59*, 617–645.

Beach, K. (1988). The role of external mnemonic symbols in acquiring an occupation. In M. M. Gruneberg, & R. N. Sykes (Eds), *Practical aspects of memory* (pp. 342–346). New York: Wiley.

Bernstein, N. A. (1967). *The coordination and regulation of movements*. Oxford: Pergamon.

Biryukova, E. V., & Bril, B. (2008). Organization of goal directed action at a high-level of motor skill: The case of stone-knapping in India. *Motor Control 12*, 181–209.

Bril, B., Rein, R., Nonaka, T., Wenban-Smith, F., & Dietrich, G. (2010). The role of expertise in tool use: Skill differences in functional action adaptations to task constraints. *Journal of Experimental Psychology: Human Perception and Performance 36*, 825–839.

Brooks, R. (1986). A robust layered control system for a mobile robot. *Journal of Robotics & Automation* 2, 14–23.

Brooks, R. (1991). Intelligence without representation. *Artificial Intelligence Journal* 47, 139–160.

Brown, J. S., Collins, A., & Duguid, P. (1989). Situated cognition and the culture of learning. *Educational Researcher* 18, 32–42.

Calvo-Merino, B., Glaser, D. E., Grèzes, J., Passingham, R. E., & Haggard, P. (2005). Action observation and acquired motor skills: an fMRI study with expert dancers. *Cerebral Cortex* 15, 1243–1249.

Chapman, S. (1968). Catching a baseball. *American Journal of Physics* 36, 868–870.

Chemero, A. (2009). *Radical embodied cognitive science*. Cambridge, MA: MIT Press.

Chen, X., Starke, S. D., Baber, C., & Howes, A. (2017). May. A cognitive model of how people make decisions through interaction with visual displays. *Proceedings of the 2017 CHI Conference on human factors in computing systems* (pp. 1205–1216). New York: ACM.

Clancey, W. J (1993). Situated action: a neuropsychological interpretation response to Vera and Simon. *Cognitive Science* 17, 87 116.

Clark, A. (1997). *Being there: Putting brain, body, and world together again*. Cambridge, MA: MIT Press.

Clark, A. (2008). *Supersizing the mind: Embodiment, action, and cognitive extension*. Oxford: Oxford University Press.

Clark, A., & Chalmers, D. J. (1998). The extended mind. *Analysis* 58, 10–23.

Collins, A., Brown, J. S., & Newman, S. E. (1988). Cognitive apprenticeship, *Thinking: The Journal of Philosophy for Children* 8, 2–10.

Cornus, S., Montagne, G., & Laurent, M. (1999). Perception of a stepping-across affordance. *Ecological Psychology* 11, 249–267

Descartes, R. (1637/2005). *Discourse on method and the meditations*, trans. F. Sutcliffe. London: Penguin.

Dreyfus, H. L. (2002). Intelligence without representation—Merleau-Ponty's critique of mental representation. *Phenomenology and the Cognitive Sciences* 1, 367–383.

Ericsson, K. A., & Simon, H. A. (1993). *Protocol analysis: Verbal reports as data*, 2nd edn. Cambridge, MA: MIT Press.

Fink, P., Foo, P., & Warren, W. (2009). Catching fly balls in virtual reality: a critical test of the outfielder problem. *Journal of Vision* 9, 14.

Freeman, W. J. (1991). The physiology of perception. *Scientific American* 264, 78–85.

Gallagher, S. (2017). Theory, practice and performance. *Connection Science* 29, 106–118.

Gallagher, S., & Zahavi, D. (2008). *The phenomenological mind: an introduction to philosophy of mind and cognitive science*, 2nd edn. New York: Routledge.

Gibbs, R. W. (2006). *Embodiment and cognitive science*. Cambridge, UK: Cambridge University Press.

Gibson, E. J. (1969). *Principles of perceptual learning and development*. New York: Prentice-Hall.

Gibson, J. J. (1977). The theory of affordance. In R. E. Shaw & J. Bransford, J. (Eds), *Perceiving, acting, and knowing: Toward an ecological psychology* (pp. 62–82). New York: Lawrence Erlbaum Associates.

Gibson, J. J. (1979). *The ecological approach to visual perception*. Boston: Houghton Mifflin.

Gigerenzer, G., & Todd, P. M. (1999). *Fast and frugal heuristics: The adaptive toolbox*. Oxford: Oxford University Press.

Gleick, J. (1992). *Genius: The life and science of Richard Feynman*. New York: Vintage.

Greeno, J. G. (1994). Gibson's affordances. *Psychological Review 101*, 336–342.

Guastello, S. J., & Gregson, R. A. (2016). *Nonlinear dynamical systems analysis for the behavioral sciences using real data.* Boca Raton, FL: CRC Press.

Hagemann, N., Schorer, J., Cañal-Bruland, R., Lotz, S., & Strauss, B. (2010). Visual perception in fencing: Do the eye movements of fencers represent their information pickup? *Attention, Perception, and Psychophysics 72*, 2204–2214.

Haken, H., Kelso, J.A.S., & Bunz, H. (1985). A theoretical model of phase transitions in human hand movements. *Biological Cybernetics*, 51, 347 356.

Høffding, S. (2014). What is skilled coping? Experts on expertise. *Journal of Consciousness Studies, 21*, 49–73.

Hoffman, R. R., Crandall, B. W., & Shadbolt, N. R. (1998). A case study in cognitive task analysis methodology: The Critical Decision Method for the elicitation of expert knowledge. *Human Factors 40*, 254–276.

Hollan, J., Hutchins, E., & Kirsch, D. (2002). Distributed cognition: Toward a new foundation for human-computer interaction. In J. Carroll (Ed.), *Human–computer interaction in the new millennium* (pp. 75–94). New York: Addison-Wesley.

Humphreys, G. W. (2001). Objects, affordances, action. *Psychologist 14*, 408–412.

Hutchins, E. (1995). *Cognition in the wild.* Cambridge, MA: MIT Press.

Jagacinski, R. J., & Flach, J. (2003). *Control theory for humans: Quantitative approaches to modelling performance.* Mahwah, NJ: Lawrence Erlbaum Associates.

Kellman, P. J., Massey, C. M., Roth, Z., Burke, T., Zucker, J., Saw, A., ... Wise, J.A. (2008). Perceptual learning and the technology of expertise: Studies in fraction learning and algebra. *Pragmatics and Cognition 16*, 356–405.

Kello, C. T., Beltz, B. C., Holden, J. G., & van Orden, G. C. (2007). The emergent coordination of cognitive function. *Journal of Experimental Psychology: General 136*, 551–568.

Kelso, J. S. (1997). *Dynamic patterns: The self-organization of brain and behavior.* Cambridge, MA: MIT Press.

Kinsella-Shaw, J. M., Shaw, B., & Turvey, M. (1992). Perceiving walk-on-able slopes. *Ecological Psychology 4*, 223–239.

Kirsh, D. (2013). Embodied cognition and the magical future of interaction design. *ACM Transactions on Computer-Human Interaction 20*, article no. 3.

Kirsh, D., & Maglio, P. (1994). On distinguishing epistemic from pragmatic action. *Cognitive Science 18*, 513–549.

Kugler, P. N., Kelso, J. A. S., & Turvey, M. (1982). On the control and coordination of naturally developing systems. In J. A. S. Kelso & J. E. Clark (Eds) *The development of movement control and coordination* (pp. 5–78), New York: Wiley.

Lakoff, G. J., & Johnson, M. (1999). *Philosophy in the flesh: The embodied mind and its challenge to Western thought.* New York: Basic Books.

Larkin, J. H. (1989). Display-based problem solving. In D. Klahr & K. Kotovsky (Eds), *Complex information processing: The impact of Herbert A. Simon* (pp. 319–341). Hillsdale, NJ: Erlbaum.

Lave, J. (1988). *Cognition in practice: Mind, mathematics, and culture in everyday life.* Cambridge, UK: Cambridge University Press.

Lave, J., & Wenger, E. (1990). *Situated learning: Legitimate peripheral participation.* Cambridge, UK: Cambridge University Press.

McBeath, M. K., Shaffer, D. M., & Kaiser, M. K. (1995). How baseball outfielders determine where to run to catch fly balls. *Science 268*, 569–573.

Mark, L. S. (1987). Eyeheight-scaled information about affordances: A study of sitting and stair climbing. *Journal of Experimental Psychology: Human Perception and Performance 13*, 361–370.

Mayes, R. T., Draper, S. W., McGregor, A. M., & Oatley, K. (1988). Information flow in a user interface: The effect of experience of and context on the recall of MacWrite screens. In D. M. Jones & R. Winder (Eds), *People and computers IV* (pp. 257–289). Cambridge, UK: Cambridge University Press.

Merleau-Ponty, M. (1945/2014). *Phenomenology of perception*. London: Routledge.

Mollo, V., & Falzon, P. (2004). Auto- and allo-confrontation as tools for reflective activities. *Applied Ergonomics 35*, 531–540.

Nemeth, C. (2003). How cognitive artifacts support distributed cognition In Acute Care. In *Proceedings of the 47th Annual Meeting of the Human Factors and Ergonomics Society* (pp. 381–385). Santa Monica, CA: Human Factors and Ergonomics Society.

Newell, K. M. (1985). Coordination, control and skill. In D. Goodman, R. B. Wilding, & I. M. Franks (Eds), *Differing perspectives in motor learning, memory and control* (pp. 295–317). Amsterdam: Elsevier.

Newell, K. M. (1986). Constraints on the development of coordination. In M. G. Wade & H. T. A. Whiting (Eds), *Motor development in children: Aspects of coordination and control* (pp. 341–361). Amsterdam: Martinus Nijhoff.

Polanyi, M. (1966). *The tacit dimension.*, Chicago: University of Chicago Press.

Proffit, D. R., & Linkenauger, S. A. (2013). Perception viewed as a phenotypic expression. In W. Prinz, M. Biesert, & A. Herwig (Eds), *Action science: Foundations of an emerging discipline* (pp. 171–179). Cambridge, MA: MIT Press.

Rosenbaum, D. D., Chapman, K. M., Weigelt, M., Weiss, D. J., & van der Wel, R. (2012). Cognition, action, and object manipulation. *Psychological Bulletin 138*, 924–946.

Rowlands, M. (1999). *The body in mind: Understanding cognitive processes*. Cambridge, UK: Cambridge University Press.

Rupert, R. D. (2009). *Cognitive systems and the extended mind*. Oxford: Oxford University Press.

Saxberg, B. V. H. (1987a). Projected free fall trajectories: I. Theory and simulation. *Biological Cybernetics 56*, 159–175.

Saxberg, B. V. H. (1987b). Projected free fall trajectories: II. human experiments. *Biological Cybernetics 56*, 177–184.

Schmidt, R. A. (1975). A schema theory of discrete motor skill learning. *Psychological Review 82*, 225.

Scribner, J. (1984). Studying working intelligence. In B. Rogoff & J. Lave (Eds), *Everyday Cognition* (pp. 9–40). Cambridge, MA: Harvard University Press.

Shapiro, L. (2011). *Embodied cognition*. New York: Routledge Press.

Shaw, R. E., & Turvey, M. T. (1981). Coalitions as models for ecosystems: A realist perspective on perceptual organization. In M. Kubovy, & J. R. Pomerantz (Eds), *Perceptual organization* (pp. 343–415). Hillsdale, NJ: Lawrence Erlbaum Associates.

Stefanucci, J. K., & Geuss, M. N. (2009). Big people, little world: The body influences size perception. *Perception 38*, 1782–1795.

Stefanucci, J. K., & Geuss, M. N. (2010). Duck! Scaling the height of a horizontal barrier to body height. *Attention, Perception and Psychophysics 72*, 1338–1349.

Steffensen, S. V., Vallée-Tourangeau, F., & Vallée-Tourangeau, G. (2016). Cognitive events in a problem-solving task: A qualitative method for investigating interactivity in the 17 Animals problem. *Journal of Cognitive Psychology 28*, 79–105.

Suchman, L. A. (1987). *Plans and situated action.* Cambridge: Cambridge University Press.

Sutton, J., McIlwain, D., Christensen, W., & Geeves, A. (2011). Applying intelligence to the reflexes: Embodied skills and habits between Dreyfus and Descartes. *Journal of the British Society for Phenomenology 42*, 78–99.

Tucker, M., & Ellis, R. (1998). On the relations between seen objects and components of potential actions. *Journal of Experimental Psychology: Human Perception and Performance 24*, 830–846.

Vallée-Tourangeau, F., Steffensen, S. V., Vallée-Tourangeau, G., & Sirota, M. (2016). Insight with hands and things. *Acta Psychologica 170*, 195–205.

Varela, F. J. (1987). Laying down a path in walking: A biologist's look at a new biology. *Cybernetic 2*, 6–15.

Varela, F., Thompson, E., & Rosch, E. (1991). *The embodied mind: Cognitive science and human experience.* Cambridge, MA: MIT Press.

Vernooij, C. A., Mouton, L. J., & Bongers, R. M. (2012). Learning to control orientation and force in a hammering task. *Zeitschrift für Psychologie 220*, 29–36.

Warren, W. H. (1984). Perceiving affordances: Visual guidance of stair climbing. *Journal of Experimental Psychology: Human Perception and Performance 10*, 683–703.

Warren, W. H., & Wang, S. (1987). Visual guidance of walking through apertures: Body-scaled information for affordances. *Journal of Experimental Psychology: Human Perception and Performance 13*, 371–383.

Wheeler, M. (2005). *Reconstructing the cognitive world: The next step.* Cambridge, MA: MIT Press.

Wilson, A. D., & Golonka, E. (2013). Embodied cognition is not what you think it is. *Frontiers in Psychology 4*(58), 1–13.

Wilson, M. (2002). Six views of embodied cognition. *Psychonomic Bulletin and Review 9*, 635–636.

Witt, K. L., & Proffit, D. R. (2008). Action-specific influences on distance perception: a role for motor simulation. *Journal of Experimental Psychology: Human Perception and Performance 34*, 1479–1492.

CHAPTER 12

ADAPTIVE EXPERTISE

KATERINA BOHLE CARBONELL AND JEROEN J. G. VAN MERRIENBOER

Introduction

INCREASINGLY, technology is used in the workplace to support decision-making of individuals (Brynjolfsson & Mcafee, 2017). As a result of this, decisions that individuals have to take without input from technology are those for which not enough quality data are available to train computers (Brynjolfsson & Mcafee, 2014; Kahneman, 2016). This occurs for rare and unpredictable situations. As a result of this change in the workplace, individuals are required to adapt effectively to unexpected situations.

To remain employable by being able to deal effectively with unexpected situations, individuals should develop adaptive expertise at work through problem solving and lifelong learning. In doing so they will be able to deal with unfamiliar situations and remain good performers. They do so by inventing new procedures and using their expert knowledge in novel ways (Hatano & Inagaki, 1986; Holyoak, 1991). The aim of this chapter is to answer the question of how the acquisition of adaptive expertise helps individuals to remain employable by being able to deal with unfamiliar problems. To achieve this aim, we will first explain the concept of adaptive expertise, and how it differs from other related concepts. In the second section, we will elaborate on how adaptive expertise is developed. In order to measure any growth in adaptive expertise, tools are needed to measure the level of adaptive expertise. This will be addressed in the third section. We will end the chapter by providing recommendations for how to teach adaptive expertise in largely formal learning settings.

What is Adaptive Expertise?

Hatano and Inagaki (1986) first described the concept of adaptive expertise, contrasting it with routine expertise. Routine expertise is the ability to perform at a high level on tasks that are representative of the domain. Adaptive expertise is defined as the ability to perform at a relatively high level in unfamiliar situations thanks to understanding why a specific procedural skill should be used in a specific situation. Unfamiliarity is defined as a task situation that an individual has not encountered previously but is within the individual's known domain. This is different to "messy" or ill-structured situations, as these situations appear to lack structure with regard to how situational aspects and relevant knowledge relate to each other (Feltovich, Coulson, & Spiro, 2013). Hatano and Inagaki (1986) used two examples to clarify what is meant with novel situations: First, the difference between gardening in the open versus gardening in a glasshouse, and second, cooking with measuring cups versus cooking without measuring cups. Hence it can be derived that the authors perceive content transfer, the transfer of a learned skill to a new situation, and context transfer, application of known procedures in a different physical surrounding as signs of adaptive expertise (Barnett & Ceci, 2002). Important for novelty to exist is that the structural features, thus aspects of the situation that are crucial for executing the tasks, are different between the original and new situation (Holyoak & Koh, 1987).

The ability to transfer performance from a familiar task to an unfamiliar task sets individuals who possess adaptive expertise apart from those who only possess routine expertise. Both routine expertise and adaptive expertise implies achieving high performance on standard tasks, those that are representative of the tasks within one's domain (Ericsson & Smith, 1991). However, changes in the task, such as different tools, computer programs, and sequences of sub tasks, require changes in application of knowledge and skill. A difference of adaptive expertise compared to relevant terms such as cognitive flexibility (given the needs of a particular ill-structured situation knowledge is being used selectively; Spiro, Collins, Thota, & Feltovich, 2003) or resilience (the ability to manage adaptive systems in order to achieve sustained adaptability; Woods, 2015) is the notion that individuals with adaptive expertise, while being confronted with an unfamiliar task, modify their conceptual understanding of the task and in the process can create new knowledge and skills (Barnett & Koslowski, 2002). They are able to engage in this transformation thanks to their conceptual understanding of domain knowledge and skills (Hatano & Inagaki, 1986). This understanding provides individuals with adaptive expertise with not just information about how to apply a skill, but also when and why to apply it.

Chi (2011) provides two arguments for the better performance of individuals with adaptive expertise in unfamiliar situations. First, those individuals have achieved a deeper understanding of the procedures or skills of their domain, which allows them to generalize their knowledge to unfamiliar situations and thus outperform individuals with routine expertise. This conceptual understanding of domain skills can be achieved at all levels of skill acquisition (Chi, 2011; Schwartz, Bransford, & Sears, 2005). Conceptual understanding entails knowing the conditions under which a procedure or skill can be applied. Secondly, adaptive expertise can also point towards a predisposition of learning during performance (Chi, 2011). The notion is that adaptive experts seek to learn from their experiences. Thus, they do not only want to complete a task, but by applying procedures and skills their secondary goal is to further their knowledge. This characterization of adaptive expertise resembles the characteristics of elite experts, experts who engage in deliberate practice. Regardless whether we perceive the source of performance of adaptive expertise as a better conceptual understanding of the procedures and skills of their domain or as a predisposition of learning, the overlapping component in both characterizations is a positive attitude towards learning (Chi, 2011).

Support for both characterizations of adaptive expertise is provided by Barnett and Koslowski (2002). They reported that adaptive experts, represented by business consultants in their study, outperformed routine experts, represented by restaurant experts, thanks to the greater number of abstract concepts the consultants possessed. Conceptually, Hatano and Inagaki (1986) seem to emphasize the second characterization of adaptive expertise, as they argue that individuals with high levels of adaptive expertise possess greater motivation to achieve a deep understanding of why specific domain-relevant procedures need to be applied. In addition, the authors argue that these individuals benefit from an environment that values learning above performance, and in which deviation from routine procedures does not carry high risk in terms of material and human costs if the desired results would not be achieved swiftly. Tentative support for this claim is reported by Han and Williams (2008).

Hatano and Inagaki (1986) also talk about the importance of problem-solving skills for the development of adaptive expertise. The presence of an unfamiliar situation presents a problem to an individual. Within the studies on adaptive expertise, several studies have indicated that problem-solving skills, such as the ability to draw analogies between familiar and unfamiliar situations, are a crucial part of adaptive expertise (Barnett & Koslowski, 2002; Bell & Kozlowski, 2008). Bell and Kozlowski (2008) demonstrated that analogical problem solving, transferring skills to unfamiliar situations, is a strong predictor of adaptive expertise. When engaging in analogical problem solving, the individual is mapping the target unfamiliar situation to a familiar source situation by looking for similarities between these two situations (Holyoak & Thagard, 1997).

In his study on problem solving of experts in history, Wineburg (1998) highlighted how persistently asking questions instead of fixating on the first idea helps to develop the proper understanding of the context and thus to reach the correct solution. These questions help to develop an understanding of the unfamiliar situation and are an indicator of learning from experience. Dane (2010) argued that domain expertise can be a potential inhibitor for problem solving as it can lead to functional fixedness (Duncker, 1945). This occurs when experts are unable to deal with unfamiliar problems because they focus on suboptimal solutions. Individuals with expertise have a large collection of scripts to rely on when solving problems. While this is important to efficiently solve routine problems, these scripts may be suboptimal in unfamiliar problems. However, the activation of a script can block the consideration of new solutions (Adelson, 1984). Based on adaptive expertise problem-solving skills it can be assumed that adaptive experts realize the inadequacy of an activated script and thus devise a new solution. Important for an adaptive expert is thus to realize that a script is suboptimal, something routine experts are less likely to do.

Bibliographic Analysis

Adaptive expertise research sits at the intersection between many different domains, such as expertise research, educational research, and cognitive science. Hatano and Inagaki (1986) had children in mind when writing their original article, building upon Piaget's theories. However, other researchers (e. g., Fisher & Peterson, 2001; Martin, Rayne, Kemp, Hart, & Diller, 2005) later expanded this work to include university students and professionals. Given the variety of domains that investigate adaptive expertise, it is interesting to analyze the knowledge sources used in adaptive expertise research in greater detail. The subsequent analysis provides a first insight into the diversity of scholarly communities investigating adaptive expertise.

To analyze the body of knowledge a co-citation analysis has been conducted, which analyzes the strength of association between two papers as perceived by the investigated scholarly community. This method explores what sources of knowledge are used in a research stream (Hall, 1998; Small, 1973; Uzzi, Mukherjee, Stringer, & Jones, 2013). Two articles are co-cited if they are cited together in a third article (Small, 1973). If a scholar cites articles A and B, then this person assumes some relationship between articles A and B. This could mean that the arguments and findings reported in articles A and B are consistent or inconsistent, or build on each other.

We used Web of Science to extract information about articles that contained relevant keywords (adaptive expert(ise), reflective expertise, cognitive readiness, switching cognitive gears, cognitively slowing down). This list of keywords is not exhaustive and has been selected from a learning perspective, in the sense of teaching adaptive expertise. Additionally, the bodies of research on, for example, resilience or transfer are very large and established. Including these keywords in the analysis will diminish the ability to perceive smaller research streams, such as adaptive expertise. The key terms *adaptive expertise* and *reflective expertise* are linked to expertise research. Cognitive readiness, switching cognitive gears, and cognitively slowing down have been added because adaptive expertise is argued to be a central attribute of cognitive readiness (O'Neil et al., 2014). We extracted meta-data of 152 articles and conducted a co-citation analysis using VOSViewer. As a threshold two articles were included as a co-cited pair if they were cited at least ten times together.

Figure 12.1 depicts the resulting co-citation pattern. Each circle represents one article. The size of the circle denotes how often an article has been cited. The smaller the distance between two articles, the more often these two have been cited together. The co-citation map reveals six clusters. A closer analysis of the articles within each cluster helps to identify the focus of research within this cluster. Based on the analyses of articles within the cluster, we conclude that the cluster 5 (red) deals with research on *child rearing*, and the cluster 6 (orange) *child rearing: social perspective*.

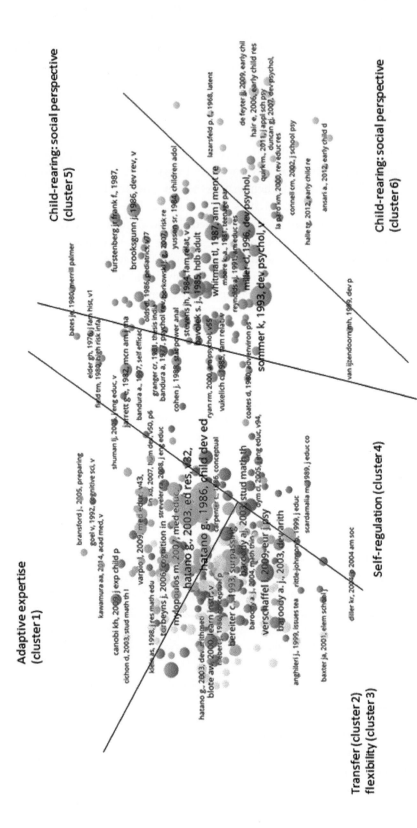

FIGURE 12.1 Co-citation of articles in the expertise literature

Articles that appear in the reference list of other articles are co-citations. Co-citations that occurred at least ten times are included. Larger circles represent articles that appear more often. Colors are mapped based on a clustering algorithm that puts articles with similar co-citation patterns next to each other.

Pinpointing the research focus of the four other clusters on the left side of the figure is more complicated. The reason is that while the articles are clustered within different groups, the distance between articles from different clusters is small. This already indicates that there is a lot of cross-citation between the research streams. In an attempt to define the clusters we named them *transfer* (cluster 2; blue), *flexibility* (cluster 3; yellow), *self-regulation* (cluster 4; purple), and *adaptive expertise* (cluster 1; green). Transfer refers to transferring something that has been learned in one situation (e. g., knowledge) to a different situation (Barnett & Ceci, 2002). This corresponds to the last three levels of transfer (near transfer, far transfer, and displacement transfer) according to Haskell (2001); it is linked to flexibility. Flexibility is the ability to select what procedures and skills need to be applied in a given situation (Spiro, Coulson, Feltovich, & Anderson, 1988) and indicates that an individual has internalized the skills and knowledge of their domain and thus achieved a deep procedural understanding of them (Chi, 2011). In order to achieve this flexibility it is necessary to regulate one's own learning activities and to reflect on one's own performance (Zimmerman, 2006). Flexibility and transfer thus require self-regulation of behavior. The concepts of transfer and flexibility are linked to adaptive expertise: Adaptive experts transfer their problem-solving abilities from familiar to unfamiliar problems (Barnett & Koslowski, 2002). They achieve this by selecting the procedures and skills that will best suit the unfamiliar situation. Self-regulation is also linked to adaptive expertise, as these are the cognitive, affective, and behavioral processes individuals use to monitor their performance (Cannon-Bowers & Bowers, 2009).

The co-citation analysis reveals that the field of adaptive expertise, viewed through an educational lens, is composed of two isolated subfields that coincide with distinct clusters (child rearing and child rearing: social perspective), and one mixed subfield combining four clusters (adaptive expertise, transfer, flexibility, self-regulation). The outcome of the co-citation analysis is dependent on the keywords. By including additional, but relevant keywords, such as adaptive performance, the analysis of the adaptive expertise research body would tap into the knowledge of other domains. While this exercise will be valuable for gaining a comprehensive view of all research about and related to adaptive expertise, it would have been too broad for our aim. The goal of this chapter is to focus on how adaptive expertise develops and can be taught. Therefore, we adopted an educational viewpoint while conducting the co-citation analysis.

Adaptive Expertise and Related Concepts

To further refine our understanding of adaptive expertise it is helpful to compare and contrast this concept with other expertise and adaptability concepts developed by other scholars. Table 12.1 provides an overview of these concepts. The list is by no means meant to be exhaustive, but aims to highlight the fuzzy boundaries between adaptive

Table 12.1 Different Conceptualizations of Expertise

Concept	Author	Domain	Characteristics
Expertise	Ericsson, Krampe, & Tesch-Romer (1993)	Cognitive psychology	• Perform at a high level on domain representative tasks.
Adaptive Expertise	Hatano & Inagaki (1986)	(Child) development, education	• Perform at a high level when executing familiar tasks. • No or few performance loss when executing unfamiliar problems. • Motivation to learn and the ability to deviate from routines are necessary for its acquisition.
Reflective Expertise	Olsen & Rasmussen (1989)	Cognitive psychology	• In familiar situations individuals apply skill-based behavior • Rule-based behavior is applied when situations are somewhat unfamiliar, but the goal that needs to be achieved is clear. • When unfamiliarity is high, individuals inspect their mental models, to create new mental models appropriate for the situation.
Professional Expertise (or flexpertise)	Van der Heijden (2000)	Industrial psychology	• Possess the knowledge, skills, and metacognition required to become and stay an expert. • Be recognized as an expert by peers (social recognition). • Be able to achieve expert performance in adjacent or radically different fields (grow and flexibility).
Elite Expertise	Chi (2011)	Cognitive psychology	• Experts, as described by Ericsson, who practice deliberate practice.
Cognitive Readiness	O'Neil, Lang, Perez, Escalante, & Fox (2014)	Cognitive psychology	• Be ready to commit to changes to deal with unfamiliar situations. • Adaptive experts understand when and why certain knowledge is appropriate.

expertise and its related concepts. For each concept we included the definition provided by the scholars who introduced the concept. Common among these different conceptualizations of expertise is the ability to perform at a high level on domain representative tasks. A high level implies executing a task with great accuracy or speed (Hatano & Inagaki, 1986). Differences among the concepts arise when further investigating what question the specific expertise concept should answer. For example, *reflective expertise*, brought forward by Olsen and Rasmussen (1989), focuses on the cognitive mechanisms that could explain how experts approach tasks with varying degrees of familiarity by reflecting on their mental models. These mental models guide the selection of behaviors. The goal of *professional expertise* (Van der Heijden, 2002) is to explain what the necessary ingredients are for individuals to gain and remain an expert in their profession. Therefore, the component of social recognition has been included. Chi (2011) describes *elite experts* as those experts who have acquired a deep conceptual understanding of the skill within their domain by intentionally seeking challenges, reflecting on their performance, and thus engaging in continuous learning. These activities are inherent when engaging in deliberate practice. Following terminology introduced by Ericsson and Lehmann (1996), Chi (2011) calls adaptive experts elite experts as they achieve elite levels of performance due to deliberate practice. Finally, *cognitive readiness* is defined by O'Neil et al. (2014) to be ready to deal with unfamiliar situations. Within this concept, the authors perceive adaptive expertise to be an attribute of cognitive readiness, stating that it entails a deep understanding of the domain. Within the concept of cognitive readiness, adaptive expertise stays close to the definition of Hatano and Inagaki (1986).

The previous description about related concepts of adaptive expertise highlights that these various concepts differ in some aspects, while sharing commonalities with others. Hence it is not possible to establish clear boundaries between the terms. For example, the concepts of *adaptive expertise* and *expertise* share the commonality of high performance in familiar tasks. The definitions of *elite experts* and *adaptive experts* both highlight an individual's motivation to learn as a driver for performance on unfamiliar tasks. *Reflective expertise* and *adaptive expertise* both consider the need to adapt an individual's mental models when change is introduced. Finally, *cognitive readiness* and *adaptive expertise* both consider that the ability to adapt is independent of the level of proficiency, but dependent on the level of familiarity with a task.

DEVELOPMENT OF ADAPTIVE EXPERTISE IN INFORMAL SETTINGS

Hatano and Inagaki (1986) write that the following ingredients are necessary for achieving adaptive expertise: Variety in practice, risk of deviation from routine, and a cultural norm favoring achieving deep understanding above achieving quick efficiency.

The first component described by Hatano and Inagaki (1986), variety in practice, concerns how often an individual observes how a certain procedure (e.g., surgical procedure of removing tonsils) leads to a specific outcome (e.g., tonsils are removed and patient fully recovered). The variety is introduced by the system, the environment, in which a procedure or skill is applied. The authors argue that if the system contains built-in randomness, individuals achieve a deeper understanding of when a procedure can be applied, and what its limits are. Variation in practice gives individuals the opportunity to observe how different aspects of the procedural skill interact with each other and what outcome they will create. Variance in practice thus implies changes in the task, work environment, and outcome. Variability in practice has been related to increased learning (Shea & Kohl, 1990; Wulf & Schmidt, 1997). According to the variability of practice hypothesis, changing parameters of a task aids in defining which features are relevant and irrelevant for a task (Paas & Van Merriënboer, 1994; Shea & Wulf, 2005). Once an individual has passed beyond the level of a novice (Schwartz et al., 2005; Shea & Wulf, 2005), variability in practice stimulates the development of an accurate and detailed schema of the procedure or skill. Adaptive expertise is developed through the observations of variance in the task and the resulting solution. In the above medical example, variation in practice could be anything that changes criteria of the patient (e.g., immune deficiency, diabetes, obesity, age of patient), the available equipment, or aspects of the surgical environment.

Hatano and Inagaki (1986) argue that risk of deviation is an influencing environmental factor when developing adaptive expertise. This factor refers to the context in which a procedure is applied. It is linked to variance of practice, as deviation refers to the individual deviating from learned procedure. This is coupled with correctly assessing the risk of failing to meet the required outcome when engaging in variation of practice. In the medical example previously mentioned, the desired outcome was achieved if two conditions were met: The tonsils were removed, and the patient fully recovered. Any deviation from practice (e.g., power outage) could result in not achieving the desired outcome by, for example, abandoning the operation. The context now becomes important in order to determine the risk of failing to perform. In case of high-risk situations, deviations from practice are avoided, eliminating the opportunities for learning and achieving adaptive expertise.

Finally, Hatano and Inagaki (1986) raise the point that individuals have reference points, which they use to determine to what degree a deep understanding is necessary and valued by their environment. This addresses the cultural norms of an individual's work environment. When achieving a deep understanding of the procedures is part of the cultural norms, the risk of deviations is lower and thus individuals engage in a higher variety of practice opportunities. However, on the other hand, if the culture values efficiency above anything else, the risk of deviating and not achieving the desired outcome is higher and thus more likely avoided. This cultural reference also indicates that the norms and values of a group determine what actions are permissible when learning and performing a skill. This social dimension is acknowledged by Grenier and Kehrhahn (2008), who argue that peers working within the same domain influence whether an individual is perceived to be an expert.

Several studies analyzed the impact variability of practice has on the development of adaptive expertise at the workplace or conducted cross-sectional studies to relate variability of practice to the level of adaptive expertise. For example, on the topic of variability of practice, Barnett and Koslowski (2002) use the variance in consultants' tasks to explain their higher levels of adaptive expertise. The authors analyzed the performance of undergraduates, restaurant managers, and consultants on the basis of a case study. The case study consisted of a problem faced by a restaurant. The optimal solution was determined by a number of professors who also possessed expertise in consulting with restaurants. The authors reported that consultants provided the most accurate solution and through further tests linked this to a greater number of general concepts they employed in their solution. This reflected their deeper understanding of business skills. The authors suggested that consultants achieved the deep level of understanding thanks to their work which confronted them with a variety of problems and tasks.

A number of studies investigated the presence and development of adaptive expertise in the healthcare sector, specifically in hospital staff. Those studies focused on adaptive expertise as a determinant for creating new knowledge in the form of work routines in order to do the work better (Mylopoulos & Regehr, 2007). While these do not aim to explain the development of adaptive expertise, they nevertheless highlight the impact a work environment can have on how to deal with unexpected situations and the level of adaptive expertise.

Mylopoulos and Scardamalia (2008) demonstrated that doctors did not recognize the potential learning opportunity for others when deviating from practice, and hence did not engage colleagues when modifying procedures to deal with unfamiliar problems. Within the same line of research, Mylopoulos and Regehr (2009) investigated intermediates' understanding of *innovation of procedures*, in the sense of creating new procedures. This definition of innovation was derived from the work of Schwartz, Bransford, and Sears (Schwartz et al., 2005) and refers to learning from the experience of applying known skills to novel situations. They reported that medical students did not perceive themselves to be fit to invent new procedures but saw their responsibilities as limited to acquisition of knowledge, thus the development of routine expertise. This study indicates that the medical domain adopts a cultural norm of efficiency above understanding (Moulton, Regehr, Mylopoulos, & MacRae, 2007). This cultural norm is understandable given the considerable risk of deviating from routines during practice in the medical domain. However, in instances in which the risk is perceived to be lower, for example during visits of low-risk patients and subsequent recording of patient data, variations from routines were communicated to novices (Varpio, Schryer, & Lingard, 2009). These variations were analyzed in interprofessional communication settings when communication procedures required the use of information and communication technology (ICT)-based electronic-patient records. Variations included abandoning the ICT tool, forcing *nonstandard entries*, and submitting to the ICT tool. Novices gained knowledge about these variations from routines but, unfortunately, did not properly manage the consequences of applying them, resulting in interprofessional communication problems.

MEASURING ADAPTIVE EXPERTISE

The assessment of learning normally focuses on evaluating the acquisition of declarative knowledge through achievement tests and skills through observations on simulations (e.g., in a skills lab or computer-controlled environment) or at work. However, to measure the level of adaptive expertise in individuals' achievement tests and observations might not suffice. Smith, Ford, and Kozlowski (1996) argue that if adaptive expertise should be measured, knowledge structures and metacognition should be analyzed. Knowledge structures can be measured through conceptual maps (for mental models) or scripts (for cognitive strategies). These knowledge structures should contain constraints about when a skill should *not* be applied (Hatano & Inagaki, 1986). These constraints act as a stop sign, alerting experts to the need to refrain from automatic execution of skills (O'Neil, Perez, & Baker, 2014). Smith, Ford, and Kozlowski (1996) also advocate analyzing the level of metacognition. However, it needs to be kept in mind that metacognition does not seem to aid in distinguishing between adaptive expertise and routine expertise (Bohle Carbonell, Könings, Segers, & van Merriënboer, 2016). While metacognition is certainly an important characteristic of expertise, too few studies have investigated its ability to distinguish between routine and adaptive expertise (Bohle Carbonell, Stalmeijer, Könings, Segers, & Van Merriënboer, 2014). If metacognition is measured, it should preferably be done in combination with the assessment of mental representations, as these would indicate adaptive experts' conceptual understanding (Olsen & Rasmussen, 1989).

Adaptive expertise has been measured through objective and subjective methods. As with expertise research, the challenge when measuring adaptive expertise objectively is to identify tasks that are representative of the domain. These are tasks that routine and adaptive experts should accomplish with the same level of performance (e. g., same speed and accuracy), whereas novices should not be able to complete the task. An example of such a task from the medical domain is *inserting a syringe*. Once the parameters of this task are known, it is possible to develop an unfamiliar task which adaptive experts can complete, but on which routine experts will fail. Regarding the example task, the parameters are *inserting, sharp object, not moving human body part, extract or insert a fluid, clean environment, no distraction*, and *lack of stress*. To transform this task into an unfamiliar task, the environment could be varied, by, for example, changing it from a hospital setting to a war zone, or changing the patient from a relatively relaxed and calm adult to an anxious and screaming toddler. However, changing the parameters is a difficult step as it is necessary to know the experiences of participants to deduce what tasks they have already encountered. As past experience provides a learning opportunity to individuals by observing how variation of actions impact outcome (Hatano & Inagaki, 1986), this should be taken into account. With increasing level of domain expertise, designing unfamiliar tasks becomes increasingly more difficult due to the large amount of experiences domain experts have assimilated. Therefore, the following suggestion for how to measure adaptive expertise objectively might be better suited for novices and intermediates than for experts. To measure the level of adaptive expertise of individuals with a great amount of experience, analyzing their conceptual understanding through assessment of their knowledge structures and collecting information about the variation of practice they encountered might be a better approach. However, the challenge here is that experts might encounter problems in articulating their knowledge properly, due to the internalization of domain knowledge (Feldon, 2006).

When using an objective measure of adaptive expertise it is best if an ideal answer exists or if in some way (e.g., through simulations) it is possible to measure the accuracy of the answer. If accuracy of the solution cannot be adequately measured, other indicators of adaptive expertise can be used and measured in combination. For instance, one could measure time to answer the question, degree of elaboration of the proposed answer in terms of steps and feasibility, and number and nature (level of abstraction) of relevant general concepts employed. Instead of asking for a complete solution it is also possible to ask individuals to list the information and steps required to solve the situation (Walker, Cordray, King, & Brophy, 2006).

Barnett and Koslowski (2002) adapted this method in one of their studies. They created a scenario of a hypothetical restaurant that faced major strategic and operational problems over a couple of years. Care was taken that the problems were plausible and domain dependent. Debriefing at the end confirmed that for all participants the problems were unfamiliar. Adaptive expertise was subsequently assessed by using a script concordance test. This test is often applied in the medical domain to assess clinical reasoning in uncertain situations by analyzing whether the expert possesses an efficient organization of domain knowledge (Charlin, Roy, Brailovsky, Goulet, & van der Vleuten, 2000). The answer provided in the script concordance test was compared with the answer provided by judges. In the case of this study, the judges were two professors whose work centered on restaurant management, who previously owned and managed restaurants, and had some experience consulting businesses, including but not limited to restaurants. The authors rewarded points for providing components of the optimal answers. They also subtracted points if participants provided answers which contradicted the optimal answers. This punished participants who provided numerous solutions, but failed to consider the appropriateness of their solutions by elaborating on them. This ensured that individuals who were creative in coming up with solutions, but lacked the ability to connect their ideas with their domain knowledge by considering their applicability, were not equated with adaptive experts. It needs to be kept in mind that experts might not always be able to properly verbalize their knowledge and skills, and demonstration of knowledge and skills is superior than articulating them. Proper scaffolding is necessary to aid participants in phrasing their understanding (Clark, Feldon, Van Merriënboer, Yates, & Early, 2007; see Chapter 17 on *A Historical Perspective on Introspection*, by Ward et al., and Chapter 19, *Incident-Based Method for Studying Expertise*, by Militello and Anders, this volume).

Wineburg (1998) used a similar method to measure adaptive expertise. He asked two historians to analyze a number of texts. While one historian was familiar with the time period from which these texts were generated, the expertise of the other historian was in a different epoch. The expert historian quickly reached a conclusion about the event mentioned in the texts. However, the non-expert needed more time. Importantly, he applied three problem-solving heuristics to correctly assess the event and come to the same conclusion as his expert colleague: He considered the source of the text to assess the value of the content of the text (sourcing), he compared the details of one text with those of another (corroborating), and finally he developed a spatial and temporal context of the event (contextualization). By applying these heuristics, the non-expert avoided simplification. These heuristics refer to strategies common among historians and their use for solving an unfamiliar problem signals the expertise status of the non-expert (Schraagen, 1993). When measuring adaptive expertise through case analysis, it is thus also possible to assess the problem-solving strategies individuals use, and investigate to what degree individuals are analyzing and rebuilding their mental model of the situation (Olsen & Rasmussen, 1989; Schraagen, 1993).

When an objective measure of adaptive expertise is not possible, subjective measures can be applied. Various instruments exist that measure an individual's ability to adapt to changes (e. g., I-ADAPT-M, Ployhart & Bliese, 2006; job adaptability inventory, Pulakos et al., 2002) Two instruments that focus specifically on adaptive expertise have been developed. Fisher and Peterson (2001) developed an instrument to measure adaptive expertise attitudes. This instrument contains four dimensions. It measures an individual's attitudes to employing multiple perspectives when solving a problem, their level of metacognition, their goals and beliefs regarding learning and expertise, and epistemological perspectives about the nature of knowledge. Example items include "When I consider a problem, I like to see how many different ways I can look at it" (multiple perspective), "I monitor my performance on a task" (metacognitive self-assessment), "Expertise can be developed through hard work" (goals and beliefs), and "Knowledge that exists today may be replaced with a new understanding tomorrow" (epistemological perspective). This instrument's internal consistency was assessed using engineering students and faculty. The authors report significant increase of adaptive expertise scores from freshman to faculty.

Bohle Carbonell et al. (2016) developed the adaptive expertise inventory. Through ten items this survey assesses adaptive expertise by measuring an individual's domain skills and innovation skills on a five-point scale. The domain skills gauge an individual's readiness to adapt whereas innovation skills tap into their ability to modify current knowledge and skills. We used a sample of graduates and professionals to validate the internal structure of the instrument. In the process of creating a valid instrument, we decided to take items referring to metacognition out. While this was conceptualized by some scholars as a defining characteristic of adaptive expertise, our literature review did not support that the level of metacognition helps to distinguish between routine expertise and adaptive expertise (Bohle Carbonell et al., 2014). As predicted, the adaptive expertise scores differ between graduates and professionals, indicating that on average professionals, possessing greater familiarity with their domain in terms of years of work experience, also achieve a greater level of adaptive expertise. In addition, the scores differed across professionals' fields. This is an important sign of its known-groups validity, as professional fields possess various amounts of naturally occurring variation, different risk structures for not reaching the desired outcome, and different cultural norms with regard to learning and performance.

Thus, the opportunity to develop adaptive expertise, without separately training for it, differs across professionals' fields.

TEACHING FOR ADAPTIVE EXPERTISE

A popular instructional design model aimed at the development of adaptive expertise in complex domains is van Merrienboer's four-component instructional design (4C/ID) model (van Merriënboer, 1997; van Merriënboer, Jelsma, & Paas, 1992; Van Merriënboer & Kirschner, 2018). The basic assumption of this model is that educational programs aimed at the development of complex skills and adaptive expertise always consist of four components:

1. *Learning tasks.* Learning tasks provide the backbone of the educational program. They can be cases, projects, professional tasks, problems, or assignments that learners work on. The learning tasks require carrying out both non-routine aspects, which are performed differently for different tasks, and routine aspects, which are performed in the same way for different tasks (van Merriënboer, 2013). The development of adaptive expertise requires an optimal balance between teaching non-routine aspects (also called *innovation*) and routine aspects (also called *efficiency*; Schwartz, Bransford, & Sears, 2005).
2. *Supportive information.* Supportive information helps students with performing the non-routine aspects of learning tasks which require problem solving, reasoning, and/or decision making. It describes how the task domain is organized (helping learners to develop *mental models* of the domain) and how problems in the domain can be approached in a systematic fashion (helping learners to develop *cognitive strategies* for performing tasks in the domain).
3. *Part-task practice.* Additional part-task practice of routine aspects is provided when a very high level of automaticity is needed for those aspects, and when the learning tasks do not provide the required amount of practice. It provides extensive repetitive practice and helps learners to develop *cognitive rules* that eventually enable the accurate, fast, and effortless performance of routine aspects of complex skills.
4. *Procedural information.* Procedural information helps students with performing the routine aspects of tasks; that is, aspects that are always performed in the same fashion. It typically has the form of *how-to* instructions given to the learner precisely when needed (i.e., just-in-time) to perform either a part-task or the routine aspects of a whole task.

The 4C/ID model prescribes instructional methods for each of the four components. A full discussion of all methods falls beyond the scope of this chapter (see van Merrienboer & Kirschner, 2018), but Table 12.2 presents for each component the

Table 12.2 Educational Components and Related Methods for Stimulating Adaptive Expertise

Educational component	Methods stimulating adaptive expertise
Learning tasks	• Variability of practice • Open learning tasks • Simulated task environments • Teaching task selection skills
Supportive information	• Guided discovery learning • Cognitive feedback
Part-task practice	• Intermix training • Teaching deliberate practice

instructional methods that explicitly aim at the development of adaptive expertise and a justification for their use. For learning tasks the methods include variability of practice, open tasks, use of simulation, and self-directed task selection; for supportive information the methods include guided discovery and cognitive feedback, and for part-task practice the methods include intermix training and teaching deliberate practice. Note that for procedural information, there are no methods explicitly aimed at the development of adaptive expertise.

With regard to learning tasks, a first instructional method for stimulating adaptive expertise is applying *variability of practice*, a concept which was originally introduced in the field of motor learning (Shea & Kohl, 1990). Paas and Van Merriënboer (1994) studied the effects of variability of practice on learning a cognitive task, namely, computer numerically controlled (CNC) machine programming. They compared low variability and high variability conditions for both solving CNC problems and studying CNC solutions (i.e., programs). They found positive effects of high variability of practice on the ability to solve novel problems, especially when it was combined with studying CNC solutions rather than solving the equivalent problems. It shows that variability of practice allows learners to generalize and abstract away from the details of each single task, yielding more abstract and general ideas that can be used when thinking up solutions for unfamiliar problems.

A second method relates to the use of *open learning tasks*, which give learners the opportunity to study, try out, and explore different solutions. In contrast to conventional tasks, which focus on performance because they require learners to come up with one single solution that transforms a given state in an acceptable goal state, open learning tasks focus on exploring possible solutions. They include, for instance, process-oriented worked examples that must be evaluated by learners (van den Boom, Paas, van Merriënboer, & van Gog, 2004; van Gog, Paas, & van Merriënboer, 2004), reverse tasks for which learners must find out for which situations given solutions are useful (Halff, 1993), and goal-free problems for which learners must explore relationships between solutions and the goals that can be reached by those solutions (Ayres, 1993; for a review of methods, see van Merriënboer & Kirschner, 2018). An example of the use of open learning tasks is provided by Martin, Rivale, and Diller (2007), who compared the effects of exploratory methods and traditional didactic lecture methods in a biomedical engineering program on the development of adaptive expertise. Their results showed that both methods yielded equivalent knowledge gains, but that students in the exploratory group demonstrated significantly greater improvement in innovative thinking abilities.

A third method relates to the use of *simulated task environments*, enabling learners to try out new approaches and learn from errors because deviations from optimal performance no longer carry risks in terms of material and human costs. We are not aware of studies that directly compare effects on the development of adaptive expertise of simulated task environments in which it is safe to make errors and real task environments in which it is not safe to make errors because they yield negative consequences. However, studies do show the importance of errors for the development of adaptive expertise. For example, Joung, Hesketh, and Neal (2006) compared the effects of two conditions for training fire incident management: One condition contained case studies in which the protagonist made errors in fire incident management, whereas case studies on the other condition did not contain any errors. A positive effect of the error condition on adaptive expertise during case study assessment was found.

A fourth method for stimulating adaptive expertise relates to giving learners the freedom to select their own learning tasks, sometimes in combination with teaching them the skills for selecting suitable tasks (e.g., Kostons, van Gog, & Paas, 2012). As an example, Corbalan, Kester, and van Merrienboer (2011) either had learners select their own inheritance tasks in the domain of genetics (learner control) or simply provided them with these tasks (system control). They predicted that the more salient the task features, the better learners can choose personally relevant and varied tasks, which would in turn enhance transfer of acquired skills. Indeed, their results show that efficiency (i.e., higher performance in combination with lower mental effort) on a transfer test, where learners had to solve novel problems, is higher for learners who selected their own tasks on the basis of salient task features than for learners who were not able to select their own tasks. Thus, giving learners the opportunity to select their own learning tasks may, under particular conditions, positively affect the ability to deal with novel problems.

With regard to supportive information, a first instructional method relates to the use of *guided discovery learning* to promote the development of deeply integrated mental models and cognitive strategies. In this approach, leading questions are used to guide the learners to the discovery of how things are organized (mental models) or how tasks are best approached in the task domain (cognitive strategies). Thus, in contrast to exploratory learning, guided discovery does not relate to actually trying out different approaches for solving problems but to deeply processing information that may help to solve these problems. Kontogiannis and Linou (2001), for example, studied guided discovery for operator training in fault diagnosis of complex systems. A guided discovery group worked with an interface containing telltale signs and received verbal discovery instructions to guide discovery of diagnostic rules; other groups worked with an interface without the telltale signs and received either verbal discovery instructions, heuristics, or plant theory. The guided discovery groups (verbal discovery instructions with or without the interface containing telltale signs) outperformed the other groups on a near transfer task. On far transfer to another plant, the guided discovery group with the telltale interface maintained superiority to the non-discovery groups and outperformed the discovery group without the telltale interface on a subset of fault scenarios. The results indicate that guided discovery learning enhances the transfer of complex skills while allowing for flexibility and adaptability in the learning environment.

A second instructional method relates to the provision of *cognitive feedback*. This should stimulate learners to critically reflect on the quality of both their personal problem-solving processes and the solutions they have found, so that they can develop more effective cognitive strategies and mental models. Its main function is thus not the detection and correction of errors, but rather the fostering of reflection by the learner. For example, Martin et al. (2005) used the *STAR.Legacy* cycle for teaching ethics in bioengineering: Students receive a challenge, generate ideas, develop multiple perspectives on those ideas, do research and revise their ideas, test their mettle, and finally go public. Compared to students who followed a standard lecture sequence, both groups learned the factual material equally well, but the *STAR.Legacy* group was more prepared to act adaptively when presented with a novel situation. An important finding was that giving students cognitive feedback and having them revise their work in light of that feedback was a critical chain in the cycle.

With regard to part-task practice, a first method for stimulating adaptive expertise is *intermix training*, in which work on learning tasks is intermixed with part-task practice, that is, practicing routines. The learning tasks are then used to create occasional *impasses* which confront the learners with situations where the routines do not work and which train them to switch from an automatic to a problem-solving mode. The learners are thus explicitly trained to *switch cognitive gears* (Louis & Sutton, 1991), to *cognitively slow down* when the situation requires it (Moulton Regehr, Lingard, Merritt, & MacRae, 2010), or to *balance efficiency and innovation* (Schwartz, Bransford, & Sears, 2005). Schwartz and Martin (2004), for example, compared in the domain of teaching statistics a tell-and-practice condition that focused on efficiency only with an invention condition that combined both innovation (i.e., finding your own solution) and efficiency (i.e., being told how to efficiently solve the problem in a subsequent lecture). They found that the invention condition outperformed the tell-and-practice condition on a transfer test, but only when learners could consult learning resources during the test. Thus, the combination of innovation and efficiency better prepared students for future learning.

Finally, teaching *deliberate practice* may help learners develop some of the lifelong learning skills that enable adaptive expertise. Ericsson and Lehman (1996) define deliberate practice as "the individualized training activities specially designed by a coach or teacher to improve *specific aspects* of an individual's performance through repetition and successive refinement" (pp. 278–279, italics added). Deliberate practice is assumed to contribute to adaptive expertise because it helps learners identify part-task practice that can help them to successively refine and fully automate routine aspects of performance, and so improve their whole-task performance on future problems or tasks (van Gog et al., 2005). Such deliberate practice can sometimes even lead to temporary drops in performance (i.e., pose learning above performance). Although we are not aware of empirical research relating the teaching of deliberate practice to adaptive expertise, an anecdote may illustrate this point. Some years ago, the golfer Tiger Woods announced that he wanted to improve one particular swing technique. He predicted a temporary drop of his position in the world rankings because *unlearning* the old swing technique and learning the new one would take a lot of time-consuming part-task practice, resulting in a temporary drop in whole-task performance as reflected in these world rankings. Indeed, this is what happened. But one year later he was number one in the world rankings again. Thus, deliberate practice helped him to be better prepared for future challenges.

Summary

This chapter presented the concept of adaptive expertise, how it develops in informal settings, how it can be measured, and how it can be explicitly taught in more formal settings. Adaptive expertise is the ability to perform at a high level in unfamiliar situations. Novelty is introduced when the environment, tools, application domain, and social context differ from what is commonly experienced. This novelty forces individuals to reassess how and why they are applying a certain skill or process. Successful solving of unfamiliar problems requires the modification of current skills and processes. A bibliographic analysis of the research field of adaptive expertise highlights that transfer, flexibility, and self-regulation are core research themes in the research field. Our further discussion of the concept of adaptive expertise described the similarities and differences among various expertise concepts.

Methods for measuring adaptive expertise can be grouped into objective and subjective measures. Objective measures focus on assessing the mental models and cognitive strategies of individuals. An important step is to design an appropriate task that can be solved adequately by adaptive experts, but not by routine experts. Indicators of adaptive expertise are accuracy of solution, time to answer, degree of elaboration, number of relevant concepts mentioned, and so forth. Simulations or case scenarios are the most common methods for objectively assessing adaptive expertise. Subjective assessment of adaptive expertise relies on individuals' perception about their behavior. Two surveys are readily available, one assessing adaptive expertise attitudes (Fisher & Peterson, 2001), whereas the other measures an individual's domain skills and innovative skills (Bohle Carbonell et al., 2016).

Adaptive expertise can be developed in informal and formal settings. First, in the natural work environment it is important that individuals are confronted with a variety of practice, low risk when deviating from routines, and work in an environment that values learning. Thus, the work environment needs to naturally present individuals with unfamiliar situations, pushing them to adapt. Secondly, adaptation might be stimulated when there is low risk for the individual, and finally adaptation might be stimulated when it is valued by peers and by the culture of the organization. These characteristics for developing adaptive expertise can also be created in more formal learning environments. The 4C/ID model (van Merrienboer & Kirschner, 2018), for example, prescribes instructional methods that stimulate the development of complex skills by using methods that stimulate the recurrent aspects of complex skills, and methods aimed at adaptive expertise for nonrecurrent aspects of complex skills.

This chapter aimed to describe what adaptive expertise is, how it can be developed, and how it can be taught. We have tried to take care to provide a foundation of what the concept of adaptive expertise entails and, importantly, how it is similar and different from other related concepts based on the current state of literature on adaptive expertise. The comparison of adaptive expertise with other related concepts highlights the fuzzy boundaries between various concepts aiming to explain how individuals deal with unfamiliar tasks. The bibliographic analysis was included to provide a more objective view of the research field and to help in this endeavor. Nevertheless, it needs to be kept in mind that the accuracy of sorting groups of articles into clusters is dependent on the algorithm, and results may diverge if different clustering methods are used. Consequently, the research field may be more or less fragmented than described in this chapter. Another limitation relates to the inability to make empirically supported claims about the development of adaptive expertise in informal learning settings. While a number of studies measuring adaptive expertise before and after a training session exists, experimental studies with pre- and post-assessments conducted on employees are lacking.

Regarding the current body of research on adaptive expertise, progress is needed with regard to further validating existing measures. This applies to objective and subjective methods for measuring adaptive expertise. While several methods for measuring adaptive expertise subjectively exist, it is not clear how sensitive they are over time, and how good they are at predicting the criterion and performance on an unfamiliar task. Thus, their test-retest reliability needs to be established. While a number of studies have tested the implication active learning strategies have on adaptive expertise development, the current chapter has provided a detailed and extensive set of activities that should promote adaptive expertise. It will be beneficial if future research investigates the relative importance of these activities for the promotion of adaptive expertise. For the teaching of adaptive expertise it is important to determine domain-specific differences in the development of adaptive expertise.

The changes of work and the increasing proliferation of knowledge and tools present challenges to individuals in the form of unfamiliar situations. However, this challenge is surmountable if individuals are stimulated to develop adaptive expertise in formal instructional settings, and in their natural work environment.

REFERENCES

Adelson, B. (1984). When novices surpass experts: The difficulty of a task may increase with expertise. *Journal of Experimental Psychology: Learning, Memory, and Cognition, 10*(3), 483–495.

Ayres, P. L. (1993). Why goal-free problems can facilitate learning. *Contemporary Educational Psychology 18*(3), 6.

Barnett, S. M., & Ceci, S. J. (2002). When and where do we apply what we learn? A taxonomy for far transfer. *Psychological Bulletin 128*(4), 612–637.

Barnett, S. M., & Koslowski, B. (2002). Adaptive expertise: Effects of type of experience and the level of theoretical understanding it generates. *Thinking & Reasoning 8*, 237–267.

Bell, B. S., & Kozlowski, S. W. J. (2008). Active learning: Effects of core training design elements on self-regulatory processes, learning, and adaptability. *Journal of Applied Psychology 93*, 296–316.

Bohle Carbonell, K., Könings, K. D., Segers, M., & van Merriënboer, J. J. G. (2016). Measuring adaptive expertise: development and validation of an instrument. *European Journal of Work and Organizational Psychology 25*, 167–180.

Bohle Carbonell, K., Stalmeijer, R. E., Könings, K. D., Segers, M., & Van Merriënboer, J. J. G. (2014). How experts deal with novel situations: A review of adaptive expertise. *Educational Research Review 12*, 14–29.

Brynjolfsson, E., & McAfee, A. (2014). *The second machine age: Work, progress, and prosperity in a time of brilliant technologies.* New York: WW Norton.

Brynjolfsson, E., & Mcafee, A. (2017). What's driving the machine learning explosion. *Harvard Business Review, The Big Idea* (July), 12–13.

Cannon-Bowers, J. A., & Bowers, C. (2009). Synthetic learning environments: On developing a science of simulation, games, and virtual worlds for training. In S. W. J. Kozlowski & E. Salas (Eds), *Learning, training, and development in organizations* (pp. 229–262). New York: Routledge.

Charlin, B., Roy, L., Brailovsky, C., Goulet, F., & van der Vleuten, C. (2000). The Script Concordance Test: A tool to assess the reflective clinician. *Teaching and Learning in Medicine 12*(4), 189–195.

Chi, M. T. H. (2011). Theoretical perspectives, methodological approaches, and trends in the study of expertise. In Y. Li & G. Kaiser (Eds), *Expertise in mathematics instruction: An international perspective* (pp. 17–39). New York: Springer.

Clark, R. E., Feldon, D. F., Van Merriënboer, J. J. G., Yates, K., & Early, S. (2007). Cognitive task analysis. In J. M. Spector, M. D. Merrill, J. J. G. van Merriënboer, & M. P. Driscoll (Eds), *Handbook of research on educational communications and technology* (pp. 577–593). Mahwah, NJ: Lawrence Erlbaum Associates.

Corbalan, G., Kester, L., & van Merriënboer, J. J. G. (2011). Learner-controlled selection of tasks with different surface and structural features: Effects on transfer and efficiency. *Computers in Human Behavior 27*(1), 76–81.

Dane, E. (2010). Reconsidering the trade-off between expertise and flexibility: A cognitive entrenchment perspective. *Academy of Management Review 35*, 579–603.

Duncker, K. (1945). On problem solving. *Psychological Monographs 58*, 270.

Ericsson, K. A., Krampe, R. T., and Tesch-Romer, C. (1993). The role of deliberate practice in the acquisition of expert performance. *Psychological Review 100*(3), 363–406.

Ericsson, K. A., & Lehmann, A. C. (1996). Expert and exceptional performance: Evidence of maximal adaptation to task constraints. *Annual Review of Psychology* 47(1), 273–305.

Ericsson, K. A., & Smith, J. (1991). Prospects and limits of the empirical study of expertise: an introduction. In K. A. Ericsson & J. Smith (Eds), *Toward a general theory of expertise: Prospects and limits* (pp. 1–38). Cambridge, UK: Cambridge University Press.

Feldon, D. F. (2006). The implications of research on expertise for curriculum and pedagogy. *Educational Psychology Review* 19(2), 91–110.

Feltovich, P. J., Coulson, R. L., & Spiro, R. J. (2013). Knowledge application and transfer for complex tasks in ill-structured domains: Implications for instruction and testing in. In D. A. Evans & V. L. Patel (Eds), *Advanced models of cognition for medical training and practice* (vol. 97, p. 213). New York: Springer Science & Business Media.

Fisher, F. T., & Peterson, P. L. (2001). A tool to measure adaptive expertise in biomedical engineering students. In *Proceedings of the 2001 American Society for Engineering Education Annual Conference, Albuquerque, NM*.

Grenier, R. S., & Kehrhahn, M. (2008). Toward an integrated model of expertise redevelopment and its implications for HRD. *Human Resource Development Review* 7, 198–217.

Halff, H. M. (1993). Supporting scenario- and simulation-based instruction: Issues from the maintenance domain. In J. M. Spector, M. C. Polson, & D. J. Muraida (Eds), *Automating instructional design: Concepts and issues* (pp. 231–248). Englewood Cliffs, NJ: Educational Technology Publications.

Hall, W. (1998). Usability studies of a remedial multimedia system. *World Wide Web Internet and Web Information Systems* 7, 207–236.

Han, T. Y., & Williams, K. J. (2008). Multilevel investigation of adaptive performance: Individual- and team-level relationships. *Group & Organization Management* 33(6), 657–684.

Haskell, R. E. (2001). *Transfer of learning: Cognition, instruction, and reasoning. Educational Psychology Series*. San Diego, CA: Academic Press.

Hatano, G., & Inagaki, K. (1986). Two courses of expertise. In H. Stevenson, H. Azuma, & K. Hakuta (Eds), *Child development and education in Japan* (pp. 262–272). New York: W. H. Freeman.

Holyoak, K. J. (1991). Symbolic connectionism: Toward third-generation theories of expertise. In K. A. Ericsson & J. Smith (Eds), *Toward a general theory of expertise: Prospects and limits* (pp. 301–335). Cambridge, UK: Cambridge University Press.

Holyoak, K. J., & Koh, K. (1987). Surface and structural similarity in analogical transfer. *Memory & Cognition* 15(4), 332–340.

Holyoak, K. J., & Thagard, P. (1997). The analogical mind. *American Psychologist* 52(1), 35–44.

Joung, W., Hesketh, B., & Neal, A. (2006). Using "war stories" to train for adaptive performance: Is it better to learn from error or success? *Applied Psychology* 55(2), 282–302.

Kahneman, D. (2016). Interview of Daniel Kahneman by Dan Pink. Philadelphia. Retrieved from https://www.danpink.com/resource/dans-interview-with-nobelist-daniel-kahneman/

Kontogiannis, T., & Linou, N. (2001). Making instructions "visible" on the interface: An approach to learning fault diagnosis skills through guided discovery. *International Journal of Human-Computer Studies* 54(1), 53–79.

Kostons, D., van Gog, T., & Paas, F. (2012). Training self-assessment and task-selection skills: A cognitive approach to improving self-regulated learning. *Learning and Instruction* 22(2), 121–132.

Louis, M. R., & Sutton, R. I. (1991). Switching cognitive gears: From habits of mind to active thinking. *Human Relations 44*(1), 55–76.

Martin, T., Rayne, K., Kemp, N. J., Hart, J. D., & Diller, K. R. (2005). Teaching for adaptive expertise in biomedical engineering ethics. *Science and Engineering Ethics 11*(2), 257–276.

Martin, T., Rivale, S. D., & Diller, K. R. (2007). Comparison of student learning in challenge-based and traditional instruction in biomedical engineering. *Annals of Biomedical Engineering 35*, 1312–1323.

Moulton, C. A., Regehr, G., Lingard, L., Merritt, C., & MacRae, H. (2010). Slowing down to stay out of trouble in the operating room: Remaining attentive in automaticity. *Academic Medicine 85*(10), 1571–1577.

Moulton, C. A., Regehr, G., Mylopoulos, M., & MacRae, H. (2007). Slowing down when you should: A new model of expert judgment. *Academic Medicine 82*(10 Suppl), S109–16.

Mylopoulos, M., & Regehr, G. (2007). Cognitive metaphors of expertise and knowledge: prospects and limitations for medical education. *Medical Education 41*, 1159–1165.

Mylopoulos, M., & Regehr, G. (2009). How student models of expertise and innovation impact the development of adaptive expertise in medicine. *Medical Education 43*, 127–132.

Mylopoulos, M., & Scardamalia, M. (2008). Doctors' perspectives on their innovations in daily practice: Implications for knowledge building in health care. *Medical Education 42*, 975–981.

O'Neil, H. F., Lang, J., Yuang-C., Perez, R. S., Escalante, D., & Fox, F. S. (2014). What is cognitive readiness? In H. F. O'Neil, R. S. Perez, E. L. Baker (Eds), *Teaching and measuring cognitive readiness* (pp. 3–23). New York, NJ: Springer.

O'Neil, H. F., Perez, R. S., & Baker, E. L. (2014). *Teaching and measuring cognitive readiness.* New York: Springer.

Olsen, S. E., & Rasmussen, J. (1989). The reflective expert and the prenovice: Notes on skill-, rule-, and knowledge-based performance in the setting of instruction and training. In L. Bainbridge & S. A. Ruiz-Quintanilla (Eds), *Developing skills with information technology* (pp. 9–33). Chichester, UK: Wiley.

Paas, F. G. W. C., & Van Merriënboer, J. J. G. (1994). Variability of worked examples and transfer of geometrical problem-solving skills: A cognitive-load approach. *Journal of Educational Psychology 86*(1), 122–133.

Ployhart, R. E., & Bliese, P. D. (2006). Individual adaptability (I-ADAPT) theory: Conceptualizing the antecedents, consequences, and measurement of individual differences in adaptability. In C. S. Burke, L. G. Pierce, & E. Salas (Eds), *Advances in human performance and cognitive engineering research* (vol. 6, pp. 3–39). Bingley, UK: Emerald Group.

Pulakos, E. D., Schmitt, N., Dorsey, D., Arad, S., Borman, W., & Hedge, J. (2002). Predicting adaptive performance: Further tests of a model of adaptability. *Human Performance 15*(4), 299–323.

Schraagen, J. M. (1993). How experts solve a novel problem in experimental design. *Cognitive Science 17*(2), 285–309.

Schwartz, D. L., Bransford, J. D., & Sears, D. (2005). Innovation and efficiency in learning and transfer. In J. P. Mestre (Ed.), *Transfer of learning from a modern multidisciplinary perspective* (pp. 1–51). Charlotte, NC: Information Age.

Schwartz, D. L., & Martin, T. (2004). Inventing to prepare for future learning: The hidden efficiency of encouraging. *Cognition and Instruction 22*(2), 129–184.

Shea, C. H., & Kohl, R. M. (1990). Specificity and Variability of Practice. *Research Quarterly for Exercise and Sport 61*(2), 169–177.

Shea, C. H., & Wulf, G. (2005). Schema theory: A critical appraisal and reevaluation. *Journal of Motor Behavior, 37*(2), 85–102.

Small, H. (1973). Co-citation in the scientific literature: A new measure of the relationship between two documents. *Journal of the American Society for Information Science, 24*(4), 265–269.

Smith, E. M., Ford, J. K., & Kozlowski, S. W. J. (1996). Building adaptive expertise: Implications for training design strategies. In M. A. Quinones & A. Ehrenstein (Eds), *Training for a rapidly changing workplace: Application of psychological research* (pp. 89–118). Washington, DC: American Psychological Association.

Spiro, R. J., Collins, B. P., Thota, J. J., & Feltovich, P. J. (2003). Cognitive flexibility theory: Hypermedia for complex learning, adaptive knowledge application, and experience acceleration. *Educational Technology 43*(5), 5–10.

Spiro, R. J., Coulson, R. L., Feltovich, P. J., & Anderson, D. K. (1988). *Cognitive Flexibility Theory: Advanced Knowledge Acquisition in Ill-Structured Domains.* Technical Report No. 441.

Uzzi, B., Mukherjee, S., Stringer, M., & Jones, B. (2013). Atypical combinations and scientific impact. *Science 342*(6157), 468–472.

van den Boom, G., Paas, F., van Merriënboer, J. J. G., & van Gog, T. (2004). Reflection prompts and tutor feedback in a web-based learning environment: Effects on students' self-regulated learning competence. *Computers in Human Behavior 20*(4), 551–567.

Van der Heijden, B. I. J. M. (2000). The development and psychometric evaluation of a multidimensional measurement instrument of professional expertise. *Journal of the European Council for High Ability 11*(1), 9–39.

Van der Heijden, B. I. J. M. (2002). Prerequisites to guarantee life-long employability. *Personnel Review, 31*(1), 44–61.

van Gog, T., Ericsson, K. A., Rikers, R. M. J. P., Paas, F., Gog, T., Ericsson, K. A., . . . Paas, F. (2005). Instructional design for advanced learners: Establishing connections between the theoretical frameworks of cognitive load and deliberate practice. *Educational Technology Research and Development 53*(3), 73–81.

van Gog, T., Paas, F., & van Merriënboer, J. J. G. (2004). Process-oriented worked examples: Improving transfer performance through enhanced understanding. *Instructional Science 32*(1/2), 83–98.

Van Merriënboer, J. J. G. (1997). *Training complex cognitive skills: A four-component instructional design model for technical training.* Englewood Cliffs, NJ: Educational Technology.

Van Merriënboer, J. J. G. (2013). Perspectives on problem solving and instruction. *Computers & Education 64*, 153–160.

Van Merriënboer, J. J. G., Jelsma, O., & Paas, F. G. W. C. (1992). Training for reflective expertise: A four-component instructional design model for complex cognitive skills. *Educational Technology Research and Development 40*(2), 23–43.

Van Merriënboer, J. J. G., & Kirschner, P. A. (2018). *Ten steps to complex learning* (3rd edn). New York: Routledge.

Varpio, L., Schryer, C. F., & Lingard, L. (2009). Routine and adaptive expert strategies for resolving ICT mediated communication problems in the team setting. *Medical Education 43*, 680–687.

Walker, J. M. T., Cordray, D. S., King, P. H., & Brophy, S. P. (2006). Design scenarios as an assessment of adaptive expertise. *International Journal of Engineering Education 22*(3), 645–651.

Wineburg, S. (1998). Reading Abraham Lincoln: An expert/expert study in the interpretation of historical texts. *Cognitive Science 22*, 319–346.

Woods, D. D. (2015). Four concepts for resilience and the implications for the future of resilience engineering. *Reliability Engineering & System Safety 141*, 5–9.

Wulf, G., & Schmidt, R. A. (1997). Variability of practice and implicit motor learning. *Journal of Experimental Psychology: Learning, Memory, and Cognition 23*(4), 987–1006.

Zimmerman, B. J. (2006). Development and Adaptation of Expertise: The role of Self-Regulatory Process and Beliefs. In K. A. Ericsson, N. Charness, P. J. Feltovih, & R. R. Hoffman (Eds), *The Cambridge handbook of expertise and expert performance* (pp. 705–722). New York: Cambridge University Press.

SECTION II

METHODS TO STUDY, TEST, ANALYSE, AND REPRESENT EXPERTISE

SECTION EDITOR:
EMILIE ROTH

CHAPTER 13

REPRESENTATIVE TEST AND TASK DEVELOPMENT AND SIMULATED TASK ENVIRONMENTS

KEVIN R. HARRIS, LINDSEY N. FOREMAN, AND DAVID W. ECCLES

INTRODUCTION

SUGGESTING that Stephen Curry is a good basketball player is an understatement. Curry, a professional player in the National Basketball Association (NBA) in the USA, is changing expectations about what can be expected for his position. He holds the record for three-pointers in a game, making 13 in one game with an impressive 76 percent success rate (13 of 17). Few other players come close to Curry's three-point dominance (Golliver, 2016). However, this dominance was by no means a given: Curry had to relearn shooting the ball during his high school sophomore year and currently often takes one thousand shots even before practice begins (Fleming, 2015).

While professional athletes, chess grandmasters, and other elite performers garner much attention for their performances, and professional associations and local communities might celebrate high-performing surgeons and teachers, understanding high-level performance requires effective methods for studying expertise and expert performers. Ideally, tasks that capture the key components of expertise can be identified, allowing expert performance to be studied under controlled conditions (Ericsson, 2004; Ericsson & Smith, 1991). Such tasks are referred to as representative tasks or representative tests (hereon, simply *representative tasks* for ease of reading). When well designed, these tasks afford researchers distinctions regarding the skill level of the performer and to identify the mechanisms underpinning expert performance within a given domain. Once the underlying mechanisms for superior performance have been elucidated and performers have been stratified based on skill, researchers can uncover the developmental steps taken by those performers to reach their current levels of performance.

Our goal for this chapter is to describe representative task development and simulated task environments and their usefulness for studying expertise and fostering the development of expertise via the design of training based on studies of experts. First, we provide a general overview of the concepts of representative tasks and simulated task environments. In the next section, we describe in more detail how representative tasks and simulated task environments can be used to achieve the twin goals of understanding expert performance and developing training on the basis of expert performers. We then present examples of recent research involving representative tasks. We finish by presenting future directions for the use of representative tasks in research and practice.

Representative Tasks and Simulated Task Environments

Before describing in detail how representative tasks and simulated task environments can be used to research and train expert performance, we provide here a more general overview of representative tasks and simulated task environments that will afford the reader an initial appreciation of these key concepts.

Representative Tasks

Representative tasks are tasks allowing researchers to "measure highly representative performance under controlled conditions that are standardized for everyone tested" (Ericsson & Ward, 2007, p. 347). Moreover, representative tasks are tasks that are central to performance in a given domain and effectively discriminate between skill levels. Ericsson (2004) noted the work of de Groot (1946/1978) as an early example of the use of a representative task to study differences in skill level. De Groot used actual in-game chess positions recorded during matches between chess grandmasters. The extracted positions were presented to chess players of various skill levels and the decision-making process of selecting the next move was studied in a controlled manner. Thus, the underpinnings of expert performance in chess were identified and extracted. In other words, the thought processes during the selection of the next move, as well as the ultimate selection decision made, for each presentation of a static individual in-game chess position are sufficient to allow researchers both to classify chess players according to ability and to study the mechanisms underpinning superior performance.

The use of static in-game chess positions is a straightforward example of a representative task because the task is an extracted moment from within the overall chess match. Other straightforward examples of representative tasks include the whole task, from beginning to completion. These whole tasks can feature time limits depending on the nature of the task. An example of a representative task without time limits is the amount of time it takes to sprint a given distance; that is, the entire task of sprinting is the representative task. An example of a representative task with time limits is the maximum amount of text that can be typed without error in one minute, where this task is a measure of typing skill (Ericsson, 2008; Ericsson & Lehmann, 1996). In contrast to this *whole task* approach, other representative tasks involve performance during more isolated subcomponents of task performance. Returning a serve during a tennis match is only one subcomponent of the overall game, but the ability to return a high-speed serve is a valid indicator of overall tennis skill (Ericsson & Ward, 2007). We suggest a framework later in this chapter for identifying and isolating the specific aspects of task performance that distinguish the observed superior performance of the top performers. However, we now introduce the concept of the simulated task environment.

Simulated Task Environments

A simulated task environment is a concept related to the concept of a representative task. Although simulated task environments are used for both research and training purposes, traditionally simulated task environments are more synonymous with training than with research. Indeed, humankind has always made use of forms of simulated environments for training: For example, consider how our ancient ancestors practiced fighting techniques or used their hunting weapons with a symbolic target representing the hunted animal. An early documented use of simulation for training in the modern era was a flight simulator known as the *Link Trainer*, which was introduced in 1922 (Grenvik & Schaefer, 2004) and patented in 1937 (Link, 1937). One benefit of simulation is that it provides opportunities to engage in a task mimicking a *real* task with fewer pressures and risks than are normally encountered during the real task. For example, a nurse learning how to perform needle injections can simulate the task by practicing with an orange or pig flesh, and a surgeon can simulate a laparoscopic procedure using hi-fidelity simulation that is virtually indistinguishable from the *real* surgical task. These examples differ in meaningful ways, the most apparent of which is the degree to which the simulations mimic the *real world*. Yet, despite these differences, both are examples of a simulated task environment.

Another benefit of simulated task environments is the ability to control the features and presentation of the simulated environment. Thus, within a research context, the easy repeatability of simulations is useful for achieving experimental control; the same stimuli can be presented to multiple participants and on multiple occasions. Within a training context, simulations can be repeated to offer opportunities for additional practice with feedback or based on additional training needs identified during previous simulated sessions. For tasks that can vary considerably across different instances of the task, simulations should be designed to present many variations on a single scenario (Harris, Eccles, & Shatzer, 2017). For example, Harris et al. advocated asking junior medical staff members to practice challenging medical decisions made by a senior colleague when a staff member was concerned that the senior colleague was making a wrong decision. In medicine, situations and colleagues are always changing, and so simulations designed to allow staff members to practice such challenges should present a variety of senior colleagues operating in a variety of situations.

Simulated task environments also can be used to train for, and undertake research in relation to, specific events and situations that do not occur at all naturally before they must be performed for real. From a research perspective, they provide unique opportunities to systematically and repeatedly examine and understand the nature of performance in these otherwise rare circumstances. From a training perspective, they offer the opportunity for performers to engage in practice in relation to the specific demands and constraints imposed by the upcoming events and situations. Examples include rare medical situations (Ericsson, Whyte, & Ward, 2007), major sports competitions (Eccles, Ward, & Woodman, 2009), military operations, and space missions, such as human Mars landings. In these examples, designers and performers go to great lengths in advance of events and situations to identify and replicate the physical and psychological demands that these events and situations will likely entail with the aim of gaining the greatest possible practice effect (Eccles et al., 2009).

Gray (2002) provided a structure for researchers and trainers to discuss simulated task environments, and outlined important considerations for their deployment. Gray provided a broad rather than precise definition of such environments, encompassing the range of simulation possibilities varying in delivery, features, and complexity. For our present purpose, the unifying element of the simulated task environment is the trainer or researcher's desire to study behavior in a task environment that is appropriate to his or her specific training and research goals (Gray, 2002). In other words, simulation can mean different things to different people, and this flexibility allows for maximizing the use of simulation by tailoring it to users' needs.

Thus, our use of the term simulated task environment allows for the inclusion of a wide range of simulation types. Let us now consider how this broad use of the term might play out in practice. An applied researcher might want to bring her work in the field into the laboratory to increase environmental control and better isolate possible answers to tightly focused research questions. For example, consider a line of research in which law enforcement officers were interviewed about how they typically handled stressful situations. A next step in this line of research might be to use a simulated task environment depicting stressful situations to capture and examine officers' thoughts and behaviors during the simulated event. In contrast, consider a laboratory-based researcher conducting basic research by measuring the thresholds at which participants' performance on standard attention span tasks suffers following standard laboratory stress induction techniques (e.g., sounds or high temperatures). This researcher might use the same, or perhaps a different, simulated task environment used to study law enforcement officers during a potential domestic violence situation to add layers of complexity by measuring the impact of traditional stressors (e.g., the addition of sounds or high temperatures) on attention in a real-world scenario (Gray, 2002). In this sense, a simulated task environment can be viewed as a potential bridge between basic and applied research that ties together the commonalities between these two approaches and offers opportunities to obtain a better overall picture of the phenomenon under investigation. Later, we explore how this flexibility is useful to researchers studying expertise and expert performers but we turn next to variations of simulated task environments.

Gray (2002) categorizes simulated task environments into five types: (a) Hi-fidelity simulations of complex systems, (b) hi-fidelity simulations of simple systems, (c) scaled worlds, (d) synthetic environments and microworlds, and (e) laboratory tasks and simulated task environments. As with the definition of simulated task environment, there is flexibility in how a particular simulated task environment is categorized and categorization ultimately depends on the research question. We now briefly discuss these categories in turn.

Hi-fidelity simulations of complex systems are perhaps the type of simulated task environment that one would envision when contemplating simulation, such as a commercial flight simulator with a cockpit identical to a real aircraft and simulated flight motions. This category of simulation attempts to capture the full complexity inherent in the real-world task being simulated. The second category, hi-fidelity simulation of simple systems, might simulate a single component of a larger complex system or fully simulate a simplistic system. An example of a single component simulation involves presenting only the navigation display screen from an aircraft cockpit and asking that the participant select the best next course of action based on this information display. Simulations of single components of a larger complex system are often used for answering tightly focused research questions (Gray, 2002); in this case, the question might concern the effect of pilot fatigue on monitoring the navigation display, for example.

The third category of simulated task environment, scaled worlds, is perhaps the most relevant to the present chapter because scaled worlds are useful in identifying functional relationships between subcomponents of a complex task and then preserving "the functional relationships in this subset while paring away others" (Gray, 2002, p. 208). Consider an example from the domain of residential roofing in which researchers created a scaled world to study roofing performance (Murphy et al., 2014). Roofers were asked to don a harness, install tar paper, install shingles, and doff a harness on a small (14 × 12 × 8 ft^3) angled roof placed at ground level. The tasks asked of the participants (e.g., installing tar paper and shingles) were functionally most important to the task of roofing and thus to the researchers, who were able to pare away other tasks deemed less important to the research question of interest (e.g., climbing the ladder). In another example, Eccles, Walsh, and Ingledew (2002a) studied navigation in the sport of orienteering, which is used by various militaries to train forces for covert maneuvers in enemy territory. While orienteering takes place in wild terrain such as hilly forests, and involves navigating while running using a map and compass, Eccles et al. focused on only the from-the-map route-planning subcomponent of orienteering. The researchers were able to pare away most of the remaining components of this larger and more complex task such that they could capture key differences in expert and novice orienteers' planning heuristics in the laboratory under controlled conditions. Thus, in a broader sense scaled worlds allow researchers to direct their finite resources to the task subcomponents of greatest interest.

Gray (2002) described a fourth category of simulated task environment called synthetic environments and microworlds, which are often computer simulations (Gonzalez, Vanyukov, & Martin, 2005) in which physical realism is exchanged with cognitive engagement (Gray, 2002; Naweed, Hockey, & Clarke, 2013). For example, in a research context, participants might be studied as they direct traffic on an artificial rail network or avoid a system malfunction by maintaining machinery at an artificial commercial brewery; in a training context, trainees might engage in practice on these artificial tasks. Synthetic environments and microworlds differ from scaled worlds in that components from several systems are combined in an artificial environment and the effects of this combination on performance are measured with the goal of generalizing the findings across other related systems. For example, researchers interested in the impact of automation on operator attention span are less concerned about translating the findings back to a specific environment, such as bicycle manufacturing, than whether the findings are generalizable across a broad range of commercial environments, from canning to automobile manufacturing. The final category of simulated task environments in Gray's taxonomy is *simple* laboratory tasks. The inclusion of these laboratory tasks again allows flexibility because it encompasses any laboratory-based simulated task, however crude, used by researchers in an attempt to answer their specific research questions.

Gray (2002) also presented three dimensions of simulated task environments: (a) tractability, (b) correspondence, and (c) engagement. Tractability refers to the degree that the simulated task environment can be managed, or controlled, by the researcher or trainer and how much training is required by the trainee prior to that individual engaging in the simulation. The second dimension is correspondence, which is the degree to which simulated task environment mimics the real world, recognizing of course that the real-world target for transfer might be a generalizable component of many systems (e.g., automation or warning systems in varied contexts) or many components within a single system (e.g., a comprehensive aircraft simulation corresponding to an actual flight environment). As we discussed earlier in relation to the definition of *simulated task environment*, what is important about correspondence is tied directly to the research question or training objective of interest. The simulated task environment need not fully mimic the actual *real-world* environment (e.g., hi-fidelity flight simulator) but the simulated task environment should allow the researcher and trainer to identify and isolate the functional relationships being studied.

The third dimension of simulated task environments is engagement, which is the degree to which the participant or trainee finds the simulated task environment to be engaging versus boring.

The Application of Representative Tasks and Simulated Task Environments for Studying Expertise and Expert Performance

Representative tasks and simulated task environments are used in the study of expertise and expert performance to (a) understand the performance of the best performers and/or (b) derive training intended to improve the performance of lesser-skilled performers. Thus, the theoretical foundation for the use of simulated task environments should be to either understand or improve an individual's performance; that is, such use should be based on "a theoretical understanding of human performance and how it is enhanced" (Harris et al., 2013, p. 6). A guiding framework specifically for developing and implementing representative tasks and/or simulated task environments is also beneficial for researchers generally (e.g., Gray, 2002), and those studying expert performance specifically (e.g., Fadde, 2009; Ward, Suss, & Basevitch, 2009).

Earlier, we presented historical examples of representative tasks, such as those involving chess. However, there are many domains of performance for which representative tasks have yet to be developed. The strategy presented in this chapter involves identifying the best performers within a domain based on their performance during representative tasks, uncovering the developmental and training pathway taken to reach this best level, and using this pathway information to improve the performance of less-skilled individuals. Thus, critical to this approach is identifying a representative task for a given domain. How, then, do we identify expert performers and/or representative tasks in domains for which representative tasks do not currently exist?

While identifying representative tasks can be challenging in some domains (Ericsson & Williams, 2007), the development of a representative task for given domain often begins with identifying expert performers on the basis of an existing objective performance hierarchy, such as the Elo rating in chess, percentage of targets correctly identified as friend or foe in the military, bowling average in bowling, and the top salesperson based on income from product sales. In other domains, experts might be identified based not on a performance hierarchy but on a performance measure critical to success in the domain, such as tissue damaged during surgery or postoperative patient survival rate. Once the best performers are preliminarily identified, *process tracing methods* can be used to capture key cognitive, behavioral, and physiological processes underpinning these high performance levels (e.g., Charness & Tuffiash, 2008). These methods include eye-tracking and other forms of attention tracing (e.g., Eccles, Walsh, & Ingledew, 2006), temporal or spatial occlusion (Williams, Ford, Eccles, & Ward, 2011), latency and motion analysis (e.g., Ward, Suss, Eccles, Williams, & Harris, 2011), think-aloud reports (e.g., Ericsson & Simon, 1993; see Chapter 17, "A Historical Perspective on Introspection," by Ward et al., this volume), and forms of interview such as the critical decision method (e.g., see Chapter 19, "Incident-Based Methods for Studying Expertise," by Militello & Anders, this volume). We now consider examples of some of these methods.

De Groot's (1946/1978) study of chess (described earlier in the section "Representative Tasks") is a classic example of the use of a process tracing method, which was think-aloud reports, to better understand the cognitive processes underpinning skilled performance on representative chess tasks. De Groot presented in-game chess positions to world-class performers and asked them to think aloud while selecting the best available next move. Analysis of the resultant verbal reports allowed de Groot to identify specific thought processes leading to superior next-move-selection decisions (Ericsson & Ward, 2007).

Research by Eccles and colleagues provides an example of how multiple studies, involving different research methods, may be used to identify representative tasks and then use them to investigate the mechanisms mediating expert performance. Eccles, Walsh, and Ingledew (2002b) first interviewed expert athletes in the sport of orienteering, which, as described earlier in the section "Simulated Task Environments", involves on-foot navigation using map and compass through a series of checkpoints in wild terrain. The aim of the study was to better understand the cognitive demands of the sport and the cognitive strategies employed by expert orienteers in the face of the those demands. The study participants reported that an important component of the sport is effectively planning upcoming navigation from the map with the aim of selecting a psychologically and physically efficient route through the terrain to the checkpoints. The participants also reported that much practice was required to become skilled at this component of the sport. Based on this information, Eccles et al. (2006) were then able to (re-)present this planning component of the sport under controlled conditions to explore, using the methods of experimental psychology, how differences in planning strategies effectively discriminated between expert and less-skilled orienteers.

Paull and Glencross (1997) used a temporal occlusion method to determine that superior tracking of baseball pitches by batters was based on their superior recognition of cues related to the pitcher's arm and early path of the ball upon release. Pitches were presented to batters using film of real pitches. Performance-critical cues in the pitcher's movement were identified by occluding (by inserting a black screen) various parts of the pitcher's movement and identifying the part of the movement that, when occluded, led to the greatest decrement in the skilled batter's performance. A related study of tennis skill used an eye-tracking method to identify that the ability to return a high-speed tennis serve was a key determinant of overall tennis skill level, and serve-return skill specifically was shown to depend on skilled players' use of movement-related cues located around the middle of the server's body (Williams, Ward, Knowles, & Smeeton, 2002). Thus, in these cases, batters with higher batting averages were better at picking up the movement cues related to the pitcher's arm during the pitch, and the higher-ranked tennis player was better at picking up movement cues during an opponent's serve; for reviews, see Eccles and Arsal (2015) and Fadde (2009). Of course, there may be multiple representative tasks for a given domain. The best tennis players also might be better than less-skilled players at returning a forehand stroke via their use of movement cues during the stroke. Nonetheless, once identified, representative tasks allow researchers to study expertise and expert performers in a controlled manner.

Recalling Gray's (2002) framework, the examples described above of the use of process tracing methods with representative tasks illustrate how identifying the mediating mechanisms underpinning performance, such as use of predictive movement cues in dynamic open skills, can help identify what task components to pare away to better understand performance on the task or to better design training. A researcher or trainer need not go beyond performance on specific representative tasks, such as returning a high-speed serve, if the task allows for the study of expert performance and classification of performers based on skill. However, there will be circumstances for which researchers will be interested in adding additional subcomponents of task performance, all the way up to presenting the whole of the task with nothing pared away, to explore expert versus less-skilled performance (Charness & Tuffiash, 2008; Ericsson & Williams, 2007).

Once identified, representative tasks can be used to further explore the underpinnings of expert performance. In a study of skilled versus inexperienced police officers, performance on 3 of 11 filmed scenarios in which the officer came under simulated threat was found to be highly discriminating, in that principal component analysis confirmed the categorization of the officers as skilled based on their performance (Ward et al., 2011). The three performance-discriminating scenarios included depictions of a boy outside a school carrying an explosive device, a hostage situation at a convenience store, and an armed man entering a school. Upon completion of each scenario, officers were asked to provide immediate retrospective reports of thinking during the scenarios. These thought data allowed the researchers to identify environmental cues used by the skilled officers as the basis for their interventions in the scenario, such as their decision to shoot a perpetrator. A follow up cue-occlusion-based study similar to the studies of sport described earlier could have been conducted as a next research step (but was not) to identify the importance, for officers' performance, of a given identified cue in the threat scenarios. In an example from the nursing domain, researchers collected think-aloud reports from experienced and novice nurses during a simulated scenario of encountering a fallen patient (Whyte et al., 2012). Analysis of the reports revealed the environmental cues used by experienced nurses to decide on their treatment response and the specific actions these nurses took to resolve the situation more successfully than the novice nurses.

The Application of Representative Tasks and Simulated Task Environments for Developing Training Based on the Study of Expertise and Expert Performance

A guiding framework for the use of representative tasks and simulated task environments to develop training expertise and expert performance is needed to maximize the effectiveness of the derived training. For example, simulations lacking structure or a guiding purpose that merely put individuals *through the motions* are ineffective, and a waste of financial and human resources (Ward, Williams, & Hancock, 2006). Without structure or guidance, individuals are left to their own assumptions about how to benefit from the training. Often these assumptions are wrong or differ widely among individuals and thus the training received is ineffective and subject to criticism. Early simulation attempts are often blamed for poor trainee progression when the culprit is actually a lack of an appropriate curriculum or design for simulation (Harris, Eccles, Ward, & Whyte, 2013). Over a decade ago, only 25 percent of US surgical residency programs with simulation centers had mandatory program requirements specifying what should be taught and how (~13 percent of total US surgical residency programs; Korndorffer, Stefanidis, & Scott, 2006) and a review of 36 surgical skills laboratories found that 12 (33 percent) did not have a documented curriculum (Kapadia, DaRosa, MacRae, & Dunnington, 2007). The approach outlined in this chapter helps provide structure during the application of training, but also during its development.

Training the specific cognitive mechanisms identified as underpinning superior performance, such as a focus on specific movement-related cues to better anticipate tennis serves, has been shown to improve the performance of less-skilled performers (e.g., Williams, Ward, Knowles, & Smeeton, 2002). However, training may not be always focused on enhancing use of an identified mediating mechanism such as coaching a tennis player to focus on the midsection of the server. When key developmental and training pathways to skilled performance in a domain have been identified already, then the emphasis is on identifying skill development practices found to be effective across performers from within these pathways. Analyses of the developmental histories of expert performers, which involve identifying for a given expert the instruction they received and practice in which they engaged over their developmental years, often reveal that experts have engaged in a wide range of types of practice activities including quite novel and unanticipated ones. These different activities can be examined empirically to identify their contribution to skill development.

Continuing to focus on chess as an example, the Soviet Union's dominance of the game was attributed to their academies' superior selection of the most promising players (Shadrick & Lussier, 2004), implying either an implicit or explicit endorsement of *natural* Soviet chess ability. However, following the breakup of the Soviet Union, analysis of the academies' techniques revealed that Soviet training went beyond traditional approaches to include exercises based on "principles of expert play that reflected the thought patterns of grandmasters" (Shadrick & Lussier, 2004, p. 2). These exercises allowed trainees to engage in deeper and more creative thinking than afforded by previous training regimens and competitions. Once the secrets of Soviet chess dominance and other grandmasters were uncovered, novices could adopt these techniques to improve their own performance. Chess grandmaster Judit Polgár's father László, who did not play chess, trained Judit and her sisters by adopting the known training methods of chess grandmasters, which included devoting much time to a specific training regimen involving feedback (Dweck, 2006).

For some time, researchers have advocated, either implicitly (Ericsson, Krampe, & Tesch-Römer, 1993) or explicitly (Williams & Ward, 2003), deriving training techniques from studies of expert performers, and the ability to improve novices' performance has been proposed as a measure of the value of dedicating resources to studies of expert performers (Charness & Tuffiash, 2008). The practice of tracing the developmental history of expert performers for training development has been labeled by different researchers as expert performance-based training (ExPerT; Ward, Suss, & Basevitch, 2009) and expertise-based training (XBT; Fadde, 2009). We now present examples of these training approaches.

In essence, the ExPerT approach was used by the Soviets when they extended the work of de Groot (1946/1978) and developed chess training exercises based on the thought processes of grandmasters (Shadrick & Lussier, 2004). ExPerT allows objective assessments of the success of training and offers a way of measuring how expert performers actually perform rather than how it is assumed they will perform and/or if the actual observed performance of the expert performer matches the way novices are trained to perform (i.e., the textbook version of performance). Observing domain performance on a representative task helps ensure that training reflects the true means to perform well within the domain. For example, Staszewski and Davison (2000) attempted to improve mine detector performance by studying an expert mine detector operator with 30 years' experience. They first confirmed the operator's high performance level by having him detect mines and then investigated the mine location methods he employed, which revealed very different methods to those outlined in the military-issued training manual. The expert's methods were then used to develop a training program. Novice mine detector operators with standard mine detection training and some active service were then trained on this program, with remarkable results. Detection performance improved beyond current levels for all types of mines but most dramatically (up to 300 percent increases) for the most difficult-to-detect mines.

As the examples in this section indicate, once representative tasks have been identified, performance on these tasks allows for confirmation of high levels of domain performance and identification of the developmental pathway taken by the performer to reach such levels of performance. The chapter opened with a description of Stephen Curry's basketball performance, which he developed by engaging in very specific self-selected ancillary drills after arriving to the NBA. These drills helped him improve further to reach his current level of performance (Davis, 2015). In one such drill, Curry dribbles in front of a wall on which differently colored flashing lights illuminate. Each color is associated with a different movement instruction that he must act on as quickly as possible (Kroichick, 2015), mimicking the rapid and varied demands of in-game dribbling and decision making. Presumably, another professional basketball player could follow Curry's training regimen and experience similar performance gains. In fact, Curry adopted a drill involving fast movements around the court with three-point shots taken from five different locations on the basis that another NBA player, who was proposed to be the best spot-up shooter in the league by one of Curry's coaches, also used this drill (Davis, 2015). Moreover, Curry's lifelong developmental history could be studied to determine the activities responsible for his skill development. It is of course an empirical question whether the techniques used by Curry would work for others, but this question could be answered by carefully designed research using the expertise-based approaches proposed here.

Recent Example Applications of Representative Tasks and Simulated Task Environments for Studying Expertise and Expert Performance

To this point, we have described the concepts of representative tasks and simulated task environments, and their application to the study of expert performance and, in turn, the design of effective training. We now present a recent example from three quite different domains of study of expert performance in which researchers have identified representative tasks and recreated these tasks under controlled conditions through careful design of experimental stimuli and environments. In doing so, these researchers have revealed important insights about the nature of expertise within and beyond these domains. In each domain, a key challenge has been to identify tasks that are central to performance in the domain and effectively discriminate between skill levels. For example, while being able to effectively shine your boots prior to playing in a soccer game might be a skill encouraged by your coach, it is not a skill central to the actual performance in the game and is unlikely to discriminate effectively between players of different skill levels. On this note, we focus first on studies in the domain of sport.

Sport

Researchers in the sports sciences have long been interested in understanding expert performance in sport, with a view, of course, in understanding how to accelerate its acquisition. A particular interest of these researchers has been in understanding expertise in *open* sports, where the environment changes independently of the individual performer, which is the case for most team sports.

Researchers identified, initially on the basis of their own task analyses and testimonies by coaches and players in these sports but later through experimentation, that one key type of task common to these sports is anticipation of changes to the environment, such as movements of teammates and opponents in team sports (Eccles & Tenenbaum, 2004; Williams et al., 2011). Subsequently, a challenge for these researchers has been to identify laboratory tasks that adequately capture and thus faithfully represent real anticipation tasks in these sports.

A study by Ward, Ericsson, and Williams (2013) provides an example of one such attempt to create and use a representative task. These researchers compared performance on laboratory-based anticipation tasks between skilled and less-skilled soccer players, where skill level was determined by playing either at the highest national level available or at an amateur high school level. The researchers created stimuli for the study by filming live games in such a way that the viewer was facing the opposing team. The film was also edited so that the film displayed the opposing team in control of the ball and moving towards the participant (Ward et al., 2013). During each trial, 10 s of the footage (described above) was played and then was ended abruptly, to show only a black screen, before, during, or after a pass was made. Participants were immediately asked to indicate, from three presented locations on the field, where the pass would end. In a second form of this anticipation task, participants were shown similar sequences of play and the footage was paused (i.e., not removed as it was in the first task) for 20 s prior to a pass, a shot on goal, or an advancing opponent player. Participants were immediately asked to generate every realistic option that could follow the video clip including the opposing team's movements on the field and actions that could be taken, such as making a pass to another player or making a shot on goal. Participants then ranked these options based on threat level in terms of scoring potential. When compared to their less-skilled counterparts, skilled players were better able to anticipate opponents' actions and their outcomes in the anticipation task, and generate more relevant response options in the assessment task. These findings demonstrate the development of a representative task allowing one to objectively distinguish between the skill levels of the players. Moreover, analysis of a play situation improves as one becomes more of an expert, which suggests that training novice performers' analytical skills may help accelerate their growth as a player.

In a similar study, researchers used video simulations of baseball pitches to compare perceptual–cognitive skills of college baseball batters identified as either higher or lower ranking performers by their coaches (Belling, Sada, & Ward, 2015). The pitches were occluded either early or later in the pitch and batters were asked to indicate the type of pitch thrown, which of nine zones the ball would pass through over the plate, and press a button when specific pitch types and zones appeared together. For both the early and late occlusion conditions, skilled batters were found to perform significantly better at recognizing pitch type, pitch destination, and *swinging* at particular pitches.

Fadde's (2016) 10-month baseball training program is an example of an application of the baseball occlusion approach. The program used an occlusion paradigm and pitch recognition software. The program was integrated into a college team's regular practice using a series of occlusion-based training regimes. The first regime presented trainees with pitches occluded either earlier or later in the pitch based on the participant's skill level, as evaluated by the program. Using the software, batters chose which pitcher, batting side, and type of drill to focus on. A second batting cage simulation occluded the simulated pitches during ball flight as the trainee hit a ball off the tee to strengthen associations between hitting and particular types of pitches. The final regimen involved a live occlusion task where the batter stood in a bullpen to receive pitches as normal but was instructed to call out the type of pitch before the pitch reached the catcher rather than swing at it. The trained team significantly improved their performance during the training period and became the top performers in their conference on all analyzed metrics, including runs scored and batting averages when compared against the batting statistics of the teams in their athletic conference. Fadde's training program suggests that the use of representative tasks to develop training based on expert performers provides an ecologically valid and effective form of training.

Researchers have used eye-tracking and collected both concurrent and retrospective verbal reports of thinking to compare the visual search and decision strategies, for instance, of novice karate athletes who had never competed and expert, internationally competitive karate athletes (Milazzo, Farrow, Ruffault, & Fournier, 2016). The karate athletes were asked to respond to an opponent's attacks that had a specific attack pattern by either launching their own counterattack or intercepting the opponent's attack. The experts reacted faster and more accurately than the novices, which was attributed to the experts identifying the pattern of attacks via their more efficient visual search strategies. This live task suggests programs can be created to encourage training novices to be more efficient with their visual search, allowing them to recognize patterns of attack and respond accordingly at a quicker pace. We now turn to examples from the domain of medicine.

Medicine

While determined to be effective for identifying expert performers (Ward et al., 2013) and developing training in sport (e.g., Fadde, 2016), representative tasks and simulated task environments are particularly useful in domains with life-or-death consequences such as medicine. As outlined earlier, researchers increasingly have been incorporating both representative tasks and simulated task environments into medicine. For example, Causer, Barach, and Williams (2014) proposed that deriving training from studies of expertise and its effects on performance could both accelerate and provide a basis for evaluating skill acquisition.

Researchers found significant differences between trained expert laparoscopic surgeons and novice medical students during a study that examined task completion time, distance traveled, speed, and curvature during simulated robotic surgery tasks

(Judkins, Oleynikov, & Stergiou, 2009). To attempt to bridge this gap in performance, the students undertook brief training in the simulated tasks of bimanual carrying, needle passing, and suture tying, where each task was representative of actual tasks involved in laparoscopic surgery. Bimanual carrying required using each hand to move pieces of rubber from one metal cap to another, while needle passing involved maneuvering a surgical needle through increasingly distant holes, and, in suture tying, two knots had to be tied into a surgical suture. The students performed significantly better than in their original performance on each task following training.

Laparoscopic simulators also have been used for training aimed at reducing errors during live surgery. One such training program involved various tasks representative of live surgery, including suturing, lifting and grasping, clipping, and ultrasonographic dissection using both hands (Ahlberg et al., 2007). After using experts in the field to establish proficiency scores for the tasks, surgical residents were randomly assigned to a training or control group which received no additional training outside of their standard curriculum, and the training group then practiced the representative tasks until they were proficient. Both groups were then evaluated during their first ten full laparoscopic procedures. Controls made three times more mistakes than trained residents and took 58 percent longer to complete their procedures. These drastic performance differences suggest that training derived from representative tasks and incorporating simulated task environments not only improves the performance of the trainee but that the improved performance translates beyond the simulated environment.

In a more recent study, McCormack, Wiggins, Loveday, and Festa (2014) compared visual cue use between consultant and staff-level experts in neonatal and pediatric care and competent non-expert medical personnel in training. Audio, visual, and eye-tracking recordings were made of the participants as they responded to an infant manikin depicting either a worsening coughing or a seizure-related emergency scenario. Experts were shown to focus on, and obtain more information from the head and face of the manikin. Experts also were shown to focus on and require shorter fixation times on the head and face areas than novices due to a greater understanding of the cues presented which allowed them to obtain more information. Novices were found to focus on the actions of the confederate nurse instead of gathering cues from the manikin and acting on them. These findings suggest that the representative task was successful at distinguishing between the performance of experts and non-experts. A logical next step might be to develop training in an attempt to improve the performance of the non-experts (e.g., Harris et al., 2013; Harris et al., 2017).

Law Enforcement, the Military, and Aviation

Representative tasks and simulated task environments also have been used to understand expertise in military, law enforcement, and closely related aviation domains. As with medicine, each of these domains commonly involves life-or-death circumstances.

Godwin et al. (2015) asked experts with actual combat experience and non-experts who had simply received risk assessment training to assess potential threats during an imagined foot patrol with the aid of photographs of actual Afghani patrol routes through a variety of environments. Expert explosive operators helped the researchers identify certain areas in each photograph as potential risk indicators, such as disturbed dirt potentially indicating an IED. The participants' task was to identify potential risks in the photographs and, if a threat was detected, halt the patrol. Eye-movement data were collected from the participants during the task. The two groups of participants were equally likely to examine the potential risk indicators but experts were quicker to make a patrol-stop decision, needed fewer returns by the eye to, and fixations on, risk indicators to make a decision, and were less likely to pronounce the situation as dangerous. The results of this study indicate that it may be worthwhile for risk assessment training programs to focus on trainees learning to make more efficient visual searches in order to make patrol-stop decisions quicker.

In a study of police officer performance, SWAT-qualified skilled officers were compared to less-skilled recruits in training during eight video-based simulations of police-related events such as a domestic disturbance or potential suicide (Suss & Ward, 2012; see also Suss & Ward, 2018) that were designed to end either violently toward the participant or peacefully. The simulation films were occluded at a turning point in each scene in which enough information had been provided to predict the conclusion. Upon occlusion, participants were asked to provide predictions about what would happen next in the scenarios and to rank these predictions in order of likelihood. Skilled officers produced significantly fewer responses that were of a higher quality than their less-skilled counterparts' and were better able to accurately predict the conclusion of the scene. Simulations such as these offer less-skilled officers an opportunity to gain some experience in a multitude of situations so that these situations will not be as novel if they are encountered on duty, leading the officers to be better able to predict and react appropriately to the situation.

Seeking to understand how expertise affected a pilot's ability to react appropriately during a failure situation, Schriver, Morrow, Wickens, and Talleur (2008) examined the attentional strategies of pilots designated as either more or less expert during 16 simulated flights, in which eight required diagnosis and rectification of an aircraft failure. Expertise was evaluated based on total flight hours, instrument flight hours, type of pilot certificate, and score on a test of aviation knowledge. The failures encountered were quite routine and included low oil pressure. The more expert pilots were found to make quicker and more accurate decisions and attended more to cues relevant to failure diagnosis and rectification. Training in which cues may be more relevant during a failure situation in the air could help novice pilots to react better in these types of situations, thereby increasing flight safety.

Future Directions and Concluding Remarks

Representative tasks and simulated task environments allow for the study of expert performance, help inform training based on the study of experts and expert performers, and provide a means for objectively measuring performance improvements. Additionally, simulated task environments provide the opportunity to train and assess performers' skills in an environment presenting significantly fewer risks than are presented in the real world, and in relation to rare events that cannot be experienced or practiced for in advance of actual performance. Importantly, training via representative tasks and simulated task environments translates into improved real-world task performance (e.g., Crochet et al., 2011).

Only recently have researchers proposed formal frameworks for using what is learned from studying expert performance on representative tasks to develop training (e.g., Ward et al., 2009) but the introduction of these frameworks has led to increasing calls, and in a variety of domains, for their application. Two authors of this chapter answered a recent call to provide a theoretical framework for use of simulation in nursing (Harris et al., 2013) and for training team performance in medicine (Harris et al., 2017), while others have suggested applying this approach to improve simulation training for individuals in medicine (Causer et al., 2014) and training perceptual–cognitive skills in athletes (Fadde, 2010).

Our hope is that the approach described in the present chapter will be adopted widely, with training in many domains benefiting as a result. Our proposed approach of directly studying expert performers to better understand performance and develop training has been criticized for emphasizing the nurture side of the nature–nurture debate (e.g., Hambrick et al., 2014; see Chapter 3, "Domain-General Models of Expertise," by Hambrick et al., this volume). However, we propose that our approach should be considered regardless of one's stance on this classic debate. We would like to make two pragmatic points. First, even if nature plays an important role in determining an individual's potential for performance, such that a given individual might be poorly gifted in some domain, the evidence is strong that high-quality training and deliberate practice are still going to be very useful, especially when extensive, in advancing that individual's performance level. Second, individuals generally underestimate their potential for adaptation following such training and practice. It is the identification of the training and practice pathways taken by currently skilled and expert performers, be they chess players, mine detection operators, or basketball athletes, that will illuminate the steps that others can take in an attempt to improve their own performance. Whether their performance improves by engaging in the proposed activities is an empirical question. Activities identified as leading to consistent and objective performance gains can be further disseminated as they are deemed valid.

References

Ahlberg, G., Enochsson, L., Gallagher, A. G., Hedman, L., Hogman, C., McClusky, D. A., . . . Arvidsson, D. (2007). Proficiency-based virtual reality training significantly reduces the error rate for residents during their first 10 laparoscopic cholecystectomies. *American Journal of Surgery* 193, 797–804.

Belling, P. K., Sada, J, & Ward, P. (2015). Assessing hitting skill in baseball using simulated and representative tasks. In *12th International naturalistic decision making conference*, 9th–12th June 2015. McLean, VA: The MITRE Corporation.

Causer, J., Barach, P., & Williams, A. M. (2014). Expertise in medicine: Using the expert performance approach to improve simulation training. *Medical Education* 48, 115–123.

Charness, N., & Tuffiash, M. (2008). The role of expertise research and human factors in capturing, explaining, and producing superior performance. *Human Factors* 50, 427–432.

Crochet, P., Aggarwal, R., Dubb, S. S., Ziprin, P., Rajaretnam, N., Grantcharov, T., . . . Darzi, A. (2011). Deliberate practice on a virtual reality laparoscopic simulator enhances the quality of surgical technical skills. *Annals of Surgery* 253, 1216–1222.

Davis, S. (2015, June). How Stephen Curry became the best shooter in the NBA. *Business Insider*. Available at https://www.businessinsider.com/stephen-curry-best-shooter-in-nba-2016-6?r=US&IR=T.

de Groot, A. D. (1978). *Thought and choice in chess* (2nd edn). The Hague, Netherlands: Mouton Publishers. (Original work published in 1946.)

Dweck, C. S. (2006). *Mindset*. New York: Random House.

Eccles, D. W., and Arsal, G. (2015). How do they make it look so easy? The expert orienteer's cognitive advantage. *Journal of Sports Sciences* 33(6), 609–615.

Eccles, D. W., Krampe, R. T., & Tesch-Römer, C. (1993). The role of deliberate practice in the acquisition of expert performance. *Psychological Review* 100(3), 363–406.

Eccles, D. W., & Tenenbaum, G. (2004). Why an expert team is more than a team of experts: A social-cognitive conceptualization of team coordination and communication in sport. *Journal of Sport and Exercise Psychology* 26(4), 542–560.

Eccles, D. W., Walsh, S. E., & Ingledew, D. K. (2002a). The use of heuristics during route planning by expert and novice orienteers. *Journal of Sports Sciences* 20, 327–337.

Eccles, D. W., Walsh, S. E., & Ingledew, D. K. (2002b). A grounded theory of expert cognition in orienteering. *Journal of Sport & Exercise Psychology* 24, 68–88.

Eccles, D. W., Walsh, S. E., & Ingledew, D. K. (2006). Visual attention in orienteers with different levels of experience. *Journal of Sports Sciences* 24, 77–87.

Eccles, D. W., Ward, P., & Woodman, T. (2009). Competition-specific preparation and expert performance. *Psychology of Sport and Exercise* 10, 96–107.

Ericsson, K. A. (2004). Deliberate practice and the acquisition and maintenance of expert performance in medicine and related domains. *Academic Medicine* 79, S70–S81.

Ericsson, K. A. (2008). Deliberate practice and acquisition of expert performance: A general overview. *Academic Emergency Medicine* 15, 988–994.

Ericsson, K. A., Krampe, R. Th., & Tesch-Römer, C. (1993). The role of deliberate practice in the acquisition of expert performance. *Psychological Review*, 100, 363–406.

Ericsson, K. A., & Lehmann, A. C. (1996). Expert and exceptional performance: Evidence of maximal adaptation to task constraints. *Annual Review of Psychology* 47, 273–305.

Ericsson, K. A., & Simon, H. A. (1993). *Protocol analysis: Verbal reports as data* (2nd edn). Cambridge, MA: MIT Press.

Ericsson, K. A., & Smith, J. (1991). Prospects and limits in the empirical study of expertise: an introduction. In K. A. Ericsson, & J. Smith (Eds), *Toward a general theory of expertise: Prospects and limits* (pp. 1-38). Cambridge, UK: Cambridge University Press.

Ericsson, K. A., & Ward, P. (2007). Capturing the naturally occurring superior performance of experts in the laboratory: Toward a science of expert and exceptional performance. *Current Directions in Psychological Science 16*, 346-350.

Ericsson, K.A., Whyte, J., & Ward, P. (2007). Expert performance in nursing: Reviewing research on expertise in nursing within the framework of the expert-performance approach. *Advances in Nursing Science 30*, E58-E71.

Ericsson, K. A., & Williams, A. M. (2007). Capturing naturally occurring superior performance in the laboratory: translational research on expert performance. *Journal of Experimental Psychology: Applied 13*, 115-123.

Fadde, P. J. (2009). Expertise-based training: Getting more learners over the bar in less time. *Technology, Instruction, Cognition and Learning 7*, 171-197.

Fadde, P. J. (2010). Look 'Ma, no hands: part-task training of perceptual-cognitive skills to accelerate psychomotor expertise. In *Proceedings of the interservice/industry technology, simulation, and education conference (I/ITSEC), Orlando, FL*.

Fadde, P. J. (2016). Instructional design for accelerated macrocognitive expertise in the baseball workplace. *Frontiers in Psychology 7*, 292.

Fleming, D. (2015, April). Stephen Curry, the full circle. *ESPN the Magazine*. Available at http://www.espn.com/espnradio/play/_/id/20228602

Golliver, B. (2016, November). Stephen Curry hits NBA record 13 three-pointers versus Pelicans. *Sports Illustrated*. Available at https://www.si.com/nba/2016/11/08/stephen-curry-golden-state-warriors-nba-three-point-record

Gonzalez, C., Vanyukov, P., & Martin, M. K. (2005). The use of microworlds to study dynamic decision making. *Computers in Human Behavior 21*, 273-286.

Godwin, H. J., Liversedge, S. P., Kirkby, J. A., Boardman, M., Cornes, K., & Donnelly, N. (2015). The influence of experience upon information-sampling and decision-making behaviour during risk assessment in military personnel. *Visual Cognition 23*, 415-431.

Gray, W. D. (2002). Simulated task environments: The role of high-fidelity simulations, scaled worlds, synthetic environments, and laboratory tasks in basic and applied cognitive research. *Cognitive Science Quarterly 2*, 205-227.

Grenvik, A., & Schaefer, J. (2004). From Resusci-Anne to Sim-Man: The evolution of simulators in medicine. *Critical Care Medicine 32*, S56-S57.

Hambrick, D. Z., Oswald, F. L., Altmann, E. M., Meinz, E. J., Gobet, F., & Campitelli, G. (2014). Deliberate practice: Is that all it takes to become an expert? *Intelligence 45*, 34-45.

Harris, K. R., Eccles, D. W., & Shatzer, J. H. (2017). Team deliberate practice in medicine and related domains: a consideration of the issues. *Advances in Health Sciences Education 22*, 209-220.

Harris, K. R., Eccles, D. W., Ward, P., & Whyte, J. (2013). A theoretical framework for simulation in nursing: Answering Schiavenato's call. *Journal of Nursing Education 52*, 6-16.

Judkins, T. N., Oleynikov, D., & Stergiou, N. (2009). Objective evaluation of expert and novice performance during robotic surgical training tasks. *Surgical Endoscopy 23*, 590.

Kapadia, M. R., DaRosa, D. A., MacRae, H. M., & Dunnington, G. L. (2007). Current assessment and future directions of surgical skills laboratories. *Journal of Surgical Education 64*, 260-265.

Korndorffer, J. R., Stefanidis, D., & Scott, D. J. (2006). Laparoscopic skills laboratories: Current assessment and a call for resident training standards. *American Journal of Surgery 191*, 17–22.

Kroichick, R. (2015, January 25). For Stephen Curry, basketball in hands is masterpiece theater. *SFGATE*. Available from https://www.sfgate.com/warriors/article/For-Stephen-Curry-basketball-in-hands-is-6037999.php.

Link, Edwin A., Jr. (1937). *U.S. Patent No. 2,099,857*. Washington, DC: US Patent and Trademark Office.

McCormack, C., Wiggins, M. W., Loveday, T., & Festa, M. (2014). Expert and competent non-expert visual cues during simulated diagnosis in intensive care. *Frontiers in Psychology 5*, 949.

Milazzo, N., Farrow, D., Ruffault, A., & Fournier, J. F. (2016). Do karate fighters use situational probability information to improve decision-making performance during on-mat tasks? *Journal of Sports Sciences 34*, 1547–1556.

Murphy, J., Akter, T., Murphy, S. A. J., & Smith-Jackson, T. (2014). Ecological validity of scaled models in construction research. In *IIE Annual Conference. Proceedings* (p. 2754). Institute of Industrial Engineers.

Naweed, A., Hockey, G. R. J., & Clarke, S. D. (2013). Designing simulator tools for rail research: The case study of a train driving microworld. *Applied Ergonomics 44*, 445–454.

Paull, G., & Glencross, D. (1997). Expert perception and decision making in baseball. *International Journal of Sport Psychology 28*, 35–56.

Schriver, A. T., Morrow, D. G., Wickens, C. D., & Talleur, D. A. (2008). Expertise differences in attentional strategies related to pilot decision making. *Human Factors 50*(6), 864–878.

Shadrick, S. B., & Lussier, J. W. (2004). *Assessment of the think like a commander training program* (No. ARI-RR-1824). Alexandria VA.: Army Research Institute for the Behavioral and Social Sciences.

Staszewski, J., & Davison, A. (2000). Mine detection training based on expert skill. *Proc. SPIE 4038, Detection and remediation technologies for mines and mine-like targets V*, 90–101.

Suss, J., & Ward, P. (2012). Use of an option generation paradigm to investigate situation assessment and response selection in law enforcement. *Proceedings of the Human Factors and Ergonomics Society Annual Meeting 56*(1) 297–301.

Suss, J., & Ward, P. (2018). Revealing perceptual-cognitive expertise in law enforcement: An iterative approach using verbal report, temporal-occlusion, and option-generation methods. *Cognition, Technology, & Work 20*(4), 585–596.

Ward, P., Ericsson, K. A., & Williams, A. M. (2013). Complex perceptual-cognitive expertise in a simulated task environment. *Journal of Cognitive Engineering and Decision Making 7*, 231–254.

Ward, P., Suss, J., & Basevitch, I. (2009). Expertise and expert performance-based training (ExPerT) in complex domains. *Technology, Instruction, Cognition and Learning 7*, 121–146.

Ward, P., Suss, J., Eccles, D. W., Williams, A. M., & Harris, K. R. (2011). Skill-based differences in option generation in a complex task: A verbal protocol analysis. *Cognitive Processing 12*, 289–300.

Ward, P., Williams, A. M., & Hancock, P. (2006). Simulation for performance and training. In K. A. Ericsson, P. Hoffman, N. Charness, & P. Feltovich (Eds), *The Cambridge handbook of expertise and expert performance* (pp.243–262). Cambridge, UK: Cambridge University Press.

Whyte IV, J., Ward, P., Eccles, D. W., Harris, K. R., Nandagopal, K., & Torof, J. M. (2012). Nurses' immediate response to the fall of a hospitalized patient: A comparison of actions and cognitions of experienced and novice nurses. *International Journal of Nursing Studies 49*, 1054–1063.

Williams, A. M., Ford, P., Eccles, D. W., & Ward, P. (2011). Perceptual-cognitive expertise in sport and its acquisition: Implications for applied cognitive psychology. *Applied Cognitive Psychology 25*, 432–442.

Williams, A. M., & Ward, P. (2003). Perceptual expertise: Development in sport. In J. L. Starkes & K. A. Ericsson (Eds), *Expert performance in sports: Advances in research in sport expertise* (pp. 219–247). Champaign, IL: Human Kinetics.

Williams, A. M., Ward, P., Knowles, J. M., & Smeeton, N. J. (2002). Anticipation skill in a real-world task: measurement, training, and transfer in tennis. *Journal of Experimental Psychology: Applied 8*, 259–270.

CHAPTER 14

DEVELOPING MASTERY MODELS TO SUPPORT THE ACQUISITION AND ASSESSMENT OF EXPERTISE

KAROL G. ROSS AND JENNIFER K. PHILLIPS

INTRODUCTION

A Mastery Model is the result of a specific process to generate a representation of development in any domain characterized by complex expertise. The Mastery Model is defined by customized key performance areas; stage profiles for each of these key areas specifying the hallmarks of performance and characterizing the progression of skill; and performance indicators for each area in five progressive levels of development. The model can support assessment and guidance for the development of a complex skill, job position, or profession. The goal of this chapter is to share the research foundation for the origin of the model, the structure of the Mastery Model, the process for development, and an example of successful application of the model. To that end the following sections address (1) the origins of this method for representing expertise, (2) the structure of the model, (3) the development process, (4) a specific example of a successful application for US Marine Corps instructors, and (5) recommendations for improvements in future development and application.

Foundations of the Mastery Model Approach

This section provides an overview of how the Mastery Model evolved and demonstrates the continuity of research that led to the current instantiation. The Dreyfus and Dreyfus (1986) general stage model of cognitive skill acquisition served as a starting point for the framework of the current Mastery Model. The original model described how individuals progressively attain knowledge and improve their performance in cognitively complex domains and hypothesized the five stages of novice, advanced beginner, competent, proficient, and expert. (See Table 14.1 for an overview of the stage model.) The general thesis of the original model is that the attainment of expertise is a function of a continuing expansion of knowledge and experiences that become integrated into easily accessible knowledge, skills, and schema attained through practice with real-life challenges, and the progression can be described by a consistent set of developmental stages across domains.

Our goal was to use this framework as the basis for documenting performance progression in different domains to support accelerating expertise and devising assessment. Since its introduction, the five-stage model had been applied to training and instruction in a number of domains, including combat aviation, nursing, education, industrial accounting, psychotherapy, and chess (Benner, 1984, 2004; Houldsworth, O'Brien, Butler, & Edwards, 1997; McElroy, Greiner, & de Chesnay, 1991). In each domain studied, the generalized stage-specific characteristics set forth by Dreyfus and Dreyfus (1986) have been confirmed.

Our initial review of the Dreyfus and Dreyfus model concluded that the five-stage model of cognitive skill acquisition represented a strong candidate as the scientific foundation for representing expertise in a progressive roadmap to meet our goal of generating approaches for assessing and accelerating expertise (Ross, Phillips, Klein, & Cohn, 2005). For example, in contrast to three-stage developmental models (such as novice, intermediate, and expert), and in contrast to expert–novice difference comparisons, the five-stage model distinguished among the intermediate stages of learning. Research in cognitive expertise has often "been conducted within a binary framework that sets novices apart from experts while ignoring any qualitative distinctions in between" (Campbell & Di Bello, 1996, p. 277). The intermediate stages are typically the most poorly understood, yet account for the performance of the vast majority of individuals in a given domain. Similarly, individuals tend to spend the most time in intermediate phases of development as learners. It follows that a five-stage model providing meaningful differentiation of intermediate levels of performance offers more value for application in training and assessment.

To study the implications of the five-stage model for supporting training interventions and assessments, we conducted a comprehensive review of the research literature associated with the Dreyfus and Dreyfus model. Findings were extracted from every

Table 14.1 Summary of Performance at Each Stage of Cognitive Skill Acquisition

Stage	Characteristics
Novice	• Lack of experience with real-world situations • Little situational perception • Rule-based or procedure-based performance • Abstract thinking without contextual anchors • No discretionary judgment
Advanced beginner	• Some experience with real-world situations, enabling recognition of recurring elements • Internalized guidelines for action based on response to limited attributes or aspects they have learned to recognize • Situational perception limited • Situational attributes and aspects are treated separately and given equal importance • Pattern recognition absent
Competent	• Sees action at least partially in terms of longer-term goals • Conscious, deliberate planning • Skilled at formulating goals and plans • Manages large amounts of information well • Standardized and routinized procedures • Plan guides performance as situation evolves and fails to adjust
Proficient	• Sees situation holistically rather than in terms of aspects • Pattern recognition • Sees what is most important in a situation • Perceives deviations from the normal pattern (anomalies) • Automatic and dynamic situational assessment based on experience • Situational factors guide performance as situation evolves • Requires analytic deliberation to reach course of action decision
Expert	• Rules or maxims entirely internalized • Focuses on only critical elements • Intuitive situational assessment based on deep tacit understanding • Intuitive recognition of appropriate decision or action • Analytic approaches used only in novel situations or when problems occur

Note: Stage descriptions are based on the original Dreyfus and Dreyfus (1986) model and were expanded through literature review.

domain-specific application of model we could access, from the expert–novice differences literature, and from relevant constructivism, cognitive development, and adult-learning sources, to produce an enhanced and extended description of cognitive skill acquisition based on empirical findings (Ross et al., 2005). We produced a refined five-stage general, descriptive model consisting of thorough descriptions of the types of knowledge and characteristics of performance at each developmental stage from the literature. The resulting expanded model of cognitive skill acquisition integrates new information into the original Dreyfus and Dreyfus model by (1) describing, in depth, the type of knowledge and the features of performance at each of the five stages in which that hypothesized model has been applied to various domains in real-world settings, and (2) by providing training implications and assessment recommendations for each stage derived from the literature (Ross et al., 2005).

Table 14.1 summarizes the characteristics of each stage as originally proposed by Dreyfus and Dreyfus and expanded by our literature review. The following stage descriptions provide an overview of our findings integrated with the original Dreyfus and Dreyfus hypothesized model. (For more information about the model, see also Ross, Phillips, & Cohn, 2009).

Novices (stage 1) have limited or no experience in situations characteristic of their domain. *Novice* does not refer to *rank beginner*. Novices may have substantial textbook or classroom knowledge, but their lack of lived experience results in a novice's understanding of the domain being largely based on rules or procedures learned absent of context (Benner, 1984; Dreyfus & Dreyfus, 1986; Glaser, 1996; McElroy et al., 1991). Therefore, performance is limited to the application of those rules, and more often than not, application is unsuccessful in realistic situational circumstances.

Once individuals move to stage 2, advanced beginner, they have enough experience to demonstrate marginally acceptable performance (Benner, 1984). They can recognize recurring elements of situations because they have experiences to use as comparison cases (Benner, 1984). Further, their experience base provides a set of guidelines for operating in the domain depending on whether recognizable attributes of situations occur (Dreyfus & Dreyfus, 1986). However, advanced beginners are limited by their inability to perceive patterns in the environment and have a tendency to prematurely jump to action. They also become easily overwhelmed because they cannot prioritize; they see every task as equally critical (Benner, 1984; Dreyfus & Dreyfus, 1986; Shanteau, 1992).

Competent performers, at stage 3 of development, are characterized by their deliberate, analytic, and intentional performance. They have acquired enough experience to understand how goals dictate appropriate actions, and as a result they are skilled at formulating plans and prioritizing tasks (Benner, 1984; Dreyfus & Dreyfus, 1986). They are also able to manage large sets of incoming information due in part to their understanding of priorities. However, because they are so reliant on structured and formulaic analysis, they tend to wed themselves to plans and fail to adjust when the situation changes (Dreyfus & Dreyfus, 1986). Their highly analytical problem-solving approach is in contrast to the agile and flexible approach seen by more advanced performers.

Once individuals reach stage 4 of development, proficient, they become less formulaic and analytical in their approach. They perceive patterns in situations and assess them holistically and intuitively, surpassing competent performers, who tend to see the situation as a set of independent attributes (Benner, 1984). Proficient performers are able to recognize when the situation has changed and the plan no longer holds up. Performance is characterized by this automatic and dynamic situational assessment ability enabled by an extensive base of experience from which to draw comparisons. However, when it comes to making decisions based on recognition of situational changes, proficient performers still require detached analysis and deliberation to reach an acceptable course of action (Dreyfus & Dreyfus, 1986; McElroy et al., 1991).

Stage 5 of development, expert, is characterized by fluid, adaptable, intuitive performance in both situational assessment and decision making. The expert focuses attention on only the critical elements of the situation, recognizes changes with immediacy, flexibly applies knowledge and experience even to novel problems, and implicitly knows what course of action will remedy the situation—rapidly mentally simulating how to implement the course of action successfully.

The literature provides support for the assertion that these stages of development are consistent across varying domains characterized by the acquisition of complex cognitive skills. The developmental timelines from novice to expert likely differ across domains, but no comparative studies have examined this question. In our work, we have seen that there are key shifts in perspective that facilitate moving to higher levels of performance which may or may not be typical across domains (Ross, Phillips, & Lineberger, 2015). During the development of US Marine Corps formal school instructors, the first focus shifts occur for advanced beginners when they (1) refocus from self to others and (2) transition from rote knowledge to the recognition of a system of domain knowledge with real-world impacts (for example, shifting from being overly concerned about how one appears to perform as an instructor to how one's overall performance impacts students). A key shift in performance midway through development, in the competent stage, is when one's repertoire of processes and procedures reaches a point of quantity and quality that domain-specific critical thinking becomes a skill. A transition in performance in the proficient stage occurs when one is sufficiently skilled to begin to contribute to the community of practice, and a systems focus is adopted. We have not examined the generalization of developmental focus shifts across domains, so this aspect of development may vary by individual, nature of the domain, or timing. However, we assert that every domain is likely to have such shifts in focus that facilitate movement to higher stages of performance.

The general stage model of cognitive skills acquisition forms the overall framework driving the Mastery Model process. However, the Mastery Model includes several elements based on that framework that enable the documentation of a roadmap to expertise for a specific domain as described in the next section.

Structure of the Mastery Model

In this section, we provide (1) an overview of the difference between developmental models (of which the Mastery Model is one type) and competency models, and (2) a description of the elements of a Mastery Model.

A stage model or developmental model generally consists of levels of progressive proficiency in a specific domain. It consists of an overall description for each stage, including some combination of a general behavioral description, specific behavioral descriptors or performance indicators, key performance areas, attitudes, abilities, skills, knowledge, and general cognitive orientation (e.g., acceptance of difference, inward focus, a heightened sense of responsibility, or improved self-awareness). Such a model may also include the key developmental tasks that must be undertaken by the learner to move to the next stage of performance, strategies to support the learner in moving to the next stage, and challenges the learner must overcome, as well as recommended assessment strategies by stage.

Competency models commonly identify key competencies and the associated knowledge, skills, and abilities required for job performance, and in some cases address the question of skill development over time. However, these models are limited in their applicability to training and assessment interventions by their overgeneralization of competencies, lack of detail in performance descriptors, and under-representation of the cognitive dimension of performance (see Shippman et al., 2000). Other problems with competency models include not operationalizing competencies so they can be observed and measured, overcoming organizational disagreements about content and face validity, too few performance indicators, lack of rater agreement for accurate measurement, and whether competencies predict job performance (Markus, Cooper-Thomas, & Allpress, 2005).

A Mastery Model is specifically designed to support training and assessment in ways competency models do not. One example of the difference is the stage profile in which a Mastery Model summarizes performance across all stages within one key performance area. A stage profile provides a holistic view of performance for a developing practitioner and supports targeting feedback and setting goals. Another key difference is the use of specific performance descriptors, using the advanced practitioners' words to the extent possible, that allow the user of the model to see performance differences at different stages through the eyes of a consolidation of experienced practitioners. For example, rather than only acknowledging that communication is a key performance area, a Mastery Model provides descriptions of that performance at every stage of development supplemented by performance indicators describing behavior at each stage. Development descriptions include cognitive, affective, and behavioral performance indicators. The key performance areas are not as general as competencies tend to be, nor are they imposed on the domain practitioners for agreement or ranking, but are derived from performance descriptions provided through interviews. Mastery Models inherently have good face validity and content validity.

The elements of the five-stage Mastery Model described in the following sections are (1) the key performance areas, which are derived from analyzing interview data; (2) stage profiles within each key performance area specifying the hallmarks of performance and characterizing the progression of skill over time; and (3) performance indicators at each stage within each performance area.

Key Performance Areas

The purpose of defining key performance areas is twofold. First, identification and description of key performance areas enable a comparison of the research findings to a job description (or when working with the military, the applicable doctrine), thereby lending face validity to performance areas that accurately reflect doctrine, and revealing important discrepancies between doctrine and practice when the model presents an alternate portrait of performance. (For example, doctrine may prescribe that first-year instructors deliver lectures they have tailored from a general template, whereas the interview-based developmental model may reveal this is not a typical performance goal in actual practice.) Second, organization of the model by performance area facilitates its application to the assessment of an individual's current performance and the identification of pertinent learning objectives at the right point in development (such as identifying that a person is exceling at learning different instructional techniques, but is not developing the ability to interact with students). An example of the 10 key performances areas for Marine instructors is shown in Table 14.2.

Table 14.2 Example of Key Performance Areas and Definitions

KPA	Definition
Instructional technique	Knowing and applying a variety of methods and strategies to secure student attention, enhance student participation, and facilitate learning, and the ability to select and adapt approaches based on learning goals and the student population.
Setting the example	The mental, physical, and character traits of an individual who embodies USMC values and ethos, demonstrates professionalism and command presence, garners respect and trust, and displays passion and commitment to the job.
Communication and delivery	Clearly, concisely, dynamically, and interactively exchanging information to transfer knowledge and promote understanding, using a combination of verbal, nonverbal, and other communication approaches.
Self-improvement	The motivation to continually increase domain knowledge and enhance instructor skills by actively seeking and engaging in a variety of knowledge and skill acquisition activities.
Developing subordinates and peers	Establishing relationships with students and peers to mentor, coach, advise, and guide their development.
Planning and preparation	Reviewing, generating, and adapting teaching materials, rehearsing for delivery, and proactively planning the administration and logistics of course delivery based on learning objectives, the role of the course within the institution's progression of instruction, and anticipated student characteristics and questions.
Learning environment	Setting and maintaining the conditions for a respectful, engaging, and motivating atmosphere that encourages active collaboration by managing time, the physical space, and student behavior.
Assessing effectiveness	Knowing and applying formal and informal assessment techniques to gauge the effectiveness of the instruction, accurately verify student knowledge, and provide performance feedback to students.
Subject matter expertise	Maintaining technical and tactical proficiency in course content and associated principles to be regarded as a credible source of information, and applying that knowledge and experience to facilitate learning.
Community of practice	Actively contributing to enhancing the collective body of instructional expertise, examining organizational practices and processes to achieve desired learning outcomes, and socializing recommendations for improving institutional and service-wide methods to meet USMC standards.

The term key performance area was chosen for the Mastery Model over the term competency to distinguish the developmental model from other performance descriptions that might be in use in the target domain. We sought to elicit descriptors that would enable us to identify all the major areas that are integral to successful performance in a broad manner and to avoid conflating those with very specific traits or competencies that are actually supporting skills.

Stage Profiles

A stage profile presents a high-level depiction of characteristics associated with each of the five stages of learning within one key performance area. Profiles support a high-level view of how performance improves over time, without requiring reading the detailed performance indicators in each performance area. Table 14.3 provides an example of a stage profile for Marine instructors for the key performance area of *developing subordinates and peers*.

Table 14.3 Stage Profile of Marine Instructor Performance for "Developing Subordinates and Peers"

Stage	Profile
Novice	Focuses on own performance as a lecturer and does not engage with students. Sees the instructor's job as transmitting information and the student's job to be hearing and retaining knowledge. Does not understand that the instructor is responsible for students' knowledge and skill acquisition as well as their development of maturity and lifelong learning habits. Does not yet understand the impact an instructor has on a student, either positive or negative, depending on the instructor.
Advanced beginner	Becomes somewhat personable and approachable in the eyes of the students. Interacts with students outside the classroom and begins to see positive effects of those interactions on student behavior in the classroom. Attempts to mentor students, but those exchanges are often more parental than thought provoking.
Competent	Possesses a full realization that the instructor role goes beyond the classroom. Cares about whether students are learning and sees it as his/her responsibility to ensure they learn. Demonstrates patience in counseling students and working with them outside of class to understand the material. Begins to mentor less-experienced instructors as well by helping them prepare for class or advising them on how to instruct a lesson.
Proficient	Relates well to individual students and is highly approachable. Builds mutually respectful relationships with students and peers. Treats individuals as people, not subordinates. Takes a personal interest in students' successes and goes out of his/her way to help them achieve. Becomes a go-to adviser for less-experienced instructors, and shares best practices and constructive evaluative feedback to improve their performance.
Expert	Builds a friendly, personable rapport with students and instructors that makes them trust and feel at ease to receive guidance and mentoring. Inspires and instills confidence. Demonstrates a genuine passion for and pride in helping others succeed. Sees it as his/her job to facilitate the development of expertise in students and instructors alike as a means of building the collective skills and abilities of the Marine Corps.

Performance Indicators

The performance indicators are bulleted lists of actions, attitudes, or behaviors an individual is likely to exhibit at each stage. The indicators connect the cognitive development that is characteristic of the developmental stage with the observable behaviors reflective of that level of cognition and degree of experiential knowledge. As such, they are the essence of the Mastery Model and the concrete, descriptive elements that inform learning goals and the development of assessment tools. Table 14.4 provides an excerpt of several key performance areas and associated performance indicators for Marine instructors as an example.

In the next section, we discuss the process for deriving the elements of the model. We provide the specific process of selecting participants, knowledge elicitation, and analysis to produce the model.

Process of Model Development

Identification of the Domain and Interview Participants

Model development starts with defining the domain of interest and assessing who the experts are and how many should be interviewed. The sponsor of the research often has very specific ideas about the domain of interest, and potential interviewees, but discussion to clarify who or what within the domain is critical ensures the researcher and sponsor agree on the intent.

The researcher should provide criteria to the sponsor of the model development for identifying an advanced performer/expert to support recruitment of a good pool of interview participants. In the development of the model for US Marine Corps instructors that is used in this chapter for illustration, the criteria included such things as years as an instructor and supervisor recommendations of highly skilled instructors. A range of organizations might be involved in providing the interviewees, and, not surprisingly, not all participants actually meet the advanced performer criteria once the researchers get into the interview process. Interviews that were planned or conducted but are not usable must be expected. Additionally, large organizations, such as the military, might request that a variety of different organizational elements be included in the interview process to represent the range of experts and to increase *buy in* from a variety of stakeholders.

Data Collection Interviews

The interview process consists of an explanation of the interview process for the participant and intended use of the outcomes; gathering of demographics; administration

Table 14.4 Excerpt of the Key Performance Areas and Performance Indicators

	Novice	Advanced beginner	Competent	Proficient	Expert
Instructional technique	Lecture is primary teaching strategy	Focuses on telling stories vice leading discussions about the experiences	Finds different strategies to get teaching points across if lecture is not working	Uses exercises or activities that make students feel they are in charge of the class	Applies different instructional strategies based on assessment of students' individual learning processes
Setting the example	Is concerned about how he/she will look in front of students	Keeps up to standards for physical fitness; sets a positive example	Begins to realize the importance of leading by example	Takes ownership and is accountable for class performance	Displays motivation to give back to the USMC and provide knowledge to students
Communication and delivery	Presents materials in an awkward, hesitant, choppy, and distracting manner	Focuses on transmitting information instead of whether students grasp the materials	Is comfortable and not nervous in front of people, as long as he/she is prepared	Balances delivery well, with less lecture time and more student participation time	Engages, motivates, and connects with students; articulates concepts in an understandable manner
Self-improvement	Struggles to integrate feedback from others to improve performance	Considers feedback about past performance when preparing for instruction	Requests and accepts critiques, feedback, and assistance from more experienced instructors	Demonstrates awareness of own weaknesses and diligently focuses on improving them	Possesses a strong sense of humility and an honest view of self
Developing subordinates and peers	Does not know enough about teaching the course to be able to give advice to other instructors	Compensates as a mentor by being parental instead of thought provoking	Shows personal investment in the students; sees the bigger picture of the instructor's role	Builds mutually respectful relationships with students	Builds teams within the instructor cadre; routinely shares new knowledge with others

(continued)

Table 14.4 Continued

	Novice	Advanced beginner	Competent	Proficient	Expert
Planning and preparation	Needs someone to physically walk him/her through the class preparations	Rehearses for class using a checklist to verify all learning objectives will be met	Has increased vision for the course; takes more ownership; is more proactive; strays from master lesson file	Creatively reorganizes materials to use classroom time more efficiently; foresees problems	Mentally simulates his/her presentation, anticipated questions, and reactions
Learning environment	Believes student is responsible for learning the material, rather than taking ownership	Ensures the class is set up and operational before students arrive for instruction; talks at students instead of interacting	Organizes the classroom so that each student can participate; focuses on the students' needs	Creates a respectful learning environment that encourages information exchange	Customizes the classroom to fit each student's personality and presentation style; makes students central to learning
Assessing effectiveness	Asks students only the required, basic checks on learning	Relies on tests and checklists to assess student understanding	Asks probing questions to look for logic flaws and gaps in student understanding	Experiments with and utilizes more qualitative assessment methods	Understands students' mental models and anticipates when the students will not understand information
Subject matter expertise	Obtained knowledge from fleet experience, but needs instructional coaching	Knows content of technical manuals, but lacks experience instructing	Understands the why behind the material and is starting to explain subject importance	Has detailed course knowledge that enables him/her to deepen student understanding	Masters the entire scope of the curriculum; knows course content and how to apply it
Community of practice	Has little understanding of instructors' additional duties	Starts to understand how elements within the program of instruction fit together	Begins to connect the program of instruction across different classes in the formal learning center	Participates in institutional reviews that provide inputs to improve overall quality	Understands a Marine's career learning process and where the current instruction fits

of the task diagram (Militello & Hutton, 1998); and conduct of the developmental progression interview technique (Phillips, Ross, Rivera, & Knarr, 2013).

Before each interview, participants are provided an overview of the project objectives and informed that the purpose of the interview is to discuss the key responsibilities and developmental progression of the specified domain. Interviews are conducted with individual practitioners by one or two researchers and normally require two hours to complete the semi-structured protocol. Group interviews are not used as the elicitation includes detailed examination of actual performance observed by the participant during their individual career. The sessions are digitally recorded and later transcribed as a means of accurately capturing all relevant information. Researchers also take field notes to facilitate the questioning and elicitation process.

Once the demographic information is captured, the researcher guides the participant in constructing a task diagram. During the task diagram portion of the interview, the researcher asks the participant to identify the four to six core task areas in the domain under study. The domain could be a skill, job, or career, for example, the skill of tactical thinking, the job of Marine Squad Leader, or the career of software developer. The participant is asked to respond based on his or her own experience versus formalized descriptions of responsibilities. The participant names these areas and provides brief explanations of each. Follow-up questions about which of these tasks are most cognitively challenging provide more opportunity to uncover areas of potential concentration for the developmental progression interview. Both the standard demographic information and the task diagram can serve to determine whether criteria for the desired expertise have been met.

Next, the researcher uses the developmental progression interview technique to collaboratively assemble a proficiency table with the participant. The researcher draws a five-column table and explains that each column represents a level of proficiency—novice, advanced beginner, competent, proficient, and expert—and explains the general nature of each. The interviewer makes it clear that the novice level of performance represents an inexperienced and/or untrained person with the potential to achieve mastery, not an individual who failed to embody good values or standards of performance, i.e., someone who is failing at the job. Participants are then asked to identify performance descriptors for each level, being reminded of the nature of each stage as the interview progresses. If necessary, tasks and areas of performance from the task diagram are used to prompt the participant to describe the nature of performance at each of the five levels of proficiency. Researchers also ask participants about activities that result in improvement, or transition, from one level of proficiency to the next.

To support identification of the point where enough interviews have been conducted, an iterative process of interviewing and incremental analysis can establish the point of data saturation, which is a typical method in qualitative research to understand when enough data have been gathered (Glaser & Strauss, 1967). Data saturation means that if the study were to be extended or repeated, it is unlikely that new information will be uncovered, because the latest interviews are no longer uncovering sufficient new information that can be used to create new and different categories of key performance areas or new performance indicators. However, the sponsor's desire for inclusion of more stakeholders and funding resources affect the decision to establish an initial target number of interviews, whether further interviews will be conducted once saturation is reached in order to include all stakeholder groups (and perhaps to compare their responses), or how many more interviews can be accomplished if saturation has not been reached.

Analysis

Two types of qualitative analyses are conducted. The first analysis identifies what is done in the domain of interest, critical tasks, and attributes, and depicts them as a set of key performance areas. The second analysis is designed to understand how a practitioner performs at different levels of proficiency, and yields performance indicators for each performance area at each stage of performance. Transcripts of the interviews and hand-drawn proficiency tables resulting from the developmental progression interview technique, which are integrated into one overall proficiency table, are the source materials for the analysis process.

To understand what tasks and attributes are associated with the target domain, the research team, typically three to six people, reviews the interview transcripts of a subset of the interview participants. The goal is to identify and extract mention of any area critical for effective performance. The team uses a card sort of the extracted items to create categories to ultimately form the set of key performance areas. The team sorts the items into like piles and then, through discussion, must agree on the name of the key performance area.

The second analysis is aimed at describing how performance areas are carried out. The research team reviews the transcripts to correct and add to the integrated proficiency table. The resulting additional descriptors of how performance is carried out are added to the initial performance indicators captured on the hand-drawn proficiency tables. A judgment of the stage in which new performance indicators fit is made, given the context provided by the interview transcript. These descriptors are *bullets* (short phrases) depicting performance indicators at each stage of development. (See Table 14.4.) A card sort is used to assign all performance indicators into key performance areas to form the initial version of the model.

In the example application described in the section "Mastery Model Application," a total of 1,752 performance items were extracted from the transcripts of the most experienced instructors and sorted into 10 finalized key performance areas. The intraclass correlation coefficient for that analysis provided evidence to conclude strong inter-rater agreement among four researchers ($r = 0.848$).

The initial model is then reviewed to deconflict the performance indicators such as by removing duplicates; re-wording statements for brevity, grammar, and precision; removing outliers that do not conform to the majority of descriptors; and removing overly vague items. Finally, researchers construct descriptions, or profiles, for each level of proficiency for each key performance area based on a summary of the performance indicators contained in the table. Validation is assumed due to the conformity of the set of descriptors obtained from the interview participants across key performance areas. In the case provided in the section "Mastery Model Application," the model was also presented to a review board of advanced instructional and management personnel and was accepted as valid based on their review after the presentation.

MASTERY MODEL APPLICATION

In this section we describe the application of the Marine Corps Instructor Mastery Model (Ross et al., 2015; Vogel-Walcutt, Phillips, Ross, & Knarr, 2015). This model details the progressive instructor development stages at Marine Corps formal schools. In this example, application of the model supported a change of training policy and supported the development of feedback-oriented assessment tools linked to the new policy for professional development practices.

Instructors and faculty advisors in the Marine Corps formal schools typically serve three-year assignments with no prior teaching experience. While they may be subject-matter experts, instruction is a distinct skill set and the training they receive varies greatly across formal schools. The requirement is to maximize performance as early as possible in a relatively short tenure across all formal schools.

Formal school managers have identified that the output of the institutional instructor development process does not sufficiently meet their needs. In response to the need to improve instructor professional development, the Marine Corps Training and Education Command is examining current practices for staff and faculty development and pursuing a collection of new or modified courses, tools, and technologies to support the professionalization of the instructor cadre throughout the Marine Corps. The specific objectives of these activities are to enhance the ability of instructors to facilitate learning in classroom, field practice, and coaching settings, and to more effectively and efficiently deliver higher-order critical thinking instruction.

The first step in the process was to develop a Marine Corps Instructor Mastery Model. The selection of interview participants was heavily influenced by the goal of including representatives from as many schools as possible, provided their instructors met the criteria defining instructors of advanced ability, resulting in a requirement from the sponsor to conduct 93 interviews as the basis for the model.

Once the development was complete, the model was presented to the Marine Corps Instructor Training and Readiness Working Group, who review and update programs of staff and faculty development to be sure they are complete, current, and aligned with changes in training and readiness requirements. As a result of the presentation and the group's subsequent review, the model was adopted in 2015 as the basis for new training policy for Marine formal school instructor development. Specifically, as a first step in optimizing and standardizing instructor performance, the 10 key performance areas were integrated by the Marine Corps Training and Education Command into the *Train-the-Trainer Training and Readiness Manual* for instructors. The purpose of a Marine Corps *Training and Readiness Manual* is to establish the standards, requirements, and policies in a manner that can guide implementation of training activities in a specific area of performance. Formalization of the instructor key performance areas as policy means that they represent the performance goals for instructor development and assessment.

To support implementation of the policy, our team developed standardized assessment tools that could be flexibly implemented by each formal school. The tools were intended to guide assessment of the key performance areas, provide developmentally oriented feedback, and provide training support. The development of the assessment tools required iterative goal and design discussions and development demonstrations with key schools representing the range of intended users.

The resulting Marine Instructor Assessment Toolkit is intended to produce a rich understanding of an instructor's current position along the path to mastery, provide insights into actions that will produce performance improvements, and produce quantitative data that can be used to provide individualized feedback, as well as examine trends over time at both individual and school levels.

The tools include an observation rubric for instructional settings, a supervisor rating form for rating holistic instructor performance, a self-reflection tool, and a situational judgment test. The tools were developed based on rating scales for each tool that were derived from the performance indicators within the Mastery Model. The observation rubric is used by experienced instructors to observe other instructors conducting regular classroom and field training sessions. Observations are meant to be conducted as part of a formal school's certification process and then as regularly as desired with the results provided to the instructor as feedback. The observation rubric only includes those key performance areas that can be observed directly. The supervisor rating form includes all 10 key performance areas and is meant to be completed once or twice annually to support formal supervision sessions with performance feedback data. The self-reflection tool is a self-assessment of performance on all 10 key performance areas designed to be used in comparison to the supervisor's assessment of one's performance. Finally, the situational judgment test is an objective evaluation of an instructor's ability to judge effective and ineffective responses to situations that can occur in classroom settings. Instructors' responses to short vignettes are compared to the responses of an expert group to assess level of proficiency on five of the key performance areas. These instructor assessment tools were field tested to gather user input at formal schools and to derive data for reliability and validity. After field testing, the tools were finalized based on a psychometric analysis and user input. The psychometric analysis provided us with information that the tools could be consistently applied and were correlated, and that the comprehensive supervisor rating form was related to levels of overall performance as shown by multiple regression analysis using a *global rating* criterion variable collected during field testing.

The validated tools have utility for supporting train-the-trainer activities in a number of ways. The profiles and performance indicators depict the nature of an individual's performance and challenges at each stage of development in a rich and nuanced manner. Formal train-the-trainer courses can be designed to target learner knowledge and experience gaps without exceeding their current capacity to learn. Similarly, the stage-specific performance descriptions can assist with observations to diagnose individual strengths and development needs across the distinct key performance areas, in service of tailoring staff and faculty development plans to individuals' unique challenges. For example, a mentoring instructor could use the model to both assess a new instructor's current skills and then help define concrete behavioral goals for that instructor to achieve in order to progress toward mastery.

Broadly speaking, the Instructor Mastery Model has supported the development of tools that can aid the selection and development of educational programs, training initiatives, sequencing of assignments, or screening of potential instructors to assess their baseline to ensure the optimal developmental progression is supported. Follow-up research should include validation that the use of the assessment tools and feedback results in the attainment of higher levels of instructor competency more rapidly.

Enhancing Mastery Model Development and Application

Based on our efforts to date, lessons learned are emerging on how to improve the development and structure of the Mastery Model to support various applications. In this section, we will review challenges to (1) model development in an emerging domain where a large cadre of experienced and high-performing subject-matter experts does not exist, (2) development of assessment instruments using the output of the Mastery Model, and (3) potential improvements in the model development and documentation process.

Model Development in Emerging Domains

In this section, we discuss the issues with working in an emerging domain that lacks a sufficient pool of expertise for creation of a model. For example, we are currently encountering this challenge as we develop a Mastery Model for cyber warfare in a specific area where experts are not pervasive.

Typically, a high level of experience is available across a large sample of practitioners in a mature domain. Currently, the expertise, processes, and procedures in the cyber area of interest are in early stages of establishing capabilities. Given the lack of a large pool of experts in the targeted population, we are modifying our approach by spending more time to clarify the requirements of performance and looking at additional sources of information.

This first step of orienting ourselves to the domain requirements as the practitioners see them also includes examining our hypothesis that the challenges are highly cognitive. This goal includes understanding the different roles in this domain to clarify which roles the Mastery Model can best describe.

Second, we are reviewing related literature. While this can be useful, one must be cautious to avoid relying too heavily on doctrine or formal job descriptions that may not accurately reflect the demands of the work. The literature has gaps in performance descriptions and is often not cognitively oriented. It is important not to seek agreement with the literature and doctrine from the expert interviews, but to gain the expert's input and then compare that to the literature to form a fuller picture from a smaller participant group.

Third, we are seeking experts outside the targeted participant group, to include domain interviews with experts from other military services or civilian government experts who have more mature capabilities in the area of interest. Following all the data collection and analysis, validation of the initial model is more critical than developing a model in a mature domain. When we include information from the literature and analogous experts, we cannot rely on data saturation in the interviews or on the process of elimination of outliers during analysis to ensure content validity.

Developing Assessment from a Mastery Model

In this section we discuss generating model-based assessment. The current model structure has a wide range of performance indicators across key performance areas. To date, we have used these indicators to develop rubrics for observation and supervisor ratings, vignette-based assessment interviews, and situational judgment tests. Some development challenges are typical of many assessment efforts, such as development of parallel forms when more than one form of an assessment is needed to support repeated testing, creating a set of tools that provide correlated rating outcomes for an individual across tools, and ensuring ratings on the tools are valid in relation to other measures of performance. Selecting among the rich pool of performance indicators to develop tools that yield scores for the different key performance areas is specific to the Mastery Model and applies to all the types of assessment we have developed.

The development of situational judgment tests has been particularly challenging. In this form of assessment, the item *stem* is in the form of a very brief vignette with several response choices that can be ranked or selected as most desirable. While having the advantages of immersing the test-taker in a cognitively challenging context, providing more objective assessment than the rubrics, and providing validation of performance, the task of aligning different levels of performance to the development of response choices is arduous. While the Mastery Model provides performance descriptors derived from a particular context, these text-based assessments require greater specificity of behaviors applicable for a particular situation or event. We have had success designing valid situational judgment tests, but the process needs to be streamlined. To capture more specific, observable behaviors during the interview, we plan to adapt the interview method in our next project. The development of situational judgment tests is more likely to be successful if the interviews elicit more critical incident data and typical responses by individuals at each stage.

Potential Improvements to the Mastery Model

The greatest challenge to application of the model is in efficient use of the performance indicators. In structuring the data into the model, we have used tables with key performance areas and subcategories arranged by stage that show all the final performance indicators. However, we have found the indicators are not grouped in a fashion that allows the user (developer of assessment) to efficiently align the indicators with measurement objectives. As users of our own models, we find that we must search a large number of performance indicators to find the discriminating behavior that suits the development of the specific assessment tool or item that we are generating.

Generally, we are considering restructuring the model to match the flow of development a user would follow to generate assessments. For example, we intend to provide, at the front of the model, a list of the key performance areas and their subcategories, as opposed to only listing and defining key performance areas as in previous models. Generic measurement objectives could be added to the model for each key performance area and subcategory to support easier use. This all-in-one reference table to the structure of the model will facilitate the conceptualization of assessments.

Another example of potentially restructuring the format of the model is to categorize the types of performance indicators within each performance area. Observable behaviors, which support text-based assessments like the situational judgment test, would be grouped together, while indicators of focus, attitudes, and what the performer can achieve at a given level would be grouped separately and made available for supervisors and trainers to use for coaching and mentoring and for use in less behaviorally oriented measurement such as self-report and rating instruments. As part of the support for assessment development, different types of assessments that can be produced by different categories of indicators can be specified in the model.

Rating rubrics typically require articulation of how a particular element of performance improves from one stage to the next; however, in some cases, performance indicators do not flow in a progressive five-stage sequence in actual mastery development. For example, if the performer masters how to break down complex material into component parts for presentation, that skill does not look different in subsequent stages but continues to be used. The performance indicator continues to be present but not to a different degree that would allow discrimination of the indicator across stages for differentiation by measurement. Altering the table format to allow some skills that have been mastered and should continue to be present in the following stages will clarify the actual progression.

Conclusion

The Mastery Model provides access to the consolidated insights of experienced practitioners. This form of expertise representation goes beyond competency models to focus specifically on performance demands and detail performance indicators by progressive stages toward mastery. As such, the model supports the generation of developmental and assessment activities intended to accelerate development.

The purpose of education and training is to move individuals from their current state of skill and knowledge to a higher state of mastery. To facilitate and assess that process, we need to understand the developmental progression in the domain of interest. However, individual professional development is a non-linear process influenced by several factors. Rate of growth depends on individual differences in motivation, dedication, and aptitude. In addition, baseline proficiency within each of the key performance areas in any domain is a result of training, education, and experience prior to first assessment based on the model. For example, an individual may exhibit stage 2 behaviors in one performance area, but stage 3 behaviors on another. It is normal for an individual to progress in one area of performance faster than in another, and therefore exhibit indicators of transitioning to a higher stage of proficiency in some areas but not others. Because individual differences create variation in developmental needs, there is no single best path to achieve development goals. Without a commonly accepted and detailed representation of the stages of development along a continuum, we lack a roadmap with which to predict and understand performance resulting from efforts to support improved performance. Therefore, by using assessment based on the performance indicators in the five-stage model, training and education efforts can target developmental activities more specifically, include more refined assessments, and provide more targeted feedback and individualized development goals. The big picture resulting from application of such a model allows the learner and instructor to understand developmental leaps and lags in the context of the individual's non-linear path toward mastery.

The development of a Mastery Model is resource intensive, but the return on investment is that the model can support multiple applications in a rich and effective manner, such as selection, placement, training for supervisors, creation of a common language about development in a specific community of practice, more targeted training and development assignments, assessment tools and processes such as developmental feedback and coaching, and growth assessment over time in comparison to a baseline.

Acknowledgments

This work was supported by the Office of Naval Research and by the U.S. Marine Corps Training and Education Command. The views and conclusions contained in this document are those of the authors and should not be interpreted as representing the official policies, either expressed or implied, of the Office of Naval Research or the US Government.

References

Benner, P. (1984). *From novice to expert: Excellence and power in clinical nursing practice*. Menlo Park, CA: Addison-Wesley.

Benner, P. (2004). Using the Dreyfus model of skill acquisition to describe and interpret skill acquisition and clinical judgment in nursing practice and education. *Bulletin of Science, Technology & Society* 24(3), 189–199.

Campbell, R. L., & Di Bello, L. (1996). Studying human expertise: Beyond the binary paradigm. *Journal of Experimental and Theoretical Artificial Intelligence* 8(3–4), 277–291.

Dreyfus, S. E., & Dreyfus, H. L. (1986). *Mind over machine: The power of human intuition and expertise in the era of the computer*. New York: The Free Press.

Glaser, R. (1996). Changing the agency for learning: Acquiring expert performance. In K. A. Ericsson (Ed.), *The road to excellence: The acquisition of expert performance in the arts and sciences, sports and games* (pp. 303–311). Mahwah, NJ: Lawrence Erlbaum.

Glaser, B. G., & Strauss, A. L. (1967). *The discovery of grounded theory: strategies for qualitative research*. Chicago: Aldine.

Houldsworth, B., O'Brien, J., Butler, J., & Edwards, J. (1997). Learning in the restructured workplace: A case study. *Education + Training* 39(6), 211–218.

McElroy, E., Greiner, D., & de Chesnay, M. (1991). Application of the skill acquisition model to the teaching of psychotherapy. *Archives of Psychiatric Nursing* 5(2), 113–117.

Markus, L., Cooper-Thomas, H. D., & Allpress, K. N. (2005). Confounded by competencies? An evaluation of the evolution and use of competency models. *New Zealand Journal of Psychology* 34(2), 117–126.

Militello, L. G., & Hutton, R. J. (1998). Applied cognitive task analysis (ACTA): A practitioner's toolkit for understanding cognitive task demands. *Ergonomics* 41(11), 1618–1641.

Phillips, J. K., Ross, K.G., Rivera, I. D., & Knarr, K. A. (2013). Squad leader mastery: A model underlying cognitive readiness interventions. *Proceedings of the Interservice/Industry Training, Simulation, and Education Conference*. Arlington VA: National Training Systems Association. Available at http://www.iitsecdocs.com/search.

Ross, K., Phillips, J., & Cohn, J. (2009). Creating expertise with technology-based training. In D. Schmorrow, J. Cohn, & D. Nicholson (Eds), *The PSI handbook of virtual environments for training and education* (vol. 1, pp. 66–79). Westport, CT: Praeger Security International.

Ross, K. G., Phillips, J. K., Klein, G., & Cohn, J. (2005). Creating expertise: A framework to guide technology-based training. Final Technical Report, Contract M67854-04-C-8035 under the Sponsorship of the Office of Naval Research and administered by PMTRASYS. Orlando FL: MARCORSYSCOM PMTRASYS.

Ross, K. G., Phillips, J. K., & Lineberger, R. E. (January, 2015). Marine Corps Instructor Mastery Model. (Approved for public release; distribution unlimited). Alexandria VA: Office of Naval Research.

Shanteau, J. (1992). Competence in experts: The role of task characteristics. *Organizational Behavior and Human Decision Processes* 53(2), 252–266.

Shippman, J. S., Ash, R. A., Carr, L., Hesketh, B., Pearlman, K., Battista, M., . . . & Prien, E. P. (2000). The practice of competency modeling. *Personnel Psychology* 53(3), 703–739.

Vogel-Walcutt, J. J., Phillips, J. K., Ross, K. G., & Knarr, K. A. (2015). Marine Corps instructor mastery model: A foundation for Marine faculty professional development. In *Proceedings of the Interservice/Industry Training, Simulation, and Education Conference*. Arlington VA: National Training Systems Association. Available at http://www.iitsecdocs.com/search.

CHAPTER 15

COMPUTATIONAL MODELS OF EXPERTISE

ALEX KIRLIK AND MICHAEL D. BYRNE

Introduction

Essentially, all models are wrong, but some are useful.
(Box & Draper, 1987)

ONE approach to the scientific study of expertise is to create and empirically evaluate computational models of psychological theories. These models, implemented in various programming languages, can be viewed as algorithms or sets of interconnected algorithms in the standard, computer science sense of the term (Sedgewick & Wayne, 2011). As such, the increasing use of computational modeling in psychological science to better understand the nature of expertise is merely one instance of the much more widespread use of algorithmic, computational techniques in science since digital computing became broadly available and cost effective to researchers in the mid-twentieth century.

Our focus in this chapter is modeling techniques and illustrative examples that are explicitly framed and presented by their creators as models of high levels of skilled or expert performance. A survey of the literature reveals no single, widely agreed-upon definition of what exactly this *expert* level is, in many cases because no objective performance measures may exist (e.g., what makes one parent or politician an expert while others merely average?). In other cases where measures may exist or can at least be created and operationalized for the purpose of a research project, a large difference may exist between the duration and nature of task experience that different researchers presume to be required for true expertise to be acquired.

For example, an experimental psychologist working within laboratory paradigms such as videogames or microworlds may be interested in describing the, often striking, differences that distinguish the first few hours of participant learning and performance and the very high or perhaps even asymptotic levels of performance that may be achieved over a period of a few weeks or perhaps a month or two of practice. In other cases, researchers using qualitative methods grounded in naturalistic observations of professionals in field studies or operational contexts may associate expertise with either seniority, job title, or displays of confidence or decisiveness that may or may not be correlated with skill (see Tetlock, 2005). The models discussed in this chapter provide instances of both types. What does, however, appear to us to be common to all these models is that they describe levels of performance that are qualitatively different in nature from the performance of novices, who exhibit behavior that is markedly inferior to that of highly practiced or experienced performers.

Chapter Overview and Organization

The remainder of this chapter focuses on providing an introduction, overview, and assessment of a broad range of computational models of expertise. We begin by contrasting computational models used primarily for the statistical analysis of data from those that provide a candidate account of the, often covert, psychological processes accounting for actually generating observable behavior, the latter of which are the focus of this chapter. A concrete example in the realm cognitive–motor skill is presented to concretize and motivate this distinction.

Next, we provide an overview of modeling approaches assuming that the basic mechanisms and knowledge representations underlying expertise are largely domain- or task-independent. The generality of these approaches is indicated in the name used to characterize them: cognitive architectures. We begin by providing a high-level overview of various cognitive architectures, and then discuss various domain- and task-dependent computational models implemented within these cognitive architectures. We end our discussion of computational modeling research leveraging the cognitive architecture approach by presenting computational models intended to capture the processes of expertise acquisition.

We then turn to briefly discussing a few computational models that are not implemented in cognitive architectures, but are instead narrower in scope in that they do not presume that every demonstration of expertise can be explained by a single theoretical framework. Instead, these models are highly domain- and task-dependent in their origins, and cover topics such as perceptual expertise, recognition-based decision making, and situated action. The chapter closes with discussions assessing the state of the science in the computational modeling of expertise, promising future directions for research, and conclusions.

Computational Models of Data versus Process

The use of computational modeling techniques in psychological science is hardly a recent development of course. Experimental and quantitative psychologists have been using computational implementations of statistical analysis and modeling techniques, such as the general linear model (GLM) and others for well over 50 years now, in software packages such as SAS, SPSS, R, and others.

We trust that the majority of readers are at least acquainted with, if not expert users of, these statistical modeling approaches, and we will have little more to say about them in this chapter. They are typically taken to be the standard techniques for the identification and examination of potential causal relations among sets of independent, dependent, and treatment variables in psychological experimentation, and also play important roles in differential psychology, the study of individual differences, psychometrics, and other areas (yet see, e.g., Gigerenzer, Krauss, & Vitouch, 2004, for a critique of some of these methods).

Instead, the style of computational modeling that is the focus of this chapter is models that are intended to simulate, at some level of fidelity, the time course of psychological activities or processes that give rise to the temporal course of observable behavior, typically in experimental or simulator-based settings. These models leverage theory that emerged from the cognitive revolution of the 1950s and 1960s, such as those viewing cognition as information processing (Neisser, 1967) and symbol manipulation (Newell & Simon, 1972), including methods for knowledge representation such as schemata, procedural (rule-based), and declarative (fact-based) knowledge.

A Motivating Example

To more concretely illustrate the contrast between these two general classes of computational models (computer-based statistical analysis techniques versus computational simulations of the temporal course of psychological activities), consider just one example from the history of the psychological study of expertise in cognitive–motor behavior. As early as 1897, Bryan and Harter performed a detailed study of both highly skilled (or expert) and unskilled telegraph operators.

With the exception of the lack of inferential statistical analysis techniques used for hypothesis testing—techniques unavailable to them at the time—the design and the reporting of their study was not unlike what one would see today: the representative selection of three groups of experimental participants ("expert," "ordinary," and "poor") from a population of "about 60" operators to which they had access from Western Union (pp. 36–37); the use of three different treatment conditions intended to focus on different aspects of the skill; the use of summary statistics such as means and variances of key press and error rates to serve as the basis for drawing comparisons; and graphs of learning curves in key press rates as a function of practice. Their paper ends with a list of eight empirical results and open questions the authors believe to be established by the data collected, notably including the flattening of the learning curve over practice. On this time-tested observation they state:

> [O]ne has to account for the great slowing down in the process of improvement. Stated otherwise, the task is to explain the nature of the changes in brain or mind which must be taking place during the period represented by the plateau, and which yet make no determinable manifestation of themselves. (p. 53)

Here, the authors are clearly pointing to a hard limit in their ability to provide a more satisfactory, explanatory, and temporally based account of learning ("the changes in brain or mind which must be taking place") based on the data and methods available to them. For the purposes of this illustration, it is crucial to note that this situation would remain unchanged had they had access to the inferential statistical methods and computer-based statistical analysis and modeling techniques available today.

Now, contrast this study of expert performance with the study of expert typists conducted by Rumelhart and Norman (1982). Unlike Bryan and Harter, whose chief contribution was the discovery of empirical regularities in expert cognitive–motor performance, such as the flattening of the learning curve among others, Rumelhart and Norman started their research with access to a sizable corpus of empirical regularities in typing, most notably information about the particular kinds of errors that both are and are not common in expert typing performance, and about inter-stroke timing regularities associated with which keys must be pressed in succession and the hands and fingers used to press those keys. Instead, their chief contribution was to pick up where Bryan and Harter were forced to stop, namely, "to explain the nature of the changes in brain or mind which must be taking place" in expert cognitive–motor performance.

To do so, or at least to provide one such theoretical explanation consistent with their data and extant findings, Rumelhart and Norman (1982) created and empirically evaluated a computational model or simulation of a skilled typist. Their goal was to create a model that was consistent with both then-current theory of knowledge representation in cognitive–motor performance and extant knowledge of expert typing error types and keystroke timing patterns. Their model took the form of an *activation-trigger-schema* (ATS) system. Each motor schema in the system represented knowledge of the rules for making various keystrokes, where "the interactions of the patterns of activation and inhibition among the schemata determine the temporal ordering for launching the keystrokes" (Rumelhart & Norman, 1982, p. 1). Although space constraints preclude providing a detailed description of their computational model, Figure 15.1 (Figure 4 from the original article) captures important aspects of their model for the purposes of this illustration.

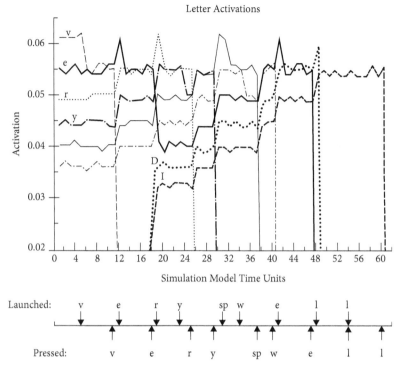

FIGURE 15.1 Temporal patterns of covert letter schema activations in the computational model for typing the string "very well" (upper graph). Timing of overt key press launches and presses along the same timeline (lower graph). See text for details.

Reproduced from Cognitive Science 6 (1), David E. Rumelhart and Donald A. Norman, Simulating a Skilled Typist: A Study of Skilled Cognitive-Motor Performance, p. 17, Figure 4, doi.org/10.1207/s15516709cog0601_1
© 1982 Cognitive Science Society, Inc.

We chose this figure to highlight the contrast between this typing research based on computational modeling and the prior research by Bryan and Harter (1897) that focused on identifying empirical regularities in cognitive–motor expertise. Specifically, note that the lower timeline in the figure (Launched, Pressed) represents overt key press timing data available to both the telegraph and typing research teams. Also note that the upper graph in Figure 15.1 can be taken to be a candidate explanation of "the nature of the changes in brain or mind which must be taking place" (Bryan & Harter, 1897, p. 53).

Providing such explanations is perhaps the most important contribution that computational modeling techniques are, arguably, uniquely suited (perhaps jointly with other emerging techniques such as fMRI) to aid in the task of investigating the mechanisms and knowledge underlying expertise. We hope that those seeking to create candidate explanations of the covert psychological processes that underlie expertise will consider including computational modeling among their methodological toolbox.

The Cognitive Architecture Approach

While there are many modeling formalisms in psychology, some of which are computational, most of them are specific to a particular domain, or even a particular task within a domain, and rarely are they models of expertise. Instead, they are generally models of typical psychology laboratory experiments; that is, models of undergraduate students with little to no prior experience with the task. However, there is a modeling approach that has been used to model both experts and the acquisition of expertise: cognitive architectures.

What are cognitive architectures? They are broad theories of human cognition based on a wide selection of human experimental data, and generally implemented as computer simulations. Young (Gray, Young, & Kirschenbaum, 1997; Ritter & Young, 2001) defines a cognitive architecture as an embodiment of "a scientific hypothesis about those aspects of human cognition that are relatively constant over time and relatively independent of task" (p. 3). One of the best descriptions of the vision for this area is presented in Newell's (1990) book *Unified Theories of Cognition*. In it, Newell argues that the time has come for cognitive psychology to stop collecting disconnected empirical phenomena and begin seriously considering theoretical unification in the form of computer simulation models. Cognitive architectures are attempts to do just this.

A cognitive architecture alone is generally not able to describe human performance on any particular task; it must be given knowledge about how to do the task. Generally speaking, a model of a task in a cognitive architecture (generally termed a *cognitive model*) consists of the architecture and the requisite knowledge to perform the specified task. This knowledge is typically based on a thorough task analysis of the target activity being modeled (for examples of task analytic methods, see Chapter 16, "Studying Expert Behavior in Sociotechnical Systems," by Salmon et al., this volume). It is increasingly common that the architecture is also connected to a complex simulation of the environment in which the task is performed. In some cases, the cognitive architecture interacts directly with the actual software that humans use to perform the task. Most cognitive architectures produce not only a prediction about performance, but in fact output actual performance. That is, they produce a time-stamped sequence of actions (e.g., mouse clicks, eye movements) that can be compared to real human performance on a task.

There are several cognitive architectures that have made appearances in the expertise literature. This includes the EPIC system of Kieras and Meyer, which has been used to model expert telephone operators (Kieras, Wood, & Meyer, 1997). The Soar architecture (Laird, 2012) has been used to model expertise in a variety of domains, from simple games to complex problem-solving.

The architecture that has supported the widest variety of models of expertise is ACT-R (Anderson & Lebiere, 1998; Anderson, 2007), which has been used to model many phenomena such as driving while dialing a mobile phone (Salvucci, 2006), aircraft maneuvering (Gluck, Ball, Krusmark, Rodgers, & Purtee, 2003) and taxiing (Byrne & Kirlik, 2005; Zemla et al., 2011), process monitoring (Gonzales, Lerch, & Lebiere, 2003), programming flight management systems (Taatgen, Huss, Dickinson, & Anderson, 2008), graph comprehension (Peebles & Cheng, 2003), searching the World Wide Web (Fu & Pirolli, 2007), and many others.

In general, models based on cognitive architectures have been successful at modeling expertise in a variety of domains. However, approaching the modeling of expertise with a cognitive architecture should not be taken lightly, as there are substantial challenges along the way. In particular, constructing models of complex tasks represents substantial time investments in multiple endeavors.

First among these is the knowledge engineering problem; in order to get a cognitive architecture to actually do anything, it must be supplied with knowledge about how to do the task. When the task is complicated, and/or situated in a complex environment, this can be particularly challenging. What do experts know, and how can that knowledge be represented in the architecture? For example, to get a cognitive architecture to taxi a simulated airplane, one must have the knowledge of how human pilots perform the task and must represent that knowledge in the language of the architecture; these languages are not unlike programming languages. Extracting the knowledge experts have about how to perform complex tasks should not be underestimated. Doing so often involves detailed task analysis, behavioral observation, and analysis of the properties of the task environment itself. All of this generally requires extensive consultation with subject-matter experts (SMEs), since cognitive modelers generally are not experts in the task domains they model.

Other challenges include the software integration problem of getting the architecture to interact with a rich simulation environment that mirrors that with which the modeled humans interact and setting free parameters. Large and complex models often have many such parameters, though in many cases these need not be fit as system default values often suffice. Exposition can also be an issue; explaining how a complex ACT-R model works and precisely which mechanisms in ACT-R provide the most predictive leverage is not always straightforward.

Despite these challenges, many modelers have been successful at producing sophisticated models of human experts in complex task domains.

Computational Models of Expertise Implemented in Cognitive Architectures

One of the first cognitive architectures to take on models of human expertise in complex domains was Soar. A model that does an excellent job of highlighting the power of cognitive architectures is NTD-Soar (Nelson, Lehman, & John, 1994). NTD stands for "NASA Test Director," who

> is responsible for coordinating many facets of the testing and preparation of the Space Shuttle before it is launched. He must complete a checklist of launch procedures that, in its current form, consists of 3000 pages of looseleaf manuals ... as well as graphical timetables describing the critical timing of particular launch events. To accomplish this, the NTD talks extensively with other members of launch team over a two-way radio ... In addition to maintaining a good understanding of the status of the entire launch, the NTD is responsible for coordinating troubleshooting attempts by managing the communication between members of the launch team who have the necessary expertise. (p. 658)

Constructing a model that is even able to perform this task at all is a significant accomplishment. Nelson, Lehman, and John were able to not only build such a model but this model was able to produce a timeline of behavior which closely matched the timeline produced by the actual NTD being modeled. That is, the ultimate result was a quantitative model of human performance, and an accurate one at that.

It is unlikely that such an effort could have been accomplished without the use of an integrated cognitive architecture. This was a Soar model that made use of other Soar models. Nelson, Lehman, and John did not have to generate and implement theory of natural language understanding to model the communication between the NTD and others, or the NTD reading the pages in the checklist, because one had already been constructed in Soar (Lewis, 1993). They did not have to construct a model of visual attention to manage the scanning and visual search of those 3000 pages, because such a model already existed in Soar (Wiesmeyer, 1992). There was still a great deal of knowledge engineering that had to go on to understand and model this complex task, but using an integrated architecture greatly eased the task of the modelers.

Soar has also been used as the basis for simulated agents in wargames (Jones et al., 1999). This model (called TacAir-Soar) participates in a virtual battlespace in which humans also participate. TacAir-Soar models take on a variety of roles in this environment, from fighter pilots to helicopter crews to refueling planes. Their interactions are more complex than simple scripted agents, and they can interact with humans in the environment with English natural language. One of the major goals of the project was to make sure that TacAir-Soar produced human-like behavior because this is critical to their role, which is to serve as part of training scenarios for human soldiers. In large-scale simulations with many entities, it is much cheaper to use computer agents than to have humans fill every role in the simulation. This effort stood somewhere between traditional AI and cognitive modeling. That is, not all of these agents were designed to necessarily exactly mimic human experts, but to at least reflect human-like time courses for behavior and show human-like limitations.

As noted previously, ACT-R has been used to model experts in a variety of domains. One of the most well-known models of expertise comes from the work of Dario Salvucci (summarized in Salvucci, 2006). Most adults in the USA have at least moderate expertise in a task that is demanding on cognitive, perceptual, and motor fronts: driving. This is not a robotics project; the ACT-R model does not actually turn the steering wheel or manipulate the pedals, but rather it communicates with the automobile simulation software. The model's primary job is to maintain lane position as the car drives down the road. Salvucci (2001) added an additional task which makes it particularly interesting: the model dials telephone numbers on a variety of mobile phone interfaces, thus examining the effects of a secondary task on expert performance. There were two factors, which were crossed: whether the telephone was dialed manually via keypad versus dialed by voice, and whether the full telephone number needed to be dialed versus a shortened *speed dial* system. The model was also validated by comparison with data from human drivers.

What both the model and the human users showed is that dialing while not driving is faster than dialing while driving, and that steering performance can be disrupted by telephone dialing. Not surprisingly, the most disruptive interface was the *full-manual* interface, where the full phone numbers were dialed on a keypad. This is due largely to the fact that dialing with the keypad requires visual guidance, causing the model (and the users) to take their eyes off the road. There was very little disruption associated with the voice interfaces, regardless of whether full numbers or speed dial was used.

This is a nice illustration of the value of cognitive architectures for a number of reasons. First, the basic driving model could simply be reused for this task; it did not have to be reimplemented. Second, the model provides an excellent quantitative fit to the human data. Third, this is an excellent example of a situation where testing human users can be difficult. Testing human drivers with interfaces that degrade driving performance is dangerous, so simulators are generally used for this kind of evaluation. However, maintaining a high-fidelity driving simulator requires a great deal of space and is quite expensive. If someone wanted to test another variant of the telephone interface, it would be much faster and cheaper to give that interface to Salvucci's model than it would be to recruit and test human drivers.

There are other domains where collection of empirical data can be difficult exactly because the potential subjects are experts, and their expertise makes them expensive (in terms of both time and dollars) to recruit. One such domain is aviation. Aviation is a broad domain, with tasks ranging from air traffic control to managing flight dynamics to navigation while taxiing. It is also a domain populated largely by people with extensive expertise. In commercial aviation, even a *novice* 737 pilot for a major airline has generally logged thousands of hours of flight, though on different equipment. Different researchers have modeled multiple aviation tasks with ACT-R; two of these will be considered here.

Many parts of commercial aviation have been automated but there are still times when pilots directly control flight dynamics. This has been studied extensively, but new technologies bring new challenges, particularly outside the commercial domain. In the late 1990s, the US military began deploying remotely piloted unmanned aerial vehicles (UAVs) on a much larger scale than before, and understanding the skill required of pilots received research attention. Gluck et al. (2003) developed an ACT-R model of UAV pilots focused on basic maneuvering. Each task required the pilot to fly seven distinct maneuvers while trying to minimize root-mean-squared deviation (RMSD) from ideal performance on altitude, airspeed, and heading. The model produced an overall performance in terms of RMSD that closely matched those produced by human pilots; however, the model showed much greater variability in some of the maneuvers than human pilots did. This is a somewhat unusual outcome in that generally models produce lower variance than the experts being modeled, which suggests that the model was lacking some knowledge that the pilots had. Still, the model generally performed similarly to the subjects, suggesting that most of the human expertise was captured by the model.

Back in the commercial aviation domain, most features of flight are now highly automated. However, there is one part of the trip that is essentially free of automation: taxiing. Taxiing a Boeing 737 at a large airport is, in practice, not entirely unlike driving an extremely large automobile around a city. There are road signs, traffic, decisions to be made about navigation, and of course physically controlling the vehicle. Based on a detailed task analysis produced by NASA and extensive consultations with SMEs, Zemla et al. (2011) were able to construct an ACT-R model of taxiing a Boeing 737 at a major US airport. While the knowledge engineering problem was substantial, dealing with the simulation environment was just as challenging. ACT-R had to be connected to a commercial flight simulator augmented with information about airport signage taken from engineering drawings of the airport surface and Google Earth satellite photos of the airport. Model performance was validated not against experimental data, but by comparison to recorded data from actual taxiing episodes collected at the same airport. Further validation was done by a kind of Turing Test; actual trajectories and model-generated trajectories were played back via the flight simulator package and expert pilots could not distinguish between the two.

Of course, not all expertise comes from the kind of rigorous and explicit training given to pilots. Some expertise comes from extensive interaction with specific environments, such as linguistic environments. This kind of expertise is reflected in Pirolli and Card's (1999; Pirolli, 2007) theory of *information foraging*. This theory is based on the idea that people forage for information in much the same way that animals forage for food. The literature on the latter type of foraging is quite rich, including equations which describe when foragers will give up on a particular *patch* of food and move on to the next one. Foraging for food is often guided by proximal cues (e.g., scent) which indicate the presence of distal content (e.g., food); analogously, information foragers may use proximal cues (e.g., visible link labels) to guide their search for information which they cannot presently access. In the spirit of the analogy, Pirolli and Card term these proximal cues information scent. That such a theory should be applicable to the Web should be fairly obvious.

This theory has spawned a number of models based on modified versions of ACT-R. The original model from Pirolli and Card (1999) was called ACT-IF (for *information foraging*). Later versions were SNIF-ACT 1.0 (Card, et al., 2001) and SNIF-ACT 2.0 (Pirolli & Fu, 2003), where SNIF stands for *scent-based navigation and information foraging*. These models use a modified form of ACT-R's spreading activation mechanism to compute information scent, which drives the model's choices as it browses through Web pages. SNIF-ACT 1.0 and later also uses a modified version of ACT-R's conflict resolution mechanism to determine when to abandon the current Web site (*patch*) and move on to a new site. SNIF-ACT 2.0 adds a further improvement, which is the use of a new measure to compute the strengths of association (and therefore spreading activation) between words. This new measure, termed point-wise mutual information (PMI), is used to represent the information structure of the linguistic environment, which is seen as the driver of expert human behavior. That is, what experts know in this domain is this structure, and relatively simple heuristics combined with knowledge of the environment combine to produce behavior. SNIF-ACT 2.0 was used to predict link selection frequency in a large ($N = 244$) Web behavior study with impressive results, accounting for 72 percent of the selection variance on one site and 90 percent of the variance on a second site.

Acquisition of Expertise in ACT-R

While these models have been able to reproduce and explain expert behavior, they are models that look at only a single point on the learning curve. What they do not model is the process of acquiring that expertise; they are not learning models.

ACT-R, however, is also capable of modeling the acquisition process. This should not be a surprise, since the originator of ACT-R, John Anderson, has been studying skill acquisition for decades. His most recent efforts (e.g., Tenison, Fincham, & Anderson, 2016; Tenison & Anderson, 2015) have been focused on modeling the acquisition of mathematical problem-solving skills. This line of research has been extensively informed by neuroscience, as subjects in these studies are also imaged via fMRI. ACT-R makes predictions about not only behavior and its time course, but also activations in specific brain regions. It turns out that as students acquire expertise, these activation patterns change, and the ACT-R models predicts these changes as a function of skill. In order to do this, the modeling is at a fine level of detail, examining specific components of the skill and identifying and modeling transitions separately for encoding, solving, and responding to problems. This research shows that different aspects of the skill are acquired at different rates and is not always smooth, but often occurs in small stages; learning is not uniform. While simple models like the power law of practice (Newell & Rosenbloom, 1981) capture skill acquisition in the aggregate, the reality at the individual level is much more complex.

This is not the only example of modeling skill acquisition with ACT-R. A particularly compelling set of results come from the work of Taatgen (2002; Taatgen & Lee, 2003). The task used in this research was the Kanfer–Ackerman air traffic control (KA-ATC) task. This task is not a real-world ATC task, but was developed with an ATC cover story to examine skill acquisition in a dynamic, game-like environment that goes beyond static math problems and requires perceptual and motor skill as well. What their data showed, and what was echoed in the ACT-R model, was again that different components of the task are also learned at different rates.

More than that, though, is that they showed that learning of the perceptual skill was a critical piece of the acquired expertise. Novices tended to spend a lot of time visually scanning the screen for information that was not relevant to the part of the task they were actually doing, while experts learned to look only at the areas they needed to and only when a specific piece of information was needed. Learning what *not* to do in a dynamic domain can be just as important as learning what to do, and ACT-R's learning mechanisms were able to capture this as well.

Both Soar and ACT-R are general-purpose theories of cognition intended to be applied across a wide variety of cognitive domains. What these models show is that not only can expertise be represented in broad cognitive architectures in multiple ways, but that the development of expertise can also be modeled. However, using cognitive architectures in this way requires specialized skill in programming the architecture and extensive understanding of the domain. While insights into expertise and detailed fits to data are possible, they are not inexpensive. There are other modeling formalisms that may be better suited, depending on the goals of the research and the level of investment available.

Domain-Specific Computational Models of Expertise

As an alternative to working within the cognitive architecture approach, some psychological scientists have instead chosen to construct computational models of expertise derived from a narrower set of theoretical prescriptions in some domain of psychological activity. In this section of the chapter, we describe a few such models drawn from theories in the domains of visual perception, recognition-based decision making, and situated cognition and action.

Computational Modeling of Perceptual Expertise

In his classic book *Cognition and Reality* (1976), the pioneering cognitive psychologist Ulric Neisser characterized perception as the most primitive and fundamental cognitive act, and spent a good portion of that book making the case for his view. Compared to the relatively much slower and mentally challenging mechanisms underlying analytical (as opposed to perceptually driven and intuitive) cognition (Hammond, Hamm, Grassia, & Pearson, 1987), or System 2 (analytical) versus System 1 (intuitive) cognition (Kahneman, 2011), it has been natural for some psychological scientists to view at least one significant aspect of the growth from novice to expert performance as due, at least in part, to an increased reliance on perceptual and other intuitive mechanisms (Kirlik, Walker, Fisk, & Nagel, 1996; Klein, 2013).

This assumption presumes, of course, that the nature of the performer's task environment provides timely and unambiguous feedback to enable perceptual or intuitive learning to become successfully adapted (Kahneman & Klein, 2009). The dynamics of the perceptual learning process that lead to perceptual expertise are still not yet well understood, as a review of the literature indicates that perceptual learning is studied at various levels of information processing and focuses on a diverse set of perceptual activities (e.g., perceptual search, discrimination, object identification, and others). According to Petrov, Dosher, and Lu (2005),

> Perceptual learning refers to performance improvements in perceptual learning tasks as a result of practice or training. Perceptual learning has been of particular interest as it may reflect plasticity at different levels of perceptual analysis—from changes in early sensory representations to higher order changes in the way these representations are used in a task. (p. 715)

The computational modeling of perceptual expertise or its development has tended to focus on one level or aspect of perceptual achievement: no comprehensive theory or computational modeling of all aspects of perceptual expertise yet exists.

There is, however, some agreement on a few distinctions between novice and expert perceptual performance that any computational model of the acquisition or characterization of perceptual expertise should exhibit. According to the review and synthesis provided by Palmeri, Wong, and Gauthier (2004), these include:

1. Novices' abilities to decompose a perceptual object or situation is task-neutral, while expert-level decompositions are highly tied to the goals to which their performance is directed.
2. While experts exhibit near transfer of their perceptual expertise better than novices, their ability to demonstrate far transfer, for example to totally novel objects or situations, remains much more limited.

This succinct list provides a nice transition to a discussion of the computational modeling research we have chosen to concretely illustrate the nature and acquisition of expertise in the perceptual domain.

Consider, for example, item 2 on contrasts between novices and experts and between near and far transfer. Perhaps the best example illustrating these contrasts was the work of Chase and Simon (1973) and by de Groot (1978) on chess masters. Two of the best known findings of this research were chess masters' highly superior ability compared to lower ranked players to recall and reconstruct meaningful chess positions (those drawn from actual master level games), yet in contrast a disappearance of this advantage when players at both the master and lesser levels were presented with positions in which the placement of the pieces had been randomly arranged—the latter, as close to one can get to far, rather than near, transfer in the chess domain.

This finding, which has been recounted in numerous textbooks, has more recently been challenged by a discovery that resulted from the development and evaluation of a model of perceptual expertise in chess called CHREST, for Chunk, Hierarchy, and REtreival Structures (de Groot & Gobet, 1996; Gobet & Simon, 2000; for a summary of the evolution of the classic approach to studying expertise, see Chapter 2, "The Classic Expertise Approach and Its Evolution," by Gobet, this volume). CHREST computationally simulates perceptual expertise acquisition in terms of the elaboration of a network of both perceptual- and action-oriented knowledge represented as chunks. As in the computational model of typing expertise by Rumelhart and Norman discussed previously, the model also includes more global forms of knowledge representation in the form of schemata, or what Gobet (2005) calls "templates" (p. 187; see this article for more detail than space constraints allow here).

As discussed in Gobet (2005), CHREST is able to mimic both learning rates and eye movement patterns in chess expertise, and also resulted in a novel discovery concerning the previously mentioned classic result on experts' abilities and limitations in achieving near and far transfer of skill. There, Gobet discusses the fact that the CHREST model's performance actually deviated from received wisdom, in that it predicted that chess masters' memory for board positions should still be slightly superior to lesser ranked players, even for random board positions. This would be a demonstration of at least some superior ability even in the far transfer case, in contrast to prior, textbook results. The CHREST model behaved in this way because it was able to recognize at least a few patterns existing solely by chance in what was believed to be a purely random board configuration. As discussed in Gobet (2005), this model-inspired prediction was subsequently confirmed by a detailed re-analysis of past data. This demonstrates that computational models may not only provide an ability to test a theory of the mechanisms underlying expertise, but they may also give rise to discoveries about the nature of expertise as well.

Computational Modeling of Recognition-Based Decision Making

Klein (1989, 2013) has provided a highly influential model of expert decision making, especially for performers in naturalistic or applied contexts. In contrast to the dominant approach to understanding decision making by most psychological scientists that focuses on the comparative evaluation of numerous decision options, the essence of Klein's recognition-primed decision (RPD) model is the ability to rapidly assess a situation in reference to a storehouse of past decision situations, and identify one or just a few promising options based on a combination of environmental cues and experiential memory.

Although the RPD model has provided a useful qualitative picture of how experts make decisions for some time, Mueller (2009) has more recently provided a computational model of at least some of the key aspects of the RPD model called the Bayesian recognitional decision model (BRDM). The BRDM focuses on "one narrow aspect of the RPD model" (p. 115) by using Bayesian techniques to infer that an optimal decision option is selected given the constraints provided by a model of the decision environment composed of its features, their base rates, and relative reliabilities and experiential knowledge. To model the acquisition of expertise, the model is fed a large corpus of previous decisions in response to similar environmental situations, and creates a set of event prototypes corresponding to environmental prototypes. Once learning is complete, the model uses largely standard Bayesian techniques to calculate a distribution of posterior decision likelihood ratios, which represent the relative strength of the evidence for various decision options. Quantitative examples are provided showing how the BRDM model is sensitive to both base rate manipulations and the differing levels of trust the decision maker has in various sensors or reporters; that is, the mechanisms or people providing source information to the expert decision maker.

Computational Modeling of Situated Cognition and Action

This research (Kirlik, Miller, & Jagacinski, 1993) involved creating and validating a computational model of expertise in situated cognition and action (Suchman, 1987). Expert performers interacted with a simulated dynamic, interactive, and uncertain environment requiring them to control both their own craft, called the scout, using joysticks and pushbuttons, along with four additional craft over which the participant exercised remote or supervisory control (e.g., fly to a specified waypoint, conduct patrol, load cargo, return to a home base). Performers were provided a top-down map display of the simulated 100-square-mile world to which activity was confined, various status displays, and a text editor for sending commands to the four supervised craft. The task was to control the activities of both the scout and the supervised craft to score points in each 30-minute session by processing valued objects that appeared on the display once sighted by scout radar.

The goal was to create a computer simulation capable of performing this challenging task, and one that would allow us to reproduce, and thus possibly explain the nature of expert task performance. We observed that expert performers seemed to be relying heavily on the external world (the interface) as its own best model (Brooks, 1991), as suggested by intimate perceptual engagement with the displays at all times. This turned us to the work of Gibson (1979), whose theory of affordances provided an account of how people might be attuned to perceiving the world functionally; in this case, in terms of actions that could be performed in particular situations.

As such, we created maps and models of the task environment in terms of the degree to which various environmental regions and objects afforded various actions such as scout locomotion (i.e., *fly-through-ability*), searching (i.e., discovering valued objects by radar), processing those objects (i.e., loading cargo, engaging enemy craft), and returning home to unload cargo and re-provision. Since performers' actions influenced the course of events experienced, they partially shaped the affordances of their own worlds.

As fully explained in Kirlik et al. (1993), this functional, affordance-based differentiation of the environment, assuming behavior was largely shaped by these affordances, provided an extremely efficient method for mimicking the behavior of our expert performers. The computational model was able to pass a Turing Test by generating behavior statistically indistinguishable from those of human experts.

Finally, we return to the robust finding that experts are often unable to verbalize rules underlying their expertise. If one does assume that expert procedural knowledge in situated activity exists in the form of *if p then q* conditionals or rules, then this computational modeling provides a different explanation of why experts may often be unable to verbalize knowledge.

Rather than placing such *if p then q* rules in the *head* of the model, we instead created perceptual mechanisms that functioned to *see* the world functionally, as affordances, which we interpret as playing the roles of the *p* terms in the *if p then q* construction. The *q* terms, in contrast, are the internal responses to assessing the world in functional terms. As such, the *if p then q* construct is distributed across the boundary of the human–environment system. This account suggests that expertise in situated cognition and action may shift from being internally rule-based to becoming based more in perceptual mechanisms, and therefore may be unable to be verbalized as these latter processes are typically covert rather than overt (cf. Greeno, 1987, on situated cognition and action).

Assessing the State of the Science

In providing an overall assessment of the state of the science in the computational modeling of expertise, it is useful to first highlight the inclusion criteria we used to select the modeling research presented in this chapter. Even though space constraints necessarily required a sparse sampling of these models to illustrate more general techniques, we nevertheless have also intentionally excluded the much broader set of computational models in the psychological literature that were not explicitly framed as models of high levels of skilled or expert performance. As such, entire classes of computational models of various psychological theories, including connectionist or neural network models (Rumelhart, McClelland, & the PDP Research Group, 1986), among many others were excluded from our review. We believe it is fair to say that the lion's share of computational models presented in the psychological literature are focused on modeling competence of one sort or another, where a match to behavioral data is the chief concern when evaluating a model's validity, rather than whether a model achieves expert-level performance at a task. Said another way, the fraction of computational modeling research being done using data collected from expert-level performers to determine model validity is quite small.

One reason this may be the case is that modeling non-expert or relatively task-naive behavior can be done by similarly non-expert or task-naive scientists. The chief requirement is expertise in cognitive modeling. However, modeling highly skilled or expert performance also requires expert knowledge of the tasks and task environments in which expertise is expressed. Neisser (1976) put the matter of predictively modeling expert performance as follows:

> What would we have to know to predict how a chess master will move his pieces, or his eyes? His moves are based on information he has picked up from the board, so they can only be predicted by someone who has access to the same information. In other words, an aspiring predictor would have to understand the position at least as well as the master does; he would have to be a chess master himself! If I play chess against the master he will always win, because he can predict and control my behavior while I cannot do the reverse. To change this situation I must improve my knowledge of chess, not of psychology. (Neisser, 1976, p. 183)

In closed worlds or domains such as performing mental arithmetic or a Tower of Hanoi puzzle, where computational modeling had its initial and impressive successes, knowledge of a task environment is not necessary either because the task is done solely internally (mental arithmetic) or because expert knowledge of task performance is readily available to the modeler, either from books (arithmetic itself) or from relatively simple task analyses (the Tower of Hanoi). Successful modeling of expertise is highly dependent on modelers having the resources to obtain expert knowledge of task environments through task analyses and collaborations with subject-matter experts.

We believe there are a number of critical ways in which computational modeling efforts have provided insight into the nature of expertise. First, most other methods have been much more qualitative. Many of the models described here provide quantitative accounts of expert performance, the development of expertise, or both. Some of these models provide fits to performance of individual experts at the sub-second level. Non-computational methods rarely achieve this. Second, along with that, many computational models demonstrate the degree to which the devil is in the details; performance improvements and expert superiority is not uniform across the entire task. Learning happens at different rates for different components of the skill, and novice–expert differences are similarly non-global. Finally, what many of these models show, again in detail and at the mechanistic level, is how in many domains expert performance is substantially a matter of adaptation to task goals and the structure of the task environment. We believe these contributions are critical to a greater understanding of the structure of expertise.

Future Research Directions

Computational models of expertise have been around for almost as long as digital computers. Over the years, the models have become increasingly sophisticated and wider ranges of modeling approaches have been used. In addition, the tasks models have been applied to have also changed, from simple static tasks that were almost entirely cognitive to complex, dynamic tasks with perceptual–motor components. We expect these trends to continue: models and modeling systems will continue to get more advanced, and the tasks modeled will also continue to increase in complexity. These are the most obvious future directions.

However, there are other developments that we believe portend things to come. Most of the research on expertise (with the possible exception of chess) have studied experts, but not necessarily experts at the most extreme end of the expertise continuum. However, recently Wayne Gray and colleagues (Gray & Lindstedt, 2017; Sibert, Gray, & Lindstedt, 2017) have been studying, and modeling, this kind of expertise. Their domain is the computer game Tetris, and they have gone to great lengths to find not just experts, but the best of the best. They hosted competitions at their university to find the superior local players and then expanded to examine players at the world championships. What they have found is that even at the most extreme levels of expertise, experts still learn, and there are discontinuities in performance—there are still dips, and then leaps ahead.

Another likely direction going forward is increasing integration with neuroscience. As neuroimaging and related techniques become more popular, there are new opportunities to marry modeling and neuroscience. Behavioral data alone are not always enough to provide adequate constraints to computational models, but the addition of neural data can help rule out models and provide increasing insight. As noted, ACT-R is well positioned in this arena, as ACT-R makes quantitative predictions about neural activity (though only in certain brain regions) and modeling of both task performance and brain activations is already underway (e.g., Tenison, Fincham, & Anderson, 2016; Tenison & Anderson, 2015). Another likely candidate for a modeling architecture that might make inroads on expertise and neuroscience is O'Reilly's Leabra architecture (O'Reilly, Hazy, & Herd, 2017), which comes from a different intellectual tradition than ACT-R, connectionism. There are certainly other directions that modeling of expertise may go, but we see these as the most likely next steps.

Conclusions

In this chapter we have provided an introduction to and overview of both foundational and contemporary research using computational modeling to further aid in the scientific understanding of human expertise. We noted the distinction between models constructed within some more global and unified psychological theory or cognitive architecture and models that are much more domain or task specific in their psychological assumptions, and have provided numerous illustrative examples of each type.

We have also provided an assessment of this overall body of research at a level we believe it can be justified given the diversity of the models discussed, one that highlights the need for extensive analysis, and even expert-level knowledge of both tasks and the environments in which expert behavior is manifest as a key requirement for successfully modeling high levels of skill or expertise. Finally, we have provided our thoughts about promising future directions for research using computational cognitive modeling, together with other emerging techniques such as neuroimaging, to provide a more fruitful approach to the study of expertise than would be available using any single psychological method or technique.

References

Anderson, J. R. (2007). *How can the human mind occur in the physical universe?* New York: Oxford University Press.

Anderson, J. R., & Lebiere, C. (1998). *The atomic components of thought.* Mahwah, NJ: Lawrence Erlbaum.

Box, G. E. P., & Draper, N. R. (1987). *Empirical model building and response surfaces.* New York: Wiley.

Brooks, R. (1991). Intelligence without representation. *Artificial Intelligence 47,* 139–159.

Bryan, W. L., & Harter, N. (1897). Studies in the physiology and psychology of the telegraphic language. *Psychological Review* 4(1), 27.

Byrne, M. D., & Kirlik, A. (2005). Using computational cognitive modeling to diagnose possible sources of aviation error. *International Journal of Aviation Psychology 15,* 135–155.

Card, S. K., Pirolli, P., Van Der Wege, M., Morrison, J. B., Reeder, R. W., Schraedley, P. K., & Boshart, J. (2001). Information scent as a driver of web behavior graphs: Results of a protocol analysis method for Web usability. In *Human factors in computing systems: Proceedings of CHI 2001* (pp. 498–505). New York: ACM.

Chase, W. G., & Simon, H. A. (1973). The mind's eye in chess. In W. G. Chase (Ed.), *Visual information processing,* (pp. 214–281). New York: Academic Press.

De Groot, A. D. (1978). *Thought and choice in chess.* (2nd English edition, 1st Dutch edition published in 1946). The Hague: Mouton.

De Groot, A. D., & Gobet, F. (1996). *Perception and memory in chess: Heuristics of the professional eye.* Assen: Van Gorcum.

Fu, W.-T., & Pirolli, P. (2007). SNIF-ACT: A cognitive model of user navigation on the World Wide Web. *Human-Computer Interaction 22,* 355–412.

Gibson, J. J. (1986). *The ecological approach to visual perception.* Hillsdale, NJ: Erlbaum. (Original work published in 1979.)

Gigerenzer, G., Krauss, S., & Vitouch, O. (2004). The null ritual: What you always wanted to know about significance testing but were afraid to ask. In D. Kaplan (Ed.), *The Sage handbook of quantitative methodology for the social sciences* (pp. 391–408). Thousand Oaks, CA: Sage.

Gluck, K. A., Ball, J. T., Krusmark, M. A., Rodgers, S. M., & Purtee, M. D. (2003). A computational process model of basic aircraft maneuvering. In F. Detje, D. Dörner, & H. Schaub (Eds), *Proceedings of the fifth international conference on cognitive modeling* (pp. 117–122). Bamberg: Universitas-Verlag Bamberg.

Gobet, F. (2005). Chunking models of expertise: implications for education. *Applied Cogntivie Psychology* 19(2), 183–204.

Gobet, F., & Simon, H. A. (2000). Five seconds or sixty: Presentation time in expert memory. *Cognitive Science 24*, 651–682.

Gonzalez, C., Lerch, F. J., & Lebiere, C. (2003). Instance-based learning in real-time dynamic decision making. *Cognitive Science 27*, 591–635.

Gray, W. D., & Lindstedt, J. K. (2017). Plateaus, dips, and leaps: Where to look for inventions and discoveries during skilled performance. *Cognitive Science 41(7)*, 1838–1870.

Gray, W. D., Young, R. M., & Kirschenbaum, S. S. (1997). Introduction to this special issue on cognitive architectures and human-computer interaction. *Human–Computer Interaction 12*, 301–309.

Greeno, J. (1987). Situations, mental models, and generative knowledge. In D. Klahr & K. Kotovsky (Eds), *Complex information processing* (pp. 285–316). Hillsdale, NJ: Lawrence Erlbaum.

Hammond, K. R., Hamm, R. M, Grassia, J. & Pearson, T. (1987). Comparison of the efficacy of intuitive versus analytical cognition in expert judgment. *IEEE Transactions on Systems, Man, and Cybernetics SMC-17(5)*, 753–770.

Jones, R. M., Laird, J. E., Nielsen, P. E., Coulter, K. J., Kenny, P., & Koss, F. V. (1999). Automated intelligent pilots for combat flight simulation. *AI Magazine 20(1)*, 27–41.

Kahneman, D. (2011). *Thinking fast and slow*. New York: Farrar, Straus and Giroux.

Kahneman, D., & Klein, G. (2009). Conditions for intuitive expertise. *American Psychologist 64(6)*, 515–526.

Kieras, D. E., Wood, S. D., & Meyer, D. E. (1997). Predictive engineering models based on the EPIC architecture for a multimodal high-performance human-computer interaction task. *ACM Transactions on Computer–Human Interaction 4*, 230–275.

Kirlik, A., Miller, R. A., & Jagacinski, R. J., (1993). Supervisory control in a dynamic uncertain environment: A process model of skilled human-environment interaction. *IEEE Transactions on Systems, Man, and Cybernetics 23(4)*, 929–952.

Kirlik, A., Walker, N., Fisk, A. D., & Nagel, K. (1996). Supporting perception in the service of dynamic decision making. *Human Factors 38(2)*, 288–299.

Klein, G. (1989). Recognition-primed decisions. In W. B. Rouse (Ed.), *Advances in man-machine systems research, Vol. 5* (pp. 47–92). Greenwich, CT: JAI Press.

Klein, G. (2013). *Seeing what others don't: The remarkable ways we gain insights*. New York: Public Affairs (Persus Books Group).

Laird, J. E. (2012). *The Soar cognitive architecture*. Cambridge, MA: MIT Press.

Lewis, R. L. (1993). *An architecturally-based theory of human sentence comprehension*. Doctoral dissertation, University of Michigan, Ann Arbor.

Mueller, S. (2009). A Bayesian recognitional decision model. *Journal of Cognitive Engineering and Decision Making 3(2)*, 111–130.

Neisser, U. (1967). *Cognitive psychology*. Upper Saddle River, NJ: Prentice Hall.

Neisser, U. (1976). *Cognition and reality*. New York: W. H. Freeman.

Nelson, G., Lehman, J. F., & John, B. E. (1994). Integrating cognitive capabilities in a real-time task. In A. R. Eiselt & K. Eiselt (Eds), *Proceedings of the sixteenth annual conference of the cognitive science society* (pp. 353–358). Hillsdale, NJ: Lawrence Erlbaum.

Newell, A. (1990). *Unified theories of cognition*. Cambridge, MA: Harvard University Press.

Newell, A., & Rosenbloom, P. S. (1981). Mechanisms of skill acquisition and the law of practice. In J. R. Anderson (Ed.), *Cognitive skills and their acquisition* (pp. 1–55). Hillsdale, NJ: Lawrence Erlbaum.

Newell, A. & Simon, H.A. (1972) *Human problem solving*. Upper Saddle River, NJ: Prentice Hall.

O'Reilly, R. C., Hazy, T. E., & Herd, S. A. (2017). The Leabra cognitive architecture: How to play 20 principles with nature and win! In S. Chipman (Ed.), *Oxford handbook of cognitive science* (pp. 91–115). Oxford: Oxford University Press.

Palmeri, T. J., Wong, A. C., & Gauthier, I. (2004). Computational approaches to the development of perceptual expertise. *Trends in Cognitive Sciences* 8(8), 378–386.

Peebles, D., & Cheng, P. C.-H. (2003). Modeling the effect of task and graphical representation on response latency in a graph reading task. *Human Factors* 45, 28–45.

Petrov, A. A., Dosher, B. A., & Lu, Z. L. (2005). The dynamics of perceptual learning: An incremental reweighting model. *Psychological Review* 112(4), 715–743.

Pirolli, P. (2007). *Information foraging theory*. New York: Oxford University Press.

Pirolli, P., & Card, S. (1999). Information foraging. *Psychological Review* 106, 643–675.

Pirolli, P., & Fu, W. (2003). SNIF-ACT: A model of information foraging on the world wide web. In P. Brusilovsky, A. Corbett, & F. de Rosis (Eds), *Proceedings of the ninth international conference on user modeling* (pp. 45–54). Johnstown, PA: Springer-Verlag.

Ritter, F. E., & Young, R. M. (2001). Embodied models as simulated users: Introduction to this special issue on using cognitive models to improve interface design. *International Journal of Human-Computer Studies* 55, 1–14.

Rumelhart, D., McClelland, J., & the PDP Research Group (1986). *Parallel distributed processing: Explorations in the microstructures of cognition*. Cambridge, MA: MIT Press.

Rumelhart, D. E., & Norman, D.A. (1982). Simulating a skilled typist: A study of skilled cognitive-motor performance. *Cognitive Science* 6, 1–36.

Salvucci, D. D. (2001). Predicting the effects of in-car interface use on driver performance: An integrated model approach. *International Journal of Human-Computer Studies* 55, 85–107.

Salvucci, D. D. (2006). Modeling driver behavior in a cognitive architecture. *Human Factors* 48, 362–380.

Sedgewick, R., & Wayne, K. (2011). *Algorithms*, 4th edn. Upper Saddle River, NJ: Pearson Education.

Sibert, C., Gray, W. D., & Lindstedt, J. K. (2017). Interrogating a feature learning model to discover insights into the development of human expertise in a real-time, dynamic decision-making task. *Topics in Cognitive Science* 9, 374–394.

Suchman, L. A. (1987). *Plans and situated actions*. New York: Cambridge University Press.

Taatgen, N. A. (2002). A model of individual differences in skill acquisition in the Kanfer-Ackerman Air Traffic Control Task. *Cognitive Systems Research* 3, 103–112.

Taatgen, N. A., Huss, D., Dicksion, D., & Anderson, J. R. (2008). The acquisition of robust and flexible cognitive skills. *Journal of Experimental Psychology: General* 137, 548–565.

Taatgen, N. A., & Lee, F. J. (2003). Production compilation: A simple mechanism to model skill acquisition. *Human Factors* 45, 61–76.

Tenison, C., & Anderson, J. R. (2015). Modeling the distinct phases of skill acquisition. *Journal of Experimental Psychology: Learning, Memory, and Cognition* 42(5), 749–767.

Tenison, C., Fincham, J. M., & Anderson, J. R. (2016). Phases of learning: How skill acquisition impacts cognitive processing. *Cognitive Psychology* 87, 1–28.

Tetlock, P. E. (2005). *Expert political judgment: How good is it? How can we know?* Princeton, NJ: Princeton University Press.

Wiesmeyer, M. (1992). *An operator-based model of covert visual attention*. Doctoral dissertation, University of Michigan, Ann Arbor.

Zemla, J. C., Ustun, V., Byrne, M. D., Kirlik, A., Riddle, K., & Alexander, A. L. (2011). An ACT-R model of commercial jetliner taxiing. *Proceedings of the Human Factors and Ergonomics Society* 55(1), 831–835.

CHAPTER 16

STUDYING EXPERT BEHAVIOR IN SOCIOTECHNICAL SYSTEMS

Hierarchical Task Analysis

PAUL M. SALMON, NEVILLE A. STANTON, GUY H. WALKER, AND GEMMA J. M. READ

Introduction

EXPERTISE is often studied naturalistically in context by examining the behavior of experts in work and everyday systems. One group of methods that is used extensively for studying behavior in context is task analysis methods. These methods are used to describe, represent, and interrogate the physical and cognitive aspects of behavior in sociotechnical systems (Stanton et al., 2013). In relation to expertise, this allows analysts to understand expert performance in terms of the physical and cognitive activities involved, which in turn is used to support the design of training programs, standard operating procedures, interfaces, and work tools and technologies. In addition, these methods are often used to study "normal performance" in terms of how work is actually performed by domain experts (as opposed to how the work is described in procedures). This aligns well with the currently popular concepts of Safety II and work as imagined versus work as done (Hollnagel, Leonhardt, Licu, & Shorrock, 2013). As a result of the increasing interest in these concepts, the ability to identify and describe how expert behavior goes beyond that described in procedures is a key requirement of human factors and safety science methods.

Within human factors, the discipline concerned with understanding and optimizing sociotechnical system performance, task analysis methods are arguably the most popular analysis approach. Indeed, their popularity is such that there are over 100 task analysis techniques available (Diaper & Stanton, 2004). In recent times, there has been a shift in the focus of task analysis methods from studying individual and team behavior to the study of the behavior of entire work systems (Salmon, Walker, Read, Goode, & Stanton, 2017). This so-called systems thinking approach involves taking the overall system as the unit of analysis, looking beyond individuals, and considering the interactions between humans and between humans and artifacts within a system. This view also encompasses factors within the broader organizational, social, or political system in which behavior takes place (Read, Beanland, Lenne, Stanton, & Salmon, 2017). Taking this perspective, behaviors emerge not from the decisions or actions of individuals but from interactions across the wider system. Expert behavior therefore can be viewed as an emergent property that is influenced by a network of interacting factors across the wider sociotechnical system. Studying expertise therefore requires analyses of both the expert behavior itself and of the network of interacting factors underpinning it.

The systems thinking philosophy has important implications for the task analysis methods that we use to understand behavior and expertise. Most important is that the overall work system itself (or as much of it as possible) is taken as the unit of analysis and that behavior is studied in the context of wider organizational, social, and political factors. For example, whilst individual physical and cognitive processes should be examined, the systemic factors influencing them should also be considered. Moreover, the focus should be on how experts interact with the wider work and social system, rather than on only the behavior of the expert themselves.

One highly popular task analysis method capable of studying behavior and expertise through a systems thinking lens is hierarchical task analysis (HTA; Annett et al., 1971; Stanton, 2006). Reflecting this, recent studies have utilized HTA in the study of expert behavior in a range of work and everyday systems (e.g., Al Hakim, Maiping, & Sevdalis, 2014; Demirel et al., 2016; Read et al., 2017). A significant strength of HTA is that it can be used to analyze the behavior of individuals, teams, organizations, and even entire systems.

The aim of this chapter is to describe and demonstrate HTA through a case study undertaken in the railway level crossing safety context. Importantly, the case study showcases how HTA can be used to study expert behavior at the so-called sharp end of systems performance as well as to study behavior at an overall sociotechnical system level. The intention is to provide analysts with practical guidance on how to study both behavior in complex sociotechnical systems and the behavior of complex sociotechnical systems.

Hierarchical Task Analysis

Hierarchical task analysis (Annett et al., 1971) is the most commonly applied task analysis method. The name is slightly misleading, however, as HTA does not focus exclusively on tasks. Rather it focuses on goals (an objective or end state) and decomposes them hierarchically to identify sub-goals and requisite operations.

HTA has its roots in the scientific management movement of the early twentieth century (Stanton, 2006). Scientific management methods were used to describe and analyze tasks in a way that supported the development of more efficient work processes. The focus was on how the work was performed physically, what was required to do the work, and how the work could be enhanced (Stanton, 2006). While this approach endured until the mid-twentieth century, HTA was developed in the 1960s in response to the changing nature of work. Cognition during work tasks was becoming more prevalent and important, which in turn created the need for methods that could describe both the physical and cognitive aspects of behavior (Annett, 2004). At the time HTA provided a significant advancement over scientific management methods which largely focused on the physical and observable aspects of behavior.

HTA works by decomposing systems and behavior into a hierarchy of goals, subordinate goals, operations, and plans. In doing so, it focuses on "what an operator... is required to do, in terms of actions and/or cognitive processes to achieve a system goal" (Kirwan & Ainsworth 1992, p. 1). Equally, however, it can be used to focus on what an organization or entire system comprising multiple organizations is required to do in terms of the actions and processes undertaken by multiple stakeholders to achieve the systems goals. It is important to note also that an *operator* may be human or non-human (e.g., system artifacts such as equipment, devices, documentation, and interfaces).

HTA outputs specify the overall goal of a particular task/scenario/organization/ system, the sub-goals to be undertaken to achieve this goal, the operations required to achieve each of the sub-goals specified, and the plans that trigger engagement and achievement of the different goals and operations. The plans component specifies the sequence, and under what conditions different sub-goals must be achieved in order to satisfy the requirements of a superordinate goal.

The human factors and safety science literature describes multiple applications of HTA undertaken in many different areas over the past five decades (see Stanton, 2006 for an overview). HTA's long-standing popularity is largely down to its flexibility—it can be used in any domain and forms the input for various additional human factors analysis methods (Salmon, Stanton, Jenkins, & Walker, 2010; Stanton et al., 2013). For example, error identification methods such as the systematic human error reduction and prediction approach (SHERPA; Embrey, 1986) and more recently the networked hazard and risk management system (NET-HARMS; Dallat et al., 2018) risk assessment method require a HTA description as their primary data input. This has resulted in HTA being applied for all manner of system design and evaluation purposes, including interface design and evaluation, job design, training program design and evaluation, error prediction and risk assessment, team task analysis, situation awareness requirements analysis, allocation of functions analysis, workload assessment, and procedure design (Stanton, 2006; Stanton et al., 2013).

Another important aspect of the method's flexibility is the ability to support different levels of analysis within sociotechnical systems. For example, HTA can be used to analyze the behavior of individuals, human–machine dyads, teams, organizations, multiorganization tasks, and even entire sociotechnical systems. The latter in particular ensures that the method will continue to be popular in light of the increasing popularity of systems thinking in human factors and ergonomics and safety science (Karsh, Waterson, & Holden, 2014; Salmon et al., 2017; Walker, Salmon, Bedinger, & Stanton, 2017).

HTA and Expertise

There are various ways in which HTA can be used in the study of expertise. These include:

- describing expert behavior in terms of the goals and physical and cognitive activities involved. These HTA descriptions can then be compared with HTAs of non-expert behavior to determine the differences between non-expert and expert behavior;
- describing work as imagined (e.g., by procedures) versus work as done by experts. These HTA descriptions are useful as they detail how experts actually get the job done as opposed to how often inaccurate standard operating procedures suggest they do;
- describing expert behavior in the context of the wider sociotechnical system to identify what factors create and influence expert behavior; and
- providing the input into various additional analysis methods that can be used to explore expertise in depth, such as error identification, situation awareness requirements analysis, training needs analysis, and allocation of functions analysis.

HTA can also be used to support the development of interventions designed either to support expert behavior or to facilitate the development of expertise. This includes using HTA to design training programs; design and represent standard operating procedures (see Stanton, Salmon, Walker, & Jenkins, 2010); and design work support tools, technologies, and interfaces.

Practical Guidance for HTA

The HTA process is relatively simple and involves collecting data about the task or system under analysis (through techniques such as observation, questionnaires, interviews with subject matter experts (SMEs), walkthroughs, user trials, and documentation review). The data are then used to decompose and describe the goals and sub-goals involved in task or system operation.

Step 1: Define Task under Analysis

The first step in applying HTA involves clearly defining the tasks or system under analysis along with any relevant analysis boundaries. For example, the analysis may be focused on individual behavior in a given context (i.e., driving a train), teamwork (i.e., between train drivers and network controllers), organizational behavior (i.e., delivery of rail transport), or the behavior of an entire sociotechnical system (i.e., operation of the railway system). It is useful practice to clearly state the analysis boundaries so that all analysts understand them and what sits outside of them.

As part of this first step the aims of the analysis should also be clearly defined. For example, for the first HTA developed as part of the rail level crossing case study described later the following was determined during step 1:

- *Task/System under analysis*: De-confliction of road and rail traffic at railway level crossings;
- *Aim of the analysis*: Describe de-confliction of road and rail traffic at railway level crossings to support error identification and the development and evaluation of new railway level crossing design concepts;
- *Analysis boundaries*: Scope of the analysis is limited to actively controlled railway level crossings in Victoria, Australia (e.g., rail level crossings with boom gates and/or flashing lights).

Step 2: Data Collection

Once the task/system under analysis and aims of the analysis are clearly defined, specific data regarding the task/system should be collected. Data should be collected regarding the goals and tasks involved, the human and non-human agents involved, the interactions between humans and non-human agents, the ordering of tasks and conditions that dictate task sequences, and information on the factors that influence behavior. A number of different approaches can be used to collect this data, including observations, concurrent verbal protocols, structured or semi-structured interviews (e.g., the critical decision method; Klein et al., 1989), questionnaires and surveys, walkthrough analysis, and documentation review (e.g., incident reports, standard operating procedures). For further information on these methods and their application the reader is referred to "Incident-Based Methods for Studying Expertise," by Militello and Anders (Chapter 19, this volume); "A Historical Perspective on Introspection: Guidelines for Eliciting Verbal and Introspective-Type Reports," by Ward, Wilson, Suss, Woody, & Hoffman (Chapter 17, this volume); and Stanton et al. (2013).

The data collection approach selected is dependent upon the aims and boundaries of the analysis as well as project constraints, such as time, number of analysts available, and access to information and personnel. For studies of expert behavior it is recommended that the following data collection approaches are used:

- Critical decision method interviews with subject matter experts;
- Task walkthroughs with SMEs providing concurrent verbal protocols;
- Documentation review (e.g., standard operating procedures, training manuals); and
- Naturalistic observations of SME performance.

For the railway level crossing case study presented later, data collection activities included documentation review, on-road studies of novice and experienced driver behavior at rail level crossings, critical decision method interviews with drivers, a diary study of road user behavior, train driver discussions and in-cab familiarization, and SME workshops involving multiple rail safety stakeholders (see Salmon et al., 2016 for a full description).

Step 3: Data Transcription and Analysis

Once the data are collected it should be transcribed as appropriate. Analysis of the data should then proceed. This can take various forms depending on what data collection approaches are used. For example, when concurrent verbal protocols are used, the data analysis process might involve coding the data to identify goals, sub-goals, tasks, operations, and performance-shaping factors. Similarly, CDM interview transcripts are typically analyzed thematically to identify goals, tasks, operations, the information used during task performance, and factors influencing behavior (e.g., Mulvihill et al., 2016).

Step 4: Determine Overall Goal

A flowchart depicting the HTA goal decomposition and description procedure is presented in Figure 16.1. The flowchart covers steps 4 to 6 of this guidance.

The analysis process begins by specifying an overall goal for the task or system under analysis. This goal is placed centrally at the top of the hierarchy and is labelled 0. For example, in the first rail level crossing case study example, the overall goal was expressed as:

0. *Safe and efficient de-confliction of road and rail traffic*

Step 5: Identify and Record Sub-goals

The next step involves decomposing the overall goal down into a series of meaningful sub-goals (four or five is normally optimal; however, this depends on the task/system under analysis and can extend beyond ten). These sub-goals should be numbered 1, 2, 3, 4, and so on. Where different sub-goals are undertaken by different actors it is useful to make a note of this within the sub-goal description. For the rail level crossing case study, the overall goal of *Safe and efficient de-confliction of road and rail traffic* was decomposed into the following sub-goals:

1. *Control vehicle (road users, train driver)*
2. *Detect presence of train (rail level crossing)*
3. *Detect presence of railway level crossing (train driver, road users)*
4. *Detect presence of road users (train driver)*
5. *Announce presence of train (rail level crossing, train driver)*
6. *Detect presence of train (road users)*
7. *Stop vehicle at railway level crossing (road user)*
8. *Deactivate rail level crossing warnings/controls (railway level crossing)*
9. *Traverse rail level crossing (train driver, road user)*

Step 6: Sub-goal Decomposition

Next, the sub-goals identified in step 5 should be broken down into further sub-goals and operations, according to the goals and tasks required. This process continues until

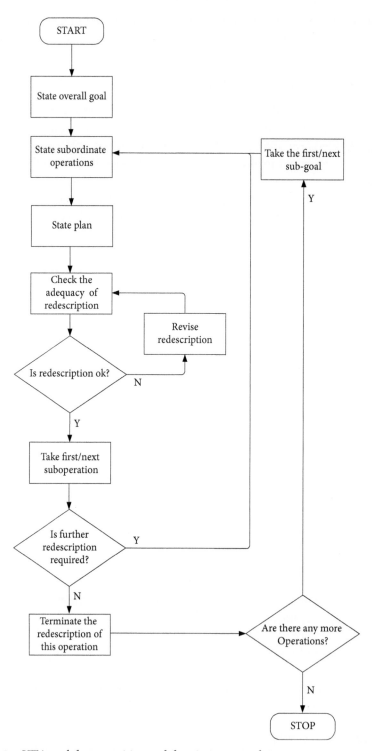

FIGURE 16.1 HTA goal decomposition and description procedure.

an appropriate set of end operations is reached or the level of granularity required for the analysis is achieved. Note, the bottom level of any branch in a HTA should always be a series of operations. Whereas everything above an operation specifies goals, operations state what actually needs to be done (e.g., check traffic light, turn wheel, change gears). Thus, operations are the actions to be made by the operator (whether human or non-human). For the rail level crossing example, sub-goal 3 *Detect presence of railway level crossing* was decomposed as follows:

3. *Detect presence of rail level crossing (train driver, road users)*
3.1. Check whistle board (train driver)
3.1.1. Look for whistle board
3.1.2. Detect whistle board
3.1.3. Interpret whistle board
3.2. Check rail level crossing road signage (road user)
3.2.1. Look for road signage
3.2.2. Detect road signage
3.2.3. Interpret road signage
3.3. Check rail level crossing road markings (road user)
3.3.1. Scan road ahead of vehicle
3.3.2. Detect rail level crossing road markings
3.3.3. Interpret rail level crossing road markings
3.4. Traverse rumble strip (road user)
3.5. Check rail level crossing (road user)
3.5.1. Scan area in front of vehicle
3.5.2. See railway level crossing

Step 7: Plans Analysis

Once all sub-goals have been fully decomposed and described, the plans that dictate the sequencing of goals, sub-goals, and operations need to be added. A simple linear plan would say "Do 1, then 2, and then 3, then EXIT." Once the plan is completed, the operator returns to the superordinate level and repeats the process for other decomposed sub-goals. There are various forms that plans can take. An overview of different plan types is presented in Table 16.1. Further, the method is flexible in that new types of plans can be added should the analysis require it.

Table 16.1 HTA plan types

Plan	Example
Linear	Do 1, then do 2, then do 3, then EXIT
Non-linear	Do 1, 2, and 3 in any order, then EXIT
Simultaneous	Do 1, then 2 and 3 together, then EXIT
Branching	Do 1, if *x* is present, then do 2 and 3, then EXIT; if *x* is not present then do 4, then EXIT
Cyclical	Do 1, then do 2, then do 3, and repeat until *X*, then EXIT
Selection	Do 1, then do 2 or 3 as required

For the railway level crossing example, the plan for *Detect presence of railway level crossing* was defined as "Plan 3: Do 3.1 (train driver) and 3.2–3.5. in any order (road user) until the presence of a railway level crossing is detected then EXIT."

Step 8: Construct HTA Diagram

Once the goals, sub-goals, operations, and plans are exhausted, the next step involves creating the HTA diagram. This is typically a tree diagram (see Figure 16.2) or a tabular representation (see Table 16.3). Various HTA software tools are available to support this process; however, drawing packages such as Microsoft Visio can also be used.

Step 9: SME Review

Once the initial draft is complete it is useful to have various SMEs review it. The HTA should then refined based on SME feedback. It is normal practice for the HTA to go through many iterations before it is finalized.

Step 10: Conduct Additional Analyses Using HTA Extension Methods

Depending on the aims of the analysis, additional methods can be used to explore the task or system under analysis further. As described earlier, examples include error identification methods (e.g., SHERPA; Embrey, 1986), risk assessment methods (e.g., NET-HARMS; Dallat et al., 2018), interface design assessment (Kirwan & Ainsworth, 1992), and allocation of functions analysis (Marsden & Kirby, 2005). Depending on the analysis aims during this step, analysts should take the HTA output and apply relevant analysis methods to it. Guidance on how to use a series of HTA-based analysis methods can be found in Stanton et al. (2013).

Step 11: Construct Task Network

A more recent extension of the HTA method involves converting the HTA into a task network depicting the relationships between high-level goals and tasks (e.g., Salmon, Lenne, Walker, Stanton, & Filtness, 2014; Stanton et al., 2013; Stanton, 2014; Stanton & Harvey, 2017). Typically, this involves taking the subordinate decomposition level of

Table 16.2 Analysis rules to support the construction of HTA-based task networks

Task network	Nodes	Relationships	Examples
	Represent high-level goals/tasks that are required during the task or system under analysis. These should be extracted from the subordinate goals level of the HTA	Represent instances where the conduct of one high-level grouping of tasks (i.e., task network node) influences, is undertaken in combination with, or is dependent on another group of tasks.	In rail level crossing system operation, the nodes "Select site for upgrade" and "Announce upgrade program" are linked because the upgrade program announcement cannot be made until sites have been selected.

the HTA and representing it in the form of a network. Table 16.2 provides a set of analysis rules to support the construction of task networks.

Step 12: Analyze Task Network

Task networks are typically analyzed using a series of quantitative network analysis metrics (e.g., Salmon et al., 2014; Stanton, 2014). Although many metrics exist, the following metrics are typically used to analyze task networks:

1. *Network density (overall network).* Network density represents the level of interconnectivity of the task network in terms of relations between tasks. Density is expressed as a value between 0 and 1, with 0 representing a task network with no connections between tasks, and 1 representing a task network in which every task is related to every other task (Kakimoto, Kamei, Ohira, & Matsumoto, 2006; cited in Walker et al., 2011). Higher density values are indicative of a well-connected network in which tasks are tightly coupled.
2. *Sociometric status (individual nodes).* Sociometric status provides a measure of how *busy* a task is relative to the total number of tasks within the task network under analysis (Houghton et al., 2006). In the present analysis tasks with sociometric status values greater than the mean sociometric status value plus one standard deviation are taken to be *key* (i.e., most connected) tasks within the network.
3. *Centrality (individual nodes).* Centrality is used to examine the standing of a task within a task network based on its geodesic distance from all other tasks in the network (Houghton et al., 2006). Central tasks represent those that are closer to the other task in the network as, for example, the interaction between one task and another task in the network would require an interaction with few other tasks (i.e., the tasks are directly related). Houghton et al. (2006) point out that well-connected nodes can still achieve low centrality values as they may be on the periphery of the network. For example, in the case of the task networks nodes with higher centrality status values are those that are closest to all other tasks in the network as they have direct rather than indirect links with them.

Analysis of the task network involves creating a matrix describing the tasks and relationships between them and then using a social network analysis software tool such as Agna to calculate the chosen network analysis metrics.

CASE STUDY EXAMPLE: STUDYING BEHAVIOR OF RAILWAY LEVEL CROSSINGS AND OF RAILWAY LEVEL CROSSING SYSTEMS

Research Context

Collisions between trains and vehicles and pedestrians at rail level crossings continue to represent a major road and rail safety issue. In Australia, for example, there were ninety-seven fatalities resulting from collisions between trains and vehicles at rail level crossings between 2000 and 2009. Further, rail level crossing fatalities account for almost half of all Australian rail fatalities (ONRSR, 2016). The problem is of a similar magnitude in the United States of America and Europe (Evans, 2011; Federal Railroad Administration, 2013).

As well as the personal and societal impacts, the financial burden of rail level crossings is significant. The annual cost of rail level crossing incidents in Australia was recently estimated to be over one hundred million dollars, taking into account human and property damage and other costs such as emergency service attendance, delays, investigation, and insurance (Tooth & Balmford, 2010). Given the combined personal, societal, and economic burden created by rail level crossing collisions, they continue to represent an area of strategic importance for road and rail safety authorities (Read et al., 2017).

Recent research has argued that rail level crossing collisions represent a systems problem (Read et al., 2013; 2017; Salmon et al., 2016; Stefanova et al., 2015). Analyses of collisions provide evidence that they have a complex web of interrelated causes spanning overall road and rail systems. For example, a systems analysis of the Kerang semi-trailer truck and train collision in 2007 identified various systemic control and feedback failures that led to diminished situation awareness on behalf of the truck driver (Salmon et al., 2013). These included a failure of the warnings to alert the driver to the presence of the train, a limited rail level crossing risk assessment process, and communications failures regarding risk and near misses at the crossing involved. Salmon et al. (2013) argued that interventions to improve rail level crossing safety should focus on strengthening controls both at the rail level crossing itself (e.g., warnings) and within the wider rail level crossing system (e.g., improved risk assessment processes).

As such, safety improvements can only be driven by the application of systems analysis methods for understanding railway level crossing operation. As part of a program of research which was undertaken to develop interventions designed to improve railway level crossing safety (Read et al., 2017), the authors used HTA to describe and analyze:

1. the interaction between vehicles and trains at rail level crossings; and
2. railway level crossing system operation.

HTA of Safe De-confliction of Road and Rail Traffic at Active Rail Level Crossings

To support the evaluation of current rail level crossing environments and the design of new rail level crossing environments, a HTA describing the interactions between vehicles and trains at actively controlled rail level crossings was developed. The intention was to develop a description of rail level crossing behavior as well as how expert on-road users (i.e., drivers) negotiate rail level crossings.

The process adopted, including data sources used, was described earlier in the practical guidance section. An extract of the HTA tree diagram is presented in Figure 16.2. The full HTA is subsequently presented in tabular form in Table 16.3.

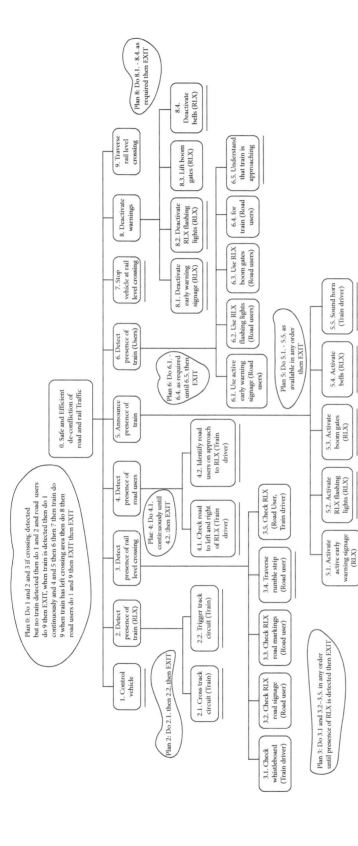

FIGURE 16.2 Rail level crossing sub-goal decomposition (note nodes without line underneath them are decomposed further in the full analysis).

Table 16.3 Safe de-confliction of road and rail traffic HTA in tabular format (RLX = Rail Level Crossing)

0. Safe and Efficient de-confliction of road and rail traffic

Plan 0: Do 1 and 2 and 3 if crossing detected but no train detected then do 1 and 2 and road users do 9 then EXIT, when train is detected then do 1 continuously and 4 and 5 then 6 then 7 then train do 9 when train has left crossing area then do 8 then road users do 1 and 9 then EXIT then EXIT

1. (Road users, train driver) Control vehicle
2. Detect presence of train (RLX)

Plan 2: Do 2.1 then 2.2 then EXIT
2.1. Cross track circuit (Train)
2.2. Trigger track circuit (Train)

3. Detect presence of rail level crossing (train driver, road users)
Plan 3: Do 3.1 and 3.2-3.5 in any order until presence of RLX is detected then EXIT
3.1. Check whistle board (train driver)
Plan 3.1: Do 3.1.1 then 3.1.2 then 3.1.3 then EXIT
3.1.1. Look for whistle board
3.1.2. Detect whistle board
3.1.3. Interpret whistle board
3.2. Check RLX road signage (road user)
Plan 3.2: Do 3.2.1 then 3.2.2 then 3.2.3 then EXIT
3.2.1. Look for road signage
3.2.2. Detect road signage
3.2.3. Interpret road signage
3.3. Check RLX road markings
Plan 3.3: Do 3.3.1 then 3.3.2 then 3.3.3 then EXIT
3.3.1. Scan road ahead of vehicle
3.3.2. Detect RLX road markings
3.3.3. Interpret RLX road markings
3.4. Traverse rumble strip
3.5. Check RLX (Road User)
Plan 3.5. Do 3.5.1 then 3.5.2 then EXIT
3.5.1. Scan area in front of vehicle
3.5.2. See RLX

4. Detect presence of road users (Train driver)
Plan 4: Do 4.1 continuously until 4.2 then EXIT
4.1. Check road to left and right of crossing
4.2. Identify road users on approach to crossing

5. Announce presence of train (RLX, signage, train driver)
Plan 5: Do 5.1-5.5 as available in any order then EXIT
5.1. Activate active early warning signage (RLX)
5.2. Activate RLX flashing lights (RLX)
5.3. Activate RLX boom gates (RLX)
5.4. Activate bells (RLX)
5.5. Sound horn (train driver)

6. Detect presence of train (road users)
Plan 6: Do 6.1-6.4 as required until 6.5 then EXIT
6.1. Use active early warning signage

Plan 6.1: Do 6.1.1 then 6.1.2 then 6.1.3 then EXIT
6.1.1. Look for signage
6.1.2. Detect flashing lights
6.1.3. Interpret flashing lights
6.2. Use RLX flashing lights
Plan 6.2: Do 6.2.1 then 6.2.2 then 6.2.3 then EXIT
6.2.1. Look for flashing light assembly
6.2.2. Detect flashing lights assembly
6.2.3. Interpret flashing lights assembly
6.3. Use RLX boom gates
Plan 6.3: Do 6.3.1 then 6.3.2 then 6.3.3 then EXIT
6.3.1. Look for rail level crossing
6.3.2. Detect rail level crossing
6.3.3. Interpret rail level crossing
6.4. Look for train
Plan 6.4: Do 6.4.1 then 6.4.2 then 6.4.3 then EXIT
6.4.1. Look for train tracks either side of RLX
6.4.2. Detect approaching train
6.4.3. Interpret approaching train
6.5. Understand that train is approaching

7. Stop vehicle at rail level crossing (road user)

8. Deactivate RLX warnings/controls
Plan 8: Do 8.1–8.4 then EXIT
8.1. Deactivate active early warning signage (RLX)
8.2. Deactivate RLX flashing lights (RLX)
8.3. Lift RLX boom gates (RLX)
8.4. Deactivate bells (RLX)

9. Traverse rail level crossing (train driver, road user)

The HTA shows the nine subordinate goals underpinning the main goal of *Safe and efficient de-confliction of road and rail traffic*. Assuming the driver maintains control of the vehicle throughout (see Walker et al., 2015 for full HTA of the driving task), the subordinate goals initially include three key detection tasks:

- The rail level crossing warning system detecting the presence of the approaching train;
- The vehicle driver and train driver detecting the presence of a rail level crossing; and
- The train driver detecting the presence of road users at the crossing.

Following its detection of the train, the rail level crossing warning system announces the presence of a train through active warnings such as flashing lights and boom gates. The onus is then placed on the driver to detect the presence of the train (either through the warnings or through seeing the train) and stop appropriately. Once the train traverses the rail level crossing, the warning system de-activates and the driver can proceed.

The subordinate goals are decomposed to reveal sub-goals and associated operations. The description of the system goals and sub-goals provided by the HTA demonstrates the amount of activity and interdependence of activity occurring at a rail level crossing. This demonstrates the flexibility of HTA beyond considering the road user perspective alone.

In relation to expertise specifically, the HTA describes the goals, sub-goals, and operations that expert users engage in when negotiating actively controlled rail level crossings. An interesting finding from the HTA is the variability in expert behavior at rail level crossings. For example, to determine that a train is approaching, expert users use the active early warning signage (if present), flashing lights, the boom gates, the train itself, or different combinations of this information depending on the crossing and its environment. Whilst this is useful as a base description of expert behavior, it can also inform the development of training and education programs as well as the design of new rail level crossing environments. For example, the HTA shows the different checks that expert users make to determine (a) the presence of a rail level crossing and (b) the presence of a train. This is useful to develop *situation awareness requirements* to inform driver training programs and education campaigns as it shows what information users require to negotiate rail level crossings safely. Indeed, the HTA becomes particularly powerful when additional naturalistic studies are used to explore aspects of behavior further. For example, as part of the same research program Salmon et al. (2013) used an on-road study to examine differences in the behavior of novice and experienced drivers at rail level crossings. They found that the experienced drivers undertook more checking behaviors, in particular checks of the railway tracks to the left and right of the roadway for approaching trains. In addition, they found that novice drivers appeared to expect warning devices (e.g., boom gates, flashing lights) at all crossings (even those that did not have them), and to engage less in checking for trains as a result. Salmon et al. (2013) concluded that this will be an issue at rail level crossings that do not provide an active warning where novice drivers might approach expecting a warning of a train but not receive one. Combining the HTA with study evidence therefore suggests that non-experts should be encouraged to check for trains as well as checking rail level crossing signage and warning devices.

As discussed, the analysis can also be used to inform more in-depth analyses of expert behavior. For example, the SHERPA error identification approach can be used to identify the errors that users of differing levels of expertise may make when negotiating rail level crossings. In addition, interviews with experts can be undertaken to identify strategies that they use to either prevent errors identified by SHERPA or detect and recover from them when they occur. This information can be used to inform the design of training programs and standard operating procedures as well as new technologies designed to enhance error tolerance. For example, Read et al. (2017) describe how SHERPA was used in conjunction with the *Safe and efficient de-confliction of road and rail traffic* HTA to identify the range of errors that drivers could make when negotiating rail level crossings. They identified a total of 92 potential errors. Around a third were action errors (e.g., driver fails to slow down, rail level crossing fails to activate early warning signage), around a third were checking errors (e.g., driver fails to look at flashing light assembly, train driver fails to look for road users either side of the crossing), and around a third were retrieval errors (e.g., driver fails to interpret flashing lights, driver misreads signage). From the HTA, the following tasks had the greatest number of potential errors associated with them:

- Detect presence of train (road user); and
- Detect rail level crossing (road user).

This suggests that driver training and education programs and design interventions should focus on enhancing drivers' capacity to detect the presence of rail level crossings and approaching trains. An interesting finding from the SHERPA analysis was that there appears to be little redundancy or fail safes present in current rail level crossing environments to cope with high criticality errors such as drivers failing to detect or interpret flashing lights.

HTA of Rail Level Crossing System Operation

As discussed in the Introduction, increasingly task analysis methods are being used to describe and understand the behavior of overall systems. In the present rail level crossing case study, HTA was also used to develop a description of rail level crossing system operation. The resulting HTA describes the goals, sub-goals, and operations required when designing, implementing, operating, and removing (i.e., grade separating) rail level crossings in Victoria, Australia. The intention was to produce a HTA that would complement the safe de-confliction of road and rail traffic rail level crossing HTA described in the previous section and would inform the identification and development of strategies designed to optimize rail level crossing safety management.

Initially, an Actormap (Rasmussen, 1997; Svedung & Rasmussen, 2002) was developed to identify which stakeholders currently contribute to rail level crossing system operation. The Actormap of the rail level crossing system in Victoria, Australia is presented in Figure 16.3. The Actormap details each level of the rail level crossing system along with the actors and organizations who share the responsibility for rail level crossing safety. While this is useful in of itself the analysis informed development of the HTA of rail level crossing system operation as it specified which actors and organizations currently work within the rail level crossing system. Development of the HTA involved identifying what goals, tasks, and operations the different actors and organizations contribute to.

The resulting HTA was subsequently converted into a task network to show the relationships between the high-level goals and to enable further analysis of the system via network analysis metrics (steps 11 and 12 in the practical guidance section). The task network for rail level crossing system operation is presented in Figure 16.4.

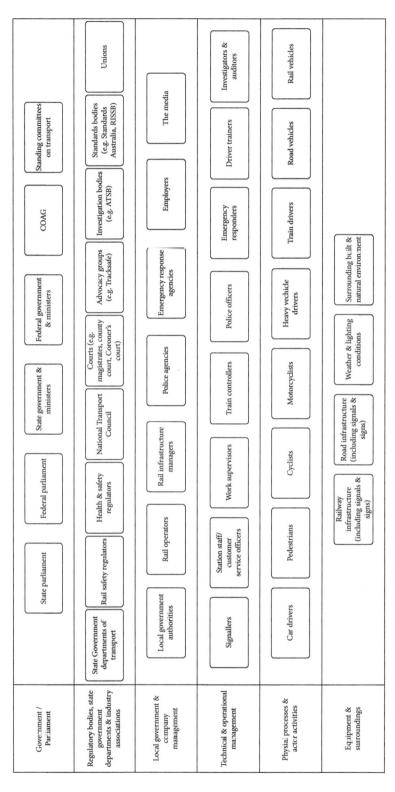

FIGURE 16.3 Actormap of the rail level crossing system in Victoria, Australia.

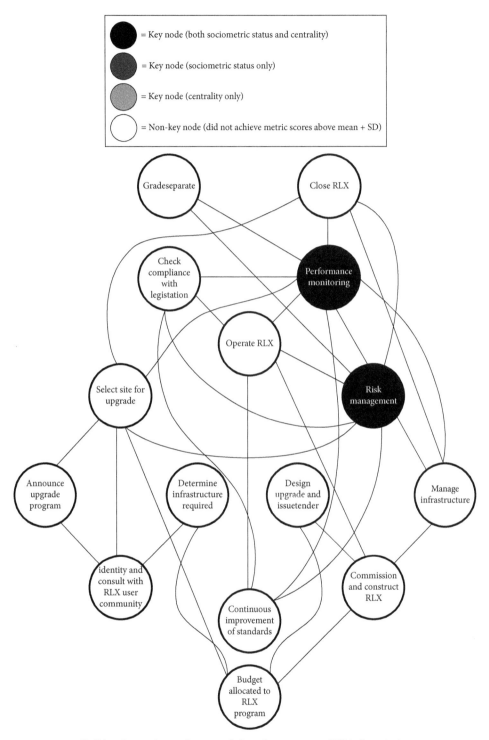

FIGURE 16.4 Rail level crossing task network based on systems HTA description.

Table 16.4 Task network analysis outputs (cells with italic text represent nodes with values above the mean + standard deviation)

Network metrics	Key nodes sociometric status (mean = 0.4, SD = 0.22)	Key nodes centrality (mean = 7.81, SD = 1.56)
Nodes = 15	*Risk management = 1.0*	*Risk management = 10.88*
Edges = 42	*Monitor performance = 0.79*	*Monitor performance = 9.98*
Density = 0.2	Operate rail level crossing = 0.5	Operate rail level crossing = 9.23
	Check compliance with legislation = 0.43	Close rail level crossing = 8.96
	Select site for upgrade = 0.43	Check compliance with legislation = 8.6
	Manage infrastructure = 0.43	Select site for upgrade = 8.83
		Manage infrastructure = 8.12

The outcomes for the task network analysis are presented in Table 16.4. For the sociometric status and centrality analysis Table 16.4 includes the values for the nodes that scored above the mean sociometric status and centrality values for the network. Nodes that achieved values above the mean + standard deviation for sociometric status and centrality are in italics. This approach has previously been used to identify key nodes within task networks (e.g., Houghton et al., 2006; Salmon et al., 2014; Stanton, 2014; Stanton, Harris, & Starr, 2016; Stanton & Harvey, 2017).

As shown in Figure 16.4, rail level crossing system operation is underpinned by a network of fifteen core sub-goals. Analysis of the task network in Table 16.2 reveals a network density of 0.2, which is indicative of a relatively loosely connected network in which many tasks are undertaken largely independent of one another. The sociometric status and centrality analyses reveal that *risk management* and *monitor performance* are the key tasks within rail level crossing system operation. The findings therefore suggest that road and rail stakeholders should attempt to strengthen their risk management and performance monitoring activities. This could be driven through the development of more comprehensive risk management processes (e.g., incorporating human factors data) and performance monitoring controls (e.g., incident and near miss reporting systems, regular behavioral assessments, and increasing data collection mechanisms at rail level crossings such as closed circuit television cameras).

Strengths and Weaknesses of HTA

While there are many benefits associated with HTA, the method is also not without its flaws. A summary of its main strengths and weakness is given in the following.

Main Strengths

- HTA requires minimal training and is relatively straightforward to apply;
- The process of constructing the HTA allows the analyst(s) to develop an in-depth understanding of the task/scenario/system under analysis;
- Various other ergonomics analysis methods can be undertaken using HTA outputs, e.g., error identification, risk assessment, interface design, allocation of functions analysis, situation awareness requirements analysis, and teamwork assessment;
- HTA is highly flexible and can be used in any domain for multiple purposes;
- HTA is relatively quick to apply compared to other systems analysis methods such as cognitive work analysis (Vicente, 1999) and EAST (Stanton et al., 2013);
- The output provides a comprehensive description of the task/scenario/system under analysis;
- HTA is capable of describing the behavior of individuals, teams, organizations, and entire systems; and
- When used for the study of expertise, HTA provides an in-depth description of expert behavior and can be used to support further in-depth analyses of expertise, e.g., error identification, situation awareness requirements analysis.

Main Weaknesses

- HTA provides mainly descriptive information rather than analytical information;
- The quality of expertise-related information produced is largely dependent on the data collection approaches used. For example, CDM interview data are extremely useful for studies of expertise, whereas observational study only is less useful (particularly regarding the cognitive aspects of expertise);
- The outputs do not provide direct design recommendations—further interpretation is required;
- HTA requires a high degree of access to the system and SME under analysis;
- HTA can become time-consuming and complex to apply when used to describe complex tasks/scenarios/systems;
- The initial data collection phase can be time consuming and resource intensive;
- The reliability of the method may be questionable in some instances. While it has been shown to demonstrate acceptable levels of validity, it has tended to score poorly in reliability testing (Stanton et al., 2013); and
- While software support exists, an optimal tool is yet to emerge.

SUMMARY

Almost half a century since it first emerged, HTA continues to represent one of the most popular task analysis and indeed ergonomics methods. While it can be used for many purposes, HTA provides a useful approach for studying expertise. Fruitful applications lie in the description and analysis of expert behavior and in examining differences between work as imagined (e.g., work procedures) and work as done (e.g., expert behavior in context). In addition, HTA acts as the input to various human factors analysis methods that provide in-depth analyses of expert behavior (e.g., error identification, situation awareness requirements analysis, allocation of functions analysis). This chapter has presented practical guidance on how to apply HTA when attempting to understand behavior within sociotechnical systems. Notably, the flexibility of HTA is such that it has been able to move with the times. In line with the increasingly relevant and popular systems ergonomics approach, HTA is capable of describing the behavior of entire sociotechnical systems as well as the behavior of individuals, teams, and organizations. As systems become more technology centric and more complex, and the requirement for systems analyses continues to expand, this flexibility will ensure that HTA remains both relevant and popular. Future areas in which HTA will no doubt be applied include some of society's most pressing issues, including artificial intelligence, automation, regulation, and cybersecurity. As such the reader is encouraged to apply HTA in their own contexts and problem spaces and to continue to expand and evolve the method.

References

Al-Hakim, L., Maiping, T., & Sevdalis, N. (2014). Applying hierarchical task analysis to improving the patient positioning for direct lateral interbody fusion in spinal surgery. *Applied Ergonomics* 45(4), 955–966

Annett, J., et al., (1971). *Task analysis*. Department of Employment Training Information Paper 6. London: HMSO.

Annett, J. (2004). Hierarchical task analysis. In D. Diaper & N. A. Stanton (Eds), *The handbook of task analysis for human-computer interaction* (pp. 67–82). Mahwah, NJ: Lawrence Erlbaum Associates,.

Dallat, C., Salmon, P. M., & Goode, N. (2018). Identifying risks and emergent risks across sociotechnical systems: The NETworked Hazard Analysis and Risk Management System (NET-HARMS). *Theoretical Issues in Ergonomics Science* 19(4), 456–482.

Demirel, D., Butler, K. L., Halic, T., Sankaranarayanan, G., Spindler, D., Cao, C., ... deMoya, M. A. (2016). A hierarchical task analysis of cricothyroidotomy procedure for a virtual airway skills trainer simulator. *American Journal of Surgery* 12(3), 475–484.

Diaper, D., & Stanton, N. A. (2004). *Handbook of task analysis in human-computer interaction*. Mahwah, NJ: Lawrence Erlbaum Associates.

Embrey, D. E. (1986). SHERPA: A systematic human error reduction and prediction approach. In *Proceedings of the International Topical Meeting on Advances in Human Factors in Nuclear Power Systems, Knoxville, Tennessee American Nuclear Society La Grange Park, Illinois 60525.*

Evans, A. W. (2011). Fatal accidents at railway level crossings in Great Britain 1946–2009. *Accident Analysis & Prevention* 43(5), 1837–1845.

Federal Railroad Administration. (2013). *Railroad safety statistics 2012 preliminary annual report*. Department of Transportation, Washington, DC.

Hollnagel, E., Leonhardt, J. Licu, T. & Shorrock, S. (2013). From Safety-I to Safety-II: A white paper. Brussels, Belgium: EUROCONTROL. Available at http://skybrary.aero/bookshelf/books/2437.pdf. Accessed 17 November 2017.

Houghton, R. J., Baber, C., McMaster, R., Stanton, N. A., Salmon, P. M., Stewart, R., & Walker, G. H. (2006). Command and control in emergency services operations: a social network analysis. *Ergonomics 49*, 1204-1225.

Independent Transport Safety Regulator. (2011). Level crossing accidents in Australia. *Transport Safety Bulletin*. Sydney: ITSR. Available from: http://www.onrsr.com.au/__data/assets/pdf_file/0020/2963/Transport-safety-bulletin-Issue-2-Level-crossing-accidents-in-Australia-August-20112.PDF. Accessed February 10, 2015.

Kakimoto, K., Kamei, Y., Ohira, M., & Matsumoto, K. (2006). Social network analysis on communications for knowledge collaboration in OSS communities. In *Proceedings of the 2nd International Workshop on Supporting Knowledge Collaboration in Software Development* (KCSD2006) (pp. 35-41). Tokyo, Japan.

Karsh, B. T., Waterson, P., & Holden, R. J. (2014). Crossing levels in systems ergonomics: A framework to support "mesoergonomic" inquiry. *Applied Ergonomics 45*(1), 45-54.

Kirwan, B., & Ainsworth, L. K. (1992). *A guide to task analysis*. London, UK: Taylor & Francis.

Klein, G. A., Calderwood, R., & Macgregor, D. (1989). Critical decision method for eliciting knowledge. *IEEE Transactions on Systems, Man, and Cybernetics: Systems 19*(3), 462-472.

Marsden, P., & Kirby, M. (2005). Allocation of functions. In: N. A. Stanton, A. Hedge, K. Brookhuis, E. Salas, & H. Hendrick (Eds), *Handbook of human factors and ergonomics methods*. London: Taylor & Francis.

Mulvihill, C., Salmon, P. M., Beanland, V., Lenne, M. G., Read, G. J. M., Walker, G. H., & Stanton, N. A. (2016). Using the decision ladder to understand road user decision making at actively controlled rail level crossings. *Applied Ergonomics 56*, 1-10.

Office of the National Rail Safety Regulator (ONRSR). (2016). *Rail Safety Report 2015-2016*. Adelaide, Australia: Office of the National Rail Safety Regulator,

Rasmussen, J. (1997). Risk management in a dynamic society: A modelling problem. *Safety Science 27*(2/3), 183-213.

Read., G. J. M., Beanland, V., Lenne, M. G., Stanton, N. A., & Salmon, P. M. (2017). *Integrating human factors methods and systems thinking for transport analysis and design*. Boca Raton, FL: CRC Press.

Read, G. J. M., Salmon, P. M., & Lenne, M. G. (2013). Sounding the warning bells: the need for a systems approach to rail level crossing safety. *Applied Ergonomics 44*, 764-774.

Salmon, P. M., Lenne, M. G., Mulvihill, C., Young, K., Cornelissen, M., Walker, G. H., & Stanton, N. A. (2016). More than meets the eye: using cognitive work analysis to identify design requirements for safer rail level crossing systems. *Applied Ergonomics 53* (Part B), 312-322.

Salmon, P. M., Lenne, M. G., Walker, G. H., Stanton, N. A., & Filtness, A. (2014). Using the Event Analysis of Systemic Teamwork (EAST) to explore conflicts between different road user groups when making right hand turns at urban intersections. *Ergonomics 57*(11), 1628-1642.

Salmon, P. M., Read, G., Stanton, N. A., & Lenné, M. G. (2013). The crash at Kerang: Investigating systemic and psychological factors leading to unintentional non-compliance at rail level crossings. *Accident Analysis and Prevention 50*, 1278-1288.

Salmon, P. M., Stanton, N. A., Jenkins, D. P., & Walker, G. H. (2010). Hierarchical task analysis versus cognitive work analysis: comparison of theory, methodology, and contribution to system design. *Theoretical Issues in Ergonomics Science* 11(6), 504–531.

Salmon, P. M., Walker, G. H., Read, G. J. M., Goode, N., & Stanton, N. A. (2017). Fitting methods to paradigms: are ergonomics methods fit for systems thinking? *Ergonomics* 60(2), 194–205.

Stanton, N. A. (2006). Hierarchical task analysis: Developments, applications, and extensions. *Applied Ergonomics* 37(1), 55–79.

Stanton, N. A. (2014). Representing distributed cognition in complex systems: how a submarine returns to periscope depth. *Ergonomics* 57(3), 403–418.

Stanton, N.A., Harris, D., & Starr, A. (2016). The future flight deck: Modelling dual, single and distributed crewing options. *Applied Ergonomics* 53, 331–342.

Stanton, N. A. & Harvey, C. (2017). Beyond human error taxonomies in assessment of risk in sociotechnical systems: a new paradigm with the EAST "broken-links" approach. *Ergonomics* 60(2) 221–233.

Stanton, N. A., Salmon, P. M., Rafferty, L., Walker, G., Baber, C., & Jenkins, D. P. (2013). *Human factors methods: A practical guide for engineering and design*, second edition. Aldershot, UK: Ashgate.

Stanton, N. A., Salmon, P. M., Walker, G. H., & Jenkins, D. P. (2010). *Human factors and the design and evaluation of central control room operations*. Boca Raton, FL: Taylor & Francis.

Stefanova, T., Burkhardt, J.-M., Filtness, A., Wullems, C., Rakotonirainy, A., & Delhomme, P. (2015). Systems-based approach to investigate unsafe pedestrian behavior at level crossings. *Accident Analysis & Prevention*, 81(0), 167–186.

Svedung, I., & Rasmussen, J. (2002). Graphic representation of accident scenarios: mapping system structure and the causation of accidents. *Safety Science* 40, 397–417.

Tooth, R., & Balmford, M., 2010. *Railway Level Crossing Incident Costing Model*. Canberra: LECG for the Rail Industry Safety and Standards Board.

Vicente, K. J. (1999). *Cognitive work analysis: Toward safe, productive, and healthy computer-based work*. Mahwah, NJ: Lawrence Erlbaum Associates.

Walker, G. H., Salmon, P. M., Bedinger, M., & Stanton, N. A. (2017). Quantum ergonomics: shifting the paradigm of the systems agenda. *Ergonomics* 60(2), 157–166.

Walker, G. H., Stanton, N. A., & Salmon, P. M. (2011). Cognitive compatibility of motorcyclists and car drivers. *Accident Analysis & Prevention* 43(3), 878–888.

Walker, G. H., Stanton, N. A., & Salmon, P. M. (2015). *Human factors in automotive engineering and technology*. Aldershot, UK: Ashgate.

CHAPTER 17

A HISTORICAL PERSPECTIVE ON INTROSPECTION

Guidelines for Eliciting Verbal and Introspective-Type Reports

PAUL WARD, KYLE WILSON, JOEL SUSS,
WILLIAM DOUGLAS WOODY, AND
ROBERT R. HOFFMAN

INTRODUCTION

> The course that an observer follows will vary in detail with the nature of the consciousness observed, with the purpose of the experiment, with the instruction given by the experimenter. Introspection is thus a generic term, and covers an indefinitely large group of specific methodical procedures. (Titchener, 1912b, p. 485)

INTROSPECTION was classically defined as a conscious, deliberate activity of reflecting on one's own thoughts and feelings. It is an inward-looking mental activity that relies on the common metaphor of *knowing as seeing*, where an individual attempts to express the conscious contents of their current or very recent flow of thought. Some psychologists of the twentieth century argued that introspection is the psychological equivalent of *inspective* methods used by natural scientists (Titchener, 1912b; Watson, 1920). A variety of specific methods have been adduced over the centuries, ranging from methods that are explicitly described as types of prayer to those that avoid the term introspection, such as verbal report;[1] originally, an effort to avoid the charge of being mentalistic and therefore not genuine science.

Some introspective methods are forms of what is today called *cognitive task analysis* (CTA). This has been described as a suite of cognitive process-tracing and knowledge elicitation methods designed to identify the perceptual, cognitive Oxford, and meta-cognitive skills, strategies, and knowledge required to perform tasks proficiently, and a means to elicit the informational requirements and contextual factors that constrain effective work (Crandall, Klein, & Hoffman, 2006). CTA has been crucial for understanding expert reasoning, representing expert knowledge, and validating computational models of expert cognition. Moreover, as one of the key tools in any CTA practitioner's toolbox, the products of expert introspection have been used successfully to develop instruction, training, and computational technologies (e.g., decision aids) that accelerate proficiency or support cognitive work.

Although the roots of introspection can be traced back to the Greek philosophers, over the past two centuries introspection has been used to study psychological phenomena. However, with only minimal qualification, the catchall term *introspection* has been used both to refer to—and to reject—a range of methods, despite practical as well as theoretical differences among them. From a philosophy-of-mind perspective, many conditions have been proposed as benchmarks for introspection. For instance, for a verbal utterance to count as introspection, it should be about (a) one's own mental events, (b) current or immediately past thinking, (c) detection of information that is consciously available directly, and (d) deliberate efforts to think about mental events (e.g., see *Encyclopedia Britannica*). Although these conditions are not universally accepted, introspection would typically involve some combination of the majority of them.

These conditions have some practical implications for studying expertise. For instance, can experts introspect about very recent as opposed to immediately past or current thoughts? How recent is very recent? Is any degree of transformation acceptable (e.g., from imagistic to verbal form; from thought sequence to a description of thinking)? Can experts introspect about knowledge related to but not part of current thinking? Can they introspect about defining features of expertise, such as tacit knowledge or perceptually automated processes (e.g., recognition-primed decisions)? For a more complete philosophical dissection of introspection we direct the reader elsewhere (e.g., Lyons, 1986; Schwitzgebel, 2016). Here we take a pragmatic and historical perspective.

This chapter consists of five parts. First, we consider specific contributions of some of the pioneers in psychology of introspective methods, highlighting key motivations and arguments that have spurred methodological evolution over the past 100+ years. Next, we review current methods of thinking aloud. Then we describe the types of verbal reports of thinking used to study expertise today. In the penultimate section, we offer some guidance on using these methods. In the concluding part, we provide a summary of the key recommendations and offer some thoughts on the future of introspection methodology

---

[1] The term *verbal report* was first introduced during the era of behaviorism to avoid any accusation that the study of *verbal behavior* was mentalistic.

that, we hope, will improve the state-of-the-science and escape the legacies of behaviorism. We begin with an abridged history of introspection.

AN ABRIDGED HISTORY OF INTROSPECTION

The Roman Catholic saint, Teresa of Ávila (1515–1582) discussed introspection as a level of prayer (i.e., reflecting on one's thoughts and beliefs with respect to one's relation to God) (Mercer, 2017). Her writings influenced the first generation of experimental psychologists. Franz Brentano and Wilhelm Wundt were amongst the first natural philosophers who advocated for a science of psychology and the use of introspection to understand the psyche. However, they approached psychology with different worldviews, and used distinct methods (Titchener, 1921). Undoubtedly, others paved the way for psychological uses of introspection, including Binet, Fortlage, Fechner, Helmholtz, James, and Stumpf (see Danziger, 1980a, 1980b). Henceforth, we will refer to the psychological use of introspection simply as introspection. We begin by examining Brentano's (1874/2009) contribution.

Franz Brentano

Brentano came to psychology from logic and theology. His psychological worldview had foundations in the active world of subjective experience rather than physiology. This perspective impacted his students (e.g., Stumpf, von Ehrenfels, Husserl) and those he taught and influenced (e.g., Freud and James) (see Broome, Harland, Owen, & Stringaris, 2012; Cohen, 2002; Huemer & Landerer, 2010; Kersten, 1969). As Titchener (1921) argued, Brentano "can claim kinship with every psychologist, of whatever school, who has approached [their] subject from the technically 'empirical' standpoint" (p. 119).

With his experiential emphasis, Brentano (1874/2009) differentiated between two types of introspective methods: *Beobachtung* (inner observation) and *Wahrnehmung* (inner perception). Inner observation involved reasoning while simultaneously observing oneself reason, whereas inner perception was an incidental awareness of one's own mental phenomena (immediately post hoc) with an awareness of the object of thought. Brentano rejected inner observation as a means to study subjective experience. He argued it was impossible to simultaneously direct full attention to the object of reasoning and to reasoning itself, at least when both acts require full attention—which would lead to an untenable degree of circularity and render introspection a pretend psychological method (see also Comte, 1798–1857, cited in James, 1890/1981, p. 188). Hence, this objection was rooted in the difficulties in splitting attention (Lyons, 1986).

Brentano (1874/2009) argued that in inner perception the introspector directs attention toward thoughts or acts just carried out and so awareness is facilitated by immediate memory for them. In essence, through his view of inner perception,

Brentano was advocating for what today might be called immediate retrospection, with the emphasis on immediate so as to avoid difficulties of memory (Lyons, 1986). Of particular relevance to the science of expertise, he asserted that "inner perception of our own mental phenomena... is the primary source of the experiences essential to psychological investigations" (1874/2009, p. 26), and one could learn the most about psychological laws by studying outstanding people and successively more and more complex phenomena. Although it is difficult to find empirical data in Brentano's writing about his use of inner perception, his perspective laid the foundations for introspective studies of expertise.

Wilhelm Wundt

In contrast with Brentano, Wundt started out as a physiologist (Titchener, 1921). "[H]is primary aim in all cases is to describe the phenomena of mind as the physiologist describes the phenomena of the living body" (p. 118). Although Wundt (1882/1906) never acknowledged Brentano's work, he was clearly influenced by the distinction between inner observation and inner perception, despite Titchener's (1921) claim that "[t]here is no middle way between Brentano and Wundt" (p. 108).

Like Brentano, Wundt (1882/1906) advocated for a psychological reformation from *introspection as inner observation* toward *introspection as inner perception*, while simultaneously discounting broad definitions of inner perception as any delayed form of retrospection. He elaborated on Brentano's focus on *immediate* retrospection, by arguing that if the time between stimulus perception (i.e., awareness of thought) and its reproduction (i.e., reporting on the thought) could be greatly reduced, then time for distortion, through reflection or memory decay, could be virtually eliminated.

Wundt was concerned with meeting the conditions of external observation (i.e., inspection) demanded by other scientists. He claimed that, in addition to eliminating memory decay, these conditions could be met by: (a) tightly controlling experimentation and stimulus presentation (using the then-new chronometric technology) (see Evans, 2000); (b) reducing the complexity of experimental tasks to the presentation of very simple stimuli (e.g., color patches), and (c) restricting inner perception to descriptions of specific qualities and intensities of the associated elementary sensations. These were the key characteristics of Wundtian introspection and of sensationism.

Wundt also argued that using an untrained introspector, who had not yet habituated to the task of perceptual awareness of the stimuli being sensed (such that it would not interfere with the target mental activity), would result in a scientifically invalid result. Therefore, he trained participants (usually graduate students or teachers) to introspect using large numbers of practice trials. In essence, Wundt was advocating for the use of a skilled introspector (and the aforementioned conditions) as a means of obtaining valid data on simple tasks, compared to Brentano who argued for valid introspection about a person's expertise, ultimately, on complex tasks (among other topics).

Edward Titchener

A student of Wundt, Titchener, used a version of Wundtian introspection to capture the sensory elements associated with complex experience. Although Titchener "emulated Wundt" (Boring, 1929, p. 504), his approach was also influenced by fellow Englishman John Stuart Mill. Like Mill, Titchener (1912a) initially used the term *self-observation* to refer to his kind of introspection but differentiated pre-critical (i.e., pre-experimental) from critical (i.e., experimental) self-observation. In doing so, he echoed Wundt's point that experimental procedures safeguarded introspection as a viable scientific method. He agreed that training should automate this activity, to the extent that the experienced introspector "*completely forgets to give subjective attention to the state of observing*" (p. 443, italics in original). Hence, in Titchener's mind, this would also overcome the issue of splitting attention (see also Lyons, 1986).

Titchener (1912b) also noted that introspection could be direct or indirect. Direct introspection is when "process and apperception[2] occur together" (p. 491). This happens when introspecting concurrently about an immediately present target or when introspecting about a target through remembered apperception (akin to instantaneous retrospection—awareness immediately *post hoc*—of the observed process; or inner perception). The caveat was that any criticisms were circumvented through both controlled experimentation and skill at introspecting. In contrast, indirect introspection is based on recollection from memory and so constituted a delayed form (i.e., retrospection).

From Titchener's (1912b) perspective, psychological descriptions of sensory processes were the true targets of introspection and the real focus of a descriptive science of psychology. He argued that the functions of perception (e.g., judgments about perceived information) were logical abstractions added *ex post facto* by the introspector via inference and, hence, were not *directly* observable. Consequently, he argued that interpretive explanations of reasoning should be separated from his targets of introspection (see the discussion about *verbal elaborations* in the section "Types of Verbal Reports of Thinking Used to Study Expertise Today"). Wundt (1905, 1913) asserted that the latter were the purview of *Völkerpsychologie* (cultural psychology) and beyond the scope of experimental introspection, albeit not beyond scientific scrutiny. Titchenerian introspection was, therefore, characterized by the critical and direct nature of reporting on sensory experience by trained introspectors, rather than interpretations of the contents of introspection.

The Würzburgers

Oswald Külpe (1893/1909), a former student of Wundt and co-founder (with Karl Marbe) of the Würzburg School, contended that all aspects of psychology (not just

---

[2] *Apperception* was first described as the immediate comprehension of a phenomenal *complex* in relation to one's knowledge. This was contrasted with *apprehension*, which was described as an immediate impression of a phenomenal complex *without attentional demand* (see Wundt, 1896/1897, p. 420).

psychophysical sensations as studied by Wundt and Titchener) could be subjected to introspective investigation. Külpe, Marbe, and other Würzburg colleagues engaged in concerted efforts to extend the field of psychology to the study of higher-order features of cognition, such as judgement, knowing, understanding, problem solving, and interpreting. Their form of introspection was pivotal in leading to a different theory and approach, termed *Denkpsychologie* (i.e., thought psychology), which moved psychology beyond the study of elementary sensations and associations or their reproduction.

Participants trained in *systematische experimentelle selbstbeobachtung* or systematic experimental self-observation (Ach, 1905, p. 23; henceforth Würzburgian introspection) described their experiences by reporting in detail, immediately after stimulus presentation and task completion, the thoughts they had experienced during the experiment. As a side note, Titchener (1912b) translated *selbstbeobachtung* as *introspection*, even though the literal translation is *self-observation*, because there was no German word for introspection.

Using systematic experimental self-observation, Marbe's students, August Mayer and Johannes Orth (1901; see also Marbe, 1901), presented participants with a simple task (e.g., verbally respond to a stimulus word with the first association that comes to mind). Immediately following, they "report[ed] all conscious events which had gone on from the moment the stimulus word was given to the end of the reaction" (see Hoffmann, Stock, & Deutsch, 1996, p. 9). Consistent with Brentano, Narziss Ach (1905) argued against concurrent reporting during the experiment because of the difficulties of trying to experience a stimulus while simultaneously monitoring one's impressions of it, which would require participants to shift their attention back and forth between the two (see Brentano's earlier criticism of inner observation).

Participants were trained to introspect impartially (immediately after task performance) about aspects of conscious thought that were noticed during the experiment, including what they had judged, grasped, and understood, and the connections made. Ach (1905) noted that this was a co-constructive process (i.e., a continuous exchange of thoughts between the experimenter and introspector, see p. 8): The experimenter restates the protocol back to the introspector (see p. 17) and asks prescriptive and undirected questions (to ensure completeness and temporal order) in an unbiased, non-suggestive manner (to avoid leading the participant and to preserve the process). This type of questioning included phrases such as: "What was going on before...?" "Were you conscious of that? "Is that the same as the previous one?" "How are they different?" "What was present at the time?" (p. 17). Experimenter–participant interactions were especially evident when the introspector was untrained. However, with practice, Ach argued that questioning should be barely noticeable and any systematic effects could be offset by coupling introspection with experimentation (Ach, 1905; see also Titchener, 1912b).

Ach further argued that this kind of interactive probing requires skill on the part of the experimenter. Importantly, it permits a full understanding of the protocol and allows the experimenter to capture the "true picture of the experience" (pp. 17–18).

Without it, Ach insisted, an "in-depth analysis would be impossible" (pp. 17–18). Hence, Würzburgian introspection was characterized by interactive but unbiased, immediately post hoc self-observation by trained introspectors, of *all* conscious events experienced during systematic experimentation, especially higher order aspects of thought.

In addition to reacting against the sensationist's view of introspection (e.g., Wundt, Titchener), the Würzburgers used introspection to challenge the then-popular view of psychology as associationism (e.g., Müller, 1911). Associationists viewed cognitive processes as a "system of diffuse reproductions" (Selz, 1924/1981, p. 25).[3] The Würzburgers, on the other hand, used their method to elucidate other types of higher-order thinking, beyond associations, including the awareness of a rule or solution principle, an intention, and the experience of remembering. Ach's (1905) use of Würzburgian introspection lead to the notion of determining tendencies,[4] a mechanism to reflect the perceived meaning of a stimulus of which one is spontaneously aware (referred to as "intuitive symbols of awareness") rather than the literal sense of a stimulus (referred to as "signs") (p. 222).

The Würzburgers' findings of different types of (imageless) thought attracted considerable attention and debate (e.g., see Ach, 1905; Angell, 1906; Binet & Simon, 1908; Stout, 1896; Titchener, 1910). Although the theoretical debate surrounding imageless thought is beyond the scope of the chapter, one of the primary concerns of some objectors was the methods used to elicit it. Wundt (1907), for instance, argued that rather than the experimenter being the real observer, in Würzburgian introspection the introspector was the observer.[5] Following Brentano and Auguste Comte, he asserted that it was impossible for the introspector to engage in observation of the self and reason about an object simultaneously. Further, he protested that the use of *unexpected* stimuli lacked scientific precision. George E. Müller (1911) posed similar criticisms. He argued that reports on complicated processes would be incomplete; post hoc observation would not counter memory loss or selectivity; and similarity (of thoughts) was questionable between post-experiment introspection and thinking during experimentation. Despite Ach's (1905) recommendation that experimenter questioning should be unbiased and non-suggestive, several other researchers rejected this practice. For instance, Wundt (1907) argued that variable questioning introduced bifurcations

---

[3] A *system of diffuse reproductions* is akin to early views of a cognitive or knowledge network, in which recurrent stimulation of a mental event activates residual associations with other events through *reproductive tendencies*—a mechanism akin to modern conceptions of *spreading activation*).

[4] Types of *determining tendencies* (which can be compared to the associationist's *reproductive tendencies*) were mechanisms of higher-order thought that included associative abstraction, determining apperception, including apperceptive fusion and substitution, and determining abstraction.

[5] Wundt's criticism of the introspector (rather than the experimenter) as the real observer—a criticism shared by Watson about the Würzburgian method (see the section "John B. Watson, Behaviorism, and Thinking Aloud")—demonstrated, in part, his narrow view of introspection, and his commitment to promoting a method that could be regarded as scientifically valid as the inspective methods used in physiology (see Titchener, 1921).

in experimentation, making it impossible to replicate. Müller (1911) asserted that questioning was negatively influential. Titchener (1912b) claimed it was dangerous.

In response to these criticisms, the Würzburgers pushed back. Karl Bühler (1908) pointed out that participants were not asked to simultaneously reason and observe as Wundt had charged, but rather to reason then reflect on that reasoning. The perceived problem of replication (due to trial-to-trial differences in questioning) was discounted given others' use of variation of stimuli or task across trials. The Würzburgers' goal was to use psychologically equitable stimuli that permitted comparison yet variation across conditions, rather than overfamiliar stimuli that rendered trial-to-trial comparison meaningless. Last, Bühler argued that using introspection to identify the conscious processes responsible for behavior was better than using *post hoc* experimenter speculation about underlying processes based on reaction times, as was Wundt and Titchener's tendency (see O'Shea & Bashore, 2012). Despite the common perception that Wundt's work was based on introspection, rich introspective data were rarely reported (e.g., in only 4 of 180 studies between 1883 and 1902). Instead, its use was to explain individual differences in objective data and provide manipulation checks (Danziger, 1980a, 1980b).

John B. Watson, Behaviorism, and Thinking Aloud

Although the opposing views persisted, this methodological debate was not fully resolved—in part due to the occurrence of World War I (see Humphrey, 1951)—even though the theoretical ideas gained some acceptance (see Woodworth, 1915). In the early 1900s behavioral psychologists in the USA challenged introspection conceptually and as a viable method of studying behavior. Watson (1913) emphatically stated that "Psychology, as the behaviorist views it, is a purely objective, experimental branch of natural science which needs introspection as little as do the sciences of chemistry and physics" (p. 158). Further, his push was to dismiss *introspectionism* from psychology entirely, including any reference to conscious processes and introspective methods of eliciting them. This alternate perspective had longevity, lasting well into the mid-twentieth century (Boring, 1953).

Watson (1924a) later clarified his position on thinking as a form of behavior. He viewed internal language activity (i.e., inner and subvocal speech) as an internalization of externally observable speech (or physical gestures that *stand for* speech). Hence, he used *thinking* as a term to capture all subvocal behavior. Ironically, given his prior dismissal of introspection, Watson (1920) contended that more could be learned about thinking *behavior*, by having individuals *think aloud* concurrently about well-defined problems.

Watson (1920) asserted that, unlike introspectionists, behaviorists collecting think aloud reports make two implicit requests of the scientific community: (a) to be treated like any other natural scientist who observes some behavior—in this case, a verbalized thought report—under stimulating conditions; and (b) not to dwell on the myriad of metaphysical questions associated with observing (because other scientists using *inspective* observation do not dwell on such matters; see Watson, 1913). Echoing Wundt's criticisms of the introspector as the observer, Watson indicated that in his method, the experimenter was the observer, not the person thinking aloud. Hence, the observed behavior (i.e., thinking out loud) was not complicated by potentially suspect introspective techniques, and understanding of that behavior was not based on the introspector's mystic self-knowledge.

Watson's (1920) goal was to generate an account of thinking based on the complexities of behavior as observed by the experimenter. Like Titchener, he argued that any conceptual meaning representation inferred from introspection was an abstraction and rationalization. However, he went further saying that such introspective "speculation serv[ed] no useful scientific purpose" (p. 103). To him, introspection was simply a popular phrase used to describe the "organized interplay of laryngeal and related muscular activity used in word responses and substitutive word responses" during thinking, which he viewed as a process of trial and error in manual learning (Watson, 1924b, p. 98).

Otto Selz, Cognitive Operations, and a Process Theory of Thinking

Despite the momentum of behaviorism in the USA, the use of Würzburgian introspection continued in Europe. Otto Selz continued this tradition through 1943, when he was deported to and, a few days later, killed in Auschwitz. His focus was on developing a new psychological theory of thinking, beyond *Denkpsychologie*. Arguably, he was the staunchest advocate of the scientific legitimacy of Würzburgian introspection (Reinert, 1981).

In the introduction to a translation of *Laws of Ordered Thinking*, Nico H. Frijda (see Selz, 1913/1981) described Selz's experimental set up as consisting of "instructional variations, unstandardized (yet not unsystematic) item series, retrospections with their potential patchiness or post hoc reconstructions, and protocol interpretation with its inherent dangers" (p. 79). The process included a reflection interval (i.e., short pause after task completion) to facilitate report completeness, which permitted assessment of agreement across elicited protocols.

Selz first used introspection to study the *transcendence problem* with Theodore Lipps (see Reinert, 1981; ter Hark, 2010), whose own interests in causal relationships was influenced by a Brentano student, Christian von Ehrenfels (1890/1988) and his theory of objects. Selz later completed his post-doctoral work with Külpe in Bonn and, hence, is known as a successor to the Würzburg school (see Herrmann & Katz, 2000). Using Würzburgian introspection, Selz (1924/1981) advocated for a process-oriented approach to thinking, viewed as a strictly ordered succession of cognitive operations (cf. James's, 1890/1981, conception of thinking as a series of transitive states).

Influenced by the work of Ach and Bühler (see Selz, 1924/1981), and by another Brentano student, Alexius Meinong (1899/1978) and his theory of higher-order relations, Selz's goal was to extend the Würzburger's research to a more complex view of thinking (see Humphrey, 1951; ter Hark, 2007). As such, he investigated associations of a peculiar kind; those in which a particular semantic relation was stated. For instance, assuming A and B were related, and when provided with A and the *relation* only, participants had to generate an option that fitted with B (e.g., A [known]–*relation* [known]–B [unknown]). Selz's introspective data indicated that, sometimes, more complex mechanisms than the associationists' *diffuse reproduction* (see footnote 3) were needed to explain reproductive thinking. He demonstrated that *Wissen* (knowledge or the process of knowing) involved an awareness of complex relational structures comprised of mental events. These structures represented the initial- and end-task state, goal awareness, and solution awareness, and were organized into what might be called a *slotted schema* (i.e., where one or more *slots* may be unknown) in current theories of cognition and expertise (e.g., Gobet & Simon, 1996).

His data suggested that for a given complex problem (as described earlier), *komplexergänzung* or *complex completion* offered a better explanation than reproductive tendencies. Completion (e.g., filling in slot B) occurred through *schematische antizipation des ziels* (schematic anticipation of the goal) in the context of the *Gesamtaufgabe* (total task); in other words, through goal-directed schematic anticipation. Today, what we call intuition (i.e., a goal-directed simple match between a stimulus and response stored in memory) is consistent with Selz's explanation of reproductive thinking (Selz, 1924/1981; see Wertheimer, 1945/1982). Simple, yet context-sensitive couplings of this kind constitute an important but not the only component of expert knowledge (see Simon, 1981).

Selz's introspective research led him to argue that schematic anticipation was the catalyzing force (much like reproductive tendencies in associationism) behind all complex cognitive operations (Selz, 1913/1981). He further argued that schematic anticipation was particularly relevant to aspects of reproductive thinking analogous to adaptation (i.e., finding new means or ways), and to adaptive thinking that was truly productive (i.e., finding new ends) (for a discussion of adaptation see Ward, Gore, Hutton, Conway, & Hoffman, 2018; see also the last five chapters in Section I, Chapters 8–12, this volume). Selz's introspective research led to him identifying two key processes involved in reproductive and productive thinking. *Mittlefindung* (means finding) was proposed as a process used when the solution is not at one's disposal but the means to find it are. This is reminiscent of current views of sensemaking (see Klein, Phillips, Rall, & Peluso, 2006). *Mittelanwendung* (means application) was proposed as a literal actioning of the found means. Moreover, Selz demonstrated that participants often vacillated between these two processes in a cumulative or subsidiary manner. De Groot (1946/1965) later noted that in tasks more complex than those studied by Selz, this often meant revising the goal (i.e., the ends) dynamically, not just adapting the ways or means (p. 76). Collectively, Selz's research on directed thinking and de Groot's subsequent research program anticipated future models of problem solving (e.g., Newell & Simon, 1972; see Simon, 1999), and more recently, the integrated model of macrocognition, which combines models of expert sensemaking and adaptive replanning (see Hoffman & Hancock, 2017; Ward et al., 2018).

In sum, Selz's use of Würzburgian introspection led him to question the associationist views of human memory organization (Simon, 1981). He asserted that thinking involved—through schematic anticipation—a number of relational matching processes that varied in degrees of complexity, sophistication, automaticity, and adaptive capacity (see also Simon, 1999/2009). These processes ranged from simple matches (as in intuition) to those that "depended on appropriate abstraction of the stimulus situation or on combination of several components of the situation" (Simon, 1981, p. 154). Moreover, Selz demonstrated that these were embedded in a dynamic interplay between means finding and means application, which often required goal restructuring. In effect, Selz's introspective experiments anticipated the memory requirements for a cognitive psychological theory by replacing associations with directed relations and schematic anticipation (see Kintsch, 1974; Simon, 1981). Arguably, the conclusions based on his introspective research foreshadowed the types of mechanisms (e.g., retrieval structures) invoked in current theories of expertise.

From Selz to Current Methods of Thinking Aloud

In addition to making significant theoretical contributions, Selz supplemented his laboratory studies with those of more complex behavior. He adapted Würzburgian introspection to study school learning and intelligence. Selz's student, Julius Bahle, adapted it further by eliciting thought reports through correspondence with musical composers to better understand song composition. However, it was Karl Duncker (e.g., 1935/1945, 1947) who adapted Würzburgian introspection in an important way to study problem solving.

Duncker was a student of Max Wertheimer (who was a student of von Ehrenfels and Stumpf and, later, a graduate of the Würzburg School) and Wolfgang Köhler (also a student of Stumpf), both co-founders of the Gestalt School of thinking in Berlin. As part of the Gestalt School, Duncker's work, like Selz's directed thinking, reacted against associationism (e.g., see Newell, 1985).

Duncker detailed his perspective on problem solving in 1926 and the associated research (1935/1945), which used verbal reporting methods, was influential on subsequent pioneers in the field (e.g., Newell, 1985; Simon, 1999). Duncker's view of thinking as a series of cognitive operations mirrored that of Selz (e.g., see Selz 1913/1981) and was reminiscent of Selz's formulation of problem solving (Selz, 1913/1981, 1924/1981; see also Frijda & de Groot, 1981; Simon, 1999/2009). Duncker's (1935/1945) perspective on problem solving was sufficiently similar to Selz's treatise to motivate Selz to write to Julius Bahle in 1935 to note this point:

> You must read Dunker's book on the psychology of productive thinking. His terms are often confessedly translations of mine. He sticks close to me even when he claims to diverge. So apparently my whole Work, parts of it somewhat watered down, is now taken over by the Berliners. On the whole, he has behaved fairly, but did not send the book to me. (cited in Simon, 1999, p. 9)

The problems Duncker (1935/1945) used in his research (i.e., the *gimlet, box* [candle], *pliers, weight* [pendulum], and *paperclip* problems; p. 86) took time to solve, which significantly delayed subsequent opportunities to introspect (for a discussion of this point, see de Groot, 1946/1965). Therefore, he opted to let participants think aloud concurrently while solving the problem. Duncker's concurrent think aloud method is difficult to distinguish methodologically from Watson's and his specification of what constituted introspection (like Watson's) was somewhat vague. It departed from Würzburgian introspection where participants reported *all conscious events experienced during the experimental task, immediately following completion of the task*. To contrast his method with introspection, Duncker suggested that where "the introspector makes *himself as thinking* the object of his attention, the subject who is thinking aloud remains *immediately directed to the problem*, so to speak allowing his activity to become verbal" (p. 2, italics added).

Duncker's (1935/1945) instructions (i.e., "Please think aloud during the experiment, so that I may hear as many of your ideas as possible, including those which you take less seriously," p. 87), however, were not too dissimilar to those of the Würzburgers (see the earlier section "The Würzburgers"). And, like the Würzburgers, Duncker (1947) embraced participants' active experiences as primary. However, the main difference noted by Duncker between methods was related to the participants' focus of attention (i.e., on the problem rather than on aspects of consciousness). But participants engaged in Würzburgian introspection were also instructed to focus on the task during experimentation, rather than on their thinking per se (see Ach, 1905).[6] Only during immediate retrospection were Würzburgian introspectors asked to report aspects of conscious thought they had noticed during task performance, thus avoiding the division of attention problem. Hence, other than the obvious difference between methods in terms of when reporting occurred (i.e., concurrently vs. *immediately* retrospectively), the difference between Duncker's method and some other introspective forms is murky at best. In summarizing Duncker's method, de Groot (1946/1965) conceded that a trade-off had been made. Although thinking aloud permitted systematic investigation of relatively lengthy sequences of thought, the cost was a "somewhat rougher, more macroscopic overall picture of the thought process" (p. 79).

---

[6] A focus on task performance, rather than on thinking per se, was facilitated through practice (i.e., at introspecting during task performance) in both Wundt and Titchener's research.

Duncker's method is usually highlighted as the first example of a concurrent think-aloud technique that is methodologically *and* conceptually consistent with one verbal reporting method commonly used to study expertise today (e.g., see Ericsson & Simon, 1980, 1993).[7] Adriaan de Groot's chess research is probably the most cited example of how the think-aloud method has been used to study expert thinking. Initially, de Groot (1946/1965) tried both Würzburgian retrospective introspection and Duncker's think-aloud technique to study chess players' move selection in different game positions. However, the extended periods of thinking time associated with chess—often several minutes at a time—meant he had to interrupt players to retrospect when using Würzburgian introspection, or risk poor reliability due to the influence of memory decay. Given these difficulties, de Groot concentrated on perfecting his use of the think-aloud technique.

Despite his conclusion that thinking aloud was less disruptive than Würzburgian introspection (because it did not require interruption), de Groot (1946/1965) noted that it was still disruptive. Thinking aloud slowed down thinking, even in self-paced tasks like chess, and increased task completion time. Instructions that directed participants to report more explicitly than normal (i.e., to make references to *objects of thinking*, rather than undecipherable utterances such as "this" and "that") further exacerbated this effect. Other potential disruptive effects were noted, such as low motivation to engage, which occurred when reporting during experimental or artificial tasks that were marginally representative of naturalistic contexts.

Like others, de Groot (1946/1965) highlighted that the incomplete nature of thinking aloud impacted the ability to capture the true course of thinking. He noted four possible causes: The (1) phase structure of thought, which likely occurred outside of consciousness, would be absent from the report; (2) speed of thought (i.e., more rapid than speech) could lead to omission of attended information; (3) exclusion of thoughts not in verbal form because transformation may disrupt thinking; and (4) intentional suppression of steps in thinking, for instance, of mistakes.

De Groot (1946/1965) used two criteria to assess completeness: (1) the participant's satisfaction with the protocol as a representation of their actual thinking sequence (perhaps a measure of reliability); and (2) the ability to follow and understand the participant's reasoning for a particular action (perhaps a measure of validity). Both criteria require the participant and experimenter to go back through the protocol in an interactive manner consistent with Ach's (1905) approach. This is frequently absent in many think-aloud practices but has been incorporated into similar reporting methods (e.g., critical decision method; CDM) through explicit restatement and follow-up probing (see Hoffman, Crandall, & Shadbolt, 1998; also see Ach, 1905).

---

[7] The same authors also provide extensive description of an equally viable immediate retrospective verbal reporting method that is not too dissimilar from Würzburgian introspection but deny that it constitutes introspection (see the section "Types of Verbal Reports of Thinking Used to Study Expertise Today").

Despite his preference for thinking aloud, de Groot did not rule out Würzburgian introspection, especially for those tasks where timing or time pressure are integral. Instead, he offered useful guidelines, such as steering participants away from describing peculiar qualities of inner experiences, toward immediately recalling the thoughts they can confidently recall having during task performance (see also Müller, 1911).

In interpreting his data, de Groot (1946/1965) argued that the elicited thought sequence did not represent an absolute truth (because it was invariably incomplete). He argued that these data should be treated as a hypothesis about the underlying cognitive process, which could be best understood in the context of a scientific theory. His preference for Selz's theory of thinking as a series of cognitive operations (cf. associationism) stemmed from the similarities in mechanisms he noted in his own research. However, he considered verbal-reporting and introspective methods valuable both for theory testing and for hypothesis formation and theory building, especially about the nature of expertise. Arguably, current methods and practices, especially those used to study expertise on both moderately complex, rule-bound tasks (e.g., chess and puzzles) and more complex work (e.g., decision making in sport and law enforcement), stem from this research.

Types of Verbal Reports of Thinking Used to Study Expertise Today

Ericsson and Simon (1980, 1993) proposed specific procedures for obtaining think-aloud and retrospective reports of thinking (see 1993, pp. 375–379), especially for studying expertise (see Ericsson & Smith, 1991). Consistent with others (e.g., de Groot, 1946/1965; Duncker, 1926; Newell & Simon, 1972; Selz, 1913), they described cognition as an ordered sequence of states (or inputs to and outputs from them) in short-term memory, which are transformed by successive, yet consciously inaccessible information processes. Since thinking states themselves are consciously accessible, they argued that these can be verbalized concurrently or immediately in retrospect. Following de Groot (1946/1965), they asserted that the elicited thought sequence provides a testable hypothesis about cognitive activity.

A crucial issue to note in following Ericsson and Simon's (1980) recommendations is that their goal was to elicit, as close as possible, a complete copy (i.e., record) of the sequence of thinking steps in which a participant engages during task performance. Accordingly, they differentiated between two types of thinking: The first we term *direct verbalizations*. Selz (1913) suggested that these are the types of verbalizations that directly capture knowledge or relational facts currently in consciousness. The second, one might think of as consciously accessible elaborations on that thinking. These are what Selz (1913) termed verbalizations about *potential* knowledge that is not currently in consciousness or attended during task performance (henceforth *verbal elaborations*).

To attain their goal, Ericsson and Simon (1980, 1993) recommended eliciting only direct verbalizations during think-aloud (and immediately retrospective) reporting. Their instructions excluded prompts to elicit verbal elaborations because this might change the course of participants' thinking, or modify the record of thinking in which the participant engaged. Hence, the primary purpose of eliciting direct verbalizations is to express the actual sequence of thinking during task performance.

Procedures for eliciting direct verbalizations differ from those used to elicit verbal elaborations. Direct verbalizations are designed to elicit only those thoughts that come to mind during task performance. Typically, instructions consist of undirected probes, such as "think aloud" or, if asking the participant to immediately retrospect, for instance: "start with the first thought you remember thinking after...." Ericsson and colleagues provided meta analytic evidence, indicating there is limited reactivity when eliciting direct verbalizations through concurrent report on non-time-constrained tasks (Fox, Ericsson, & Best, 2011). In other words, thinking aloud concurrently does not really alter task performance compared to when not thinking aloud. However, others have noted reactivity as a potential concern when verbally reporting during complex tasks; most notably de Groot (1946/1965). For example, in a study of experienced drivers on a test track, Salmon et al. (2017) demonstrated that while no differences were observed in general driving behavior, participants braked more gradually on approach to an intersection and accelerated away more smoothly when thinking aloud concurrently compared to when not (see also Gagné & Smith, 1962; Russo, Johnson, & Stephens, 1989).

In contrast to direct verbalizations, verbal elaborations include explanations, generalizations, and assumptions about (thinking during) task performance, as well as descriptions, summaries, and self-analyses of thinking or performance. They are often elicited via more directed questioning (i.e., directed towards a specific event, issue, or thought) than direct verbalizations (which, contrary to their name, use undirected methods of probing to elicit thinking directly). For instance, probes designed to elicit verbal elaborations may direct participants to verbalize thoughts, such as assessments made, options generated, goals pursued, and priorities managed. They might also include asking for a description of (e.g., "Tell me how you did that"), or explanation about task performance (e.g., "What are you trying to do?"), or an interpretation of their performance (e.g., "How did that go?"). Probes that produce verbal elaborations may include those that encourage generalization to the current situation from other incidents (e.g., "Do you usually do that?") or from another specific incident (e.g., "Were you reminded of any previous experience?").

Perhaps the best examples of instructions to elicit verbal elaborations are the cognitive probes used in the *deepening* phase of the CDM (e.g., as illustrated in Crandall et al., 2006, see p. 79). Incidentally, *deepening* was based on the *progressive deepening* phase that de Groot identified in the think-aloud reports he elicited, where participants reinvestigated part of the task problem more deeply and specifically (e.g., to revise a plan, or to explore alternatives). The key difference is that in de Groot's method, the decision to reinvestigate occurred as part of the natural course of the participant's thinking without directed and specific probing from the experimenter (although, this might also be led by the experimenter during the restatement process). In contrast, in the CDM, the investigator scaffolds the participant's thinking at key decision points (cf. point of reinvestigation) by attempting to help them unpack their thoughts using relevant directed probes. Importantly, per Ach's (1905) recommendation, in the CDM the goal is to minimize suggestive questions beyond the specific focus of the probe (for further discussion of the CDM, see Chapter 19, "Incident-Based Methods for Studying Expertise," by Militello and Anders, this volume).

Verbal elaborations are useful aspects of communication in everyday life, helping individuals externalize and translate thoughts (that might otherwise be considered garbled) into coherent form. As a product of elicitation, they provide content, for instance, for creating an expert knowledge repository, understanding an expert's motivations for action, or developing representations of expert mental models that provide the basis for employing expert knowledge, skills, and strategies. Importantly, verbal elaborations are useful for making such models explicit, for instance, in building concept maps with experts—which is often a long, involved, and co-constructive thinking-aloud process (see Hoffman, LaDue, Mogil, Roebber, & Trafton, 2017). Last, verbal elaborations are indispensable in terms of conducting a priori task analyses to flesh out a problem space (for more information on task analytic methods, see Chapter 16, "Studying Expert Behavior in Sociotechnical Systems: Hierarchical Task Analysis," by Salmon et al., this volume). A problem space is an explicit representation of a well-defined problem, stated in terms of the problem's initial, intermediate, and end states (which includes the correct, best, or alternative solutions), against which researchers can determine the viability of a particular thought sequence elicited through direct verbalization (i.e., to determine whether thinking contains a viable problem solution). The process of comparing if/how a direct verbalization *fits* with a verbally elaborated problem space is a method of analyzing verbal report data often referred to as *protocol analysis* (see Ericsson & Simon, 1993).

Verbalizations may also occur without being solicited (experts often like to explain their thinking spontaneously) but, depending on the goal of the research, this may or may not be helpful. For instance, instructions to elicit a direct verbalization (e.g., sequence of thinking during task performance) might, instead, elicit a verbal elaboration (e.g., description of how to do it, explanation for why it was done) or vice versa. Hence, the art of eliciting a verbal report lies in the ability to elicit the desired kind of verbalization, know when to engage in divergent thinking and when to converge, distinguish between types of verbalizations reported, and determine whether they meet the stated goal.

Moreover, determining the validity of a verbal report is key. As others have noted, unskilled participants performing arbitrary problem-solving tasks may offer implicit theories, culturally derived social rules, or irrelevant generalizations from past experiences, or generate causal hypotheses that could, but do not necessarily, explain their behavior post hoc (see Nisbett & Wilson, 1977). Hence, any information elicited through introspective methods, whether for the purposes of building a theory, testing it, developing an intervention, or testing its effectiveness, should be put to a strong test both inside and outside the laboratory (e.g., via experimental manipulation; field observations, etc.); that is, against all possible alternative explanations, not just the preferred one (see Feynman, 1974). This means determining: (a) whether the contents of a verbal report are generalizable beyond the specific incident from which they were elicited; and (b) whether they offer a *unique* means to explain the observed pattern of behavior, or are just one of a number of alternative and equally plausible explanations.

One curious off-shoot in the history of introspection is the tendency of some to deny that their method constitutes introspection. Titchener's (1912b; see opening quote of this chapter) view of introspection was broad, even though his own method was relatively narrow. The Würzburgers' view was broader still. With only a spurious explanation of differences in attentional focus, Duncker rejected the idea that his version of thinking aloud was introspection. Watson rejected the term introspection because of its association with mentalistic concepts. Ericsson and colleagues argued that direct verbalizations do not constitute introspection—which they reserved for the elicitation of verbal elaborations—because direct verbalizations are qualitatively different from verbal elaborations and, they argued, the latter are tantamount to inner observation (see Ericsson & Fox, 2011). In their review of introspection, Ericsson and Crutcher (1991) argued that systematic introspection (cf. systematic *experimental* introspection) writ large was not possible for the reasons already cited against inner observation (but also cited criticisms similar to those aimed at the Würzburgers—such as inaccessible thought processes, divided attention, and memory decay).

The historical review presented here clearly indicates that each of these narrow views is overly simplistic, that all of the concepts are fuzzy, and that each of the researchers may have been somewhat revisionist in their outlook. Historically, the pre-behaviorist philosophical jargon was replaced by the purist jargon of the behaviorists, which in turn was up-ended by the information-processing jargon of the 1970s and 1980s. With each wave of major change in outlook and jargon, introspective methods have remained remarkably similar and contingent on the task. This is not to say that key methodological distinctions have fallen by the wayside and are irrelevant to expertise. However, some generally accepted norms should come with clear caveats and the associated assumptions noted explicitly. Hopefully, we can learn from historical precedents and consider these when using introspective-type methods in future.

Guidance on Collecting Verbal Reports of Thinking

Several recommendations can be made based on the historical literature about how to collect viable introspective-type reports and how to consider their validity. First, we recommend specifying the goal of eliciting a verbal report. If, like Ericsson and Simon, the goal is to elicit a record of a sequence of thoughts, then instructions that restrict elicitation to direct verbalizations are recommended. Internal validity would be determined by the extent to which those verbalizations reflect the actual thinking sequence. While elicitation of direct verbalizations has been common practice, this has often been dogmatic. For instance, some researchers have excluded verbal elaborations simply because past recommendations have suggested doing so, or because of lingering concerns about interference and memory decay. We encourage a more open and thoughtful approach, especially if pursing an alternate goal. All methods have strengths and limitations. Just because a method might have limitations, it does not mean that one should avoid using it to see where the data might lead.

If the goal is, for instance, to gain a better understanding of the underlying knowledge representation that supports expert thinking, then verbal elaborations (and direct verbalizations) would be useful, if not necessary. The caveat is that any verbal report be verifiable, that is, treated as a testable hypothesis and validated against one's goal. While direct verbalizations may be valid in terms of capturing a sequence of thinking, they may be invalid, for instance, in terms of capturing the rationale for generating a particular course of action.

As stated earlier, verbal elaborations are especially useful in conducting a priori task analyses, and for explicitly specifying potential problem spaces (see the section "Types of Verbal Reports of Thinking Used to Study Expertise Today"). Likewise, computational models of expert reasoning often require prior specification of a repository of interconnected knowledge on which the model draws during computation (e.g., Kintsch, 1988; see also Chapter 15, "Computational Models of Expertise," by Kirlik & Byrne, this volume). This kind of knowledge has to come from somewhere (e.g., through verbal elaboration). Hence, although direct verbalizations might support eliciting an actual thought sequence, verbal elaborations may be more useful in conducting more comprehensive task analyses, and in building better representations and models of expertise. Knowing what you are eliciting and why you are eliciting it, however, is key!

A defining feature of expertise is the meta-cognitive nature of expert thinking, including the use of self-explanations (e.g., Glaser & Chi, 1988; see also Chi, Bassok, Lewis, Reimann, & Glaser, 1989; Ward et al., 2018). If people, including experts, were unable to explain their reasoning and knowledge to other people, such as students, trainees, or mentees, then the evolution of human intellect may have been considerably hindered! When participants explain to themselves as part of their normal work, these explanations are integral to understanding thinking and representing the underlying knowledge structure that supports successful work. However, this is different from an explanation that is for the benefit of someone else (e.g., the experimenter) or justifying thinking or actions post hoc. Both are different from an explanation by one person to another (e.g., a team mate) as a means, for instance, of making sense of a situation collaboratively. Each is a rich source of data but are useful for different purposes, and serve as means for achieving different ends. If the goal is *not* to trace an actual thought sequence then the type of verbalization probably does not matter. But the context, rationale, and/or motives should be considered, the data interpreted accordingly, and, like any hypothesis, the resulting data should be put to the test.

Following goal specification, we recommend conducting a complete task analysis to flesh out the expert problem space and the important information cues and relationships, using both introspective and other methods (e.g., see each of the other chapters in Section II, this volume). Most importantly, we recommend that researchers engage in extensive practice at eliciting verbal reports using a variety of methods of structured interviewing and task reflection. Practice will help the participant provide the desired kind of report, and the experimenter develop their understanding of the utility and suitability of different probes for eliciting different kinds of verbalizations. It will also help both researchers and introspectors detect and differentiate between types of verbalizations (e.g., spotting a justifying explanation versus a self-explanation). The interested reader can find concrete examples of the types of guidance used for eliciting verbal elaborations and knowledge elicitation techniques more broadly in Crandall et al. (2006). For direct verbalizations, see Ericsson and Simon (1993).

Historically, most researchers have recommended that the time between an activity and reporting should be minimized to avoid contamination from memory (e.g., forgetting, generalizing). In some instances, especially where time pressure is an issue, retrospection may be preferred. Although the original instructions for direct verbalization were developed for use in simple, non-time-constrained tasks (e.g., numerical puzzles, chess; see Ericsson & Simon, 1980), variations of these instructions have been used in complex domains that change rapidly, and where uncertainty and time pressure are high (e.g., Ward, Suss, Eccles, Williams, & Harris, 2011; Ward, Ericsson, & Williams, 2013).

For example, in a study of law enforcement decision making, skilled and rookie officers were presented with 30- to 90-second simulated scenarios, including high- and low-frequency, lethal and non-lethal incidents (see Ward et al., 2011). Participants responded by deciding whether to de-/escalate use of force and by responding accordingly. Officers were asked immediately after randomly pre-selected trials to "Verbally report the first thought you can remember thinking as the scenario began." Additional undirected prompts were used to ensure participants continued to verbalize their thinking (e.g., "What was your next thought immediately after that?"). Participants first received instructions on how to give direct verbalizations immediately in retrospect, and engaged in introspection practice with feedback. Per Müller's (1911) recommendation (see also Ericsson & Simon, 1993), participants were instructed to recall only those thoughts they could confidently remember having.

Our goal was to test alternative accounts of expertise, specifically about option generation during different phases of decision making. Hence, the verbal data were coded post hoc as being about situational events (*assessment phase of decision making*) or personal courses of action (*intervention phase of decision making*). The verbal report data revealed that skilled officers generated a higher number of options (than rookie officers) during both decision-making phases, and the strategies they employed were characteristic of both intuitive and apperceptive thinking (see footnote 2), which were supported by differences in deliberative thinking.

In some domains, there may be extended periods of time between the target activity and an opportunity to elicit a verbal report about that activity (e.g., a fighter pilot during an intense dogfight). We have noted that experts often retain detailed and vivid knowledge of previously encountered tough cases, some of which they may have encountered months or years earlier. They think about such cases over and over, reasoning about what they did and why, and what they might have done differently. In such cases, some researchers have advocated for concurrent collection of verifiable sources of data (e.g., observer's record; black-box data; audio/video film recording) against which retrospection can be compared (e.g., Crandall et al., 2006; Ericsson & Ward, 2007; Flanagan, 1954). These data not only help verify subjectively recalled events from memory but provide additional data to support the development of a task analysis or cognitive model.

In the absence of any verifiable record of events, especially when reporting is delayed substantially, memory scaffolding techniques can be used to help facilitate accurate recall, such as those used in the CDM (see Flanagan, 1954; Hoffman et al., 1998; Klein, Calderwood, & Clinton-Cirocco, 1986; see also Militello and Anders, this volume). Scaffolding methods include asking for recollections of a specific incident rather than generalizations across events, or of an incident that is particularly salient to the introspector (e.g., they were a key decision maker, their decision affected the outcome). Likewise, experimenter restatement using a participant's own language, phrasing, and/or syntax (i.e., from the participant's recollection of the timeline) can cue recall of specific memories during subsequent elicitation.

In an example of this delayed approach, Harris, Eccles, Freeman, and Ward (2017) adapted the CDM (to include more undirected probing) to examine how officers prepared for, coped with, and made decisions under threat-of-death stress when *on the street*. After introspection training (similar to Ward et al., 2011), officers were asked to briefly describe a recent stressful work-related event in which they: (a) were a key decision maker, and their decisions affected the event outcome; (b) unholstered their service weapon; and (c) were required to adapt to the situation (Harris et al., 2017, p. 1114).

Following the CDM procedures (see Crandall et al., 2006), while one researcher generated an event timeline, another elicited an incident overview from the officer. After restatement, the verified sub-events in the timeline were used as probes to aid participant recall. In a final pass through the timeline, undirected probes were used to elicit thoughts or feelings of stress and anxiety, followed by directed probes to aid recall of "any attempt to control these feelings; what effect an attempt had on the stress/anxiety experienced; and when the stress experienced was reduced to a more normal level" (Harris et al., 2017, p. 1114). The officers experienced a wide range of events and coped with stress predominantly via problem-focused strategies (i.e., centered around planning), which were adapted based on the available context. Officers rarely used an emotion-focused approach to cope with in-event stress or anxiety. Instead, they deferred dealing with their feelings until after the event.

The extent to which the experimenter should actively intervene when using introspective-type methods has been one of greatest points of disagreement over the years. Examples of intervention include restatement of the recalled timeline back to the participant (see Ach, 1905; de Groot 1946/1965), or using specific probes to unpack a particular thought (e.g., Crandall et al., 2006; Hoffman et al., 1998; Klein et al., 1986). Those who advocated against intervention warned of its potential influence on what is reported, presumably because it introduced personal prejudices (e.g., see Müller, 1911). Arguably, influence is only problematic if it results in an inaccurate or dishonest report. A more complete, accurate, and honest report would be beneficial. To protect against negative influence, however, some have recommended that data elicited through questioning be differentiated from those derived spontaneously, or questioning be replaced with experimental manipulations to resolve ambiguities. However, the contention that intervention via questioning is dangerous and unscientific (e.g., Titchener, 1912b; Wundt, 1907) is grounded in the dogmatic perspective that the *true experiment* is the gold standard in research. From our perspective the standard should not be to hold objectivism above subjectivism, or internal above external validity, but to ask researchers using any method or research design to employ them with greater veracity. One can aspire to obtain external and internal validity simultaneously and veraciously, albeit with some tradeoff (see Stokes, 1997).

Arguably, experimenter intervention in the form of both restatement and experimenter questioning (e.g., Ach, 1905; Crandall et al., 2006; de Groot, 1946/1965; Hoffman et al., 1998) helps guarantee completeness and accuracy of the thought report. Boren and Ramey (2000) echoed this sentiment indicating that intervention was especially helpful in complex task environments, such as usability testing. They argued that Ericsson and Simon's approach was overly restrictive and needed to be modified to be more communicative so that the researcher can help scaffold introspection of actual thinking. This view of knowledge elicitation as a collaborative and co-constructive process has long been recognized in computer science (see Agnew, Ford, & Hayes, 1997; Ford & Adams-Webber, 1992; Ford, Cañas, & Coffey, 1993), where interactive introspective-type methods are common place (e.g., cognitive walkthrough, heuristic evaluation).

Differences in recommendations about the type of communication and questions posed, typically, have been manifest in the directedness of questioning. *Undirected*, albeit incident-based, methods of questioning have been proposed to avoid being suggestive or leading participants to talk about information they would not otherwise have mentioned (e.g., Ach, 1905; Ericsson & Simon, 1980, 1993). Others have advised, depending on the context, beginning with undirected probes, followed by progressive waves of increasingly directed probing (e.g., Crandall et al., 2006; Hoffman et al., 1998; Klein et al., 1986). For instance, one might use undirected probes to elicit a timeline of events, then use markers on the timeline to cue more detailed memory recall, and then use more directed probes about specific types of thinking (i.e., identified a priori though other methods, e.g., task analyses, experimentation, verbal reports). Others have tested information elicited through directed probes using a subsequent simulation-based interview (e.g., Militello & Hutton, 1998; see also Militello and Anders, this volume). This method presents hypothetical situations (based on claims made in prior interviews) in scenario form as a means for eliciting knowledge that would dis/confirm the kinds of expert thinking and strategies elicited previously through directed probing. In most instances, researchers would agree that elicitation is a skilled behavior that requires practice, and an unbiased yet interactive approach permits a more complete understanding of the participant's experience being reported (e.g., see Ach, 1905).

Experts often think aloud while conducting their tasks. This often happens during mentoring (see Chapter 44, "Learning at the Edge," by Petushek et al., this volume), but there are also domains in which the primary task involves concurrent verbalization (e.g., the coroner's autopsy process). If interference were rampant and thoroughly destructive of any empirical value that such data might have, humankind would, again, be remitted to the *extinct* category of species. Even more to the point, when a concurrent verbalization influences reasoning ("Wait a minute, I just thought of something"), that need not be branded as interference, without thinking twice about the dogmatic implication. The historical baggage of the term *interference* needs to be abandoned.

In instances where thinking aloud does not interfere with concurrent work, one strategy is to use observation, followed by a combination of uninterrupted, concurrent think-aloud reports and probing; the order, frequency, and directedness of which would depend on the stated goal. In instances where thinking aloud disrupts performance, one could use a verifiable record of the event (e.g., observation, video recording) to first build a timeline retrospectively, then use identified timeline markers to probe the participant. Without any verifiable record, we might use undirected probes to build a timeline demarked with key events, followed by restatement (e.g., collaborative elicitation) to help ensure completeness, reliability, and accuracy of the elicited report thus far (including order of events). Once undirected probing and restatement have been exhausted, one can follow up by using more directed probing (ideally, based on some prior analyses of performance on the same task). Where possible, we recommend subsequent testing of elicited information via the use of multiple methods. For instance, this can be done by collecting additional elicitation, via the presentation of hypotheticals, and/or the subsequent use of follow-up experimentation, simulation, modeling, or naturalistic studies, including observation. And like any good handbook recommendation, repeat as necessary until a more complete representation is formed.

In some instances, there is insufficient time to follow multiples steps (as outlined above) and so some form of abridged introspective method or rapidized CTA needs to be implemented (see Hoffman et al., 2014). This may mean omitting steps in the process (e.g., creating a timeline beforehand) and/or identifying the method that will produce the largest yield in the shortest time. In another recent study of law enforcement officer decision making, Suss and Ward (2018; see also Suss, Belling, & Ward, 2014) used an abridged undirected retrospective verbal-reporting method immediately after participants engaged in a simulated scenario in which they made a decision or prediction. Participants were only given very brief instructions on how to verbally

Table 17.1 The types of probes and specific questions used by Suss, Belling, and Ward (2014) to probe participants during video-stimulated recall

Probe type	Probe
Cues heeded	What were you paying attention to at this point?
Evaluations and inferences	You said you were paying attention to X. What did that mean to you at this point? What was your understanding of the situation at this point?
Anticipations	At this point what did you think would happen in the next few seconds?
Decisions and responses	What course(s) of action were you considering at this point? What led you to consider that option, or reject other options? Was there a rule that you followed that led you to consider that option?"
Prior knowledge and experience	What prior knowledge or personal experience influenced your thinking at this point?

Source: Data from Suss, J., Belling, P., & Ward, P. (2014), Use of cognitive task analysis to probe option-generation in law enforcement, *Proceedings of the Human Factors and Ergonomics Society Annual Meeting 58*(1), 280–284.

report retrospectively. Retrospective reporting was then followed by probing during video-stimulated recall.

During video-stimulated recall, participants were instructed to re-watch the scenario in which they had just participated, to pause it when they could recall having a particular thought, and then respond to a set of undirected and directed probes (see Table 17.1). These were designed to elicit verbal elaborations (because direct verbalizations had already been elicited via immediate retrospective report).

In comparison to the abridged retrospective method, introspection via probing with stimulated recall yielded the greatest return, and was most informative in testing and generating hypotheses about skilled officer performance. However, we should note that the choice of probes used was based on two sources of information: The immediate retrospective verbal reports of thinking from the police study discussed earlier (i.e., Ward et al., 2011) and several similar studies employing introspective and cognitive task analytic methods (e.g., Suss, 2013; Suss & Ward, 2012, 2013).

Concluding Remarks

The history of introspection provides a springboard from which to plan future studies using introspective methods. It provides insights into both how (and how not) to execute particular techniques for studying expertise and how these methods can advance theory underpinning expertise (e.g., Selz, 1924/1981). We encourage the reader to consider both in designing their next study and, in particular, in advancing the state of the science on expertise.

In sum, there are many factors to consider when using introspective-type methods. Arguably the primary factors should be to consider the primary goal of elicitation and to gain much practice in both elicitation and introspection. Like any skill, it takes time to acquire skill in introspecting itself and in eliciting introspective-type verbal reports. Appropriately worded instructions will help elicit the intended type of verbalization or information. Proficient CTA practitioners should be able to (a) differentiate between general and incident-specific introspections; (b) select the appropriate follow-up, un/directed probe to further unpack thinking; (c) understand the impact of using different types of probes on verbalization; and (d) spot the difference between reports given, such as between summaries of thinking and actual thought sequences, and between explanations, self-explanations, and emergent explanations. Most report types that follow the guidelines presented here generally result in useful data that are dependent on the study goal.

Last, the skilled experimenter and CTA practitioner should be sensitive to the contextual constraints associated with the incident being introspected, and to the goals of the stakeholder—the end user of the product derived from introspective data. This means asking questions like: Are sufficient resources available? Are there any organizational constraints? Are experts available to be interviewed? How much time do they have? Is there time to provide introspection training (of any kind) to participants? How might concurrent reporting influence reasoning or performance? What is the expected time lag between activity and reporting? Perhaps most importantly, how does the stakeholder intend to use the data: To build a better description of an expert's thinking? To develop theory? To build a reliably predictive model? To design an intervention or support technology? Each contextual constraint, stakeholder input, and desired outcome should help specify the goal of using a specific bout of a particular type of introspection, and should drive the development of instructions and the choice of approach to be employed.

Thinking Out of the Box

There has always been a push in psychology for our science to be more like the hard sciences. Arguably, this push has crept in to the study of experts (see Chapter 49, "The 'War' on Expertise," by Klein et al., this volume). Historically, this has often resulted in the use of a reductionist approach and valuing the laboratory experiment as the gold standard. Overly reductionist approaches have been described as a *psychology of the organism* rather than, arguably, what psychology should be: a *psychology of the organism–environment interaction* (Brunswik, 1956). In turn, this has often manifested itself in the use of introspective methods that are also reductive (e.g., elementary sensationism; direct verbalizations only), at the cost of understanding higher-order thinking as it occurs in context. And, as is typical in a tribe (but *should be* atypical in science), "disdain has routinely been expressed by a diverse range of scientists for those 'in the other camp', whose position, purpose, and methods have been described by those holding contrary views as having little scientific or societal value" (Ward, Belling, Petushek, & Ehrlinger, 2017; p. 18; see also Gigerenzer, 2004).

Brunswik argued that the hard sciences, with their strict experimental methods, may not be a good analog for psychology. Instead, he argued that geography, with its emphasis on observational and descriptive methods, might be a better model. Perhaps this view should be extended to the study of expertise, or at least some balance struck. If, as Brunswik noted, such methods were good enough for Darwin, and as Brentano argued that it had worked for some areas of physics (e.g., astronomy), one could reasonably argue it should be good enough for those studying expertise. The obvious caveat is that when verification of ideas generated through observation, including self-observation, is possible, this is rigorously pursued. We would add that researchers should strive to both make observations in natural environments and test in artificial ones and vice versa. No single context should hold precedence if we are ever to develop a fuller understanding of expertise (see Hoffman, 2018). Despite the propensity to elevate one methodological approach over another (e.g., Banaji & Crowder, 1989), the complementarity between naturalism and experimentalism has long been appreciated (e.g., by Darwin) and has recently been reiterated (see Klein et al., 2003).

Historically, the view that psychology should be about studying *increasingly complex* phenomena (rather than *simpler and simpler* phenomena, as is the case in an overly reductive approach) has been shared by some of the earliest pioneers of introspective methods. For instance, Brentano (1874/2009) suggested that:

> Scientists have begun to pay very special attention to the method of psychology. In fact, you could say that no other general theoretical sciences are as noteworthy and instructive in this regard as psychology, on the one hand, and mathematics, on the other... Mathematics considers the most simple and independent phenomena, psychology those that are most dependent and complex. Consequently, mathematics reveals in a clear and understandable way the fundamental nature of all true scientific investigation... Psychology alone, on the other hand, demonstrates all the richness to which scientific method lends itself, by seeking to adapt itself to successively more and more complex phenomena. (p. 21)

One could argue that studying individuals who are best adapted to increasingly complex phenomena, such as experts operating in complex environments, could be viewed as the ultimate application of psychology. The key question, given the aforementioned criticisms of some methods, is whether particular forms of introspection are best suited to moving our understanding of expertise forward. The answer depends on one's goal, not on the prevailing dogma.

In agreement with Franz Brentano and Bertrand Russell (1872–1970), we see much value in engaging in intellectual dissent and fielding opposing viewpoints regarding the value of differing methods, even when those views are biased, one-sided, or tribal. Our hope, however, is that others also view the psychology of expertise through Brentano's lens: as one target domain for studying *more complex phenomena*. As we have seen, introspective methods have been especially instructive, not just for describing expert thinking but for understanding its context-sensitive nature, and the organism–environment interactions that underpin expertise. More than 140 years on, however, we agree with Brentano (1874/2009):

> More light would undoubtedly be shed if the psychological method itself were more clearly known and more fully developed. In this respect there remains much to be done, for only with the progress of the science does a true understanding of its method gradually develop. (p. 22)

We hope that our current exposition of introspective methods goes some way to providing some semblance of clarity in this regard, and helps others leverage these useful, yet often demonized methods to study future experts in a thoughtful and productive manner. For those who are still skeptical of introspective and similar qualitative methods, we point them to the McNamara fallacy—the erroneous bias towards making decisions based on quantifications at the expense of other important observations—eloquently summarized by Daniel Yankelovich (1972):

> The first step is to measure whatever can be easily measured. This is OK as far as it goes. The second step is to disregard that which cannot be measured or give it an arbitrary quantitative value. This is artificial and misleading. The third step is to presume that what can't be measured easily isn't very important. This is blindness. The fourth step is to say that what can't be measured doesn't really exist. This is suicide.

References

Ach, N. K. (1905). *Über die willenstätigkeit und das denken*. Göttingen: Vandenhoeck & Ruprecht.

Agnew, N. M., Ford, K. M., and Hayes, P. J. (1997). Expertise in context: Personally constructed, socially selected, & reality-relevant? In P. J. Feltovich, K. M. Ford, & R. R. Hoffman (Eds), *Expertise in context* (pp. 219–244). Cambridge, MA: MIT Press.

Angell, J. R. (1906). Studies in psychology. *Journal of Philosophy, Psychology and Scientific Methods* 3(26), 637–643.

Banaji, M. R., & Crowder, R. G. (1989). The bankruptcy of everyday memory. *American Psychologist* 44(9), 1185–1193.

Binet, A. & Simon, T. (1908). The development of intelligence in children. *L'Annee Psychologique* 14, 1–94.

Boren, T., & Ramey, J. (2000). Thinking aloud: Reconciling theory and practice. *IEEE Transactions on Professional Communication* 43(3), 261–278.

Boring, E. G. (1929). *A history of experimental psychology*. New York: Appleton-Century.

Boring, E. G. (1953). A history of introspection. *Psychological Bulletin* 50, 169–188.

Brentano, F. (2009). *Psychology from an empirical standpoint*. London: Taylor & Francis. (Original work published 1874.)

Broome, M. R., Harland, R., Owen, G. S., & Stringaris, A. (2012). Franz Brentano (1838-1917). Editors' introduction. In M. R. Broome, R. Harland, G. S. Owen, & A. Stringaris (Eds), *The

Maudley reader in phenomenological psychiatry (pp. 3–5). New York: Cambridge University Press.

Brunswik, E. (1956). *Perception and the representative design of psychological experiments* (2nd edn). Oakland, CA: University of California Press.

Bühler, K. (1908). Antwort auf die von W. Wundt erhobenen einwande gegen die methode der selbstbeobachtung an experimentell erzeugten Erlebnissen. *Archiv fur die Gesamte Psychologie* 12, 93–123.

Chi, M. T. H., Bassok, M., Lewis, M., Reimann, P., & Glaser, R. (1989). Self-explanations: How students study and use examples in learning to solve problems. *Cognitive Science* 13, 145–182.

Cohen, A. (2002). Franz Brentano, Freud's philosophical mentor. In G. Van de Vijver & F. Geerardyn (Eds), *The pre-psychoanalytic writings of Sigmund Freud* (pp. 88–100). London: H. Karnac.

Crandall, B., Klein, G., & Hoffman, R. R. (2006). *Working minds*. Cambridge, MA: MIT Press-Bradford.

Danziger, K. (1980a). The history of introspection reconsidered. *Journal of the History of the Behavioral Sciences* 16(3), 241–262.

Danziger, K. (1980b). Wundt's psychological experiment in the light of his philosophy of science. *Psychological Research* 42, 109–122.

De Groot, A. (1965). *Thought and choice in chess*. The Hague, Netherlands: Mouton. (Original work published 1946.)

Duncker, K. (1926). A qualitative (experimental and theoretical) study of productive thinking (solving of comprehensible problems). *Pedagogical Seminary* 33, 642–708.

Duncker, K. (1935). On problem solving. *Psychological Monographs* 58(5), 1–113. Translated by L. S. Lees. (Original work published 1945; *Zur Psychologie des produktiven Denkens*.)

Duncker, K. (1947). Phenomenology and epistemology of consciousness of objects. *Philosophy and Phenomenological Research* 7, 505–542.

Ericsson, K. A., & Crutcher, R. J. (1991). Introspection and verbal reports on cognitive processes—Two approaches to the study of thinking: A response to Howe. *New Ideas in Psychology* 9(1), 57–71.

Ericsson, K. A., & Fox, M. (2011). Thinking aloud is not a form of introspection but a qualitatively different methodology: Reply to Schooler (2011). *Psychological Bulletin* 137(2), 351–354.

Ericsson, K. A., & Simon, H. (1980). Verbal reports as data. *Psychological Review* 87(3), 215–251.

Ericsson, K. A., & Simon, H. (1993). *Protocol analysis: Verbal reports as data* (rev. edn). Cambridge, MA: MIT Press.

Ericsson, K. A., & Smith, J. (1991) Prospects and limits of the empirical study of expertise: an introduction (pp. 1–38). In K. A. Ericsson & J. Smith (Eds), *Toward a general theory of expertise: Prospects and limits*. Cambridge, UK: Cambridge University Press.

Ericsson, K. A., & Ward, P. (2007). Capturing the naturally-occurring superior performance of experts in the laboratory: Toward a science of expert and exceptional performance. *Current Directions in Psychological Science* 16(6), 346–350.

Evans, R. B. (2000). Psychological instruments at the turn of the century. *American Psychologist* 55, 322–325.

Feynman, R. P. (1974). Cargo cult science: Some remarks on science, pseudoscience, and learning how not to fool yourself. Caltech's 1974 commencement address. *Engineering and Science* 37(7), 11–13.

Flanagan, J. C. (1954). The critical incident technique. *Psychological Bulletin* 51(4), 327–358.

Ford, K. M., & Adams-Webber, J. R. (1992). Knowledge acquisition and constructivist epistemology. In R. R. Hoffman (Ed.), *The psychology of expertise: Cognitive research and empirical AI* (pp. 121–136). New York: Springer.

Ford, K. M., Cañas, A. J., & Coffey, J. (1993). Participatory explanation. *Presented at the FLAIRS 93: Sixth Florida Artificial Intelligence Research Symposium* (pp. 111–115). Ft. Lauderadale, FL: Institute for Human and Machine Cognition.

Fox, M., Ericsson, K. A., & Best, R. (2011). Do procedures for verbal reporting of thinking have to be reactive? A meta-analysis and recommendations for best reporting methods. *Psychological Bulletin* 137(2), 316–44.

Frijda, N. H., & de Groot, A. D. (Eds) (1981). *Otto Selz: His contribution to psychology*. The Hague: Mouton Publisher.

Gagné, R. M., & Smith Jr, E. C. (1962). A study of the effects of verbalization on problem solving. *Journal of Experimental Psychology* 63(1), 12.

Gigerenzer, G. (2004). Mindless statistics. *Journal of Socio-Economics* 33, 587–606.

Glaser, R., & Chi, M. T. H. (1988). Overview. In M. T. H. Chi, R. Glaser, & M. J. Farr (Eds.). *The nature of expertise* (pp. xv–xxviii). Hillsdale, NJ: Lawrence Erlbaum.

Gobet, F., & Simon, H. A. (1996). Templates in chess memory. A mechanism for recalling several boards. *Cognitive Psychology* 31(1), 1–40.

Harris, K. R., Eccles, D. W., Freeman, C., & Ward, P. (2017). "Gun! Gun! Gun!": An exploration of law enforcement officers' performance under stress during actual events. *Ergonomics* 60(8), 1112–1122.

Herrmann, T., & Katz, S. (2000). Otto Selz and the Würzburg school. In L. Albertazzi (Ed.), *The dawn of cognitive science: Early European contributors* (pp. 225–235). Dordrecht: Kluwer.

Hoffman, R. R. (2018). Macrocognition: A commentary on "Combining the strengths of naturalistic and laboratory decision making research to seek optimal level of fuzz" by Art Markman. *Journal of Applied Research in Memory & Cognition* 7(1), 23–25.

Hoffman, R. R., Crandall, B., & Shadbolt, N. (1998). Use of the critical decision method to elicit expert knowledge: A case study in the methodology of cognitive task analysis. *Human Factors* 40(2), 254–276.

Hoffman, R. R., & Hancock, P. A. (2017). Measuring resilience. *Human Factors* 59(14), 564–581.

Hoffman, R. R., LaDue, D. S., Mogil, H. M., Roebber, P. J., & Trafton, G. (2017). *Minding the weather: How expert forecasters think*. Cambridge, MA: MIT Press.

Hoffman, R. R., Ward, P., Feltovich, P. J., DiBello, L., Fiore, S. M., & Andrews, D. (2014). *Accelerated expertise: Training for high proficiency in a complex world*. New York: Psychology Press.

Hoffmann, J., Stock, A., & Deutsch, R. (1996). The Würzburg school. In J. Hoffmann and A. Sebald (Eds), *Cognitive Psychology in Europe: Proceedings of the Ninth Conference of the European Society for Cognitive Psychology* (pp. 147–172). Lengerich: Pabst Science Publishers.

Humphrey, G. (1951). *Thinking: An introduction to experimental psychology*. London: Methuen.

Huemer, W., & Landerer, C. (2010). Mathematics, experience, and laboratories: Herbart's and Brentano's role in the rise of scientific psychology. *History of the Human Sciences* 23, 72–94.

James, W. (1981). *The principles of psychology.* Cambridge, MA: Harvard. (Original work published 1890.)

Kersten, F. (1969). Franz Brentano and William James. *Journal of the History of Philosophy* 7, 177–191.

Kintsch, W. (1974). *The representation of meaning in memory.* Hillsdale, NJ: Lawrence Erlbaum.

Kintsch, W. (1988). The use of knowledge in discourse processing: A construction-integration model. *Psychological Review* 95, 163–182.

Klein, G. A., Calderwood, R. & Clinton-Cirocco, A. (1986). Rapid decision making on the fire ground. *Proceedings of the Human Factors and Ergonomics Society Annual Meeting* 30, 576–580.

Klein, G., Phillips, J. K., Rall, E. L., & Peluso, D. A. (2006). A data/frame theory of sensemaking. In Robert R. Hoffman (Ed.), *Expertise out of context: Proceedings of the Sixth International Conference on Naturalistic Decision Making* (pp. 113–155). Boca Raton, FL: Taylor and Francis.

Klein, G., Ross, K. G., Moon, B. M., Klein, D. E., Hoffman, R. R., & Hollnagel, E. (2003). Macrocognition. *IEEE Intelligent Systems: Human Centered Computing* (May/June), 81–85.

Külpe, O. (1909). *Grundriss der psychologie; auf experimenteller grundlage dargestellt.* Leipzig: Engelmann. (Original work published in 1893.)

Lyons, W. (1986). *The disappearance of introspection.* Cambridge, MA: MIT Press.

Marbe, K. (1901). *Experimentell-Psychologische Untersuchungen über das Urteil, eine Einleitung in die Logik.* Leipzig: Engelmann.

Mayer, A., & Orth, J. (1901). Zur qualitativen untersuchung der association. *Zeitschrift für psychologie und physiologie der sinnesorgane* 26, 1–13.

Meinong, A. (1978). Über gegenstände höherer ordnung und deren verhältnis zur inneren wahrnehmung. *Zeitschrift für Psychologie und Physiologie der Sinnesorgane* 21, 182–272. (Original work published 1899.)

Mercer, C. (2017). Descartes' debt to Teresa of Ávila, or why we should work on women in the history of philosophy. *Philosophical Studies* 174(10), 2539–2555.

Militello, L. G., & Hutton, R. J. B. (1998). Applied cognitive task analysis (ACTA): A practitioner's toolkit for understanding cognitive task demands. *Ergonomics* 41(11), 1618–1641.

Müller, G. E. (1911). *Zur analyse der gedächtnistätigkeit und des vorstellungsverlaufes, I.* Leipzig: Verlag von Ambrosius Barth.

Newell, A. (1985). Duncker on thinking: An inquiry into progression in cognition. In S. Koch & D. Leary (Eds), *A Century of Psychology as Science: Retrospections and Assessments* (pp. 392–419). New York: McGraw-Hill.

Newell, A., & Simon, H. A. (1972). *Human problem solving.* Englewood Cliffs, NJ: Prentice-Hall.

Nisbett, R. E., & Wilson, T. D. (1977). Telling more than we can know: Verbal reports on mental processes. *Psychological Review* 84(3), 231–259.

O'Shea, G., & Bashore, T. R., Jr. (2012). The vital roles of *The American Journal of Psychology* in the early and continuing history of mental chronometry. *American Journal of Psychology* 125, 435–448.

Reinert, G. (1981). In memoriam Otto Selz (spoken upon the occasion of the post-humous award of the Wilhelm Wundt plaquette in 1971). In N. H. Frijda, & A. D. de Groot (1981) (Eds), *Otto Selz: His contribution to psychology* (pp. 13–19). The Hague: Mouton Publisher.

Russo, J. E., Johnson, E. J., & Stephens, D. L. (1989). The validity of verbal protocols. *Memory & Cognition* 17(6), 759–769.

Salmon, P. M., Goode, N., Spiertz, A., Thomas, M., Grant, E., & Clacy, A. (2017). Is it really good to talk? Testing the impact of providing concurrent verbal protocols on driving performance. *Ergonomics* 60(6), 770–779.

Schwitzgebel, E. (2016). Introspection. In E. N. Zalta (Ed.), *The Stanford Encyclopedia of Philosophy* (Winter 2016 edn). Stanford, CA: Metaphysics Research Lab. Retrieved January 11, 2018 from https://plato.stanford.edu/entries/introspection/

Selz, O. (1913). Über die gesetze des geordneten denkverlaufs. Eine experimentelle untersuchung. Stuttgart. Reproduced (in part) in N. H. Frijda, & A. D. de Groot (Eds) (1981), *Otto Selz: His contribution to psychology* (pp. 76–106). The Hague: Mouton Publisher.

Selz, O. (1924). *Die gesetze der produktiven und reproduktiven geistestätigkeit kurzgefasste darstellung*. Bonn. Reproduced in N. H. Frijda, & A. D. de Groot (Eds) (1981), *Otto Selz: His contribution to psychology* (pp. 21–75). The Hague: Mouton Publisher.

Simon, H. A. (1981). Otto Selz and information-processing psychology. In N. H. Frijda & A. de Groot (Eds), *Otto Selz: His contribution to psychology* (pp. 147–163). The Hague: Mouton Publisher.

Simon, H. A. (1999). Karl Duncker and cognitive science. *From Past to Future* 1(2), 1–11. (Republished in 2009).

Stokes, D. E. (1997). *Pasteur's quadrant—Basic science and technological innovation*. Washington, DC: Brookings Institution Press.

Stout, G. F. (1896). *Analytic Psychology* (Vols 1 & 2). London: Swan Sonnenschein.

Suss, J. (2013). Using a prediction and option generation paradigm to understand decision making. Published Doctoral Dissertation, Michigan Technological University.

Suss, J., Belling, P., & Ward, P. (2014). Use of cognitive task analysis to probe option-generation in law enforcement. *Proceedings of the Human Factors and Ergonomics Society Annual Meeting* 58(1), 280–284.

Suss, J., & Ward, P. (2012). Use of an option generation paradigm to investigate situation assessment and response selection in law enforcement. *Proceedings of the Human Factors and Ergonomics Society Annual Meeting* 56(1), 297–301.

Suss, J., & Ward, P. (2013). Investigating perceptual anticipation in a naturalistic task using a temporal occlusion paradigm: A method for determining optimal occlusion points. *Proceedings of the Human Factors and Ergonomics Society Annual Meeting* 57(1), 304–308.

Suss, J., & Ward, P. (2018). Revealing perceptual-cognitive expertise in law enforcement: An iterative approach using verbal report, temporal-occlusion, and option-generation methods. *Cognition, Technology, & Work* 20(4), 585–596.

ter Hark, M. (2007). Popper, Otto Selz and Meinong's gegenstanstheorie. *Archiv für die Geschichte der Philosophie* 89, 60–78.

ter Hark, M. (2010). The psychology of thinking before the cognitive revolution: Otto Selz on problems, schema, and creativity. *History of Psychology* 13, 2–24.

Titchener, E. B. (1910). *A textbook of psychology*. New York: Macmillan.

Titchener, E. B. (1912a). Prolegomena to a study of introspection. *American Journal of Psychology* 23, 427–448.

Titchener, E. B. (1912b). The schema of introspection. *American Journal of Psychology* 23, 485–508.

Titchener, E. B. (1921). Brentano and Wundt: Empirical and experimental psychology. *American Journal of Psychology* 32, 108–120.

von Ehrenfels, C. (1988). Über "gestaltqualitäten". *Vierteljahrsschrift für wissenschaftliche philosophie 14*, 249–292. (Original work published 1890.)

Ward, P., Belling, P., Petushek, P., & Ehrlinger, J. (2017). Does talent exist? A re-evaluation of the nature–nurture debate. In J. Baker, S. Cobley, J. Schorer, and N. Wattie (Eds.), *Routledge handbook of talent identification and development in sport* (pp. 19–34). London: Routledge.

Ward, P., Ericsson, K. A., & Williams, A. M. (2013). Complex perceptual-cognitive expertise in a simulated task environment. *Journal of Cognitive Engineering and Decision Making 7*, 231–254.

Ward, P., Gore, J., Hutton, R., Conway, G., & Hoffman, R. (2018). Adaptive skill as the *conditio sine qua non* of expertise. *Journal of Applied Research in Memory and Cognition 7*(1), 35–50.

Ward, P., Suss, J., Eccles, D. W., Williams, A. M., & Harris, K. R. (2011). Skill-based differences in option generation in a complex task: A verbal protocol analysis. *Cognitive Processing: International Quarterly of Cognitive Science 12*, 289–300.

Watson, J. B. (1913). Psychology as the behaviorist views it. *Psychological Review 20*, 158–177.

Watson, J. B. (1920). Is thinking merely the action of language mechanisms? *British Journal of Psychology 11*, 87–104.

Watson, J. B. (1924a). The place of kinaesthetic, visceral and laryngeal organization in thinking. *Psychological Review 31*, 339–347.

Watson, J. B. (1924b). The unverbalized in human behavior. *Psychological Review 31*, 273–280.

Wertheimer, M. (1982). *Productive thinking* (enlarged edition). Chicago: University of Chicago Press. (Original work published 1945.)

Woodworth, R. S. (1915). A revision of imageless thought. *Psychological Review 22*, 1–27.

Wundt, W. (1897). *Outlines of psychology*. Leipzig: Engelmann. (Original work published 1896)

Wundt, W. (1906). *Die aufgaben der experimentellen psychologie. Unsere Zeit*. Reprinted in *Essays* (2nd edn). Leipzig: Engelmann. (Original work published 1882.)

Wundt, W. (1905). *Völkerpsychologie: Eine untersuchung der entwicklungsgesetze—sprache, mythus und sitte*. Leipzig: Verlag von Wilhelm Englcmann.

Wundt, W. (1907). Über ausfrageexperimente und über die methoden zur psychologie des denkens. *Psychologische Studien 3*, 301–360.

Wundt, W. (1913). *Elemente der Völkerpsychologie: Gurndlinien einer psychologischen entwicklungsgeschichte der menschheit*. Leipzig: Alfred Kröner Verlag.

Yankelovich, D. (1972). *Corporate priorities: A continuing study of the new demands on business*. Stamford, CT: Yankelovich.

CHAPTER 18

CLOSE-TO-PRACTICE QUALITATIVE RESEARCH METHODS

SARAH YARDLEY, KAREN MATTICK,
AND TIM DORNAN

INTRODUCTION

THIS chapter argues that qualitative methods make an important and distinct contribution to the study of professional expertise. It describes various methodologies, encouraging readers already familiar with qualitative research to extend their repertoires. Schön's theory that expert practice "emerges" in unpredictable situations is central to our argument (Schön, 1987). While the strength of quantitative methodologies is that they allow researchers to measure changes and conduct experiments, the need to predetermine variables limits the power of these methodologies to explore indeterminate aspects of expert practice. Qualitative methodologies, which make fewer a priori assumptions, can explore the emergent expertise of professionals as they carry out complex tasks in authentic settings. This chapter highlights *close-to-practice* qualitative methods. It describes several such methodologies, illustrating the different research questions these can answer and explaining how they can explore the messiness and complexity of *real-world* expertise. It then presents qualitative studies which have changed our own understanding of expertise. In making the case for qualitative research, it uses a range of examples from medicine and medical education because these research disciplines have enthusiastically adopted close-to-practice qualitative methodologies. Specific examples are included to help readers visualize how qualitative methodologies might play out in their own disciplines. Finally, the chapter discusses key considerations in undertaking qualitative research and future directions for the interested reader.

Why is Qualitative Research Needed?

This section outlines three arguments for qualitative research into expertise: a *complexity* argument, a *social and professional* argument, and a *special and different* argument.

The Complexity Argument

Expertise is complex because practitioners must make difficult judgments based on imperfect information under time pressure. Medicine has championed qualitative expertise research because doctors routinely base significant decisions on short interactions with patients who are diverse (e.g., rich or poor, native or immigrant, articulate or inarticulate), and whose illnesses have emotional and psychological as well as physical dimensions. They treat presumed rather than *proven beyond doubt* diagnoses, balancing the likelihoods of competing diagnoses and of different interventions leading to more or less favorable outcomes. They are not alone in this. Experts in all fields make, communicate, and justify tough decisions, while managing the expectations of their clients about the likelihood of unpredictable outcomes.

Cognitive psychology has made important contributions to the study of expertise. Researchers have described two types of knowledge: analytical knowledge (Norman et al., 2007) and experiential knowledge. Novices draw heavily on analytical knowledge, gathering information systematically and relying on their limited practical experience to tell them how to proceed. Experts, in contrast, draw heavily on experience and use internal narratives to recognize patterns, draw inferences, and make predictions (Charlin et al., 2007; Schmidt et al., 1990). Expert doctors, for example, draw on their repertoires of *illness scripts* to rule diagnoses in or out, hone their questioning, and choose investigations that can confirm, refine, or refute working diagnoses (Groves et al., 2003). Expert knowledge is qualitatively different from novices' knowledge. Not only do experts understand concepts better but they have a wealth of experience of applying those concepts in practice. Experts can identify the difference between uncertainty about their own knowledge, uncertainty about the limits of knowledge, and uncertainty about which of the former is the true source of uncertainty in any specific circumstance (Fox, 1957).

Quantitative psychological studies of expertise, however, tend to focus on tightly bounded decision making and diagnostic reasoning (Norman 2005). They test expertise under controlled conditions, often with only one right answer. These can reliably measure the knowledge obtained by different people under different training conditions but have a limited capacity to predict how experts will perform in the messy real world. Other quantitative approaches, while making important contributions to practice development, have similar limitations. The evidence-based practice movement valorizes quantitative over *narrative* knowledge (Dornan, Peile, & Spencer, 2008).

Competency-based education (ten Cate, 2016) shifts the emphasis from educating students to think and act independently to training students to achieve predetermined and measurable learning outcomes. Researchers and professional leaders thereby decontextualize and simplify expertise in order to conduct experiments on it and regulate professional performance. Expert practitioners, however, are more at home in messy situations than experimentalists. This allows them to work in more intuitive and individual ways than these pedagogies recognize.

Schön (1983, 1987) used the term *messy* to describe the complex, indeterminate situations that expert practitioners routinely encounter. He uses the term *technical rationality* to describe the science, evidence, and competences, which emanate from the *high hard ground of universities*. Expert practitioners, according to Schön, practice in *swampy lowlands* where technical rationality alone is insufficient. They make messy problems amenable to technically rational solutions by virtue of their ability to frame problems. The writings of other theorists (Billett, 2014; Billett & Bruner, 1986; Eraut, 2004) reinforce Schön's view that professional expertise is often tacit, personal to practitioners, and context-specific.

Quantitative experimentalists might contest our use of the term *complexity* to describe this argument. That is because in vitro research filters out misleading and emotionally salient features of practice, which experts must resolve before they even start to make diagnostic or therapeutic choices. Researchers have a responsibility to *represent complexity well* (Regehr, 2010). This calls for methodologies that can provide valid knowledge without first filtering out contextual features that create uncertainty and indeterminacy. Qualitative methodologies are better suited than quantitative ones in situations that depend on human choices and where values and feelings defy measurement.

The Social and Professional Argument

Experts often work within large and complex organizations, which employ members of different professions, each with their own hierarchy. Here, acknowledging, interpreting, and navigating social contexts is critical to expert judgment. As Bull, Mattick, and Postlethwaite (2013) put it: "This context is far more than the stage on which decision-making happens: it shapes what kinds of decisions are possible and what counts as good decision-making" (p. 402).

Notions of expertise, evidence, and competence that focus attention on knowledge and skills tend to conceptualize practitioners individualistically, rather than as members of social groups or teams. Further, they focus on cognitive attributes rather than on the integrated performance of mind and body in social settings. Medicine has recently re-learned the lesson (known since Hippocrates) that, while doctors must be technically skilled, expert doctors are not solely technicians. Medicine is, essentially, a social practice. Context is a determining feature of expertise. Surgery used to be cited as a counter-example but we now know that technically competent surgeons who lack the social skills to create a collaborative climate in operating theatres have poor patient outcomes (Lingard et al., 2008; McCulloch et al., 2009; Mishra et al., 2008). Research has to address both the social and technical dimensions of expertise, and how the two interact (Papoutsi et al., 2017).

The social argument acknowledges, also, the exercise of power in practice settings. Consider a general practitioner treating a patient who believes their cough is due to pneumonia rather than asthma and will not resolve without antibiotics. Conceding to this patient's wish will contribute to the global epidemic of antibiotic-resistant bacteria. Negotiating a safer outcome requires the doctor to manage the expectations and concerns of patient and carers, who may disagree between themselves. The doctor has to *gatekeep* society's collective health from what the patient may view as the lowly status of a generalist. Precisely the same action by a specialist may be accepted with demur. Expertise research tends to be conducted from a professional standpoint, which bypasses some of this social complexity.

Democratization of knowledge by the Internet is adding to this complexity and changing the nature of specialist as well as generalist expertise in all fields of practice. Experts no longer have privileged access to knowledge. They critically appraise, synthesize, and communicate information to help clients make informed choices. This further perturbs professional–public relationships, already troubled by social mistrust of expertise (see Chapter 4, "Studies of Expertise and Experience," by Collins & Evans, this volume). People empowered by the wider availability of information expect responses to be tailored to their unique needs. Additionally, increasing numbers of stakeholder groups are scrutinizing expert practice, each from a different perspective. A rich understanding of the social environment of practice is needed. This poses research questions like "What do clients expect from safety-critical services funded by general taxation?" and "How do service users access and interpret information before consulting experts?" These are not questions that lend themselves well to quantitative enquiry.

The Special and Different Argument

A third contribution of qualitative research is to explore how an understanding of expertise gained in one setting can be transferred to other settings. This requires in-depth understanding of the nature of work in those settings, differences between their social contexts, and how these interact. A prime example is the unfavorable comparison often made between medicine and civil aviation. Without denying that medicine could be safer, the comparison ignores difference in the indeterminacy of practice in those two settings. A pilot can decide it is too unsafe to fly or eject unsuitable passengers from their plane. The consultant in charge of an emergency department with ambulances lined up outside does not have that luxury. The net result is that the very notions of safety and risk are much more open to interpretation in medicine. A more suitable comparator than aviation might be military personnel who are trained to focus on a common goal but achieve this through semiautonomous functioning and on-the-spot decision making. Such comparisons are reliant on a rich understanding of the realities of practice in a given setting and detailed consideration of the degree to which certain approaches are context-specific or might provide useful insights to another setting, with appropriate tailoring.

Our three arguments share the assumption that expertise involves a far wider range of human capabilities than just cognitive ones. For example, a large multisite ethnographic study of accountability in healthcare (Aveling, Parker, & Dixon-Woods, 2016) concluded that, while moral responsibility is inherent in professional practice, individuals and systems are not mutually exclusive but interacting. Experts need a fine-tuned moral compass and the ability to navigate complex social situations where power is at play as well as intellectual and psychomotor skills. They have to be tolerant of ambiguity and have a capacity to withhold action or act in the face of uncertainty, based on a fine balance of risks and benefits. Qualitative research can help us to understand this complexity.

What is (Different about) Qualitative Research?

Research Topics and Questions

Qualitative research can build a rich picture of phenomena exploring subtle social dynamics and interactions between individuals and employing organizations, as is often the case in expertise development. In such situations, researchers are trying to understand what is happening, when, and why. This chapter focuses particularly on close-to-practice qualitative research because this captures the complexity of expertise as it plays out dynamically in authentic contexts. Researchers may choose to examine one or more of several levels of practice, from the activities of individual professionals, to local settings, or to the entire organization or system in which they work. Researchers typically seek out different stakeholders whose different perspectives and explanations contribute to an in-depth analysis of the practice. The exploratory and explanatory nature of qualitative research makes it better suited to some topics than others and that is reflected in the phraseology of research questions. Closed research questions (inviting yes/no answers), questions about the quantity of something, and comparative questions are generally unsuited to qualitative research, yet inexperienced researchers often ask them. As a rule of thumb, *how* and *why* questions are more appropriate than *whether*, *which*, and *how often* questions. Imagine you are an expertise researcher wishing to understand the expertise of professional musicians. A suitable research question might be: "How do musicians reconcile their wish to build successful careers with the need to protect themselves from locomotor problems caused by overuse?" Or considering front-line lawyers working with under-served populations, who must make sparing use of specialist lawyers charging high fees: "How do generalists experience their responsibilities to serve individual clients well whilst making judicious use of specialist services?" Or an emergency physician who must prioritize care for sick patients who exceed the capacity of their hospital to admit them: "How do front-line practitioners prioritize acutely ill patients for hospital admission when the demand for beds outstrips capacity?" Qualitative research is well placed to answer these types of question because of the scope it allows experts to raise issues that would have been challenging for researchers to anticipate (Brooks & King, 2017; Bryman, 2015; Cohen, Mannion, & Morrison, 2001; Corbin & Strauss, 2008).

Qualitative research can also illuminate and explain quantitative findings. Depending on the topic and research questions, it may be appropriate to use multiple qualitative methods or combinations of qualitative and quantitative methods. The term *mixed methods* usually means combining quantitative and qualitative methods in a single-study design. The term *multiple methods* describes combinations of either quantitative or qualitative methods (see Yardley, Brosnan, & Richardson, 2013 for an example of multiple methods). Terms such as *an embedded qualitative study* are used to describe the collection of qualitative data from participants in a, usually larger, quantitative study. The embedded qualitative study may be designed to clarify quantitative findings, or explain participants' experience and why something did or did not work (e.g., McLachlan et al., 2015; McLellan et al., 2016). Another way qualitative and quantitative research can synergize is when a qualitative interview study identifies constructs to include in a measurement scale. In this case, qualitative and quantitative methodologies are combined sequentially rather than concurrently. Schifferdecker and Reed (2009) provide a helpful typology of categories of mixed methods research designs.

Ontology and Epistemology

Differences between qualitative and quantitative research are far deeper than a choice of whether to use numbers or words to represent more or less the same thing. Numbers are only as good as the a priori choices researchers have made about what to measure. Words give research participants far greater agency and can illuminate realms that neither researcher nor participant could have contemplated a priori. Quantitative and qualitative research represent different research paradigms, with different assumptions about reality and the nature of knowledge. Qualitative researchers' starting assumption is that there is no single external reality or, if there is, then it can never be fully and completely defined—as human nature is to view it through a complex set of pre-existing beliefs and experience. Phenomena, therefore, exist as conceptualized by different people. Different researchers, or even the same researcher on different occasions, can apprehend them in different but equally valid ways. Subjectivity is inherent to qualitative research and, if deliberatively applied to qualitative analysis, an asset rather than a problem. Rather than proving the existence of phenomena and measuring how those phenomena vary, qualitative research provides rich representations of phenomena (Crotty, 1998).

Critiques of interview-based research have questioned whether participants *tell the truth*, whether they give partial accounts of events (intentionally or otherwise), and whether what they say reflects what they do (Hammersley, 2005). These concerns arise because people who are thinking within one paradigmatic set of assumptions apply those assumptions to a different paradigm. If one believes in a single reality that can be measured and known, then the assumptions underpinning qualitative research are problematic. On the other hand, if one believes reality or truth for human beings can only be partially known through their personal perspectives, then qualitative research gives access to multiple understandings of reality—and quantitative research lacks richness. Qualitative researchers are careful not to use leading questions or value-laden language that would make it hard for interview participants to speak freely. They may increase the validity of their work by grounding interview conversations in the realities of practice (such as in close-to-practice qualitative research) to avoid social desirability biases that may occur in less contextualized conversations (Miles & Huberman, 1994; Saks & Allsop, 2007).

Theory and Methodology

Qualitative research is not a single methodology but a family of methodologies that share an ability to find meaning in ill-structured situations. Qualitative research is made rigorous by purposefully choosing a theory of knowledge that is suited to the research question. That is what the term *methodology* means. There are many theories of knowledge and qualitative methods, allowing researchers to answer a wide range of research questions, or consider things from a wide range of perspectives. For example, there are methods that allow researchers to interpret interview or focus group transcripts subjectively and develop a theory of the phenomenon of interest by alternately interpreting data collected before and gathering fresh data (e.g., constructivist grounded theory; Watling & Lingard, 2012). Other methods allow researchers to interpret and report peoples' lived experiences by listening deeply to how subjects speak of their experiences and interpreting the phenomenon from participants' perspectives (e.g., interpretive phenomenology; King & Horrocks, 2010). Yet others examine how one person's choice of words, use of metaphors, and grammatical constructions exercise power over another person (e.g., discourse analysis; Hodges, Kuper, & Reeves, 2008). Focus groups, interviews, constant comparative analysis, and so on contribute to qualitative research but they are not, of themselves, qualitative research. Using those methods within a declared methodology to answer an appropriate research question, however, is qualitative research (Silverman, 2005).

It is critical that qualitative research shows coherence between the overarching aim, research question(s), theoretical perspective, and methodology. A defining feature of any piece of high-quality qualitative research is that it has an explicit conceptual orientation. In other words, theory helps researchers conduct their work in rigorous ways and arrive at valid conclusions. There are two ways in which theory does this. *Subject matter theories* provide ways of thinking about a topic. And *methodological theories* help researchers arrive at valid conclusions by defining what constitutes knowledge. Our argument that human activities are inseparable from the contexts in which they take place, for example, reflects our orientation toward social theory. As a result, some of our work draws on sociocultural theories, including activity theory (Johnston & Dornan, 2015). Analyzing textual data about complex medical expertise from the perspective of social theory would help a researcher build on a corpus of widely accepted knowledge and relate their findings to the findings of other researchers. Regarding methodological theories, quantitative research is more straightforward than qualitative research because it regards phenomena as existing or not existing. If they do exist and are worthy of investigation, they are measured. Core tenets like representativeness, objectivity, reliability, and probability define quantitative rigor and validate conclusions drawn from measurement.

The relativist nature of qualitative inquiry opens new possibilities for researching expertise. A realist approach (more aligned with quantitative research) would require a clear definition of expertise, which would help researchers operationalize a priori constructs to measure how those constructs differ under different conditions. A relativist approach (more aligned with qualitative research) makes it possible to take the stance: "I can't define expertise but I know it when I experience it; let me find instances of it, explore them, and see how that allows me to think about expertise in new ways." It is possible, moreover, to choose a qualitative methodology that serves a specific purpose. Grounded theory could explain how experts match available solutions to the expectations of different clients. Phenomenology could richly describe how clients experience expert practitioners. And discourse analysis could analyze how experts encourage clients either to exercise autonomy or to do just what the expert wants.

The exploratory and inherently uncertain nature of qualitative research, moreover, allows researchers to find out how phenomena are enacted in social settings. In a series of our own studies, for example, we came to a startling conclusion. Whereas curricula teach young doctors how to choose medications for patients in hospital, there are many occasions when prescriptions are more the product of an idiosyncratic set of circumstances than the choice of any one individual (McLellan et al., 2015; Mattick, Kelly, & Rees, 2014—see later in this chapter, Papoutsi et al., 2017). The requisite expertise for young doctors, then, is to manage the social context of practice as much as it is to make rational therapeutic choices. This has significant implications for curricula and clinical services.

Exemplar Close-to-Practice *Qualitative Research Methods*

This section presents approaches to qualitative research that ensure qualitative data are collected close-to-practice and are, therefore, relatively true to the contextualized practice under scrutiny.

One approach is ethnography (Hammersley & Atkinson, 2007). Different academic disciplines and practice traditions use the term in different ways, sometimes far removed from its original meaning. Ethnography involves immersive study of people, their cultures, their customs, and their habits, and how these differ between populations or ethnic groups. Ethnographers observe, sometimes as *flies on the wall*, sometimes as participants, how social groups live their lives. Ethnography typically involves participant observation whereby researchers observe extensively in a given setting (e.g., a professional workplace), shadowing participants within that setting (receptionists, managers, accountants, and practitioners, for example) and interviewing people whose experiences might illuminate the researcher's interpretation. Ethnography can provide a rich understanding of the context in which expertise develops and the individuals and activities contributing to that development.

A narrative approach is underpinned by the assumption that people make sense of their experiences through narrative (stories). Retelling those stories can yield data that are close-to-practice (Riessman 2008). Participants in expertise research would be asked to recount significant events in their experience of practice, and provide real examples and events, typically as a story with a beginning, middle, and end. How participants select events to describe and sequence their stories gives researchers deep insight into the practice.

Another way of ensuring research data are contextualized to practice is to incorporate visual stimuli from workplaces into interviews. Video reflexivity is one such approach, whereby video recordings of clinical practice (either real or simulated) are used as a stimulus for evaluating practice and promoting behavior change of individual practitioners or healthcare teams. For example, Gough, Yohannes, and Murray (2016) videoed final-year physiotherapy students working through simulated practice scenarios and then interviewed them individually to explore their perspectives on that action. Stimulated recall is another approach. Bull et al. (2013) used field notes they had made while observing junior doctors' ward-based practice to stimulate recall of events for discussion in subsequent research interviews. Walking interviews, when the research participant and interviewer walk through the relevant environment while undertaking the interview, are another approach. In a walking interview the researcher can experience something of the participants' daily routines and use artifacts and environments as prompts for discussion. Dube, Schinke, Strasser, and Lightfoot (2014) used a participant led *guided walk* to explore the lived experiences of medical students undertaking placements by moving through their environment with them while discussing their experiences, thereby generating data from context-rich interactions. A hypothetical example in expertise development would be to video-record a healthcare team making a complex decision and then ask participants to review, prompted by the video tape, and discuss their various contributions to that process.

Researchers may also give research participants tasks. In photo- or object-elicitation interviews, for example, participants bring an image or object which they see as relevant to the topic under discussion (Kronk, Weideman, Cunningham, & Resick, 2015). Graphic elicitation interviews incorporate a drawing task within an interview; for example, to draw out a timeline, or relationships between groups within an organization (Bagnoli, 2009). Many of these approaches can be applied within group interviews or focus groups, as well as individual interviews (Kvale & Brinkman, 2009).

Data analysis procedures are equally varied. Again, these must be congruent with the researcher's explicitly stated theoretical perspective and approach to data collection. For example, a researcher might thematically analyze qualitative data, seeking patterns across and within data sets. Narrative analysis examines research participants' stories in order to understand how characters and events contribute to the richness of the rich story (see Riessman, 2008). A question well suited to thematic analysis might be "What do early career professionals regard as important for expertise and why?"; a question well suited to narrative analysis might be "How do early career professionals describe their experiences of leadership development?" If thematic analysis is chosen, then the research may stop at the point of having identified and described component themes, with exemplar quotes to illustrate each theme. They may, alternatively, undertake an interpretative analysis and/or develop a new theory or a three-dimensional model or framework.

As the body of published qualitative studies grows, researchers are increasingly synthesizing *meta-interpretations* from multiple primary investigations addressing a single topic. Readers are referred to a recent journal article (Gough & Thomas, 2016) which describes diverse approaches to systematic reviews of research. In addition, there is a proliferation of qualitative metasynthesis methodologies, which are increasingly refined and sophisticated. Secondary analysis of existing qualitative datasets is another emerging approach. This is attractive because it is resource-efficient, although researchers may face methodological problems and must consider any ethical issues raised by potentially reusing data without participants' explicit consent (Hinds, Vogel, & Clarke-Steffen, 1997; Yardley, Watts, Pearson, & Richardson, 2014).

QUALITATIVE STUDIES THAT HAVE CHANGED OUR UNDERSTANDING OF EXPERTISE

Here, we describe qualitative studies that have changed our own understanding of expertise. Although our selection is from medicine, their theoretical implications are wider.

Establishing the Value of Qualitative Research

Long before medical education was an identifiable research field, turning novice medical students into expert doctors drew the attention of sociologists. Two large sociological studies are credited with introducing qualitative research (including, but not limited to, classical ethnography) to the study of medical expertise. First published was *The Student-Physician* (Merton, Reader, & Kendall, 1957), soon followed by *Boys in White* (Becker, Geer, & Strauss, 1961). The importance of these studies is that they recognized the importance of social and professional factors in being (and becoming) a doctor, and made innovative use of qualitative research to explore them.

American society's increasing interest in professionalization led Merton and colleagues to investigate *the making of the medical man.* They reconceptualized the medical school as a "social environment in which the professional culture of medicine is variously transmitted to novices" (Preface, p. vii). They sought to understand how novices developed the attitudes, self-image, and values of expert doctors. They investigated students' career choices, development of attitudes, preferences for different types of patients, and hence types of practice. A series of qualitative studies took a broadly social constructionist angle, positioning students as subjects within social environments. Ethnographic data were generated by observing students and making field notes, and interviewing them. This research identified tensions arising from the pressure to accommodate expanding scientific knowledge into an already overfilled curriculum. It showed how medical students socialized into the powerful position of doctors by modeling on faculty and senior doctors. Students were not just seeking to know what doctors know but to be what doctors are.

Boys in White sought to explain student culture, focusing on social interactions in medical school and students' changing perspectives as they progressed through their first year. This study showed how students transitioned from novices towards expert by being subversive. For example, they rapidly changed from trying to *learn everything* to recognizing this was impossible and developing strategies to meet faculty demands, while accommodating the knowledge that real-life practice could not be fully anticipated through academic study. Becker et al. were more explicit than Merton et al. about how they conducted their research. Their starting point was "to discover what medical school did to medical students other than giving them a technical education" (p. 17). The theoretical orientation of their studies was toward collective social action; in particular, the concept of symbolic interaction and the study of phenomena that caused tension among their medical student subjects. This led the research team to rely heavily on participant observation as a method for data generation, complemented by other qualitative methods such as interviewing. *Boys in White* further exposed the importance of social interaction and the importance of recognizing expertise as an identity as well as a knowledge base and set of practical and technological skills. These findings are not unique to medicine, or even to the so-called professions, vocations, or scientific disciplines. The expertise, often called know-how or work-arounds, gained in workplaces may not always be formally recognized but is crucially important for understanding how workers function and handle uncertainty (for more recent further examples see the work of Stephen Billett, 2014, and for examples outside of medicine see Scott, 1998). Together these studies changed our understanding of expertise development by drawing attention to social and cultural influences that shape expertise in a person, challenging the view that expertise could solely be conceptualized as scientific and technical knowledge, and identifying the importance of considering organizational and interpersonal interactions in studies of expertise development rather than simply measuring individual attainment.

Qualitative Research Addressing Contemporary Concerns

The concept of safety is dominating thinking in health and other professions. Pursuing an appropriate course of action is equated with practicing *safely*. Avoidable harm caused by poor performance of technical procedures or omissions of simple protective actions has rightly caused outrage. But while medicine and other professions need to reduce preventable error, the idea that all risk is predictable and preventable is contentious. It is true that risk may be predictable and preventable in determinate situations, but it cannot, by definition, be wholly preventable in indeterminate ones. Applying the blanket assumption of simplicity to complex situations will, according to the law of unintended consequences, do harm as well as good—and maybe even more of the former than the latter. Years of experience help experts choose the least unsafe course of action in indeterminate, safety-critical situations but they may not understand how they do so, let alone explain this to novices. Organizations typically put in place mass mandatory training to prevent harms such as this. This may do no more than drive uncertainty underground. What may be more effective is the precise opposite—to legitimize uncertainty.

As part of this safety debate, researchers have sought to understand how issues of safety relate to the development of expertise, for example in how the process of decision making is conducted. The use of qualitative research methods has facilitated investigation of *real-life* practice, achieving a level of dynamic understanding of situated expertise to direct future education interventions for the development of expertise, increasing the likelihood of real-world impact. For example, in healthcare, prescribing errors related to antibiotics are common. Doctors have to weigh up significant potential consequences arising from either under- or over-prescription for individual patients and populations. Judgements about when and what to prescribe are often required without full clinical information such as laboratory test results. Patient pressure to prescribe can contribute to the challenges. Mattick et al. (2014) undertook a narrative interview study of junior doctors at two UK hospitals. The doctors were asked to tell stories about their experiences of prescribing antibiotics, in order to understand what types of experiences they have and how they make sense of these experiences. The researchers drew on social constructionist epistemology and used multiple analytical approaches to answer the research questions, including framework analysis (a type of thematic analysis; Ritchie, Spencer, Bryman, & Burgess, 1994) and in-depth narrative analysis. They found that the decision to prescribe, and if so what to prescribe, is much more complicated and socially mediated than has been recognized. Junior doctors reported significant variability in local practices, received seemingly conflicting advice from senior staff, and felt they rarely received feedback on the consequences of their decisions. The authors' use of constructionist methodologies showed that social hierarchies and expectations were coming into conflict with evidence-based practice when doctors prescribed antibiotics. The problem was not so much that doctors lacked expertise as that they found it hard to decide which expertise should have most weight when making decisions. This finding suggested that informal practice-based opportunities to discuss decision making and receive feedback were more likely to reduce errors than provide information.

Other studies have used alternative frameworks such as activity theory, a sociocultural theory that provides a framework for studying the influence of informal learning in workplaces on medical practice. McLellan et al. (2015) studied how medical students learn to prescribe in a study informed by cognitive psychology, sociocultural theory, and systems thinking. Participants kept audio diaries over a two-week period and participated in minimally structured qualitative interviews. The researchers also observed practice and conducted short in-situ interviews with participants who had particular contributions to make. Grounded theory analysis demonstrated a complex interplay between individual students and social dimensions of learning: learners needed to be situated in the right environment and exposed to meaningful learning opportunities (including active engagement in the process) if they were to develop their own internal cognitive-based expertise for future prescribing tasks. That is, the ability to develop expertise was dependent on unquantifiable but essential social factors. Newly qualified doctors in the UK are required to demonstrate their prescribing abilities in decontextualized, simulated assessments. Qualitative research by Mattick et al. (2014) and McLellan et al. (2015) explains this does not prevent prescribing mishaps and suggests potential solutions.

Studies of workplace-based and vocational learning have shown how pedagogical processes can support the development of expertise (Yardley, Teunissen, & Dornan, 2012) but there are tensions between what is best for learners and what is best for patients. Returning to the aviation metaphor, people would like to be treated by expert doctors, just as they would like to be flown by expert pilots, and know that an expert engineer designed their plane. The snag is that health facilities may have come into existence because of historical accident rather than purposeful design. Simulation can, as in aviation, improve doctors' expertise in more clearly defined and predictable situations (see Chapter 13, "Representative Test and Task Development and Simulated Task Environments," by Harris, Foreman, & Eccles, this volume). But practicing medicine can be more like the Battle of Britain than piloting a long-haul scheduled flight so simulation has limitations, at least for some important aspects of clinical practice. Unless patients submit to the training of novices, including whatever risk is entailed in that, we cannot educate excellent doctors. Despite all the advances in medical education over the past half century, newly qualified doctors are novices in the *practice* of medicine to an even greater extent than they were in earlier times.

De Feijter, de Grave, Dornan, Koopmans, and Scherpbier (2011) sought to understand medical student perceptions of patient safety during the transition from undergraduate to postgraduate training. This transition represents a significant shift in clinical responsibility. The researchers conducted a thematic analysis of data from focus groups with final year medical students and used activity theory as their analytical framework. This identified contradictions between their need to learn and the need for patients to receive safe care. This study identified interconnections between identity and expertise and showed the importance of considering environmental factors that might cause someone to act beyond their expertise. In some cases, students would delay or defer seeking help in situations beyond their expertise because they did not perceive the risk of harm to patients to outweigh the risks to themselves of *not managing* the situation. This is all the more concerning when one considers it alongside the findings of Bull et al. (2013), whose qualitative study of junior doctors' decision making found that this group were often enacting *inherited* decisions, without participating or understanding how their seniors (the experts) had reached their conclusions. Furthermore, other studies (including those highlighted earlier) have shown that junior doctors are influenced by cultural rules and the personal preferences of seniors, where etiquette can dictate that one should not question their expertise. This illustrates that, despite our arguments earlier that medical expertise is not entirely cognitively based or technical, the embodiment of expertise can itself be used as a form of power (Charani et al., 2013). The senior doctors who see any attempt to influence their prescribing as an affront to their autonomy may in fact be choosing to work outside a system designed to support their expertise (Aveling, Parker, & Dixon-Woods, 2016).

Qualitative Research Enriching the Conceptualization of What Makes an Expert

Throughout this chapter, we have argued that experts make socially situated complex clinical judgements and that qualitative research can uncover processes of expert judgement as well as mechanisms of developing this expertise. In a further example, Cristancho, Lingard, Forbes, Ott, & Novick (2016) draw on their study of surgeons' experiences in complex situations to question the idea that experts are simply better at problem solving. These authors used systems engineering theory to explore problem definition (rather than problem solving) in authentic clinical judgements. They started from the premises that person and context were inseparable, and that what emerges as a judgement is an act of choice. Using the concept of a *rich picture* (whereby surgeons were asked to produce and describe a picture or diagram of complex operations) alongside observations and interviews the researchers were able to investigate how problem definition occurred and was refined by different people interacting during surgical operations and to explore different participant perspectives on these interactions. Part of the analysis process involved a gallery walk whereby two researchers walked around a room with the pictures on display and discussed, compared, and contrasted the different pictures through conversation and written memos. What they found was that problem definition was a live evolving process of making sense of *what is going on* that experts responded to flexibly, adapting their choices and behaviors. Hence, expertise was not knowing what solution to deliver but knowing how to respond to a changing complex situation (see Chapter 51, "Reflections on the Study of Expertise and Implications for the Future of the Field," by Ward et al., this volume).

Key Considerations in Qualitative Research

We hope we have persuaded readers of the potential for qualitative methods, particularly those that involve observation within authentic workplace settings, to contribute to research into expertise development. This section aims to highlight some key considerations when researching expertise using qualitative methods: quality, research ethics, and researcher training and development.

Quality in Qualitative Studies

It is clearly important to ensure that qualitative research is high quality, in order to maximize its contribution to knowledge and theory and to underpin the development and evaluation of various initiatives. Key features of quality are likely to include a compelling description of the problem needing research, a clear aim and/or research question(s), details of the research team and their theoretical framework, a clear and detailed justification and account of the methodological approach, detailed results that are clearly derived from the methodological approach and address the original research question, and conclusions that do not overstate the findings. However, appropriate criteria with which to judge the quality of qualitative research is an ongoing debate, partly due to the significant diversity of qualitative methodologies which do not lend themselves to a *one size fits all* approach. It is challenging to envisage a common list of features of high-quality qualitative research that are applicable across the diverse range of methodologies that sit under this umbrella term. There are also understandable challenges when concepts from quantitative research, such as validity and reliability, are transferred to qualitative research, given the different underpinning assumptions and worldview. A useful summary of the debate can be found in Hammersley & Atkinson (2007). Despite the ongoing discussion, sets of criteria for qualitative research are increasingly available and, in some cases, being made a requirement for publication in peer-reviewed journals. One example is the COnsolidated criteria for REporting Qualitative research (COREQ), a 32-item checklist for interviews and focus groups with three domains (research team and reflexivity, study design, analysis and findings). The main aim of COREQ is to improve the quality of reporting of qualitative research, which enables the reader to make a better-informed judgment of quality, but implicit within it are some assumptions about quality in qualitative research. For example, COREQ items prompt the appraiser to consider whether the following aspects are reported: "What experience or training did the researcher have?", "What are the important characteristics of the sample?", "Were themes identified in advance or derived from the data?", and "Was there consistency between the data presented and the findings?" Other checklists, such as the Critical Appraisal Skills Program (CASP) checklist for qualitative research, are more explicit in the link between the checklist and judgment of quality of qualitative research, asking the appraiser to make judgments such as "Was the research design appropriate to address the aims of the research?" and "Have ethical issues been taken into consideration?"

Ethical Aspects of Qualitative Research

The protection of participants and other stakeholders is a key concern, even when the focus of the research is on professionals rather than those arguably more in need of protection (such as patients within the healthcare context). Research which involves access to the workplace setting for the purpose of data collection poses specific considerations. Gatekeepers who can introduce the researcher to key staff and advocate for their project will help enormously with buy in for and recruitment to the research project. These relationships may require substantial investment, however, and the research will need to ensure that gatekeepers are familiar with research ethics principles, for example to ensure that research participants are free from any coercion or undue influence. Where field notes or video are being made in workplace, appropriate informed consent will be required and confidential treatment of information including good data storage to provide assurance about security will be important. The researcher may need to undergo checks, such as the Disclosure and Barring Service checks in the UK for previous criminal convictions, and have an institutional sponsor, for example from their university. They may also need to undergo specific training, for example relating to safety in that environment, or to ensure an understanding of the relevant legislative or ethical frameworks that apply. There may also be multiple research ethics approval processes, for example to ensure scrutiny from the university perspective (where the researcher may be employed) and the workplace perspective (where the participants may be employed), often making this stage more involved than for some other kinds of research. A useful discussion of the principles of ethical research can be found in the Economic and Social Research Council guidelines (ESRC, 2015).

Researcher Development and Support

Conducting high-quality qualitative research also requires expertise. Many researchers and practitioners from scientific disciplines are more familiar with quantitative than qualitative research. Before developing their qualitative research expertise, researchers may have studied a wide range of disciplines. Therefore, qualitative research teams are often multidisciplinary, each contributing their own expertise and insights to the research at hand. This has substantial potential benefits but may also highlight a range of development needs. A researcher will need development opportunities to keep abreast of developments in their own discipline, in order to remain a *content expert*. However, they will also need a broader understanding of research methodologies and a good understanding of the context in order to see the niche for their research. Furthermore, communicating the nuances of their methodological approach to the wider research team, and key stakeholders who may have a different paradigmatic stance, may require the researcher to invest in their communication capabilities.

Conclusions, Future Directions, and Recommendations

Expertise is inherently linked to the context in which experts work, as we have explained. Therefore, it is critical to understand the context and be open to unexpected findings about that context in order to truly understand how expert judgment works in the real world. This is not a static picture either: the context is changing as we have shown. Changing times call for new types of expertise, and experts who are flexible and adaptable and who understand the need to learn at all stages of a professional career. Expertise may be less about deep knowledge and technical skill, but more about making judgements in complex situations. There are probably no short-cuts to this kind of expertise but the importance of exposure to complex problems in situ in the workplace will be critically important to build experience. Employers will need to value and support those on the journey toward expertise and understand the challenges that this journey will involve. Failure to do this will lead to provision of suboptimal professional practice and the loss of talented young people from the professions. The application of qualitative methodologies to the study of expertise provides researchers with tools to explore and explain contextualized and social practices. Qualitative methods are used to provide answers to *how* and *why* questions and to disentangle the impact of different interactions occurring in complex and uncertain situations. When people are involved, the use of qualitative methods can illuminate how the human drive to create meaning from experience shapes attitudes and behaviors. Although close-to-practice qualitative research is not without its challenges, the rewards of undertaking this kind of research usually make up for the effort invested.

In this chapter, we have outlined the possibilities and benefits to be gained, demonstrating applications within medicine and medical application. We have demonstrated that the complexity of expertise arises because it is socially situated and mediated through human interactions in workplaces and elsewhere. Those in other fields also dealing with complexity, uncertainty, and the need for expert judgment under pressure will recognize the potential for transfer to their own fields, although tailoring may be required.

To summarize, qualitative research allows us to develop understanding of when, how, and why the social environment impacts on the translation of cognitive and technical aspects of knowledge and skills into embedded practice, providing a means to identify targeted solutions to the challenges of transitioning from novice to expert as well as the development of new expertise following scientific advancements. For disciplines and professions to engage with the challenges outlined in this chapter (which makes a strong case for the need for qualitative research), a critical first step will be the development of awareness and appreciation of the potential of qualitative research amongst a larger proportion of researchers and practitioners. We advise those previously unfamiliar with qualitative methods to start by expanding their reading further with reference to texts focusing on practical guidance for the use of these methods in the design and conduct of future work and methodologies that have proved fruitful in researching professional practice and workplaces (Hager, Lee, & Reich, 2012; Layder, 1997; Malloch, 2013; Mason & Dale, 2010; see full reference list also). Subsequent to this, formal education and training in methodologies, and qualitative methods with application through cross-disciplinary networks and collaborations holds substantial potential for realizing the added value qualitative methods can offer to our understanding of expertise.

References

Aveling, E., Parker, M., & Dixon-Woods, M. (2016). What is the role of individual accountability in patient safety? A multi-site ethnographic study. *Sociology of Health and Illness 38*, 216–232.

Bagnoli, A. (2009) Beyond the standard interview: The use of graphic elicitation and arts-based methods. *Qualitative Research 9*, 547–570.

Becker, H., Geer, B., & Strauss, A. (1961). *Boys in white*. New York: Transaction.

Billett, S. (2014). *Mimetic learning at work: Learning in the circumstances of practice*. New York: Springer.

Billett, S., & Bruner, J. (1986). *Actual minds, possible worlds*. Cambridge, MA: Harvard University Press.

Brooks, J., & King, N. (2017). *Applied qualitative research in psychology*. London: Palgrave.

Bryman, A. (2015). *Social research methods*, 5th edn. Oxford: Oxford University Press.

Bull, S., Mattick, K., & Postlethwaite, K. (2013). "Junior doctor decision making: isn't that an oxymoron?" A qualitative analysis of junior doctors' ward-based decision-making. *Journal of Vocational Educational Training 65*, 402–421.

Charani, E., Castro-Sanchez, E. C., Sevdalis, N., Kyratsis, Y., Drimright, L., Shah, N., & Holmes, A. (2013). Understanding the determinants of antimicrobial prescribing within hospitals: the role of "prescribing etiquette." *Clinical Infectious Diseases* 57, 188–196.

Charlin, B., Boshuizen, H. P. A., Custers, E. J., & Feltovich, P. J. (2007). Scripts and clinical reasoning. *Medical Education* 41, 1148–1184.

Cohen, L., Mannion, L., & Morrison, K. (2001). *Research methods in education*, 5th edn. London: Routledge.

Corbin, J., & Strauss, A. (2008). *Basics of qualitative research*, 3rd edn. London: Sage.

Cristancho, S., Lingard, L., Forbes, T., Ott, M., & Novick, R. (2016). Putting the puzzle together: The role of "problem definition" in complex clinical judgement. *Medical Education* 51, 207–214.

Crotty, M. (1998). *The foundations of social research: Meaning and perspective in the research process*. London: Sage.

De Feijter, J., de Grave, W., Dornan, T., Koopmans, R., & Scherpbier, A. (2011). Students' perceptions of patient safety during the transition from undergraduate to postgraduate training: an activity theory analysis. *Advances in Health Science Education* 16, 347–358.

Dornan, T., Peile, E., & Spencer, E. (2008). On "evidence." *Medical Education* 42, 232–234.

Dube, T., Schinke, R., Strasser, R., & Lightfoot, N. (2014). Interviewing in situ: employing the guided walk as a dynamic form of qualitative inquiry. *Medical Education* 48, 1092–1100.

Economic and Social Research Council (ESRC) (2015). *ESRC framework for research ethics*. London: ESRC.

Eraut, M. (2004). Informal learning in the workplace. *Studies in Continuing Education* 26, 247–273.

Fox, R. (1957). Training for certainty. In R. Merton, G. Reader, & P. Kendall (Eds), *The student-physician* (pp. 207–230). Cambridge, MA: Harvard University Press.

Gough, D., & Thomas, J. (2016). Systematic reviews of research in education: Aims, myths and multiple methods. *Review of Educational Research* 4, 84–102.

Gough, S., Yohannes, A., & Murray, J. (2016). Using video-reflexive ethnography and simulation-based education to explore patient management and error recognition by pre-registration physiotherapists. *Advances in Simulation* 1, 9.

Groves, M., O'Rourke, P., & Alexander, H. (2003). The clinical reasoning characteristics of diagnostic experts. *Medical Teacher* 25, 308–313.

Hager, P., Lee, A., & Reich, A. (2012). *Practice, learning and change: Practice-theory perspectives on professional learning*. London: Springer.

Hammersley, M. (2005). Ethnography: Potential, practice, and problems: Criticism of Interview-based Qualitative Research. *Qualitative Research Methodology Seminar Series*. Available from http://eprints.ncrm.ac.uk/24/. Accessed 8 May 2017.

Hammersley, M., & Atkinson, P. (2007). *Ethnography: Principles in practice*. London: Routledge.

Hinds, P., Vogel, R., & Clarke-Steffen, L. (1997). The possibilities and pitfalls of doing a secondary analysis of a qualitative data set. *Qualitative Health Research* 7, 408–424.

Hodges, B., Kuper, A., & Reeves, S. (2008). Discourse analysis. *British Medical Journal* 337, 570–572.

Johnston, J., & Dornan, T. (2015). Activity theory: Mediating change in medical education. In J. Cleland, & S. Durning (Eds), *Researching medical education*. Edinburgh: ASME.

King, N., & Horrocks, C. (2010). *Interviews in qualitative research*. London: Sage.

Kronk, R., Weideman, Y., Cunningham, L., & Resick, L. (2015). Capturing student transformation from a global service-learning experience: The efficacy of photo-elicitation as a qualitative research method. *Journal of Nursing Education* 54, s99–s102.

Kvale, S., & Brinkman, S. (2009). *Interviews: Learning the craft of qualitative research interviewing.* London: Sage.

Layder, D. (1997). *Modern social theory: Key Debates and new directions.* London: Routledge.

Lingard, L., Regehr, G., Orser, B., Reznick, R., Baker, G., Doran, D., . . . Whyte, S. (2008). Evaluation of a preoperative checklist and team briefing among surgeons, nurses and anesthesiologists to reduce failures in communication. *Archives of Surgery* 143, 12–17.

McCulloch, P., Mishra, A., Handa, A., Dale, T., Hirst, G., & Catchpole, K. (2009). The effects of aviation-style non-technical skills training on technical performance and outcome in the operating theatre. *British Medical Journal: Quality & Safety* 18, 109–115.

McLachlan, S., Mansell, G., Saunders, T., Yardley, S., van der Windt, D., Brindle, L., . . . Little, P. (2015). Symptom perceptions and help-seeking behavior prior to lung and colorectal cancer diagnoses: A qualitative study. *Family Practitioner* 32, 568–577.

McLellan, L., Dornan, T., Newton, P., Williams, S., Lewis, P., Steinke, D., & Tully, M. (2016). Pharmacist-led feedback workshops increase appropriate prescribing of antimicrobials. *Journal of Antimicrobial Chemotherapy* 71, 1415–1425.

McLellan, L., Yardley, S., Norris, B., de Bruin, A., Tully, M. & Dornan, T. (2015). Preparing to prescribe: how do clerkship students learning in the midst of complexity? *Advances in Health Science Education* 20, 1339–1354.

Malloch, M. (2013). *The Sage handbook of workplace learning.* London: Sage.

Mason, J., & Dale, A. (2010). *Understanding social research: Thinking creatively about method.* London: Sage.

Mattick, K., Kelly, N., & Rees, C. (2014). A window into the lives of junior doctors: narrative interviews exploring antimicrobial prescribing experiences. *Journal of Antimicrobial Chemotherapy* 69, 2274–2283.

Merton, R. K., Reader, G., & Kendall, P. L. (1957). *The student-physician: Introductory studies in the sociology of medical education.* Cambridge, MA: Harvard University Press.

Miles, M. B., & Huberman, A. (1994). *An expanded sourcebook: Qualitative data analysis*, 2nd edn. Thousand Oaks, CA: Sage.

Mishra, A., Catchpole, K., Dale, T., & McCulloch, P. (2008). The influence of non-technical performance on technical outcome in laparoscopic cholecystectomy. *Surgical Endoscopy* 22, 68–73.

Norman, G. (2005). Research on clinical reasoning: past history and current trends. *Medical Education* 39, 418–427.

Norman, G., Young, M. & Brooks, L. (2007). Non-analytical models of clinical reasoning; the role of experience. *Medical Education* 41, 1140–1145.

Papoutsi, C., Mattick, K., Pearson, M., Brennan, N., Briscoe, S., & Wong, G. (2017). Social and professional influences on antimicrobial prescribing for doctors in training: a realist review. *Journal of Antimicrobial Chemotherapy* 72, 2418–2430.

Regehr, G. (2010). It's NOT rocket science: rethinking our metaphors for research in health professions education. *Medical Education* 44, 31–39.

Riessman C. (2008) *Narrative methods for the human sciences.* Thousand Oaks, CA: Sage.

Ritchie, J., Spencer, L., Bryman, A. & Burgess, R. G. (1994). Qualitative data analysis for applied policy research. In A. Bryman & R. G. Burgess (Eds), *Analyzing qualitative data* (pp. 173–194). London: Routledge.

Saks, M., & Allsop, J. (2007). *Researching health: Qualitative, quantitative and mixed methods.* London: Sage.

Schifferdecker, K., & Reed, V. (2009). Using mixed methods research in medical education: basic guidelines for researchers. Medical Education 43, 637–644.

Schmidt, H. G., Norman, G. R., & Boshuizen, H. P. A. (1990). A cognitive perspective on medical expertise: theory and implications. *Academic Medicine 10,* 611–621.

Schon, D. A. (1983). *The reflective practitioner.* New York: Basic Books.

Schon, D. A. (1987). *Educating the reflective practitioner.* San Francisco: Jossey-Bass.

Scott, J. (1998). *Seeing like a state: how certain schemes to improve the human condition have failed.* London: Yale University Press.

Silverman, D. (2005). *Doing qualitative research,* 2nd edn. London: Sage.

Ten Cate, O. (2016). Competency-based medical education and its competency frameworks. In M. Mulder (2016) *Competence-based vocational and professional education* (pp. 903–929). Basel: Springer.

Watling, C., & Lingard, L. (2012). Grounded theory in medical education research: AMEE Guide No. 70. *Medical Teacher 34,* 850–861.

Yardley, S., Brosnan, C., & Richardson, J. (2013). Sharing methodology: a worked example of theoretical integration with qualitative data to clarify practical understanding of learning and generate new theoretical development. *Medical Teacher 35,* e1011–e1019.

Yardley, S., Teunissen, P., & Dornan, T. (2012). Experiential learning: AMEE Guide No. 63. *Medical Teacher 34,* e102–e115.

Yardley, S., Watts, K., Pearson, J. & Richardson, J. (2014). Ethical issues in the reuse of qualitative data: perspectives from literature, practice and participants. *Qualitative Health Research 24,* 102–113.

CHAPTER 19

INCIDENT-BASED METHODS FOR STUDYING EXPERTISE

LAURA G. MILITELLO AND SHILO ANDERS

Introduction

INCIDENT-BASED interview methods have been used in many settings to better understand expert decision making and other macrocognitive activities. In this chapter we discuss four types of incident-based interview methods, examine analysis strategies, and provide examples of products resulting from this type of research and their influence on applications. We consider practical issues associated with the use of incident-based interview methods and discuss future directions.

Incident-based interview methods include retrospective and concurrent accounts of what the expert senses, attends to, and takes meaning from in the context of a specific incident. These methods provide an important counterpoint to experimental methods that employ a more controlled and objective view of expert behavior, but require inference on the part of the investigator to understand cognitive activity. Incident-based interview methods, in contrast, provide a first-person perspective as the interviewee shares his or her perception and framing of the situation, actions considered, actions taken, and the underlying rationale.

Many alternative interview strategies have been used to unpack expertise: the most well known include hierarchical task analysis (Annett, 2004); goals, operators, methods, and selection rules (GOMS) (Card, Moran, Newell, 1983; Kieras, 1988); and think-aloud protocols (Watson, 1920, Ericsson & Simon, 1998) (for an overview of some of these methods, see the other chapters in Section II, this volume). These techniques focus on articulating and representing routine, common, or optimal workflow, goals, and strategies, while highlighting the cognitive aspects of work.

Incident-based interview methods, in contrast, examine challenging, non-routine incidents. The goal is to understand cognitive work as it occurs, acknowledging the ambiguity and uncertainty present in cognitively complex domains. Rather than beginning with a hierarchical structure to frame the conversation, the semi-structured nature of the incident-based interviews allows the interviewer the flexibility to explore common, but unsatisfying responses such as "it depends," "I just knew," and "something didn't seem right" to discover dependencies, exceptions, and sometimes subtle cues that influence the way that experts understand a situation and react.

Some investigators use less formal, unstructured interviews and focus groups to unpack expertise (Cooke, 1994). Although these approaches are less likely to be documented in the academic literature, it is not uncommon to receive requests from professionals who started down this type of interview path, asking questions such as "What are the tough decisions you make?" and "How do you decide what to do when faced with [insert example from domain of interest]?" Although this direct approach has intuitive appeal, the difficulty that humans have in introspecting about their own cognitive processes has been well documented. (See Ericsson & Simon, 1984; Hoffman & Militello, 2008, pp. 156–165; Nisbett & Wilson, 1977 for a thorough discussion of the limits of self-report.) In fact, the limitations of some forms of introspection (e.g., inner observation, see Chapter 17, "A Historical Perspective on Introspection," by Ward et al., this volume) are often overgeneralized to include the types of retrospective interviews often used in incident-based interview methods, making researchers reluctant to use interviews of any type to explore human cognition.

To overcome these limitations, incident-based interview methods focus on a real-lived experience, asking the interviewee to recall what they saw, heard, smelled, and touched. Humans are much better at reporting accurately what they sensed and what meaning it had for them than at describing their own thought processes. Furthermore, people are better at recalling details when asked about incidents related to their area of expertise than when placed in novel situations and asked to complete tasks for which they have little interest or engagement. In fact, studies show that incident-based interviews with experienced practitioners evoke more accurate and detailed accounts than other types of interviews (Crandall, Klein, & Hoffman, 2006).

Types of Incident-Based Interview Methods

There are many variants on incident-based interview methods; in fact, they tend to be tailored to each project. The four types of methods described here have common elements and have been adapted for use in different contexts and by different researchers.

Critical Decision Method

Perhaps the most well-known retrospective interview technique for unpacking expertise is the critical decision method (Crandall, Klein, & Hoffman, 2006). This technique involves four sweeps through an incident (Figure 19.1). During the first sweep, incident identification, the interviewer and interviewee work together to select an incident to explore. At this point typically two to three incidents are described at a very high level (i.e., the interviewer may ask the interviewee to "tell me about the incident in 100 words or less"). The goal is to choose an incident that required skill or expertise on the part of the interviewee and is relevant to the goals of the study. During the second sweep, timeline verification, the interviewer and interviewee co-create a timeline of the incident depicting the sequence of major events from the interviewee's perspective, including rough estimates of time intervals. It works best if the timeline is drawn on a large sheet of paper or whiteboard so both can use it as a frame of reference through the rest of the interview. The third sweep, deepening, is the phase in which the interviewer begins to probe each shift in the interviewee's assessment of the situation, critical cues, points of confusion, strategies to manage uncertainty, and actions taken. The third sweep generally takes the longest as the interviewer seeks to understand the incident from the perspective of the interviewee. In the fourth sweep, hypotheticals, the

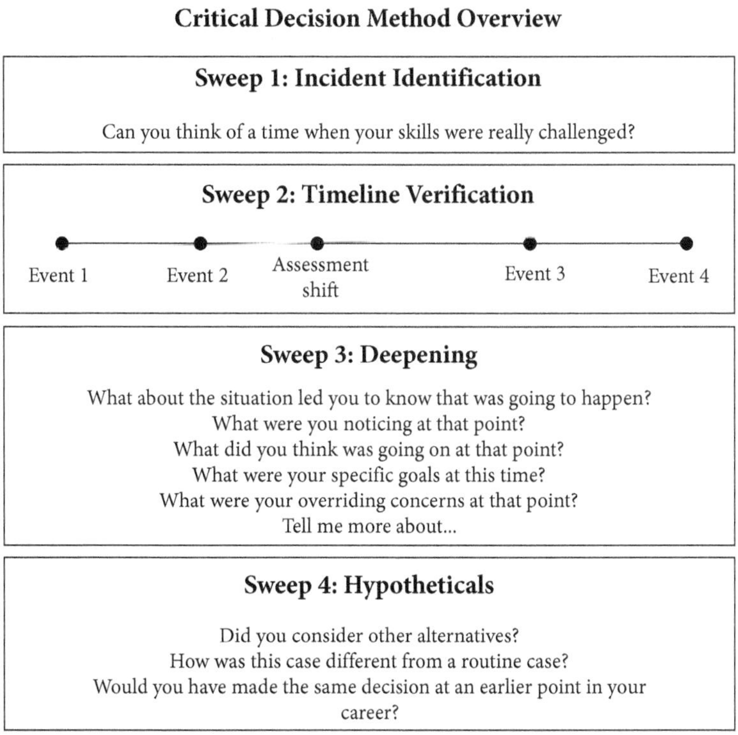

Critical Decision Method Overview

Sweep 1: Incident Identification

Can you think of a time when your skills were really challenged?

Sweep 2: Timeline Verification

Event 1 — Event 2 — Assessment shift — Event 3 — Event 4

Sweep 3: Deepening

What about the situation led you to know that was going to happen?
What were you noticing at that point?
What did you think was going on at that point?
What were your specific goals at this time?
What were your overriding concerns at that point?
Tell me more about...

Sweep 4: Hypotheticals

Did you consider other alternatives?
How was this case different from a routine case?
Would you have made the same decision at an earlier point in your career?

FIGURE 19.1 Four sweeps of the critical decision method.

interviewer may ask hypothetical questions, such as what an inexperienced person might have done in that same situation or how the situation might have ended differently if it happened at a different time or in a different context. Critical decision method interviews generally require a 1.5- to 2-hour session with each interviewee.

Nine years after the publication of the first article describing the critical decision method, a study reported that the critical decision method had been used in over 30 research projects for a diverse set of domains (Klein, Calderwood, & Macgregor, 1989). At the time this chapter was written, Google Scholar indicated that the seminal article has been cited 1123 times, and a book on this technique, *Working Minds* (Crandall, Klein, & Hoffman, 2006), has been cited 1092 times. The method has been used many times to explore expertise across a broad range of domains.

Knowledge Audit

The knowledge audit (Klein & Militello, 2005; Militello & Hutton, 1998) is another variation on the critical decision method developed for use by practitioners who do not have a background in cognitive psychology, qualitative research, or related fields. One of the challenges of conducting a critical decision method interview is recognizing a promising incident to explore and knowing what aspects of the incident to probe deeply in the third sweep. In the knowledge audit interview, the interviewer elicits shorter vignettes of real-lived incidents that illustrate cognitive challenges. The interviewer asks about cues and strategies used in each incident, as well as what would make this incident difficult for someone with less experience. Probes were designed to elicit vignettes related to specific aspects of expertise (e.g., Chi, Feltovich, & Glaser, 1981; Klein & Hoffman, 1993), including diagnosing and prediction, situation assessment, pattern recognition, heuristics, improvising, metacognition, recognizing anomalies, and compensating for automation limitations. A more general probe is termed the *scenario from Hell*, in which the interviewee is asked to describe what they would include in a training scenario designed to show how difficult a job can be. The knowledge audit interview provides a more general survey of the cognitive challenges associated with a specific domain of study as well as important cues and strategies related to each. An interview composed of a series of short vignettes is easier to manage than an in-depth exploration of a single incident.

The knowledge audit interview was used in a recent study of experienced software developers (Alarcon et al., 2017). The goal of the study was to develop a descriptive model of how software programmers assess the trustworthiness of code when considering whether to reuse existing code. The authors chose to collect a broader collection of vignettes using the knowledge audit rather than exploring a small set of cases in depth using the critical decision method because there are so many types of code and contexts in which reuse is considered. Probes focusing on diagnosing and predicting, situation assessment, pattern recognition, and anomaly detection were adapted for the software domain. For example, to elicit incidents related to noticing anomalies, authors asked: "When you think about [the code the interviewee has already described], what did you notice that made you think something was not quite right?" The corpus of vignettes collected allows for examination of trustworthiness cues and strategies across a broad range of programming contexts.

Simulation Interview

Simulation interviews (Militello & Hutton, 1998) are a variant of the critical decision method and can be used either as a retrospective or concurrent interview technique. For this interview, the interviewer provides the incident, which creates additional complexity prior to the interview but allows for a more streamlined interview. The incident can be presented in written form or embedded in a higher-fidelity simulator facility and generally focuses on a challenging situation. Depending on the setting, the interviewer may stop the simulation at key decision points and present a series of cognitive probes (concurrent interview), or the interviewer may wait until the interviewee has experienced the entire simulated scenario and then present the cognitive probes for each major event in the scenario (retrospective). In either case, the cognitive probes are similar to those used in the critical decision method sweeps three and four. The following are offered as potential starting points, and the interviewer is encouraged to elaborate and adapt as needed:

- What actions, if any, would you take at this point in time?
- What do you think is going on here? What is your assessment of the situation at this point in time?
- What pieces of information led you to this situation assessment and these actions?
- What errors would an inexperienced person likely make in this situation?

These probes are designed to encourage the interviewee to articulate what information s/he is attending to, what meaning s/he takes from the information available, what actions s/he would likely take in this situation, and what aspects of the situation might be challenging for a novice.

The simulation interview has been used to effectively explore differences in expert and novice practitioners. For example, Dominguez (2001; Dominguez et al., 2004) used video records of laparoscopic cholecystectomy procedures as a foundation for interviews with surgeons of different experience levels. The video was stopped at key points in the procedure and each participant was asked to describe his/her assessment and likely actions at that point in time, allowing for comparison of cues noticed, assessments, and intended actions across experience levels. Seagull and Xiao (2001) used a similar strategy, collecting eye-tracking data during tracheal intubation to create a recorded event that could be used to simulate the experience for other anesthesiologists. The resulting video depicted the first-person perspective of the incident with cross-hairs overlaid onto the video record to indicate the specific point-of-gaze of the anesthesiologist. The recorded events were then used as the basis for interviews with multiple anesthesiologists. In this case, in addition to reporting on their own assessment and actions, interviewees were asked to speculate on why the anesthesiologist in the recorded incident focused on specific cues at different points in the scenario and how the information gleaned was likely used by the anesthesiologist in the context of the incident.

The simulation interview explores how different people interpret and react to the same scenario. This is useful in comparing responses from different groups (i.e., experts versus novices). It is also valuable in domains for which security concerns make it difficult for interviewees to discuss the details of lived incidents.

Cued-Retrospective Interviews

In cued-retrospective interviews, the interviewee reviews a record taken from his/her perspective during a recent incident. For example, Omodei, McLennan, and Wearing (2005) used a small camera mounted to the helmet of Melbourne Metropolitan Fire Brigade Station officers to create a video and audio record of an incident. The subsequent interviews consisted of two sweeps through the incident. During the first sweep, termed psychological re-immersion, the interviewee reviewed the recording, and the interviewer encouraged him/her to relive and describe all cognitive events that occurred during the incident, uncensored for relevance or appropriateness. During the second sweep, the interviewer asked focused questions to fill in gaps and highlight the cognitive aspects of performance (Omodei, McLennan, & Wearing 2005; Omodei, Wearing, & McLennan, 1998).

In a variation on this technique, Russ and her colleagues (Russ et al., 2017) interviewed pharmacists and primary care clinicians about medication errors that were caught very quickly or even before an adverse event occurred. Because quickly identified errors and near-misses happen routinely, specific incidents are difficult to recall. To overcome this challenge, Russ and her team developed a *Nice Catch!* card that could be carried in the pocket of a lab coat and asked study participants to write down a few details about an incident involving a potential medication issue soon after it occurred. Cards were collected at the end of each day, and an interview was scheduled within 1 month of the incident. The act of recording a summary of the incident immediately following the event has been shown to increase recall accuracy (Gabbert, Hope, & Fisher, 2009). During the interview, the interviewer made the *Nice Catch!* card available as a memory aid and also provided the clinician access to his/her notes and other patient data available in the electronic health record to further support recall of the incident.

Cued-retrospective interviews are notionally distinct from the simulation interview in that the focus is on a real-world incident experienced by the interviewee that has been recorded, while simulation interviews are based on created incidents. However, the two techniques are easily merged as a study participant may be asked to experience a simulated incident and then to take part in a cue-retrospective interview using a record of the participant's perspective during the simulated incident.

Analyzing Data from Incident-Based Interviews

A number of approaches exist for analyzing qualitative data (for a discussion of traditional qualitative analysis methods, see Chapter 18, "Close-to-Practice Qualitative Research Methods," by Yardley, Mattick, & Dornan, this volume). For the purposes of this chapter, we divide them into decompositional and integrative approaches.

Decompositional Analysis

Decompositional approaches focus on coding segments or excerpts of transcripts to explore themes and to obtain frequency counts of specific cues, strategies, approaches, and/or concepts. These segments may be coded at a highly granular utterance level, with each utterance containing only a few words, or in larger segments, such as sentences or paragraphs with related content. With decompositional approaches, researchers may begin by independently reviewing a small number of transcripts (e.g., one or two) and identifying candidate codes of interest based upon research objectives. Alternatively, codes may be based on an existing framework or coding scheme developed in prior research. Developing candidate codes of interest could be deductive (i.e., coding categories that link to consensus cues for sepsis such as fever and elevated heart rate), or inductive (i.e., coding categories that are hypotheses about themes that begin to emerge across incidents involving patients suspected of having sepsis), or some combination. Analysts code individual transcripts using the preliminary codebook and meet to discuss after each transcript until agreement has been established. The analysts proceed iteratively until no new themes emerge and the codebook is deemed stable. Concurrent with these iterations the codebook is refined as analysts create a definition for each code and provide examples from the transcripts. The resulting codebook constrains the analysis to codes contained therein and also serves as a training resource for new coding team members. Other researchers or new additions to the research team can utilize the codebook to learn the coding scheme and establish common ground with other coders.

In some cases, researchers choose to assess inter-rater agreement using an inferential statistic such as Cohen's Kappa to explore how reliably coders apply the codes. If agreement is lower than expected, retraining and/or recoding may be necessary for further analysis of the data. Alternatively, the research team may decide that exploration of differing viewpoints and interpretations across a multidisciplinary analysis team is preferred. In this case, coders independently code each transcript and then meet to discuss different interpretations until consensus is reached. The consensus approach allows the team to explore and retain multiple interpretations of the same data and is preferred for exploratory studies in which new discoveries and insights are desired.

One common framework for guiding the coding process is grounded theory (Glaser & Strauss, 1967; see Yardley et al., this volume, for discussion of additional theoretical and methodological frameworks for analyzing qualitative data). Grounded theory is an inductive research method that codes or groups concepts that researchers or analysts find in qualitative data. These concepts and categories are then used as the basis for generating new theory, in contrast to the more traditional research model of testing existing theory with newly collected data.

After coding is complete, upward abstraction of coded data may be used to identify insights and implications. Details that are specific to the context of a setting are abstracted to describe the underlying strategies and performance criteria relevant across settings. Coded data allow for comparison across participants and support examination of patterns and idiosyncrasies. For example, in a project investigating the barriers and facilitators to the use of computerized clinical reminders, Saleem et al. (2005) collected and coded observational and opportunistic interview data from 90 healthcare practitioners. Researchers identified 30 emergent codes from those data. These codes were subsequently upwardly abstracted to five barriers and four facilitators of clinical reminder use, which then served as the basis for the redesign of the reminders.

In many cases, analysis of coded data includes frequency counts and minimal, if any, aggregation of the codes into higher-level concepts or categories. Frequency counts may be used to support hypothesis testing or as a component of data exploration compatible with grounded theory. For example, a recent study of health information needs of pregnant women and their caregivers coded all mentions of information needs during a semi-structured interview (Robinson et al., 2017). For this analysis, the team used a hierarchical taxonomy that was previously documented in the literature for consumer health-related information needs to organize the data and explore potential emergent themes (Shenson, Ingram, Colon, & Jackson, 2015). At least two research team members reviewed all needs independently and coded them according to information need category and hierarchical category. After substantial training, Cohen's Kappa was used to calculate agreement, and acceptable agreement was achieved.

Analysis of these coded data revealed that 100 participants (71 pregnant women, most with a fetal anomaly, and 29 caregivers) reported 1052 information needs about their pregnancy, of which almost 30% remained unmet. The hypothesis that pregnant women and their caregivers (e.g., spouse, mother) would have different types of information needs was not supported by these data. The majority of the information needs for both pregnant women and caregivers focused on understanding the medical information received from healthcare providers, such as understandable explanations of diagnoses, survivability associated with particular diagnoses, the anticipated benefits of recommended procedures, and what specific surgical procedures entail (Robinson et al., 2017).

Integrative analysis

Integrative approaches look across incidents for insights, rather than focusing on coded segments. This type of analysis requires a focus on entire incidents, so for representations such as timeline summaries, text-based vignettes are often a starting place. Timelines or vignettes are reviewed and analyzed for common elements as well as for insights about complex macrocognitive skills and strategies. For example, in a study of emergency department pediatricians assessing patients for sepsis, a comparison of cues mentioned in incidents related by residents (i.e., physicians with less than 4 years of experience and working under the supervision of an attending physician) versus experienced clinicians uncovered no expert–novice differences. A review of incident summaries, however, revealed that the inexperienced resident physicians related clear and detailed lists of cues, but did not tend to describe the implications of the cues or how they related to actions taken. Experienced clinician narratives, in contrast, described cues in the context of a hypothesis and the tests and interventions administered to confirm or disconfirm the presence of infection or other suspected condition. Without context elements, expertise was difficult to discriminate when examining frequency counts of coded data. In contrast, reviewing entire incidents revealed important differences in cue utilization and sensemaking (Patterson et al., 2016).

Decision requirements tables are often created to examine the key decisions in each incident. In reviewing each incident, investigators note key decisions, as well as critical cues, strategies, and common errors associated with each. By linking these important contextual elements to the key decision, investigators are able to combine common elements and note important differences and conflicts while retaining the link to the original incident. This type of analysis highlights opportunities for better supporting expertise via improved system design, new technology, and training.

As part of a study to articulate cognitive support for pararescue jumpers, decision requirements tables were created for a variety of incidents (Militello, Sushereba, Branlat, Bean, & Finomore, 2015). Table 19.1 depicts a decision requirements table created based on an incident recounted by a pararescue jumper in the US Air Force (described in Sine, 2013). In this incident the interviewee was called to rescue a pilot who had ejected from an A-4 Skyhawk aircraft and landed in a tree in dense jungle in the Philippines. The pararescue jumper expected to be able to hike to the tree and climb up, but the dense jungle made landing an aircraft nearby impossible, and the approaching nightfall made it too risky for a long hike. The pararescue jumper improvised a rescue strategy. He contacted a nearby helicopter that would be able

Table 19.1 Decision requirements table summarizing key decisions in rescuing a pilot who ejected in the Philippines jungle

Decision	Cues	Strategies	Why difficult?
1. Initial planning	• Downed pilot in a tree • Hanging from parachute	• Land nearby • Use tree climbing apparatus to rescue pilot	• Did not know the tree was in dense jungle. • Time pressure made it difficult to come up with a workable contingency plan.
2. Re-planning	• Could not land nearby (5 miles to dry creek bed) • Nightfall was approaching—would not be able to find the tree in the dark • Helicopter was too big to hover at the required altitude	• Contact nearby Navy helicopter, as they could hover at required altitude • Lower self from helo cable to tree and swing in to rescue pilot • Avoid focusing on all the reasons why this plan might not work	• Lots of uncertainty (not sure this plan will work) • Have not done this type of rescue before • Time pressure (it's getting dark, and pilot is panicking because his parachute keeps slipping) • Comms not compatible—would not be able to communicate with Navy helicopter pilot
3. Rescue	• Dense tree canopy • Downed pilot cannot help; tangled in parachute cords and has been hanging so long has lost sensation in legs	• Swing from helicopter cable to get close enough to pilot • Use knife to cut entangled parachute cords • Avoid panicking by focusing on immediate task • Use hand signals to communicate with Navy helicopter pilot	• Do not have equipment designed for this type of rescue.
4. Treat pilot after rescue	• The pilot had been in the parachute for hours • Hanging from a parachute for a length of time can cause serious problems, even death • Pilot's legs were numb	• Treat with O_2 and IV fluids.	• Can only do limited assessment in helicopter

to hover at the required altitude, and then lowered himself from the helicopter cable to swing into the tree. After finding both the pilot and himself tangled in the pilot's parachute cords, he was able to cut them both free and hoist the pilot into the helicopter where treatment for his injuries was administered. Similar tables were created for each incident collected in this study. Looking across incidents, cognitive requirements intended to inform the design of cognitive support for pararescue were abstracted (Militello et al., 2015).

Researchers may choose to employ a decompositional coding approach in conjunction with a more integrative analysis approach. For example, in a study that used observations and interviews to develop the requirements and design of a tool to support a burn Intensive Care Unit (ICU), physicians and staff employed both a decompositional and integrative analysis approach. Researchers used week-long data collection visits to observe and interview ICU personnel. The initial data analysis consisted of coding the interview transcripts from site visits using a grounded theory approach in which researchers developed themes related to communication and coordination among practitioners. Then researchers conducted additional data collection, eliciting critical incidents. A second round of data analysis focused on developing a decision requirements table using both the critical incidents and the themes developed during the grounded theory phase (Nemeth et al., 2016). This type of analysis was advantageous in this case because the research team used a decompositional coding approach to first understand and quasi-quantify the qualitative data in order to identify existing gaps and focus incident elicitation in the next stage of data collection. This provided a valuable complement to the later integrative analysis. These analyses enabled the specification of design requirements for the development of a technology system to improve patient care.

Regardless of the analysis approach used, the number of people involved in the coding process may vary greatly, with as few as two coders to teams of them, depending on the amount of data, the specificity of coding to be done, expertise level needed to understand the data, and time constraints for the project. Key elements of analyzing incident-based data include tailoring the approach to the resources available and goals of the project, applying the approach systematically, and carefully documenting the approach used including any mid-analysis corrections.

Products

Products of studies using incident-based methods include observations about expert–novice differences, critical cue inventories, descriptive models, cognitive requirements, and workflow diagrams. These products are commonly used to inform the design of technology, training, and systems. Brief case studies illustrating different products are discussed.

Expert–Novice Differences and Critical Cue Inventories

In the study of expert–novice differences in sepsis recognition by emergency department pediatricians, findings changed the way simulation-based training was administered. Investigators used the critical decision method to elicit a set of incidents from residents and experienced clinicians. Exploration of expert–novice differences supported the anecdotal notion that residents exhibit skill at recording and reporting cues, but their narratives lack a first-person perspective and evidence of sensemaking and hypothesis testing seen in incidents recounted by experienced clinicians. These findings, in combination with a critical cue inventory derived from the incidents (Table 19.2), led to the design of a new approach to scenario-based training. The incidents provided the building blocks of improved scenarios in terms of critical cues and realistic situational factors that residents may not experience during their residency. The findings also influenced trainers to create a setting where the learner was in the role of decision maker and must experience, recognize, and interpret cues on his/her own. In more traditional training, it had been customary to present learners with radiographic findings or distal perfusion by simply reporting the results. Findings from this study prompted trainers to present the learner with raw data to interpret and to use for forming hypotheses and choosing appropriate actions (Patterson et al., 2016).

Table 19.2 Critical cues for assessing distal perfusion

Distal Perfusion			
Skin color	Extremities	Temperature	Other
Pale	Mottling, especially lower extremities	Cold	Delayed cap refill
Paleish gray	Hands were mottled	Mottled and warm	Decreased peripheral perfusion
Pasty	Extremities were cold	Sweating	Poor peripheral pulses
Pallor (African American patient)	Pale extremities	Perfusion was warm	Vasoconstricted
Yellowish	Nose was yellowish		Pulses were thready
No nice flush on cheeks			
Mottled			
Flushed			
Purple			
Reticulated pattern			

Reproduced from Mary D. Patterson, Laura G. Militello, Amy Bunger, Regina G. Taylor, Derek S. Wheeler, Gary Klein, and Gary L. Geis, Leveraging the critical decision method to develop simulation-based training for early recognition of sepsis, *Journal of Cognitive Engineering and Decision Making* 10(1), p. 47, Table 4, doi.org/10.1177/1555343416629520 Copyright © 2016 by Human Factors and Ergonomics Society.

Descriptive Models

In the study of software developers using the knowledge audit (Alarcon et al., 2017), findings informed a descriptive model of software trustworthiness (Figure 19.2). Specifically, investigators found that experienced software developers consider the performance, reputation, and transparency of software as they determine whether it can be reused or adapted for a new software product. Furthermore, their criteria for acceptable levels of uncertainty and risk are influenced by environmental factors such as customer needs and requirements, organizational and resource constraints, and the consequence of failure. The goal of this study was to develop a descriptive model that would serve as a foundation for future experiments to explore the contributions of specific elements of the descriptive model and to increase our understanding of presenting code in a way that appears trustworthy to programmers (Alarcon et al., 2017).

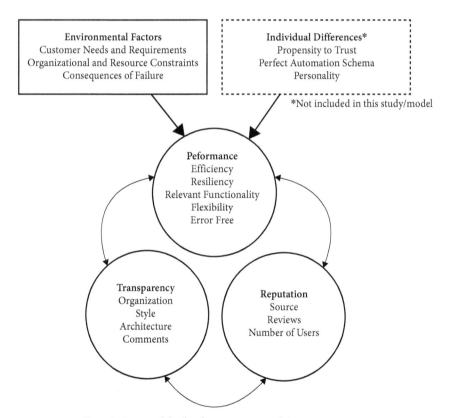

FIGURE 19.2 Descriptive model of software trustworthiness.
Reproduced from Gene M. Alarcon, Laura G. Militello, Patrick Ryan, Sarah A. Jessup, Christopher S. Calhoun, and Joseph B. Lyons, A descriptive model of computer code trustworthiness, *Journal of Cognitive Engineering and Decision Making* 11(2), pp, 116, Figure 2, doi.org/10.1177/1555343416657236
Copyright © 2017 by Human Factors and Ergonomics Society.

Cognitive Requirements

Cognitive requirements are commonly used to guide the design of training and technology to support expertise. In a study of how primary care clinicians track and manage colorectal cancer testing for their patients, a combination of observations and critical decision method interviews were used to articulate six cognitive requirements: (1) determine whether the patient is in screening or surveillance mode, (2) obtain a big picture perspective of the patient's testing history, (3) know where the patient is in the screening cycle, (4) consider conditions or medication that have implications for testing, (5) assess and monitor a patient's individual risk for colorectal cancer, and (6) educate and inform patients. These informed the design of a one-page visualization, centered on a timeline display providing data about testing history (Figure 19.3) (Militello et al., 2016). Existing user interfaces require review and mental

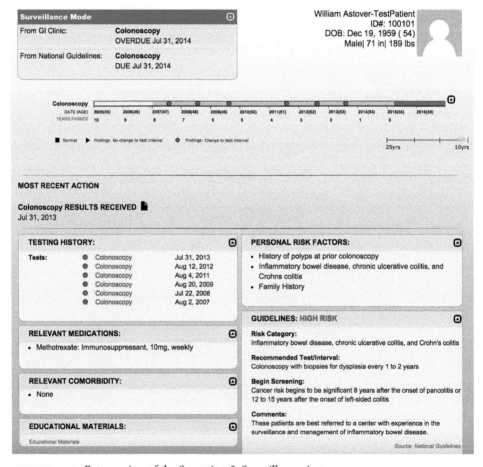

FIGURE 19.3 Beta version of the Screening & Surveillance App.

Reproduced from Laura G. Militello, Jason J. Saleem, Morgan R. Borders, Christen E. Sushereba, Donald Haverkamp, Steven P. Wolf, and Bradley N. Doebbeling, Designing colorectal cancer screening decision support: A cognitive engineering enterprise, *Journal of Cognitive Engineering and Decision Making* 10(1), p. 83, Figure 6, doi.org/10.1177/1555343416630875 Copyright © 2017 by Human Factors and Ergonomics Society.

integration of notes, lab reports, and gastroenterology reports in order to build a mental timeline, which creates unnecessary cognitive load and competes with other patient care tasks, such as assessing and interacting with the patient.

Practical Issues

Pragmatic issues such as determining whom to interview, obtaining an appropriate sample, limited time with interviewees, concerns about faulty memory or fabrication, and managing difficult interviewees can be challenges for studies using incident-based methods.

Whom to Interview

Determining whom to interview is an important consideration in framing a study. Retrospective methods such as the critical decision method and knowledge audit work best with experienced practitioners because they have an extensive experience base to draw from for discussion. These methods have been used successfully with less experienced practitioners; however, it is not uncommon for a less experienced interviewee to struggle to recall a relevant incident. We recommend oversampling when conducting retrospective interviews with less experienced practitioners.

If the goal of the project is to study expert strategies, or expert–novice differences, the investigator must determine how to define expertise for the domain of study. While some research suggests that domain-general aptitudes may positively influence the acquisition of expertise, applied research is generally focused on domain-specific expertise (Hoffman, 1998; cf. Chapter 3, "Domain-General Models of Expertise: The Role of Cognitive Ability," by Hambrick et al., this volume). In some domains, such as chess, standard rankings used to assess skill can be used to determine expertise level. For many domains, however, no objective measures of skill are available. In this case, years of experience and peer nominations are generally used to select experts. Therefore, in most cases, it is best to work with subject-matter experts to determine meaningful criteria for identifying experts in a specific domain.

How Many People to Interview

Determining how many people to interview is another challenge. The sample size is largely dependent on project goals, project resources, and access to study participants. In many early cognitive task analysis studies, limited funding, time, and/or access to people with the needed expertise led to an emphasis on strategies for obtaining insights from in-depth interviews with a small sample of interviewees having specialized experience. In projects with these types of constraints, we recommend including at least three participants, but prefer to include six to eight highly experienced practitioners (Militello & Hutton, 1998). Often after six interviews, we begin to hear common themes, suggesting a reduced likelihood of new insights with additional interviews. If project resources allow a larger sample, we recommend analyzing the first six to eight interviews to identify gaps in understanding and hypotheses for further exploration in future interviews *before* conducting a second set of interviews.

For more extensive studies, there may be a need to interview multiple types of participants. For example, study goals may include comparison of participants with different levels of experience, or related but distinct types of expertise working to accomplish a shared task (i.e., physicians, nurses, and respiratory therapists). In this case, we recommend at least three participants in each group. If time and resources allow, six to eight participants in each group is preferred.

It is also worth noting that in recent years, we have seen an emphasis on larger sample studies. Grant reviewers, particularly in healthcare domains, may require sample sizes of 50, or even 100, to include participants from multiple sites, different levels of expertise, and/or different roles. Proposers may be asked to include power analyses to justify proposed sample sizes. We suspect this emphasis on large-sample studies is an overgeneralization from the traditional multisite randomized controlled trials used to assess the safety of new medications. While large-sample studies and power analyses are appropriate for experiments, the value of this approach for studies using in-depth qualitative interviews is less clear. Studies using incident-based methods tend to be exploratory. Even those involving comparisons of different experience levels or roles tend to be descriptive in nature rather than experimental; the focus is generally on hypothesis generation rather than hypothesis testing. These large-sample studies tend to shift the emphasis from a careful analysis of a small set of high-quality, in-depth, semi-structured interviews to highly standardized interviews and efficient analysis techniques.

What Types of Incidents to Go after

Incident-based methods focus on exploring challenging incidents. In interviews using retrospective methods (i.e., critical decision method, knowledge audit), highly emotional (e.g., perhaps someone died or justice was not served) incidents might be the first to come to the interviewee's awareness. It is best to elicit incidents in which the interviewee's skills were particularly challenged or in which a less experienced person may not have been as successful, rather incidents that are memorable because of the emotional content.

Another consideration when using retrospective methods is how recent the incident should be (for a discussion of this issue, see Ward et al., this volume). Certainly, participants are likely to experience less memory decay with recent incidents, and recent incidents are less likely to have been rehearsed with changing details over time. Nonetheless, we have found that sometimes a highly experienced practitioner will remember an incident earlier in his/her career that changed his/her mental model or approach for managing a specific type of challenge. Unpacking the circumstances, cues, and factors that led to the new insight can be quite valuable.

For simulation interviews, incidents may be recordings of real-life events. In this case, incidents that represent challenges relevant to study goals are selected. In some cases, investigators create a scenario, often beginning with a small set of critical decision method or knowledge audit interviews, or even published accounts of incidents in the domain of interest. Elements from the real-world examples are used to create plausible scenarios that highlight cognitive challenges such as goal conflicts, uncertainty, and important perceptual discriminations.

Limited Interview Time

In some cases, experts are willing to participate in interviews but have limited time available. Critical decision method interviews can take up to 2 hours to explore a single incident in depth. However, interviewees in highly scheduled work domains, such as primary care physicians, may be available for only 30–45 minutes over lunch or during administrative time between patients. In these constrained settings, methods such as the knowledge audit or simulation interview are more practical than the critical decision method or cued retrospective interviews. The use of vignettes in the knowledge audit allows for shorter, focused interviews. The simulation interview may be more concise as it eliminates the need for the interviewer and interviewee to spend time selecting an appropriate incident. Other strategies for making the most of limited interview time include providing background about the study and collecting informed consent and demographic information before the interview session. We have explored the possibility of scheduling two 45-minute interview sessions with each interviewee but have never been able to overcome the associated scheduling challenges. We recommend interview sessions of 60 minutes or more whenever possible.

Avoiding Faulty Memories and Fabrication

When conducting interviews, faulty memories and fabrication are always concerns (Nisbett & Wilson, 1977). Indeed, incident-based methods were developed specifically to overcome some of these limitations. In addition to focusing on the participant's lived experiences, the ways in which questions are asked can support accurate, detailed accounts. For example, we avoid questions that begin with *why*, as these tend to encourage interviewees to generate a plausible explanation that may or may not link to the incident under investigation (i.e., see Ericsson & Simon, 1984; for an alternative argument about the value of asking for verbal elaborations such as explanations, see Ward et al., this volume). We try to bring the interviewee back to the incident to stay grounded in real-world examples and avoid discussion of general approaches, optimal conditions, and idealized strategies. When an interviewee responds with "it depends," the focus is on further exploring the dependencies rather than forcing a choice to determine which option would be optimal. The interviewer explores the influence of specific factors in the incident of study, and also how the incident may have played out differently if those factors had been different. Incident-based interviews are characterized by an attempt by the interviewer to understand the interviewee's first-person perspective. By taking this stance, the interviewer conveys to the interviewee that it is acceptable to discuss initial perspectives that do not make sense in retrospect, and that it is acceptable to note aspects of the incident for which uncertainty remains or memory is not clear. The interviewer is in the role of learning from the interviewee; the interviewee is not being evaluated or compared to other interviewees during this session.

Difficult Interviews

The objective of the interview is not simply to keep the conversation going or to collect interesting stories. Understanding the cognitive challenges and the cues and strategies experienced practitioners use to manage them are the real measures of success for an incident-based interview. It sounds simple enough, but managing the interview can be difficult, even for an experienced interviewer. Some interviewees will be reluctant participants and have little to say. Some may have a lecture style of communication that is difficult (if not impossible) to redirect into discussing an actual incident. Sometimes, an interviewee is recommended by peers or supervisors as an expert but does not have experiences relative to the study goal. For all of these reasons, we recommend coming into each interview with at least one contingency plan. If your initial plan is to use the critical decision method, prepare some knowledge audit probes, or bring along a scenario that will allow you to shift to a simulation interview if needed. If the interviewee cannot be shaken from the lecture s/he prepared, consider whether it may be useful to understand how learners are trained in this domain and whether this lecture is a relevant example. Our goal is to be flexible enough to learn something relevant from each interview. Sometimes, the main insight is that we need to adapt our interviewing strategy.

Future Directions

We anticipate that incident-based methods will continue to be a valuable resource when investigating expertise in complex environments. As recording technology becomes more affordable and portable, there will be increased opportunities to leverage audio, video, eye gaze, and even keystroke/mouse click data to recreate incidents. These can be used to create a timeline prior to the interview, as a memory aid during the interview, or as a means to compare retrospective accounts to recorded events. Refining and extending existing strategies for leveraging these technologies for use in cued-retrospective interviews and simulation interviews will be an important step forward. Records of lived events will provide fertile material for better understanding the strengths and limits of self-report, and the effect of different questioning and cueing strategies.

If the trend toward large-sample studies using incident-based methods in healthcare continues, meaningful and efficient analysis of many incidents often collected by distributed, multisite data collection teams will continue to be a challenge. Existing software designed to support qualitative analysis can be useful in tracking and coding transcripts (e.g. DeDoose, NVivo, MaxQDA). However, these software tools tend to have limited support for comparing codes across coders to facilitate consensus coding, and provide even less support for thematic analysis and upward abstraction to extract meaning and insights from the data. Better tools to support these analyses are needed. Equally, if not more important, is the need for strategies for analysis of these large data sets. Traditional approaches for looking across incidents obtained in small-sample studies using timeline summaries and incident vignettes do not easily scale up. A small analysis team of two to four people can conduct in-depth examination of six to eight incidents, exploring different representations and discussion of insights and implications from each. When analyzing 100 incidents, this task becomes quite daunting. Discussion of what skills should be represented in the analysis team, how to frame and structure the analyses, and what portions of the process require consensus or inter-rater agreement are important future directions.

Although decompositional analysis techniques such as coding of excerpts continues to be considered the most rigorous approach by many reviewers, important insights can be obscured by examining decontextualized components of incidents. A continuing challenge for studies of expertise using incident-based methods is to make the case for the value of integrative analysis approaches, to articulate rigorous strategies for this type of analysis, and to increase the visibility of integrative analysis approaches in the scientific literature so that they can be applied more broadly.

SUMMARY AND CONCLUSION

Incident-based methods are powerful tools for studying expertise in context. They provide insight into expert cognition that cannot be obtained with other methods. These methods facilitate description of expertise in the context of work and aid in developing testable hypotheses. They can be used to uncover cognitive requirements to support acquisition of skill via decision support design and training. Findings can also be used to inform the design of tools to support expertise and descriptive models to inform basic research.

Incident-based interview methods complement behavioral measures that show what an expert can do, using a description of what is going on in the interviewee's head. Incident-based methods provide insight into what the interviewee attends to, how the interviewee makes sense of the information available, what the interviewee expects, what surprises the interviewee (i.e., violated expectancies), and how confident or worried the interviewee was in a particular situation (i.e., risk assessments, prioritization). Each interview provides one individual's first-hand perspective. Incident-based techniques are designed to increase recall accuracy and elicit rich details. Analysis techniques may be decompositional (i.e., coding) or integrative (i.e., comparison of vignettes or timelines), or include some combination of both depending on the goals of the study and resources available.

Future directions for incident-based methods include leveraging technologies that facilitate recordings of real-world incidents, developing strategies for analyzing large data sets, and better articulating rigorous methods for conducting integrative analyses of incidents.

References

Alarcon, G. M., Militello, L. G., Ryan, P., Jessup, S. A., Calhoun, C. S., & Lyons, J. B. (2017). A descriptive model of computer code trustworthiness. *Journal of Cognitive Engineering and Decision Making* 11(2), 107–121.

Annett, J. (2004). Hierarchical task analysis (HTA). In N. A. Stanton, A. Hedge, K. Brookhuis, E. Salas, & H. W. Hendrick (Eds), *Handbook of human factors and ergonomics methods* (pp. 33–41). Boca Raton, FL: CRC Press.

Card, S. K., Moran, T. P., & Newell, A. (1983). *The psychology of human–computer interaction*. Hillsdale, NJ: Lawrence Erlbaum.

Chi, M. T. H., Feltovich, P. J., & Glaser, R. (1981). Categorization and representation of physics problems by experts and novices. *Cognitive Science* 5(2), 121–152.

Cooke, N. J. (1994). Varieties of knowledge elicitation techniques. *International Journal of Human-Computer Studies* 41(6), 801–849.

Crandall, B., Klein, G. A., & Hoffman, R. R. (2006). *Working minds: A practitioner's guide to cognitive task analysis*. Cambridge, MA: MIT Press.

Dominguez, C. O. (2001). Expertise and metacognition in laparoscopic surgery: A field study. *Proceedings of the Human Factors and Ergonomics Society Annual Meeting* 45(17), pp. 1298–1302.

Dominguez, C. O., Flach, J. M., & Dunn, M. (2004). The conversion decision in laparoscopic surgery: Knowing your limits and limiting your risks. In K. Smith, J. Shanteau, & P. Johnson (Eds), *Psychological investigations of competence in decision making* (pp. 7–39). Cambridge, UK: Cambridge University Press.

Ericsson, K. A., & Simon, H. A. (1984). *Protocol analysis*, rev edn. Cambridge, MA: MIT Press.

Ericsson, K. A. & Simon, H. A. (1998). How to study thinking in everyday life: Contrasting think-aloud protocols with descriptions and explanations of thinking. *Mind, Culture, and Activity* 5(3), 178–186.

Gabbert, F., Hope, L., & Fisher, R. P. (2009). Protecting eyewitness evidence: Examining the efficacy of a self-administered interview tool. *Law and Human Behavior* 33, 298–307.

Glaser, R., & Strauss, A., (1967) *The discovery of grounded theory: Strategies for qualitative research*. New Brunswick, NJ: AldineTransaction.

Hoffman, R. R. (1998). How can expertise be defined? Implications of research from cognitive psychology. In W. Faulkner, J. Fleck, & R. Williams (Eds), *Exploring expertise* (pp. 81–100). London: Palgrave Macmillan.

Hoffman, R. R., & Militello, L. G. (2008). *Perspectives on cognitive task analysis: Historical origins and modern communities of practice*. New York: Taylor and Francis.

Kieras, D. E., (1988). Towards a practical GOMS model methodology for user interface design. In M. Helander (Ed.), *Handbook of human–computer interaction* (pp. 135–158). Amsterdam: Elsevier.

Klein, G. A., Calderwood, R., & Macgregor, D. (1989). Critical decision method for eliciting knowledge. *IEEE Transactions on Systems, Man, and Cybernetics* 19(3), 462–472.

Klein, G. A., & Hoffman, R. R. (1993). Seeing the invisible: Perceptual-cognitive aspects of expertise. In M. Rabinowitz (Ed.), *Cognitive science foundations of instruction* (pp. 203–226). Hillsdale, NJ: Lawrence Erlbaum.

Klein, G., & Militello, L. (2005). The knowledge audit as a method for cognitive task analysis. In H. Montgomery, R. Lipshitz, B. Brehmer (Eds) *How professionals make decisions* (pp. 335-342). Mahwah, NJ: Lawrence Erlbaum.

Militello, L. G. & Hutton, R. J. B. (1998). Applied cognitive task analysis (ACTA): A practitioner's toolkit for understanding cognitive task demands. *Ergonomics* 41(11), 1618–1641.

Militello, L. G., Saleem. J. J., Borders, M. R., Sushereba, C. E., Haverkamp, D., Wolf, S. P., & Doebbeling, B. N. (2016). Designing colorectal cancer screening decision support: A cognitive engineering enterprise. *Journal of Cognitive Engineering and Decision Making* 10(1), 74–90.

Militello, L. G., Sushereba, C. E., Branlat, M., Bean, R., & Finomore, V. (2015). Designing for military pararescue: Naturalistic decision-making perspective, methods, and frameworks. *Journal of Occupational and Organizational Psychology* 88, 251–272.

Nemeth, C., Anders, S., Strouse, R., Grome, A., Crandall, B., Pamplin, J., ... Mann-Salinas, E. (2016). Developing a cognitive and communications tool for burns intensive care units clinicians. *Military Medicine* 181(supp 5), 205–213.

Nisbett, R. E., & Wilson, T. D. (1977). Telling more than we can know: Verbal reports on mental processes. *Psychological Review* 84, 231–259.

Omodei, M., Wearing, A., & McLennan, J. (1998). Integrating field observation and computer simulation to analyse command behaviour. In *Proceedings of the NATO RTO workshop on the human in command*.

Omodei, M. M., McLennan, J., & Wearing, A. J. (2005). How expertise is applied in real-world dynamic environments: Head mounted video and cued recall as a methodology for studying routines of decision making. In T. Betsch & S. Haberstroh (Eds), *The routines of decision making* (pp. 271–288). New York: Psychology Press.

Patterson, M. D., Militello, L. G., Bunger, A., Taylor, R. G., Wheeler, D. S., Klein, G., & Geis, G. L. (2016). Leveraging the critical decision method to develop simulation-based training for early recognition of sepsis. *Journal of Cognitive Engineering and Decision Making* 10(1), 36–56.

Robinson, J. R., Anders, S., Novak, L. L., Simpson, C. L., Holroyd, L. E., & Jackson, G. P. (2017). Health-related needs of pregnant women and their caregivers. In *Proceedings of American Medical Informatics Association Annual Conference.* Washington DC.

Russ, A. L., Militello, L. G., Glassman, P. A., Arthur, K. J., Zillich, A. J., & Weiner, M. (2017). Adapting cognitive task analysis to investigate clinical decision-making and medication safety incidents. *Journal of Patient Safety*, online ahead of print, May 3. DOI: 10.1097/PTS.0000000000000324

Saleem, J. J., Patterson, E. S., Militello, L., Render, M. L., Orshansky, G. & Asch, S. M. (2005). Exploring barriers and facilitators to the use of computerized clinical reminders. *Journal of American Medical Informatics Association* 12(4), 438–447.

Seagull, F. J., & Xiao, Y. (2001). Using eye-tracking video data to augment knowledge elicitation in cognitive task analysis. *Proceedings of the Human Factors and Ergonomics Society Annual Meeting* 45(4), pp. 400–403.

Shenson, J.A., Ingram, E., Colon, N., Jackson, G.P. (2015). Application of a consumer health information needs taxonomy to questions in maternal-fetal care. *AMIA Annual Symposium Proceedings*, 1148–56.

Sine, W. F. (2013). *Guardian angel: Life and death adventures with pararescue, the world's most powerful commando rescue force.* Philadelphia: Casemate Publishers.

Watson, J. B. (1920). Is thinking merely the action of language mechanisms? *British Journal of Psychology* 11, 87–104.

CHAPTER 20

COGNITIVE WORK ANALYSIS

Models of Expertise

CATHERINE M. BURNS

INTRODUCTION

COGNITIVE work analysis (CWA) emerged in the late 1980s as a possible way of looking at human work. CWA came from an engineering perspective and developed from the need to have an approach that could handle complex operational problems in safety critical systems. Much of the analytical work underlying CWA had developed from analyses proposed by Jens Rasmussen while at the Risø National Lab in Denmark (Vicente, 1997) and was brought together as a new tool for looking at human work by Kim Vicente (1999). CWA differed from hierarchical task analysis, or user-centered design approaches popular at the time, by including an analysis of work constraints (the abstraction hierarchy), strategies, socio-organizational analysis, and the skills–rule–knowledge framework.

The contexts that CWA was directed toward were complex and safety critical systems. The work of operators in these systems could often not be addressed solely through interview and observational methods, and an approach that included a fundamental analysis of the domain of work of the operators added insight to the understanding of cognitive work. It was also recognized early on that highly experienced operators understood their work differently from novice operators. However, for safe operations, it was advantageous to progress the expertise of novice operators as quickly as possible. This desire meant that the strategies, knowledge, and competencies of highly experienced operators needed to be understood. For this reason, CWA has always incorporated models of expertise within its analysis.

A full CWA has five phases of analysis. Most commonly these phases progress from work domain analysis, control task analysis, strategies analysis, social organizational

Table 20.1 CWA Phases and their associated models of knowledge

CWA phase	Model of knowledge
Work domain analysis	Abstraction hierarchy
Control task analysis	Decision ladder
Strategies analysis	Information flow maps or narrative
Social organizational analysis	(No particular model)
Worker competency analysis	Skills–rule–knowledge model

analysis, and work competency analysis. Within each of these analyses, there are particular models of knowledge that are used (Table 20.1). In the next sections, we will discuss these models and how they represent expert knowledge structures or behavior. Examples of these representations are given in later sections of this chapter. For more information how to conduct CWA, consider Burns, Enomoto and Momtahan (2008), Burns and Hajdukiewicz (2004), and Vicente (1999).

A CWA is developed from a concentrated period of studying human work in a particular environment. A CWA analyst will spend time talking to expert operators and novice operators, observing performance on the job or in simulations, talking to trainers on what expertise they aim to develop, and developing an understanding of the fundamental relationships that need to be known. While this might seem daunting, analysts who work regularly in the same context (e.g., process operations, or healthcare) quickly get a strong knowledge of the fundamentals of the domain and can quickly focus their analysis on the particular nuances of a new environment. In the following sections, the basics of the CWA analyses will be described, to provide background for the discussion on expertise.

Work Domain Analysis

Work domain analysis is often the first phase of CWA. From a theoretical perspective, work domain analysis provides the first level of constraint on human work, that being that human work is constrained by the laws and relationships of their work context. In many cases, these are physical constraints, but analyses of modern intentional systems may also include constraints of value, purpose, and proper regulatory functioning in the work domain description. The work domain analysis has typically used the abstraction hierarchy as its model of constraints. The abstraction hierarchy is a model of functional purpose achievement, with purposes at the highest level, constrained by the physical laws and the capabilities of the domain components. In this way, the abstraction hierarchy is often called a *means–end* structure. This structure is important in that means–ends relations are functional, and the relation itself is critical information to the knowledge structure. A means–end relation describes how one element (a means) helps to achieve a function (an end). Ontologically, the means–end relations build into a structure that usually has five levels with purpose at the top and component capabilities at the bottom. The intermediate levels describe functional processes and physical laws. An abstraction hierarchy is often informed by manuals, trainers, and fundamental knowledge about the design of the work problem. For examples and guidance on how to conduct work domain analysis consider Naikar (2013).

Control Task Analysis

Control task analysis is usually the second phase of CWA. The control task analysis is the first phase of CWA that explicitly considers human information processing. A control task is a category of task in a work domain that is so fundamental that it represents a particular mode and context of operation. There are often many smaller tasks that can be modeled within a control task. Examples of control tasks in a process environment may be startup, normal operations, and shutdown. The control task analysis uses the decision ladder as its knowledge model. The decision ladder is a template model of human information processing, on which different processing routes can be represented. The model is often used to show the different processing routes between novices and experts, different contexts, or different resources. A decision ladder is often informed by observations and interviews of operators as they perform control tasks. For examples on how to use the decision ladder consider Jenkins et al. (2010), McIlroy and Stanton (2015), and Naikar, Moylen, and Pearce (2006). The use of decision ladders to describe expertise is discussed later in this chapter.

Strategy Analysis

The strategy analysis is often the third phase of CWA. Strategy analysis is conducted by interviews with users and operators, focusing on when they shift to alternative routes. Common strategies include various problem reasoning strategies (deduction, search), or strategies for different workload contexts. Analysts should interview both novices and experts and spend some time focusing on the different approaches novices and experts may take. Novices will often focus on techniques and knowledge from training, while experts focus on heuristics and experience to drive their strategies. For examples of strategies analysis, consider Kilgore, St-Cyr, and Jamieson (2009) or Ashoori and Burns (2012).

Social Organizational Analysis

Social organization analysis may be positioned as the fourth analysis but, in reality, could be understood at any point past the work domain analysis. Social organizational analysis has been conducted with a wide variety of models, depending on the direction of the analysis. Similarly, this analysis can be conducted in a variety of ways, depending on the scale of social factors and effects that the analyst is looking to examine. Kilgore, St-Cyr, and Jamieson (2009) and Ashoori, Burns, Momtahan, and d'Entremont (2014) are some examples of social organizational analysis in CWA.

Worker Competency Analysis

Worker competency analysis uses the skills–rules–knowledge framework to understand individual competencies to succeed in the work environment. Understanding novice to expert development has always been a core part of this analysis. One of the fundamental principles of this analysis is that experts can use experience to develop heuristics or rules, therefore bypassing knowledge-based decision making in situations that they recognize. This principle is compatible with the suggestions of recognition-primed decision making, which argues that expert decision making takes advantage of the ability to quickly recognize a situation and its course of action, as opposed to engaging in more analytical processes. For examples of worker competency analysis consider McIlroy and Stanton (2011) or Ashoori and Burns (2012).

In the next sections, I discuss these knowledge models in more detail, focusing on how expertise is shown through the models. Expertise in this chapter means expert skill and knowledge. This may or may not be correlated with the length of time spent in a role. Instead, an expert is considered to be a person with deep knowledge and experience that allows them to demonstrate significantly different and more effective behaviors than novices.

The Abstraction Hierarchy: Model of Expert Reasoning

CWA grew from a context where the technology of nuclear power was evolving quickly, and in some ways dangerously. This environment led to the development of the Risø National Lab, and eventually to Jens Rasmussen as the head of the electronics department (Vicente, 1997). This department was focused on hardware reliability but soon learned that, even with reliable hardware, the human reliability in the system needed to be considered, if the overall system was going to be reliable. Rasmussen's team learned that while experienced teams would often perform well under normal, expected conditions when faced with unanticipated situations, they would make errors and poor decisions.

In this context, Rasmussen began to study the strategies of experienced troubleshooters, to understand how they found problems. He started by seeking to understand the performance of experts, learning what they did well, as opposed to studying novices to see where they went wrong. (This has analogs to the recent interest in Safety II; see Hollnagel, 2014. Hollnagel argued that it is important to study when things go right, as opposed to when things go wrong.) In the strategies of these expert troubleshooters, Rasmussen identified three core approaches: topographic search, functional search, and search by fault evaluation. Two of these approaches, functional search and search by fault evaluation, influenced the structure of the abstraction hierarchy. Functional search refers to the operator's ability to interpret improper function and reason down to the component level. Search by fault evaluation is the reverse path: by observing changes in component responses, the operator must use their mental models of the system to determine function. These search strategies developed the *How/Why?* structure of the abstraction hierarchy (Vicente, 1999).

The abstraction hierarchy is, from its foundations, a model of expert deductive reasoning. The deductive analysis is built into the structure. To be useful, an abstraction hierarchy is built to reflect the actual working constraints of the system and practitioners building these hierarchies must take time to investigate experts, trainers, and manuals to build these hierarchies as correctly as possible. Figure 20.1 shows this idea that abstraction hierarchy is a model of how the system works and experts will have clarity on these relationships.

The abstraction hierarchy is also useful for modeling large work domains and showing how various operators can collaborate in that workspace. Models of collaboration typically build a broader abstraction hierarchy that encompasses the scope of the team, then develops models of individual expertise zones. Examples of these models are seen in the literature (e.g., Burns, Torenvliet, Scott, & Chalmers, 2009; Hajdukiewicz, Vicente, Doyle, Milgram, & Burns, 2001; and the team models of Ashoori et al., 2014). The areas where expertise overlaps are often of concern as support for collaboration may be required, and tools in this space must work for experts from different domains, who may have different requirements or languages.

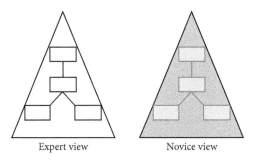

FIGURE 20.1 Differences between the expert and novice understanding of system relationships as modeled in the abstraction hierarchy. Experts see the relationships more clearly.

Figure 20.2 shows the situation where these people may have shared goals (on the left), and work on different components of the domain, or different goals and have shared work components (on the right). We have seen models on the left where operators have distinct work areas, contributing to a common goal. Many healthcare models (like Hajdukiewicz et al., 2001) have the shared work component structure where the patient is a shared component while people with different expertise work to improve their health. In this case, each expertise has its own particular goals; e.g., a surgeon may be correcting a problem, while the patient's physician tries to work on the problem from a pharmaceutical treatment perspective. Note, the overall goal may still be to improve patient health, but they may have individual purposes specific to their disciplines. These models are dependent on the scope of the domain chosen; with a broad domain scope, overall purposes will overlap. Some models have very complex overlaps between components and goals, such as Ashoori et al. (2014).

A limited example of this complexity is shown in Figure 20.3. Heart failure and kidney failure are common comorbidities that influence the progression of each disease. The reduction in blood flow in heart failure can cause kidney damage due to backed-up blood in the kidneys and reduced oxygenation of the kidneys. Similarly, when the kidneys begin to fail, the body increases blood pressure to provide more blood to the kidneys, damaging the heart further (Silverberg, Wexler, Blum,

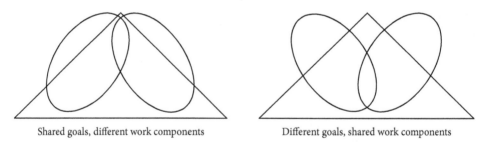

Shared goals, different work components   Different goals, shared work components

FIGURE 20.2 Modeling different domains of expertise on the same work domain.

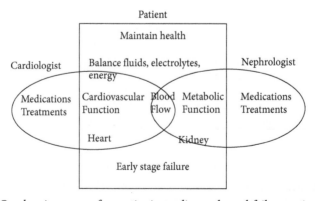

FIGURE 20.3 Overlapping areas of expertise in cardiac and renal failure patients.

Schwartz, & Iaina, 2004). The diseases are so closely linked that heart failure is present in 65–70 percent of end-stage renal patients. Patients may see both a cardiologist and nephrologist, but it is vitally important that the two professionals work compatibly. Both are likely prescribing various medications, treatments, and lifestyle changes. Both professionals are experts in the particular specialties, but a good solution for the patient requires that each physician works on their problem while accommodating the problem of the other physician. Solutions that preserve both cardiac and renal function as well as possible must be sought. It should be noted by the reader that Figure 20.3 is simplified to show the overlap. Both physicians share goals of retaining as much function as possible for the patient, and both would be consulting with the patient on treatments, lifestyle changes, and dietary changes that could help their disease.

The Decision Ladder: Differences in Processing

The decision ladder may be the phase of CWA where a novice–expert difference can be modeled most clearly. The decision ladder is a template of human information processing, from alert to decision making, to execution. With increasing expertise, operators can shunt through the ladder, moving quickly from information to execution. This corresponds with the rule-based behavior as discussed in the skills–rules–knowledge (SRK) model. Skill-based behavior is driven by perception and motor patterns, connecting perception to actions closely. Rule-based knowledge is driven by heuristics and experience, and allows people to quickly assess a situation and select an appropriate action. Knowledge-based behavior is driven by analytical thinking, where people must reason about the situation. Knowledge-based behavior is cognitively effortful, but with expertise, people can move from using knowledge-based behavior to rule-based behavior in many situations.

In the decision ladder, the rules, as they become developed with expertise, allow the shunts to occur. Where rules are not developed, people must move through the full ladder, assessing and interpreting their goals (Figure 20.4).

As an example of where changes in experience could be seen influencing processing, Figure 20.5 shows how physicians interacted with advanced practice nurses (Burns, Enomoto, & Momtahan, 2008). In familiar situations, the physicians would quickly cross from their observations (on the left of the ladder) to the required task (on the right of the ladder) and communicate the task to the nurses. In unusual situations though, they would move through the upper parts of the decision ladder, while they tried to assess a more complicated case, or determine a less standard treatment approach. With experience and a wider range of cases, this behavior becomes less frequent.

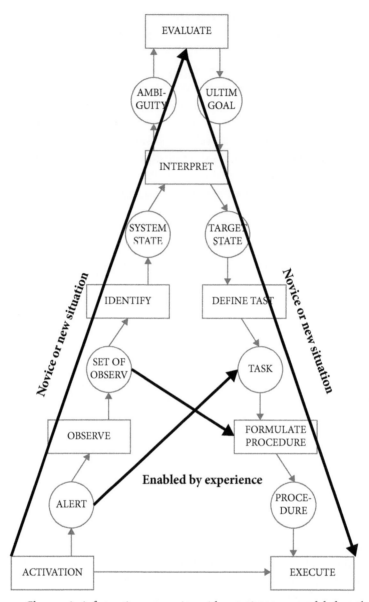

FIGURE 20.4 Changes in information processing with experience, as modeled on the decision ladder. Experts use experience to shortcut the process and move quickly to actions.

STRATEGIES ANALYSIS: LOOKING AT DIFFERENCES

In strategies analysis, the various ways that people accomplish tasks are revealed. A strategies analysis will look at how people shift approaches with workload, context,

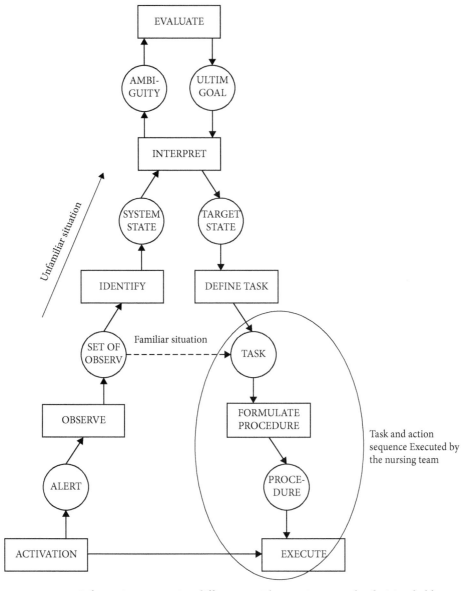

FIGURE 20.5 Information processing differences with experience on the decision ladder.

and experience differences, or in some cases a combination of these triggers (Figure 20.6). Expertise differences often provide a clue to how a design can be developed to encourage more expert strategies. Novices may not have discovered the strategies that experts have, and some operators may have experienced different contexts that influence their strategy choice.

We found examples of expertise differences in strategies in an exploration of cognitive work by advanced practice nurses performing cardiac triage (Burns et al., 2008). In this study we found five strategies for choosing triage questions: Open-ended,

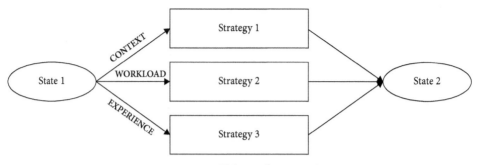

FIGURE 20.6 Triggers for strategies.

Table 20.2 Experience and how it influences strategies

Patient statement: "I have chest pain..."

Strategy	Next question	Experience needed to question
Open-ended	Describe your pain to me.	Triggered by context and experience in social interaction
Standardized	When did it start? (onset)	Triggered by rules taught in education
Topographical	Do you have any pain in your shoulders?	Triggered by rule-based experience of associations
Hypothesis and test	Is it relieved by nitroglycerin?	Triggered by experience with knowledge-based behavior
Ruling out possibilities	Do you have any problems breathing?	Triggered by rules of experience with commonly seen problems

standardized, topographical, hypothesis and test, ruling out possibilities. Table 20.2 shows how these strategies would affect the nurse's triage direction in response to an initiating concern "I have chest pain." All of these strategies are triggered by various kinds of experience, and some may be triggered by experience that has been consolidated into rule-based behavior. In particular, the standardized questioning took advantage of a mnemonic that had been taught to some nurses in their training, "OLDCART," where each letter initiated a new question. Ruling out possibilities was demonstrated more with experience, and directed towards eliminating commonly seen problems. This is also an experienced process but relies on having seen a wide number of cases, to be used effectively (Table 20.2). These are the two rule-based strategies. Hypothesis and test is a strategy that relies on experience to generate targeted hypotheses, as well as knowledge of what the appropriate test for that hypothesis would be. This strategy requires an experienced but highly deductive process and demonstrative of knowledge-based behavior. The topographical strategy is triggered by known associations such as anatomy and pain referral patterns. The open-ended strategy is triggered by context and experience in social interaction and was largely used when the nurse anticipated her patient to be articulate and able to describe the situation.

The intention behind performing a strategies analysis is to uncover the various possible strategies, and ensure that a design to support these operators will both encourage the discovery of new strategies and ensure that known strategies can be executed efficiently.

Worker Competencies

The SRK model is probably most widely recognized as a model of expertise. First, the model itself, as defined by Rasmussen, is a model of *skilled human operators*. Sometimes the model is misinterpreted as a learning model whereby people progress from skills to rules to knowledge. In fact, the model is intended quite differently. A skilled operator demonstrates all three components—skills, rules, and knowledge—with fluid and near-nvisible shifts between the three. In many cases all three components work together, with skilled sensory or control behavior, well-developed rules to guide behavior, and knowledge seeking occurring simultaneously.

Sometimes it is misunderstood that the *knowledge-based* behavior is actually a result of expertise. Highly experienced performers can avoid the conscious problem solving of knowledge-based behavior by taking advantage of stored rules or procedures. With greater expertise, we would expect operators to have larger stores of these rules, and be able to work in rule-based behavior for greater percentages of their time (Figure 20.7). Knowledge-based behavior is controlled and analytical. Goals are established, plans considered, and responses tested. Knowledge-based behavior occurs in unanticipated situations or situations requiring new problem solving to resolve.

Worker competencies analysis outlines the skills and competencies that the job requires to be done well. The SRK model is used to describe the skill-, rule-, and knowledge-based competencies. A reasonable interpretation of the worker competencies analysis would be that newer workers would be developing these competencies

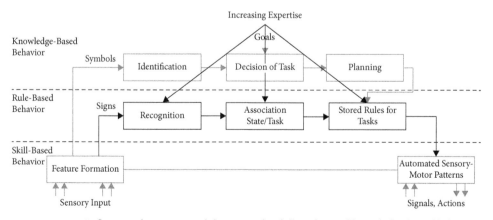

FIGURE 20.7 Influence of expertise and design on the skill-, rule-, and knowledge-based behavior.

and might need support, and experienced workers would have deep competencies, particularly in the skill- and rule-based behaviors. Recently, Ashoori et al. (2014) extended the worker competencies to describe social competencies important for effective teamwork.

Designing to Support Expertise

CWA has often been used for design, through either a formal ecological interface design process (EID: Burns & Hajdukiewicz, 2004) or less formal design guidance. The advantage of CWA is that it identifies the information important for expert behavior. In this way, CWA can advance users from a novice to expert state more quickly. We will discuss later, in the Future Directions section, how EID is relatively passive in how it encourages the development of expertise. It is quite possible EID could be improved to take a more active role in expertise development. Two key areas where EID can have a clear impact on improving expertise are by understanding fundamental relationships better, and by helping users to develop heuristics. These two opportunities are discussed in the next sections.

Understanding Fundamental Relationships Better

The abstraction hierarchy has often been used as a design tool in the approach of EID. If the hierarchy is used as a design tool, as in EID, it must be as correct as possible as the users of the design will learn from the display information. Research has shown that people using ecological displays can improve their mental model of the system they are working on over time, gradually absorbing the expertise embedded in the displays (Christoffersen, Hunter, and Vicente, 1996). This idea of developing expertise and improving knowledge through the display is a fundamental premise of EID. Novices may not understand these relationships as well, but a good design can help to make these relationships clearer for them, and over time, they will also learn these relationships well. We can modify Figure 20.1 now considering that the appropriate use of technology can make the relationships that experts understand well be visible to novices (Figure 20.8).

Building Heuristics

Two of the models, the decision ladder and the strategies analysis, identified the use of heuristics by experts. Heuristics are seen in the decision ladder where experts can shunt more directly to execution, in the strategies analysis where experts may be able to access

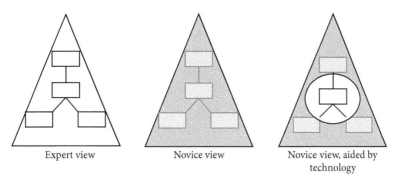

FIGURE 20.8 How technology can improve the understanding of relationships by novices.

different strategies, and in the SRK model where experts can reduce the need for knowledge-based decision making. A heuristic strategy used by experts could show up in all three of these models. The difference would be in the context and the understanding of that strategy. In the decision ladder, we can see how that strategy aids information processing and what stages of processing are involved. A design can take the knowledge of experts and deliberately show these heuristic pathways to novices (Figure 20.9). Pathways may be built into the design, or another approach, like intelligent coaching, may be able to suggest the pathway to the users.

In the strategy analysis, we can see how that strategy is triggered, and when and why it might be used when there are alternative strategies. In the SRK model, we see how experience with strategies improves the competency of expert workers. The takeaway is that expert strategies can be some of the most effective additions to a design. The design should show the strategy pathway, and the triggers for execution of the strategy appropriately (Figure 20.10). Designs that contain information on strategies will present novice operators with more options, and also remind experienced operators of strategies they may not have used lately.

FUTURE DIRECTION: DEVELOPING EXPERTISE

While cognitive work analysis provides ways of thinking about and eliciting expertise, there is not much design guidance on how to develop expertise. EID takes a passive position, assuming that if we show people how the domain works correctly, then they will learn and adapt. The research suggests that this does happen, and EID has shown good performance in many situations (Vicente, 2002).

EID though was developed for a worker situation. A worker is trained, employed, and incentivized to learn to do their job well. They are exposed to their work environment in largely uninterrupted fashion for several hours, usually on a daily basis. There are many contexts, however, where expertise and performance must be

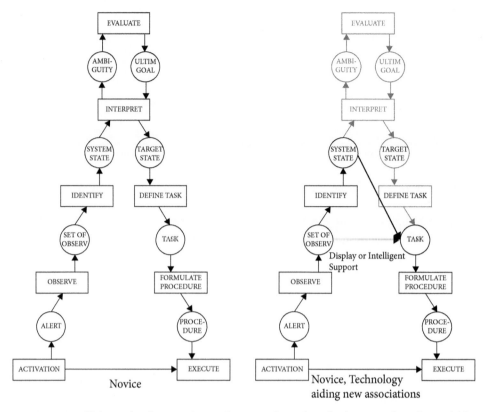

FIGURE 20.9 Using technology to give novices experience-based rules to reach action quickly.

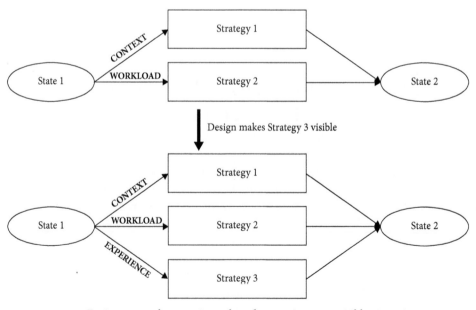

FIGURE 20.10 Design can make experience-based strategies more visible to novices.

developed outside of these conditions. A very good example of this situation occurs in health care, where a patient must monitor and treat their health at home. This person may have low expertise in treating their disease, is not incentivized in the direct manner of employment, and may have many diverse goals and interruptions throughout the day. There is a gap in the CWA and EID literature on how to train or to develop expertise in an active and deliberate manner.

Currently, we are exploring the use of persuasive design as a complement to EID and CWA as a method for triggering and motivating the development of expertise (Burns, Sadat Rezai, & St. Maurice, 2018). Persuasive design incorporates various design elements to encourage various levels of behavior change. We have used this design, in conjunction with a CWA approach to encourage better record-keeping behaviors (St. Maurice, 2017) and blood pressure management (Sadat Rezai & Burns, 2014), and to encourage fitness behaviors (Sadat Rezai, Chin, Bassett-Gunter, & Burns, 2017). For example, in the case of record-keeping, expert strategies of updating records daily were identified from the CWA. This strategy of daily updates was persuaded by including a count of the number of records kept up to date (pushing the strategy observed by experts), in the context of behaviors of other users in the team (taking advantage that competition and quality were identified as values during the work domain analysis).

The hypothesis behind the complementary use of persuasive design with CWA is to be deliberate and strategic about encouraging behavior change. With the appropriate information displayed to operators (as in the EID approach) we know that, over time, they will begin to shift their behavior and behave more like experts. With the persuasive approach, we deliberately target certain changes in behavior, like moving from information to action in the decision ladder, or an alternative strategy from the strategy analysis (Figure 20.11). Beyond just making this pathway more visible, the EID approach, now we incentivize and persuade the operator to consider the new pathway.

The behavior change can be triggered in many different ways in accordance with the theories and research of persuasive design. Some common design manipulations may include making performance goals more visible, showing progress towards goals, or showing behavior of one team member in relation to other team members.

To date we have explored persuasive design techniques in the contexts of encouraging fitness behavior and encouraging medical data entry. In both situations, adding

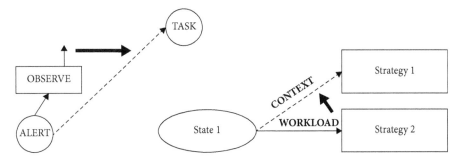

FIGURE 20.11 Deliberating changing behavior, using CWA as a foundation.

the persuasive technique was effective at creating behavior change. The persuasive approach was used to train and develop the patterns of the best performers in the particular context. As a cautionary note, when choosing to persuade a user along a certain behavior you must have substantial confidence and reasoning to suggest that that behavior is an important one to increase.

Summary

CWA is an approach that was founded on the respect for highly expert behavior and the recognized need to develop expertise. In this chapter we have looked at how CWA describes expert behavior and can inform the design of information systems to foster expert performance. CWA can provide many good requirements from domain relationships, rule-based behaviors to develop, and strategies that can help operators develop greater expertise. Combining CWA and EID with a design approach like persuasive design may be able to motivate the development of expertise further, or more quickly.

References

Ashoori, M., & Burns, C. M. (2012). *Team CWA Part 2: Strategies and competencies.* White paper, October 2012. Available on researchgate at 10.13140/RG.2.2.28790.09281.

Ashoori M., Burns C. M., Momtahan K., & d'Entremont B. (2014) Using team cognitive work analysis to reveal healthcare team interactions in a labour and delivery unit. *Ergonomics* 57(4), 973–986.

Burns, C. M., Enomoto, Y., & Momtahan, K. (2008). A cognitive work analysis of cardiac care nurses performing teletriage. In A. Bisantz, & C. M. Burns (Eds), *Applications of cognitive work analysis* (pp. 149–174). Mahwah, NJ: Lawrence Erlbaum and Associates,

Burns, C. M. & Hajdukiewicz, J. R. (2004). *Ecological interface design.* Boca Raton, FL: CRC Press.

Burns, C. M., Sadat Rezai, L., & St. Maurice, J. (2018). Understanding the context for health behavior change with cognitive work analysis and persuasive design. *Hawaii International Conference for Systems Sciences.*

Burns, C. M., Torenvliet, G., Scott, S. & Chalmers, B. (2009). Work domain analysis for establishing collaborative work requirements. *Proceedings of the 53rd Annual Meeting of the Human Factors and Ergonomics Society,* pp. 314–318.

Christoffersen K., Hunter C. N., & Vicente K. J. (1996) A longitudinal study of the effects of Ecological Interface Design on skill acquisition. *Human Factors* 38(3), 523–541.

Hajdukiewicz, J. R., Vicente, K. J., Doyle, D. J., Milgram, P., & Burns, C. M. (2001). Modeling a medical environment: An ontology for integrated medical informatics design. *International Journal of Medical Informatics* 62, 79–99.

Hollnagel, E. (2014). *Safety I and safety II: The present and future of safety management.* Dorchester, UK: Ashgate.

Jenkins, D. P., Stanton, N. A., Salmon, P. M., Walker, G. H., & Rafferty, L. (2010). Using the decision ladder to add a formative element to naturalistic decision-making research. *International Journal of Human-Computer Interaction* 26(2–3), 132–146.

Kilgore, R. M., St-Cyr, O., & Jamieson, G. A. (2009). From work domains to worker competencies: a five phase CWA. In A. M. Bisantz & C. M. Burns, *Applications of cognitive work analysis* (pp. 15–47). Boca Raton, FL: CRC Press.

McIlroy, R. C., & Stanton, N. A. (2011). Getting past first base: Going all the way with Cognitive Work Analysis. *Applied ergonomics* 42(2), 358–370.

McIlroy, R. C., & Stanton, N. A. (2015). A decision ladder analysis of eco-driving: the first step towards fuel-efficient driving behaviour. *Ergonomics* 58(6), 866–882.

Naikar, N. (2013). *Work domain analysis: Concepts, guidelines, and cases*. Boca Raton, FL: CRC Press.

Naikar, N., Moylan, A., & Pearce, B. (2006). Analysing activity in complex systems with cognitive work analysis: concepts, guidelines and case study for control task analysis. *Theoretical Issues in Ergonomics Science* 7(4), 371–394.

Rasmussen, J., & Jensen, A. (1974). Mental procedures in real-life tasks: A case study of electronic trouble shooting. *Ergonomics* 17(3), 293–307.

Sadat Rezai, L., & Burns, C. M. (2014). Using cognitive work analysis and a persuasive design approach to create effective blood pressure management systems. *HFES 2014 International Symposium on Human Factors and Ergonomics in Health Care*, vol. 3, pp. 36–43.

Sadat Rezai, L., Chin, J., Bassett-Gunter, R., & Burns, C. M. (2017). Developing persuasive health messages for a behaviour-change-support-system that promotes physical activity. *Persuasive Technology XII, Amsterdam, April 3–6*.

Silverberg, D., Wexler, D., Blum, M., Schwartz, D., & Iaina, A. (2004). The association between congestive heart failure and chronic renal disease. *Current Opinion in Nephrology and Hypertension* 13(2), 163–170.

St. Maurice, J. (2017). *Improving data quality in primary care: Modelling, measurement, and the design of interventions*. PhD thesis, University of Waterloo.

Vicente, K. J. (1997). A history of cognitive engineering research at Risø (1962–1979). *Proceedings of the human factors and ergonomics society 41st Annual Meeting*, 210–215.

Vicente, K. J. (1999). *Cognitive work analysis: Toward safe, productive, and healthy computer-based work*. Mahwah, NJ: Erlbaum and Associates.

Vicente, K. J. (2002). Ecological interface design: Progress and challenges. *Human Factors* 44(1), 62–78.

CHAPTER 21

REFLECTIONS ON THE PROFESSIONAL PRACTICE OF KNOWLEDGE CAPTURE

BRIAN MOON

> I suggest that we may look upon these myths, these ideas and theories, as some of the most characteristic products of human activity. Like tools, they are organs evolving outside our skins. They are exosomatic artefacts. Thus we may count among these characteristic products especially what is called "human knowledge"; where we take the word "knowledge" in the objective or impersonal sense, in which it may be said to be contained in a book; or stored in a library; or taught in a university... This allows us to think of knowledge produced by men as analogous to the honey produced by bees: the honey is made by bees, stored by bees, and consumed by bees; and the individual bee which consumes honey will not, in general, consume only the bit it has produced itself: honey is also consumed by the drones which have not produced any at all (not to mention that stored treasure of honey which the bees may lose to bears or beekeepers). It is also interesting to note that, in order to keep up its powers to produce more honey, each working bee has to consume honey, some of it usually produced by other bees.
>
> —Karl Popper, *Objective Knowledge* (Oxford University Press, 1972)

INTRODUCTION

THE goal of this chapter is to describe the landscape of the practice of capturing cognitive performance and expertise—*knowledge capture* for brevity. The notion of capturing things connotes several activities:

1. to take and hold someone as a prisoner especially by using force;
2. to describe or show someone or something in a very accurate way by using writing, painting, film, etc.; and
3. to get and put information into a form that can be read or used by a computer (Merriam-Webster, 2004).

Capturing cognitive performance and expertise can reflect each of these connotations, depending on one's perspective. For the grizzled veteran who has spent a career gaining experience and *learning the hard way*, the very notion of capturing expertise may feel very much like the first, at least at the outset. For the professional whose role it is to gather stories and models to inform the design or evaluation of a new capability, capturing cognitive performance can feel quite akin to the work of an artist, journalist, or historian. For the computer scientist seeking next-generation human–machine interfaces, capturing expertise may ultimately be a stepping stone—albeit, a wet, slippery mossy kind of stone—to the holy grail of machine readable, human-derived knowledge. Each of these perspectives matter in knowledge capture industry, and each can play a role when knowledge is targeted for capture.

This chapter targets connotation 2, the work of the professional knowledge capturer. In describing such work, the perspectives of 1 and 3 will also be considered, for every knowledge capture attempt must include a holder of knowledge—typically an expert—and is performed for some ultimate end—even if that end is simply preservation. The remainder of the Introduction lays out the scope of and uses for knowledge capture, briefly reviews the origins of knowledge capture, exploring guidance from related fields, and concludes with some illustrative stories of knowledge capture from the field.

The next section focuses on the praxis of knowledge capture. Protocols and methods for knowledge elicitation, analysis, and representation have been described in detail elsewhere (e.g., Crandall, Klein, & Hoffman, 2006; see also Chapter 17, "A Historical Perspective on Introspection," by Ward et al., and Chapter 19, "Incident-Based Methods for Studying Expertise," by Militello & Anders, both this volume). This section offers insights into the components common to all capture methods—structure and probing questions. It also explores how the burden of capturing is shared across the parties involved, how purpose guides execution, and how constraints can shape practical developments in approaches to knowledge capture.

The penultimate section reviews some of the requirements toward and challenges in becoming a professional knowledge capturer. The chapter concludes by looking at future directions for the profession.

A bit of context may serve to ground the chapter. My own pathway in becoming a knowledge capturer includes academic training in the naturalistic social sciences, including a year interviewing and observing members of new religious movements (cults); a stint conducting unstructured interviews in support of criminal defense teams defending clients charged with capital crimes, facing the penalty of death; introduction to and practice of cognitive task analysis (CTA) through working with mentors Gary Klein and Robert Hoffman; and the practical application of CTA methods in my own company offering expertise management (Moon, Baxter, & Klein, 2015) and human-centered design services. The views reflected in this chapter are inextricably linked to this practitioner-oriented pathway.

Scope of and Uses for Knowledge Capture

For the purposes of this chapter, knowledge capture is defined broadly as any attempt to capture knowledge people have gained through professional experience. Professional knowledge may include knowledge about technical matters and practices, organizational operations, historical events in which knowledge has been applied successfully or otherwise, and personnel and their behavior. *Knowledge* is also defined quite broadly, to include the models people develop to make sense of and explain the world (Klein, Moon, & Hoffman, 2006), the purposeful application of such models—i.e., skills—and facts, recollections, opinions, perspectives, hypotheses, and their understandings of their selves and others. Knowledge, then, is the stuff that people have in their mind that they use for their work, and all of it is potential fodder for knowledge capture.

Capture covers activities that help people articulate their knowledge, and activities that help organize, analyze, and represent for use by others. All capture activities may happen simultaneously, or they may be conducted in phases.

To scope the focus of the chapter, attempts at knowledge capture are limited to *direct engagement* with the holders of knowledge, including interviewing methods, observational techniques, and review of work artifacts. A knowledge capturer can talk to people, watch them, and look at what they produce. Each of these approaches can yield useful knowledge about their work, validate findings from the other methods, and be blended to represent the captured knowledge. They all have their limitations. Each requires their own set of guidelines, and all are typically governed by a set of rules of engagement.

Organizations may use knowledge capture when they fear that knowledge may be lost, the results of which could prove to be at least costly and, in some cases, disastrous (DeLong, 2004). In such projects, knowledge capture may be housed under some variant of organizational development or human resources component. The immediate goal for these projects may simply be knowledge preservation. Knowledge capture may also be called upon where organizations seek the solution to some problem, or to improve upon ongoing performance of the organization.

The nature of any given knowledge capture project helps to consider which knowledge to capture, and which knowledge capture approach to take. Where solutions to workplace problems and/or large-scale improvements are sought, knowledge capture may need to be quite comprehensive. Knowledge may need to be captured from a range of performers, at a variety of points in the workflow, and using the entire toolkit. Klinger (2007) provided one of the best-known examples of this use of knowledge capture that ultimately resulted in demonstrable improvements for a nuclear power organization. Where organizations seek to retain the knowledge of a retiring expert, knowledge capture will likely be most efficient using interviewing methods, augmented with artifact reviews. The key methodological challenge concerns which aspects of the experts' vast knowledge to hone in on (Moon & Kelley, 2010).

This chapter draws no methodological distinctions between the various purposes of knowledge capture. Methodologically, there is nothing different about capturing knowledge for the purpose of knowledge preservation as for engineering problem solutions. Knowledge is knowledge, methods are methods. Purpose certainly affects when to deploy methods and what to capture and represent—it does not necessitate differences in how to deploy methods or imply differences in the nature of cognitive performance and expertise.

Knowledge Capture as a Professional Practice

It is difficult to put a fence around the flock of professional knowledge capture practitioners. The skilled journalist and careful historian can offer deep insight into cognitive performance and expertise. Knowledge capturers come to the profession from a variety of pathways, having a wide range of experiences, and providing a host of services based in knowledge capture. I have worked with professional knowledge capturers with educational backgrounds in administrative, library and military sciences, cognitive psychology and sociology, and software engineering, to name a few. These professionals have developed, used, and refined diverse toolkits for knowledge capture, and their service and tool offerings are equally as diverse. To varying degrees, these professionals capture cognitive performance and expertise.

Hoffman and Militello (2012) have traced the historical origins of one strand of knowledge capture practice, that of cognitive task analysis. Readers would do well to review their volume to gain an appreciation of the various threads in twentieth-century psychology that gave rise to the need to formalize methods for understanding the macrocognitive processes and functions that professionals use to achieve cognitive work.

An important theme for the development of CTA can be found in the injection of an *anthropological point of view* into the study of work and the retention of *real-world context* in the exploration of cognitive performance—i.e., a naturalistic approach. Adoption of this naturalistic stance brings methodological imperatives:

> [Knowledge capture] is a tough job requiring a high order of careful and honest probing, creative yet disciplined imagination, resourcefulness and flexibility in study, pondering over what one is finding, and a constant readiness to test and recast one's views and images of the area... [It cannot be properly achieved by relying on] a picture of [the worker's] world derived from a few scattered observations of it. (Blumer, 1986; pp. 21–47)

It is this stance that first and foremost defines the professional knowledge capturer. Helping other professionals articulate their performance of cognitive work is indeed a tough job. The best work is marked by direct, careful, and probing examination. The knowledge capture methods described in the next section are aids to this examination. But because each expert is different, each project seeking different goals, the best professional practitioners are characterized by their creative yet disciplined imagination, resourcefulness, and flexibility in the application of these aids and the managing of their knowledge capture targets.

Illustrative Stories

A few short stories can help to illustrate how professional knowledge capture can play out and can drive home the need for a naturalistic stance. In each of these stories, I was retained to capture knowledge from professionals, for different purposes and under different project goals. And each brings into high relief challenges a knowledge capturer can face.

The Unwilling Expert

A chemical company faced the retirement of their most senior analyst. Throughout his career, the analyst had been responsible for analyzing unknown and complex substances, using analytic techniques such as mass spectrometry that also evolved in complexity and analytic power. Earlier in his career, the expert worked as a lone wolf, handling all aspects of the analysis workflow: receipt and preparation of samples, running of analytic procedures, data analysis, and reporting. He often collaborated with a colleague in a sister discipline, working the hardest problems the company faced. Their efforts earned them global renown in their respective fields.

The sponsor's request was to *capture this expert's knowledge*—not an uncommon request from managers and executives who fear lost knowledge and search for any solution to mitigate it. As often happens, the request came quite late—the expert was due to retire in a couple of months—and preparations were rushed. The ultimate goal was to help accelerate the achievement of expertise in the other professionals that had joined the now growing laboratory. On the eve of the start of the project, the expert approached me to express, rather forcefully, opposition to the entire effort. His belief—which was not unfounded—was that the other personnel needed to learn the business the *hard way, get their hands dirty*, just as he had done all those years. He was not going to participate in any knowledge capture activities.

That is a tough way to start a knowledge capture project. Thinking creatively, we all switched gears. I spent a day using knowledge capture methods with the personnel whose roles surrounded the expert, exploring how they work with him and homing in on the aspects of his cognitive performance that they did not understand. As the picture of their work emerged, it became apparent that they would likely never reach his level of expertise. As their workload had increased, the laboratory had been reorganized to assign personnel to segments of the overall workflow. No one person saw an entire problem through from start to finish. Moreover, the hardest problems were siphoned off to the expert and his colleague while the rest of the staff worked routine procedures against easier problems. Collectively, we conveyed this picture to the expert, and expressed their desire to watch him work a difficult problem. He somewhat begrudgingly agreed, and as he did,

I (carefully) introduced probing questions that targeted his cognitive performance. From the looks on the rest of the staff's faces, it was clear as he articulated his assumptions and considerations that we were collectively capturing knowledge. Yet this story taught me that not all knowledge desires to be captured.

The Useless Method

A company in the utilities industry faced retirements across their workforce. Their primary concern was with their most senior engineers who had helped to design, construct, and maintain the 40+-year-old fleet of nuclear power plants. Succession planning and knowledge sharing was deeply embedded in the corporate culture, and management had funded a large-scale project of knowledge capture and sharing, including the development of a portal through which current and future engineers could learn from these esteemed engineers.

Our plan was to exercise a range of methods for knowledge capture, several of which are described in the next section. Among these was a method for which I had become well known for using—applied concept mapping (Moon et al., 2011a). A goal for the project was to capture the engineers' technical, managerial, marketing, and organizational knowledge in concept maps, which colleagues of mine had been doing successfully with similar professionals for several years (Coffey, Eskridge, & Sanchez, 2004).

The experts were identified, a methodological guide was developed, and appointments for sessions were arranged. When I first met with one expert, however, it quickly became apparent that our plan was worthless. This expert was one of the most well-regarded engineers not only in the company, but throughout the industry. He had quite literally been involved in the field since the dawn of the nuclear age. He had also immigrated to the United States from Japan as a young adult. The idea of co-creating a diagram as complex as a concept map—about the history and decision making involved during the development of a dynamic control mechanism for nuclear power—with an elderly engineer who spoke English as a second language... Needless to say, I never made a single concept map with this expert. This story taught me that not all knowledge capture methods are applicable all the time.

Bubbling over with Cognitive Performance

Robert Hoffman and I were invited to conduct a survey of healthcare facilities with the goal of elucidating critical challenges in delivering care using electronic health records. The sponsor of the project sought to create a sweeping survey, and arranged for the participation of seven facilities throughout the system. Our teammates on the effort thankfully handled the logistics and would accompany us to the facilities. Together, we set about conducting knowledge capture with 60 healthcare professionals—a number that is frankly unheard of for any knowledge capture project. This was, in many respects, a dream project. Plenty of participants who held knowledge about healthcare focused on one of the most interesting and perplexing aspects of modern healthcare.

At our first facility, the plan for scheduling and initiating our knowledge capture sessions began to become apparent. At each facility, our team was provided a point of contact, who was to arrange interviewing and observation sessions with members of the staff, given an overarching schedule of *days onsite*. The points of contact at each facility, as we started to learn, had different ideas about the purpose and requirements of our visit. Moreover, sessions were often scheduled for convenience of the organization—i.e., whoever was available—not necessarily for evidence of cognitive performance. While we met with many providers and nurses, whose cognitive performance was primary, their levels of experience were quite varied. Some were new to the profession and to the facility, so their experience was limited. We also met with information technology technicians and hospital administrators, whose insight into the cognitive work of healthcare providers was second-hand.

We sought to elicit examples from their experience that could illustrate their cognitive performance and, ideally, their expertise. Our approach was to employ the critical decision method described in the next section. With lesser experienced personnel, this was of course a challenge. But it also proved challenging with career professionals. Given their vast experience in which every patient appointment could be considered a case and our broad focus, asking these professionals to recall a case in which they exercised their expertise was like asking a farmer to find a strand of hay in a haystack. This story taught me that *cognition in the wild* (Hutchins, 1995) needs a good safari guide to keep from getting lost in the reservation, and that too much cognition can be a difficult thing to wrangle, let alone capture.

These stories offer a flavor for the context in which knowledge capture occurs and some of the extra-methodological factors that must be considered. The next section focuses on the methodology of knowledge capture.

Praxis

Effective and efficient knowledge capture requires methods and techniques. Proctor (2017) offers a listing of many, with advantages and disadvantages; Burge (2017) provides an extensive overview and classification of knowledge elicitation tools.

For the methods and techniques that enable *direct engagement* with cognitive performers, it is useful to break them down into two components: structure and probing questions. Reviewing these components separately enables insight into the interaction between capturer and holder, how purpose can drive the execution of a method, and how adjustments to either can be the keys to success or failure.

Structure

The structure of the engagement refers to the steps, phases, and/or sweeps that are to be executed when conducting a knowledge capture. Structure enables the capturer to know where the engagement is heading, if not the exact path it will take, by providing a guide as to activities that need to be accomplished and the order for achieving them. It enables efficiency by giving the capturer a direction to go once an activity has been achieved.

Knowledge capture methods range in their level of structure. Some knowledge capture protocols provide a specific set of steps, to be followed in a specific order, and these ensure that a basic set of knowledge will be captured. Such structure may capture elements of cognitive performance, but likely will stifle the digging necessary to explore expertise. Other knowledge capture methods—such as observations—are virtually structure-less, bound only by the time of arrival and departure. In these cases, questions can take the primary role for effective knowledge capture, and efficiency may fall.

The role of structure can be explored in three methods: cognitive interviewing, applied concept mapping, and the critical decision method.

Cognitive Interviewing: Emergent Structuring

The term *cognitive interviewing* has been used in several fields. For example, in forensics and public health, cognitive interviewing is a "set of discrete techniques [with] synergies with unstructured qualitative interviewing" that provides a context for a knowledge holder to recall their experiences (Waddington & Bull, 2007, p. 2).

When used in a knowledge capture context, cognitive interviewing uses knowledge elicitation probe questions that are designed to elicit a descriptive account of the cognitive functions and processes required to make decisions and perform complex tasks, and builds an emergent structure as the interviewing progresses. Cognitive interviewing can uncover the cues, expectancies, goals, strategies, and typical actions taken by domain practitioners. Importantly, cognitive interviewing can focus on aspects of cognitive work influenced and enabled by technologies. For variants of this approach see Ward et al. (this volume) and Militello and Anders (this volume).

Applied Concept Mapping: Iterative Structuring

Applied concept mapping (ACM) is the application of concept mapping to problem solving in the workplace (Moon et al., 2011a). Concept maps are organized sets of propositions, expressed as *concept—relation—concept*, that capture knowledge. An example concept map describing ACM is provided in Figure 21.1. ACM has been used as a knowledge capture technique in domains as varied as weather forecasting and silk weaving (Hoffman & Beach, 2013), power production, and consumer packaged goods (Moon et al., 2011a).

A major advantage in using this technique lies in the fact that executing the technique enables the possibility of the concurrence of articulation, organization, and representation of knowledge, though these activities may also need to be executed in sequence. Crandall, Klein, and Hoffman (2006) provide a structured technique for co-creating concept maps with knowledge holders. One researcher acts as an elicitor while the other acts as the mapper and captures the participant's statements in the concept map, which is projected on a screen for all to see. The elicitor prompts the expert to attempt to answer the focus questions by asking for key concepts, using probing questions, and encouraging the knowledge holder to expand on their reflections. Doing so provides the fodder to start structuring the knowledge holder's thoughts.

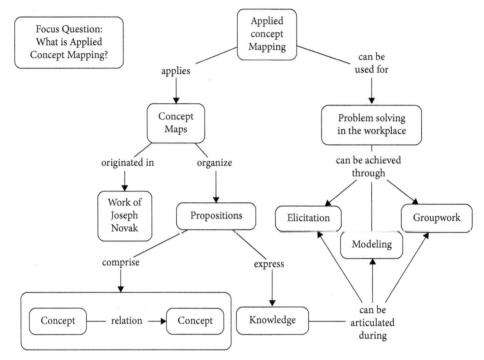

FIGURE 21.1 Concept map of applied concept mapping

The practicalities of knowledge capture, such as funding, can often require that the roles of facilitator and concept mapper be executed by a single facilitator/mapper. My colleagues and I (Moon, Johnston, Rizvi, & Dister, 2016) have described a structure that streamlines the Crandall et al. approach. Managing the dual roles while attempting to maintain direct engagement with the knowledge holder can prove difficult, at best. We recommended a revised structure, shown in Table 21.1.

The main contrast is the collapsing of steps 3–6 into a single step 3, which is executed by the mapper *after* the knowledge capture session. The mapper uses the parking lot-arranged concepts captured during the session, linking them together *based on* the recorded statements from the knowledge holder, which were provided in their natural, spoken language. The draft map(s) are then provided back to the knowledge holder for review, which can be done in a format most comfortable for the knowledge holder.

This example of structure revision demonstrates a valuable lesson in knowledge capture. Structure should be considered more like Legos than rebar. Knowledge capturers should rework structure where necessary, or where opportunity suggests that a structure change may yield practical benefits.

When created with the aid of software, other revisions to maps can enhance their value as a knowledge capture product. They can be linked to other maps and digital knowledge products. Eskridge and Hoffman (2012) developed a technique for creating and using concept map-like diagrams to enable *opportunistic computation*. The diagrams they describe include components captured from humans that describe their

Table 21.1 Comparing recommended and modified method for ACM used for knowledge capture

Crandall et al. (2006)	Executed with knowledge holder?	Moon et al. (2016)	Executed with knowledge holder?
Step 1: Select the domain and focus	Yes	Step 1: Select the domain and focus	Typically, though both may be provided prior to the knowledge capture session
Step 2: Set up the "parking lot" and arrange the concepts	Yes	Step 2: Set up the "parking lot" and arrange the concepts—while audio recording	Yes
Step 3: Begin to link the concepts	Yes	Step 3: Link the concepts, refine the concept map, look for new relations and cross-links, and further refine the concept map, build the knowledge model, if appropriate	No
Step 4: Refine the concept map	Yes	Step 4: Knowledge holder review of concept map(s); revise as appropriate	Possibly, or knowledge holder reviews alone
Step 5: Look for new relations and cross-links, and further refine the concept map	Yes		
Step 6: Build the knowledge model	No		

knowledge, marked up with components that a computer can read—inching ever closer to that holy grail of machine readable, human-derived knowledge.

Critical Decision Method: Time-Tested Structuring

The critical decision method (CDM) is a knowledge capture method that enables knowledge holders to articulate experiences they have lived through and thereby describe their cognitive performance in exquisite detail (for in-depth discussion of this method, see Militello and Anders, this volume). For practitioners of CTA, it is one of, if not the, most widely used methods. Indeed, for many, CTA might as well be spelled CDM.

Structurally speaking, CDM is described as *multiple-pass event retrospection guided by probe questions* (Hoffman, Crandall, & Shadbolt, 1998). The multiple passes typically number four and are:

1. Incident selection, during which one or more descriptions of incidents related to the project goals are briefly described, with the one holding the most promise selected for continuation;

2. Decision/judgment/assessment identification and ordering, during which the major punctuations in the incident that shaped the outcome are specified and temporally ordered;
3. Deepening, during which probing questions prompt the knowledge holder to recall their cognitive performance and experience in detail; and
4. What-if-ing, curing which hypothetical questions challenge the knowledge holder to consider alternative cognitive performance and outcomes.

Variations on the CDM structure have been offered. Stanton, Salmon, and Walker (2005) suggest nine required steps—the extra steps pertaining to the setup for the method and analysis of the articulations of the knowledge holder. They also exclude the What-if-ing pass, which, of the four typical passes, can be considered the most expendable.

CDM has been fruitfully used across the globe, in hundreds of domains, by scores of professional knowledge capturers. Its wide use is testament to the utility of its structure. While experienced knowledge capturers may blend the sweeps—particularly 2 and 3—or discard 4 under time pressure, the CDM's time-tested structure offers a high degree of control over the direction of a knowledge capture session.

Questions

Questions are the shovels, trowels, pick-axes, and brushes that enable careful digging. Regardless the structure, carefully selected questions are the *sine qua non* of knowledge capture. This is an important point. Some methods are little more than questions (e.g., observations, cognitive interviewing). Some methods inextricably weave structure and method (e.g., CDM). But even methods that are primarily structure and technique (e.g., ACM) are best served with a healthy side dish of good questions. The professional knowledge capturer puts forth questions with the intent of scaffolding knowledge holders to review and put into expression their experience. The ideal expression is verbal or even diagrammatic, but nonverbal responses to questions suggest a lot—often that another question is necessary or appropriate.

No knowledge capture method can specify which questions should be asked or when—there is an emergent quality to any question. Probing questions are selected and pitched based on the goals of a project, the type and context of domain, and the experience level of the knowledge holder. As knowledge emerges within and across interviews, probing questions are asked that hold promise for revealing additional, relevant details about the knowledge are of interest. In real time, they are selected based on the knowledge shared by the holder, particularly the *red flags* that suggest that value can be gained by probing deeper. Sometimes the red flags are obvious—"I just know how to do that from experience." Sometimes they are elusive—"It's hard to describe." They can be straightforward decision points and judgments. Red flags also start flying when inconsistencies, gaps, conflicts, and leaps are offered by the knowledge holder, not for some devious purpose but as a consequence of the knowledge holder attempting to organize their own knowledge or recall a story. Mismatches could be from within the story, or to the *official* protocols, or across the mental models of other performers. They might also emerge as knowledge capturers consider the implications of the holder's knowledge for their own mental models. It is perfectly reasonable and appropriate to probe based on one's own disconnects—there is where expertise will be exposed.

The best questions should feel, to the knowledge holder, like the creation of a conversation. They are relevant but probing, inviting without leading, insightful yet inquisitive. The tone of questions may even take on a critical bent, particularly as disconnects form the basis of the question. To get at the point of a question, the knowledge capturer may need to circle back to a line of inquiry, repeat questions, or reframe them, perhaps with the benefit of additional knowledge articulated by the holder. For the knowledge capturer, questions are tools to get at cognitive performance and context to which it is responsive. Like all tools, they require experience to use well.

The following two tables suggest a set of questions that can be fashioned into the emergent conversation. They draw on and are expansions of questions offered by Militello and Hutton (1998) and Crandall et al. (2006). Table 21.2 presents a set of generic questions that a knowledge capturer may use to get at cognitive performance.

Table 21.2 Cognitive performance questions

Cognition	
Cues	What were you seeing? Hearing? Smelling? What did you think you should be seeing, hearing, smelling? What are the most critical things you watch out for?
Analogues	Did this remind you of anything? Had you seen this elsewhere, such as in training?
Typicality	Is this what you would usually do? Are trained to do? Is this what everyone else does? What stuck out to you? What seemed unusual?
Information needs	What did you need to know then? Where did you get the information? What do you ignore?
Situation assessment	If you had to describe the situation to someone at this point, what would you say? What did you think was happening when you first arrived?
Mental models	Describe how you thought it worked or should have worked. What were you thinking would happen next? What did you think was going on?
Big picture	What else was going on that you were tracking? If you could track something else, what would it be?
Options	What could you have done? What did you want to do, but could not?
Rationale	Why? Why not? Why did you pick that and not the other thing? Were you following a particular rule or approach?
Improvising	Since you could not do that, what did you do? Where could you cut corners?

Table 21.3 Context questions

Context	
Goals	What were you trying to accomplish? Did others have different goals? Had your goals changed at this point?
Task	What are the steps/phases/stages of this task? What are the subtasks?
Order	What happened next? Did you skip or add a step, or think about skipping or adding?
Products	What products are you required to produce, or produce just because? Where do you fall in the production line?
Time pressure	What was driving the schedule or pace? What concerned you about time? If you needed more time, what could you do to get it?
Challenges	Why is this so difficult to do? Where might someone without your experience have trouble? In what ways do you work around the challenges?
Constraints	What kept you from doing what you would have wanted to do? What interrupts you?
Errors	What kinds of mistakes do you see in less-experienced people? What mistakes do you still make?
Self-monitoring	Which parts of your performance were you tracking? What tells you that you are on or off from where you should be?
Feedback	What forms does your feedback take? Who notices your performance?

Table 21.3 presents a set of generic questions that a knowledge capturer may use to get at the contexts of cognitive performance.

Depending on the purpose for the knowledge capture, another set of context questions may be useful for digging into the social-technical context. Table 21.4 offers a generic set for these.

Capture as a Shared Burden

Knowledge capture inherently involves two roles—the capturer and the holder. Each plays a specific role in what can be a complex duet conditioned by rapport building, goal-seeking, attention management, and just plain hard thinking. It can be difficult to discern which role is leading which. Structure suggests that the capturer leads, but without content provided by the holder, structure is lost.

Knowledge capture methods impose varying levels of cognitive burden on each. CDM shifts the burden to the capturer. With CDM, knowledge holders are charged

Table 21.4 Socio-technical context questions

Technology	
Aids	What would you have wanted to have at that moment?
	Do you have any special aids you use?
Fitness	Did you get what you needed from your technology?
	What shape was your technology in?
Requirements	Is this a required tool?
	Did you lose any capabilities with the introduction of this required tool?
Changes	What would you change if you could?
Teamwork	
Teamwork	What are the roles you are supposed to work with?
	Which roles were not working?
	Where were the bottlenecks?
Monitoring	Which parts of other performers' performance do you track?
	Do you provide feedback to other performers, based on your monitoring activities?
Makework, workarounds, kludges, surprises	
Makework	Do you have any repetitive, mundane tasks that you must do just to make a system work?
	Do you know of other's experiences with such tasks?
Workarounds	Do you have ways of working that you have developed in order to work around some part of your system?
Kludges	Are there parts to your system that you have had to force-fit together, in order to make things work?
	In what ways is that arrangement vulnerable?
Surprises	Have you ever been surprised by your technology?
Information management	
Needs	What information do you need?
	Where did you get the information?
	Where else might you have gotten it?
Sharing	To whom do you pass information?
	How do you know when and if they need it?

with recalling their experience. While recall of intimate and specific details that may have lied dormant for some time can certainly be a challenge, CDM only requires recall of experience—not active reflection on that experience. The burden is on the knowledge capturer to reflect on the experience to determine its implications and where to take the engagement next.

For ACM, the cognitive burden shifts to the knowledge holders. Make no mistake—in conducting an ACM session, a knowledge capturer is executing many skills simultaneously, including management of the knowledge holder, executing the protocol, and in many cases, using software (Moon et al., 2011b). Yet the experience is still more cognitively burdensome on the knowledge holders than with the CDM because they must reflect on their cognitive performance, figure out its accurate and appropriate expression, and guide its organization. They are doing much more work than telling a story as with CDM.

Cognitive interviewing, with its blend of recall and reflection questions and its virtual lack of structure, plays out on a mostly level field. Knowledge capturers must figure out where to start and where to go next as the knowledge holders work to respond to the probing. Anecdotally, it is safe to say that this method most frequently elicits from knowledge holders their own question: "Are you getting what you want?" The negotiation can feel like a wander. The burden falls primarily on the knowledge capturer to find opportunities to expose fruitful cognitive performance.

The shared burden also involves the starting and end points of all knowledge capture sessions and projects. Knowing where to start the knowledge capture activities is a major key to ensuring effective and efficient work (Moon & Kelley, 2010). There are many starting places for a knowledge capture session, such as general and specific goals of the project or suggested topics from a sponsor or other stakeholder. How far to pursue any engagement is determined in part by practicalities—e.g., schedules and available time of the busy expert—and methods—e.g., completion of the four sweeps in a CDM. Completion is also an emergent quality of the engagement. Knowledge capture sessions should start to wrap up when the knowledge holder can provide no more details or suggest any new directions. But some topics and methods could be pursued *ad infintium*. When to call a concept map *complete* is a judgment call—branches of the map can always be extended and new maps can always be started. Comprehensiveness is a negotiated goal.

A final note on the shared burden concerns the originality of the captured content. Many knowledge capturers work across domains. Domain unfamiliarity offers several advantages. Chief among these is the observation that some questions that may seem *basic* to domain practitioners may, in fact, be sources of deep knowledge and/or conflict across practitioners, and can prompt valuable capture around assumptions and losses of common ground (Klein et al., 2005). But unfamiliarity can also be a source of inefficiency. Particularly early in knowledge capture projects, the knowledge capturer is placed in a constant state of wondering, "Is this *new* to the organization, or just new to me?" A bad outcome for a session is the re-capture of knowledge that is already well known or even worse well documented. Knowledge holders and capturers share the burden of ensuring that the focus of the capture is predominately around elusive knowledge.

Analysis and products

To serve the ultimate purpose, knowledge capture activities must produce a tangible artifact. Conversely, the ultimate purpose effects the execution of the knowledge capture activities in several ways. Initially, the purpose should be considered in the selection of methods. If the goal of a project is to improve the decision skills of the next generation of cognitive performers by developing a scenario-based training program, then producing a set of concept maps will not provide the necessary fodder for developing the scenarios. CDM will produce the skills and scenarios. If the goal is to preserve expertise for reference during the future execution of a set of tasks, then a set of stories derived from execution of CDM, while potentially useful, will require significant effort to extract, organize, and represent decision support. ACM can yield a ready-made reference product by the end of a session.

Purpose also determines the level of analysis to be applied to the outputs of a knowledge capture session. Knowledge preservation projects often limit analysis to validation of the captured knowledge by other stakeholders. Engineering- and learning-oriented projects require significant analysis to extract requirements, design seeds, and evaluation criteria. The intended analysis level can, in turn, effect selection of methods. Concept maps offer opportunities for semi-automated merging of knowledge capture sessions (Harter & Moon, 2011), which can provide an efficient pathway for determining differences across the levels of cognitive performance leading to expertise (Moon, Ross, & Phillips, 2010). CDM and cognitive interviewing can get at these same analytic concepts, but require a greater level of analysis.

How to capture the tangible artifacts can also be determined by the purpose. Concept maps can be a final product, though their value as a product may be determined by the comfort of users with diagrams (Moon, 2016). Their overall efficiency for knowledge capture is greatly enhanced by their immediate availability as a refined artifact. But as noted in Table 21.1, post-session organization may be necessary from draft maps and recordings. CDM outputs require, at a minimum, organization of the recalled story, which may proceed from notes, recordings, and/or transcripts. Some knowledge capture service providers have used video recording as a means of producing tangible artifacts: "The best way is to get the workers to talk about—or even demonstrate—what they do, how they do it, and how they know when it's done right, and capture that conversation on video" (Lundy, 2014, p. 1). Video recording may be used to capture interviews, or simply observe performance. While video and audio recording undoubtedly can enhance text or diagram representations, the post-production requirements and their additional effort must always be considered.

As these preceding sections suggest, capturing knowledge at the individual and organizational level requires a high level of professional skill and judgment. The next section explores this professionalism and challenges to its development.

Professionalism

Professional knowledge capture requires two critical elements: knowledge about cognitive performance, and facility with the cognitive performance of capturing knowledge.

Successful knowledge capture builds on results of decades of knowledge capture that have resulted in the understanding of expertise covered other places in this volume. To find expertise, a knowledge capturer must know what to look for. It is one thing to interview a performer about their work and to capture aspects of the work that enable task completion. It is another to hear in an expert's description of their cognitive performance the key pointers that suggest here is an area of the performance where a great deal of expertise is wrapped up and needs to be detangled. The professional knowledge capturer must be familiar with the generic models of expertise—e.g., models of macrocognition (Klein et al., 2006)—and features of *franchise expert*-level cognitive performance (Hoffman et al., 2011) as a starting point. Methods alone have not enabled me to help experts articulate their expertise in domains as varied as nuclear engineering, helicopter resupply, baby diaper and cheese culture production, cybersecurity, and roadside bomb defeat. But in each of these varied domains, the methods have enabled the search for how expert cognitive performance is developed and demonstrated.

Professional knowledge capturers must also develop a facility with the cognitive performance of knowledge capture. This performance involves simultaneously tracking and managing: the efficient execution of method; the careful consideration of expert responses for their influence on the direction of the interview and implications for the wider effort; the art of conversation; the restraint of listening to capture but not to learn (i.e., to compare what is being articulated to one's own beliefs); the patience to trust that the methods will yield something of value; the flexibility to consider new lines of inquiry without disrupting trains of thought or destroying efficiency; among other things. It is fair to say that not everyone is cut out for such complexity.

Having trained scores of people in the methods of knowledge capture, I have been struck by the differences across learners and how these differences have expressed themselves. During one such training I conducted in Beijing, I introduced two people to ACM and CDM, as their company was exploring how to develop organic capabilities in knowledge capture. Neither person was a native English speaker—their first languages were Chinese and Hindi. Our targets for the knowledge capture spoke still another language, Japanese. One person had a bit more professional experience than the other, but both worked in product development and they were familiar with the topics chosen for this pilot effort. Both received the same training program I delivered. Over the course of a week, we gradually shifted from my leading the knowledge capture interviews to their leading, with the goal of having each try to lead both methods. As we did, it became apparent that they were both equally skilled at *one* of the methods. One learner took very quickly to ACM, crafting elegant propositions and mastering their organization almost as rapidly as the target performers offered them. This learner, however, struggled with CDM. The idea of structuring the engagement while selecting questions based on the performer's responses felt very discomforting. The second learner, in contrast, showed a facility with CDM from the start. This learner felt at ease with the give and take of the method, and lead engagements that were fluid and productive. But the challenge of crafting propositions on the fly while guiding the performer to co-create a concept map felt very unnatural to this learner. After several attempts, this learner declared an intention to never use this method. The experience demonstrated to me that knowledge capture methods are not universally usable—other, mostly ill-defined skills are required for efficient and effective knowledge capture.

There is irony in the situation that the practice of helping experts articulate their expertise involves aspects of expertise that are difficult to articulate. Many who practice knowledge capture know what enables the acceleration of expertise (Hoffman et al., 2013). Experts develop through gaining experience with hard problems, getting direct and timely feedback from seasoned mentors, and exercising through deliberate practice. Developing expertise in the practice of knowledge capture is, by these criteria, an elusive pathway. Learning about knowledge capture methods is certainly part of the pathway. But development of expertise in knowledge capture cannot be achieved in a classroom. Practitioners should face unwilling or at least suspicious experts, try methods that do not seem to work for the situation, analyze large volumes of findings, and attempt to represent and convey knowledge for others to use.

Connecting apprentices to mentors is a necessary but not sufficient requirement for encouraging professionalization. Experience is the crucial element. By the time I started my career in knowledge capture, I had already conducted scores of interviews with criminal defendants and their families and acquaintances, in literal life-and-death circumstances. Yet few knowledge capture practitioners see so many opportunities to hone their skills. Ericsson's *10,000-hour* rule of thumb is virtually unheard of for knowledge capturers—meeting that threshold would require engaging with hundreds of experts. Regarding ACM, it is conceivable to reach a comparable threshold. I have suggested that ACM expertise could reasonably be considered achieved after *10,000 propositions* (Moon, 2016).

FUTURE DIRECTIONS

Professional knowledge capture as described is a nascent field. The past several decades saw the coalescence of several fields of practice and academic traditions, resulting in handbooks describing useful techniques, articles exploring methodological implications, workshops and courses offering training, hundreds of studies using the methods, and scores of companies and practitioners setting out shingles to provide knowledge capture services. Given the societal needs for mitigating lost knowledge (Moon, 2014) and the demonstrable benefits of capturing expertise for engineering (Cooke & Durso, 2007) and training (Hoffman et al., 2013), the practice of capturing cognitive performance and expertise will likely remain a sought-after field. Yet the field will need to address two key challenges to ensure that the supply continues to satisfy the demand.

Developing Knowledge Capture Skills

In many ways, the knowledge capture community of practice faces the very problem it purports to help others mitigate. Those who have codified, evaluated, and evolved knowledge capture methods are themselves in the twilights of their careers. Mostly through apprenticeship relationships, they have passed along lessons learned and provided the critical feedback to help the following generation develop their skills. Through self-selection, market competition, and the pressures of delivering quality, usable products, the community remains relatively small. Opportunities to hone skills remain even more narrow.

Enabling skill development will require large-scale training circumstances that provide opportunities to try out many methods in many contexts and get direct feedback. The notion of a knowledge capture academy to provide the context and curriculum for such a circumstance seems unlikely. The apprentice model will likely continue to be the ad hoc solution for developing skills. But it requires capable, available, and professional mentors who can provide experience-guided feedback.

Because knowledge capture includes aspects of human interaction that are challenging to *train*, criteria to select promising capturers should also be made more explicit and incorporated in the skill development process. Handling the more delicate aspects of performance can sometimes prove challenging for any mentor. But to tell an aspiring knowledge capturer that they are not developing rapport with knowledge holders is akin to telling them they are not likable or awkward. Descriptions of such skills may be best used for selection.

Gaining Efficiencies

Brain dump. Mind meld. Matrix download. These are the metaphors knowledge capturers often hear used to describe their work. They are typically used in jest, but also subtly convey a desire held by knowledge holders and the sponsors of knowledge capture activities—make it quick. Projects may be scoped by time, and everyone wants to know how long the knowledge capture is going to take. The concerns are legitimate and the response from knowledge capturers has traditionally been the same:

> all of these methods take time. An interview using the critical decision method typically takes just a couple of hours but can take all day, or even longer. The data must then be coded and recoded, meaning that additional time is required to transform the data into representations.
> (Zachary, Hoffman, Crandall, Miller, & Nemeth, 2012, p. 61)

The knowledge capture community continuously seeks opportunities to hasten the process. Kelley, Sass, and Moon (2013) demonstrated the evolution of a knowledge capture program that sought to offer several knowledge capture services based on the organization's needs and available time. Moon et al. (2015) demonstrated the critical necessity of scoping the knowledge topics of interest. Knowing the focus questions or having a set of examples to spark recall upfront is the single best programmatic factor for introducing efficiencies into knowledge capture sessions.

The application of methods, too, can be a source for efficiencies. Efficiencies in the protocols for ACM that help manage the efficiency *of the expert's time* were suggested earlier. The steps all remain necessary, but it may not always be necessary to execute them with the knowledge holder.

Another potential source of efficiencies may lay in the development of knowledge capture methods that directly target models of cognitive performance—i.e., macrocognition (Klein et al., 2003). Descriptions of cognitive performance in many domains—yielded through knowledge capture—have led to generic models of cognitive performance and expertise that hold across domains. Probing questions can reflect these models. Moon, Wei, and Cox (2004) suggested translation of macrocognitive models into criteria against which technologies that claim to enable expert performance can be judged. Similarly, macrocognitive models could be translated into knowledge capture methods that more directly target expert performance. For example, Klein et al. (2005) identified features of joint activity that define how high-performing teams develop and maintain common ground. A *common ground audit* translates the features into probing questions that get directly at them, for example:

"Describe methods for and examples of ways that you and your teammates:

- Establish your initial calibration, for example at the beginning of a shift
- Provide and receive clarifications and reminders
- Update others about changes that occurred outside of view or when they were otherwise engaged
- Monitor other team members to gauge whether common ground is breaking down
- Signal a potential loss of common ground
- Repair the loss of common ground."

Such an approach shifts a great deal of the shared burden to the knowledge holder, but suggesting common examples can help spark their own experience. Subjecting other macrocognitive models to similar treatment should yield methods that cut more directly to cognitive performance.

Technological advances will likely also enable efficiencies. Speech-to-text offers the potential to reduce the overall level of effort and to move more rapidly to usable products. Rapid conversion of knowledge capture products into knowledge *transfer* experiences will also be enhanced by technology. Klein's ShadowBox (Klein et al., 2015) has demonstrated this capability for CDM outputs, and Moon's Sero! learning assessment platform (Moon et al., 2016) has similarly demonstrated a capability using ACM outputs. Applications such as these will expand the ever-growing utility of capturing cognitive performance and expertise.

Regardless of applications and innovations, the future of knowledge capture will no doubt continue to balance the structure and questions that lie at the heart of knowledge capture, and seek a productive balance of effort between knowledge capturer and holder.

References

Blumer, H. (1986). *Symbolic interactionism: Perspective and method*. Los Angeles: University of California Press.

Burge, J. (2017). *Knowledge Elicitation Tool Classification*. Available at https://web.cs.wpi.edu/~jburge/thesis/kematrix.html

Coffey, J. W., Eskridge, T., & Sanchez, D. P. (2004). A case study in knowledge elicitation for institutional memory preservation using concept maps. In *Concept maps: Theory, methodology, technology: Proceedings of the first International Conference on Concept Mapping* (pp. 151–158). Servicio de Publicaciones de la Universidad Pública de Navarra.

Cooke, N. J., & Durso, F. (2007). *Stories of modern technology failures and cognitive engineering successes*. Boca Raton, FL: CRC Press.

Crandall, B., Klein, G., & Hoffman, R. (2006). *Working minds: A practitioner's guide to cognitive task analysis*. Cambridge, MA: MIT Press.

DeLong, D. W. (2004). *Lost knowledge: Confronting the threat of an aging workforce*. New York: Oxford University Press.

Eskridge, T. C., & Hoffman, R. (2012). Ontology creation as a sensemaking activity. *IEEE Intelligent Systems* 27(5), 58–65.

Harter, A., & Moon, B. (2011). Common lexicon initiative: A concept mapping approach to semiautomated definition integration. In B. Moon, R. Hoffman, J. Novak, & A. Cañas (Eds), *Applied concept mapping: Capturing, analyzing, and organizing lnowledge* (pp. 131–150). New York: CRC Press.

Hoffman, R., & Beach, J. (2013). Lessons learned across a decade of knowledge modeling. *JETT* 4(1), 85–95.

Hoffman, R., Ziebell, D., Feltovich, P., Moon, B., & Fiore, S. (2011). Franchise experts. *IEEE Intelligent Systems* 26(5), 72–77.

Hoffman, R. R., Crandall, B., & Shadbolt, N. (1998). Use of the critical decision method to elicit expert knowledge: A case study in the methodology of cognitive task analysis. *Human Factors* 40(2), 254–276.

Hoffman, R. R., & Militello, L. G. (2012). *Perspectives on cognitive task analysis: Historical origins and modern communities of practice*. New York: Psychology Press.

Hoffman, R. R., Ward, P., Feltovich, P. J., DiBello, L., Fiore, S. M., & Andrews, D. H. (2013). *Accelerated expertise: Training for high proficiency in a complex world*. New York: Psychology Press.

Hutchins, E. (1995). *Cognition in the wild*. Cambridge, MA: MIT Press.

Kelley, M., Sass, M., & Moon, B. (2013). Maturity of a nuclear-related knowledge management solution. Presented at the *Annual Meeting of the American Nuclear Society*, June 16–20.

Klein, G., Borders, J., Wright, C., & Newsome, E. (2015). An empirical evaluation of the ShadowBox™ training method. In *12th International Conference on Naturalistic Decision Making*, McLean, VA.

Klein, G., Feltovich, P. J., Bradshaw, J. M., & Woods, D. D. (2005). Common ground and coordination in joint activity. *Organizational simulation* 53, 139–184.

Klein, G., Moon, B., & Hoffman, R. (2006). Making sense of sensemaking 2: A macrocognitive model. *IEEE Intelligent Systems* 21(5), 88–92.

Klein, G., Ross, K., Moon, B., Klein, D., Hoffman, R., & Hollnagel, E. (2003). Macrocognition. *IEEE Intelligent Systems* 18(3), 81–85.

Klinger, D. (2007). Too many cooks. In N. J. Cooke & F. Durso (Eds), *Stories of modern technology failures and cognitive engineering successes*. Boca Raton, FL: CRC Press.

Lundy, J. (2014). Using video to capture workforce knowledge. Morgan Hill, CA: Aragon Research.

Merriam-Webster. (2004). *Merriam-Webster's collegiate dictionary* (11th edn). Springfield, MA: Merriam-Webster.

Militello, L. G., & Hutton, R. J. (1998). Applied cognitive task analysis (ACTA): a practitioner's toolkit for understanding cognitive task demands. *Ergonomics 41*(11), 1618–1641.

Moon, B. (2014). The right way to maintain expertise in health care. *Hospitals and Health Networks Daily Magazine*. American Hospital Association.

Moon, B. (2016). Learning, creating, and using Cmaps: Successes and challenges for concept maps as facilitative tools in corporations. Keynote presentation to the *2016 International Conference on Concept Mapping*. Available at http://cmc.ihmc.us/cmc2016papers/BrianMoon-Keynote-CMC2016.pdf

Moon, B., Baxter, H. and Klein, G., (2015). Expertise management: Challenges for adopting naturalistic decision making as a knowledge management paradigm. *International Conference on Naturalistic Decision Making* 2015, McLean, VA.

Moon, B., Hoffman, R., Eskridge, T. & Coffey, J. (2011b). Skills in Concept Mapping. In B. Moon, R. Hoffman, J. Novak, & A. Cañas (Eds), *Applied Concept mapping: Capturing, analyzing, and organizing knowledge* (pp. 23–46). New York: CRC Press.

Moon, B., Hoffman, R., Novak, J. & Cañas, A. (2011a). *Applied concept mapping: Capturing, analyzing, and organizing knowledge*. New York: CRC Press.

Moon, B., Johnston, C., Rizvi, S., & Dister, C. (2016). Eliciting, representing, and evaluating adult knowledge: A model for organizational use of concept mapping and concept maps. In *International Conference on Concept Mapping* (pp. 66–82). Berlin: Springer International.

Moon, B., & Kelley, M. (2010). Lessons learned in knowledge elicitation with nuclear experts. Invited paper. *Seventh American Nuclear Society international topical meeting on nuclear plant instrumentation, control and human–machine interface technologies*, Las Vegas, Nevada, November 7–11.

Moon, B., Ross, K., & Phillips, J. (2010). Concept map-based assessment for adult learners. In A. Cañas, J. Sánchez, & J. Novak (Eds), *Concept maps: Making learning meaningful. Proceedings of the Fourth International Conference on Concept Mapping* (pp. 1–8). Viña del Mar, Chile.

Moon, B., Wei, S., & Cox, D. (2004). Cognitive impact metrics: Applying macrocognition during the design of complex cognitive systems. *Proceedings of the Human Factors and Ergonomics Society Annual Meeting 48*(3), 473–477.

Popper, K. R. (1972). *Objective knowledge: An evolutionary approach*. Oxford: Oxford University Press.

Proctor, R. (2017). *Knowledge elicitation methods and their major advantages and disadvantages*. Available at http://www3.psych.purdue.edu/~rproctor/ACM_tables.pdf

Stanton, N., Salmon P, Walker G, et al. (2005) Cognitive task analysis methods. In *Human factors methods: A practical guide for engineering and design* (pp. 77–108). Aldershot, UK: Ashgate;

Waddington, P. A. J., & Bull, R. (2007). Cognitive interviewing as a research technique. *Social Research Update 50*, 1–4.

Zachary, W., Hoffman, R., Crandall, B., Miller, T., & Nemeth, C. (2012). "Rapidized" cognitive task analysis. *IEEE Intelligent Systems 27*(2), 61–66.

CHAPTER 22

STRESS, SKILLED PERFORMANCE, AND EXPERTISE

Overload and Beyond

GERALD MATTHEWS, RYAN W. WOHLEBER, AND JINCHAO LIN

Introduction

SKILLS are executed and acquired in a variety of stressful contexts. Case studies provide dramatic examples of stress apparently impairing performance. For example, the 2009 crash of Air France 447 resulted from several interlocking factors: tropical storms, the icing up of air-speed sensors, autopilot disconnection, and a series of crew errors (Wise, Rio, & Fedouach, 2011). The inexperience of the co-pilot, deficiencies in cockpit displays, and lack of training for crisis events compounded the situation. The most egregious human error was a co-pilot's persistent pulling back on the stick to climb, an action that led to loss of airspeed and an eventually fatal stall. The cockpit voice recording left no doubt about the stress experienced by the men or about their failure to comprehend the developing crisis. It is plausible that stress impaired their judgment and situation awareness. Events of this kind support a narrative of operator overload applicable to a variety of domains. Stress in settings including transportation and combat may lead to drastic outcomes, but overload may also provoke undramatic but significant loss of performance in contexts such as office work and education.

Contrary to the narrative that stress overloads cognitive processing, the modest stressors of the laboratory often fail to impair performance (Hockey, 1997). In the military context, trained, motivated warfighters maintain high performance standards in combat (Stokes & Kite, 1994). The overload metaphor also neglects the extent to which people actively regulate stress, through finding ways to reduce task demands or otherwise cope with external pressures. The Lazarus and Folkman (1984) transactional theory emphasizes that stress is a dynamic process that unfolds over time, mediated by cognitive processes including appraisal and coping. Skill acquisition is similarly dynamic, so that we expect relationships between stress and performance to change as the learner becomes increasingly expert.

It is questionable too whether there is any general relationship between stress and skilled performance. Skidding on ice while driving, failing a critical exam, and operating on a patient close to death may all be highly stressful, but each context differs in the specific stress factors impacting the performer, the skills to be executed, and the motives of the performer. The relevance of laboratory studies of stressors such as loud noise and high temperatures to real life is also debatable. Also, selection for skill differs sharply across contexts: many more can drive a car than fly a military fighter plane.

In this chapter, expertise is defined as competence in the cognitive and/or psycho-motor skills central to accomplishing performance goals across a range of applied domains. The chapter will explore the utility and limitations of the overload metaphor in understanding expert performance under stress across multiple applied domains. We will introduce a simplified overload model that expresses domain-general principles. If stress overloads working memory (WM) or attention, both immediate performance and capacity for skill learning may be impaired. We also survey some of the complexities of the stress literature, including individual difference factors, and the empirical evidence on stress and performance. We then turn to methodological issues, especially techniques for measuring stress in the performance context, and the challenge of the divergence of different stress indices. The final part of this chapter reviews studies of stress, skilled performance, and expertise in four domains of application: test anxiety, sports performance, surgery, and vehicle driving. We will examine the extent to which a general overload model captures the main empirical findings and supports practical interventions.

A Standard Capacity Model for Stress and Skilled Performance

Stress and Overload: The Standard Capacity Model

Figure 22.1 illustrates how stressor impacts on skilled performance may be conceptualized in terms of cognitive overload, together with some challenges to the overload perspective (in italics). We will call this the Standard Capacity Model (SCM), representing an attempt to capture common assumptions about overload. External stressors tend to produce overload, either directly as task demands or indirectly through reducing the processing capacity available for managing demands. One challenge is that stressor impacts depend on appraisal and coping, rather than objective task demands

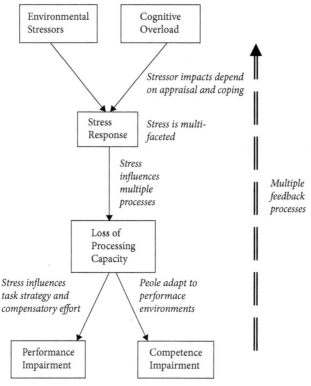

FIGURE 22.1 The Standard Capacity Model: an outline model for overload as a driver of stress response, competence impairment, and performance impairment. Text in italics indicates possible challenges to the assumptions of the model.

(Matthews, 2001). Stress typically correlates with workload, but people may experience disproportionate stress or, conversely, an enjoyable challenge (Matthews et al., 2002). In addition, stress responses have multiple psychological and physiological facets that have differing impacts on information-processing and performance (Matthews, 2016).

The stress response is associated with loss of functional processing capacity (Hancock & Matthews, 2015). Capacity loss impacts skilled performance in two ways. In the short term, stress may impair immediate performance through overload of attention or WM, so that performance falls short of actual competence. For example, well-prepared students may fail exams because their anxiety interferes with processing of the test questions (Zeidner, 1998). Also, people seek to cope actively with stressors, and the strategy chosen may exacerbate or mitigate stressor impact (Hockey, 1997). In the long term, stress may impair learning and the acquisition of competence during extended training and practice. However, performance consequences of chronic stress may not be straightforward as people acquire various adaptive and maladaptive coping skills over time (Wells & Matthews, 2015). The dynamic nature of the interplay between task demands, stress, and performance is itself a major challenge to specifying general relationships between stress and expertise.

Mechanisms for Performance Impairment in Novices and Experts

As illustrated in Figure 22.1, cognitive overload elevates stress, which in turn impairs immediate performance as well as reduces functional information-processing capacity. Thus, overload may have both direct effects on performance (increased needs for capacity) and indirect effects mediated by stress (loss of available capacity). Researchers typically assume that processing is regulated by a limited-capacity supervisory executive that requires attentional resources, WM, and/or cognitive control processes, functions that are vulnerable to overload when task demands are high (Baddeley, 2012; Ilkowska & Engle, 2010).

Stress and anxiety effects on WM are well established: a recent meta-analysis (Moran, 2016) demonstrated a moderate effect size for the anxiety—WM relationship across 177 samples. The effect appeared to be fairly domain-general. Moran (2016) also confirmed the sensitivity of dual-task performance to anxiety, and adverse effects of experimental stress manipulations, e.g., threat of electric shock, on WM. These findings suggest a direct impact of stress consistent with SCM. Loss of WM might reflect both the distracting effects of cognitive, emotional, and physical stress responses, as well as reduction in availability of capacity due to impairment of key brain systems including prefrontal cortex (Ilkowska & Engle, 2010).

The SCM also indicates how the level of operator expertise might moderate stressor impacts. Broadly, if expertise leads to reduction in the processing capacity required for performance, the operator should become less prone to stress responses, and less vulnerable to impairments in performance and learning if stress does develop. The SCM applies to both competence—the person's maximal performance under ideal conditions—and actual performance under stress, which may fall short of competence. Expertise reflects increasing competence as the person transitions from the early cognitive stage to well-practiced autonomous skill execution through practice. In Anderson's (1987, 2007) cognitive skill theory, WM is critical for the proceduralization of skill. At the early cognitive stage, productions operate on knowledge held in WM. At later stages, as domain-specific procedures are developed, the role of WM diminishes; automatic skill execution is essentially pattern recognition.

WM limitations provoke several types of error (Anderson, 1987). Accurate feedback on error responses is critical for proceduralization; WM failures may lead to inaccurate rules being encoded into productions. If stress leads to loss of WM and increased errors, proceduralization will be delayed. Figure 22.2 (top) shows schematically performance change according to the power law of practice (Anderson, 1983). Task practice under stressful conditions is associated with delayed acquisition of competence, relative to no-stress conditions.

WM limitations also lead to errors in the firing of well-formed procedures (Anderson, 1987). Thus, situational stressors that impair WM may lead to transient performance deficits. Stressor impacts can be described in terms of attentional resource theory (Norman & Bobrow, 1975). Figure 22.2 (center) shows the hypothetical performance-resource function (PRF) that links resource availability to performance. Performance is data-limited as it approaches floor and ceiling levels, but it is resource-limited over the middle part of the range. Hence, skill may be most vulnerable to disruption by stress when the person is attending moderately hard, but is not fully deploying attentional capacity or WM.

Finally, competence and performance deficits associated with stress may interact. Stress slows proceduralization in the chronically stressed learner, preserving the vulnerability of performance to WM impairment. Thus, the shape of the PRF will differ in higher- and lower-stress individuals who have acquired some degree of expertise through extended practice. The lower-stress person is less vulnerable to WM loss, and hence to stress (Figure 22.2, bottom). To the extent that experts are less dependent on WM, they will also be less vulnerable to stress-induced performance deterioration, as some of the applied research suggests (Wickens, 1996).

Figure 22.2 also illustrates how expertise may influence vulnerability to task stress. Cognitive overload elicits stress (Matthews, 2001, 2016); hence, operator vulnerability to experiencing stress will parallel vulnerability of performance to stress impacts. Increasing expertise should reduce stress vulnerability (Figure 22.2, top), but stress-induced delay in proceduralization will maintain stress vulnerability (Figure 22.2, bottom). There is a substantial literature on the roles of appraisal and coping in generating stress responses in performance environments (e.g., Staal, Bolton, Yaroush, & Bourne, 2008) to which we cannot do justice here. Generally, growing expertise should lead to appraisals of increasing performance competence, a larger repertoire of coping skills, and consequently less stress vulnerability.

Limitations of the SCM

Even in terms of the foundational theories of performance, the SCM has several limitations. First, PRFs are hypothetical and difficult to determine empirically (Matthews, Davies, Westerman, & Stammers, 2000), limiting the predictive utility of the model. Second, the rate of proceduralization depends on the extent to which the S–R mappings supporting the skill are varied or consistent across trials; consistency promotes proceduralization (Ackerman, 1987; Anderson, 1987). Thus, stress effects will depend on the nature of the skill acquired. Vulnerability of stress to performance deficits will persist in highly practiced individuals in varied- but not consistent-mapping tasks. Third, strategies for coping with stressors are also skills that can be acquired and potentially proceduralized over time (Wells & Matthews, 2015). Coping skills themselves represent an aspect of expertise.

Thus far, theory suggests an entirely benign view of expertise. The expert gains doubly from reduced stress response and reduced impact of stress on performance. However, while the expert may sometimes enjoy serene invulnerability, various factors complicate associations between expertise, stress, and performance.

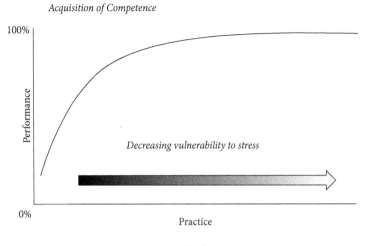
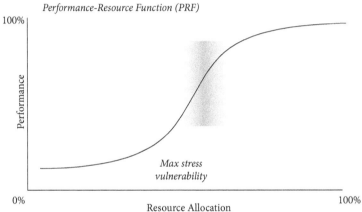
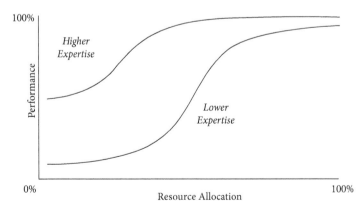

FIGURE 22.2 Stress vulnerability as a function of competence and resource allocation.

- *Complexity of stress.* Stress is more than a general psychobiological fight-or-flight response. Qualitatively different external stressors have differing impacts on stress responses and performance (Matthews, 2016; Matthews et al., 2000). Some stressors are intrinsic to the task performed, such as cognitive overload and time pressure. Others reflect extraneous factors such as environmental stressors (e.g., noise), working conditions, and social pressures. The person's own negative self-appraisals may also generate stress. Both subjective and physiological stress response metrics are varied and often poorly intercorrelated (Matthews, 2016; Matthews, Reinerman-Jones, Abich, & Barber, 2015). Multiple biological and cognitive mechanisms may mediate stressor effects on performance.
- *Underload as a stressor.* Typically, stress accompanies high workload, but the monotony of underload may also induce stress associated with fatigue (Hancock & Warm, 1989). Vigilance tasks also provoke stress (Warm, Matthews, & Finomore, 2008). If skill acquisition drives down workload, the expert may be more prone to stress of this kind than the novice. Some occupational studies suggest that mastering a task to the level that it requires little conscious thought promotes stress in the form of boredom (Loukidou, Loan-Clarke, & Daniels, 2009). Consistent tasks conducive to automatization of skill are most likely to produce this downside to expertise.
- *Dynamic adaptation.* As expertise grows, multiple factors change, requiring a dynamic perspective on performance. In a longitudinal study of practice effects on WM, Matthews and Campbell (2010) showed that task-induced distress attenuated across time, but the diminished response resulted from both lower post-task distress and increasing, anticipatory pre-task distress. Distress reflects not only decreasing workload with growing expertise, but also changes in regulating negative affect. In real-life, increasing expertise is often paired with increasing task demands. A student driver might work up from practice in an empty parking lot to managing progressively more difficult traffic conditions as skills improve. Decreasing stress response might be obscured by increasing task difficulty.
- *Changes in awareness and insight.* Experts and novices differ in their awareness of threat in complex ways. To the extent that skill acquisition represents substitution of conscious, cognitive performance with unconscious, autonomous performance (Anderson, 1987), the novice has generally greater awareness of performance, and hence vulnerability to stress associated with appraisals of performance failure. However, expertise may be associated with increased knowledge of potential threats and risks. Novice drivers are poorer on objective tests of hazard perception (Wetton, Hill, & Horswill, 2011) and they also overestimate their own driving skills (Glendon et al., 1996), factors that may mitigate stress, but lead to risk-taking behaviors. Conversely, expert awareness of performance may be associated with "choking under pressure" (Beilock & Carr, 2001) as further discussed later.

Especially in real-world settings, there is no simple relationship between stress and expertise, and stress measures are not directly informative about expertise. Novices may be prone to stress associated with cognitive overload and perceived lack of self-efficacy. However, experts may also show high stress response due to factors including stress-vulnerable personality, incidental stressors and life events extraneous to the task, or poor system/interface design. Experts may also be pushed to work at the limits of their expertise. A fire chief might select the most experienced member of the firefighting team to enter a burning building; experts may also voluntarily seek challenging encounters to test their skills.

Mechanisms for Stressor Impacts on Performance

Stressors have multiple, complex effects on the person, and it is difficult to attribute performance changes to any single mechanism. Matthews (1997) outlined a cognitive science perspective that differentiates three distinct levels of explanation associated with stress biology, *virtual* information processing based on cognitive architecture parameters, and goal-driven self-regulation based on interpretation of the task environment and its personal relevance. Repeated practice of a skill in a stressful environment will change processes at all three levels. Chronic exposure to stress changes neural responses (Ulrich-Lai & Herman, 2009), proceduralization of skill influences cognitive architecture (Anderson, 2007), and the person's attitudes, motivations, and coping skills also change with expertise (Matthews, 1999). Differentiating multiple levels of explanation suggest different lines of research on skilled performance, as we will now outline.

The Psychobiological Perspective

Stress is traditionally understood as a biological response to external stressors, captured by notions such as Walter Cannon's fight-or-flight response and Hans Selye's General Adaptation System. Traditional arousal theory invokes the inverted-U function said to relate arousal to performance, but it is oversimplified and lacks utility for performance prediction (Hancock & Ganey, 2003; Matthews et al., 2000). Modern views are more nuanced (Ulrich-Lai & Herman, 2009). Multiple brain systems regulate stress response, including the sympathetic-adrenomedullary and slower-acting hypothalamo-pituitary-adrenocortical (HPA) systems.

Mechanisms for acute stress effects on performance include impairments in functioning of prefrontal cortex and other regions supporting WM and attentional control (Arnsten, 2009; Okon-Singer, Hendler, Pessoa, & Shackman, 2015). Binding of cortisol onto glucocorticoid and mineralocorticoid receptors in the brain influences information-processing more widely (Vogel & Schwabe, 2016). With repeated stress exposures, the HPA habituates faster than sympathetic responses (Schommer, Hellhammer, & Kirschbaum, 2003).

Well-learnt skills may then be more resistant to acute stress than those requiring extensive cognitive mediation (Wickens, 1996), due to decreased HPA sensitivity as well as decreased prefrontal mediation of performance. Indeed, as in the case of Air France 447 (Wise et al., 2011), stress may encourage regression to overlearned or initially learned habits (Wickens, 1996).

Stress also influences neural learning mechanisms. Simplifying considerably, sympathetic arousal and HPA activation enhance learning of material related to the stressor, as in traumatic memory, but impair learning of neutral material and retrieval processes (Vogel & Schwabe, 2016). Different studies find that stressors delivered following learning both enhance and impair memory consolidation (Trammell & Clore, 2014). Stress also promotes a shift from reliance on a declarative hippocampus-dependent memory system to a procedural striatum-dependent memory system (Schwabe & Wolf, 2013). This shift may be an adaptive response to impairment in the cognitively demanding declarative system.

The Cognitive-Psychological Perspective

Cognitive-psychological perspectives on stress specify general limitations on processing, associated with attentional resources, WM, or executive control of performance. WM is especially important for endogenous, top-down attention and maintaining task sets and goals (Okon-Singer et al., 2015). According to Hancock and Warm (1989), stress impairs performance because it disrupts the person's normal capacity to adapt physiologically and psychologically to changes in task demand. Both underload and overload have this disruptive effect, characterized as a loss of attentional resources. If experts require fewer resources than novices, experts should be less vulnerable to both over- and underload.

Studies of skilled performance identify additional cognitive processes beyond capacity loss that are sensitive to stress (see Staal, 2004; Wickens, 1996, for reviews). Stress may produce impairments in long-term memory (especially in encoding and maintenance), fine motor skills and manual dexterity, and judgment and decision-making. Stressed operators are also vulnerable to attentional narrowing, including perceptual neglect of peripheral cues and ignoring alternative interpretations of a situation. Wickens (1996) attributed to cognitive narrowing the failures of the Three Mile Island nuclear power plant operators to correctly diagnose the developing crisis.

Thus, the SCM is oversimplistic, given that multiple cognitive processes may contribute to impairments in competence and performance. Indeed, multiple-resource theory identifies separate pools of capacity for different aspects of processing (Wickens, 2008), a perspective incorporated in more recent versions of Anderson's (2007) ACT-R model. The SCM also neglects performance effects that reflect efforts to compensate for loss of capacity, which may include shedding low-priority tasks and relying on familiar routines.

The Self-Regulative Perspective

The third level of explanation attributes stress effects to its impact on self-regulation and choice of strategies for maintaining a sense of personal competence and adequacy. People may increase effort to compensate for loss of processing efficiency (Eysenck & Calvo, 1992), reduce effort if the task is perceived as excessively difficult (Matthews & Campbell, 2009), or lower task demands by changing the task environment. In terms of the Lazarus and Folkman (1984) transactional theory, stressors influence the operator's appraisals of the task, and choice of coping strategy (Matthews, 2001; Szalma, 2009). For example, driver stress elicits both strategies that directly reduce safety, such as confrontive coping, and emotion-focused strategies that reduce safety indirectly, by diverting attention from monitoring the traffic environment to processing personal concerns (Matthews, 2002). Experimental studies of stressors show that WM capacity predicts greater flexibility in use of emotion-regulation strategies (Hofmann, Friese, Schmeichel, & Baddeley, 2011).

This brief review of theory reinforces the point that attempts to develop a general account of the relationship between stress and expertise succeed only in broad-brush terms. Instead, the focus should be on how context-specific interactions between operator and task environment are reflected in stress response.

Individual Differences

Individuals vary in both immediate performance quality and acquisition of competence (Ackerman, 1987; Szalma, 2009). Each of the paths shown in Figure 22.1 is moderated by a range of individual difference factors, including personality factors. The *Big Five* neuroticism trait is reliably associated with subjective stress in performance environments, and more weakly with physiological stress indices and performance (Matthews, Deary, & Whiteman, 2009). Specialized traits associated with resilience, such as hardiness and grit, also predict stress response, although effects vary across different stressors (Matthews, Lin, & Wohleber, 2017a). For example, using a simulation of UAV operation, Wohleber et al. (2015) found that grit predicted both subjective and EEG response to negative feedback, but not to elevated task demand. Cognitive resilience refers to maintenance of cognitive performance under stress (Matthews et al., 2017a; Staal et al., 2008). As the SCM predicts, individuals higher in WM capacity maintain performance better under stress, although generalization of this effect to field settings is uncertain (Buszard, Masters, & Farrow, 2017). Traits for negative affectivity such as neuroticism and anxiety also influence attention, WM, and executive control of processing (Eysenck & Derakshan, 2011). It can be difficult to discriminate performance loss due to neuroticism amplifying subjective stress from impairment due to neuroticism moderating the impact of stress on performance (Matthews, 2008). Personality effects may also be mediated by coping: Delahaij, van Dam, Gaillard, and

Soeters (2011) showed that military recruits possessing more adaptive coping styles performed better on stressful military exercises.

Little research has addressed long-term individual differences in learning and skill acquisition under stress. Traits predict various real-life skills (Matthews, 1999) but the dependence of these associations on stress processes is uncertain.

Assessment of individual differences is central to several application domains, including test anxiety, sports anxiety, and driver stress. Stress vulnerability traits interact with situational pressures to influence both state stress response and performance impairment. For example, trait-anxious students respond to examination pressures with higher state anxiety and poorer test performance (Zeidner, 1998). Contextualized traits such as test anxiety often predict performance better than general personality dimensions in their respective domains. The individual-differences approach is of value because controlled manipulation of situations cannot easily be effected in field settings; e.g., researchers typically have no control over examination stressors. However, we cannot infer causal influences on performance from correlational data. For example, it can be hard to gauge whether trait anxiety effects on performance are directly mediated by state anxiety, whether trait anxiety moderates the impact of state anxiety on performance, or whether elevated state anxiety is a consequence of the person's awareness of performance deterioration.

Measures of Stress Response

There are two motivations for measuring stress in studies of expertise. The first is to assess stress as a dependent variable, whether in experimental or naturalistic designs. Studies of acute impacts focus on how manipulations of task demand and extraneous stressors impact response. For example, in driving simulation studies, we could confirm that manipulations such as level of congestion, weather conditions, and time pressure actually influenced stress outcomes (Matthews et al., 2011). From a training perspective, we could investigate how instructional interventions influenced stress, or how stress response changed as expertise increased.

A second aim is to investigate stress-related constructs as independent variables that influence performance, or as variables that mediate the impact of external stressors on performance. For example, state anxiety may directly impact performance (Spielberger, 1972). Testing causal hypotheses is difficult in the case of stress because, as previously discussed, stress measures index multiple mechanisms. For example, an anxiety-performance association might be variously attributed to neural stress processes, impairment of attention, or coping efforts focused on mental withdrawal (Matthews, 2001). Generally, a case for causal influence is built from multiple lines of evidence, combining evidence from experimental and correlational studies. Structural equation modeling is useful for formal testing of causal models (Matthews & Campbell, 2010).

Methods for assessing stress response break down into subjective and objective methods. Subjective measures such as stress symptom checklists and questionnaires are easy to administer, but vulnerable to response bias. Objective methods are primarily psychophysiological in nature, although there is interest in behavioral assessment of anxiety and other negative emotions through facial expressions or posture (e.g., Calvo & Miguel-Tobal, 1998).

Subjective Measures

The simplest subjective measures assess basic dimensions of mood or emotion, such as positive affect (Gray & Watson, 2007). Scales also assess subjective responses beyond mood: e.g., loss of interest and worry are important components of test anxiety (Zeidner, 1998). The Dundee Stress State Questionnaire (DSSQ: Matthews et al., 2002) assesses primary dimensions linked to mood, motivation, and cognition, as well as three higher level factors of task engagement, distress, and worry. These factors characterize operator response to a variety of complex tasks including surface and aerial vehicle operation, command-and-control (C2), and tactical decision-making (Matthews, 2016).

In the expertise context, subjective measures define the impact of task factors and stressors on the operator. For example, assessing subjective stress and emotion response can identify features of the human–computer interface (HCI) responsible for user discomfort (Brave & Nass, 2008). Subjective measures can inform about whether learning environments are overly stressful and whether acquisition of expertise mitigates stress. Subjective measures also support performance prediction. Studies utilizing the DSSQ suggest that task engagement predicts performance on tasks requiring focused and sustained attention, whereas distress predicts performance in multitasking scenarios (Matthews, Reinerman-Jones, Abich, & Kustubayeva, 2017b).

Subjective scales have two major limitations. First, especially in high-stakes applied settings, respondents may be consciously or unconsciously motivated to distort their responses, for example to project competence by minimizing stress. Second, subjective measures are typically administered following the task; they are not suitable for monitoring stress continuously during performance.

Psychophysiological Measures

The apparent simplicity of the fight-or-flight response is misleading. There are multiple physiological responses to stressors which do not necessarily converge (Fahrenberg & Foerster, 1982; Matthews et al., 2015). Here, we provide only an outline of leading measures (see Cacioppo, Tassinary, & Berntson, 2007, for a review).

Sympathetic Activation

Sympathetic activity increases perspiration and hence electrical conductance of the skin. Measures of electrodermal activity (EDA) linked to stress include tonic conductance and phasic response to specific stimuli. Other sympathetic indices include muscle activity, control of eye movements, and pupillary response (e.g., Tichon, Wallis, Riek, & Mavin, 2014). Disruption of gaze control by stress may directly impact task performance when visuospatial attention is critical, as in airplane piloting (Vine et al., 2015).

Cardiac Activity

Increased heart rate reflects sympathetic and parasympathetic influences on the baroreflex which controls blood pressure (Fairclough & Mulder, 2012). Other metrics derived from the electrocardiogram (ECG) include heart rate variability (HRV) and its spectral frequency components, blood pressure (mean and variability), and respiratory sinus arrhythmia (RSA), which is linked to respiration rate. ECG measures are most useful in expertise research when physical activity can be controlled, as in studies of cognitive skill.

Central Nervous System Activity

The electroencephalogram (EEG) is used to assess general cortical arousal using measures of spectral power in delta, theta, alpha, and beta bands. Higher arousal is associated with higher-frequency power increases and lower-frequency decreases (Borghini et al., 2014; Craig & Tran, 2012). Frontal theta also indexes effort (Cavanagh & Frank, 2014). Borghini et al. (2013) trained novices on a flight simulation task and found an inverted-U shaped relationship between practice session and frontal theta. They suggest that effort increases over time initially, but further practice builds expertise which reduces the need for effort.

Hormonal Indices

The measures thus far discussed index acute stress response over time intervals ranging from microseconds (EEG) to a few seconds (EDA). Hormonal responses attributed to the HPA operate over longer time spans of minutes. Salivary cortisol is considered a direct HPA index, though various physiological influences limit its validity (Hellhammer, Wüst, & Kudielka, 2009). Studies in military and sports contexts show performance deficits linked to cortisol; Lautenbach, Laborde, Achtzehn, and Raab (2014) found that cortisol response predicted drop in serving performance in tennis players. Other hormones used as stress indices include salivary amylase (Valentin et al., 2015) and dehydroepiandrosterone (DHEA: Lieberman et al., 2016).

Psychophysiological measures require technical expertise to use and may be intrusive to the performer. In the performance context, different metrics may fail to converge, suggesting that they assess different aspects of stress and cognitive workload (Matthews et al., 2015). Field studies may require ambulatory recording which raises additional challenges (Sharma & Gedeon, 2012). Further guidance may be found in Fairclough's (2009) review of the validity of psychophysiological measures in the HCI context.

Use of Stress Measures in Expertise Studies

Given the multiplicity of objective and subjective stress measures, choosing one or more for a study of expertise is challenging. Use of multiple measures to delineate patterns of response may be more informative about the nature of the stress process than any single measure (Abich, Reinerman-Jones, & Matthews, 2017; Matthews & Reinerman-Jones, 2017). Modern conceptions of validity, enshrined in the AERA/APA/NCME (2014) standards, emphasize making an evidence-based case that a measure is suitable for a specified purpose. Purposes for stress measures in expertise studies include the following:

1. Evaluating learning environments: whether different settings for skill acquisition, such as real or virtual environments, influence learner stress.
2. Evaluating task demands: how changes to tasks influence stress. For example, we might vary memory load, or in applied settings, design features such as interfaces.
3. Evaluating counter-measures to stress: how effective are interventions for mitigating stress.
4. Evaluating individual differences: whether personality and other factors predict stress response to the task.
5. Evaluating predictors of performance: which stress measures are diagnostic of performance competence or fitness for duty.

Application Domains

We review four representative domains in which stress and expertise are studied: academic test performance, sports, surgery, and vehicle driving. For each domain, research supports an appraisal of the validity of the SCM, limitations of the capacity perspective, and implications for interventions to enhance performance. These domains differ in the cognitive and/or motor skills that support expertise, the external stressors that may impact performance, the extent to which performers are typically selected for expertise, and the stress response measures commonly used.

Academic expertise is a purely cognitive skill, whereas expertise in the other domains requires integration of cognitive and motor skills. Overload is a ubiquitous stressor, but each domain imposes its own unique pressures. Test and sports performance require attainment of socially defined goals that can be highly rewarded. Stakes in surgery and driving include preserving one's own life (driving only) and those of others. Assessment of anxiety is common to all domains but focus on other subjective factors is more context-dependent; for example, anger is especially salient in driver stress. Psychophysiological stress measures are used extensively in vehicle driving research (Borghini et al., 2014), but less so in other domains. A cross-domain limitation is that most studies investigate short-term associations between stress and performance. Longitudinal studies examining the interplay between stress and expertise over time are generally lacking.

Test Anxiety

Stress factors

In formal academic education, students acquire cognitive expertise as both declarative and procedural knowledge (Anderson, 1983). Students' ability to acquire and use this expertise is threatened by stress in the form of anxiety (Cassady, 2010a). Test anxiety is defined as the subjective, physiological, and behavioral responses that accompany concern about possible negative consequences of failure in an evaluative situation (Zeidner, 1998, 2007). Related constructs include evaluation anxieties linked to specific subjects including math, computing, and foreign language learning. Research typically centers on individual differences. Anxiety is a common experience, but students differ in their vulnerability to intense anxiety and performance impairments.

The multiple facets of acute anxiety include cognitive, affective, and somatic symptoms; typically, cognitive measures such as worry are the most reliably associated with performance deficits (Zeidner & Matthews, 2005). Cassady and Finch (2015) reported elevations of cortisol and sympathetic arousal during tests but the functional significance of these responses remains to be elucidated.

Performance Effects

A meta-analysis (Richardson, Abraham, & Bond, 2012) reported a correlation of -.24 between test anxiety and academic performance. Test anxiety is also quite consistently but modestly associated with a variety of performance indices in lab studies (Zeidner & Matthews, 2005). High anxiety is especially detrimental in low working-memory individuals (Owens, Stevenson, Hadwin, & Norgate, 2014), consistent with the SCM. Math and computer anxieties are similarly associated with performance impairments, including reduced WM (e.g., Chang & Beilock, 2016). Studies of online learning suggest computer anxiety may be associated with lower grades (Vician & Davis, 2003) and poorer computer-mediated communication (Fuller, Vician, & Brown, 2016). From a more dynamic perspective, deficits in WM and executive functioning may be a risk factor for the development of test anxiety (O'Donnell, 2017).

Researchers typically identify a performance deficit: high anxiety interferes with attention, WM, and retrieval from memory, so that the test score underestimates the student's actual expertise (Sarason, Sarason, & Pierce, 1995). Sarason et al.'s (1995) model is consistent with the SCM. Anxious testees anticipate failure and experience self-critical thoughts, which divert attention away from processing test questions. The model also incorporates feedback processes. Anxious testees are aware of their own distraction, which leads to further negative self-referent thoughts, and amplifies anxiety. Consistent with the SCM, relationships between anxiety and performance are moderated by environmental stressors. Experimental manipulations that increase cognitive load (e.g., time pressure) or critical self-appraisals (e.g., negative feedback) strengthen test anxiety effects (Zeidner, 1998).

Challenges to the SCM

Test anxiety–performance correlations do not necessarily reflect acute impacts of anxiety on attention and/or WM. Naveh-Benjamin (1991) suggested that a third variable—poor study skills—underlies the association. Students who are poorly prepared for a test experience realistic test anxiety, but it is the lack of effective study that is the causal influence on outcome. In addition, test anxiety may reflect various self-serving strategies for managing academic stress. Self-handicapping students use test anxiety as an excuse for poor grades, and perfectionistic over-strivers suffer from exaggerated, unrealistic performance expectations (Zeidner, 1998). Cassady (2010b) emphasizes that test anxiety influences prior preparation and the period following the exam, in which the student must decide how to modify their study habits according to their success or failure. The diagnostic significance of test anxiety thus varies for different individuals. For some students, the SCM is approximately correct, and measures to lower state anxiety in the exam will improve performance. For other individuals, anxiety signals a need to examine their study habits, attitudes to academic achievement, and motivations.

Test anxiety may also influence the acquisition of the knowledge that supports academic expertise. Learning processes vulnerable to anxiety include failure to organize and elaborate material effectively at encoding, deficient reading comprehension, greater retention loss over time, and impaired information retrieval (Zeidner, 1998). Rather little research directly demonstrates reduced competence in anxious students (as opposed to acute performance deficits). Naveh-Benjamin, McKeachie, Lin, and Lavi (1997) tested the relationship between course knowledge and test anxiety in students at intervals of up to 7 years after course completion. Students tested soon after completion showed typical test-anxiety deficits—less knowledge and poorer organization of information. However, over longer retention intervals, the advantage of low test anxiety disappeared. Naveh-Benjamin et al. (1997) concluded that test-anxious students do not have a learning deficit. Instead, they have a retrieval deficit during the period they are exposed to academic pressures. A recent study of knowledge retention during an introductory human anatomy course also failed to find any test anxiety effect (Fournier, Couret, Ramsay, & Caulkins, 2017). It is questionable whether anxiety influences academic competence, but further studies are needed.

Applications

Interventions divide into those that seek to mitigate anxiety directly, using cognitive, behavioral, and cognitive–behavioral stress management techniques, and those that aim to build study or test-taking skills (Cassady, 2010b; Von Der Embse, Barterian, & Segool, 2013). In addition, a supportive test environment can sometimes be created, e.g., by relaxing time limits, allowing open books, or utilizing adaptive testing (Zeidner, 2007). From the perspective of the SCM, alleviating state anxiety should reduce the gap between performance and competence in students prone to situational anxiety (Zeidner & Matthews, 2005).

A meta-analysis (Ergene, 2003) found that a range of interventions was generally effective in reducing test anxiety. Techniques including systematic desensitization and cognitive restructuring had large effect sizes (Cohen's $d > 0.8$), but skills training was associated with a modest d of 0.28. The analysis was directed toward anxiety as an outcome rather than academic performance, for which skills training might be more effective. However, it does suggest that skills deficits play only a limited role in test anxiety, although they could be important for individual students. Alternatively, skills-based interventions might be improved. For example, collaborative two-stage testing which includes group discussions as well as individual assessment may foster cooperative learning and mitigate test anxiety effects, although evidence from several studies is mixed (Fournier et al., 2017). Cognitively based methods for supporting test-anxious students include summary writing tasks (Mok & Chan, 2016) and retrieval practice via use of practice tests and quizzes (Agarwal et al., 2014).

Sports Performance

Stress Factors

Research on sports expertise has primarily focused on stress expressed as anxiety; mental fatigue and anger responses are also potentially important. As with test anxiety, there is an evaluative element to sports anxiety (Zeidner & Matthews, 2005). Competition raises the specter of failure and potentially strong and vocal disapproval by significant others including teammates, coaches, and spectators (Woodman & Hardy, 2001). Sarkar and Fletcher (2014) identified additional non-evaluative competitive stressors including injury and the fear of injury. Organizational stressors include poor support from coaches and the player's club, as well as issues surrounding equipment, facilities, and travel. Personal stressors include balancing commitment to sport with personal relationships, and relocation-related pressures.

Stressors differ somewhat for elite and non-elite performers. Indeed, Fletcher, Hanton, Mellalieu, and Neil (2012) found that elite performers reported a higher incidence of organizational stressors, including issues with income and funding, media attention, and a lack of participation in decision-making processes. Thus, in the sports context, exceptional expertise does not necessarily protect against stress. The benefits of superior skills must be balanced against the costs of higher expectations and responsibility, and harsher criticism of perceived performance failings.

Performance Effects

Both observational and laboratory studies demonstrate detrimental effects of anxiety. For example, penalty-taking in soccer is considered vulnerable to anxiety because the player is expected to score, failure is highly visible, and the consequences of failure may be disastrous. Jordet, Hartman, Visscher, and Lemmink (2007) showed that penalty misses in international competitions increased with the importance of the penalty, taken as an indicator of stress. Jordet and Hartman (2008) analyzed video of real penalty attempts. Avoidance behaviors linked to anxiety, such as preparing the shot quickly, predicted missed penalties. In controlled studies using experienced soccer players as participants, anxiety is induced by using instructions to instill a sense of competitive pressure. Such instructions enhance state anxiety, especially in high trait-anxious players, and produce a higher incidence of penalty misses (Horikawa & Yagi, 2012). Other skills dependent on a discrete response lend themselves to similar methodologies, such as basketball free throws, golf putting, and archery.

Discrete responses are easier to investigate than dynamic skills such as an American football quarterback improvising a play under pressure when pass protection has broken down. Nieuwenhuys and Oudejans (2017) proposed an integrated model of anxiety and perceptual–motor performance, which distinguishes three levels of processing vulnerable to anxiety. In addition to impairing attention, anxiety may also bias negatively the interpretation of ambiguous information and affect physical response through increased muscle tension, and directing responses away from a stressor. These different levels of control have different effects on perception, selection, and action aspects of movement behavior.

Meta-analyses of relationships between anxiety measures and sports performance suggest that, overall, the relationship is decidedly modest (Craft et al., 2003; Woodman & Hardy, 2003). Nevertheless, anxiety may sometimes overload attentional resources or WM (Wilson, 2008). Supporting the SCM perspective, detrimental effects of anxiety seem to decrease with skill level (Jones & Swain, 1995), although competitive experience plays an additional role (Hanton et al., 2007). WM also moderates anxiety effects. Williams, Vickers, and Rodrigues (2001) found that anxiety, manipulated via competitive pressure, had stronger impacts on processing efficiency on the higher WM version of a table tennis task. Similarly, participants low in WM capacity showed greater performance impairment in handgun shooting when an evaluative threat was imposed (Wood, Vine, & Wilson, 2016).

In fact, the version of SCM most often cited in the sports literature is Eysenck and Calvo's (1992) processing efficiency theory. Anxiety impairments in processing efficiency may be offset by compensatory effort, which maintains processing effectiveness. Some studies show that anxiety is associated with increased effort but no overall change in performance quality (Wilson, 2008). Theory may also be tested using psychophysiological measures including HRV, pupillary response, and EEG (evoked potentials), although results are mixed (Wilson, 2008).

Loss of processing efficiency reflects impairments in top-down attentional control (Eysenck & Derakshan, 2011), a mechanism that may account for effects of anxiety on the "quiet eye," the final period of fixation on a relevant object prior to initiating a movement such as putt or free throw. In sports, experts initiate the quiet eye earlier than novices, and maintain it for longer (Mann, Williams, Ward, & Janelle, 2007). Anxiety disrupts the attentional control necessary to maintain the quiet eye, specifically via failure to maintain the fixation (Vine, Lee, Moore, & Wilson, 2013).

The relevance of the SCM to the long-term development of expertise is unclear. Higher WM speeds acquisition of simple motor skills, reflecting its involvement in processing error feedback, but there is no evidence that WM facilitates acquisition of complex motor skills, including in sports (Buszard et al., 2017).

Training prepares the performer for the stress of competition. Studies of tasks including handgun shooting, basketball free throws, dart throwing, and returning badminton serves have confirmed that exposing performers to mild anxiety supports better subsequent performance under simulated competitive anxiety (e.g., Alder, Ford, Causer, & Williams, 2016; Oudejans & Pijpers, 2010). Lawrence et al. (2014) demonstrated the advantages of training under anxiety in studies of golf putting and rock climbing. Training under anxiety led to a performance decrement when the person transferred to a non-anxious control condition. Thus, consistent with the specificity principle, performance is optimized when the training environment matches the target environment for performance.

Challenges to the SCM

As for test anxiety, high anxiety may reflect unrealistic perfectionism or self-handicapping (Uphill, 2016). Contrary to expectation, higher anxiety may sometimes benefit sports performance; Parfitt and Hardy (1993) reported a positive, linear association between cognitive anxiety and basketball rebound shooting. Task demands play a role; shooting is a low WM task, which may reduce its vulnerability to cognitive interference from worry. More fundamentally, athletes differ in their perceptions of whether anxiety symptoms are debilitative or facilitative of performance. Some athletes interpret their anxiety symptoms constructively as a signal to increase effort and attentional focus, whereas others lose focus through becoming distracted by symptoms (Neil, Hanton, Mellalieu, & Fletcher, 2011). Swain and Jones (1996) found that ratings of this directionality of effect predicted a composite measure of basketball performance. Similarly, the individual zones of optimal functioning (IZOF) model (Ruiz, Raglin, & Hanin, 2017) proposes that athletes differ in the level of anxiety that is optimum for performance. Both too little and too much anxiety relative to the individual's optimum produce performance deficits.

Elite athletes do not necessarily experience lower levels of anxiety than non-elite performers, but they are more likely to interpret anxiety as facilitative rather than debilitating (Mellalieu, Hanton, & Fletcher, 2006). Elite performers build confidence and a sense of control over symptoms through emotion-regulation strategies acquired through exposure to competition, including mental rehearsal and positive self-talk (Mellalieu et al., 2006). These perspectives do not negate the SCM—worry can still overload attention—but they do suggest that adaptive motivational factors can override harmful effects of anxiety.

Another challenge to the SCM comes from the phenomenon of choking under pressure. In the sports context, choking describes a substantial performance failure with major consequences (Hill, Hanton, Matthews, & Fleming, 2010). An example often cited is Belgian golfer Jean van de Velde. With a three-shot lead at the final hole in the 1999 British Open, he hit a series of inept shots on his way to a triple bogey seven, eventually losing the championship in a play-off. Choking is also distinct from another affliction of golfers, the *yips*, associated with involuntary muscle contractions that impair motor skill execution, which can be accentuated by stress (Clark, Tofler, & Lardon, 2005). Episodes of choking challenge the SCM because they suggest that expertise may not protect against serious stress-related impairments. In addition, choking may reflect the disruptive effects of self-conscious awareness of automatic skill execution, rather than any overload of attention (Baumeister, 1984).

Debate continues over whether choking can be attributed to overload or to self-consciousness. Choking may also be mediated by disruption of quiet eye fixation (Vine et al., 2013). Studies showing that lower WM is related to poorer performance in pressure situations tend to support an overload interpretation (Wilson, 2008). For example, Bijleveld and Veling (2014) showed that lower WM players were relatively likely to lose the decisive set in competitive tennis matches. However, whether these performance effects qualify as "choking" is moot (Buszard et al., 2017).

Explanations based on self-consciousness assume that either close monitoring and/or controlling well-learned, proceduralized skills disrupts performance (Beilock & Carr, 2001; Hill et al., 2010). (The simplistic model of Figure 22.2 assumes that conscious attention is merely irrelevant to proceduralized skill.) Manipulating self-focus impaired performance on golf putting and soccer dribbling tasks, relative to a distraction manipulation that focused attention externally (Beilock, Carr, MacMahon, & Starkes, 2002). Both overload and self-consciousness mechanisms may operate in different circumstances (DeCaro et al., 2011, Hill et al., 2010). Further progress requires a deeper understanding of the complex ways in which automatic and controlled processes interact with one another (Toner, Montero, & Moran, 2015), as well as the numerous factors that moderate choking (Hill et al., 2010).

Applications

The SCM suggests that interventions for stress should reduce anxiety and other stress responses in order to limit diversion of attentional capacity away from task processing. The majority of studies of stress management (reviewed by Rumbold, Fletcher, & Daniels, 2012) report multimodal interventions that may include treatments to reduce anxiety and arousal, restructure cognitions adaptively, and enhance motivation and goal-setting. Most studies showed beneficial effects on stress symptoms, although Rumbold et al. (2012) found that only about 50 percent of studies had positive effects on both stress and performance.

Beyond simple stress reduction, interventions are guided by some of the subtleties of the relationships between stress and skilled performance identified in the research literature. For example, the distinction between debilitative and facilitate anxiety suggests treatments to induce more constructive appraisals of anxiety symptoms. Indeed, nonelite athletes tend to rely on relaxation whereas elite performers use strategies such as goal setting, imagery, and self-talk to reinterpret their anxiety symptoms as facilitative (Mellalieu et al., 2006). Empirically supported techniques for constructive emotion regulation include imagery routines, building self-confidence via cognitive restructuring, positive self-talk, and setting goals for process and performance (Uphill, 2016).

Mitigating choking requires countering the self-conscious awareness of skill execution that may disrupt expert performance. A focus on implicit learning could minimize the explicit learning that may foster awareness. Studies of table tennis and basketball shooting suggest that performers who learn by visuomotor analogy are more resistant to pressure than those receiving explicit instruction (Lam, Maxwell, & Masters, 2009). However, implicit training limits athletes' ability to self-correct technical errors and to deploy their explicit knowledge (Hill et al., 2010). Redirecting attentional focus also supports interventions for choking. Distraction techniques include music, secondary tasks irrelevant to the skill, and skill-relevant secondary tasks such as saying "hit" as a golf ball is struck (Mesagno, Geukes, & Larkin, 2015). More evidence on the practical effectiveness of such methods is needed (Mesagno et al., 2015). Moore, Vine, Freeman, and Wilson (2013) showed that quiet eye training improved pressure putting in novices. Interestingly, the effect was mediated by more positive appraisals, suggesting a transactional perspective on training.

Surgery

Stress Factors

The practice of medicine involves numerous stressors, encountered by physicians, nurses, paramedics, and others. Stress derives partly from occupational factors including the level of support provided by the practice, and the social challenges of interacting with patients (Horner, Matthews, & Yi, 2012). Stress also results from the cognitive and physical demands of clinical work itself. The role of practitioner expertise has been studied in most depth in surgeons. Specific task-related stress factors encountered by this group include multitasking, time pressure, and evaluation threat (Poolton et al., 2011). Demands of the fine motor control required for laparoscopic surgery may be a stressor (Klein et al., 2012). The SCM suggests that ergonomic interventions that lower attentional demands should also mitigate stress impacts: evidence from studies of robotic-assisted laparoscopic surgery supports this hypothesis (Hurley et al., 2015; Klein et al. 2012). Use of simulation methods in training may confer comparable benefits (Stefanides et al., 2017).

Performance effects

A systematic review (Arora et al., 2012) examined field and simulation studies of surgical procedures. Surgical work tended to increase both subjective stress and electrocardiac and electodermal stress indices. One study reported elevation of cortisol on surgery days. Performance measures in the studies reviewed were primarily measures of productivity, such as time taken for tasks or number of knots tied in laparoscopic procedures. Generally, factors that increased stress also elicited poorer performance, although evidence was lacking on whether stress correlated directly with performance. There was limited evidence that stress impaired aspects of *non-technical performance* such as communication, leadership, and decision making. As in sports, mental rehearsal prior to performing surgery may mitigate stress and enhance performance (Cocks, Moulton, Luu, & Cil, 2014). Thus, like star athletes, expert surgeons acquire expertise in regulating stress in parallel with acquiring proceduralized knowledge that is resistant to stress.

Studies of surgery show clearer evidence for expertise moderating stress response than most other task domains. Arora et al. (2012) cited four studies that confirmed lower stress response in experienced than in novice surgeons, evident in measures including self-report, electocardiac response, and skin conductance. Whether reduced stress translates into more effective performance requires further investigation. Gallagher, Satava, and O'Sullivan (2015) provide a clinical perspective on the importance of attentional capacity in surgery that suggests expert performance may be more resilient. Expert surgeons may have more spare capacity than novices to deal with the additional cognitive demands imposed by unexpected events such as intraoperative bleeding and instrument failure. Similarly, experts may be less vulnerable to seemingly minor distractions such as recalling whether a particular piece of equipment is available, or whether a nurse will need to be prompted to hand over an instrument. However, more research is needed to substantiate novice–expert performance differences specific to stressful environments.

Challenges to the SCM

The capacity model appears quite well-supported in the surgery context, but close scrutiny of studies reveals some complications. For example, Wetzel et al. (2010) compared groups of low- and high-experience surgeons who performed a procedure to remove plaque from the carotid artery using a realistic simulation including a full operating theater team. Performances during non-crisis and crisis scenarios were compared. The crisis scenario involved severe adverse reactions by the patient and stress factors such as technical challenge, time pressure, the need for multitasking, and poor assistance. The crisis scenario elevated self-report stress but did not affect psychophysiological indices including electocardiac response and salivary cortisol. Consistent with the SCM an interactive effect of surgeon experience and scenario type was obtained: only more experienced surgeons maintained performance in the crisis scenario. However, the stress indices did not predict performance. Coping did predict performance, but there was no effect of expertise on coping. Thus, while expertise conferred the expected performance advantage under stress, the study does not indicate the source of the advantage.

Applications

In the training context, proceduralization of knowledge (Gallagher et al., 2015) and general stress management interventions (Ignacio et al., 2016) should support effective performance under stress, especially in novices. Crewther et al. (2016) monitored subjective and psychophysiological stress measures during a 3-week laparascopic surgery training program. Over time, stress declined and performance improved; gains of both types were retained 4 weeks after the cessation of training. Harmful effects of stress during training were demonstrated by Flinn et al. (2016). Medical students learned laparascopic pattern cutting. They showed elevated psychological and physiological stress and performed more poorly when criticized by an expert surgeon.

Additional stress-mitigation techniques include training mental skills such as imagery and thought management, but it is unclear whether the resulting stress reduction translates into preservation of performance skills under stress (Anton et al., 2016; Anton, Bean, Hammonds, & Stefanidis, 2017). Focusing more directly on performance, surgeons may benefit from quiet eye fixation, as in sports (Moore et al., 2013). Training in quiet eye enhanced surgical knot tying during an anxiety induction (Causer et al., 2014).

Vehicle Driving

Stress factors

Driving requires attentionally demanding skills for controlling the vehicle, monitoring for hazards, and exercising judgment and decision making. Common stress symptoms include anxiety, anger, and fatigue, especially when driving conditions are demanding (Gulian et al., 1990). Driver stress is understood transactionally, reflecting the interaction between personality characteristics associated with stress vulnerability and external stressors such as poor visibility and congestion (Matthews, 2002). Plausibly, stress response is associated with loss of processing capacity, performance impairment, and increased vulnerabilty to crashes.

Performance Effects

Stress is a known risk factor for crash likelihood: divorce and financial problems are among several stress factors implicated in risk (Ge et al., 2014; Rowden, Matthews, Watson, & Biggs, 2011). Life stress spills over into driver stress when task demands are high, e.g., in high traffic congestion (Hennessy, Wiesenthal, & Kohn, 2000). Driver safety may also be impacted by emotion-inducing events on the road, and by mental health conditions including anxiety and depression (Cunningham & Regan, 2016).

Inattention frequently contributes to crashes, although failures of attention take different forms (Beanland et al., 2013). Stress has weaker effects on attention than intoxication, fatigue, and distraction, but is nevertheless a significant safety factor (Beanland et al., 2013). Stress is also associated with higher levels of self-reported errors and attentional lapses, although the validity of self-reports is debatable (Rowden et al., 2011). Drivers with lower STM capacity are vulnerable to failures of hazard perception under conditions of high cognitive load (Wood, Hartley, Furley, & Wilson, 2016). Thus, stress may impair attentional functioning, which leads to driver error and to crash vulnerability, consistent with the SCM.

Driver behavior research has employed various psychophysiological stress measures, primarily indexing sympathetic activity. Cortisol assays demonstrate chronic stress in professional drivers (Taylor & Dorn, 2006). EDA is sensitive to stressful events in both real (Healey & Picard, 2005) and simulated driving (Lanatà et al., 2015). EEG, ECG, and eye movements are all sensitive to task demands during vehicle driving and in related contexts such as piloting an airplane (Borghini et al., 2014). Hazard detection during driving is also associated with electrodermal and cardiac responses (Kübler et al., 2014).

The role of stress in acquisition of skills in student and expert drivers requires further investigation. Initial exposure to driving can be stressful (Lee & Winston, 2016), but experts typically report similar or higher levels of stress than novices (Banuls Eseda, Carbonell Vaya, Casonoves, & Chisvert, 1997; Useche, Serge, & Alonso, 2015), and stress is minimally related to age and expertise in qualified drivers (Matthews, Desmond, Joyner, & Carcary, 1997). The graduated nature of driver training may regulate stress response. So too may the tendency for inexperienced drivers to lack awareness of their shortcomings in skill (Gregersen, 1996).

Challenges to the SCM

One challenge to the SCM is that driver stress takes multiple forms that differ in their impacts on information processing on safety. Different dimensions of stress vulnerability, including aggression, dislike of driving, and fatigue proneness, differ in their associations with performance criteria (Matthews et al., 1997). Dislike of driving predicts state anxiety but not crash risk (Matthews, 2002; Qu et al., 2016). The attentional costs of anxiety may be compensated by adaptive stress outcomes such as speed reduction and lower confidence in driving competence (Matthews, 2002). Most drivers show the *above-average effect* of overestimating their skills and safety relative to other drivers, but individuals high in dislike of driving are more realistic in their self-appraisals, which may encourage a cautious driving strategy (Wohleber & Matthews, 2016).

Another challenge is that driving simulation data show that drivers high in dislike show greater performance deficits under lower rather than higher workload conditions, contrary to prediction from a resource model (Matthews, 2002). Low workload may encourage anxious drivers to divert attention to their personal concerns, although they can reorient attention externally when necessary to deal with increased demands.

In fact, anger and aggression are the elements of driver stress response that most consistently predict loss of safety (Wickens, Mann, Ialomiteanu, & Stoduto, 2016). Frustration promotes dangerous confrontive coping behaviors intended to threaten or punish other drivers (Emo, Matthews, & Funke, 2016). Confrontive behaviors such as gesturing and swearing may involve loss of task-directed attention, but often risk results directly from aggressive maneuves such as tailgating and risky passing. Such strategies primarily reflect maladaptive coping with perceived interpersonal stressors, such as hostile behaviors of others, and thus are beyond the scope of the SCM.

Applications

Stress management techniques have targeted professional groups such as bus drivers who are especially stress-vulnerable (Machin & Hoare, 2008; Taylor & Dorn, 2006). Machin (2003) developed situational exercises for bus drivers that had respondents evaluate different ways of coping with realistic stressful work scenarios. However, anxiety reduction is only a partial solution to safety issues. Indeed, interventions that elevate self-confidence excessively may be counterproductive (Wohleber & Matthews, 2016). Training higher order skills in situation awareness, hazard perception, and risk management enhances aspects of driver performance while also lowering self-confidence (Isler, Starkey & Sheppard, 2011). Such methods might be adapted for stress mitigation. In regard to aggressive driving, Deffenbacher (2016) reviews various cognitive, relaxation, and behavioral interventions that are effective in reducing driver anger, although more evidence is needed on their effects on skill execution.

Conclusions

Stress-as-overload provides a simple and appealing narrative for performance impairments. The idea that stress reduces functional capacity for processing captures an important aspect of stress effects, especially for the acute impact of severe stressors. Expertise protects against both overload-induced stress and loss of capacity, so that experts should be better able to maintain performance under pressure. However, the overload model (the SCM) is overly general. Multiple processes contribute to attention, WM, and executive control, and more granular theory is needed. The overload model is also incomplete. Additional factors including underload, strategic compensation, and changes in task appraisal and motivation play important roles in mediating performance effects. Applied studies reveal further processes beyond overload such as strategic use of anxiety for psychological defense (test anxiety), self-conscious choking under pressure (sports), and angry aggression (driving) that may impact skill execution. The complexity of the psychological science is paralleled by complexity in methodology, especially in regard to accommodating multiple stress measures. In consequence, the SCM has deficiencies as the basis for practical interventions to support expertise. Stress management techniques may contribute to enhancing attention and working memory under stress. However, domain-specific interventions such as training study skills (test anxiety), emotion-regulation skills for competition (sports), and anger management (driving) may be as effective.

This analysis suggests several future directions. Simplistic, general stress models, whether biological or cognitive, have serious limitations. Future research needs to develop basic science that accommodates the many forms of stress as well as contextually grounded accounts of the interplay between stressors, neurocognitive processes, and skill acquisition and execution within individual domains of application.

More sophisticated theoretical accounts call for parallel advances in methodology to measure and differentiate the multiple neural, cognitive, strategic, and social processes that together generate *stress*. Practical interventions for stress must go beyond simple stress management. Personalized treatments that support the individual's acquisition of stress-regulation skills appropriate to the performance context may be especially promising. All of these aims are challenged by technological advancements, which introduce novel stressors, cultural and generational shifts in perceptions of technology, and possible computer-mediated strategies for mitigating stress. Indeed, stress is increasingly a moving target, and rapid development of theory and methodology is required to anticipate and mitigate stress effects on the new domains of expertise that will emerge this century.

References

Abich J., IV, Reinerman-Jones, L., & Matthews, G. (2017). Impact of three task demand factors on simulated unmanned system intelligence, surveillance, and reconnaissance operations, *Ergonomics 60*, 791–809.

Ackerman, P. L. (1987). Individual differences in skill learning: An integration of psychometric and information processing perspectives. *Psychological Bulletin 102*, 3–27.

AERA, APA, & NCME (2014). *Standards for educational and psychological testing*. Washington, DC: AERA.

Agarwal, P. K., D'Antonio, L., Roediger, H. L., McDermott, K. B., & McDaniel, M. A. (2014). Classroom-based programs of retrieval practice reduce middle school and high school students' test anxiety. *Journal of Applied Research in Memory and Cognition 3*, 131–139.

Alder, D., Ford, P. R., Causer, J., & Williams, A. M. (2016). The effects of high-and low-anxiety training on the anticipation judgments of elite performers. *Journal of Sport and Exercise Psychology 38*, 93–104.

Anderson, J. R. (1983). *The architecture of cognition*. Hillsdale, NJ: Lawrence Erlbaum.

Anderson, J. R. (1987). Skill acquisition: Compilation of weak-method problem situations. *Psychological Review 94*, 192–210.

Anderson, J. R. (2007). *How can the human mind occur in the physical universe?* New York: Oxford University Press.

Anton, N. E., Bean, E. A., Hammonds, S. C., & Stefanidis, D. (2017). Application of mental skills training in surgery: A review of its effectiveness and proposed next steps. *Journal of Laparoendoscopic & Advanced Surgical Techniques 27*, 459–469.

Anton, N. E., Howley, L. D., Pimentel, M., Davis, C. K., Brown, C., & Stefanidis, D. (2016). Effectiveness of a mental skills curriculum to reduce novices' stress. *Journal of Surgical Research 206*(1), 199–205.

Arnsten, A. F. (2009). Stress signalling pathways that impair prefrontal cortex structure and function. *Nature Reviews Neuroscience 10*, 410–422.

Arora, S., Atkin, G., Mathur, P., Patel, B., & Elton, C. (2012). Training tomorrow's surgeons: The rule of '4s'. *International surgical congress of the Association-of-Surgeons-of-Great-Britain-and-Ireland (ASGBI)* (pp. 61–62). London: Wiley-Blackwell.

Baddeley, A. (2012). Working memory: Theories, models, and controversies. *Annual Review of Psychology 63*, 1–29.

Banuls Esada, R., Carbonell Vaya, E., Casonoves, M., & Chisvert, M. (1997). Different emotional responses in novice and professional drivers. In T. Rothengatter & E. Carbonell Vaya (Eds), *Traffic and transport psychology: Theory and application* (pp. 343–352). Amsterdam: Pergamon.

Baumeister, R. F. (1984). Choking under pressure: self-consciousness and paradoxical effects of incentives on skillful performance. *Journal of Personality and Social Psychology 46*, 610–620.

Beanland, V., Fitzharris, M., Young, K. L., & Lenné, M. G. (2013). Driver inattention and driver distraction in serious casualty crashes: Data from the Australian National Crash In-depth Study. *Accident Analysis & Prevention 54*, 99–107.

Beilock, S. L., & Carr, T. H. (2001). On the fragility of skilled performance: What governs choking under pressure? *Journal of Experimental Psychology: General 130*, 701–725.

Beilock, S. L., Carr, T. H., MacMahon, C., & Starkes, J. L. (2002). When paying attention becomes counterproductive: Impact of divided versus skill-focused attention on novice and experienced performance of sensorimotor skills. *Journal of Experimental Psychology: Applied 8*, 6–16.

Bijleveld, E., & Veling, H. (2014). Separating chokers from nonchokers: Predicting real-life tennis performance under pressure from behavioral tasks that tap into working memory functioning. *Journal of Sport and Exercise Psychology 36*, 347–356.

Borghini, G., Arico, P., Astolfi, L., Toppi, J., Cincotti, F., Mattia, D., . . . & Babiloni, F. (2013). Frontal EEG theta changes assess the training improvements of novices in flight simulation tasks. In *Engineering in Medicine and Biology Society (EMBC), 2013 35th Annual International Conference of the IEEE* (pp. 6619–6622). IEEE.

Borghini, G., Astolfi, L., Vecchiato, G., Mattia, D., & Babiloni, F. (2014). Measuring neurophysiological signals in aircraft pilots and car drivers for the assessment of mental workload, fatigue and drowsiness. *Neuroscience & Biobehavioral Reviews 44*, 58–75.

Brave, S., & Nass, C. (2008). Emotion in human-computer interaction. In A. Sears and J. A. Jacko (Eds), *The human–computer interaction handbook: Fundamentals, evolving technologies and emerging applications* (pp. 77–92). New York: Taylor & Francis.

Buszard, T., Masters, R. S., & Farrow, D. (2017). The generalizability of working-memory capacity in the sport domain. *Current Opinion in Psychology 16*, 54–57.

Cacioppo, J. T., Tassinary, L. G., & Berntson, G. (Eds). (2007). *Handbook of psychophysiology*. New York: Cambridge University Press.

Calvo, M. G., & Miguel-Tobal, J. J. (1998). The anxiety response: Concordance among components. *Motivation and Emotion 22*, 211–230.

Cassady, J. C. (2010a). (Ed.) *Anxiety in schools: The causes, consequences, and solutions for academic anxieties*. New York: Peter Lang.

Cassady, J. C. (2010b). Test anxiety: Contemporary theories and implications for learning. In J. C. Casady (Ed.), *Anxiety in schools: The causes, consequences, and solutions for academic anxieties*. New York: Peter Lang.

Cassady, J. C., & Finch, W. H. (2015). Using factor mixture modeling to identify dimensions of cognitive test anxiety. *Learning and Individual Differences 41*, 14–20.

Causer, J., Harvey, A., Snelgrove, R., Arsenault, G., & Vickers, J. N. (2014). Quiet eye training improves surgical knot tying more than traditional technical training: a randomized controlled study. *American Journal of Surgery 208*, 171–177.

Cavanagh, J. F., & Frank, M. J. (2014). Frontal theta as a mechanism for cognitive control. *Trends in Cognitive Sciences 18*, 414–421.

Chang, H., & Beilock, S. L. (2016). The math anxiety-math performance link and its relation to individual and environmental factors: A review of current behavioral and psychophysiological research. *Current Opinion in Behavioral Sciences 10*, 33–38.

Clark, T. P., Tofler, I. R., & Lardon, M. T. (2005). The sport psychiatrist and golf. *Clinics in Sports Medicine 24*, 959–971.

Cocks, M., Moulton, C. A., Luu, S., & Cil, T. (2014). What surgeons can learn from athletes: Mental practice in sports and surgery. *Journal of Surgical Education 71*, 262–269.

Craft, L. L., Magyar, T. M., Becker, B. J., & Feltz, D. L. (2003). The relationship between the Competitive State Anxiety Inventory-2 and sport performance: A meta-analysis. *Journal of Sport and Exercise Psychology 25*, 44–65.

Craig, A., & Tran, Y. (2012). The influence of fatigue on brain activity. In G. Matthews, P. A. Desmond, C. Neubauer, & P. A. Hancock (Eds), *Handbook of operator fatigue* (pp. 185–196). Aldershot, UK: Ashgate Press.

Crewther, B. T., Shetty, K., Jarchi, D., Selvadurai, S., Cook, C. J., Leff, D. R., ... Yang, G.-Z. (2016). Skill acquisition and stress adaptations following laparoscopic surgery training and detraining in novice surgeons. *Surgical Endoscopy 30*, 2961–2968.

Cunningham, M. L., & Regan, M. A. (2016). The impact of emotion, life stress and mental health issues on driving performance and safety. *Road & Transport Research 25*, 40–50.

DeCaro, M. S., Thomas, R. D., Albert, N. B., & Beilock, S. L. (2011). Choking under pressure: Multiple routes to skill failure. *Journal of Experimental Psychology: General 140*, 390–406.

Deffenbacher, J. L. (2016). A review of interventions for the reduction of driving anger. *Transportation Research Part F: Traffic Psychology and Behaviour 42*, 411–421.

Delahaij, R., van Dam, K., Gaillard, A. W., & Soeters, J. (2011). Predicting performance under acute stress: The role of individual characteristics. *International Journal of Stress Management 18*, 49–66.

Emo, A. K., Matthews, G., & Funke, G.J. (2016). The slow and the furious: Anger, stress and risky passing in simulated traffic congestion. *Transportation Research Part F: Traffic Psychology and Behaviour 42*, 1–14.

Ergene, T. (2003). Effective interventions on test anxiety reduction: A meta-analysis. *School Psychology International 24*, 313–328.

Eysenck, M. W., & Calvo, M. G. (1992). Anxiety and performance: The processing efficiency theory. *Cognition & Emotion 6*, 409–434.

Eysenck, M. W., & Derakshan, N. (2011). New perspectives in attentional control theory. *Personality and Individual Differences 50*, 955–960.

Fahrenberg, J., & Foerster, F. (1982). Covariation and consistency of activation parameters. *Biological Psychology 15*, 151–169.

Fairclough, S. H. (2009). Fundamentals of physiological computing. *Interacting with Computers 21*, 133–145.

Fairclough, S. H., & Mulder, L. M. J. (2012). Psychophysiological processes of mental effort investment. In R. Wright & G. Gendolla (Eds), *Motivational perspectives on the cardiovascular response* (pp. 61–76). Washington, DC: APA.

Fletcher, D., Hanton, S., Mellalieu, S. D., & Neil, R. (2012). A conceptual framework of organizational stressors in sport performers. *Scandinavian Journal of Medicine & Science in Sports 22*, 545–557.

Flinn, J. T., Miller, A., Pyatka, N., Brewer, J., Schneider, T., & Cao, C. G. (2016). The effect of stress on learning in surgical skill acquisition. *Medical Teacher 38*, 897–903.

Fournier, K. A., Couret, J., Ramsay, J. B., & Caulkins, J. L. (2017). Using collaborative two-stage examinations to address test anxiety in a large enrollment gateway course. *Anatomical Sciences Education 10*, 409–422.

Fuller, R. M., Vician, C. M., & Brown, S. A. (2016). Longitudinal effects of computer-mediated communication anxiety on interaction in virtual teams. *IEEE Transactions on Professional Communication 59*, 166–185.

Gallagher, A. G., Satava, R. M., & O'Sullivan, G. C. (2015). Attentional capacity: an essential aspect of surgeon performance. *Annals of Surgery 261*, e60–e61.

Ge, Y., Qu, W., Jiang, C., Du, F., Sun, X., & Zhang, K. (2014). The effect of stress and personality on dangerous driving behavior among Chinese drivers. *Accident Analysis & Prevention 73*, 34–40.

Glendon, A. I., Dorn, L., Davies, D. R., Matthews, G., & Taylor, R. G. (1996). Age and gender differences in perceived accident likelihood and driver competences. *Risk Analysis 16*, 755–762.

Gray, E. K, & Watson, D. (2007). Assessing positive and negative affect via self report. In J. A. Coan & J. J. B. Allen (Eds), *Handbook of emotion elicitation and assessment* (pp. 171–183). New York: Oxford University Press.

Gregersen, N. P. (1996). Young drivers' overestimation of their own skill—an experiment on the relation between training strategy and skill. *Accident Analysis & Prevention 28*, 243–250.

Gulian, E., Glendon, A. I., Matthews, G., Davies, D. R., & Debney, L. M. (1990). The stress of driving: A diary study. *Work and Stress 4*(1), 7–16.

Hancock, P. A., & Ganey, H. C. (2003). From the inverted-U to the extended-U: The evolution of a law of psychology. *Journal of Human Performance in Extreme Environments 7*, 5–14.

Hancock, P. A., & Matthews, G. (2015). Stress and attention. In E. Risko, A. Kingstone, & J. Fawcett (Eds), *The handbook of attention* (pp. 547–568). Boston: MIT Press.

Hancock, P. A., & Warm, J. S. (1989). A dynamic model of stress and sustained attention. *Human Factors 31*, 519–537.

Hanton, S., Cropley, B., Neil, R., Mellalieu, S. D., & Miles, A. (2007). Experience in sport and its relationship with competitive anxiety. *International Journal of Sport and Exercise Psychology 5*, 28–53.

Healey, J. A., & Picard, R. W. (2005). Detecting stress during real-world driving tasks using physiological sensors. *IEEE Transactions on Intelligent Transportation Systems 6*, 156–166.

Hellhammer, D. H., Wüst, S., & Kudielka, B. M. (2009). Salivary cortisol as a biomarker in stress research. *Psychoneuroendocrinology 34*, 163–171.

Hennessy, D. A., Wiesenthal, D. L., & Kohn, P. M. (2000). The influence of traffic congestion, daily hassles, and trait stress susceptibility on state driver stress: An interactive perspective. *Journal of Applied Biobehavioral Research 5*, 162–179.

Hill, D. M., Hanton, S., Matthews, N., & Fleming, S. (2010). Choking in sport: A review. *International Review of Sport and Exercise Psychology 3*, 24–39.

Hockey, G. R. J. (1997). Compensatory control in the regulation of human performance under stress and high workload: A cognitive-energetical framework. *Biological Psychology 45*, 73–93.

Hofmann, W., Friese, M., Schmeichel, B. J., & Baddeley, A. D. (2011). Working memory and self-regulation. In K. D. Vohs & R. F. Baumeister (Eds), *Handbook of self-regulation: Research, theory, and applications* (pp. 204–225). New York: Guilford Press.

Horikawa, M., & Yagi, A. (2012). The relationships among trait anxiety, state anxiety and the goal performance of penalty shoot-out by university soccer players. *PloS One 7*, e35727.

Horner, R. D., Matthews, G., & Yi, M. (2012). A conceptual model of physician work intensity: Guidance for evaluating policies and practices to improve healthcare delivery. *Medical Care* 50, 654–661.

Hurley, A. M., Kennedy, P. J., O'Connor, L., Dinan, T. G., Cryan, J. F., Boylan, G., & O'Reilly, B. A. (2015). SOS save our surgeons: Stress levels reduced by robotic surgery. *Gynecological Surgery* 12, 197–206.

Ignacio, J., Dolmans, D., Scherpbier, A., Rethans, J.-J., Chan, S., & Liaw, S. Y. (2016). Stress and anxiety management strategies in health professions' simulation training: a review of the literature. *BMJ Simulation and Technology Enhanced Learning* 2, 42–46.

Ilkowska, M. & Engle, R. W. (2010). Trait and state differences in working memory capacity. In A. Gruszka, G. Matthews, & B. Szymura (Eds), *Handbook of individual differences in cognition: Attention, memory, and executive control* (pp. 295–320). New York: Springer.

Isler, R. B., Starkey, N. J., & Sheppard, P. (2011). Effects of higher-order driving skill training on young, inexperienced drivers' on-road driving performance. *Accident Analysis & Prevention* 43, 1818–1827.

Jones, G., & Swain, A. (1995). Predispositions to experience debilitative and facilitative anxiety in elite and nonelite performers. *The Sport Psychologist* 9, 201–211.

Jordet, G., & Hartman, E. (2008). Avoidance motivation and choking under pressure in soccer penalty shootouts. *Journal of Sport and Exercise Psychology* 30, 450–457.

Jordet, G., Hartman, E., Visscher, C., & Lemmink, K. A. (2007). Kicks from the penalty mark in soccer: The roles of stress, skill, and fatigue for kick outcomes. *Journal of Sports Sciences* 25, 121–129.

Klein, M. I., Warm, J. S., Riley, M. A., Matthews, G., Doarn, C., Donovan, J. F., & Gaitonde, K. (2012). Mental workload and stress perceived by novice operators in the laparoscopic and robotic minimally invasive surgical interfaces. *Journal of Endourology* 26, 1089–1094.

Kübler, T. C., Kasneci, E., Rosenstiel, W., Schiefer, U., Nagel, K., & Papageorgiou, E. (2014). Stress-indicators and exploratory gaze for the analysis of hazard perception in patients with visual field loss. *Transportation Research Part F: Traffic Psychology and Behaviour* 24, 231–243.

Lam, W. K., Maxwell, J. P., & Masters, R. (2009). Analogy learning and the performance of motor skills under pressure. *Journal of Sport and Exercise Psychology* 31, 337–357.

Lanatà, A., Valenza, G., Greco, A., Gentili, C., Bartolozzi, R., Bucchi, F., . . . & Scilingo, E. P. (2015). How the autonomic nervous system and driving style change with incremental stressing conditions during simulated driving. *IEEE Transactions on Intelligent Transportation Systems* 16, 1505–1517.

Lautenbach, F., Laborde, S., Achtzehn, S., & Raab, M. (2014). Preliminary evidence of salivary cortisol predicting performance in a controlled setting. *Psychoneuroendocrinology* 42, 218–224.

Lawrence, G. P., Cassell, V. E., Beattie, S., Woodman, T., Khan, M. A., Hardy, L., & Gottwald, V. M. (2014). Practice with anxiety improves performance, but only when anxious: evidence for the specificity of practice hypothesis. *Psychological Research* 78, 634–650.

Lazarus, R. S., & Folkman, S. (1984). *Stress, appraisal and coping*. New York: Springer.

Lee, Y. C., & Winston, F. K. (2016). Stress induction techniques in a driving simulator and reactions from newly licensed drivers. *Transportation Research Part F: Traffic Psychology and Behaviour* 42, 44–55.

Lieberman, H. R., Farina, E. K., Caldwell, J., Williams, K. W., Thompson, L. A., Niro, P. J., . . . & McClung, J. P. (2016). Cognitive function, stress hormones, heart rate and nutritional status during simulated captivity in military survival training. *Physiology & Behavior 165*, 86–97.

Loukidou, L., Loan-Clarke, J., & Daniels, K. (2009). Boredom in the workplace: More than monotonous tasks. *International Journal of Management Reviews 11*, 381–405.

Machin, M. A. (2003). Evaluating a fatigue management program for coach drivers. In L. Dorn (Ed.), *Driver behaviour and training* (pp. 75–83). Aldershot, UK: Ashgate.

Machin, M. A., & Hoare, P. N. (2008). The role of workload and driver coping styles in predicting bus drivers' need for recovery, positive and negative affect, and physical symptoms. *Anxiety, Stress, & Coping 21*, 359–375.

Mann, D. L., Williams, A. M., Ward, P., & Janelle, C. M. (2007). Perceptual–cognitive expertise in sport: A meta-analysis. *Journal of Sport & Exercise Psychology 29*, 457–478.

Matthews, G. (1997). An introduction to the cognitive science of personality and emotion. In G. Matthews (Ed.), *Cognitive science perspectives on personality and emotion* (pp. 3–30). Amsterdam: Elsevier.

Matthews, G. (1999). Personality and skill: A cognitive-adaptive framework. In P. L. Ackerman, P.C. Kyllonen, & R. D. Roberts (Eds), *The future of learning and individual differences research: Processes, traits, and content* (pp. 251–270). Washington, DC: APA.

Matthews, G. (2001). A transactional model of driver stress. In P. A. Hancock & P. A. Desmond (Eds), *Stress, workload and fatigue* (pp. 133–163). Mahwah, NJ: Lawrence Erlbaum.

Matthews, G. (2002). Towards a transactional ergonomics for driver stress and fatigue. *Theoretical Issues in Ergonomics Science 3*, 195–211.

Matthews, G. (2008). Personality and information processing: A cognitive-adaptive theory. In G. J. Boyle, G. Matthews, & D. H. Saklofske (Eds), *Handbook of personality theory and testing*: Vol. 1: *Personality theories and models* (pp. 56–79). Thousand Oaks, CA: Sage.

Matthews, G. (2016). Multidimensional profiling of task stress states for human factors: A brief review. *Human Factors 58*, 801–813.

Matthews, G., & Campbell, S. E. (2009). Sustained performance under overload: Personality and individual differences in stress and coping. *Theoretical Issues in Ergonomics Science 10*, 417–442.

Matthews, G., & Campbell, S. E. (2010). Dynamic relationships between stress states and working memory. *Cognition and Emotion 24*, 357–373.

Matthews, G., Campbell, S. E., Falconer, S., Joyner, L., Huggins, J., Gilliland, K., ... & Warm, J. S. (2002). Fundamental dimensions of subjective state in performance settings: Task engagement, distress and worry. *Emotion 2*, 315–340.

Matthews, G., Davies, D. R., Westerman, S. J., & Stammers, R. B. (2000). *Human performance: Cognition, stress and individual differences*. London: Psychology Press.

Matthews, G., Deary, I. J., & Whiteman, M. C. (2009). *Personality traits*, 3rd edn. Cambridge, UK: Cambridge University Press.

Matthews, G., Desmond, P. A., Joyner, L. A., & Carcary, B. (1997). A comprehensive questionnaire measure of driver stress and affect. In E. Carbonell Vaya & J. A. Rothengatter (Eds), *Traffic and transport psychology: Theory and application* (pp. 317–324). Amsterdam: Pergamon.

Matthews, G., Lin, J., & Wohleber, R. (2017a). Personality, stress and resilience: A multifactorial cognitive science perspective. *Psychological Topics 26*, 139–162.

Matthews, G., & Reinerman-Jones, L (2017). *Workload assessment: How to diagnose workload issues and enhance performance*. Santa Monica, CA: Human Factors and Ergonomics Society.

Matthews, G., Reinerman-Jones, L., Abich, J., IV, & Kustubayeva, A. (2017b). Metrics for individual differences in EEG response to cognitive workload: Optimizing performance prediction. *Personality and Individual Differences 118*, 22–28.

Matthews, G., Reinerman-Jones, L. E., Barber, D. J., & Abich, J. (2015). The psychometrics of mental workload: Multiple measures are sensitive but divergent. *Human Factors 57*, 125–143.

Matthews, G., Saxby, D. J., Funke, G. J., Emo, A. K., & Desmond, P. A. (2011). Driving in states of fatigue or stress. In D. Fisher, M. Rizzo, J. Caird, & J. Lee (Eds), *Handbook of driving simulation for engineering, medicine and psychology* (pp. 29-1–29-11). Boca Raton, FL: Taylor and Francis.

Mellalieu, S. D., Hanton, S., & Fletcher, D. (2006). A competitive anxiety review: Recent directions in sport psychology research. In S. Hanton & S. D. Mellalieu (Eds), *Literature reviews in sport psychology* (pp. 1–45). Hauppauge, NY: Nova Science.

Mesagno, C., Geukes, K., & Larkin, P. (2015). Choking under pressure: A review of current debates, literature, and interventions. In S. Mellalieu & S. Hanton (Eds), *Contemporary advances in sport psychology: A review* (pp. 148–174). Abingdon, UK: Routledge.

Mok, W. S. Y., & Chan, W. W. L. (2016). How do tests and summary writing tasks enhance long-term retention of students with different levels of test anxiety? *Instructional Science 44*, 567–581.

Moore, L. J., Vine, S. J., Freeman, P., & Wilson, M. R. (2013). Quiet eye training promotes challenge appraisals and aids performance under elevated anxiety. *International Journal of Sport and Exercise Psychology 11*, 169–183.

Moran, T. P. (2016). Anxiety and working memory capacity: A meta-analysis and narrative review. *Psychological Bulletin 142*, 831–864.

Naveh-Benjamin, M. (1991). A comparison of training programs intended for different types of test-anxious students: Further support for an information-processing model. *Journal of Educational Psychology 83*, 134–139.

Naveh-Benjamin, M., McKeachie, W. J., Lin, Y. G., & Lavi, H. (1997). Individual differences in students' retention of knowledge and conceptual structures learned in university and high school courses: The case of test anxiety. *Applied Cognitive Psychology 11*, 507–526.

Neil, R., Hanton, S., Mellalieu, S. D., & Fletcher, D. (2011). Competition stress and emotions in sport performers: The role of further appraisals. *Psychology of Sport and Exercise 12*, 460–470.

Nieuwenhuys, A., & Oudejans, R. R. (2017). Anxiety and performance: Perceptual-motor behavior in high-pressure contexts. *Current Opinion in Psychology 16*, 28–33.

Norman, D. A., & Bobrow, D. B. (1975). On data-limited and resource-limited processes. *Cognitive Psychology 7*, 44–64.

O'Donnell, P. S. (2017). Executive functioning profiles and test anxiety in college students. *Journal of Psychoeducational Assessment 35*, 447–459.

Okon-Singer, H., Hendler, T., Pessoa, L., & Shackman, A. J. (2015). The neurobiology of emotion–cognition interactions: Fundamental questions and strategies for future research. *Frontiers in Human Neuroscience 9*, 58.

Oudejans, R. R., & Pijpers, J. R. (2010). Training with mild anxiety may prevent choking under higher levels of anxiety. *Psychology of Sport and Exercise 11*, 44–50.

Owens, M., Stevenson, J., Hadwin, J. A., & Norgate, R. (2014). When does anxiety help or hinder cognitive test performance? The role of working memory capacity. *British Journal of Psychology 105*, 92–101.

Parfitt, G., & Hardy, L. (1993). The effects of competitive anxiety on memory span and rebound shooting tasks in basketball players. *Journal of Sports Sciences 11*, 517–524.

Poolton, J. M., Wilson, M. R., Malhotra, N., Ngo, K., & Masters, R. S. W. (2011). A comparison of evaluation, time pressure, and multitasking as stressors of psychomotor operative performance. *Surgery 149*, 776–782.

Richardson, M., Abraham, C., & Bond, R. (2012). Psychological correlates of university students' academic performance: A systematic review and meta-analysis. *Psychological Bulletin 138*, 353–387.

Rowden, P., Matthews, G., Watson, B., & Briggs, H. (2011). The relative impact of occupational stress, life stress, and driving environment stress on driving outcomes. *Accident Analysis and Prevention 44*, 1332–1340.

Ruiz, M. C., Raglin, J. S., & Hanin, Y. L. (2017). The individual zones of optimal functioning (IZOF) model (1978–2014): historical overview of its development and use. *International Journal of Sport and Exercise Psychology 15*, 41–63.

Qu, W., Zhang, Q., Zhao, W., Zhang, K., & Ge, Y. (2016). Validation of the Driver Stress Inventory in China: Relationship with dangerous driving behaviors. *Accident Analysis & Prevention 87*, 50–58.

Rumbold, J. L., Fletcher, D., & Daniels, K. (2012). A systematic review of stress management interventions with sport performers. *Sport, Exercise, and Performance Psychology 1*, 173–193.

Sarason, I. G., Sarason, B. R., & Pierce, G. R. (1995). Cognitive interference: At the intelligence-personality crossroads. In D. H. Saklofske & M. Zeidner (Eds), *International handbook of personality and intelligence* (pp. 91–112). New York: Plenum.

Sarkar, M., & Fletcher, D. (2014). Psychological resilience in sport performers: a review of stressors and protective factors. *Journal of Sports Sciences 32*, 1419–1434.

Schommer, N. C., Hellhammer, D. H., & Kirschbaum, C. (2003). Dissociation between reactivity of the hypothalamus-pituitary-adrenal axis and the sympathetic-adrenal-medullary system to repeated psychosocial stress. *Psychosomatic Medicine 65*, 450–460.

Schwabe, L., & Wolf, O. T. (2013). Stress and multiple memory systems: From "thinking" to "doing." *Trends in Cognitive Sciences 17*, 60–68.

Sharma, N., & Gedeon, T. (2012). Objective measures, sensors and computational techniques for stress recognition and classification: A survey. *Computer Methods and Programs in Biomedicine 108*, 1287–1301.

Spielberger, C. D. (1972). Anxiety as an emotional state. In C. D. Spielberger (Ed.), *Anxiety: Current trends in theory and research*, Vol. 1 (pp. 23–49). London: Academic Press.

Staal, M. A. (2004). *Stress, cognition, and human performance: A literature review and conceptual framework* (NASA Tech. Memorandum 212824). Moffett Field, CA: NASA Ames Research Center.

Staal, M. A., Bolton, A. E., Yaroush, R. A., & Bourne, L. E., Jr (2008). Cognitive performance and resilience to stress. In B. Lukey, & V. Tepe (Eds), *Biobehavioral resilience to stress* (pp. 259–299). London: Francis & Taylor.

Stefanidis, D., Anton, N. E., Howley, L. D., Bean, E., Yurco, A., Pimentel, M. E., & Davis, C. K. (2017). Effectiveness of a comprehensive mental skills curriculum in enhancing surgical performance: Results of a randomized controlled trial. *American Journal of Surgery 213*, 318–324.

Stokes, A. F., & Kite, K. (1994). *Flight stress: Stress, fatigue and performance in aviation*. Aldershot, UK: Avebury Aviation.

Swain, A., & Jones, G. (1996). Explaining performance variance: The relative contribution of intensity and direction dimensions of competitive state anxiety. *Anxiety, Stress, and Coping* 9, 1–18.

Szalma, J. L. (2009). Individual differences in human–technology interaction: incorporating variation in human characteristics into human factors and ergonomics research and design. *Theoretical Issues in Ergonomics Science* 10, 381–397.

Taylor, A. H., & Dorn, L. (2006). Stress, fatigue, health, and risk of road traffic accidents among professional drivers: The contribution of physical inactivity. *Annual Review of Public Health* 27, 371–391.

Tichon, J. G., Mavin, T., Wallis, G., Visser, T. A., & Riek, S. (2014). Using pupillometry and electromyography to track positive and negative affect during flight simulation. *Aviation Psychology and Applied Human Factors* 4, 23–32.

Toner, J., Montero, B. G., & Moran, A. (2015). The perils of automaticity. *Review of General Psychology* 19, 431–442.

Trammell, J. P., & Clore, G. L. (2014). Does stress enhance or impair memory consolidation? *Cognition & Emotion* 28, 361–374.

Ulrich-Lai, Y. M., & Herman, J. P. (2009). Neural regulation of endocrine and autonomic stress responses. *Nature Reviews Neuroscience* 10, 397–409.

Uphill, M. (2016). Anxiety in sport: Are we any closer to untangling the knots? In A. M. Lane (Ed.), *Sport and exercise psychology*, 2nd edn (pp. 50–75). London: Routledge.

Useche, S., Serge, A., & Alonso, F. (2015). Risky behaviors and stress indicators between novice and experienced drivers. *American Journal of Applied Psychology* 3, 11–14.

Valentin, B., Grottke, O., Skorning, M., Bergrath, S., Fischermann, H., Rörtgen, D., ... & Rossaint, R. (2015). Cortisol and alpha-amylase as stress response indicators during pre-hospital emergency medicine training with repetitive high-fidelity simulation and scenarios with standardized patients. *Scandinavian Journal of Trauma, Resuscitation and Emergency Medicine* 23, 31.

Vician, C., & Davis, L. R. (2003). Investigating computer anxiety and communication apprehension as performance antecedents in a computing-intensive learning environment. *Journal of Computer Information Systems* 43, 51–57.

Vine, S. J., Lee, D., Moore, L. J., & Wilson, M. R. (2013). Quiet eye and choking: Online control breaks down at the point of performance failure. *Medicine & Science in Sports & Exercise* 45, 1988–1994.

Vine, S. J., Uiga, L., Lavric, A., Moore, L. J., Tsaneva-Atanasova, K., & Wilson, M. R. (2015). Individual reactions to stress predict performance during a critical aviation incident. *Anxiety, Stress, & Coping* 28, 467–477.

Vogel, S., & Schwabe, L. (2016). Stress in the zoo: Tracking the impact of stress on memory formation over time. *Psychoneuroendocrinology* 71, 64–72.

Von Der Embse, N., Barterian, J., & Segool, N. (2013). Test anxiety interventions for children and adolescents: A systematic review of treatment studies from 2000–2010. *Psychology in the Schools* 50, 57–71.

Warm, J. S., Matthews, G., & Finomore, V. S. (2008). Workload and stress in sustained attention. In P. A. Hancock & J. L. Szalma (Eds), *Performance under stress* (pp.115–141). Aldershot, UK: Ashgate.

Wells, A., & Matthews, G. (2015). *Attention and emotion: A clinical perspective* (classic edition). New York: Psychology Press.

Wetton, M. A., Hill, A., & Horswill, M. S. (2011). The development and validation of a hazard perception test for use in driver licensing. *Accident Analysis & Prevention* 43, 1759–1770.

Wetzel, C. M., Black, S. A., Hanna, G. B., Athanasiou, T., Kneebone, R. L., Nestel, D., ... & Woloshynowych, M. (2010). The effects of stress and coping on surgical performance during simulations. *Annals of Surgery* 251, 171–176.

Wickens, C. D. (1996). Designing for stress. In J. Driskell & E. Salas (Eds), *Stress and human performance* (pp. 279–295). Hillsdale, NJ: Lawrence Erlbaum.

Wickens, C. D. (2008). Multiple resources and mental workload. *Human Factors* 50, 449–455.

Wickens, C. M., Mann, R. E., Ialomiteanu, A. R., & Stoduto, G. (2016). Do driver anger and aggression contribute to the odds of a crash? A population-level analysis. *Transportation Research Part F: Traffic Psychology and Behaviour* 42, 389–399.

Williams, A. M., Vickers, J., & Rodrigues, S. (2001). The effects of anxiety on visual search, movement kinematics, and performance in table tennis: A test of Eysenck and Calvo's processing efficiency theory. *Journal of Sport and Exercise Psychology* 23, 438–455.

Wilson, M. (2008). From processing efficiency to attentional control: a mechanistic account of the anxiety–performance relationship. *International Review of Sport and Exercise Psychology* 1, 184–201.

Wise, J., Rio, A. F., & Fedouach, M. (2011). What really happened aboard Air France 447. *Popular Mechanics*, 6. Available at http://www.popularmechanics.com/technology/aviation/crashes/what-really-happened-aboard-air-france-447-6611877. Accessed on June 2, 2017.

Wohleber, R. W., & Matthews, G. (2016). Multiple facets of overconfidence: implications for driving safety. *Transportation Research Part F: Traffic Psychology and Behaviour* 43, 265–278.

Wohleber, R. W., Matthews, G., Reinerman-Jones, L. E., Panganiban, A. R., & Scribner, D. (2015). Individual differences in resilience and affective response during simulated UAV operations. *Proceedings of the Human Factors and Ergonomics Society Annual Meeting* 59, 751–755.

Wood, G., Hartley, G., Furley, P. A., & Wilson, M. R. (2016). Working memory capacity, visual attention and hazard perception in driving. *Journal of Applied Research in Memory and Cognition* 5, 454–462.

Wood, G., Vine, S. J., & Wilson, M. R. (2016). Working memory capacity, controlled attention and aiming performance under pressure. *Psychological Research* 80, 510–517.

Woodman, T., & Hardy, L. (2001). Stress and anxiety. In R. Singer, H. Hausenblas, & C. Janelle (Eds), *Handbook of sport psychology*, 2nd edn (pp. 290–318). New York: Wiley.

Woodman, T., & Hardy, L. (2003). The relative impact of cognitive anxiety and self-confidence upon sport performance: A meta-analysis. *Journal of Sports Sciences* 21, 443–457.

Zeidner, M. (1998). *Test anxiety: The state of the art*. New York: Plenum.

Zeidner, M. (2007). Test anxiety in educational contexts: Concepts, findings and future directions. In P. A. Schutz & R. Pekrun (Eds), *Emotion in education* (pp. 165–184). San Diego, CA: Academic Press.

Zeidner, M., & Matthews, G. (2005). Evaluation anxiety. In A. J. Elliot & C. S. Dweck (Eds), *Handbook of competence and motivation* (pp. 141–163). New York: Guilford Press.

SECTION III

DOMAINS AND APPLICATIONS

SECTION EDITOR:
JULIE GORE

CHAPTER 23

EXPERTISE IN STEM DISCIPLINES

DAVID F. FELDON, SOOJEONG JEONG,
AND JOANA FRANCO

Introduction

THE development of expertise in science, technology, engineering, and mathematics (STEM) is a topic of great interest in modern policy contexts. Enhancing STEM workforce capacity expands the potential of these fields to both enhance human knowledge about the natural world and drive the development of new technologies that provide strong economic benefits (Rothwell, 2013). However, despite broad consensus that STEM expertise is necessary for the realization of these benefits, empirical evidence of how the development of such expertise can be supported and enhanced is limited. In part, this is due to a bifurcated view of the indicators and mechanisms of expertise development in these fields. On the one hand, research into individuals' development of research skills and the practices of STEM experts has existed largely within the cognitive and psychological sciences. On the other, the training, evaluation, and consumption of research does not occur separately from disciplinary and academic communities of practice that are social in nature. As such, the broader social and cultural contexts of universities—especially their graduate programs—and of the STEM disciplines themselves play a major role in the development of experts' identities, expectations, and beliefs as they relate to the definition of expertise and expert performance (Gardner, 2010; Holley, 2009). Thus, for the purposes of this chapter, we define expertise in STEM as the mastery of the knowledge and skills necessary to produce new knowledge that meets or exceeds the standards of rigor appropriate to the disciplinary context in which it is situated.

This chapter describes the major areas of research targeting expertise in STEM disciplines, with a specific emphasis on science and engineering. These emphases reflect the relative proportion of literature developed on expertise within the specific disciplines, although some of the literature described engages samples that draw participants from across STEM disciplines. In synthesizing relevant studies, the chapter draws from four major types of research: cognitive psychology, the psychology of science, sociology and anthropology, and educational research. As such, there is an active attempt to interweave the psychological framing of expertise as a property of individuals who possess an extensive base of knowledge and skills with the social framing of expertise as a transactive recognition of valued attainment and stature in relevant fields. To this end, we first address the role that disciplines and fields play in defining the domain of expertise. Second, we discuss cognitive models and mechanisms of reasoning in science and engineering. Third, we examine the criteria by which expertise is identified within STEM fields from anthropological and sociological perspectives. Fourth, we discuss research on the factors that specifically influence the development of STEM expertise by focusing on both educational findings related to learning progressions and the process of socialization that engages the interactions between graduate students and faculty in the modern university context. We close the chapter with a summary of unresolved issues in the study of STEM expertise and specific questions for future research.

Importance of Recognizing Distinct STEM Disciplines

By definition, expertise is domain specific (Ericsson & Smith, 1991). Thus, meaningfully engaging expertise in STEM disciplines requires recognition of discipline as an essential component of domain. Disciplinary identity establishes which skills, knowledge, assumptions, and evaluative criteria are central to the recognition and development of expertise. Secondly, field of study or practice also contributes to the establishment of domain boundaries by anchoring disciplinary work to a specific phenomenon around which knowledge has accumulated. For example, one individual may be described as an expert chemist (i.e., discipline) whose research focuses on pollution in the atmosphere (i.e., field), and another may be an expert mathematician with research in the same field. While each would apply their own disciplinary tools and frameworks, they might each call upon overlapping bases of knowledge stemming from past research on atmospheric pollution.

STEM disciplines bear both similarities and differences to one another. For example, in science, as in engineering, empirical (typically quantitative) evidence and logical inference are used and valued as the primary basis for conducting their work. However, each of these disciplines has its own goals (Feldon, Hurst, Rates, & Elliott, 2013). For instance, scientists attempt to generate empirically supported principles that provide generalizable and parsimonious explanations for an entire class of phenomena (Blackburn, 1999). In contrast, engineers attempt to design targeted solutions to specific problems within localized constraints that span both physical and social parameters (Koen, 1985), such that optimal solutions are unlikely to be universally generalizable. Thus, solving the nature of outstanding performance or achievement differs by discipline, because "maximal adaptation to task constraints" (Ericsson & Lehmann, 1996, p. 273) will manifest differently.

The definition of discipline traditionally engages both structural and cultural components (Lattuca, 2001). Structurally, becoming an expert entails mastering the norms and practices that describe a discipline's subject matter and its accepted methodologies. Both are inherently grounded in the discipline's view of knowledge or *epistemic culture* (Knorr-Cetina, 1997), which cannot be separated from the social and historical dimensions of the disciplinary community and its norms. A group of scholars organized around a domain of expertise shares research questions, inquiry methods, and problem-solving approaches that configure the intellectual undertakings of that discipline (Chubin, 1983; Kuhn, 1962; Price, 1965). For example, epistemological assumptions in cellular and molecular biology heavily favor experimental designs that require nuanced implementations of control conditions (e.g., use of both positive and negative controls; Gross & Mantel, 1967; Rates & Feldon, 2014). These differ from other disciplines in which positive and negative controls may not be differentiated (e.g., physics; Galison, 1997) or in which controlled experiments may not be considered viable for contributing meaningful knowledge contributions (e.g., climatology; Navarra, Kinter, & Tribbia, 2010).

In many cases, disciplinary knowledge and experience allow for the creation of solutions to novel problems. However, there are cases in which the solution to a given problem lies beyond the search space established by a single discipline's task constraints. In some cases, Lattuca (2001) described such interdisciplinarity as outreach from one discipline to others (i.e., informed disciplinarity) or as a convergence between different disciplines around a common problem (i.e., synthetic interdisciplinarity). In other cases, questions inherently cross traditional disciplinary lines (i.e., transdisciplinarity), or arise independent of an established disciplinary context (i.e., conceptual interdisciplinarity). Although such efforts are challenging and not always successful, results can realize major gains in understanding complex problems (Rhoten & Parker, 2004). For example, interdisciplinary work including physicists, chemists, and engineers at the Joint Institute for Laboratory Astrophysics produced the Bose–Einstein condensate (originally theorized in the 1920s), for which they earned a Nobel Prize in 2001 (Aldhous, 2003; Cech & Rubin, 2004). Notwithstanding, it is currently unclear how domains of expertise might be conceptualized within an interdisciplinary context.

Mechanisms of Reasoning in STEM Disciplines

The psychology of science has a long history of studying the mechanisms of reasoning that are common to various STEM disciplines. Efforts to characterize these mechanisms have primarily focused on two levels of analysis: fundamental cognitive processes and central tasks that meet disciplinary criteria for building disciplinary knowledge. The basic cognitive processes include mental simulation, analogical reasoning, causal thinking, and generating inductive and deductive inferences (Dunbar & Fugelsang, 2005), which occur in the context of criterial disciplinary tasks for developing knowledge claims such as formulating productive research questions, generating testable hypotheses, making observations, collecting and analyzing data, presenting findings, drawing conclusions, and building models and theories (Zimmerman, 2000).

Mental simulation serves as a form of forward reasoning (cf. Chi, Feltovich, & Glaser, 1981) that is documented across many STEM disciplines (Anzai & Yokoyama, 1984; Ball & Christensen, 2009; Brown, 2011). Leveraging their extensive knowledge of relevant prior research, findings, principles, and theories, experts draw inferences about the necessary properties of an unknown component necessary to the solution of a problem, reasoning through likely outcomes when encountering uncertainty or ambiguity (Christensen & Schunn, 2009). Studies have documented such strategies with scientists when designing experiments where established protocols are not known (Dhillon, 1998; Schraagen, 1993) and analyzing data when the results obtained are unclear or inconsistent with expectations (Trickett & Trafton, 2007). Similarly, during product development, engineers are likely to mentally simulate the functions or features of a developing product and the interactions that end-users might have with them to reduce the extent of uncertainty in making subsequent design decisions (Christensen & Schunn, 2009).

To identify the role of mental simulation under uncertainty in engineering design, Christensen and Schunn (2009) captured five months of a professional design team's meetings on video as they worked to develop new features for a product. Reflecting about nine hours of conversation, the transcripts of these meetings were analyzed by identifying verb phrases that reflected mental operations pertinent to the design effort. The researchers then classified all segments ($n = 6,171$) according to whether they reflected mental simulation. If segments were determined to indicate mental simulation, they were also coded according to whether simulation was applied to the functional aspects of the product or end-user behaviors. Segments were also coded for indicators of uncertainty on the part of the speaker (e.g., "maybe," "not certain"). Statistical analyses indicated that mental simulation occurred significantly more often immediately following an expression of uncertainty (17 percent) than the base rate for expressions of uncertainty (8 percent). Further, the frequency of segments indicating uncertainty were significantly higher than the base rate of uncertainty before (17 percent) and during (13 percent) mental simulation than following the completion of a simulation event (11 percent; not significantly different than base rate).

Analogical reasoning is another core mental strategy used in both scientific investigation and engineering problem-solving (Ball & Christensen, 2009; Chan & Schunn, 2015; Clement 1988; Holyoak, 2005; Nersessian, 1984; Robinson, 1998). For example, when constructing new hypotheses or designing experiments, scientists retrieve prior relevant knowledge or experience from long-term memory. They then evaluate the similarities between hypotheses or experiments previously used in other studies (i.e., source) and those currently targeted by analogical mapping (Dunbar, 1997; Holyoak, 2005; Klahr & Simon, 1999). Due to the association between knowledge base (past experiences and domain knowledge) and analogical thinking, expert scientists tend to make more extensive use of analogies during scientific problem-solving than novice scientists (Dunbar, 2000; Hmelo-Silver, Nagarajan, & Day, 2002). In addition, scientists frequently use analogies when they explain or present a certain concept or result to others. However, these explanatory analogies tend to be made to relatively different areas or domains (i.e., distant analogies) compared to analogies used in the phases of hypothesis formation and experiment design.

In engineering design contexts, three types of analogies are typically used during design problem-solving: *explanation*, *problem identification*, and *solution generation* (Ball & Christensen, 2009). Specifically, explanatory analogies play a critical role for engineering designers to explain their ideas and solutions to others, and the distance between the target and the source of these analogies is relatively larger than with the other two types of analogies. In addition, when finding potential problems in a novel design or generating particular design solutions, engineers use previous designs or existing exemplars as analogous sources (Christensen & Schunn, 2009).

Dual Space Framework for Scientific Discovery

According to general theories of human problem-solving, a problem space is composed of a set of initial problem states, a goal state (i.e., target solution), and operators for moving from one state to another in the navigation from initial to goal state (Newell & Simon, 1972). Problem-solving thus refers to an activity that occurs by search in a problem space to seek a route successfully connecting the initial state to the goal state. Employing this metaphor, Klahr and Dunbar (1988) proposed "a general model of scientific reasoning, that can be applied to any context in which hypotheses are proposed and data is collected," known as "scientific discovery as dual search (SDDS)" (p. 32).

Klahr and Dunbar (1988; Klahr, 2000) consider hypothesis formation and experimental design to be the two major aspects of scientific discovery as a complex problem-solving process. To solve problems, scientists thus engage heuristic searches in two distinct but interacting problem spaces. Hypothesis formation is the construction and validation of theory through obtained findings, and experimental design is the design of empirical procedures to generate data capable of supporting or disconfirming hypotheses.

The SDDS model specifies that the hypothesis space includes the current knowledge base related to the relationships among the variables in the domain and all possible hypotheses generated either from that knowledge or from prospective experimental findings. Similarly, the experiment space includes the current hypotheses and all possible experiments that can be carried out to test the hypotheses. Scientific investigation is then conducted by simultaneously searching these two spaces. The SDDS framework also distinguishes three primary components that direct search in these two problem spaces: *search hypothesis space* (i.e., formulating a completely specified hypothesis that yields the most likely explanation or prediction), *test hypothesis* (i.e., evaluating the obtained hypotheses through experiments), and *evaluate evidence* (i.e., determining the appropriateness of the cumulative data to accept or reject the hypotheses). As such, progress in navigating the hypothesis problem space entails successful navigation of the experiment problem space to yield valuable evidence to determine the viability of a hypothesis. Likewise, progress in navigating the experiment space is dependent upon solutions from the hypothesis space to target variables for isolation and inform decisions about optimal methods for generating informative data.

Decision-Based Design in Engineering

Although varying definitions of engineering thinking and practice exist, many researchers characterize engineering design as a deliberate decision-making process constrained by both physical and social constraints (Chen & Wassenaar, 2003; Dym, Agogino, Eris, Frey, & Leifer, 2005; Koen, 1985; National Research Council, 2001). The *decision-based design* (DBD) framework, originally coined by Hazelrigg (1998), assumes that engineering design is a rational process of selecting design alternatives against highly situated criteria, which follows "the rule that the preferred decision is the option whose expectation has the highest value" (p. 656). This DBD process basically involves two major phases: *alternative generation* (i.e., creating all potential product design options) and *alternative selection* (i.e., selecting the best optimal one) (Chen & Wassenaar, 2003).

In terms of generating alternatives, the set of potential design options is inherently extremely large or infinite, because the set of possible design configurations and dimensions for each configuration is unlimited (Hazelrigg, 1998). Engineers might use different strategies to create possible design alternatives during this phase, such as analogies (Keeney, 2004; Keller & Ho, 1988). In addition, when evaluating and choosing optimal alternatives, engineers encounter situations of uncertainty and risk, because they cannot exactly foresee how their designs will be actually implemented as products. Thus, engineers often use modeling or simulation to proximate the functions and behavior of their designs (Christensen & Schunn, 2009; Koen, 1985).

RECOGNITION OF EXPERT PERFORMANCE IN STEM FIELDS

Ericsson and Smith (1991) suggested that expertise can be defined as superior performance that is shown to be consistent and reproducible for particular skills or knowledge of a domain. However, studies of expert physicists have found widely discrepant and inconsistent applications of domain knowledge to problem solving (Cooke & Breedin, 1994; Reif & Allen, 1992). Further, expertise is often operationalized in terms of the attainment of credentials and the number of years of training or work experience in the domain. This experience-based approach tends to depend largely on peer evaluation or social recognition as a way of identifying experts in a particular discipline (Ericsson & Towne, 2010). The identification of experts is, thus, challenging because experience-based judgments of expertise do not always correspond to both field knowledge and problem-solving performance (Ericsson & Lehmann, 1996). In addition, "criteria may differ from one field to another, and they may be loosely and even inconsistently applied from one case to another" (Sternberg, 1997, p. 158).

Scientists' activities are verified and judged by fellow researchers in their disciplines via peer-reviewed conference papers and journal publications. Indeed, Latour and Woolgar (1979) argued that developing skills in laboratory work "is only a means to the end of publishing a paper" (p. 71), and it is the paper that is the ultimate contribution to the scientific community. Through the peer review process, work products deemed to be of high quality allow scientists to receive peer recognition that leads to various benefits for their further work (Allison, Long, & Krauze, 1982). Specifically, scientific recognition

> of one's contributions and consequent collegiate reputation ... is the key currency of the open science reward system. To this are tied the academic researcher's material rewards, such as salary and job tenure, and access to the human resources and physical facilities that scientists typically need to produce published results.
> (David, 1994, p. 70; see also Ehrenberg, Zuckerman, Groen, & Brucker, 2009)

Yet, the social recognition that scientists achieve may not precisely reflect their actual performance or contributions, especially in a team science environment (Feldon, Maher, & Timmerman, 2010). Different factors other than a scientist's actual skills are involved in the process of how scientists gain recognition and reputation over the course of their careers within the scientific fields.

The Matthew Effect

Merton (1968, 1988) originally developed the idea of cumulative advantage in the career development of scientists, in which early recognition of scientific accomplishments leads disproportionately to future recognition. Analyzing interview data from Nobel Prize laureates, he found a considerably skewed distribution of recognition and credit for scholarly accomplishments in science that favored those who published earliest on a topic, regardless of the actual significance of the contribution to the collective knowledge of the field. This Matthew effect posits that "eminent scientists get disproportionately great credit for their contributions to science while relatively unknown scientists tend to get disproportionately little credit for comparable contributions" (Merton, 1968, p. 57). For example, the better recognized scientist will get most of the credit for collaborative work (e.g., co-authored papers) irrespective of his or her actual contribution. Similarly, if two scientists generate the same scientific discovery independently, the more established of the two scientists will get more recognition for the discovery. Zuckerman (1992) also confirmed this inequality in recognition and evaluation for scientific achievement by demonstrating that Nobel Prize recipients are more likely to become major candidates for other awards, because the Nobel Prize seems to increase the reputation of those awards and reduce the likelihood of making the wrong choice of award winners. Thus, advantage (e.g., invitations to collaborate, access to research grant dollars) accumulates over time for those who are initially successful, resulting in both greater opportunities to publish and greater recognition. Such processes account for a robust underlying social stratification in science (Allison, Long, & Krauze, 1982; Allison & Stewart, 1974; Cole & Cole, 1973) that undercut Ericsson and Smith's (1991) requirement that domains of expertise appropriate for rigorous study must provide equitable access to demonstrate outstanding performance.

A similar effect can be found in the career trajectories of early career scientists. Many studies have examined the effects of the institutions where scientists are trained and the eminence of their advisors on their academic careers, including productivity and recognition (e.g., Sheltzer & Smith, 2014; Zuckerman, 1992). Crane (1965), for instance, found that scientists who received their graduate training at a more prestigious university were likely to be more productive than those trained at a less prestigious university. The eminence of the graduate school from which a doctoral degree was obtained was a stronger predictor of a scientists' later productivity than that of their current academic affiliation. In addition, scientists who were formerly students of prestigious advisors are likely to be more productive than those who were students of lesser known advisors. Further, obtaining a postdoctoral research position at one of a very limited number of highly prestigious laboratories yields a higher rate of entry into faculty positions within major research universities (Sheltzer & Smith, 2014).

Such a correlation between young scientists' home laboratories and their high performance and recognition can be accounted for by a "joint process of self-selection and selective recruitment" (Zuckerman, 1992, p. 157). That is, talented students interested in becoming scientists are discriminating in choosing prominent universities and elite advisors conducting significant work in the field. Simultaneously, prominent universities and advisors also select these talented individuals as their trainees. Further, because high prestige graduate schools tend to have relatively greater material and human resources, students in these schools are better positioned to demonstrate early success as a scientist (Merton, 1988). Similarly, the elite advisors transfer to their students not only their knowledge and skills but also the scientific standards, values, social connections, and self-confidence that are necessary to make significant advances in science (Gopaul, 2016). These students are, therefore, prepared well for elite status throughout their training and demonstrate higher rates of publication in excess of the small initial advantages they may demonstrate (Green & Bauer, 1995; Paglis, Green, & Bauer, 2006).

Unfortunately, such conditions create a challenging environment for establishing expertise in absolute comparative terms (cf. Ericsson & Smith, 1991).

Bibliometric Analysis

Bibliometric analyses provide a means for assessing product-based, rather than reputational, indicators of expertise using publication and citation analyses to more objectively assess scientists' scholarly performance (Long, Plucker, Yu, Ding, & Kaufman, 2014). Using this approach, Simonton (1997, 2004) proposed the model of creative productivity, which is based on a Darwinian perspective that views the generation of human creative thinking and products as a *blind-variation-and-selective-retention process* (Campbell, 1960; Simonton, 1999). The variation–selection process implies that the ultimate impact or success of a new scientific discovery is more likely to be assessed retrospectively than prospectively due to the substantially blind nature of the variation–selection process. In other words, individual scientists cannot necessarily predict which combinations of ideas and information will have greater impact over time, so stronger and weaker variations in individual scientific outputs are randomly distributed among scientists and within the careers of individual scientists (Simonton, 1997). Further, the selection process is also considered to be blind because it typically runs at multiple levels simultaneously, with authors selecting publication outlets, reviewers examining scholarly merits, and editors prioritizing areas of scientific work without prior knowledge of the decisions others in each category will make. For example, individual scientists who select which of their articles to submit to a given journal or conference cannot foresee what topics or how many rival manuscripts will be submitted. Similarly, reviewers offer their comments independently of one another, such that one reviewer's opinion cannot influence that of another. Moreover, emerging technologies and findings in the field enhance the unpredictability of scientific products or articles that will ultimately have a high impact in the future.

Using bibliometric analysis, Simonton (1997, 2004) argues that scientific awards or honors may not predict the recipients' later success or the impact of their work, because the award process is a type of a variation–selection process that is blind and random. Rather, a scientist's productivity (i.e., total number of published articles) is the strongest predictor of the ultimate influence (i.e., citation rate) of a certain article. The variation–selection model assumes that if the distribution of quality across publications (as judged by citation rate) is random, the probability of large numbers of citations is constant across all publications. Therefore, the only way to increase the number of widely cited publications is to increase the total number of publications produced.

Conceptualizing expertise in this framework indicates that the bibliometric strength of a given scholar is a marker of their expertise. However, Simonton's (1997, 2004) model does not link directly to cognitive processes that would tie it to most of the literature on expertise. Prior scholarship on creativity by Campbell (1960) and Sweller (2009; Sweller & Sweller, 2006) suggests that, similar to Simonton's model, creative cognitive processes represent a random generate-and-test model, in which navigation of a problem space occurs as a random walk. Under this assumption, the likelihood of obtaining a goal state is thus dependent on the size of the problem space, with the probability of attaining the goal state increasing as the possible search space decreases. If knowledge of relevant scientific concepts, principles, and strategies constrains the realm of possible solutions and solution paths, then experts with more extensive or more optimally organized knowledge in relation to the problem would be searching smaller problem spaces. As such, an expert advantage would be maintained, even within a context of stochastic processes driving scientific discovery (Feldon et al., 2013).

Factors Contributing to the Development of Expertise in Scientific Research

Educational interventions that can facilitate the development of expertise in STEM disciplines have been extensively studied at multiple levels of schooling. Yet, few efforts have directly linked childhood experiences specifically to expertise attainment. Retrospective studies of STEM professionals indicate that the presence of role models and early interest play an important role (e.g., Chakraverty & Tai, 2013). Likewise, schooling experiences tend to shape trajectories toward or away from advanced study of STEM topics based on both the extent to which students perceive themselves as compatible with a STEM identity (e.g., Archer et al., 2010, 2012) and the quality of academic preparation for advanced study in relevant topics, including mathematics (e.g., Arcidiacono, Aucejo, & Hotz, 2016, Wang, 2013).

Progressions of Learning

Recent attention to longitudinal progressions of science learning has focused efforts in K–12 research around the identification of optimized sequences of content instruction that are most likely to yield successively more sophisticated ways of thinking about central disciplinary concepts (Duncan & Hmelo-Silver, 2009; Wilson, 2009). These learning progressions facilitate students' development of frameworks that can sustain complex tasks such as extended hypothesis testing and modeling relevant data (National Research Council, 2007). For example, Schwarz and colleagues (2009) developed a scientific learning progression for elementary through middle grades that builds fundamentals of scientific practice, encompassing two dimensions: (1) modeling as a generative process to aggregate and synthesize information into scientific knowledge and (2) modeling as a dynamic, iterative process of revision and modification as new data and understandings are obtained. Beginning at the middle school level, other progressions develop science argumentation skills. These deepen the emphasis on domain-specific argumentation skills and critical analysis to begin the transition to authentic scientific practices that entail using understanding of theoretical concepts to shape novel inquiry (Berland & McNeill, 2010; Osborne et al., 2016).

Research to understand and support progressions of scientific knowledge and skills has also emerged at the level of graduate training. Through extensive interviews with faculty supervisors of graduate students, Kiley (2009; Kiley & Wisker, 2009) identified several foundational concepts that commonly served as substantial barriers to further progress until mastered. These threshold concepts are such that, "once grasped, [they] lead to a qualitatively different view of the subject matter" (Kiley & Wisker, 2009, p. 432). Threshold concepts are, therefore, likely to be transformative, integrative, irreversible, troublesome, and bounded (Meyer & Land, 2003, 2005). Given these characteristics, threshold concepts entail a sequential structure in which each serves as a barrier to further expertise development until that threshold is crossed (Roberts, 2016). With each threshold crossed, it becomes possible to conceptualize and develop skills that target more nuanced aspects of disciplinary practice in ways that would not be accessible to a student who had not yet crossed (Urquhart, Maher, Feldon, & Gilmore, 2016).

Overall, these threshold concepts center around major competencies for advancing disciplinarily acceptable arguments and knowledge claims. These include the full comprehension and effective use of theoretical frameworks, constructing coherent arguments or theses, and developing conceptual models as coherent representations of findings. Theoretical frameworks appropriately situate a research study or argument within the broader context of the field and discipline. Theses present research evidence with explicit links to the significance or implications of the research findings. Conceptual models synthesize empirical findings with prior results from the primary literature to present coherent, supported hypotheses around the phenomenon of interest.

Through analysis of STEM graduate students' scholarly writing over time, Timmerman, Feldon, Maher, Strickland, and Gilmore (2013) similarly found that the effective use of primary literature (as entailed in Kiley & Wisker's (2009) threshold concepts of theoretical framework and conceptual model) and the ability to generate disciplinarily appropriate testable hypotheses (as entailed in Kiley & Wisker's threshold concept of constructing a disciplinary argument or thesis) systematically preceded the development of other research skills. Only when students' scores in these areas reached a certain level of proficiency did scores in other areas (e.g., data analysis, identifying limitations of a study design) increase.

Further data suggest that these progressions of skill development are nonlinear in nature. As with the Matthew effect (Merton, 1968) described previously, small initial differences in STEM graduate students' skill levels at the beginning of an academic year increase over time, creating a widening gap in research skills (Feldon, Maher, Roksa, & Peugh, 2016). The mechanisms driving this trend are not yet clear. It is possible that small initial advantages over other students attract the notice of supervising faculty, leading to greater recognition and increased access to mentoring and research opportunities (Green & Bauer, 1995). However, Feldon and colleagues documented only very minor differences in the reported experiences of participants (e.g., quality of relationship with faculty advisor, participation in supervised research, publishing opportunities) in both the stronger and weaker skill groups. A possible alternative explanation with support from their data suggests that students in the stronger group may be more proactive in constructing extended meaning from their experiences and more inclined to reinvest effort as tasks become easier (cf. Bereiter & Scardamalia, 1993).

Socialization

Socialization theory posits that graduate students training to become a researcher undergo a process of acculturation into their discipline as they join their research teams and academic departments (Gardner, 2010; Weidman, 2010). It means that students establish both formal and informal interactions with those in the disciplinary and graduate communities and, as a result, start understanding their dynamics and incorporating their values and practices in an active and self-evaluative manner (Weidman et al., 2001).

One way in which socialization occurs is through the interactions of a graduate student and his or her faculty mentor. Cognitive apprenticeship is such a pervasive pedagogical approach that it has been labeled a signature feature of doctoral training (Golde, Conklin Bueschel, Jones, & Walker, 2009). For example, a study exploring doctoral students' motivation to graduate showed that the more collegial the relationship with their faculty advisor, the higher students' satisfaction and motivation (Mason, 2012). Further, it is through engaging in supervised research activities that students build their self-efficacy beliefs and identity as science researchers (Holley, 2009). However, Maher, Gilmore, Feldon, and Davis (2013) found that in contrast to ideal mentoring practice, which assumes progression in the complexity of assignments and subsequent feedback over time, students tended to report a constant level of heavy cognitive demands related to research productivity with limited quantity and depth of feedback (for more information on cognitive apprenticeship and other examples of mentoring, see Chapter 44, "Learning at the Edge: The Role of Mentors, Coaches, and Their Surrogates in Developing Expertise," by Petushek et al., this volume).

Heavy reliance on the mentorship model assumes that advisors occupy an ideal position from which to view students' performance in STEM research, provide feedback, and craft opportunities to help students develop their expertise through deliberate practice (Ericsson, Krampe, & Tesch-Römer, 1993). Unfortunately, empirical tests of this assumption do not endorse its full acceptance. In a study of intact pairs of STEM graduate students and their faculty mentors, students' written research proposals were scored and compared against mentors' statements about the skill strengths and weaknesses they observed in their students. These perspectives predicted student performance at no better than chance and often were unrelated to or directly contradicted students' self-perceptions of strengths and weaknesses (Feldon, Maher, Hurst, & Timmerman, 2015).

This study (Feldon et al., 2015) recruited 81 students pursuing research-intensive graduate degrees in STEM disciplines and elicited from them sole-authored research proposals early in the fall semester. These students were then asked to revise and resubmit their proposals (again as sole authors) late in the following spring semester. Pairs of domain experts trained in the use of a previously validated rubric scored these research proposals, with focal aspects including students' abilities to set the proposed research in the context of the field, frame productive research questions and testable hypotheses (as appropriate), design valid and appropriate studies, and identify the limitations and delimitations of prospective findings (Feldon et al., 2011).

In addition to the scores from the rubrics, the research team interviewed all participants late in both the fall and spring semesters and asked them to identify their strengths and weaknesses as a researcher. Independently, the researchers also interviewed the faculty whom participating students identified as their research mentors shortly after the fall and spring student interviews. In the faculty mentor interviews, each individual was asked to articulate the strengths and weaknesses of the identifying student in terms of their research skills. For both students and their faculty mentors, researchers kept the interview prompts intentionally vague to avoid biasing participant responses by focusing them on specific skills for which they may or may not have held specific assessments for the student in question.

Emergent themes from all interviews yielded sixteen codes related to research knowledge and skills. Of those, nine aligned directly to skills assessed by the rubric used to score research proposals. Statements of matched faculty–student pairs were compared to assess levels of agreement across all themes. In turn, statements from student and faculty interviews, respectively, that aligned with one or more of the nine rubric-identified skills were compared to scores attained on those skills for each submission of the research proposal. For statements describing a specific skill as a strength by one member of the student–faculty pair, researchers compared the statement to the assessment offered by the other member of the pair as well as the scores attained through rubric scoring. For each skill, a score at or above the sample mean was considered to be a strength, and a score below the sample mean was considered to be a weakness.

Results from these analyses indicated that in more than 75 percent of cases, students and their faculty mentors were focused on completely unrelated skills. However, when both members of a pair identified the same skill, they disagreed on whether it was a strength or weakness in approximately half of those cases. Further, for most skills, neither the student nor the faculty mentor was able to predict performance as assessed by the rubric scores at a rate greater than chance. The one exception to this trend was graduate students' self-assessments of their ability to utilize primary literature to frame a study, which concurred with rubric scores in 67 percent of cases from the spring proposals and interviews.

These findings reflecting limited insight by faculty mentors into the skill development of their students may be attributable to the evolution of university-based STEM research practices, in which the pace and volume of work required of faculty is increasing (Anderson et al., 2011; Austin & McDaniels, 2006; Johnson, Lee, & Green, 2000). It is also increasingly common for such endeavors to be team-based (Charlesworth, Farrall, Stokes, & Turnbull, 1989; Delamont & Atkinson, 2001; Knorr-Cetina, 1999; Parry, 2007). These conditions have given rise to a model of cascading mentorship (Golde et al., 2009), in which supervising faculty work most extensively with postdoctoral researchers, who in turn mentor senior graduate students. These senior students then support junior graduate students, who become primarily responsible for supervising undergraduate researchers. As such, faculty mentors may directly observe a students' work only rarely.

The lack of mentorship guidance and feedback requires successful STEM students to be independent learners. In this context, it is likely that academic self-regulation—i.e., "the degree to which students are metacognitively, motivationally, and behaviorally active participants in their own learning processes" (Zimmerman, 2013, p. 137)—becomes important to the development of graduate students' success, because the system within which graduate students develop expertise offers little direct externally imposed regulation. Even though research in self-regulated learning (SRL) has not heavily targeted graduate students, some studies do indicate that higher levels of expertise are accompanied by higher levels of self-regulation (Artino & Stephens, 2009; Cleary & Zimmerman, 2001; Kitsantas & Zimmerman, 2002). Further, Zimmerman and Campillo (2003, p. 238) affirm that novices and experts have very different profiles of self-regulation competency, in that "experts display greater use of hierarchical knowledge when formulating strategic solutions, greater use and self-monitoring of strategies, more accurate self-evaluation, and greater motivation than novices."

FUTURE DIRECTIONS

This chapter highlights a number of the major areas of research in efforts to understand how expertise manifests and develops in STEM disciplines. Due to the dependence of expert performance on both the cultural dimensions of scholarly productivity within disciplines and the ability of individuals and teams to solve complex problems, understanding the ways in which experts realize their contributions is highly challenging and defies some of the long-established criteria for the empirical study of expertise. For example, to contribute to a general theory of expertise, Ericsson and Smith (1991) specify that individual experts suitable for study must demonstrate stability in the characteristics that lead to outstanding performance. In this regard, they specifically state that this criterion excludes those who excel in games of chance. However, the most prestigious forms of social recognition of outstanding performance in science, such as a Nobel Prize, can be awarded for a single major discovery that builds on the accumulated work of many scholars and may have required the participation of experts from many domains. Further, the bibliometric analyses offered by Simonton (1997, 2004) and the model of cognitive creativity established by Campbell (1960) and Sweller (2009; Sweller & Sweller, 2006) explicitly rely on stochastic processes to generate scientific achievements that impact the progress of the field. In addition, substantial research has documented gender biases favoring men that skew the relationship between the scholarly contributions made and the recognition of those contributions (Feldon, Peugh, Maher, Roksa, & Tofel-Grehl, 2017; Lincoln, Pincus, Koster, & Leboy, 2012).

Another challenge pertains to the availability of "a larger group of other individuals (a 'control' group of sorts) who have experienced similar opportunities to make contributions or to achieve" (Ericsson & Smith, 1991, p. 2). Given the documented social advantages of individuals based on early, minor contributions to an area of research (Merton, 1968), demonstration of small advantages over peers in skill at an early stage of training (Gopaul, 2016; Green & Bauer, 1995), and receiving initial training from more prestigious university faculty (Crane, 1965; Sheltzer & Smith, 2014; Zuckerman, 1992), it becomes difficult to argue that all scientists receive "similar opportunities." It is likewise difficult to establish consistent definitions of domains of expertise as fields evolve and interdisciplinarity becomes a more prominent mechanism for solving challenging scientific problems (Lattuca, 2001; Rhoten & Parker, 2004).

Further, it is typically considered necessary to be able to "account for the acquisition of the characteristics and cognitive structures and processes that have been found to mediate the superior performances of experts" (Ericsson & Smith, 1991, p. 12). Nonetheless, the conventional mechanisms by which expertise is assumed to be developed during graduate training have failed to yield consistent results. For example, the mentoring practices employed by university faculty supervising graduate students in STEM are frequently documented to be subpar (Maher et al., 2013) and students whose research skills develop at faster and slower rates report largely identical mentoring experiences (Feldon et al., 2016). Likewise, with the rise of team science models, tracking the flows of mentorship becomes even more challenging for a cascading model of mentorship from senior peers (Golde et al., 2009). As pointed out previously, these phenomena suggest that increased attention to the role of self-regulated learning in the preparation of STEM experts may be warranted.

While addressing these issues with understanding the development and function of expertise in STEM disciplines will prove challenging, they are not necessarily intractable. Substantial progress has been made in parsing out the sequence of development for necessary skills (Kiley & Wisker, 2009; Timmerman et al., 2013), as well as in identifying discrete interventions and activities that lead to demonstrably greater research skill development during graduate training (e.g., coupling teaching and research experiences, Feldon et al., 2011; coauthoring research publications between students and faculty, Feldon, Shukla, & Maher, 2016). While such findings are encouraging, they represent only one aspect of the social and cognitive duality of STEM expertise, and it is understanding this complex dynamic that may ultimately yield the most complete picture of the phenomenon.

We recommend several practices that have the potential to substantially advance the field. First, insofar as it is possible, we suggest evaluating developing expertise using performance-based metrics (e.g., rubric-assessed, sole-authored samples of scholarly writing). While this is not practical for gathering data from leading experts in the field due to publishing norms in team science environments, it can be successfully applied at the level of graduate training. Such data may provide a mechanism for longitudinal modeling in which skill acquisition patterns during training may be predictive of subsequent bibliometric indicators of field influence. Further, assessments that do not rely on written products would also be informative for those individuals who are developing STEM expertise outside of an academic context (e.g., industry). Second, comprehensive studies that link performance and experiential data from individuals, programmatic and instructional practices at their training institutions and laboratories, and subsequent bibliometric performance data are likely to draw more informative conclusions about the constellations of factors that lead to the development of both demonstrable and socially recognized expertise in the field. Third, the dynamic nature of STEM disciplines and interdisciplinarity must be accounted for in two ways: (1) as domains in which performance across multiple individuals can be compared, and (2) as social contexts that drive productivity choices (e.g., valuing holding patents over publishing journal articles) such that bibliographic analyses may not fully credit the expertise or level of influence on a field.

Acknowledgments

The authors gratefully acknowledge the support of the U.S. National Science Foundation. This material is based upon work supported under Grants 0723686, 1242369, and 1431234. Any opinions, findings, and conclusions or recommendations expressed in this material are those of the authors and do not necessarily reflect the views of the National Science Foundation. We also thank the following colleagues for their collaboration related to this work: Joanna Gilmore, Michelle Maher, James Peugh, Christopher Rates, Josipa Roksa, Alok Shenoy, Kathan Shukla, Chongning Sun, and Briana Timmerman.

References

Aldhous, P. (2003). Atomic physics: Rocky Mountain high. *Nature* 423(6943), 915–916.

Allison, P. D., Long, J. S., & Krauze, T. K. (1982). Cumulative advantage and inequality in science. *American Sociological Review* 47(5), 615–625.

Allison, P. D., & Stewart, J. A. (1974). Productivity differences among scientists: Evidence for accumulative advantage. *American Sociological Review* 39(4), 596–606.

Anderson, W. A., Banerjee, U., Drennan, C. L., Elgin, S. C. R., Epstein, I. R., . . . Warner, I. M. (2011). Changing the culture of science education at research universities. *Science* 331(6014), 152–153.

Anzai, Y., & Yokoyama, T. (1984). Internal models in physics problem solving. *Cognition and Instruction* 1(4), 397–450.
Archer, L., DeWitt, J., Osborne, J., Dillon, J., Willis, B., & Wong, B. (2010). "Doing" science versus "being" a scientist: Examining 10/11-year-old schoolchildren's constructions of science through the lens of identity. *Science Education* 94(4), 617–639.
Archer, L., DeWitt, J., Osborne, J., Dillon, J., Willis, B., & Wong, B. (2012). Science aspirations, capital, and family habitus: How families shape children's engagement and identification with science. *American Educational Research Journal* 49(5), 881–908.
Arcidiacono, P., Aucejo, E., & Hotz, V. (2016). University differences in the graduation of minorities in STEM fields: Evidence from California. *American Economic Review 106*, 525–562.
Artino, A., & Stephens, J. (2009). Academic motivation and self-regulation: A comparative analysis of undergraduate and graduate students learning online. *Internet and Higher Education 12*, 146–151.
Austin, A., and McDaniels, M. (2006). Preparing the professoriate of the future: Graduate student socialization for faculty roles. In J. C. Smart (Ed.), *Higher education: Handbook of theory and research* (Vol. 21, pp. 397–456). The Netherlands: Kluwer Academic.
Ball, L.., & Christensen, B. (2009). Analogical reasoning and mental simulation in design: two strategies linked to uncertainty resolution. *Design Studies 30*(2), 169–186.
Bereiter, C., & Scardamalia, M. (1993). *Surpassing ourselves: An inquiry into the nature and implications of expertise*. Chicago: Open Court.
Berland, L., & McNeill, K. (2010). A learning progression for scientific argumentation: Understanding student work and designing supportive instructional contexts. *Science Education* 94(5), 765–793.
Blackburn, S. (1999). *Think: A compelling introduction to philosophy*. London: Oxford University Press.
Brown, J. (2011). *The laboratory of the mind: Thought experiments in the natural sciences* (2nd edn). London: Routledge.
Campbell, D. (1960). Blind variation and selective retention in creative thought as in other knowledge processes. *Psychological Review 67*, 380–400.
Cech, T., & Rubin, G. M. (2004). Nurturing interdisciplinary research. *Nature 11*(12), 1166–1169.
Chakraverty, D., & Tai, R. (2013). Parental occupation inspiring science interest: Perspectives from physical scientists. *Bulletin of Science, Technology, & Society 33*, 44–52.
Chan, J., & Schunn, C. (2015). The impact of analogies on creative concept generation: Lessons from an in vivo study in engineering design. *Cognitive Science 39*(1), 126–155.
Charlesworth, M., Farrall, L., Stokes, T., & Turnbull, D. (1989). *Life among the scientists*. Melbourne: Oxford University Press.
Chen, W., & Wassenaar, H. (2003). An approach to decision-based design with discrete choice analysis for demand modeling. *Journal of Mechanical Design 125*(3), 490–497.
Chi, M., Feltovich, P., & Glaser, R. (1981). Categorization and representation of physics problems by experts and novices. *Cognitive Science 5*(2), 121–152.
Christensen, B., & Schunn, C. (2009). The role and impact of mental simulation in design. *Applied Cognitive Psychology 23*(3), 327–344.
Chubin, D. (1983). *Sociology of sciences: An annotated bibliography on invisible colleges, 1972–1981*. New York: Garland.
Cleary, T., & Zimmerman, B. (2001). Self-regulation differences during athletic practice by experts, non-experts, and novices. *Journal of Applied Sport Psychology 13*(2), 185–206.

Clement, J. (1988). Observed methods for generating analogies in scientific problem solving. *Cognitive Science* 12(4), 563–586.

Cole, J., & Cole, S. (1973). *Social stratification in science*. Chicago: University of Chicago.

Cooke, N., & Breedin, S. (1994). Constructing naive theories of motion on the-fly. *Memory and Cognition* 22, 474–493.

Crane, D. (1965). Scientists at major and minor universities: A study of productivity and recognition. *American Sociological Review* 30(5), 699–714.

David, P. (1994). Positive feedbacks and research productivity in science: Reopening another black box. In O. Granstrand (Ed.), *Economics and Technology* (pp. 65–89). Amsterdam, Netherlands: Elsevier.

Delamont, S., & Atkinson, P. (2001). Doctoring uncertainty: Mastering craft knowledge. *Social Studies Od Science* 31(1), 87–107.

Dhillon, A. (1998). Individual differences within problem-solving strategies used in physics. *Science Education* 82, 379–405.

Dunbar, K. (1997). How scientists think: On-line creativity and conceptual change in science. In T. B. Ward, S. M. Smith, & J. Vaid (Eds), *Creative thought: An investigation of conceptual structures and processes* (pp. 461–493). Washington, DC: American Psychological Association.

Dunbar, K. (2000). How scientists think in the real world: Implications for science education. *Journal of Applied Developmental Psychology* 21(1), 49–58.

Dunbar, K., & Fugelsang, J. (2005). Scientific thinking and reasoning. In K. J. Holyoak & R. Morrison (Eds), *Cambridge handbook of thinking and reasoning* (pp. 705–726). Cambridge, UK: Cambridge University Press.

Duncan, R., & Hmelo-Silver, C. (2009). Learning progressions: Aligning curriculum, instruction, and assessment. *Journal of Research in Science Teaching* 46(6), 606–609.

Dym, C., Agogino, A., Eris, O., Frey, D., & Leifer, L. (2005). Engineering design thinking, teaching, and learning. *Journal of Engineering Education* 94(1), 103–120.

Ehrenberg, R., Zuckerman, H., Groen, J., & Brucker, S. (2009). *Educating scholars: Doctoral education in the humanities*. Princeton, NJ: Princeton University Press.

Ericsson, K., Krampe, R., & Tesch-Römer, C. (1993). The role of deliberate practice in the acquisition of expert performance. *Psychological Review* 100(3), 363–406.

Ericsson, K., & Lehmann, A. (1996). Expert and exceptional performance: Evidence of maximal adaptation to task constraints. *Annual Review of Psychology* 47(1), 273–305.

Ericsson, K., & Smith, J. (1991). *Toward a general theory of expertise: Prospects and limits*. New York: Cambridge University Press.

Ericsson, K., & Towne, T. (2010). Expertise. *Wiley Interdisciplinary Reviews: Cognitive Science* 1(3), 404–416.

Feldon, D., Hurst, M., Rates, C., & Elliott, J. (2013). Innovation in science, technology, engineering, and mathematics (STEM) disciplines: Implications for educational practices. In L. Shavinina (Ed.), *The International handbook of innovation education* (pp. 359–371). New York: Taylor & Francis/Routledge.

Feldon, D., Maher, M., Hurst, M., & Timmerman, B. (2015). Faculty mentors, graduate students, and performance-based assessments of students' research skill development. *American Educational Research Journal* 52(2), 334–370.

Feldon, D., Maher, M., Roksa, J., & Peugh, J. (2016). Cumulative advantage in the skill development of STEM graduate students: A mixed methods study. *American Educational Research Journal* 53, 132–161.

Feldon, D., Shukla, K., & Maher, M. (2016). Faculty-student coauthorship as a means to enhance STEM graduate students' research skills. *International Journal for Researcher Development* 7(2), 178–191.

Feldon, D. F., Maher, M. A., & Timmerman, B. E. (2010). A call for performance-based data in the study of STEM Ph.D. education. *Science* 329(5989), 282–283.

Feldon, D. F., Peugh, J., Maher, M. A., Roksa, J., & Tofel-Grehl, C. (2017). Effort-to-credit gender inequities of first-year PhD students in the biological sciences. *CBE-Life Sciences Education* 16(1), ar4.

Feldon, D. F., Peugh, J., Timmerman, B. E., Maher, M., Hurst, M., Strickland, D. C., ..., Stiegelmeyer, C. (2011). Graduate students' teaching experiences improve their methodological research skills. *Science* 333(6055), 1037–1039.

Galison, P. (1997). *Image and logic: A material culture of microphysics*. Chicago: University of Chicago Press.

Gardner, S. (2010). Contrasting the socialization experiences of doctoral students in high- and low-completing departments: A qualitative analysis of disciplinary contexts at one institution. *Journal of Higher Education* 81, 61–81.

Golde, C., Conklin Bueschel, A., Jones, L., & Walker, G. (2009). Advocating apprenticeship and intellectual community: Lessons from the Carnegie Initiative on the Doctorate. In R. G. Ehrenberg & C. V. Kuh (Eds), *Doctoral education and faculty of the future* (pp. 53–64). Ithaca, NY: Cornell University Press.

Gopaul, B. (2016). Applying cultural capital and field to doctoral student socialization. *International Journal of Researcher Development* 7, 46–62.

Green, S., & Bauer, T. (1995). Supervisory mentoring by advisers: Relationships with doctoral student potential, productivity, and commitment. *Personnel Psychology* 48, 537–562.

Gross, A., & Mantel, N. (1967). Negative controls in screening experiments. *Biometrics* 23, 285–295.

Hazelrigg, G. (1998). A framework for decision-based engineering design. *Journal of Mechanical Design* 120, 653–658.

Hmelo-Silver, C., Nagarajan, A., & Day, R. (2002), "It's harder than we thought it would be": A comparative case study of expert-novice experimentation strategies. *Science Education* 86, 219–243.

Holley, K. (2009). Animal research practices and doctoral student identity development in a scientific community. *Studies in Higher Education* 34(5), 577–591.

Holyoak, K. (2005). Analogy. In K. J. Holyoak & R. Morrison (Eds), *Cambridge handbook of thinking and reasoning* (pp. 117–142). Cambridge, UK: Cambridge University Press.

Johnson, L., Lee, L., & Green, B. (2000). The PhD and the autonomous self: Gender, rationality, and postgraduate pedadgogy. *Studies in Higher Education* 25(2), 135–147.

Keeney, R. (2004). Stimulating creative design alternatives using customer values. *IEEE Transactions on Systems Man and Cybernetics Part C Applications and Reviews* 34(4), 450–459.

Keller, L., & Ho, J. (1988). Decision problem structuring: Generating options. *IEEE Transactions on Systems, Man, and Cybernetics* 18(5), 715–728.

Kiley, M. (2009). Identifying threshold concepts and proposing strategies to support doctoral candidates. *Innovations in Education and Teaching International* 46(3), 293–304.

Kiley, M., & Wisker, G. (2009). Threshold concepts in research education and evidence of threshold crossing. *Higher Education Research & Development* 28(4), 431–441.

Kitsantas, A., & Zimmerman, B. (2002). Comparing self-regulatory processes among novice, non-expert, and expert volleyball players: A microanalytic study. *Journal of Applied Sport Psychology* 14(2), 91–105.

Klahr, D. (2000). *Exploring science: The cognition and development of discovery processes.* Cambridge, MA: The MIT Press.

Klahr, D., & Dunbar, K. (1988). Dual space search during scientific reasoning. *Cognitive Science 12*(1), 1–55.

Klahr, D., & Simon, H. (1999). Studies of scientific discovery: Complementary approaches and convergent findings. *Psychological Bulletin 125*(5), 524–543.

Knorr-Cetina, K. (1997). What scientists do? In T. Ibáñez & L. Íñiguez (Eds), *Critical social psychology.* London: Sage Publications.

Knorr-Cetina, K. (1999). *Epistemic cultures: How the sciences make knowledge.* Cambridge, MA: Harvard University Press.

Koen, B. (1985). *Definition of the engineering method.* Washington, DC: American Society for Engineering Education.

Kuhn, T. (1962). *The structure of scientific revolutions.* Chicago: University of Chicago Press.

Latour, B., & Woolgar, S. (1979). *Laboratory life.* Beverly Hills, CA: Sage.

Lattuca, L. (2001). *Creating interdisciplinarity: Interdisciplinary research and teaching among college and university faculty.* New York: Vanderbilt University Press.

Lincoln, A. E., Pincus, S., Koster, J. B., & Leboy, P. S. (2012). The Matilda effect in science: Awards and prizes in the US, 1990s and 2000s. *Social Studies of Science 42*, 307–320.

Long, H., Plucker, J., Yu, Q., Ding, Y., & Kaufman, J. (2014). Research productivity and performance of journals in the creativity sciences: A bibliometric analysis. *Creativity Research Journal 26*(3), 353–360.

Maher, M., Gilmore, J., Feldon, D., & Davis, T. (2013). Cognitive apprenticeship and the supervision of science and engineering research assistants. *Journal of Research Practice 9*(2), Article M5. Retrieved from http://jrp.icaap.org/index.php/jrp/article/view/354/311

Mason, M. (2012). Motivation, satisfaction, and innate psychological needs. *International Journal of Doctoral Studies 7*, 259–277.

Merton, R. (1968). The Matthew effect in science. *Science 159*(3810), 56–63.

Merton, R. (1988). The Matthew effect in science, II: Cumulative advantage and the symbolism of intellectual property. *ISIS 79*, 606–623.

Meyer, J. H. F., Land, R. (2003). Threshold concepts and troublesome knowledge: Linkages to ways of thinking and practising within the disciplines. In C. Rust (Ed.), *Improving student learning: Improving student learning theory and practice—Ten years on* (pp. 412–424). Oxford: Oxford Centre for Staff and Learning Development.

Meyer, J., & Land, R. (2005). Threshold concepts and troublesome knowledge (2): Epistemological considerations and a conceptual framework for teaching and learning. *Higher Education 49*, 373–388.

National Research Council (2001). *Theoretical foundations for decision making in engineering design.* Washington, DC: National Academy Press.

National Research Council (2007). *Taking science to school: Learning and teaching science in grades K–8.* Washington, DC: National Academy Press.

Navarra, A., Kinter, J., & Tribbia, J. (2010). Crucial experiments in climate science. *Bulletin of the American Meteorological Society 91*, 343–352.

Nersessian, N. (1984). *Faraday to Einstein: Constructing meaning in scientific theories.* Dordretch, Netherlands: Martinus Nijhoff.

Newell, A., & Simon, H. (1972). *Human problem solving.* Englewood Cliffs, NJ: Prentice-Hall.

Osborne, J., Henderson, J., MacPherson, A., Szu, E., Wild, A., & Yao, S. (2016). The development and validation of a learning progression for argumentation in science. *Journal of Research in Science Teaching* 53(6), 821–846.

Paglis, L., Green, S., & Bauer, T. (2006). Does adviser mentoring add value? A longitudinal study of mentoring and doctoral student outcomes. *Research in Higher Education* 47(4), 451–476.

Parry, S. (2007). *Disciplines and the doctorate*. Dordrecht, the Netherlands: Springer.

Price, D. (1965). Networks of scientific papers. *Science* 149, 510–515.

Rates, C., & Feldon, D. (2014). *Threshold concepts within doctoral biology programs*. Paper presented at the 2014 meeting of the American Educational Research Association. Philadelphia, PA: April 3, 2014.

Reif, F., & Allen, S. (1992). Cognition for interpreting scientific concepts: A study of acceleration. *Cognition & Instruction* 9, 1–44.

Rhoten, D., & Parker, A. (2004). Risks and rewards of an interdisciplinary research path. *Science* 306(5704), 2046.

Roberts, R. (2016). Understanding the validity of data: a knowledge-based network underlying research expertise in scientific disciplines. *Higher Education* 72(5), 651–668.

Robinson, J. A. (1998). Engineering thinking and rhetoric. *Journal of Engineering Education* 87(3), 227–229.

Rothwell, J. (2013). *The hidden STEM economy*. Washington, DC: Brookings.

Schraagen, J. (1993). How experts solve a novel problem in experimental design. *Cognitive Science* 17(2), 285–309.

Schwarz, C., Reiser, B., Davis, E., Kenyon, L., Achér, A., et al. (2009). Developing a learning progression for scientific modeling: Making scientific modeling accessible and meaningful for learners. *Journal of Research in Science Teaching* 46(6), 632–654.

Sheltzer, J., & Smith J. (2014). Elite male faculty in the life sciences employ fewer women. *Proceedings of the National Academy of Sciences* 111(28), 10107–10112.

Simonton, D. (1997). Creative productivity: A predictive and explanatory model of career trajectories and landmarks. *Psychological Review* 104, 66–89.

Simonton, D. (1999). Creativity as blind variation and selective retention: Is the creative process Darwinian? *Psychological Inquiry* 10(4), 309–328.

Simonton, D. (2004). *Creativity in science: Chance, logic, genius, zeitgeist*. Cambridge, UK: Cambridge University Press.

Sternberg, R. (1997). Cognitive conceptions of expertise. In P. Feltovich, K. Ford, & R. Hoffman (Eds), *Expertise in context* (pp. 149–162). Menlo Park, CA: American Association for Artificial Intelligence Press.

Sweller, J. (2009). Cognitive bases of human creativity. *Educational Psychology Review* 21(1), 11–19.

Sweller, J., & Sweller, S. (2006). Natural information processing systems. *Evolutionary Psychology* 4, 434–458.

Timmerman, B., Feldon, D., Maher, M., Strickland, D., & Gilmore, J. (2013). Performance-based assessment of graduate student research skills: Timing, trajectory, and potential thresholds. *Studies in Higher Education* 38(5), 693–710.

Trickett, S., & Trafton, J. (2007). "What if . . .": The use of conceptual simulations in scientific reasoning. *Cognitive Science* 31(5), 843–875.

Urquhart, S., Maher, M., Feldon, D., & Gilmore, J. (2016). Factors associated with novice graduate student researchers' engagement with primary literature. *International Journal of Researcher Development 7*, 141–158.

Wang, X. (2013). Why students choose STEM majors. *American Educational Research Journal 50*, 1081–1121.

Weidman, J. (2010). Doctoral student socialization for research. In S. Gardner & P. Mendoza (Eds.), *On becoming a scholar: Socialization and development in doctoral education* (pp. 29–44). Sterling, VA: Stylus.

Weidman, J., Twale, D., & Stein, E. (2001). *Socialization of graduate and professional students: A perilous passage?* (ASHE-ERIC Higher Education Report, 28, No. 3). Washington, DC: Association of the Study of Higher Education.

Wilson, M. (2009). Measuring progressions: Assessment structures underlying a learning progression. *Journal of Research on Science Teaching 46*(6), 716–730.

Zimmerman, B. (2013). From cognitive modeling to self-regulation: A social cognitive career path. *Educational Psychologist 48*, 135–147.

Zimmerman, B., & Campillo, M. (2003). Motivating self-regulated problem solvers. In J. Davidson & R. Sternberg (Eds.), *The psychology of problem solving* (pp. 233–262). Cambridge, UK: Cambridge University Press.

Zimmerman, C. (2000). The development of scientific reasoning skills. *Developmental Review 20*(1), 99–149.

Zuckerman, H. (1992). The scientific elite: Nobel laureates' mutual influences. In R. Albert (Ed.), *Genius and eminence*, 2nd edn (pp. 157–170). Oxford: Pergamon Press.

CHAPTER 24

A COGNITIVE EXAMINATION OF SKILL AND EXPERTISE IN WORD GAMES AND PUZZLES

SHANE T. MUELLER

Introduction

EXPERTISE in word games requires deep and rich knowledge of language, along with the ability to deal with specialized procedures, strategies, and rules that constitute the game or puzzle. However, language itself has sometimes been considered an expert skill (see Ericsson & Kintsch, 1995), which suggests that we all may already possess the core knowledge for expertise in these contexts. To the extent that some individuals have demonstrated exceptional performance, word game play may involve specific skills that are built on the core of an already-existing linguistic knowledge expertise. Thus, examining expert word game play may help identify boundaries of current theories on expertise, especially in terms of how existing knowledge and expertise is adapted to new problems.

In this chapter, I will examine examples, data, and records on expertise in word games. I will first propose an organizing taxonomy of word games, and use this to discuss a variety of games from a cognitive perspective. Next, I will discuss expert abilities in word games, focusing on SCRABBLE and crossword. Then, I will catalog some of the expert–novice differences that have been examined in an array of related cognitive skills. Finally, I will document aspects of how experts in these domains practice, including how they improve over time. Overall, these aspects will help document the underlying skills involved, and identify ways in which expertise in word games manifests.

A Cognitive Taxonomy of Word Games and Puzzles

There are several ways to conceptually organize word games and puzzles, usually based on the rules of play. For example, the *Oxford Guide to Word Games*, Volume 2 (Augarde, 2003), includes twenty-seven chapters, each describing a different form or set of forms, including many historical ones. More recently, Shortz and Huang (2012) described twenty-two puzzle types, and Selinker and Snyder (2013) published a puzzle creators guidebook that provided step-by-step instructions for creating many modern word games and puzzles. Although they come in a variety of forms, word games and puzzles appear to be distributed around two foci that map onto the cognitive psychology of language and memory: surface structure (including spelling, sound, and other surface representations) and deep structure (meaning). To illustrate this, I asked several professional puzzle creators to rate nineteen word puzzles and games along six dimensions (letter, phonology, meaning, logic, speed/fluent thinking, lateral/creative thinking, and logic), and created a multidimensional scaling (MDS) visualization for the solution (Figure 24.1).

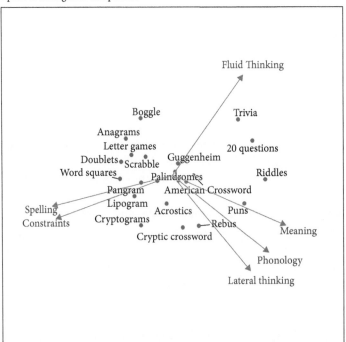

FIGURE 24.1 Multidimensional scaling space of classes of word games, and underlying psychological processes, as rated by a group of word game constructors.

Figure 24.1 shows that one basic division among games and puzzles is whether they focus on meaning or spelling, which is the primary left–right axis in the MDS solution. Each of the other primary dimensions can be mapped into this space. For example, the *constraint* dimension is similar to spelling, because most games appear to place constraints primarily on letters. Similarly, the phonology dimension is similar to the meaning dimension, suggesting that maybe pure sound-based constructions such as puns need to have meaning incorporated. In general, games near the top of the space are thought to require faster, fluid thinking, whereas those closer to the bottom require more creative *outside-the-box* thinking.

A number of games are primarily letter-based, where knowledge of words and letters is paramount. These include competitive board games like SCRABBLE and Boggle, and related parlor games like doubles (creating a chain of words between two related words by changing one letter at a time), palindromes (words and phrases that read the same forward and backward), pangrams (sentences that use all of the letters of an alphabet), and more esoteric games like lippograms (sentences or larger works that avoid using a particular character) and univocalics (sentences using only one vowel). These last three cluster together near the center of the space perhaps because they require some high-level planning based on grammar and meaning. American-style crossword and related puzzles, including the generally more challenging acrostic (in which select letters form a longer phrase) and the simpler Guggenheim (in which one names elements of a category that start with a specific set of letters), also are located more centrally. The remaining games fall into a larger unstructured group that are often less constrained, including puns, riddles, cryptograms, and the like.

The primary meaning–orthography dimension also maps on to how previous researchers have described word games, including Nickerson (1977; 2011) for crosswords, and Schulman (1996) for more general puzzle creation and solving. Furthermore, Mueller and Thanasuan (2013) and Thanasuan and Mueller (2014, 2016) described the primary associative representations used to solve word games like crossword puzzles as involving these same two pathways.

We may expect expertise in word games to involve core cognitive knowledge centered in these foci, as well as an ability required for both: declarative memory retrieval from remote cues. Thus, it is informative to establish the particular ways in which expert performance in word games differs from non-experts and, if possible, understand why most adults—who have strong language and knowledge—are not able to perform similarly.

Expert Ability in Word Games

For many of the word games I described in the previous section, there are individuals who might be considered experts. This is well known in domains such as trivia, SCRABBLE, and crossword (which have all have tournaments and competitions).

But many of the more esoteric puzzles in Figure 24.1 also have individuals of exceptional abilities. For example, E. V. Wright is famous for his 50,000-word lipogrammatic novel *Gadsby* (Wright, 1939) that did not use the letter "e," and several other authors have created similar works. Members of *Oulipo* (see Augarde, 2003, Chapter 27) explore esoteric word games, developing extensive skill in contrived and constrained writing techniques, including palindromes, lipograms, univocalics, and other forms. Palindromes have a small community with regular publications and a world-championship competition that has been held twice,[1] and there are several regularly held pun contests.[2] Each of these communities likely involve individuals of incredible skill, but whose work is obscure because the community is small.

Two domains of word game experts have been studied in the psychological literature most broadly: SCRABBLE and crossword players. Given the different focus of these games, we might expect different pathways to expertise, especially when focusing on the core memory retrieval processes that support expertise.

SCRABBLE Expertise

The competitive SCRABBLE community has a long history, which Fatsis (2001) documented in a best-selling book. By far the most frequent goal of researchers studying SCRABBLE players is to understand the cognition and neuroscience of visual word form identification (e.g., Cansino, Ruiz, & López-Alonso, 1999; Halpern & Wai, 2007; Hargreaves, Pexman, Zdrazilova, & Sargious, 2012; Perea, Marcet, & Gomez, 2016; Protzner et al., 2016; van Hees, Sayffarth, Pexman, Cortese, & Protzner, 2017); although some have established covariates with age or other cognitive skill (e.g., Halpern & Wai, 2007; Toma, Halpern, & Berger, 2014); and some have looked more specifically at the expertise involved (Halpern & Wai, 2007; Tuffiash, Roring, & Ericsson, 2007; van Hees et al., 2016). One lesson is that the advantages shown by experts appear to be primarily related to the word forms, and not about deeper lexical meaning or semantic associations. For example, Protzner et al. (2016) found that SCRABBLE experts exhibited better performance on a number of linguistic measures, including word fluency, anagram solution ability, and lexical decision time, and that they were more likely than non-players to use brain regions associated with vision (and not word meaning) in the lexical decision task.

The fact that the locus of SCRABBLE is on spelling is not surprising, but it sheds light on some interesting facts about competitive and expert SCRABBLE players.

---

[1] The world palindrome competition involves events such as developing a palindrome on a theme with specific letter constraints within a time limit. Remarkable submissions to the 2013 competition included a palindomic self-referential sonnet by Martin Clear, following classic rhyme structures (see Zimmer, 2013).

[2] In the United States, the O Henry Pun-off World Championship has been held since the 1970s; in the United Kingdom, the UK Pun Championships are held as part of the Leicester Comedy Festival.

For example, Zimmer (2015) reports that Nigel Richards won the 2015 French-language SCRABBLE championship, although he did not speak French.[3] This is not an isolated case—Thailand has a crossword game association, and Brand's Crossword Game King's Cup is hosted there each year for international competitors in SCRABBLE. Although English-language speakers have often won, Thai players are frequently runners-up or winners. Thai players also frequently appear in the highest divisions of the World SCRABBLE tournaments, while numerous sources have reported that these players typically have very limited English language skills (e.g., Wachtell, 2004).[4]

Some understanding of top SCRABBLE play can be gained by examining game point totals during competition. Typical advice is that amateur two-player games will have scores in the 200-point range, whereas competitive players rarely score below 300 points in a game, and 400–450 points are good tournament scores. At a recent tournament (2017 Niagara Falls International Open) with about 2400 head-to-head games, the average per-player score was 411 points, with a range of 162 to 716. The winner, whose record was 21–1 over the event, averaged 499 points per game. Since there are 102 tiles in the SCRABBLE set, and the average game earned 822 total points, this amounts to about 8 points per tile played (tiles average 1.83 points each). Thus, unlike novices, whose play is primarily constrained by simply finding a word, expertise in this domain is not just letter-based, but involves finding high-scoring words, not being satisfied with an easy-to-generate low-scoring word, and incorporating other factors (multipliers and 50-point bonuses) that lead to higher scores. This may help explain how Richards was able to excel in French SCRABBLE so easily—much of his skill is about factors outside of language.

Crossword Expertise

The crossword community and methods used by that community to develop crossword skill have been described by a number of authors (Amende, 2002; Reynaldo, 2007; Romano, 2005). In fact, crossword involves two distinct games—the American-style crossword that typically involves extensive crossing-letters and straightforward but remote clues, and the cryptic-style, which involves fewer crossing constraints, but multiple hints provided by rule-based constraints in the clue. I will focus mostly on American-style crossword expertise, although I will consider some of the research on cryptic experts later in this chapter.

---

[3] Richards is a competitive English-language SCRABBLE player, and claims to have first picked up a French dictionary nine weeks before the competition.

[4] Popular press coverage includes 2016 stories "Thailand fields Scrabble masters who don't really understand English," available at http://www.wionews.com/world/thailand-fields-scrabble-masters-who-dont-really-understand-english-1580; and "Thais master Scrabble without speaking English at King's Cup," available at http://www.abc.net.au/news/2016-07-01/thais-master-scrabble-without-speaking-english/7561666

Unlike SCRABBLE, crossword players do not typically compete head-to-head, but the community centers on daily puzzles and the annual American Crossword Puzzle Tournament (ACPT). At tournaments, performance is usually measured in a way that combines accuracy and time, with each letter mistake incurring a large time cost.

Development of expertise in crossword play requires a different set of skills than SCRABBLE. Certainly, semantic associations, trivia, and general knowledge play an important role. But orthographic constraints matter as well, leaving questions regarding the relative importance of each. Later, I will discuss how many top players became competitive in a relatively short period of time after they began to play seriously. This is somewhat similar to Nigel Richards' French SCRABBLE play, which might suggest that crossword experts have a distinct advantage in the same abilities required for SCRABBLE play: visual pattern completion, perhaps supported by highly developed visual word-form networks (e.g., Perea et al., 2016). According to this hypothesis, players may tend to *see* the answer in the grid based on word fragments, bypassing the semantic clue route completely.

In support of this, letter constraints can often narrow down a word to just a few possibilities, and AI systems that solve crossword puzzles (e.g., Ginsberg, 2011) rely heavily on constraints from crossing letters. Moreover, Thanasuan and Mueller (2014) showed that experts solving real puzzles tended to use *more* letters to solve clues than novices did, which might suggest a greater reliance on orthographic information.

However, there are many good reasons to believe this is not the case, and that the advantage experts have lies primarily in their semantic-route access. For example, the computational models developed by Thanasuan and Mueller (2014) rely heavily on semantic information to solve puzzles, and orthographic-route information is typically not sufficient to solve puzzles well (for a review of some of this research, see Chapter 15, "Computational Models of Expertise," by Kirlik & Byrne, this volume). We concluded that orthographic-route retrieval is used only when just a few letters are missing. Logically, this make sense because when no letters are available to solve a clue, only semantic information can be used, and this is exactly the type of solving that experts are much better at than novices.

This suggests that crossword play shares much in common with semantic knowledge games like trivia. In support of this, I have found many top players are also frequent trivia players. For example, at the 2014 crossword tournament, the organizers informally polled the competitors about who had competed on the television program *Jeopardy!*. Out of the roughly 600 participants in the room, approximately 50 raised their hand (and many of them also acknowledged they were Jeopardy! champions). As a final example, in 2006 Ken Jennings, who holds one of the longest winning streaks on the Jeopardy! program, competed in the American crossword puzzle tournament. He won the Rookie division and placed thirty-seventh overall, suggesting his knowledge gave him a strong advantage to developing expertise in crossword solving. This all suggests that crossword skill is strongly supported by semantic-route memory search ability, and is associated with fast recall of general knowledge. The cognitive underpinnings may involve better encoding, better memory search from remote cues, reduced interference, or a combination of these.

Expert Crossword Solution Times as a Function of Difficulty

Time to solve a puzzle is the common measure used for players to judge their performance. Most top players will be able to solve any standard puzzle completely (or with very few missing or erroneous letters) without any outside help. Consequently, time-to-solve is a reasonable and pure measure of performance. For lesser players (even very experienced ones), several factors may impede using solution times as a measure of problem-solving ability; this includes using external aids such as a search engine or Wikipedia, the greater likelihood of mistakes, the use of the crossword software to give letter hints, and so on. Thus, comparing solution times across expertise groups may be imperfect but informative.

To examine solution times, it is useful to look at *New York Times* (*NYT*) crossword puzzles, as they are widely solved, their form is well understood, and they have a systematic difficulty variation. *NYT* puzzles generally progress in difficulty throughout the week, starting on Monday with a very straightforward puzzle, increasing in difficulty through Thursday, which will often include themes or trickery such as rebuses that make for greater challenge. These 15 × 15 puzzles typically contain up to 78 clues, and only occasionally involve answers that are 15 letters long. This changes qualitatively for Friday and Saturday puzzles, which typically do not have the themes involved in earlier puzzles, but will have more 15-letter answers, and fewer total clues. Sunday puzzles are generally similar in difficulty to Thursday puzzles (and again may include themes), but are larger (21 × 21, with a maximum of 140 clues), for about twice as many cells (441 vs 225), and so are difficult to compare directly to solution times of other puzzles.

Several sources record the speed at which players solve *NYT* puzzles. Currently, a group of top players record their daily solution times on a public website.[5] Although each player typically records times for a subset of about seven different puzzles, most typically solve the *NYT* puzzle. Both time and completion criteria are self-determined, but players of this caliber rarely make mistakes or require hints, and typically use software for solving that records their play times exactly.

I examined solution times for a group of fifty-five puzzles from the summer of 2017, which involved a total of 299 solutions among the group. During this time period, six distinct experts recorded times, and each puzzle solution time was recorded by, on average, 5.5 individuals.

Figure 24.2 shows the distribution of solving times across the week. Each day consists of around five to six solutions per player for seven to eight puzzles. As the week progresses, the mean solution time increases from 139 s (2.3 minutes) on Monday to 309 s (5.2 min) on Saturday, increasing by a factor of 2.2, but very little increase occurs

---

[5] http://dandoesnotblog.blogspot.com/

after Thursday. For these experts, although there is substantial overlap, Friday puzzles were on average solved a bit faster than Thursday puzzles. This suggests that although Friday puzzles may be intended to be more difficult than Thursday puzzles for average solvers, the differences (more long words and fewer themes and tricks) do not present a greater difficulty for top players. Leban (2016) performed a detailed analysis of solution times from the puzzazz.com website, and found that this Thursday/Friday reversal was systematic for a wider range of player skill. This flattening at the end of the week may stem in part from higher attrition (which Leban also established)—fewer players start or complete these puzzles, and so do not appear in the time records.

Overall, note that for a solution time of 139 s, this amounts to roughly one clue every two seconds, and filling more than one letter cell per second. This does not approach the limits of typical typing speeds (60 wpm is roughly five characters per second), but it is clear that for experts, very little contemplation or reasoning can be involved, and answers must primarily involve fast retrieval from memory and letter-based pattern completion. Later week puzzles roughly double the time needed (309 s for Saturday puzzles), but is still a frenetic pace of around one clue every four seconds, and one letter completed every second.

Puzzazz.com website lets users solve various daily puzzles on their tablets, phones, or computers, and records a leaderboard for those completing the puzzle, showing the fastest 100 players to solve each puzzle, along with an average solution time from all of their customers. To compare the best players with more typical enthusiasts, I collected

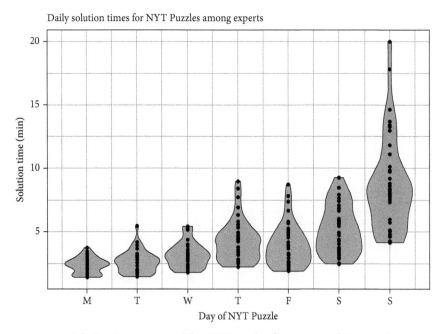

FIGURE 24.2 Solution times reported for *NYT* puzzles for a group of six top players over a period of 55 days.

Puzzazz website records on the same puzzles, including the 10th, 50th, and 100th place, and the reported average solution time across all players. The intercorrelations of these measures are all above 0.97, and the average and best expert solution time also predicts the average Puzzazz player solution time extremely well ($r = 0.938$ and 0.948). This indicates that aspects of puzzle difficulty impact all players similarly. However, the slopes of the relationship indicate that the average solution time for Puzzazz players was about ten times as long as the fastest expert, and Puzzazz players are typically enthusiasts who play frequently and are engaged enough to pay subscription fees to solve the puzzle.

This factor of 10 is modulated somewhat by the difficulty of the puzzle, as shown in Figure 24.3. Here, the lowest two lines indicate solution times for the fastest and average expert, and the shallow slope of these lines in comparison to the others shows relative insensitivity to difficulty. The ratio between average player and the best player starts at 8.99 on Monday, but grows to 10.8, 10.5, 10.8, 12.9, 11.4, and 10.7 on subsequent days. This indicates that, relatively speaking, experts are less impacted by increased difficulty than are average players.

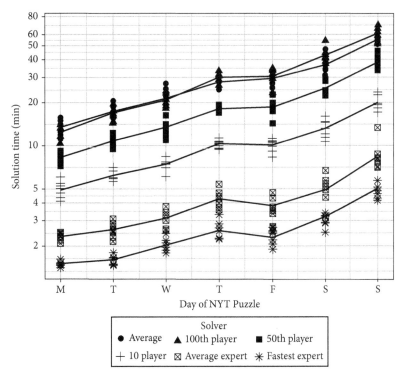

FIGURE 24.3 Different measures of puzzle solution times for fifty-five puzzles on a log-transformed time scale. As puzzles increase in difficulty (as the week progresses), increased time is required.

Strategy versus Memory Skills in Crossword Experts

It appears that much of the skill involved in word games like crossword is about memory search, activation, and retrieval from a rich semantic network. Nevertheless, there are a number of other skills, abilities, and strategies that are learned and develop with practice and expertise.

Some of these involve the low-level mechanics of play, including how to type or write letters, and how to move one's eyes efficiently and not lose place. Others are more strategic, such as how to start a puzzle (some like to always start with a clue they are 100 percent sure of; others pick a corner and start from there), how to move through the grid, how to guess, skip answers, or queue multiple answers. Thanasuan and Mueller (2014) concluded that both low-level motor skills and aspects of gameplay strategy could increase speed in some circumstances, but this is only effective if the player has enough knowledge to play well already. Many of these higher level skills allow a good player to play faster, but would not allow a poor player to play well. For example, one piece of advice a top ACPT player gave about how to solve the puzzle is "Never let your pencil stop moving." Using this strategy, a player may need to be very daring about looking ahead to the next clue while writing, and this type of strategy may help a very good player reduce their time by a few seconds, but the advice is useless to a novice. Some aspects of expert strategy include the following:

- Experts have the ability to solve clues with fewer letters than novices, but they tend to solve the puzzle in a way such that they solve clues that have more filled-in letters. Their grid-fill is more deliberate, building on sections that are already solved.
- Deliberate solving also promotes place-keeping, and helps players avoid missing a cell accidentally.
- Some players will use working memory strategies to improve speed. For example, a small block of three to four down clues will be next to one another in the clue listing. Some good players will try to look at and solve all of them in their head BEFORE writing the solutions.
- Speed can be improved by reducing error-checking. At tournaments, a common mistake is to leave an answer blank accidentally. I once observed a top player at a tournament who completed the puzzle, and immediately realized he had made a mistake. He remembered that he had guessed on a single letter of an early clue, recognized it immediately, but decided to wait to correct his mistake at the end of the puzzle in order to save time. However, he forgot to make the correction.

- Daring strategies are common enough that top players have innovated meta-strategies for error flagging. The *Ripstein mark* is attributed to an ACPT champion Ellen Ripstein, and it is a mark made to indicate a guess so that it can later be checked or corrected.
- When solving for speed on paper, top players will sometimes use the finger of one hand and their pen or pencil to mark current location in the grid and clue set, avoiding the constant need for visual search.
- Many players will typically start at 1-across as their first clue. However, some have developed other strategies. Some players report searching the clues for one that they have 100 percent confidence they are correct on, and build on that. For tournament puzzles, solvers have reported suspecting the puzzle author will try something deliberately misleading on the 1-across clue, and so may avoid starting there because of it.

These expert strategies are informative, but most would be bad advice for novices. They may help a very good player, but for a novice player they would not be helpful and may in fact hurt performance.

Cognitive Skills Associated with Expertise in Word Games

Much of what is known about the cognitive skills involved in word games comes from studying college students or other non-expert populations. Forshaw (1994) notes that even for anagrams, which are understood better than any other wordplay game, little is known about expertise. A small body of research has attempted to determine whether cognitive skills differ between novice and expert populations, and these have sometimes found differences between groups. These results can be difficult to interpret, because they are usually correlational—we cannot tell whether the differences stem from the expertise, or the expertise arises because of the differences. Tables 24.1 through 24.3 summarize some of the results across three different styles of word games.

Overall, the results are quite variable. There are few measures that consistently show differences between experts and comparison groups, in part because of variability in selection criteria for expert groups, and in part because comparison groups differed widely—including age-matched controls, comparable enthusiasts, young non-players, and experts in other word games. Several studies used correlational comparisons within a single group of players. In general, differences have been reported in tasks that are closely linked to the expert domain, especially when comparison groups had little experience in word games. For example, fluency measures were often better for experts than control groups; lexical decision tasks were sometimes better, but many linguistic and memory tasks showed no differences. This suggests that future research on expert–novice differences should consider carefully criteria for inclusion, as well as appropriate control groups. Furthermore, the source of the differences cannot be established unless true random assignment and longitudinal practice studies are conducted. Even if differences were found and were consistent, we cannot determine whether they lead to better play, or are caused by extensive play.

Table 24.1 Cognitive skills associated with cryptic crossword skill, as identified by past research

Source	Comparison	Key findings result
Underwood, et al. (1994)	Expert vs intermediate solvers	Word generation: better Anagrams: better Word stem completion: faster High-frequency words access: faster Meaning of unusual words: better Lexical decision: no difference Synonym judgment: no difference AH4-verbal: no difference AH4-spatial: no difference Vocabulary: better
Underwood et al. (1988)	Experts vs non-matched non-puzzlers	Generate associations: no difference primed lexical decisions: no difference
Fine & Friedlander (2013)	Experts vs matched non-experts	AH4 Problem solving: Relationships: no difference Directions: experts better Verbal analogies: experts better Numerical series: experts better Non-verbal: no difference
Friedlander & Fine (2016)	Experts vs ordinary enthusiasts	Need for cognition: no difference
Forshaw (1994)	Correlation across range of abilities in elderly players	Homophone detection: faster Fixed-length word generation: better Stem completion: better Trigram fluency: better Word fragments: better Anagrams: better Homophone judgment: better Ambiguous sentences: n.s. Word length judgment: n.s. Letter detection: n.s. Word from definition: n.s. Meaning check: n.s. Choice RT: n.s.

Table 24.2 Cognitive skills associated with SCRABBLE expertise, as identified by past researchers

Source	Comparison	Key findings result
Tuffiash et al. (2007)	Matched non-elite players	SCRABBLE Moves: better Digit-symbol: n.s. UN- Letter fluency: better FAS letter fluency: n.s. 4-letter noun fluency: n.s. Vocabulary: n.s. Reading: n.s. Anagrams: better
Hargreaves et al., (2012)	Age-matched controls	Greater COWAT word fluency Higher anagram accuracy Faster lexical decision task Smaller effect of vertical presentation Smaller effect of concreteness Vocabulary: No difference
Protzner et al. (2016)	Age-matched controls	In fMRI studies using lexical decision task, experts used brain regions associated with working memory and visual perception.
Van Hees et al. (2016)	Age-matched controls	Greater COWAT word fluency Greater anagram accuracy
Perea et al. (2016)	Non-age-matched controls	Lexical decision task with transposed letters had smaller time cost
Halpern & Wai (2007)	Novices	Extended vocabulary: better Lexical decision: n.s. Mental rotation: n.s. Paper folding: faster Shape memory: better
Toma et al. (2014)	Young novices	Mental rotation: n.s. Symmetry span: n.s. Analogies: better Reading span: better
Toma et al. (2014)	American crossword	Mental rotation: n.s. Analogies: n.s. Reading span: n.s. Symmetry span: n.s.

Table 24.3 Cognitive skills associated with American crossword ability

Source	Comparison	Key findings result
Toma et al. (2014)	Young novices	Mental rotation: n.s. Symmetry span: n.s. Analogies: better Reading span: better
Toma et al. (2014)	SCRABBLE	Mental rotation: n.s. Analogies: n.s. Symmetry span: n.s. Reading span: n.s.
Hambrick et al. (1999)	Correlations in older enthusiast population in structural equation model	General knowledge (Study 1,2,3,4) Puzzle experience (1,2,3,4) Word retrieval (Study 1,2) Speed (Study 2,3) Reasoning (Study 4)

Practice among Word Game Experts

The time, extent, nature, and purpose of practice has rarely been the sole focus of research on expert word game players, but it has often been reported as part of demographic surveys in tests of expert–novice differences. Thus, reports of practice are linked to samples recruited to form some comparison with a non-expert group, and not to establish practice routines and commitments of the typical elite player.

Practice among SCRABBLE Experts

Studies with SCRABBLE players (Table 24.2) have recruited participants from varying levels of competitiveness, and so our conclusions about practice depend on sampling and inclusion criteria, which will tend to include high-performing enthusiasts that might not be considered true experts.

Halpern and Wai (2007) noted that a sample of competitive SCRABBLE players dedicate 4.6 hours/week to deliberate practice, and reported playing SCRABBLE on average 211 days per year, with a total lifetime average of 1,904 hours and ranging up to 14,872. Similarly, Tuffiash et al. (2007) found that across a number of measures of activities related to SCRABBLE play and practice, their group of experts typically spent considerable more time in serious study and competitive play than average players with the same years of experience. Elite players reported a sum total of 20.6 hours/week in SCRABBLE-related activities, whereas average tournament-players reported 16.3

hours. Also, elite SCRABBLE players reported only 0.1 hours of this time was involved in other kinds of word games, in comparison to 4.0 hours of the average players time. Tuffiash et al. also used these reports to estimate lifetime estimates of time spent in study (5084 hr) and play (6043 hr) for elite players, which contrasted to 1318 and 2379 hours, respectively, for average competitors. Toma et al. (2014) reported that elite SCRABBLE players reported playing on average 138 days in the prior year, while spending around 4.7 hours/week studying word lists. In addition, van Hees et al. (2016) reported that their participants with SCRABBLE expertise reported playing on average 6.8 hours/week.

Overall, this suggests expert SCRABBLE practice often involves extensive deliberate study of word lists—nearly half of the preparation time. There are several specific word types that SCRABBLE players study:

- *Short words.* Experts are likely to know every two-letter (about 100) and three-letter words (about 1000), and have high familiarity with the 4-letter words (about 4000).
- *Special cases.* Many SCRABBLE guides provide word lists based on specific letters that have high payoff (J, Q, Q without U, Z, etc.),
- *Bingos.* Because of the point bonus for using all seven letters (a *Bingo*), experts study seven-letter words extensively. We tested one nationally competitive SCRABBLE player in our laboratory on word-stem completion problems that required him to generate six-letter words that fit certain patterns. In the post-study interview, he told us that despite the fact that seven-letter words would have been objectively more difficult, he would have done better because he studied them extensively.
- *Bingo stems.* There are clusters of 6 letters that will form many different words in combination with almost any additional letter. This gives many options if a player can retain those letters, and so knowing all the words that can be made will offer a competitive advantage. For example, the letters AEINST form the base for seventy-three different seven-letter words, and there are many other six-letter combinations that form more than fifty seven-letter words. The SCRABBLE study guide identifies the top stems as TISANE, SATIRE, RETAIN, ARSINE, and SENIOR.

In summary, expert SCRABBLE play typically involves long-term study and practice, with a self-reported commitment of 5–20 hours/week. Competitive players rely on deliberate practice outside of play—especially in the form of studying and memorizing word lists. Deliberate practice is often focused on information that will give the player options and point advantages.

Practice in American-style Crossword Players

Practice for crossword players primarily (and almost exclusively) involves playing the game. Toma et al. (2014) established that in contrast to a comparison of elite

SCRABBLE players who spent 4.1 hours/week studying word lists, only one elite crossword player (out of thirty-one) reported any deliberate study of words, and this was for just 30 minutes/week. Although little deliberate practice is reported by experts, there are variations of play that experts report doing that either challenges themselves or helps develop skill. One top player reported that he would solve using only the down clues (or across clues), eliminating the benefit of crossing letters and thus strengthening ability to solve with clues alone and fewer letter clues. Relatedly, the acrostic puzzle is a variation that provides similar crossword clues but has fewer crossing letters, and experts have reported using them to help challenge their abilities.

Perhaps as important is daily, regular, and long-term commitment to solving. For example, Toma et al. (2014) examined a group of competitive crossword players and estimated that they averaged solving puzzles 309 days in the past year. Moreover, Moxley, Ericsson, Scheiner, and Tuffiash (2015) examined how long-term practice was associated with points earned at the 2005 American Crossword Puzzle Tournament. Points earned were negatively correlated with years of pre-tournament experience ($r = -0.24$), indicating that those who started competing after playing longer earned fewer points); years post-tournament experience ($r = 0.32$, indicating those who had competed more earned more points), and total number of hiatuses from playing ($r = -0.21$, indicating those who had taken more hiatuses from regular play earned fewer points). Interestingly, measures of total practice time over lifetime and current practice were not as strongly predictive, with total estimated word study producing a correlation of $r = 0.1$ and current hours of weekly practice producing $r = -0.09$, indicating those who spent more time practicing in the past year actually did slightly worse. This is probably a consequence of the fact that the very best players spend relatively less time playing because they solve so quickly. In this sample, mean play per week was 7.4 hours. In contrast, for the six experts whose data appear in Figures 24.2 and 24.3, players typically recorded solution times for five to seven puzzles per day, but typically under 30 minutes a day (the mean solution time across experts was 4.23 minutes, which often amounted to around 20 minutes per day playing time). Players who are likely to earn fewer points at the ACPT may spend more time playing, even while solving fewer puzzles.

These values contrast with similar numbers reported by Hambrick, Salthouse, and Meinz (1999), who studied older adults recruited as crossword enthusiasts. Across studies 2–4, participants reported averages between 4.7 and 5.4 hours/week playing crossword puzzles—around the total time reported by Moxley et al. (2015) regarding ACPT players—but at the same time reporting playing only between 5.2 and 6.5 puzzles per week—about an hour per puzzle.

Although time alone is not a great predictor of performance, number of puzzles solved may be better. In my lab, we conducted a survey to assess practice among the 2013 ACPT competitors, and asked two questions of the ninety-two players we recruited: what their best finish (in terms of rank) was at a recent ACP tournament, and how often they played crossword puzzles. Table 24.4 shows this relationship. Results show that for top-50

Table 24.4 Responses from 92 participants recruited from ACPT contestants

What is your best or current ACPT Rank?	How often do you play			
	Usually several a day	Almost every day	At least once a week	Occasionally
Top 10	5	1	1	0
Top 25	4	0	0	0
Top 50	1	2	0	0
Top 200	7	3	2	2
Below top 200	8	10	1	1
Not previously competed	12	20	9	3

Note: Participants were asked to identify their best or current finishing rank, and how often they played crossword puzzles.

players, they almost universally play multiple puzzles per day. For lesser players, they tend to report playing fewer and less often. A Spearman rank–order correlation showed that these two ratings were significantly related ($\rho = 0.28$, $p = 0.007$).

Elite Players' History of Tournament Play Prior to Making ACPT Finals

Finally, among ACPT competitors, there is a history of young players and players new to competitive crossword play joining the elite ranks after just a few years of competition. One example is Dan Feyer, who Mueller and Thanasuan (2013) and Grady (2010) documented began playing seriously in 2008, had finished fourth by the 2009 tournament, won in 2010, and has been the dominant player ever since. During his initial three years, he estimated solving more than 6000 puzzles per year. His is not an isolated case, and we can see evidence of fast rises by examining published tournament play records.

Generally, the top 25 point-earners at the ACPT complete all puzzles, make almost no mistakes, and are primarily separated by completion times. If we consider the top 25 finishers between 2000 and 2017, on average 1.4 per year were *Junior* competitors (age 25 or under), 0.5 were *rookies* (having never competed before), and 5.7 were classified as *Division B* or lower—having not won a prize during the previous three tournaments. Thus, around 25 percent of the top 25 appear to be relatively new to highly ranked finishes. This is even clearer when I examined the prior competition records for players who had finished in the top 3 (and thus gone on to the live playoff) for the first time during the years 2002–2018, as shown in Figure 24.4. For these ten players, the median prior competitions was 6, but in fact only two players who started as non-juniors

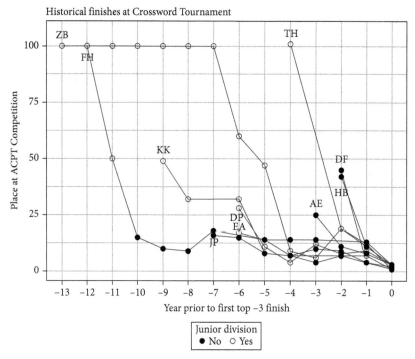

FIGURE 24.4 Ranks achieved during prior tournaments by the nine players who finished in the top 3 places during the years 2002–2017 at the ACPT. Unknown finishes outside the top 100 are truncated to 100.

began outside the top 25, and only one player spent more than a year outside the top 50 (ZB, who began competing when he was 14). Excluding ZB, the median debut of future top players was 1028 points behind the top 3 (where the lowest point total wins). In contrast, the median player averaged 3297 points behind the top 3, which was more than three times as large. Only competitors EA and JP finished in the top 25 in their first appearance—and JP was a competitive trivia player prior to competing in the ACPT.

This suggests that when the players who later became elite started competing, they were typically well ahead of average competitors, had a steep increase to near-top level, and then sometimes took several years of maintenance or steady increase until they finished in the top 3. This last phase is spent mostly in the top 25—where mostly speed distinguishes players. It may be that moving up in ranks at this level is more reliant on procedural skill and strategy tuning, whereas the sharp rise comes from rich expansion of relevant semantic knowledge. Thus, the best players start good, get better very quickly, but sometimes take longer to achieve their top play, which is consistent with recent evidence in some other fields of expertise (Hambrick et al., 2013; Lombardo & Deaner, 2014; Wai, 2014).

Practice among Cryptic Solvers

Practice strategies for cryptic solvers have some similarities to American-style crossword solvers. Friedlander and Fine (2009) surveyed cryptic crossword solvers about their solving style, focus, and practice. All respondents were highly skilled, began solving on average at age seventeen, had typically been solving cryptics thirty-five to forty years (depending on skill and focus category), and reported spending on average eight hours a week solving. Similarly, Friedland and Fine (2016) found that respondents spent on average seven hours/week solving, and this did not differ between ordinary and expert solvers. These numbers are a bit larger than those reported by American-style enthusiasts, within the time reported by SCRABBLE competitors.

Forshaw (1994) examined a group of older adults ranging in experience solving the cryptic crossword and found that the better players had been playing thirty-two years (in comparison to twenty-six years of poorer players), indicating that all players had spent substantial time playing. This may indicate an interesting interplay between processing-intensive tasks often associated with fluid intelligence (which is thought to decline with age) and verbal knowledge often associated with crystallized intelligence (which is maintained or increases; see Park et al., 2002). Cryptics certainly involve both types of reasoning, but most expert players appear to have substantial years of practice, indicating that the strategies and knowledge earned over long periods of time play an important role in their expertise.

Summary of Practice among Word Game Experts

Across several studies covering both SCRABBLE and crossword, reports indicate that most elite players recruited to be in psychological studies have typically spent years or decades studying their word game of choice, and often spend close to an hour a day playing or practicing. Serious crossword players appear to spend less time per week playing than serious SCRABBLE competitors spend preparing, but SCRABBLE players devote only about half of their preparation to play, and spend the other half in deliberate study, which is rare for crossword players. Daily time spent playing is a somewhat imperfect measure of expertise for crossword players, because experts can solve the same number of puzzles much faster than lesser players. This suggests that for crossword play, initial superior abilities for some individuals allow for much more efficient practice, resulting in a steeper learning curve to the top of their ultimate abilities.

SUMMARY AND CONCLUSIONS

I have identified several consistent phenomena about word game expertise that helps to paint a picture about the nature of these skills:

- Expertise in letter-based word games can be achieved without being able to understand the language of play, and training can be fast for existing experts.
- Top crossword players may be ten times faster than enthusiasts who play as often or as long as the best players.
- Elite crossword players tend to have started competitive play above average, had a steep increase lasting 1–2 years, followed sometimes by a slower approach to their peak.
- Solution times between expert and non-expert crossword players tend to be highly correlated across puzzles, indicating their skill is highly constrained by the game difficulty.
- Strong overlap appears to exist for trivia and crossword games, insofar as they involve semantic associations as primarily memory retrieval route.

Challenges to Theories of Expertise

Word games and puzzles involve, at their core, normal human language abilities with non-typical memory cues and constraints. The nature of the cues and constraints forms the rules of different word games, and the ability to manage these rules forms a core aspect of expertise in the game. Superior play appears to involve substantial strategic knowledge, specific linguistic knowledge, and also perceptual-motor skills that allow the knowledge to be applied to the word game context, which are likely to be well accounted for by any account of expertise. However, other aspects of expertise reasonably arise from other sources, including general language expertise common to most adult humans and transfer from other similar word game domains (trivia to crossword, English to French SCRABBLE, SCRABBLE to anagramming, etc.). Furthermore, most top crossword players appear to have begun competing when they were already very good, underwent a steep increase to near-top performance, and then spent a few years improving incrementally to their peak.

Thus, models of expertise may need to account for at least five sources of performance advantages: (1) improvement from deliberate practice; (2) transfer from similar games and tasks; (3) support from general language expertise; (4) individual differences in specific mnemonic skills that may be best considered traits; and (5) a caveat that developmental changes through adolescence mean that few players have reached their peak before their early twenties, and many have done so at much older ages. This notional model mirrors diathesis-stress models of mental disorder, and might be considered a predisposition-opportunity model of expertise. In this account, individual differences in some low-level cognitive abilities related to memory encoding or retrieval may be a key that enables better initial performance. This, together with general language expertise and transfer from similar domains, permits time-efficient practice of both declarative knowledge and skill, and produces a fast rise toward peak performance. Future research will benefit from examining how these factors mutually support one another, for both expertise in word games and other expert domains that have similar properties.

References

Amende, C. (2002). *The crossword obsession: The history and love of the world's most popular pastime.* New York: Berkley.

Augarde, T. (2003). *The Oxford guide to word games*, second edition. Oxford: Oxford University Press.

Cansino, S., Ruiz, A., & López-Alonso, V. (1999). What does the brain do while playing Scrabble?: ERPs associated with a short–long-term memory task. *International Journal of Psychophysiology 31*(3), 261–274.

Fatsis, S. (2001). *Word freak: Heartbreak, triumph, genius, and obsession in the World of competitive Scrabble players.* New York: Penguin Books.

Fine, P., & Friedlander, K. (2013). Cryptic crossword expertise and fluid intelligence. In *Proceedings of the International Symposium on Performance Science* (pp. 415–420). European Association of Conservatoires (AEC).

Forshaw, M. J. (1994). *Expertise and ageing: The crossword-puzzle paradigm.* Thesis, University of Manchester.

Friedlander, K. J., & Fine, P. A. (2016). The grounded expertise components approach in the novel area of cryptic crossword solving. *Frontiers in Psychology 7*, 567.

Friedlander, K., & Fine, P. (2009). Expertise in cryptic crossword performance: An exploratory survey. In *Proceedings of the International Symposium on Performance Science* (pp. 279–284). Auckland.

Ericsson, K. A., & Kintsch, W. (1995). Long-term working memory. *Psychological Review 102*(2), 211–245.

Ginsberg, M. L. (2011). DR. FILL: Crosswords and an implemented solver for singly weighted CSPs. *Journal of Artificial Intelligence Research 42*(1), 851–886.

Grady, D. (December 6, 2010). Across and down, the wizard who is fastest of all. Retrieved from https://www.nytimes.com/2010/12/07/science/07profile.html.

Halpern, D. F., & Wai, J. (2007). The world of competitive Scrabble: Novice and expert differences in visuopatial and verbal abilities. *Journal of Experimental Psychology: Applied 13*(2), 79–94.

Hambrick, D. Z., Oswald, F. L., Altmann, E. M., Meinz, E. J., Gobet, F., & Campitelli, G. (2013). Deliberate practice: Is that all it takes to become an expert? *Intelligence 45*, 34–45.

Hambrick, D. Z., Salthouse, T. A., & Meinz, E. J. (1999). Predictors of crossword puzzle proficiency and moderators of age–cognition relations. *Journal of Experimental Psychology: General 128*(2), 131–164.

Hargreaves, I. S., Pexman, P. M., Zdrazilova, L., & Sargious, P. (2012). How a hobby can shape cognition: visual word recognition in competitive Scrabble players. *Memory & Cognition 40*(1), 1–7.

Leban, R. (2016). How hard is this puzzle anyway? Levering user analytics to measure and understand the difficulty of crossword puzzle. Available from http://www.puzzazz.com/puzzle-difficulty-index

Lombardo, M. P., & Deaner, R. O. (2014). You can't teach speed: sprinters falsify the deliberate practice model of expertise. *PeerJ 2*, e445.

Moxley, J. H., Ericsson, K. A., Scheiner, A., & Tuffiash, M. (2015). The effects of experience and disuse on crossword solving. *Applied Cognitive Psychology 29*(1), 73–80.

Mueller, S. T., & Thanasuan, K. (2013). A model of constrained knowledge access in crossword puzzle players. In *Proceedings of the 12th International Conference on Cognitive Modeling (casing).* Ottawa: Concordia University.

Nickerson, R. S. (1977). Crossword puzzles and lexical memory. *Attention and Performance VI, Proceedings of the Sixth International Symposium on Attention and Performance* (pp. 699–718). Stockholm, Sweden.

Nickerson, R. S. (2011). Five down, absquatulated: Crossword puzzle clues to how the mind works. *Psychonomic Bulletin & Review 18*(2), 217–241.

Park, D. C., Lautenschlager, G., Hedden, T., Davidson, N. S., Smith, A. D., & Smith, P. K. (2002). Models of visuospatial and verbal memory across the adult life span. *Psychology and Aging 17*(2), 299.

Perea, M., Marcet, A., & Gomez, P. (2016). How do Scrabble players encode letter position during reading? *Psicothema 28,* 7–12.

Protzner, A. B., Hargreaves, I. S., Campbell, J. A., . . . Pexman, P. M. (2016). This is your brain on Scrabble: neural correlates of visual word recognition in competitive Scrabble players as measured during task and resting-state. *Cortex 75,* 204–219.

Reynaldo, A. (2007). *How to conquer the New York Times crossword puzzle: Tips, tricks and techniques to master America's favorite puzzle.* New York: St. Martin's Press.

Romano, M. (2005). *Crossworld: One man's journey into America's crossword obsession.* New York: Broadway Books.

Schulman, A. (1996). The art of the puzzler. In M. P. Friedman & E. C. Carterette (Eds), *Cognitive ecology: Handbook of perception and cognition,* Vol. 16, 2nd edn (pp. 293–321). San Diego: Academic Press.

Selinker, M., & Snyder, T. (2013). *Puzzlecraft: The ultimate guide on how to construct every kind of puzzle.* New York: Puzzlewright Press.

Shortz, W., & Huang, W.-H. (2012). *Will Shortz's puzzle master workout.* New York: Seven Footer Press.

Thanasuan, K., & Mueller, S. T. (2014). Crossword expertise as recognitional decision making: an artificial intelligence approach. *Frontiers in Psychology 5,* 1018.

Thanasuan, K., & Mueller, S. T. (2016). Investigating and simulating the effect of word fragments as orthographic clues in crossword solutions. In D. Reitter & F. E. Ritter (Eds), *Proceedings of the 14th International Conference on Cognitive Modeling (ICCM 2016).* University Park, PA: Penn State.

Toma, M., Halpern, D. F., & Berger, D. E. (2014). Cognitive abilities of elite nationally ranked SCRABBLE and crossword experts. *Applied Cognitive Psychology 28,* 727–737.

Tuffiash, M., Roring, R. W., & Ericsson, K. A. (2007). Expert performance in SCRABBLE: Implications for the study of the structure and acquisition of complex skills. *Journal of Experimental Psychology: Applied 13*(3), 124–134.

Underwood, G., MacKeith, J., & Everatt, J. (1988). Individual difference in reading skill and lexical memory: The case of the crossword puzzle expert. In M. M. Gruneberg, P. E. Morris, & R. N. Sykes (Eds), *Practical aspects of memory: Current research and issues 2.* Chichester, UK: Wiley.

Underwood, G., Deihim, C., & Batt, V. (1994). Expert performance in solving word puzzles: From retrieval cues to crossword clues. *Applied Cognitive Psychology 8*(6), 531–548.

van Hees, S., Pexman, P. M., Hargreaves, I. S., . . . Protzner, A. B. (2016). Testing the limits of skill transfer for Scrabble experts in behavior and brain. *Frontiers in Human Neuroscience 10.*

van Hees, S., Seyffarth, S., Pexman, P. M., Cortese, F., & Protzner, A. B. (2017). An ERP investigation of vertical reading fluency in Scrabble experts. *Brain Research 1667,* 1–10.

Wachtell, D. (Aug 18, 2004). Word up: You don't have to know English to play Scrabble. Dispatches: Notes from Different Corners of the World. *Slate.* Available at http://www.

slate.com/articles/news_and_politics/dispatches/features/2004/word_up/you_dont_have_to_know_english_to_play_scrabble.html

Wai, J. (2014). Experts are born, then made: Combining prospective and retrospective longitudinal data shows that cognitive ability matters. *Intelligence 45*, 74–80.

Wright, E. V. (1939). *Gadbsy: 50,000 word novel without the letter "E."* Los Angeles: Wetzel. Available from https://archive.org/details/Gadsby

Zimmer, B. (2013). *Anointing the crossword and palindrome champions. Word routes: Exploring the pathways of our lexicon.* Available from https://www.visualthesaurus.com/cm/wordroutes/anointing-the-crossword-and-palindrome-champions/

Zimmer, B. (2015, Aug). A non-French speaker won the French language Scrabble championship. How is that possible? Lexicon Valley, Slate. Available from http://www.slate.com/blogs/lexicon_valley/2015/08/05/international_scrabble_the_size_of_the_dictionary_depends_on_how_words_get.html

CHAPTER 25

MUSICAL EXPERTISE

JENNIFER MISHRA

Introduction

IN the field of musical expertise, there is a dichotomy of thinking. On the one hand, there is a widespread perception in the general population that expert musicians have innate talent, or giftedness,[1] beyond ordinary abilities (e.g., Davis, 1994; Gagne, 1999; Scripp, Ulibarri, & Flax, 2013). Talent, as part of the vernacular in the field of music, is usually assumed to be a stable trait—one is either born with musical talent or not. Music aptitude tests popular in the early to mid-twentieth century, such as the *Seashore Tests of Musical Talent* (Seashore, Lewis, & Saetveit, 1919) and the *Music Aptitude Profile* (Gordon, 1965), attempted to find children who had this musical talent (for reviews, see Grashel, 2008; Humphreys, 1998). On the other hand, there is a very real feeling that ability in music comes from a disciplined work ethic. It would be unacceptable, even for those considered talented, not to practice. In fact, those who are considered talented are expected to practice all the more (Haroutounian, 2002; Scripp et al., 2013).

The trend coming into the twenty-first century has been to favor work ethic over talent (Mosing & Ullen, 2016), though by no means has the idea of musical talent been abandoned (Gagne, 2013). The mission statement of the National Association for Music Education (NAfME), America's primary music education organization, promotes music making by all children—not just for the talented few (NAfME, 2017). Popular musical pedagogies, such as the one developed by Shinichi Suzuki, are based on the premise that all children can learn to perform music at a high level, just as all children can learn to speak their native language[2] (Suzuki, 1996). Cognitive and neurological research supports musical ability as a fundamental human trait (e.g., McPherson & Hallam, 2009; Scripp et al., 2013; Sloboda, 1996) that can be developed

---

[1] For the purposes of this chapter, *talented* will be used synonymously with *gifted*, though there are those who argue for a differentiation with talent being developed from innate gifts (Gagne, 2013).

[2] The Suzuki pedagogy is termed *talent education*, which may be somewhat confusing in this context.

in all but a small minority (1.5 percent) of the population who possess the rare neurological condition *amusia* (Peretz, 2016). Even infants show high levels of musical ability (e.g., Hallam, 2015; Sloboda, 1996; Trehub, 2001). The ability to hear music, perform music (even if in informal settings like the shower or while driving), and make musical judgments (e.g., same/different, recognition of errors) is common in the general population.

The theory of expert performance as defined by Ericsson and colleagues (Ericsson, Krampe, and Tesch-Römer, 1993) is attractive in the current philosophical climate, favoring acquired over innate musical ability. In this theory, deliberate practice is the central requirement to achieving performance skill in music, putting it firmly within the control of the musician. Musicians who attentively work at musical production (e.g., performing on an instrument, singing, composition) can become experts, regardless of whether they are considered *talented*, a compelling idea that provides hope and motivation for aspiring musicians.

This is a democratic view of ability—all are created equal. But casual observation brings this view into question. Some children do indeed seem more adept at picking up musical skills than others. There is also the question of physical limitations, though there are notable examples of performers overcoming physical disadvantages (e.g., Evelyn Glennie, world renowned deaf percussionist).

Researchers continue to discuss the relative merits of nature versus nurture in the development of musical expertise (Hallam, 2015; McPherson & Williamon, 2015). However, recently, the field of music has begun exploring the position that deliberate practice is necessary, but not sufficient, to create musical experts (Campitelli & Gobet, 2011).

This chapter will begin by describing the foundational role music has played in the development of the theory of expert performance. The question of what constitutes a *musical expert* will then be considered, looking both at how expertise has been defined and how it potentially could be defined in future research. A discussion of deliberate practice follows, including how this is operationally defined in musical research. The chapter will end with a consideration of other variables, such as genetics, that could interact with deliberate practice to develop expertise in the musical domain.

MUSIC'S ROLE IN THE DEVELOPMENT OF THE THEORY OF EXPERTISE

Music is one of the foundational domains used to develop the theory of expert performance. In the seminal study by Ericsson et al. (1993), violinists of varying (perceived) performance ability were asked to estimate amounts of deliberate practice since first starting to play their instrument, and to track current practicing patterns. From these estimates of practice, the authors argued for a direct connection between expertise in music and accumulated practice time over approximately 10 years of study.[3] These results form part of the underlying support for the popular "10-year, 10,000 hour" rule, widely reported in the press, for achieving expertise.[4] Further, Ericsson et al. (1993) argue that the motor advantages (coordination) of expert pianists were domain specific rather than general. This domain specificity is central to the theory of expertise development. From the evidence collected for the 1993 paper, the authors began to seek connections with other domains to develop a widespread theory of expert performance, and music remains one of the most studied domains in the field of expertise (Bilalić, Langner, Campitelli, Turella & Grodd, 2015; Hambrick et al., 2014).

Several elements make music well suited for studying expertise. Musical ability can be psychometrically (e.g., music aptitude) and physically (e.g., keystroke response time) measured. Music has a system of intensive training that can be measured, and levels of mastery that can be operationally defined (e.g., acceptance into a music school, progression to more difficult repertoire). Children begin the supervised study of music at a young age, and can be tracked through the hypothesized 10 years required to achieve expert levels. However, age and expertise are not necessarily as closely linked in music as they are in domains such as sports, where physical development limits practice and achievement (Ericsson et al., 1993).

Defining Music Experts

Colloquially, there is a broad view of what constitutes an expert in the area of musical performance. Expertise can be equated with everything from good, or competent musicianship, to virtuosity, to household fame. In music, there is no agreed-upon ranking or hierarchical system for defining an expert, like there is in chess or some sports, or a universal barrier beyond which a musician is considered an expert.

Operational Definition of Musical Expertise

Musical performance expertise is generally considered to develop along a continuum. Researchers often include study participants who have varying levels of expertise rather

---

[3] Because of the research design, correlations were not computed, but a significant difference in the amount of reported practice time between the two higher ability groups and the lower ability group was found. Ericsson and colleagues argued for a correlation stating: "Our framework predicts a monotonic relation between the current level of performance and the accumulated amount of deliberate practice for individuals attaining expert performance" (p. 378).

[4] Malcolm Gladwell popularized this rule in his 2008 book *Outliers*, based on research by Ericsson et al. (1993) and Chase and Simon (1973). Recently, this rule has been called into question and is at most now considered a guideline.

than a group with no performing experience at all. The most common ways of assessing musical expertise include amount of formal instruction (i.e., years of formal private music lessons), group membership (e.g., music major; Ollen, 2006), or some form of performance evaluation (e.g., an audition).

In 2007, Ericsson and colleagues defined a musical expert as someone who performs consistently above the level of his or her peers, achieves concrete results (e.g., winning auditions, competitions), and has the ability to replicate superior performance in a laboratory setting (Ericsson, Prietrula, & Cokely, 2007). This definition appears to apply to the undisputed musical experts (i.e., those who perform at a professional level), but it is difficult to translate into a developmental definition of musical performance expertise.

Because the field of musical expertise research has not adopted a consistent definition or hierarchical model of expertise, operational definitions in the research can be somewhat inconsistent. For instance, a graduate student in music might be considered an expert in one study (Kopiez & Lee, 2008) while in another he or she would be considered a novice (Repp, 1997). Similarly, a musician who has as many as 15 years of formal study may be termed a musical *amateur* because they are not pursuing a music degree at a university (Krampe & Ericsson, 1996). Researchers, and those who interpret research, must be cognizant of the inconsistency in which the term *musical expert* has been used, thus limiting the ability to compare findings from one study with another.

One solution would be to adopt a definition or hierarchical system developed for another domain. Hoffman (1998) proposed a hierarchy based on medieval guilds with levels from *Naivete* ("One who is totally ignorant of a domain") to *Master* ("expert who is also qualified to teach at the lower level... [and] regarded by the other experts as being 'the' expert"). Considering Hoffman's hierarchy alongside how musical expertise has been operationally defined in the literature (as well as colloquially), a musical performance expertise hierarchy may include levels such as: *student musician* and *developing expert* along the expertise continuum. A student musician has formal instruction on an instrument and is focused on learning basic technical skills (i.e., instrumental control) from a teacher. This level is distinguished from a *non-performer*,[5] a person who has not intensively studied an instrument. A developing expert has demonstrated their skills through some sort of audition or public performance, but continues to work on technical concerns under the supervision of a teacher (e.g., a music major at a university). An expert publicly performs concert repertoire and can prepare this repertoire with little or no supervision from a teacher (i.e., shows musical independence). This is a preliminary proposal of what a hierarchy of musical expertise might look like. Future consideration is needed to develop a consistent definition of musical expertise and a hierarchy that makes sense in the field of music.

---

[5] Though the term "non-musician" is used frequently in the research literature in this way, though as discussed earlier in the chapter humans are, by nature, musical. I prefer the term "non-performer."

Table 25.1 Proposed musical performance expertise hierarchy

Non-performer	Someone who has not systematically studied an instrument (including voice)
Student musician	A person who has systematically studied or is currently studying an instrument and is focused on learning basic technical skills (i.e., instrumental or vocal control) from a teacher.
Developing expert	A musician who has demonstrated their skills through some sort of audition or public performance, but continues to work on technical concerns under the supervision of a teacher (e.g., a music major at a university).
Expert musician	A musician who publicly performs concert repertoire and can prepare this repertoire with little or no supervision from a teacher (i.e., shows musical independence).
Influential expert	A high-achieving musical expert who sets the standard for the field and influences other experts.

Proposed in Ericsson et al. (1993) is the idea of *eminent performer*, one that "goes beyond expert mastery of available knowledge and requires an important and innovative contribution to the domain" (p. 370). This type of expertise changes the domain. When considering creativity, Csikszentmihalyi (1996) distinguished *creativity* (with a lower-case "c") and *Creativity* (with an upper-case "C"), with the former referencing everyday creativity and the latter changing the field. Using this model, experts might be individuals who have abilities beyond the general population and Experts could be those who achieve high status as professional musicians, or musicians who set the standard for the field (e.g., members of the New York Philharmonic, Yo-Yo Ma). Another way to conceptualize this idea is to term outstanding musical experts who change or teach the field as *influential experts* (see Table 25.1).

Regardless of the definitions or hierarchy adopted by the field of music performance research, a greater level of consistency is necessary when defining musical expertise. The definitions should be broad enough to encompass both the performance tradition that predominates in Western classical music conservatory tradition and the other musical traditions and other types of musical expertise.

Other Musical Experts

Research in musical expertise has been generally limited to musicians functioning within the Western classical tradition. Further, the research has focused almost exclusively on performers within this tradition. However, there exist musical experts who do not perform. McPherson and Hallam (2009) list eight distinct types of musical talents that can be developed through practice, only one of which is performing: improvising, composing, arranging, analyzing, performing, appraising, conducting, and music teaching. Though some underlying skills may exist across these types of musical expertise, they are distinct. As Gjerdingen (2003) colorfully puts it:

> Many researchers in the psychology of music have been troubled by the facile equation of the degree of musical comprehension with the number of years of music lessons. By that measure, an elderly fan of symphonic music who had never taken lessons but listened to 8 hours of classical radio daily for 50 years would be a classical-music "novice" whereas a 12-year-old student with 5 years of piano lessons and a passion for the music of Britney Spears and Christina Aguilera would be a classical-music "expert." (p. 491)

The focus on performance may be understandable. The general population appears to equate musical ability with performance ability (Hallam & Prince, 2003), and performance generally underlies the Western conservatory tradition. Instruction in musical skills such as conducting and arranging generally begin later for performers already somewhat advanced in their studies. In other words, performance is a gateway skill for many of the other skills. However, it is erroneous to assume that musical expertise *equals* musical performance.

Further, music performance itself constitutes different skills (e.g., sight reading, improvisation[6]), and musicians vary in their skill levels. Within a group of expert musical performers, some will be expert sight readers and some will be expert improvisers, but not necessarily both. Some musicians are expected to sight read or improvise, while others are not (Lehmann & Davidson, 2002). Solitary practice, termed *deliberate practice*, does not necessarily enhance specific musical skills such as sight reading (Kopiez & Lee, 2006, 2008; Lehmann & Ericsson, 1996). While Ericsson defines expert performance as "typically displayed in solo performance of a piece that the musician has extensively studied beforehand" (p. 287), sight reading, the initial playing of a piece from notation before practice begins, is often used to assess performance ability (see Platz, Kopiez, Lehmann, & Wolf, 2014).

Expertise is also influenced by other factors such as genre. Western classical music has been the standard for research in the field of musical expertise, but musicians outside this tradition often learn music in a very different setting. For instance, it is possible for a non-classical musician (e.g., a fiddler or a pop musician) to have few, if any, years of *formal instruction*, i.e., one-on-one training with a private instructor. They may instead have a great deal of informal training that comes from self-instruction or enculturation. A non-classical musician might be unlikely to pass an audition into a music school if the requirement is to perform in the genre of classical music—just as a classical performer would likely be unable to perform non-classical music at an expert level without specialized training. There is some evidence that musicians across genres may share views on practice (Hallam & Shaw, 2002) and report a similar number of practice hours per week (Creech et al., 2008; Gruber, Degner, & Lehmann, 2004). Musicians, though, may differ on their perceptions of the relevance of other musical skills or activities (e.g., importance of playing music by ear, engaging in music for fun, practicing with others; Papageorgi, 2014).

---

[6] Sight reading is the ability to perform from musical notation without previously practicing the music. Improvisation is the creation of music during performance.

Deliberate Practice

Research into the development of expertise fits squarely into the domain of music education, which is concerned with increasing the musical abilities of children as they move along a continuum towards becoming musical experts. By and large, this research has centered on practicing.

Defining Deliberate Practice

In Ericsson et al. (1993), deliberate practice is distinguished from other types of musical activities and is defined as: (a) structured, (b) having the primary purpose of improving skills, and (c) requiring effort and attention. Instruction and feedback also play a role. Furthermore, deliberate practice is not expected to be enjoyable or immediately rewarding. Ericsson and Lehmann (1999) defined deliberate practice as follows:

> a structured activity, often designed by teachers or coaches with the explicit goal of increasing an individual's current level of performance. In contrast to work and play, it requires the generation of specific goals for improvement and the monitoring of various aspects of performance. Furthermore, deliberate practice involves trying to exceed one's previous limit, which requires full concentration and effort. (p. 695)

Sloboda's definition (1996) is altogether simpler, referring to deliberate practice as effortful and structured.

The theory of expertise development encompasses both quality and quantity, and estimates that 10,000 hours of quality (deliberate) practice is needed to achieve expertise. Many researchers have shown that the amount of deliberate practice in music directly relates to levels of achievement (Ericsson, 2006; Ericsson et al., 1993; Jabusch, Alpers, Kopiez, Vauth, & Altenmuller, 2009; Jørgensen, 2002; O'Neill, 1999; Platz, Kopiez, Lehmann, & Wolf, 2014; Ruthsatz, Detterman, Griscom, & Cirullo, 2008; Sloboda, Davidson, Howe, & Moore, 1996) though others have found that the quality of practice is more important (Miksza, 2007; Williamon & Valentine, 2000). The quality of practice is sometimes forgotten in music, as the number of hours is concrete while quality is more difficult to characterize.

The Williamon and Valentine (2000) study is an example of how research in the area of musical performance expertise is often conducted. In this study, performers were recommended by private music teachers and grouped by performance ability using the Associated Board of the Royal Schools of Music (ABRSM[7]). Performers were given a new piece of music to practice and then perform. Williamon and Valentine used

---

[7] The ABRSM has a well-developed system of grading musicians based on yearly performance evaluations; however, it is not commonly used in the United States, so this ranking system is not universal.

published music, but other researchers have music especially composed for the research study. It is essential that the music be novel for all participants and sufficiently challenging for the performers. Performers were allowed to practice the music in any way they wish which is common unless a particular practice treatment is being tested. The performers were also allowed unlimited practice time though practice time may be restricted. Williamon and Valentine recorded all practice sessions and analyzed these both qualitatively (through interviews with performers) and quantitatively (e.g., measuring practice time and frequency). A mixed-methodology approach is common in the field, though researchers differ in what variables are measured and how the recordings are qualitatively analyzed. Finally, performances were recorded and scored by judges using a researcher-developed rating scale.

Solitary Practice

A number of researchers have focused on conceptually defining deliberate practice, but less attention has been given to creating an operational definition—exactly what deliberate practice looks like in a musical setting. Most researchers have relied on Ericsson's original operational definition of deliberate practice, i.e., solitary practice.

After tracking expert violinists' musical and non-musical activities throughout the week, Ericsson and colleagues (1993) quickly focused their attention on how much time these musicians reported practicing alone, an activity reported by the violinists as requiring a high level of effort and as being highly relevant to achieving musical expertise. At the same time, other musical activities (taking lessons, solo or group performance, and practicing with others) were also reported as effortful and highly relevant. However, the researchers dismissed these activities, primarily because the performer was not considered to be solely in control of the activity. It is unclear why this mattered. Perhaps because Ericsson and colleagues did not consider activities other than solitary practice, performance activities other than practicing alone have been largely marginalized or ignored in the discussion of deliberate practice. Based on this foundational research, these activities could be as important as solitary practice to the development of musical experts (Moore, Burland, & Davidson, 2003).

Even restricting the definition of deliberate practice to solitary practice, musicians still understand that practicing alone does not necessarily equate to the attentive, goal-directed practice that constitutes deliberate practice (Lehmann & Davidson, 2002), which is much more difficult to measure. Hypothetically, two musicians could spend similar amounts of time practicing alone, but one may practice in an effective and efficient way while the other could have a more superficial attitude towards practice. These two individuals would not be expected to progress towards musical expertise in a similar manner. Ericsson's (2014) quote about athletic expertise applies equally to music:

> it is not the total number of hours of practice that matters, but a particular type of practice that predicts the difference between elite and sub-elite athletes. (p. 94)

Hours are easy to track, but what goes on inside a musician's mind when practicing is not.

Effective Musical Practice

In the development of the theory of musical expertise, deliberate practice in music has been addressed somewhat superficially. In reality, solitary practice is not, in itself, a deliberate practice strategy. Musicians have at their disposal dozens of practice strategies when they practice alone, some more effective, effortful, and deliberate than others. For instance, musicians may vary tempos (slow practice), target specific errors, isolate musical elements (e.g., pitch, rhythm), or play one part or hand in isolation. There are even non-playing practice strategies, including listening to a recording of the piece being practiced, singing, or studying a full score. Unfortunately, mindless repetition is one of the most common, though inefficient, strategies used by novice musicians (Austin & Berg, 2006; Barry & Hallam, 2002; Carter & Grahn, 2016; Renwick & McPherson, 2000).

A number of researchers have investigated the relative effectiveness of various musical practice strategies (e.g., Hallam et al., 2012; Miksza, 2007; Miksza & Tan, 2015; Zhukov, 2009), and have found that musicians vary in their ability to use effective practice strategies, to self-regulate learning, and to define practice goals (Byo & Cassidy, 2008; Hatfield, 2016; Jørgensen, 2004, 2008; McPherson & Renwick, 2001). The use of more effective practice strategies appears to be related to expertise (Barry & Hallam, 2002) with musical experts reporting the use of a wide variety of effective practice strategies (Hallam, 1995; Hallam et al. 2012; McPherson, Davidson, & Faulkner, 2012). There are individual preferences amongst the experts when it comes to practice strategies, and the type and scope of the strategy may vary, depending on the music being practiced (Mishra & Fast, 2015).

While there may be differences between expert musicians, there are also patterns. Hallam (1995) interviewed freelance musicians about their practice routines. There was considerable variation in responses, but there were also some similarities that allowed Hallam to group the musicians as either analytic holists or intuitive serialists. Analytic holists listened to recordings, cognitively analyzed the structure of the music, and sought underlying meanings and connections between musical elements. The intuitive serialists rejected cognitive analysis and resisted external influences including listening to recordings, letting their understanding of the piece emerge naturally through the playing of the piece. When learning technically challenging music, the musicians either used a repetitious approach, playing slowly with a metronome and systematically increasing the tempo, or an analytic approach, changing rhythms or other musical elements.

Case studies have been key in determining which practice strategies are effective for experts (e.g., Ginsborg & Chaffin, 2010). To conduct this type of research, undisputed experts (such as members of a professional symphony orchestra) are interviewed about their practice strategies (e.g., Mishra & Fast, 2015) and their practice may be recorded for later qualitative and/or quantitative analysis (e.g., Chaffin & Imreh, 2001). The case-study approach allows for depth, but assumes that individual experts are rare and unique entities. However, expert musicians are not rare; thousands of indisputably top musical experts exist. Many conclusions concerning effective practice based on case studies need verification with a broader selection of musical experts.

What constitutes effective practice strategies that could be termed *deliberate* remains a central research question (e.g., Hallam et al., 2012; Miksza, 2007; Zhukov, 2009). Some researchers continue to study expert performers, under the assumption that strategies used by the experts are, by definition, effective. Others focus on novice musicians' practice behaviors, and how teachers direct student practice. Research at this stage is for the most part foundational: investigating novice musicians' reported and actual use of various practice strategies (e.g., Byo & Cassidy, 2008; Miksza, 2007) or surveying teachers about the perceived effectiveness of different practice strategies (e.g., Barry, 2007). Though the musical profession as a whole agrees that effective practice strategies lead to expert performance abilities, the research has not yet established a cause-and-effect link over time.

10,000 Hours

The 10,000 hours estimated by Ericsson et al. (1993) to achieve performance expertise has been widely accepted by the field of music. However, there is variability in the amount of practice expected across instruments (Jørgensen, 1997, 2002). Pianists practice more than string players, who practice more than wind players. This may be down to the technical aspects of the instruments or the relative difficulty of the repertoire. On this point alone, 10,000 hours can only be taken as a guide; reality is much more complex. In fact, in the field of music there is no agreed-upon standard of musical practice (Jørgensen, 2004; Kostka, 2002).

The 10,000-hour goal is based largely on estimates of long-term practice and how these relate to some form of musical achievement. To conduct this type of research, musicians of varying abilities are asked to retroactively estimate their lifetime practice sometimes with the help of prompts that help musicians remember various musical activities (in the form of either a questionnaire or interview questions). These data can be supplemented and verified with practice diaries kept during the course of the research study. Using self-reports of practice time is difficult because it relies on the musicians' honesty.

Enjoyment and Motivation

Perhaps a minor point, but one that may be related to motivation and retention is that deliberate practice, as described by Ericsson et al. (1993), is inherently unenjoyable. There is no explanation given in Ericsson et al.'s 1993 article to clarify on what this assumption is based. Ericsson's own survey does not indicate exceptionally low levels of pleasure for the effortful and relevant musical activities. It is perplexing to understand why musicians might spend as much as three hours a day for ten years doing something that they do not find enjoyable. Indeed, flow theory (Csikszentmihalyi, 1990) would indicate that musicians pursuing an attentive, goal-direct activity, which they can control, and receiving feedback (all part of the original description of deliberate practice and also required for flow state) would likely enjoy the activity. In fact, one could argue that musicians who are not encountering flow during practice might not be fully engaged in deliberate practice.

Csikszentmihalyi and colleagues (Csikszentmihalyi, Rathunde, & Whalen, 1996) have discussed the connection between flow and talent development in teenagers, and the connection between enjoyment and musical practice has some support from research. Butkovic, Ullen, and Mosing (2015) found that music-specific flow was the strongest predictor of practice, and other researchers have supported the positive connection between flow experiences (or another measure of enjoyment) and practice (Ginsborg, 2014; Miksza & Tan, 2015; O'Neill, 1999). Understanding factors, such as motivation, that underly deliberate practice may be as important as understanding the role that deliberate practice itself plays in developing experts.

This role of pleasure in deliberate practice has not received a great deal of direct attention in music. The theory of flow is accepted alongside deliberate practice and while I would argue that these theories are not incompatible, there has been little written on the link between these two theories.

Differences between Experts and Novices

Central to the theory of expertise development is that experts in a domain are expected to process and remember stimuli from their domain of expertise differently than novices, but are not expected to have a global increase in processing or memory. As with other domains, musical experts have been found to think and behave differently from musical novices. Some researchers have documented the differences in aural processing and abilities (e.g., Glushko, Steinhauer, DePriest, & Koelsch, 2016; Pallesen et al., 2010); others have focused on memory or cognitive processing (e.g., Sloboda 1976; Strait, Kraus, Parbery-Clark, & Ashley, 2010; Zuk, Benjamin, Kenyon, & Gaab, 2014). As with chess players, expert musicians appear to remember music better, but only for expected musical patterns. Memory is not necessarily better for a random series of pitches (Sala & Gobet, 2017; Schulze, Dowling, & Tillman, 2011). More recent research has focused on neuroprocessing and anatomical differences related to musical experience (e.g., Brown, Zatorre, & Penhune, 2015; Kraus & Chandrasekaran, 2010; Moreno & Bidelman, 2014; Schlaug, 2009; Scripp et al., 2013). For some skills, expert brains show increased activation in intensity, or more or even different areas of the brain are activated (e.g., Wong & Gauthier, 2010). For other skills, expert brains show a systematic deactivation (e.g., Berkowitz & Ansari, 2010; Limb and Braun, 2008).

The Role of Genetics

Recent research has focused on the question of whether deliberate practice is, in and of itself, sufficient to develop musical expertise. In the theory put forth by Ericsson et al. (1993), deliberate practice is the key. With enough motivation and attention, anyone can become an expert. However, casual observation does not necessarily support this; most people who want to become experts do not (Sternberg, 1996) and researchers have not always found the expected correlations between accumulated practice and ability (Hallam, 1998, 2013; Hallam et al., 2012; Hambrick et al., 2014). Some people seem to have to work harder to attain the same level of musical performance as others (Hambrick et al., 2014; Sloboda et al., 1996), and there appears to be a limit or plateau that is reached by some musicians, which no amount of practice seems able to overcome. In his differentiated model of giftedness and talent, Gagne (1999, 2009, 2013) attributes these observations to genetic *giftedness*.

While prodigies and musical savants may appear to be, on the face of it, evidence for the role of genetics in expertise development, this is not necessarily the case. Prodigies, however early they emerge, do not *spontaneously* emerge. There is practice and hard work behind their skills. Some children begin studying music at the age of three and devote many hours to learning. These types of children often demonstrate a *rage to master*, an almost obsessive desire to learn the skills of the domain (Winner, 2014). Practice is involved in their development and continues to be a central component (Hallam, 2015; McPherson & Hallam, 2009). Some have argued that even prodigies such as Mozart required 10 years to become proficient and may have practiced for 10,000 hours (Hayes, 1981; Schonberg, 1970; Scripp et al., 2013). Savants may have a genetic predisposition for certain characteristics that enhance learning in a domain such as music, but without practice, these skills will not necessarily develop into expertise (Hallam, 2015; Heaton, 2009).

One confounding element mentioned by Ericsson et al. in their 1993 article, and expanded upon by Malcolm Gladwell in his book *Outliers* (2008), is that not all children receive similar resources in developing expertise in a particular domain. In essence, some children receive more encouragement than others, sometimes because they are perceived to be *talented* by a parent or teacher (whether or not they actually are more precocious than other children) (McPherson, 2009; McPherson & Davidson, 2002), sometimes because of other external factors (Gladwell, 2008; Scripp et al., 2013). It becomes unclear whether the children are truly more talented than their peers or advantaged because of the additional instruction received. This cycle was described by Bloom (1985)[8] and still holds true in the domain of music (Gembris & Davidson, 2002; Howe, Davidson, Moore, & Sloboda, 1995; Kemp & Mills, 2002). This makes it difficult to separate perceived talent from practice. Further, while there are children who appear

---

[8] A musical case study was presented by Sosniak in Bloom's book (Sosniak, 1985).

to develop faster, this is not always an indication of future abilities (Howe et al., 1995; Howe, 1999).

Recent studies, some using twins, have attempted to tease out the relative effects of environment and genetics. A number have found that both contribute to musical expertise (Coon & Carey, 1989; Hambrick & Tucker-Drob, 2015 Mosing & Ullen, 2016; Mosing, Madison, Pedersen, Kuja-Halkola, & Ullen, 2014; Ruthsatz et al., 2008; Tan, McPherson, Peretz, Berkovic, & Wilson, 2014; Vinkhuyzen, van der Sluis, Posthuma, & Boomsma, 2009). One idea emerging from this line of research is that there may be a genetic predisposition to practice (Butkovic et al., 2015; Hambrick & Tucker-Drob, 2015; Mosing et al., 2014) driven by factors such as personality and motivation (Burland & Davidson, 2002; Kemp & Mills, 2002).

Further, there is reason to expect that genetics and practice are not the sole contributors to musical expertise development. A number of meta-analyses have found that deliberate practice explains some, but not all, of the variance in level of performance (Hambrick et al., 2014; Macnamera, Hambrick & Oswald, 2014; Meinz & Hambrick, 2010; Mosing et al., 2014; Platz et al., 2014). Genetics may have a role, as do intelligence and working memory (Hambrick & Meinz, 2012; Mosing & Ullen, 2016), but there are other factors. Hambrick et al. (2014) proposed a few characteristics that might be important in the development of expertise, including the age that instruction began, working memory capacity, personality (especially elements that impact motivation), genetics, and IQ. Results vary on how important deliberate practice is, but this research suggests the presence of other still as yet unidentified factors at work in the development of expertise.

Conclusion

Music has played a pivotal role in the development of the theory of expert performance with much of the foundational research conducted with musical experts (Ericsson et al., 1993). This chapter explored the role music played in the development of the theory and issues arising from the original research that have not yet been fully addressed (e.g., forms of deliberate practice other than solitary practice). The definitions of musical expertise and deliberate practice in music were then explored to better understand the research that has been conducted in the field and highlight remaining questions. The chapter ended with a consideration of other variables, such as genetics, that could interact with deliberate practice to develop expertise in the musical domain.

Current thinking is that deliberate practice is important, but insufficient to explain expertise. Researchers are currently exploring the effects of genetics, revisiting the nature versus nurture debate. However, the development of musical experts may be more complex than either the amount of deliberate practice or genetics can explain. While researchers appear to have taken up polarizing positions, the development of musical expertise is likely much more complex than either position can explain. Rather than expertise resulting from either genetics *or* deliberate practice, it is likely both play a role. Future researchers should continue to explore the interaction between genetics and practice, but also be open to additional factors (e.g., motivational, educational) that may impact expertise development.

As mentioned earlier in this chapter, there is a need for a consistent definition of *musical expert* specifically addressing the developmental assumptions that permeates the literature. Currently, it is difficult to compare the results of research studies in which musical experts are differently defined. The hierarchy proposed in this chapter (see Table 25.1) may serve as a starting point, but additional attention must be given to this important issue.

Research conducted to date on musical expertise is almost exclusively focused on Western classical music performers. This provides a foundation for future research with performers in non-Western classical traditions and non-performance-based experts. Future research needs to broaden the definition of musical expert to include experts from traditions other than Western classical performance, since it is possible that some assumptions underlying deliberate practice do not hold for such musicians. However, finding similarities across genres and in the non-performance domains would strengthen the theory of expertise development.

REFERENCES

Austin, J. R., & Berg, M. H. (2006). Exploring music practice among sixth-grade band and orchestra students. *Psychology of Music* 34(4), 535–558.

Barry, N. H. (2007). A qualitative study of applied music lessons and subsequent student practice sessions. *Contributions to Music Education* 34, 51–65.

Barry, N. H., & Hallam, S. (2002). Practice. In R. Parncutt & G. E. McPherson (Eds), *The science and psychology of music performance* (pp. 151–165). Oxford: Oxford University Press.

Berkowitz, A. L., & Ansari, D. (2010). Expertise-related deactivation of the right temporoparietal junction during musical improvisation. *Neuroimage* 49, 712–719.

Bilalić, M., Langner, R., Campitelli, G., Turella, L., & Grodd, W. (2015). Editorial: Neural implementation of expertise. *Frontiers in Psychology* 9, 545.

Bloom, (1985). *Developing talent in young people*. New York: Ballantine Books.

Brown, R. M., Zatorre, R. J., & Penhune, V. B. (2015). Expert music performance: Cognitive, neural, and developmental bases. *Progress in Brain Research* 217, 57–86.

Burland, K., & Davidson, J. W. (2002). Training the talented. *Music Education Research* 4, 121–140.

Butkovic, A., Ullen, F., & Mosing, M. A. (2015). Personality related traits as predictors of music practice: Underlying environmental and genetic influences. *Personality and Individual Differences* 74, 133–138.

Byo, J. L., & Cassidy, J. W. (2008). An exploratory study of time use in the practice of music majors: Self-report and observation analysis. *Update: Applications of Research in Music Education* 27(1), 33–40.

Campitelli, G., & Gobet, F. (2011). Deliberate practice: Necessary but not sufficient. *Current Directions in Psychological Science 20*, 280–285.

Carter, C. E., & Grahn, J. A. (2016). Optimizing music learning: Exploring how blocked and interleaved practice schedules affect advanced performance. *Frontiers in Psychology 7*, 1251.

Chaffin, R., & Imreh, G. (2001). A comparison of practice and self-report as sources of information about the goals of expert practice. *Psychology of Music 29*, 39–69.

Chase, W. G., & Simon, H. A. (1973). The mind's eye in chess. In W. G. Chase (Ed.), *Visual information processing* (pp. 215–281). San Diego, CA: Academic Press.

Coon, H., & Carey, G. (1989). Genetic and environmental determinants of musical ability in twins. *Behavior Genetics 19*, 183–193.

Creech, A., Papageorgi, I., Duffy, C., Morton, F., Hadden, E., Potter, J., . . . & Welch, G. (2008). Investigating musical performance: Commonality and diversity amongst classical and non-classical musicians. *Music Education Research 10*(2), 215–234.

Csikszentmihalyi, M. (1990). *Flow: The psychology of optimal experience*. New York: Harper & Row.

Csikszentmihalyi, M. (1996). *Creativity: Flow and the psychology of discovery and invention*. New York: Harper Perennial.

Csikszentmihalyi, M., Rathunde, K., & Whalen, S. (1996). *Talent teenagers: The roots of success and failure*. Cambridge, UK.: Cambridge University Press.

Davis, M. (1994). Folk music psychology. *The Psychologist 7*(12), 537.

Ericsson, K. A. (2006). The influence of experience and deliberate practice on the development of superior expert performance. In K. A. Ericsson, N. Charness, P. Feltovich, & R. R. Hoffman (Eds.), *The Cambridge handbook of expertise and expert performance* (pp. 685–706). Cambridge, UK: Cambridge University Press.

Ericsson, K. A. (2014). Why expert performance is special and cannot be extrapolated from studies of performance in the general population: A response to criticisms. *Intelligence 45*, 81–103.

Ericsson, K. A., Krampe, R. T., & Tesch-Romer, C. (1993). The role of deliberate practice in the acquisition of expert performance. *Psychological Review 100*(3), 363–406.

Ericsson, K. A., & Lehmann, A. C. (1999). Expertise. In M. A. Runco & S. R. Pritzker (Eds), *Encyclopedia of creativity* (vol. 1, pp. 695–707). San Diego, CA: Academic Press.

Ericsson, K. A., Prietrula, M. J., & Cokely, E. T. (2007). The making of an expert. *Harvard Business Review 85*(7–8), 114–121.

Gagne, F. (1999). Nature or nurture? A re-examination of Sloboda and Howe's (1991) interview study on talent development in music. *Psychology of Music 27*, 38–51.

Gagne, F. (2009). Building gifts into talents: Detailed overview of the DMGT 2.0. In B. MacFarlane & T. Stambaugh (Eds), *Leading change in gifted education: The festschrift of Dr. Joyce VanTassel-Baska* (pp. 61–80). Waco, TX: Prufrock Press.

Gagne, F. (2013). The DMGT: Changes within, beneath, and beyond. *Talent Development & Excellence 5*, 5–19.

Gembris, H., & Davidson, J. W. (2002). Environmental influences. In R. Parncutt & G. E. McPherson (Eds), *The science and psychology of music performance* (pp. 17–30). New York: Oxford University Press.

Ginsborg, J. (2014). Focus, effort, and enjoyment in chamber music: Rehearsal strategies of successful and "failed" student ensembles. *International Symposium on Performance Science*. Retrieved from https://www.researchgate.net/publication/239590237_Focus_effort_

and_enjoyment_in_chamber_music_Rehearsal_strategies_of_successful_and_failed_student_ensembles

Ginsborg, J., & Chaffin, R. (2010). Preparation and spontaneity in performance: A singer's thoughts while singing Schoenberg. *Psychomusicology 21*(1&2), 1–22.

Gjerdingen, R. (2003). Review of the book *What to listen for in rock: A stylistic analysis*. *Music Perception 20*, 491–497.

Gladwell, M. (2008). *Outliers*. Toronto, Canada: ExecuGo media.

Glushko, A., Steinhauer, K., DePriest, J., & Koelsch, S. (2016). Neurophysiological correlates of musical and prosodic phrasing: Shared processing mechanisms and effects of musical expertise. *PLoS ONE 11*(5): e0155300.

Gordon, E. (1965). *Musical aptitude profile*. Boston: Houghton Mifflin.

Grashel, J. (2008). The measurement of musical aptitude in 20th century United States: A brief history. *Bulletin of the Council for Research in Music Education 176*, 45–49.

Gruber, H., Degner, S., & Lehmann, A. C. (2004). Why do some commit themselves in deliberate practice for many years—and so many do not? In M. Radovan & N. Dordevic (Eds), *Current issues in adult learning and motivation 7th adult education colloquium* (pp. 222–235). Ljubljana: Slovenian Institute for Adult Education.

Hallam, S. (1995). Professional musicians' orientations to practice: Implications for teaching. *British Journal of Music Education 12*, 3–19.

Hallam, S. (1998). The predictors of achievement and dropout in instrumental tuition. *Psychology of Music 26*, 116–132.

Hallam, S. (2013). What predicts level of expertise attained, quality of performance, and future musical aspirations in young instrumental players? *Psychology of Music 41*, 267–291.

Hallam, S. (2015). Musicality. In G. E. McPherson (Ed.), *The child as musician: A handbook of musical development*, 2nd edn (pp. 67–80). Oxford: Oxford University Press.

Hallam, S., & Prince, V. (2003). Conceptions of musical ability. *Research Studies in Music Education 20*, 2–22.

Hallam, S., Rinta, T., Varvarigou, M., Creech, A., Papageorgi, I., Gomes, T., & Lanipekun, J. (2012). The development of practising strategies in young people. *Psychology of Music 40*(5), 652–680.

Hallam, S., & Shaw, J. (2002). Constructions of musical ability. *Bulletin of the Council for Research in Music Education 153/154*, 102–108.

Hambrick, D. Z., & Meinz, E. J. (2012). Working memory capacity and musical skill. In T. P. Alloway (Ed.), *Working memory: The connected intelligence*. London: Psychology Press.

Hambrick, D. Z., Oswald, F. L., Altmann, E. M., Meinz, E. J., Gobet, F., & Campitelli, G. (2014). Deliberate practice: Is that all it takes to become an expert? *Intelligence 45*, 34–45.

Hambrick, D. Z., & Tucker-Drob, E. M. (2015). The genetics of music accomplishment: Evidence for gene-environment correlation and interaction. *Psychonomic Bulletin & Review 22*, 112–120.

Haroutounian, J. (2002). *Kindling the spark: Recognizing and developing musical talent*. Oxford: Oxford University Press.

Hatfield, J. L. (2016). Performing at the top of one's musical game. *Frontiers in Psychology 7*, 1356.

Hayes, J. R. (1981). *The complete problem solver*. Philadelphia, PA: Franklin Institute Press.

Heaton, P. (2009). Assessing musical skills in autistic children who are not savants. *Philosophical Transactions of the Royal Society B: Biological Sciences 364*(1522), 1443–1447.

Hoffman, R. R. (1998). How can expertise be defined? Implications of research from cognitive psychology. In R. Williams, W. Faulkner, & J. Fleck (Eds), *Exploring expertise* (pp. 81–100). Edinburgh: University of Edinburgh Press.

Howe, M. J. A. (1999). Prodigies and creativity. In R. J. Sternberg (Ed.), *Handbook of creativity*. New York: Cambridge University Press.

Howe, M. J. A., Davidson, J. W., Moore, D. G., & Sloboda, J. A. (1995). Are there early childhood signs of musical ability? *Psychology of Music 23*, 162–176.

Humphreys, J. T. (1998). Musical aptitude testing: From James McKeen Cattell to Carl Emil Seashore. *Research Studies in Music Education 10*(1), 42–53.

Jabusch, H. C., Alpers, H., Kopiez, R., Vauth, H., & Altenmuller, E. (2009). The influence of practice on the development of motor skills in pianists: A longitudinal study in a selected motor task. *Human Movement Science 28*(1), 74–84.

Jørgensen H. (1997). Time for practising? Higher level music students' use of time for instrumental practising. In H. Jørgensen & A. C. Lehmann (Eds), *Does practice make perfect? Current theory and research on instrumental music practice* (pp. 123–129). Oslo, Norway: Norges Musikkhøgskole.

Jørgensen, H. (2002). Instrumental performance expertise and amount of practice among instrumental students in a conservatoire. *Music Education Research 4*, 105–119.

Jørgensen H. (2004). Strategies for individual practice. In A. Williamon (Ed.), *Musical excellence: Strategies and techniques to enhance performance* (pp. 85–104). Oxford: Oxford University Press.

Jørgensen, H. (2008). Instrumental practice: Quality and quantity. *Musiikkikasvatus: The Finnish Journal of Music Education 11*, 8–18.

Kemp, A. E., & Mills, J. (2002). Musical potential. In R. Parncutt & G. E. McPherson (Eds), *The science and psychology of music performance.* (pp. 3–16). New York: Oxford University Press.

Kopiez, R., & Lee, J. I. (2006). Towards a dynamic model of skills involved in sight reading music. *Music Education Research 8*(1), 96–120.

Kopiez, R., & Lee, J. I. (2008). Towards a general model of skills involved in sight reading music. *Music Education Research 10*, 41–62.

Kostka, M. J. (2002). Practice expectations and attitudes: A survey of college-level music teachers and students. *Journal of Research in Music Education 50*, 145–154.

Krampe, R. T., & Ericsson, K. A. (1996). Maintaining excellence: Deliberate practice and elite performance in young and older pianists. *Journal of Experimental Psychology: General 125*, 331–359.

Kraus, N., & Chandrasekaran, B. (2010). Music training for the development of auditory skills. *Nature Reviews: Neuroscience 11*, 599–605.

Lehmann, A. C., & Davidson, J. W. (2002). Taking an acquired skills perspective on music performance. In R. Colwell & C. Richardson (Eds), *The new handbook of research on music teaching and learning* (pp. 542–560). New York: Oxford University Press.

Lehmann, A. C., & Ericsson, K. A. (1996). Performance without preparation: Structure and acquisition of expert sight-reading and accompanying performance. *Psychomusicology 15*(1–2), 1–29.

Limb, C. J., & Braun, A. R. (2008). Neural substrates of spontaneous musical performance: An FMRI study of jazz improvisation. *PLoS One 3*(2), e1679.

Macnamera, B. N., Hambrick, D. Z., & Oswald, F. L. (2014). Deliberate practice and performance in music, games, sports, education, and professions: A meta-analysis. *Psychological Science 25*, 1608–1618.

McPherson, G. E. (2009). The role of parents in children's musical development. *Psychology of Music* 37(1), 91–110.

McPherson, G. E., & Davidson, J. W. (2002). Musical practice: Mother and child interactions during the first year of learning an instrument. *Music Education Research* 4(1), 141–156.

McPherson, G. E., Davidson, J. W., & Faulkner, R. (2012). *Music in our lives: Rethinking musical ability, development and identity*. Oxford: Oxford University Press.

McPherson, G. E., & Hallam, S. (2009). Musical potential. In S. Hallam, I. Cross, & M. Thaut (Eds), *The Oxford handbook of music psychology* (pp. 433–448). New York: Oxford University Press.

McPherson, G. E., & Renwick, J. M. (2001). A longitudinal study of self-regulation in children's musical practice. *Music Education Research* 3, 169–186.

McPherson, G. E., & Williamon, A. (2015). Building gifts into musical talents. In G. E. McPherson (Ed.), *The child as musician: A handbook of musical development*, 2nd edn (pp. 340–360). Oxford: Oxford University Press.

Meinz, E. J., & Hambrick, D. Z. (2010). Deliberate practice is necessary but not sufficient to explain individual differences in piano sight-reading skill: The role of working memory capacity. *Psychological Science* 21, 914–919.

Miksza, P. (2007). Effective practice: An investigation of observed practice behaviors, self-reported practice habits, and the performance achievement of high school wind players. *Journal of Research in Music Education* 55, 359–375.

Miksza, P., & Tan, L. (2015). Predicting collegiate wind players' practice efficiency, flow, and self-efficacy for self-regulation: An exploratory study of relationships between teachers' instruction and students' practicing. *Journal of Research in Music Education* 63, 162–179.

Mishra, J., & Fast, B. (2015). Practising in the new world: A case study of practising strategies related to the premiere of contemporary music. *Music Performance Research* 7(1), 65–80.

Moore, D. G., Burland, K., & Davidson, J. W. (2003). The social context of musical success: A developmental account. *British Journal of Psychology* 94(4), 529–549.

Moreno, S., & Bidelman, G. M. (2014). Examining neural plasticity and cognitive benefit through the unique lens of musical training. *Hearing Research* 308, 84–97.

Mosing, M. A., Madison, G., Pedersen, N. L., Kuja-Halkola, R., & Ullen, F. (2014). Practice does not make perfect: No causal effect of music practice on music ability. *Psychological Science* 25, 1795–1803.

Mosing, M. A., & Ullen, F. (2016). Genetic influences on musical giftedness, talent, and practice. In G. E. McPherson (Ed.), *Musical prodigies: Interpretations from psychology, education, musicology* (pp. 156–167). Oxford: Oxford University Press.

National Association for Music Education. (2017). Strategic plan. Available from http://www.nafme.org/about/mission-and-goals/

Ollen, J. E. (2006). *A criterion-related validity test of selected indicators of musical sophistication using expert ratings*. Unpublished doctoral dissertation, Ohio State University, Columbus.

O'Neill, S. (1999). Flow theory and the development of musical performance skills. *Bulletin of the Council for Research in Music Education* 141, 129–134.

Pallesen, K. J., Brattico, E., Bailey, C. J., Korvenoja, A., Gjedde, A., & Carlson, S. (2010). Cognitive control in auditory working memory is enhanced in musicians. *PLoS ONE* 5(6): e11120.

Papageorgi, I. (2014). Developing and maintaining expertise in musical performance. In I. Papageori & G. Welch (Eds), *Advanced musical performance: Investigations in higher education learning* (pp. 303–318). New York: Routledge.

Peretz, I. (2016). Neurobiology of congenital amusia. *Trends in Cognitive Sciences* 20, 857–867.

Platz, F., Kopiez, R., Lehmann, A. C., & Wolf, A. (2014). The influence of deliberate practice on musical achievement: A meta-analysis. *Frontiers in Psychology 5*, 646.

Renwick, J. M., & McPherson, G. E. (2000). I've got to do my scale first!: A case study of a novice's clarinet practice. In C. Woods, G. B. Luck, R. Brochard, F. Seddon, & J. A. Sloboda (Eds), *Proceedings of the sixth international conference on music perception and cognition.* Keele: Keele University, Department of Psychology.

Repp, B. (1997). Expressive timing in a Debussy prelude. *Musicae* Sientiea 1(2), 257–268.

Ruthsatz, J., Detterman, D., Griscom, W. S., & Cirullo, B. A. (2008). Becoming an expert in the musical domain: It takes more than just practice. *Intelligence 36*, 330–338.

Sala, G., & Gobet, F. (2017). Experts' memory superiority for domain-specific random material generalizes across fields of expertise: A meta-analysis. *Memory & Cognition 45*, 183–193.

Schlaug, G. (2009). Music, musicians, and brain plasticity. In S. Hallam, I. Cross, & M. Thaut (Eds), *The Oxford handbook of music psychology* (pp. 197–207). New York: Oxford University Press.

Schonberg, H. C. (1970). *The lives of great composers.* New York: Norton.

Schulze, K., Dowling, W. J., & Tillman, B. (2011). Working memory for tonal and atonal sequences during a forward and a backward recognition task. *Music Perception: An Interdisciplinary Journal 29*, 255–267.

Scripp, L., Ulibarri, D., & Flax, R. (2013). Thinking beyond the myths and misconceptions of talent: Creating music education policy that advances music's essential contribution to twenty-first-century teaching and learning. *Arts Education Policy Review 114*, 54–102.

Seashore, C. E., Lewis, D., & Saetveit, J. G. (1919). *Seashore measures of musical talent.* New York: Psychological Corporation.

Sloboda, J. A. (1976). Visual perception of musical notation: Registering pitch symbols in memory. *Quarterly Journal of Experimental Psychology 28*, 1–16.

Sloboda, J. A. (1996). The acquisition of musical performance expertise: Deconstructing the "talent" account of individual differences in musical expressivity. In K. A. Ericsson (Ed.), *The road to excellence: The acquisition of expert performance in the arts and sciences, sports, and games* (pp. 107–126). Mahwah, NJ: Lawrence Erlbaum.

Sloboda, J. A., Davidson, J. W., Howe, M. J. A., & Moore, D. G. (1996). The role of practice in the development of performing musicians. *British Journal of Psychology 87*, 287–309.

Sosniak, L. A. (1985). One concert pianist. In B. S. Bloom (Ed.), *Developing talent in young people* (pp. 68–89). New York: Ballantine Books.

Sternberg, R. J. (1996). The costs of expertise. In K. A. Ericsson (Ed.), *The road to excellence: The acquisition of expert performance in the arts and sciences, sports, and games* (pp. 347–354). Mahwah, NJ: Lawrence Erlbaum.

Strait, D. L., Kraus, N., Parbery-Clark, A., & Ashley, R. (2010). Musical experience shapes top-down auditory mechanisms: Evidence from masking and auditory attention performance. *Hearing Research 261*, 22–29.

Suzuki, S. (1996). *Nurtured by love: The classic approach to talent education.* Miami, FL: Summy-Birchard.

Tan, Y. T., McPherson, G. E., Peretz, I., Berkovic, S. F., & Wilson, S. J. (2014). The genetic basis of music ability. *Frontiers in Psychology 5*, 658.

Trehub, S. E. (2001). Musical predispositions in infancy. *Annals of the New York Academy of Sciences 930*, 1–16.

Vinkhuyzen, A. A. E., van der Sluis, S., Posthuma, D., & Boomsma, D. I. (2009). The heritability of aptitude and exceptional talent across different domains in adolescents and young adults. *Behavior Genetics 39*, 380–392.

Williamon, A., & Valentine, E. (2000). Quantity and quality of musical practice as predictors of performance quality. *British Journal of Psychology 91*, 353–376.

Winner, E. (2014). The rage to master: The decisive role of talent in the visual arts. In K. A. Ericsson (Ed.), *The road to excellence: The acquisition of expert performance in the arts and sciences, sports, and games* (pp. 271–301). Mahwah, NJ: Lawrence Erlbaum.

Wong, Y. K., & Gauthier, I. (2010). A multimodal neural network recruited by expertise with musical notation. *Journal of Cognitive Neuroscience 22*, 695–713.

Zhukov, K. (2009). Effective practising: A research perspective. *Australian Journal of Music Education 1*, 3–12.

Zuk, J., Benjamin, C., Kenyon, A., & Gaab, N. (2014). Behavioral and neural correlates of executive functioning in musicians and non-musicians. *PLoS ONE 9*(6): e99868.

CHAPTER 26

SKILLED ANTICIPATION IN SPORT

Past, Present, and Future

A. MARK WILLIAMS, BRADLEY FAWVER,
DAVID P. BROADBENT, COLM P. MURPHY,
AND PAUL WARD

Introduction

In many complex and dynamic situations, including high-performance sport, performers at all levels of proficiency must deal with uncertainty by processing and integrating prior knowledge with relevant cues from the environment in order to facilitate an appropriate response (e.g., Ward, Williams, & Hancock; 2006; Williams, Ford, Hodges, & Ward, 2018; Williams & Jackson, 2019). Of the several perceptual-cognitive skills important for successful performance, the ability to anticipate what will happen next is a crucial factor underpinning successful decision making and response execution (e.g., Williams, Casanova, & Toledo, in press). In this chapter, we review current knowledge and understanding of how skilled anticipation occurs in sport.

In the first half of this chapter, we have two aims. First, we present a succinct historical overview of research that has examined how experts anticipate effectively and efficiently in sport. This overview spans at least four decades of research, dating back to seminal work in the late 1970s (e.g., Jones & Miles, 1978; Salmela & Fiorito, 1979). Second, we highlight advances over the past decade or so in our knowledge and understanding of the processes underpinning expert anticipation, including how

---

Skilled Anticipation in Sport: Past, Present and Future by A. Mark Williams, B. Fawver, D. P. Broadbent, C. P. Murphy, and Paul Ward

The author (Paul Ward's) affiliation with The MITRE Corporation is provided for identification purposes only, and is not intended to convey or imply MITRE's concurrence with, or support for, the positions, opinions or viewpoints expressed by the author.

various sources of information interact to guide anticipation in dynamic sport environments (for a review of the role of anticipation in other dynamic work contexts, see Suss & Ward, 2015). We then turn our focus to current lines of enquiry and look forward, rather than backward, in an effort to present some new avenues for research that could provide fresh impetus to the field. We suggest potential directions for future work and highlight areas where the current state of knowledge remains limited. It is not our intention to present an all-encompassing review, but rather to focus on a few select areas of research where, ultimately, we feel significant progress can be made relatively quickly. Moreover, we delimit our scope mostly to research that has attempted to identify the processes and mechanisms underpinning superior anticipation (i.e., the ability to predict the outcome of an opponent's action prior to the execution of that action or the consequences being evident). Consequently, we largely avoid discussion of the equally interesting topic area associated with how anticipation may be trained across domains (for recent reviews of this latter area, see Broadbent, Causer, Williams, & Ford, 2014; Loffing, Hagemann & Farrow, 2017).

The First Few Decades of Research: A Restricted Focus on Identifying Postural Cues and Efforts to Enhance Methods and Measurement Sensitivity

Although there were some tentative efforts to examine skilled anticipation in athletes during the late 1960s (e.g., Enberg, 1968), the pivotal work was undertaken almost a decade later. In a landmark study in the domain of tennis, Jones and Miles (1978) used a film-based method to occlude potentially important stimuli (i.e., they prevented viewing of the stimuli by replacing a dynamic image with a blank screen) after a given point in the action sequence. Jones and Miles (1978) presented participants with film from a first-person perspective of an opponent's tennis serve. In separate trials, the film was occluded at various time points before, at, or after the tennis ball made contact with the racket during the serve. Participants were required to anticipate where the tennis ball would land in his/her service box by marking a response on a scaled, schematic representation of the court. This technique is not unlike subsequent methods developed in other domains, such as the situation awareness global assessment technique (SAGAT), which, in part, have been used to assess an individual's skill at using current information to predict future situational actions or events (e.g., Endsley, 1988, 1995; see Suss & Ward, 2015).

Jones and Miles (1978) used the *temporal occlusion* method to examine whether skilled and less-skilled tennis players could anticipate the outcome of the serve using only the information available prior to the point of occlusion, such as the information emanating from an opponent's body shape. This method restricts participants from anticipating the outcome based on information that is only available later in the action sequence (i.e., after the point of occlusion), such as information from ball flight. The authors reported that expert players were better than novices at anticipating serve direction at the earliest point of occlusion (i.e., 42 ms before ball–racket contact), whereas no differences in judgment accuracy were observed when the film was occluded 336 ms after ball–racket contact, when the initial part of ball flight was visible. These data indicate that a key difference between skilled and less-skilled athletes is the ability to recognize and interpret information from the body shape of an opponent in advance of a key event such as ball–racket contact, which provides the basis for accurately anticipating the future outcome. A myriad of publications subsequently followed using the temporal occlusion method, with comparable conclusions being reported in sports as diverse as squash (Abernethy, 1990), badminton (Abernethy & Zawi, 2007), baseball (Moore & Müller, 2014), cricket (Abernethy & Russell, 1984), soccer (Williams & Burwitz, 1993), volleyball (Wright, Pleasants, & Gomez-Meza, 1990), and field hockey (Starkes, 1987).

At the same time, researchers began to explore alternative methods for examining skilled anticipation in the domain of sport. Several researchers presented similar first-person perspective film of an opponent (e.g., an opposing tennis player making a serve) and had participants respond virtually (e.g., by mimicking a service return or moving toward the direction of the ball). The key difference of this *response time* method—compared to temporal occlusion—was that occlusion of the stimuli was initiated by the participant's own response (e.g., by pressing a key on a keyboard, or by using movement-sensing technology capable of registering initiation of a full body response). The participant knew that the act of physically responding would essentially eliminate further access to the stimuli. The goal was to determine how quickly a participant could accurately (and occasionally, how confident they were) respond when they, rather than the experimenter, controlled the availability of information (Williams, Davids, Burwitz, & Williams, 1994). On average, skilled players have been shown to respond more quickly than their less-skilled counterparts, but do not differ in response accuracy despite relying on earlier arising information (see Helsen & Starkes, 1999; Williams & Davids, 1998). A disadvantage with this method is that subtle trade-offs between speed and accuracy can make interpreting the findings difficult and, in the real-world setting, athletes are not required to respond quickly and accurately, but rather to respond accurately in the time available (Alain & Proteau, 1980) and in line with their own action capabilities (Dicks, Davids, & Button, 2010).

Over the ensuing decade, the focus shifted toward a perceived need to embrace more representative task designs that employed more life-like response modes in the performance environment, driven in part by the upsurge in interest in alternative theoretical frameworks grounded in ecological psychology (Araujo & Kirlik, 2008; Williams, Davids, & Williams, 1999). There were efforts to replicate the film-based occlusion technique in-situ using liquid-crystal occlusion goggles (e.g., Farrow & Abernethy, 2003), which are specialized glasses that are triggered using a range of electronic sensors, such as infrared beams, pressure-sensitive pads, and optoelectronic motion analysis systems (see Müller & Abernethy, 2006; Oudejans & Coolen, 2003; Starkes, Edwards, Dissanayake, & Dunn, 1995). These devices are designed to detect the onset of movement (e.g., infrared beams, pressure-sensitive pads) or quantify the motion of the participant using reflective markers attached to various parts of the body (e.g., optoelectronic motion capture systems). In experimental designs using this technology, vision can be quickly and selectively occluded linking the liquid crystal glasses to the movements made by the participant. However, despite advances in methods, sensing technology, and measurement sensitivity, the main conclusions emanating from this body of work remained largely unchanged.

A shortcoming with the approaches presented above is that only the key time period for information extraction was identified, rather than the crucial sources of information itself. In order to ascertain what information is extracted, and at what time, these methods must be integrated with other experimental manipulations, or cognitive process methods, to verify hypotheses about the value of particular information sources. Several researchers began using complementary approaches such as the spatial or event occlusion techniques or eye movement recording that provided opportunities for experimental manipulation and the use of process-tracing techniques (Williams & Ericsson, 2005). In the spatial occlusion method, different information sources are dynamically occluded (i.e., masked), usually for the entire duration of the clip. For example, cues thought to be informative, such as the hips of the penalty taker in a soccer penalty kick or the server's arm and the racket in tennis, are removed from view, for instance by superimposing a mask (e.g., black box) over the respective cues or digitally editing the information by replacing the foreground with background. The decrement in performance when various postural cues are occluded provides support for the claim that body position is a relevant source of information that contributes to successful anticipation (e.g., Causer, Smeeton, & Williams, 2017; Jackson & Mogan, 2007; Müller, Abernethy, & Farrow, 2006; Williams & Davids, 1998).

The recording of gaze behavior has also been used as a process measure to provide additional evidence of the information sources that are visually fixated on during anticipation (Williams, Janelle, & Davids, 2004). The assumption is that experts fixate on specific body regions in an effort to extract information from that source to guide anticipation. Overall, this body of work has shown that experts fixate on different (more informative) areas of the display compared to their less-skilled counterparts, typically using the visual system in a more efficient and effective manner (Mann, Williams, Ward, & Janelle, 2007). However, findings suggest that the specific attentional strategy employed by the performer is influenced by various task, situation, and context-specific constraints, and strategies have been shown to vary markedly across domains (for a review, see Ward et al., 2006). For example, the most effective gaze behaviors in team ball games vary based on the number of players and the playing area involved (Vaeyens, Lenior, Williams, Mazyn, & Philippaerts, 2007), the distance of the ball from the player (Roca, Ford, McRobert, & Williams, 2013), the strategic intention of the player (Helsen & Starkes, 1999; Williams & Davids, 1998), and the levels of fatigue (Casanova et al., 2013) and anxiety (Vater, Roca, & Williams, 2015; Williams & Elliott, 1999) experienced.

A shortcoming with the eye movement recording method is that it only identifies the position of the fovea in space, which may not be directly related to the information actually used given the well-reported tendency of expert athletes in particular to *anchor* their gaze on a given stimulus (somewhat independent of the importance of that specific area) in order to detect relevant information via the para-fovea and the visual periphery (Williams et al., 1999). It is recommended that a number of complementary methods are used (e.g., film occlusion, verbal reports) in conjunction with eye movement recording in an effort to better identify what information is used and how this is picked up by the visual system (Mann & Savelsbergh, 2015; Williams & Ericsson, 2005).

Several researchers have embraced alternative methods to conventional film-based approaches, for instance, by using point-light displays and stick-figure images to examine the specific information sources used by skilled athletes when attempting to anticipate an outcome based on an opponent's postural information or movement pattern (e.g., Abernethy, Gill, Parks, & Packer, 2001; Abernethy & Zawi, 2007; Shim, Carlton, & Kwon, 2006; Ward, Williams, & Bennett, 2002). This body of work has highlighted the importance of biological motion, particularly the relative motions between limbs, during anticipation when compared to the use of background or structural information. Moreover, sophisticated methods have been used to isolate and identify the specific information used during anticipation via statistical modeling (e.g., using principal component analyses) and/or experimental manipulations (i.e., occluding, neutralizing, or exaggerating information from one or more limbs) of the information presented (Bourne, Bennett, Hayes, Smeeton, & Williams, 2013; Cañal-Bruland & Williams, 2010; Huys et al., 2009; Smeeton & Huys, 2011). Skilled and less-skilled participants have been shown to use different sources of information to guide anticipatory behavior such as critical postural cues like the throwing arm, hip, and trunk in handball (Bourne et al., 2013). Moreover, there is evidence to suggest that skilled participants use a more global perceptual approach such that information pick-up is distributed over a larger part of the opponent's body rather than on more local and isolated cues (Huys et al., 2009).

The New Millennium: A Focus on Identifying Perceptual–Cognitive Skills Other Than Postural Cue Usage and Examining How These Interact in Dynamic Environments

An overemphasis on the importance of postural cues during the first two or three decades of research in this field led to a rather restricted and limited view of anticipation. Perhaps some 90 percent of published papers during this early period focused on the perceptual cues used by experts. Although a few published reports examined whether superior anticipation was based on skill in recognizing patterns of information (e.g., Allard, Graham, & Paarsalu, 1980; Garland & Barry, 1991; Williams & Davids, 1995) or whether athletes assigned specific probabilities to any given event as a means of dealing with apparent uncertainty (e.g., Alain & Girardin, 1978; Alain & Proteau, 1978), only recently has there been a more substantive focus on the role of perceptual–cognitive skills other than cue usage in anticipation (Cañal-Bruland & Mann, 2015).

Following the early work of Allard and colleagues (1980), researchers have attempted to better identify the nature of the information athletes use to recognize evolving sequences of play, such as the environmental structure imbued in the semantic and structural relations between elements that make up a pattern. Following research in chess that examined skilled recognition (Goldin, 1978, 1979) and recall of patterns (e.g., Chase & Simon, 1973), similar methods have been used to examine the relationship between skill and pattern familiarity in sport. For example, Williams, Hodges, North, and Barton (2006) used a recognition task to examine differences between skilled and less-skilled soccer players. Participants were first shown a set of short (e.g., 3–10 s) film sequences of both structured (i.e., real offensive sequences of play) and unstructured (i.e., random arrangements of players) soccer match-play, and subsequently asked whether they recognize from the first set either new (i.e., not previously seen) or old (previously seen) structured and unstructured stimuli. Skilled players were more accurate than their less-skilled counterparts in recognizing structured sequences of soccer play, but not unstructured sequences. The authors interpreted these findings as being due to the skilled individual's superior representation of domain-specific knowledge (North & Williams, 2008).

The superior pattern recognition performance of skilled compared with less-skilled players has been demonstrated in numerous team sports, including soccer (Ward & Williams, 2003; Williams & North, 2009), field hockey (Smeeton, Ward, & Williams, 2004), and basketball (Gorman, Abernethy, & Farrow, 2012; 2013). In an attempt to explain these findings post hoc, Williams et al. (2006) suggested that Dittrich's (1999) interactive encoding model may offer a plausible explanation for the skilled-based differences found for pattern recognition. Dittrich (1999) proposed that during biological motion perception, skilled performers initially extract low-level structural motion information and temporal relationships between features of relevant stimuli before matching those features to high-level semantic concepts or templates stored in memory in an effort to facilitate recognition. A prediction of the model is that as the performer develops more elaborate cognitive processes there is a transition from low-level to higher-level processing. Williams et al. (2006) suggested that in sports like soccer, low-level information might be reflected in the positions and movements of teammates, opponents, and the ball, as well as pitch/field markings which both skilled and less-skilled athletes can process. However, the authors proposed that, compared to skilled athletes, less-skilled individuals are unable to detect important relational information and have fewer higher-level semantic concepts or templates, constraining them to employ more distinctive surface features when making such judgments.

In an effort to further examine the processes underpinning superior recognition, researchers have manipulated film stimuli presented in these tasks and represented them as point-light displays (e.g., North, Williams, Hodges, Ward, & Ericsson, 2009). The assumption is that skilled players are not recognizing the superficial characteristics of the individuals or elements in the scene, or even characteristics of their behavior (such as postural changes, or behavioral gestures), but rather the information germane to point-light displays, such as positions of key elements, and the absolute and relative motions that exist between individual elements. In this research, rather than represent parts of the body (e.g., joint centres) using points of light, entire players (as well the ball) are represented as individual points of light moving against a black background which included only the pitch or field markings. North et al. (2009) demonstrated that skilled players made superior recognition judgments under both point-light and film conditions, implying that the information sources used by skilled participants is largely reflected in the key relative motions between players and the crucial, higher-order strategic information that is conveyed, as opposed to using more superficial background and structural information to recognize sequences (see North, Ward, Ericsson, & Williams, 2011). Subsequently, researchers have attempted to isolate the relational elements conveyed in core aspects of the pattern. Recent findings (e.g., North, Hope, & Williams, 2017; Williams, North, & Hope, 2012) seem to suggest that accurate pattern recognition may be based on the positioning of only a few select pieces of information (i.e., particular players in the field of play), but this information changes over time, emerges at relatively discrete moments as the pattern of play unfolds, and is not continuously available throughout each action sequence (North & Williams, 2008).

Another perceptual–cognitive skill that has attracted interest of late, having been largely ignored following the seminal work of Alain and colleagues, is the focus on how experts make situational assessments about unfolding events to facilitate anticipation. This work was inspired, initially, by Alain and Proteau (1978), whose focus was on whether athletes make decisions by first assigning actual probabilities to likely events as a means to deal with uncertainty. Ward and colleagues (e.g., Ward & Williams, 2003; Ward, Ericsson, & Williams, 2013) presented skilled and less-skilled soccer players with film sequences involving offensive sequences of play from the perspective of a defender (see also Belling, Suss, & Ward, 2015). These film sequences were paused and players were asked to highlight the situational options (i.e., courses of action) that were available to the opposing player in possession of the ball. Options were ranked in regards to their offensive threat, as rated by a panel of subject-matter experts (i.e., expert soccer coaches). Skilled players were more accurate in their predictions, and were more accurate in identifying task-relevant options, ignoring irrelevant options, and were better at rank ordering those options in a manner consistent with the subject-matter experts. In a subsequent study, Ward et al. (2013) demonstrated that greater prediction accuracy was associated with superior situational assessments. Although few options are generated, and better options are typically generated first, having a better representation of the potential alternative courses of action that an opponent might take is positively associated with superior anticipation of what that opponent will actually do next.

Perhaps one of the most significant areas of advancement over the past decade has been the shift away from considering anticipation as being based on a single perceptual–cognitive skill toward a greater emphasis on examining the importance of complex interactions between many different skills. It is now well established, as the preceding parts of the chapter demonstrate, that skilled athletes: (1) are able to use postural cues from an opponent early in an action (i.e., advanced cue utilization skill); (2) are more accurate in recognizing environmental structure in evolving sequences of play (i.e., pattern recognition skill); and (3) are better at identifying alternative options and assessing their relative threat (situational assessment skill). Just as cue-based information changes dynamically in sport environments and becomes more or less predictive of the actual outcome over time, the relative influences of these skills are likely to be different from one moment to the next as a function of complexity of the contextual constraints (Suss & Ward, 2015; Williams, 2009).

In an effort to explore how the importance of these skills differ as a function of the constraints of the task, Roca et al. (2013) presented skilled and less-skilled soccer players with offensive sequences of play (i.e., the opposing team was moving in possession of the ball toward the receiving team's goal) where the ball was either near or far away from the observer/defender. In the near task, the offensive pattern of play commenced in the half of the field in which the participant was situated, whereas, in the far task, the sequence of play started with the ball in the other half of the pitch (approximately, double the distance away from the participant). Visual gaze data and retrospective verbal reports of thoughts were taken as players attempted to anticipate which action their opponent would perform next (i.e., pass, shoot, or dribble the ball). Less-skilled players did not alter their gaze significantly across the near and far task manipulation, often fixating on the ball, the opponent in possession, or the ball flight movement under both task conditions. In contrast, skilled participants employed more fixations of shorter durations to a greater number of different areas of the field (i.e., opponents, teammates, and areas of free space) under the far task condition, compared to in the near task, where fixations were fewer and focused on their own half of the field. In addition, skilled players spent more time fixating on teammates, opponents, and areas of free space in the far task, compared to the near task in which they spent more time fixating on the player in possession of the ball. The strategy adopted by the skilled players in the far task likely helped them identify structural relations between features such as the players and the ball (North et al., 2011), whereas in the near task, the skilled players focused on the player in possession of the ball, perhaps facilitating the pick-up of postural cues (Mann et al., 2007).

In an attempt to examine the interactions between the different perceptual–cognitive skills, and to supplement the gaze behavior data, verbal reports were elicited from participants and coded relative to their use of advance postural cues, pattern recognition, and situational probabilities. These data are presented in Figure 26.1. The data indicate that skilled players used the aforementioned perceptual–cognitive skills

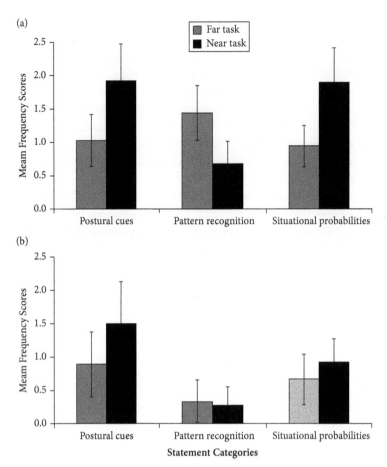

FIGURE 26.1 Mean frequency scores by verbal statement and task condition for (a) skilled and (b) less-skilled players.

Adapted from André Roca, Paul R. Ford, Allistair P. McRobert, and A. Mark Williams, Perceptual-cognitive skills and their interaction as a function of task constraints in soccer, *Journal of Sport and Exercise Psychology*, 35 (2), pp. 144–155, Figure 3, doi.org/10.1123/jsep.35.2.144 © 2013, Human Kinetics.

differently across the two task conditions. In the far task, skilled players articulated a more substantive role for pattern recognition during anticipation, whereas in the near task there was greater reliance on using advanced cue utilization and situational assessment to anticipate what the opponents would do. These findings demonstrate that changes in task constraints and context alter the influence of the different sources of information and corresponding perceptual–cognitive skills used to guide anticipatory behavior.

In summary, following the early work on anticipation in the late 1970s, researchers were overly pre-occupied with assessing the ability of skilled athletes to use advance visual cues from the postural orientation of an opponent, across a myriad of sports (Mann et al., 2007). While this work was on the whole innovative, with progressively more sophisticated methods and measures being used, this focus did lead to a fairly narrow conceptualization of anticipation, overemphasizing the importance of biological motion perception at the expense of extending knowledge of the role of high-level cognitive processes in anticipation. Although some researchers had, to some extent, simultaneously focused on these higher-level processes, only more recently has this work gathered any degree of momentum, with greater efforts being made to better identify the role of pattern recognition and the use of situational assessment in anticipation. It may be that the interaction between these different perceptual–cognitive skills is what is most important during anticipation, rather than the manner in which these skills are used in isolation. In the final part of the chapter, we highlight more recent research that has demonstrated how anticipation involves much more diverse and abstract cognitive processes than initially thought, with these being greatly influenced by the specific context under which such judgments are made.

A Contemporary Perspective: Presenting a Broader Context for the Study of Skilled Anticipation

An increasing number of researchers are now focusing attention on the role of context and how it may be used in conjunction with postural information to facilitate anticipation. In one of the earliest studies that allowed inferences to be made about the type of information being used to anticipate the intentions of an opponent, Abernethy et al. (2001) demonstrated that skilled squash players were able to accurately anticipate an opponent's shot as early as 620 ms in advance of the opponent striking the ball. Since accuracy levels were significantly above chance, one could hypothesize that these players were not using information from the racket and ball that are only available immediately before, at, or after ball–racket contact, but instead used other sources that are readily available in that early time window (i.e., >600 ms prior to ball–racket contact). This hypothesis was recently supported using a detailed video-based analysis of over 3,000 rallies involving tennis players who had achieved a top-10, world-ranking (Triolet, Benguigui, Le Runigo, & Williams, 2013). The authors identified four distinct time windows in which different sources of information is required to respond correctly to an opponent's stroke (see Figure 26.2).

In Figure 26.2, windows 1, 2, and 3 represent anticipatory judgments, while window 4 represents decisions that were reactive, in the sense that the movements the players made were thought to be based on ball flight information (as evidenced by 100 percent accuracy levels). If we assume that the typical visuo-motor delay period (i.e., the time needed for the performer to initiate a response to the actual stimulus) is between 100 and 200 ms, in windows 1, 2, and 3 players were thought to be responding based on

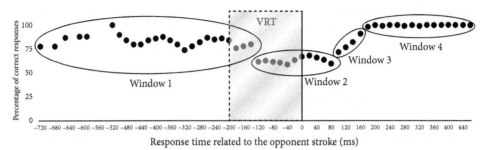

FIGURE 26.2 Percentage of correct responses to the left or right relative to response time. $t = 0$ ms (solid vertical black line) represents the moment of the opponent's ball–racket contact. The shaded area with a dotted outline represents typical visual response time (VRT).

Adapted from Celine Triolet, Nicolas Benguigui, Cyrille Le Runigo, and A. Mark Williams, Quantifying the nature of anticipation in professional tennis, *Journal of Sports Sciences*, 31 (8), pp. 820–830, Figure 3, doi.org/10.1080/02640414.2012.759658 © 2013, Taylor & Francis. Reprinted by permission of the publisher (Taylor & Francis Ltd, http://www.tandfonline.com).

information detected at or before the moment when the opponent struck the ball (depicted as $t = 0$ in Figure 26.2). The observable anticipatory behavior (i.e., those responses initiated during temporal windows 1, 2, or 3) was apparent on only 13.4 percent of the shots played, with response accuracy significantly decreasing during these time windows. The possibility exists that in tennis there is a tendency for players not to commit too early to a shot, because the potential cost of anticipating incorrectly (i.e., moving in the opposite direction to the actual shot) may outweigh the benefits (i.e., more time to prepare a return shot) achieved via early anticipatory responses. Moreover, the data suggested that an early anticipatory response was likely to be employed more as a defensive strategy, and used only when a player was under severe time constraints and in danger of losing the rally, rather than an offensive strategy, in an effort to gain an advantage over an opponent. It appears that alongside making anticipatory judgments in high-performance sport, players may conduct some sort of cost–benefit analysis to evaluate the time available to respond and the likelihood of making a correct response.

Perhaps the most interesting finding from the task analysis undertaken by Triolet et al. (2013) is that approximately one-third of the observable anticipation behaviors occurred at least 140 ms before the opponent has struck the ball (window 1), with high levels of accuracy achieved on these occasions. Given the length of the aforementioned visuo-motor delay period, the implication is that these judgments are made prior to the availability of postural cues relating to stroke direction (i.e., > 340 ms before ball–racket contact). These data, notably recorded from real-world situations involving elite performers, demonstrate that effective anticipation can occur with reduced kinematic information and potentially highlighted the need for researchers to broaden their focus to examine the role of non-kinematic (contextual) sources of information during anticipation (Cañal-Bruland & Mann, 2015). The paradox that exists in our field is that while the study by Triolet and colleagues (2013) suggests that only a minority of anticipatory responses are seemingly based on the pick-up of postural cues around the moment of ball–racket contact, the majority of published reports on anticipation in sport has focused on this issue. There have been limited attempts to identify how accurate anticipation is accomplished well ahead of ball–racket, or ball–foot, contact.

In an effort to verify the ability of skilled tennis players to anticipate accurately in the absence of postural cues from an opponent, Murphy and colleagues (2016) presented skilled and less-skilled players with animated sequences of tennis rallies in which they removed access to all background and structural information, including cues from the opponent's posture and racket. The court markings and the ball were retained and the players were represented as colored cylinders. The authors reported that while a decrement in anticipation accuracy was observed in the animated compared with a film condition, both groups of players were able to respond at levels above chance, with the skilled players maintaining their superiority over their less-skilled counterparts across conditions. Players were able to effectively use contextual information based on the positioning of the players on the court, the relative movements of the players and ball, shot sequencing, and the angles between the players and court markings, to anticipate an opponent's next shot. In follow-up work, Murphy, Jackson, & Williams (2018) utilized various manipulations to the animated sequences, and players were able to extract information from the immediately preceding sequence of shots played to more effectively anticipate the direction of the opponent's shot. Moreover, the authors observed that information emanating from the movements and positioning of players, and to a lesser extent the ball's flight path, was helpful during anticipation. It is likely that in some instances this information is sufficient, and may be the only relevant information available to the athlete, whereas in other instances information arising later from the opponent's postural orientation as ball–racket contact approaches is either used exclusively or may simply confirm the veracity of predictions made well ahead of ball–racket contact.

A body of work has now identified several non-kinematic sources of information available to athletes, including prior knowledge about opponents' action tendencies (Gredin, Bishop, Broadbent, Tucker, & Williams, 2018; Mann, Schaefers, & Cañal-Bruland, 2014; Navia, van der Kamp, & Ruiz, 2013), action patterns associated with certain game scores (Farrow & Reid, 2012; Gray, 2002; Runswick, Roca, Williams, Bezodis, & North, 2018c), court positioning (Loffing & Hagemann, 2014; Loffing, Sölter, Hagemann, & Strauss, 2016; Murphy et al., 2018), and sequences of action outcomes (Loffing, Stern, & Hagemann, 2015; Murphy et al., 2018). While our knowledge of what sources of contextual information may be used in conjunction with postural cues to facilitate anticipation is increasing, an important question that has received limited attention is how the information available to a performer at any given time is integrated to inform anticipation.

One of the first attempts to shed light on how these processes interact was reported by Loffing and Hagemann (2014). Skilled and novice tennis players were presented with dynamic point-light images of an opponent hitting cross-court and down-the-line shots. The lateral positioning of the opponent and the moment of occlusion was systematically varied such that the interaction between contextual information (i.e., court positioning), postural information, and expertise could be examined. When trials were occluded at the moment of ball–racket contact, participants were not influenced by the opponent's court positioning, relying almost entirely on the pick-up of biological motion information for anticipation. Conversely, when trials were occluded 800 ms prior to ball–racket contact, when relatively few relevant postural cues would be available, participants' expectancies of shot outcomes were strongly influenced by contextual information picked up from the opponent's court positioning. In particular, skilled participants, but not novices, displayed a shift from a reliance on contextual information picked up from an opponent's court positioning to the use of biological motion information as more of this information became available.

In a more recent study, Runswick et al. (2018a) examined how skilled and less-skilled cricket batters integrate various types of contextual information with the pick-up of postural information when anticipating the ball location of a bowler using a film-based simulation. A novel approach was used which necessitated that participants rated verbally the relative importance of the different sources of information at various time points during the anticipation process. Participants had access to contextual information in the form of the game score (number of overs bowled, runs scored, and wickets taken) and the field setting (i.e., position of surrounding players). This information was displayed on-screen (the field settings were displayed using a schematic representation) prior to seeing the bowler. The clip selected from each bowler was then occluded at four different time points, creating four conditions: (i) *no kinematic information*—participants were only exposed to contextual information (the game situation and field setting) and received no kinematic information (the bowler was not shown); (ii) *mid-run*—the trial was occluded at the mid run-up point, defined as the frame midway between the initiation of the bowler's run-up and the moment they released the ball; (iii) *pre-release*—the trial was occluded immediately prior to ball release; and (iv) *ball flight*—the clip was occluded after 80 ms of ball flight. The amount of contextual information presented was constant across viewing conditions. Skilled batters recorded higher accuracy scores than less-skilled batters across conditions, even at the earliest occlusion condition, when only context-specific information was available in the absence of any postural information from the bowler.

The verbal report data, presented in Figure 26.3, indicate that skilled batters use different sources of information when compared to their less-skilled counterparts across the four conditions. When the information sources were classified into verbalizations relating to contextual, kinematic, or ball flight information, it was observed that the reliance on kinematic information progressively increased from the mid-run to the pre-release and ball flight conditions (excluding the no kinematic condition). This increase in utilization of kinematic information is apparent despite the amount of contextual information presented remaining consistent across conditions. Moreover, this shift is highlighted more clearly in Figure 26.4, where the importance of the different sources of information are collapsed into kinematic and contextual information sources.

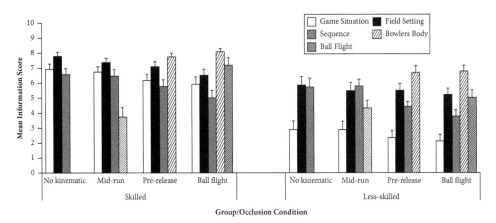

FIGURE 26.3 Mean total information scores for each category for skilled and less-skilled groups across the four occlusion conditions (SE).

Data from Runswick, O., Roca, A., Williams, A. M., Bezodis, N. E., McRobert, A. P., & North, J. (2018a), The impact of contextual information and a secondary task on anticipation performance: An interpretation using cognitive load theory, *Applied Cognitive Psychology* 32(2), 141–149.

While several sources of contextual information have been identified as contributing to skilled anticipation, and researchers have now started to make in-roads into demonstrating how various sources of contextual and postural information interact in the lead-up to a critical event, our understanding of how these sources of information are used and integrated by skilled performers remains limited. In one exception, McRobert, Ward, Eccles, and Williams (2011) examined how skilled and less-skilled cricket batters are influenced by having multiple exposures to the same bowler. They used a film-based simulation task, where batters executed a stroke in response to each delivery from a bowler as well as recording gaze behaviors and think-aloud verbal reports of thoughts. Under the high-context condition, the batters viewed a number of 6-ball overs from bowlers, whereas under a low-context condition, only a single delivery was observed from each bowler in a random order. Participants were more accurate under the high- than the low-context condition, suggesting that they were able to benefit from the additional context. Also, significant changes were observed in the underlying thought processes employed for all batters as an increase in evaluation and future planning related statements were found under the high-context condition, compared to the low-context condition, which elicited more lower-level cognition statements (e.g., factual information). In contrast to the less-skilled batters, high-skilled batters demonstrated reduced mean fixation duration employed under the high-context condition compared to the low-context condition, with the authors concluding that this reduction may represent an increased efficiency in picking up relevant cues when contextual information builds up over a series of deliveries from one bowler. It appears that skilled performers are able to quickly adapt their cognitive processes by using situation-specific context to create a cognitive model *on the fly* that enables them to shift from a more reactive to a more anticipative mode of function.

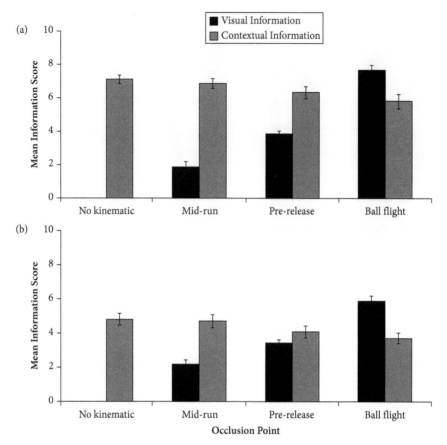

FIGURE 26.4 Mean total information scores for (a) skilled and (b) less-skilled batters for kinematic/visual (bowler's body and ball flight) and contextual (sequence, game situation, and field setting) information sources across the four occlusion conditions.

Data from Runswick, O., Roca, A., Williams, A. M., Bezodis, N. E., McRobert, A. P., & North, J. (2018a), The impact of contextual information and a secondary task on anticipation performance: An interpretation using cognitive load theory, *Applied Cognitive Psychology* 32(2), 141–149.

Although performers may utilize a variety of information sources during anticipation, these sources may not necessarily be congruent with one another, or indeed the outcome of the task. Using the same cricket-batting task described previously, Runswick, Roca, Williams, McRobert, and North (2018b) used a modification of the Posner cost–benefit method to manipulate the accuracy of the contextual information such that it could be either congruent or incongruent with the eventual outcome. As shown in Figure 26.5, when the outcome (i.e., ball landing position) was congruent with the contextual information presented prior to the bowler commencing his delivery, errors rates were low. In contrast, if these two sources of information were incongruent the error margins increased significantly. Moreover, an interaction between congruence and skill level was observed. The skilled group performed better

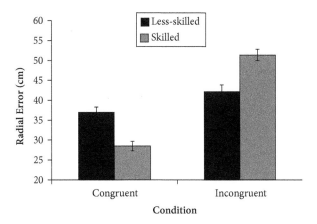

FIGURE 26.5 Anticipation accuracy as a function of radial error (cm) for skilled and less-skilled groups under congruent and incongruent conditions (SE).

Data from Runswick., O., Roca, A., Williams, A. M., McRobert, A., & North, J. (2018b), *The temporal integration of information during anticipation*, Psychology of Sport and Exercise 37, 100–108.

than the less-skilled group when contextual and kinematic information were congruent, whereas their accuracy scores were lower than less-skilled participants when the available information was incongruent.

It has been reported that contextual information regarding an opponent's action tendencies can positively and negatively influence anticipation based on the level of congruence in volleyball (Loffing et al., 2015), baseball (Gray, 2015), handball (Mann et al., 2014), and soccer (Navia et al., 2013). This finding is in line with earlier research using a cost–benefit method to examine the effects of prior knowledge on simple reaction time (e.g. Posner, Nissen, & Ogden, 1978). However, recently Gredin et al. (2018) used a video-based, 2-versus-2 soccer anticipation task to examine how skilled and less-skilled soccer players utilize and integrate contextual information, regarding the opponent's action tendencies, with postural cue information. While prior contextual information did influence the skilled players' online expectations and allocation of visual attention to more context-rich areas of the display (i.e., opponent off the ball), they were able to integrate confirmatory or conflicting postural information in the final stages of the action with the contextual information to update their pre-held expectations, to enhance their anticipatory judgments on congruent actions, but most importantly to maintain performance on actions incongruent with the opponent's action tendencies (Gredin et al., 2018).

Back to the Future: New Directions

Overall, the more recent programs of work reviewed up to this point illustrate the importance of context or other higher-level cognitive processes during anticipation in sport.

In future, researchers need to improve our understanding of how different perceptual–cognitive skills interact and are used dynamically in time-constrained performance domains. Moreover, we echo the recommendations made by Cañal-Bruland and Mann (2015) that greater efforts are needed to identify the nature of contextual information, how it is shaped by circumstances, and how contextual information or situational assessment interacts with non-contextual information and other perceptual–cognitive skills, such as cue utilization and pattern recognition. In addition, we suggest that more research is needed to ascertain how different types of information are prioritized and, most likely, how this may change as a function of changes in contextual demands and constraints.

Cañal-Bruland and Mann (2015) suggest that increasing our knowledge and understanding in this area should contribute to the development of an overarching framework that incorporates the relative influence of different sources of kinematic and non-kinematic information and a temporal understanding of how these different sources of information integrate together in dynamic environments to facilitate anticipation. While the research presented earlier in this chapter potentially bring us a step closer to this goal, it should be noted that the development of such a framework is no small feat and there is still much research required before this is possible. Several factors need to be taken into consideration such as different types of context and whether it is situation-specific (Runswick et al., 2018c) as well as additional environmental factors such as the impact of performance pressure on the integration and processing of the various sources of information (Cocks, Jackson, Bishop, & Williams, 2015; Broadbent, Gredin, Rye, Williams, & Bishop, 2018). Moreover, when researchers have proposed a framework for explaining anticipation in the past, these have been constrained to a single sport (e.g., see Müller & Abernethy, 2012). However, when attempting to propose a theoretical model with testable hypotheses that helps explain and predict anticipation behavior across all sports, the challenges are increased significantly due to differences in the knowledge structures employed as a function of task constraints within each domain.

Finally, although we have refrained from discussing the value of training programs intended to improve anticipation in this chapter, research is needed to explore the implications of a more multifaceted approach to anticipation. Scientists who have explored how anticipation may be trained have generally used a fairly reductionist approach, focusing on the testing and training of a single perceptual–cognitive skill. Moreover, and in keeping with the historical precedence for research in the field, the vast majority of published reports focusing on the training of anticipation have focused exclusively on how to improve the ability of athletes to pick up postural cues. While the body of literature on perceptual training remains comparatively small, findings strongly support the value of this type of training (Broadbent et al., 2014; Loffing et al., 2017). However, we express some words of caution. The well-documented efficacy of training programs that improve the ability to make use of perceptual cues may be overstated. The strength of our existing conclusions is based on the very narrow manner in which researchers have typically assessed transfer, such as examining whether the training effect that is classically seen on a laboratory-based test of anticipation transfers to a field-based version of the same test. The problem of course is that neither test situation replicates the real-world context. The key question to address is whether the training effects generally seen under relatively context-poor conditions remain when performance is assessed under the context-rich conditions seen during the actual competitive situation (e.g., Broadbent, Ford, O'Hara, Williams, & Causer, 2017) or when individuals are required to adapt to the types of complexity that push, or even extend beyond the bounds of their current competence (for extensive reviews, see Hoffman et al., 2014; Ward, Gore, Hutton, Conway, & Hoffman, 2018).

To summarize, in this chapter we have presented a historical overview of research in the area of anticipation in sport over the past four decades. Table 26.1 provides a simplified overview of the three time periods outlined in this chapter regarding research on anticipation in sport, including the methods used and the key findings. The early focus was on identifying the key postural cues that athletes picked up from an

Table 26.1 A simplified historical overview of research on anticipation in sport, including the methods used and the key findings

	Methods	Key findings
Focus of early research up to 2000	Video-based and in-situ representative tasks - Temporal occlusion - Response time - Spatial occlusion - Eye movement recording - Point-light displays	This body of research demonstrated the ability of expert athletes to pick up and utilize postural cues emanating from an opponent in advance of action execution.
Shift in focus for research in the past two decades	Video-based and in-situ representative tasks - Eye movement recording - Temporal occlusion - Point-light displays - Retrospective verbal reports	This body of work demonstrated the important role of pattern recognition, situational assessment, and the interaction of these with advanced postural cue utilization.
Focus of recent and future research	Match analysis, video-based, and in-situ representative tasks - Animations (removing postural cues) - Eye movement recording - Temporal occlusion - Inclusion of contextual information (game status, field positioning, prior knowledge of opponent action tendencies)	This body of work demonstrated the important role of non-kinematic (contextual) sources of information such as prior knowledge about opponents' action tendencies, action patterns associated with certain game scores, court positioning, and sequences of action outcomes. The focus of future work is to understand more fully how all the different sources of information interact with one another to facilitate efficient and effective anticipation

opponent ahead of a crucial event such as ball–racket or foot–ball contact. It is now well reported that skilled athletes are far superior to their less-skilled counterparts at picking up postural information from opponents to facilitate anticipation. Although more research is needed to identify which cues are most crucial, and when these are extracted, across sports, it is unlikely that such descriptive work will continue to add much to current knowledge and understanding. More recently, there has been an increased focus on identifying the importance of other perceptual–cognitive skills, such as pattern recognition and situation assessment, to make rank-order judgments, as well as how each of these skills may vary in importance from one moment to the next. We suggest that these later areas of study remain potentially fruitful ones for the field.

Most recently, researchers have embraced a more holistic approach to anticipation by trying to identify how low-level (e.g., pick-up of kinematic information) and high-level (e.g., context) cognitive processes interact during anticipation in dynamic, time-constrained environments (e.g., Gredin et al., 2018). In contrast to this recent work, we have focused too little effort over the years on identifying the broader role of context in framing and shaping anticipation behavior. If the goal is to better explain rather than simply describe behavior, then we need to accept that anticipation is a much more multifaceted and complex process than originally thought and that it involves the integration of numerous perceptual–cognitive processes and mechanisms working together in a dynamic and flexible manner. Our ability to develop new methods that enable us to effectively measure and ultimately to enhance anticipation behavior in sport, and other professional domains, is dependent on our ability to adequately deal with this challenge.

REFERENCES

Abernethy, B. (1990). Expertise, visual search, and information pick-up in squash. *Perception* 19(1), 63–77.

Abernethy, B., Gill, D., Parks, S., & Packer, S. (2001). Expertise and the perception of kinematic and situational probability information. *Perception* 30(2), 233–252.

Abernethy, B., & Russell, D. (1984). Advance cue utilisation by skilled cricket batsmen. *Australian Journal of Science and Medicine in Sport* 16(2), 2–10.

Abernethy, B., & Zawi, K. (2007). Pickup of essential kinematics underpins expert perception of movement patterns. *Journal of Motor Behavior* 39(5), 353–367.

Alain, C., & Girardin, Y. (1978). The use of uncertainty in racquetball competition. *Canadian Journal of Applied Sport Sciences* 3, 240–243.

Alain, C., & Proteau, L. (1978). Étude des variables relative au traitement de l'information en sport de raquette. *Canadian Journal of Applied Sport Sciences* 3, 27–35.

Alain, C., & Proteau, L. (1980). Decision-making in sport. In C. H. Nadeau, W. R. Halliwell, K. M. Newell, & G. C. Roberts (Eds), *Psychology of motor behavior and sport* (pp. 465–477). Champaign, IL: Human Kinetics.

Allard, F., Graham, S., & Paarsalu, M. E., (1980). Perception in sport: Basketball. *Journal of Sport Psychology* 2, 14–21.

Araujo, D., & Kirlik, A. (2008). Towards an ecological approach to visual anticipation for expert performance in sport. *International Journal of Sport Psychology* 39(2), 157.

Belling, P., Suss, J., & Ward, P. (2015). Advancing theory and application of cognitive research in sport: Using representative tasks to explain and predict skilled anticipation, decision-making, and option-generation behavior. *Psychology of Sport and Exercise* 16(1), 45–59.

Bourne, M., Bennett, S. J., Hayes, S. J., Smeeton, N. J., & Williams, A. M. (2013). Information underpinning anticipation of goal-directed throwing. *Attention, Perception, & Psychophysics* 75(7), 1559–1569.

Broadbent, D. P., Causer, J., Williams, A. M., & Ford, P. R. (2014). Perceptual-cognitive skill training and its transfer to expert performance in the field: Future research directions. *European Journal of Sport Science* 15(4), 322–331.

Broadbent, D. P., Ford, P. R., O'Hara, D. A., Williams, A. M., & Causer, J. (2017). The effect of a sequential structure of practice for the training of perceptual-cognitive skills in tennis. *PloS One* 12, e0174311.

Broadbent, D. P., Gredin, N. V., Rye, J. L., Williams, A. M., & Bishop, D. T. (2018). The impact of contextual priors and anxiety on performance effectiveness and processing efficiency in anticipation. *Cognition and Emotion*, 1–8. DOI: 10.1080/02699931.2018.1464434.

Cañal-Bruland, R., & Mann, D. L. (2015). Time to broaden the scope of research on anticipatory behaviour: A case for the role of probabilistic information. *Frontiers in Psychology* 6, 1518.

Cañal-Bruland, R., & Williams, A. M. (2010). Recognizing and predicting movement effects. *Experimental Psychology* 57(4), 320–326.

Casanova, F., Garganta, J., Silva, G., Alves, A., Oliveira, J., & Williams, A. (2013). Effects of prolonged intermittent exercise on perceptual-cognitive processes. *Medicine & Science in Sports & Exercise* 45(8), 1610–1617.

Causer, J., Smeeton, N. J., & Williams, A. M. (2017). Expertise differences in anticipatory judgements during a temporally and spatially occluded task. *PloS One* 12(2), e0171330.

Chase, W. G., & Simon, H. A. (1973). Perception in chess. *Cognitive Psychology* 4(1), 55–81.

Cocks, A. J., Jackson, R. C., Bishop, D. T., & Williams, A. M. (2015). Anxiety, anticipation and contextual information: A test of attentional control theory. *Cognition and Emotion* 30, 1037–1048.

Dicks, M., Davids, K., & Button, C. (2010). Individual differences in the visual control of intercepting a penalty kick in association football. *Human Movement Science* 29(3), 401–411.

Dittrich, W. H. (1999). Seeing biological motion: Is there a role for cognitive strategies? In A. Braffort, R. Gherbi, S. Gibet, J. Richardson, & D. Teil (Eds), *Gesture-based communication in human-computer interaction* (pp. 3–22). Berlin: Springer-Verlag.

Enberg, M. L. (1968). Assessing perception of object directionality in tennis. Doctoral dissertation, Purdue University. Dissertation Abstracts International, 29, 806A.

Endsley M. R. (1988). Situational awareness global assessment technique (SAGAT). In *Proceedings of the National Aerospace and Electronics Conference* (pp. 789–795).

Endsley, M. R. (1995). Measurement of Situation Awareness in Dynamic Systems. *Human Factors* 37(1), 65–84.

Farrow, D., & Abernethy, B. (2003). Do expertise and the degree of perception—action coupling affect natural anticipatory performance? *Perception* 32, 1127–1139.

Farrow, D., & Reid, M. (2012). The contribution of situational probability information to anticipatory skill. *Journal of Science and Medicine in Sport* 15, 368–373.

Garland, D. J., & Barry, J. R. (1991). Cognitive advantage in sport: The nature of perceptual structures. *American Journal of Psychology 104*(2), 211–228.

Goldin, S. E. (1978). Memory for the ordinary: Typicality effects in chess memory. *Journal of Experimental Psychology: Human Learning and Memory 4*, 605–616.

Goldin, S. E. (1979). Recognition memory for chess positions: Some preliminary research. *American Journal of Psychology 92*(1), 19–31.

Gorman, A. D., Abernethy, B., & Farrow, D. (2012). Classical pattern recall tests and the prospective nature of expert performance. *Quarterly Journal of Experimental Psychology 65*(6), 1151–1160.

Gorman, A. D., Abernethy, B., & Farrow, D. (2013). Is the relationship between pattern recall and decision-making influenced by anticipatory recall? *Quarterly Journal of Experimental Psychology 66*(11), 2219–2236.

Gray, R. (2002). Behavior of college baseball players in a virtual batting task. *Journal of Experimental Psychology: Human Perception and Performance 28*(5), 1131–1148.

Gray, R. (2015). The moneyball problem: What is the best way to present situational statistics to an athlete? *Proceedings of the Human Factors and Ergonomics Society Annual Meeting 59*, 1377–1381.

Gredin, N. V., Bishop, D., Broadbent, D., Tucker, A., & Williams, A. M. (2018). Experts integrate explicit contextual priors and environmental information to improve anticipation efficiency. *Journal of Experimental Psychology: Applied 24*(4), 509–520.

Helsen, W. F., & Starkes, J. L. (1999). A multidimensional approach to skilled perception and performance in sport. *Applied Cognitive Psychology 13*(1), 1–27.

Hoffman, R. R., Ward, P., Feltovich, P. J., DiBello, L., Fiore, S. M., & Andrews, D. (2014). *Accelerated expertise: Training for high proficiency in a complex world.* New York: Psychology Press.

Huys, R., Cañal-Bruland, R., Hagemann, N., Beek, P. J., Smeeton, N. J., & Williams, A. M. (2009). Global information pickup underpins anticipation of tennis shot direction. *Journal of Motor Behavior 41*(2), 158–171.

Jackson, R. C., & Mogan, P. (2007). Advance visual information, awareness, and anticipation skill. *Journal of Motor Behavior 39*, 341–351.

Jones, C., & Miles, T. (1978). Use of advance cues in predicting the flight of a lawn tennis ball. *Journal of Human Movement Studies 4*(4), 231–235.

Loffing, F., & Hagemann, N. (2014). On-court position influences skilled tennis players' anticipation of shot outcome. *Journal of Sport & Exercise Psychology 36*, 14–26.

Loffing, F., Hagemann, N., & Farrow, D. (2017). Perceptual-cognitive training: The next piece of the puzzle. In J. Baker, S. Cobley, J. Schorer, & N. Wattie (Eds), *Routledge handbook of talent identification and development in sport* (pp. 207–220). New York: Routledge.

Loffing, F., Sölter, F., Hagemann, N., & Strauss, B. (2016). On-court position and handedness in visual anticipation of stroke direction in tennis. *Psychology of Sport and Exercise 27*, 195–204.

Loffing, F., Stern, R., & Hagemann, N. (2015). Pattern-induced expectation bias in visual anticipation of action outcomes. *Acta Psychologica 161*, 45–53.

McRobert, A. P., Ward, P., Eccles, D. W., & Williams, A. M. (2011). The effect of manipulating context-specific information on perceptual-cognitive processes during a simulated anticipation task. *British Journal of Psychology 102*, 519–534.

Mann, D. L., & Savelsbergh, G. J. P. (2015). Issues in the measurement of anticipation. In J. Baker & D. Farrow (Eds), *Routledge handbook of sport expertise* (pp. 166–175). Oxon: Routledge.

Mann, D. L., Schaefers, T., & Cañal-Bruland, R. (2014). Action preferences and the anticipation of action outcomes. *Acta Psychologica 152*, 1–9.

Mann, D., Williams, A. M., Ward, P., & Janelle, C. M. (2007). Perceptual-cognitive expertise in sport: A meta-analysis. *Journal of Sport & Exercise Psychology 29*, 457–478.

Moore, C.G., & Müller, S. (2014). Transfer of expert visual anticipation to a similar domain. *Quarterly Journal of Experimental Psychology 67*(1), 186–196.

Müller, S., & Abernethy, B. (2006). Batting with occluded vision: An in situ examination of the information pick-up and interceptive skills of high-and low-skilled cricket batsmen. *Journal of Science and Medicine in Sport 9*(6), 446–458.

Müller, S., & Abernethy, B. (2012). Expert anticipatory skill in striking sports: A review and a model. *Research Quarterly for Exercise and Sport 83*, 175–187.

Müller, S., Abernethy, B., & Farrow, D. (2006). How do world-class cricket batsmen anticipate a bowler's intention? *Quarterly Journal of Experimental Psychology 59*(12), 2162–2186.

Murphy, C. P., Jackson, R. C., Cooke, K., Roca, A., Benguigui, N., & Williams, A. M. (2016). Contextual information and perceptual-cognitive expertise in a dynamic, temporally-constrained task. *Journal of Experimental Psychology: Applied 22*, 455–470.

Murphy, C. P., Jackson, R. C., & Williams, A. M. (2018). The role of contextual information during skilled anticipation. *Quarterly Journal of Experimental Psychology 71*(10), 2070–2087.

Navia, J. A., van der Kamp, J., & Ruiz, L. M. (2013). On the use of situational and body information in goalkeeper actions during a soccer penalty kick. *International Journal of Sport Psychology 44*, 234–251.

North, J. S., Hope, E., & Williams, A. M. (2017). Identifying the micro-relations underpinning familiarity detection in dynamic displays containing multiple objects. *Frontiers in Psychology 8*, 963.

North, J. S., Ward, P., Ericsson, K. A., & Williams, A. M. (2011). Mechanisms underlying skilled anticipation and recognition in a dynamic and temporally constrained domain. *Memory 19*(2), 155–168.

North, J. S., & Williams, M. A. (2008). Identifying the critical time period for information extraction when recognizing sequences of play. *Research Quarterly for Exercise and Sport 79*(2), 268–273.

North, J. S., Williams, A., Hodges, N., Ward, P., & Ericsson, K. K. (2009). Perceiving patterns in dynamic action sequences: Investigating the processes underpinning stimulus recognition and anticipation skill. *Applied Cognitive Psychology 23*(6), 878–894.

Oudejans, R., & Coolen, B. (2003). Human kinematics and event control: on-line movement registration as a means for experimental manipulation. *Journal of Sports Sciences 21*(7), 567–576.

Posner, M. I., Nissen, M. J., & Ogden, W. C. (1978). Attended and unattended processing modes: The role of set for spatial location. In H. L. Pick & I. J. Saltzman (Eds), *Modes of perceiving and processing information* (pp. 137–157). Hillsdale, NJ: Lawrence Erlbaum.

Roca, A., Ford, P. R., McRobert, A. P., & Williams, A. M. (2013). Perceptual-cognitive skills and their interaction as a function of task constraints in soccer. *Journal of Sport & Exercise Psychology 35*(2), 144–155.

Runswick, O., Roca, A., Williams, A. M., Bezodis, N. E., McRobert, A. P., & North, J. S. (2018a). The impact of contextual information and a secondary task on anticipation performance: An interpretation using cognitive load theory. *Applied Cognitive Psychology 32*(2), 141–149.

Runswick, O. R., Roca, A., Williams, A. M., Bezodis, N. E., & North, J. S. (2018c). The effects of anxiety and situation-specific context on perceptual-motor skill: A multi-level investigation. *Psychological Research 82*(4), 708–719.

Runswick, O. R., Roca, A., Williams, A. M., McRobert, A., & North, J. S. (2018b). The temporal integration of information during anticipation. *Psychology of Sport and Exercise 37*, 100–108.

Salmela, J. H., & Fiorito, P. (1979). Visual cues in ice hockey goaltending. *Canadian Journal of Applied Sport Sciences 4*(1), 56–59.

Shim, J., Carlton, L. G., & Kwon, Y-H. (2006). Perception of kinematic characteristics of tennis strokes for anticipating stroke type and direction. *Research Quarterly for Exercise and Sport 77*, 326–339.

Smeeton, N. J., & Huys, R. (2011). Anticipation of tennis-shot direction from whole-body movement: The role of movement amplitude and dynamics. *Human Movement Science 30*(5), 957–965.

Smeeton, N. J., Ward, P., & Williams, A. M. (2004). Do pattern recognition skills transfer across sports? A preliminary analysis. *Journal of Sports Sciences 22*(2), 205–213.

Starkes, J. L. (1987). Skill in field hockey: The nature of the cognitive advantage. *Journal of Sport Psychology 9*(2), 146–160.

Starkes, J. L., Edwards, P., Dissanayake, P., & Dunn, T. (1995). A new technology and field test of advance cue usage in volleyball. *Research Quarterly for Exercise and Sport 66*(2), 162–167.

Suss, J., & Ward, P. (2015). Predicting the future in perceptual-motor domains: Perceptual anticipation, option generation and expertise. In R. R. Hoffman, P. A. Hancock, M. Scerbo, & J. L. Szalma (Eds), *Cambridge handbook of applied perception research* (pp. 951–976). New York: Cambridge University Press.

Triolet, C., Benguigui, N., Le Runigo, C., & Williams, A. M. (2013). Quantifying the nature of anticipation in professional tennis. *Journal of Sports Sciences 31*, 820–830.

Vaeyens, R., Lenoir, M., Williams, A. M., Mazyn, L., & Philippaerts, R. M. (2007). The effects of task constraints on visual search behavior and decision-making skill in youth soccer players. *Journal of Sport & Exercise Psychology 29*(2), 147.

Vater, C., Roca, A., & Williams, A. M. (2015). Effects of anxiety on anticipation and visual search in dynamic, time-constrained situations. *Sport, Exercise, and Performance Psychology 5*(3), 179.

Ward, P., Ericsson, K. A., & Williams, A. M. (2013). Complex perceptual-cognitive expertise in a simulated task environment. *Journal of Cognitive Engineering and Decision Making 7*(3), 231–254.

Ward, P., Gore, J., Hutton, R., Conway, G., & Hoffman, R. (2018). Adaptive skill as the *conditio sine qua non* of expertise. *Journal of Applied Research in Memory and Cognition 7*(1), 35–50.

Ward, P., & Williams, A. M. (2003). Perceptual and cognitive skill development in soccer: The multidimensional nature of expert performance. *Journal of Sport and Exercise Psychology 25*(1), 93–111.

Ward, P., Williams, A. M., & Bennett, S. J. (2002). Visual search and biological motion perception in tennis. *Research Quarterly for Exercise and Sport 73*(1), 107–112.

Ward, P., Williams, A. M., & Hancock, P. A. (2006). Simulation for performance and training. In K. A. Ericsson, N. Charness, P. Feltovich, & R. R. Hoffman (Eds), *Cambridge handbook of expertise and expert performance* (pp. 243–262). Cambridge, UK: Cambridge University Press.

Williams, A.M. (2009). Perceiving the intentions of others: How do skilled performers make anticipation judgments? In M. Raab, J. G. Johnson, & H. R. Heekeren (Eds), *Progress in brain research, mind and motion: The bidirectional link between thought and action* (Vol. 174, pp. 73–83). Amsterdam: Reed Elsevier.

Williams, A. M., & Burwitz, L. (1993). Advance cue utilization in soccer. In T. Reilly, J. Clarys, & A. Stibbe (Eds), *Science and Football II* (pp. 239–243). London: E & FN Spon.

Williams, A. M., Casanova, F., & Toledo., I. (in press) Anticipation. In V. Zeigler-Hill and T. K. Shackelford (Eds), *Encyclopedia of personality and individual differences*. Berlin: Springer-Verlag: Berlin.

Williams, A. M., & Davids, K. (1998). Visual search strategy, selective attention, and expertise in soccer. *Research Quarterly for Exercise and Sport* 69(2), 111–128.

Williams, A. M., Davids, K., Burwitz, L., & Williams, J. G. (1994). Visual search strategies in experienced and inexperienced soccer players. *Research Quarterly for Exercise and Sport* 65(2), 127–135.

Williams, A. M., Davids, K., & Williams, J. G. (1999). *Visual perception and action in sport*. London: E & FN Spon.

Williams, A. M., & Elliott, D. (1999). Anxiety, expertise, and visual search strategy in karate. *Journal of Sport & Exercise Psychology* 21(4), 362.

Williams, A. M., & Ericsson, K. A. (2005). Perceptual-cognitive expertise in sport: Some considerations when applying the expert performance approach. *Human Movement Science* 24, 283–307.

Williams, A. M., Ford, P., Hodges, J., & Ward, P. (2018). Expertise in sport: Specificity, plasticity and adaptability. In K. A. Ericsson, N. Charness, R. Hoffman, & A. M. Williams (Eds), *Cambridge handbook of expertise and expert performance*, 2nd edn (pp. 653–674). Cambridge, UK: Cambridge University Press.

Williams, A. M., Hodges, N. J., North, J. S., & Barton, G. (2006). Perceiving patterns of play in dynamic sport tasks: Investigating the essential information underlying skilled performance. *Perception* 35(3), 317.

Williams, A. M., & Jackson, R. J. (2019). *Anticipation and decision-making in sport*. London: Routledge.

Williams, A. M., Janelle, C. M., & Davids, K. (2004). Constraints on the search for visual information in sport. *International Journal of Sport and Exercise Psychology* 2(3), 301.

Williams, A. M., Murphy., C. P., Broadbent, D. P., & Janelle, C. J. (2018). Anticipation in sport. In T. Horn (Ed.), *Advances in sport psychology*, third edn (pp. 229–246). Champaign, IL: Human Kinetics.

Williams, A. M., & North, J. S. (2009). Identifying the minimal essential information underpinning pattern recognition. In, D. Araujo, H. Ripoll, & M. Raab (Eds), *Perspectives on cognition and action* (pp. 95–107). Hauppauge, NY: Nova Science.

Williams, A. M., North, J. S., & Hope, E. R. (2012). Identifying the mechanisms underpinning recognition of structured sequences of action. *Quarterly Journal of Experimental Psychology* 65, 1975–1992.

Williams, A.M., & Davids, K. (1995). Declarative knowledge in sport: A by-product of experience or a characteristic of expertise? *Journal of Sport and Exercise Psychology* 17(3), 259–275.

Wright, D. L., Pleasants, F., & Gomez-Meza, M. (1990). Use of advanced visual cue sources in volleyball. *Journal of Sport and Exercise Psychology* 12(4), 406–414.

CHAPTER 27

DIAGNOSTIC REASONING AND EXPERTISE IN HEALTH CARE

VIMLA L. PATEL, DAVID R. KAUFMAN, AND THOMAS G. KANNAMPALLIL

INTRODUCTION

MEDICAL cognition names a body of work pertaining to the study of cognitive structure and processes, such as perception and action, memory, comprehension, problem solving, decision making, and knowledge representation in medical practice or in tasks representative of medical practice. Diagnostic reasoning research describes a form of qualitative and quantitative inquiry that examines the cognitive processes underlying medical decisions. Researchers study individuals at various levels of experience (from novices to experts) and in different clinical roles in medical settings, including medical students, physicians, nurses, and biomedical scientists.

Medical problem solving, diagnostic reasoning, and decision making are all terms used in a growing body of literature that examines how clinicians understand biomedical information, solve clinical problems, and make patient-care decisions. The study of diagnostic reasoning underlies much of medical cognition, and has been the focus of considerable research in cognitive science and artificial intelligence as applied to medicine.

Diagnostic reasoning involves an inferential process for making diagnostic and therapeutic decisions or understanding the pathology of a disease process. On the one hand, medical reasoning is basic to all higher-level cognitive processes in medicine, such as problem solving and medical text comprehension. On the other hand, the structure of medical reasoning is, itself, the subject of considerable scrutiny.

Generally, experts (in knowledge-based domains such as medicine, physics, and chess) are known to have superior memory skills in recognizing patterns and expert problem solvers use knowledge-based reasoning strategies (i.e., a heuristic search) in solving problems in their domain of expertise (Anzai, 1991; Chase & Simon, 1973; Larkin, McDermott, Simon, & Simon, 1980; Patel & Groen, 1991a, 1991b). These characteristics can be reliably tested. This is contrasted with goal-directed reasoning of a novice such as a medical student, working from a hypothesis regarding an unknown to the given (known information). Here, expertise is more about *knowing that* as well as knowing *how* (actions).

An *expert* can be defined in terms of achieving a certain level of performance, as exemplified by an Elo rating in chess, or by virtue of being certified by a sanctioned licensing body, as is characteristic of medicine. In a complex domain such as medicine, expertise is not a monolithic entity, there is considerable variation and specialization, and individuals typically perform at an expert level, only within a narrow-specialized context. However, the definition of expertise with the use of professional certifications is not often precise. Years of experience is also not necessarily the best criterion for measuring expertise. Although there is a correlation between experience and expert performance, the degree of specialized knowledge and skill is an important factor. As a working definition, medical expertise is defined as either generic expertise (medical-certified doctor such as an MD) or specialized expertise (a doctor with a certification in a specialty such as endocrinology).

The domains of medicine and health care are in a state of rapid transformation. This has been partially spurred by new developments in biological and computer-based technologies, as well as by societal influences such as budget cuts, changing population demographics, and the prevalence of certain diseases. Medical and health sciences curricula have been under great pressure to adapt to the rapid growth of knowledge and also to meet the changing needs of society. It is widely recognized that the didactic lecture alone is inadequate for conveying the range of knowledge and skills required to support medical practice. Instead, modern health information technology (HIT) plays an important role in clinical practice and in training programs, and has vastly changed the task of clinicians, requiring a different nature of expertise.

It has been demonstrated that superior expert performance is mediated by highly structured and richly interconnected domain-specific knowledge (Feltovich, Johnson, Moller, & Swanson, 1984; Schmidt & Boshuizen, 1993). Their knowledge bases are hierarchical and densely interconnected, which allows new pieces of information to become well integrated. For example, expert cardiologists are routinely called upon to integrate clinical findings at various levels of aggregation, from biochemical abnormalities evidenced in blood tests to perturbations at the system level to clinical manifestations as expressed in the patient's complaints. Given that cognition is constrained by the limitations of human working memory, experts' well-organized domain knowledge enables them to circumvent some of these limitations. Clinicians are increasingly provided with external memory and decision support aids to offset human limitations. These support systems are provided via the use of HIT, which help health professionals to reduce their cognitive load, often generated by having to multitask and to deal with constant interruptions, during clinical practice. In addition, patients play an important role in health care decisions, requiring them to work in team problem solving and decision-making situations. Expansion of scientific and clinical knowledge is growing exponentially, and this necessitates the health professionals to constantly update their knowledge base by translating and integrating such knowledge for their practices. All these create challenges to not only acquire, but also maintain a certain level of expertise in health care.

This chapter begins with a brief historical perspective on expertise research within the medical domain, followed by an evolving perspective on diagnostic reasoning in the context of expertise (and methods of investigation). Some recent research that shed light on the nature of expertise taking into account the complexity of medical domain is discussed next. The role of support technologies, including technology-mediated diagnostic reasoning and its role in expertise research, is discussed subsequently, followed by conclusion and future directions about expertise research for better understanding of diagnostic reasoning in health care.

History and Evolving Perspectives on Diagnostic Reasoning

Diagnostic reasoning in medicine can be understood as the cognitive processes involved in categorizing a set of signs and symptoms into diagnostic concepts (i.e., a diagnosis or diagnostic category) (Feltovich, Spiro, & Coulson, 1993). It also serves to identify a body of scholarly literature in disparate fields including medical cognition (Patel & Kaufman, 2013), decision making (Dawes & Kagan, 1988), and medical education and medical artificial intelligence (Clancey & Shortliffe, 1984). The nature of diagnostic expertise has been the subject of research for more than fifty years (Rimoldi, 1961). Foundational work on diagnostic reasoning was carried out over a period of time from the late 1950s through the early 1970s. This includes influential work by several researchers (e.g., De Dombal, Leaper, Staniland, McCann, & Horrocks, 1972; Feinstein, 1973a, 1973b; Kleinmuntz, 1968; Ledley, Lusted, & Ledley, 1959; Rimoldi, 1961), whose objectives included improving reasoning in the context of medical education and formalizing decision-making activities, and developing systems to assist the diagnostic process, described in the early research on medical education (Groen & Patel, 1985; Norman, 2005; Norman et al., 1996; Patel, Arocha, & Kaufman, 1994).

Medical cognition refers to a research area that began with the influential work of investigators like Elstein and colleagues (Elstein, Shulman, & Sprafka, 1978). The work was steeped in theories and methods from cognitive and decision sciences. Studies in medical cognition include studies of perception, comprehension, reasoning, decision making, and problem solving (Patel, Arocha, & Kaufman, 2001). There have been a number of reviews and edited texts that summarize this body of research (e.g., Higgs, 2008). Given the differences in theory and methodology, it is useful to group studies of medical cognition into a decision-making and judgment perspective, whereby decisions are contrasted with a normative model, based on probability theory, indicating optimal choices under conditions of uncertainty (Dawes & Kagan, 1988), and a problem-solving approach, whereby the focus is on describing the cognitive processes in reasoning tasks (Ericsson & Simon, 1993). This review predominantly focuses on research in the problem-solving tradition.

A parallel distinction can be made between a systematic and analytic form of reasoning and a more intuitive form that emphasizes pattern recognition. A common thread in early work on medical cognition was that clinical reasoning can be construed as a rigorous logical process necessitating prerequisite knowledge and skills for competent performance (Feinstein, 1973a, 1973b). However, studies found that clinicians' reasoning was often flawed (De Dombal et al., 1972) and less experienced clinicians lacked the reasoning acumen of seasoned physicians (Kleinmuntz, 1968). On the other hand, expert reasoning in medicine has been construed as a heuristically driven process (Patel & Groen, 1991a). In recent years, dual-process theory (Kahneman, 2003) has been invoked as a mechanism to explain these differences (Croskerry, 2009). Croskerry employs this distinction to explain different facets of clinical reasoning and decision making. The intuitive, non-empirical approach draws heavily on the experience of the decision maker. Seasoned clinicians recognize patterns in the information presented and act accordingly (Croskerry, 2009). The decision is said to be recognition-primed (Klein, 1989). On the other hand, the analytical approach is more rigorous and resource intensive (more intellectually demanding) and involves systematic testing of hypotheses.

The dual-process theory captures an important distinction in diagnostic-reasoning research. In their pioneering work, Elstein and colleagues characterized the process of diagnostic reasoning as hypothetico-deductive (Elstein et al., 1978). This involves the rapid generation of a small set of hypotheses to explain the patient's problem, which are then tested against incoming information (e.g., new laboratory results). The initial hypotheses serve to reduce the problem space and enable clinicians to selectively attend, process, and classify findings. In this framework, data are interpreted as positive, negative, or noncontributory to the various hypotheses. In a series of studies, the approach did not differentiate between physicians, who were judged as exceptional diagnosticians, and those who were not designated as such. The model of reasoning was very influential in medical cognition and medical education research. It gave further credence to the belief that the hypothetico-deductive model both is descriptive of diagnostic reasoning and defines a normative standard and skill set that should be taught to medical students.

The work of Elstein and colleagues was strongly influenced by information-processing theories and the seminal work of Newell and Simon (1972). Much of the pioneering tasks in the early years of cognitive science were done in knowledge-lean tasks such as the Tower of Hanoi or Cryptarithmetic problems (e.g., Newell & Simon, 1972), where the latter is a type of mathematical puzzle in which the digits are replaced by letters of the alphabet, to solve an addition problem. Although individuals can vary substantially in terms of skilled performance, the rules are fairly simple and do not require much in the way of prior knowledge.

Elstein and colleagues' work similarly emphasized process rather than knowledge. In the subsequent decade, a convergence of factors including the development of expert systems led cognitive researchers to investigate how differences in knowledge impact clinical reasoning. The emergence of medical AI (particularly, expert systems technology) paralleled the early work on medical cognition. The focus was on the development of methods and implementations of expert systems that, to a certain extent, approximate medical experts' reasoning.

AI in medicine and medical cognition mutually influenced each other in a number of ways, including (1) providing a basis for developing formal models of competence in problem-solving tasks; (2) elucidating the structure of medical knowledge and providing important epistemological distinctions; and (3) characterizing productive and less-productive lines of reasoning in diagnostic and therapeutic tasks. Gorry (1973) conducted a series of studies comparing a computational model of medical problem solving with the actual problem-solving behavior of physicians. This analysis provided a basis for characterizing a sequential process of medical decision making, one that differs in important respects from early diagnostic computational systems based on Bayes theorem (Ledley et al., 1959), which describes the relations between conditional probabilities and provides a normative rule for updating belief in light of evidence.

Pauker, Gorry, Kassirer, and Schwartz (1976) drew on Gorry's earlier work to develop the Present Illness Program (PIP), a program designed to take the history of a patient with edema. Questions guiding this research, including the nature and organization of expert knowledge, were of central concern to both developers of medical expert systems and researchers in medical cognition. The development and refinement of the program was partially based on studies of clinical problem solving. Medical expert consultation systems, such as Internist (Miller, Pople, & Myers, 1984) and MYCIN (Shortliffe, 2012), introduced ideas about knowledge-based reasoning strategies across a range of cognitive tasks. MYCIN, in particular, had a substantial influence on cognitive science. It contributed several advances (e.g., representing reasoning under uncertainty) in the use of production systems as a representation scheme in a complex, knowledge-based domain. MYCIN also highlighted the difference between medical problem solving and the cognitive dimensions of medical explanation. Clancey's research (Clancey, 1985; Clancey & Letsinger, 1982) on GUIDON and NEOMYCIN was particularly influential in the evolution of new models of medical cognition. Clancey reconfigured MYCIN in order to utilize it to teach medical students about meningitis and related disorders. NEOMYCIN was based on a more psychologically plausible model of medical diagnosis, and raised a host of questions regarding experts' knowledge-based strategies.

Patel and Groen (1986) studied knowledge-based solution strategies of expert cardiologists, examining their pathophysiological explanations and diagnostic reasoning strategies in solving a complex clinical problem. The results revealed a notable difference between participants who correctly diagnosed the problem and those who misdiagnosed the clinical case. Clinicians, who accurately diagnosed the problem, employed a forward-oriented (data-driven) reasoning strategy, using patient data to move towards a complete diagnosis as a sequence of inferences, reasoning from data to hypothesis. In contrast, participants who did not accurately reach a diagnosis (partial or misdiagnosis) employed a backward or hypothesis-driven reasoning strategy. Patel and Groen (1986) were able to show that experts most often reason using forward-driven inferences in a familiar domain. Contrary to the findings by Elstein and colleagues (1978) hypothetico-deductive reasoning did not figure prominently in their solution strategies.

An explanation for the apparently conflicting findings is that forward reasoning is used when a physician has significant domain knowledge regarding the problem at hand. In contrast, they resort to backward reasoning when their available knowledge is insufficient to solve the problem. In other words, their domain knowledge is not adequate to seamlessly process the presented information. When the patient problem is familiar to a clinician, it can invoke a process of pattern recognition that rapidly partitions the problem space and may instantiate a single candidate hypothesis. This is in contrast to a hypothetico-deductive process, which is cognitively effortful, and more memory taxing as it presupposes that one must simultaneously consider multiple hypotheses against the available evidence. In a series of studies comparing clinicians at various levels of expertise and medical problems of variable complexity, Patel and Groen studied the conditions under which either mode predominated in diagnostic reasoning tasks (Patel & Groen, 1991a; Patel, Groen, & Arocha, 1990).

As seen in the research so far, experts have traditionally been identified based on differences in search strategies (including their directionality of reasoning). The two basic forms of reasoning are deductive and inductive. However, reasoning in the *real world* does not fit cleanly into these basic types. For this reason, a third form of reasoning, termed *abductive reasoning* (Peirce, 1955), has been recognized as best capturing the generation of clinical hypotheses, where deduction and induction are intermixed. Abductive reasoning is a data-driven process and also dependent on domain knowledge.

Here, the processes of hypothesis generation and testing, described by Stefanelli and Ramoni (1992), can be characterized in terms of four types of processes: abstraction, abduction, deduction, and induction. Following this, hypotheses that could account for the current situation are related through a process of *abduction*, characterized by a *backward flow* of inferences across a chain of directed relations that identify those initial conditions from which the current abstract representation of the problem originates. This process provides tentative solutions to the problem by way of hypotheses. For example, knowing that disease A will causes symptom b, by abduction one will try to identify the explanation for b, while through deduction one will forecast that a patient affected by disease A will manifest symptom b; both inferences use the same relation along two different directions. The major feature of induction is, therefore, the ability to rule out those hypotheses whose expected consequences turn out to be not in agreement with the patient problem. There is no way to logically confirm a hypothesis, but we can only disconfirm it in the presence of contrary evidence. These types of clinical reasoning in medicine are described by Patel and Ramoni (1997), where experts and less than experts are shown to have specific characteristic in patterns of diagnostic reasoning. There is a support from AI literature where the cognitive model of reasoning was validated against a generic system containing abductive reasoning capabilities (Stefanelli & Ramoni, 1992).

The model presented in the previous paragraph can be used to explain the medical diagnostic process and role of expertise, and there is enough empirical evidence to add support to the model. First, seasoned clinicians (experts) are selective in the data they collect (*abstraction*), focusing on relevant data for generating diagnostic hypotheses, while ignoring other less-relevant data, as found in Patel and Groen (1986) and later supported in other studies. Successful clinicians focus on the fewest and closely related pieces of data and are better able to integrate these pieces of data into a coherent explanation for the problems (for example, see Patel & Groen, 1991a, 1991b). Second, typically physicians generate a small set of hypotheses very early (*abduction*), often as soon as the first pieces of data become available, as was first shown by Elstein, Shulman, and Sprafka (1978) and elaborated by Feltovich et al.'s (1984) logical competitor set. Third, as originally shown by Elstein, Shulman, and Sprafka (1978), physicians sometimes make use of the hypothetico-deductive process (*deduction*). Cues in the clinical case lead to the generation of a few selected hypotheses, and then each hypothesis is evaluated for consistency with the cues (*induction*). Mixture of such reasoning strategies are most often used in a complex real-world clinical environment.

As described earlier, decision making in natural environment differs in important respects from laboratory-based decisions. Naturalistic decision making (NDM) is concerned with the study of cognition in *real-world* work environments that are often dynamic. The majority of this research combines conventional protocol-analytic methods with innovative methods designed to investigate reasoning and behavior in realistic settings (Rasmussen, Pejtersen, & Goodstein, 1994; Woods, 1993). The investigation of decision making in the work context necessitates an extension of the cognitive framework based on knowledge structures, processes, and skills, to include variables such as stress, time pressure, and fatigue as well as communication patterns that affect team performance.

Among the issues that have been investigated in NDM are understanding how decisions are jointly negotiated and updated by clinicians with different areas of expertise (e.g., pharmacology, respiratory medicine); the complex communication process; role of technology in mediating decisions and its impacts on reasoning; and the sources of error generated in the decision-making process.

Patel, Kaufman, and Magder (1996) studied decision making in a medical intensive care unit (ICU) with the purpose of describing jointly negotiated decisions, communication processes, and the development of expertise. In real-time observational data collection, ICU decision making was found to be characterized by a rapid, serial evaluation of options leading to immediate action, where reasoning is schema-driven in a forward direction toward action with minimal inference or justification (where explanation and justification interferes with utility). However, when the patient outcome is not consistent with the original proposed hypothesis, then the initial decision comes under scrutiny. This strategy can result in a brainstorming session in which the team retrospectively evaluates and reconsiders the decision and considers possible alternatives. Various patterns of reasoning are used to evaluate alternatives in these *brainstorming* sessions, including probabilistic reasoning, diagnostic reasoning, and biomedical causal reasoning, as shown in Figure 27.1, supporting decision making in clinical settings necessitates even better understanding of communication patterns and its relationship to error monitoring in the context of expertise.

The concepts of explicit and implicit knowledge requires some discussion because a key to understanding the role of implicit knowledge in expert performance lies in the relationship between formally acquired knowledge and informally acquired knowledge (Sternberg & Horvath, 1999). We make the following claims: (1) explicit knowledge and implicit knowledge are two separate forms of knowledge, which are effected by different mechanisms and acquired through different experiences; (2) the equivalent to these two types of knowledge in the medical domain are domain-specific knowledge, which is explicitly acquired, and practical clinical knowledge, which is partially learned through experience in hospitals and other health care settings; (3) the successful utilization of implicit knowledge rests upon the acquisition of well-formed, biomedical knowledge structures; and (4) implicit knowledge is situated. That is, it becomes available in routine situations of practice and can only be acquired with practice. It is important to note that implicit knowledge can be made explicit and this is a critical part of the learning process.

COMPLEXITY AND EXPERTISE IN DIAGNOSTIC REASONING

Error Detection and Recovery

As discussed in this chapter, experts exhibit superior knowledge organization, solution strategies, performance efficiency, a highly refined ability to recognize and integrate the critical features of problems, and an improved ability to predict the consequences of decisions taken. However, there also exists compelling evidence that experts are by no means immune to error. Intuitively, it makes sense to believe that experts make fewer errors than medical trainees. However, researchers in other domains, such as aviation, have abandoned the goal for zero tolerance, focusing instead on the development of strategies to enhance the ability to recover from the generated errors (Amalberti, 2001). Given the evidence from other domains as well as evidence of error recovery observed in our prior ethnographic studies (Cohen, Blatter, Almeida, & Patel, 2007; Kubose,

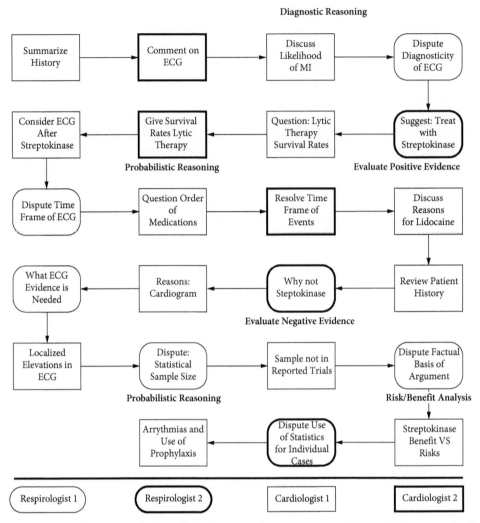

FIGURE 27.1 Reasoning during clinical review, showing pattern of negotiations typical of complex clinical care practice.

Adapted from Patel, V. L., Kaufman, D. R., and Magder, S. A., 'The acquisition of medical expertise in complex dynamic environments,' in K. A. Ericsson, ed., *The road to excellence: The acquisition of expert performance in the arts and sciences, sports and games*, pp. 127–165 © Laurence Erlbaum Associates, Inc., 1996.

Patel, & Jordan, 2002), we perceived a need to explore further the concept of promoting error recovery.

In our model, error is viewed as something that cannot be eliminated, but can be negotiated in a complex environment (Patel, Cohen, Murarka, et al., 2011). However, mechanisms of error detection and recovery in complex clinical settings are poorly understood, and the roles of distribution of attention, expertise, and team members in a complex health care system is inherently unpredictable, as this distribution is an emergent property of a workflow that is directed by circumstance and patient need (Franklin, Robinson, & Zhang, 2014; Kannampallil, Cohen, Kaufman, & Patel, 2014; Patel, Kaufman, & Cohen, 2014; Patel, Zhang, Yoskowitz, Green, & Sayan, 2008). In Figure 27.2, we present a model of error detection and recovery during a process of diagnostic reasoning in a complex health-care setting. While the detection of error requires significant cognitive resources, recovery is an even more complex process, where there is a reconstruction and reassessment of the original erroneous action to develop and implement remedial measures. Our studies show that error detection and recovery play a central role in the development of clinical expertise. This process is important for distributed patient care environment, where clinical teams and the environment can create opportunities to learn iteratively on the job.

The error boundaries in Figure 27.2 provide opportunities for us to design appropriate interventions where *near misses* can be used for the learning phase to recognize how such errors occur (Croskerry, 2003). However, interventions to address the *error recovery* phase will require knowledge.

Errors are contextualized within the clinical workflow and a multimethod approach is used to characterize the workflow, which include human-intensive observation, as well as laboratory-based controlled studies of error management, both individually and in teams (Patel, Cohen, Batwara, & Almoosa, 2011; Patel, Cohen, Murarka, et al., 2011; Patel, Kaufman, et al., 2014). They are supplemented by technological tools to mediate rapid and consistent explanation of workflow activities. Automated approaches in which the movements of multiple team members are monitored using sensor-based technology such as radio frequency identification (RFID) tags are shown to contribute new insights into clinical workflow (Kannampallil et al., 2014; Vankipuram, Kahol, Cohen, & Patel, 2009; Vankipuram, Traub, & Patel, 2017), including the characterization of team aggregation and dissemination as emergent properties of the entire system. The research described in this section provide details of how studies of error detection and correction as a function of expertise can be conducted in health care domains.

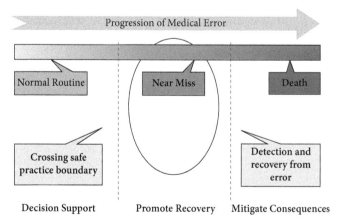

FIGURE 27.2. A framework for error detection and correction in a clinical environment with identification of unsafe situations, and recovering from them.

We conducted a pilot in situ study of expertise and team decision making to understand the mechanisms by which errors are detected and corrected in this dynamic, high velocity, time-pressured environment, and to investigate the strategies for error management and risk assessment as a function of expertise. Using participant observation and shadowing techniques, an expert in cardiothoracic ICU in a busy academic hospital detected and recovered from 75 percent of the *errors* made during the 10-hour period. On further analysis, we found that 70 percent of the errors identified required expert knowledge for recovery, and were likely to have serious consequences if undetected. Resident physicians detected fewer errors than experts subjects, and corrected a smaller proportion of these (61 percent) errors. A senior-level medical student detected the least number of errors, and those that were corrected were mostly routine errors that did not require detailed medical knowledge (Kubose et al., 2002).

The fast pace of decision making by experts, combined with a high level of confidence, meant that the mistakes were generated quickly and often. However, due to the attending physician's expert knowledge and ability to evaluate the problem situation, generated errors were rapidly identified and corrected.

With our promising pilot results, we began studies of expertise and error detection and recovery in semi-naturalistic environment in a more formal manner. We showed that teams detected and corrected more errors than under the individual experimental condition, and teams worked collaboratively to manage these errors (Patel, Cohen, Batwara, et al., 2011). But, with an increase in problem elaboration (and discussion), teams generated new errors, not all of which were detected or corrected. There appears to be two different processes related to error detection and error correction, where the former is related to judgment (requiring general knowledge) and the latter to action (requiring practical knowledge). These results reflect a delicate relationship between delivering competent and safe patient care (real-time *patient* rounds, usually at the bedside) and learning from errors (*teaching rounds*, which involve *near miss* case discussions after the fact), suggesting that both real-time problem solving and post-hoc explanation of near misses are necessary to meet the needs for teaching trainees on the job and providing competent patient care to understand what experts do (Patel, Kannampallil, & Shortliffe, 2015).

Extensions of this study in a real-time naturalistic setting (in an ICU) showed that collaboration among team members in practice resulted in significantly better recovery from errors (i.e., two-thirds of the errors generated were corrected compared to in either the individual condition or semi-naturalistic team situation) (Patel, Shine, & Almoosa, 2014). The clinical environment provided some form of safety net for error recovery, which was not available during individual problem solving or in a semi-naturalistic situation. Understanding the limits and failures of human decision making by experts and novices is important for building robust decision-support systems to manage the boundaries of risk of error in decision making for experts and trainees. Detection and correction of potential error is an integral part of cognitive work in complex, critical care settings.

EXPERTISE AND DEVIATIONS FROM STANDARD PROTOCOL

Standardized clinical protocols are important for consistent and safe practices. Nevertheless, complex clinical environments are highly dynamic in nature and often require clinicians, confronted with non-standard situations, to adjust and deviate from standard protocol. Some of these deviations are errors which can result in harmful outcomes. On the other hand, some of the deviations can be innovations, creating new potential solution paths. These deviations are dynamic adjustments to the protocols made by clinicians to adapt the current operational conditions, which can result in greater accuracy and efficiency. However, we know very little about the underlying cognitive processes related to errors and innovations.

In a study by Kahol and colleagues (Kahol, Vankipuram, Patel, & Smith, 2011), the authors investigated the extent to which deviations are classified as errors and innovations, as a function of expertise in a trauma ICU. Using ethnographic observations, trauma cases were observed, and data were analyzed using measures that included customized activity-error-innovation ontology, activity timestamps, and the expertise of the team members. The authors found that the experts' deviations from the standard trauma protocol were a combination of errors and innovations, whereas the novices' deviations were mostly errors (Kahol et.al, 2011). Memory is the term given to the structures and processes involved in the storage and subsequent retrieval of information. How the formation and use of retrieval structures to encode and retrieve information develop is a function of expertise. It is possible that the existence of these retrieval structures (a well-known psychological concept) among experts and top-down information processing allow for adaptive time critical thinking (i.e., backtracking strategies prominent in learning) that supports innovation in experts (Ericsson & Kintsch, 1995). This could be supplemented by the information filtering for critical information, supported by retrieval structures (Patel & Kannampallil, 2015). On the other hand, novices are driven by a bottom-up reasoning mechanism, and with limited support from not so well-developed retrieval structures and filtering. As a result, they are overwhelmed by data, resulting in errors. As expertise develops, there is less reliance on clinical protocols, and the task is shifted more toward evaluations of patients' condition against prior knowledge of similar conditions.

An example of such an expert adaptive situation was seen when a patient was admitted to the Emergency Room (ER) at a Canadian hospital with severe chest pain. One important option for stabilizing the patient was to give him an intravenous drug called streptokinase. But, according to the then-current clinical guideline, this drug was contraindicated for patients who were likely to require subsequent surgery. Given the age of the patient and other clinical factors, the physician decided to administer the drug, recognizing that the patient might require urgent surgery at a later time. The rationale for this decision was that this patient could not be stabilized in the ER without administration of the drug, making it unrealistic to avoid the drug simply because there was a possibility of future surgery. The patient was saved, and the clinical guidelines have now been updated. However, there was substantial controversy surrounding this case at the time because the physician explicitly violated the existing guideline, even though he was aware of that guidance.

Adherence to standard protocols or guidelines is important, but effective training helps to ensure that physicians have the ability to recognize situations that require a creative adaptive solution, and conditions where a standard approach may not correct. Such educational approaches will be critical for a comprehensive evaluation of necessary cognitive skills that can recognize conditions under which deviations from a protocol may be necessary. This will require further expert–trainee studies on the acquisition of development of such skills, and to develop a model such that can adequately inform the training process for medical students and other health care trainees.

Technology-Mediated Diagnostic Reasoning and Expertise

Clinical settings are often characterized as complex environments where the limited attentional resources of clinicians are in high demand. Such attentional demands are exacerbated in critical care settings such as ICUs and emergency care environments where clinical activities evolve at a very rapid pace. In addition, modern clinical settings are equipped with health information technologies aimed at improving and supporting clinical activities. Primary among the HIT solutions is an integrated electronic health record (EHR).

HIT as a Mediating Artifact

Given the current focus on providing cognitive support to clinicians via technology, and federal incentives on the use of EHRs in clinical practice, there has been considerable debate regarding their impact, including its influence on clinicians. Although there have been positive reports regarding the success of EHR in providing decision support, alleviating the memory and cognitive load, confirmatory evidence is still lacking (Garg et al., 2005).

However, technology mediates human performance, transforming how individuals and groups interact with technology. In cognitive science, this is referred to as the representational effect—where different representations of a common abstract structure can result in different representational efficiencies, task complexities, and behavioral outcomes (e.g., Zhang & Norman, 1994).

Reports and initiatives from the Office of National Coordinator for HIT and the Institute of Medicine (now, the National Academy of Medicine) have characterized the importance of aligning HIT within the social fabric of the health care environment. Achieving such an alignment requires, at least in part, understanding of the limits of human cognition and performance (Institute of Medicine, 2011). Similarly, a recent report from the American Medical Informatics Association (AMIA) echoed the importance of better understanding the technical, cognitive, organizational, and policy implications of improving the interface between humans and HIT (Bloomrosen et al., 2011).

The question then becomes the following: what are the challenges at the *human–HIT interface*? To answer this question, one must consider HIT (and more specifically, EHRs) as a cognitive artifact—"an artificial device that maintain[s], display[s] or operate[s] on information in order to serve a representational function and that affect[s] human cognitive performance" (Norman, 1991, p. 17). Considering an EHR as a cognitive artifact can help in characterizing the outcomes that arise from interactions—both positive and negative. For example, what are the causes of unanticipated errors and adverse events that arise as a result of HIT use? How can such adverse events be mitigated?

In the rest of this section, we review the role of human cognition in interactions with a range of HIT applications. We describe how clinical reasoning and decision making are affected by HIT, and the role of expertise in such interactions. The review is informed by a variety of topical areas ranging from human factors, health cognition, biomedical informatics, to HIT research. The insights gained can help in addressing the challenges of medical errors, their cause and characteristics, and efforts to mitigate them. In addition to being instrumental in the design of tools, understanding the mediating role of HIT can also help hospital administrators and organizational leaders in improving the quality and safety of patient care.

Considering the EHR as a cognitive construct helps in characterizing it as an external tool assisting physicians in performing their clinical activities. Given the collaborative nature of clinical settings, EHRs coordinate clinician activities within a distributed environment potentially minimizing cognitive demands, serving as a memory aid (Hutchins, 1995), facilitating task offloading (e.g., simplifying tedious manual tasks) (Zhang & Norman, 1994), providing decision making and reasoning support (Patel, Kushniruk, Yang, & Yale, 2000), and supporting routine clinical workflow. However, it also does not always function in the expected ways, leading to unintended consequences and errors (e.g., Ash, Berg, & Coiera, 2004; Koppel et al., 2005).

HIT and Its Effects on Diagnostic Reasoning

The EHR is a mediating artifact and it profoundly shapes the interactions, decision making, and reasoning behaviors of clinicians. One of the earliest studies that investigated the role of EHRs in changing the process of reasoning was by Patel and colleagues (2000). The central thesis of these studies was that EHR was a cognitive artifact *mediating* the interactions and that its impact could be measured by cognitive methods. In a study of users of a computerized patient record system, Kushniruk and colleagues (1996) evaluated the changes in clinicians' decision-making and reasoning strategies. Based on their evaluation using a combination of data collection methods, they found that the strategies evolved over time: as clinicians became accustomed to the interface and the interactions, their actions were increasingly driven by the presentation of information on the interface—i.e., their interactions were based on a screen-based strategy for understanding the patient context and making their assessments and plans regarding patient care. Such an approach resulted in more comprehensive consideration of recorded information, but it also promoted something of a confirmatory bias in which clinicians ignored information that was not aligned with their operative diagnostic hypothesis.

Follow-up studies evaluated how such changes in reasoning processes occurred. For example, Patel and colleagues (2000) tracked the changes in physicians' information retrieval and knowledge organization from a paper-based health record to an EHR. They found that information was organized differently by the physicians on these two records: narrative for paper form and discrete and sparse information for the electronic form. Different information organization results in the application of different reasoning strategies for making decisions. The authors found that paper-based records prompted physicians to use a data-driven reasoning strategy—where the focus was on gathering information from the patient to make the diagnostic decision. In contrast, EHRs supported a hypothesis-driven strategy, where the physicians relied on decision support prompted by the EHRs to narrow the field of search (such as using the cardiology filter for patient problems related to chest pain). The impact of technology was so profound, that after the EHRs were removed from the clinical workplace, the clinicians used the organization similar to the EHRs in working with paper-based forms, showing residual effect of technology on knowledge organization.

Other researchers have evaluated how information organization on interfaces have changed the nature of reasoning, and thereby, diagnostic outcomes. Zeng and colleagues (Zeng, Cimino, & Zou, 2002) organized clinical information around clinical concepts; such an organization, they argued, potentially reduced the information overload on the clinicians. Their evaluation study revealed that concept-based organization resulted in more efficient information retrieval and, in general, improved the overall diagnostic and reasoning process. Other representations, such as the use of temporal narratives, have also found changes in the clinician reasoning process (Plaisant et al., 1998). Studies have also evaluated specific interface elements on the quality and efficacy of the reasoning process. For example, Bauer and Guerlain (2011) evaluated the effect of graphical presentation of results in a critical care setting. Their presentation, based on sparklines, a minimalist graphical representation, had effects on the efficiency and quality of diagnostic accuracy.

Additionally, the organization of information on an interface also affects its retrieval and use. In a study exploring the nature of information retrieved by physicians in a real-world information search task, Kannampallil and colleagues found differences in the nature of information retrieved from electronic sources and paper-based records, thereby affecting the interpretive process of reasoning (Kannampallil et al., 2013). They found that the information retrieved from electronic records was at a higher cognitive level (e.g., findings) than those retrieved from paper-based records, which were basic clinical data elements (e.g., observation). Although the authors did not investigate the specific effects of expertise, the fact that information retrieved from the EHR included higher-level cognitive elements potentially introduces challenges to clinicians with lesser levels of expertise.

In modern EHRs, computerized physician order entry (CPOE) systems are used to order and track medications. As reported extensively in the literature, medication ordering, dispensing, and administration process using a CPOE is fraught with errors (e.g., Koppel et al., 2005). The reasons for CPOE-based errors can be attributed a number of factors; a primary cause for such errors is the information and cognitive overload that is at these interfaces. Errors included fragmented presentation of information across screens, improper organization of categories, and usability challenges. Such inconsistencies often affect the reasoning process of the clinician when creating medication orders, leading to several types of errors: ordering the wrong drug, order on the wrong patient or encounter, and improperly composed orders.

In a study investigating the effects of the interface elements and organization on the reasoning and decision-making process of clinicians, Horsky and colleagues (2006) provide a detailed description of the source, progression, and effects of errors. Their report was based on a case study of a potassium chloride (KCl) overdose. Based on interviews, EHR usage logs and task analysis, the authors investigated how the overdose errors occurred over a seven-day period. The CPOE interface provided order forms for critical-care drip (saline-medication bags hung over the bed) and IV bolus orders that were similar, but had different calculation modalities. IV bolus doses are usually specified by dosage, and IV drip administration is usually specified by volume. The IV administration refers specifically to the size of the IV bag and not the volume. The interface did not provide options for the order review; as a result, the task of dose calculation was complicated by its organization on the interface. One can trace the causal influences of the interface organization on the reasoning process of the clinician (human factors). First, there was a mismatch regarding clinician expectations of how dosage was to be calculated (i.e., by volume vs by time to complete a saline-medication bag flow). This mismatch initiated the primary error. Second, there were limited, if any, checks on the process for the erroneous entry (i.e., minimal decision support to block out the range of values). The entire process resulted in errors by a number of clinicians, compounding over seven days, ending in a massive overdose of KCl. This suggests that interface elements can influence clinician decisions, compromising patient safety, and appropriate cognitive support is instrumental in avoiding such CPOE-based errors.

Another topical area of exploration within the use of technology and reasoning is the visual search and diagnosis process (see review by Norman, Coblentz, Brooks, & Babcook, 1992). Much of the literature has focused on visual diagnosis with slide or radiology images. More recently, the effects of expertise have been explored with a focus on digital slide images. Brunye and colleagues investigated the interactive process of zooming, panning, and other interactive action for the interpretation of whole slide digital images (Brunyé, Mercan, Weaver, & Elmore, 2017). They found differences in the efficiency of interaction and accuracy of interpretation based on the expertise and clinical interpretation.

Clinical Decision Support and Expertise

Another related research domain that has grown along with the development of EHRs is that of clinical decision support (CDS) (Greenes, 2011). Research on CDS has focused on developing and evaluating tools that can help clinicians in functioning with *efficiency*. Towards this end, much of the work has focused on developing rules that can aid clinicians in diagnosing and treating clinical conditions. In other words, CDS applications act as an external surveillance mechanism, functioning as an *expert clinician* overseeing certain clinical actions. Most CDS applications have focused on addressing condition-specific and patient-specific aspects for improving care and care processes, specifically focusing on supporting clinician actions within their workflow (Berner, 2009).

In general, diagnostic decision support has been prominent in three areas: specific clinical conditions (e.g., sepsis), specialty areas (e.g., asthma), and images (e.g., whole slide images). Across all of these topical areas, the focus is to support the clinician's tasks by aligning with the clinical workflow, in a quick and timely manner (Bates et al., 1998).

Studies on the effect of decision support on reasoning and diagnostic decision making have been conducted in several clinical domains. For example, Friedman et al. (1999) investigated how clinicians' diagnostic reasoning and hypotheses were affected by the use of a CDS system. Based on their evaluation they found that the effect of the CDS was more prominent among clinicians with lower clinical expertise (medical students) than among those with significantly more expertise (residents, fellows, and attending physicians). Other studies have investigated the role of CDS tools for diagnostic accuracy (Berner, 2009), for interpretation of EKG (e.g., Tsai, Fridsma, & Gatti, 2003), and for diagnostic decision making in an emergency room (Graber & VanScoy, 2003).

In summary, much of the work on the effects of CDS has focused on how it can potentially alleviate some of the limitations of clinical cognition. In other words, CDS has been designed and used with a focus on overcoming and avoiding some of the challenges of the limitations of expertise in certain domains. Although CDS applications have been discussed as a mechanism of cognitive support, their evaluation has focused primarily on their effectiveness, cost effects, and treatment outcomes, and not on their effect on reasoning or strategies of clinical care.

As HIT has been adopted in most clinical settings, new challenges have been identified at the human–machine interface. Although the confirmatory evidence regarding the effect of HIT on patient safety and clinical efficiency is still open to debate, the importance of considering the cognitive aspects of care is widely acknowledged (Patel & Kannampallil, 2014, 2015; Patel, Kaufman, & Kannampallil, 2013).

With the almost central role of HIT in most clinical environments, its effects on reasoning and decision making have come under increased scrutiny. With clinicians ranging in the level of expertise, HIT use has also varied significantly. Additional focus on how HIT changes work activities and practices leading to unintended consequences has also been discussed. Much of the prior research has characterized how HIT has changed the reasoning process, and modified the ways in which clinicians gather, organize, and utilize clinical information for their decision-making activities. Furthermore, several threads of research have attempted to characterize the tools and technologies that can help clinicians with lesser expertise navigate the clinical decision-making process.

CONCLUSIONS AND FUTURE DIRECTIONS

The study of medical expertise, drawing on cognitive theories and methods, dates back more than 40 years. The early years focused first on characterizing process, and later in understanding the nature of expert (and novice) knowledge, and its impact on performance including memory, comprehension, and reasoning. Research has demonstrated that medical expertise was not a simple construct, and its development was not characterized by linear growth in skills and knowledge. Specific measureable characteristics of medical expertise were delineated in studies of expert–novice differences in laboratory-based settings.

Newer research in the past couple of decades or more has moved toward carefully monitored real-world studies. This has been enabled by new technology, which has facilitated more precise in situ methods of data collection and analysis. Findings from these studies contribute to our knowledge of medical experts' performance as they change as a function of domain complexity, and as the generated errors get corrected (recovered), before they impact patient care. These newer studies also shed light on the *error* generation process, where all expert *errors* are not real errors, and experts' deviations from standard practice, producing novel explanations or innovations. Novices' deviations from the standard invariably lead to errors. This line of research on expertise and error monitoring and patient safety has been extended to study the mechanism of how such errors are generated and corrected with team-based practices, where subtasks are distributed among team members. The expertise model being considered here for the future appears to have a more collaborative nature with decision making shared with other team members and the artifacts.

In the past two decades, research has also shifted toward understanding the mediating role of technology and cognition that is stretched across teams of clinical practitioners. This latter work did not typically employ the contrastive methods of performance common to studies of expertise.

Many of these research studies on medical performance have been situated in applied contexts. The question is how can we leverage the decades of basic science research within the expertise tradition to inform our understanding of applied cognition in health care? More recent studies have used a combination of both laboratory-based and naturalistic studies such that the specific questions raised in the studies in natural clinical environment can be brought to be investigated in a controlled laboratory setting (bed to the bench side). The findings from these studies can be brought back to the clinics (bench to the bedside). This combination method increases our capabilities to create generalizable knowledge that can lead to scientific contributions as well as knowledge that serves to improve health care practices.

Another particularly important issue is whether basic cognitive constructs can illuminate facets of the mediation of HIT. For example, how does different EHR configurations impact memory, cognitive load, and mental models? Similarly, how can we study the distribution of knowledge across individuals within a team? How does individual expertise contribute to team performance, and conversely, what elements cannot be predicted by individual expertise? How can we characterize adaptive expertise in relation to the changing landscape in health care with regards to technology, precision medicine, and the ever-expanding role of the patient as an agent in the decision-making process? We believe that fundamental studies in psychology and cognitive science in general can still provide general guiding principles to study these issues, but they need to be combined with field studies, which serves to illuminate different facets and contextualize the phenomena observed in laboratory studies.

Finally, we need to find ways to train novice physicians or medical students to move more efficiently and effectively toward expert behavior without just mimicking the experts. More recent move toward wide use of simulations in training and education represents a significant development in addressing this concern. We can now leverage technology such as computer-based simulations, virtual reality, and serious games. However, it is important that the content of the simulation have ecologically valid information derived from real-world practice.

In our view, there is much to be learned about the future by capitalizing on an understanding gained from 40 years of research into medical expertise and diagnostic reasoning. This can be coupled with an appreciation of the changing landscape of health care and the ability to project what a future health care system will bring.

References

Amalberti, R. (2001). The paradoxes of almost totally safe transportation systems. *Safety Science* 37(2), 109–126.

Anzai, Y. (1991). Learning and use of representations for physics expertise. In K. A. Ericsson and J. Smith (Eds), *Towards a general theory of expertise: Prospects and limits* (pp. 64–92). New York: Cambridge University Press.

Ash, J. S., Berg, M., & Coiera, E. (2004). Some unintended consequences of information technology in health care: the nature of patient care information system-related errors. *Journal of the American Medical Informatics Association* 11(2), 104–112.

Bates, D. W., Boyle, D. L., Rittenberg, E., Kuperman, G. J., Ma'Luf, N., Menkin, V.,... Tanasijevic, M. J. (1998). What proportion of common diagnostic tests appear redundant? *American Journal of Medicine* 104(4), 361–368.

Bauer, D. T., & Guerlain, S. (2011). Improving the usability of intravenous medication labels to support safe medication delivery. *International Journal of Industrial Ergonomics* 41(4), 394–399.

Berner, E. S. (2009). *Diagnostic error in medicine: introduction.* Heidelberg: Springer Netherlands.

Bloomrosen, M., Starren, J., Lorenzi, N. M., Ash, J. S., Patel, V. L., & Shortliffe, E. H. (2011). Anticipating and addressing the unintended consequences of health IT and policy: A report from the AMIA 2009 Health Policy Meeting. *Journal of the American Medical Informatics Association* 18(1), 82–90.

Brunyé, T. T., Mercan, E., Weaver, D. L., & Elmore, J. G. (2017). Accuracy is in the eyes of the pathologist: The visual interpretive process and diagnostic accuracy with digital whole slide images. *Journal of Biomedical Informatics* 66, 171–179.

Chase, W. G., & Simon, H. A. (1973). Perception in chess. *Cognitive Psychology* 4(1), 55–81.

Clancey, W. J. (1985). Heuristic classification. *Artificial Intelligence* 27(3), 289–350.

Clancey, W. J., & Letsinger, R. (1982). *NEOMYCIN: Reconfiguring a rule-based expert system for application to teaching.* Stanford, CA: Department of Computer Science, Stanford University.

Clancey, W. J., & Shortliffe, E. (Eds) (1984). *Readings in medical artificial intelligence: The first decade.* Reading, MA: Addison-Wesley.

Cohen, T., Blatter, B., Almeida, C., & Patel, V. L. (2007). Reevaluating recovery: perceived violations and preemptive interventions on emergency psychiatry rounds. *Journal of the American Medical Informatics Association* 14(3), 312–319.

Croskerry, P. (2003). The importance of cognitive errors in diagnosis and strategies to minimize them. *Academic Medicine* 78(8), 775–780.

Croskerry, P. (2009). A universal model of diagnostic reasoning. *Academic Medicine* 84(8), 1022–1028.

Dawes, R. M., & Kagan, J. (1988). *Rational choice in an uncertain world.* San Diego, CA: Harcourt Brace Jovanovich.

De Dombal, F., Leaper, D., Staniland, J. R., McCann, A., & Horrocks, J. C. (1972). Computer-aided diagnosis of acute abdominal pain. *British Medical Journal* 2(5804), 9–13.

Elstein, A. S., Shulman, L. S., & Sprafka, S. A. (1978). *Medical problem solving an analysis of clinical reasoning.* Cambridge, MA: Harvard University Press.

Ericsson, K. A., & Kintsch, W. (1995). Long-term working memory. *Psychological Review* 102(2), 211.

Ericsson, K. A., & Simon, H. A. (1993). *Protocol analysis.* Cambridge, MA: MIT Press.

Feinstein, A. (1973a). An analysis of diagnostic reasoning. II. The strategy of intermediate decisions. *Yale Journal of Biology and Medicine* 46(4), 264.

Feinstein, A. (1973b). An analysis of diagnostic reasoning. I. The domains and disorders of clinical macrobiology. *Yale Journal of Biology and Medicine* 46(3), 212.

Feltovich, P. J., Johnson, P. E., Moller, J. H., & Swanson, D. B. (1984). LCS: The role and development of medical knowledge in diagnostic expertise. In W. J. Clancey and S. H. Shortliffe (Eds), *Readings in medical artificial intelligence: The first decade* (pp. 275–319). Reading, MA: Addison-Wesley.

Feltovich, P. J., Spiro, R. J., & Coulson, R. L. (1993). Learning, teaching, and testing for complex conceptual understanding. In N. Frederiksen, R. J. Mislevy, & I. I. Bejar (Eds), *Test theory for a new generation of tests* (pp. 181–217). Hillsdale, NJ: Lawrence Erlbaum.

Franklin, A., Robinson, D. J., & Zhang, J. (2014). Characterizing the nature of work and forces for decision making in emergency care. In V. L. Patel, D. R. Kaufman, & T. Cohen (Eds), *Cognitive informatics in health and biomedicine: Case studies on critical care, complexity, and errors* (pp. 127–145). London: Springer.

Friedman, C. P., Elstein, A. S., Wolf, F. M., Murphy, G. C., Franz, T. M., Heckerling, P. S., . . ., Abraham, V. (1999). Enhancement of clinicians' diagnostic reasoning by computer-based consultation: a multisite study of 2 systems. *JAMA: Journal of the American Medical Association 282*(19), 1851–1856.

Garg, A. X., Adhikari, N. K., McDonald, H., Rosas-Arellano, M. P., Devereaux, P., Beyene, J., . . ., Haynes, R. B. (2005). Effects of computerized clinical decision support systems on practitioner performance and patient outcomes: a systematic review. *Journal of the American Medical Association 293*(10), 1223–1238.

Gorry, G. A. (1973). Computer-assisted clinical decision making. *Methods of Information in Medicine 12*(1), 45–51.

Graber, M., & VanScoy, D. (2003). How well does decision support software perform in the emergency department? *Emergency Medicine Journal 20*(5), 426–428.

Greenes, R. A. (2011). *Clinical decision support: the road ahead.* London: Academic Press.

Groen, G. J., & Patel, V. L. (1985). Medical problem solving: some questionable assumptions. *Medical Education 19*, 95–100.

Higgs, J. (2008). *Clinical reasoning in the health professions.* Amsterdam: Elsevier Health Sciences.

Horsky, J., Gutnik, L., & Patel, V. L. (2006). Technology for emergency care: cognitive and workflow considerations. *AMIA Annual Symposium Proceedings 2006*, 344–348.

Hutchins, E. (1995). *Cognition in the wild.* Cambridge, MA: MIT Press.

Institute of Medicine (2011). *Health IT and patient safety: Building safer systems for better care.* Washington, DC: Institute of Medicine of the National Academies.

Kahneman, D. (2003). A perspective on judgment and choice: mapping bounded rationality. *American Psychologist 58*(9), 697–720.

Kahol, K., Vankipuram, M., Patel, V. L., & Smith, M. L. (2011). Deviations from protocol in a complex trauma environment: Errors or innovations? *Journal of Biomedical Informatics 44*(3), 425–431.

Kannampallil, T., Cohen, T., Kaufman, D. R., & Patel, V. (2014). Re-thinking complexity in the critical care environment. In V. L. Patel, D. R. Kaufman, & T. Cohen (Eds), *Cognitive informatics in health and biomedicine: Case studies on critical care, complexity, and errors* (pp. 343–356). London: Springer.

Kannampallil, T. G., Franklin, A., Mishra, R., Almoosa, K. F., Cohen, T., & Patel, V. L. (2013). Understanding the nature of information seeking behavior in critical care: implications for the design of health information technology. *Artificial Intelligence in Medicine 57*(1), 21–29.

Klein, G. A. (1989). *Strategies of decision making.* Yellow Springs, OH: Klein Associates.

Kleinmuntz, B. (1968). The processing of clinical information by man and machine. In B. Kleinmuntz (Ed.), *Formal representation of human judgment* (pp. 149–186). New York: Wiley.

Koppel, R., Metlay, J. P., Cohen, A., Abaluck, B., Localio, A. R., Kimmel, S. E., & Strom, B. L. (2005). Role of computerized physician order entry systems in facilitating medication errors. *Journal of the American Medical Association 293*(10), 1197–1203.

Kubose, T. T., Patel, V. L., & Jordan, D. (2002). Dynamic adaptation to critical care medical environment: error recovery as cognitive activity. In *Proceedings of the Twenty-fourth annual conference for the cognitive science society* (pp. 43–45).

Kushniruk, A. W., Kaufman, D. R., Patel, V. L., Lévesque, Y., & Lottin, P. (1995). Assessment of a computerized patient record system: a cognitive approach to evaluating medical technology. *MD Computing: Computers in Medical Practice* 13(5), 406–415.

Larkin, J., McDermott, J., Simon, D. P., & Simon, H. A. (1980). Expert and novice performance in solving physics problems. *Science 208*, 1335–1342.

Ledley, S., Lusted, L. B., & Ledley, R. S. (1959). Reasoning foundations of medical diagnosis. *Science 130*(3366), 9–21.

Miller, R. A., Pople, H. E., & Myers, J. D. (1984). Internist-I, an experimental computer-based diagnostic for general internal medicine. In W. J. Clancey and S. H. Shortliffe (Eds), *Readings in medical artificial intelligence: The first decade*. Reading, MA: Addison-Wesley.

Newell, A., & Simon, H. A. (1972). *Human problem solving* (Vol. 104, No. 9). Englewood Cliffs, NJ: Prentice-Hall.

Norman, D. A. (1991). Cognitive artifacts. In J. M. Carroll (Ed.), *Designing interaction: Psychology at the human–computer interface* (pp. 17–38). New York: Cambridge University Press.

Norman, G. (2005). Research in clinical reasoning: past history and current trends. *Medical Education* 39(4), 418–427.

Norman, G. R., Brooks, L. R., Cunnington, J. P., Shali, V., Marriott, M., & Regehr, G. (1996). Expert-novice differences in the use of history and visual information from patients. *Academic Medicine* 71(10), S62–S64.

Norman, G. R., Coblentz, C. L., Brooks, L., & Babcook, C. (1992). Expertise in visual diagnosis: a review of the literature. *Academic Medicine* 67(10), S78–83.

Patel, V., Arocha, J. F., & Kaufman, D. R. (1994). Diagnostic reasoning and expertise. *Psychology of Learning Motivation* 31, 137–252.

Patel, V., Arocha, J. F., & Kaufman, D. R. (2001). A primer on aspects of cognition for medical informatics. *Journal of the American Medical Informatics Association* 8(4), 324–343.

Patel, V., Cohen, T., Batwara, S., & Almoosa, K. F. (2011). Teamwork and error management in critical care. In V. L. Patel, D. R. Kaufman, & T. Cohen (Eds), *Cognitive informatics in health and biomedicine: Case studies on critical care, complexity, and errors* (pp. 59–90). London: Springer.

Patel, V. L., Cohen, T., Murarka, T., Olsen, J., Kagita, S., Myneni, S., . . . , Ghaemmaghami, V. (2011). Recovery at the edge of error: debunking the myth of the infallible expert. *Journal of Biomedical Informatics* 44(3), 413–424.

Patel, V. L., & Groen, G. J. (1986). Knowledge based solution strategies in medical reasoning. *Cognitive Science* 10(1), 91–116.

Patel, V. L., & Groen, G. J. (1991a). Developmental accounts of the transition from medical student to doctor: some problems and suggestions. *Medical Education* 25(6), 527–535.

Patel, V. L., & Groen, G. J. (1991b). The general and specific nature of medical expertise: A critical look. In K. A. Ericsson and J. Smith (Eds), *Towards a general theory of expertise: Prospects and limits* (pp. 63–125): New York: Cambridge University Press.

Patel, V. L., Groen, G. J., & Arocha, J. F. (1990). Medical expertise as a function of task difficulty. *Memory & Cognition* 18(4), 394–406.

Patel, V. L., & Kannampallil, T. (2014). Human factors and health information technology: current challenges and future directions. *Yearbook of Medical Informatics* 9(1), 58.

Patel, V. L., & Kannampallil, T. G. (2015). Cognitive informatics in biomedicine and healthcare. *Journal of Biomedical Informatics* 53, 3–14.

Patel, V. L., Kannampallil, T. G., & Shortliffe, E. H. (2015). Role of cognition in generating and mitigating clinical errors. *British Medical Journal: Quality & Safety* 24(7), 468–474.

Patel, V. L., & Kaufman, D. R. (2013). Cognitive Science and Biomedical Informatics. In E. H. Shortliffe & J. J. Cimino (Eds), *Biomedical informatics: Computer applications in health care and biomedicine*, 4th edn (pp. 133–185). New York: Springer-Verlag.

Patel, V. L., Kaufman, D. R., & Cohen, T. (2014). Complexity and errors in critical care. In V. L. Patel, D. R. Kaufman, & T. Cohen (Eds), *Cognitive informatics in health and biomedicine: Case studies on critical care, complexity, and errors* (pp. 1–13). London: Springer.

Patel, V. L., Kaufman, D. R., & Kannampallil, T. G. (2013). Diagnostic reasoning and decision making in the context of health information technology. *Reviews of Human Factors and Ergonomics* 8(1), 149–190.

Patel, V. L., Kaufman, D. R., & Magder, S. A. (1996). The acquisition of medical expertise in complex dynamic environments. In K. A. Ericsson (Ed.), *The road to excellence: The acquisition of expert performance in the arts and sciences, sports and games* (pp. 127–165). Mahwah, NJ: Laurence Erlbaum.

Patel, V. L., Kushniruk, A. W., Yang, S., & Yale, J. F. (2000). Impact of a computer-based patient record system on data collection, knowledge organization, and reasoning. *Journal of the American Medical Informatics Association* 7(6), 569–585.

Patel, V. L., & Ramoni, M. (1997). Cognitive models of directional inference in expert medical reasoning. In K. Ford, P. Feltovich, & R. Hoffman (Eds), *Human and machine cognition*. (pp. 67–99). Hillsdale, NJ: Lawrence Erlbaum.

Patel, V. L., Zhang, J., Yoskowitz, N. A., Green, R., & Sayan, O. R. (2008). Translational cognition for decision support in critical care environments: a review. *Journal of Biomedical Informatics* 41(3), 413–431.

Patel, V., Shine, A., & Almoosa, K. F. (2014). Error recovery in the wilderness of ICU. In V. L. Patel, D. R. Kaufman, & T. Cohen (Eds), *Cognitive informatics in health and biomedicine: Case studies on critical care, complexity, and errors* (pp. 1–13). London: Springer.

Pauker, S. G., Gorry, G. A., Kassirer, J. P., & Schwartz, W. B. (1976). Towards the simulation of clinical cognition: taking a present illness by computer. *American Journal of Medicine* 60(7), 981–996.

Peirce, C. S. (1955). *Philosophical writings of Peirce*. Ed. by Justus Buchler. New York: Dover.

Plaisant, C., Mushlin, R., Snyder, A., Li, J., Heller, D., & Shneiderman, B. (1998). LifeLines: Using visualization to enhance navigation and analysis of patient records. *Proceedings of the AMIA Symposium 1998*, 76–80.

Rasmussen, J., Pejtersen, A., & Goodstein, L. P. (1994). *Cognitive systems engineering*. New York: Wiley.

Rimoldi, H. J. (1961). The test of diagnostic skills. *Academic Medicine* 36(1), 73–79.

Schmidt, H. G., & Boshuizen, H. P. (1993). On acquiring expertise in medicine. *Educational Psychology Review* 5(3), 205–221.

Shortliffe, E. (2012). *Computer-based medical consultations: MYCIN* (Vol. 2). Amsterdam: Elsevier.

Stefanelli, M., & Ramoni, M. (1992). Epistemological constraints on medical knowledge-based systems. In D. A. Evans and V. L. Patel (Eds), *Advanced models of cognition for medical training and practice* (pp. 3–20). London: Springer.

Sternberg, R. J., & Horvath, J. A. (Eds). (1999). *Tacit knowledge in professional practice: Researcher and practitioner perspectives.* Hove, UK: Psychology Press.

Tsai, T. L., Fridsma, D. B., & Gatti, G. (2003). Computer decision support as a source of interpretation error: the case of electrocardiograms. *Journal of the American Medical Informatics Association 10*(5), 478–483.

Vankipuram, M., Kahol, K., Cohen, T., & Patel, V. L. (2009). Visualization and analysis of activities in critical care environments. *AMIA Annual Symposium Proceedings 2009,* 662–666.

Vankipuram, A., Traub, S. J., & Patel, V. L. (2017). Clinical workflow visualization: Representation of clinician activity from location tracking data. In *AMIA Annual Symposium, Washington, DC.* Bethesda, MD: American Medical Informatics Association.

Woods, D. D. (1993). Process tracing methods for the study of cognition outside of the experimental psychology laboratory. In G. A. Klein, J. Orasanu, R. Calderwood, & C. E. Zsambok (Eds), *Decision making in action: Models and methods* (pp. 228–251). Norwood, NJ: Ablex.

Zeng, Q., Cimino, J. J., & Zou, K. H. (2002). Providing concept-oriented views for clinical data using a knowledge-based system. *Journal of the American Medical Informatics Association 9*(3), 294–305.

Zhang, J., & Norman, D. A. (1994). Representations in distributed cognitive tasks. *Cognitive Science 18*(1), 87–122.

CHAPTER 28

FIRE FIGHTING AND EMERGENCY RESPONDING

MARK WIGGINS, JAIME AUTON, AND MELANIE TAYLOR

INTRODUCTION

EMERGENCIES are non-routine, often fast-paced, dynamic situations that, left unchecked, impact adversely the safety and/or security of the broader population and/or the operating environment. They range in scale from events that have the potential to affect millions, to events that affect a single individual, and their onset can be triggered by the behavior of operators, and/or by environmental events, including failures of equipment.

In this context, expertise involves a recognition of the onset of emergency conditions, the rapid synthesis of relevant information, often from disparate sources, and the capacity to generate viable options, taking account of what is often a dynamic and uncertain environment (Klein, 1997; Kowalski-Trakofler, Vaught, & Scharf, 2003). This chapter examines the components of expertise that enable the successful management of firefighting and similar emergency situations, including a detailed and accurate mental model, and the rapid discrimination and interpretation of task-related cues. Underpinning these components is a continuous process of information acquisition and interpretation that enables the anticipation and response to changes in the system state.

The complex, dynamic, and uncertain nature of emergency situations means that the study of expertise in this context is particularly problematic. For example, in evaluating expert performance, exposure to naturalistic environments not only imposes threats to the safety of operational personnel, but the nature of the stimuli to which they are exposed is impossible to control. As a result, the outcomes associated with the management of emergency situations may not necessarily reflect levels of expertise, since events may have occurred that were completely outside the range of expectations.

To illustrate the problem, consider the events of February 6, 2009, when a warning was issued throughout Southeastern Australia, to the effect that catastrophic wildfire conditions were expected the following day (Whittaker, Haynes, Handmer, & McLennan, 2013). Despite preparations, the speed and ferocity of the conflagration of wildfire fronts was largely underestimated. Despite a great deal of expertise in the management of wildfires, the result was indecision and a delay, the outcome of which was the loss of 172 lives and approximately 300 dwellings during what became known as *Black Saturday* (Handmer & O'Neill, 2016).

It is the unpredictable character of emergencies, in both their onset and their trajectory, that creates challenges for explorations of expertise. The alternative is to examine expertise in the context of simulations, over which a degree of experimental control can be exercised. In this case, the difficulty lies in ensuring that the key features on which experts rely to formulate assessments of the situation are both available and presented in the form in which they occur in reality (Cohen-Hatton & Honey, 2015). Shanteau (1992) has argued that even minor changes to the presentation of information can result in a deterioration in the performance of personnel who would otherwise be regarded as experts.

Apart from the environment within which the nature of expertise can be explored, there is also a need, in context of emergency management, to consider how expertise is evaluated. For example, traditional approaches to the study of expertise target outcomes as a key measure of performance. In the context of mathematics or similar disciplines, the *correct* answer tends to be unequivocal. However, in emergency situations, there may be a range of responses, each of which results in a positive outcome. The question is whether a positive outcome necessarily equates to a *correct* response.

In experimental environments that recreate a historical event, the progression of a scenario can be modeled as it occurred (Lewandowsky & Kirsner, 2000). This enables a potentially more objective assessment of performance, since the outcomes are known. However, even in this case, there remains a level of uncertainty, since a slightly different response on the part of the emergency responder might result in an outcome that departs from the actual events and, therefore, is difficult to model accurately.

To further complicate the problem of expert assessment in emergency situations are the windows of opportunity within which a response may be more or less appropriate. From an empirical perspective, comparative analyses become impossible, since different emergency responders are effectively engaging with different stimuli at different periods within a scenario. The consequence is a lack of generalizability and a tendency towards more descriptive analyses.

Descriptive Approaches to the Assessment of Expertise in Emergencies

Given the difficulties in recreating simulations with sufficient fidelity, together with the dynamic nature of the environments, the research methodology of choice within the context of emergency environments has tended to be descriptive. Where some researchers have focused on post-event descriptions (e.g., Frye & Wearing, 2011; Klein, Calderwood, & Clinton-Cirocco, 2010), others have undertaken descriptive research in situ (e.g., Lewandowsky & Kirsner, 2000). In each case, the aim has been to identify common strategies and to infer from this information, underlying cognitive and perceptual strategies that are likely to be involved.

Post-Hoc Descriptive Approaches

Post-hoc analyses can target both successful and unsuccessful outcomes, and can be undertaken through interviews with the personnel involved, or through analyses of investigative reports. Where the personnel involved are available, cognitive task analyses, incorporating cognitive interviews, tend to be the preferred methodology (Mendonca, Beroggi, & Wallace, 2001). This strategy enables explorations of a wide range of cognitive, perceptual, and behavioral strategies that might have been engaged in response to the events (e.g., Adams & Ericsson, 2000). Importantly, it also enables the integration between the progression of events and the responses to the events. Collapsing across groups enables the identification of consistent responses, enabling inferences to be drawn concerning the differences between expert and non-expert performance (e.g., Schubert, Denmark, Crandall, Grome, & Pappas, 2013).

The outcomes of cognitive interviews amongst emergency responders have yielded a number of important outcomes concerning the nature of expertise, including an extended period of information acquisition (Ash & Smallman, 2010), the use of non-compensatory decision strategies (Klein et al., 2010), a reliance on a detailed and sophisticated mental model (Okoli, Weller, & Watt, 2014), and the application of a process whereby the outcomes of a response are reviewed and the information integrated to form a revised and updated mental model (Okoli, Watt, Weller, & Wong, 2016; Patterson et al., 2016).

The difficulty associated with cognitive interviews relates to the dependence upon memory and the potential for the rationalization of behavior (Condie, 2012). To some extent, this can be overcome by undertaking interviews as soon as possible after an event, and/or by using a particularly memorable event as the basis for the interview. The latter is referred to as a critical incident analysis and has been used both as a research tool and as an investigative tool following an incident (Hoffman, Crandall, & Shadbolt, 1998; O'Hare, Wiggins, Williams, & Wong, 1998; see also Chapter 19, "Incident-Based Methods," by Militello and Anders, this volume).

Importantly, cognitive interviews, incorporating a critical incident analysis, can be used to target the precursors to successful and/unsuccessful outcomes. By contrast, post-hoc analyses based on accident or incident reports tend to target the precursors to unsuccessful performance and, by inference, establish the foundations for success. To illustrate the process, consider the events of February 9, 2015, when TransAsia Flight 235 experienced an engine failure immediately after takeoff from Taipei Songshan

Airport. As a twin-engine aircraft, the pilots were required to shut down the errant engine on the ATR 72 to reduce aerodynamic drag and enable the aircraft to climb out safely on the second engine. In the confusion, the captain shut down the functioning engine, thereby reducing thrust and causing an aerodynamic stall at low altitude, and from which recovery was impossible (Aviation Safety Council, 2016).

Identifying and shutting down an engine, particularly in an emergency, requires the rapid recognition and interpretation of task-related information, the activation of the appropriate mental model which, in turn, initiates appropriate behavior. However, in the absence of a well-established, precise, and accurate mental model, the interpretation of the information becomes cognitively demanding and time-consuming (McCutchen, 2011). The path of least resistance in the case of a time-constrained emergency is to *guess*, and this may explain the misdiagnosis on the part of the captain.

Like the captain of TransAsia Flight 235, the first officer aboard British Midland Flight 92 misdiagnosed the engine that had failed during the initial stages of a flight from Heathrow to Belfast on January 8, 1989 (Besnard, Greathead, & Baxter, 2004). The pilots had recently transitioned from the Boeing 737-300 series aircraft to the Boeing 737-400 series and there were differences in both the representation of the information on the flight deck and the design of the aircraft systems. Their experience and understanding of these systems were limited to a short course during which they had been introduced to the features of the new aircraft. In the absence of a detailed mental model, the key information was overlooked, and the error ensued (Besnard et al., 2004).

The similarities between the cases can be used to underscore the requirements of expert performance. In the case of TransAsia Flight 235 and British Midlands Flight 92, the lack of a sophisticated and easily accessible mental model may have contributed to the outcomes and, by association, constitutes a basic requirement for successful performance. However, as with post-hoc assessment more generally, causal inferences are impossible to test in the absence of further analysis.

In Situ Descriptive Approaches

To complement and, in some cases, provide the triggers for post-event cognitive interviews, data can be recorded during the course of an event. A number of different types of data can be collected, but the miniaturization and increasing reliability of wearable technologies has enabled the collection of a range of physiological and behavioral data. Eye tracking technology, including fixations and saccades, offers behavioral data pertaining to the sources of information, and the preference for different types of information during a sequence of events (Omodei, McLennan, & Wearing, 2005; Vickers & Lewinski, 2012). Saccades also enable assessments of the transition between different sources of information, thereby establishing a potential hierarchy and the framework for inferences pertaining to the application of specific cognitive skills (McCormack, Wiggins, Loveday, & Festa, 2014).

Physiological data, including measures of exertion, have been employed to differentiate expert from non-expert performance, and are often complemented with subjective assessments that are recorded at specific intervals (Perroni et al., 2010). Like eye-tracking data, these descriptive data provide the basis for inferences concerning underlying cognitive and perceptual strategies that are engaged by different cohorts. The main difficulty in using physiological data involves accounting for individual differences in physiological functioning, especially where there are marked differences in age and health that coincide with differences in expertise (Holmér & Gavhed, 2007).

Despite inherent difficulties associated with in situ assessments of expertise in emergency conditions, a number of notable observations have been reported, including experts' reliance on patterns of cues to diagnose a situation while maintaining an oversight of multiple activities (Flin, Slaven, & Stewart, 1996; Schubert et al., 2013), and their capacity to adapt their approach, depending upon the complexity and changing nature of a situation (Gunnarsson & Stomberg, 2009).

Experimental Approaches to the Assessment of Expertise in Emergencies

One of the most significant issues associated with descriptive approaches to the study of expertise concerns the difficulties in establishing cohort differences between groups. This is due, in part, to the lack of experimental control that can be exercised over stimuli to which personnel are responding. The idiosyncratic nature of emergencies is such that even within domains, the types of situations that emergency personnel confront are quite different, and therefore embody quite different features. For example, the behavior of fires in metropolitan areas differs significantly from the behavior of wildfires due, in large part, to the nature of the combustible materials. However, even within the context of an event, the conditions can change quickly and unpredictably to a point where different emergency responders are dealing with different environmental conditions often at the same time.

To exercise greater levels of experimental control and establish cohort-level differences between experts and non-experts, carefully designed simulations of differing levels of fidelity have been employed. At the highest levels of fidelity, mock scenarios are engaged where emergency responders are required to manage simulated events in environments where they are likely to occur (Cha, Han, Lee, & Choi, 2012). Common scenarios include post-disaster management such as aircraft accidents and law enforcement tactical response operations such as riot control (Vincent, Sherstyuk, Burgess, & Connolly, 2008). In each case, the intention is to recreate the environmental conditions that are likely to confront emergency responders.

Although high fidelity scenarios offer an opportunity for engagement with realistic environments, they necessarily embody a storyline within which there are decision points, such as which patients to triage in the case of a disaster recovery scenario, or whether to advance or defend a position in the case of a tactical response scenario. Inevitably, the responses at these decision points will differ between emergency responders, thereby impacting the course of the scenario and the comparative analyses that are possible between individuals (Comfort, Ko, & Zagorecki, 2004).

Enabling descriptions of behavior at a cohort level necessarily requires fewer decision paths so that emergency responders are interacting with comparable situations. One solution has been the development of scenarios within which there are fewer decision points. However, the options at decision points can also be constrained so that emergency responders *travel* along similar paths and are thereby confronted by similar conditions (Gonzalez, Vanyukov, & Martin, 2005).

Although the constraints embodied within scenarios potentially engender a lack of realism, fidelity can be maintained, to an extent, through the careful design of stimuli that mimic the nature and behavior of the conditions that occur in reality. For example, Omodei and Wearing (1995) developed a computer-based representation of fire behavior, referred to as Fire Chief. Although the representation of the fire was limited to a series of icons, the behavior of the fire was designed to be consistent with the behavior of an actual fire front. Since the stimuli to which emergency responders were exposed was relatively consistent, the outcomes enabled cohort-level comparisons.

Referred to as a *microworld*, these simulated environments enable the introduction of changes to the environment systematically and consistently (Malakis, Kontogiannis, & Kirwan, 2010). Differences in the behavior of users can be described, and the underlying cognitive strategies inferred. The outcomes of this type of descriptive research have been used to explain the causes of real-world events where the consequences appeared otherwise inexplicable. For example, in 1994, fourteen experienced firefighters lost their lives fighting a wildfire on Storm King Mountain in Colorado when they were caught between two fire fronts. The fire *spotted* beneath their location and then burnt rapidly uphill toward their position at an estimated six to nine feet per second (Butler et al., 1998).

This effect, of a fire burning rapidly uphill, is somewhat counterintuitive, and the outcomes of research involving microworlds demonstrated that a proportion of firefighters held assumptions about the movement of fire that were incorrect. They inferred on this basis that they held an incorrect or incomplete mental model of the passage of fire and that this was likely due to their lack of task-related experience.

Although the use of simulations, including microworlds, offer the capacity to exercise a degree of experimental control over both the conditions to which emergency responders are exposed and the problem-solving path that is activated, the extent to which the behavior of participants represents behavior in reality is the subject of some debate. For example, Chapman, Nettelbeck, Welsh, and Mills (2006) failed to establish differences in the performance of experts and non-experts in team-based simulated firefighting scenarios using a networked version of the Fire Chief microworld. This was interpreted as potential evidence of a lack of criterion validity for the use of microworlds in exploring the operational performance.

There is no doubt that there are limitations concerning the use of microworlds as surrogates for real-world environments. For example, Chapman et al. (2006) employed Fire Chief as part of a complex, team-based exercise, using multiple agencies that were dispersed geographically. Although this type of scenario represents the environment within which the management of large-scale fires occurs, it perhaps constitutes an overestimation of the extent to which microworlds can accurately represent reality. Therefore, there remains some debate as to how and when to use microworlds to investigate human performance, especially in the dynamic, multi-agent environments that characterize the management of emergency operations.

Clearly, the difficulties associated with the use of microworlds to examine human performance in emergency situations reflect the broader issue of maintaining experimental control while ensuring a level realism that allows for cohort-based assessments in environments that are true to life. Part of the difficulty lies in the expectation that microworlds can represent the totality of behavior in these environments. An alternative approach is to use techniques such as microworlds more judiciously to test specific theoretical propositions that are purported to differentiate expert from non-expert performance.

Inferential Approaches to the Study of Expertise in Emergency Situations

While descriptive approaches to the analysis of expertise in emergency conditions have proven useful, establishing the cognitive and perceptual mechanisms that enable expert performance is problematic after an event (Ericsson, Whyte, & Ward, 2007). In the absence of a theoretically driven research question that is advanced a priori, explanations of performance constitute speculation that need to be evaluated empirically. Consequently, while more empirical approaches to the assessment of expertise have been applied in the context of emergency management, the techniques necessarily require experimental designs that enable the isolation of variables. Arguably, the consequence of this reductionist approach is a strategy that targets specific features of expertise to the exclusion of other dimensions of the construct, and the development simulated scenarios that are designed to engage these features.

The application of the experimental approach is intended to enable the assessment of expertise systematically, thereby enabling the construction of a model of expert performance. This process has generally occurred by differentiating expert from non-expert responses using the expert–novice experimental paradigm (Shanteau, 1988).

Differences that emerge between groups are used to infer capabilities associated with expert performance that are yet to be acquired or established amongst non-experts.

Expert–novice differences in the context of emergency management include differences in the capability to ignore extraneous information (Hutton & Klein, 1999), anticipate change in the external environment (Cellier, Eyrolle, & Marine, 1997; Cioffi, 2000), manage multiple activities (Leprohon & Patel, 1995), generate a limited number of possible diagnoses (Pelaccia et al., 2015), cue utilization and discrimination (Reischman & Yarandi, 2002), selective attention (Gegenfurtner, Lehtinen, & Säljö, 2011), and the storage and retrieval of knowledge (Case, Harrison, & Roskell, 2000).

In combination, these outcomes characterize expertise in emergency conditions as dependent upon a comprehensive mental model that constrains information acquisition to a relatively limited number of highly diagnostic features, thereby reducing the demands on information processing and enabling the rapid acquisition and integration of new information as it emerges (Ozel, 2001). While this is broadly consistent with the outcomes of descriptive approaches, mental models are difficult to evaluate using cross-sectional experimental designs, since expertise needs to be defined a priori, and attributing causality remains problematic.

Attributions of Expertise in Emergency Conditions

In emergency environments where the progression of events is highly dynamic and uncertain, and where the outcomes can be outside the direct control of operational personnel, establishing expertise based on the performance of emergency responders can be difficult. There are many variables, including the behavior of other personnel, and the availability of reliable information, that will impact the success of a strategy. Consequently, differences in performance might reflect the levels of support provided to an emergency responder, rather than differences inherent in the level of expertise.

In experimental designs, expertise is normally defined a priori using one or more of a range of variables, including the number of encounters, the years of experience accumulated, and/or the level of seniority achieved (Breckwoldt et al., 2012; Lewandowsky & Kirsner, 2000). Although there is clearly an association between measures of experience and exposure, and the progression to expertise, there are likely to be individual differences in the rate of skill acquisition per unit of exposure or experience (Bayouth, Keren, Franke, & Godby, 2013; Ericsson & Ward, 2007). As a result, categorizing levels of expertise on the basis of descriptors may minimize any differences between groups that might otherwise be apparent. This may explain results where no differences in performance are apparent.

An alternative to the use of descriptors is to use a measure of performance that is characteristic of expertise but which does necessarily require interaction with a fully developed, realistic scenario. For example, the outcomes of both descriptive and experimental research in the context of emergency management suggest that experts possess highly developed and well-structured mental models of the environment (Klein, 2008; Lewandowsky, Dunn, Kirsner, & Randell, 1997). Importantly, these mental models are not directly observable and their existence and architecture has been inferred, either through careful observations of performance or through post-event discussions and interviews.

Although mental models are not directly observable, the immediate outcomes of a highly developed mental model are observable in the context of cue utilization (Okoli, Weller, & Watt, 2014; Wiggins, 2014). Cue utilization is a process dependent upon pre-existing associations in memory that, in combination, provide the patterns or schemas that underpin mental models (Braley & Johnson, 1963; Rouse & Morris, 1986). Differences in cue utilization have been reported between experts and novices' responses to emergency scenarios, suggesting that the utilization of cues might be used to reflect the characteristics of an underlying mental model, thereby enabling a more accurate assessment of levels of expertise (Okoli et al., 2014).

Previous approaches to the assessment of cue utilization have required the identification of the specific cues on which an accurate assessment of a situation might be based (e.g., Causer, Barach, & Williams, 2014; Okoli et al., 2016). In the case of laboratory tasks, this has been achieved by carefully constructing scenarios so that the critical information necessary to resolve a situation is embodied within one or more key features (Lagnado, Newell, Kahan, & Shanks, 2006). The identification of the key features and, thus, the successful resolution of the problem are presumed dependent upon a level of skilled performance characteristic of expertise.

Under field conditions, where situations tend to be dynamic and unpredictable, it can be difficult to identify with certainty the specific features on which a successful outcome will depend. One approach involves seeking the consensus of subject-matter experts who review scenarios and then identify key features, and the events or objects with which they are associated (Hirsch, Corey, & Martell, 1998). Ideally, this universal set of features could be used as a *gold standard* against which to assess the mental models of non-experts (Hoffman, 1998).

Since cue-based relationships, like the mental models on which they are based, are idiosyncratic, attempts to identify a universal set of features on which all experts agree have proven problematic (e.g., Feeley & De Turck, 1995). This is especially the case in the complex, time-constrained environments that characterize emergency conditions. While a limited number of universal features will inevitably attract the attention of a cohort of experts, their interpretation and application are based on the presumption that the learning trajectories that precede expert performance are sufficiently consistent to ensure that cues, and patterns of cues, are acquired and interpreted similarly.

In fact, the progression of learning is markedly idiosyncratic due to both the capabilities of different learners and the environments to which they are exposed (Ackerman, 1988). The consequence is a repertoire of feature–event/object or cue-based relationships that are also idiosyncratic, even though emergency responders may share a similar mental model. This presents difficulties where assessments of expertise are based on performance involving the acquisition of information derived from prescribed environmental features. What constitutes a cue for one person may be simply represent an innocuous feature to another.

Although there are differences between experts' reliance on, and interpretation of, specific environmental features, the similarities that exist between mental models ensure that the characteristics of information remain similar. For example, given a set of related problems to sort, experts will tend to create a greater number of categories than non-experts, reflecting a greater level of discrimination (Shanteau, Weiss, Thomas, & Pounds, 2002). This capacity for discrimination is presumed indicative of a greater repertoire of more precise mental models that enable the superior performance of experts.

Underpinning a greater repertoire of mental models is the need to identify and respond to cue-based associations at a faster rate and with greater accuracy than less experienced practitioners (Abernethy & Russell, 1987). For example, in diagnosing complex scenes, experts need relatively less time and fewer features to be able to draw accurate conclusions (Kioumourtzoglou, Kourtessis, Michalopoulou, & Derri, 1998). Similarly, they require fewer cues to anticipate accurately the future state of a system (Abernethy, 1990). In combination, this evidence confirms a highly developed relationship between cues and mental models which provides the basis for subsequent responses.

The prescriptive role of mental models is further evident in patterns of information acquisition in response to ambiguous scenarios. In constructing an assessment of a situation, experts adopt a more systematic approach to the acquisition of information (Wiggins & O'Hare, 1995; Wiggins, Stevens, Howard, Henley, & O'Hare, 2002). This effect is demonstrated by presenting practitioners with a problem, the task-related information for which is listed in a random sequence. Non-experts tend to acquire information in the sequence in which it is presented, even though the sequence is randomized (Loveday, Wiggins, Harris, Smith, & O'Hare, 2013). Experts tend to impose a structure to the acquisition of information, consistent with the application of pre-existing model or strategy (Wiggins & O'Hare, 1995). Characterizing this structure allows comparisons between cohorts and the establishment of a normed database against which the performance of individual practitioners can be assessed.

In preparing for emergencies, responders will participate in a range of simulations that are intended to provide a breadth of experiences. This continuous process assures the internalization of feature–event/object relationships to form cues and reduces cognitive demands so that responses are both rapid and accurate when necessary (Chevalier, Dauvier, & Blaye, 2018; Gigerenzer & Goldstein, 1996). However, there is evidence to suggest that there are individual differences in the rate at which this internalization occurs (Wiggins, Brouwers, Davies, & Loveday, 2014). Where exposure to a small number of scenarios may be sufficient for the acquisition of cues amongst some practitioners, others need more frequent, and/or a greater variety of experiences to reach a similar level of capability (Brouwers, Wiggins, Griffin, Helton, & O'Hare, 2017).

Measures of cue utilization are now being employed across a range of emergency and non-emergency domains (Loveday, Wiggins, Harris et al., 2013; Loveday, Wiggins,

Searle, Festa, & Schell, 2013; Loveday & Wiggins, 2014). The technique involves the application of number of different tasks that, in combination, assess different aspects of cue utilization in response to task-related stimuli (Wiggins, 2015). Scores on each of the tasks are standardized and compared against population norms, which overcomes the difficulties associated with subjective assessments. Performance is assessed across different dimensions, including response latency, accuracy, and variance in judgements (Wiggins et a., 2014).

Developing, Maintaining, and Losing Expertise in Managing Emergencies

If a repertoire of sophisticated and precise mental models forms the basis of expertise in managing emergencies successfully, the development and maintenance of these models are critically important. Where explicit mental models, such as the progression of a fire up a slope, can be articulated, implicit models that refer to the heat and the color of flames are more difficult to impart (Mathews et al., 1989). Typically, the acquisition of implicit mental models requires a process of internalization where learners are actively involved and experience a range of conditions from which they can develop a robust mental model and the cues with which they are associated.

Active involvement, together with exposure to a breadth of experiences, is a particular challenge in the context of emergency management. Typically, learners will act as close observers of more experienced practitioners and, at some stage, will be considered sufficiently competent to manage the situation under observation and, eventually, in a leadership role (Ford & Schmidt, 2000; Williams-Bell, Kapralos, Hogue, Murphy, & Weckman, 2015). However, the unpredictable nature of emergency situations, together with their frequency and the contexts to which learners are exposed, is often such that the breadth of exposure is necessarily constrained.

Constraining the breadth of exposure can limit the rate at which skills are acquired and limit the extent to which those skills acquired can be generalized to other tasks, providing the basis for adaptive expertise (Ford & Schmidt, 2000). Emergencies, by definition, are non-routine, and differ in geographic location, environmental conditions, the progression of the emergency, and the resources available. Therefore, the management of emergencies requires the capacity to adapt existing knowledge and skills quickly and accurately from one context to another.

At the operational level, the opportunity afforded by a focus on key cues lies in minimizing cognitive demands to those features that offer the greatest diagnosticity (Brouwers et al., 2017; Frye & Wearing, 2016). Features perceived as less diagnostic are not processed and, therefore, the cues acquired may be activated across a range of contexts. This capacity for adaptation is characteristic of expertise whereby an expert practitioner recognizes the fundamental behavior of a fire, irrespective of the specific context. Where the context does impact the behavior of a fire, a new mental model that pertains specifically to those features is generated. For example, the presence or absence of eucalyptus trees would activate different mental models as to the behavior of a wildfire since these trees contain an oil that, when heated, releases a flammable gas that enables a fire to travel across the tree canopy (Bowman, Wilson, & Hooper, 1988).

Since the capacity for generalization is a characteristic of expertise, improving the rate at which expertise is acquired will ensure that practitioners are appropriately equipped to respond to emergencies in different contexts (Jaarsma, Jarodzka, Nap, Merrienboer, & Boshuizen, 2014). This requires an environment in which learners can engage with a range of relevant feature–event/object relationships in the form in which they will be encountered in the operational context (Joung, Hesketh, & Neal, 2006; Lewandowsky et al., 1997; Thiele, Baldwin, Hyde, Sloan, & Strandquist, 1986). Actively engaging in their interpretation, together with the provision of timely, targeted feedback will enable the development of a greater repertoire of more precise mental models and the associated feature–event/object relationships that cue their activation (Ward, Suss, & Basevitch, 2009). However, where combinations of cues emerge in a situation that are associated with complementary mental models, it provides the trigger for the emergence of hybrid models, reflecting a level adaptation to what would otherwise constitute a novel situation. This process is illustrated in Figure 28.1.

In emergency management, learning is best achieved through simulation that, to the greatest extent possible, represents the characteristics inherent in the operational context, including the time constraints and uncertainty (Ford & Schmidt, 2000). Simulation embodies a level of flexibility that enables repeated exposure to feature–event/object associations together with the provision of feedback, thereby creating an environment conducive to their embodiment as cues. The intention is to contribute to the range and variety of possible cue-based associations in memory, facilitating the development, retention, and subsequent refinement of a repertoire of mental models sufficient to respond to a variety of situations.

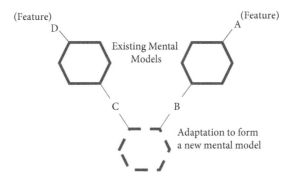

FIGURE 28.1 The process by which combinations of feature–event/object relationships in the form of cues provide a trigger to enable the construction of new mental models, reflecting adaptive expertise.

Cue-based learning appears to be more effective following the acquisition of an explicit, rudimentary mental model, usually by the intermediate stage of skill acquisition (Wiggins & O'Hare, 2003; Wiggins et al., 2014). Performance prior to this level tends to be governed by exemplars or cases in memory that are used to direct behavior (Leake, 1996; Nosofsky, 1991). Therefore, exposure to cue-based associations at the intermediate level offers the opportunity to recast and refine pre-existing associations and thereby develop distinct mental models but with potentially greater precision (Wiggins & Bollwerk, 2006; Wiggins et al., 2002). Together with a breadth of cue-based experiences, it is possible that a number of cue pathways are developed that, although similar, activate different mental representations.

Just as cue-based associations are constantly being formed, reinforced, or refined in memory, they can also be discarded where they are no longer accurate or where the association has been supplanted by a feature–event/object association with greater diagnosticity. However, cues may also lose their sensitivity in the absence of reinforcement, and this is particularly problematic for emergency operations where experience with actual emergency conditions can be sporadic. In the absence of recent experience, otherwise expert practitioners may fail to respond to task-related features that are immediately evident to practitioners with less, but recent exposure to the operational context.

While there is evidence to suggest that cue-based associations decline in the absence of exposure (e.g., Arthur, Bennett, Stanush, & McNelly, 1998), what remains unclear is whether there are differences in the rates at which these associations decline, depending upon the levels of expertise. It might be reasoned that the inherent strength of association between feature and events/objects, together with a well-developed mental model, shields against their loss in the absence of exposure. However, this has yet to be demonstrated empirically as has the role of individual differences in this process.

Future Directions

Despite the increasing importance of effective and efficient responses to emergency conditions, there remains a dearth of research concerning the nature of expertise in this context, and the means through which expertise is achieved and maintained. The research available to date comprises largely cross-sectional experimental designs, descriptive accounts, and/or a reliance on experimental contexts that capture only some aspects of the complex and dynamic situations that confront emergency responders. The result is a largely theoretically driven account that, although useful, requires evaluations using longitudinal research designs in environments that take account of the operational environment.

In undertaking research in the emergency conditions, one of the principal challenges lies in determining suitable measures of performance. Work in emergencies is often undertaken in teams, with any number of emergency responders contributing to the overall outcome. Isolating the performance of individual operators in this context is challenging and needs to draw on research initiatives involving expert teams. Specifically, multi-level modeling needs to be undertaken in the context of emergency conditions, where performance can be assessed at both the team and individual levels.

In addition to the challenges associated with differentiating the performance of individuals from that of teams, responses to emergencies are also complicated by the dynamic and uncertain nature of the conditions that are experienced. Consequently, the successful resolution of emergency situations is not always the most effective nor accurate means of assessing performance. More sensitive measures are needed that are also capable of distinguishing performance at different stages of the response to an emergency.

The lack of a valid and reliable means of assessing the performance of emergency responders also contributes to difficulties in identifying genuine subject-matter experts for both research and practice purposes. At a research level, subject-matter experts enjoy an important role in establishing a comparative group for cross-sectional research designs and in the construction of scenarios for both experimental and descriptive research initiatives (e.g., Okoli, Watt, & Weller, 2017). The critical contribution of subject-matter experts to expertise-related research is such that the misclassification of a competent practitioner as an expert will have significant ramifications for the conclusions drawn and conceptualization of expertise more generally. While some work has been undertaken into the use of cue utilization as a barometer of expertise, additional research is required to establish normed databases and ensure the validity and reliability of the metric.

The impact of cue utilization on the construction and maintenance of mental models is particularly important in the context of emergency management given the demand to adapt to different environmental conditions rapidly and accurately. Understanding the mechanisms by which this adaptation does or does not occur will inform both training initiatives and on-the-job decisions concerning the deployment of resources.

At a pragmatic level, research involving emergencies needs to be undertaken in environments that are sufficiently realistic and that enable a degree of experimental control over the presentation of stimuli (features). Access to suitable populations is also necessary, and preferably over an extended period. This will allow for the application of more sophisticated and robust measures and causal relationships to be inferred between variables.

Conclusion

The prospect of climate change, coupled with the development of increasingly interdependent and large-scale industrial systems, means that system failures in the future are likely to have a much larger and more immediate effect on a greater proportion of the population than has previously been the case. Therefore, the role of emergency responders is likely to become increasingly important in the future to safeguard the population and/or ensure the continuity of services.

Since emergencies are often characterized by a rapid onset, multiple changes in the system state, and a considerable degree of uncertainty, timely and accurate responses are critical to initially contain the situation, minimize disruption, and then effect an optimal resolution in the circumstances. This capability demands a level of expertise and has prompted a range of research initiatives to investigate both the characteristics of expertise in the context of emergencies and the means by which expertise is achieved. The broader goal has been to provide the foundation for strategies that will facilitate the acquisition of expertise amongst emergency responders.

While there has been a great deal of research concerned with characterizing expertise in emergencies, the dynamic and uncertain nature of the environment has presented difficulties for researchers, particularly in explaining the causal relationships between variables. Much of the research to date has relied upon descriptive approaches that provide the basis for more experimental strategies. However, more experimental strategies depend upon the exercise of a degree of experimental control, and access to large cohorts ideally, over an extended period.

Although there are a number of well-articulated theoretical models of expertise, establishing the veracity of these models requires a comparative evaluation against performance metrics. Where some environments are conducive to the collection of performance metrics, reliable and objective outcome metrics are difficult to isolate in emergency response. Cue utilization offers a potential solution to this problem, although further evaluation is necessary.

Despite the challenges associated with research in emergency environments, understanding expertise in this context remains an important goal not simply to ensure that responses to emergencies are timely and appropriate, but to ensure the safety of those emergency responders on which society depends. It constitutes a unique role, and society has a responsibility more generally to ensure that emergency responders are optimally equipped to manage the range of situations that they are likely to confront: in effect, that they possess the knowledge, skills, and capabilities of experts.

References

Abernethy, B. (1990). Anticipation in squash: Differences in advance cue utilization between expert and novice players. *Journal of Sports Sciences* 8(1), 17–34.

Abernethy, B., & Russell, D. G. (1987). Expert-novice differences in an applied selective attention task. *Journal of Sport Psychology* 9(4), 326–345.

Ackerman, P. L. (1988). Determinants of individual differences during skill acquisition: Cognitive abilities and information processing. *Journal of Experimental Psychology: General* 117(3), 288.

Adams, R. J., & Ericsson, A. E. (2000). Introduction to cognitive processes of expert pilots. *Journal of Human Performance in Extreme Environments* 5(1), 5.

Arthur, W., Jr., Bennett, W., Jr., Stanush, P. L., & McNelly, T. L. (1998). Factors that influence skill decay and retention: A quantitative review and analysis. *Human Performance* 11(1), 57–101.

Ash, J., & Smallman, C. (2010). A case study of decision making in emergencies. *Risk Management* 12(3), 185–207.

Aviation Safety Council. (2016). *Aviation Occurrence Report: TransAsia Airways Flight GE235, ATR72-212A, Loss of control and crashed into the Keelung River, three nautical miles east of Songshan Airport, Taipei, 4 February, 2015*. Report Number ASC-AOR-16-06-001. Taipei, Taiwan: Aviation Safety Council.

Bayouth, S. T., Keren, N., Franke, W. D., & Godby, K. (2013). Examining Firefighter Decision-Making: How Experience Influences Speed in Process and Choice. *International Fire Service Journal of Leadership and Management* 7, 51–60.

Besnard, D., Greathead, D., & Baxter, G. (2004). When mental models go wrong: co-occurrences in dynamic, critical systems. *International Journal of Human-Computer Studies* 60(1), 117–128.

Bowman, D. M. J. S., Wilson, B. A., & Hooper, R. J. (1988). Response of Eucalyptus forest and woodland to four fire regimes at Munmarlary, Northern Territory, Australia. *Journal of Ecology* 76(1), 215–232.

Braley, L. S., & Johnson, D. M. (1963). Novelty effects in cue acquisition and utilization. *Journal of Experimental Psychology* 66(4), 421–422.

Breckwoldt, J., Klemstein, S., Brunne, B., Schnitzer, L., Arntz, H. R., & Mochmann, H. C. (2012). Expertise in prehospital endotracheal intubation by emergency medicine physicians—comparing "proficient performers" and "experts." *Resuscitation* 83(4), 434–439.

Brouwers, S., Wiggins, M. W., Griffin, B., Helton, W. S., & O'Hare, D. (2017). The role of cue utilisation in reducing the workload in a train control task. *Ergonomics* 60(11), 1500–1515.

Butler, B. W., Bartlette, R. A., Bradshaw, L. S., Cohen, J. D., Andrews, P. L., Putnam, T., & Mangan, R. J. (1998). Fire behavior associated with the 1994 South Canyon Fire on Storm King Mountain, Colorado. Research Paper RMRS-RP-9. Washington, DC: United States Department of Agriculture.

Case, K., Harrison, K., & Roskell, C. (2000). Differences in the clinical reasoning process of expert and novice cardiorespiratory physiotherapists. *Physiotherapy* 86(1), 14–21.

Causer, J., Barach, P., & Williams, A. M. (2014). Expertise in medicine: using the expert performance approach to improve simulation training. *Medical Education* 48(2), 115–123.

Cellier, J. M., Eyrolle, H., & Mariné, C. (1997). Expertise in dynamic environments. *Ergonomics* 40(1), 28–50.

Cha, M., Han, S., Lee, J., & Choi, B. (2012). A virtual reality-based fire training simulator integrated with fire dynamics data. *Fire Safety Journal* 50, 12–24.

Chapman, T., Nettelbeck, T., Welsh, M., & Mills, V. (2006). Investigating the construct validity associated with microworld research: A comparison of performance under different management structures across expert and non-expert naturalistic decision-making groups. *Australian Journal of Psychology* 58(1), 40–47.

Chevalier, N., Dauvier, B., & Blaye, A. (2018). From prioritizing objects to prioritizing cues: a developmental shift for cognitive control. *Developmental Science* 21(2), doi: 10.1111/desc.12534.

Cioffi, J. (2000). Recognition of patients who require emergency assistance: A descriptive study. *Heart & Lung: The Journal of Acute and Critical Care* 29(4), 262–268.

Cohen-Hatton, S. R., & Honey, R. C. (2015). Goal-oriented training affects decision-making processes in virtual and simulated fire and rescue environments. *Journal of Experimental Psychology: Applied* 21(4), 395–406.

Comfort, L. K., Ko, K., & Zagorecki, A. (2004). Coordination in rapidly evolving disaster response systems: The role of information. *American Behavioral Scientist* 48(3), 295–313.

Condie, J. (2012). Beyond rationalisations: Improving interview data quality. *Qualitative Research in Accounting & Management* 9(2), 168–193.

Ericsson, K. A., & Ward, P. (2007). Capturing the naturally occurring superior performance of experts in the laboratory: Toward a science of expert and exceptional performance. *Current Directions in Psychological Science* 16(6), 346–350.

Ericsson, K. A., Whyte IV, J., & Ward, P. (2007). Expert performance in nursing: reviewing research on expertise in nursing within the framework of the expert-performance approach. *Advances in Nursing Science* 30(1), E58-E71.

Feeley, T. H., & DeTurck, M. A. (1995). Global cue usage in behavioral lie detection. *Communication Quarterly* 43(4), 420–430.

Flin, R., Slaven, G., & Stewart, K. (1996). Emergency decision making in the offshore oil and gas industry. *Human Factors* 38(2), 262–277.

Ford, J. K., & Schmidt, A. M. (2000). Emergency response training: strategies for enhancing real-world performance. *Journal of Hazardous Materials* 75(2), 195–215.

Frye, L., & Wearing, A. (2011). The central mountain fire project: achieving cognitive control during bushfire response. *Journal of Cognitive Technology* 16(2), 33–44.

Frye, L. M., & Wearing, A. J. (2016). A model of metacognition for bushfire fighters. *Cognition, Technology & Work* 18(3), 613–619.

Gegenfurtner, A., Lehtinen, E., & Säljö, R. (2011). Expertise differences in the comprehension of visualizations: A meta-analysis of eye-tracking research in professional domains. *Educational Psychology Review* 23(4), 523–552.

Gigerenzer, G., & Goldstein, D. G. (1996). Reasoning the fast and frugal way: models of bounded rationality. *Psychological Review* 103(4), 650–669.

Gonzalez, C., Vanyukov, P., & Martin, M. K. (2005). The use of microworlds to study dynamic decision making. *Computers in Human Behavior* 21(2), 273–286.

Gunnarsson, B. M., & Stomberg, M. W. (2009). Factors influencing decision making among ambulance nurses in emergency care situations. *International Emergency Nursing* 17(2), 83–89.

Handmer, J., & O'Neill, S. (2016). Examining bushfire policy in action: Preparedness and behaviour in the 2009 Black Saturday fires. *Environmental Science & Policy* 63, 55–62.

Hirsch, K. G., Corey, P. N., & Martell, D. L. (1998). Using expert judgment to model initial attack fire crew effectiveness. *Forest Science* 44(4), 539–549.

Hoffman, R. R. (1998). How can expertise be defined? Implications of research from cognitive psychology. In J. Fleck, W. Faulkner, & R. Williams (Eds), *Exploring expertise* (pp. 81–100). Palgrave Macmillan UK.

Hoffman, R. R., Crandall, B., & Shadbolt, N. (1998). Use of the critical decision method to elicit expert knowledge: A case study in the methodology of cognitive task analysis. *Human Factors* 40(2), 254–276.

Holmér, I., & Gavhed, D. (2007). Classification of metabolic and respiratory demands in fire fighting activity with extreme workloads. *Applied Ergonomics* 38(1), 45–52.

Hutton, R. J., & Klein, G. (1999). Expert decision making. *Systems Engineering* 2(1), 32–45.

Jaarsma, T., Jarodzka, H., Nap, M., Merrienboer, J. J., & Boshuizen, H. (2014). Expertise under the microscope: Processing histopathological slides. *Medical Education 48*(3), 292–300.

Joung, W., Hesketh, B., & Neal, A. (2006). Using "war stories" to train for adaptive performance: Is it better to learn from error or success? *Applied Psychology, 55*(2), 282–302.

Kioumourtzoglou, E., Kourtessis, T., Michalopoulou, M., & Derri, V. (1998). Differences in several perceptual abilities between experts and novices in basketball, volleyball and waterpolo. *Perceptual and Motor Skills 86*(3), 899–912.

Klein, G. (1997). Developing expertise in decision making. *Thinking & Reasoning 3*(4), 337–352.

Klein, G. (2008). Naturalistic decision making. *Human Factors: Journal of the Human Factors and Ergonomics Society 50*(3), 456–460.

Klein, G., Calderwood, R., & Clinton-Cirocco, A. (2010). Rapid decision making on the fire ground: The original study plus a postscript. *Journal of Cognitive Engineering and Decision Making 4*(3), 186–209.

Kowalski-Trakofler, K. M., Vaught, C., & Scharf, T. (2003). Judgment and decision making under stress: an overview for emergency managers. *International Journal of Emergency Management 1*(3), 278–289.

Lagnado, D. A., Newell, B. R., Kahan, S., & Shanks, D. R. (2006). Insight and strategy in multiple-cue learning. *Journal of Experimental Psychology: General 135*(2), 162.

Leake, D. B. (1996). Experience, introspection and expertise: Learning to refine the case-based reasoning process. *Journal of Experimental & Theoretical Artificial Intelligence 8*(3–4), 319–339.

Lewandowsky, S., Dunn, J. C., Kirsner, K., & Randell, M. (1997). Expertise in the management of bushfires: Training and decision support. *Australian Psychologist 32*(3), 171–177.

Lewandowsky, S., & Kirsner, K. (2000). Knowledge partitioning: Context-dependent use of expertise. *Memory & Cognition 28*(2), 295–305.

Leprohon, J., & Patel, V. L. (1995). Decision-making strategies for telephone triage in emergency medical services. *Medical Decision Making 15*(3), 240–253.

Loveday, T., & Wiggins, M. W. (2014). Cue utilization and broad indicators of workplace expertise. *Journal of Cognitive Engineering and Decision-Making 8*, 98–113.

Loveday, T., Wiggins, M. W. Harris, J., Smith, N., & O'Hare, D. (2013). An objective approach to identifying diagnostic expertise amongst power system controllers. *Human Factors 55*, 90–107.

Loveday, T., Wiggins, M. W., Searle, B. J., Festa, M., & Schell, D. (2013). The capability of static and dynamic features to distinguish competent from genuinely expert practitioners in pediatric diagnosis. *Human Factors 55*, 125–137.

McCormack, C., Wiggins, M. W., Loveday, T., & Festa, M. (2014). Expert and competent nonexpert visual cues during simulated diagnosis in intensive care. *Frontiers in Psychology 5*, 949.

McCutchen, D. (2011). From novice to expert: Implications of language skills and writing-relevant knowledge for memory during the development of writing skill. *Journal of Writing Research 3*(1), 51–68.

Malakis, S., Kontogiannis, T., & Kirwan, B. (2010). Managing emergencies and abnormal situations in air traffic control (part I): Taskwork strategies. *Applied Ergonomics 41*(4), 620–627.

Mathews, R. C., Buss, R. R., Stanley, W. B., Blanchard-Fields, F., Cho, J. R., & Druhan, B. (1989). Role of implicit and explicit processes in learning from examples: A synergistic effect. *Journal of Experimental Psychology: Learning, Memory, and Cognition 15*(6), 1083.

Mendonca, D., Beroggi, G. E., & Wallace, W. A. (2001). Decision support for improvisation during emergency response operations. *International Journal of Emergency Management* 1(1), 30–38.

Nosofsky, R. M. (1991). Tests of an exemplar model for relating perceptual classification and recognition memory. *Journal of Experimental Psychology: Human Perception and Performance* 17(1), 3–27.

O'Hare, D., Wiggins, M., Williams, A., & Wong, W. (1998). Cognitive task analysis for decision centred design and training. *Ergonomics* 41, 1698–1718.

Okoli, J., Watt, J., & Weller, G. (2017). Towards the classification of fireground cues: A qualitative analysis of expert reports. *Journal of Contingencies and Crisis Management* 25(4), 197–208.

Okoli, J., Watt, J., Weller, G., & Wong, W. B. (2016). The role of expertise in dynamic risk assessment: A reflection of the problem-solving strategies used by experienced fireground commanders. *Risk Management* 18(1), 4–25.

Okoli, J., Weller, G., & Watt, J. (2014). Eliciting experts' knowledge in emergency response organizations. *International Journal of Emergency Services* 3(2), 118–130.

Omodei, M. M., McLennan, J., & Wearing, A. J. (2005). How expertise is applied in real-world dynamic environments: Head mounted video and cued recall as a methodology for studying routines of decision making. In T. Betch & S. Haberstroh (Eds), *The routines of decision making* (pp. 271–288). Mahwah, NJ: Lawrence Erlbaum.

Omodei, M. M., & Wearing, A. J. (1995). The Fire Chief microworld generating program: An illustration of computer-simulated microworlds as an experimental paradigm for studying complex decision-making behavior. *Behavior Research Methods* 27(3), 303–316.

Ozel, F. (2001). Time pressure and stress as a factor during emergency egress. *Safety Science* 38(2), 95–107.

Patterson, M. D., Militello, L. G., Bunger, A., Taylor, R. G., Wheeler, D. S., Klein, G., & Geis, G. L. (2016). Leveraging the critical decision method to develop simulation-based training for early recognition of sepsis. *Journal of Cognitive Engineering and Decision Making* 10(1), 36–56.

Pelaccia, T., Tardif, J., Triby, E., Ammirati, C., Bertrand, C., Charlin, B., & Dory, V. (2015). Insights into emergency physicians' minds in the seconds before and into a patient encounter. *Internal and Emergency Medicine* 10(7), 865–873.

Perroni, F., Tessitore, A., Cortis, C., Lupo, C., D'artibale, E., Cignitti, L., & Capranica, L. (2010). Energy cost and energy sources during a simulated firefighting activity. *Journal of Strength & Conditioning Research* 24(12), 3457–3463.

Reischman, R. R., & Yarandi, H. N. (2002). Critical care cardiovascular nurse expert and novice diagnostic cue utilization. *Journal of Advanced Nursing* 39(1), 24–34.

Rouse, W. B., & Morris, N. M. (1986). On looking into the black box: Prospects and limits in the search for mental models. *Psychological Bulletin* 100(3), 349.

Schubert, C. C., Denmark, T. K., Crandall, B., Grome, A., & Pappas, J. (2013). Characterizing novice-expert differences in macrocognition: an exploratory study of cognitive work in the emergency department. *Annals of Emergency Medicine* 61(1), 96–109.

Shanteau, J. (1988). Psychological characteristics and strategies of expert decision makers. *Acta Psychologica* 68(1), 203–215.

Shanteau, J. (1992). Competence in experts: The role of task characteristics. *Organizational Behavior and Human Decision Processes* 53(2), 252–266.

Shanteau, J., Weiss, D. J., Thomas, R. P., & Pounds, J. C. (2002). Performance-based assessment of expertise: How to decide if someone is an expert or not. *European Journal of Operational Research* 136(2), 253–263.

Thiele, J. E., Baldwin, J. H., Hyde, R. S., Sloan, B., & Strandquist, G. A. (1986). An investigation of decision theory: what are the effects of teaching cue recognition? *Journal of Nursing Education* 25(8), 319–324.

Vickers, J. N., & Lewinski, W. (2012). Performing under pressure: Gaze control, decision making and shooting performance of elite and rookie police officers. *Human Movement Science* 31(1), 101–117.

Vincent, D. S., Sherstyuk, A., Burgess, L., & Connolly, K. K. (2008). Teaching mass casualty triage skills using immersive three-dimensional virtual reality. *Academic Emergency Medicine* 15(11), 1160–1165.

Ward, P., Suss, J., & Basevitch, I. (2009). Expertise and expert performance-based training (ExPerT) in complex domains. *Technology, Instruction, Cognition and Learning* 7(2), 121–145.

Whittaker, J., Haynes, K., Handmer, J., & McLennan, J. (2013). Community safety during the 2009 Australian "Black Saturday" bushfires: An analysis of household preparedness and response. *International Journal of Wildland Fire* 22(6), 841–849.

Wiggins, M. W. (2014). The role of cue utilisation and adaptive interface design in the management of skilled performance in operations control. *Theoretical Issues in Ergonomics Science* 15(3), 283–292.

Wiggins, M. W. (2015). Cues in diagnostic reasoning. In M. W. Wiggins and T. Loveday (Eds), *Diagnostic expertise in organizational environments* (pp. 1–13). Aldershot, UK: Ashgate.

Wiggins, M. W., & Bollwerk, S. (2006). Heuristic-based information acquisition and decision making among pilots. *Human Factors* 48(4), 734–746.

Wiggins, M., Brouwers, S., Davies, J., & Loveday, T. (2014). Trait-based cue utilization and initial skill acquisition: implications for models of the progression to expertise. *Frontiers in Cognition* 5, 541.

Wiggins, M., & O'Hare, D. (1995). Expertise in aeronautical weather-related decision making: A cross-sectional analysis of general aviation pilots. *Journal of Experimental Psychology: Applied* 1(4), 305–320.

Wiggins, M., & O'Hare, D. (2003). Weatherwise: Evaluation of a cue-based training approach for the recognition of deteriorating weather conditions during flight. *Human Factors* 45(2), 337–345.

Wiggins, M., Stevens, C., Howard, A., Henley, I., & O'Hare, D. (2002). Expert, intermediate and novice performance during simulated pre-flight decision-making. *Australian Journal of Psychology* 54(3), 162–167.

Williams-Bell, F. M., Kapralos, B., Hogue, A., Murphy, B. M., & Weckman, E. J. (2015). Using serious games and virtual simulation for training in the fire service: a review. *Fire Technology* 51(3), 553–584.

CHAPTER 29

EXPERTISE IN AVIATION

CHRISTOPHER D. WICKENS
AND FREDERIC DEHAIS

Introduction

In 2009, Captain Chelsey "Sully" Sullenberg landed a crippled aircraft, loaded with 150 passengers and no power following a bird strike, on the Hudson River in New York. The incredible skill shown by Captain Sullenberg and his co-pilot Jeffrey Skyles was credited with saving the lives of all on board in what could have been a near total disaster. Other examples of skilled piloting saving lives abound, including the landing of another totally crippled airline with no steering capability on the runway at Sioux Falls, Iowa. Here, because the runway surface is far less forgiving than a river surface, many lives were lost; but equally many were saved. Again, the incredible skill and expertise of the pilot, Captain Hanes, and his crew were credited for this disaster management.

These two disaster management responses reveal many different aspects of aviation expertise. Certainly, the finely tuned stick and rudder flying skills of Captain Sullenberg, reflecting his perceptual–motor coordination skill, permitted him to maintain a dangerously unstable aircraft on its critical glide slope to hit the water at precisely the correct angle and at high speed, so that its nose or wings did not penetrate the water and invite a catastrophic upset. For Captain Hanes, his calm communications skills, his ability to harness all the resources of both his cabin crew, an extra skilled pilot on board, and the services of air traffic control were all critical in accomplishing the life-saving actions. These factors combine as a set of skills known as crew resource management, which was first documented in the social psychology literature (Foushee, 1984; Salas et al., 2010). In between these skills supporting two disparate research areas in applied psychology (i.e., perceptual–motor coordination and social psychology) lies the expert decision-making skills of pilots, exhibited in the examples above in the choice of how and where to land.

In this chapter, we will describe the nature of expertise in aviation, both on the flight deck and in air traffic control (we here define both as *aviation professional*).

We investigate what changes occur in these professionals with learning, and how an expert differs from a novice. In doing so, we first define what we mean by *expertise* in aviation, considering two alternative approaches: the amount of flight experience and the level of proficiency. Next, we provide a clear description of the different psychological skills required of the aviation professional and their relationship to more fundamental information-processing abilities. Then we review the research that has distinguished between levels of proficiency, typically novice from expert aviation professionals, or has evaluated the change in skill differences as learning takes place. Finally, we briefly describe research that has tried to adopt novel training strategies to accelerate the trajectory of skill development.

Defining Expertise

There is an important distinction between the two different ways in which expertise has been defined in aviation. On the one hand, it is easy to define expertise a priori, in terms of the amount of flight experience. This is typically characterized by the total number of hours spent piloting an aircraft (henceforth, flight hours). In some cases, the number of flight hours in visual flight (i.e., when the ground is in sight, visual meteorological conditions or VMC) is differentiated from those spent under instrument flight conditions (instrument meteorological conditions or IMC; i.e., when the ground is not in sight and the pilot must rely totally on navigational instruments). These two conditions are the basis for two types of pilot ratings or certifications. With only a visual flight rating (VFR), the pilot cannot fly under IMC. With an instrument flight rating, (IFR), the pilot can fly under both VMC and IMC, and is considered more proficient. Finally, sometimes experience is also defined in terms of flight qualifications. In particular, those qualified to fly only general aviation are considered less proficient than those also certified to fly commercial transport aircraft, as run by the airline industry.

The second distinction of aviation professional expertise is simply how *proficient* the professional is at his or her task. Although we could credit the proficient management of individual incidents (such as those described at the outset) as being representative of an expert level of proficiency, we cannot necessarily say that this was a consequence of their many hours of both instrument and visual flight, nor whether these efforts would be repeatable and, therefore, truly expert. Indeed, National Transportation and Safety Board accident reports are replete with similar examples but of tragic accidents attributable to the errors of pilots who had many years of experience, such as the crash of an Eastern Airlines Jet into the Everglades (Dismukes, Berman, & Loukopoulos, 2010; Wiener, 1977). As is true in many other domains, years of experience do not guarantee a high level of proficiency (Ericsson & Charness, 1997), particularly with a skill as critically dependent on fluent decision making as is the case with aviation. As has been previously noted, decision making is a task often ill suited for learning from experience (Einhorn & Hogarth, 1978).

Aviation Tasks

The Pilot's Tasks

Conventionally, the work of a pilot has been categorized into four tasks considered primary.

Aviate

This is the standard perceptual–motor *stick-and-rudder* task required to keep the plane airborne. It involves selecting the right combination of *airspeed* via the throttle, *pitch* (nose pointed upward or downward) via controls on the wings and the elevator on the tail, and *bank angle* via the ailerons located on the wings to assure that the air flow is greater over the wings than beneath them. This differential flow in turn creates a partial vacuum above the wings, which literally *lifts* the aircraft upward toward the sky. There is greater lift when the aircraft is flying faster through the air, and when the wings are level. When this lift is lost, the plane will *stall* and start to fall toward the earth. In addition to maintaining lift (and preventing stall), these controls with the stick and throttle serve to change the aircraft's *attitude* (bank and pitch) in a way that can direct it to different vectors or *3D trajectories* through the air (e.g., turn, climb, descend, accelerate, or decelerate). These changes in turn are the basic building blocks of navigation as described in the next section, "Navigate."

What makes aviating so complex, and requires a great amount of skill development, is the fact that all of these axes are cross-coupled (i.e., each action has both primary and secondary effects). For example, when a pilot banks to turn the plane, it can also start to *slide* downward, hence both losing altitude and gaining airspeed. And when the pilot pitches the airplane upward to climb, it will also lose airspeed. Such cross-coupling of primary and secondary effects requires a great deal of mental integration. Also, with larger aircraft, there is a greater lag between when a control is implemented, and when the plane starts to change the controlled variable. The dynamics of manual control systems are such that dealing with lags requires mental prediction which is cognitively quite demanding (Wickens, 1986; 2003; 2007; Wickens & Gopher, 1977). Failure to aviate well is often a precursor to a *loss of control* (LOC) accident—the most lethal kind of accident in both commercial and general aviation. Failure to properly aviate has been identified as the causal factor for half the accidents of general aviation pilots in the UK during a 6-year period (Taylor, 2014).

Navigate

While aviating can put the plane on a trajectory to establish where is it going, navigating determines the precise trajectory (e.g., a climb to 5000 feet, a heading of 270°) and precise targets (e.g., a fix over a certain point on the ground, at a particular altitude, at a point in particular time). Airborne navigation can be particularly challenging because of the number of attributes to be controlled (heading, climb, altitude, position, speed in four dimensions), and targets may be specified in various frames of reference (Wickens, 1999; Wickens, Vincow, & Yeh, 2005). For example, a navigational command may be given in a world-referenced frame (fly 250°), but exercising it may require thinking in ego-referenced terms (I must turn 40° from my current heading). The cognitive demands of transforming between different frames of reference (mental rotation) are familiar any time we are driving southward, negotiating complex intersections, while consulting a north-up map (Aretz & Wickens, 1992; Wickens, 1999; Wickens et al., 2005). These complexities are amplified in the three dimensions of the airspace.

The close relationship between aviating and navigating has been described as a form of mental calculus—every bit as challenging as learning to intuitively apply integral and differential calculus computations(!!)—because the parameters controlled in aviating are generally integrated over time to establish the navigational goals (Wickens, 2007). One should note that both of these tasks (aviate and navigate) are, in many aircraft, supported by various forms of automation and technology. Autopilots can relieve the pilot of many aspects of aviating. Likewise, new navigational systems and some advanced displays can alleviate many of the mental transformations of navigating. However, such systems are always susceptible to failure, and it seems reasonable to assume that pilot skills at flying the aircraft *by hand* should always be practiced. Also training a pilot to fly should always proceed through the sequence of hand flying, before learning the capabilities of automation.

Communications

This refers primarily to voice communications between the pilot and air traffic control (Mosier et al., 2013). Verbal *protocols*—precise *readback* of controller's instructions and precise *hearbacks* by the controller (who must assure that the pilot reads back precisely what the controller said)—must be maintained during their communications. This activity also requires tuning of the radios to different frequency channels in order to deal with different controllers along the flight path. The challenge of precise communications is amplified by two factors. First, so much of it is accomplished by voice, and by the auditory modality which is so susceptible to short-term memory forgetting (Gateau, Ayaz, & Dehais, 2018; Gateau, Durantin, Lancelot, Scannella, & Dehais, 2015; Helleberg & Wickens, 2003; Latorella, 1996). Second, much of it involves numbers, which, if confused, can have catastrophic consequences (e.g., confusing or mis-remembering a heading, altitude, and airspeed command of 320, 25, 350 respectively, as, for example, 350, 25, 320).

Systems Management

This refers to assuring the proper mechanical and electrical status of all on-board systems, such as power, fuel, and engine functioning, as well as assuring the correct functioning of navigation and aviating systems. Systems management primarily involves monitoring and awareness of the system status, whereas the prior three tasks also require the performance of specific, and often skilled, actions. However, system management can also escalate rapidly into required diagnosis, which involve action-driven troubleshooting and systems corrections when things go wrong. The diagnosis sequence sometimes terminates in crisis decision making regarding what to do with the crippled airplane.

Mission Tasks

In addition to the previous four primary tasks, many flights also require a *mission-oriented* task. An aerial photographer must photograph, a fire tanker must drop its load of fire retardant precisely, and of course almost all military aircraft have a mission-critical combat objective. These are often added to the four requirements of basic flying.

Task hierarchy

The four primary tasks of aviate, navigate, communicate, and systems management (ANCS) are listed in the that order because it is generally considered that these tasks are hierarchical and pilots should adhere to this ordering of prioritization when tasks conflict. For instance, unless the pilot keeps the plane in the sky (aviate), he or she cannot accomplish any of the tasks below it in the hierarchy; and, for safety reasons, the pilot should always know where the plane is traveling and where it is, regarding navigational goals and hazard avoidance, before engaging in communication or systems management. The ANCS hierarchy needs to be flexible because certain tasks may temporarily demand top priority (Schutte & Trujillo, 1996). These task management skills, referred to as cockpit task management (CTM; Funk, 1991), are in themselves complicated and constitute important features of expert piloting. We discuss these features of expert piloting later in the sections on "Task Management" and "Expertise in Aviation." Next we discuss the tasks in which expert air traffic controllers engage.

Air Traffic Controller Tasks

The controller can have a work environment that is in many ways every bit as demanding as that of the pilot, even though it contains only two, rather than five major task categories (Wickens, Mavor, & McGee, 1997). These tasks can be defined as the following.

Maintaining Separation

The number one safety priority of the controller is to keep planes from colliding, in the air, or on the airport surface. To ensure that this is the case requires that the controller keeps each aircraft outside of a *protected zone* around each other one, a sort of cylindrical *hockey puck* of a designated vertical and circular extent (this may vary in its size, depending on the region of the airspace). To accomplish this, controllers are required to have an extensive amount of 3D visualization skills, particularly those involving spatial prediction. If one plane is flying toward another's protected zone, a command issued to the pilot to avoid penetration must be issued well before the zone is penetrated. And such proactive control requires the skill of making predictions (Boudes & Cellier, 2000).

Maintaining Flow

An individual plane does not require controllers to direct it from start to destination. However, in the commercial airspace it is necessary that controllers maintain the collective flow between airports. This means they must continually control departures and arrivals. The controllers maximize the efficiency of travel, and hence minimize delays, by managing the fundamental components of navigation (speed, heading, and altitude) of an entire stream of aircraft.

Efficiency goals, such as *maintaining flow* of air traffic, and safety goals, such as *maintaining separation* between aircraft, are somewhat conflicting. In a busy airspace, efficiency can be best preserved by tightly packing planes together. This can occur only up to a limit where separation standards are not violated. But where that boundary is is difficult to establish. The balance between these competing forces is a key characteristic of controller proficiency. Unlike the pilot's clear definition of a task hierarchy, the relative priority for maintaining wide separation versus flow is not as easily determined given its conflicting nature.

Communications

Communications is a task shared equally by pilots and controllers. Although it may be argued that since the controller is generally in charge of issuing voice instructions, comprehension skill demonstrated during this exchange is most vital for the pilot, who after all is doing the majority of listening.

Cognitive or Non-technical Skills

As in other societal-technical domains of high complexity such as medicine or power plant management, so the skills in aviation may be divided into the so-called "technical skills," related specifically to the task of controlling the aircraft, and non-technical skills of a more generic nature.

At a slightly different level of description from the aviation task-oriented skills discussed in the previous sections, and that are clearly articulated in pilot and controller training manuals, are a set of what we define as cognitive, or non-technical skills that can often differentiate better from poorer performing aviation professionals. We discuss four of these fundamental skills as follows.

Task Management

We referred briefly to task management skills earlier in the context of the ANCS priority hierarchy on the flight deck of cockpit task management (Chou, Madhavan, &

Funk, 1996). But good task management skills go beyond adhering to this hierarchy to include such skills as:

- Effectively and flexibly moving a lower hierarchy task toward the top in case of an emergency (Schutte & Trujillo, 1996). For example, an engine overheat may suddenly bring systems management to a greater priority than navigating.
- Switching with sufficient frequency between tasks to avoid some form of cognitive tunneling on a task deemed high priority (e.g., troubleshooting an engine failure) at the expense of other safety critical tasks (e.g., altitude or airspeed monitoring; Wickens & Alexander, 2009).
- Resisting dealing with a low priority interruption while in the middle of performing a higher priority task (Latorella, 1996).
- Remembering to return to a temporarily deferred task (Loukopolis, Dismukes, & Barshi, 2009).

Failures of appropriate task management in any of these four types have been attributed as the cause of aircraft and ATC mishaps (Dismukes, Berman, & Loukopolous, 2010; Loukopolous, Dismukes, & Barshi, 2009; Wickens, 2003). We examine later the extent to which these improve with expertise.

Decision Making and Diagnostic Skills

Since the pioneering work of Jensen (1982) and Wiener (1977), aviation professionals have realized the vulnerability of pilot and controller performance to making inappropriate or non-optimal decisions and judgments that lead to accidents (Sicard, Taillemite, Jouve, & Blin, 2003). For every good decision in crisis, such as that made by Captain Sullenberger to try to land on the river, there are countless poor decisions. These include examples such as the decision of a pilot who is not instrument rated, to try to *fly through* bad weather (Wiggins & O'Hare, 1995) in order to reach a final destination at all costs. A study conducted by the French Safety Board revealed that such erroneous behavior (*perseveration*) was responsible for more than 40 percent fatalities in general aviation (see Dehais, Tessier, Christophe, & Reuzeau, 2010). This trend to persist in hazardous decision making also occurs in commercial aviation. For instance, 97 percent of all unstabilized approaches end up with a decision to land (Curtis & Smith, 2013). These typically lead to satisfactory outcomes (potentially via good luck), but are occasionally the source of tragic fatalities, such as the decision by a pilot to take off with ice remaining on the wings at Washington National Airport in 1987.

The study of cognition, and decision making in particular, has provided a great deal of knowledge on how decision skills are learned (e.g., for reviews, see Hoffman et al., 2014; Suss & Ward, 2015), and it is often assumed that such skills will naturally develop with *time on task* or years of flight experience. Yet certain specific characteristics that we will discuss later render the acquisition of skilled pilot judgment a slow and sometimes unreliable process.

Situation Awareness

Numerous aviation accidents have been directly attributed to a loss of situation awareness (SA), where a pilot or controller fails to maintain the big picture of what is going on, fails to notice changes in the environment or to predict the implications of the evolving situation (Durso & Alexander, 2010; Hopkin, 1999; Jones & Endsley, 1996; Wickens, 2002). SA (or its complement, the loss of SA: LSA) can apply to almost any dynamic feature of the aerospace environment, including the awareness of:

- Aircraft attitude (pitch and roll);
- Navigational and geographical information;
- Personnel (e.g., "What is my captain doing/thinking?"); and
- System and automation state.

Orthogonal to these four examples of dynamic processes to be aware of lie three components or *levels* of SA (Endsley, 1995, 1999).

1. *Noticing* that things have changed, a process heavily dependent on visual scanning (level 1 SA);
2. *Understanding* the meaning of changes, a process heavily dependent on prior knowledge (level 2 SA); and
3. *Predicting* the implications of the change, a cognitive activity quite dependent on working memory, a process dependent upon both prior knowledge and working memory (level 3 SA).

While these categories map onto the pilot's task hierarchy, the skill of maintaining SA is thought to be more general than the particular knowledge and skill set associated with any one of the ANCS tasks. Effort has been invested in trying to assess and train SA skills that might benefit any and all of the tasks we've discussed.

Crew Resource Management

Some skills emerge uniquely in a team, whether between pilot and co-pilot on the flight deck, or between flight deck personnel and ATC or the cabin crew. These have been termed crew resource management (CRM), which includes communications skills. However, these go well beyond adhering to the strict communications protocols described earlier, to include effective communications in emergencies where there may be no protocol and the harnessing of effective teamwork strategies, including what has been described as team situation awareness (Stanton, Salmon, Walker, Salas, & Hancock, 2017). As with the other non-technical skills described earlier, airspace incidents and accidents provide plentiful examples where CRM was both effective in *saving the day* (e.g., the Sioux City crash) or was found wanting, leading to confusion, uncertainty, and often disaster (e.g., the crash of a commercial airliner in the Florida Everglades, when all three personnel on the flight deck concentrated on a potential landing gear failure, and did not pay attention to altitude) (see O'Hare & Roscoe, 1990).

EXPERTISE IN AVIATION

Taken together, both technical and non-technical skills define a powerful array, which should differentiate the expert and highly experienced pilot or controller from one who either has little experience or performs poorly at the task. But is this intuition correct? In the following we review the experimental and descriptive data for the different categories of skills that indicate the extent to which these intuitions are correct, and the more specific features that differentiate levels of expertise. To the extent that these skills do differentiate across levels of proficiency, we examine the success of training strategies targeted directly at the four non-technical skills to establish whether the trajectory of expertise can be shortened.

In the following section we review the literature on professional aviation expertise. Our operational definition of expertise is high proficiency in performing aviation tasks. As such, proficiency may be assessed by instructor ratings, objective quality of performance on subtasks or often, as is the case particularly of pilot judgment, of decision outcomes that are either labeled *good* (e.g., turning back in deteriorating weather) or *poor* (e.g., flying on into a storm). At the extreme, mishaps and accidents are often the consequence of poor performance. Increasing *expertise* implies better outcomes, and the *expert* is one who consistently produces the best outcomes.

Our review seeks to understand three main factors that may contribute to outcome quality:

- Flight or ATC *experience* and certification or type ratings;
- Natural cognitive or psychomotor *abilities*; these may sometimes extend to personality types or measures of cognitive style; and
- Certain *cognitive strategies* typically learned through aviation experience; prominent among these which we will discuss are ocular-motor or visual scanning strategies.

The four main non-technical skills described in the section "Cognitive or Non-technical Skills" often mediate the relationship between the three main causal variables and the outcome. For example, an aviator may demonstrate good outcomes because she possesses superior situation awareness (Trapsilawati, Wickens, Cheun, & Qu, 2017). However, we note that in many cases these non-technical skills represent the final outcome of aviation performance assessment. Finally, we consider two phenomena well known to develop with experience, whether through deliberate practice or simply on-the-job training (e.g., Patrick, 2006): the decreased attention demands of performance, known to occur with a phenomenon termed *automaticity* (e.g., Fisk, Ackerman, & Schneider, 1987), and an increase in knowledge (e.g., Simon & Gilmartin, 1973). We refer to these two phenomena as the *known signatures of expertise*.

The Research

In the following pages we first discuss research supporting two of the most prominent features of expertise in general, automaticity and knowledge. We then consider the findings of general changes in performance that result as expertise develops before considering in depth how these are expressed in the non-technical skills of improved situation awareness, better decision making, and improved task management and resource management skills. A final research section is devoted to explicit efforts to train these non-technical skills.

Known Signatures of Expertise

Automaticity and Spare Capacity

It has been long known that increasing practice on a task, whether deliberate or just through repeated performance, reduces the attention/resource demands of the task (Fitts & Posner, 1967). This is one feature of a characteristic we refer to as automaticity (Fisk, Ackerman, & Schneider, 1987; Schneider & Shiffrin, 1977). One consequence of automaticity is to make available more attentional capacity for use on other tasks, which often results in improved time sharing. This phenomenon was well illustrated by Damos (1978), who demonstrated that flight instructors with greater flying experience performed better on a secondary task while flying than student pilots (novices).

There is also some evidence that individual differences in cognitive ability can play a role in making available spare attentional resources. In particular working memory (WM) span needed, for instance, in flight communications has been shown to vary broadly across people in general (e.g., Engle, 2002) and pilots specifically (e.g., Morrow et al., 2003). Hence those with greater WM span should demonstrate greater proficiency on other tasks while engaged in tasks that place a considerable demand on working memory.

It is important to note, however, that individuals develop automaticity better on some tasks than others or, more specifically, some individuals develop automaticity better when learning under some conditions than others. In particular, classic work by Schneider and Shiffrin (1977; Fisk, Ackerman, & Schneider, 1987) showed that only tasks for which there is high *consistency* between the mapping of events in the world and appropriate actions can develop total automaticity (and hence demand no resources). An example of such consistency is flight control without turbulence, in which a given action on the stick will be guaranteed to produce nearly the identical aircraft response every time given the same conditions (e.g., aircraft cargo weight and temperature/humidity/pressure) and within the same aircraft dynamics type.

However, unlike flight control (the *aviate* tasks), many other aviation tasks do not provide the opportunity to acquire skills that have a consistent mapping between events, actions, and their consequences, particularly many decision tasks. This is because the appropriate decision may depend heavily on the context or conditions in which the decision is required, which can vary markedly as would be expected in any complex domain. For example, the decision to fly on or turn back may depend not only on the weather, but also the fuel remaining or conditions at the different airports. Furthermore, appropriate decisions also depend on future conditions. Not only can information about the current situation be incomplete, future conditions are often imperfectly predictable, and hence provide an inconsistent context to which pilots must actively adapt (for a description of adaptive skill, see Ward, Gore, Hutton, Conway, & Hoffman, 2018). In short, many aviation decisions simply cannot be automatized and will always be resource-demanding, independent of the level of pilot expertise.

Knowledge

Just as reduced resource demand in consistently mapped tasks is a signature of experience-producing expertise, so also is the increase in knowledge about aviation. Such increase allows both more rapid and accurate retrieval of facts and procedures with experience. In air traffic control, the amount of time spent at a particular facility is one of the greatest predictors of expertise, allowing the controller to understand all the particular quirks and features of the relevant airspace (Seamster, Redding, & Kaempf, 1997; Wickens et al., 1997). Studies have revealed that not only does the amount of knowledge increase with experience, the qualitative nature of knowledge organization becomes more sophisticated (Sherry & Polson, 1999).

Expertise in General Flight Performance

Hardy and Parasuraman (1997) argued that domain independent knowledge (i.e., used in domain-general cognitive functioning) and pilot's characteristics (i.e., domain-specific expertise) collectively determine general flight performance. Whereas most of the empirical studies report a close relationship between flight experience and basic flying skills, several cognitive ability factors have been inferred to mediate this relation such as time-sharing ability (Tsang & Shaner, 1995), speed of processing (Taylor et al., 1994), attention (Knapp & Johnson, 1996), and both psychomotor ability and general intelligence (g) (Carretta & Ree, 2003). Yakimovitch et al. (1994) were among the very first to use a method to investigate these complex interactions between individual, expertise, and flight performance. Their approach administered a battery of cognitive tests (e.g., CogScreen-AE) and showed that it was predictive of flight parameter violation under real flight conditions. Following this approach of test battery correlations with flight performance, Taylor and colleagues (2000) found that speed of processing, working memory, visual associative memory, motor coordination, and tracking abilities explained 45 percent of the variance of the flight simulator performance.

More recent studies have been able to replicate and expand these findings regarding processing speed (Kennedy et al., 2013; Tolton, 2014; Van Benthem & Herdman, 2016), working memory (Causse, Dehais, Arexis, & Pastor, 2011a; Causse, Dehais, & Pastor, 2011b; Tolton, 2014; Van Benthem & Herdman, 2016), and tracking ability (Tolton, 2014), as well as visual attention allocation, cognitive flexibility (Van Benthem & Herdman, 2016), and logical reasoning (Causse et al., 2011b). These were all positively correlated with to the ability to maintain flightpath and keep control of the aircraft; i.e., the task of aviating.

In contrast to the previous studies, however, Causse et al. (2011b) did not find any relationship between basic reaction time speed and flight performance, a conclusion echoing that of Carretta and Ree (2003). Also, Johnston and Catano (2013) reported only limited success of cognitive ability tests to predict success in Canadian military aviation training. In reviewing pilot selection test batteries Damos (1996) concluded that, collectively, these results showed only a limited ability to predict expert flight performance. In sum, although some studies show a positive correlation between cognitive ability and expert performance, the correlations are low, although sometimes significant ($p < 0.05$) between $r = 0.15$ and $r = 0.40$ (see Causse, Matton, & Del Campo, 2012). We now turn to the sources of expertise in the four more specific non-technical tasks.

Situation Awareness

As we might expect, each level of SA (noticing, understanding, and predicting) in experts depends on different types of skills and abilities (Sohn & Doane, 2004; Wickens, 2007). At level 1, visual scanning strategies are a major component (Wickens et al., 2008). At level 2, knowledge is critical for understanding (Sohn & Doane, 2004). And at level 3, because of the critical cognitive demands of projecting, there is a vital role for working memory as well as knowledge. We now describe each of these levels in more detail.

Visual Scanning and Level 1 SA

The monitoring of the flight parameters on the flight deck is a key issue for flight safety. The National Transportation Safety Board (NTSB) and the International Civil Aviation

Organization (ICAO) state that deficiencies in monitoring were a causal factor in most of the major recent civilian accidents (NTSB, 2013; UK Civilian Aviation Authority, 2013). Jones and Endsley (1996) determined that 755 of air force aircraft mishaps resulting from LSA resulted from the breakdown of level 1 SA. Indeed, the volume of information that needs to be dynamically processed can overwhelm human operators and lead to poor situation awareness with regards to primary flight parameters. Several eye-tracking studies have revealed that more experienced pilots have developed specific scanning strategies, different from novices, to ensure better awareness and flying performance (Kim, Palmisano, Ash, & Allison, 2010; Kirby, Kennedy, & Yang, 2014; Li, Chiu, & Wu, 2012; Ottati, Hickox, & Richter, 1999; Robinski & Stein, 2013; Fox, Merwin, Marsh, McConkie, & Kramer, 1996). See also the comprehensive review of pilot scanning studies by Ziv (2016), and Peissl and Wickens (2018). As an example, a majority of studies revealed that experienced pilots exhibited shorter dwell times on the instrument displays but checked them more frequently (Bellenkes, Wickens, & Kramer, 1997; Kasarskis, Stehwien, Hickox, Aretz, & Wickens, 2001; Kramer, Tham, Konrad, Wickens, & Lintern, 1994; Li, Chiu, Kuo, & Wu, 2013; Sullivan, Yang, Day, & Kennedy, 2011; Tole, Stephens, Vivaudou, Ephrath, & Young, 1983). These finding suggest that expertise is associated with more efficient visual scanning, that is, greater skill at extracting relevant information in a shorter amount of time.

Some explanations may rely on the more experienced pilot's qualitatively different visual search pattern than novices and the proficiency of expert's mental model of their aircraft dynamics. For example, Bellenkes et al. (1997) found that more experienced pilots scanned predictive instruments more than novices, as if they were *looking ahead* of the aircraft to support level 3 situation awareness. Wickens et al. (2008) noted that those pilots who showed greater adherence to the prescriptions of an optimal priority-driven scan model were more proficient in detecting possible traffic conflicts, hence linking differences in level 1 SA to expertise in the navigational component (hazard avoidance) of pilot performance. Schriver, Morrow, Wickens, and Talleur (2008) found that more experienced pilots, during simulated in-flight system failures, both made better decisions and spent more time fixating on more relevant instruments. Here there was a link between experience and expertise in decision making, mediated by level 1 SA. The scanning patterns of more experienced pilots has also been shown to be more robust and less affected by increased workload and stress (Tole et al., 1983) and also to be more flexible to adapt to contingencies (Bellenkes et al., 1997).

Despite evidence that pilots improve their visual scanning with practice, some eye tracking studies have pointed out the importance of inter-individual differences with regard to visual abilities and level 1 situation awareness. For example, a study in a motion simulator revealed that dwell time on the landing gear indicator, but not flight experience, was predictive of the ability to detect an auditory landing gear alarm (Dehais et al., 2012, 2014). This study is in line with previous research demonstrating that the detection of unexpected events might be compromised by inadequate scanning and focused attention (Alexander & Wickens, 2006). Li et al. (2012) found that specific scanning strategies led pilots to have a better situation awareness of a hydraulic failure independently of the level of experience. Moreover, their study revealed that dwell time on relevant instruments such as airspeed strongly mediated experience to optimize fuel consumption and flight duration.

Individual differences in scanning strategies have also been found to discriminate *good* versus *poor* flight performance during critical flight phases such as landing (Gray, Navia, & Allsop, 2014). Lefrancois, Matton, Gourinat, Peysakhovich, and Causse, (2016) found that pilots who had an inadequate dwell time on the attitude indicator were more likely to face an unstabilized approach and had to perform a go-around. Correspondingly, Reynal, Rister, Scannella, Wickens, and Dehais (2017) observed that higher dwell time on the attitude indicator and lower fixation time on the navigation display was associated with poor awareness, a destabilized approach, and the resulting necessity to perform a go-around. Regarding this latter flight phase, crews in which the co-pilot spent more time glancing on the speed indicator exhibited better flightpath management (Dehais, Behrend, Peysakhovich, Causse, & Wickens, 2017).

Our review failed to identify studies in air traffic control where individual differences in scanning strategies were specifically associated with performance differences. The only study located (Hasse, Grasshoff, & Bruder, 2012) revealed null results. Part of the challenge here is the more ill-defined (compared to the cockpit) designation of specific areas of interest at the controller work station for quantifying controller scanning, and for determining optimal scan strategies.

Understanding and Predicting: Levels 2 and 3 SA

In reviewing this literature, we combine research on levels 2 and 3 SA, because some researchers do not distinguish between them, and in any case, the borderline between what is happening (level 2) and what will happen (level 3) is a very fuzzy one. To illustrate, in a collision avoidance situation in which the pilot is in a state of predicted conflict (understanding current state), this state is indistinguishable from saying that the pilot *will* collide (prediction), if a maneuver is not initiated. Sohn and Doane (2004) found that experience (flight hours) was a strong predictor of aircraft state awareness, a finding consistent with that of Bellenkes et al. (1997), who examined the frequency of looking at predictive instruments. It certainly makes sense that greater experience provides better knowledge and a better mental model of flight dynamics, hence allowing the pilot to more easily seek and absorb incoming information from appropriate sources. Sohn and Doane also found that SA was higher for pilots who had higher visualization skills and a greater capacity of long-term working memory (LTWM; Ericsson & Kintch, 1995; see Chapter 2, "The Classic Expertise Approach and Its Evolution," by Gobet, this volume). This latter construct—described as a retrieval structure-based mechanism that permits the limits of working memory to be circumvented—lies at the intersection of working memory and long-term memory. Those who have developed LTWM skills have essentially developed the ability to rapidly access and retrieve material regarding changing state, even if that material is not being actively rehearsed.

Sulistyawati, Wickens, and Chui (2011) observed that spatial ability and general working memory capacity predicted situation awareness in air force fighter pilots, and Carretta, Perry, and Ree (1996) observed a corresponding correlation of working memory capacity with the situation awareness of fighter pilots as rated by their superiors. Importantly, Carretta and Ree (2003) found that psychomotor performance and personality tests did not predict higher rated SA for their military sample.

Of these studies, Sulityawati et al. (2011) explicitly distinguished predictors of level 2 from level 3 SA, observing that only level 3 expertise was uniquely predicted by reasoning and logic ability. Furthermore, they found that pilots with high level 3 SA performed better in simulated air-to-air combat scenarios, but level 2 SA was not a significant predictor of performance here. Endsley and Bolstadt (1994) found that levels 2 and 3 SA of fighter pilots correlated with spatial skills, perceptual speed, and pattern-matching ability.

In air traffic control, Durso, Blekely, and Dattel (2006) examined the speed and accuracy of simulated traffic management. The better performing novices (i.e., those developing greater expertise) showed higher cognitive abilities of spatial working memory, perceptual closure, and need for cognition. Furthermore, a test of controller SA developed by the authors added to the prediction of performance above and beyond those cognitive abilities. Unlike Sohn and Doane (2004), their test was not explicitly designed to test LTWM theory (as with Sohn & Doane, 2004), but the authors drew similar conclusions.

In conclusion, increases in memory capacity, both working memory and LTWM are associated with higher situation awareness in aviation environments. In the case of LTWM, this is consistent with the findings that experts in other domains possess greater LTWM skills (e.g., see Suss & Ward, 2015, 2018; Ward et al., 2013). In at least one study, higher SA was found to be directly associated with increased aviation flight performance.

Decision and Judgment

Good aviation judgment depends on good situation awareness but is distinct from it in its focus on the specific choice or action to be generated and then taken on the basis of a dynamic situation assessment. Ward et al. (2013) referred to decision making and the situation assessment process (i.e., sensemaking) as two reciprocal sides of the same dynamic system (e.g., Neisser, 1976; for a model of how these two processes work collectively to bring about adaptive performance, also see Ward et al., 2018). Hence, we might expect the quality of aviation decision making to be driven by some but not all of the factors that drive SA. We might also expect that decision making is driven by factors that are uniquely related to the choice process, such as generating options for action, based on past experience and then choosing between them.

Experimental studies of expertise in pilot judgment and decision making have generally taken two forms: Discriminating those who make *good* decisions from those who make poor decisions in a particular context (such as the poor decision to continue the flight in poor weather), or assessing the characteristics of good pilot judgment in general (rather than in a particular context). In the former case, we assume that greater expertise is associated with those who make the better decision.

Good vs Poor Decisions

Several investigations have looked specifically at the tendency to make a poor judgment and fly on into deteriorating weather conditions, when the better choice is to turn back given the pilot's lack of qualifications. Several studies have indeed found that greater experience (i.e., more flight hours) supports better decision making in this context (Goh & Wiegmann, 2001; Hunter, Martinussen, Wiggins, & O'Hare, 2011; Johnson & Wiegmann, 2015; Wiegmann, Goh, & O'Hare, 2002; Wiggins & O'Hare, 1995; 2003). However, Wiggins and Bollwerk (2006) found better decision making to be more highly correlated with recency in flight experience (i.e., past 60 days) than with overall flight hours. A major reason experience provided an advantage is that those with more experience tend to employ the strategies of seeking and interpreting cues in the weather environment. These cognitive activities (seeking, interpreting) are closely related to, if not the same as, the expert's advantage in levels 1 and 2 SA, respectively.

Johnson and Wiegmann (2015) qualified that it was not purely the amount of flight hours that distinguished expert pilots making good decisions from those at a lower level of proficiency making poorer decisions. Rather it was the amount of time actually flying in poor weather that predicted decision quality. This implies a note of caution in simply using overall hours as a proxy for experience. Goh and Wiegmann (2001) observed that those who chose to continue flight when it was inappropriate to do so provided higher ratings of their own skills and judgments, a finding that might implicate a greater degree of overconfidence. We note, however, that none of the researchers from this class of studies examined the influence of individual differences in, for instance, cognitive ability on judgment and decision making.

To complement research on the decision to fly on into bad weather, a second type of research discriminating good from poor choices has examined the choice to continue with a landing in ill-advised circumstances. Several behaviors describe the kind of factors that help avoid such poor decision making. For instance, as described earlier, those who did not choose to continue under suboptimal conditions (i.e., chose to go around) employed more efficient scanning strategies (e.g., higher number of and shorter fixations on information that permitted insight into upcoming events; see Reynal et al., 2017). Good decision makers had higher working memory capacity and greater attentional flexibility (Causse et al., 2011a), were better at risk assessment (Hunter et al., 2011), and exhibited lower impulsivity (Behrend, Dehais, & Koechlin, 2017; Causse et al., 2011b). Subsequent research has also shown that when monetary incentive and uncertainty were manipulated in the helicopter landing decision task, risky decision makers exhibited lower activation of the prefrontal areas (i.e., dorsolateral prefrontal cortex and anterior cingulate cortex) than good decision makers. These areas signify rationality (Causse et al., 2013). In a related study, Adamson et al. (2014) found that lower activity in the caudate nucleus was associated with higher landing decision accuracy under instrument meteorological condition. Unfortunately, however, how these differences are moderated by individual differences in expertise remains relatively uncharted, specifically when examining performance on an aviation task (cf. Hunter et al., 2011). Further, none of the studies of landing decisions appear to have associated decision quality with the differences in experience (e.g., number of flight hours).

Overall Decision Quality

The more general approach to measuring expertise in pilot decision making, going beyond a particular decision (e.g., to land or go-around), is illustrated by two classes of studies. As an example of the first type, Stokes and his colleagues developed a pilot judgment trainer/evaluator simulator called MIDIS, which presented various decision scenarios to pilots and employed skilled flight instructors to evaluate their choices of action in terms of decision quality (Barnett et al., 1987; Stokes, Kemper, & Marsh, 1992; Stokes et al., 1987; Wickens, Stokes, Barnett, & Hyman, 1993). Barnett et al. (1987) observed that within a cohort of more experienced pilots, better decisions were made by those with higher working memory capacity. Stokes et al. (1992) found that experts generally made more optimal choices than novices, for example turning back when it was appropriate to do so. Results from the other two MIDIS studies in which pilots with different levels of experience (i.e., more vs less flight time) were compared found either no difference in decision quality between groups or ambiguous results (i.e., differences on some metrics but not others; Stokes et al., 1987).

As noted earlier, Schriver et al. (2008) examined experience differences in pilot decision making following in-flight failures in a simulator, a skill heavily dependent upon diagnostic ability and cue seeking. As a consequence of seeking different cues (e.g., oil pressure indicator, airspeed) by those at a lower level of proficiency, more experienced pilots' decisions were superior in both speed and accuracy.

The second type of general decision quality study has examined non-experimental aspects of data. Rebok, Qiang, Baker, McCarthy, and Li (2005) studied a large number of *violations* by air taxi pilots, where a violation was defined as an intentional decision to not follow or deviate from aviation rules. The authors found that fewer violations (i.e., *bad decisions*) were committed by pilots with more than 5000 flying hours. However, among this group, there was no tendency for violation rate to decrease with additional flight experience.

In a related study, Hunter (2006) measured pilots' perception of perceived risk of different flight scenarios, as a function of their flight certification category, and observed that pilots in more advanced categories (e.g., transport pilots) who were, therefore, more experienced generally perceived lower risk. It is not clear whether such pilots simply have greater confidence in their judgment because of greater proficiency, or rather, perhaps have greater overconfidence in their abilities. This finding potentially echoes that of Goh and Wiegmann (2001) and is consistent with other decision-making research, which suggests that that higher levels of decision-making experience often fosters increasing levels of overconfidence (Kahneman, 2011; Wickens, Hollands, Banbury, & Parasuraman, 2012).

Finally, a pessimistic view of the relationship between experience and judgment quality was offered by the data of McKinney (1993), who examined the quality of professional's decision outcomes. Professional fighter pilots' quality of decision making following air force aircraft mishaps was analyzed by two experienced pilots. No differences in quality were observed as a function of years of experience. Furthermore, McKinney's data revealed that those pilots flying in the leadership position exhibited poorer quality decisions, a finding McKinney attributes in part to greater overconfidence of those leads and in part to their lack of habit of soliciting information from others.

The absence of an experience effect is consistent with two other observations made by decision scientists on the effects of experience on decision making in other contexts. Einhorn and Hogarth (1978) have identified relatively poor learning (i.e., limited improvement with experience) with respect to their decision making because the feedback is often delayed. Furthermore, feedback in many real-world contexts is often misleading in an uncertain world because poor decision sometimes produce (luckily) good outcomes, and vice versa (i.e., good decisions can have poor effects). Following the earlier research of Shanteau (1992), Kahneman (2011; Kahneman & Klein, 2009) examined the characteristics of professions in which experience does not produce expertise in (i.e., better) decision making and judgment, and argued that the unpredictable aspects of such environments, and the resulting limited or challenging opportunities to learn from feedback, are primary reasons for this disconnect (for similar effects in healthcare, see Ericsson, 2004; Ericsson, Whyte, & Ward, 2007). Given the complexities of many aviation environments, it is understandable that this environment may, sometimes, result in only a loose association between expertise and experience.

As reviewed earlier, few researchers have directly examined these kinds of decisions as a function of cognitive ability differences. Those that have have observed that any such differences are not reliably predictive of the choices made by pilots. The ability to predict expertise in decision making on the basis of experience appears to grow increasingly problematic as we move from specific in-context decisions (e.g., flying through bad weather, inappropriate landings, or inaccurate diagnosis of system failures) to more generic context-independent evaluations of the process. The effect of experience therefore appears to be to support seeking out relevant perceptual cues (e.g., Schriver et al., 2008), not necessarily a reliance on superior capacities.

Cockpit Task Management

Surprisingly little research has been carried out on experience and expertise in aviation cockpit task management (Chou, Madhavan, & Funk, 1996; Loukopolis et al., 2009).

The focus of this research is distinct from scanning, in that it defines how one switches attentional resources between tasks, rather than how the eyeball switches between sources of information. It is distinctive from much attention research in that it examines attention's sequential, rather than its parallel properties (Wickens, Gutwiller, & Santamaria, 2015). In one study Wickens and Raby (1991) imposed sudden high workload on advanced student pilots doing a landing approach. The distinguishing feature of expertise between those performing best (in terms of flight path deviations) and those performing worst was that the former group scheduled higher priority tasks at more appropriate or optimal times. Similar findings in a flight simulator study were reported by Laudemann and Palmer (1995). Neither study, however, examined differences in experience nor correlated task management strategies with individual differences in cognitive ability.

Although basic research has correlated differences in basic attentional functioning with attention management strategies in multitasking environments, this has not been done in an aviation context, let alone with consideration of level of proficiency in this domain. Future aviation expertise research should examine this issue further.

Crew Resource Management

Crew resource management defines the coordination among teams (pilot, copilot, ATC) in, typically, dealing with in-flight emergencies or unpredicted events. While crew experience and CRM are beyond the scope of the current chapter (see Salas et al., 2010), one study of voice communications is directly relevant to both the non-technical skill of CRM and to the fourth task on the pilots' ANCS hierarchy: Communications. In this study, Morrow et al. (2003) found that experienced pilots showed better recall of typical ATC communications information than did non-pilots. However, there was little evidence that within the pilot population, increasing experience led to better recall. But for all groups, superior performance on communications recall was, expectedly, associated with greater working memory capacity and greater spatial ability. The latter benefit can be attributed to the fact that although numerical symbolic information was communicated, its interpretation was in terms of spatial locations and trajectories within the 3D airspace.

Training and Creating Expertise in Non-technical Aviation Skills

If experts perform better than novices in non-technical skills, an important issue is whether there are particular training modules available in these skills to *shortcut the trajectory* from novice to expert aviation professional. We examine whether their application has documented any success via positive transfer to more in-context aviation tasks. Here the evidence is scant and mixed. Regarding crew resource management, a host of CRM training programs have been adopted by nearly all of the American and European airlines, but a meta-analysis by Salas et al. (2006) concluded little consistent evidence for positive transfer. Regarding cockpit task management, Gopher, Weil, and Baeket (1994) have demonstrated some success of using the Space Fortress multitask video game to develop attention management skills that appeared to improve the chances of Israeli Air Force pilots to become qualified for the highest proficiency fighter pilot slots. Regarding decision making, other researchers (e.g., Walmsley & Gilbey, 2017) have developed specific aviation decision training modules (e.g., MIDIS: Stokes et al., 1987), but these have not been documented to transfer to better decision making in more remote contexts beyond the MIDIS simulator itself.

Regarding situation awareness, Endsley and Robertson (2000) and Endsley and Garland (2000) have discussed the need for such training but very few studies have proposed and evaluated indirect training for this important non-technical skill. For example, two PC-based situation awareness training programs were, respectively, implemented for navy cadets (Strater et al., 2004) and general aviation pilots (Bolstad, Endsley, Costello, & Howell, 2010). These studies failed to report strong evidences of situation awareness improvement. Only two situation awareness modules programs were developed for airlines pilots (Hoermann et al., 2003) and airline student pilots (Gayraud, Matton, & Tricot, 2017). The results of these two studies disclosed better performance and situation awareness scores for the experimental versus the control group when confronted with scenarios in simulated flights. The findings of these studies should encourage more research to define and test training solutions to enhance situation awareness ability and other non-technical aviation skills such as scanning as a means to potentially accelerate the acquisition of expertise.

Conclusions

In conclusion, expertise in aviation is certainly multidimensional, a complex mix of experience, abilities, and strategies, and this situation reflects the complex mix of technical and non-technical skills required. It is important to note too that within the field, there is not the same sort of competition to identify *the best* pilot that exists in other domains of performance (the exception being competition in university aviation flight teams). Although military aviation training and selection does sometimes refer to those selected to be fighter pilots as "the elite," it is hardly fair to characterize them, as a class, as experts because their skills for high proficiency may be quite different from those demanded by the transport pilot, or helicopter pilot.

A further challenge for defining expertise in aviation arises from the emerging dominance of flight deck automation. This tends to level the playing field of flying skills, except on those rare and unexpected occasions when automation fails (Wickens,

2009) and here, the better performing pilot or controller (i.e., the *expert*) may, ironically, be the one who has greater proficiency performing without automation, so that graceful recovery can be accomplished.

Ultimately one can argue that what truly defines expertise in aviation is the guarantee of safety. But this is an exceedingly difficult commodity to assess, perhaps defined by wise decisions to avoid unsafe conditions and possessing the non-technical skills to escape those conditions should they unexpectedly be thrust on the pilot. But experience in making the former decision, as we saw, does not necessarily lead to competence, and the occurrence of the latter decision is, fortunately, quite rare but as a consequence is hard to reliably assess.

Hence, we see a strong need for research to continue to examine these differences in expertise defined by aviation proficiency, where and however they can be found, and correlate them with abilities, strategies (i.e., maintaining SA, scanning), and indeed experience, to continue compiling these in a systematic way. We hope that this chapter has provided a foundation for this effort.

References

Adamson, M. M., Taylor, J. L., Heraldez, D., Khorasani, A., Noda, A., Hernandez, B., & Yesavage, J. A. (2014). Higher landing accuracy in expert pilots is associated with lower activity in the caudate nucleus. *PloS One* 9(11), e112607.

Alexander, A. L., & Wickens, C. D. (2006). Integrated hazard displays: Individual differences in visual scanning and pilot performance. *Proceedings of the Human Factors and Ergonomics Society Annual Meeting 50*, 81–85.

Aretz, A. J., & Wickens, C. D. (1992). The mental rotation of map displays. *Human Performance 5*, 303–328.

Barnett, B., Stokes, A., Wickens, C. D., Davis, T., Rosenblum, R., & Hyman, F. (1987). A componential analysis of pilot decision-making? *Proceedings of the 31st Annual Meeting of the Human Factors Society*. Santa Monica, CA: Human Factors Society.

Boudes, N., & Cellier, J.-M. (2000). Accuracy of estimations made by air traffic controllers. *International Journal of Aviation Psychology 10*, 207–225.

Behrend, J., Dehais, F., & Koechlin, E. (2017, September). Impulsivity modulates pilot decision making under uncertainty. In *The Human Factors and Ergonomics Society Europe* (HFES 2017) (p. 1–12).

Bellenkes, A. H., Wickens, C. D., & Kramer, A. F. (1997). Visual scanning and pilot expertise: the role of attentional flexibility and mental model development. *Aviation, Space, and Environmental Medicine 68*, 569–579.

Bolstad, C., Endsley, M. R., Costello, A. M., & Howell, C. D. (2010). Evaluation of computer-based situation awareness training for general aviation pilots. *International Journal of Aviation Psychology 20*(3), 269–294.

Carretta, T. R., Perry, D. C., & Ree, M. J. (1996). Prediction of situational awareness in F-15 pilots. *International Journal of Aviation Psychology 6*, 21–41.

Carretta, T., & Ree, M. J. (2003). Pilot selection methods. In M. Vidulich & P. Tsang (Eds), *Principles and practices of aviation psychology*. Mahwah, NJ: Lawrence Erlbaum.

Causse, M., Dehais, F., Arexis, M., & Pastor, J. (2011a). Cognitive aging and flight performances in general aviation pilots. *Aging, Neuropsychology, and Cognition* 18(5), 544-561.

Causse, M., Dehais, F., & Pastor, J. (2011b). Executive functions and pilot characteristics predict flight simulator performance in general aviation pilots. *International Journal of Aviation Psychology* 21(3), 217-234.

Causse, M., Matton, N., & Del Campo, N. (2012). Cognition and piloting performance: offline and online measurements. In: *Human Factors and Ergonomics Society Europe Chapter Annual Meeting in Toulouse*, France.

Causse, M., Péran, P., Dehais, F., Caravasso, C. F., Zeffiro, T., Sabatini, U., & Pastor, J. (2013). Affective decision making under uncertainty during a plausible aviation task: An fMRI study. *NeuroImage* 71, 19-29.

Chou, C., Madhavan, D., & Funk, K. (1996). Studies of cockpit task management errors. *International Journal of Aviation Psychology* 6, 307-320.

Curtis, B., & Smith, M. (2013). *Why are go-around policies ineffective?* Missisauga, Canada: Presage Group.

Damos, D. L. (1978). Residual attention as a predictor of pilot performance. *Human Factors* 20, 435-440.

Damos, D. L. (1996). Pilot selection batteries: Shortcomings and perspectives. *International Journal of Aviation Psychology* 6, 199-209.

Dehais, F., Behrend, J., Peysakhovich, V., Causse, M., & Wickens, C. D. (2017). Pilot flying and pilot monitoring's state awareness during go-around execution in aviation: A behavioral and eye-tracking study. *International Journal of Aerospace Psychology* 27(1-2), 15-28.

Dehais, F., Causse, M., Régis, N., Menant, E., Labedan, P., Vachon, F., & Tremblay, S. (2012). Missing critical auditory alarms in aeronautics: Evidence for inattentional deafness? *Proceedings of the Human Factors and Ergonomics Society Annual Meeting* 56, 1639-1643.

Dehais, F., Causse, M., Vachon, F., Régis, N., Menant, E., & Tremblay, S. (2014). Failure to detect critical auditory alerts in the cockpit evidence for inattentional deafness. *Human Factors* 56(4), 631-644.

Dehais, F., Tessier, C., Christophe, L., & Reuzeau, F. (2010). The perseveration syndrome in the pilot's activity: guidelines and cognitive countermeasures. In P. Palanques, J. Vanderdonckt, & M. Winkler (Eds), *Human error, safety and systems development* (pp. 68-80). New York: Springer.

Dismukes, R. K., Berman, B. A., & Loukopoulos, L. D. (2010). *The limits of expertise: Rethinking pilot error and the causes of airline accidents.* Aldershot, UK: Ashgate.

Durso, F., & Alexander, A. (2010). Managing workload and situation awareness in aviation systems. In E. Salas & D. Maurino (Eds), *Human factors in aviation.* New York: Academic Press.

Durso, F., Bleckley, M. K., & Dattel, A. R. (2006). Does situation awareness add to the validity of cognitive tests? *Human Factors* 48, 721-733.

Einhorn, H. J., & Hogarth, R. M. (1978). Confidence in judgment: Persistence of the illusion of validity. *Psychological Review* 85, 395-416.

Endsley, M. R. (1995). Toward a theory of situation awareness in dynamic systems. *Human Factors* 37, 32-64.

Endsley, M. R. (1999). Situation awareness in aviation systems. In D. Garland, J. Wise, & V. D. Hopkin (Eds), *Handbook of aviation human factors.* Mahwah, NJ: Lawrence Erlbaum.

Endsley, M. R., & Boldstadt, C. (1994). Individual differences in pilot situation awareness. *International Journal of Aviation Psychology* 4, 241-264.

Endsley, M. R., & Garland, D. J. (2000). Pilot situation awareness training in general aviation. *Proceedings of the Human Factors and Ergonomics Society Annual Meeting 44*(11), 357–360.

Endsley, M. R., & Robertson, M. M. (2000). Training for situation awareness in individuals and teams. Situation awareness analysis and measurement. In M. R. Endsley & D. J. Garland (Eds), *Situation awareness analysis and measurement* (pp. 349–365). Mahwah, NJ: Lawrence Erlbaum.

Engle, R. W. (2002). Working memory capacity as executive attention. *Current Directions in Psychological Science 11*(1), 19–23.

Ericsson, K. A. (2004). Deliberate practice and the acquisition and maintenance of expert performance in medicine and related domains. *Academic Medicine 79*(10, suppl), S70–S81.

Ericcson, K. A., & Charness, N. (1997). Cognitive and developmental factors in expert performance. In P. Feltovich, K. Ford, & R. Hoffman (Eds), *Expertise in context* (pp. 3–41). Cambridge, MA: MIT Press.

Ericsson, K. A., & Kintsch, W. (1995). Long-term working memory. *Psychological Review 102*, 211–245.

Ericsson, K. A., Whyte, J., & Ward, P. (2007). Expert performance in nursing: Reviewing research on expertise within the framework of the expert-performance approach. *Advances in Nursing Science 30*(1), E58–E71.

Fisk, A. D., Ackerman, P. L., & Schneider, W. (1987). Automatic and controlled processing theory and its applications to human factors problems. In P. A. Hancock (Ed.), *Human factors psychology* (pp. 159–197). Amsterdam: Elsevier.

Fitts, P. M., & Posner, M. A. (1967). *Human performance.* Pacific Palisades, CA: Brooks Cole.

Foushee, H. C. (1984). Dyads and triads at 35,000 ft: Factors affecting group process and aircrew performance. *American Psychologist 39*, 885–893.

Fox, J., Merwin, D., Marsh, R., McConkie, G., & Kramer, A. (1996). Information extraction during instrument flight: An evaluation of the validity of the eye-mind hypothesis. *Proceedings of the Human Factors and Ergonomics Society Annual Meeting 40*(2), 77–81.

Funk, K. (1991). Cockpit task management. *International Journal of Aviation Psychology 1*, 271–286.

Gateau, T., Ayaz, H., & Dehais, F. (2018). In silico versus over the clouds: On-the-fly mental state estimation of aircraft pilots, using a functional near infrared spectroscopy-based passive-BCI. *Frontiers in Human Neuroscience 12*, 187.

Gateau, T., Durantin, G., Lancelot, F., Scannella, S., & Dehais, F. (2015). Real-time state estimation in a flight simulator using fNIRS. *PLoS ONE 10*(3), e0121279.

Gayraud, D., Matton, N., & Tricot, A. (2017). Efficiency of a situation awareness training module in initial training pilot. *Presented at ISAP'17, Dayton, Ohio, USA.*

Goh, J., & Wiegmann, D. (2001). Visual flight rules (VFR) into adverse weather. *International Journal of Aviation Psychology. 11,* 259–279.

Gopher, D., Weil, M., & Bareket, T. (1994). Transfer of skill from a computer game trainer to flight. *Human Factors 36,* 387–405.

Gray, R., Navia, J. A., & Allsop, J. (2014). Action-specific effects in aviation: What determines judged runway size? *Perception 43*(2–3), 145–154.

Hardy, D. J., & Parasuraman, R. (1997). Cognition and flight performance in older pilots. *Journal of Experimental Psychology Applied 3,* 313–348.

Hasse, C., Grasshoff, D., & Bruder, C. (2012). Eye-tracking parameters as a predictor of human performance in the detection of automation failures. In D. de Waard, K. Brookhuis, F. Dehais, C. Weikert, S. Röttger, D. Manzey, . . . P. Terrier (Eds), *Human factors: A view from an integrative perspective* (pp. 133–144). Proceedings HFES Europe Chapter Conference Toulouse.

Helleberg, J. R., & Wickens, C. D. (2003). Effects of data-link modality and display redundancy on pilot performance: An attentional perspective. *International Journal of Aviation Psychology 13*, 189–210.

Hoermann, H.-J., Soll, H., Banbury, S., Dudfield, H., Blokzijl, C., Bieder, C., Paries, J., ... Polo, L. (2003). *ESSAI Enhanced safety through situation awareness integration in training: A validated recurrent training programme for situation awareness and threat management* (ESSAI/NLR/WPR/WP6/2.0).

Hoffman, R. R., Ward, P., Feltovich, P. J., DiBello, L., Fiore, S. M., & Andrews, D. (2014). *Accelerated expertise: Training for high proficiency in a complex world*. New York: Psychology Press.

Hopkin, V. D. (1999). Air traffic control automation. In D. Garland, J. Wise, and V. D. Hopkin (Eds), *Handbook of aviation human factors*. Mahwah NJ: Laurence Erlbaum.

Hunter, D. (2006). *Risk perception in general aviation pilots. International Journal of Aviation Psychology 16*, 135–144.

Hunter, D. R., Martinussen, M., Wiggins, M., & O'Hare, D. (2011). Situational and personal characteristics associated with adverse weather encounters by pilots. *Accident Analysis & Prevention 43*(1), 176–186.

Jensen, R. S. (1982). Pilot judgment: Training and evaluation. *Human Factors 24*, 61–74.

Johnson, C., & Wiegmann, D. (2015). VFR into IMC: Using simulation to improve weather-related decision making. *International Journal of Aviation Psychology 23*, 63–76.

Johnston, P., & Catano, V. (2013). Investigating the validity of previous fying experience in predicting initial and advanced military pilot training performance. *International Journal of Aviation Psychology 23*, 227–244.

Jones, D., & Endsley, M. (1996). Sources of situation awareness errors in aviation. *Aviation, Space and Environmental Medicine 67*, 507–512.

Kahneman, D. (2011). *Thinking fast and slow*. New York: Farrar, Strauss and Giroux.

Kahneman, D., & Klein, G. A. (2009). Conditions for intuitive expertise: A failure to disagree. *American Psychologist 64*, 515–524.

Kasarskis, P., Stehwien, J., Hickox, J., Aretz, A., & Wickens, C. (2001). Comparison of expert and novice scan behaviors during VFR flight. *Presented at the Proceedings of the 11th International Symposium on Aviation Psychology*, Columbus, OH: Dept. of Aerospace Engineering, Applied Mechanics, and Aviation, Ohio State University.

Kennedy, Q., Taylor, J., Heraldez, D., Noda, A., Lazzeroni, L. C., & Yesavage, J. (2013). Intraindividual variability in basic reaction time predicts middle-aged and older pilots' flight simulator performance. *Journals of Gerontology Series B: Psychological Sciences and Social Sciences 68*(4), 487–494.

Kim, J., Palmisano, S. A., Ash, A., & Allison, R. S. (2010). Pilot gaze and glideslope control. *ACM Transactions on Applied Perception (TAP) 7*(3), 18.

Kirby, C. E., Kennedy, Q., & Yang, J. H. (2014). Helicopter pilot scan techniques during low-altitude high-speed flight. *Aviation, Space, and Environmental Medicine 85*(7), 740–744.

Knapp, C., & Johnson, R. (1996). F-16 Class A mishaps in the US Air Force, 1975–93. *Aviation, Space, and Environmental Medicine 67*, 777.

Kramer, A., Tham, M., Konrad, C., Wickens, C., & Lintern, G. (1994). Instrument scan and pilot expertise. *Proceedings of the Human Factors and Ergonomics Society Annual Meeting 38*, 36–40.

Latorella, K. A. (1996). Investigating interruptions—An example from the flightdeck. *Proceedings of the Human Factors and Ergonomics Society Annual Meeting 40*, 249–253.

Laudeman, I. V., & Palmer, E. A. (1995). Quantitative measurement of observed workload in the analysis of aircrew performance. *International Journal of Aviation Psychology 5*, 187–198.

Lefrancois, O., Matton, N., Gourinat, Y., Peysakhovich, V., & Causse, M. (2016). The role of pilots' monitoring strategies in flight performance. Presented at EAAP'16, Portugal.

Li, W.-C., Chiu, F.-C., Kuo, Y., & Wu, K.-J. (2013). The investigation of visual attention and workload by experts and novices in the cockpit. In *International Conference on Engineering Psychology and Cognitive Ergonomics* (pp. 167–176). London: Springer.

Li, W.-C., Chiu, F.-C., & Wu, K.-J. (2012). The evaluation of pilots' performance and mental workload by eye movement. *Proceedings of the 30th European Association for Aviation Psychology Conference*, Sardinia, Italy, 24–28 September 2012.

Loukopoulos, L., Dismukes, R. K., & Barshi, E. (2009). *The multi-tasking myth*. Burlington, VT: Ashgate.

McKinney, E. (1993). Flight leads and crisis decision making. *Aviation Week and Environmental Medicine 64*, 359–362.

Morrow, D. G., Ridolfo, H. E., Menard, W. E., Sanborn, A., Stine-Morrow, E. A. L., Magnor, C., ... Bryant, D. (2003). Environmental support promotes expertise-based mitigation of age differences in pilot communication tasks. *Psychology and Aging 18*, 268–284.

Mosier, K., Rettenmaier, P., McDearmid, M., Wilson, J., Mak, S., Raj, L., & Orasanu, J. (2013). Pilot-ATC communications conflicts: implications for NexttGen. *International Journal of Aviation Psychology 23*, 213–226.

Neisser, U. (1976). *Cognition and reality: Principles and implications of cognitive psychology*. New York: Freeman.

National Transportation Safety Board. (2013). Crash during a nighttime nonprecision instrument approach to landing UPS Flight 1354 airbus A300-600, N155UP Birmingham, Alabama August 14, 2013 (NTSB/AAR-14/02 PB2014- 107898). Washington, DC: NTSB.

O'Hare, D. & Roscoe, S. (1990). *Flightdeck performance: The human factor*. Ames, IA: Iowa State University Press.

Ottati, W. L., Hickox, J. C., & Richter, J. (1999). Eye scan patterns of experienced and novice pilots during visual flight rules (VFR) navigation. *Proceedings of the Human Factors and Ergonomics Society Annual Meeting 43*, 66–70.

Patrick, J. (2006). Training. In P. Tsang & M. Vidulich (Eds), *Principles and practices of aviation psychology*. Mahwah, NJ: Lawrence Earlbaum.

Peissl, S., & Wickens, C. (2018). Eye tracking measures in aviation: a review of the literature. *International Journal of Aerospace Psychology 28*(3–4), 98–112.

Rebok, G., Qiang, Y., Baker, S., McCarthy, M. & Li, G. (2005). Age, flight experience and violation risk in mature commuter and air taxi pilots. *International Journal of Aviation Psychology 15*, 363–374.

Reynal, M., Rister, F., Scannella, S., Wickens, C. D., & Dehais, F. (2017). Investigating pilot's decision when facing an unstabilized approach: an eye tracking study. *Presented at International Symposium of Aviation Psychology*, Dayton, Ohio, USA.

Robinski, M., & Stein, M. (2013). Tracking visual scanning techniques in training simulation for helicopter landing. *Journal of Eye Movement Research 6*(2), 1–17.

Salas, E., Shuffler, M., & Diazgranados, D. (2010). Team dynamics at 35,000 Ft. In E. Salas & D. Maruino (Eds), *Human factors in aviation*. New York: Academic Press.

Salas, E., Wilson, K., Burke, C., & Wightman, D. (2006). Does CRM training work? An update, extension and some critical needs. *Human Factors 48*, 392–412.

Schneider, W., & Shiffrin, R. M. (1977). Controlled and automatic human information processing I: Detection, search, and attention. *Psychological Review 84*, 1–66.

Schriver, A. T., Morrow, D. G., Wickens, C. D., and Talleur, D. A. (2008). Expertise differences in attentional strategies related to pilot decision making. *Human Factors 50*, 864–878.

Schutte, P. C., & Trujillo, A. C. (1996). Flight crew task management in non-normal situations. *Proceedings of the Human Factors and Ergonomics Society Annual Meeting 40*, 244–248.

Seamster, T., Redding, R., & Kaempf, G. (1997). *Applied cognitive task analysis in aviation*. Aldershot, UK.: Averbury.

Shanteau, J. (1992). Competence in experts: The role of task characteristics. *Organizational Behavior and Human Decision Processes 53*, 252–266.

Sohn, Y. W., & Doane, S. M. (2004). Memory processes of flight situation awareness: Interactive roles of working memory capacity, long-term working memory, and expertise. *Human Factors 46*, 461–475.

Sicard, B., Taillemite, J., Jouve, E., & Blin, O. (2003). Risk propensity in commercial and military pilots. *Aviation, Space, and Environmental Medicine 74*, 879–881.

Sherry, L., & Polson, P. (1999). Shared mental models of flight management system vertical guidance. *International Journal of Aviation Psychology 9*, 139–154.

Simon, H. A., & Gilmartin, K. (1973). A simulation of memory for chess positions. *Cognitive Psychology 5*, 29–46.

Stanton, N., Salmon, P., Walker, G., Salas, E., & Hancock, P. (2017). State-of-science: situation awareness in individuals, teams and systems. *Ergonomics 60*(4), 449–466.

Stokes, A., Kemper, K., & Marsh, R. (1992). Time stressed flight decision making: a study of expert and novice aviators. *Aviation Research Lab Tech report ARL-93-1/INEL 93-1*. Savoy: University of Illinois Institute of Aviation.

Stokes, A., Wickens, C. D., Davis, T., Jr., Barnett, B., Rosenblum, R., & Hyman, F. (1987). A study of pilot decision making using MIDIS—A microcomputer-based flight decision training system. *Proceedings of the Fourth International Symposium on Aviation Psychology*. Columbus, OH.

Strater, L. D., Reynolds, J. P., Faulkner, L. A., Birch, D. K., Hyatt, J., Swetnam, S., & Endsley, M. R. (2004). PC-based training to improve infantry situation awareness. *Proceedings of the Human Factors and Ergonomics Society Annual Meeting 48*(3), 668–672.

Sulistyawati, K., Wickens, C. D., & Chui, Y. P. (2011). Prediction in situation awareness: Confidence bias and underlying cognitive abilities. *International Journal of Aviation Psychology 21*, 153–174.

Sullivan, J., Yang, J. H., Day, M., & Kennedy, Q. (2011). Training simulation for helicopter navigation by characterizing visual scan patterns. *Aviation, Space, and Environmental Medicine 82*(9), 871–878.

Suss, J., & Ward, P. (2015). Predicting the future in perceptual-motor domains: Perceptual anticipation, option generation and expertise. In R. R. Hoffman, P. A. Hancock, M. Scerbo, R. Parasuraman, & J. L. Szalma (Eds), *Cambridge handbook of applied perception research* (pp. 951–976). New York: Cambridge University Press.

Suss, J., & Ward, P. (2018). Revealing perceptual-cognitive expertise in law enforcement: An iterative approach using verbal report, temporal-occlusion, and option-generation methods. *Cognition, Technology, & Work 20*(4), 585–596.

Taylor, A. (2014). UK General Aviation accidents: increasing safety through improved training. Master Thesis, University of Leeds.

Taylor, J., O'Hara, R., Mumenthaler, M. S., & Yesavage, J. (2000). Relationship of CogScreen-AE to flight simulator performance and pilot age. *Aviation Space and Environmental Medicine* 71(4), 373–380.

Taylor, J., Yesavage, J., Morrow, D., Dolhert, N., Brooks III, J., & Poon, L. (1994). The effects of information load and speech rate on younger and older aircraft pilots' ability to execute simulated air-traffic controller instructions. *Journal of Gerontology* 49, 191–200.

Tole, J. R., Stephens, A. T., Vivaudou, M., Ephrath, A. R., & Young, L. R. (1983). Visual scanning behavior and pilot workload. NASA Contractor Report 3717.

Tolton, R. G. (2014). Relationship of individual pilot factors to simulated flight performance. Master of Health Sciences, University of Otago.

Trapsilawati, F., Wickens, C., Cheun, H., & Qu, X. (2017). Transparency and conflict resolution automation reliability in air traffic control. In P. Tsang, M. Vidulich, & J. Flach (Eds), *Proceedings of the 2017 International Symposium on Aviation Psychology* (pp. 419–424).

Tsang, P., & Shaner, T. (1995). Age, expertise, structural similarity, and time-sharing efficiency (of pilots and non-pilots). *Proceedings of the Human Factors and Ergonomics Society Annual Meeting* 39(2), 124–128.

UK Civilian Aviation Authority. (2013). *Loss of control action group, monitoring matters—Guidance on the development of pilot monitoring skills* (CAA Paper 2013/02). London: UK CAA.

Van Benthem, K., & Herdman, C. M. (2016). Cognitive factors mediate the relation between age and flight path maintenance in general aviation. *Aviation Psychology and Applied Human Factors* 6, 81–90.

Walmsley, S., & Gilbey, A. (2017). Debiasing visual pilots' weather-related decision making. *Applied Ergonomics* 65, 200–208.

Ward, P., Ericsson, K., and Williams, A. M. (2013). Complex perceptual–cognitive expertise in a simulated task environment. *Journal of Cognitive Engineering and Decision Making* 7(3), 231–254.

Ward, P., Gore, J., Hutton, R., Conway, G., & Hoffman, R. (2018). Adaptive skill as the *conditio sine qua non* of expertise. *Journal of Applied Research in Memory and Cognition* 7(1), 35–50.

Weignmann, D. Goh, J., & O'Hare D. (2002). The role of situation assessment and flight experience in pilots' decisions to continue visual flight rules flight into adverse weather. *Human Factors* 44, 187–197.

Wickens, C. D. (1986). The effects of control dynamics on performance. In K. R. Boff, L. Kaufman, & J. P. Thomas (Eds), *Handbook of perception and performance* (Vol. II, pp. 39-1/39–60). New York: Wiley.

Wickens, C. D. (1999). Frames of reference for navigation. In D. Gopher & A. Koriat (Eds), *Attention and performance XVI* (pp. 113–144). Orlando, FL: Academic Press.

Wickens, C. D. (2002). Situation awareness and workload in aviation. *Current Directions in Psychological Science* 11(4), 128–133.

Wickens, C. D. (2003). Aviation tasks and actions. In P. Tsang & M. Vidulich (Eds), *Principles and practices of aviation psychology*. Mahwah, NJ: Lawrence Erlbaum.

Wickens, C. D. (2007). Aviation. In F Durso (Ed.), *Handbook of applied cognition*. Chichester, UK: Wiley.

Wickens, C. D. (2009). The psychology of aviation surprise: an 8 year update regarding the noticing of black swans. In J. Flach & P. Tsang (Eds), *Proceedings 2009 Symposium on aviation psychology*. Dayton, OH: Wright State University.

Wickens, C. D., & Alexander, A. L. (2009). Attentional tunneling and task management in synthetic vision displays. *International Journal of Aviation Psychology 19*(2), 182–199.

Wickens, C. D., & Gopher, D. (1977). Control theory measures of tracking as indices of attention allocation strategies. *Human Factors 19*, 249–366.

Wickens, C. D., Gutzwiller, R., & Santamaria, A. (2015). Discrete task switching in overload: A meta-analyses and a model. *International Journal of Human-Computer Studies 79*, 79–84.

Wickens, C. D., Hollands, J., Banbury, S., & Parasuraman, R. (2012). *Engineering psychology and human performance* (4th edn). Upper Saddle River, NJ: Pearson.

Wickens, C. D., McCarley, J. S., Alexander, A. L., Thomas, L. C., Ambinder, M. & Zheng, S. (2008). Attention-situation awareness (A-SA) model of pilot error. In D. Foyle & B. Hooey (Eds), *Human performance models in aviation* (pp. 213–239). Boca Raton, FL: Taylor & Francis.

Wickens, C. D., Mavor, A., & McGee, J. (1997). *Flight to the future*: Washington, DC: National Academy Press.

Wickens, C. D., & Raby, M. (1991). Individual differences in strategic flight management and scheduling. In R. Jensen (Ed.), *Proceedings of the sixth international symposium on aviation psychology*. Columbus, OH: Ohio State University.

Wickens, C. D., Stokes, A., Barnett, B., & Hyman, F. (1993). The effects of stress on pilot judgment in a MIDIS simulator. In O. Svenson & A. J. Maule (Eds), *Time pressure and stress in human judgment and decision making* (pp. 271–292). New York: Plenum Press.

Wickens, C. D., Vincow, M., & Yeh, M. (2005). Design applications of visual spatial thinking: The importance of frame of reference. In A. Miyaki & P. Shah (Eds), *Handbook of visual spatial thinking*. New York: Oxford University Press.

Wiener, E. L. (1977). Controlled flight into terrain accidents: System-induced errors. *Human Factors 19*, 171–181.

Wiggins, M., & Bollwerk, N. (2006). HFJ Heuristic-based information acquisition and decision making among pilots. *Human Factors 48*, 734–746.

Wiggins, M., & O'Hare, D. (1995). Expertise in aeronautical weather-related decision making. *Journal of Experimental Psychology: Applied 1*, 305–320.

Wiggins, M., & O'Hare, D. (2003). Expert and novice perception of static in-flight images. of weather. *International Journal of Aviation Psychology 13*, 173–184.

Yakimovitch, N., Strongin, G., Go'orushenko, V., Schroeder, D., & Kay, G. (1994). Flight performance and CogScreen test battery in Russian pilots. *Aviation, Space, and Environmental Medicine 65*, 443.

Ziv, G. (2016). Gaze behavior and visual attention: A review of eye tracking studies in aviation. *International Journal of Aviation Psychology 26*(3–4), 75–104.

CHAPTER 30

UNCOVERING EXPERTISE FOR SAFE AND EFFICIENT PERFORMANCE IN RAILROAD OPERATIONS

EMILIE M. ROTH, ANJUM NAWEED, AND JORDAN MULTER

INTRODUCTION

THE success of a railroad depends on the safe movement of goods and people. The phrase "train wreck" encapsulates the damage to a railroad's reputation when an accident occurs. Train collisions and derailments can cause significant harm to people, property, and the environment. Although train accidents are decreasing in frequency, railroads receive negative attention in the media when they occur. In complex systems and high hazard industries like railroad, human expertise is critical to its safe and effective operation. Understanding how railroad employees acquire, make use of, and maintain expertise can enable stakeholders to foster this critical resource. This chapter provides an overview of the growing literature on expertise in railroad operations and how it contributes to overall safety and productivity. Our definition of expertise corresponds with Hoffman et al. (2014), and with Hoffman (1998), where an employee is highly regarded by his or her peers and shows "consummate skill, economy of effort, and can deal with rare or tough cases" (p. 86).

The chapter is informed by research conducted across multiple continents including: Australia, Europe, and the United States. While there are idiosyncratic differences in railroad operations across countries, there is surprising overlap in how railroad employees acquire, use, and sustain expertise. Furthermore, while the rate of change in railroad industries in different countries varies, they face a common set of challenges.

The chapter reviews research primarily relating to locomotive engineers, but also covers other railroad positions. We begin by providing an overview of railroad

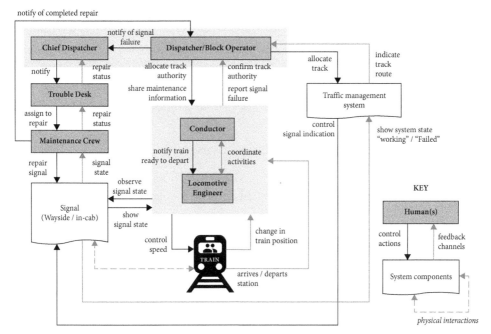

FIGURE 30.1 Overview of the primary roles and interactions of frontline workers in passenger rail operations.

operations and how expertise fits into the railroad enterprise. We then introduce substantive methods used to uncover expertise in railroad research, followed by a review of the variety of types of expertise that railroad workers exhibit that contribute to safe and efficient operation. This includes cognitive strategies used by individuals to avoid and overcome error traps as well as teamwork strategies that contribute to overall system safety and resilience. We follow with a discussion of the impact of ongoing technological changes on the requirements for expertise, and end with speculation on the likely changes in railroad technologies and nature of expertise in the coming decades.

Frontline Workers in Rail Operations

Railroad operation is a complex, dynamic environment that depends on the contributions of a large organization. Figure 30.1 provides an overview of the role of primary actors in the passenger context, though positions and responsibilities are largely similar in freight. Key frontline workers include the locomotive engineer (train driver) and the conductor (train guard) that collectively are referred to as the train crew.[1] The locomotive engineer operates the train, while the conductor is in charge of the train

---

[1] Different terms are used to refer to the same railroad position in different countries. For simplicity, we will use locomotive engineer, conductor, dispatcher, block operator, and roadway worker in the

and, among other things, couples/decouples cars and manages paperwork. The dispatcher (rail traffic controller) is responsible for managing track authority, deciding who gets access and for how long to particular sections of track. In Europe and Australia, there may also be block operators (signalers/area controllers) who manually control the signals and switches for a portion of a railroad's territory. Roadway workers inspect and maintain the track infrastructure and signal system. Effective communication and coordination of work across these multiple crafts is required for smooth and safe operation.

Changing Landscape

Changes in technology have impacted how work is performed in railroad operations, the expertise that is required, and the way that employees acquire knowledge and skills. Introduction of new technologies has reduced physical demands (e.g., using powered switches instead of manual switches) while increasing cognitive demands (Roth & Multer, 2009). The locomotive engineer must allocate attention to displays inside the cab that provide information about the operation of the train while also attending to what is happening outside the cab and managing communications over radio channels.

Technology has also reduced the number of people needed to do the same work and changed the distribution of work across positions. Freight trains that operated with up to seven employees in the mid-twentieth century now operate with one or two. The change from steam locomotives to diesel and electric eliminated jobs like the fireman's (who provided the fuel).[2]

There have also been changes in the background of individuals hired. Since the inception of railroading in the late nineteenth century, and until recently, children followed their parents into the business. Consequently, they were familiar with the language and culture of railroading. Today, new employees tend to be hired *off the street* with no background in railroad operations.[3]

Further, while employees started in less safety critical positions in the past, gaining valuable knowledge and experience along the way, in many cases this is no longer possible because those positions no longer exist. For example, in the USA, a dispatcher would first learn railroad operations as a block operator in the field. They physically manipulated switches and could see and directly interact with trains operating over their territory via radio. This experience provided important grounding to understand the context and constraints in which other employees and equipment operated before

---

remainder of the chapter. These terms are commonly used in the USA. Alternative terms used in other countries are indicated in parentheses.

[2] The title, however, has endured (a second driver is still referred to as a fireman in some countries).

[3] New employees without railroad background would correspond to a naïve in the scheme developed by Hoffman (1998).

taking on the role of dispatcher, whose interaction with field operations is mediated through computer displays and radio communications. In the USA, many block operator positions were eliminated as increased automation enabled dispatchers to control signals and switches remotely from a centrally located dispatch center.[4] This removed a path for employees to gain first-hand experience working in the field. A dispatcher lacking this knowledge may have diminished appreciation of what train crews will see and how they may respond that could put the train at risk of an accident.

These shifts have altered opportunities for gaining expertise while placing greater emphasis on cognitive skill. In today's environment, new employees take from several months to more than a year of initial training. For crafts like locomotive engineer or dispatcher, it typically takes 5–7 years to be considered expert by their peers.

The next section reviews the types of methods that have been used to uncover railroad worker expertise.

Unearthing Expertise in Railroad Operations: Research and Practice

Numerous methods have been used to unearth the nature of work, and the basis of expertise in railroad operations. These vary from controlled laboratory experiments using quantitative methods to answer-focused research questions, to field studies using qualitative methods to derive richer accounts of expertise. As widely recognized, knowledge elicitation of expertise in complex systems can, however, be difficult as much of it is tacit and resistant to conscious introspection (Cullen & Bryman, 1988). For example, Branton (1979) described the underlying skills in railroad driving as *enactive*, implying the difficulty of extracting knowledge accumulated through routine behavior. For this reason, innovative qualitative research methods have been areas of particular focus.

Interviews and Focus Groups

Interviews, focus groups, and other participative methods are widely used to elicit knowledge in naturalistic settings (Bisantz, Roth, & Watts-Englert, 2015; Cooke, 1994; Shadbolt, 2005). This makes sense given that the information is from individuals who are experts of their own tasks. However, as expertise can be very tacit, the validity of interviewing in terms of completeness and accuracy of the data elicited has long been criticized (Nisbett & Wilson, 1977; for reviews of these issues see Chapter 19,

---

[4] Block operator (signaler) positions have not been eliminated to the same extent in Europe and Australia.

"Incident-Based Methods for Studying Expertise," by Militello and Anders; and Chapter 17, "A Historical Perspective on Introspection," by Ward et al., both this volume). Consequently, studies of railroad worker expertise typically couple interviews with other methods such as field observations to obtain a more comprehensive picture (for a review of methods that employ field observation, see Chapter 18, "Close-to-Practice Qualitative Research Methods," by Yardley et al., this volume).

Focus groups are a good way of addressing criticisms about completeness of the data in that they reflect the knowledge of more than one individual (Cooke & Schvaneveldt, 1988; Grabowski, Massey, & Wallace, 1992). They allow points of disagreement to be uncovered and discussed, and overcome ethical hurdles of topics that are perceived to be taboo by providing a *safe* context for sharing opinions while maintaining the anonymity of individual responders when reporting data (e.g., Naweed et al., 2017). While they require skilled facilitation, focus groups have the further advantage that individuals end up eliciting knowledge from each other, enriching the output. For these reasons focus groups are widely used in railroad applications (e.g., Roth, Multer, & Raslear, 2006; Roth, 2009; Safar, Multer, & Roth, 2015).

Field Observations

Field observations can be used to verify what an expert actually does, while minimally interfering with the task itself (Bisantz, Roth, & Watts-Englert, 2015). They can help unearth manifestations of expert work in the railroad domain that may be missed through interviews and focus groups (e.g., Naweed & Moody, 2015; Roth & Patterson, 2005). Field observation data may be free form (e.g., handwritten notes documenting the activities observed). Alternatively, observational protocols can also be very structured, involving a predefined coding scheme for what information is to be recorded. One example from the railroad industry is an adaptation of the line of sight audit methodology, which involves observing and documenting specific behaviors and contextual cues (Naweed & Dawson, 2016).

In the railroad industry field observations are often coupled with interviews and focus groups to provide multiple converging methods that leverage each other and overcome the limitations of any single method (e.g., Roth, Malsch, & Multer, 2001; Wreathall, Roth, Bley, & Multer, 2007).

Process-Tracing, Retrospection, and Harnessing Creativity

In many cases, interviews and observations are simply an engagement platform for other methods to access underlying expertise. Sometimes, the *when* is also as important as the *how*. A good reflection of this is the family of verbal protocol process-tracing techniques, undertaken at the same time as the task of interest (Elstein et al., 1978; see also Ward et al., this volume), which connect thought processes, problem-solving strategies, and information requirements in situ. Verbal protocol methods have been used in a variety of railroad contexts, for example with train crews (Naweed & Balakrishnan, 2014). The critical decision method, a structured methodology for having experts discuss a specific past incident they experienced that provides insight into their expertise (Klein, Calderwood, & MacGregor, 1989; see also Militello and Anders, this volume), has also been used with considerable success in rail (e.g., Naweed, 2014; Naweed & Balakrishnan, 2014).

Sometimes, elicitation of deeper types of expertise can be harnessed through bench-top simulation exercises. One such method called the scenario invention task technique (Naweed, 2015) has been developed and used extensively with train crews in Australia and New Zealand (e.g., Naweed, 2013; Naweed & Rainbird, 2015; Naweed, Rainbird, & Chapman, 2015; Naweed, Rainbird, & Dance, 2015; Rainbird & Naweed, 2016), as well as in other domains like aviation (e.g. Naweed & Kingshott, 2019). In this method, experts are asked to draw a picture illustrating elements of a challenging railroad *scenario* (e.g., one that can lead to an accident). This method aims to externalize expertise by having experts draw pictures to reveal knowledge that might not be easily verbalized. The theory underpinning this method relates to multiplicity of dimensions, meaning that daily visual and sensory experiences may not always be reduced to language (Eisner, 2008, as quoted by Bagnoli, 2009). Including non-linguistic dimensions may therefore allow access to layers of experience that are not easily put into words (Bagnoli 2009; Gauntlett, 2007).

Figure 30.2 shows example data collected using the scenario invention task during a focus group with locomotive engineers who created challenging scenarios involving breaches of movement authority (e.g., driving through stop signals, referred to as *signal*

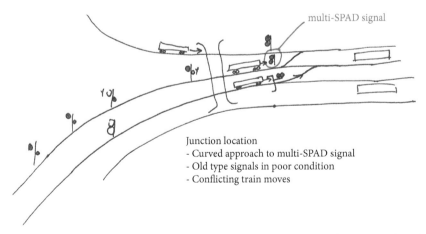

FIGURE 30.2 Example data from the scenario invention task. Handwriting replaced with typescript. Created by an expert locomotive engineer with 11 years' driving experience. Drawing depicts a signal passed at danger (SPAD) scenario.

Used with permission from Anjum Naweed.

passed at danger or SPAD). The drawing depicts the experts' understanding of their occupational environment in terms of its dynamism, opacity, and complexity, and illustrates how the design of the system and its various actors are perceived to behave.

Having interviewees provide hand-generated drawings, as in the scenario invention task, also helps generate rapport, which improves the fluidity and efficiency of the process. It can also help experts gain greater awareness of the dimensions of their expertise, their limitations, and the consequences of their own actions.

Task Analysis and Other Systematic Approaches

Methods in the task analysis family are also widely used in railroad settings (Rose, Bearman, & Naweed, 2013; Tichon, 2007; see also Chapter 16, "Studying Expert Behavior in Sociotechnical Systems," by Salmon et al., this volume). Research and practice in railroad operations have also adopted cognitive work analysis methods (CWA; Rasmussen, et al., 1994; Vicente, 1999; see also Chapter 20, "Cognitive Work Analysis," by Burns, this volume). CWA enables investigation of expertise relating to the domain as well as the tasks, strategies, and socio-organizational structures and their constraints. CWA has been used in a variety of ways in rail, illustrating for example demands of the work domain, strategies, communications, social relationships, learning and development, and inter-team differences of railroad employees (Farrington-Darby, Wilson, Norris, & Clarke, 2006; Roth, 2009; Salmon et al., 2015). It has also been used to identify the specific features of the rail domain required to produce an effective simulator (Naweed, Hockey, & Clarke, 2013). There are also systemic approaches that have been used to examine expertise in the context of accident analysis in rail, including HFACs (Baysari, McIntosh, & Wilson, 2008; Wiegmann & Shappell, 2003), Accimap (Chen, Zhao, & Zhao, 2015; Rasmussen, 1997), CREAM (Hollnagel, 1998; Phillips & Sagberg, 2014), FRAM (Belmonte, Schön, Heurley, & Capel, 2011; Hollnagel, 2012), and STAMP (Leveson, 2004; Ouyang, Hong, Yu, & Fei, 2010; Read, Naweed, & Salmon, 2019).

Experimental Research, Simulators, and Quantitative Performance Measures

More quantitative studies have also sought to uncover what expertise looks like in the rail domain. Many of these are lab-based and have tracked various aspects of expertise and performance, for example, how it is that certain individuals may fail to learn from threatening or aversive experiences in train driving (Hickey & Collins, 2017). Simulated task environments have also been used to investigate train crew information requirement, showing, for example, that performance can be impacted by predictive information (e.g., Einhorn, Sheridan & Multer, 2005; Naweed, Hockey, & Clarke, 2009; for a review of research using these methods in other domains, see Chapter 13,

"Representative Test and Task Development and Simulated Task Environments," by Harris et al., this volume). It has also been used to identify tasks and how they are shared by locomotive engineers in two-driver operations (Naweed, Balakrishnan, & Dorrian, 2018).

Quantitative data collection methods have also been used in real-world contexts. For example, locomotive data loggers have been used to understand locomotive engineer driver behavior (Dorrian, Hussey, & Dawson, 2007). Data loggers reflect expertise in the form of managing braking, throttle regulation, and fuel consumption in response to dynamic changing operational needs. Studies have also employed eye-trackers to monitor eye movements of locomotive engineers during actual train runs. These have been useful in identifying the effect of specific environmental features (e.g., location and pattern of wayside signals) on visual behavior, as well as the role that expectations play in guiding eye gaze (e.g., Luke, Brook-Carter, Parkes, Grimes, & Mills, 2006).

Because these studies are conducted in the actual work context, they are less likely to have the issues with motivation that are possible in simulators (e.g., *buy-in* or engagement with the simulator as in with the real world), but the knowledge that you are being monitored is inescapable and may affect performance in the real train (i.e., the Hawthorne effect).

Finally, attempts have been made to develop quantitative models of train crew performance with the aim of generating more precise quantitative predictions of performance parameters (e.g., monitoring pattern, time to respond) and likelihood of error (e.g., Moray, Groeger, & Stanton, 2017). These types of models tell us that expertise in the railroad domain is shaped by the complexity in the operational context. They are useful in explaining and predicting how aspects of the environment (e.g., multiple, ambiguously placed signals) can lead to performance difficulty and error (e.g., passing a stop signal) as well as how elements of expertise (e.g., experience-based knowledge of the terrain) allow experienced locomotive engineers to anticipate and avoid potential problems (e.g., begin to brake early in anticipation of an upcoming stop signal that is not yet visible).

Exhibited Expertise Contributing to Safe and Efficient Operation

Branton (1993) highlighted three important principles of person-centered ergonomics which tell us much about how systems operate from the perspective of expertise. They are:

1. the need to consider the complete individual;
2. the need to consider purposes (goals) rather than causes of actions; and

3. the need to understand the values and philosophical bases upon which individual behavior within the system rests, including how the system and its functions are conceptualized.

These principles highlight the importance of adopting both a systems perspective and a first-person perspective taken from the view of domain practitioners themselves (i.e., their perceptions, values, and goals) when trying to understand the role of expertise in railroad operations.

Figure 30.3 provides a system decomposition description of railroad operations that is informed by these principles. The figure synthesizes some of the major findings related to locomotive engineer expertise (Naweed, 2014; Roth & Multer, 2009; Multer, Safar, & Roth, 2015; Safar, Multer, & Roth, 2015). The findings are organized around inter-related layers that influence locomotive engineer expertise in railroad operations, including the physical equipment (locomotive and cars), the psychological and physical characteristics of the train crew, the physical terrain and railroad infrastructure, the goals and purposes of the organization, human values and philosophies, including social and cultural norms, and weather that can impact visibility and track conditions.

In the next section, we provide an overview of how the demands of the railroad system shape expertise. This is followed by more detailed discussions of select hallmarks of expertise in railroad operations.

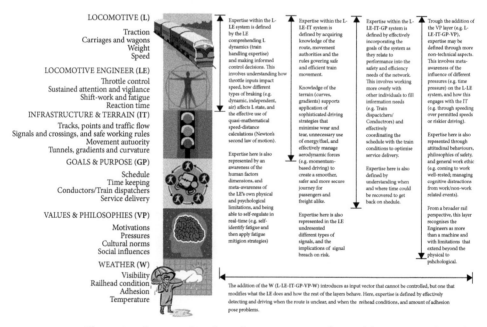

FIGURE 30.3 The various layers within the rail system creating demand for expertise from the railroad driving perspective.

Used with permission from Anjum Naweed.

Overview of How Rail System Demands Shape Expertise

One of the distinguishing characteristics of the railroad system is that trains *roll* on guided track. For this reason, train driving is essentially predicated by throttle and braking manipulations. This is a key difference to other transportation systems. For example, while automobiles can maneuver around an obstacle to avoid a collision, this is not possible in the case of a train. Further, because braking distances to stop can be long, by the time a locomotive engineer sees an obstacle, it may be too late to stop in time. This places a high premium on being able to anticipate the territory ahead and take appropriate control action (e.g., begin to slow down in anticipation of an upcoming stop sign). Weather can also complicate performance, by reducing visibility (placing a greater premium on knowledge of terrain and anticipatory behavior) as well as reducing track adhesion, placing greater demands on train braking and control strategies.

Another unique aspect of train operation is that it is controlled by *movement authority*. Movement authority directs whether a train can enter a portion of track and at what speed. Movement authority is provided by dispatchers via radio, or more typically, electronically via external wayside signals. Wayside signaling systems employ signal aspects (i.e., the pattern and color of signal lights) to indicate whether the train must stop, or can continue, and if so, at what maximum speed. The signals are usually presented in a progression (e.g., clear, indicating that can travel up to a maximum speed; caution, indicating that the train should slow down and prepare to stop at the next signal; and stop, indicating that the train must stop prior to reaching the signal) to give the locomotive engineers the time they need to react, correct their speed, and brake appropriately. Figure 30.4 illustrates some examples of how traditional rail signals are designed. While they vary in color, and configuration, signal aspect systems use the same principle for driving safely on railways—a multi-aspect sequencing to provide the human with early indication of cautionary and stop signals.

More than any other transportation mode, expertise in railroad operations relies on a good understanding of movement authority, safe working, and appreciation of other dynamic control tasks.

Train driving is also contextually bound by performance goals and service delivery requirements—in particular, trains need to meet a schedule. A hallmark of locomotive engineer expertise is the ability to manage the train schedule and associated time pressure, while continuing to operate safely. This includes knowing where and how to make up time.

Finally, railroad operations are round the clock, typically 24/7/365. As a consequence, for many of its employees, it is a shift-working environment, which means that people must apply multiple strategies to manage and compensate for inherent cognitive limitations (e.g., fatigue, mental over- or underload) born from the associated human factors (circadian disruption, time on task). A great deal of locomotive engineer expertise relates to strategies for preventing and managing fatigue, as well as preventing and managing distraction (Filtness & Naweed, 2017; Naweed, 2013; Roth & Multer, 2009; Safar, Multer, & Roth, 2015).

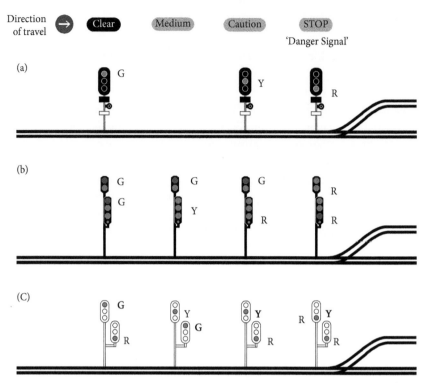

FIGURE 30.4 Examples of multi-aspect signaling design. Tracks (a) and (b) illustrate three- and four-aspect signaling conventions and track (c) illustrates a variation on color arrangements.

Reprinted from *Safety Science*, 76, Anjum Naweed, Sophia Rainbird, and Craig Dance, Are you fit to continue? Approaching rail systems thinking at the cusp of safety and the apex of performance, pp. 101–10, Figure 1, doi.org/10.1016/j.ssci.2015.02.016. Copyright © 2015 Elsevier Ltd. All rights reserved.

Prediction and Anticipation

Prediction (and anticipation) of system state underlies expert performance in railroad operations, which is a common feature of expertise across domains (Suss & Ward, 2015). Locomotive engineers must react to information situated beyond their visual field. This, in turn, requires forming accurate expectations of upcoming events (e.g., sharp curves, railroad crossings, locations where trains will meet or pass each other) (Naweed, 2013; 2014; Roth & Multer, 2009). With increasing experience train crews develop route knowledge, including physical layout of the track, the type and location of signals, and the pattern of other rail traffic, that allows them to operate on expectations (Naweed, Rainbird, & Dance, 2015). This works well when expectations are met (e.g., when they come up to a signal it is clear, same as it has been every day) but can lead to error with safety consequences (going through a stop signal) when expectations are violated (Naweed, 2013; Safar, Multer, & Roth, 2015). A good example of an expectation-driven error is a *signal misread* scenario where a locomotive engineer reads and follows a signal set for a different line (i.e., the wrong signal). An expert locomotive engineer may have intimate knowledge of their own signal's location and knowledge of the existence of multiple signals placed in close proximity. However, the nature of rail driving, and the fact that signal aspects are an independent and dynamic feature means that they may easily see what they expect to see (e.g., a signal reading "proceed" that is for a different track) instead of what the signal intended for them actually shows (e.g., "stop").

Like most other surface transportation modes, rail also tends to operate in an open environment, which means that the task is susceptible to unpredictable threats, such as transgressions from others (e.g., rail level crossing users; pedestrians at stations; roadway workers operating near the track). For train crews, this places a premium on monitoring outside the cab to identify and respond to unanticipated situations as well as monitoring radio communications, to develop and maintain an awareness of the dynamic changes in the environment. For dispatchers, it means a need to maintain awareness of and rapidly communicate situations with potential safety consequences (roadway workers operating near the track; hazards on the track; non-routine train movements) over radio (Roth, Malsch, & Multer, 2001).

As the discussion above makes apparent, railroad operations are, by their very nature, a perceptual–cognitive exercise filled with demands for sustained attention, multiple sources of distraction, and potential error traps. This means that the expertise and underlying development of skill in this domain requires active strategies. These include strategies for developing and maintaining situation awareness as well as meta-cognitive strategies for managing attention, combating fatigue, and preventing memory lapses. They are described in more detail in the next sections, "Maintaining Situation Awareness" and "Meta-cognitive Strategies Compensating for Cognitive Performance Threats."

Maintaining Situation Awareness

Beyond core infrastructure (e.g., rail tracks), the information guiding when and where to move is dynamically changing. This means that locomotive engineers must constantly monitor the environment, assess the situation and project what is likely to occur next. For example, they need to maintain awareness of other trains, roadway workers, and possible trespassers (unauthorized individuals on or near the track) in their general area and anticipate what they are likely to do. This is important for ensuring safe operation. One of the best ways to conceptualize the challenges here, and by extension the requisite skills and expertise, is through theories of applied attention and situation awareness.

To draw on but one of these theories, Endsley (1995) has defined situation awareness according to the ability to perceive and comprehend appropriately, and to project ahead. Figure 30.5 extends the decomposition in Figure 30.3 with a conceptual model of various dimension of expertise in the context of the locomotive engineer. It shows the interplay between the extant knowledge base (e.g., route knowledge) and dynamic

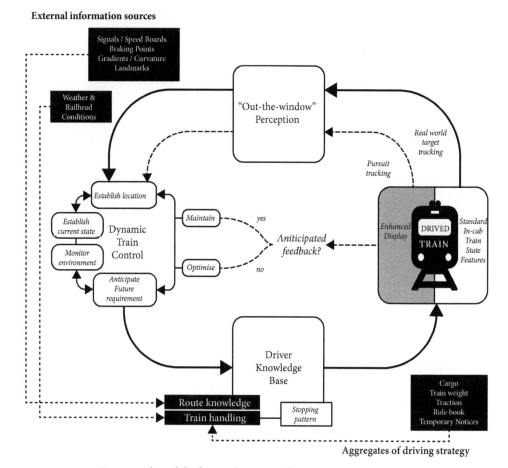

FIGURE 30.5 Conceptual model of train-driving performance.
Reprinted from *Applied Ergonomics*, 45(3), Anjum Naweed, Investigations into the skills of modern and traditional train driving, pp. 462–470, Figure 5, doi.org/10.1016/j.apergo.2013.06.006
Copyright © 2013 Elsevier Ltd and The Ergonomics Society. All rights reserved.

train control (i.e., monitoring the environment, establishing current state, and anticipating future requirements). Situation awareness is built and maintained by drawing on route knowledge developed from prior experiences and information extracted from external monitoring processes. Railroad workers engage in active processes to extract information about the locations, activities, and intents of others working in their area. For example, railroad workers, including train crews, dispatchers, and roadway workers monitor radio communication directed at others to extract information of relevance to themselves. By listening to radio communications, expert train crews are better able to maintain awareness of the locations, activities and likely future actions of other trains, roadway workers, and trespassers in their immediate surroundings (Roth & Multer, 2009).

Meta-cognitive Strategies Compensating for Cognitive Performance Threats

Cognitive limitations in performance are invariably a part of the human condition, irrespective of skill level or experience—this includes the inability to maintain attention for extended periods of time, to disengage and *switch* attention from thoughts or activities that interfere with their focus, and memory lapses. Characteristics of the work environment can contribute to these effects. Is the job engaging or monotonous, and does the mind wander? Does the work induce anxiety, time pressure, or stress, and is it possible to be thrown from a state of abject boredom to unequivocally high anxiety at the drop of a hat? For railroad operations, and indeed for many other occupational settings, the answers to all these questions may be a resounding "yes." Therein lie various threat factors that need to be managed to ensure performance is sustained and safe. To work around a changing dynamic, experts in railroad operations display a variety of meta-cognitive strategies to manage and compensate for these inherent cognitive performance threats.

One category of meta-cognitive strategies observed in experts are strategies intended to manage distractions. The railroad environment is replete with external distractions ranging from radio communication with dispatchers, to malfunctioning doors, to non-informative audio alarms that suddenly trigger, to non-work-related conversations with conductors. The locomotive engineer's attention can be drawn to these external distractions, as well as to internal (i.e., cognitive) preoccupation with say, a recent argument they had with their spouse, or what they will be doing that evening. Increasingly, rail research is starting to show that cognitive distraction from internal sources (sometimes referred to as mind wandering) is a growing issue (Safar, Multer, & Roth, 2015). Based on current rail safety research, time pressure (and the ensuing heightened anxiety) can also be viewed as a form of distraction that contributes to train driver errors (Hickey & Collins, 2017; Naweed, 2013).

In the context of train driving, research has shown that train crews devise strategies to maintain an attentive state and minimize the impact of distraction on performance. For example, they may visually inspect the train (e.g., wheels, bogies, air reservoir, fuel gauge, wagon conditions) to engage themselves with the work before settling into the shift. They *get themselves in the game* so to speak.[5] Expert locomotive engineers have also developed strategies to avoid losing focus during monotonous period of travel. For example, Branton (1993) described the use of *finger pointing procedures* (e.g., physically pointing and calling out signal aspects when seeing them) as a way of enhancing vigilance.

With increased experience, locomotive engineers also develop attention management and goal prioritization skills that enable them to ignore competing demands for

---

[5] Durso et al., Glover (2015) have posited a *head-in-the-game* hypothesis as a possible explanation for the fact that railroad incidents often happen early in a shift or after a long period of being off, suggesting that it may take time on the job to become mentally situated in the work context and thus avoid *error traps*.

attention during safety critical periods to focus on safe driving (Safar, Multer, & Roth, 2015). For example, they will delay responding to radio requests if they are approaching a complex portion of track or signal (Naweed, Rainbird, & Chapman, 2015). They are also better able to prioritize safety over on-time performance in the face of time pressure.

Locomotive engineers also self-monitor to evaluate the extent to which emotional distraction is likely to impede their capacity to work. Experienced engineers will self-assess their emotional and physical state prior to taking on a shift, and call in sick if necessary (Filtness & Naweed, 2017).

Expert locomotive engineers have also developed strategies for managing fatigue. Managing operator fatigue in railroad operations is an age-old issue, an ongoing debate, and a very *wicked problem* in the sense that it is linked with cultural challenges and contradictory requirements that are often incomplete (e.g., shift-swapping between drivers is not always tracked against fatigue risk management tools), hidden (e.g., employees do not report fatigue for fear it will trigger a medical assessment), and difficult to recognize (e.g., employees decide to work because they feel obliged to even if they feel tired) (Chapman & Naweed, 2015; Naweed, 2016; Rittel & Weber, 1973). While mental fatigue is a hallmark of prolonged time-on-task, many railroad workers operate in shifts, further contributing to fatigue due to ineffective sleep regulation. Regulatory bodies and railroads manage fatigue formally in a number of ways, principally by attempting to control work and traveling hours, creating reporting processes, and education and awareness. In practice, however, these do not always work, and drivers use their own informally developed strategies to combat fatigue. In recent research (Filtness & Naweed, 2017), expert railroad drivers reported use of biomechanical (e.g., standing up or kneeling in cab or taking a power nap when at station) and biochemical (ingestion of caffeine, energy drinks, etc.) strategies to manage their fatigue, with more sophisticated cognitive strategies including reprioritizing goals (e.g., drive slower). The biggest indicator of expertise and meta-awareness of the task requirements was self-assessment and self-regulation of fatigue; for example, sick leave was often used to manage fatigue, though given cultural concerns, reported officially as physical ill-heath.

Expert locomotive engineers have also developed strategies to avoid memory lapses. These informal strategies include use of physical objects as aids to memory (Naweed, Rainbird, & Chapman, 2015). For example, when stopped at a station experienced engineers often use external aids to remind them to check the signal aspect prior to leaving the station. This avoids the commonly observed error of locomotive engineers taking off as soon as they get the OK from the conductor, forgetting to first check that they still have a clear signal. Example strategies include centering the reverser (analogous to placing a car in neutral) or applying the park brake (analogous to a handbrake). These create a different cue–response link to remind the expert that they are outside their train movement schema, thus preventing the triggering of an inappropriate automated response. Train crews have also used placement of physical objects such as keys, coins, or similar objects on their dashboard when driving to remind them of where they are. Other examples include a re-appropriation of the metal gooseneck of the microphone in the cab or the blinds, so that erect/downward positions mean different things. These examples of offloading memory demands onto physical objects in the environment are reminiscent of the findings of Hutchins (1995; Hollan, Hutchins & Kirsch, 2000) with respect to distributed cognition in cockpits.

Expert locomotive engineers also used biomechanical strategies (e.g., *kneeling* or *standing when driving*), as a reminder. This change in stance (while prohibited in some organizations) was considered a highly reliable reminder that something beyond them had changed (e.g., that they were about to come to a stop light, or a work zone where they needed to monitor for roadway workers that may be working around the track). Train crews also reported using physical gestures to acknowledge a signal had been sighted, for example, the aforementioned finger pointing, qualifying its role as both a vigilance aid and a memory aid. Epizeuxis (repeating a word over and over for emphasis) was also harnessed by experts as a strategy for ensuring that they do not forget an important piece of information by maintaining it in short term memory through repeated rehearsal (e.g., calling out "stop signal" repeatedly as a way to make sure that they don't forget that the next upcoming signal is a *stop*) and was reported to help with accurate recollection. Interestingly, the Rail Safety & Standards Board (2008) recommended a similar strategy for responding to high risk states (e.g., approaching a stop signal) which they referred to as *risk-based commentary*.

Expert strategies sometimes operate at the psychomotor level. A good example of this is what has been observed to alleviate musculoskeletal strain from sustained push forces. In a study of light rail operations (Naweed & Moody, 2015) a throttle control lever on a train exceeded the recommended duration, and was considered the main cause for incidence of musculoskeletal injury in a group of drivers. During cab observations, frequent changes in the hand positions of experts who operated this lever were observed, though the same drivers did not display these movements in conscious demonstrations, suggesting that this type of expertise was a kind of tacit knowledge that is not easily articulated.

Importantly these strategies are not explicitly taught but rather informally developed and disseminated by train crews.

More Than the Sum of Its Parts: Strategies for Operating as Cooperating Teams

Railroad operations fundamentally involves teamwork (Morgan, Olson, Kyte, Roop, & Carlisle, 2006; Roth, Multer, & Raslear, 2006; TRB, 2011). This includes activities within a train crew, as well as activities of the broader distributed team of rail workers (e.g., train crews, dispatchers, roadway workers) that actively coordinate work across time and space. Below we summarize results of a series of studies conducted to explore expert teamwork strategies and their contribution to safe and efficient railroad operation (Rosenhand, Roth, & Multer, 2012; Roth, Malsch, & Multer, 2001; Roth & Multer, 2007, 2009).

Team Work Processes in Train Crews

One set of studies explored teamwork processes employed by freight train crews, consisting of a locomotive engineer and a conductor (Rosenhand, Roth, & Multer, 2012; Roth & Multer, 2009). Expert crews exhibited the characteristics of high-performing teams that have been shown across industries (Salas, Diazgranados, & Lazzara, 2011; Salas, Sims, & Burke, 2005). These include mutual performance monitoring (to catch and correct errors) and active support of each other's activities (e.g., fill in knowledge gaps, collaboratively problem-solve; point out potential risks and how they can be mitigated).

Conductors and locomotive engineers operated as a tightly coupled cooperative team. While each had a distinct set of formal responsibilities, they jointly contributed to the set of cognitive activities required to operate the train safely and efficiently. Significantly, these teamwork activities went beyond the requirements of formal operating rules and were not explicitly covered in training.

Conductors and engineers served as an extension of *eyes and ears* for each other, jointly monitoring outside the window, listening to the radio, and catching and communicating information that the other may have missed. They also extended each other cognitively, filling in knowledge gaps, reminding each other about upcoming tasks, and contributing jointly to problem-solving and decision-making situations that arise. For example, conductors participated jointly with locomotive engineers in planning activities to perform along the route and identifying and mitigating potential risk. Locomotive engineers served a similar support role for conductors.

A key contributor to successful performance was the ability to establish and maintain *common ground* (a shared understanding) with respect to goals, planned activities, potential risks, and ways to mitigate them (Klein et al., 2005). Maintaining common ground reduced the potential for miscommunication, facilitated work, and enhanced overall safety. Train crews stressed the importance of effective cab communication and job briefing skills for developing and maintaining this common ground, such as the importance of *rolling* job briefs, conducted immediately prior to performing challenging tasks. They fostered a shared understanding of task objectives, potential hazards in the environment, and actions that needed to be taken to mitigate risk. Discussing potential risks enabled less experienced individuals to better recognize hazards. Equally important, the very act of discussing the potential hazards served as a form of sensitization and rehearsal, enabling the conductor and engineer to be better prepared.

These results complement research conducted by Naweed and colleagues examining one- versus two-person train crew operation (Naweed, Every, Balakrishnan, & Dorrian, 2014). They found that a second person in the locomotive cab was considered helpful "if they talked in a timely or on-task way, the driver was new and still needed guidance, the difficulty in the task was very high (i.e., from fatigue, visibility issues, and/or track complexity) or if the main driver was more extroverted and liked company" (p. 5).

Teamwork Processes in Distributed Teams

Studies have also examined expert practices across the larger distributed team that is made up of dispatchers, train crews, and roadway workers (Roth, Malsch, & Multer, 2001; Roth & Multer, 2007; Roth, Multer, & Raslear, 2006). The distributed teams engaged in active cognitive and collaborative processes to develop and maintain common ground with respect to each other's location, activities, and intentions. These included active strategies for extracting relevant information by *listening in* on radio communications directed at others, allowing them to identify information that had a bearing on achieving their own goals or maintaining their safety. It also enabled them to recognize situations where information in their possession was relevant to the performance or safety of others.

The studies also identified informal cooperative practices intended to foster common ground, redundancy checks intended to reduce the possibility of error, and proactive actions intended to level workload and facilitate work across the distributed organization. Figure 30.6 summarizes these informal proactive strategies. They are described in more detail in Roth, Multer, and Raslear (2006). In combination, they contribute to more reliable performance and enhanced safety by enabling train crews

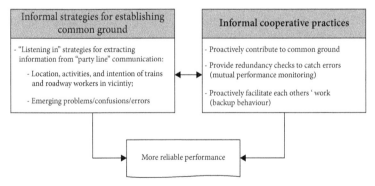

FIGURE 30.6 Distributed teams actively work to develop and maintain common ground contributing to more reliable performance and increased safety (material adapted from Roth, Multer, & Raslear, 2006).

Adapted from *Organization Studies*, 27(7), Emilie M. Roth, Jordan Multer, and Thomas Raslear, Shared situation awareness as a contributor to high reliability performance in railroad operations, pp. 967–987, doi.org/10.1177/0170840606065705 Copyright © 2006, SAGE Publications.

and roadway workers to be aware of the locations, activities, and intentions of each other, improving overall efficiency and avoiding potential accidents (Roth, Multer and Scott, 2009).

One example of a proactive communication intended to enhance safety is alerting roadway workers of unusual or unexpected conditions that may pose danger. Roth et al. (2006) found that dispatchers called roadway workers to alert them of unscheduled trains, or trains running at a different time, track, or direction than usual. As one dispatcher put it, "I let them know what my plan is so that they are not startled" (p. 976). This call is not required by policy. It exemplifies the informal *safety net* created by cooperative strategies of railroad workers.

As in the case of train crews, these informal strategies went beyond the requirements of formal rules and procedures. They were referred to as *courtesies* to emphasize their voluntary nature. Nevertheless, these cooperative strategies contribute to overall efficiency and resilience to error of railroad operations.

One interesting question that was not addressed by this work is where these informal cooperative practices emerge from. It is not clear what causes them to emerge and how they are learned from one railroad worker to the next. The reason these are important questions is that prosocial behavior may arise from, and be reinforced by, aspects of the broader sociotechnical system. If changes are made without awareness of what influences these prosocial behaviors, we may inadvertently disrupt or discourage these behaviors, resulting in less safe systems.

Impact of New Technologies on Expert Strategies

The section on changing landscape highlighted recent-past changes that have impacted the requirements for railroad expertise. In this section we look to currently emerging trends in railroad technology that promise to alter requirements for expertise further (Naweed, 2014; Roth & Multer, 2007; Roth & Multer 2009; Roth et al., 2013).

New display and automation technology promise to dramatically alter the expertise required to operate trains and the distribution of roles and responsibilities in the modern train cab (Naweed, 2014; Rosenhand, Roth, & Multer, 2012; Roth & Multer, 2009). Train technology trends in Europe, Australia, and the USA fall into two main categories. One class provides enhanced visualizations to increase awareness of train speed limits and enable crews to better balance the goals of on-time performance, energy efficiency, ride quality, and safety. For example, the Australian Energymeiser system (Howlett & Pudney, 2000) displays energy consumption and recommends train speed trajectory to minimize energy consumption while maintaining on-time performance (See Figure 30.7a). The locomotive engineer remains in control with complete discretion whether to follow the speed trajectory recommended by Energymeiser. The

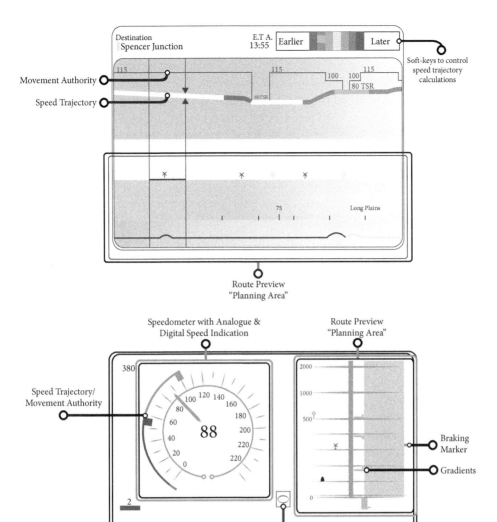

FIGURE 30.7 Schematic of the displays in the (a) Energymeiser system and the (b) European Train Control System (ETCS) and callouts (from Naweed, 2014). Energymeiser provides a route preview, information on movement authority (speed limits), and a recommended speed trajectory that, if followed, would make most efficient use of energy. Importantly, Energymeiser involves no speed enforcement and the locomotive engineer is free to deviate from the recommended speed trajectory. ETCS provides a preview of the upcoming route together with a marker indicating when to brake (Braking Marker). It also provides speed guidance in the form of a band depicting upper and lower acceptable speed bounds around the speedometer. This band, marked "Speed Trajectory/Movement Authority," has different color ranges that guide speed changes. In this example, the permitted speed is 100 mph and current locomotive speed is 88 mph. The locomotive engineer will reduce the speed of the locomotive to 55 mph, the new upcoming permitted speed, which is conveyed by the darker shaded band and rectangular maximum limit marker. This reduction in speed will happen in concert with gradual reduction of the lighter shaded band and rectangular maximum limit marker, leaving behind only the darker shaded band. Automatic braking will occur if the locomotive engineer exceeds the speed boundary.

Reprinted from Applied Ergonomics, 45(3), Anjum Naweed, Investigations into the skills of modern and traditional train driving, pp. 462–470, Figure 1, doi.org/10.1016/j.apergo.201306.06.006 Copyright © 2013 Elsevier Ltd and The Ergonomics Society. All rights reserved.

second class of systems deploy increased automation for more active train protection. These systems place active constraints on train handling. Examples are the European Train Control System (ETCS) and the Positive Train Control System (PTC) being implemented in the USA. An example ETCS display (CENELEC, 2005) is shown in Figure 30.7b. Both ETCS and PTC apply automatic braking to stop the train if it is projected to exceed upcoming speed restrictions or pass stop signals. Since train crews are penalized if an automatic brake is applied, these systems enforce speed and braking regimen.

ETCS and Energymeiser are currently in use in intercity passenger and freight operations (Naweed, 2014). Similar systems are being field tested in the USA as well (Roth, Rosenhand, & Multer, 2013). Studies have been conducted in both the USA and Europe examining the impact of new in-cab technologies on train crew performance and requirements for expertise, specifically ETCS in Europe (Naweed, 2014) and PTC in the USA. (Roth & Multer, 2009).

Roth and Multer (2009) conducted interviews and observations at sites where PTC technologies were introduced. They found that while some cognitive demands were reduced (e.g., some PTC systems reduced memory demands by presenting route preview), new cognitive demands emerged, and changes in train control strategies were required. For example, PTC systems employ conservative braking algorithms designed to ensure ability to stop under restrictive assumptions (e.g., heavy train or slippery track). This meant that under most conditions, train crews needed to initiate braking at an earlier point than they would choose on their own. With experience, locomotive engineers developed strategies for staying within the braking requirements of the PTC system while still operating as efficiently as possible. One example was to delay manual braking, coming as close as possible to the trigger for an automatic "penalty" brake, without exceeding that point, to maintain efficient train operation.

Increases in information and alerts provided by in-cab displays also impacted attention management and monitoring strategies. Locomotive engineers needed to develop new strategies for managing the competing attentional demands between monitoring train control inside the cab and monitoring hazards such as trespassers, motor vehicles approaching grade crossings, and objects on the track outside the cab.

Interviews with train crews exercising the PTC system also indicated a need to guard against becoming overly reliant on the new train control technologies (Roth & Multer, 2009). Locomotive engineers expressed concern that they may come to rely on the PTC system to alert them of upcoming speed and authority limits and to automatically stop the train should they fail to do so themselves. They worried that they may erroneously continue to rely on the PTC system to stop the train even in cases where the PTC system is not operating. Train crews may not realize or may forget that the PTC system has failed (or is off) and is thus no longer providing the level of support they are expecting. They also expressed concern of their ability to take over manually should the system become unavailable. Even if they are initially trained for manual operation, they may lose skill due to lack of practice, and thus may not be able to perform tasks as well when PTC is not available as they would have had it never been installed. This implies a need for more careful design and evaluation of new forms of automation to ensure that train crews become immediately aware when the automation is not functioning, and regular refresher training to ensure that train crews preserve their manual operation expertise.

Depending on the design, automated train control systems can also impact established teamwork strategies, particularly if the display is placed where only the locomotive engineer can see it. This can disrupt shared understanding of the current situation and reduce the ability of the second individual in the cab to provide a redundancy check.

Very similar results have been observed in studies of European automated braking technologies (Naweed, 2014). Naweed found that systems such as ETCS that provide upcoming route information and speed-based instruction resulted in locomotive engineers having less need to monitor the environment, and placed more emphasis on in-cab monitoring than traditional train operation. Further, the task of speed control was converted into a *pursuit tracking* task where the goal is to maintain the speedometer reading within the upper and lower control bounds provided by the ETCS display.

Importantly, Naweed (2014) noted that while locomotive engineers "drove to the display, the task of maintaining and optimizing speed-error was found to actually require some careful planning and tentative adjustments if train speed were not to be exceeded" (p. 469). This finding is consistent with the results of the studies of US PTC systems. They highlight that while new technologies can alter the nature of locomotive engineer expertise, they do not eliminate the need for expert skills and strategies.

FORWARD THINKING: WHAT MAY BE THE NATURE OF EXPERTISE IN RAILROAD OPERATIONS IN COMING DECADES?

The review of expertise in railroad operations provided in this chapter highlights the role of people in compensating for system limitations, contributing to system resilience, and enhancing overall safety. Increasingly, research of people working in highly regulated environments, including rail operations, is showing that expert decision making is a fluid and ongoing process with preventive strategies that *flex* in response to evolving demands (O'Keeffe, Tuckey, & Naweed, 2015). The railroad industry, which is highly rule-based with very specific procedures around safe working, can have rules that are very rigid, implying that there is one best response rather than several suitable options. In contrast, our findings summarized here highlight that decision strategies of train crews are inherently flexible, and flex in response to evolving events, such that they become a means for assessing risk and a practical mechanism for reducing it by taking account of the wider sociocultural environment. They may also be used to manage or minimize averse mental health consequences, such as stress or exhaustion (Naweed, O'Keeffe, & Tuckey, 2016).

Questions arise as to what roles people will take on in railroad operation of the future. In a series of international workshops across Australia and New Zealand, hundreds of stakeholders representing the entire distributed railroad system were asked what the future would look like for their operations. These workshops were designed to envision a future where there were no safety incidents and everything ran smoothly (Naweed & Rainbird, 2014). Figure 30.8 depicts a newspaper headline and illustration of the year 2024 that was created by one group. The detail underneath the headline included the following:

> Following the successful completion of extensive trials, [rail organization X] announced they are introducing a train operating system where [train crew are] directly linked to the train system via wireless telepathy [...] the group, using advances in stem cell research, were able to develop a link to connect neuro receptors to an onboard computer to control onboard traction systems.

This projection of the future of rail operations, while fanciful, is nevertheless important for staging prospective answers to the titular question of this concluding section.

Figure 30.8 is very telling. It provides a story about the intimate (literally cybernetic) connection between the operator and the train. It speaks to distributed cognition, and the more central role that technology plays within that. Is it science fiction, or science eventuality?

While cybernetic connections may be a distant vision whose merits are debatable, we anticipate that the trend toward greater machine autonomy and the shift in roles and responsibilities of people in the system will continue to accelerate. While the figure depicts a person in the cab, it is more likely that human control will increasingly be exercised remotely (whether via *wireless telepathy* or not). Where railroads now often operate with two people in the cab, we see a growing shift to single-person cabs, and in the not-too-distant future, to "no-person" cab operation.

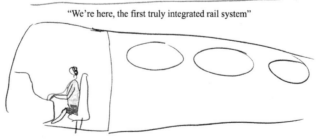

FIGURE 30.8 Newsletter headline and illustration created during a Future Inquiry Workshop in the railroad setting. Handwriting has been substituted with typescript.

Unpublished data from Naweed, A., & Rainbird, S., *Managing and mitigating SPAD risk in rail operations* (Tech. Rep. R2.116), Central Queensland University, 2014. Used with permission from Anjum Naweed.

We also expect railroad systems to become more complex and opaquer than ever, even as the role of the person becomes more relevant than ever before. Figure 30.8 speaks to tight integration, but past experience suggests that attempts at integration create new challenges. As more automated functions are introduced (e.g., energy management and automated safety response functions), how will these subsystems interact, and will that happen in ways that facilitate understanding, anticipation, and control by the human? What are the implications for the learning curve to gain expertise, and what new knowledge and strategies will be required to maintain engagement and supervisory control?

Addressing these questions will be critical to ensure that the people in the system are able to adapt to evolving demands and contribute to system resilience and overall safety.

References

Bagnoli, A. (2009). Beyond the standard interview: the use of graphic elicitation and arts-based methods. *Qualitative Research* 9(5), 547–570.

Baysari, M. T., McIntosh, A. S., & Wilson, J. R. (2008). Understanding the human factors contribution to railway accidents and incidents in Australia. *Accident Analysis & Prevention* 40(5), 1750–1757.

Belmonte, F., Schön, W., Heurley, L., & Capel, R. (2011). Interdisciplinary safety analysis of complex socio-technological systems based on the functional resonance accident model: An application to railway traffic supervision. *Reliability Engineering & System Safety* 96(2), 237–249.

Bisantz, A., Roth, E. M., & Watts-Englert, J. (2015). Study and analysis of complex cognitive work. In J. R. Wilson & S. Sharples (Eds), *Evaluation of human work*. Boca Raton, FL: CRC Press.

Branton, P. (1979). Investigations into the skills of train-driving. *Ergonomics* 22(2),155–164

Branton, R. (1993). *Person-centred ergonomics: A Brantonian view of human factors*. London: Taylor and Francis.

CENELEC (2005). European Standard e Railway applications e Communication, signalling and processing systems e European Rail Traffic Management Systeme Driver-Machine Interface e Part 4: Data entry for the ERTMS/ETCS/GSM-R systems Comité Européen de Normalisation Electrotechnique. (Vol. CLC/TS50459e4:2005.)

Chapman, J., & Naweed, A. (2015). Health initiatives to target obesity in surface transport industries: Review and implications for action. *Evidence Base* 2, 1–32.

Chen, L., Zhao, Y., & Zhao, T. (2015). An AcciMap Analysis on the China-Yongwen Railway Accident. In *Engineering Asset Management-Systems, Professional Practices and Certification* (pp. 1247–1253). Cham: Springer.

Cooke, N. J. (1994). Varieties of knowledge elicitation techniques. *International Journal of Human-Computer Studies* 41(6), 801–849.

Cooke, N. J., & Schvaneveldt, R. W. (1988). Effects of computer programming experience on network representations of abstract programming concepts. *International Journal of Man-Machine Studies* 29(4), 407–427.

Cullen, J., & Bryman, A. (1988). The knowledge acquisition bottleneck: A time for reassessment? *Expert Systems* 5(3), 216–225.

Dorrian, J., Hussey, F., & Dawson, D. (2007). Train driving efficiency and safety: examining the cost of fatigue. *Journal of Sleep Research 16*(1), 1–11.

Durso, F. T., Gregg, S., Ferguson, A., Kazi, S. McDonald, J. M., & Glover, K. (2015). Human factors of run through switches in US rail operations. In *19th Triennial Congress of the International Ergonomics Association*: Melbourne, Australia.

Einhorn, J., Sheridan, T., & Multer, J. (2005). *Preview information in cab displays for high-speed locomotives: Human factors in railroad operations* (DOTVNTSC-FRA-04-06). Cambridge, MA: Department of Transportation, John A. Volpe National Transportation Systems Center.

Eisner, E. (2008). Art and knowledge. In J. G. Knowles & A. L. Cole (Eds), *Handbook of the arts in qualitative research: Perspectives, methodologies, examples and issues* (pp. 71–81). London: Sage.

Elstein, A. S., Shulman, L. S., & Sprafka, S. A. (1978). *Medical problem solving: An analysis of clinical reasoning*. Cambridge, MA: Harvard University Press.

Endsley, M. R. (1995). Toward a theory of situation awareness in dynamic systems. *Human Factors 37*(1), 32–64.

Farrington-Darby, T., Wilson, J. R., Norris, B. J., & Clarke, T. (2006). A naturalistic study of railway controllers. *Ergonomics 49*(12–13), 1370–1394.

Filtness, A. J., & Naweed, A. (2017). Causes, consequences and countermeasures to driver fatigue in the rail industry: The train driver perspective. *Applied Ergonomics 60*, 12–21.

Gauntlett, D. (2007). *Creative explorations. New approaches to identities and audiences*. London: Routledge.

Grabowski, M., Massey, A. P., & Wallace, W. A. (1992). Focus groups as a group knowledge acquisition technique. *Knowledge Acquisition 4*(4), 407–425.

Hickey, A. R., & Collins, M. D. (2017). Disinhibition and train driver performance. *Safety Science 95*, 104–115.

Hoffman, R. R. (1998). How can expertise be defined: Implications of research from cognitive psychology. In R. Williams, W. Faulkner, & J. Fleck (Eds), *Exploring expertise* (pp. 91–100). New York: Macmillan.

Hoffman, R. R., Ward, P., Feltovich, P. J., DiBello, L., Fiore, S. M., & Andrews, D. H. (2014). *Accelerated expertise: Training for high proficiency in a complex world*. New York: Psychology Press.

Hollan, J., Hutchins, E., & Kirsch, D. (2000). Distributed cognition: Toward a new foundation for human-computer interaction research. *ACM Transactions on Computer-Human Interaction 7*, 174–196.

Hollnagel, E. (1998). *Cognitive reliability and error analysis method (CREAM)*. Oxford: Elsevier Science.

Hollnagel, E. (2012). *FRAM, the functional resonance analysis method: Modelling complex socio-technical systems*. London: Ashgate.

Howlett, P., & Pudney, P. (2000). Energy-efficient driving strategies for long-haul trains. Paper presented at the *CORE 2000 Conference on Railway Engineering*, Adelaide, Australia.

Hutchins, E. (1995). How a cockpit remembers its speeds. *Cognitive Science, 19*, 265–288.

Klein, G. A., Calderwood, R., & MacGregor, D. (1989). Critical decision method for eliciting knowledge. *IEEE Transactions on Systems, Man, & Cybernetics 19*(3), 462–472.

Klein, G., Feltovich, P. J., Bradshaw, J. M., & Woods, D. D. (2005). Common ground and coordinating joint activity. In W. B. Rouse & K. R. Boff (Eds), *Organizational simulation*. New York: Wiley.

Leveson, N. G. (2004). A new accident model for engineering safer systems. *Safety Science* 42(4), 237–270.

Luke, T., Brook-Carter, N., Parkes, A. M., Grimes, E., & Mills, A. (2006). An investigation of train driver visual strategies. *Cognition, Technology and Work 8*, 15–29.

Moray, N., Groeger, J., & Stanton, N. (2017). Quantitative modelling in cognitive ergonomics: predicting signals passed at danger. *Ergonomics 60*, 206–220.

Morgan, C. A., Olson, L. E., Kyte, T. B., Roop, S. S. and Carlisle, T. D. (2006). *Railroad crew resource management (CRM): Survey of teams in the railroad operating environment and identification of available CRM training methods*. Report DOT/FRA/ORD-06/10. Federal Railroad Administration, U.S. Department of Transportation.

Multer, J., Safar, H., & Roth, E. (2015). Understanding how signals passed at danger occur through the lens of sociotechnical systems. In *5th International Rail Human Factors Meeting (London, UK)*.

Naweed, A. (2013). Psychological factors for driver distraction and inattention in the Australian and New Zealand rail industry. *Accident Analysis and Prevention 60*, 193–204.

Naweed, A. (2014). Investigations into the skills of modern and traditional train driving. *Applied Ergonomics 45*, 462–470.

Naweed, A. (2015). The "scenario invention task" (SIT): An innovative method for harnessing natural human creativity. *19th Triennial Congress of the International Ergonomics Association*. Melbourne, AU (9–14 August).

Naweed, A. (2016). The case of the crooked clock and the distracted driver. *Narrative 24*(2), 212–221.

Naweed, A., & Balakrishnan, G. (2014). Understanding the visual skills and strategies of train drivers in the urban rail environment. *Work 47*, 339–352.

Naweed, A., Chapman, J., Allan, M., & Trigg, J. (2017). It comes with the Job: Work organization, job design, and self-regulatory barriers to improving the health status of train drivers. *Journal of Occupational and Environmental Medicine 59*(3), 264–273.

Naweed, A., & Dawson, D. (2016). *Mulgrave mill cane rail safety risk study*. Cairns, QLD: MSF Sugar.

Naweed, A., Every, D., Balakrishnan, G., & Dorrian, J. (2014). One is the loneliest number: Exploring the role of the second driver in Australian rail operations. *Ergonomics Australia 1*, 1–8.

Naweed, A., Hockey, B., & Clarke, S. (2009). Enhanced information design for high speed train displays: Determining goal set operation under a supervisory automated braking system. *Human Factors, Security and Safety 189*, 202.

Naweed, A., Hockey, G. R., & Clarke, S. D. (2013). Designing simulator tools for rail research: The case study of a train driving microworld. *Applied Ergonomics 44*(3), 445–454.

Naweed, A., & Kingshott, K. (2019). Flying off the handle: Affective influences on action tendencies in real-world aircraft maintenance engineering scenarios. *Journal of Cognitive Engineering and Decision Making*, 13(2), 81–101. DOI: 10.1177/1555343418821507

Naweed, A., & Moody, H. (2015). A streetcar undesired: Investigating ergonomics and human factors issues in the driver-cab interface of Australian trams. *Urban Rail Transit 1*(1), 1–10.

Naweed, A., O'Keeffe, V.J., & Tuckey, M.R. (2016). The art of train driving: Flexing the boundaries to manage risk within an inflexible system. *Eat Sleep Work 1*, 68–72.

Naweed, A., & Rainbird, S. (2014). *Managing and mitigating SPAD risk in rail operations* (Tech. Rep. R2.116). South Australia, AU: CQUniversity.

Naweed, A., & Rainbird, S. (2015). Recovering time or chasing rainbows? Exploring time perception, conceptualization of time recovery, and time pressure mitigation in train driving. *IIE Transactions on Occupational Ergonomics & Human Factors 3*(2), 91–104.

Naweed, A., Rainbird, S., & Chapman J. (2015). Investigating the formal countermeasures and informal strategies used to mitigate SPAD risk in train driving. *Ergonomics 58*, 883–896.

Naweed, A., Rainbird, S., & Dance, C. (2015). Are you fit to continue? Approaching rail systems thinking at the cusp of safety and the apex of performance. *Safety Science 76*, 101–110.

Naweed, A., Balakrishnan, G., & Dorrian, J. (2018). Going Solo: Hierarchical Task Analysis of the Second Driver in "Two-up" (multi-person) Freight Rail Operations. *Applied Ergonomics 70*, 202–231.

Nisbett, R. E., & Wilson, T. D. (1977). Telling more than we can know: Verbal reports on mental processes. *Psychological Review 84*(3), 231–259.

O'Keeffe, V. J., Tuckey, M. R., & Naweed, A. (2015). Whose safety? Flexible risk assessment boundaries balance nurse safety with patient care. *Safety Science 76*, 111–120.

Ouyang, M., Hong, L., Yu, M. H., & Fei, Q. (2010). STAMP-based analysis on the railway accident and accident spreading: Taking the China–Jiaoji railway accident for example. *Safety Science 48*(5), 544–555.

Phillips, R. O., & Sagberg, F. (2014). What did you expect? CREAM analysis of hazardous incidents occurring on approach to rail signals. *Safety Science 66*, 92–100.

Rail Safety & Standards Board (2008). *Good practice guide on cognitive and individual risk factors* (RS/232 Issue 1).

Rainbird, S., & Naweed, A. (2016). Signs of respect: Embodying the train driver-signal relationship to avoid rail disasters. *Applied Mobilities 1*(2), 50–66.

Rasmussen, J. (1997). Risk management in a dynamic society: A modelling problem. *Safety Science 27*(2–3), 183–213.

Rasmussen, J., Pejtersen, A. M., & Goodstein, L. P. (1994). *Cognitive systems engineering.* New York: Wiley.

Read, G., Naweed, A., & Salmon, P. (2019). Complexity on the rails: A systems-based approach to understanding safety management in rail transport. *Reliability Engineering & System Safety 118*, 352–365.

Rittel, H. W., & Weber, M. M. (1973). Dilemmas in a general theory of planning. *Policy Sciences 4*, 155–169.

Rose, J., Bearman, C., & Naweed, A. (2013). Using task analysis to inform the development and evaluation of new technologies. In C. Bearman, A. Naweed, J. Dorrian, J. Rose, & D. Dawson (Eds), *Evaluation of rail technology: A practical human factors guide* (pp. 129–169). Surrey, UK: Ashgate.

Rosenhand, H., Roth, E., & Multer, J. (2012). *Cognitive and collaborative demands of freight conductor activities: Results and implications of a cognitive task analysis.* (DOT/FRA/ORD-12/13). Cambridge, MA: US DOT Volpe National Transportation Systems Center. Available at https://rosap.ntl.bts.gov/view/dot/9681

Roth, E. M. (2009). Understanding cognitive strategies for shared situation awareness across a distributed system: An example of strategies analysis. In A. M. Bisantz & C. M. Burns (Eds), *Applications of cognitive work analysis* (pp. 129–147). Boca Raton, FL: CRC Press.

Roth, E. M., Malsch, N., & Multer, J. (2001). *Understanding how train dispatchers manage and control trains: Results of a cognitive task analysis.* Washington, DC: US Department of Transportation, Federal Railroad Administration. (DOT/FRA/ORD-01/02). Available at http://ntl.bts.gov/lib/33000/33600/33672/33672.pdf

Roth, E., & Multer, J. (2007). *Communication and coordination demands of railroad roadway worker activities and implications for new technology.* Washington, DC: US Department of

Transportation, Federal Railroad Administration (DOT/FRA/ORD-07/28). Retrieved from http://www.fra.dot.gov/eLib/details/L01602

Roth, E. M., & Multer, J. (2009). *Technology implications of a cognitive task analysis for locomotive engineers.* Washington, DC: US Department of Transportation, Federal Railroad Administration (DOT/FRA/ORD-09/03). Available at https://www.fra.dot.gov/Elib/Document/381

Roth, E. M., Multer, J., & Raslear, T. (2006). Shared situation awareness as a contributor to high reliability performance in railroad operations. *Organization Studies* 27(7), 967–987.

Roth, E. M., Multer, J., & Scott, R. (2009). Understanding and contributing to resilient work systems. In C. P. Nemeth, E. Hollnagel, & S. Dekker (Eds), *Resilience engineering perspectives volume 2: Preparation and restoration.* Burlington, VT: Ashgate.

Roth, E. M., & Patterson, E. S. (2005). Using observational study as a tool for discovery: Uncovering cognitive and collaborative demands and adaptive strategies. In H. Montgomery, R. Lipshitz, & B. Brehmer (Eds), *How professionals make decisions* (pp. 379–393). Mahwah, NJ: Erlbaum.

Roth, E. M, Rosenhand, H., & Multer, J. (2013). *Using cognitive task analysis to inform issues in human systems integration in railroad operation.* Washington, DC: US Department of Transportation, Federal Railroad Administration, Office of Research and Development (DOT/FRA/ORD-13/31).

Safar, H., Multer, J., & Roth, E. (2015). *An investigation of passing stop signals at a passenger railroad.* Washington, DC: US Department of Transportation, Federal Railroad Administration, Office of Research and Development (DOT/FRA/ORD-15/25).

Salas, E. Diazgranados, D., & Lazzara, E. (2011). Promoting teamwork when lives depend on it: What matters in the railroad industry? In *Teamwork in U.S. railroad operations* (pp. 10–26). Irvine, CA: Transportation Research Board of the National Academies.

Salas, E., Sims, D., E., & Burke, C. S. (2005). Is there a "Big Five" in teamwork? *Small Group Research* 36, 555–599.

Salmon, P. M., Lenne, M. G., Read, G. J. M., Mulvihill, C. M., Cornelissen, M., Young, K. L., ... & Stevens, N. (2015). Beyond the crossing: a cognitive work analysis of rail level crossing systems. *Procedia Manufacturing* 3, 2921–2928.

Shadbolt, N. R. (2005). Eliciting expertise. In J. R. Wilson and N. E. Corlett (Eds), *Evaluation of human work* (pp. 185–218). London: Taylor & Francis.

Suss, J., & Ward, P. (2015). Predicting the future in perceptual-motor domains: Perceptual anticipation, option generation and expertise. In R. R. Hoffman, P. A. Hancock, M. Scerbo, & J. L. Szalma (Eds), *Cambridge handbook of applied perception research* (pp. 951–976). New York: Cambridge University Press.

Tichon, J. G. (2007). The use of expert knowledge in the development of simulations for train driver training. *Cognition, Technology & Work* 9, 177–187.

TRB (2011). Teamwork in U.S. railroad operations (E-159). Irvine, CA: Transportation Research Board of the National Academies.

Vicente, K. (1999). *Cognitive work analysis: Toward safe, productive, and healthy computer-based work.* Mahwah, NJ: Lawrence Erlbaum.

Wiegmann, D. A., & Shappell, S. A. (2003). *A human error approach to aviation accident analysis: The human factors analysis and classification system.* Burlington, VT: Ashgate.

Wreathall, J., Roth, E., Bley, D., & Multer, J. (2007). *Human factors considerations in the evaluation of processor-based signal and train control systems.* Washington, DC: US Department of Transportation, Federal Railroad Administration (DOT/FRA/ORD-07/07). Available at http://www.fra.dot.gov/eLib/details/L01620

CHAPTER 31

THE CYBER DOMAINS
Understanding Expertise for Network Security

ROBERT THOMSON

INTRODUCTION

ONE of the challenges in writing a book chapter on cyber expertise is that cyber—unlike chess—isn't well-defined: there is no consistent and agreed-upon set of knowledge, skills, and attributes that make up the cyber domain(s). There are no rules, no game-boards, and few constraints. There is little state-of-the-art research in cyber expertise, not because impactful work is not being conducted, but because the field is progressing so quickly that research findings can become obsolete before they are even published. Cyber expertise research is like astronomy: what we perceive is actually a snapshot of the past.

The cyber domain(s) impact(s) every facet of modern life from the electricity that powers our homes to the financial networks that underlie our economy. Cyber is more than just the set of electronic devices connected to a network; it also encompasses how these devices and networks interact. This interaction goes beyond a hierarchy of networks, but includes how these networks touch people (e.g., the cyber–social) and the real world (e.g., the cyber–physical). To be successful, the people who operate within this field need a combination of technical skills, domain-specific knowledge, and social intelligence. They, like the networks they operate, must also be secure, trustworthy, and resilient. Defining the knowledge, skills, attributes, and other characteristics is not as simple as defining a group of technical skills on which people can be trained; the complexity of the cyber domain(s) makes this a unique challenge.

Defining expertise is especially difficult because outside of the few established first-generation professionals who are considered experts by virtue of being the original architects of the field, there is no standard definition for cyber expertise. Research in the cyber domain has generally operationalized expertise using questionnaires, peer identification, or self-selection (Rajivan, Moriano, Kelley, & Camp, 2017). Questionnaires usually divide relative novices from *experts* by using one or more of the following criteria: years of experience, job title, technical competency, and range of competencies (see Ben-Asher & Gonzalez, 2015).

Classification in this manner likely underspecifies expertise and oversimplifies where people lie along the spectrum from relative novice to expert (see Hoffman, 1998). Instead, cyber expertise should be operationalized along a continuum defined by the representational reorganization and schematization of knowledge structures with the proceduralization of reasoning that comes from practice, training, and experience in one's domain, finally culminating in the accurate and rapid deployment of attentional and reasoning processes during complex decision-making (e.g., Hoffman, 1998; Hoffman, Ward, DiBello, Feltovich, & Fiore, 2014). This is not to argue that these three factors are sufficient for expertise, but that these factors provide the preconditions for the kinds of ontological shifts and schematization of knowledge that appear to be the hallmarks of experts (Chase & Simon, 1973). Some individuals may never become experts no matter their training or experience.

In the context of the cyber domain(s), there has yet to be sustained research into cyber expertise at this level of cognitive fidelity. Instead, extant research has focused on examining technical skills that may correlate with future job performance in information technology or as a cyber operator. While jobs in information technology may be filled by individuals who are responsible for setting up the network infrastructure, cyber operators maintain the security of that network. Before delving into these details, let's look at the evolution of the cyber domain.

The structure of this chapter is as follows: I begin by describing the historical and continuing evolution of the cyber domains. Next, I review current literature on cyber expertise and provide several case studies. Third, I argue for a more cognitively inspired definition of cyber expertise focusing on schematization of knowledge and the role of team-based expertise. Finally, I forecast the evolution of the cyber domains and present candidate research paradigms to further investigate cyber expertise.

Evolution of the Cyber Domains

The historical use of the term *cyber* (as short for cybernetics) arose in the 1940s to describe the study of communication and control in both living systems and machines. The vernacular use of the term *cyber* (as a short form for cyberspace) is attributed to novelist William Gibson in 1982 to describe electronic communication and virtual reality. When people think of cyber in terms of cybersecurity, popular film references, such as to *Wargames* (1983), *Hackers* (1995), and the *Matrix* trilogy (1990s), bring forth images of hackers taking down security systems with a few keystrokes, a swipe of the hand, or even by concentration alone.

Perhaps the first major application of cyber expertise was Alan Turing's development of the cryptanalysis machine in WWII to decrypt the Nazi's Enigma code. By the 1960s, computers were being used for more complex calculations, and by the 1970s systems were being networked together over long distances (using phone lines). With the development of the TCP/IP protocol in 1982 and Tim Berners-Lee's concept of the World Wide Web implemented between 1989 and 1991, the modern Internet became accessible to the public. The 2000s further saw the rise of smartphones, wireless networking, and social media. The 2010s saw the proliferation of smart devices inside the home (the Internet of Things; IoT). In 2016, nearly 81 percent of people in the developed world use the Internet, and there were 6.4 billion connected devices (International Telecommunication Union, 2016).

As the number and uses for connected devices grow, the complexity of cyber infrastructure grows exponentially. Each new device and new software application can be connected to innumerable others, and each adds new vulnerabilities. Planned obsolescence causes older devices and software to become more vulnerable as developers focus their resources on current technologies. The potential for breakthroughs in quantum computing technology may render much extant encryption useless, leaving our sensitive information exposed. Furthermore, artificial intelligence can rapidly automate both the detection and exploitation of vulnerabilities. Finally, the rise of *big* data analytics provides new outlets for socially engineering our behavior at a massive scale.

The most popular term for someone who exploits a vulnerability in a computer system is a *hacker*, although the cyber community usually uses the term *adversary*. The term *hacker* was coined in the 1960s at MIT where extremely skilled computer programmers would push the limits of software and hardware. In the 1970s to mid-1980s a select group of hackers discovered a method for making free long-distance phone calls by using pure sound tones to confuse AT&T's automated call system. These early phone hackers were known as phreakers.

By the end of the 1980s and with the approaching rise of the modern Internet, the first malware (computer viruses) began infecting systems. The 1988 Morris computer worm (named after Robert Tappan Morris, now an MIT professor) exploited a Unix operating system vulnerability in an attempt to measure the number of computers on the fledgling Internet. A worm is a form of malware software that makes a copy of itself to spread to another computer system. Due to a flaw in its design, the worm went from a harmless experiment to the first major denial-of-service (DoS) attack by overwhelming networked computers with an exponentially increasing number of requests and by installing numerous software services that would continuously slow computers down. DoS attacks flood a website or computer with requests, eventually overwhelming its ability to handle said requests and thus taking it offline.

The severity and variability of cyber attacks steadily increased throughout the 1990s and 2000s, causing computer security, such as anti-virus software and built-in firewalls, to become a mainstream feature of operating systems as a means to defend against such attacks. The most popular kind of malware was the *bot-net* (i.e., robot network)—a network of malware-infected computers that allow hackers to remotely use the machine to infect more machines and launch distributed DoS attacks. In 2007, Estonia was victim of a large DoS attack that took government and banking services temporarily offline, and was purported to be politically motivated.

In 2010, the Stuxnet computer worm exhibited physical consequences when its software specifically targeted the Siemens centrifuges used in an Iranian nuclear facility, causing them to blow apart with the apparent goal of derailing Iran's nuclear program. Stuxnet resulted in the public realization that cyberattacks could affect the physical world and were not limited to just destroying or exfiltrating information or temporarily taking a network offline.

Finally, the recent rise in ransomware has added a new wrinkle in cybersecurity. Ransomware is a scenario whereby malware encrypts information on a target system, rendering it inaccessible until a ransom payment is made. In May 2017, the WannaCry ransomware targeted older Windows machines, infecting over 200,000 machines worldwide and infecting parts of the United Kingdom's National Health Service. WannaCry exploited a Windows operating system vulnerability in how Windows shares files and then encrypted the contents of vulnerable computers, making them unusable until a decryption key is sent. Infected computers demanded that a *ransom* be paid in bitcoin, which is a new, relatively untraceable, online form of currency.

In summary, cyber attacks not only have an online component, but also extend to social and physical interactions depending on the particular type of attack and goal. A small group of adversaries can do substantial damage to our infrastructure, and it only takes one vulnerability to get a foothold in a computer system or network. Before delving into recent research in cyber expertise, let us briefly review the work roles and role of education in the cyber domains.

Work Roles in the Cyber Domains

To develop an effective workforce, one method is to instill a set of knowledge, skills, and abilities which provide an ontology of work roles. Several such ontologies exist, and while they have been developed to support US government hiring requirements, they represent the most well-documented rostering of work roles in the cyber domains. The Department of Homeland Security's National Initiative for Cybersecurity Careers and Studies (NICCS) developed a Cybersecurity Workforce Framework (Newhouse, Keith, & Witte, 2016) to provide government and industry with a baseline for common positions. This collection includes nine work-role categories, 31 specialty areas, and over 1000 types of knowledge, skills, and abilities. Major categories are described in Table 31.1.

Securely provision roles revolve around the more traditional information technology field including software developers, computer programmers, and network architects. The *operate and maintain* roles include system administrators, knowledge management, and security analysts. The *oversee and govern* roles include managerial roles, and

Table 31.1 Cybersecurity Workforce Framework

Work-role category	Description
Securely provision	Conceptualizes, designs, and builds secure information technology (IT) systems, with responsibility for aspects of systems and/or networks development.
Operate and maintain	Provides the support, administration, and maintenance necessary to ensure effective and efficient IT system performance and security.
Oversee and govern	Provides leadership, management, direction, or development and advocacy so the organization may effectively conduct cybersecurity work.
Protect and defend	Identifies, analyzes, and mitigates threats to internal IT systems and/or networks.
Analyze	Performs highly specialized review and evaluation of incoming cybersecurity information to determine its usefulness for intelligence.
Collect and operate	Provides specialized denial and deception operations and collection of cybersecurity information that may be used to develop intelligence
Investigate	Investigates cybersecurity events or crimes related to IT systems, networks, and digital evidence.

Reproduced from William Newhouse, Stephanie Keith, Benjamin Scribner, and Greg Witte, National Initiative for Cybersecurity Education (NICE) Cybersecurity Workforce Framework, *NIST Special Publication*, 800-181, p. 14, © National Institute of Standards and Technology, 2016.

those working in cyber law, policy development, and education. The *protect and defend* roles include cyber analysts (operators) and network defenders. The *analyze, collect, and operate* and *investigate* roles all encompass the broad field of digital forensics and tend to be government or law enforcement positions (Caulkins, Badillo-Urquiola, Bockelman, & Leis, 2016).

The current common understanding of technical aspects of the cyber domain is often viewed separately from the social aspects occurring in the domain. A limitation of the NICCS Workforce Framework is that, of the 1060 types of knowledge, skills, and aptitudes, fewer than 1 percent describe social fit or teamwork. This means that the framework paints an incomplete picture of workforce proficiency (Seong, Kristof-Brown, Park, Hong, & Shin, 2015). Cultivating talent in the cyber domains involves recognizing that the people drawn to this domain may have distinctive social psychological traits and tendencies that make them uniquely suited to excel in this space. The development of any cyber workforce that omits the social aspect of human behavior on the network neglects a critical component of the cyber domain. Cyber defense would be aided by an understanding of human behavior and how it introduces risk to the network (Asgharpour, Liu, & Camp, 2007; Arachchilage & Love, 2013; Fontenele & Sun, 2016; Pfleeger & Caputo, 2012). Convincing users to engage in best practices is a skillset that relies more on social skill and persuasion than technical skill (Shillair et al., 2015). Similarly, offensive cyber operations are often contingent upon exploiting known human behaviors (e.g., phishing attacks; many attacks start with someone opening an infected e-mail).

This section has briefly described current efforts toward delineating work roles and the knowledge, skills, and abilities that comprise cyber expertise. While current efforts capture many of the technical aspects of work roles, more work needs to be done to capture social requirements for effective cyber operators. In the next section, I briefly examine how new cyber operators are educated.

Cyber Education and Training

The Department of Homeland Security's NICCS and the National Security Agency have sponsored National Centers of Academic Excellence in Cyber Defense and Cyber Operations, and have identified over 200 colleges and universities in the United States whose cyber curricula align with the cybersecurity knowledge, skills, and abilities in their Cybersecurity Workforce Framework. In addition to classroom education, there are numerous certifications such as the Certified Information Systems Security Professional (CISSP), whose hallmark is its insistence on re-certification every three years with continuing professional education requirements.

The truth is that many institutions have jumped on the *cyber* bandwagon without really understanding the requirements for developing competent cyber operators. There is limited understanding of either the cognitive processes involved in cyber operations or the kinds of training that promote the development of expertise in this domain. Baseline knowledge, skills, and abilities do go a long way toward developing initial capabilities that can be honed into more specialized skills that are organizationally dependent. However, attempting to develop these key baselines without first defining the organizational environment will likely only result in a limited ability to produce an effective cyber workforce (Cable & Parsons, 2001; Seong, Kristof-Brown, Park, Hong, & Shin, 2015). Furthermore, developing the knowledge, skills, and abilities needed across teams would arguably provide greater fidelity on the types of teams needed to build an effective cyber workforce (Rajivan, 2014; Rajivan, Janssen, & Cooke, 2013).

While cybersecurity education is burgeoning, relatively few high school students are exposed to cybersecurity. Where students are exposed to cyber, it only happens relatively late in their education (Frank, 2016). A consequence of promoting cyber education is that we reduce the overall attack surface for malicious actors by improving cyber *hygiene*—keeping data safe by using strong passwords and keeping computer security up to date—across groups that can be differentially targeted with techniques such as phishing attacks and ransomware.

In recent years, cybersecurity-themed competitions (known as capture-the-flag competitions) have emerged as an engaging way to educate middle and high school students about potential career paths in computer science and related cybersecurity fields (Bell, 2014; Capalbo, 2011; Chapman, Burket, & Brumley, 2014; Cowan, 2003;

Dodge, 2007; Eagle, 2004; Hoffman, 2005; Wagner, 2004; Werther, 2011). In these competitions, individuals or teams solve computer security challenges that vary in type, including but not limited to digital forensics, cryptographic methods, software reverse engineering, web security, and network traffic analysis. In each challenge, the aim is to produce the best or desired outcome such as defend a network against an adversary, identify the likely threat, or decide whether to *sandbox* an adversary or to deny access to them.

As an example, the PicoCTF platform was a competition hosted in 2013 and 2014 targeting middle and high school students with varying levels of skill (Chapman, Burket, & Brumley, 2014). PicoCTF was designed with gradated gameplay to bootstrap students by providing challenges with advancing levels of difficulty beginning with critical thinking skills to final, more-advanced problems that can be technically challenging even to cybersecurity experts. Supporting all levels of difficulty, PicoCTF used an integrated story-driven gamified environment to scaffold early learners, with the option of a text-based viewer for experienced players. These competitions (and similar internal efforts) have been used by companies to promote training their existing workforce and as a task in the interview phase for assessing prospective employees.

The PicoCTF 2013 competition featured 57 challenge problems that spanned five main categories: forensics ($n = 16$), cryptography ($n = 8$), reverse engineering ($n = 9$), script exploitation ($n = 13$), and binary exploitation ($n = 11$). In the forensics challenges, students searched for hidden data in images, network traffic, and file systems. The cryptography challenges required students to decipher text and audio messages encoded with classic ciphers as well as more modern encryption schemes. Reverse engineering problems involved understanding the behavior of compiled, obfuscated, or cryptic program code. In web and script exploitation challenges, students were required to attack web applications using common techniques such as database *Structured Query Language* (SQL) injection. Finally, binary exploitation problems required students to use vulnerabilities in complied programs (e.g., buffer overflow, format string, and return-oriented programming attacks) to gain control of a target system's operating system.

While capture-the-flag competitions have been effective in engaging students, high-scoring teams have tended to be from schools in districts with high socioeconomic status, where resources were in place for students to already have experience with these concepts. Effective use of mentors may provide a means for offsetting this discrepancy by providing a baseline opportunity for all interested participants. Gains on outcome measures from youth mentoring have been shown to be considerable (22 percentile points) when the targeted group is disadvantaged, there is a good fit between mentors and tasks, and when mentors and a youth have a similarity of interest (Levina & Nidiffer, 1996). In a recent study examining factors that had the highest impact on determining whether to pursue a career in cybersecurity, the most influential aspects included the availability of internships, cybersecurity competitions, and mentorship (Chen & Cotoranu, 2013). (For more information on the effect of mentoring, see

Chapter 44, "Learning at the Edge: The Role of Mentors, Coaches, and Their Surrogates in Developing Expertise," by Petushek et al., this volume.)

This section described the current state of education initiatives to promote the development of cyber expertise. Schools are struggling to adapt to the changing cyber landscape, while the gamification of cyber education has brought both awareness and general expertise to a broad range of ages and skills. The following section will address the complexities of measuring expertise in the rapidly evolving cyber domains.

THE COMPLEXITIES OF CYBER EXPERTISE

The domains of cyber expertise encompass an interdisciplinary blend of computer science, mathematics, economics, law, psychology, and engineering, at a minimum. While it may seem that a cyber expert needs be a technological polyglot and polymath, I instead argue that successful cyber operators are those with overlapping expertise who will best operate within a niche defined by a set of competencies. Given the rapidly changing technologies within the cyber domains, it may be impossible for individuals to develop in-depth expertise that remains relevant for longer than a few months. In this sense, there may be no such thing as an expert cyber operator, but instead an operator has expertise in domains that best fit particular work roles (Cook, 2014).

In this section, I will focus primarily on defensive cyber operations. Offensive cyber operations are usually methodically planned interventions requiring a blend of long-term strategic and deceptive thought. The offensive domain is legally constrained to government teams, and is generally performed in classified environments. As such, a review of offensive techniques and specific offensive expertise is generally beyond the scope of this chapter. Defensive operations tend to fall into two broad categories: *information technology managers* that set up and maintain the infrastructure of their networks, and *defensive cyber operators* whose work roles include detecting intrusions, performing digital forensics, and mitigating damage to infrastructure.

Defending a network requires thinking through the vulnerabilities as though one were going to attack the network (Baker, 2016). The success of a cyber attack is determined by the adversaries' ability to find vulnerabilities and exploit some elements of the system, including the weakest link, the human part (Pfleeger & Caputo, 2012; Rid & Buchanan, 2015). With new vulnerabilities constantly emerging, cyber operators require a life-long commitment to learning to constantly stay abreast of new technologies and potential new attack vectors which adversaries may exploit. Cyber operators also need to exhibit strong situational awareness (Dutt, Ahn, & Gonzalez, 2013; Jajodia, Liu, Swarup, & Wang, 2010), including juggling information such as over the health of the network, historical and current network activity, and performing a continual assessment of risk (Mahoney et al., 2010; Shin, Son, Khalil, & Heo, 2015).

Cyber operators need to have good mental flexibility, attention, and pattern-matching abilities as well as a good knowledge of what constitutes anomalous behavior in network traffic logs (Baker, 2016; Ben-Asher & Gonzalez, 2015; Champion, Jariwala, Ward, & Cooke, 2014). They should possess considerable skill and knowledge regarding particular operating systems, and particular skills in using a range of analytical tools for such things as network scanning, network mapping, vulnerability analysis, and malware analysis. Operators work with a set of software tools to sort through network traffic logs and visualize the flow of information in order to detect and attribute potenital intrusions. In essence, skilled operators are able to determine whether any network activity is anomalous in a high-noise environment, and judge whether to pass this information up to a superior to investigate further (D'Amico & Whitley, 2008; Genge, Kiss, & Haller, 2015).

In terms of specific tasks, skilled operators examine a large number of alerts and network events across multiple computer screens with the goal of identifying threats while minimizing false alerts (D'Amico & Whitley, 2008). To address the massive incoming information, they generally do a quick first pass to identify potential anomalous behavior. Those alerts considered non-threatening are deemed *white-listed* after the first pass and are rarely reviewed again. A second pass examines the remaining potential anomalies in more detail to determine the positive alerts from the false alarms. These positive alerts are then passed up to their supervisor for additional investigation and possible remediation. The complexity of this task is evidenced by its distributed nature: multiple operators use differing tools to detect clues to potential intrusions where these clues may be perceived at different locations on the network, by different operators, at different times, and may be classified differently by each operator based on their own experiences and judgment. Unfortunately, the cognitive mechanisms by which operators accurately identify anomalies have not been fully studied.

I argue that operators require two kinds of expertise in this context to be successful: domain-specific knowledge and situated knowledge (Cook, 2014; Goodall, Lutters, & Komlodi, 2004, 2009). Domain-specific knowledge includes formal education and training as well as informal practice with the tools and processes in the field. Unfortunately, this domain-specific knowledge is generally insufficient to support the recognition of anomalies in an operational environment because network traffic is heavily tied to the many valid operations within an environment, and what may be considered potentially malicious in one context could be normal in another (Yurcik, Barlow, & Rosendale, 2003). Instead, operators need local situated knowledge to understand the operation of the local network topology, the goal of the network application (e.g., company internal network; industrial control system), and what constitutes acceptable network traffic. This requirement for situated knowledge means that intrusion detection requires substantial operational expertise that goes beyond traditional formal education requirements. For this reason, many operators will team with local information technology specialists who impart their situated knowledge.

Armed with an understanding of the complexities involved in the work of defensive cyber operators, I will next examine several research paradigms to review the current state of research into cyber expertise. Two topics have been studied in some detail: cyber situational awareness and teamwork in intrusion detection.

Current Cyber Research

Before delving into specific research, a few commonalities and limitations need to be identified. In many cases, the research presented in the following is based on data from college students as opposed to actual cyber operators. This is not a criticism of the researchers' methodology, because there is limited access to actual cyber operators. Operators may be limited in their ability to disclose their tactics and teamwork, and their work locations are generally restricted. Of the research that has involved actual cyber operators, these operators tended to be from the private sector and the findings might not generalize to their military counterparts. In general, military cyber protection teams (CPTs) are better staffed, are better funded, and accord their operators time to keep abreast of recent developments in software, vulnerabilities, and remediation. On the other hand, much of the private sector, except maybe the largest conglomerates, are: (1) desperately attempting to staff up to their security requirements, (2) unable to afford meeting their security requirements, or (3) not aware of their security requirements (Srinidhi, Yan, & Tayi, 2015). As there is relatively more access to private sector cyber operators, most of our evaluation of current research is based on researchers' perception of private sector cyber defense and not the strategies of military CPTs.

Cyber Situational Awareness in Intrusion Detection

One of the most studied tasks is that of intrusion detection. In intrusion detection, the goal of the operator is to review and classify incoming alerts and network logs. This dynamic and uncertain environment tends to blend long periods of sustained attention looking at computer screens with periods of high cognitive load once an alert is detected. The act of compiling and fusing multiple sources of data (e.g., network traffic logs and alerts) to detect patterns (e.g., potential attacks) has been called cyber situational awareness (for recent meta-analyses see Franke & Brynielsson, 2014; Liu, Jajodia, & Wang, 2017; and Onwubiko & Owens, 2011). In the intelligence domain, a similar procedure has also been called sensemaking (Klein, Moon, & Hoffman, 2006).

Studies examining cyber situational awareness do not tend to use experimental manipulation, but instead rely on a range of other cognitive task analysis (CTA) methods, including observation coupled with directed interviews and questionnaires, usually performed over one or more scenarios (Liu, Jajodia, & Wang, 2017; Onwubiko &

Owens, 2011). The CTA methods used generally involve walking a team of experts through hypothetical scenarios of dilemma (e.g., a potential network intrusion; some taken from real-world prior examples) to find a common set of tasks, knowledge, and strategies that are used to complete the scenario. Goodall et al. (2009) used a form of CTA to interview 12 cyber operators of varying expertise who were all familiar with the open-source intrusions in the detection system Snort (Roesch. 1999). While they did not investigate how operators gained their expertise, they did identify that the requirement for situated expertise made intrusion detection a challenging task to transfer their expertise to other tasks in the cyber domains.

Champion, Jariwala, Ward, and Cooke (2014) administered a questionnaire (Study 1) and performed a cognitive task analysis (Study 2) to investigate the contribution of informal education to developing cybersecurity expertise. Perhaps unsurprisingly, 69 of 82 professionals reported that informal education supplementation was a necessary prerequisite for employment in cyber domains. Furthermore, 40 percent of professionals felt that job experience was the highest factor in positive performance rather than degree of knowledge/education (12 percent). Many professionals anecdotally reported that those receiving on-the-job training and mentoring exhibited the highest performance benefits as measured by future career success. Similarly, Asgharpour, Liu, and Camp (2007) found that operators who subjectively rated themselves with higher levels of expertise tended to have more competencies, which were more diverse than those with less self-professed expertise.

Mahoney et al. (2010) relied on a single subject matter expert to conduct a CTA on a set of five scenarios, including detecting a malicious insider threat attempting to penetrate a system. Mahoney et al. did not study expertise directly, but did attempt to decompose the job of cyber operators into tasks which could be analyzed in more detail in future research. Similarly, D'Amico and colleagues interviewed 41 cyber professionals across the commercial sector and the Department of Defense (D'Amico, Whitley, Tesone, O'Brien, & Roth, 2005). They identified several key task stages during intrusion detection including initial *triage* to weed out false positives, *escalation* by correlating patterns and trends in prior incidents with multiple and historical data sources, and finally multiple techniques for *attribution*. They also grouped their analytical processes into several broad stages in situation awareness including perception, comprehension, and projection (threat assessment). Perception included data acquisition and initial triage, comprehension involved situation assessments through escalation and correlation across multiple data sources, and projection included attribution by categorizing confirmed incidents in order to identify common features which can be used to both remediate the current threat and prevent future threats. Unfortunately, D'Amico et al. did not provide any quantification of situational assessments with which to justify their classifications.

In most of these CTA studies, only indirect measures of cognitive performance were collected, and expertise was not operationalized. Likewise, contrasts across different levels of proficiency were not made and so it is difficult to discern whether these are discriminating features of expertise. Some studies have modeled situation awareness computationally; however, the results of this research are generally not verified against human performance (Dutt, Ahn, & Gonzalez, 2011, 2013). Moreover, such studies assume that expertise is a function of accumulated instances in memory and that there is no cognitive re-organization that occurs through the processes of gaining expertise. In addition, there is a tendency to pre-define *expert models* as those with more instances in memory (i.e., more relevant knowledge). A limitation of these approaches is that they do not model the process of learning or otherwise acquiring expertise, and therefore are limited in their ability to make prescriptive conclusions about the nature of cyber expertise. Hence, only broad conclusions can be drawn from these studies. Still, much of this research provides an important first step in determining the right tasks for future researchers to use in operationalizing cyber expertise and in attempting to delineate the important cognitive processes involved in skilled cyber sensemaking and decision making. Future research that fruitfully operationalizes different kinds of expertise and critically analyzes past conclusions is required, something echoed in Rajivan and Cooke (2018).

In one of the first studies measuring human expertise at different skill levels on intrusion detection, Ben-Asher and Gonzalez (2015) developed a series of scenarios in which participants defended the network of a retail company. They simulated a stereotypical small corporate network including a public web server, private file server, and several computer workstations. This network simulated a firewall to prevent unwanted Internet connections from gaining access to the private file server and workstations. During a scenario, participants perceived a series of twenty network events (including seven that were malicious) presented on a simplified Snort-like interface (see Figure 31.1). A new event was displayed on the interface every ten seconds. The participants' goal was to determine whether each event was malicious and, at the end of the scenario, whether the series of events was representative of an ongoing attack. Participants could change their mind at any time about whether an event was malicious.

Events were presented in the form of logs from the intrusion detection system and a description of the event and an alert as to whether the event was suspicious. This included the network component name (e.g., web server, file server), active process

Is threat	ID	Alert	Description
☐	1		The web server is running ftpd and httpd services. The traffic is 3.3 Mbps between internet and web server, 3.3 Mbps between web server and file server, and 3.3 Mbps between web server and workstation.
☐	2	ftpd has started running on web server	The web server is running ftpd and httpd services. The traffic is 3.3 Mbps between internet and web server, 3.3 Mbps between web server and file server, and 3.3 Mbps between web server and workstation. An ftpd operation has been executed.
☐	3		The workstation is running a user process. The traffic is 3.3 Mbps between file server and workstation, and 3.3 Mbps between workstation and web server.

FIGURE 31.1 Sample scenario from Security Intrusion Detection System game.

Reprinted from *Computers in Human Behavior*, 48, Noam Ben-Asher and Cleotilde Gonzalez, Effects of cyber security knowledge on attack detection, pp. 51–61, doi.org/10.1016/j.chb.2015.01.039. Copyright © 2015 Elsevier Ltd.

(e.g., http, ftp), amount of network traffic between components (0 Mbps, 3.3 Mbps, 6.7 Mbps, and 10 Mbps), and whether an operation was accessed on the network component. In each scenario, normal traffic on the network was denoted as 3.3 Mbps. Participants were informed that the intrusion detection system may provide erroneous alerts such as a *false alarm* (i.e., incorrectly an alert when the event was benign) or a *miss* (i.e., have no alert to an event that was malicious). Although participants were not told, an attack (i.e., an event that was actually malicious) was operationally defined as *in progress* (or having occurred) when both an alert was presented *and either* there was irregular network traffic *or* an operation was executed (i.e., the system was designed to have no *misses*). In total there were five scenarios: website defacement, detection of a password sniffer/backdoor, DoS, stealing confidental data, and no attack present.

Ben-Asher and Gonzalez (2015) recruited 55 participants from the Carnegie Mellon University general student population as the novice group. In addition, 20 security professionals were recruited as the expert group. Each participant completed a questionnaire examining their domain knowledge including an understanding of technical terms and intrusion detection, and were asked demographic information about years of experience and understanding of network operation and network security including intrusion detection systems. The novice group completed each of the five scenarios twice ($N=10$) while the expert group only completed three randomly selected scenarios. They did not operationalize the professionals' expertise into different levels (e.g., journeyman, expert, or master).

Interestingly, there were no significant differences between the expert group and novice group in detecting whether a scenario was malicious (68 vs. 67 percent correct, 12 vs. 14 percent false positive, respectively). Further, the expert group was no more confident in their result (3.35 vs. 3.12 on a 1–5 low to high confidence Likert scale). The expert group was significantly more accurate at both correctly identifying the malicious events within each scenario (i.e., hits: 55 percent vs. 44 percent; $\chi(1, N = 3829) = 15.651$, $p < .001$) and correctly identifying benign events (i.e., correct rejections: 85 percent vs. 82 percent; $\chi(1, N = 8371) = 15.068$, $p = .024$). The novice group and expert group also generated different explanations to account for their cyber-attack decisions. Novice group participants tended to generate event-driven explanations (e.g., sustained increased network congestion) based on the Snort-like tool they used, while expert group participants tended to rely on more complex goal-directed explanations of the network behavior (e.g., an adversary wanting to use a workstation to install malicious code on a webserver). A similar pattern of behavior was also reported by Goodall, Lutters, and Komlodi (2004).

There are several lessons to be learned by examining this study. It is quite possible that many of the professionals were not experts, and were not very experienced in intrusion detection specifically. Although the *experts* were recruited from security jobs, the authors did not report to confirm their level of expertise or determine any individual differences, for instance, in experience on these type of tasks. Without operationalizing their expertise, it is possible that the actual proficiency of the so-called *experts* varied markedly and ranged between, for instance, junior apprentice and senior expert. The potential many gradations of expertise included in this study may further explain the subtle differences between the groups' performance. Therfore, rather than suggest that there is a clear expert–novice difference, the main take-away message from this study may be that cyber professionals who possess more technical experience appear to be somewhat better able to adapt to cyber environments—that are likely to have some similarity to their own work environments—than those with less technical proficiency (Thompson, Rantanen, & Yurcik, 2006). In the next section, we review the research on expertise in teams.

Teamwork in Intrusion Detection

While we have reviewed research examining individual cyber operators performing intrusion detection, in reality there are many roles and requirements, from the actual detection of an intrusion all the way through to remediation and attribution of the adversary. To offset this complexity, many operators will need to work in teams. As previously discussed, government cyber operators have the resources to work in properly staffed teams. If they need expertise in the Linux operating system, they can consult their team member with that expertise. In the private sector and other defense agencies, there is less teamwork and more working in silos (i.e., an operator with a supervisor with a fairly vertical chain of command). An issue with many current private sector teams of cyber analysts is that they tend to operate as a group of similarly talented individuals rather than as a group with diverse work roles and backgorunds (Champion et al., 2012).

Recent research has identified that cybersecurity teams are better able to solve complex tasks than individual analysts, potentially due to the distribution of expertise among analysts (Rajivan, 2014; Rajivan et al., 2013; Rajivan & Cooke, 2018; Rajivan, Janssen, & Cooke, 2013). For instance, performance on incident triage was highest with a diverse group of heterogeneous talents as opposed to a team with members of similar background and skills (Rajivan, Janssen, & Cooke, 2013).

Rajivan (2014) developed an experimental environment called CyberCog to measure collaboration during simulated network defense. CyberCog is a three-person synthetic environment that presents a series of security alerts to participants whose goal as a team was to correlate these alerts between each other and to categorize these alerts using multiple sources of information. These sources included network logs, a database of vulnerabilities, and a news feed. Thirty teams of three novice participants completed two scenarios. Participants received training to become familiar with computer networks, develop an understanding of how adversaries can compromise a network, and learn to communicate with others to share information. After conducting a familiarization trial, participants performed two test trials. During these trials, each participant was provided with a report containing eight security alerts generated during a suspected large-scale network intrusion attack, with each participant receiving a report from a different part of the network. In each report, four of the alerts were from *shared attacks*, where the same attack was occurring in multiple network locations. Two of these alerts were common to two of three team members, while two were available to all three team members. In addition, two alerts were from *unique attacks*, which were different attacks occurring at separate network locations but were part of a larger-scale attack pattern. This evidence required participants to pool their information and infer that the multiple attacks constituted a pattern. Finally, two alerts were from *isolated attacks*, which were different attacks occurring at separate network locations but were not part of the larger-scale attack pattern. Participants were not made explcitly aware that there were different kinds of alerts in each report.

Participants were instructed to discuss and pool their knowledge regarding the evidence available to each of them in order to detect incidents that were part of a large-scale attack, which included multiple types of attacks across several network locations. They were to report all of the attacks that were a part of the large-scale attack, but to not report isolated alerts not deemed part of the larger attack. Interestingly, participants spent over 60 percent of their time discussing evidence from shared attacks and only 15–17 percent of their time discussing evidence from unique attacks. A potentially more optimal solution would have been for participants to discuss events that were uniquely available to each team member (i.e., *unique* and *isolated attacks*), as this required the most complex information pooling to determine that the attacks were part of the larger overall attack pattern. In our experience, integrating disparate and isolated events are crucial when detecting large-scale attacks with multiple incoming and temporally extended kinds of network penetrations. Operators analyzing different parts of the network may find corroboratory evidence pointing to a common set of attacks reflecting a unique attack pattern.

While the literture on teamwork examines the role of individual knowledge-sharing and provides evidence for the right kinds of knowledge to share, more research needs to be done to determine how differing levels of expertise and areas of specialization influence knowledge sharing, and how this expertise may influence the use and efficacy of visualizations of network event data. Furthermore, integrating this research with the prior literature on cyber work roles may provide insight into more optimal team configurations and methods for optimizing efficacy.

In the last section, I will identify several challenges of past research and provide a roadmap for the future of research in cyber expertise. I will also recommend several projects that will hopefully spur new research and fill gaps in existing paradigms.

Future Directions for Cyber Expertise Research

A running thread of this chapter has been that cyber expertise research is in its infancy. There have been relatively few articles published discussing what constitutes cyber expertise, and of those, robust measures of skill, proficiency, and operational definitions of expertise were not utilized. Furthermore, research reports tend to fall into the trap of dichotomizing expertise into novice or expert rather than relying on far richer and useful proficiency scales that are available in the literature on expertise studies (see Hoffman, 1998) Assumptions are often made that professionals with some general cybersecurity experience are experts (or are treated as an expert sample) in other cyber domains without verifying their proficiency. In nearly all studied domains, genuine expertise is only achieved after many years of experience in one's actual job. One does not graduate from a program, cyber or otherwise, and qualify as expert (Hoffman et al., 2014).

Another challenge is that there is no single definition of what constitutes *cyber* as a domain or set of domains. I argue that this is due, in part, to the rapidly changing landscape of the cyber domains; technical research is often superseded by the time it is published. More research needs to be devoted to identifying sufficiency conditions for good cyber operators looking not only at the current workforce, but at future workforce requirements. In many cases, the current cadre of senior cyber operators have expertise focused on the past decade of cyber operations as opposed to the next decade. Finally, expertise research needs to go beyond technical skills to look at underlying aptitudes and social fit.

While I will make several recommendations for specific research projects, the general framework for research has already been addressed by Rajivan and colleagues (2013). They argued that we need to integrate what we know from cognitive task analyses into more realistic laboratory studies using synthetic task environments. This research should be supplemented with cognitive modeling to understand the underlying cognitive representations and conceptual changes that occur during the accumulation of cyber expertise.

Currently, there are no longitudinal studies of cyber operators. This is quite challenging outside the military domain as it is hard to identify potential candidates. Forming an alliance with the US Cyber National Mission Force to track recruits from the start of their cyber training through their military careers should be a high priority. However, first we need to understand what to track and determine whether there are ways of investigating tracking civilian operators. One such method of tracking would be to engage with cyber challenge competitions such as PicoCTF (Chapman, Burket, & Brumley, 2014) and to help develop challenge problems which not only measure technical ability, but also underlying cognitive competencies. A first such effort was conducted by Henshel and colleagues (2016), who measured questionnaire pre-test with actual team-based cyber exercise data. Requiring participants to login and use a supplied environment would afford researchers the most information as all interactions with the envrionment could be examined.

Another area for future research is to create realistic synthetic environments that are ecologically representative and ecologically salient, and in which cyber operations can be studied in detail. Much like simulation in aviation, surgery, and healthcare, numerous training environments of this type, called cyber ranges, have been developed and are becoming more popular. Cyber ranges offer valuable opportunities for operators to hone their skills but this may require cyber researchers to develop collaboration with those companies developing or using this technology in order to keep up with technological change. In a similar vein to the collaborations that have occurred at cyber competitions, this kind of collaboration would permit researchers and practioners to conduct experiments and capture observational data on the cyber operators using these environments (Trent, Hoffman, & Lathrop, 2016). One such way of doing this would be to integrate and extend tools and scenarios such as those used in Rajivan's (2014) CyberCog into these training environments. Of import is to go beyond Rajivan's three-player team to study realistic compositions of cyber operations teams of various skills.

Using immersive synthetic environments, it would be possible to run many smaller studies including multiple fast-fail pilot tasks (Lathrop, Trent, & Hoffman, 2016). Instead of trying to run fewer large and difficult-to-control experiments, using these ranges we could run a campaign of many smaller experiments and allow the complexity to fall naturally out of integrating these smaller studies together using meta-analyses.

A series of experiments which could be designed include determining the cognitive abilities utilized in the tasks identified in cognitive task analyses. This includes verifying that the tasks involved are reasonably unique tasks. An understanding of these underlying cognitive competencies would provide the necessary details to determine which tasks and work roles naturally align due to cognitive overlap. With this more robust understanding of the underlying competencies involved in work roles, it may be possible to further identify whether existing taxonomies of work roles and team composition are optimally designed and whether a standard baseline of skills and abilities for cyber operations could be developed. Tying this into different levels of expertise, it may be possible to track any changes (e.g., conceptual realignment) that occur throughout operators' careers.

Beyond understanding the conceptual realignment (if any) that occurs during the accumulation of experience is the understanding of how cyber professionals use their tools, and whether differences in expertise influence the kinds of information and degree of automation that tools should provide. In our experience and echoed by career cyber officers (Lathrop et al., 2016), there are many open-source tools (e.g., Snort and Bro) and plugins for these tools including ways of visualizing the incoming streams of data. To our knowledge, there have been no published usability studies or other human factors research on these visualizations. While Liu and colleagues (2017) recently released an edited volume that describes theory and models identifying some of the right kinds of information to present to operators, they do not operationalize how to optimally present this information to users, or to present different kinds of information to different users in the same team.

Another reason to focus on these tools and visualizations is that, at least in the military setting, many cyber operators have only a few years of experience. They are most likely the ones in need of decision support as they are not yet experts. Decision-support tools not only involve visualizations, but also include artificial intelligence. A better understanding of the cognitive underpinnings of cyber expertise would allow for the development of cognitive models which approximate these cognitive abilities. In the context of intelligence analysis and malware identification, I have previously adapted a cognitive model into a malware identification tool that was licensed by a large security provider (Thomson, Lebiere, Bennati, Shakarian, & Nunes, 2015).

Based on the research reviewed above, and my own experience of operating in this domain, I conclude with five assertions about the attributes needed to develop cyber expertise and build successful cyber teams. Cyber operators must:

1. Be systematic thinkers, able to conceptualize systems in their entirety as well as understanding internal variations within systems (D'Amico & Whitley, 2008).
2. Be able to work well in teams. The cyber domain is too broad, complex, and rapidly changing to be able to depend on a single individual (Choo, 2011; Krawczyk et al., 2013; Liu, Jajodia, & Wang, 2017).
3. Possess technical aptitude and curiosity that may not necessarily be captured by certificates or other traditional skill identifiers. Experience is one of the most prized attributes, and cyber training should include substantial on-the-job training comparable to the trades (Cook, 2014).
4. Have an insatiable need to solve puzzles and problems.
5. Be able to bridge the technical communications divide and be able to communicate effectively to non-expert decision makers (Dawson & Thomson, 2018).

Acknowledgements

I would like to thank Drs. Prashanth Rajivan, Aryn Pyke, Stoney Trent, Robert Hoffman, and MAJ Jessica Dawson for their comments and support on this chapter.

Disclaimer

The views presented and opinions expressed in this chapter are those of the author and do not represent the Department of Defense, the Department of the Army, or the United States Government.

References

Arachchilage, N. A., & Love, S. (2013). A game design framework for avoiding phishing attacks. *Computers in Human Behavior* 29(3), 706–714.

Asgharpour, F., Liu, D., & Camp, L. J. (2007). Mental models of computer security risks. In S. Dietrich, & R. Dhamija (Eds), *International conference on financial cryptography and data security*. Berlin: Springer.

Baker, M. (2016). *Striving for effective cyber workforce development*. Pittsburgh, PA: Software Engineering Institute—Carnegie Mellon University.

Bell, S. V. (2014). A longitudinal study of students in an introductory cybersecurity course. In *Proceedings of the 121st Annual ASEE Conference and Exposition*, Indianapolis, IN.

Ben-Asher, N., & Gonzalez, C. (2015). Effects of cyber security knowledge on attack detection. *Computers in Human Behavior* 48(1), 51–61.

Cable, D. M., & Parsons, C. K. (2001). Socialization tactics and person-organization fit. *Personnel Psychology* 54(1), 1–23.

Capalbo, N. R. (2011). RTFn: Enabling cybersecurity education through a mobile capture the flag client. In *The 2011 international conference on security and management*, Las Vegas, NV.

Caulkins, B. D., Badillo-Urquiola, K., Bockelman, P., & Leis, R. (2016). Cyber workforce development using a behavioral cybersecurity paradigm. In C. Connelly, A. Brantly, R. Thomson, N. Vanatta, P. Maxwell, & D. Thomson (Eds), *International conference for cyber conflict US*. West Point, NY: Army Cyber Institute.

Champion, M., Jariwala, S., Ward, P., & Cooke, N. J. (2014). Using cognitive task analyis to investigate the contribution of informational education to developing cyber security expertise. *Proceedings of the Human Factors and Ergonomics Society Annual Meeting* 58(1), 310–314.

Champion, M., Rajivan, P., Cooke, N. J., & Jarwala, S. (2012). Team-based cyber defense analysis. In *2012 IEEE international multidisciplinary conference on cognitive methods in situation awareness and decision support, CogSIMA 2012* (pp. 218–221). Piscataway, NJ: IEEE Press.

Chapman, P., Burket, J., & Brumley, D. (2014). PicoCTF: A game-based computer security competition for high school students. In *2014 USENIX Summit on Gaming, Games, and Gamification in Security Education*, San Diego.

Chase, W. G., & Simon, H. A. (1973). Perception in chess. *Cognitive Psychology* 4(1), 55–81.

Chen, L. C., & Cotoranu, A. (2013). *Enhancing the interdisciplinary curriculum in cybersecurity by engaging high-impact educational practices*. Pace University, Thinkfinity Center for Innovative Teaching, Technology and Research. Pace University Digital Commons.

Choo, K. K. (2011). The cyber threat landscape: Challenges and future research directions. *Computers & Security* 30(8), 719–731.

Cook, M. (2014). *Cyber acquisition professionals need expertise (But they don't necessarily need to be experts)*. Fort Belvoir, VA: Defense Acquisition University.

Cowan, C. (2003). Defcon capture the flag: Defending vulnerable code from intense attack. In *DARPA Information Survivability Conference and Exhibition*, Las Vegas, NV.

D'Amico, A., & Whitley, K. (2008). The real work of computer network defense analysts. In G. Conti, J. R. Goodall, & K. L. Ma (Eds), *Proceedings of the Workshop on Visualization for Computer Security* (pp. 19–37). Berlin: Springer.

D'Amico, A., Whitley, K., Tesone, D., O'Brien, B., & Roth, E. (2005). Achieving cyber defense situational awareness: A cognitive task analysis of information assurance analysts. *Proceedings of the Human Factors and Ergonomics Society* 49(3), 229–233.

Dawson, J., & Thomson, R. (2018). The future cybersecurity workforce: Going beyond technical skills for successful cyber performance. *Frontiers in Psychology* 9, 744.

Dodge, R. C. (2007). Phishing for user security awareness. *Computers & Security* 26(1), 73–80.

Dutt, V., Ahn, Y., & Gonzalez, C. (2011). Cyber situation awareness: Modeling the security analyst in a cyber-attack scenario through instance-based learning. *Data and Applications Security and Privacy* 25, 280–292.

Dutt, V., Ahn, Y.-S., & Gonzalez, C. (2013). Cyber situation awareness: Modeling detection of cyber attacks with instance-based learning theory. *Human Factors* 55(3), 605–618.

Eagle, C. (2004). *Capture-the-flag: Learning computer security under fire*. Monterey: Naval Postgraduate School.

Fontenele, M., & Sun, L. (2016). Knowledge management of cyber security expertise: An ontological approach to talent discovery. In *International conference on cyber security And protection of digital services*. London: IEEE.

Frank, C. E. (2016). Early undergraduate cybersecurity research. *Journal of Computing Sciences in Colleges* 32(1), 46–51.

Franke, U., & Brynielsson, J. (2014). Cyber situational awareness: A systematic review of the literature. *Computers & Security 46(1)*, 18–31.

Genge, B., Kiss, I., & Haller, P. (2015). A system dynamics approach for assessing the impact of cyber attacks on critical infrastructure. *International Journal of Critical Infrastructure Protection* 10, 3–17.

Goodall, J. R., Lutters, W. G., & Komlodi, A. (2004). I know my network: Collaboration and expertise in intrusion detection. In J. Herbsleb & G. Olson (Eds), *Proceedings of the 2004 ACM conference on computer supported cooperative work* (pp. 342–345). New York: Association for Computing Machinery.

Goodall, J. R., Lutters, W. G., & Komlodi, A. (2009). Developing expertise for network intrusion detection. *Information Technology & People* 22(2), 92–108.

Henshel, D. S., Deckard, G. M., Lufkin, B., Buchler, N., Hoffman, B., Rajivan, P., & Collman, S. (2016). Predicting proficiency in cyber defense team exercises. In *Military communications conference IEEE* (pp. 776–781). Baltimore: IEEE.

Hoffman, L. J. (2005). Exploring a national cybersecurity exercise for universities. *IEEE Security and Privacy* 3(5), 27–33.

Hoffman, R. R. (1998). How can expertise be defined? Implications of research from cognitive psychology. In R. Williams, W. Faulkner, & J. Fleck (Eds), *Exploring expertise* (pp. 81–100). London: Palgrave Macmillan.

Hoffman, R. R., Ward, P., DiBello, L., Feltovich, P., & Fiore, S. M. (2014). *Accelerated expertise: Training for high proficiency in a complex world*. Boca Raton, FL: Taylor and Francis/CRC Press.

International Telecommunication Union (2016). *ICT Facts and Figures*. Geneva, Switzerland: International Telecommunication Union.

Jajodia, S., Liu, P., Swarup, V., & Wang, C. (2010). *Cyber situational awareness*. New York: Springer.

Klein, G., Moon, B., & Hoffman, R. (2006). Making sense of sensemaking 1: Alternative perspectives. *IEEE Intelligent Systems* 21(4), 70–73.

Krawczyk, D., Bartless, J., Kantarcioglu, M., Hamlen, K., & Thuraisingham, B. (2013). Measuring expertise and bias in cyber security using cognitive and neuroscience approaches. In K. Glass, R. Colbaugh, A. Sanfillippo, A. Kao, M. Gabbay, C. Corley, ... A. Yaghoobi (Eds), *2013 IEEE International Conference on Intelligence and Security Informatics* (pp. 364–367). Piscataway: IEEE.

Lathrop, S. D., Trent, S., & Hoffman, R. R. (2016). Applying human factors research towards cyberspace operations: A practitioner's perspective. In D. Nicholson (Ed.), *Advances in Human Factors in Cybersecurity* (pp. 281–293). Gewerbestrasse, Switzerland: Springer.

Levina, A., & Nidiffer, J. (1996). *Beating the odds: How the poor get to college*. San Francisco: Institute of Education Sciences.

Liu, P., Jajodia, S., & Wang, C. (2017). *Theory and models for cyber situation Awareness.* Gewerbestrasse, Switzerland: Springer.

Mahoney, S., Roth, E., Steinke, K., Pfautz, J., Wu, C., & Farry, M. (2010). A cognitive task analysis for cyber situational awareness. *Proceedings of the Human Factors and Ergonomics Society Annual Meeting* 54(4), 279–283.

Newhouse, B., Keith, S. S., & Witte, G. (2016). *NICE Cybersecurity Workforce Framework.* Gathersburg, MD: National Institute of Standards and Technology.

Onwubiko, C., & Owens, T. J. (2011). *Situational awareness in computer network defense: Principles, methods and applications.* Hershey, PA: Information Science Reference.

Pfleeger, S. L., & Caputo, D. D. (2012). Leveraging behavioral science to mitigate cyber security risk. *Computers & Security* 31(4), 597–611.

Rajivan, P. (2014). *Information pooling bias in collaborative cyber forensics.* Doctoral dissertation, Arizona State University.

Rajivan, P., & Cooke, N. (2018). Information pooling bias in collaborative security incident analysis. *Human Factors* 60(5), 626–639.

Rajivan, P., Champion, M., Cooke, N., Jariwala, S., Dube, G., & Buchanan, V. (2013). Effects of teamwork versus group work on signal detection in cyber defense teams. In D. D. Schmorrow & C. M. Fidopiastis (Eds), *AC/HCII* (pp. 172–180). Berlin: Springer.

Rajivan, P., Janssen, M. A., & Cooke, N. J. (2013). Agent-based model of a cyber security defense analyst team. *Proceedings of the Human factors and Ergonomics Society Annual Meeting* 57, 314–318.

Rajivan, P., Moriano, P., Kelley, T., & Camp, J. (2017). Factors in an end user security expertise instrument. *Information and Computer Security* 25(2), 190–205.

Rid, T., & Buchanan, B. (2015). Attributing cyber attacks. *Journal of Strategic Studies* 38(1), 4–37.

Roesch, M. (1999). Snort-lightweight intrusion detection for networks. In *Proceedings of the 13th USENIX Conference on System Administration* (pp. 229–238). Seattle, WA: USENIX Association.

Seong, J. Y., Kristof-Brown, A. L., Park, W. W., Hong, D. S., & Shin, Y. (2015). Person-group fit: Diversity antecedents, proximal outcomes, and performance at the group level. *Journal of Management* 41(4), 1184–1213.

Shillair, R., Cotten, S. R., Tsai, H. Y., Alhabash, S., LaRose, R., & Rifon, N. (2015). Online safety begins with you and me: Convincing internet users to protect themselves. *Computers in Human Behavior* 48(1), 199–207.

Shin, J., Son, H., Khalil, R., & Heo, G. (2015). Development of a cybersecurity risk model using Bayesian Networks. *Reliability Engineering and System Safety* 134(1), 208–217.

Srinidhi, B., Yan, J., & Tayi, G. K. (2015). Allocation of resources to cyber-security: The effect of misalignment of interest between managers and investors. *Decision Support Systems* 75(1), 49–62.

Thompson, R. S., Rantanen, E. M., & Yurcik, W. (2006). Network intrusion detection cognitive task analysis: Textual and visual tool usage and recommendations. *Proceedings of the Human Factors and Ergonomics Society Annual Meeting* 50(5), 669–673.

Thomson, R., Lebiere, C., Bennati, S., Shakarian, P., & Nunes, E. (2015). Malware identification using cognitively-inspired inference. In *Proceedings of the 24th Annual Behavior Representation in Modeling and Simulation Conference.* Washington, DC: Springer.

Trent, S., Hoffman, R. R., & Lathrop, S. (2016). Applied research in support of cyberspace operations: Difficult but critical. *Cyber Defense Review.*

Wagner, P. J. (2004). Designing and implementing a cyberwar laboratory exercise for a computer security course. *ACM SIGCSE Bulletin* 36(1), 402–406.

Werther, J. Z. (2011). Experiences in cyber security education: The MIT Lincoln laboratory capture-the-flag exercises. In *Proceedings of the 4th Confrence on cyber security experimentation and test* (pp. 2–12). San Francisco: USENIX Association.

Yurcik, W., Barlow, J., & Rosendale, J. (2003). Maintaining perspective on who is the enemy in the security systems administration of computer networks. In *ACM CHI Workshop on System Administrators Are Users* (pp. 345–347). New York: Association for Computing Machinery Press.

CHAPTER 32

EXPERTISE IN INTELLIGENCE ANALYSIS

MICHAEL P. JENKINS AND
JONATHAN D. PFAUTZ

INTRODUCTION

To conduct an intelligence analysis (IA), analysts perform multiple perceptual and cognitive tasks to understand the past, present, and/or future states of some real-world environment or situation. They attempt to *illuminate the future* (Taylor, 2005) and provide *actionable intelligence* (Scholtz et al., 2006) by compiling, reviewing, and processing as much information as they can. The purpose of IA is to formulate interpretations of often ambiguous, substantive problems based on both available information and analysts' experience (Tam, 2009). IA is unlike many other analytical processes in that making a decision and taking action based on that decision is not the immediate goal (Hutchins, Pirolli, & Card, 2004). Instead, analysts filter, validate, associate, and summarize massive amounts of interrelated information relevant to a real-world scenario and send the results to one or more consumers, in some cases without even knowing who those consumers are (who are often remote). The consumer then takes action (and faces the consequences). The real-world scenario being studied must be considered on a dynamic and comprehensive scale (Mangio & Wilkinson, 2008)—"the world is literally its province" (Platt, 1957, p. 16). As time passes during an analyst's investigation, the situation will change, therefore requiring they anticipate these changes or continually review and revise their situation assessment to maintain an accurate representation of the state of the world. Because of the important nature of IA work and how difficult and complicated it is to generate accurate representations, IA has drawn considerable attention from researchers seeking to identify what makes IA tasks challenging and what defines an expert in the field.

Although the tasks and cognitive processes that analysts carry out have remained relatively unchanged over the past half-century (as this chapter will illustrate), research on the challenges, complexities, and constraints that make performance difficult has evolved, providing a better understanding of the key characteristics that define expertise in this field. The goal of this chapter is to highlight the unique characteristics of expert IA practitioners, compared to other domains. We highlight a range of past research that characterizes the domain, including research that focuses on workflow as well as other perspectives that focus on individual analyst information processing. These characterizations allow us to highlight different elements of expert behaviors that emerge throughout the IA process. We conclude by detailing the implications of these different elements of expert behaviors on emerging research directions and the design of new assistive technologies to augment analysts executing modern and next generation IA.

INTELLIGENCE ANALYSIS OVERVIEW

Hughes and Schum (2003) present one possible framing of the IA domain, noting the IA process is always composed of three parts: (1) hypotheses generation; (2) evidence gathering and evaluation; and (3) generation and evaluation of arguments linking evidence and hypotheses. These components emerge within any robust analysis regardless of the specifics of the task, domain of application, or an individual's level of expertise. This perspective has been widely accepted.

Predating this perspective, but still in alignment, Kent (1965) presented one of the earliest attempts to define a model of the process of IA. This seven-stage process characterized a flexible yet encompassing workflow that moves from the appearance (Stage 1) and understanding (Stage 2) of the substantive problem, to the collection (Stage 3) and evaluation (Stage 4) of data to support generation of a new hypothesis to address the problem as the analyst understands it (Stage 5), to evaluating and testing one or more competing hypotheses (Stage 6), to the final stage (Stage 7)—generating an artifact to communicate the analyst's key findings and operational recommendations. This workflow underlies subsequent models and characterizations presented over forty years later. For example, one of the most widely referenced characterizations of the cognitive processes and challenges of IA is presented by Heuer (1999) in his seminal book, *Psychology of Intelligence Analysis*. The steps in the IA process described in this work include: (1) defining the problem; (2) generating hypotheses; (3) collecting information; (4) evaluating hypotheses; (5) selecting the most likely hypothesis; and (6) ongoing monitoring and integration of new information. The similarities between these two characterizations are readily apparent despite Heuer's being based on decades of experience and research that culminated with his seminal publication thirty-five years later. However, due to their similarity, they both have the same two shortcomings, which arise from their task-based characterization of IA: they fail to consider how the steps required to perform IA tasks align with human capabilities, and they fail to describe the cognitive processes and strategies of expert analysts.

In the early 2000s, there was a shift in the research community toward investigating these capabilities and cognitive processes of expert analysts to better understand what makes IA expertise difficult to obtain and maintain (e.g., see Badalamente & Greitzer, 2005; Elm et al., 2004; Elm et al., 2005; Greitzer, 2005; Grossman, Woods, & Patterson, 2007). This shift was largely motivated by: (1) advances in intelligence, surveillance, and reconnaissance (ISR) capabilities (i.e., sensor improvements that enabled capture of larger volumes of high-resolution data); (2) improved digitization and broader sharing of data (i.e., improvements in global network transmission efficiencies, data compression algorithms, and storage mechanisms); and (3) an increase in conflicts around the world that lack historical precedent. As a result, intelligence analysts were faced with situations where they had increasing volumes of information to process and decreasing timelines to produce results, without any increase in staff or resources to support their work. At the time, the US Army was also transitioning to a *reachback model* for IA support, where commanders in the field issued information requirements (IRs) that were routed to remotely located intelligence shops with specialist analysts. This shift introduced latency into the process that compounded the issue of fixed critical deadlines for intelligence delivery. The reachback model also introduced ambiguities for the analysts, who now had to decipher IR requests routed to them from a remote customer with whom communications were limited (i.e., there was an increased level of uncertainty with respect to the commander's intent and the operational context). While the anticipated benefits were compelling (e.g., more detailed expertise could be focused on a given problem, rapid retasking, and staff reallocation to new requests and priorities), this shift made requests for clarification challenging and costly and added pressure to delivery timelines.

As a result, the intelligence community and stakeholders sought new processes, training, and technologies to support and/or enhance analyst performance. This motivated the research community to refocus and adopt new methods for characterizing the challenging dimensions of IA. Methods began to look more holistically at the limitations and affordances of analysts, supporting technology, operating environment, and cognitive tasks being executed within the work domain. Characterizations better suited for identifying and framing technology requirements that support modern IA started to appear. As these characterizations matured and were validated, they provided the foundation with which to highlight and define expertise in the field.

STUDIES OF EXPERTISE IN INTELLIGENCE ANALYSIS

Expertise in IA is a fluid construct that increasingly focuses on the meta-cognitive characteristics of an individual (e.g., strategies an individual adopts to apply a skill, awareness of self-limitations, ability to appropriately calibrate confidence in the result of a skill-based outcome) than their cognitive (e.g., technical; cf., meta-cognitive) skills and knowledge. These meta-characteristics heavily influence the expertise construct, and their applicability can change based on the context of a given analysis.

There is currently no singular field of study, no critical domain of knowledge, and no single technical specialization[1] that will produce an expert intelligence analyst. This is not unlike the work of government policy professionals (especially, generalists) as described by Conway and Gore (see their Chapter 48, "Framing and Translating Expertise for Government and Business," this volume). Therefore, to understand what meta-characteristics of an individual result in expertise, we must understand the metacognitive *and* cognitive requirements of IA that are most salient in a given operational and analytical context. In this chapter, we will not only establish a foundation for expertise, but will also review the methods applied over the past two decades to extract these requirements in an attempt to reveal which factors contribute to IA expertise.

To investigate these IA requirements, methods that largely originated for studying supervisory control systems were employed. This has included, for instance, Rasmussen's (1981, 1983) seminal work on mental models, his human performance models of power plant operators during diagnosis, and the plume of research that grew into the field of *cognitive systems engineering* (see Hollnagel & Woods, 2006; Rasmussen, Pejtersen, & Goodstein, 1994; Woods & Roth, 1988). The cross-application of these methodologies was appropriate given the many similarities that IA shares with supervisory control domains (e.g., complex interconnected systems of study, high degree of risk of catastrophic failures, distal view of system status indicators resulting in uncertainty, short decision timeline). However, as Patterson, Woods, Tinapple, & Roth (2001) described, there are important distinctions between the two domains: in IA, analysts are responsible for monitoring a mix of technological and sociocultural systems whereas those in domains where work can be described as a supervisory control task engage primarily with intentionally engineered systems. Patterson, Roth, and Woods (2001) support this distinction, pointing out that supervisory control tasks support operator interventions with alarms that draw operator attention to relevant events, while IA tasks do not.

Despite these and other differences, cognitive task analysis (CTA) methods such as cognitive work analysis (CWA), applied cognitive task analysis (ACTA), critical decision method (CDM), and the critical incident technique (CIT), among others, have successfully been adapted and applied to the study of IA (for further information on

---

[1] Intelligence analysts do require a basic minimum technical specialization. For example, communications intelligence (COMINT) analysts focus on analysis of textual messages or voice communications and as a result must have the technical skills to understand how to decompose that type of information. Likewise, as Clark (2016) highlighted, there are different models of skill expertise (e.g., mathematics, text, visual models, targets) that are required for different analyst roles. Our discussion here, however, assumes the individual has the core technical skills to understand the type of information they are tasked with analyzing, as other chapters in this *Handbook* cover the development of core technical skills of this nature.

these methods, see Chapter 19, "Incident-Based Methods of Studying Expertise," by Militello & Anders, this volume; and Chapter 20, "Cognitive Work Analysis: Models of Expertise," by Burns, this volume). Across these investigations, the consensus is that, from a cognition viewpoint, IA is about performing inferential analysis to determine the "best explanation for uncertain, contradictory, and incomplete data" (Patterson, Roth, & Woods, 2001, p. 225). As Hajdukiewicz, Burns, Vicente, & Eggleston (1999) point out, the results of IA represent the best but not the most certain explanation for a situation given dynamic, incomplete, potentially deceptive (also Halpern, 2001), and, as a result, *uncertain* evidence (Patterson, Roth, & Woods, 2001). Given the abductive nature of IA, a successful analysis process must compare analyst hypotheses to consider many competing explanations for the available data, eventually converging on one (or several) best explanations justified by available evidence (Heuer, 1999).

To understand the structures, constraints, and complexities in IA that make the development of expertise challenging, we must first understand what defines success. A natural metric of success is an analysis that results in 100 percent valid and complete explanation of the substantive problem. However, given the characteristics of the domain, including high levels of uncertainty and missing information, this standard is rarely achieved. Instead, success in IA is often characterized by convergence (Grossman et al., 2007). Elm et al. (2005) characterized effective convergence within the IA process as "a stable balance of applying broadening and narrowing" (p. 2) "to focus on a reduction of the problem towards an answer" (p. 1). Given this definition, expert analysts' efforts are typically characterized by multiple cycles of broadening (i.e., divergence) and narrowing (i.e., convergence), based on available resources (e.g., time, information, cognitive capacity) to avoid premature closure (e.g., cognitive fixation) and arrive at a set of final hypotheses that best explain the substantive problem.

To better understand this broadening and narrowing cycle, Roth et al. (2010) conducted a CTA of the intelligence analysis process with the help of twenty-two Army intelligence analysts. Their results showed that analysts engage in three primary tasks during this cycle. These tasks, all of which eventually narrow the analysis, include: (1) down collect—acquiring information relevant to the information request or question being answered (narrowing the volume of data for consideration); (2) conflict and corroboration—exploring and analyzing collected data to uncover unexplained relationships (narrowing the interrelationships of the data to identify salient and anomalous features); and (3) hypothesis exploration—developing hypotheses and narratives using collected and corroborated data to fill in the gaps posed by the unexplained relationships (narrowing toward a solution to the substantive problem). In addition, expert analysts also engage in broadening cross-checks, preventing premature closure by considering alternative hypotheses (Heuer, 1999; Roth et al., 2010). By combining these broadening cross-checks with the functions designed to narrow in on the best information, Elm et al. (2005) produced a model of IA that can serve as a general construct for tracking cognitive processes and development of expertise.

Now that we have defined success, ultimately, as a process that produces converging evidence, and have identified the broadening and narrowing process that achieves it, the next step to understand the cognitive tasks and limitations of IA is to characterize the cognitive and metacognitive processes employed in IA work. To date, however, research into these IA processes has focused primarily on cognitive processes as a tool to characterize and identify challenges within the IA domain. One of the first attempts to understand the processing in an analyst's mind during IA was presented by Katter, Montgomery, and Thompson (1979) as a model of low-level information processing components handling a flow of incoming information. This model is similar to the widely accepted information processing model of decision making presented by Wickens and Hollands (2000), but without the components of choice and execution. These missing components do not exist in Katter et al.'s model because the goal of analysis is to gain an understanding of a situation, not to take action. The models are similar because decision making also requires this situational awareness prior to (and after) action. Katter et al.'s model of information processing helps characterize expertise requirements in IA by describing a three-step cyclical process. First, initial processing of information from the environment is preattentive. Preattentive processes are automatic, and rapidly and accurately compare raw sensory information against stored information patterns (see Treisman & Gelade, 1980, for more information on preattentive processing theory). Second, initial meanings are formed that remain outside analyst awareness (based on triggered patterns of information which subsequently form the basis for generating awareness and guiding attention). Third, central cognitive functions take over processing in a continuous cycle of comparing incoming information against a network of information in the analyst's memory and then updating memory based on any matches or associations in the memory network (in a manner similar to that described by Anderson, 1983). As more information is perceived, the cycle continues, i.e., additional information is perceived, understood, and integrated, to increase the analyst's understanding of the situation.

This model illustrates how IA is similar to the early stages of decision making where humans attempt to first understand the problem, then continuously refine and expand their understanding until they can draw on analogous experiences to generate a potential solution. This generation of potential solutions is equivalent to developing multiple hypotheses to explain the IA problem. This comparison is valuable because it allows the extensive body of literature on human decision making to be applied to IA. For example, limitations of human processing capabilities such as short-term memory limits (Cowan, 2001; Miller, 1956) that have been empirically studied in psychology can also be applied to IA. Cowan (2001) demonstrated that raw short-term memory capacity is around 4 ± 1 pieces of information when considering both storage and processing. However, skilled individuals (who are often experts) acquire and employ strategies to *group* related information in to patterns (e.g., chunks, retrieval structures, slotted schemas). That is, these individuals practice cognitive strategies in which they relate pieces of information, circumventing short-term memory limitations by mentally representing relevant patterns as a single unit of information, often by associating pieces of information with representations already stored in long-term memory. These expert memory-based mechanisms (sometimes referred to as long-term working memory (see Ericsson & Kintsch, 1995; Ward, Ericsson, & Williams, 2013) can extend short-term working memory capacity (see also Neath & Surprenant, 2003, Tulving & Craik, 2000; for more information see Chapter 2, "The Classic Expertise Approach and Its Evolution," by Gobet, this volume).

However, as the volume of available and potentially relevant information continues to increase, analysts are more likely to encounter information overload (e.g., fan effect, see Anderson, 1974; Drucker, 2006). This type of information overloading may increase the likelihood that analysts will adopt satisficing strategies, such as heuristics (e.g., see Simon, 1956) or kluges (e.g., see Koopman & Hoffman, 2003); sometimes inelegant but often workable and effective shortcuts. In the context of IA, these can short-circuit the broadening-and-narrowing process and help analysts converge more quickly on what they believe is the best hypothesis. Much like experts operating in other complex environments, it is inevitable that intelligence analysts will resort to using such *heuristics* given the frequent ambiguity, uncertainty, and time pressure that characterize complex domains in general, and IA specifically. A key distinction between expert and non-expert IA analysts therefore becomes the ability to apply the heuristics best suited to the context of the problem, available information, goals of the analysis, and even the analyst's own strengths and weaknesses. Unfortunately, especially among non-experts, these same heuristics can lead to negative biasing (Kahneman, Slovic, & Tversky, 1982). These biases have been observed in IA tasks in both empirical and operational settings (Heuer, 1999), and significantly increase the chance of a flawed analysis. Accounting for biases is one of the major challenges facing analysts en route to becoming experts who are trying to perform robust and successful analyses.

Combining the surface-level models of IA presented by Heuer (1999) and Kent (1965), the primary functional tasks presented by Roth et al. (2010), the framework of broadening and narrowing developed by Elm et al. (2005), and the low-level model of information processing that occurs during IA from Katter et al. (1979), we can safely conclude that all of these descriptions support the belief that the overall cognitive task of IA is to develop a situational understanding of some real-world environment or situation. Both the tasks of IA and the cognitive processes of analysts support the development of this situation awareness (SA). SA is defined by Endsley (1988) as "the perception of the elements in the environment within a volume of time and space, the comprehension of their meaning and the projection of their status in the near future" (pp. 97–101). Analysts with highly developed SA around the information being analyzed can produce more accurate hypotheses of the substantive problem. To help understand how an analyst's SA evolves, Endsley (1996) presents three distinct levels of SA (see Table 32.1).

To *illuminate the future*, an analyst must achieve a state of awareness where he/she can accurately predict the state of the future within the contextual bounds of his/her substantive problem and analytical focus (i.e., Level 3 SA). Understanding the process of achieving this level of awareness, combined with the previous understanding of the stages of IA, will help to reveal the structures, constraints, and complexities in IA that make achieving expertise challenging.

Table 32.1 Endsley's (1999) three distinct levels of situational awareness a person can achieve, most often in sequence

Level	Expert behavior/skill
1	Perceiving critical factors in the environment
2	Understanding what those factors mean, particularly when integrated together in relation to the person's goals
3	Understanding what will happen in the near future

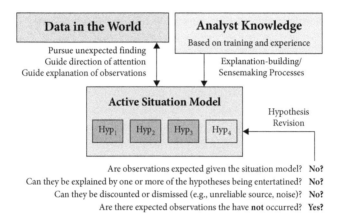

FIGURE 32.1 Overview of cognitive processes involved in generating situational awareness.

Adapted from Jonathan Pfautz, Ted Fichtl, Sean Guarino, Eric Carlson, Gerald Powell, and Emilie Roth, "Cognitive Complexities Impacting Army Intelligence Analysis," *Human Factors and Ergonomics Society Annual Meeting Proceedings*, 50 (3), pp. 452–456, doi.org/10.1177/154193120605000351
Copyright © 2006, © SAGE Publications

Pfautz et al. (2006) provide an overview of the cognitive processes involved in generating SA within IA (see Figure 32.1). This model was based on a CTA and in-depth interviews of intelligence analysts. In it situational awareness is developed by integrating the analyst's prior knowledge with incoming information from the world. During this process of developing SA (i.e., sensemaking), the analyst begins with an initial understanding of the situation and determines whether the incoming information meets expectations. If it does, then the information is integrated into the situation model as a new piece of evidence. If it does not, then the analyst either ignores the information (e.g. writing it off as unreliable) or revises their mental model to incorporate the disparate piece of information. Because this process is largely driven by the analyst's expectations, it is normally a top-down cognitive process. The analyst knows how to direct their attention appropriately to either fill in the gaps or explain the information away. Pfautz et al. (2006) also suggested that bottom-up processes exist similar to those of Katter et al.'s (1979) model, where bottom-up processes interpret and integrate incoming information into the analyst's understanding. A challenge in describing and attempting to synthesize the results of all of this research on the IA domain is the lack of a consistent approach to framing human cognitive and perceptual processes. This challenge is not fundamental to cognitive engineering processes (such as CTA, CWA, etc.), but rather to the ongoing problem of identifying how best to interpret general human behavior, establish task and perceptual performance, and incorporate increasingly researched neurocognitive processes into a cohesive form. This broader epistemological problem is as apparent in this summary of approaches to understanding expertise in IA as it is in other domains. Overall, Pfautz et al. concluded that the analyst carries out the sensemaking process by attempting to explain and integrate newly acquired pieces of information (and updated pieces of existing information) based on expectations driven by the analyst's current SA, which is motivated by their efforts to vet hypotheses currently under consideration (i.e., to drive the convergence process to closure).

If we accept that sensemaking is the primary cognitive task of IA, we can frame expertise by focusing on those elements that make the broadening and narrowing cycle that drives the sensemaking process challenging. Pinelli et al. (1993) and Pirolli (1993, 2006) developed a notional model of analyst sensemaking based on a CTA that provides arguably the most inclusive framework for IA. It integrates the previously discussed models and components into a multilevel model characterized by varying degrees of information structure and analyst effort. This model has sixteen different stages, from initial search and filtering of information, to building cases to support hypotheses, to the final presentation of a product to the end consumer(s). While these stages are presented in an ordered fashion, each stage presents an opportunity to revert back to previous stages by combining top-down and bottom-up processes, similar to the Elm et al. (2005) model of narrowing and broadening. The bottom-up processes of the framework incorporate new information into the case for the analyst's current hypotheses or motivate the creation of new hypotheses.

Viewing IA expertise from the perspective of it being a process, an expert analyst can be defined as one who can fluidly progress through the process of continuous and balanced narrowing and broadening en route to organizing and providing structure to information. This then enables that expert analyst to achieve a level of SA with well-calibrated responses to information requests given available time, resources, and information. Adopting this broad perspective of expertise, we can leverage Pirolli's (2006) model to characterize the skills that define expertise. Consequently, in the following section we characterize expert behavior for each stage of Pirolli's model of sensemaking in IA.

Pirolli's (2006) Model of Sensemaking in IA

Search and Filter. In this stage the analyst searches for data and filters it for relevance. This incorporates the data collection stage of the Kent (1965) and Heuer (1999) models, the down collect process of Elm et al. (2005) and Roth et al. (2010), and development of the ability to perceive critical factors in the information environment (Endsley, 1996; Level 1 SA), and is guided by the prior knowledge expectations described in Pfautz et al. (2006). Success in this stage is driven by the ability to know which *sources* have the greatest potential for producing relevant information, how to access those sources, and how to efficiently identify relevant information amidst any noise. Expert behavior is likely to come simply with experience as analysts learn what sources are available and begin to build trust in those that are most reliable and produce the most salient relevant information.

Read and Extract. In this stage analysts review the data to extract pieces of supporting or disconfirming evidence. This stage incorporates the data evaluation component (Stage 4) of the Kent (1965) and Heuer (1999) models, the conflict and corroboration process of Elm et al. (2005) and Roth et al. (2010), further refinement of analysts' ability to perceive critical factors in the information environment (Level 1 SA), and understanding what those factors mean in relation to the substantive problem (early Level 2 SA), and is guided by expectations based on the analyst's active hypotheses in terms of Pfautz et al.'s (2006) model. This stage can also trigger new hypotheses based on evidence that fails to meet the expectations of active hypotheses. Success in this stage is efficiently reviewing information and extracting evidence aligned (confirming or disconfirming) with an open question or hypothesis. Expert behavior is balancing multiple active hypotheses and questions and effectively comparing evidence against each in parallel.

A more interesting aspect of expert behavior is when analysts can dynamically and appropriately calibrate their trust in a given piece of information based on surrounding uncertainties (e.g., pedigree, staleness). An expert analyst maintains a self-awareness of their trust in available sources, and actively seeks to continuously confirm/disconfirm and update that level of trust over time. In contrast, a novice analyst may rely on the same sources based on an initial level of trust, which may have resulted from a previous analysis where it was found to be reliable at that time.

Schematize. In this stage, analysts organize incoming evidence into small-scale stories or narratives to prepare the raw data for integration. This stage is not directly called out in other IA models, because it is characterized by data organization, which these models do not explicitly address. The focus of this stage—organizing the data for integration with an active hypothesis or supporting narrative—is in many ways a framing effort. Success in this stage is identifying coherent relationships among evidence and active hypotheses or narratives. Expertise requires that the analyst self-evaluate whether a narrative they are constructing has a high degree of coherence (i.e., it is internally consistent with what the analyst knows and believes) and correspondence (i.e., the analyst's internal model is accurately aligned to the real world) (Hammond, 2000).

Build a Case. In this stage, analysts integrate supporting evidence into existing hypotheses, generate new hypotheses, or eliminate hypotheses when information does not support them. This stage incorporates the hypothesis generation and evaluation/comparison stages (Stages 5 and 6) of the Kent (1965) and Heuer (1999) models, the hypothesis exploration process of the Elm et al. (2005) and Roth et al. (2010) models, and the integration of information to support or refute active hypotheses (Level 2 SA), while also beginning to predict future outcomes (Level 3 SA) when there is convergence of information. At a minimum, success in this stage is forming feasible (and ideally, well-supported) hypotheses given available information. Expert behavior occurs when an analyst can generate multiple competing hypotheses (see Heuer's, 1999, analysis of competing hypotheses (ACH) method for IA). These competing hypotheses may contradict each other. Assessing each competing narrative creates a more robust analysis that attempts to avoid premature closure by considering a wider range of plausible alternatives. Experts also know when to narrow versus broaden their analysis. They must maintain awareness of available information (how it will evolve, how much can it be feasibly considered given resource constraints, and how to prioritize it for analysis), deadlines for a given analysis (to avoid broadening to the point where reasonable conclusions cannot be vetted given available time and resources), and when to broaden an analysis. Knowing when to broaden is arguably one of the most challenging skills for an analyst because it requires a meta-analysis of the active hypotheses, mental model, and SA of a given problem space to determine what may be flawed or biased. Expert analysts continuously critique their own assessments, assumptions, hypotheses, and narratives, while knowing when it is appropriate to take action and delve deeper into validation or exploration of alternatives, without crippling their ability to make progress. As Hetzler and Turner (2004) point out, this is an exercise in satisficing, where prematurely focusing on a piece of evidence or hypothesis can lead to unduly weighted evidence or conclusions.

Tell a Story. In this stage, analysts organize and prepare evidence and hypotheses to create a robust and supported narrative that addresses the original substantive problem that motivated the analysis. Success in this stage is aligning the competing hypotheses and their supporting evidence into a deliverable, artifact, or response that effectively answers the question. Expert behavior is positioning the analysis, evidence, and conclusions in a way that enables commanders and staff to make decisions and take actions. Expert analysts understand that this is achieved by incorporating context into their tasks, looking beyond the substantive problem as stated and taking into account implicit but relevant characteristics of the operational scenario. An effectively contextualized deliverable facilitates decision making. In essence, expertise is understanding the target audience's motivations and needs for requesting a piece of intelligence, then tailoring a response to meet those needs. For example, the commander who requests data on the length of a runway, but really wants to know whether a transport aircraft could land there would be better served by the analyst who can understand why commanders ask that question and can inject appropriate context when crafting their response. Therefore, expertise in IA relates to the analyst's familiarity with the area of operations and their relationship with the requesting individual and other stakeholders who will rely on their intelligence report. This socio-organizational expertise is often reflected by knowing how (i.e., what communications channels are effective/appropriate), knowing when to act (i.e., recognizing when a request becomes stale, relevant data needs updating, and recommended intelligence loses its value), and knowing what questions to ask to clarify a substantive question and ground it in the operational need (i.e., aligning to the commander's intent).

With our discussion of critical, bottom-up driven processes complete, we now turn to those driven by top-down processes. All of the top-down driven processes complement the bottom-up processes. These top-down processes are characterized by expectation-driven searching, based on the analyst's current hypotheses or prior experience. An expert analyst effectively draws analogies from prior experience, even in cases where relationships may not be readily apparent. For intelligent analysts, this is one of the most critical expert-level skills. Intelligence analysts are not focused on mastery of a rigid and discrete process, but rather how to successfully navigate and progress at appropriate times through a process, given changing circumstances and context of the domain and problem. For the top-down processing aspects of this process, an expert analyst not only possesses a range of experiences (which can only occur with time), but also effectively recalls those experiences, identifying emerging patterns that relate to one or more active substantive problems they are working, and drawing analogies between those patterns, the problem, active (and potentially previously rejected) hypotheses, and available evidence. These analogies must also be formed in a way that facilitates further action, whether that is targeted narrowing (e.g., "I've seen a situation similar to this one before so I know what to look for") or broadening (e.g., "given what we know, I can think of four different outcomes that are plausible and worth exploring given similar scenarios I've seen in the past").

The Pirolli and Card (2005) sensemaking model incorporates a low-level view of the cognitive processes used during sensemaking, leaving the primary analysis portion (i.e., the deeper cognitive work) largely unexplored. For example, the role of memory in driving expectations (described by Pfautz et al., 2006), creating new hypotheses, and integrating incoming pieces of information is largely left unexplained. In psychology, theories such as Schank and Abelson's (1977) theory on mental scripts or theories on the use of schemata (e.g., Rumelhart & Normal, 1985; Rumelhart, Smolensky, McClelland, & Hinton, 1986) describe how individuals use memories and past experiences to guide expectations and integrate new information. Domains such as reading or language comprehension offer theories of how individuals construct mental models of situations described by a writer or speaker (Bower & Morrow, 1990; Ericsson & Kintsch, 1995; Kintsch, 1988, 1998). These models are referred to in the IA literature, such as in Mangio and Wilkinson (2008), who cite Davis's (1992) characterization of the use of mental models in IA as a "distillation of the analyst's cumulative factual and conceptual knowledge into a framework for making estimative judgments on a complex subject" (p. 13) (see Heuer, 1999, and Hutchins, Pirolli, & Card, 2006, for additional references on schemata, mental models, and scripts in IA).

These referenced theories were only recently adapted to sensemaking. Klein, Phillips, Rall, and Peluso (2007) adapted and extended these theories to complement the model of cognitive sensemaking processes collectively integrated into Pirolli et al.'s model. They proposed their model which further characterized the stages of analysis that take place after information is initially perceived and filtered, when higher level cognitive behaviors are invoked (e.g., knowledge-based behaviors within Rasmussen's (1983) skills-, rules-, and knowledge-based behavior taxonomy).

Klein et al.'s (2007) data-frame theory (DFT) of sensemaking explains the cognitive processes that occur when a person is consciously attempting to understand events. This conscious attempt at understanding is the primary cognitive task of intelligence analysts; therefore, this theory presents a well-adapted model for, collectively speaking, providing a more complete and detailed picture of the mental processes that take place (and therefore the expertise required) during the substantive analysis within IA. This theory is based on the idea of mental frames—"explanatory structure[s] that define entities by describing their relationships to other entities" (Klein et al., 2007, p. 118). The concept of mental frames rests on a large body of previous evidence, especially as related to expert performance (see Chase & Simon, 1973a; 1973b; Chi et al., 1981; Irwin-Zarecka, 2007; Minsky, 1975; Mou & McNamara, 2002).

The DFT suggests that sensemaking is characterized by a cycle of information that determines the frames to employ, and frames that determine the information to notice. Klein et al. point out that, in this cycle, neither the information nor the frame comes first; rather, "the data elicit and help to construct the frame; the frame defines, connects, and filters the data" (p. 118). Frames are defined by a series of anchor points along with a much larger set of missing information. For example, most people have a frame for understanding how an automobile works: an engine uses gasoline to turn the wheels. However, there is a great deal of information missing from this frame, such as the information a mechanic has in their frame of how each component of an automobile works. These incomplete frames, aka mental models, represent just-in-time frames of understanding that people access and update when needed (Klein et al., 2007). For example, analysts monitoring suspicious activities in a city may from past experience know that a bomb-making network will most often involve a financier, a materials supplier, a bomb technician, and triggermen. They may even have a sense of where these parties will be located and how information will be passed between them based on their initial frame or mental model of a bomb-making network; however, they do not activate this frame until some piece of information activates the frame to guide their attention and actions.

For a more comprehensive review of expertise from the perspective of mental frames, schemes, and models, see Chapter 9, "Macrocognitive Models of Expertise," by Hutton (this volume). With respect to IA, the robustness of an analyst's repertoire of frames is a key characteristic that affects the difficulty they will encounter with new requests for information or novel pieces of data. A robust set of mental frames provides a larger number of anchor points (i.e., key nodes in a frame that allow analysts to quickly activate them when data are found that aligns to what is expected for that key node), which allow the analyst to identify more alternatives and draw analogies from these alternatives even when faced with a substantive problem beyond their scope of technical or regional expertise (Taylor, 2005). Drawing analogies is critical to formulating and exploring competing hypotheses; thus, more distinct frames, which are most often developed over time through varied experiences, is a strong characteristic of expertise in IA (as Hutton, this volume, highlights is true of other domains as well). When analysts have limited knowledge about the operational context, availability of resources, or domain, they are more likely to experience difficulty identifying patterns in the data (i.e., frames that can be used to align the data into a coherent narrative). These patterns form the basis of new hypotheses or mental models to generate understanding (i.e., level 2 SA) or the ability to project forward in time (i.e., level 3 SA) and effectively respond to a substantive problem or information request (Pfautz et al., 2006; Pirolli & Card, 2005).

Our discussion thus far has focused on past efforts to model the cognitive processes engaged during IA and extracting characteristics of expertise. Self-awareness and robust past experiences are common throughout these characteristics. While these characteristics illustrate some factors that underlie expertise in IA, we can also consider other challenges that arise during cognitive processes. Most notably, working memory overload can occur during the IA (and more general sensemaking) process. Factors causing overload include the uncertainty, complexity, and dynamic nature of the data, the need to maintain awareness of multiple analysis layers (e.g., sources, intelligence consumers, one's model; see Taylor, 2005), continually consider different explanations, and ask, "What am I missing?" As a result, the cognitive processes described are almost always used in situations where analysts must apply simplifying information processing strategies. With access to observe and better approaches to model these strategies (see Rasmussen's (1981) decision ladder, and its many variations, e.g., Cummings, 2003; Hoffman & McCloskey, 2013), we can provide insight into expert performance (i.e., intentionally adopting heuristics based on context to shorten the path to convergence without negatively biasing analysis) versus non-expert performance (i.e., analysts who may unintentionally or intentionally adopt heuristics that result in the negative biases presented by Kahneman et al., 1982).

Multiple CTA efforts (Hutchins et al., 2006; Pfautz et al., 2006) support the case that even expert analysts may fall prey to the inappropriate application of simplifying strategies—the use of some heuristic is inevitable given their high cognitive workload demands (Endsley, 1996). When employing these heuristics, analysts often introduce biases into their analysis process (Heuer, 1999; Hutchins et al., 2006; Johnston, 2005; Mangio & Wilkinson, 2008; Patterson, Woods, Tinapple, & Kuperman, 2001; Patterson, Woods, Tinapple, & Roth, 2001; Pirolli, 2006; Pirolli & Card, 2005; Roth et al., 2010; Tam, 2009; Taylor, 2005). While experts can often apply heuristics that result in positive outcomes (i.e., appropriate biases given the analysis context), the appearance of negative biases may still arise, and do so frequently for novices. These common negative biases are presented in Table 32.2.

These biases have also been observed in other domains where sensemaking and decision-making processes are stressed. To avoid these biases during analysis, Heuer (1999) implied that expert analysts are self-conscious about their reasoning process so they can understand when they are using these simplified heuristics; however, an increased awareness of biases does little on its own to help analysts avoid these biases

Table 32.2 Cognitive biases with potential for negative analysis impacts most commonly encountered by intelligence analysts

Bias	Summary
Selectivity	Characterized by information being selectively recalled as a function of how salient the information is to the individual analyst.
Absence of evidence	Characterized by a failure to recognize and incorporate missing data into judgments of abstract problems (e.g., the "what am I missing?" question).
Confirmation bias	Characterized by the tendency to seek and interpret information in way as to confirm existing beliefs or hypotheses.
Overconfidence	Characterized by overconfidence in one's own hypotheses being correct when in fact most of the time they are wrong.
Oversensitivity to consistency	Characterized by a tendency to place too much reliance on small samples or the inability to discern multiple reports from the same source information.
Discredited evidence	Characterized by the persistence of beliefs or hypotheses even after evidence fully discrediting those hypotheses has been perceived.

during analysis. They need targeted strategies to counteract their tendencies to adopt and employ them. For this reason, many technology solutions have focused on augmenting human cognitive capabilities by offloading and/or externalizing analyst reasoning, assumptions, and active mental models of competing hypotheses. Pfautz et al. (2006) and Heuer (1999) both suggest that any method or tool which can help to better structure information, challenge the analyst's assumptions, or explore alternative interpretations—without adding a cognitive burden—will significantly increase the robustness and validity of an analyst's performance. Finally, Tam (2009) focuses on technology-based support to augment analyst capabilities and bring them to an expert level of performance and provides a list of analyst skills and sub-skills that are believed to help successfully conduct robust analyses and avoid many biases. This list of analyst skills and sub-skills, which complement the meta-characteristics of expertise in intelligence analysis (and even includes a self-regulation skillset that aligns with our assertions on expertise in IA), includes the non-technical skill areas listed in Table 32.3.

By combining the characteristics identified through prior cognitive work analyses (e.g., experts are self-aware of their own reasoning strategies, strive to evaluate multiple competing hypotheses and avoid premature closure, and continually revisit and revise their trust in information sources) and the models of IA from cognitive task and information processing domain analyses, a more complete picture of expertise is revealed. The skills and sub-skills of an individual analyst that can contribute to expertise are largely knowledge-based or *soft skills* (see Table 32.4 for a summary) that typically require more experiential learning to master. This insight supports our assertion that, unlike expertise in many other domains and fields of work, IA expertise is a fluid construct centered on the meta-characteristics of an individual (e.g., awareness of skill limitations and strategies to calibrate one's work based on that awareness) versus

Table 32.3 Non-technical skill and sub-skill areas that help intelligence analysts conduct robust analyses and avoid common decision-making biases

Skill	Sub-skills
Interpretation	• Categorization • Decoding significance • Clarifying meaning
Analysis	• Examining ideas • Identifying arguments • Analyzing arguments
Evaluation	• Assessing claims • Assessing arguments
Inference	• Querying evidence • Conjecturing alternatives • Drawing conclusions
Explanation	• Stating results • Justifying procedures • Presenting arguments
Self-regulation	• Self-examination • Self-correction

Reproduced from 2nd Lt. Chin Ki Tam, Behavioral and psychosocial considerations in intelligence analysis: A preliminary review of literature on critical thinking skills, Interim report, May 2008–October 2009. Report ID: AFRL-RH-AZ-TR-2009-0009, Air Force Research Lab, Mesa, AR, Human Effectiveness Directorate, p. 17, http://www.dtic.mil/dtic/tr/fulltext/u2/a502215.pdf. Accessed April 2018.

domain-relevant or technical skills and knowledge. However, there is currently no singular field of study, no critical domain of knowledge, and no single technical specialization that will produce an IA expert. This synthesis of a considerable body of prior research, while acknowledging current research efforts, emphasizes a critical concern: studies of IA are not unified (coming from multiple fields of study), nor is there currently an epistemological approach to aggregating knowledge about this domain.

IMPLICATIONS

The domain of IA and the sensemaking process more generally have received considerable international attention over the past two decades from researchers hoping to understand the challenges facing analysts and why IA expertise has been so hard to define. These investigations were motivated by complexities resulting from rapid increases in technology (especially in access to data), to increasing operational tempos (further constraining individual and socio-organizational analytic processes), as well as reductions to IA staff, shifting stakeholder organizational structures (that moved

Table 32.4 Summary of areas of expertise in intelligence analysis and exemplifying expert behaviors of each

Area of expertise	Expert behavior/skill
Source selection and evaluation	Continually revisits and recalibrates trust in potential sources over time by seeking corroborating information from alternative sources. Maintains set of active evidence to support a source's trustworthiness.
Analysis of competing hypotheses	Generates and effectively maintains multiple different potential explanations for anomalies or interconnected events in the data to avoid premature closure. Seeks evidence that will confirm or disconfirm competing hypotheses, even after a single hypothesis has significant supporting evidence.
Validating coherence and correspondence of hypotheses	Evaluates hypotheses for both internal logic and consistency (coherence), and alignment with evidence on the true state of the world (correspondence).
Broadening	Appropriately engages in broadening activities while balancing an awareness of available information and its metadata, deadlines for a given analysis, and feasibility of active hypotheses.
Narrowing	Appropriately engages in narrowing activities while continuously critiquing one's own assessments, assumptions, hypotheses, and narratives to know when confirmation (or disconfirmation) should be sought and when alternatives should be explored
Audience analysis	Awareness and understanding of the intelligence requestor's and consumer's motivations and needs for requesting a piece of intelligence, and then tailoring a response given that understanding
Maintaining wealth of experience	Readily accessing robust repertoire of mental frames based on past experiences to draw appropriate analogies
Meta-cognition	Monitoring and criticizing one's reasoning processes to understand when simplifying heuristics or assumptions are being applied to constrain a problem or advance an analysis

different types of analysts both closer and farther away from decision-makers). The previous sections of this chapter highlight a considerable amount of scholarly work on the challenges in IA that experts must overcome. However, the ongoing changes to the domain emphasize the need for future work. For example, the impact of emerging technology on IA practices, and the concept of *expertise* in this domain both need to be addressed. The advent of new data sources and processing techniques require expert analysts to have additional technical knowledge and skills, and trust in technological innovations to support correct interpretation of data. *Trust* will influence analysts' decisions to rely on technology, data sources, and the data itself, which may change the how, and the extent to which risk is communicated to decision-makers. Future research is needed to further understand the implications of rapidly emerging technologies—such as the emerging class of automated analytics support systems (e.g., information fusion systems, trends in *big data* analytics, and machine-learning solutions). The increase in data availability, along with the increase in machine learning, has created a demand for transparency and explainability. In terms of IA expertise, the shifting technology landscape introduces questions about how best to characterize (and to continue to conduct applied research across) both current and future levels of proficiency as analysts progress from apprentice to journeymen to expert. For example, characteristics that define IA expertise may shift from IA-relevant cognitive activities (e.g., explicit hypothesis definition, knowing when to narrow/broaden search) alone to supporting these with the skills required to rapidly adapt to new technological enhancements that are introduced in to the domain. For example, while iteratively broadening and narrowing from a particular hypothesis about formulating a plan, an analyst may also need a deep understanding of how and why multiple disparate and (potentially haphazardly integrated) information fusion systems recommend conflicting courses of action.

Compounding such challenges, new technology is provided within an acquisition process that fails to acknowledge the prior research on analyst processes (detailed to some extent in this chapter) and therefore typically overlooks expert and novice analysts as well as the demands of the analyst's socio-organizational structure. Achananuparp, Park, Proctor, Wania, and Zhou (2005) provided one overview of several technology-based capabilities specifically designed to aid the analyst. However, Achananuparp et al.'s summary is now over a decade old and focused on technologies that existed or were under development at the time of their publication. These technologies were intended to provide support for a range of intelligence and sensemaking challenge areas, including collaboration (see Rodriguez et al., 2005; Wang et al., 2005), information retrieval (see Badia, 2004; Srihari, Crist & Niu, 2005), information visualization (see Keahey & Cox, 2005; Rasmussen & Jensen, 1974), knowledge management (see Ong et al., 2001; Pettersson, 2001), language translation (see Diekema et al., 2005), memory and multitasking (see Hillman, 1998; Kaplan et al., 2005; Risch et al., 1996), and overcoming bias (see Alonso & Li, 2005). Despite the high standard of scholarship of much of this research, it highlights the aforementioned challenge; technological capabilities relevant to IA are increasingly available (and accelerating in their capabilities and potential impact on society at a rate consistent with Moore's Law) (e.g., see Kurzweil, 2005). Yet, the need to align these capabilities with clearly identified cognitive and perceptual processes inherent to IA remains a challenge. This highlights the need for a new kind of IA expertise, as mentioned earlier—the ability to understand new technologies across the analytic process, and rapidly incorporate, use, and evaluate (and discard appropriately) new capabilities. Traditional skill- or knowledge-based IA expertise (e.g., hypothesis generation, evaluation, and comparison) will not disappear, but a new cognitive burden is being introduced, and percolates throughout the entire IA process: from calibrating trust in a new data source, to being able to explain the features extracted by a massive-scale data analysis technique, to being able to defend the results of a machine-learned statistical prediction. These tasks are increasingly part of the definition of *expertise* in intelligence analysis.

Critically, the prior work to understand the cognitive and perceptual tasks in IA remains an important foundation for future work. Where and when should human–machine interactions be provided for new technologies, and how should these interactions be designed to facilitate analyst goals (e.g., establish and maintain an accurate representation of the system's internal model of the operating environment to understand its limitations and affordances) and allow dynamic, appropriate calibration of analyst trust in the system (e.g., providing explainable artificial intelligence and machine learning so analysts can develop trust not only in source information, but in the reasoning mechanisms that are acting upon it to reach system outputs/conclusions)?

The challenge of coping with, and optimally exploiting new technology, is not unique to IA, and appears in related reasoning and decision-making domains where technology is being introduced in an effort to improve the efficiency, accuracy, robustness, and/or resiliency of outcomes. Across these domains, and as exemplified by ongoing developments in IA, the need to project and anticipate changes in how we describe *expertise* will remain an important challenge for the research community. The same technology that provides new capabilities has the potential to also support new methods for rigorous study of this problem, e.g., to capture and study where and how fielded technology (in IA and elsewhere) is employed, and what user–technology interaction patterns emerge. Specific opportunities include: (1) supporting all stages of system engineering and the broader process of technology development to develop tools for modern or future IA; (2) creating new lines of scientific inquiry for researchers aiming to characterize cognitive/perceptual processes used in IA; and (3) projecting how IA expertise must evolve if a new technology-enabled capability succeeds (e.g., how analyst trust can be appropriately calibrated to the rate of success). Such promises can only be realized if a notion of expertise in IA is the basis for exploring each opportunity to transform the state of the practice.

For example, consideration must be given to the interaction cycles that will occur with the introduction of a new technology to understand what inputs will be required by the analyst and what system outputs will result from those supervisory or directive inputs. In that vein, consideration must be given to what skills will be required for analysts to achieve their goals, use the system as intended, and interpret the results to foster an appropriate level of reliance. However, beyond these likely more straightforward technical skills, consideration must also be given to how the beneficial heuristics that expert analysts employ may result in new negative biases and outcomes. While there is a body of evidence on how to overcome negative biases that result from common heuristics faced by non-experts, there is a gap to date on characterizing how to ensure that the negative effects of the biases imparted by the use of shortcutting heuristics are lessened, rather than compounded, by introducing new technology that is intended to make processes more efficient.

The effects of new technology on the definition of expertise go beyond the individual analyst, to the organization within which they perform their tasks. Organizational definitions of expertise are likely to change, and allocation of tasks across the members of the IA community will be important for future research, as much as understanding the individual's understanding of how to best employ and/or trust new technology. The socio-organizational structure and dimensions of IA are another area ripe for additional research and understanding. As this chapter demonstrated, there has been significant effort and truly profound findings related to characterization of the individual analyst. However, a gap exists to characterize how the differing organizational structures of modern IA (e.g., integrated IA cells, remote IA reachback centers, centralized IA reporting repositories) influence the dynamics of expertise and how dimensions of expertise translate across these socio-organizational structures. This need is further amplified by an increased number of joint and coalition operation centers, where distributed intelligence is distributed not only within and across US military branches, but also within and across coalition and non-coalition bounds (e.g., IA for space battle management command and control (BMC2)). As illustrated in one of the recent case studies presented in McChrystal, Collins, Silverman, & Fussell (2015), the individuals who will succeed in these embedded coalition and cross-service/-organization roles are those with exceptional interpersonal skills (e.g., ability to influence without authority, ability to empathize with the perspective and goals of others, ability to adapt to norms of different organizational cultures). This presents yet another area for additional research, turning our focus from that of the individual analyst and his/her internal cognitive processes, to studies of the larger socio-organizational structure of modern and next-generation IA.

Conclusions

Significant opportunities exist to develop technologies to aid intelligence analysts, as well as individuals preforming similar sensemaking tasks. There is a significant and well-structured body of research that spans the cognitive and perceptual processes, meta-cognitive processes, and dynamic socio-organizational aspects of performance on the range of challenging tasks inherent to IA. This chapter has provided an overview of the methods and conceptual models that have been developed to establish an understanding of expertise in IA. Decades of work have established an understanding of individual analyst tasks as well as their cognitive processes, heuristics, and biases. Still, the IA domain, as it continues to evolve, remains a ripe area for future research.

As in other domains, a challenge remains to design and implement capabilities that go beyond attempts to automate or expedite isolated tasks and procedures of individual analysts, and instead provide support that is adaptive to the highly contextualized and dynamic nature of IA. This requires consideration of how characterizations of expertise must not only reflect tasks associated with core analytic tasks, but also the need to assess and utilize increasingly capable technological systems. Over 50 years have passed since Kent's (1965) seminal work characterizing IA, and despite the significant advances in our understanding of how to characterize IA expertise, the evolution of the domain continues to generate exciting and compelling research opportunities. These are not only focused on what constitutes human expertise in a world where technological innovation is constantly accelerating, but also on how human expertise can (and should) interoperate with emerging technologically driven ways to make sense of a wide range of data. IA, given its potential impacts on decisions that span from strategic policy to tactical decisions, should remain at the forefront of research into how expertise is developed, evolves, and is applied in critical, highly impactful situations.

References

Achananuparp, P., Park, S. J., Proctor, J. M., Wania, C., & Zhou, N. (2005). *An analysis of tools that support the cognitive processes of intelligence analysis*. Philadelphia: T. Hewett, Drexel University.

Alonso, R., & Li, H. (2005). Combating cognitive biases in information retrieval. *Proceedings of the first international conference on intelligence analysis*. McLean, VA: MITRE Corp.

Anderson, J. (1974). Retrieval of propositional information from long-term memory. *Cognitive Psychology* 6(4), 451–474.

Anderson, J. R. (1983). A spreading activation theory of memory. *Journal of Verbal Learning and Verbal Behavior* 22(3), 261–295.

Badalamente, R. V., & Greitzer, F. L. (2005). Top ten needs for intelligence analysis tool development. *Proceeding of the first annual conference on intelligence analysis methods and tools*. McLean, VA: MITRE Corp.

Badia, A. (2004). Interactive query languages for intelligence tasks. In H. Chen, R. Moore, D. D. Zeng, & J. Leavitt (Eds), *Proceedings of intelligence and security informatics: Second symposium on intelligence and security Informatics* (pp. 114–124). Lecture Notes in Computer Science, vol 3073. Berlin/Heidelberg: Springer.

Bower, G. H., & Morrow, D. G. (1990). Mental models in narrative comprehension. *Science* 247, 44–48.

Chase, W. G., & Simon, H. A. (1973a). The mind's eye in chess. In W. G. Chase (Ed.), *Visual information processing* (pp. 215–281). New York: Academic Press.

Chase, W. G., & Simon, H. A. (1973b). Perception in chess. *Cognitive Psychology* 4(1), 55–81.

Chi, M. T. H., Feltovich, P. J., & Glaser, R. (1981). Categorization and representation of physics problems by experts and novices. *Cognitive Science* 5(2), 121–152.

Clark, R. M. (2016). *Intelligence analysis: a target-centric approach*. London: CQ Press.

Cowan, N. (2001). The magical number 4 in short-term memory: A reconsideration of mental storage capacity. *Behavioral and Brain Sciences* 24, 87–185.

Cummings, M. (2003). Designing decision support systems for revolutionary command and control domains. Dissertation, School of Engineering and Applied Science, University of Virginia.

Davis, J. (1992). Combating mindset. *Studies in Intelligence* 36(5), 13–18.

Diekema, A., Hannouche, J., Ingersoll, G., Oddy, R., & Liddy, E. (2005). Arabic information retrieval. *Proceedings of the first international conference in intelligence analysis*. McLean, VA: MITRE Corp.

Drucker, S. M. (2006). Coping with information overload in the new interface era. *Proceedings of the CHI2006 workshop "What is the Next Generation of Human Computer Interaction?,"* Montreal Canada.

Elm, W., Cook, M., Greitzer, F. L., Hoffman, R. R., Moon, B., & Hutchins, S. G. (2004). Designing support for intelligence analysis. *Proceedings of the Human Factors and Ergonomics Society 48th Annual Meeting*. New Orleans, LA.

Elm, W., Potter, S., Tittle, J., Woods, D. D., Grossman, J. & Patterson, E. (2005). Finding decision support requirements for effective intelligence analysis tools. *Proceedings of the Human Factors and Ergonomics Society 49th Annual Meeting 49*, 297–301.

Endsley, M. R. (1988). Design and evaluation for situation awareness enhancement. *Human Factors and Ergonomics Society Annual Meeting Proceedings 32*, 97–101.

Endsley, M. R. (1996). Automation and situational awareness. In R. Parasuraman & M. Mouloua (Eds), *Automation and human performance: Theory and applications* (pp. 163–182). Mahwah, NJ: Lawrence Erlbaum.

Ericsson, K. A., & Kintsch, W. (1995). Long-term working memory. *Psychological Review 102*, 211–245.

Greitzer, F. L. (2005). Toward the development of cognitive task difficulty metrics to support intelligence analysis research. *Fourth IEEE Conference on Cognitive Informatics, 2005 (ICCI 2005)* (pp. 315–320). IEEE.

Grossman, J. B., Woods, D. D., & Patterson, E. S. (2007). Supporting the cognitive work of information analysis and synthesis: A study of the military intelligence domain. *Proceedings of the Human Factors and Ergonomics Society 1*, 348–352.

Hajdukiewicz, J. R., Burns, C. M., Vicente, K. J., & Eggleston, R. G. (1999). Work domain analysis for intentional systems. *Human Factors and Ergonomics Society Annual Meeting Proceedings 43*, 333–337).

Halpern, J. Y. (2001). Conditional plausibility measures and Bayesian networks. *Journal of Artificial Intelligence Research 14*, 359–389.

Hammond, K. R. (2000). Coherence and correspondence theories in judgment and decision making. In T. Connolly, H. R. Arkes, & K. R. Hammond (Eds), *Cambridge series on judgment and decision making. Judgment and decision making: An interdisciplinary reader* (pp 53–65). New York: Cambridge University Press.

Hetzler, E., & Turner, A. (2004). Analysis experiences using information visualization. *IEEE Computer Graphics and Applications 24*(5), 22–26.

Heuer, R. J. J. (1999). *The psychology of intelligence analysis*. Washington DC: Center for the Study of Intelligence.

Hillman, D. V. (1998). Inquest: A prototype intelligence tool. *Proceedings of the first international conference on industrial and engineering applications of artificial intelligence and expert systems* (vol. 1, pp. 147–156).

Hoffman, R. R., & McCloskey, M. J. (2013). The macrocognitive decision ladder. *Proceedings of the Human Factors and Ergonomics Society Annual Meeting 57*(1), 245–249.

Hollnagel, E., & Woods, D. D. (2006). *Joint cognitive systems: Patterns in cognitive systems engineering*. CRC Press: Boca Raton, FL.

Hughes, F. J., & Schum, D. A. (2003). Preparing for the future of intelligence analysis: Discovery—proof—choice. *Joint Military Intelligence College*. Unpublished Manuscript.

Hutchins, S. G., Pirolli, P. L. & Card, S. K. (2004). A new perspective on use of the critical decision method with intelligence analysts. In R. R. Hoffman (Ed.), *Expertise out of context: Proceedings of the 6th international conference on naturalistic decision making*. Mahwah, NJ: Lawrence Erlbaum.

Hutchins, S. G., Pirolli, P. L., & Card, S. K. (2006). What makes intelligence analysis difficult? A cognitive task analysis of intelligence analysts. In P. L. Pirolli (Ed.), *Assisting people to become independent learners in the analysis of intelligence* (pp. 6–52). Palo Alto, CA: Palo Alto Research Center.

Irwin-Zarecka, I. (2007). *Frames of remembrance: The dynamics of collective memory*. Piscataway, NJ: Transaction Publishers.

Johnston, R. (2005). *Analytic culture in the US intelligence community: An ethnographic study*. Washington, DC: US Government Printing Office.

Kahneman, D., Slovic, P., & Tversky, A. (1982). *Judgment under uncertainty: Heuristics and biases*. New York: Cambridge University Press.

Kaplan, R., Crouch, R., King, T., Tepper, M., & Bobrow, D. (2005). A note-taking appliance for intelligence analysts. *Proceedings of the first international conference in intelligence analysis*. McLean, VA: MITRE Corp.

Katter, R. V., Montgomery, C. A., & Thompson, J. R. (1979). *Human processes in intelligence analysis: Phase I overview* (p. 74). Virginia: Army Research Inst. for the Behavioral and Social Sciences.

Keahey, T. A., & Cox, K. C. (2005). VIM: A framework for intelligence analysis. *Proceedings of the first international conference on intelligence analysis*. McLean, VA: MITRE Corp.

Kent, S. (1965). Special problems of method in intelligence work. In S. Kent (Ed.), *Strategic intelligence for American world policy* (pp. 159–179). Hamden, CT: Archon Books.

Kintsch, W. (1988). The role of knowledge in discourse comprehension construction-integration model. *Psychological Review* 95, 163–182.

Kintsch, W. (1998). *Comprehension: A paradigm for cognition*. New York: Cambridge University Press.

Klein, G. A., Phillips, J. K., Rall, E. L., & Peluso, D. A. (2007). A data-frame theory of sensemaking. In R. R. Hoffman (Ed.), *Expertise out of context* (pp. 113–155). Mahwah, NJ: Lawrence Erlbaum.

Koopman, P., & Hoffman, R. R. (2003). Work-arounds, make-work, and kludges. *IEEE Intelligent Systems* 18(6), 70–75.

Kurzweil, R. (2005). *The singularity is near*. London: Penguin Group.

McChrystal, G. S., Collins, T., Silverman, D., & Fussell, C. (2015). *Team of teams: New rules of engagement for a complex world*. London: Penguin.

Mangio, C. A. & Wilkinson, B. J. (2008). Intelligence analysis: Once again. Paper presented at the *International Studies Association 2008 Annual Convention*. San Francisco, CA.

Miller, G. A. (1956). The magical number seven, plus or minus two: some limits on our capacity for processing information. *Psychological Review* 63(2), 81–97.

Minsky, M. (1975). A framework for representing knowledge. In P. Winston (Ed.), *The psychology of computer vision* (pp. 211–277). New York: McGraw-Hill.

Mou, W., & McNamara, T. P. (2002). Intrinsic frames of reference in spatial memory. *Journal of Experimental Psychology: Learning, Memory, and Cognition* 28(1), 162–170.

Neath, I., & Surprenant, A. M. (2003). *Human memory: An introduction to research, data, and theory* (2nd edn). Belmont, CA: Wadsworth.

Ong, H.-L., Tan, A.-H., Ng, J., Pan, H. & Li, Q.-X. (2001). FOCI: Flexible organizer for competitive intelligence. *Proceedings of the tenth international conference on information and knowledge management* (CIKM '01) (523–525). New York: ACM Press.

Patterson, E., Roth, E. M., & Woods, D. D. (2001). Predicting vulnerabilities in computer-supported inferential analysis under data overload. *Cognition, Technology & Work* 3, 224–237.

Patterson, E. S., Woods, D. D., Tinapple, D., & Kuperman, G. G. (2001). *Aiding the intelligence analyst: From problem definition to design concept exploration*. Columbus, OH: Ohio State University.

Patterson, E. S., Woods, D. D., Tinapple, D., & Roth, E. M. (2001). Using cognitive task analysis (CTA) to seed design concepts for intelligence analysts under data overload. *Proceedings of the Human Factors and Ergonomics Society 45th Annual Meeting 45*, 439–443.

Pettersson, U. (2001). Creating an intelligence system at the Swedish national financial management authority. *Competitive Intelligence Review 12*(2), 20–31.

Pfautz, J., Fichtl, T. F., Guarino, S., Carlson, E., Powell, G., & Roth, E. M. (2006). Cognitive complexities impacting army intelligence analysis. *Human Factors and Ergonomics Society Annual Meeting Proceedings 50*: 452–456.

Pinelli, T. E., Bishop, A. P., Barclay, R. O., & Kennedy, J. M. (1993). The information-seeking behavior of engineers. In A. Kent & C. M. Hall (Eds), *Encyclopedia of library and information science* (vol. 52, pp. 167–201). New York: Marcel Dekker.

Pirolli, P. (1993). Towards a unified model of learning to program. In E. Lemut (Ed.), *Cognitive models and intelligent environments for learning programming* (pp. 34–48). Berlin, Heidelberg: Springer.

Pirolli, P. (2006). *Assisting people to become independent learners in the analysis of intelligence*. Pal Alto, CA: Palo Alto Research Center.

Pirolli, P. L. & S. K. Card (2005). The sensemaking process and leverage points for analyst technology as identified through cognitive task analysis. *Proceedings of the first international conference on intelligence analysis*. McLean, VA: MITRE Corp.

Platt, W. (1957). *Strategic intelligence production: Basic principles*. New York: Frederick A. Praeger.

Rasmussen, J. (1981). Models of mental strategies in process plant diagnosis. In J. Rasmussen, & W. B. Rouse (Eds), *Human detection and diagnosis of system failures* (pp. 241–258). New York: Plenum Press.

Rasmussen, J. (1983). Skills, rules, knowledge: Signals, signs, and symbols, and other distinctions in human performance models. *IEEE Transactions on Systems, Man and Cybernetics 13*, 257–266.

Rasmussen, J., & Jensen, A. (1974). Mental procedures in real-life tasks: A case study of electronic troubleshooting. *Ergonomics 17*, 293–307.

Rasmussen, J., Pejtersen, A. M., & Goodstein, L. P. (1994). *Cognitive systems engineering*. New York: Wiley.

Risch, J. S., May, R. A., Dowson, S. T., & Thomas, J. J. (1996). A virtual environment for multimedia intelligence data analysis. *IEEE Computer Graphics and Applications 16*(6), 33–41.

Rodriguez, A., Boyce, T., Lowrance, J., & Yeh, E. (2005). Angler: Collaboratively expanding your cognitive horizon. *Proceedings of first international conference on intelligence analysis* (pp. 215–216). McLean, VA: MITRE Corp.

Roth, E. M., Pfautz, J. D., Mahoney, S. M., Powell, G. M., Carlson, E. C., Guarino, S. L., ... Potter, S. S. (2010). Framing and contextualizing information requests: Problem formulation as part of the intelligence analysis process. *Journal of Cognitive Engineering and Decision Making 4*(3), 210–239.

Rumelhart, D. E., & Norman, D. A. (1985). The representation of knowledge. In A. M. Aitkenhead, & J. M. Slack (Eds), *Issues in cognitive modeling* (pp. 15–62). Hillsdale, NJ: Lawrence Erlbaum.

Rumelhart, D. E., Smolensky, P., McClelland, J. L., & Hinton, G. E. (1986). Schemata and sequential thought processes in PDP Models. In J. L. McClelland, D. E. Rumelhart, & P. R. Group (Eds), *Parallel distributed processing* (vol. 2, pp. 7–57). Cambridge, MA: MIT Press.

Schank, R. C., & Abelson, R. P. (1977). *Scripts, plans, goals, and understanding: An inquiry into human knowledge structures.* Hillsdale, NJ: Lawrence Erlbaum.

Scholtz, J., Morse, E., & Potts Steves, M. (2006). Evaluation metrics and methodologies for user-centered evaluation of intelligent systems. *Interacting with Computers* 18(6), 1186–1214.

Simon, H. A. (1956). Rational choice and the structure of the environment. *Psychological Review* 63(2), 129–138.

Srihari, R. K., Li, W., Crist, L., & Niu, C. (2005). Intelligence discovery portal based on corpus level information extraction. *Proceedings of the first international conference on intelligence analysis.* McLean, VA: MITRE Corp.

Tam, C. K. (2009). *Behavioral and psychosocial considerations in intelligence analysis: A preliminary review of literature on critical thinking skills.* Interim report, May 2008–October 2009, Report ID: AFRL-RH-AZ-TR-2009-0009, Air Force Research Lab, Mesa, AR, Human Effectiveness Directorate, p. 17. Available at http://www.dtic.mil/dtic/tr/fulltext/u2/a502215.pdf. Accessed April 2018.

Taylor, S. M. (2005). The several worlds of the intelligence analyst. *Proceedings of the first international conference on intelligence analysis.* McLean, VA: MITRE Corp.

Treisman, A. M., & Gelade, G. (1980). A feature-integration theory of attention. *Cognitive Psychology* 12, 97–136.

Tulving, E., & Craik, F. I. M. (2000). *The Oxford handbook of memory.* Oxford: Oxford University Press.

Wang, R., Kogut, P., Zhu, S., Leung, Y., & Yen, J. (2005). Semantic web enabled collaborative agents for supporting analyst teams. Presented at the *first international conference on intelligence analysis.* McLean, VA: MITRE Corp.

Ward, P., Ericsson, K. A., & Williams, A. M. (2013). Complex perceptual-cognitive expertise in a simulated task environment. *Journal of Cognitive Engineering and Decision Making* 7(3), 231–254.

Wickens, C. D. & Hollands, J. G. (2000). *Engineering psychology and human performance* (3rd edn). Englewood Cliffs, NJ: Prentice Hall.

Woods, D. D. & Roth, E. M. (1988). Cognitive systems engineering. In M. Helander, T.K. Landauer, & P.V. Prabhu (Eds), *Handbook of human-computer interaction* (pp. 3–43). Amsterdam: Elsevier.

CHAPTER 33

EXPERTISE IN LAW ENFORCEMENT

JOEL SUSS AND LAURA BOULTON

Introduction

Law enforcement is a profession that encompasses many specialties. Over their careers, officers may acquire skill in one or more roles, including undercover work, human trafficking, patrol, sex crime, cybercrime, tactical operations (e.g., special weapons and tactics or SWAT), investigation (e.g., forensic, homicide, theft), hostage negotiation, counterterrorism, canine operations, crisis intervention, instructor (e.g., firearm, defensive tactics, use of force), missing persons, and surveillance. But compared to others domains in which expertise can be readily defined and objectively measured (e.g., chess), there is no universally accepted way to describe or measure expertise in law enforcement.

Our aim in this chapter is to provide an overview of expertise research in law enforcement and guidance for those planning on conducting research themselves. We begin by describing conceptualizations of expertise in law enforcement, and offer a working definition. We then identify challenges facing researchers interested in conducting expertise research in law enforcement, and advise how to overcome those challenges. Next, we provide concrete examples of how research in this domain has been conducted. In doing so, we cover a broad range of methods and highlight the subtleties that researchers new to the domain should consider when designing and conducting expertise research. Following that, we describe key insights from the literature that cover the spectrum of expertise in law enforcement. Notably, the review provides an international perspective of this topic, presenting findings from research conducted around the world. Finally, we suggest directions for future research.

Conceptualizations of Expertise in Law Enforcement

The extent to which policing, in general, is a craft versus a science has been widely debated in the policing and criminal justice literature (e.g., Bayley & Bittner, 1984; Willis & Mastrofski, 2014). Those who view policing strictly as a craft assume that officers learn largely through experience on the job; formal education and training have little perceived value. Wilson (1968) encapsulates this perspective:

> The patrolman is neither a bureaucrat nor a professional, but a member of a *craft*. As with most crafts, his has no body of generalized, written knowledge nor a set of detailed prescriptions as to how to behave—it has, in short, neither theory nor rules. Learning in the craft is by apprenticeship, but on the job and not in the academy.... And the members of the craft, conscious of having a special skill or task, think of themselves as set apart from society, possessors of an art that can be learned only by experience, and in need of restrictions on entry into their occupation. But unlike other members of a craft—carpenters, for example, or newspaperman—the police work in an apprehensive or hostile environment producing a service the value of which is not easily judged. (p. 283)

On the other hand, there has been a long push toward professionalization of policing. This includes the requirement for formalized basic training at a police academy and academic qualifications (College of Policing, 2016; Marshall, 2015). The very existence of field training programs, during which a rookie officer accompanies and is mentored by an experienced officer, is an acknowledgment that academy training alone is not sufficient for officers to work independently (for a review of mentoring, see Chapter 44, "Learning at the Edge," by Petushek, Arsal, Ward, Hoffman, & Whyte, this volume).

The judicial system often presumes that trained police officers possess expertise, but critics have questioned the evidential basis for this presumption (Lvovsky, 2017). Traditionally, police have been assumed to possess special expertise that they developed through their work, which civilians cannot understand (Bayley, 2016). Bayley notes, however, that many of the public are now questioning such assumptions, and are seeking proof that such special expertise exists. In the United States, police actions continue to be scrutinized due to high-profile incidents, such as the shooting of unarmed black men. In legal circles, there has been much debate about the validity of police officers' intuitions, hunches, and gut instincts (e.g., as a basis for forming suspicions about who to stop and frisk), and whether these abilities—if they indeed exist—constitute expertise (Alschuler, 2007; Fulford, 2011; Segal, 2012; Taslitz, 2010; Worrall, 2013). There is often no evidence or rational basis provided in support of these intuitions, but they are the *lingua franca* of law enforcement.

In 2004 the US National Institute of Justice, Federal Bureau of Investigation, and American Psychological Association held a workshop with researchers, psychologists, and police experts from around the world to examine the nature and influence of intuition in law enforcement. The report on the workshop noted the lack of consensus over how to define intuition, with comparisons made to *complex pattern recognition*, *complex emergent processes*, the *sense of dread* that came with some police calls and not others, and the *hairs standing up on the back of your neck* (American Psychological Association, 2004, p. 3). Although the workshop identified pertinent research questions related to intuition, to the best of our knowledge no police-related intuition research was specifically funded or published. More recently, however, the role of intuition in law enforcement has been examined in the UK in comparison to more analytical modes of thinking (Akinci & Sadler-Smith, 2013).

Among police officers themselves, experts have been referred to as the *5 percenters* (Force Science News, 2006), and their close cousins, the *10 percenters* (Savelli, 2010). These are essentially officers in the right-hand tail of a hypothetical performance distribution: those who are known or perceived to be better than most officers. This conceptualization of expertise normally addresses street smarts, tactical proficiency, and a warrior mindset. Force Science News (2006) describes some of the characteristics of 5 percenters, which include the ability to detect subtle, but important, cues and the ability to quickly select the appropriate level of force needed to resolve a situation.

Measuring police performance and expertise, however, has been problematic. In law enforcement, performance is often evaluated via appraisals (e.g., supervisor and peer ratings of performance; Love, 1983). These evaluations typically focus on broader measures of performance (e.g., number of arrests) rather than on skill evaluation (e.g., Shane, 2010). They have also been criticized as being ritualistic and meaningless (Manning, 2008).

In response to these criticisms, more principled approaches have been suggested. For example, situational judgment tests have been used to elicit tacit knowledge and identify expertise (Taylor et al., 2013). In the field of forensics, expertise has been conceptualized as a forensic examiner's reliability and biasability (Dror, 2016). In the realm of use-of-force situations, researchers have taken steps to develop metrics for performance (Vila, 2014; see also Wollert, 2008). Nonetheless, assessing performance and measuring expertise is difficult because even experts may not agree about what constitutes criterion performance.

A Working Definition

These different conceptualizations of expertise in law enforcement suggest that it is indeed a difficult concept to define. In light of this, we offer a working definition: Expertise in law enforcement is characterized by the ability to adaptively apply one's skills, knowledge, and attributes to novel and complex (e.g., uncertain, time-pressured, dangerous) situations and environments. We hope that this definition, which draws on notions of cognitive readiness, adaptive expertise, and accelerated expertise

(Ericsson, 2014; Hoffman et al., 2014; O'Neil, Lang, Perez, Escalante, & Fox, 2014; Ward, Gore, Hutton, Conway, & Hoffman, 2018), is useful to expertise researchers and those responsible for training law enforcement officers (see also Chapter 12, "Adaptive Expertise," by Bohle Carbonell & van Merrienboer, this volume).

Challenges to Conducting Expertise Research in Law Enforcement

There are many challenges facing researchers interested in conducting expertise research in law enforcement. These challenges range from those generally encountered by researchers wishing to conduct research in applied settings to those that are specific to the domain of law enforcement. In this section, we describe some of the challenges and suggest ways to overcome them (see Table 33.1).

The Cultural Divide between Police and Researchers

Although expertise researchers may be enthusiastic about conducting research in law enforcement, police leaders and operational personnel may be less enthusiastic about collaborating with researchers (e.g., Cockbain & Knutsson, 2014; Cordner & White, 2010; Fyfe & Wilson, 2012; Murji, 2010; Rojek, Martin, & Alpert, 2014). For example, law enforcement agencies may be concerned that researchers will be critical and try to portray police in a negative light (e.g., racially biased, prone to aggression). A related issue stems from the perspective of legal liability: If research reveals that current training is substandard, the agency may feel compelled to change policy, procedure, and/or practice, regardless of whether it has the resources and motivation to do so. As described in Table 33.1, there are several things that researchers can do to overcome these barriers. In our experience, the most helpful is to demonstrate an understanding of the challenges faced by police in their daily work, and to spend time getting to know police. In the USA, many law enforcement agencies conduct regular citizen police academies, which aim to educate the public about police work and improve police–community relations (Lee, 2016). Participating in a citizen police academy is a good way for researchers to become more familiar with law enforcement and meet officers who can facilitate research collaboration. Researchers should also be prepared to tout the benefits of law enforcement–academic collaborations, which include bringing novel perspectives and ideas, as well as improved policies and procedures (Burkhardt et al., 2017; Guillaume, Sidebottom, & Tilley, 2012; Hansen, Alpert, & Rojek, 2014).

Table 33.1 Challenges, barriers and strategies for conducting expertise research in law enforcement

Challenge	Barriers	Strategies to overcome barriers
Cultural divide	Distrust Communication Relevance of research Legal liability to disclose findings	Get to know the police and the challenges they face Use the International Association of Chiefs of Police guide Align research to each individual forces' research agenda Pracademic (practitioner-academics) collaboration
Psychologist stereotype	Association with screening tests, mental illness, psychoanalysis and profiling Unawareness of breadth of psychology	Explain intent during initial contact Explicitly state that not there to judge Leverage officer familiarity with existing police research
Ethical approval	Increased risk Participation coercion or pressure	Research around existing training activities Letters of support from law enforcement agencies Brief officers directly without presence of superior officers Check if force has own ethics committee and/or procedures
Access to officers	Union representation Limited time Physical and mental demands Non-distinct groups (experts vs. novices) Limited number of experts	Adhere to ethical code and practices of anonymity Offer incentives Screen participants to determine grouping Thoroughly consider research method and design Pilot and be prepared (i.e. researchers and kit)
Funding	Lack of research funding Lack of law enforcement agency funding	Work with larger forces Identify other sources of funding
Access to video stimuli	Expense Suitability of content for research purposes No control of stimuli as variables	Acquire access to law enforcement simulator Design own video stimuli Role players with weapon handling experience
Live role play scenarios	Personal protective equipment interference Poor reproducibility and lack of experimental control Risk of injury	Laser-based systems Measures to mitigate actor/role players variance (i.e. practice) Compliance with standardised safety procedure training

"Oh No, Not the Psychologists!"

The academic foundation for many expertise researchers is the field of psychology. When meeting others, it is common for expertise researchers to describe themselves as *psychologists* or *psychological researchers*. Even if researchers refer to themselves using different terms, other cues may point toward an affiliation with psychology (e.g., academic department name on business cards and email signatures). Police officers generally distrust academics, among other groups in society (Van Maanen, 1978) and are wary of psychologists in particular (Max, 2000; White & Honig, 1995).

From the perspective of police, *psychology* is most commonly associated with clinical or forensic psychology. That is, police may be unaware of the breadth of psychology as a field of study; they may not be familiar with expertise research, (applied) cognitive psychology, human factors, cognitive engineering, and naturalistic decision making. In light of this, we suggest that expertise researchers explain—during their initial contact with law enforcement agencies—what their general intent is (e.g., "I am interested in understanding what makes your top performers so good. How do they see situations differently than other officers?"). We have also found it helpful to state, explicitly, that we are *not* clinical psychologists. An additional piece of advice is to leverage police officers' familiarity with existing, accessible police research. For example, in the United States many police trainers are familiar with the work of the Force Science Institute (http://www.forcescience.org)—a research, consulting, and training organization that conducts and publishes police-related human performance research. In our experience, demonstrating awareness of the Force Science Institute and its work has helped law enforcement agencies better understand our goals and the type of research we seek to conduct.

Ethics Approval for Research

Before conducting research, researchers must typically obtain ethics approval from their institution. Expertise studies in law enforcement may involve experimental tasks with risk. For example, tasks might incorporate live-fire shooting, the use of paint-marking ammunition (e.g., http://simunition.com, https://utmworldwide.com) during role-play scenarios, exposure to oleoresin capsicum (i.e., pepper) spray, defensive tactics using electric shock knives (e.g., http://www.shocknife.com), high-speed driving, intense physical activity, and the use of a shootback cannon during video-based simulations. Aside from the potential for physical injury, there is also the possibility that participants will experience high levels of stress and anxiety. In fact, some studies specifically aim to manipulate the level of stress and anxiety. Given these factors, it is understandable that institutional review boards and research ethics committees will scrutinize applications.

Researchers can facilitate the ethics process by foreshadowing and pre-emptively addressing common concerns. For instance, one way to deal with the risk of injury is to plan research around existing training activities (i.e., activities that would be occurring whether the researchers were present or not), and using those opportunities for the purpose of collecting data. We also find it helpful to attach a letter of support from the collaborating law enforcement agency, stating that the planned experimental tasks are in fact routine training activities conducted under the guidance of experienced instructors, and listing the safety precautions that are in place for such training.

Another reasonable concern is that due to the hierarchical nature of law enforcement organizations, with their inherent power imbalances, personnel may be coerced by supervisors, either explicitly or implicitly, into participating in research. For example, an officer may feel that they will get passed over for a promotion or transfer opportunity if they refuse to participate in a research study. Such coercion would contravene the ethical mandate that participation in research be voluntary. Concerns about coercion can be addressed by stating that the researcher will be given the opportunity to address officers directly and solicit their participation, without a superior officer present. During this time, researchers should take due care to explain the voluntary nature of research participation.

Gaining Access to Officers as Research Participants

Even when law enforcement agencies are willing to support research, another hurdle that must be overcome is the logistics of gaining access to officers who could participate. One issue that may arise is that officers may be represented by a union that has concerns about whether data collected (e.g., about officers' decision-making abilities or shooting accuracy) could later be used as evidence against officers. We typically respond to this concern by stating that, as researchers, we: (a) adhere to an ethical code that places a strong emphasis on protecting the welfare of research participants, and (b) will keep participants' study-related information confidential to the extent permitted by law.

Ideally, the collaborating law enforcement agency will allow officers to participate in research during their scheduled shifts or mandatory training time. However, this is not always possible; many law enforcement agencies are understaffed and cannot afford to take officers away from their regular duties or their highly structured training schedule. Recruits can also be simply exhausted due to the intense academic and physical demands of their training. Additionally, larger regional training academies may not be in a position to directly facilitate access to potential participants. Such academies serve many law enforcement agencies, with each class comprising a few students from each agency. While at the academy, recruits are paid by their employing agency, which often imposes strict limitations on activities that fall outside of the mandatory training (e.g., because there are no funds available to pay for overtime). An alternative is to offer an incentive (e.g., gift voucher) for officers to participant when they are not on duty.

Note too, that not all recruits enter training with limited skills. Trainees with prior military experience might be expected to perform better under stress, and have higher levels of shooting and defensive tactics skills. Additionally, in some jurisdictions, officers who transfer from another department must complete recruit training again (even if they already have several years of law enforcement experience). Researchers would be wise to screen their samples for such individuals.

Another issue for researchers to consider is that due to the nature of expertise, there may be relatively few highly experienced officers, or *experts*, on the force. For example, there are relatively few highly experienced, tactically trained officers. They are often on call, in training, or operationally deployed, making it difficult to gain access to them. Subsequently, researchers should be prepared to have only limited access to domain experts and select their research methods accordingly. Researchers should thoroughly pilot test all procedures before collecting data with law enforcement personnel, and have a sufficient number of trained researchers on hand to ensure that data collection proceeds smoothly and efficiently.

Lack of Research Funding

Specific funding for expertise studies in law enforcement is difficult to come by. Individual law enforcement agencies typically do not have the financial resources to fund research studies, although they may be willing to support research by providing access to participants, facilities, equipment, and instructors/safety personnel. Whilst *in-kind* contributions like these, if costed, can add up to large investments and should not be underestimated, some form of financial contribution is usually necessary for the successful completion of research projects. An exception can occur for larger agencies; for example, the New York Police Department commissioned a report on firearm training (Rostker et al., 2008). From time to time, national funding bodies present requests for proposals that could encompass expertise research in law enforcement. In the United States, such agencies include the National Institute of Justice, National Science Foundation, and Department of Defense (e.g., Office of Naval Research, Army Research Lab). Other potential funders include insurance pool programs that provide insurance coverage to municipalities and police departments (see Aveni, 2008). Such programs have a vested interest in reducing payouts due to poor training and negligence. The US Federal Law Enforcement Training Centers (FLETC), although not a direct funder of research, has an Applied Research Branch which has produced research on police performance (e.g., Federal Law Enforcement Training Center, 2004). FLETC has mechanisms for establishing collaborations with universities, and also offers internship opportunities for undergraduate and graduate students interested in human performance research through the Department of Homeland Security.

Access to Representative Video Stimuli

For researchers interested in perceptual and decision-making expertise, video-based stimuli can provide a greater degree of ecological validity than pictorial or text-based stimuli. However, obtaining access to representative stimuli can be challenging. One solution is to acquire a law enforcement judgment-and-decision-making simulator, which includes a library of videos scenarios. Although there are several companies producing such simulators (e.g., http://www.lasershot.com, http://www.meggitttrainingsystems.com, http://www.milorange.com, http://www.cubic.com, http://www.titraining.com, and http://www.virtra.com), the simulators are relatively expensive, placing them beyond the reach of many researchers. Note, though, that several simulator companies do offer their video scenario library as a stand-alone product.

Even if such video stimuli can be obtained, researchers still need to evaluate the stimuli and determine whether any are suitable for their research purpose (i.e., depict truly representative tasks). To gain more control over the stimuli, some researchers have produced their own video stimuli for research (e.g., Aveni, 2008; James, Klinger, & Vila, 2014; Johnson et al., 2014). Depending on the complexity of the scenarios to be filmed, researchers should be aware of factors they might need to control while filming multiple trials. These factors include the actor's facial expression, arrangement of clothing, speed of physical actions, and furtive movements/glances (see Aveni, 2008). Employing police officers to play the role of suspects can enhance the realism of stimuli, as officers are more familiar with typical behaviors exhibited by civilians in police–citizen encounters, and better at deploying weapons. If trained actors are used, researchers should either employ actors with weapons-handling experience or provide instruction in weapons handling prior to filming.

Researchers should ensure that, in addition to scenarios that require officers to respond using force (e.g., shoot), they also include *don't shoot* or *no-threat* scenarios. This is necessary, as participants could exhibit a response bias (e.g., always respond by shooting). The appropriate ratio of threat to no-threat stimuli is an open question, and depends on the research goal. A common standard is to present equal proportions of threat and no-threat stimuli (Correll et al., 2007; James, James, & Vila, 2016; Nieuwenhuys, Savelsbergh, & Oudejans, 2012). Other researchers have used a higher proportion of threat stimuli (e.g., Johnson et al., 2014; Suss & Ward, 2018; Ward, Suss, Eccles, Williams, & Harris, 2011).

Using Live Role-Play Scenarios

In law enforcement, live role-play scenarios (also known as reality-based or force-on-force scenarios) are considered to be the most realistic type of training available for the development of tactical skills. Researchers have used live role-play scenarios as a basis for investigating human performance (e.g., Brisinda et al., 2014; Federal Law

Enforcement Training Center, 2004). In these scenarios, officers interact with *suspects*, who act out pre-defined roles. Officers are typically equipped with specially modified weapons that fire paint-marking rounds. Depending on the scenario, the suspect may be similarly armed, armed with a different weapon (e.g., replica knife), or unarmed. Because the marking rounds are designed to inflict a pain penalty, both the suspect and the officer must don protective equipment, including face masks. This can hamper verbal communication and also hides the suspect's face, obscuring potential cues of an impending attack. Laser-based systems that provide feedback via electric shock have been developed to overcome this issue (e.g., http://www.stressvest.com), but are less common.

Another issue is the reproducibility of the suspect's actions across participants. It is impossible to achieve the same level of experimental control using live-role player scenarios compared to video stimuli. If the goal is to maximize internal validity (i.e., identify cause-and-effect relationships), researchers who use live-role player scenarios to present the same scenario to multiple participants should take steps to train the actors to produce the same movement repeatedly. This can involve extensive drilling and practice so that the actors are given the opportunity to refine their actions, so as to minimize the variability of their body posture, facial expressions, verbal communication (i.e., content, tone, volume), and physical actions between trials/participants. In contrast, when the goal is to assess the efficacy of training programs, purposefully introducing stimulus variability is likely to enhance adaptability in operational settings (see concept-case coupling and case-proficiency scaling principles in Ward et al., 2018).

Illustrative Examples of How Research Has Been Conducted in the Context of Expertise in Law Enforcement

We now turn to provide selected, illustrative examples of research accomplishments to date, including key details of methods used. We focus on expertise studies related to decision making in tactical situations. We cover a variety of cognitive task analysis methods, including retrospective incident-based interviews, gaze tracking, temporal occlusion, and option generation. Generally, the studies described in this chapter all have the common aim of identifying strategies and/or skills that separate expert from novice decisional performance, with a view to making evidence-based recommendations for accelerating the acquisition of decision-making skill. For examples utilizing other methods, readers are referred to studies on driving (Crundall, Chapman, Phelps, & Underwood, 2003; Crundall, Chapman, France, Underwood, & Phelps, 2005), intoxication judgments (Langenbucher & Nathan, 1983; Pisoni & Martin, 1989), and shooting performance (Landman, Nieuwenhuys, & Oudejans, 2016).

Use of Retrospective, Incident-Based Report Methods for Understanding Cognition during Critical Incidents

To identify the cognitive processes and strategies that support superior decision-making performance during armed confrontations, Boulton and Cole (2016) conducted critical decision method interviews with UK firearms officers. Experienced firearms officers had at least 10 years of specialized experience; less-experienced firearms officers had 3 or less years of experience in the role. During the interviews, officers were asked to walk through a *challenging* and non-routine armed confrontation that they had experienced. Note that the critical decision method interview protocol has been successfully used by several researchers to develop insight and understanding of the cognitive processes, skill, and strategies used during critical decision-making circumstances (e.g., Harris, Eccles, Freeman, & Ward, 2017; Klein, Klein, Lande, Borders, & Whitacre, 2015).

In line with guidelines and recommendations for conducting critical decision method interviews (Crandall, Klein, & Hoffman, 2006), multiple sweeps were made through the recalled incidents: (i) initial free recall of incident by the participant, (ii) interviewer restatement to establish consistency, (iii) incident timeline creation and identification of decision points, (iv) decision point probing, and finally (v) hypothetical probes (see Chapter 19, "Incident-Based Methods for Studying Expertise," by Militello & Anders, this volume). The researchers used a script to ensure that the order and content of the probe questions was standardized across all officers. A large pad of paper was used to draw the timeline for each incident; this then served as a visual aid to identify and examine key decision points using the probe questions. The interviews lasted between 1 and 2 hours each; each interview was audio recorded and transcribed for analysis.

During analysis, the researchers read the transcripts multiple times noting repeated themes and cognitive issues (e.g., cue recognition, situation assessment). The transcripts were then inductively coded for repeated ideas, which were reviewed and grouped into themes and subthemes. This process was iterative and involved multiple revisions. To increase validity of the analysis process, qualitative data analysis software NVivo 10 (QSR International) was used to create a transparent and *auditable footprint* (Sinkovics & Alfoldi, 2012) of the analysis. To demonstrate the veracity of the research method, analysis, and conclusions, qualitative assessments were tested for inter-rater reliability using Cohen's kappa to quantify the level of consistency among two independent raters who coded 30 percent of the data. Data were consolidated into a decision requirements table which was used to represent key decisions and to organize recalled cues, strategies, and practices associated with expertise, as well as to identify specific challenges, potential pitfalls, and errors typically associated with inexperience.

The results highlighted the importance of adaptability as a defining feature of expert decision making in the context of armed confrontation. This key finding was used to make recommendations for police firearms training which seeks to accelerate the development of adaptive decision-making skills (i.e., mental modelling, sense-making, and cognitive flexibility) through the systemic exposure of trainees to a variety of scenarios, including *worst-case* scenarios, that cannot be solved through standard operating procedures.

Expertise Differences in Visual Attention during Shoot/Don't-Shoot Scenarios

Vickers and Lewinski (2012) investigated differences in performance and visual attention between less-experienced and experienced tactical team members. The research incorporated live role-player scenarios, handguns adapted to fire paint-marking ammunition, and a mobile eye tracker. The scenario was set in a government office, with the armed officer providing security inside. A female receptionist sat at a desk and was approached by a male who wanted to resolve an issue with his passport. The officer wore the eye tracker and stood seven meters from the desk, facing the receptionist (i.e., observing the male from behind). The officer was confined to that location and instructed to resolve any threat using their handgun (i.e., they were not allowed to approach the male). The male became more and more agitated and began arguing loudly with the receptionist. Approximately 50 seconds after entering the office, the male spun around to face the officer while drawing either a handgun or a cell phone. The officer responded by either shooting at the male or inhibiting a shooting response. After an initial practice *gun* trial, each officer completed four gun trials and two cell phone trials in a randomized order.

Officers' gaze behavior and physical responses were recorded synchronously using the vision-in-action system (Vickers, 2007). Performance analysis focused on the initial 7 seconds—when the male role player entered the scene—and on the final 7 seconds leading up to the shoot/don't-shoot decision. In the final 7 seconds, for example, the researchers identified three phases of observable response (i.e., draw firearm, hold firearm, aim/fire). The onset and duration of each phase was determined via the video recording, and then analyzed using experience as a between-participants factor. The analysis of gaze behavior focused on the final six fixations leading up to the shoot/don't-shoot decision (i.e., when the male role player spun around). Fixations were coded by location (i.e., assailant's weapon/cell phone, other locations on assailant's body, fixations that were not on assailant, officer's firearm/sights).

Vickers and Lewinski's (2012) results were used to make specific recommendations to police firearms training in terms of target fixation, weapon alignment, and the type of conditions training should be conducted under (i.e., high levels of pressure and anxiety). Based on these findings, they suggested that changes in police training consistent with their recommendations would contribute to better decision making and performance in less-experienced officers.

Use of Temporal-Occlusion and Option-Generation Methods

Suss and Ward (2012, 2018) used video scenarios from a police judgment and decision-making simulator (http://www.milorange.com) to examine skill-based differences in anticipation and response ability. The researchers first identified candidate video clips of high- and low-frequency law enforcement situations in which there was sufficient context with which to anticipate the outcome. Then, each video was edited to end (i.e., black screen) at a point where it was possible, in theory, to correctly anticipate the outcome. Less-experienced and experienced police officers observed each clip. To elicit anticipation options, officers responded to the question, "What could happen next on the screen in the next few seconds?" After listing their anticipation option(s), they assigned likelihood ratings to each option (likelihood ratings totaled 100), and rated how threatening each option was if it were to occur next (each option was rated independently on a scale ranging from 0 to 100). After generating anticipation options, officers were prompted to generate response options: "How could you respond in the next few seconds?" Officers then rated the likelihood with which they would pursue each option, and how good each option was for their own personal safety.

After gathering the data, each option was classified by a subject-matter expert as being either relevant or irrelevant with respect to the specific situation. The analysis focused on experience-based differences in the number and type (relevant/irrelevant) of options generated, and the relationship between the number of options generated in the prediction and response phases (i.e., Does generating more, or more relevant, prediction options result in the generation of better responses options?).

As the other two studies did, the results from this research were used to make specific recommendations for improving the decision-making skills and performance of less-experienced officers. These recommendations suggested that to improve prediction accuracy and lead to better response options, training should be designed to reduce officers' focus on irrelevant options rather than aiming to increase the generation of relevant options.

KEY INSIGHTS: TASK-SPECIFIC EXPERTISE

The general public, perhaps driven by TV depictions of police, often see police officers as superheroes, imbued with decision-making abilities that exceed those of the average citizen. Therefore, a general perception remains that police officers are *expert* decision makers more broadly, and expected to behave expertly based on their intuition and hunches (Alschuler, 2007; Segal, 2012). This is despite findings that demonstrate that police are susceptible to decision biases, just as non-police officers are (Ask & Granhag, 2005; Fahsing & Ask, 2013; Taslitz, 2010).

This perception also appears to be evident in academia, where some researchers have compared police officers as an *expert* group to the general public as a *control* group. Others have relied on years of police experience as a proxy for expertise (e.g., Fulford, 2011), rather than defining expertise based on evidence of performance reproducibility (Ericsson & Ward, 2007). However, law enforcement is a profession encompassing many tasks that require a combination of skills of which someone could have expertise in. Therefore, it is difficult to determine that there is such a thing as *general* law enforcement expertise.

Instead, expertise in law enforcement often encompasses skill specialization. For example, officers' decision-making processes may vary depending on their specific role (e.g., investigative vs tactical and/or strategic). Reflecting this, we should remain clear on the area of expertise that we are examining in order to ensure the officer's experience within their role evidences their status as an expert within that specific subfield. With this in mind, a review of the key insights into task-specific expertise within law enforcement was conducted to identify how academia, police, and the justice system talk about and address expertise. Some key research findings regarding selected types of policing expertise (i.e., decision making, tactical skills, conflict resolution and social skill, visual perception and observation) will be discussed here.

Decision Making

Studies have examined a variety of different applications to decision-making expertise in law enforcement. Some have focused on identifying psychophysiological indices of expert versus novice performance in deadly force judgment and decision making (Johnson et al., 2014); others have explored how expertise impacts police officers' assessment of operational situations (Baber & Butler, 2012). Although there is a big difference between tactical (fast) and strategic/investigative (slow) operational tasks, the common thread in both types of law enforcement roles is the need for appropriate decision making.

Comparing the decisional processes and strategies underlying the performance of expert and novice British firearms officers during armed confrontations, Boulton and Cole (2016) highlighted the importance of adaptability in terms of the flexible application of experiential knowledge, strategies, and skills in response to situational demands, to expertise in this context. With their extensive domain experience, expert officers were better able to: (i) categorize incidents; (ii) recognize anomalies; (iii) be aware of, and quickly adapt to, the dynamic environment; and (iv) use their training automatically. Compared to the flexible experiential-based decisions of expert officers, novice officers reported a more sequential and linear process of tactical decision making that involved extended verbalizations and continued conscious processing throughout. Girodo's (2007) experimental study also revealed that generally, tactical experts (e.g., tactical force leaders) reasoned analytically and deliberatively compared to the more reactive and procedural reasoning exhibited by non-expert tactical force officers.

A between-groups (expert vs novice) comparison of EEG/physiological response during high-fidelity deadly force judgment and decision-making simulations conducted by Johnson et al. (2014) found that not only did police officers and military personnel (who Johnson et al. defined as experts) have a pass rate significantly higher than that of untrained civilians, but that this difference was also reflected in physiological responses. Heart rate acceleration from rest during the scenario was significantly greater in the expert group and this was suggested to be linked to more responsive threat detection.

Despite findings of cognitive processing and physiological response differences between experienced and less-experienced officers in firearms situations, such differences have not been conclusively found to result in superior performance. When comparing shooting behavior in simulated firearms environment, Ho (1994; 1997) found that rookies consistently displayed better judgment and shooting accuracy than veterans, in that: (i) rookies had a higher survival rate than veterans when responding to life-threatening situation, and (ii) rookies showed better judgment than veterans in averting or withholding fire when confronting a harmless suspect during high-risk encounters. Such surprising results may reflect the more advanced and up-to-date training rookies received compared to veterans in the sample (Doerner & Ho, 1994). In a quasi-experimental, between-participants contrast of handgun-shooting skill in police recruits with differing shooting experience, Lewinski, Avery, Dysterheft, Dicks, and Bushey (2015) concluded that trained officers had no advantage over intermediate shooters and only a small advantage over novices.

Studies which explore racial bias in police shoot/don't-shoot decisions have typically compared a police sample, as an expert group, to a civilian control sample. Generally, these studies have found a reduced racial bias in the decisions of officers compared to that in civilians; however, they also identify important complexities within this finding that is impacted by situational features (Correll, Hudson, Guillermo, & Ma, 2014; Luini & Marucci, 2015; Sim, Correll, & Sadler, 2013). Correll et al. (2014) compared police officers and civilians on a first-person shooter task designed to examine racial bias toward African-Americans and found that the police officers were faster, more accurate, and less racially biased in terms of errors; however, officers still exhibited a bias toward shooting unarmed African-American males. These findings suggest that expertise enabled officers to minimize behavioral consequences of stereotypes (i.e., learned to over-ride a pre-potent response) via exercise of cognitive control. The final part of this study examined whether training could reduce racial bias responses and found that, although promising results indicated training did eliminate racial bias under controlled conditions, this bias could re-emerge under conditions of high cognitive demand. This has important implications for training, indicating that it is crucial for police to train in situations that replicate operational settings (i.e., intense video or live-action simulation that induce high arousal). Sim et al. (2013) also found that when compared with lay participants, police officers generally showed less racial bias in laboratory-based shooter simulations (Correll et al., 2007); however, when the training context or operational experience reinforced officers' association between African-Americans and danger, training did not seem to attenuate bias.

Another task in which decision-making expertise has been investigated is urban house-clearing operations. Harris-Thompson, Wiggins, and Ho (2006) employed cognitive task analysis techniques (see Crandall et al., 2006; Hoffman & Militello, 2008) to identify the critical decisions and cues used to assess these dynamic and potentially dangerous situations. The cues were classified into four types: environmental assessment (e.g., noticing whether hinges on doors open inward or outward), threat assessment (e.g., watching the suspect's hands), situational assessment (e.g., time of day), and team assessment (e.g., hearing a team member call for backup). Experts' performance appeared to rely on their ability to rapidly switch between following standard operating procedures, on the one hand, and making recognition-primed decisions (Klein, 1989) based on the dynamic, unfolding events, on the other.

In a slower-paced decisional police environment, Baber and Butler (2012) compared novice and expert crime scene examiners' search strategies in simulated crime scenes using concurrent verbal reports and head-mounted video recordings (for a discussion of introspective-type and verbal reports, see Chapter 17, "A Historical Perspective on Introspection," by Ward, Wilson, Suss, Woody, & Hoffman, this volume). Baber and Butler found that although both groups paid attention to the likely modus operandi of the perpetrator (in terms of possible actions taken), the experts paid more attention to objects with evidential value based on consideration of the potential future analysis and actions that can be taken. Therefore, expertise in the specific area of crime scene investigation lays in the selective search strategies toward objects of evidential importance and involve predictive mental modeling.

Conflict Resolution and Social Skills

Researchers have also sought to identify the social skills that lead to successful, positive police–public interactions. Sun (2003) examined the behavioral differences between police field-training officers, who provide on-the-job mentoring to rookies, and their comparable colleagues in handling interpersonal conflicts. Sun found that field-training officers performed a greater number of supportive actions than non-field-training officers throughout their encounters with citizens. Similarly, Klein and colleagues' (Klein et al., 2015; Lande & Klein, 2016) *Good Stranger* research concluded that expertise in managing civilian encounters without creating hostility was most significantly predicted by officers' ability to build trust. In turn, the ability to take another's perspective and gauge prudent risk significantly predicted the ability to build trust. Furthermore, the researchers identified several pathways for acquiring a Good Stranger frame, including observing role models, peer pressure, becoming more effective at gaining civilian cooperation, and recognizing the problems created by failing to build trust.

Armed and Unarmed Use-of-Force Tactics

In terms of tactical skill expertise, research has predominantly explored the utility of training to increase or accelerate skilled performance. For instance, Renden, Landman, Savelsbergh, and Oudejans (2015) found that engagement in martial arts training benefited defensive tactics (i.e., hand-to-hand combat) performance under threatening conditions and suggested that this improvement reflected the development of anticipation skills and ability to counterattack. Staller and Abraham (2016) interviewed expert self-defense instructors about the characteristics of optimal training environments. The main themes included understanding the nature of violent attacks, learning and teaching how to solve problems, achieving a balance between realism and safety in training, and providing trainees with opportunities for deliberate practice. Developing expertise in defensive tactics is especially important in countries where police are routinely unarmed, and in jurisdictions that encourage the use of non-lethal force when safe for officers.

Firearms proficiency is lacking in US law enforcement, where officers typically receive only 50 hours training and/or training that lacks efficacy (Charles & Copay, 2003; see also Lewinski et al., 2015). Biggs, Cain, and Mitroff (2015) not only found shooting error to be negatively correlated with the cognitive ability to inhibit an initiated response, but also that active response-inhibition training reduced error. Therefore, Biggs et al.'s findings indicate that there is potential to improve shooting performance and thus increase or accelerate shooting expertise via cognitive training. Charles and Copay (2003) conducted a repeated-measures comparison of inexperienced shooters before and after basic law-enforcement firearms training. They found that marksmanship skill significantly improved after training, and that participants were significantly quicker to load, reload, unload, and clear malfunctions after the course. Furthermore, they suggested that specificity of training relative to performance requirement was crucial to skill attainment and that scenario-based training that included an element of stress was most effective. Other tactical research related to firearms compared the efficacy of different room-entry techniques used by police when searching buildings (Blair & Martaindale, 2014).

Research into use-of-force (e.g., defensive tactics, shooting) skill perishability and retention suggests that there is some scientific basis for providing explicit knowledge of results to trainees early on in training, but then decreasing this as skills become automatic (Angel et al., 2012). This paper concluded that the critical factors that influence skill retention are the characteristics of the individual, the nature of the task, and the nature of the training that should expose trainees to as many different situations as possible to promote knowledge, skill transfer, and adaptability (Angel et al., 2012; Boulton & Cole, 2016).

Visual Perception and Observation Skills

Some literature suggests that expert performers' advantage is due to their perceptual abilities and observation skills. Generally, experienced operators make eye movements toward expected goal-relevant areas of the scene or an increased frequency of fixations on goal-relevant information (Crundall & Eyre-Jackson, 2017; Howard, Troscianko, Gilchrist, Behera, & Hogg, 2013). Vickers and Lewinski (2012) investigated differences in performance and visual attention between experienced and less-experienced firearms officers during a shoot/don't-shoot decision-making scenario. Compared to the less-experienced officers, experienced officers (a) drew their guns earlier in the scenario, (b) fired at the assailant less frequently under the cell phone condition, (c) shot before the assailant did on a greater percentage of trials under the gun condition, and (d) hit the assailant more frequently under the gun condition. Analysis of the eye-tracking data revealed that experienced officers fixated on more locations on the assailant where a weapon could be concealed and more of the experienced officers fixated on the assailant's weapon or cell phone than did less-experienced officers. Compared to the experienced officers, more of the less-experienced officers fixated on their own weapon (e.g., sights), and on non-weapon locations on the assailant and off the assailant, suggesting a difference in officers' weapon focus and the role of optimal gaze control when under extreme pressure and threat which could be utilized in firearms training.

Crundall and Eyre-Jackson (2017) conducted an independent group comparison of criminal activity prediction from CCTV clips between police officers and a control group. Signal detection analysis revealed that the police officers were marginally more accurate than the control group at detecting imminent criminal activity or anti-social behavior, and that the police were better than the control group at identifying the type of crime about to happen. Based on these results, Crundall and Eyre-Jackson suggested that the benefit of expertise in this task lies in the ability to direct visual attention to the most relevant locations in the footage, at the most appropriate time.

Along similar lines, Koller, Wetter, and Hofer (2016) examined the ability to use nonverbal behavior to detect imminent baggage theft at an international airport. Civilians, police recruits, inexperienced officers, experienced officers, and criminal investigators observed video clips of actual baggage theft incidents that were temporally occluded prior to the theft. At the point of occlusion, participants indicated the individual(s) they anticipated would commit theft. Signal detection analysis revealed that criminal investigators—who were most familiar with the thieves' modus operandi—exhibited better sensitivity than all other groups, except experienced officers. Experienced officers were, in turn, more sensitive than civilians and recruits, but not inexperienced officers.

Observation is crucial to certain aspects of law enforcement work and investigation. Exploring the impact of expertise on incident report-writing skills, Vredeveldt, Knol, and van Koppen (2017) found that surveillance detectives provided more accurate incident reports that may serve as evidence in court than both untrained civilians and uniformed police officers, suggesting that specialized detectives on surveillance teams are more observant of the crime-relevant aspects of an incident. In a related surveillance task, Stainer, Scott-Brown, and Tatler (2013) found that trained CCTV operators spent most of their time searching on the single-scene spot-monitor, rather than spending a lot of time viewing the multiplex wall, suggesting a selective approach based on crime likelihood prediction. Damjanovic, Pinkham, Clarke, and Phillips (2014) examined the ability of experienced officers with extensive riot control experience to identify threats within the context of emotional and neutral faces. The experienced officers showed enhanced detection for threatening faces and greater degree of inhibitory control over angry face distractors, compared to trainee officers and civilians.

Future Directions

Societies are justly concerned about appropriate use of force by police. Academic research—particularly expertise research—has had relatively little direct impact on law enforcement training practices. We believe, therefore, that expertise researchers can make significant contributions by establishing training methods that are accessible, easily implemented, and directly relevant to police work, and whose efficacy is empirically supported. Here, we propose ideas based on our experience working with, and observing, law enforcement.

Police officers must excel at perceptual discrimination under stress (e.g., real firearm vs replica firearm, armed undercover officer vs armed suspect; see Band, Ray, Wollert, & Norris, 2016). Employing methods from visual cognition and sport science, researchers should identify whether skill exists in these perceptual discrimination tasks. If expertise is identified, researchers should examine the cognitive mechanisms that underlie skilled performance, and develop and evaluate training methods to improve novices' ability. Ericsson's expert performance approach (Ericsson & Ward, 2007) has been proposed as a guide to such an endeavor (see Ward, Suss, & Basevitch, 2009). Similarly, researchers should assess whether highly trained officers excel at anticipating a suspect's actions (i.e., whether the suspect is drawing a weapon vs a non-weapon from concealment). This line of research could leverage temporal- and spatial-occlusion methods that have been used to investigate anticipation in sport, and identify the cues used by experts.

A related line of research would seek to develop gamified, web-based tools that provide opportunities to deliberately practice perceptual–cognitive skills. This would be particularly useful during police academy training, during which trainees often have periods of downtime (e.g., when instructors are running trainees individually through role-play scenarios). A web-based training tool would allow trainees to engage in deliberate practice of otherwise difficult-to-train skills, without requiring the presence of an instructor. Recently, researchers have created a platform to support domain-specific cognitive training that incorporates expert feedback (e.g., Klein & Borders, 2016).

Another challenge for expertise researchers is to identify how to optimally integrate different training modalities (e.g., static live fire, dynamic force-on-force, interactive video simulation, defensive tactics, less than lethal) to create adaptive experts. It is relatively common for police instructors to specialize in either defensive tactics or firearms, creating training *silos* (Force Science, 2016). Trainees, however, need to be able to integrate all of these skills in use-of-force situations, possibly transitioning from aiming their firearm to using physical combat skills in response to a suspect's actions. Although recent research has described an approach for designing integrated training (Staller, Bertram, & Körner, 2017), there is a lack of empirical evidence that details when, how much, and what type of force-on-force training (e.g., using paintball-marking ammunition) should be introduced during training. Note that researchers have investigated debriefing and feedback techniques for force-on-force training (e.g., Phelps, Strype, Le Bellu, Lahlou, & Aandal, 2018; Sjöberg & Karp, 2012).

Finally, expertise researchers should evaluate the relative efficacy and efficiency of different types of training for improving performance. Recently, training designed to "reduce psychological threat perception and improve physiological control" was found to improve Finnish tactical officers' decision making (Andersen & Gustafsberg, 2016, p. 6; also see Andersen et al, 2015; Shipley & Baranski, 2002). This training incorporated mental imagery and breathing control. An interesting question is whether this type of training—or perceptual–cognitive skill training—results in better performance given equal training time, and which type of training is easier to deliver and more readily accepted by trainees.

To achieve these goals, we encourage researchers to engage with the law enforcement community and learn about the many challenges it faces in developing adaptive experts. Researchers without a background in law enforcement should consider partnering with *pracademics*: researchers interested in studying their own work domain (Huey & Mitchell, 2016). Alternatively, researchers should establish collaborations with practitioners (see International Association of Chiefs of Police, n.d.). A collaborative approach can improve the quality of the research; skilled researchers can increase the reliability of the findings, whilst the subject-matter expertise of practitioners can increase the validity and applicability of the findings to the real world. A particularly fruitful collaboration has been that between the Dutch national police and researchers at VU University Amsterdam (e.g., Nieuwenhuys, Caljouw, Leijsen, Schmeits, & Oudejans, 2009; Nieuwenhuys et al., 2012). Finally, researchers should ensure that the research they conduct will produce value for the practitioners (Rynes, Bartunek, & Daft, 2001). Offering practitioner-friendly summary reports of the research or conducting workshops to disseminate findings may more effectively communicate the impact of the research than sharing the resulting academic journal articles.

Acknowledgments

We have both been fortunate to conduct research with several law enforcement agencies. We acknowledge the administrators and officers who welcomed us to the inner sanctum of police training, and shared their knowledge, expertise, perspectives, and ideas with us. Without those visionaries, we could not have pursued our passion to make scientific contributions to law enforcement.

REFERENCES

Akinci, C., & Sadler-Smith, E. (2013). Assessing individual differences in experiential (intuitive) and rational (analytical) cognitive styles. *International Journal of Selection and Assessment 21*, 211–221.

Alschuler, A. W. (2007). The upside and downside of police hunches and expertise. *Journal of Law, Economics & Policy 4*, 115–117.

American Psychological Association. (2004). *The nature and influence of intuition in law enforcement: Integration of theory and practice.* Retrieved from http://www.apa.org/about/gr/science/advocacy/2004/intuition.aspx

Andersen, J. P., & Gustafsberg, H. (2016). A training method to improve police use of force decision making. *SAGE Open 6*(2).

Andersen, J. P., Papazoglou, K., Koskelainen, M., Nyman, M., Gustafsberg, H., & Arnetz, B. B. (2015). Applying resilience promotion training among special forces police officers. *SAGE Open 5*(2).

Angel, H., Adams, B. D., Brown, A., Flear, C., Mangan, B., Morten, A., & Ste-Croix, C. (2012). *Review of the skills perishability of police "use of force" skills.* HumanSystems Incorporated: Guelph, Canada. Retrieved from https://www.publicsafety.gc.ca/lbrr/archives/cnmcs-plcng/cn28923-eng.pdf

Ask, K., & Granhag, P. A. (2005). Motivational sources of confirmation bias in criminal investigations: The need for cognitive closure. *Journal of Investigative Psychology and Offender Profiling 2*, 43–63.

Aveni, T. J. (2008). *The MMRMA deadly force project: A critical analysis of police shootings under ambiguous circumstances.* Retrieved from http://www.theppsc.org/Research/V3.MMRMA_Deadly_Force_Project.pdf

Baber, C., & Butler, M. (2012). Expertise in crime scene examination: Comparing search strategies of expert and novice crime scene examiners in simulated crime scenes. *Human Factors 54*, 413–424.

Band, D., Ray, R., Wollert, T., & Norris, W. (2016). *Responses to encounters: Uniformed officer responses to encounters with plain clothes officers* (Report FLETC-ARB-01-2016). Glynco, GA: Applied Research Branch, Federal Law Enforcement Training Center. Retrieved from https://www.fletc.gov/sites/default/files/ARB_%20Newsletter-2016.pdf

Bayley, D. H. (2016). The complexities of 21st century policing. *Policing 10*, 163–170.

Bayley, D. H., & Bittner, E. (1984). Learning the skills of policing. *Law and Contemporary Problems 47*(4), 35–59.

Biggs, A. T., Cain, M. S., & Mitroff, S. R. (2015). Cognitive training can reduce civilian casualties in a simulated shooting environment. *Psychological Science 26*, 1164–1176.

Blair, J. P., & Martaindale, M. H. (2014). *Evaluating police tactics: An empirical assessment of room entry techniques.* Oxford: Anderson Publishing.

Boulton, L., & Cole, J. (2016). Adaptive flexibility: Examining the role of expertise in the decision making of authorized firearms officers during armed confrontation. *Journal of Cognitive Engineering and Decision Making 10*, 291–308.

Brisinda, D., Venuti, A., Cataldi, C., Efremov, K., Intorno, E., & Fenici, R. (2014). Real-time imaging of stress-induced cardiac autonomic adaptation during realistic force-on-force police scenarios. *Journal of Police and Criminal Psychology 30*, 71–86.

Burkhardt, B. C., Akins, S., Sassaman, J., Jackson, S., Elwer, K., Lanfear, C., ... Stevens, K. (2017). University researcher and law enforcement collaboration. *International Journal of Offender Therapy and Comparative Criminology 61*, 508–525.

Charles, M. T., & Copay, A. G. (2003). Acquisition of marksmanship and gun handling skills through basic law enforcement training in an American police department. *International Journal of Police Science and Management 5*, 16–30.

Cockbain, E., & Knutsson, J. (Eds.). (2014). *Applied police research: Challenges and opportunities*. Abington, Oxon, UK: Routledge.

College of Policing. (2016). *Policing Education Qualifications Framework: Frequently asked questions* (C305I0216). London: College of Policing. Retrieved from http://www.college.police.uk/What-we-do/Learning/Policing-Education-Qualifications-Framework/Documents/PEQF_faqs_final_290116.pdf

Cordner, G., & White, S. (2010). The evolving relationship between police research and police practice. *Police Practice and Research 11*, 90–94.

Correll, J., Hudson, S. M., Guillermo, S., & Ma, D. S. (2014). The police officer's dilemma: A decade of research on racial bias in the decision to shoot. *Social and Personality Psychology Compass 8*, 201–213.

Correll, J., Park, B., Judd, C. M., Wittenbrink, B., Sadler, M. S., & Keesee, T. (2007). Across the thin blue line: Police officers and racial bias in the decision to shoot. *Journal of Personality and Social Psychology 92*, 1006–1023.

Crandall, B. W., Klein, G. A., & Hoffman, R. R. (2006). *Working minds: A practitioner's guide to cognitive task analysis*. Cambridge, MA: MIT Press.

Crundall, D., Chapman, P., France, E., Underwood, G., & Phelps, N. (2005). What attracts attention during police pursuit driving? *Applied Cognitive Psychology 19*, 409–420.

Crundall, D., Chapman, P., Phelps, N., & Underwood, G. (2003). Eye movements and hazard perception in police pursuit and emergency response driving. *Journal of Experimental Psychology: Applied 9*, 163–174.

Crundall, D., & Eyre-Jackson, L. (2017). Predicting criminal incidents on the basis of non-verbal behaviour: The role of experience. *Security Journal 30*(3), 703–716.

Damjanovic, L., Pinkham, A. E., Clarke, P., & Phillips, J. (2014). Enhanced threat detection in experienced riot police officers: Cognitive evidence from the face-in-the-crowd effect. *Quarterly Journal of Experimental Psychology 67*, 1004–1018.

Doerner, W. G., & Ho, T. (1994). Shoot—don't shoot: Police use of deadly force under simulated field conditions. *Journal of Crime and Justice 17*(2), 49–68.

Dror, I. E. (2016). A hierarchy of expert performance. *Journal of Applied Research in Memory and Cognition 5*, 121–127.

Ericsson, K. A. (2014). Adaptive expertise and cognitive readiness: A perspective from the expert-performance approach. In H. F. O'Neil, R. S. Perez, & E. L. Baker (Eds), *Teaching and measuring cognitive readiness* (pp. 179–197). Boston: Springer.

Ericsson, K. A., & Ward, P. (2007). Capturing the naturally occurring superior performance of experts in the laboratory: Toward a science of expert and exceptional performance. *Current Directions in Psychological Science 16*, 346–350.

Fahsing, I., & Ask, K. (2013). Decision making and decisional tipping points in homicide investigations: An interview study of British and Norwegian detectives. *Journal of Investigative Psychology and Offender Profiling 10*, 155–165.

Federal Law Enforcement Training Center. (2004). *Survival scores research project*. Glynco, GA: Author.
Force Science. (2016, April 21). "Strong but wrong" training: A Force Science lead instructor shares his thoughts on fatal mistakes common to police training. Retrieved from https://www.calibrepress.com/2016/04/strong-but-wrong-training/
Force Science News. (2006). How are 5%ers created? By "effortful study," new report says. *Force Science News 50*. Retrieved from https://www.forcescience.org/2006/08/how-are-5ers-created-by-effortful-study-new-report-says/
Fulford, T. R. (2011). Writing scripts for silent movies: How officer experience and high-crime areas turn innocuous behavior into criminal conduct. *Suffolk University Law Review 45*, 497–522.
Fyfe, N. R., & Wilson, P. (2012). Knowledge exchange and police practice: Broadening and deepening the debate around researcher–practitioner collaborations. *Police Practice and Research 13*, 306–314.
Girodo, M. (2007). Personality and cognitive processes in life and death decision making: An exploration into the source of judgment errors by police special squads. *International Journal of Psychology 42*, 418–426.
Guillaume, P., Sidebottom, A., & Tilley, N. (2012). On police and university collaborations: A problem-oriented policing case study. *Police Practice and Research 13*, 389–401.
Hansen, J. A., Alpert, G. P., & Rojek, J. J. (2014). The benefits of police practitioner–researcher partnerships to participating agencies. *Policing 8*, 307–320.
Harris, K. R., Eccles, D. W., Freeman, C., & Ward, P. (2017). "Gun! Gun! Gun!": An exploration of law enforcement officers' decision-making and coping under stress during actual events. *Ergonomics 60*, 1112–1122.
Harris-Thompson, D., Wiggins, S. L., & Ho, G. (2006). *Using cognitive task analysis to develop scenario-based training for house-clearing teams*. Fairborn, OH: Applied Research Associates. Retrieved from http://cradpdf.drdc-rddc.gc.ca/PDFS/unc92/p532703.pdf
Ho, T. (1994). Individual and situational determinants of the use of deadly force: A simulation. *American Journal of Criminal Justice 18*, 41–60.
Ho, T. (1997). Police use of deadly force and experience: Rookie v. veteran. *Criminal Justice Studies 10*, 127–141.
Hoffman, R. R., & Militello, L. (2008). *Perspectives on cognitive task analysis: Historical origins and modern communities of practice*. New York: Psychology Press.
Hoffman, R. R., Ward, P., Feltovich, P. J., DiBello, L., Fiore, S. M., & Andrews, D. H. (2014). *Accelerated expertise: Training for high proficiency in a complex world*. New York: Psychology Press.
Howard, C. J., Troscianko, T., Gilchrist, I. D., Behera, A., & Hogg, D. C. (2013). Suspiciousness perception in dynamic scenes: A comparison of CCTV operators and novices. *Frontiers in Human Neuroscience 7*, 441.
Huey, L., & Mitchell, R. J. (2016). Unearthing hidden keys: Why pracademics are an invaluable (if underutilized) resource in policing research. *Policing 10*, 300–307.
International Association of Chiefs of Police. (n.d.). Establishing & sustaining law enforcement-researcher partnerships: Guide for researchers. Arlington, VA: Author. Retrieved from https://www.theiacp.org/sites/default/files/all/d-e/EstablishingSustaingLawEnforcement-ResearchPartnershipsGuideforResearchers.pdf

James, L., James, S. M., & Vila, B. J. (2016). The reverse racism effect. *Criminology & Public Policy* 15, 457–479.

James, L., Klinger, D., & Vila, B. (2014). Racial and ethnic bias in decisions to shoot seen through a stronger lens: Experimental results from high-fidelity laboratory simulations. *Journal of Experimental Criminology* 10, 323–340.

Johnson, R. R., Stone, B. T., Miranda, C. M., Vila, B., Lois, J., James, S. M., . . . Berka, C. (2014). Identifying psychophysiological indices of expert versus novice performance in deadly force judgment and decision making. *Frontiers in Human Neuroscience* 8, 512.

Klein, G. A. (1989). Recognition-primed decisions. In W. B. Rouse (Ed.), *Advances in man-machine systems research* (vol. 5, pp. 47–92). Greenwich, CT: JAI Press.

Klein, G., & Borders, J. (2016). The ShadowBox approach to cognitive skills training: An empirical evaluation. *Journal of Cognitive Engineering and Decision Making* 10, 268–280.

Klein, G. A., Klein, H. A., Lande, B., Borders, J., & Whitacre, J. C. (2015). Police and military as good strangers. *Journal of Occupational and Organizational Psychology* 88, 231–250.

Koller, C. I., Wetter, O. E., & Hofer, F. (2016). "Who's the thief?" the influence of knowledge and experience on early detection of criminal intentions. *Applied Cognitive Psychology* 30, 178–187.

Langenbucher, J. W., & Nathan, P. E. (1983). Psychology, public policy, and the evidence for alcohol intoxication. *American Psychologist* 38, 1070–1077.

Lande, B., & Klein, G. A. (2016). Moving the needle: The science of good police-citizen encounters. *The Police Chief* 83(3), 28–33.

Landman, A., Nieuwenhuys, A., & Oudejans, R. R. D. (2016). The impact of personality traits and professional experience on police officers' shooting performance under pressure. *Ergonomics* 59, 950–961.

Lee, T. L. (2016). Tennessee citizen police academies: Program and participant characteristics. *American Journal of Criminal Justice* 41, 236–254.

Lewinski, W. J., Avery, R., Dysterheft, J., Dicks, N. D., & Bushey, J. (2015). The real risks during deadly police shootouts: Accuracy of the naïve shooter. *International Journal of Police Science & Management* 17, 117–127.

Love, K. G. (1983). Empirical recommendations for the use of peer rankings in the evaluation of police officer performance. *Public Personnel Management* 12, 25–32.

Luini, L., & Marucci, F. (2015). Prediction–confirmation hypothesis and affective deflection model to account for split-second decisions and decision-making under pressure of proficient decision-makers. *Cognition, Technology & Work* 17, 329–344.

Lvovsky, A. (2017). The judicial presumption of police expertise. *Harvard Law Review* 130(8), 1995–2081.

Manning, P. K. (2008). Performance rituals. *Policing* 2, 284–293.

Marshall, A. (2015, September 29). Police work is changing, so officers must get the recognition they deserve. *The Guardian*. Retrieved from http://www.theguardian.com/public-leaders-network/2015/sep/29/police-officers-crime-qualifications

Max, D. J. (2000, December 3). The cop and the therapist. *New York Times Magazine*, 94–98. Retrieved from http://partners.nytimes.com/library/magazine/home/20001203mag-max.html

Murji, K. (2010). Introduction: Academic–police collaborations—beyond "two worlds." *Policing* 4, 92–94.

Nieuwenhuys, A., Caljouw, S. R., Leijsen, M. R., Schmeits, B. A. J., & Oudejans, R. R. D. (2009). Quantifying police officers' arrest and self-defence skills: Does performance decrease under pressure? *Ergonomics* 52, 1460–1468.

Nieuwenhuys, A., Savelsbergh, G. J. P., & Oudejans, R. R. D. (2012). Shoot or don't shoot? Why police officers are more inclined to shoot when they are anxious. *Emotion* 12, 827–833.

O'Neil, H. F., Lang, J., Perez, R. S., Escalante, D., & Fox, F. S. (2014). What is cognitive readiness? In H. F. O'Neil, R. S. Perez, & E. L. Baker (Eds), *Teaching and measuring cognitive readiness* (pp. 3–24). New York: Springer.

Phelps, J. M., Strype, J., Le Bellu, S., Lahlou, S., & Aandal, J. (2018). Experiential learning and simulation-based training in Norwegian police education: Examining body-worn video as a tool to encourage reflection. *Policing* 12(1), 50–65.

Pisoni, D. B., & Martin, C. S. (1989). Effects of alcohol on the acoustic-phonetic properties of speech: Perceptual and acoustic analyses. *Alcoholism: Clinical and Experimental Research* 13, 577–587.

Renden, P. G., Landman, A., Savelsbergh, G. J. P., & Oudejans, R. R. D. (2015). Police arrest and self-defence skills: Performance under anxiety of officers with and without additional experience in martial arts. *Ergonomics* 58, 1496–1506.

Rojek, J., Martin, P., & Alpert, G. P. (2014). *Developing and maintaining police-researcher partnerships to facilitate research use: A comparative analysis.* New York: Springer.

Rostker, B. D., Hanser, L. M., Hix, W. M., Jensen, C., Morral, A. R., Ridgeway, G., & Schell, T. L. (2008). *Evaluation of the New York City Police Department firearm training and firearm-discharge review process.* Santa Monica: RAND Corporation.

Rynes, S. L., Bartunek, J. M., & Daft, R. L. (2001). Across the great divide: Knowledge creation and transfer between practitioners and academics. *Academy of Management Journal* 44, 340–355.

Savelli, L. (2010). Ten percenters: The cops who really make a difference! Retrieved from http://policelink.monster.com/training/articles/147694-ten-percenters-the-cops-who-really-make-a-difference#comment_form

Segal, J. (2012). All of the mysticism of police expertise: Legalizing stop-and-frisk in New York, 1961–1968. *Harvard Civil Rights-Civil Liberties Law Review* 47, 573–616.

Shane, J. M. (2010). Performance management in police agencies: A conceptual framework. *Policing: An International Journal of Police Strategies & Management* 33, 6–29.

Shipley, P., & Baranski, J. V. (2002). Police officer performance under stress: A pilot study on the effects of visuo-motor behavior rehearsal. *International Journal of Stress Management* 9, 71–80.

Sim, J. J., Correll, J., & Sadler, M. S. (2013). Understanding police and expert performance: When training attenuates (vs. exacerbates) stereotypic bias in the decision to shoot. *Personality and Social Psychology Bulletin* 39, 291–304.

Sinkovics, R., & Alfoldi, E. A. (2012). Facilitating the interaction between theory and data in qualitative research using CAQDAS. In G. Symon & C. Cassell (Eds), *Qualitative organizational research: Core methods and current challenges* (pp. 109–131). London: Sage.

Sjöberg, D., & Karp, S. (2012). Video-based debriefing enhances reflection, motivation and performance for police students in realistic scenario training. *Procedia—Social and Behavioral Sciences* 46, 2816–2824.

Stainer, M. J., Scott-Brown, K. C., & Tatler, B. (2013). Looking for trouble: A description of oculomotor search strategies during live CCTV operation. *Frontiers in Human Neuroscience* 7, 615.

Staller, M., & Abraham, A. (2016). "Work on your problem-solving": Krav maga experts' views on optimal learning environments for self-defence training. *International Journal of Coaching Science 10*(2), 91–113.

Staller, M. S., Bertram, O., & Körner, S. (2017). Weapon system selection in police use-of-force training: Value to skill transfer categorisation matrix. *Salus Journal 5*(2), 1–15.

Sun, I. Y. (2003). A comparison of police field training officers' and nontraining officers' conflict resolution styles: Controlling versus supportive strategies. *Police Quarterly 6*, 22–50.

Suss, J., & Ward, P. (2012). Use of an option generation paradigm to investigate situation assessment and response selection in law enforcement. *Human Factors and Ergonomics Society Annual Meeting Proceedings 56*, 297–301.

Suss, J., & Ward, P. (2018). Revealing perceptual–cognitive expertise in law enforcement: An iterative approach using verbal-report, temporal-occlusion, and option-generation methods. *Cognition, Technology & Work 20*(4), 585–596.

Taslitz, A. E. (2010). Police are people too: Cognitive obstacles to, and opportunities for, police getting the individualized suspicion judgment right. *Ohio State Journal of Criminal Law 8*, 7–78.

Taylor, T. Z., Elison-Bowers, P., Werth, E., Bell, E., Carbajal, J., Lamm, K. B., & Velazquez, E. (2013). A police officer's tacit knowledge inventory (POTKI): Establishing construct validity and exploring applications. *Police Practice and Research 14*, 478–490.

Van Maanen, J. (1978). The asshole. In P. K. Manning & J. Van Maanen (Eds), *Policing: A view from the street* (pp. 221–238). Santa Monica, CA: Goodyear.

Vickers, J. N. (2007). *Perception, cognition, and decision training: The quiet eye in action.* Champaign, IL: Human Kinetics.

Vickers, J. N., & Lewinski, W. (2012). Performing under pressure: Gaze control, decision making and shooting performance of elite and rookie police officers. *Human Movement Science 31*, 101–117.

Vila, B. (2014). *Developing a common metric for evaluating police performance in deadly force situations in the United States, 2009–2011.* Ann Arbor, MI: Inter-university Consortium for Political and Social Research. Retrieved from http://doi.org/10.3886/ICPSR33141.v1

Vredeveldt, A., Knol, J. W., & van Koppen, P. J. (2017). Observing offenders: Incident reports by surveillance detectives, uniformed police, and civilians. *Legal and Criminological Psychology 22*, 150–163.

Ward, P., Gore, J., Hutton, R., Conway, G. E., & Hoffman, R. R. (2018). Adaptive skill as the conditio sine qua non of expertise. *Journal of Applied Research in Memory and Cognition 7*, 35–50.

Ward, P., Suss, J., & Basevitch, I. (2009). Expertise and expert performance-based training (ExPerT) in complex domains. *Technology, Instruction, Cognition and Learning 7*, 121–145.

Ward, P., Suss, J., Eccles, D. W., Williams, A. M., & Harris, K. R. (2011). Skill-based differences in option generation in a complex task: A verbal protocol analysis. *Cognitive Processing 12*, 289–300.

White, E. K., & Honig, A. L. (1995). The role of the police psychologist in training. In M. I. Kurke & E. M. Scrivner (Eds), *Police psychology into the 21st century* (pp. 257–277). Hillsdale, NJ: Lawrence Erlbaum.

Willis, J. J., & Mastrofski, S. D. (2014). Pulling together: Integrating craft and science. *Policing 8*, 321–329.

Wilson, J. Q. (1968). *Varieties of police behavior: The management of law and order in eight communities*. Cambridge, MA: Harvard University Press.

Wollert, T. (2008). The right kind of feedback. *Federal Law Enforcement Training Center Journal* 6(1), 44–50.

Worrall, J. L. (2013). The police sixth sense: An observation in search of a theory. *American Journal of Criminal Justice* 38, 306–322.

CHAPTER 34

MILITARY EXPERTISE

J. D. FLETCHER AND DENNIS KOWAL

Introduction

As chapters in this volume and elsewhere suggest, the value of expertise is as evident in empirical research as it is in casual observation (e.g., Ericsson, Charness, Feltovich, & Hoffman, 2006; Gorman, 1990; Hoffman, 1992; Hoffman et al., 2014). Expertise and solve basic problems quickly, reliably, and accurately. To an appreciable extent, this level of ability is also expected of competent journeyman performers. Expertise, however, implies performance beyond that. It indicates an ability to successfully solve uncommon, unusually difficult, and/or strategic problems that others cannot. It is manifested by knowledge, skill, and intuition needed but not necessarily assured by managerial level, years of experience, or, in the present case, military rank (Feltovich, Prietula, & Ericsson, 2006).

Expertise includes an ability to perform what Wetzel-Smith and Wulfeck (2010) identify as *incredibly complex tasks*. They describe these as broad, multifaceted, abstract, codependent, and nonlinear tasks that require a large repertoire of patterns and pattern-recognition capabilities for their solution. Similarly, Mayer and Wittrock (1996) describe expertise as an ability to deal with problems that present too much data, too many options, and unknown levels of risk. Clark and Wittrock (2000) suggest that experts need X-ray vision to see past superficial features and symptoms to identify and their solutions. Sternberg and Hedlund (2002) found that expertise involves tacit knowledge—the latent knowledge that few experts can fully articulate but that enables them to solve unique and especially complex problems.

Research to assess experts' cognitive powers of speed, memory, and intelligence usually reveals higher than average abilities, but intellectual preeminence among experts is rare. Moreover, expert capability is almost always restricted to a single, specific domain. For instance, Ericsson and Lehmann (1996) determined that measures of intellect do not predict attainment of expertise and that expertise in any activity rarely transfers beyond that activity. Personal characteristics of individual experts are more likely to be based on interest in a particular topic, an intuitive sense of their talent for an area of activity, enjoyment of activities that develop expertise, or other factors not associated with general intellectual capability.

THE VALUE OF DEVELOPING MILITARY EXPERTISE

The value of expertise elsewhere due to the exposure of individuals performing military operations to physical, if not lethal, harm—often while operating complex, highly advanced technologies. For example, Chatham and Braddock (2001) reported that a training system developed by Wetzel-Smith and Wulfeck (2010) to develop expertise in sonar operators could expand a submarine's effective search area by a factor of 10.5. In effect, a single submarine with sonar operators trained using this system could provide the sonar surveillance of ten submarines. This capability, intended to develop expertise in operating advanced, complex technology, substantially reduces exposure to lethal threats from opposing forces. The monetary and operational return from investment in this training system and the expertise it provides is of significant value to the navy and its submariners in particular.

Military training has a long history of simulation used to develop expertise under the well-founded and often-verified assumption that practice with feedback will produce competence, if not expertise (e.g., Ericsson, 2006). Simulation was applied from the beginning to relieve the high cost of operating aircraft in basic pilot training and, later, for engaging in air combat (e.g., Andrews & Bell, 2000). Early on, much of that training involved a single pilot operating a simulator—a form of simulation that is described as *virtual* because it involves an aircraft simulator sitting on the ground. Because of its expense, there was limited use of *live* simulation (aircraft in the air) to train combat pilots in a manner comparable to that long used in ground warfare field exercises. That approach changed with the war in Vietnam.

During that war, roughly 1965–1973, the US Navy and US Air Force flew aircraft of comparable capabilities. In fact, many of the aircraft used were exactly the same, armed with the same weapons. In the first four years of air-to-air combat, both the Navy and the Air Force achieved nearly identical loss-exchange ratios of North Vietnamese to US aircraft. On average about 2.4 North Vietnamese aircraft were downed for every US Navy aircraft lost. The ratio was about 2.2 for the US Air Force. However, that ratio diminished rapidly for the Navy. In the first half of 1968, its loss-exchange ratio dropped to less than one-to-one—nine enemy aircraft shot down with the loss of ten Navy aircraft (Armed Forces Journal, 1974).

For a variety of reasons, there was a hiatus in the air war from late 1968 to 1970. During that time, the Navy initiated a live simulation training program, now identified as Top Gun. It used highly instrumented force-on-force combat engagements in live simulations to enhance pilot performance in air-to-air combat. Promising Navy pilots were pitted against the Navy's most proficient combat pilots who had been trained in enemy tactics and were flying Mikoyan-and-Gurevich- (MiG) aircraft.

Following each simulated engagement there was an after-action review where the lessons were not presented didactically by observers but facilitated and drawn from the trainees themselves (Morrison & Meliza, 1999). Progress and development of expertise were determined, not by who spoke first or loudest, but by evidence based on the instrumentation, which, among other technologies, included extensive videography. These engagements were played, reviewed in detail, and replayed until the Navy student flyers reached a level of expertise that matched or exceeded that of the experts training them.

The value of this expertise was determined in 1970, when the air war resumed, providing results from what might be described as a natural experiment. Navy pilots, flying the same aircraft as their Air Force counterparts but trained by Top Gun engagement simulations, were found to be achieving a loss-exchange ratio of 12.5:1, whereas the ratio for Air Force pilots remained at 2:1. Costs to provide this training were high, but the value in return from this investment in creating and developing expertise far exceeded its costs as found in analysis by Fletcher and Chatham (2010).

The value of this approach in developing military expertise was further reinforced by the subsequent, rapid development of Blue Flag by the Air Force, which was modeled on the Navy's Top Gun. The US Army followed later with the development and use of MILES (multiple integrated laser engagement simulation), which used lasers and laser detectors in land warfare training. This approach was extended to provide live simulation combining armor, infantry, and combat air support in California's National Training Center desert.

In further pursuit of military expertise, the US Defense Advanced Research Program Agency (DARPA) developed simulation networking (SIMNET). This training technology was initially intended to develop expertise in armor operations with an inexpensive, networked gaming environment for tank commanders and gunners. It was soon expanded to a global simulation capability that networked simulated land, air, and sea engagements to develop the expertise of participants at geographically diverse installations training together in simulated, joint force-on-force operations. In effect, a tank simulator in California being attacked by a helicopter simulator in Alabama could call for air support from a jet simulator in Germany.

Finally, it should be noted that about 80 percent of military activity involves combat support such as logistics, maintenance, and medical care. The expertise of technicians in these areas is essential in most fields of work, but especially in the military where technical expertise is critical. A notable difference for military technicians is that they may have to perform their work rapidly, under lethal threat. The expertise of military technicians is similar to that of their civilian counterparts but by civilian technicians.

With the need and requirements for military technicians in mind, DARPA invested in the development and demonstration of a digital tutor to apply the processes used by human tutors in one-on-one training. This tutor was intended to illustrate the value of machine intelligence used to provide tutorial problem solving similar to that of one tutor, in this case a computer program, tailoring instruction to the needs and abilities of each learner. The program was intended to accelerate acquisition of information technology (IT) expertise by newly enlisted Navy sailors.

After sixteen weeks of training with the DARPA tutor, the novice sailors were found in extensive IT problem-solving exercises to have developed levels of expertise substantially superior to other new sailors who had received thirty-five weeks of standard IT classroom training and to sailors with an average of nine years' IT experience in the fleet. In both cases, the results exceeded four standard deviations in favor of the tutor (Fletcher & Morrison, 2014). By reducing the need for on-the-job training and experience, the monetary value of this training far exceeded the cost of its design and development (Cohn & Fletcher, 2010). The operational value (i.e., improved effectiveness of naval operations) of the tutor is more difficult to quantify due to the variety and vicissitudes of naval combat, but it is likely to return equal, if not greater, value.

These results were replicated with military veterans with little or no prior IT experience who received eighteen weeks of training for civilian employment with the tutor (Fletcher, 2017). Nearly all were subsequently hired at levels ordinarily reserved for IT technicians with three to five years of experience, or more. Again, the monetary return to the government far exceeded the cost of developing the tutor and using it to provide technical training.

MILITARY DECISION-MAKING

There have always been warriors. After the dawn of cultural modernity, about the past 60,000 years, some individuals have always assumed special responsibilities for protecting the family, tribe, city-state, or nation against existential threats. Their expertise has long been valued and considered essential. Their decision making occurs at every level ranging from individuals and small teams to high levels of command responsible for thousands of individuals. Their expertise is needed in both combat and combat support activities such as logistics, maintenance, troubleshooting, and repair. Individuals from corporals to the highest ranks must make and carry out decisions that affect every level of command. These occasions include the command, coordination, and control of missions involving a full range of land, sea, and air assets, some of which may be unfamiliar to the decision maker who must nonetheless assign their missions.

Through the years, perhaps beginning with Sun Tzu's *Art of War* (e.g., 1971), many countries have developed principles of war. A review of commonalities among principles established for today's militaries suggests areas of requisite expertise. As Holz,

O'Hara, and Keesling (1994)—among others—found, there is considerable agreement about these principles. They include the ability to:

- Establish and maintain moral influence and morale ("That which causes the people to be in harmony with their leaders" (Sun Tzu, 1971, p. 64);
- Establish clear objectives and maintain attention to them;
- Ensure economy of effort;
- Maintain the offensive;
- Establish communication and cooperation among assets;
- Ensure unity of command;
- Devise offense and prepare defense for surprise and the unexpected; and
- Ensure security to whatever degree is possible.

Commonalities among these principles and across the years suggest competencies that military leaders and scholars consider essential. This list may be modified or adjusted to address the missions of specific military organizations, but it provides a basis for developing and assessing military expertise.

Today, as in times past, the military faces an operating environment characterized by volatility, uncertainty, complexity, ambiguity, and lethal threat (Stiehm, 2002). Success in dealing with the fog of war is a prime indicator of military expertise. Military leaders at all levels must make sense of the chaos they confront, rapidly assess and re-assess conditions, determine a relevant course of action, and execute it. They must continually examine the heuristics and intuitive judgements underlying their decisions, the assumptions they are making, the likelihood of deception in what they perceive, and the probable consequences of their actions.

Military expertise, as other forms of expertise, has been described as applying critical and creative thinking to understand, visualize, and solve complex, ill-structured problems and develop approaches to solve them (US Army FM 5-0, 2010). Instead of one doctrinal solution for a range of military problem categories, current approaches acknowledge the need to assess each situation and recognize that any solution will be unique to the conditions under which it is applied. Expertise in this environment requires framing a problem and continually reframing it as conditions change and alternative courses of action must be applied.

However, the tempo of combat decisions may render these steps infeasible. Leaders at all levels must often apply intuitive or *recognition-primed decision making*. These situations require decision making based on rapid, implicit assessment of the situation supported by pattern recognition developed through training, practice, and experience (Klein, Orasanu, Calderwood, & Zsambok, 1993). Leaders from corporals to generals must often decide, perform, act, and react rapidly, despite the value added by a more deliberate and carefully planned approach (e.g., Moxley, Ericsson, Charness, & Krampe, 2012).

Military expertise therefore requires a capability to quickly understand a unique, unfamiliar situation under the most stressful of conditions. On the basis of their expertise, military leaders must determine actions to counter and defeat opposing leaders endeavoring to do the same to them. As Kahneman and Klein agreed (2006), this approach to decision making emphasizes pattern recognition based on factors such as knowledge, experience, skill, and cognitive readiness, which, in turn requires capabilities for situation awareness, creativity, and adaptiveness.

Because intuitive decision making involves subjective assessments and emotion intertwined with complex cognitive process, military decision making must, as in other areas of expertise, be reviewed by post action reflection to determine conflict or agreement with prior knowledge. This review is often provided by after action reviews such as those performed following training exercises such as Top Gun and in operational warfare (Morrison & Meliza, 1999; Youngblut, et al. 2010). Accepting new ideas in these reviews is as important as recognizing and verifying those already established.

Categories of Military Expertise

Military expertise is typically keyed to three categories of decision making and action common to all military missions: tactical, operational, and strategic (e.g., Bellamy, 1985; Rogers, 2006). These three levels are often viewed as rising from tactical to strategic in accord with military rank. However, each has become increasingly relevant at all levels of command, as described by Krulak's (1999) *strategic corporal*, in the irregular operating environments increasingly characteristic of today's military missions. These environments require all service members to balance local, regional, and national economic, religious, and social concerns of combatants and non-combatants while achieving mission objectives. They may further require different military services to work together, or *jointly*, to coordinate and exercise their specific capabilities (Youngblut et al., 2010).

In accord with Drucker's (e.g., 2006) often-cited distinction, managerial expertise at all three levels concerns doing things right, but also doing the right things. Tactical expertise is especially required and manifest at the front lines of military engagements where problem solving may demand immediate and urgent attention—doing things right, but also not doing the wrong things, which may alienate the local, or even national, population. Although tactical expertise often addresses short-term objectives, it is essential for both operational and strategic success (i.e., winning wars as well as battles). Prominent military historians, such as Marshall (1947/1978) and Hart (1944), have emphasized that success, ranging from local conflicts to full theaters of action, is typically determined by the accumulation of small tactical engagements. Military planning at all levels must be founded on assessments of tactical expertise and estimated probabilities of success.

Operational expertise is a matter of extending tactical expertise to a wider range of military operations. Operational decision making involves logistical and unit decision making to ensure that all elements (e.g., land, air, sea, and logistical support) needed to prevail in an engagement or campaign are coordinated, communicating, informed, and cooperating with full understanding of the commander's intentions. In today's irregular mission environment, operational design must additionally and correctly account for the constraints and opportunities presented by civilian populations. Operational expertise must then balance analysis of the mission with its goals. It must establish objectives and standards for success while fusing local political aims with military objectives. It connects the details of tactics with the goals of strategy.

Tactical and operational expertise is critical in determining the outcome of battles. Even Napoleon Bonaparte could err in carrying them out. As von Clausewitz, who was present at the Battle of Waterloo, reported, Bonaparte reassigned his operational assets too soon after they had defeated Blucher's Prussian forces (von Clausewitz, 1976/1984). The Prussians were then able to reassemble, continue their advance, and arrive in time to reinforce the British, Belgian, and Dutch forces in repelling Napoleon's critical and nearly successful attack on their right flank. The Duke of Wellington observed that the Battle of Waterloo was "nearest run thing you ever saw" (Keegan, 1987, p. 103).

Strategic expertise stands somewhat separate from tactical and operational issues, but it is nonetheless essential. In Drucker's terms, it focuses on doing the right things. For instance, the World War II Japanese attack on Pearl Harbor is often viewed as a tactical and operational triumph. However, and as some senior Japanese staff feared, it turned out to be a strategic disaster. It served to mobilize US economic and technical potential and focus it on defeating Japan.

Strategic expertise requires full consideration of national assets, economic and otherwise, friendly and opposition, while engaging and enlisting assets from allied nations and neutral bystanders, whose economic, cultural, and religious sensitivities must be considered and taken into account. Further, these decisions may be required at any level—again emphasizing the need to do the right things as well as doing them right. Military expertise requires Krulak's corporals, as well as generals, to make strategic decisions while pursuing tactical and operational objectives.

Most military organizations provide post-graduate education for their officers and senior enlisted personnel as a way to enhance their capabilities and, eventually, develop tactical, operational, and strategic expertise. This education may be offered through graduate education courses in civilian colleges and universities or through professional military education programs supported by either single-service or all-service (*joint*) schools. These schools focus on a particular military specialty such as industrial resources, logistics, international security affairs, applied strategy, or, simply, on warfighting as performed by specific military services. Nonetheless, all three levels—tactical, operational, and strategic—must be integrated in this education.

Finally, military decision making and doctrine must continually adapt to rapidly changing circumstances. Critical, creative thinking is needed to understand, visualize, and describe complex, ill-structured problems and then develop approaches for solving them.

In addition to physical courage, military expertise requires exercise of bureaucratic courage in what are notoriously rigid and hierarchical organizations to challenge existing procedures and doctrine. As Machiavelli (1513/1952) long ago noted, change is difficult to pursue, perilous to conduct, and uncertain in its success. Nonetheless, it is a recognized, although not always rewarded, element of military expertise.

It is rare to find expertise at all three levels of expertise in a single commander. An example is the often-cited contrast of Patton and Eisenhower. As discussed by Atkinson (2002), among others, Patton had extensive battlefield experience. His combination of intellect and tactical aggressiveness made him ideal (except for his impatience with logistics) for battlefield command in World War II given his mastery of both tactics and military leadership. However, many commentators contrast Patton's warfighting expertise with that of Eisenhower, who became a master strategist, ideally suited to be the supreme commander in World War II Europe.

A twenty- to thirty-year military career may be insufficient to develop the repertoire of pattern recognition and intuition required for expertise. Competence seems achievable, but expertise requires a more intense approach. Relevant, battlefield experience as well as classroom education is needed to develop the foundation needed for a full range of military tactical, operational, and strategic expertise (e.g., Atkinson, 2002; Keegan, 1987).

Hoffman et al. (2014) determined that acquisition of expertise through education and training was in our reach, if not our grasp. Digital tutoring has been found to accelerate and develop tactical and conceptual expertise in technicians (e.g., Fletcher & Morrison, 2014; Kulik & Fletcher, 2016). It remains to be seen whether the adjustments and flexibility of digital tutoring can produce similar problem-solving expertise across the full range of military tactical, operational, and strategic decision making by military personnel.

Expertise of this sort may be particularly elusive in today's irregular, asymmetric military environment with its increasingly sophisticated and effective weapons on all sides and corporals as capable of creating international incidents as generals. To a considerable extent, today's military personnel must be able to foresee the implications of their actions across a range of possibilities that, especially in today's operational environment, is wider than those considered by earlier generations.

Another aspect of military expertise involves the requirement to develop and maintain the readiness of military command units (e.g., squads, squadrons, ships, and divisions) needed to successfully complete a full range of assigned missions and campaigns with all the exigencies and constraints that inevitably arise. Readiness of modern military organizations is usually assessed in four basic categories:

- *Materiel*: Are there enough "systems," such as aircraft, tanks, trucks, radios, and radars?
- *Equipment*: Are there enough spares, supplies, consumables, and similar items?
- *Personnel*: Are there enough training-certified people with the necessary skills at requisite levels of certification?

- *Training*: Has each military unit, from squads up through divisions and corps, completed all requisite training events, such as field exercises, firing range exercises, command and control exercises, and maneuver exercises on the ground, in the air, and at sea?

Military commands have long used these categories to assess their readiness to perform missions. Expertise within military units can be essential in ensuring that all four requirements are met. However, these consist of easily measurable items, making competence as well as expertise, more accessible. Military expertise is now seen to require an additional, cognitive component.

MILITARY EXPERTISE AS A COGNITIVE CAPABILITY

Most of the discussion in the previous section concerns behavior we might find in military expertise. However, another view concerns the cognitive traits of individuals who are considered to be expert in the performance of military activities, an aspect that may be as important as their behavior.

Traits are less easily measured, more difficult to shape through instruction, and more abstract than behavior. They concern issues less commonly considered in personnel selection and training despite their relevance and applicability across a full range of operational activity. In this context, Napoleon's famous question "I know he is a good general, but is he lucky?" presaged what Kahneman and Klein (2006), Ross, Shafer, and Klein (2006), and others viewed as a repertoire of subtle and intuitive cues that stochastically link to specific outcomes and that characterize expertise.

An individual may therefore be knowledgeable and skilled but not able to perform at an expert level when dealing with the intense environment of warfare. It is therefore important to measure performance, not just knowledge and skill. However, we cannot call a war just to assess levels of military expertise. This is the *criterion problem* discussed by Hiller (1987). In effect, we must train individuals and groups of individuals without the full benefit of criterion-referenced assessment. Instead, and as discussed earlier, we measure and evaluate units for their readiness. Given the lethal and destructive nature of combat operations, we can be glad the criterion problem exists. Still, assessing the preparedness of military forces for combat engagements beyond matters of materiel, equipment, personnel, and training remains a compelling and significant challenge for students of human behavior.

Rasmussen's (1983) often-cited model of expertise concerns the trait or ability to assess an unfamiliar situation and choose a course of action quickly using only a few points of data. This expertise is developed through experience, after an individual has built up a foundation of knowledge (Norman, 2006).

In effect, there are no general measures of expertise in the military, except outcomes in operational settings that differ substantially from those confronted by experts in other fields. They do not concern monetary market share or earnings, which are central to expertise in other fields. The brutal nature of military operations is not easily translated into other professions' requirements, standards, and outcomes.

Rasmussen's model of military expertise consists of four components with a fifth, more recent, component (cognitive readiness) added here:

- *Military-technical*: Applying military assets to accomplish missions based on the knowledge, skills, and abilities provided in the career development process and training of all military personnel.
- *Moral-ethical*: Accomplishing missions with regard to basic human rights.
- *Political-cultural*: Understanding and operating in a multicultural, complex world. This component includes the task of understanding other branches and allied forces that may be included, as well as the context and culture of their own military services along with the combatant and non-combatant culture(s) within the area of operation.
- *Leader/human development*: Recruiting, developing, and inspiring other military professionals.
- *Cognitive readiness*: Preparing for the inevitable surprise and unexpected events that arise in performing missions at every level (O'Neil, Perez, & Baker, 2014).

Much is understood about preparing individuals for the jobs, tasks, and missions that they are likely to perform along with the environments and situations in which they may be required. This understanding provides a foundation for the tasks, conditions, and standards that define objectives for training and education in the military and, for that matter, across a wide range of activity. The issue addressed by cognitive readiness is preparing individuals for surprise, i.e., situations, activities, and requirements that cannot be anticipated in advance.

The cognitive readiness literature applies varying definitions and uses a variety of terms of art (O'Neil, Perez, & Baker, 2014). The following definitions are representative:

- Cognitive readiness is the psychological (mental) and sociological (social) knowledge, skills, and attitudes (KSAs) that individuals and team members need to sustain competent performance and mental well-being in the dynamic, complex, and unpredictable environments of military operations (Bolstad, Cuevas, Babbitt, Semple, & Vestewig, 2006).
- Cognitive readiness is meta-preparedness. It is to prepare individuals for surprise—the unexpected responses, tasks, missions, and environments that cannot be anticipated in advance, but are inevitable in military operations and frequently make the difference between operational success and failure (Fletcher & Wind, 2014; Morrison & Fletcher, 2002).

Unanticipated tactics, new technological capabilities, novel applications of existing technologies, and the element of surprise are notoriously characteristic of combat engagements. Even in non-combat military operations (e.g., peacemaking, peacekeeping, humanitarian relief, and crisis management), unpredictability is a certainty. Dealing with surprise and the unexpected is, then, an essential aspect of military expertise.

Definitions of cognitive readiness appear to agree on three characteristics that may be abstracted from research and discussions on the subject:

1. Cognitive readiness is a predictor of cognitive performance.
2. Cognitive readiness requires tolerance of the uncertain, demanding, and stressful environment in which military operations take place. Other, more easily measured attributes have been shown to affect performance, particularly in high-stress situations. However, cognitive readiness attempts to identify less commonly considered cognitive qualities in measurable terms that enable further research and development for military expertise.
3. Cognitive readiness requires both an individual's level of cognitive readiness and preparation for the current situation. An individual can have an optimum level of cognitive readiness for a particular military operation, but it may not be optimal for all operations. Cognitive readiness at the expert level involves a more compete repertoire of roles, responsibilities, and operations than those assumed or found at other levels.

As a cognitive quality, then, traits associated with military expertise transcend the usual issues of readiness to emphasize items such cognitive capabilities as adaptability, creativity, pattern recognition, teamwork, communication, interpersonal skills, resilience (or *grit*), critical thinking (thinking about others' thinking), and metacognition (thinking about one's own thinking). This view is supported by von Clausewitz's (1976/1984) observation that quick intuitive judgment and determination are more important for successful military leadership than brilliance or reflection in thought and mind—"a strong mind rather than a brilliant one" in Howard and Paret's translation.

The specific elements of cognitive readiness remain at issue. These elements are necessary for education, training, and, eventually, assessment of cognitive expertise and readiness. Overall, three issues under question are whether an element of cognitive readiness can be reliably measured (are we measuring anything?), can it be validly measured (are we measuring the right thing?), and can it be improved by training or education? Some elements such as adaptability, creativity, metacognition, and situation awareness have satisfied these issues, as found in research separately reviewed by both O'Neil, Lang, and Perez (2014) and Fletcher and Wind (2014). Others, such as interpersonal skills, resilience (or *grit*), and critical thinking remain in question, while still others such as pattern recognition, leadership, and emotional control remain to be examined.

Research on cognitive readiness is continuing. Its value is widely accepted, but whether components reasonably suggested as relevant for cognitive readiness can be measured with sufficient precision and then improved through instruction requires additional research.

Final Word

Requirements for military expertise have remained relatively stable in human history, with occasional introductions of new technology such as chariots, gun powder, and the long bow (van Creveld, 1991), for literally thousands of years. However, the pace with which new technologies have been appearing over the past 200 years, along with their lethality, is increasing and accelerating rapidly (Hammes, 2004). A corresponding need for military organizations to acquire, apply, and counter these technologies as rapidly as they emerge has affected the character and development of military expertise, much as it has done in other areas of expertise.

Moreover, today's irregular warfare environment involves more than nationally organized forces (Youngblut et al., 2010). Non-state actors are acquiring sophisticated weapons and associated operational capabilities once expected only of organized national militaries. Operations to stabilize undergoverned regions and manage internal instability, once thought to be *operations other than war*, are now understood and expected to include intense, organized combat. The boundaries of tactical, operational, and strategic levels of command and conflict continue to blur and often fail to correspond to military hierarchies or to well-understood national interests. Requirements for military expertise must evolve as rapidly as the challenges themselves.

Nevertheless, the strategic level remains a primary concern for high military command, the operational level remains the focus of theater commands, and the tactical level remains the focus of field commands (Rogers, 2006). Further, it is now recognized that every level of command should consider strategic issues, including analysis of the situation, estimating friendly and enemy capabilities and limitations, the concerns and needs of non-combatant, neutral populations, and devising effective courses of action.

Decision-making at every level of command must include tactics and operations as well as strategy. Tactical and operational decisions must be constantly re-evaluated given incomplete, contrary information in a constant state of flux. A key for success is the expertise needed to adapt rapidly to changing situations and to exploit transient opportunities rather than strict adherence to predetermined approaches. The ability to adapt and exploit these opportunities requires expertise and extraordinary judgment, a fluid and comprehensive grasp of the situation, knowing what must be done, and finally how to proceed.

In some respects, military expertise appears to be similar to expertise in other areas. As in all decision making, military expertise requires individuals to see past superficial features of problems and situations and identify the underlying concepts and principles that lead to successful courses of action (Mayer & Wittrock, 1996). Despite their differences and approaches to research on the cognitive processes of decision making, Kahneman and Klein (2006) agreed that expertise in any activity requires a large repertoire of relevant, accessible cues that are stochastically linked to problem-solving patterns that may provide specifically targeted outcomes. These patterns may be applied consciously or unconsciously, intuitively, or with deliberation, to determine the course or courses of action most likely to succeed.

However, military expertise and warriors themselves appear to be unique due to the intensity, volatility, uncertainty, and lethality of the environment in which core military decision making takes place. John Keegan, a prominent military historian who has spent his career teaching soldiers, observed that "Soldiers are not as other men, [their values and skills are those of] a very ancient world, which exists in parallel with the everyday world, but does not belong to it." (1993, p. xvi).

Expertise in business may similarly involve efforts to thwart or remove opposition, and its decisions may also be made in volatile, uncertain, and ambiguous circumstances. But threats, even existential threats, to business organizations or individuals in them rarely include levels of physical harm and lethality that are characteristic of decision making at the core of military expertise. Lethal threats also occur in non-military activities such as firefighting and police work, but without the intensity and large numbers of participants characteristic of military decision making. Personal attributes of military expertise, such as physical and mental courage, self-sacrifice, and loyalty remain as relevant as ever.

References

Andrews, D. H., & Bell, H. (2000). Modeling and simulation applications to training. In S. Tobias and J. D. Fletcher (Eds), *Training and retraining: A handbook for business, industry, government, and the military* (pp. 357–384). New York: Macmillan Reference.

Armed Forces Journal (May, 1974). "You fight like you train" and TOP GUN crews train hard. *Armed Forces Journal International* 25–27.

Atkinson, R. (2002). *An army at dawn.* New York, NY: Henry Holt.

Bellamy, C. (1985). Trends in land warfare: The operational art of the European theater. *Defense Yearbook.* London: Brassey's Defense Publishers.

Bolsted, C. A., Cuevas, H. M., Babbitt, B. A., Semple, C. A., & Vestewig, R. E. (2006). Predicting cognitive readiness of military health teams. *Proceedings of the International Ergonomics Association 16th World Congress on Ergonomics.* Burlington, MA: Elsevier.

Chatham, R. E., & Braddock, J. V. (2001). *Training superiority and training surprise.* Washington, DC: Defense Science Board, Department of Defense.

Clark, R., & Wittrock, M. (2000). Psychological principles in training. In S. Tobias & J. D. Fletcher (Eds), *Training and retraining: A handbook for business, industry, government, and the Military* (pp. 51–84). New York: Macmillan Reference.

Cohn, J., & Fletcher, J. D. (2010). What is a pound of training worth? Frameworks and practical examples for assessing return on investment in training. *Proceedings of the InterService/Industry Training, Simulation and Education Annual Conference.* Arlington, VA: National Training and Simulation Association.

Drucker, P. F. (2006). *The practice of management*. New York: Harper Business.
Ericsson, K. A. (2006). The influence of experience and deliberate practice on the development of superior expert performance. In K. A. Ericcson, N. Charness, P. J. Feltovich, & R. R. Hoffman (Eds), *The Cambridge handbook of experts and expert performance* (pp. 683–722). New York: Cambridge University Press.
Ericsson, K. A., Charness, N., Feltovich, P. J., & Hoffman, R. R. (Eds) (2006). *Cambridge handbook of expertise and expert performance*. New York: Cambridge University Press.
Ericsson, K. A., & Lehmann, A. C. (1996). Expert and exceptional performance: Evidence of maximal adaptation to task. *Annual Review of Psychology 47*, 273–305.
Feltovich, P. J., Prietula, M. J., & Ericsson, K. A. (2006). Studies of expertise from psychological perspectives. In K. A. Ericsson, N. Charness, P. J. Feltovich, & R. R. Hoffman (Eds), *The Cambridge handbook of expertise and expert performance* (pp. 41–67). New York: Cambridge University Press.
Fletcher, J. D. (2017). The value of digital tutoring and accelerated expertise for military veterans. *Educational Technology Research and Development 65*, 679–698.
Fletcher, J. D., & Chatham, R. E. (2010). Measuring return on investment in military training and human performance. In P. E. O'Connor & J. V. Cohn (Eds), *Human performance enhancements in high-risk environments* (pp. 106–128). Santa Barbara, CA: Praeger/ABC-CLIO.
Fletcher, J. D., & Morrison, J. E. (2014). *Accelerating development of expertise: A digital tutor for navy technical training* (D-5358). Alexandria, VA: Institute for Defense Analyses. (DTIC AD1002362)
Fletcher, J. D., & Wind, A. P. (2014). The evolving definition of cognitive readiness for military operations. In H. F. O'Neil, Jr., R. S. Perez, & E. L. Baker (Eds), *Teaching and measuring cognitive readiness* (pp. 25–52). New York: Springer.
Gorman, P. F. (1990). *The military value of training* (IDA Paper P-2515). Alexandria, VA: Institute for Defense Analyses. (DTIC/NTIS ADA 232 460.)
Hammes, T. X. (2004). *The sling and the stone*. St. Paul, MN: MBI Publishing.
Hart, H. B. L. (1944). *Thoughts on war*. London: Faber & Faber.
Hiller, J. H. (1987). Deriving useful lessons from combat simulations. *Defense Management Journal 23*, 29–33.
Hoffman, R. R. (1992). *The Psychology of Expertise: Cognitive Research and Empirical AI*. New York: Springer-Verlag.
Hoffman, R. R., Ward, P., Feltovich, P. J., DiBello, L., Fiore, S. M., & Andrews, D. H. (2014). *Accelerated expertise: Training for high proficiency in a complex world*. New York: Psychology Press.
Holz, R. F., O'Hara, F., & Keesling, W. (1994). Determinants of effective unit performance at the National Training Center: Project overview. In R. F. Holz, J. H. Hiller, & H. McFann (Eds), *Determinants of effective unit performance: Research on measuring and managing unit training readiness*. Alexandria, VA: US Army Research Institute for the Behavioral and Social Sciences. (DTIC/NTIS No. ADA 292 342.)
Kahneman, D., & Klein, G. (2006). Conditions for intuitive expertise: A failure to disagree. *American Psychologist 64*, 515–526.
Keegan, J. (1987). *The mask of command*. New York: Viking Penguin.
Klein, G. A., Orasanu, J., Calderwood, R., & Zsambok, C. E. (Eds) (1993). *Decision making in action: Models and methods*. Norwood, NJ: Ablex.
Krulak, C. C. (1999). The strategic corporal: Leadership in the three block war. *Marines Magazine 83*(1).

Kulik, J. A., & Fletcher, J. D. (2016). Effectiveness of intelligent tutoring systems: A meta-analytic review. *Review of Educational Research 86*, 42–78.

Machiavelli, N. (1513/1952). *The Prince.* New York: New American Library.

Marshall, S. L. A. (1947/1978). *Men against fire: The problem of battle command in future war.* Gloucester, MA: Peter Smith.

Mayer, R. E., & M. C. Wittrock (1996). Problem solving transfer. In D. C. Berliner & R. C. Calfee (Eds), *Handbook of educational psychology* (pp. 47–62). New York: Macmillan.

Morrison, J. E., & Fletcher, J. D. (2002). Cognitive Readiness. In H. Florian (Ed.), *Military pedagogy* (pp. 165–187). Frankfurt am Main, Germany: Peter Lang.

Morrison, J. E., & Meliza, L. L. (1999). *Foundation of the after action review* (IDA Document D-2332). Alexandria, VA: Institute for Defense Analyses.

Moxley, J. H., Ericsson, K. A., Charness, N., & Krampe, R. T. (2012). The role of intuition and deliberate thinking in experts' superior tactical decision-making. *Cognition 124*, 72–78.

Norman, G. (2006). Building on experience: The development of clinical reasoning. *New England Journal of Medicine 355*, 2251 2252.

O'Neil, H. F., Perez, R. S., & Baker, E. L. (Eds) (2014). *Teaching and measuring cognitive readiness.* New York: Springer.

O'Neil, H. F., Lang, J. Y.-C., Perez, R. S., Escalante, D., & Fox, F. S. (2014). What is cognitive readiness. In H. F. O'Neil, Jr., R. S. Perez, & E. L. Baker (Eds), *Teaching and measuring cognitive readiness* (pp. 3–24). New York: Springer.

Rasmussen, J. (1983). Skills, rules, and knowledge: Signals, signs, and symbols, and other distinctions in human performance. *IEEE Transactions on Systems, Man, and Cybernetics 13*, 257–266.

Rogers, C. J. (2006). Strategy, operational design, and tactics. In J. C. Bradford (Ed.), *International Encyclopedia of Military History.* New York: Routledge.

Ross, K. G., Shafer, J. L., & Klein, G. (2006). Professional judgments and "naturalistic decision making." In K. A. Ericsson, N. Charness, P. J. Feltovich, & R. R. Hoffman (Eds), *Cambridge handbook of expertise and expert performance* (pp. 403–420). New York: Cambridge University Press.

Sternberg, R., & Hedlund, J. (2002). Practical intelligence, g, and work psychology. *Human Performance 15*, 143–60.

Stiehm, J. (2002). *Military education in a democracy.* Philadelphia, PA: Temple University Press.

Sun Tzu (c.500 BC/1971). *The art of war.* New York: Oxford University Press.

US Army FM 5-0 (2010). *The operations process.* Washington, DC: Headquarters, Department of the Army.

Van Creveld, M. (1991). *Technology and war from 2000 B.C. to the present.* New York: Macmillan, The Free Press.

von Clausewitz, C. (1976/1984). *On war.* Translated and edited by M. Howard and P. Paret (1942). Princeton, NJ: Princeton University Press.

Wetzel-Smith, S. K., & Wulfeck, W. H. (2010). Training incredibly complex tasks. In P. E. O'Connor & J. V. Cohn (Eds), *Performance enhancement in high risk environments* (pp. 74–89). Westport, CT: Praeger.

Youngblut, C., Adesnik, D., Fletcher, J. D., Kowal, D. M., Numrich, S. K, & Hartman, F. E. (2010). *Cognitive readiness for irregular warfare* (IDA Document NS D-4280). Alexandria, VA: Institute for Defense Analyses.

CHAPTER 35

EXPERTISE IN BUSINESS
Evolving with a Changing World

LIA A. DIBELLO

SETTING THE SCENE OF EXPERTISE IN BUSINESS

IN order to write a chapter on expertise in business, my team and I have had to become attuned to the ways that business has changed and therefore, how business expertise plays a role in today's society. By looking at business expertise, we have also learned a great deal about expertise in general and its role in human adaptation to challenging environments. Perhaps for this reason business *expertise* has become important to those outside the expertise research community.

When it comes to business specifically, what we have learned from our work with over 7000 people from 53 different companies is that highly effective teams that understand the challenges before them and efficiently deploy their resources might be more critical *experts* than individuals. In fact, we suspect that our ways of accelerating expertise are so effective at helping companies because they actually create highly expert teams. The individual members do gain in measurable expertise—and there is no question that there is value in that—but with teams there are marked changes in the reshuffling of resources, and this is what has the impact when the teams return to work.

In this chapter I will discuss our view of expertise and why we think business is an important domain for expertise research, and review what we have learned from helping our clients by providing some illustrative cases. It has been challenging to summarize a 25-year career in a few pages and choose business cases that non-experts in business can follow; my intention is to share some insights from these experiences.

Why is Business Expertise Important Now?

There is no question that higher levels of expertise are required among business professionals than in times past. In recent years, we estimate that senior leaders must reinvent the value of their firms or change strategy as often as every 18 months, making senior management assessment an important feature of responsible corporate governance (e.g., Stamoulis 2009; Kaplan & Minton, 2012). In contrast, in decades past, a single approach could work for an entire career. More recently, we are seeing that mid- and low-level professionals are expected to have a deeper understanding of business fundamentals, and especially forces involved in value creation.

The emergence of *expertise* in business is driven in part by the relatively recent increased pace and complexity of business itself, which in turn, is driven by the influx of advanced information technology. This trend began transforming the nature of business in the early 1980s and gaining full momentum in the late 1990s. This new economy, typified by global information and communication technologies, has enabled exponential growth and innovation from unexpected sources, increased competition for markets, and a vast interconnectedness. And yet, our work indicates that while the expertise of business leaders is complex, and largely intuitive, it may not be a black box we once thought it was.

Our View of Expertise

Workplace Technology Research Inc. (WTRI) studies and assesses business expertise of all kinds and helps companies re-align the expertise of their senior teams to keep pace with changes in the marketplace through services and products it has developed over the past twenty-five years.

The research on the nature of *intuitive expertise* that most influences our notions of business expertise probably began in earnest in the 1980s with work by Robert Glaser, Micki Chi, Robert Hoffman, and others (Chi, Feltovich, & Glaser, 1981; Chi, Glaser, & Farr, 1988), and continuing (e.g., Chi, 2006; Ericsson, 2003, 2016).

In the 1980s Gary Klein conducted pioneering research on expertise in dynamic areas, or areas of knowledge that change quickly and where effective actions require understanding the dynamic underlying principals defining the domain itself (Klein, 1989). This framework was well suited for looking at business experts in action and helped us with developing assessments and with mapping the problem spaces of business experts (e.g., Klein 1993, 1998, 2000).

In general, dynamic domains require that the same cognitive mechanisms come into play as that in other kinds of expertise, but manifest in different decision-making capabilities. Very broadly speaking, these capabilities were:

1. Developing an intuitive understanding of the domain that constitutes a gestalt shift, resulting in seeing things that non-experts cannot see;

2. Being able to assess and address situations not cognitively available to non-experts;
3. A way of thinking based on *first principles*, rather than the result of memorizing a set of facts or concepts; and
4. The ability to *forward simulate* eventualities in a domain that is rapidly changing and take appropriate actions proactively.

The question is how do these capabilities manifest in business experts?

Our view of expertise is that it comes about as a result of stage-like development in the sense of Piaget (e.g., Piaget 1977) but in a domain-specific way. In this sense, we are like others who have looked at expertise using Piagetian or constructivist models (e.g., Bickhard & Campbell, 1996; Campbell & DiBello, 1996; Spiro, Feltovich, Jacobson, & Coulson, 1992). However, we differ from other genetic epistemologists in that our work incorporates Vygotsky's theory of *scientific concept* (Vygotsky, 1978). Vygotsky was a contemporary of Piaget but did not share Piaget's view that higher levels of cognitive capability were the result of epistemological development. Rather, he believed that knowledge was acquired through an elaborate process of shared understanding through shared activity. His notion of *scientific concepts* in today's terms is systems of knowledge created collectively within a field through social agreement and formalized through language and symbolic representation, such as formulas or written language. *Understanding* develops dialectically, as a result of activity, usually with others.

In other words, expertise develops through a series of stages. However, *what* develops and what constitutes an expert is guided by the nature of the domain itself. As such, experts *enter* into a way of understanding the world already in place for other experts; i.e., the elements of the domain itself apply an adaptive pressure to select specific cognitive processes or capabilities that come into play as mastery develops (see DiBello, 2002 DiBello & Missildine, 2011, 2010, 2012; DiBello, Missildine, & Struttmann, 2010).

OUR METHODS AND WHAT WE LEARNED

Although our aim was to understand the nature of expertise and the cognition of experts, our opportunities for getting that data tended to be via distressed organizations. Over the years this led to a set of services and assessment products that allowed us to assess levels of expertise among executives and build solutions that accelerated skill performance outcomes.

Assessing Expertise

We had developed a number of ways of assessing the working mental models of business professionals using knowledge elicitation instruments very similar to those designed by others who research expertise (e.g., Crandall, Klein, & Hoffman, 2006; Hoffman & Lintern, 2006; Hoffman et al., 2014; Hoffman, Shadbolt, Burton, & Klein, 1995; Schraagen, 2009; see also Chapter 19, "Incident-Based Methods for Studying Expertise," by Militello & Anders, and Chapter 17, "A Historical Perspective on Introspection," by Ward et al., both this volume).

After several years using different methods, we noticed that highly talented business performers are very similar to each other. That is, they have *decoded* the domain of business much like chess masters had decoded chess (see Chapter 2, "The Classic Expertise Approach and Its Evolution," by Gobet, this volume). This led us to realize that—as a domain—business itself is an orderly closed system of relations between principles, and that so-called intuitive experts in business have an implicit grasp of this. If this is the case, it would follow that all highly skilled business experts should recognize each other and share a common mental model of these principles, although they may manifest differently in different industries. Since business has been evolving in recent years, this model would have to be very robust, running deeper than recent societal changes.

In a study funded by the National Science Foundation conducted over four years (NSF Award ENG 9548631) we found that those who show considerable and consistent talent in business have such a mental model, shown in predictable ways of making use of information. Leveraging our relationships with companies and clients, we had unusual personal access to study a large number of highly placed leaders. Doing in-depth studies of talented business leaders who were repeat successes—even in very challenging markets—and who had risen to very high positions (such as chairmen of large corporations) and maintained that level of position even as business itself has grown much more complex, we discovered a shared mental model among all of them. Like experts in other domains, business experts rely not on greater analysis or greater information, but *better ways of structuring or organizing their knowledge* (in the sense of Ericsson et al., 2006; Ross et al., 2004). Further, a business expert differs in the *manner in which he or she looks at the business landscape, particularly with regard to what is linked with what and what is important to manipulate.* For those interested in the details of the model, the results of that research are published in-depth elsewhere (DiBello, Lehmann, & Missildine, 2011).

This research led to a standardized instrument for measuring business expertise along a stage-like continuum. At this point, it is in an online form that automates much of the scoring and analysis. The participant is assigned a case relevant to his or her industry. Expertise is measured by how accurately he or she predicts the future outcomes of decisions by examining several years of a company's case material, one year at a time, predicting the subsequent year's events, then getting the opportunity to read that material to see how accurate their predictions were, doing the same thing again, and so on. Since the scenarios are based on actual companies where the outcomes are known and the material is actual raw material such as financial reports or letters from the company CEO, the task is to see whether people can see what *experts* have seen in the same material and make the same predictions. Over the years, these instruments have proven to be extremely reliable predictors of expertise for both executives and—more important to our clients—executive teams, and are routinely used by companies for succession planning, executive coaching, and executive committee coaching. They are also useful for observing what is *broken* in the thinking of an executive or a team of executives. What follow is an example of how the instrument was used to help a *broken* team.

An Example of a Broken Executive Team and One Expert

Having a useful model of an ideal business expert that has stood the test of time for ten years has been useful in helping companies find out what is lacking in their executives. Comprehensive business expertise rarely occurs in one person; more likely, it occurs among an executive committee or team, and for the most part, it is more useful to ensure that—as a team—an executive committee is working as an expert.

One of our most dramatic cases was a financial services company in Europe which was badly hurt in the sub-prime mortgage crisis in 2007–2008. This company is not a *bank* in the normal sense, but the kind of services company that has other banks or countries as its customers. Being on the brink of implosion was a crisis not only for the company, but for its country. The executive committee had a misunderstanding of mortgage risk and reinsured enormous holdings of extremely risky subprime mortgages. In order to survive they had to borrow $3 billion overnight with an interest rate of 12 percent, with pre-payment penalties of 44 percent. We used one of our online assessments to elicit the working mental model of the executive committee and their staffs. About 70 people were evaluated. Our assessment revealed that the entire issue was caused by a failure to anticipate a market crisis that was obvious to many experts outside the company. Unfortunately, it is common for an entire executive committee to share a blind spot that can cascade down through major decisions.

The assessment used here was a version of our online business simulation in which each participant must solve a business mystery by predicting what will happen over the course of a business's life in one-year intervals after being given a year's worth of data at a time and *judge* the actions of the executives in the case. We chose an actual *case* that was similar to this company's but which encountered its challenges in a different time of history. In contrast to their company, the company in the *case* had not made the same mistakes. After removing all identifying information—including dates—we presented each person with five years of material, one year at a time, and asked them to examine the company's data, decisions, plans, and declarations and predict the outcomes of the following year and judge the actions of the executives. We use a scoring scheme designed to evaluate the presence or absence of our model of an *ideal* business expert. We also ask the participants to choose the *clues* in the material that drove their thinking. Each *clue* is like DNA or fingerprints at a crime scene. An expert will not only see that it is significant, but *what* it signifies. Therefore, if the participants predict that the executives in the case do not see that a market downturn is coming and choose the clues that in fact are an early warning sign of that fall, we can see that they are attending to the right clues in the *market* portion of the model. Each question and each *clue* is coded using an association matrix that links back to the model. From this we can generate charts of a person's mental model compared to an ideal *expert*, and a *heat map* showing by color where good clues were identified and where there are blind spots (black) where clues were overlooked.

Figure 35.1 shows some material that a participant would examine. The question and clue selection screens are all online and involve easy clicks. The participant can stop mid-stream, save their work, and come back to it later. The idea is to replicate the actual decision process of busy business executives.

In real life, the banking executives were having a particularly difficult crisis as a team. They were having serious disagreements with their CEO about how to recover. Some were working behind the scenes to have him removed by the board of directors. A unique feature of our instrument is that it can show how a *team* is thinking. When we mapped the executive team's results—as a team—on the dimensions of expertise critical to their recovery, we found that they were at risk for making the same mistake again due to a shared blind spot in the area of finance and risk management. The only person on the team who could mentally simulate the eventualities accurately was the CEO with whom they were violently disagreeing and whom they were trying to depose.

Figure 35.2 shows the teams' results on this important dimension. This chart shows what they predict will happen in the case versus what actually happened in the real case. They did not see the impending disaster (the executives in the case did) and did not agree with what the executives in the case did to prepare.

Figures 35.3 and 35.4 show the CEO's predictions on the same instrument. As can be seen, he predicted exactly what happened in the case. On other dimensions of business

FIGURE 35.1 Material from CAAT.

EXPERTISE IN BUSINESS 813

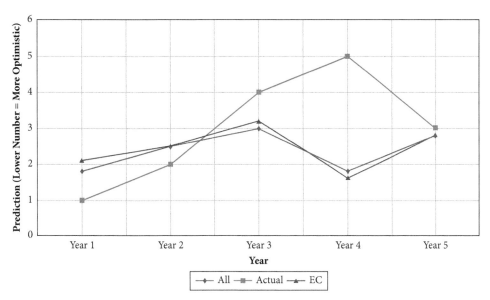

FIGURE 35.2 Financial risk prediction. This shows that all the executive committee members underestimated the financial risk coming in years 3 and 4.

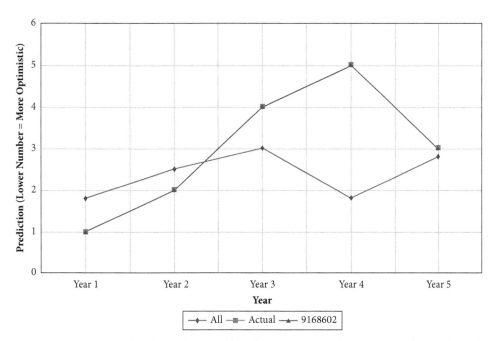

FIGURE 35.3 Financial risk prediction. This shows the CEO (9168602) predicted the risk exactly.

Category	Item				
Growth Model	Organic Growth	0	0	0	0
	Acquired Growth	1	1	1	0
	Data Mining	1	1	1	2
Strategy	Clinical Trials	0	0	0	0
	Therapeutic Areas	1	0	2	1
	Int. & Ext. Partnering	2	0	1	2
	Regulatory approval	1	1	0	0
	Financial Perf. Company	0	1	0	1
	Perf. System	3	1	0	0
	Mkt Val & WACC	1	0	0	0
Custom Layer	Scientific competency	0	0	2	1
	Patient Centric	1	1	1	2
	Collaborations	1	1	1	1
	Timely Performance	0	1	2	2
	Governance & Risk	0	0	1	1

Key		
	1	Expert Selected, You Selected
	0	Expert Selected, You Did Not Select
	1	Expert Did Not Select, You Selected
	0	Expert Did Not Select, You Did Not Select

FIGURE 35.4 A "heat map" showing the blind spots in the participant's reasoning.

expertise, they were very impressive and quite homogeneous. Figure 35.4 shows a heat map example illustrating how the blind spots can be identified.

This study is interesting on two fronts. It shows how easily a team can develop a common *theory of the crime* over time, even when it is not working, perhaps explaining how whole companies can sometimes implode. But it also shows how complex business has become and how the insights of one person can be difficult to communicate. Fortunately, using these data, the executives' advisors were able to convince the executive committee that—as a team—they needed to re-distribute responsibility and put those with the deepest insight into the capital risk environment on the front line during the recovery. As a committee, they could function as an *expert* but only if they knew where the different elements of their expertise resided.

This company did recover, enjoying a doubled stock price in less than a year, but it was by developing a shared understanding of how they were going to divide up running the business and yet keep all the parts linked in the way they ran the committee meetings. In the end, they followed the insights of the one lone wolf in one area, abandoning their preferred notions of what would work while they came up to speed on what he was seeing.

Achieving Expertise with Rehearsal

WTRI *fixes* a company that is underperforming or failing in the market with custom-designed emulations of our clients' companies which allow companies to *rehearse* the probable future facing their companies, somewhat like a *war game* used in the military. The context of these emulations is a simulation of the world economy with the emulated company and its competitors in it. The idea originally was to help companies adapt to changes in the business world, usually having to do with implementing enterprise information technology, but we realized early on that we are actually helping individuals and teams develop new capabilities for addressing competitive issues in a dynamic economy. Expertise comes from experience, and specifically from experience with difficult and complex problems in which one can link the decision made with the outcome that occurred. It is generally believed that it can take many years of mistakes and reflection to develop enough exemplars to see the pattern between decisions and the outcomes that occurred. Most corporate professional development plans of pro-motability assume five to ten years before managers can navigate complexity. In today's business world, this learning curve is too long.

With *rehearsals*, *expertise* can develop quickly under the right conditions. In fact, the European financial company discussed in the previous section was turned around with a rehearsal in a very short time, although the rehearsal was extremely complex. Rather than chronological time, cycles of trial and error against a highly visible goal with granular feedback can accelerate the development of expertise. We believe that we are actually accelerating cognitive reorganization of existing procedural knowledge or propelling new levels of expertise with compressed cycles of experience, or both. We are exploring both hypotheses in our lab.

In our rehearsals, we found that compressing time and providing instant feedback could result in five to ten years' worth of expertise in a few short days. If the emulated company resembled the *real* company in a reality-analogous way (from the point of the view of the participants and their experience), the transfer back to the real workplace was immediate and profound (DiBello et al., 2010).

In most cases (see Figure 35.5), we built actual physical models of the company emulated the actual enterprise technologies and the worldwide distribution and economy.

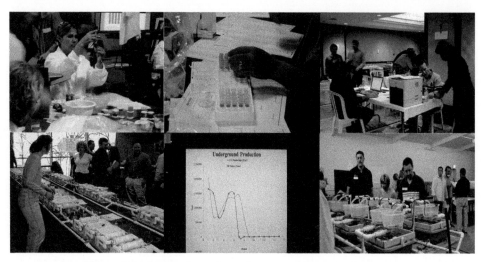

FIGURE 35.5 Photos from face-to-face rehearsal events.

There are four critical components to the rehearsals that turned out to be important to accelerating expertise.

1. Time compression;
2. A rich unfolding future that has a pattern;
3. A non-negotiable goal with multiple ways available to achieve it; and
4. A rich history that is similar to the past the participants have experienced but not identical.

Participants run through the rehearsal twice, experiencing several cycles of failure and detailed feedback before they hit on a solution that will work. The first day of the rehearsal does not go well; when a major organizational breakdown is present in real life, participants invariably fall back on default behaviors and *fail* in a way that replicates their current problems in the real workplace. At the start of the second round of the rehearsal, participants are given feedback based on various performance metrics from the activities of the previous day. Detailed data are provided that show the ideal versus the actual performance. The participants discuss among themselves three issues: (1) what went wrong, (2) what we will do differently this time, and (3) how will we know it is working. On Day 2 they are given an opportunity to rework and develop novel solutions. As such, the solutions they devise on Day 2 reflect workable strategies to bring back into the real workplace. Because various teams can coordinate across multiple functional areas with dynamic, real time feedback in a consequence-free environment, the rehearsal allows old systems to be discarded and new models emerge.

Over the years, rehearsals have become very powerful tools; almost without exception the financial performance achieved in the rehearsal after two rounds is achieved by the actual company within a few months after the team implemented the rehearsed approach (e.g., see Bower, 2004 DiBello & Missildine, 2011, 2012; Hoffman et al., 2014).

We became well known for these events because of their impact on companies. Working with our interdisciplinary team of economists, artists, developers, and cognitive scientists working out of our lab in San Diego, we were able to turn around the performance of scores of companies radically in a very short time frame with predictable and measurable financial results. With support from investors and basic research support from the National Science Foundation, we developed a number of enabling technologies to make it easier to design powerful rehearsals more quickly.

We estimate the delivered value to companies when the rehearsed outcomes are replicated in companies to be in the billions. However, the biggest benefit of these events is what we learned about the nature of business expertise and how it might be changing. The following section describes an example of a success that was financially modest (this was a relatively small client for us) but cognitively very striking, and compared to many of our cases, a relatively easy business to understand for illustration purposes.

The Midwest Foundry

The Midwest Foundry Company (MFC) is an old foundry that makes all sorts of iron castings, including small machine parts and large-scale castings such as cast iron engine blocks for freight train locomotives. However, over the past several decades, the foundry business has changed drastically. Many items once cast in iron are now fabricated using other means, and small castings are outsourced offshore at greatly reduced cost. However, there is still modest demand for domestically made very large (multi-ton) castings, which are normally not made in large quantities. MFC had facilities capable of making castings weighing as much as 30,000 pounds, with their average casting being 5,000 pounds.

When sales in 2000 were down 11 percent—the general perception was that the company was losing money because it was losing sales. As a result, the owner and the management entered into increasingly risky agreements in order to attempt to retain customer business. These were viewed as marketing programs designed to secure additional business. However, many times prices charged to desirable customers were below cost of production for MFC. As such, the more castings they made, the more money they lost.

MFC's real opportunity resided in their close proximity to their customers and their ability to make very large castings, but they needed to look at the opportunity differently than they had been. They needed to offer greater value rather than lower prices. Customers would pay a premium for something they need quickly and which is hard to get. Large castings are often used in other products that are not bought in large quantities but which, when needed quickly, are highly profitable for the seller. Therefore, price cuts were not necessary for items customers are desperate to have.

Problems on the Shop Floor

Just as the process of casting has not changed much over the centuries, foundry workers and managers tend to be second- or third-generation foundry men. Most of the workers had lived in the Mid-west for generations and still farmed the land they were born on as a way to subsidize their living. The workers and many of the managers were also not financially knowledgeable enough to know how their foundry was doing or what they could do to contribute to its survival. Notions like *breaking even* and *covering overhead* were not part of their daily language on the shop floor. They were efficient and skilled mold makers, pourers, and finishers, not businessmen. They tended to make as many molds as they could—whether there were orders for them or not. This created a widening disconnect between the day-to-day operations and the larger goals of the company.

There was also a disconnect between the different functional areas. The life of a casting begins with the pattern. A pattern is made and handed off to the mold makers. Large-scale casting molds are made in three pieces. The true craft of large casting involves getting as much of the intricacy of the final piece to be a part of the casting by controlling the flow of the molten metal in such a way that you get uniform filling and cooling. As such, mold makers cannot simply follow a set of directions; rather, the real expertise for the mold makers is in understanding how the molten metal will flow into the mold. This requires a complex placement of holes and *chills* that control the proper moving and cooling of the molten metal. With large items, the placement of the holes and chills and the skill in making each individual mold is as important as the design of the pattern.

A near non-existent tracking process led to a six-week lead time from the time a customer placed an order to the time of delivery, even though much of that time, the casting may have been sitting in a pile of uncleaned inventory. Six weeks was about the amount of time needed to get through the twenty-foot pile that resulted from excess inventory. Once an item was selected for cleaning, the focus of the workers' attention was on doing a stellar job of preparing the casting for the customer. Again, the workers proved superior at this task as others did at making the molds. A great deal of artisan skill and knowledge accumulated over generations was involved in preparing the final casting. However, many of these were not due for shipment while others that were already late could not be found.

This process led to a less than 30 percent on time delivery. In addition, because the molds were not cleaned with any priority, any flaws in the castings were only revealed when the part got through the finisher. Flaws in the castings could be a problem with either the pouring process or the original pattern. Understanding the nature of the flaw requires visual inspection by mold makers. Yet, often a casting was not scrapped until quite a bit of labor had gone into *knocking off* the sand mold to reveal the flaw. Since this happened in another building entirely, the molding and pouring personnel may not know that rework had to be done until weeks after the casting left their control.

Simply, MFC could not afford to continue with the same work patterns that had made them successful in the past. Business used to move more slowly; on-time delivery was not as important as price and inventory practices. Workers had decades of expertise as mold makers, pourers, or finishers. They had developed their priorities and impressive decision-making skills in response to environmental pressures at a time when quality was difficult to obtain and timing was not as important.

The Rehearsal

For MFC, we developed a rehearsal to develop expertise in on-time delivery and value pricing. MFC's rehearsal was a two-day, simulation-based intervention, in which participants engage with a *reality analogous* version of the foundry's mold making, pouring, an ERP system, as well as its profile of customers (see Figure 35.6 to get a sense of what the *rehearsal* looks like as well as pictures of the actual foundry floor).

Organizing Forces and Leading Activities

Before designing the mock work environment, we needed to understand what kinds of thinking supported their current practices. We asked the workers to develop an *as is* map of their process, outlining how they see the workflow from pattern to delivery, and subsequently develop a *to be* map. During the rehearsal it emerged that their *as is* map did not include on-time delivery, and their *to be* map seemed to be an extension of the current way of thinking—they simply created new ways to push more product through the pipeline, without realizing the new bottlenecks that this would create. In addition,

FIGURE 35.6 Molds in inventory on the foundry floor vs those in the rehearsal.

for the mold makers, cleaning and shipping were not part of their process maps. Activities not involved in actually making the casting were not the object of their thinking and decision making. Therefore, entire parts of the process, such as handing off the job, trucking the unfinished casting to the cleaning facility, and storing were not represented on their maps and were not objects of their attention. We needed to get them to think about these missing things.

If their old ways of thinking had emerged in response to adapting to one environment, we needed to create an environment with new goals that were highly visible. As such, the rehearsal design began with a set of outcomes to which they would be held accountable:

1. Customer orders must be shipped on time; anything late had to still be delivered, but there would be no payment for it;
2. A lower cost per item budget;
3. Reduce scrap rate goals; and
4. A specific revenue goal with specific profit margin that had to be met by the game's end, with reasonable interim milestones along the way showing progress.

We constructed a miniature foundry in which they had to make molds from miniature *patterns*, prepare them for pouring, and generally route the product through the process laid out on a series of large table tops. We did not replicate the entire foundry in miniature; rather, we created an environment that exaggerated what was important to work on and downplayed what was not. In that sense, it is a physical caricature of the decision space.

The rehearsal design ideally represents those aspects of the work environment where critical decisions are made and places them in the context of the new outcomes. For MFC, we replicated pattern retrieval, mold making, and casting cleanup, because these are the points in the process where decisions are made and things can go awry. We used a polymer sand that, when mixed with water, makes a temporary mold used for making plaster of Paris castings. This also replicated the fragility of the real molds; within 30–45 minutes, these polymer molds start to fall apart (real-life sand molds start degrading within days). They are supposed to be used immediately.

The process of making molds engaged their default, routine work activities, but in the rehearsal, these activities were placed in the context of the primary decision points that were tied specifically to high-level outcomes for which they were held accountable. On the shop floor, mold making alone is the final measure of their performance. In the rehearsal, mold making and routing are part of a larger process of on-time delivery, cost reduction, reduced scrap rate, and meeting profit margins.

While making molds was easy for them, in the context of new requirements the task became more complicated. For example, there are about eight steps to making most castings. If you are late with step 2, you are already *too late*, even if the casting is not yet due to the customer. Thinking about the lateness of a casting by step 2 was not part of their thinking. Time had not been thought about in that way before in the context of their work. To develop this kind of thinking, we gave them tools offering visibility into the effect of accumulating lateness. One tool was a MRP (material requirements planning systems)-generated list of internal deadlines (by customer order) for internal customers (i.e., the eight steps in the process) such as *making the cores*, the *pouring room*, or *finishing facility*, which allowed to them see that delay in one process had a domino effect on other parts of the process, even if it is weeks ahead of the planned ship date. We required that the MRP router (a printed form with all the steps and interim due dates) travel with the job (pattern, mold, and casting) and each person had to check off their step and when they did it. The due date (or game period, in this case) was shown for each step, not simply the due date at step 8. Thus, at every hand-off, they could see the accumulating lateness. In addition, the customers (played by staff facilitators) did not always accept orders that were late and certainly did not pay for them. MFC had had a MRP system for nearly four years, but they had never used it except to track customer information, such as shipping addresses, customer item numbers, and sales contacts.

Results in the Rehearsal

The first day of the rehearsal, the MFC workers replicated many of the dysfunctions that had occurred in real life. Less than 30 percent of the orders were on time, and the financial picture by the end of Day 1 was nearly identical to their real finances. All the molds and castings were beautifully done, but always late. As the day progressed, we tracked their performance and projected charts using an LCD projector. The projected charts were updated every few minutes. Even though they could see where they were headed, they could not come up with a solution. In order to get the *customers* to take the castings, they cut prices below cost and tried to make up the losses with volume. Even though this group had developed a careful *to-be* plan and process map, this was entirely abandoned during the rehearsal. The phones keep ringing and the suppliers kept sending bills. We stayed in role as unhappy customers throughout the day, allowing the numbers to speak for themselves.

On the second day, we reset all of the numbers and gave them an opportunity to meet privately and devise a new approach. The second rehearsal was an opportunity to rehearse a new strategy. All of the same parameters were in place. They had new molds to make, and were held accountable to the same top-line revenue goals, but the customers and orders were different. Nearly everyone showed up early and each team was rearranging their tabletop work floor before we got there. When they saw us, they asked for their data again, and asked for it in different forms. They were given a prep period of a couple hours to figure out what went wrong the day before. When the rehearsal started again, they had to come up with a plan to fix the problems that had occurred the day before.

Very early into Day 2, they began asking the customers (facilitators in role) how important was it that the castings be on time. They wanted to know if we would pay a premium if we got guaranteed on-time delivery. We said we would pay full price for on-time delivery and a slightly higher price for a shorter lead time. They took this information back to their huddle. They also rearranged their tables and examined their data from the day before, and began strategizing how they would move the casting quickly from stage to stage with as little sitting around as possible. They also discussed developing ways to know where a given order was and if it was late. A sample of dialogue illustrates:

> D: Okay, but when you're done with making that, you can't just say "okay I did my part" and make a little pile at your elbow. *I am waiting for you to be done with that so I can do my piece.* You have to let me know it's ready or give it to me, because if you get it to me later than I need it, it's *late*, period.
> T: Right, and the whole team doesn't get paid.
> L: Yes, we don't get paid. Everybody got that! You don't move your stuff, we don't get paid!!!!! So maybe we should move you guys to be closer together and come up with a way to signal that you're done.
> B: Listen up you guys. I just got the customer to agree to a higher price with a shorter lead time. Can we do that? Let's look at the routing and see if we need all the time on there.
> T: Has anybody figured out the break-even by piece? Are we charging enough? We need to look at the prices; the more stuff we made yesterday, the more money we lost.

It should strike the reader that long-term experience of the workers has not changed, but their thinking is changing, both individually and collectively. In fact, it is their experience that makes the new level of expertise possible, even though before, it was the very thing holding them back. By rearranging the context in which their expertise is deployed, new goals will select ways of thinking and develop capabilities that make meeting those goals possible. The design of the rehearsal only focused the participants on accountability for outcomes, not on the best way to get there. This stimulated participants to begin to find new ways of thinking about the same things.

On the second day of the rehearsal, *on-time* was incorporated as part of *excellent casting* and *more throughput* was transformed into the idea of *more of what we need; less of what nobody is buying*. By the end of Day 2, they had designed a pull system as opposed to the push system they had previously had in place. This made the MRP routings and reports as central objects in the decision landscape. Important to note, all the *unachievable* goals of Day 1 were achieved easily on Day 2.

Results in the real world. After 9 months, they were on time nearly 82 percent of the time and returned to profitability. As a result, they reduced their scrap rate from 13 to 9 percent. They began to turn a profit, focusing on premium pricing and on-time delivery and gradually retired commodity products.

A few months later the company confronted a crisis. The owner was involved in a lawsuit with the bank; he was accused of using bank loans secured for capital improvements to make acquisitions of other businesses. The bank took possession of the company and the owner suddenly passed away. This was very discouraging news since the workforce had made considerable progress in understanding the key to profitability in today's market. Although the plant never actually shut down, the workforce struggled to meet commitments. They found it hard to pay suppliers with assets being seized by the bank. After trying to negotiate with the bank, some of the managers who had participated in the rehearsal sought out and found a financial backer. This backer bought the company on an "assets only" basis (he did not assume the debt), which allowed the workforce to continue with their program of improvement while the investor acted largely as a silent partner. The idea was that they would repay him with interest from the profits.

Two-Year Follow-up

On our return visit after two years, most of the same people we had worked with were still there. Orders were being *pulled* through the system in an orderly fashion, on time, at a low cost, with little scrap. They had achieved 100 percent on time to the customer and had shortened their lead-time from six weeks to two weeks and negotiated premium prices from customers for shorter lead times. They had reduced their scrap rate down to 2.9 percent, well below the 6 percent industry average. The entire customer service department was eliminated, profits were consistent, and for a foundry, the margins were higher than is typical.

We found that they were still using a version of tools developed in the rehearsal, but had gone beyond them, innovating new technologies. This was remarkable for a workforce that had been computer illiterate two years before. As the industry changed again, the workforce modified the information tools to support better decisions. Each team of workers gets information each morning with all orders and all offsets, and a detailed version of accountabilities for the whole company. The larger goals and complexities of the entire process were made transparent to workers in all parts of the process. In other words, every worker has visibility of his or her work in the context of the whole business and all workers have access to the tracking technologies.

As stated earlier, two years prior there was typically a 16-foot high by 40 foot in diameter *pile* of sand-encased castings to be cleaned in the cleaning area. Cranes were used to move the multi-ton castings in order to find out what was there and find those that were due to be cleaned in time for shipping. During our follow-up visit, there were only two relatively sand-free castings that had just arrived for cleaning. Within 20 minutes, cleaners who were expecting them had them moved them to cleaning booths.

In summary, the workers had gone well beyond what they learned from the rehearsal. What they developed in their rehearsal was a first principles understanding of the dynamic nature of their business and an ability to continually adapt and change their strategy as the business changed.

Scaling Rehearsals for Education

Businesses are more alike than different, and scalable educational applications in the form of *generic* rehearsals may have the same expertise accelerating benefit as our labor-intensive and expensive face-to-face rehearsals.

We are in the early stages of exploring ways to bring expertise in business to a much broader audience. Recognizing that society does need scalable solutions to meet the need for increased expertise in business, for the past two years we have focused on scalable versions of our approach for the explicit purpose of developing expertise in both individuals and teams. We are developing virtual world *rehearsals* on a proprietary platform. Unlike other virtual world platforms, this is built specifically for complex business scenarios. Our partners in the early work are the National Science Foundation, IBM Research, and the Project Management Institute among others. We are currently focused on two areas that capture much of business expertise but are not as complex as running a whole company. These are mergers and acquisitions scenarios and complex project management for business. Below is a summary of a project management study which was our first attempt to test a scaled acceleration of expertise for small teams and individuals in a business domain,

After extensive research on what made an *expert* project manager, we made the virtual world versions of our rehearsals as complex as our physical face-to-face emulations (see Figure 35.7). In one sense it was easier; the enabling technologies that make the experience rich and complex can be attached to the virtual worlds permanently and enrich an immersive experience. Further, the worlds are quite large. Some of our agricultural and mining worlds are fifty square miles or two miles deep and can be experienced with a virtual reality headset. In single-player mode, however, the other *players* are robots, and the programming can be challenging.

We added detailed feedback using dynamic dashboards (see Figure 35.8) which update instantly and automatically and which can be accessed via a button in the user's heads-up display.

In general, the multiplayer virtual world rehearsals performed exactly as the face-to-face versions, differing only in that they did not involve travel by the participants, could

FIGURE 35.7 Screenshots of the project management "worlds."

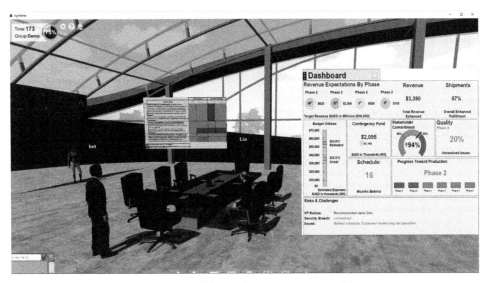

FIGURE 35.8 Example of a dashboard for tracking progress inworld.

be stored in the cloud, and could be conducted over a number of weeks instead of two days. The learning impact was the same even with the increased time between inworld sessions as long as the iterative grammar was preserved. Teams still got to complete the entire rehearsal twice and invariably made all their mistakes the first time through. We were less hopeful about the single-player versions of the product, where the participant is essentially on their own playing against and with robotic avatars. However, here too, the impact held up.

In a world-wide test in 2016, 203 participants tested a single-player version of a product we built for a one of the largest project management certification organizations in the world. All were volunteers. By our data collection deadline date, 46 had completed two sessions by completing the 60 or more tasks involved to finish (eventually 78 finished).

A complicated scoring scheme developed over two years with funding from the National Science Foundation was used to measure learning. The details are beyond the scope of this chapter. In sum, it involved five levels of expertise evidenced by levels of difficulty in the project in addition to challenging events that acted as both embedded cognitive probes and obstacles that had to be overcome.

As can be seen in Tables 35.1, 35.2, and 35.3, the participants scored at a level higher on all types of scorable tasks between time one and time two. To put this in context, this kind of change is normally seen in "real life" after three to five years of experience. We were both surprised and encouraged by these results.

This study showed that the basic principles proven to accelerate expertise—namely, cognitive reorganization through iterative trial and error with embedded feedback—held up in a single-player application with volunteers working alone, relatively unguided, and remotely. In general, the product resulted in learning benefits much better than anticipated—especially for a product still in development—and much

Table 35.1 Level of Performance on All Activities between Time One and Time Two

Group	N	Av ratio time one	Av ratio time two	df	Sig.
<60 events	46	0.433	0.591	45	$p < 0.0001$
>60 events	73	0.377	0.523	72	$p < 0.0001$

Student's T analysis of the differences between time one and time two change in level-up scores on all scoreable activities, and includes all activities, challenges, and core events.

Table 35.2 All Challenges; Comparison between Time One and Time Two

Group	N	Av score time one	Av score time two	df	Sig.
Challenges	46	0.761	1.054	45	$p < 0.001$

Student's T analysis of the differences between time one and time two change in level-up scores on all challenges.

Table 35.3 Performance on Challenges Unique to Time Two

Group	N	Av score time one	Av score time two	df	Sig.
Challenges	46	0.35	0.44	45	$p < 0.04$

Student's T analysis of the differences between time one and time two change in level-up scores on unique challenges that functioned as cognitive transfer indicators.

better for something that was completely voluntary. We were also able to show that fundamental cognitive mechanisms must be involved; because this method had been heavily tested with groups prior to this study, it could be argued that *individuals* had not necessarily undergone a change in their ways of thinking. In this study, all participants were working on their own and showed measurable changes in expertise in universal and consistent ways.

Beyond proving the robustness of the method, there are other implications. The strong results show promise not only for this product, but for the potential to deliver accelerated expertise educational methods more widely than thought possible for professional development purposes.

Conclusion and Further Work

Much of the work we have done—using the understanding of business expertise we possess to devise ways to assess and help our clients—has showed us what we don't know and what future research must address. We have touched on a number of themes throughout the chapter.

From the evidence we have collected it appears that the new *expert* in business may not be an individual at all, but rather a high-performing, highly efficient team, with team-level self-awareness of who has what piece of the skill puzzle in addressing complex challenges.

Further, all business employees probably need to be closer to an expert than in the past, if only to add value as a member of a high-performing team. Pushing the envelope on accelerated learning can only help more people feel secure that they will be able to find a meaningful and rewarding role in the future workforce. The response to products we are experimenting with now show us that scalable, remotely accessible, and affordable accelerated skill development solutions need to be widely available.

It is becoming more obvious that lifelong learning—even within the domain of business—will be necessary for all people. Perhaps the rehearsal has shown us that there is a fundamental mechanism involved in the development of expertise that comes into play, and that what we now call intuitive expertise is actually an innately human adaptive response to challenging environments.

REFERENCES

Bickhard, M. H., & Campbell, R. L. (1996). Developmental aspects of expertise: Rationality and generalization. *Journal of Experimental and Theoretical Artificial Intelligence 8*, 399–417.

Bower, B. (2004). Reworking Intuition: business simulations spark rapid workplace renovations. *Science News 23*, 263–264.

Campbell, R. L., & Di Bello, L. A. (1996). Studying human expertise: Beyond the binary paradigm. *Journal of Experimental and Theoretical Artificial Intelligence 8*, 277–291.

Chi, M., Feltovich, P., & Glaser, R. (1981). Categorization and representation of physics problems by experts and novices. *Cognitive Science 5*(2), 121–152.

Chi, M. T. H. (2006). Laboratory studies of expertise. In K. A. Ericsson, N. Charness, P. Feltovich, & R. Hoffman (Eds), *Cambridge handbook of expertise and expert performance* (pp. 167–184). New York: Cambridge University Press.

Chi, M. T. H., Glaser, R., & Farr, M. L. (Eds.) (1988). *The nature of expertise*. Hillsdale, NJ: Lawrence Erlbaum.

Crandall, B., Klein, G., & Hoffman, R. R. (2006). *Working minds: A practitioner's guide to cognitive task analysis*. Cambridge, MA: MIT Press.

DiBello, L. (2002). Solving the problem of employee resistance to technology by reframing the problem as one of experts and their tools. In E. Salas & G. Klein (Eds), *Linking expertise and naturalistic decision making* (pp. 71–94). Mahwah, NJ.: Lawrence Erlbaum.

DiBello, L., Lehmann, D., & Missildine, W. (2011). Using a unique profiling tool. In K. Mosier & U. Fischer (Eds), *Informed by knowledge: Expert performance in complex situations* (pp. 261–274). New York: Taylor & Francis.

DiBello, L., & Missildine, W. (2010). Information technologies and intuitive expertise: A method for implementing complex organizational change among New York City Transit Authority's bus maintainers. *Cognition, Technology & Work 12*, 61–75.

DiBello, L., & Missildine, W. (2011). The future of immersive instructional design for the global knowledge economy: A case study of an IBM project management training in virtual worlds. *International Journal of Web-based Learning and Training Technologies* 6(3), 14–34.

DiBello, L., & Missildine, W. (2012). The future of immersive instructional design for the global knowledge economy: A case study of an IBM project management training in virtual worlds. In N. Karacapilidis, M. Raisinghani, & E. Ng (Eds), *Web-based and blended educational tools and innovations* (pp. 115–135). Hershey, PA: Idea Group Inc.

DiBello, L., Missildine, W., & Struttmann, M. (2010). Intuitive expertise and empowerment: the long-term impact of simulation training on changing accountabilities in a biotech firm. *Mind, Culture & Activity* 16, 11–31.

Ericsson, K. A. (2003). The search for general abilities and basic capacities: Theoretical implications from the modifiability and complexity of mechanisms mediating expert performance. In R. J. Sternberg & E. L. Grigorenko (Eds), *The psychology of abilities, competencies, and expertise* (pp. 93–125). Cambridge, MA: Cambridge University Press.

Ericsson, K. A., Charness, N., Feltovich, P. J., & Hoffman, R. R. (Eds). (2006). *Cambridge handbook of expertise and expert performance*. New York: Cambridge University Press.

Ericsson, K. A. (2016). *Peak: Secrets from the new science of expertise.* Boston, MA: Houghton Mifflin Harcourt.

Hoffman, R. R., & Lintern, G. (2006). Eliciting and representing the knowledge of experts. In K. A., Ericsson, N. Charness, P. Feltovich, & R. Hoffman (Eds), *Cambridge handbook of expertise and expert performance* (pp. 203–222). New York: Cambridge University Press.

Hoffman, R. R., Shadbolt, N. R., Burton, A. M., and Klein, G. (1995). Eliciting knowledge from experts: A methodological analysis. *Organizational Behavior and Human Decision Processes* 62, 129–158.

Hoffman, R. R., Ward, P., Feltovich, P. J., DiBello, L., Fiore, S. M. & Andrews, D. (2014). *Accelerated expertise: Training for high proficiency in a complex world*. Boca Raton, FL: Taylor and Francis.

Kaplan, S. N., & Minton, B. A. (2012). How has CEO Turnover Changed? *International Review of Finance* 12(5), 57–87.

Klein, G. (1989). Recognition-primed decision making. In W. B. Rouse (Ed.), *Advances in man-machine system research* (vol. 5, pp. 47–92). Greenwich, CT: JAI Press.

Klein, G. (1993). A recognition primed decision (RPD) model of rapid decision making. In G. Klein, J. Orasanu, R. Calderwood, & C. E. Zsambok (Eds), *Decision making in action: Models and methods* (pp. 138–147). Norwood, NJ: Ablex.

Klein, G. (1998). *Sources of power: How people make decisions.* Cambridge MA: MIT Press.

Klein, G. (2000). Cognitive task analysis of teams. In J. M. Schraagen, S. F. Chipman, & V. L. Shalin (Eds), *Cognitive task analysis* (pp. 417–430). Mahwah, NJ: Lawrence Erlbaum.

Piaget, J. (trans. A. Rosin) (1977). *The development of thought: Equilibration of cognitive structures.* New York: Viking.

Ross, K. G., Lussier, J. W., & Klein, G. (2004). From recognition-primed decision making to decision skills training. In S. Haberstroh & T. Betsch (Eds), *Routines of decision making* (pp. 327–342). Mahwah, NJ: Lawrence Erlbaum.

Schraagen, J. M. (2009). Designing training for professionals based on subject matter experts and cognitive task analysis. In K. A. Ericson (Ed.), *Development of professional expertise* (pp. 157–179). Cambridge, UK: Cambridge University Press.

Spiro, R. J., Feltovich, P. J., Jacobson, M. J., & Coulson, R. L. (1992). Cognitive flexibility, constructivism, and hypertext: Random access instruction for advanced knowledge acquisition in ill-structured domains. In T. Duffy and D. Jonassen (Eds), *Constructivism and the technology of instruction* (pp. 57–75). Hillsdale, NJ: Lawrence Erlbaum.

Stamoulis, D. (2009). *Senior executive assessment. A key to responsible corporate governance.* Malden MA: Wiley-Blackwell.

Vygotsky, L. S. (1978). *Mind in society: The development of higher psychological processes.* Cambridge, MA: Harvard University Press.

CHAPTER 36

TEAMWORK IN SPACEFLIGHT OPERATIONS

UTE FISCHER AND KATHLEEN MOSIER

Apollo 13: An Example of Teamwork in Spaceflight Operations

"Houston, we've had a problem" (Test Division Apollo Spacecraft Program Office, 1970, p. 160). This now famous statement by Jim Lovell, the Commander of Apollo 13, marked the beginning of an extraordinary rescue mission. Apollo 13 was 200,000 miles away from Earth when an oxygen tank in the spacecraft's service module exploded, resulting in the depletion of the crew's supply of electricity, light, water, and oxygen.

To get the Apollo 13 crew back home safely required much individual expertise and ingenuity, but above all, it required exceptional teamwork both within mission control and between flight controllers and the crew.

Gene Kranz, the lead Flight Director for Apollo 13, orchestrated the rescue effort. He and his team of flight controllers were working the shift in the Mission Control Center at Johnson Space Center when Lovell's message came in. Updates from the crew—meter readings and warning light indications—and system reports from flight controllers presented a confusing picture. "Everything we knew about our spacecraft," Kranz recalls, "all that we had learned about the design, precluded the kind of massive failure we were seeing" (Kranz, 2000, p. 313). While flight controllers dedicated to the craft's life support systems assisted the space crew in their troubleshooting, others backed them up and eased their workload. For instance, they monitored the signal strengths of the craft's antennas and kept track of the craft's attitude, and when critical changes occurred, they called for appropriate crew action to ensure uninterrupted communication and usable readings from the craft's gyroscope. Mission control's centralized decision making and communication ensured an orderly interaction with the crew (Vessey, 2014). Flight controllers made their recommendations on crew actions to the Flight Director who, insofar as he agreed, had the Capsule Communicator (CAPCOM), an astronaut, relay the instructions to the crew.

Fifteen minutes into the crisis, the nature and scope of the problem became clear at last when Lovell reported that the craft was "venting something out (...) into space. It's a gas of some sort" (Kranz, 2000, p. 163). At that moment, it became apparent that an explosion in the service module had destroyed the space craft's cryogenics and fuel cells, and that the mission had become one of survival (Lovell, 1975). As the oxygen supply in the spacecraft's command module kept decreasing, crew and mission control agreed to move the crew into the lunar module and to use it to propel the command/service module so that its remaining resources could be saved for reentry. The next major question mission control faced was how to bring Apollo 13 back to Earth. Kranz and Glynn Lunney, his flight director colleague, worried that an immediate abort would curtail their options, given they didn't know whether the craft's main engine had been damaged in the explosion. They wanted more time to plan for reentry and thus favored the long way around the Moon. Confident that mission control and crew had the expertise to solve resource limitations on the spacecraft, they made the case for this option in discussions with other flight directors and the chief of flight operations, Chris Kraft, who ultimately agreed. Kranz then handed over command to the other flight directors and their teams of flight controllers to man mission control around the clock while he and his team went to work on detailing the return procedures for Apollo 13. Several issues had to be planned for—maneuver procedures to speed up the return; an integrated checklist for the reentry phase; use of spacecraft resources; management of lunar module resources; and a master plan to lay out the steps required for reentry and their timing. Subteams of flight controllers together with astronauts and engineers worked out solutions and tested them in simulations, all over the course of three days. On April 17, 1970 at 1:08 p.m. CT Apollo 13 splashed down successfully in the Pacific. As Kranz (2000) recounts: "Our crew was home. We—crew, contractors, controllers—had done the impossible. The human factor had carried the day" (p. 337).

Spaceflight operations—thankfully—tend not to be as dramatic as Apollo 13. Nonetheless, even though it is an extreme example, Apollo 13 brings to the fore the central role teamwork plays in any space mission. While the success of space missions unquestionably depends on the technical expertise of individuals—of astronauts and mission controllers alike—their complexity requires superior teamwork as well. In this chapter, we will show that teamwork in spaceflight operations is a multiteam effort requiring the coordination and collaboration not only of individuals within a team (mission control or space crew) but importantly also between the teams. We will discuss the strategies and procedures these expert teams have established to ensure common task and team models, as well as to facilitate their communication and joint performance. We also will address the teamwork challenges of future space exploration, and we will describe efforts to mitigate these problems. Our chapter begins with a discussion of the teams (mission control and space crew) that make up the multiteam system, highlighting important features of their (intra)teamwork. In the final section, we raise questions and methodological issues that research will need to address to support the collaboration between space crews and mission control during future long-duration exploration missions.

Mission Control and Space Crew as a Multiteam System

Conducting space missions is a complex task. Knowledge and expertise are distributed across co-located and remote team members so that communication and collaboration among not only individuals but also teams of experts—the crew in space and flight controllers on the ground in mission control—are essential to successful performance. This form of teamwork has been characterized as a multiteam system (MTS; Mathieu, Marks, & Zaccaro, 2001) to emphasize the interdependencies that exist between component teams.

As part of a multiteam system, crew and mission control rely on each other for information and resources. They cannot complete their respective tasks without successful task performance by the other team, and mission goals are achieved through a concerted effort. This interdependence between the teams is most visible during extravehicular activities (EVAs) when one or two crewmembers are outside their spacecraft to conduct specific tasks, such as maintenance or repair work. This situation is depicted in Figure 36.1 with the right-hand panel showing two crewmembers as they perform a maintenance EVA on the International Space Station; the left-hand panel captures the concurrent activities in the Mission Control Center (MCC) at the Johnson

FIGURE 36.1 (Left) An overview of the flight control room in the Johnson Space Center's Mission Control Center (MCC) as flight controllers support extravehicular activity (EVA) by astronauts on the International Space Station (ISS). (Right) Two astronauts during an EVA conducting maintenance activities outside the ISS.

Images courtesy of NASA.

Space Center in Houston, Texas. During an EVA, controllers in MCC fulfill critical monitoring and managing functions (Miller, 2017). EVA ground personnel monitor the performance of the EV (extravehicular) crewmembers and keep track of the task timeline. They watch the crew's limited consumables (e.g., O_2, CO_2, etc.), assess their health, and ensure that the communication and power systems supporting the EV crew are configured accurately and are functioning properly. One dedicated position, the EVA flight controller, leads and coordinates all EVA activities in MCC while communication from MCC with the EV crew is channeled through the ground intravehicular operator (ground IV) who assists the EV crew with procedures and checklists and relays timeline changes to them.

Missions to the International Space Station add an additional layer of complexity to the MCC/space crew multiteam system insofar as they are carried out jointly by teams from different space agencies residing in different countries (e.g., NASA in the USA, Roscosmos in Russia, the European Space Agency with centers across Europe, and JAXTA in Japan). In this chapter, we will focus exclusively on the US multiteam system consisting of NASA astronauts and NASA mission control. Before we discuss teamwork as it relates to the MCC/space crew multiteam system, we need to take a closer look at team processes in its component teams.

Teamwork in the Mission Control Center

The current focus of MCC operations at NASA Johnson Space Center is on missions to the International Space Station (ISS). Flight controllers in MCC are trained to become experts in specific ISS systems (e.g., power and communication systems; Vessey, 2014). They must learn not only how that system functions and how to troubleshoot it, but also the impact that abnormalities or malfunctions will have on other systems on the ISS. Additionally, they train to pool their expertise effectively with other flight controllers. For example, a flight controller may be assigned to work in a *backroom* during off-nominal events to support the *frontroom* controllers and highly experienced system specialists who have complementary expertise (Kranz, 2000). This operational setup, which has been characterized as distributed supervisory coordination (Caldwell, 2001; 2005), means that subject-matter expertise within and across ISS systems is distributed among several flight controllers in two different locations (frontroom and backroom), and requires complex coordination and collaboration among these experts. Other team members in MCC include the Flight Director, who is the formal leader and decision maker in MCC, and the CAPCOM, who manages communications between MCC and the astronauts (Vessey, 2014).

In addition to subject-matter expertise, flight controllers exhibit the other five dimensions of expertise described by Garrett, Caldwell, Harris, and Gonzalez (2009; see also Onken & Caldwell, 2011): situational context expertise, or the ability to assess a situation by gathering information and to know how the situation impacts goals; expert identification expertise, related to transactive memory (DeChurch & Mesmer-Magnus, 2010), or the awareness of who knows what and when to ask for data or information from them; communication expertise, or the ability to communicate clearly with team members; and two dimensions of expertise around the technology used in MCC—interface tools and information-flow path. These dimensions come into play during different operational events and phases of events. For example, detection of an anomaly may be guided by a controller's subject-matter expertise; in contrast, isolation of that anomaly and creating a hypothesis about its cause requires teamwork among controllers and may involve all dimensions of expertise (Onken & Caldwell, 2011).

MCC expertise is honed using complex high-fidelity simulations that emulate events and configurations found in actual missions. This enables extensive practice on both frequently occurring tasks and low-occurrence events such as system malfunctions. Effective MCC task coordination depends on the quality of team members' taskwork (i.e., actions directly related to accomplishing their assigned task), teamwork (i.e., behavior in support of their collaboration), and pathwork (i.e., capabilities concerned with the flow of essential computer and vehicle sensor data). Both taskwork and teamwork entail extensive knowledge sharing, and challenges for particular situational contexts (operations, problems, failures) include knowing what information to share, and how and when information should be shared or requested—for example, when to call upon the knowledge of backroom specialists (Caldwell, 2005).

A variety of methods (observation, critical incident interview, and operator feedback) and contexts (live shuttle operations, ISS missions, and training simulations) have been used to understand team coordination in mission control (for reviews of these and other methods, see Section II, "Methods to Study, Test, Analyze, and Represent Expertise," this volume). Research has focused on how information is shared and how actions are coordinated across consoles—i.e., between flight controllers with different technical expertise and responsibilities—and between the front- and backroom flight controllers who share the same technical expertise.

Fiore, Wiltshire, Oglesby, O'Keefe, and Salas (2014), for example, conducted semi-structured critical incident interviews to investigate MCC problem-solving processes around a complex problem—the failure of the ISS main bus switching unit. This particular problem requires MCC team collaboration so that relevant data are identified and pooled and MCC team members come to a shared problem model, and generate and execute solutions. The goals of this naturalistic study were to identify the team processes involved in collaborative problem solving by MCC team members and to determine whether they were consistent with the Macrocognition in Teams Model (Fiore, Smith-Jentsch, Salas, Warner, & Letsky, 2010). The analysis of the interview data revealed that as predicted by the model, most of the behaviors mentioned reflected efforts by these experts to build a common knowledge base and a shared problem model. In so doing they drew on position-specific knowledge of individual controllers as well as relied on shared domain knowledge. Specifically, MCC team members were sensitive to the time constraints of their teammates and recognized when to disseminate information to others, how to transform information into *actionable knowledge*, and when to consult subject-matter experts for relevant system knowledge.

Mission control requires the integration of multiple data points into meaningful patterns and involves continuous tasks (e.g., monitoring a system malfunction) as well as long-term plans that have to be carried out across multiple shifts, possibly until the end of a mission. Patterson and Woods (2001) conducted field observations to understand the processes involved in handovers between outgoing and incoming MCC personnel. When mission controllers change work shifts, the handoff meetings are initiated by incoming controllers and conducted in an interactive fashion. Incoming personnel gather information on current mission events or significant occurrences in the recent past prior to their meeting with their outgoing counterparts (Patterson & Woods, 2001). They look through relevant documents, such as the flight log, and join their outgoing colleagues at their console, viewing the monitors and listening to ongoing communications. These actions serve several purposes. Incoming controllers develop an understanding of the task environment, its current status, how it evolved, and how it might develop. Their interpretations likely facilitate interactions with their outgoing colleagues because aspects of the task environment can be presumed to be shared knowledge and to require no further discussion. Handoff meetings can thus focus on verifying the accuracy of the incoming controller's situation model and on providing additional detail (Durso, Crutchfield, & Harvey, 2007), and as a result promote a collaborative review and check (Patterson & Woods, 2001). Additionally, incoming controllers' interpretations and expectations may serve as a check on outgoing controllers' assumptions regarding a situation so that errors are identified and prevented from escalating into problems (Patterson & Woods, 2001).

The interactive nature of their discussion provides controllers with the opportunity to detect and correct knowledge gaps or misconceptions concerning mission-critical aspects, to remind one another of important issues, and to build shared mental models of the task environment, including off-nominal events and their implications, as well as plans and their associated assumptions. Handoffs between incoming and outgoing controllers do not follow a predefined script and rely on everyday conversational norms. To ensure the completeness and accuracy of these briefings and to coordinate activities with other controllers on the shift, a cross-checking process is invoked: incoming backroom controllers brief incoming frontroom controllers via a public voice loop communication system on the content of their individual handoff meetings with the colleagues they are replacing (Patterson & Woods, 2001).

In addition to direct dissemination through handovers, information is indirectly transmitted via voice loop groupware technology. This technology facilitates MCC teamwork by enabling flight controllers not only to communicate with other people in MCC, but also to monitor the conversations and activities of others—thus enhancing their situation awareness and providing the opportunity to detect instances that do not match their expectations, deviations, errors, or omissions on tasks (Caldwell, 2005; Patterson, Watts-Perotti, & Woods, 1999). An important facet of controller expertise is the ability to remain peripherally attentive to these voice loops and to shift attention to a communication channel when appropriate, for instance when something is discussed that will have a direct impact on the controller's subsystem. Not surprisingly, the Flight Director's channel has been found during simulated training missions to have higher utilization across missions and phases of missions than other channels, in particular the air–ground and ground controller loops (Wang & Caldwell, 2003).

One of the most important teamwork functions facilitated by voice loops is anticipation. When controllers monitor ongoing activities, they can anticipate events or problems and synchronize communications and actions with other controllers. They are aware of other controllers' workload, can judge their progress, and discern when to interrupt them. Moreover, they can take advantage of other controllers' data transfer as it evolves into integrated, event-level information, enabling them to quickly ascertain the impact of anomalies in systems related to their own subsystem (Patterson et al., 1999).

Teamwork in Space Crews

During the early years of spaceflight crews consisted of two or three astronauts. In current ISS operations, astronauts are members of a multinational team of six. Crewmembers are highly motivated and capable of adapting to stress and changing conditions. They are achievement-oriented, are intelligent, and frequently have an advanced degree in engineering, the biological and physical sciences, or mathematics, and/or professional experience in these fields (Landon, Vessey, & Barrett, 2016; Vessey, 2014). The selection process for astronaut candidates is highly competitive and rigorous and once selected, they will be in training for 5 to 10 years before their first mission. Astronaut training traditionally has emphasized technical expertise but in recent years, training in team skills and cultural competencies has been included partly in response to reports of interpersonal tension and conflict during Mir and ISS missions, and partly in preparation for future long-duration exploration missions.

While crewmembers have a common passion for space exploration and share many personality characteristics, they bring different technical expertise to a mission and thus fulfill different roles. They also have different ranks. One crewmember is the commander who during the flight

> has onboard responsibility for the vehicle, crew, mission success and safety of flight. The pilot assists the commander in controlling and operating the vehicle, and may assist in the mission experiments and payloads and their operations. Mission specialist astronauts work with the commander and the pilot and (...) are trained in the details of the orbiter onboard systems, as well as the operational characteristics, mission requirements and objectives, and supporting equipment/systems for each of the experiments conducted on their assigned missions. Mission specialists perform EVAs (extravehicular activity), operate the remote manipulator system, and are responsible for payloads and specific experiment operations.
>
> (Erickson, 2010, p. 281)

The importance of leadership to the cohesion and effectiveness of teams that work under isolated and extreme conditions has been well established in surveys of Antarctic science team members (Wood et al., 2005), crewmembers in space analogs (Kanas, Weiss, & Marmar, 1996) and astronauts during Mir/Shuttle missions (Kanas, 2005; Palinkas, 2001) and on the ISS (Kanas et al., 2006). Surprisingly, Kanas and colleagues (2006) observed that crewmembers' ratings of team cohesion were related only to perceived leader support. There was no significant link between perceptions of a leader's task role and team cohesion. The researchers suggest that this finding may have been a function of the small crew size on these missions. With only three crewmembers who had specialized knowledge and responsibilities, the task role of the leader seemed less important than his/her supportive role. Both task and supportive leader roles are addressed in the NASA Crew Office list of team competencies, as is the situational nature of leadership (Barrett, Holland, & Vessey, 2015).

Team dynamics are an important issue for crewmembers. In pre-mission surveys, ISS astronauts indicated that maintaining positive relationships with teammates would be their highest priority during the mission (Stuster, 2016). As is evident in their journal entries, astronauts "actively worked to maintain interpersonal harmony by cooperating, avoiding certain topics in conversation, and other sincere acts of comradeship" (Stuster, 2016, p. 33). Interpersonal friction or tension was rarely mentioned in the journals. This finding is consistent with research by Kanas and colleagues (2006). Their surveys of astronauts and cosmonauts did not reveal any significant decline in team cohesion over the course of a mission. Likewise, crewmembers' daily logs included few reports of negative experiences. However, Kanas et al. (2006) noted that almost half of the negative incidents referred to interpersonal (intra-crew and crew–ground) issues.

While team cohesion, collaboration, and conflict are less of a concern for the relatively short-duration missions on the ISS, they pose a significant risk for future long-duration space exploration (LDSE) missions. A mission to Mars will present challenges to crewmembers no one has experienced before. A vivid description of what lies ahead is provided by Salas et al. (2015):

> Imagine living and working in a small, confined space with five other teammates for over a year. Your team needs to complete a series of scientific experiments and perform other rigorous tasks, eventually exploring a distant location in a dangerous, even life-threatening mission. If you are successful, you will then spend 6 months "commuting" home in the same confined quarters and challenging conditions. During this assignment, headquarters cannot provide you with quick advice or coaching, because there is up to 20-minute communication delay (one-way), but you still need to coordinate as a team with people back at headquarters. From a personal perspective, during these 2 to 3 years, you cannot see Earth, feel gravity, or spend time with your family. And if you or any of your teammates are having a bad day, you cannot simply go out for a walk or call in sick. (pp. 200–201).

Mars missions will involve considerably more crew autonomy than is the case in current spaceflight operations. The move to more crew autonomy will require that some of the expertise currently held by flight controllers will have to be shared by crewmembers. This shift will result in more diverse crews as mission-relevant expertise will be distributed among their members. As crews act more self-reliantly, the parameters of their teamwork also will change. This issue became clear in semi-structured interviews Mesmer-Magnus and colleagues (Mesmer-Magnus, Carter, Asencio, & DeChurch, 2016) conducted with astronauts and members of their support teams (psychologists, trainers, operations managers, flight directors, and engineers). The discussions suggested that teamwork during LDSE missions will be considerably more fluid than in current missions. For example, an overlooked facet of crew performance during current missions is that crewmembers alternate between working independently and collaborating with others on tasks. The need to switch between individual and collective work will likely be more pronounced during LDSE as crewmembers will have more distributed expertise and unique responsibilities. Likewise, crewmembers will switch between (sub)teams, collaborating with different crewmembers as necessitated by different tasks. Fluid work structures and team memberships create complex, multilevel, and dynamic influences on team cohesion that measurements will need to address.

These insights are reflected in current efforts to advance new methodologies for assessing team cohesion and coordination during LDSE missions. The goal is to develop measures that capture team cohesion as a "dynamic, multi-faceted, and multilevel phenomenon" (Salas et al., 2015, p. 202) and that can be automated to provide crewmembers with timely feedback and guidance. The latter requirement is particularly important for LDSE to enable crews to be self-regulating. For example, a team of organizational psychologists and engineers is developing a high-precision wearable sensor system that can capture multimodal data indicative of teamwork interaction and process dynamics, such as intensity of physical movement, distance and changes in distance between team members, heart rate, and vocal characteristics (duration, interval, and intensity of vocalization; Kozlowski, Chang, & Biswas, 2013). Crewmembers wear a badge with a sensor array and a receiver/server that continuously records data and distributes them via a web interface for viewing in real time. Currently the badges are being field-tested in space analog environments. Ultimately the goal is to identify benchmark patterns characteristic of a well-functioning team to which the incoming data stream can be compared and, if an anomaly is detected, countermeasures to restore cohesiveness can be triggered (Kozlowski, Chao, Chang, & Fernandez, 2016). Another research approach (Miller, Wu, Schmer-Galunder, Rye, & Ott, 2011; Wu, Rye, Miller, Schmer-Galunder, & Ott, 2013) focuses on the automated analysis of crewmembers' written communications (log entries) using linguistic indicators of team effectiveness (Fischer, McDonnell, & Orasanu, 2007) to characterize team dynamics (e.g., who talks to whom), group identification (e.g., ingroup vs outgroup), emotional valence (positive/negative), and politeness. By tracking the social dynamic of a crew over time, sudden and unanticipated shifts can be flagged and used to initiate appropriate interventions.

Effective teamwork in space (among crewmembers) and on the ground (among mission controllers) are but two ingredients of successful space missions. As the safe return of Apollo 13 so vividly demonstrates, much of the success of human spaceflight hinges on communication and cooperation between space crews and mission control.

Collaboration between Mission Control and Space Crew

As with any collaborative work, successful collaboration between mission control and a space crew depends not only on members' technical expertise but also on skilled teamwork. Team processes and competencies that have been associated with multiteam effectiveness are communication, shared team and task models, and leadership (Kanas & Manzey, 2008; Mathieu et al., 2001).

Communication between mission control and crew is usually passed through one position in MCC, the Capsule Communicator, with the Flight Director and other flight controllers following the conversation over the space-to-ground voice channel. This centralized approach to mission control–crew communication has several benefits. The CAPCOM as information hub ensures that communication proceeds in an orderly fashion and thus facilitates maintenance of common ground within mission control and between mission control and crew. A second advantage of this communication process is that it affords multiple layers of checks. Flight controllers provide input not to the CAPCOM but to the Flight Director who needs to give approval before a message is transmitted to the crew, and other flight controllers partake in the decision making over a voice loop. A further advantage stems from the fact that CAPCOMs typically are members of the astronaut corps. Since they share knowledge and expertise with the crew and understand their perspective, they are uniquely suited to foster common task and team models between crew and mission control; i.e., a shared understanding of how and when teams need to coordinate with each other, and each team's capabilities, resources, expertise, and task responsibilities and constraints. They are *boundary spanners* who build connections and trust between teams (Landon, 2017). As one astronaut noted in a diary entry

> Capcoms can really make your day. But there's a big difference between being upbeat and being seriously upbeat... if a capcom is always saying "great job" or "awesome work" it only goes so far. The best capcoms are the ones who are to the point, give you the info you need succinctly, and sound like they know what you're doing... a sense of trust builds between the capcom and the crew and positive remarks from the capcom have more meaning.
>
> (Stuster, 2016, p. 28)

Moreover, in their capacity as astronauts, CAPCOMs represent the crew in mission control and may speak for them when events prevent mission control from consulting with the crew. One such instance is provided by Kranz (2000) in his account of the deliberations preceding the decision on the return path for Apollo 13. "In the scramble to secure the command module, we didn't have a chance to brief the crew or even get their opinion on the return path" (p. 318). Nonetheless, he felt that with the CAPCOM present they had a representation, that "Lousma, as their representative, would speak out if needed" (p. 318).

Several conventions governing ground–crew communication aim to support shared mental models and facilitate team coordination (Uhlig, Mannel, Fortunato, & Illmer, 2015). As mentioned earlier, all operational communications between CAPCOM and the crew are conducted on a specific voice channel and everybody in mission control is required to attend to the conversation. Daily planning conferences between crew and mission control take place twice a day—at the beginning and at the end of the crew's work day—to review or preview their schedule, and discuss mission-relevant events and issues on-board or on the ground. These conference calls ensure that team members have a shared understanding of upcoming tasks, task progress, and the task environment. Detailed procedures that have been worked out in advance enable crewmembers to perform many assigned tasks without having to communicate with mission control. However, some crew-to-ground communications are mandated. These include progress updates—mainly to inform mission control that a task has been completed—as well as the transfer of information that is otherwise not accessible to mission control, such as readings on devices not integrated with the ISS's data management system. Some of these communications facilitate coordinated actions between the teams. For example, the crewmembers need to let mission control know that they have reached a particular step in a procedure, and before they can continue, they need to wait for mission control to execute some action (e.g., change a system configuration) and to OK the crew's next step. Operational communications between crew and mission control are highly formalized, consisting of a call sign to identify the addressee and caller and the specification of the voice channel, followed by a succinct statement; more complex information is presented in meaningful *chunks*. Messages are promptly acknowledged, with either a generic phrase, such as "copy," or a shortened repeat of the message content. These conventions aid mutual understanding by ensuring that an addressee attends to the message (call sign), that messages do not overly tax his/her working memory (brief messages or chunking of content), that a message has been received and understood ("copy"), and that a misunderstanding is caught right away (repeat of message content).

While communication conventions support team members' evolving task and situation understanding, timeline and task procedures encapsulate the more static aspects of team members' shared knowledge. In current space operations, the multi-team system (MTS) of space crew and mission control "is driven from the ground" (Vessey, 2014, p. 141) with mission control in a leadership and monitoring role while the crew takes action as directed by or in consultation with ground. Mission planning is accomplished by mission control typically a year prior to launch and results in a detailed schedule for crew activities such as science experiments or EVAs that crewmembers are expected to adhere to during the actual mission (Kortenkamp, 2003).

Likewise, expertise on ISS systems resides with flight controllers, as does responsibility for any data management. The crew therefore needs to rely on mission control for processing system data and for catching and troubleshooting irregularities, as well as for making any changes to system settings. The Flight Director is in charge of operations in MCC, and is also the de facto leader of the MTS comprised of mission control and crew. When missions go according to plan, crew–MCC collaboration is governed by the mission timeline and standard operational procedures. MTS leadership by the Flight Director is called for when events occur that require adaptive collective responses. In these situations, the Flight Director is the person who specifies what needs to be done, by which team or teams, and when actions must be accomplished and completed.

Gene Kranz's leadership during the Apollo 13 crisis provides an excellent example of MTS leadership. He handed over the Control Room to his flight director colleague, Glynn Lunney, and assembled his team of controllers, program officers, and engineers in a separate room to *work the problem*. Together they defined the problem space. In Kranz's words: "thinking out loud so that everyone understood the options, alternatives, risks, and uncertainties of every path" (p. 321). He then assigned the most experienced controllers to lead the problem-solving efforts in their areas of expertise.

> My three leads will be Aldrich, Peters, and Aaron. Make sure everyone, and I mean everyone, knows the mandate I'm giving them. Aldrich will be the master of the integrated checklist for the reentry phase. He will build the checklist for the CSM [command/service module] from the time we start power-up until the crew is on the water. John Aaron will develop the checklist strategy and has the spacecraft resources. He will build and control the budgets for the electrical, water, life support, and any other resources to get us home. Whatever he says goes. He has absolute veto authority over any use of our consumables. Bill Peters will focus on the lunar module lifeboat. There are probably a lot of things we have not considered and he will lead the effort on how to turn a two-man, two-day spacecraft into one that will last for four days with three men. Whatever any of these three ask of you, you will do. (Kranz, 2000, p. 320)

While the teams of controllers and engineers around Aldrich, Peters, and Aaron worked on their assigned tasks, Kranz served as their liaison to Lunney and the other controllers in MCC who, in turn, were the link to the Apollo 13 crew. In addition to managing the task, Kranz was also instrumental in motivating team members, in making them believe "that this crew is coming home" (Kranz, 2000, p. 321).

NASA carefully prepares astronauts and flight controllers for their joint work during missions. Spaceflight resource management courses train important team competencies that are reinforced during simulated missions (Rogers, 2010). The collaboration between mission control and the space crew is further aided by communication conventions and task procedures as well as a centralized leadership structure. However, spaceflight presents formidable challenges to MCC–space crew collaboration that may lead to inadequate communication and cooperation between the two teams.

Challenges to MCC–Space Crew Collaboration

Any remote collaboration is challenging. Because team members are spatially separated, they live in different environments and their interactions lack the immediacy present in co-located teams. Spaceflight quite literally magnifies the distance between team members and the different worlds they inhabit. This impression is captured in Figure 36.2 which depicts the International Space Station orbiting above Earth. It is thus not surprising that potential friction between space crew and mission control has been identified as a threat to the success of space operations. As Ball and Evans (2001) report, "[c]osmonauts and astronauts alike agreed that the most challenging interpersonal problems were not among the crewmembers but, rather, were between the crewmembers in space and the mission controllers on the ground" (p. 140). Many of these problems seem to reflect differences in the task models of crew and mission control. For example, Stuster (2010; 2016) analyzed personal journals maintained by 20 astronauts on ISS missions and noted that the main sources of work-related stress and frustration that crewmembers reported were unrealistic time estimates made by mission planners, or procedures that did not sufficiently account for their perspective, as the following quotes illustrate:

> Today was a hard day. Small things are getting to me. I am tired. I think that the ground is scheduling less time for tasks than before. So, there is very little, if any fat left in the schedule for me to use to catch up on little things during the day. (Stuster, 2010, p. 10)

FIGURE 36.2 The International Space Station (ISS) as seen from the space shuttle *Atlantis*.

Image courtesy of NASA.

> The procedures are not very clear nor written in an order that makes sense for the work we are doing. They also have outdated pictures for equipment we are no longer using. In addition, they do not clearly state which steps in the procedures we will be running.
>
> (Stuster, 2016; p. 61)

The spatial distance between space crew and mission control may also lead to psychological closing and information filtration by crewmembers toward ground support. This phenomenon was noted by Gushin and colleagues (Gushin et al., 1997) in their analysis of crew-to-ground communications in two isolation studies lasting 90 and 135 days. They found that, over time, crewmembers talked less often with flight controllers and addressed fewer work-related themes. Crewmembers' interactions with mission control may also suffer from displacement of within-crew conflict to flight controllers. Displacement is operationalized in terms of an increase in reported negative mood accompanied by a drop in perceived outside support, and was observed during Shuttle/Mir missions as well as ISS missions (Kanas et al., 2007; Kanas et al., 2006). By portraying mission control as unsupportive or hostile, crewmembers were apparently able to deflect internal problems, at least temporally (Kanas & Manzey, 2008).

As missions travel further from Earth, delays in communication with mission control will be unavoidable. During long-duration missions and missions beyond low Earth orbit, space–ground communications will involve delays up to 20 minutes one way, a reality that poses a significant challenge to remote team communication and coordination and ultimately to mission safety and success. In an experiment conducted on the ISS, Kintz and colleagues (Kintz, Chou, Vessey, Leveton & Palinkas, 2016) introduced a communication delay of 50s while crew and flight controllers collaborated on off-nominal tasks. Team members' perceptions of their communication quality were significantly lower for asynchronous than for synchronous flight segments. The question of how best to support crew–mission control communication during asynchronous conditions was taken up by Fischer and colleagues (Fischer, Mosier, & Orasanu, 2013; Fischer & Mosier, 2014, 2015), using an integrated approach of laboratory studies and research in space analogs that included as participants current astronauts and flight controllers, astronaut-like professionals (i.e., individuals who in terms of education, personality characteristics, and age were comparable to members of the astronaut corps), and non-expert adults. Research identified communication problems associated with transmission delays (50s and 300s), underlying cognitive mechanisms, and strategies that facilitated remote collaboration under asynchronous conditions and supported team effectiveness. Analyses of the communication between remote team members revealed that transmission delays—irrespective of length—degraded the communication process and made it more difficult for team members to establish common ground. Problems included step-ons (i.e., a voice message that a remote team member had sent 50s or 300s prior was received while transmitting a radio message), disrupted message sequence (i.e., related messages by different team members, such as an answer to a question, were not adjacent but separated as other communications intervened), and outdated information (i.e., critical information was received too late). Team members exacerbated these problems by failing to adapt their communication behavior to the delayed conditions and by applying instead expectations and conventions of everyday (i.e., synchronous) discourse, causing unnecessary communications or misunderstandings. For example, team members often expected an immediate response and misinterpreted the time lag as a communication problem, or mistook a remote member's communication received immediately after their own transmission as a response to it. Well-performing teams adopted adaptive strategies, such as announcing the specific time at which to expect a transmission, or specifying the topic of a message. However, they did not consistently adhere to these strategies, especially when workload was high. These strategies were the starting point for the development of a communication protocol—a structured template to facilitate remote collaboration under time-delayed conditions. It consists of four segments (call sign, topic, message, closing) and specifies their content and organization to address the major challenges of asynchronous communication—time, conversational thread, and transmission efficiency. The protocol was implemented in several space analog studies to assess its feasibility. Astronauts and astronaut-like professionals showed high acceptance of the communicational protocol and rated it as highly effective in supporting their interactions with ground support (Fischer & Mosier, 2016).

In future exploration missions, space crews will need to manage tasks more autonomously than in current operations, although they will continue to be part of the MTS composed of members in space and in mission control. Introducing crew autonomy into the design of future space operations will impact the interdependencies that currently exist between crew and mission control teams; most notably it will involve a change in how responsibilities are going to be distributed between crew and mission control and this shift, in turn, will be associated with different information, action, and coordination requirements by the teams.

For instance in an experiment by Frank and colleagues (Frank et al., 2013), crew autonomy enabled by automated monitoring systems put the crewmembers into the role of "*doers*, responsible for performing most of the procedures associated with their assigned activities, and completing troubleshooting procedures in response to system failures and medical emergencies" while the role of flight controllers in mission control "was more supportive, advising, and guiding crewmembers as they went about their activities" (p. 3). In comparison to present-day conditions, crewmembers reported a reduction in workload during autonomous mission operations. Flight controllers, in contrast, experienced both "workload benefits and penalties" (p. 16), apparently because although workload was reduced when they were less directly involved in tasks, they spent more time monitoring systems in an attempt to understand the crew's actions. Complementary findings are reported by Kanas and colleagues (Kanas et al., 2010), who implemented crew autonomy in several space-analog studies. Mood ratings revealed that while space crews generally enjoyed working autonomously, flight controllers experienced confusion about their role when a space crew had task autonomy. These findings suggest that crew autonomy could potentially lead to a misalignment of the task and team models held by space crewmembers and flight controllers. Moreover, the long communication delays associated with space exploration may further exacerbate this problem.

Future Research

We have come a long way since the Apollo 13 incident. Astronauts live on the International Space Station for months at a time, and the teamwork between them and MCC continues to exemplify a highly successful collaboration. However, space operations will undergo radical changes as NASA implements long-duration exploration missions to Mars and other interplanetary destinations. In these missions, space crews will have to function more autonomously than in current ISS operations because communication delays due to an increased distance to Earth will eliminate the possibility of real-time assistance from MCC. Nonetheless, given the complexity of future spacecraft systems MCC–space crew collaboration will remain important to the safety and success of missions. This prediction raises a number of important research questions. Foremost we need to address what organizational changes to the current multiteam system of space crew and mission control are required to support long-duration space exploration. For example, should MCC abandon its traditional control function and act in a support capacity in long-duration space exploration? This shift, in turn, would require changes to the current (MCC-driven) leadership structure (Landon, et al., 2016). One possibility is that future space operations will involve shared leadership between MCC and the space crew; a structure in which leadership functions are distributed among the teams (Burke, Fiore, & Salas, 2003). Research is needed to determine whether and how shared leadership should be implemented; in particular, when it should be adopted and how best to allocate responsibility (e.g., tied to specific tasks, or to tactical versus strategic issues). A related question is how crew autonomy should be introduced (Vessey, 2014). Is it effective in terms of both mission safety and MTS performance to give space crews high autonomy throughout LDSE missions, or should crew autonomy be granted gradually in the course of a mission as a function of communication delay? Additionally, we need to specify the information- and knowledge-sharing requirements for effective MCC–space crew collaboration during autonomous operations, and develop team training and communication procedures that ensure common task and team models by space crew and MCC during crew autonomy.

To support this kind of research, appropriate analogs that simulate conditions of long-duration exploration missions are needed as well as participants that belong to or are representative of the target populations, that is, astronauts and flight controllers. Some natural extreme environments, such as Antarctic outposts, have been tapped as research venues similar to space operations—they are remote; inhabitants typically stay for a long period, are confined to the premises, and must wear special protective clothing to go outside of the facility; and communication with support personnel is limited. Other analog environments, such as NASA's HERA (Human Experimental Research Analog) at Johnson Space Center, and NEEMO (NASA Extreme Environment Mission Operations) in the Florida Keys, as well as HI-SEAS (Hawaii Space Exploration Analog and Simulation), an experimental analog on the slopes of the Mauna Loa volcano in Hawaii, offer more control than natural environments with respect to experimental manipulations, schedules, and tasks, but these research venues are limited in terms of the participant pool and/or duration of missions. Interestingly, the ISS has been considered as a research analog for LDSE missions; however, the real-time work occurring in the ISS limits its availability and flexibility for experimental research efforts.

A formidable difficulty for space research arises from the fact that there are logistically few opportunities per year to conduct a long-duration simulation. As a result, multiple studies with different and potentially conflicting research foci must be integrated and coordinated in the same simulation study. NASA has adopted this practice for its space analogs, so that researchers are required to adapt their work to a multi-study environment and many researchers collect data from the same simulation. Of course, multi-study simulations introduce their own problems associated with *test burnout* due to an excessive number of surveys and other measures. Researchers must recognize and respect limits to the number of measurements that crews can be expected to provide due to time limitations and workload.

Lastly, the required research inevitably involves a small sample size, limiting the possibilities for statistical data analysis. This highlights the need to collect data across simulations and to share data among researchers to boost statistical power. Again, NASA has begun to address this need by grouping simulations into *campaigns* that include several similar *missions*, and by including data-sharing agreements as part of the requirements to conduct research in the analogs. It has become clear that in order to develop a valid and viable perspective of teamwork in space operations, researchers must be as collaborative as the MTS teams they seek to understand.

Acknowledgments

This work was supported by National Aeronautics and Space Administration (NASA) Grants NNX12AR19G and NNX16AM16G to the Georgia Institute of Technology with sub-awards to Dr. Mosier. The views expressed in this work are those of the authors and do not necessarily reflect the opinion of NASA or Georgia Tech.

References

Ball, J. R., & Evans, C. H. (2001). *Safe passage: Astronaut care for exploration missions.* Washington, DC: National Academy Press.

Barrett, J. D., Holland, A. W., & Vessey, W. B. (2015, April 23). Identifying the "right stuff": An exploration-focused astronaut job analysis. Paper presented at the 30th Annual

Conference of the Society for Industrial and Organizational Psychology, Philadelphia, PA. Available at: https://ntrs.nasa.gov/search.jsp?R=20140011498

Burke, C. S., Fiore, S., & Salas, E. (2003). The role of shared cognition in enabling shared leadership and team adaptability. In J. Conger & C. Pearce (Eds), *Shared leadership: Reframing the hows and whys of leadership* (pp. 103–122). Thousand Oaks, CA: Sage Publishers.

Caldwell, B. S. (2001). *Information exchange and knowledge sharing flows in Mission Control Center operations.* Summer Faculty Fellowship Program Technical Report. Houston: NASA Johnson Space Center.

Caldwell, B. S. (2005). Multi-team dynamics and distributed expertise in mission operations. *Aviation, Space, and Environmental Medicine* 76(6), II, B145–B153.

DeChurch, L. A., & Mesmer-Magnus, J. R. (2010). The cognitive underpinnings of effective teamwork: A meta-analysis. *Journal of Applied Psychology* 95, 32–53.

Durso, F. T., Crutchfield, J. M., & Harvey, C. M. (2007). The cooperative shift change: An illustration using air traffic control. *Theoretical Issues in Ergonomics Science* 8(3), 213–232.

Erickson, L. K. (2010). *Space flight: History, technology, and operations.* Retrieved from http://ebookcentral.proquest.com.

Fiore, S. M., Smith-Jentsch, K. A., Salas, E., Warner, N., & Letsky, M. (2010). Towards an understanding of macrocognition in teams: Developing and defining complex collaborative processes and products. *Theoretical Issues in Ergonomic Science* 11, 250–271.

Fiore, S. M., Wiltshire, T. J., Oglesby, J. M., O'Keefe, W. S., & Salas, E. (2014). Complex collaborative problem-solving processes in mission control. *Aviation, Space, and Environmental Medicine* 85(4), 456–461.

Fischer, U., McDonnell, L., & Orasanu, J. (2007). Linguistic correlates of team performance: Toward a tool for monitoring team functioning during space missions. *Aviation, Space and Environmental Medicine* 78(5), II, B86–95.

Fischer, U., & Mosier, K. (2014). The impact of communication delay and medium on team performance and communication in distributed teams. *Proceedings of the Human Factors and Ergonomics Society Annual Meeting* 58, 115–119.

Fischer, U. & Mosier, K. (2015). Communication protocols to support collaboration in distributed teams under asynchronous conditions. *Proceedings of the Human Factors and Ergonomics Society Annual Meeting* 59, 1–5.

Fischer, U. & Mosier, K. (2016). *Protocols for asynchronous communication in space operations: Communication analyses and experimental studies.* Atlanta, GA: Georgia Institute of Technology (Final Report NASA Grant NNX12AR19G).

Fischer, U., Mosier, K., & Orasanu, J. (2013). The impact of transmission delays on mission control—space crew communication. *Proceedings of the Human Factors and Ergonomics Society Annual Meeting* 57, 1372–1376.

Frank, J., Spirkovska, L., McCann, R., Wang, L., Pohlkamp, K., & Morin, L. (2013). Autonomous mission operations. *IEEE Aerospace Conference Proceedings*, March 2–9. Big Sky, Montana.

Garrett, S. K., Caldwell, B. S., Harris, E., & Gonzalez, M. C. (2009). Six dimensions of expertise: A more comprehensive definition of cognitive expertise for team coordination. *Theoretical Issues in Ergonomics Science* 10, 93–105.

Gushin, V. I., Zaprisa, N. S., Kolinitchenko, T. B., Efimov, V. A., Smirnova, T. M., Vinokhodova, A. G., & Kanas, N. (1997). Content analysis of the crew communication with external communicants under prolonged isolation. *Aviation, Space and Environmental Medicine* 68, 1093–1098.

Kanas, N. (2005). Interpersonal issues in space: Shuttle/Mir and beyond. *Aviation, Space, and Environmental Medicine 76*(6, II), B127–134.

Kanas, N., & Manzey, D. (2008). *Space psychology and psychiatry*, vol. 22. Berlin: Springer Science & Business Media.

Kanas, N., Salnitskiy, V. P., Boyd, J. E., Gushin, V. I., Weiss, D. S., Saylor, S.,... Marmar, C. (2007). Crewmember and mission control personnel interactions during International Space Station missions. *Aviation, Space and Environmental Medicine 78*(6), 601–607.

Kanas, N. A., Salnitskiy, V. P., Ritsher, J. B., Gushin, V. I., Weiss, D. S., Saylor, S. A.,... Marmar, C. R. (2006). Human interactions in space: ISS vs. Shuttle/Mir. *Acta Astronautica 59*, 413–419.

Kanas, N., Saylor, S., Harris, M., Neylan, T., Boyd, J., Weiss, D. S.,... Marmar, C. (2010). High versus low crewmember autonomy in space simulation environments. *Acta Astronautica 67*, 731–738.

Kanas, N., Weiss, D. S., & Marmar, C. R. (1996). Crewmember interactions during a MIR space station simulation. *Aviation, Space, and Environmental Medicine 67*(10), 969–975.

Kintz, N., Chou, C.-P., Vessey, W. B., Leveton, L. B., & Palinkas, L. A. (2016). The impact of communication delays to and from the International Space Station on self-reported individual and team behavior and performance: A mixed-methods approach. *Acta Astronautica 129*, 193–200.

Kortenkamp, D. (2003). A day in an astronaut's life: Reflections on advanced planning and scheduling technology. *IEEE Intelligent Systems 3/4*, 8–11.

Kozlowski, S. W. J., Chang, C.-H., & Biswas, S. (2013). *Measuring, monitoring, and regulating teamwork during long duration missions*. Washington, DC: National Aeronautics and Space Administration.

Kozlowski, S. W. J., Chao, G. T., Chang, C.-H., & Fernandez, R. (2016). Team dynamics: Using "big" data" to advance the science of team effectiveness. In S. Tonidandel, E. King, & J. Cortina (Eds), *Big data at work: The data science revolution and organizational psychology* (pp. 272–309). New York: Routledge.

Kranz, G. (2000). *Failure is not an option: Mission Control from Mercury to Apollo 13 and beyond*. New York: Simon & Schuster.

Landon, L. (2017). "Houston, this is Station...": Mission-critical communication for teams. *Re: Work*. Available at: https://rework.withgoogle.com/blog/nasa-communication-teams/

Landon, L. B., Vessey, W. B., & Barrett, J. D. (2016). *Evidence report: Risk of performance and behavioral decrements due to inadequate cooperation, coordination, communication, and psychosocial adaptation within a team*. Houston, TX: Human Research Program Behavioral Health and Performance, National Aeronautics and Space Administration Lyndon B. Johnson Space Center.

Lovell, J. A. (1975). "Houston, we've had a problem"—A crippled bird limps safely home. In A. M. Cortright (Ed.), *Apollo Expeditions to the Moon*. Washington, DC: Scientific and Technical Information Office, National Aeronautics and Space Administration. Available from: https://history.nasa.gov/SP-350/ch-13-1.html.

Mathieu, J. E., Marks, M. A., & Zaccaro, S. J. (2001). Multiteam systems. In N. Anderson, D. Ones, H. K. Sinangil, & C. Viswesvaran (Eds), *International handbook of work and organizational psychology*, vol. 2 (pp. 289–313). London: Sage.

Mesmer-Magnus, J. R., Carter, D. R., Asencio, R., & DeChurch, L. A. (2016). Space exploration illuminates the next frontier for teams research. *Group & Organization Management 4*(5), 595–628.

Miller, C., Wu, P., Schmer-Galunder, S., Rye, J., & Ott, T. (2011). AD ASTRA: Automated detection of attitudes and states through transaction recordings analysis. Washington, DC: National Aeronautics and Space Administration.

Miller, M. J. (2017). *Decision support system development for human extravehicular activity*. Unpublished dissertation. Georgia Institute of Technology, Atlanta, GA.

Onken, J. D., & Caldwell, B. S. (2011). Problem solving in expert teams: Functional models and task processes. *Proceedings of the Human Factors and Ergonomics Society Annual Meeting 55*(1), 1150–1154.

Palinkas, L. (2001). Psychosocial issues in long-term space flight. *Gravitational and Space Bulletin 14*(2), 25–33.

Patterson, E. S., Watts-Perotti, J., & Woods, D. D. (1999). Voice loops as coordination aids in space shuttle mission control. *Computer Supported Cooperative Work (CSCW) 8*(4), 353–371.

Patterson, E. S., & Woods, D. D. (2001). Shift changes, updates, and the on-call architecture in space shuttle mission control. *Computer Supported Cooperative Work (CSCW) 10*(3), 317–346.

Rogers, D. G. (2010). Crew resource management: Spaceflight resource management. In B. Kanki, R. Helmreich, & J. Anca (Eds), *Crew resource management* (pp. 301–315). San Diego: Academic Press.

Salas, E., Tannenbaum, S. I., Kozlowski, S. W. J., Miller, C. A., Mathieu, J. E., & Vessey, W. B. (2015). Teams in space exploration: A new frontier for the science of team effectiveness. *Current Directions in Psychological Science 24*(3), 200–207.

Stuster, J. (2010). *Behavioral issues associated with long-duration space expeditions: Review and analysis of astronaut journals experiment 01-E104 (Journals): Final report*. Houston, TX: NASA/TM-2010-216130.

Stuster, J. (2016). *Behavioral issues associated with long-duration space expeditions: Review and analysis of astronaut journals experiment 01-E104 (Journals): Phase 2 Final report*. Houston, TX: NASA/TM-2016-218603.

Test Division Apollo Spacecraft Program Office (1970). *Apollo 13: Technical air-to-ground voice transcription*. Houston, TX: National Aeronautics and Space Administration Manned Spacecraft Center.

Uhlig, T., Mannel, T., Fortunato, A., & Illmer, N. (2015). Space-to-ground communication for Columbus: A quantitative analysis. *Scientific World Journal 2015* 1–7. Available from: http://dx.doi.org/10.1155/2015/308031

Vessey, W. B. (2014). Multiteam systems in the spaceflight context: Current and future challenges. In M. L. Shuffler, R. Rico, & E. Salas (Eds), *Pushing the boundaries: Multiteam systems in research and practice* (pp. 135–156). Bingley, UK: Emerald.

Wang, E., & Caldwell, B. (2003). Human information flow and communication pattern in NASA Mission Control system. *Proceedings of the Human Factors and Ergonomics Society Annual Meeting 47*, 11–15.

Wood, J. Schmidt, L., Lugg, D., Ayton, J., Phillips, T., & Shepanek, M. (2005). Life, survival, and behavioral health in small closed communities: 10 years of studying isolated Antarctic groups. *Aviation, Space, and Environmental Medicine 76*(6, II), B89–93.

Wu, P., Rye, J., Miller, C., Schmer-Galunder, S., & Ott, T. (2013). Non-intrusive detection of psycho-social dimensions using sociolinguistics. In *Proceedings of the 2013 IEEE/ACM International Conference on Advances in Social Networks Analysis and Mining* (pp. 1337–1344). New York: ACM.

CHAPTER 37

DEVELOPING OPERATOR EXPERTISE ON NUCLEAR POWER PRODUCTION FACILITIES AND OIL & GAS INSTALLATIONS

†MARGARET CRICHTON, SCOTT MOFFAT, AND LAUREN CRICHTON

Introduction

NUCLEAR power plants and offshore oil & gas installations share many common aspects; they both require continuous operations (twenty-four hours a day, seven days a week), operators work on a shift cycle, and they involve high risk and complexity. Emergencies on such facilities are extremely rare, yet, should an emergency occur, operators are expected to be able to respond effectively and accurately. This requires plans, procedures, and suitably trained and experienced personnel to be able to make decisions, communicate, work in ad hoc teams established under an emergency command and control structure, and manage periods of high stress. The aim of this chapter is to describe research undertaken to date examining how operators develop expertise in routine operations, and then make the shift to responding to non-routine atypical or emergency events. Topics such as identifying the knowledge, skills, and abilities required by operators in routine events, through to suggesting which training techniques or experiential learning methods can assist operators to make this shift will be discussed.

The chapter starts by briefly explaining the processes required by both nuclear power production facilities (NPPF) and offshore oil & gas (OO&G) installations. Next, the knowledge, skills, and abilities required by operators in routine operations are described along with methods used to investigate operator cognition. The third section discusses preparedness for managing non-routine or emergency events in terms of procedures, plans, and guidance, as well as the social and cognitive skills in these two sectors. The applicability of simulation techniques for training operators to manage emergencies is referred to in the following section. The final section outlines the challenges to be overcome in examining expertise to support operation safety and reliability.

Continuous Process Operations on Nuclear Power Plant and Offshore Oil & Gas Installations

Both NPPF and OO&G installations are complex, highly interconnected systems (see definition by Dekker, 2011) that require continuous processing, and involve operators from various disciplines working in shifts who collaborate and cooperate to ensure safe and reliable operations. The processes and tasks are heavily proceduralized, although it is recognized that not all decisions can be covered by procedures, as there will always be situations that do not match prescribed procedures, or that rapid decisions are required. Such circumstances require that individuals draw on their expertise to assess risks and carry out the necessary actions (Hopkins, 2011).

Incidents in both settings are rare but the consequences can be severe. Safety is paramount due to the potential consequences of incidents and operators must be highly trained and competent for routine and non-routine operations. This section describes each of these industries and interconnected processes required for safe operations.

Nuclear Power Production Facility Operations

The process of generating nuclear power occurs by releasing energy from atoms (NIA, 2017). In nuclear power production, Uranium (e.g., U-235) is used as the fuel. In the fission process, neutrons split the atoms of uranium with the resultant release of energy (i.e., heat). This energy is then transferred to a coolant such as carbon dioxide which heats water and generates steam which is used to make electricity through turning turbines. The electricity is then transported through the national grid network and made available to homes and businesses.

The size and complexity of a NPPF necessitates a highly structured organization (Apostolakis, 1999), with standardized work processes designed to achieve the goal of safe power production. This requires coordination of activities by maintenance and engineering technicians, control room operators, health physicists, quality control personnel, and reliability engineers (Gaddy & Wachtel, 1992). The control room is the hub of activity of the NPPF, and control room personnel are licensed to manage the duties of monitoring and controlling the plant. In the UK, operators on an NPPF abide to a strict competency framework, i.e., controlled by a set of license conditions, one of which is specific to training and competency of personnel involved in the operation of the reactors. This ensures that all personnel are suitably qualified experienced personnel and trained (SQEPT—as commonly used in the nuclear industry). In addition, all operational staff need to be able to manage an emergency or upsetting conditions requiring coordination of activities (ONR, 2016), especially cognitive tasks under stress (IMC, 1999). Crews do not always comprise the same people and it is common practice for the crews to learn to adapt to new individuals joining the crew (Gaddy & Wachtel, 1992).

Offshore Oil & Gas Installation Operations

Locating and retrieving hydrocarbons (oil & gas) from underground reservoirs is a complicated process. As described by Devold (2006), it starts with the facilities and processes required for exploration (locating possible sources of hydrocarbons) and results in upstream, which includes production and stabilization (i.e., bringing the hydrocarbons to the surface and then to onshore facilities). At each stage in this process, an interaction of technology, systems, and people is essential. Operators must be suitably trained and qualified to operate and maintain the technology, with support from the systems and procedures in place.

In the exploration phase of upstream, drilling a well is a highly uncertain, risky, and dynamic process (see Crichton, 2017). A drilling team is comprised of multiple personnel, including the operator, the contractor, and third party operators/specialists, as shown in Figure 37.1. The operator decides how the well will be drilled and creates the drilling plan, and the contractor carries out the plan, with support from specialist companies.

Team members may be geographically dispersed (Lauche, Crichton, & Bayerl, 2009) with the well operations personnel being located on the actual installation, and other team members based onshore possibly at various sites globally. Each team member has their own contribution to make to the team's overall performance in terms of specific role and responsibilities. Referring to multidiscipline teams, the US Chemical Safety and Hazard Investigation Board (2016) report stated that:

> While the well operations crew members often get credit for making decisions and taking direct action to conduct the drilling activities, a number of management and engineering personnel play a role in the decision-making/action-taking process through various means, such as providing leadership instruction, guidance, and technical analysis of the well. (p. 22)

Distributed drilling team

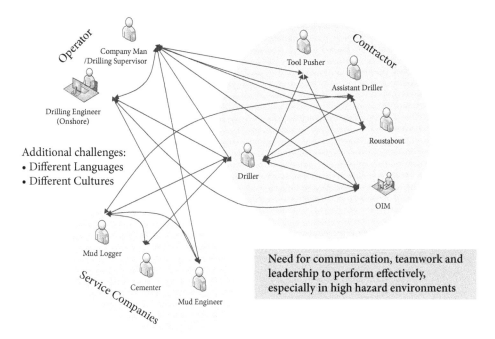

FIGURE 37.1 Distributed drilling team.

© People Factor Consultants Ltd (L. Crichton), 2015.

Although many decisions will have been made during the planning phase, others need to be made dynamically during the operations, particularly in response to ambiguous or inconclusive data from the well.

The required knowledge, rules, and skills (KRSs) of operators in both NPPF and OO&G installations have been highlighted following incidents such as Chernobyl (Reason, 1987), Three Mile Island (Three Mile Island Special Inquiry Group, 1980), Piper Alpha (Cullen, 1990), and Deepwater Horizon (Hopkins, 2012; US Chemical Safety and Hazard Investigation Board, 2016), especially the effects of decision making under stress. KRSs are defined as shown in Figure 37.2.

The question arises of how expertise is developed from routine operations and how it transfers to, and is adapted in, non-routine situations. In the next section, we first examine the required KRSs and how they relate to expertise in routine operations. In the subsequent section, we examine how these need to be adapted for non-routine situations.

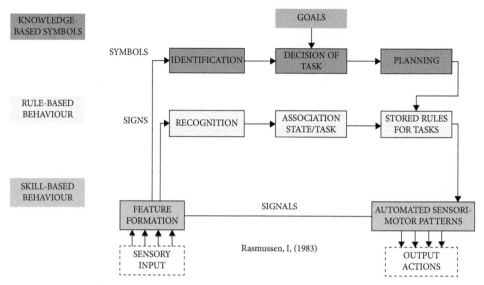

FIGURE 37.2 Definition of knowledge, skills, rules.

Data from Rasmussen, I., Skills, rules, knowledge, signals, signs, and symbol and other distinctions in human performance modelling, *IEEE Transactions on Systems, Man and Cybernetics* 13(3), pp. 257–267, doi: 10.1109/TSMC.1983.6313160, 1983.

Identifying Expertise in Routine Operating Conditions

Sonnentag (2000) provided two conceptualizations of expertise. One based on experience, the other based on performance. Many studies have shown the positive correlation between duration of experience and level of attained skill; i.e., the more experience a person has at a particular task, the better they perform. On the other hand, others have shown that experience alone may be insufficient to attain high levels of performance (e.g., see Ericsson, 2004; Ericsson, Whyte, & Ward, 2007). Sonnentag demonstrated that higher performing individuals may be selected or self-select to perform the associated tasks, such that individuals with comparable lengths of experience may differ in their levels of attained performance. Moroever, she indicated that some experts may be different in their ability to adapt (see also Ward, Gore, Hutton, Conway, & Hoffman, 2018); Chapter 12, "Adaptive Expertise," by Carbonell & van Merrienboer, this volume).

Based on the work of Hatano and Inagaki (1986), Sonnentag (2000) defined routine expertise as a *high level of fast, accurate, and automatic performance in the face of well-known tasks* (p. 257). Adaptive expertise, on the other hand, refers to flexibility and adaptability to new problems, based on a deep understanding of the domain and the task, thus allowing operators to create novel responses or procedures. More recently, Carbonell, Stalmeijer, Konings. Segers, & Merrienboer (2014) examined adaptive expertise in education, and concluded that individual and environmental factors are influential, such as cognitive and analogical problem solving and past experiences, but that knowledge representation is key, along with the opportunity to learn new tasks and manage novel situations. However, the effects of factors such as autonomy and task variability in the workplace, and personality, remain unclear. It is precisely these issues that make the study into the development of expertise so fascinating, and simultaneously frustrating. More frequently studies are comparing novices and experts in particular tasks or contexts (Masakowski & Hardinge 2000; Randel, Pugh, & Reed, 1996). On the other hand, applied research is seldom straightforward, and different methodologies are required. For example, one method for observing a descriptive behavioral framework has been implemented to assist observations and feedback (van Avermaete & Kruijsen, 1998).

Following up the routine/adaptive expertise question, this section focuses on the development of expertise under routine operating conditions in high hazard settings. In both nuclear and oil & gas settings, many activities at the operational level are highly proceduralized. Procedures are designed and implemented to reduce the complexity of the processes and minimize the potential for human error (Carvalho, 2006). Decisions made by operational personnel must balance safety and production, with regulations or rules and procedures defining the boundaries of decision making (Hayes, 2012). However, compliance to predefined procedures limits decisions made on professional judgement, and can reduce cognitive flexibility (Hayes, 2012; see also Carvalho, 2006).

On an OO&G installation, a safety management system (HSE, 2013) outlines the work schedule prepared in advance that lists the activities to be carried out, including risk assessments, permits to work, and procedures. No work should be undertaken unless clearly described and understood, and personnel must be assessed as competent and adequately supervised. Organizations typically implement a competence framework that describes the training and assessments that employees must complete to allow them to carry out specific tasks, with performance being assessed against specific identifiable standards defined in the competence framework. These may comprise technical qualifications, duration of experience, and previous roles held.

Defining someone as being an *expert* is straightforward if external criteria against which to compare exist (Shanteau, Weiss, Thomas & Pounds, 2002). While such criteria typically relate to readily discernable technical capabilities, important capabilities are frequently omitted from criteria or are unspecified. Identifiable standards may also require compliance with procedures, and someone who successfully applies and completes a standard operating procedure (SOP) is considered to be competent. Hence, one view of expertise is associated with such compliance.

More challenging to assess are the qualities associated with *human capital*, which refers to the technical qualifications, competencies, and experience of individuals, along with a range of personal and interpersonal qualities (Griffin et al., 2014). These additional qualities include detecting weak signals of potential malfunctions (i.e., situation awareness), and being able to forestall system disturbances before they become emergencies (i.e., decision making). Developing expertise in these skills depends upon identifying the strategies used by experts in the domain (Klein, 1997). Learning the skills then leads to the next step of evaluating competence in these skills.

Under the license arrangements in the UK, NPPF must encompass the philosophy of defense in depth (HSE, 1994), as part of which only suitably qualified and experienced persons are allowed to perform any duties which may affect the safety of operations on the site (ONR, 2016). A report by the International Atomic Energy Authority (IAEA) stated that overall nuclear power production safety and performance is improved through enhancing the performance of individual employees (IAEA, 2001). Although a combination of organizational factors, equipment, and human performance improvement mechanisms has been adopted, historically, less focus had been placed on human competency. To overcome such a shortfall, the IAEA proposed that competencies, such as communication, teamwork, leadership, and adaptability, are specifically trained, with the aim of increasing trust in organizations and ensuring that personnel understand and value their own contribution to the organization and that of others.

The focus of much research in NPPF has predominantly been operator tasks, especially by control room operators, who have two main objectives, namely, to keep the plant running efficiently and without interruptions in service, and to prevent any threats to public safety. As such, their responsibilities include monitoring of plant status for potential malfunctions and examining the cause should a malfunction occur (Meister, 1995).

Monitoring of activities under routine operating conditions requires a high level of cognitive effort (Mumaw, Roth, Vicente, & Burns, 2000). Following a series of field observation studies and interviews into how nuclear power plant control room operators carry out the crucial task of monitoring plant states, Mumaw et al. (2000) concluded that challenges arise in identifying subtle abnormal indications against a background of *noise*, for example, unnecessary or nuisance alarms. Control room operators therefore construct strategies for highlighting the salience of situational changes, which in turn supports their active problem solving. With respect to implications for training from these observational studies, Mumaw et al. (2000) found that *good operators rely extensively on knowledge-driven monitoring instead of rote procedural compliance* (p. 53), suggesting that training to enhance knowledge and expertise should play a larger role in training and licensing rather than the current focus on procedural compliance. Cognitive task analyses of control room operators' activities indicated that on 80 percent of occasions a recognitional strategy was used by control room supervisors for making decisions in micro-incidents, whereas a normative strategy was used for 20 percent of decisions (Carvalho, dos Santos, & Vidal, 2005). Decisions were made based on pattern recognition and condition–action rules, and relied heavily on information exchange to construct individual and shared situation awareness. The condition–action rules were derived from experience and training as opposed to purely implementing standard operating procedures.

Carvalho (2006) emphasized how control room operators rely on verbal exchanges to maintain individual and shared situation awareness, thus achieving system safety and efficiency. Operators are faced with the challenge of overcoming issues with human–computer interaction and displays of relevant data, along with background noise. Condition–action responses are again predominantly applied by experts when responding to minor or potential system malfunctions, and can manage situations that do not match the prescribed procedures.

In the OO&G setting, competence and expertise are similarly crucial in that decisions made at operational to strategic levels have implications for safety and efficiency. Economically, annual losses made in the oil industry can amount to US $30 billion due to poor decision outcomes that can be attributable to the attention paid to technical and economic factors while the important aspects of individual expertise have been overlooked (Malhotra, Lee, & Khurana, 2004). When drilling wells, uncertainty plays a major role, but the decision-making procedures followed have typically leaned towards a normative decision-making process, which is more appropriate for making decisions under certainty. Malhotra et al. (2004) argue that decisions made under uncertainty, rather than those made in accordance with the organizational hierarchy, should be made by the people with expertise. Optimal decisions are achieved when decision making types and processes are understood (Mackie, Begg, Smith, & Welsh, 2010), in that the correct process is tailored to the type of decision and that real-world decision making is acknowledged for decisions characterized by uncertainty. This suggestion was previously made by Begg, Bratvold, and Campbell (2003) and, hence, there needs to be more recognition in OO&G about the impact of human cognition, including working memory limitations, when dealing with complexity.

Even with extensive planning prior to operations beginning, it is important to describe how decisions can fail when faced with uncertainty, unexpected events, and surprises. This can be done through peer reviews, risk assessments, and statements of requirements, which culminate in the preparation of a detailed drilling programme. In uncertain and complex situations team members need to be sufficiently experienced to be able to adapt dynamically to shifting operational demands (Thorogood & Crichton, 2013). This requires the integration of a high degree of operational discipline with NTS. Operational discipline includes a standardized chain of command and decision-making procedures, clear decision-making authority, and demonstrated competence, which is combined with the implementation of NTS, along with a recognition of the effects of cognitive biases.

The International Association of Oil & Gas Producers (IOGP) published a report identifying individual non-technical skills relevant to seventeen safety critical roles in drill crew, including drillers, assistant drillers, tool pushers, and drilling supervisors (IOGP 501, 2014). Interviews were held with thirty-three representatives across these roles with the data being analyzed to identify the NTS required in routine and non-routine work-related situations. The resultant macrocognitive categories (situation awareness, decision making, communication, teamwork, leadership, and performance shaping factors—stress and fatigue) underpin training and debriefing to enhance behaviors in the workplace.

Due to the geographic dispersion of drilling team members, there has been an increase in OO&G in the introduction of collaborative environments where real-time data from the installation is monitored by experts who can liaise closely with the drilling crew on-site to make decisions (Israel et al., 2015). This requires not only communication hardware and software, such as data voice and video conferencing programs, but also necessitates effective interactions between the different experts (Munro, 2008). Lauche, Sawaryn, and Thorogood (2009) examined the work processes involved in the introduction of real-time onshore operation centers (OOC) by a UK-based drilling team by interviewing thirteen team members as well as conducting a longitudinal attitudes study with thirty-three onshore and offshore participants. The study concluded that social and cognitive skills are essential for collaboration in such computer-mediated settings.

To improve NTS behaviors, training courses are frequently associated with interventions such as crew resource management (Kanki, Helmreich & Anca, 2010), with behavioral marker frameworks being developed to assess the competence and capabilities of the individuals. In terms of assessing NTS, the NOTECHS (nontechnical skills) descriptive behavioral framework in aviation provides a tool against which to define and evaluate the implementation of nontechnical behaviors, knowledge, and skills by cockpit crew members (van Avermaete & Kruijsen, 1998). Similar systems have been introduced in other settings, especially in medicine (anaesthetists nontechnical skills (ANTS), Fletcher et al., 2003; nontechnical skills for surgeons (NOTSS), Yule et al., 2008). However, IOGP report 501 recommends that any formal assessment of NTS should be separate from the actual training programme. This is primarily because NTS are so new to the industry that personnel need time to practice the skills, and also because formal assessments require a defined system of training assessments with competent training assessors and assessment programs in place. On the other hand, debriefings and feedback can be provided by qualified and experienced observers following training exercises so that individuals and teams can receive objective feedback about strengths and weakness in their NTS performance. This reinforces the NTS classroom training by using the same phrases and definitions, and encourages operators to reflect on their responses and actions. Once the language of NTS becomes more familiar, and a formal competency framework is established, then formal assessments should take place.

This section has reviewed both NPPF and OO&G installations to describe the relevance of expertise, specifically addressing the application and adaptation of cognitive and social interactions by individuals and teams working in high hazard environments. The next section builds on these foundations by expanding these interactions in the context of non-routine events.

Identifying Expertise in Nonroutine or Emergency Conditions

The previous section discussed expertise in terms of demonstrating NTS under routine conditions; however, in high hazard environments such as OO&G installations and NPPF, it is essential that on-site personnel can manage a nonroutine event or emergency. In a review of leaders and teams for incident command, Flin (1996) proposed that although high hazard industries may have documented and regulated plans and preparations in place for managing emergencies, this may be insufficient to guarantee effective incident management. Cognitive and social skills such as managing stress, decision making, and teamwork are, however, essential.

In line with the focus on decision making in non-routine situations, a key component of high reliability organizations (HRO) is the successful containment of unexpected events (HSE, 2011) which includes *allowing people with expertise, irrespective of rank, to make important safety-related decisions in emergencies* (HSE, 2011; p. vi). According to Weick & Sutcliffe (2007), deference to expertise during nonroutine, or emergency situations, differs from decision making during routine events. In routine events, decision making is hierarchical and has a clear differentiation of responsibilities, whereas decisions required during emergencies should be made by individuals with expertise and not related purely to their hierarchical position. The concept of expertise in relation to nonroutine events appears to fulfil the definition of adaptive expertise where individuals adapt existing knowledge to new technology or processes (Saetren & Laumann, 2015). Operators in both OO&G installations and NPPF may need to make decisions in novel events or under dynamically changing conditions. As personnel in these settings prepare for the *possibility* of a nonroutine event rather than the *probability* that a nonroutine event might occur, training provides the most suitable method through which to gain experience and consequently expertise.

To ensure that any complex system can be operated as safely as possible in all situations, all operators need to be trained to enable them to deal effectively with any possible emergency or unfamiliar situation (Ainsworth & Marshall, 1994/5). Coping with an emergency or a serious incident on an NPPF requires fluent and coordinated team activity which includes decision making and problem solving, all under extreme stress. The nuclear industry has recognized that training is a key factor in ensuring adequate and reliable performance during infrequent accident scenarios, whereby operators' intervention may be crucial in mitigating the consequences of the event. As a result, the nuclear industry devotes a considerable amount of time and effort to the training of personnel to deal with a wide range of situations, many of which are incredible, infrequent, or unlikely (Ainsworth & Marshall, 1994/5).

In the UK, regulations set down by the HSE (1994) describe general arrangements for minimizing the effects on the general public of a nuclear accident, and specify that a site emergency plan must be established to deal with nuclear accidents where a release

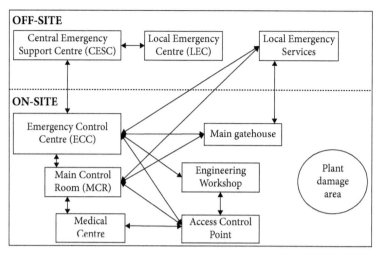

FIGURE 37.3 Site emergency plan.

Reproduced from Carthey, Jane, "Communication and decision making in nuclear emergencies: A field study," unpublished PhD thesis, 1998.

of radioactivity may occur or with any other emergency on the site. The plan also sets out the emergency response organization to be established in response to a nonroutine incident (see Figure 37.3 for an example) and details responsibilities for making assessments, giving advice, and making and implementing decisions.

On-scene decision-makers are located at the key command centers (as shown in Figure 37.3; Carthey, 1998) and, under the emergency plans, are expected to manage nonroutine incidents. Severe accident management scenarios were reviewed by Mumaw, Roth, and Schoenfeld (1993) to identify the types of complexity that can arise, the potential for human error, and the cognitive skills required to manage severe accidents, which include the ability to monitor or detect, to interpret the current state of the plant, to determine implications and establish a goal, to plan, and to control the situation.

Crichton, Flin, and McGeorge (2005) used a card-sorting task (Rugg & McGeorge, 1997) to identify the cognitive structures that assist incident commanders to assess situations and make decisions as comprising: availability of procedures, uncertainty, typicality of the decision, and advice from others. The transfer of expertise based on normal operational role to an emergency role and decision-making strategies selection were also examined (Crichton, McGeorge, & Flin, 2007). Using a dual-task methodology of an ecologically valid computer-based, context-specific decision-making task (single condition) combined with a visuospatial memory task to create a high working memory load (dual-task condition), participants were found to draw on stored knowledge, recognition, and interpretation of situational cues, and create analogies to pre-existing condition–action rules. Knowledge, especially in the form of previous comparable episodes was found to be the best predictor of decision-making expertise. Episodes themselves were frequently based on experiential training for decision making in emergencies, which will be discussed further in the section "Techniques for Developing Expertise in High Hazard Environments."

Teams on NPPF and OO&G installations are made up of individual experts, and the team itself must function as an expert team (Salas, Cannon-Bowers, & Johnston, 1997) in that it performs safely and adaptively. Team members must exhibit the social and cognitive skills to effectively function within their own team, and interact with other teams. Emergency teams are typically ad hoc; that is, they are made up of available personnel and may not be established teams. The teams may also be extended by specialists from other relevant sections within the organization and even by members of external agencies. Following a series of interviews with members of a UK nuclear emergency response organization, the necessary nontechnical skills were identified as decision making and situation assessment, as well as communication, teamwork, and stress management (Crichton & Flin, 2004). Leaders in a nuclear emergency response organization need cognitive, social, and interpersonal abilities and also need to adopt a different leadership style than that used in routine operations (Schumacher, Kleinmann, & Melchers, 2011) in that a more directive style is required to manage emergencies.

The OO&G industry has experienced a number of high-profile incidents over the past forty years, as listed in Crichton, Lauche, and Flin (2008),[1] and more recently the Deepwater Horizon tragedy in 2010 (Hopkins, 2012). Following the Piper Alpha disaster in 1988, the official inquiry specifically highlighted the command and control capabilities of the Offshore Installation Manager (Cullen, 1990) commenting that:

> the failure of the OIMs (Offshore Installation Managers) to cope with the problems they faced on the night of the disaster clearly demonstrates that conventional selection and training of OIMs is no guarantee of ability to cope if the man himself is not able in the end to take critical decisions and lead those under his command in a time of extreme stress.
>
> (para 20.59)

Subsequently, OIMs and Deputy OIMs in the UK now undergo training and assessment in their major emergency control and management skills (OPITO, 2017) that includes competences such as assessing situation and taking effective action, managing individual and team performance, and dealing with stress in self and others.

The ubiquitous availability of SOPs in the OO&G industry may be appropriate for routine situations, but standard operating procedures do not necessarily provide the answer in emergencies (Skriver & Flin, 1996). Skriver (1998) interviewed experienced OIMs and concluded that intimate knowledge of managing emergencies gathered through experience and training, as well as knowledge and understanding of the safety management systems on their installations, provides a basis for emergency decision making. This concurs with the idea that the ability to draw on existing knowledge and skills and adapt them to be applied in a novel situation underpins expertise.

---

[1] The Piper Alpha disaster in 1988 is, to date, the largest industrial accident in the UK. A total of 167 people lost their lives. Lord Cullen mentioned that the command and control capabilities of the Offshore Installation Manager (Cullen, 1990) was unable to cope with the situation.

The Deepwater Horizon tragedy in 2010 resulted in the death of eleven workers. Nontechnical skills such as situation awareness, communication, and leadership were identified as key to the development and managing the situation.

Incident management skills had been identified following an investigation into an actual novel nonroutine incident in 2003 which arose for a drilling team in the Gulf of Mexico (Crichton et al., 2008). Semi-structured interviews, based on the critical decision method (Hoffman, Crandall, & Shadbolt, 1998), were undertaken with seven key decision-makers in the incident command structure. The interviewees described the skills and organizational issues that emerged while managing this incident. By coincidence these participants had only six weeks earlier undergone an emergency exercise where the scenario included many of the features that occurred during the actual incident. The interviews highlighted particular challenges that differed between the exercise and the incident. Many of these arose during the recovery phase of the incident rather than the response phase, and included stressors on decision making such as uncertainty, novelty, time pressure and communication failures, the difficulties in anticipating future states, considering the consequences of decisions and actions, and the effects of fatigue on performance. Pre-existing knowledge and condition–action decisions were found to be applied during the first two days of the incident (i.e., the *response phase*), whereas more analytical decisions were required along with input from technical experts from Day 3 onwards. In terms of training for incident command, the participants commented that although an exercise provides knowledge and understanding of the incident management system and processes, there was little opportunity to experience making decisions under stress. It is this aspect in particular where additional experiences and training should be used to enhance expertise.

NTS have subsequently been emphasized in the inquiry by the US Chemical Safety and Hazard Board who address the competencies, with respect to NTS, of the drill team members involved in the Macondo/Deepwater Horizon incident (USCSHB, 2016). The report comments that traditional training typically focuses on managing conditions based on plans (rules, procedures, and policies). However, it adds that this "does not guarantee error-free performance" and that "expertise is required to recognise when the unexpected is present or may arise" (vol. 3, p. 67). The report includes examples from the management of the escalating situation on the Deepwater Horizon to identify the specific NTS that would be required to manage such an event.

As this section has illustrated, although research into the knowledge and skills that underpin experience has been undertaken, it is the ability to implement those aspects during unexpected or novel incidents that defines expertise. This then raises the issue of how knowledge and skills can be enhanced to facilitate their adaptation.

Techniques for Developing Expertise in High Hazard Environments

A number of different training techniques exist that can, as Klein (1998) stated, help to train novices not to think like experts but to learn like experts, for example, learning how to notice patterns, pick up anomalies, and take the big picture (see Section IV of this *Handbook*, "Developing, Accelerating, and Preserving Expertise"). Experiential learning (Kolb, 1984), whereby learners observe and reflect, try out, and learn from again, is a recognized form of gaining experience and expertise. However, this can be limited by access to a variety of opportunities, receiving feedback in the workplace, and, of course, increased potential risks when learning in hazardous environments (Klein,1998). For example, it would be highly dangerous to practice some of the tasks carried out on both NPPF and OO&G installations during real-time operations; thus, simulation allows people to practice to NIA's high level of realism without the worry of causing harm to either themselves or the plant or installation (Crichton, Moffat, & Crichton, 2017; Hoffman et al., 2014; Stanton, 1996). To overcome these concerns, high hazard industries increasingly rely upon simulator-based training that provides a safe, controlled, repeatable, directed learning environment that accelerates expertise (Owen, Mugford, Follows, & Plummer, 2006). A major advantage of simulation is the opportunity to test out and adapt knowledge and skills in scenarios that can range in complexity and novelty, especially to practice managing nonroutine scenarios. Observations of personnel managing complex and unpredictable tasks can provide support to trainers that operator reactions under simulated conditions will represent those in a real-time environment (Stanton, 1996).

Simulator-based training to enhance NTS ranges from low-fidelity methods, such as pen and paper or computer games, up to high-fidelity, virtual reality simulators (e.g., Crichton, 2017); however, it is not necessarily the case that the higher the fidelity, the better the learning. The benefits of the simulation depend on training design, outcomes, strategies, measurable learning, and feedback (Beaubien & Baker, 2004; Salas, Cannon-Bowers, & Blickensderfer, 1997). How closely the simulation matches the tasks being simulated, particularly the replication of the cognitive processes used in the real world (Kozlowski & DeShon, 2004; see also Chapter 40, "Learning with Zeal: From Deliberate Practice to Deliberate Performance," by Fadde & Jalaeian, this volume), is crucial. Although focusing primarily on teams, Salas, Reyes, and McDaniel (2018) emphasizes the psychological fidelity and safety necessary.

High-fidelity simulators tend to look and operate similarly to the technology they represent, and they are used in a number of different industries to train people when it would be risky in real life to engage in in situ training (Flin, O'Connor, & Crichton, 2008). On an NPPF, full-scope simulators are used to train shift or control room crews (Lindauer, 2012). As previously mentioned, the role of the control room crew is to monitor and control nuclear power production, which involves following procedures and plans, identifying malfunctions, and if necessary shutting down the plant. A high-fidelity control room simulator therefore provides the opportunity to increase the number of experiences, especially on tough cases, under simulated real-world conditions. Decisions can be made, team interactions observed, and stress induced through manipulation of the scenario details.

More recently, high-fidelity drilling floor simulators have been developed to enhance the capabilities of individuals and teams when preparing for drilling wells. For example, a major international oil & gas company created a training intervention, integrating simulator-based training exercises to enhance the capabilities of team members in both technical and nontechnical skills (Crichton et al., 2017). Behaviors were observed and feedback was provided at the end of each exercise to facilitate self and team reflection on performance and, as required, improvements over the course of the series of exercises.

Low-fidelity simulation techniques can be extremely cost-effective, easy to establish, and less resource intensive than higher fidelity simulations, yet result in enhanced cognitive expertise through a focus on psychological fidelity. One particularly effective low-fidelity simulation is that of tactical decision games (TDGs) that have been used to practice decision making under complex, dynamic, and stressful conditions to prepare for the unexpected (Klein, 1997; Schmitt & Klein, 1996; Woltjer, Trnka, Lundberg, & Johansson, 2006). TDGs have been developed in NPPF, in particular to train on-scene emergency response team members (Crichton, Flin, & Rattray, 2000). Scenarios in this type of low-fidelity simulation include ambiguous, incomplete, or inconsistent information which allows alternative interpretations of the situation, and encourages participants to discuss their situation assessment and learn vicariously from others.

In sum, the risks and potential consequences of errors in high-hazard environments demand that operators are knowledgeable and competent, and can draw on their expertise to manage unexpected novel circumstances. Simulation, low or high fidelity, provides a safe environment through which individuals and teams can test out their skills and enhance their expertise.

Overcoming Current and Future Challenges in Developing Expertise in High-Hazard Environments

Safety is paramount in terms of the public and the environment, as well as the personal safety of operators working on nuclear facilities and oil & gas installations. The approach to ensuring such safety has traditionally been addressed through the implementation of safety management systems, with associated rules and procedures which have been developed to assure the reduction of risks and compliance with rules. Increasingly, the expertise of operators is being acknowledged, and their cognitive and social skills are being identified, trained, and in many cases, assessed.

Conducting practical research in applied settings is recognizably difficult, and even more so when examining performance in high-risk uncertain environments. Gaining access to nuclear facilities can itself be a challenge as visitors must adhere to the site rules and regulations to ensure their physical safety and must receive clearance to visit the site in the first place. Carrying out research on an oil & gas installation is even more difficult as approval must be gained from the organization for researchers to visit the installation, who must either have undergone certificated survival training or receive special dispensation for a certain period to cover the duration of the visit, particularly as the travel itself can be dangerous and involves a round-trip helicopter journey.

To collect data in operational settings the methods to be used must be accurately identified, with clear planning as to how these methods will be implemented. The work environment should be understood, especially interactions between humans and technology, but the technology should not constrain research activities, or overshadow the basic needs of personnel performing their roles (Stanton, Salmon, Walker, Baber, & Jenkins, 2013). Data collection methods such as interviews and observations raise additional concerns including accessing the personnel during their work time which may distract them from their duties, leading to employers being less than supportive of the research study. Indeed, as Stanton et al. (2013) comment, some organizations may not readily agree to personnel being observed in the workplace and therefore do not give authorization.

Such considerations become ever more complex when attempting to research the topic of expertise in operational environments, especially in terms of social and cognitive skills, and a variety of studies that have been conducted to identify expertise have been described in this chapter. The most frequently used methods would include questionnaires (such as attitude questionnaires: Gregorich, Helmreich, & Wilhelm, 1990), interviews (e.g., the critical decision method: Hoffman et al., 1998), and simulator-based observations (see Crichton, 2017), yet each of these methods suffer limitations in terms of use in applied research as described earlier.

Moreover, as the development of expertise in high-hazard settings is so tightly associated with the ability to adapt to nonroutine events, identifying the attributes of expertise during actual nonroutine events can be problematic. Klein (1998) describes how his company studied how people make decisions under time pressure, a project sponsored by the US Army Research Institute for the Behavioral and Social Sciences, which underpinned the Recognition Primed Decision Making model. Due to the challenges of conducting such research with battle commanders, the research was undertaken with firefighters, and it was later confirmed that firefighters use the same strategies as military commanders.

Observations of simulator-based and table-top or full-scale emergency exercises should be designed to, as much as reasonably possible, expose participants to the types of environments and problems with which they might be confronted in a real-life event. Psychological fidelity (Kozlowski & DeShon, 2004) is crucial in that cognitive tasks should be equivalent to those that would be encountered in the real world. However, aspects such as increased levels of stress, time pressures, novelty, timescale, and resource limitations differ between an exercise and a real-life emergency. Participants in exercises may report experience of what could be called *exercisitis* where they expect complications to arise and are therefore exhibit expectation bias, where they actively look for potential *tricks* and for escalations (Crichton & Kelly, 2012). In contrast, in a real incident, there can be less opportunity to think ahead due to dealing with the current aspects of the incident.

So where does this leave us in relation to developing expertise in high-hazard environments in future research? There are so many questions yet to be answered: Do we look to an individual's years of experience to determine whether they are an expert, or do we look at the level of performance they achieve in a particular task? In terms of variety of experience, if someone is particularly good at one task, does this necessarily translate to other tasks, even if in the same setting? Is the old adage of *the more you practice, the better you get* applicable in all situations? How can some people adapt their experience to novel situations and perform well, especially under stress? How do we measure this combination of task, duration of experience, and level of performance to determine expertise?

Just because something is difficult does not mean that it should be ignored. In contrast, such a challenge gives researchers the opportunity to think of innovative ways or methods for resolving the problem. Due to the difficulties of conducting applied research in high-hazard environments, innovative or novel techniques can be designed, particularly with ever-increasing technology advances. Simulator-based scenario exercises provide a useful method through which to manipulate variables to examine expertise, but, as explained earlier, these need to be specifically designed and tested. Existing methods will always provide a basis, but the enticing opportunity for researchers is to consider how any findings can be implemented in the real world to enhance the development of expertise and consequently to improve the effectiveness and primarily the safety of individuals and teams operating in hazardous environments. From high physical fidelity (full-scale simulators) through drills or paper-based exercises, and low-fidelity psychological fidelity, the aim is to encourage participants to practice, test out KRSs, and learn from their experiences. Reflecting on their own performance along with directed feedback enhances confidence, competence, and capabilities.

Acknowledgments

We would like to dedicate this chapter to Professor Margaret Crichton, the founder of People Factor Consultants who died in September 2018. Her outstanding leadership directed all our applied human factors projects and her academic expertise enabled an appreciation and deeper understanding of the theoretical basis of the discipline among her staff. She is sorely missed, and her professional legacy is the enthusiasm PFC carries forward.

We would also like to thank all those who assisted with the writing of this chapter, in particular Mr. Terry Kelly for all his guidance on operations on nuclear power production facilities and emergency response arrangements.

References

Ainsworth, L. K., & Marshall, E. C. (1994/5). Training for infrequent and stressful scenarios (Generic Nuclear Safety Review (Human Factors Technical Area)). HF/GNSR/08. Unpublished report prepared for the Industry Management Committee (IMC), Human Factors Technical Area, UK.

Apostolakis, G. E. (1999). Organisational factors and nuclear power plant safety. In J. Misumi, B. Wilpert, & R. Miller (Eds), *Nuclear safety: A human factors perspective*. London: Taylor & Francis.

Beaubien, J. M., & Baker, D. P. (2004). The use of simulation for training teamwork skills in health care: How low can you go? *Quality and Safety in Health Care* 13, i51–i56.

Begg, S. J., Bratvold, R. B., & Campbell, J. M. (2003). Shrinks or quants: Who will improve decision-making. SPE 84238-MS. Paper presented at SPE annual technical conference and exhibition, 5–8 October, Denver, Colorado.

Carbonell, K. B., Stalmeijer, R. E., Konings, K. D., Segers, M. & van Merrienboer, J. J. G. (2014). How experts deal with novel situations: A review of adaptive expertise. *Educational Research Review* 12, 14–29.

Carthey, J. (1998). Communication and decision making in nuclear emergencies: A field study. Unpublished PhD thesis (Confidential), University of Birmingham, Birmingham, UK.

Carvalho, P. V. R. (2006). Ergonomic field studies in a nuclear power plant control room. *Progress in Nuclear Energy* 48, 51–69.

Carvalho, P. V. R., dos Santos, I. L., & Vidal, M. C. R. (2005). Nuclear power plant shift supervisor's decision making during microincidents. *International Journal of Industrial Ergonomics* 35, 619–644.

Crichton, M., & Flin, R. (2004). Identifying and training non-technical skills of nuclear emergency response teams. *Annals of Nuclear Energy* 31(12), 1317–1330.

Crichton, M., Flin, R., & McGeorge, P. (2005). Decision making by on-scene incident commanders in nuclear emergencies. *Cognition, Technology & Work* 7(3), 156–166.

Crichton, M., Flin, R., & Rattray, W. A. (2000). Training decision makers—tactical decision games. *Journal of Contingencies and Crisis Management* 8(4), 208–217.

Crichton, M. & Kelly, T. (2012). Developing emergency exercises for hazardous material transportation: process, documents and templates. *Journal of Business Continuity and Emergency Planning* 6(1), 32–46.

Crichton, M., Lauche, K., & Flin, R. (2008). Learning from experience: Incident management team leader training. In J. M. Schraagen, L. Militello, T. Ormerod, & R. Lipshitz (Eds), *Naturalistic decision making and macrocognition*. Aldershot, UK: Ashgate.

Crichton, M., McGeorge, P., & Flin, R. (2007). Decision making by operational incident commanders: Decision strategy selection. In M. Cook, J. Noyes, & Y. Masakowski (Eds), *Decision making in complex environments*. Aldershot, UK: Ashgate.

Crichton, M. T. (2017). From cockpit to operating theatre to drilling rig floor: Five principles for improving safety using simulator-based exercises to enhance drilling team cognition. *Cognition, Technology & Work* 19(1), 73–84.

Crichton, M. T., Moffat, S., & Crichton, L. M. (2017). Developing a team behavioral marker framework using observations of simulator-based exercises to improve team effectiveness: A drilling team case study. *Simulation and Gaming*, online February 1.

Cullen, The Lord (1990). *The public inquiry into the Piper Alpha disaster* (vols I and III (Cm 1310)). London: HMSO.

Dekker, S. (2011). *Drift into failure: From hunting broken components to understanding complex systems*. Burlington, VT: Ashgate.

Devold, H. (2006). *Oil and gas production handbook. An introduction to oil and gas production, transport, refining and petrochemical industry* (3rd edn). Oslo, Norway: ABB Oil and Gas.

Ericsson, K. A. (2004). Deliberate practice and the acquisition and maintenance of expert performance in medicine and related domains. *Academy of Medecine* 79(10, suppl), 70–81.

Ericsson, K. A., Whyte, J., & Ward, P. (2007). Expert performance in nursing: Reviewing research on expertise within the framework of the expert-performance approach. *Advances in Nursing Science 30*(1), 58–71.

Fletcher, G., Flin, R., McGeorge, P., Glavin, R., Maran, N., & Patey, R. (2003). Anaesthetists' non-technical skills (ANTS): Evaluation of a behavioral marker system. *British Journal of Anaesthesia 90*, 580–588.

Flin, R. (1996). *Sitting in the hot seat: Leaders and teams for critical incident management.* Chichester, UK: Wiley.

Flin, R., O'Connor, P., & Crichton, M. (2008). *Safety at the sharp end: A guide to non-technical skills.* Ashgate, UK: Aldershot.

Gaddy, C. D., & Wachtel, J. A. (1992). Team skills training in nuclear power plant operations. In R. W. Swezey & E. Salas (Eds), *Teams: Their training and performance.* Norwood, NJ: Ablex.

Gregorich, S., Helmreich, R., & Wilhelm, J. (1990). The structure of cockpit management attitudes. *Journal of Applied Psychology 75*, 682–690.

Griffin, M. A., Hodkiewicz, M. R., Dunster, J., Kanse, L., Parkes, K. R., Finnerty, D., . . . Unsworth, K. L. (2014). A conceptual framework and practical guide for assessing fitness-to-operate in the offshore oil and gas industry. *Accident Analysis and Prevention 68*, 156–171.

Hatano, G., & Inagaki, K. (1986). Two courses of expertise. In H. Stevenson, J. Azuma, & K. Hakuta (Eds), *Child development and education in Japan* (pp. 262–272). San Francisco: Freeman.

Hayes, J. (2012). Use of safety barriers in operational safety decision making. *Safety Science 50*, 424–432.

Hoffman, R. R., Crandall, B., & Shadbolt, N. (1998). Use of the critical decision method to elicit expert knowledge: A case study in the methodology of cognitive task analysis. *Human Factors 40*(2), 254–276.

Hoffman, R. R., Ward, P., Feltovich, P. J., DiBello, L., Fiore, S. M., & Andrews, D. H. (2014). *Accelerated expertise: Training for high proficiency in a complex world.* New York: Psychology Press.

Hopkins, A. (2011). Risk-management and rule-compliance: Decision-making in hazardous industries. *Safety Science 49*, 110–120.

Hopkins, A. (2012). *Disastrous decisions. The human and organizational causes of the Gulf of Mexico blowout.* Australia: CCH.

HSE. (1994). *Arrangements for responding to nuclear emergencies.* Sudbury: HSE.

HSE. (2011). *High reliability organizations: A review of the literature.* Research Report RR899. Available at: http://www.hse.gov.uk/research/rrpdf/rr899.pdf

HSE. (2013). *Managing for health and safety.* Report HSG65. Available at: http://www.hse.gov.uk/pubns/books/hsg65.htm.

IAEA. (2001). *A systematic approach to human performance improvement in nuclear power plants: Training solutions.* Report IAEA-TECDOC-1204. Vienna, Austria.

IMC. (1999). *Acute stress in the nuclear industry: Practical intervention methods—Summary report* (Summary report HF/GNSR/5041; Annex to report DERA/CHS/MID/CR990154/1.0). Confidential report prepared for the Industry Management Committee (IMC). Risley: IMC.

IOGP. (2014). *Crew resource management for well operations teams,* Report 501. International Association of Oil and Gas Producers.

Israel, R., Mason, C., Whiteley, N., Gibson, K., Dobson, D., & Andresen, P. A. (2015). Well advisor—integrating real-time data with predictive tools, processes and expertise to enable more informed operational decisions. Paper presented at *SPE/IADC drilling conference and exhibition*, 17–19 March, London.

Kanki, B., Helmreich, R., & Anca, J. (Eds). (2010). *Crew resource management* (2nd edn). San Diego: Academic Press.

Klein, G. (1997). Developing expertise in decision making. *Thinking and Reasoning* 3(4), 337–352.

Klein, G. (1998). *Sources of power: How people make decisions*. Cambridge, MA: MIT Press.

Kolb, D. A. (1984). *Experiential learning*. New York: Prentice-Hall.

Kozlowski, S., & DeShon, R. P. (2004). A psychological fidelity approach to simulation-based training: Theory, research and principles. In S. G. Schiflett, L. Elliot, E. Salas, & M. D. Coovert (Eds), *Scaled worlds: Development, validation and application*. Aldershot, UK: Ashgate.

Lauche, K., Crichton, M., & Bayerl, P. S. (2009). Tactical decision games: Developing scenario-based training for decision-making in distributed teams. In *Proceedings of the 9th naturalistic decision making conference, London, June*. Available at: http://dl.acm.org/citation.cfm?id=2228107&picked=prox. Accessed August 15, 2017.

Lauche, K., Sawaryn, S. J., & Thorogood, J. L. (2009). Human factors implications of remote drilling operations: A case study from the North Sea, SPE-99774-PA, Society of Petroleum Engineers SPE Drilling & Completions,. 24(1), 7–14. DOI: 10.2118/99774-PA

Lindauer, E. (2012). Appendix 4—Simulator training for nuclear power plant control room personnel. In A. Alonso (Ed.), *Infrastructure and methodologies for the justification of nuclear power programs* (pp. 934–949). Oxford: Woodhead Publishing Series in Energy.

Mackie, S. I., Begg, S. H., Smith, C., & Welsh, M. B. (2010). Human decision making in the oil and gas industry. SPE 131144-MS. Paper presented at *SPE Asia Pacific Oil and Gas Conference and Exhibition*, 18–20 October, Brisbane, Queensland, Australia.

Malhotra, V., Lee, M. D., & Khurana, A. (2004). Decisions and uncertainty management: Expertise matters. SPE 88511-MS. Paper presented at the *SPE Asia Pacific oil and gas conference and exhibition, 18–20 October, Perth, Australia*.

Masakowski, Y. R., & Hardinge, N. (2000, 26-28 May). Cognitive task analysis of decision strategies: What are the performance differences between USN and UK Navy novices and experts? Paper presented at the *Fifth conference on naturalistic decision making*.

Meister, D. (1995). Cognitive behaviour of nuclear reactor operators. *International Journal of Industrial Ergonomics* 16, 109–122.

Mumaw, R. J., Roth, E. M., & Schoenfeld, I. (1993). Analysis of complexity in nuclear power severe accident management. *Proceedings of the Human Factors and Ergonomics Society Annual Meeting* 37(4), 377–381.

Mumaw, R. J., Roth, E. M., Vicente, K. J., & Burns, C. M. (2000). There is more to monitoring a nuclear power plant than meets the eye. *Human Factors* 42(1), 36–55.

Munro, M. (2008). Real-time access to expertise: Improving decision making and collaboration through unified communications. SPE 112253-MS. Paper presented at the *Intelligent energy conference and exhibition, 25–27 February, Amsterdam, The Netherlands*.

NIA (Nuclear Industry Association). (2017). Nuclear energy facts. Available at https://www.niauk.org/wp-content/uploads/2017/02/nia_factsbook_2017.pdf. AccessedApril 28, 2017.

ONR (Office for Nuclear Regulation). (2016). *Licence condition handbook*. Bootle, UK: ONR.

OPITO (2017). *OIM controlling emergencies. Code 7025*. Available at: http://www.opito.com/media/downloads/oim-controlling-emergencies-standard-rev-4-revision-effective-from-01-june-2017.pdf. Accessed April 25, 2017.

Owen, H., Mugford, B., Follows, V., & Plummer, J. L. (2006). Comparison of three simulation-based training methods for management of medical emergencies. *Resuscitation* 71, 204–211.

Randel, J. M., Pugh, H. L., & Reed, S. K. (1996). Differences in expert and novice situation awareness in naturalistic decision making. *International Journal of Human–Computer Studies* 45, 579–597.

Rasmussen, I. (1983). Skills, rules, knowledge, signals, signs, and symbol and other distinctions in human performance modelling. *IEEE Transactions on Systems, Man, and Cybernetics SMC-13*(3), 257–266.

Reason, J. (1987). The Chernobyl errors. *Bulletin of the British Psychological Society* 40, 201–206.

Rugg, G., & McGeorge, P. (1997). The sorting techniques: A tutorial paper on card sorts, picture sorts and item sorts. *Expert Systems* 14(2), 80–94.

Sætren, G. B., & Laumann, K. (2015). Effects of trust in high-risk organizations during technological changes. *Cognition Technology & Work* 17, 131–144.

Salas, E., Cannon-Bowers, J. A., & Blickensderfer, E. L. (1997). Enhancing reciprocity between training theory and practice: principles, guidelines, and specifications. In J. K. V. E. Ford, S. W. J. Kozlowski, K. Kraiger, E. Salas, & M. S. Teachout (Eds), *Improving training effectiveness in work organizations*. New York: Psychology Press.

Salas, E., Cannon-Bowers, J. A., & Johnston, J. H. (1997). How can you turn a team of experts into an expert team? Emerging training strategies. In C. Zsambok & G. A. Klein (Eds), *Naturalistic decision making* (pp. 359–370). Hillsdale, NJ: Lawrence Erlbaum.

Salas, E., Reyes, D. L., & McDaniel, S. H. (2018). The science of teamwork: Progress, reflections and the road ahead. *American Psychologist* 71(4).

Schmitt, J. F., & Klein, G. (1996). Fighting in the fog: Dealing with battlefield uncertainty. *Marine Corps Gazette* 80(August), 62–69.

Schumacher, S., Kleinmann, M., & Melchers, K. G. (2011). Job requirements for control room jobs in nuclear power plants. *Safety Science* 49, 394–405.

Shanteau, J., Weiss, D. J., Thomas, R. P., & Pounds, J. C. (2002). Performance-based assessment of expertise: How to decide if someone is an expert or not. *European Journal of Operational Research* 136, 253–263.

Skriver, J. (1998). Emergency decision making on offshore installations. Unpublished PhD, University of Aberdeen, Aberdeen.

Skriver, J., & Flin, R. (1996). Decision making in offshore emergencies: Are Standard Operating Procedures the solution? SPE-35940-MS. Paper presented at the *SPE conference on health, safety and environment in oil and gas industry*, New Orleans, Louisiana.

Sonnentag, S. (2000). Expertise at work: experience and excellent performance. In C. L. Cooper and I. T. Robertson (Eds), *International review of industrial and organizational psychology* (vol. 15). London: Wiley.

Stanton, N. (1996). Simulators: A review of research and practice. In N. Stanton (Ed.), *Human factors in nuclear safety* (pp.117–140). London: Taylor & Francis.

Stanton, N. A., Salmon, P. M., Walker, G. H., Baber, C., & Jenkins, D. P. (2013). *Human factors methods. A practical guide for engineering and design* (2nd Ed). Aldershot, UK: Ashgate.

Thorogood, J. L., & Crichton, M. (2013). Operational control and managing change: The integration of non-technical skills with workplace procedures. SPE-163489-MS. In *Proceedings of the SPE/IADC conference*, Amsterdam, The Netherlands.

Three Mile Island Special Inquiry Group, NRC (1980). *Human factors evaluation of control room design and operator performance at Three Mile Island-2*, vol. 1 (Final Report NUREG/CR-1270-V-1). Washington, DC: US Department of Commerce, National Technical Information Service.

US Chemical Safety and Hazard Investigation Board (2016). *Drilling rig explosion and fire at the Macondo well. Investigation Report*, vol. 3. Report 2010-10-I-OS. Washington DC: USCSB.

van Avermaete, J. A. G., & Kruijsen, E. A. C. (1998). *NOTECHS: The evaluation of non-technical skills of multi-pilot aircrew in relation to the JAR-FCL requirements*. EC NOTECHS project final report: CR 98443, Amsterdam.

Ward, P., Gore, J., Hutton, R., Conway, & Hoffman, R. R. (2018). Adaptive skill as the *conditio sine qua non* of expertise. *Journal of Applied Research in Memory and Cognition* 7(1), 35–50.

Weick, K. E., & Sutcliffe, K. M. (2007). *Managing the unexpected: Resilient performance in an age of uncertainty* (2nd edn). San Francisco: Jossey-Bass.

Woltjer, R., Trnka, J., Lundberg, J., & Johansson, B. (2006). Role-playing exercises to strengthen the resilience of command and control systems. In *Proceedings of the 13th European Conference on cognitive ergonomics: Trust and control in complex socio-technical systems, Zurich, Switzerland* (pp. 71–78).

Yule, S., Flin, R., Maran, N., Rowley, D., Youngson, G., & Paterson-Brown, S. (2008). Surgeons' non-technical skills in the operating room: Reliability testing of the NOTSS behavior rating system. *World Journal of Surgery* 32, 548–556.

CHAPTER 38

EXPERTISE IN WEATHER FORECASTING

DAPHNE S. LADUE, PHAEDRA DAIPHA,
REBECCA M. PLISKE, AND
ROBERT R. HOFFMAN

INTRODUCTION

THE term *weather forecasting* is often used to refer to the forecast of parameters such as temperature, humidity, and wind over a period of hours from the current time to a week or so in the future. But weather forecasting encompasses many different weather processes and time scales ranging from seasonal climate prediction to the few minutes prior to tornado formation. The approaches to creating a forecast at different points along this spectrum of lead time vary substantially, as do the motivations for creating the forecast. Seasonal climate forecasting is motivated, for example, by large losses to America's $2B apparel industry when winter coats failed to sell during a warm winter (Barbaro, 2007). Changes in climate have also impacted Coca-Cola's sourcing and use of water to make its beverage products and scarcity of cotton for production of Nike's athletic clothing (Davenport, 2014). Weather has significant economic impacts in the more traditional hours-to-days timeframe as well, with estimates of $4.2B losses in the airline industry each year from weather-related delays (National Research Council, 2010), and 1.5M crashes, 7,400 fatalities, and $42B in economic losses on our nation's highway system (National Academy of Sciences, 2010). Since 1980 the United States has experienced 233 weather disasters costing at least $1B each (CPI adjusted to 2017), with a total of $1.5 trillion in losses (National Centers for Environmental Information, 2018). Loss of life is also notable, with an average of 585 direct, weather-related fatalities annually between 2002 and 2016 (National Weather Service, 2016).

Weather forecasting is a highly technical domain with many advanced technological tools and datasets. While known equations of motion and meteorological principles govern the atmosphere, atmospheric properties are undersampled, leading to forecast inaccuracies; nonlinear effects can grow quickly and dramatically affect a forecast (a notion made famous by Gleick's popular book, *Chaos: Making a New Science* in 1987). Forecasters supplement their meteorological knowledge with pattern recognition and heuristic rules based on experience.

This chapter demonstrates how four different focus points for studying weather forecasting complement each other to provide insights about expertise:

- Organizational contexts impact learning and development of expertise;
- Weather situations require forecasters to cope with uncertainty and risk;
- Individual differences are manifested in differing forecasting styles and strategies; and
- The diversity of weather situations requires forecasters to be facile at a great many specialized workflows.

The work analysis techniques that have been engaged to study these things can be roughly described as ethnographic but are also inspired by particular disciplinary lenses, including education; science, technology, and society studies; and cognitive systems engineering. These particular studies span across a continuum of studying forecasting expertise from the contextualist/organizational level to the cognitive/individual level of analysis. All of the discipline-based research approaches recognize the procedural and task-driven nature of expertise and, therefore, primarily favor the depth afforded by studies of weather forecasters at work. We begin at the *contextualist* pole of the continuum by discussing research on how people become proficient forecasters, and the real-world process of uncertainty management during forecast production. We then shift to the *cognitivist* pole to look at how to scale proficiency in forecasting, and the differences in reasoning between novice and expert forecasters.

Educational, Training, and Mentoring Issues in Learning to Forecast

In most US undergraduate meteorology or atmospheric science programs students are encouraged to practice forecasting but are not explicitly taught how; the overall perspective of the field is that the job of the university is not to prepare students for any *particular* meteorology job (Baum, 1975). The point is made that meteorologists are individuals whose work involves weather; they are not necessarily forecasters. While many individuals with a bachelor's degree pursue forecasting as a profession, they might instead work in software development, environmental or air quality monitoring, risk or catastrophe modeling, education, or a variety of other specific professions that interface with meteorology (Bureau of Labor Statistics, 2018).

LaDue (2011) conducted a literature search on the field of meteorological education, which yielded few discussions of learning how to forecast. Most importantly, no studies addressed learning after the receipt of a degree in meteorology, where the majority of learning how to forecast—and develop expertise—takes place. This finding motivated an in-depth study of how working forecasters had learned how to forecast.

Method

LaDue conducted in-depth ethnographic interviews with 11 forecasters (eight males, three females) from a range of locations and forecasting responsibilities. Participants were chosen based on factors by which learning might vary:

- Type of forecast (public forecasts, hydrologic forecasts, agricultural forecasts, marine forecasts, aviation forecasts); time period of the forecast (ranging from minutes to seasons in the future);
- Locale for which forecasts are made (e.g., tropical marine environment vs. inland mid-latitude locations); and
- Employment (seven worked in the public vs. four in the private sector).

Because expertise studies suggest that expertise often develops over a long period, forecasters were sampled across a range of time-in-service (i.e., from 1 to 17 years).

Interviews probed for information about learning that had occurred over the past year, a strategy typically used in studies of learning of professionals (e.g., Fox, Mazmanian, & Putnam, 1989). The main interview questions blended the strategy of asking about critical incidents (Dunn & Hamilton, 1986) with changes in practice (Fox et al., 1989). Follow-up questions included when the participant first started attempting to forecast; their formal education; their preferred way to learn; ways in which their learning strategies had changed over time; the role of social interactions in their learning; and any barriers to learning that they may have encountered.

Analysis procedures included coding interviews line-by-line, comparing codes within and across interviews, and exploring how codes relate to build an understanding of the phenomenon being studied (Charmaz, 2006; Corbin & Strauss, 2008; Lincoln & Guba, 1985; Strauss & Corbin, 1998). Results from these different analytical methods were triangulated through follow-up interviews, discussions with individuals responsible for forecaster learning in both the public and private sectors, and consistency with findings of other studies discussed later in this chapter.

Results

The resulting major categories, with two to five subcodes each, included conditions for learning, long-term learning strategies, short-term learning strategies, initiation/

culmination of learning efforts, and expression of the self and relation to others. The central category, which related to all the major categories, was a forecaster's sense of professional identity. No attempt was made to determine a forecaster's expertise level; thus although the term *novice* is used to describe the results, the term *expert* is not. The majority of participants were at least *proficient*, meaning they had responsibility to create forecasts without help; three were interns.

Figure 38.1 shows how learning events were triggered at different points in forecasters' careers (gray boxes at the top). The most common trigger for learning among beginning forecasters was their general lack of skill. All but one forecaster reported being ill-prepared to forecast on their own after graduating with their degree. In the public sector forecasters do not have forecast duties right away. Three of the four private sector forecasters worked for companies that provided either formal training or focused mentoring; the fourth said he was shown his desk and told to start forecasting. When questioned he reiterated, "I had no formal training there. They just, boom. They said, go ahead." Forecasters with some experience can find themselves at a novice stage if they begin forecasting for a new weather regime or if constructs they had not learned in school are commonly used. After 1.5–3 years' experience, forecasters learned most often after being surprised when a forecast they thought would be good did not verify. As one forecaster said, "Nine out of ten times it's gonna verify. There's that one out of ten when it doesn't... That's when you learn and improve as a forecaster." By the middle of their career, forecasters were rarely surprised, and learning tended to be for cases where no one knew the answer. Some forecasters with higher experience levels worked to advance the state of the art by either researching tough cases themselves or carefully delineating and publishing such cases for others to research, whereas those

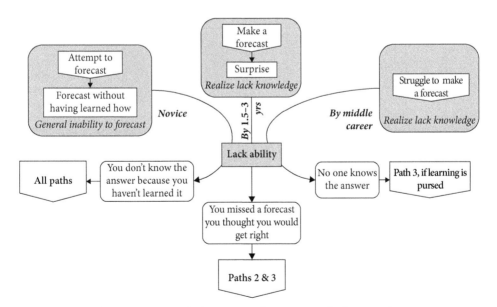

FIGURE 38.1 Common triggers for learning at three stages of a forecaster's career.

who were not learning simply accepted some level of forecast errors as inevitable and their cause unsolvable. The pentagons at the sides and bottom of Figure 38.1 indicate which learning path was most common for each situation: novices on the left, early career down the middle, and middle career on the right. Shaded boxes with a pentagon inside distinguish three starting points. Rectangles with sharp edges indicate a coding category from the data. Rectangles with rounded edges indicate an explanation that connects boxes. Ending pentagons for each path indicate where this diagram is connected to in Figure 38.2.

Figure 38.2 illustrates how the learning of the complex aspects of forecasting cannot be easily achieved by learning from books or similar resources. These paths represent forecaster descriptions of more complex thinking about a specific situation, either as individuals or through forecast discussions. These learning paths were independent of time in service, though path 1, for example, was more common among beginning forecasters. Gray boxes indicate codes with variation built into the theory.

Path 1 flows down the left side of Figure 38.2 and shows an inexperienced forecaster being taught by experienced forecasters, without prompting. Stories that fit here illustrated very efficient learning: a forecaster was not prepared to do the job, or some aspect of the job, and someone reached out to teach them what they needed to learn. All but one participant described mentoring interactions that could be described

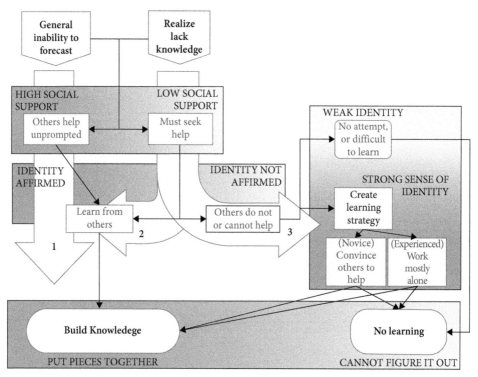

FIGURE 38.2 The elements and relationships representing a preliminary theory of how meteorologists learn to forecast.

in this way (for a discussion of mentoring more broadly, see Chapter 44, "Learning at the Edge," by Petushek et al., this volume). These interactions ranged from one experienced forecaster mentoring the new forecaster to a mix of interactions with several older forecasters. Participants highly valued these interactions, through which a deeper conceptual understanding of weather processes could be quickly gained. Path 2 is very similar, though the help was actively sought. Other forecasters either did not realize that the forecaster needed help or simply did not help unprompted.

Participants reported that when help was readily given, the forecaster learned. All participants experienced learning by seeking help. Importantly, they felt affirmed through these interactions, such as when an older forecaster responded to one younger forecaster's question, "You're seeing something, you're picking it up... let me elaborate a little bit and tell you what is causing this." A forecaster with considerable experience with tropical weather forecasting stopped "pretending," and "[asked] somebody" who then complemented him for asking what a midlatitude weather instability construct meant.

Paths 1 and 2 describe cases in which knowledge was available locally that others knew and *could* provide. If a forecaster did not or could not find help from a more experienced forecaster (low social support) their attempts to learn appeared to be moderated by how strongly they identified as a forecaster: The weaker their identity, the less likely they were to rise to the challenge. Forecasters with a particularly strong sense of identity went, at times, to great lengths to learn. These stories are described by Path 3 in Figure 38.2. These were the most significant learning events, where forecasters were conscious of having to create a strategy in order to learn. They were often successful, though sometimes the causes behind a missed forecast remained a mystery.

Path 3 has within it several subpaths, and learning was not always the outcome. Younger forecasters created strategies to link what they had learned in school with how to apply that knowledge to forecasting. One participant reported she had been told to "go through these modules and they should help you a lot," but convinced colleagues to sit down with her afterward to show her how to apply what she had learned. Another participant reported that she had found displaced real-time cases overwhelming because she was still learning how to discriminate features of interest from the noise in radar data. The forecaster with no training program reported spending significant time outside his work hours to learn all he could through publically available training resources, books, and journals that dealt with weather in a practical way. Forecasters who were still building their expertise also described learning through this path, and creating new ways to display data that might help keep them from missing important signals.

Other forecasters were building upon their ability to do their job or extending the science. About half of the stories involved reference to other forecasters, regardless of time-in-service. All but one forecaster described their learning strategy as one that they had created themselves. These learning events were motivated by curiosity or ongoing irritations with their poor forecasts, eventually leading to extensive learning efforts. One forecaster reported that he had to create strategies and extend the science of forecasting because his type of forecast, seasonal climate applied to agriculture, was in its infancy. Verifying a forecast was not possible until several months after the issuance of the forecast, requiring him to develop strategies to help him remember his reasoning: He took careful notes on how he had used data, and subsequently assessed how well the forecasting strategy had worked.

The knowledge required for forecasting the weather is extensive and complex. Thus, it is difficult to learn how to forecast without someone helping the learner to think through complex weather processes. The fact that the majority of learning episodes involved social interaction led LaDue to specifically ask for instances of learning that were entirely asocial. To this, one forecaster said, "It's hard to think of any specific examples right now... in the process of [investigating] I start [talking to] people and seeing what they have seen." This response was one variation on a consistent theme expressed in the comments by all of the participants, that they all incorporated a strategy involving others at some point in their learning. It is clear that developing forecasting expertise is a highly social activity.

Proficiency at forecasting is not just a continuous process of learning, it is also a continuous process of adaptive problem solving, as discussed in the next section.

How Weather Forecasters Cope with Uncertainty

Science, technology, and society studies (STS) is an interdisciplinary field established in the early 1970s that, crossing sociology, history, and anthropology, draws attention to the sociotechnical foundations of scientific knowledge production (for a review of the field, see Collins & Evans, 2002; see also Chapter 4, "Studies of Expertise and Experience," by Collins & Evans, this volume). The main focus of STS research is on *expertise in the making*, or on the background of arrangements and practices that make expertise possible (e.g., Cambrosio, Limoges, & Hoffman, 1992; Dreyfus, 1972; Latour, 1987). Expertise, in this sense, is understood as a robust configuration of humans, technologies, institutions, and nature, typically standardized and blackboxed out of view such that expertise appears to be embodied by the expert. STS can be thought of as complementary to cognitive psychology by regarding knowledge and expertise acquisition as a sociotechnical achievement (Daipha, 2015b; Fine, 2007; Henderson, 2016).

Studying forecasting expertise as a distributed group process requires studying it in action, in the actual settings and conditions of weather forecasting operations, for protracted periods of time. STS studies consist of immersive ethnographies of forecasting life, ranging from 1 up to 5 years of field observations, along with in-depth qualitative interviews and archival research. In addition to eliminating observer effect, long-term immersion in the culture and discourse of an expertise domain affords STS researchers *interactional expertise* (Collins & Evans, 2007): although they do not possess the expertise to produce a skillful weather forecast, they can follow the process and relay back what transpired to the satisfaction of any seasoned forecaster. Long-term immersion also affords STS researchers the ability to analytically push past the common sense accounts of their informants and, just as importantly, to gain a longitudinal understanding of the effects of technological innovations and organizational upheavals on weather forecasting knowledge and performance standards. To probe the cultural and institutional foundations of expertise, ethnographies of weather forecasting operations often feature a comparative research design, such as a comparison among forecast offices or among forecast producers and consumers. One such STS study, by Daipha (2010, 2015a, 2015b), took the forecasting task as its unit of analysis and thus has the potential to be integrated with other efforts in cognitive psychology to characterize and evaluate expertise.

Method

Daipha conducted an immersive ethnography of forecasting operations in one of 23 Weather Forecast Offices of the National Weather Service Eastern Region, hereafter referred to as "Neborough." Fieldwork at the Neborough office occurred in two waves: 14 months from 2003 through 2004 and another 8 months in 2008. After an initial intensive 3-month familiarization period, ethnographic visits settled into a weekly 4-day schedule designed to overlap with forecasting shifts as well as major weather events (for methodological details, see Daipha, 2015b: 16ff.). Data analysis was performed iteratively with data collection, and adhered to grounded theory techniques similarly as that described earlier for the study about forecaster learning (Charmaz, 2006; Strauss & Corbin, 1998). To incorporate forecasters' lived experiences and further contextualize field observations, in-depth semi-structured interviews averaging 2 hours in length were conducted with all but one of the 20-member staff toward the end of the 2004 and 2008 fieldwork waves. The analysis also benefited from unlimited access to all facets of Neborough forecasting operations, including paperwork, chat room discussions, office meetings, training workshops, and outreach events.

In order to triangulate and generalize beyond case-specific results, the following six-part strategy was implemented. First, the study of a major operational controversy, which Daipha was fortunate to observe unfold on the ground, proved critical for making visible the normally hidden cultural, infrastructural, and institutional processes that sustain and regulate National Weather Service (NWS) systems of expertise. Second, broader conclusions about NWS forecasting practice were elicited by leveraging insights from an earlier comparative ethnography of NWS office culture in the US Midwest (Fine, 2007). Third, ethnographic interviews were conducted at the NWS Eastern Region Headquarters. In addition to conducting interviews with the director, deputy director, and seven department heads, Daipha sat in on the daily regional weather briefing and a senior staff meeting. These interviews were intended to ascertain the external validity of results to all 23 forecast offices of the region. Fourth, to explore the generalizability of results across the weather industry, Daipha conducted interviews with eight broadcast meteorologists stationed in Neborough's biggest media market as well as brief interviews with two private forecasters from Meteorlogix, Inc. during their visit at the Neborough office.

Fifth, Daipha conducted 59 interviews in two Neborough fishing communities with one or more commercial fishers participating in each interview. Given that scientific credibility hinges on how its claims of expertise fare downstream by science consumers (Gieryn, 1999), it was critical to examine how weather forecasting is shaped by—and, in turn, shapes—forecast use and needs. Finally, Daipha also administered a small-scale survey to Neborough recreational boaters ($N=110$). This choice of forecast user community was determined by the fact that, unlike most other publics, mariners directly and consistently evaluated NWS forecasting expertise because only the NWS issued publicly available marine forecasts.

Results

Figure 38.3 summarizes study results for the process of expert weather forecasting at the NWS, while Figure 38.4 presents a more general conceptual schematic for the process of expert decision making. Note that the unit of analysis is the decision-making task at hand, which provides the empirical context for prospective action. In the case of expert weather forecasting at the NWS, the particular forecasting task *frames* decision-making action. The problem to be solved is the weather situation at hand, and the objective is a good weather forecast, in accordance with specific, NWS-mandated criteria of accuracy and timeliness. As this model makes clear, weather forecasting skill cannot be properly understood without reference to its institutional and organizational context. Operating in the institutional environment of the NWS, forecasters have equipped themselves with particular notions of what counts as a weather risk, particular resources and heuristics for predicting such a risk, and a particular sense of what constitutes a good weather prediction. Tellingly, Daipha's study found that NWS forecasting operations are characterized by an exceptionally strong culture of apprenticeship: one must rise through the NWS operational ranks to become senior forecaster; general forecasters are always paired with a senior forecaster during a shift, and forecasting staff have ongoing training requirements on new or improved tools and technologies. Notwithstanding the variability in forecasters' stock of heuristics and techniques depending on their early career training, socialization, and mentorship experiences (Daipha, 2010), being an effective NWS forecaster effectively means to master forecasting the NWS way.

While the importance of the institutional environment cannot be overstated, it is within the microcontext of the forecasting task that weather forecasting proficiency is forged and displayed. The indeterminacy of weather dynamics and the need for situational awareness resist the trend toward rationalization and standardization.

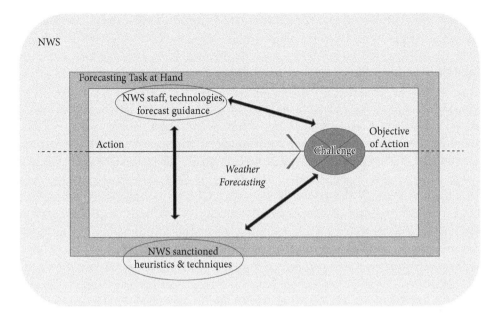

FIGURE 38.3 Conceptual schematic of the process of expert weather forecasting.
Adapted from Phaedra Daipha, *Masters of Uncertainty: Weather Forecasters and the Quest for Ground Truth*, p. 119, © 2015, University of Chicago Press.

To maintain situational awareness of the evolving weather situation, NWS forecasters have had to cultivate an omnivorous appetite for information, routinely using their own bodies as an additional weather instrument. Such adaptive strategies of *disciplined improvisation* converge into different organizational forms and systems of expertise depending on the temporal risk profile of the task at hand (Daipha, 2015b: 138ff.). The threat of a rapidly evolving weather risk such as a tornado, for example, requires all eyes on deck and restructures the forecasting staff into a *severe weather warning team* that includes a storm coordinator, several forecasters on radar duty, a mesoscale analyst, and a ham radio coordinator. Conversely, the threat of a blizzard requires paying deference to model guidance and, being a long game, is handled by consecutive regular forecasting shifts.

Notably, and regardless of the specifics of the task at hand, forecasting skill is distributed across physical, virtual, and interpersonal space—both in practice and by design. Throughout the weather forecasting profession, skepticism as to whether uncertainty management should, or indeed can, be a process that begins and ends in the mind is quite palpable, with the hub of cognitive activity prominently reassigned to the forecasting workstation instead. To capture expert weather forecasting in action, Daipha therefore closely analyzed NWS forecasters' *screenwork* (that is, their processing of information via screen devices) and traced when, which information bits mattered, and how. Conceptually, the distillation of weather complexity was theorized as a process of collage: as a mental modeling process but also a *digital* process of assembling, appropriating, superimposing, juxtaposing, and blurring available information (Daipha, 2015a). In turn, conceptualizing the distillation of complexity as the art of collage offered a useful shorthand for the process of skill acquisition, given that mastering a skill hinges upon one's ability to evaluate available information in progressively more contextual, intuitive, and holistic terms (Dreyfus, 2004).

After years of observing NWS forecasters with varying experience, ranging from a newly hired intern with a master's in meteorology to a 20-year veteran, Daipha's study thus concluded that what distinguishes experts is not just a deeper causative understanding of the weather situation at hand but rather a superior practical creative ability to produce a meaningful forecast gestalt out of available information pieces. Figure 38.4 proposes an extension of this model to the process of expert decision making more broadly. Accordingly, decision making takes place at the iteratively concretized, co-constitutive interface of one's applicable stock of knowledge, available resources, and a given objective (be it a weather forecast, a medical treatment plan, or a terrorism risk assessment). And decision-making expertise is characterized by proficiency in the art of collage; namely, by a masterful practical ability to fashion an adaptive solution to the challenge at hand by assembling, appropriating, superimposing, and juxtaposing available information without getting overwhelmed by guidance or the complexity of the situation.

These findings were further buttressed by the analysis of the interviews and survey with Neborough mariners. Results revealed little appreciable difference in how professionals and lay persons arrive at a forecast decision, with Neborough fishers relying on multiple sources of information (e.g., the NWS forecast, weather charts, private sector forecasts, personal weather observations, etc.) to decide whether tomorrow is going to

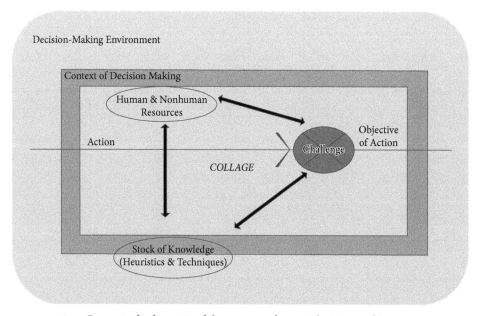

FIGURE 38.4 Conceptual schematic of the process of expert decision making.

be fishable. Furthermore, fishers were less concerned about improvements in NWS forecasting skill, instead requesting more frequent information updates and access to uncertainty information (Daipha, 2015b). Neborough mariners must remain flexible and alert to the evolving context of action if they are to successfully accomplish their task; they cannot afford to blindly heed professional advice. Treating the NWS forecast as one of many sources of weather guidance, they integrate information into a dynamic weather prediction that is meaningfully suited to their needs. Indeed, this study found some evidence that the more expert at distilling meteorological information Neborough fishers are, the more they hope for an inaccurate NWS forecast, to keep the competition at bay.

As we have explained, cognitivist and contextualist views and methodologies are complementary. LaDue examined forecaster learning in the organizational context and Daipha examined the production and consumption of expert weather forecasting in real-world settings. We now move a bit closer to the cognitivist pole to address the question of how to scale proficiency, and specifically what counts as individual expertise at forecasting, while retaining an appreciation for the fact that forecasting is always "in context."

Reasoning Styles and Strategies

Cognitive psychology research on forecasting expertise is motivated by recent efforts to improve models of expertise by linking it to naturalistic decision making (NDM; e.g., Salas & Klein, 2001). Research has employed methods of cognitive task analysis (CTA) to study forecaster knowledge and reasoning. In 1996, Pliske and her colleagues studied expertise in weather forecasting in the US Air Force (Pliske et al., 1997). Although the Air Force had recently made a large investment in new weather technologies to support forecasting, performance of their weather forecasters was perceived as declining rather than improving. Pliske et al.'s research was guided by the NDM perspective (Pliske & Klein, 2003), which studies how people make difficult decisions under time pressure and uncertainty within the context of team and organizational constraints. NDM researchers study experts to identify proficient reasoning practices in order to develop training or systems support to improve the performance of non-experts. NDM researchers define expertise by peer judgments as well as by quantitative performance measures (Kahneman & Klein, 2009).

Method

Pliske et al. (1997) used two specific CTA methods: the critical decision method (CDM; Crandall, Klein, & Hoffman, 2006) and the knowledge audit (KA; Militello & Hutton,

1998) (for a description of these methods see Chapter 19, "Incident-Based Methods for Studying Expertise," by Militello & Anders, this volume). The CDM scaffolds the expert in detailed recounting of a previous experience that was particularly difficult. Participants were asked to describe a time when they made a forecast and other people thought they were wrong, but it turned out they were right. After a good example was identified, the interviewer led the interviewee back over the incident to probe for salient cues, situational assessments, options evaluated, and other information regarding the incident. The KA utilizes probe questions that focus on the key aspects of reasoning that are known to typify expertise: situation awareness, perceptual skills, metacognitive skills, and anomaly recognition. One probe that was particularly useful for studying expertise in weather forecasting was the KA probe about relevant technology:

> Novices usually believe whatever the equipment tells them; they do not know when to be skeptical. Have there been times when the equipment pointed in one direction but your own judgment told you to do something else? Were there times when you had to rely on experience to avoid being led astray by the equipment?
> (Militello & Hutton, 1998, p. 1622)

In addition to conducting the CDM and KA interviews, Pliske and colleagues observed the forecasters' work environment and recorded the number of personnel on duty, the types and frequency of requests for weather information, and the types of equipment available.

Teams of two researchers (one note taker and one interviewer) interviewed 22 Air Force forecasters who worked at six different weather stations at a variety of geographic locations (e.g., Ohio, Florida, Alabama, Texas, and Colorado). The number of years of experience varied from 4 months to 21 years (median = 11 years). Most of the participants had not obtained a meteorology undergraduate degree, instead completing concentrated training by the Air Force.

Results

Analyses were based on the detailed notes made by the interview team and transcriptions of sections of interviews in which the participant described their forecasting process (see also Pliske, Crandall, & Klein, 2004). The four researchers who had conducted the interviews used an interactive card-sorting process. Working independently, they sorted the names of the forecasters into categories based on their perceptions of similarity of the forecasters' reasoning styles or approaches. After the initial sorting, the researchers compared and discussed their individual categories to achieve a consensus categorization scheme. The categories that emerged from the sorting process corresponded closely to the levels of the development of expertise described by Dreyfus and Dreyfus (1986). According to their stage model, the performance of novices is quite limited and is based on incomplete, rule-based knowledge. After they gain competency, they are able to become proficient performers who are able to integrate more abstract rules with more well-developed perceptual skills. However, when they are *experts* they no longer rely on analytic rules and demonstrate the highest level of performance characterized as being more fluid and flexible than the performance of those who are considered less proficient. The characteristics defining the categories are shown in Table 38.1.

The data summarized in Table 38.1 indicate that only two of the Air Force weather forecasters fell at the highest level of proficiency. Furthermore, there was not a clear linear relationship between years of experience and level of expertise. Additionally, 6 of the 22 forecasters were not included in Table 38.1 because the researchers could not reach consensus on category assignment.

Table 38.1 Results of card sort of characteristics of US Air Force weather forecasters

Category	Number of forecasters	Years of experience	Characteristics	Level of expertise
Disengaged	4	2–12	Limited knowledge base of rules; unmotivated to improve	Novice
Procedure-based mechanics	3	5–11	Limited knowledge of rules; always used same sources of information in same sequence; locally proficient; lacked motivation to improve	Advanced beginner
Procedure-based observers	3	2–8	Knowledge of rules insufficient to construct a useful mental model; lacked understanding of weather as a global system; keen observational skills; motivated to improve	Competent
Rule-based scientists	4	12–21	Extensive knowledge base of rules; high-level pattern recognition skills; integrated a wide variety of information sources; constructed a complete and useful mental model; used an analytic reasoning style	Proficient
Intuitive-based scientists	2	11–15	Used highly visual, dynamic mental models; high-level pattern recognition skills; flexible use of information sources depending on problem of the day; did not think in terms of rules	Expert

Data from Pliske et al. (1997) *Understanding skilled weather forecasting: Implications for training and the design of forecasting tools*, Technical Report 1L/HR-CR-1997-0003 for Armstrong Laboratory: Brooks AFB, TX.

In order to gain additional insight into the cognitive skills involved in expert weather forecasting, Pliske et al. (1997) conducted small group interviews with 13 NWS forecasters who had been assembled to predict the weather for the 1996 Olympic games in Atlanta, Georgia. The number of years of forecasting experience for this group varied from 3 to 25 years. Many of these forecasters described a forecasting process in which they initially generate a visual mental model of the atmosphere and then examine multiple computer models to check for similarities and differences with their own mental model. They also described how they would identify the problem of the day to focus their information-gathering activities and avoid information overload. Most of these forecasters described flexible and fluid strategies that would place them at the highest level of expertise in the Dreyfus and Dreyfus (1986) model.

Pliske et al. (1997) conducted content analyses to identify cognitive activities used by weather forecasters (e.g., noticing patterns, meaning making, etc.) and to determine critical factors in the development of expertise. They identified three factors as important for the development of expertise in weather forecasting: formal training opportunities, on-the-job training opportunities, and opportunities for feedback on specific forecasts. A major difference between the Air Force and NWS forecasters was that the Air Force forecasters had only 35 weeks of formal classroom training but the NWS forecasters had (at a minimum) completed a bachelor's level degree in meteorology. Only the most highly skilled Air Force weather forecasters mentioned they had been fortunate to have a mentor early in their careers to provide much needed on-the-job training. In contrast, the NWS requires forecasters to serve as intern forecasters for at least 2 years before qualifying to produce official forecasts. Most surprising to Pliske and colleagues was the lack of timely feedback provided to the Air Force weather forecasters about their weather forecasts. The Air Force weather forecasters stated that they typically only received feedback about specific forecasts when they had failed to issue a critical warning. These feedback sessions were perceived more as a type of punishment than an opportunity to learn.

Pliske et al. (1997) discovered that it was very difficult to identify genuine experts in the domain of weather forecasting due to the lack of operationally defined criteria of optimal performance. One might think it would be very straightforward to identify expert weather forecasters simply by discerning which forecasters are most often correct in their predictions. However, validating the accuracy of a forecast depends on how one defines *correct*. If a forecaster predicts fog that will lift by 0900 and it lifts at 0800, was that forecast wrong? Many forecasters reported that they were not judged (by peers or supervisors) as making bad forecasts as long as they had a sound rationale for their forecast. Forecasters often expressed their belief that weather is still hard to predict given the current state of knowledge of the atmosphere and current technology.

Pliske and colleagues examined forecasting in context to address the question of what counts as cognitive proficiency and expertise in forecasting. Moving a bit further out on the contextualist-cognitive continuum, Hoffman, Coffey, Ford, and Novak

(2006) drilled down on the claim that individual forecasters possess extensive knowledge and the claim that forecasting can be captured by a specific, single model of *the* forecasting task.

Modeling the Knowledge and Reasoning of Expert Forecasters

Researchers in the areas of cognitive systems engineering and NDM have developed methods for helping experts express their tacit and explicit knowledge, and for describing their reasoning strategies. Applying these methods to map the knowledge of domain expert weather forecasters has been no exception. Studies have included military, civilian, and private sector forecasters. The topics of studies range from reasoning in the warning task (Hahn, Rall, & Klinger, 2003) to forecaster adaptation of their knowledge and work processes (Hoffman, LaDue, Mogil, Roebber, & Trafton, 2017). Risk communication studies have examined the mental models of weather warning professionals during hurricane or flash flooding scenarios (Bostrom et al., 2016; Morss, Demuth, Bostrom, Lazo, & Lazrus. 2015). Reasoning strategies of warning forecasters have also used eye tracking to identify productive and less useful data interrogation strategies (Bowden, Heinselman, Kingfield, & Thomas, 2015).

The representation of weather forecasters' mental models can be achieved through CTA (Klein & Hoffman, 2008). What is lost in expert system-wide analytical breadth is gained in expert forecaster-specific analytical depth. Studies of the severe weather warning task at the US NWS thus trace the development of a forecaster's mental model of the storm as the decision-making task unfolds; and they point to dramatic regional and seasonal differences in the mental models of experts from different parts of the country (e.g., Hahn et al., 2003). Yet, importantly, cognitive task analyses of weather forecasting operations also underscore the limitations of mental models to fully capture the workings of expertise. Knowledge is inherently dynamic, for many reasons. Reasoning strategies and methods morph, if only because the technology is continually advancing. Thus, it is understood that individual models of either knowledge or reasoning are transient and brittle to change of context. When weather forecasters achieve expertise, they do not get locked into a rigid mental framework of how a storm will develop but, instead, are able to discerningly adapt or even replace their existing mental model in line with unusual or surprising information.

Hoffman et al. (2006) conducted an extensive study of forecasting at the Naval Meteorology and Oceanography Training Facility at the Pensacola Naval Air Station. The research involved cognitive interviews similar to those utilized by Daipha and LaDue, looking at the forecasting context and forecaster training. In addition, there was a concerted effort to create an objective proficiency scale, spanning the trainee, journeyman, and expert levels of proficiency and relying on performance data as well

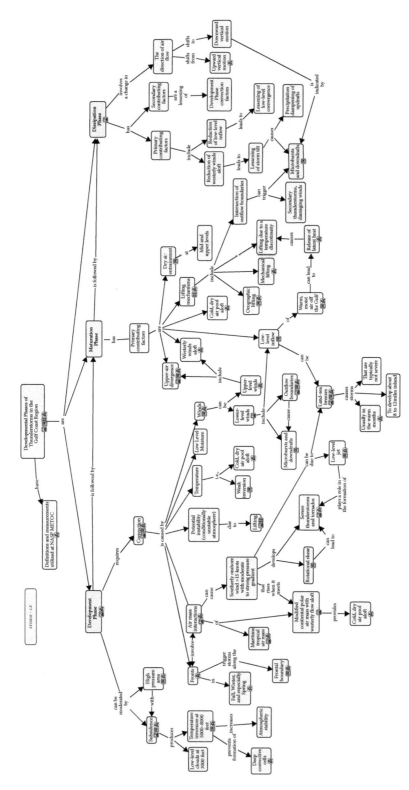

FIGURE 38.5 A conceptual diagram of expert knowledge about thunderstorms in the US Gulf Coast region.

as career interviews. Figure 38.5 illustrates the extent of expert knowledge by drilling down on the topic of the developmental phases of thunderstorms in the Gulf Coast region of the US. The icons under the nodes in the diagram connect each hyperlink to another conceptual diagram that goes into detail about the concept that is named in the node. As one might surmise from this, the full attempt to model expert knowledge resulted in many dozens of conceptual diagrams, covering all sorts of weather, weather technologies (e.g., radar), seasonal regimes, and other elements to forecasting weather in the US Gulf Coast region. Hoffman et al. (2006) estimated that the conceptual mapping of expert knowledge broadly would take literally hundreds of conceptual diagrams. This finding highlights the urgency and importance of capturing the knowledge of the *boomer generation* of senior forecasters who are close to retirement (see Hoffman & Hanes, 2003).

Figures 38.6 and 38.7 show models of two forecasters' process for the *forecasting problem of the day*. These are related to Daipha's Figures 38.3 and 38.4, which contextualize forecasting for *the problem at hand*. Indeed, the models depicted in Figures 38.6 and

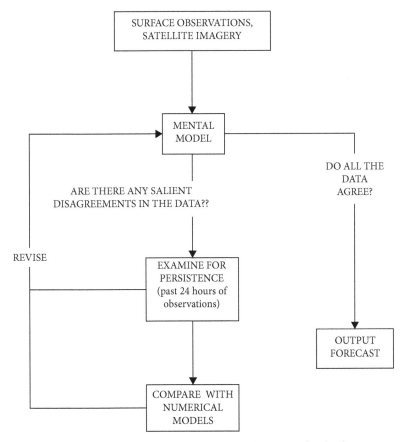

FIGURE 38.6 A conceptual model of a junior journeyman's process for the *forecasting problem of the day*.

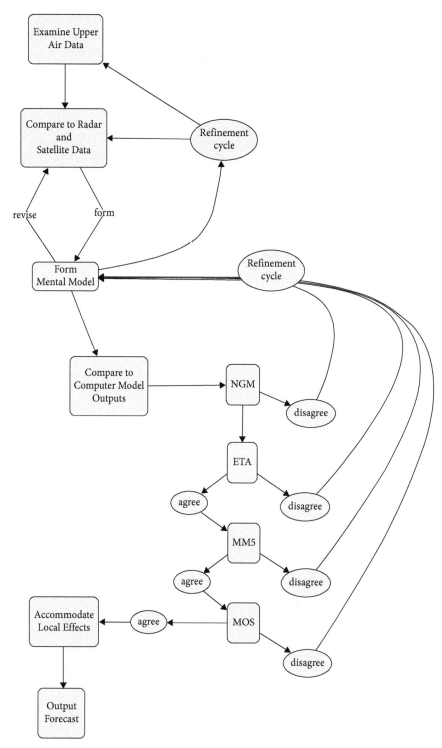

FIGURE 38.7 A conceptual model of an expert's process for the *forecasting problem of the day*.

38.7 can be seen as individualist/cognitive specifications on Daipha's contextualization. And at the same time, Figures 38.6 and 38.7 illustrate differences between less proficient (junior journeyman) and more proficient (senior expert) reasoning. Journeymen (Figure 38.6) were more likely to over-rely on the output of the computer models and are more likely to create their forecast by starting with the previous one. Experts (Figure 38.7) relied more on their own process of mentally integrating and sensemaking the data, an ability to engage their heuristics concerning local effects and trends, and an awareness of the limitations of individual computer models (NGM, ETA, MM5, and MOS are all computer model outputs).

Conclusions

The inductive generation of the conceptual categories of weather forecasters (the career interviewing conducted by LaDue), the analysis of organizational cultures via ethnographic interviews (the work of Daipha), the use of structured interviews for proficiency scaling (the work of Pliske), and the generation of knowledge models (the work of Hoffman) are all variations on a theme of empirical research on professional activity in context. One of the strengths of these methods is their potential for invigorating the study of expertise by revealing the importance of training, organizational context, work experience, and situational complexity.

The study of weather forecasters' use of technology, their self-learning, their mechanisms for dealing with uncertainty and risk communication, their knowledge and reasoning, their achievement of expertise by training and mentoring in context, and their adaptation to particular forecasting challenges, all draw expertise studies in critically important directions.

Human–Machine Interdependence

Weather forecasting, by its very nature, is an act of prospective reasoning. Pliske et al.'s (1997) disengaged forecaster (novice) relies on rule-based methods and is often outperformed by weather forecasting models. Models have made such advances they have lulled forecasters into a lowered state of vigilance at inopportune times (Bosart, 2003). Expert forecasters identified by Pliske et al. as intuitive-based scientists instead are known to engage in anticipatory modeling of future possibilities (Andra, Quoetone, & Bunting, 2002; Klein 1998). They are thus prepared and able to provide critical corrections and decision support to users of the forecast. The act of prospection strongly shapes an expert's cognition, risk perception, affect, and action (Baumeister & Vohs, 2016).

The resulting state is one of a human–computer interdependence that could be better exploited in the domain of weather forecasting. For example, the US National Weather Service hesitates to dramatically alter forecasts when numerical guidance provides drastically different solutions run-to-run (e.g., cold sunny skies forecasted for an area is shown as a snowstorm in the next model run). They instead favor a deliberate stepping toward the new solution in subsequent forecast products, for in many cases the model forecast changes again (Hitchens & Brooks, 2014).

Conveying Uncertainty

The limited predictability of weather sometimes results in these divergent solutions until weather systems are better sampled over the continental United States. Forecasters could be provided more tools to reveal the underlying causes of divergent solutions in model forecasts, thus enabling expert forecasters to anticipate an appropriate range of future possibilities. Only then can forecasters provide expert decision support to states, communities, and various types of emergency planning groups who perform best when they have accurate information on high-impact weather a week or more in advance.

Meteorological prospection is amenable to research in forecast offices, even if largely hidden to outside observers by deterministic forecast products that do not express uncertainty. That said, after many high-impact weather events wherein forecast uncertainty was not conveyed adequately (a recent example is described in Samenow, 2015), the US National Weather Service is increasingly unmasking its uncertainty in forecast products (Novak, Bright, & Brennan, 2008) and is creating entirely new forecast products that embrace uncertainty (Karstens et al., 2015). Current and future technologies could easily parse forecasts as more complex, user-defined, and situationally dependent forecasts (Rothfusz, Karstens, & Hilderbrand 2014). Little is known about what private sector companies do for their clients but we surmise they are also engaging in such advances according to their clients' needs.

Putting Forecasting Expertise in Context

The study of weather forecasting enables, if not encourages, simultaneous study of expertise in interdependent groups, and this has at least two interesting qualities. First, it allows the study of the interactions and interdependencies of expert and non-expert judgments, as well as how each group grants, or not, a judgment of expertise on members of the other group. For example, Daipha (2015b) studied the forecaster–mariner dyad for marine forecasting, while LaDue and colleagues conducted high-fidelity simulations of the forecaster–emergency manager–broadcast meteorologist triad to study the potential impacts of a new severe weather warning paradigm for tornadoes, wind, lightning, and hail (LaDue et al., 2017). Second, weather forecasting simultaneously encompasses organizationally sanctioned predictions wherein individual judgments are tempered (e.g., the coordination between NWS forecast offices and national centers) and disorganized predictions, where various actors in the weather enterprise sometimes make differing forecasts that cause confusion (e.g., Mersereau, 2015).

Translating the Results of Expertise Studies into Training Applications

Studies of weather forecasting expertise have shown that without comprehensive educational and mentoring programs designed to facilitate the achievement of proficiency, trainees can fail to achieve expertise (the work of Pliske and colleagues). Without an operationalized understanding of how expertise develops, each professional's learning becomes strongly mediated by the peculiar, unique social interactions in individual forecasting offices (LaDue's poorly supported forecasters) and institutional, cultural, and technological context in which they work (Daipha's work). On the other hand, some weather forecasters not only achieve expertise but actively enhance the state of the science. How can more of the latter, and less of the former, be widely achieved?

The meteorology community has proven to be resourceful, scraping together efforts such as the University Corporation for Atmospheric Research's world-wide education and training program (http://www.comet.ucar.edu) and many small learning projects. These provide any weather forecaster who possesses a strong identity with free or small-fee resources with which to build their expertise even if they have poor social support and widely varying local expertise from which to learn. A reliance upon these resources, however, points to systemic neglect and dependency on the gumption of a sufficient number of weather forecasters so as to maintain some level of value for the weather forecasting enterprise. In order to properly facilitate the building of expertise there must be sustained efforts to help individual weather forecasters with resources supporting their perceived learning needs. Critically important, there must also be efforts to help forecasters identify learning needs they may not have recognized for themselves.

Weather forecasting is a challenging job to learn, with many mental models of processes necessary to account for particular types of weather being forecast, as well as the time scales upon which that type of weather evolves. We have shown that some forecasters may be predisposed to a reasoning style that allows them to somewhat independently learn how to manipulate knowledge so as to apply it to novel problems, and that a strong sense of passion and personal identity as a forecaster enable a capacity to thrive in almost any environment. Even these professionals, however, clearly benefit from strong mentorship by senior forecasters who have achieved expertise and/or an environment that promotes training and upskilling. All forecasters learn and perform better and more efficiently in this way.

Further, expertise itself involves an ability to flexibly apply concepts to problems for which there is no clear, best approach to a solution (Schön, 1987). Techniques for operationalizing this skill into educational programs are advanced by Schön himself, van Merriënboer and Kirschner (2007), the National Academy's Institute of Medicine, and its basis in medical education literature (Davis, Barnes, & Fox, 2003). Techniques for capturing and making this complex knowledge accessible to younger forecasters have been attempted on a regional scale (Hoffman et al., 2006), but programs like these suffer systemic underfunding, leading to a limited set of knowledge models and scant hard evidence of their efficacy.

Finally, these research insights on the temporal and contextual properties of weather forecasting expertise can be used to inform the development and use of operational weather forecasting tools and guidance (e.g., Heinselman et al., 2015). It is critically important to conduct high-fidelity testing of current and developmental systems to discover and mitigate potential operational challenges. The National Oceanic and Atmospheric Administration currently does this to some extent through several test beds and a proving ground facility (http://www.testbeds.noaa.gov). This testing, however, must extend across temporally varying weather event types, with multiple offices, with partners and publics included, if we are to achieve a holistic understanding of how meteorological intuitive and deliberative thinking would interact in practice, under different time constraints, in particular sociopolitical contexts. Only then can the resulting tools transcend particular forecast challenges and support weather forecasters across their spectrum of expertise development.

REFERENCES

Andra, D. L., Quoetone, E. M., & Bunting, W. F. (2002). Warning decision making: The relative roles of conceptual models, technology, strategy, and forecaster expertise on May 3, 1999. *Weather and Forecasting 17*, 559–566.

Barbaro, M. (2007, December 2). Meteorologists shape fashion trends. *The New York Times*. Retrieved from https://www.nytimes.com/2007/12/02/business/02weather.html/

Baum, W. A. (1975). The roles of universities and weather services in the education of meteorological personnel. *Bulletin of the American Meteorological Society 56*(2), 226–228.

Baumeister, R. F., & Vohs, K. D. (2016). Introduction to the Special Issue: The science of prospection. *Review of General Psychology 20*(1), 1–2.

Bosart, L. F. (2003). Whither the weather analysis and forecasting process? *Weather and Forecasting, 18* 520–529.

Bostrom, A., Morss, R. E., Lazo, J. K., Demuth, J. L., Lazrus, H., & Hudson, R. (2016). A mental models study of hurricane forecast and warning production, communication, and decision-making. *Weather, Climate, and Society 8*, 111–129.

Bowden, K. A., Heinselman, P. L., Kingfield, D. M., & Thomas, R. P. (2015). Impacts of phased array radar data on forecaster performance during severe hail and wind events. *Weather and Forecasting 30*, 389–404.

Bureau of Labor Statistics (2018). Atmospheric scientists, including meteorologists. *Occupational outlook handbook*. Retrieved from https://www.bls.gov/ooh/Life-Physical-and-Social-Science/Atmospheric-scientists-including-meteorologists.htm

Cambrosio, A., Limoges, C., & Hoffman, E. (1992). Expertise as a network: A case study of the controversies over the environmental release of genetically modified organisms. In N. Stehr & R. V. Erickson (Eds.), *The culture and power of knowledge* (pp. 341–361). Berlin: Walter de Gruyter.

Charmaz, K. (2006). *Constructing grounded theory: A practical guide through qualitative analysis*. Thousand Oaks, CA: SAGE.

Collins, H. & Evans, R. (2002). The third wave of science studies: Studies of expertise and experience. *Social Studies of Science 32*, 235–296.

Collins, H. & Evans, R. (2007). *Rethinking expertise*. Chicago: University of Chicago Press.

Corbin, J., & Strauss, A. L. (2008). *Basics of qualitative research*, 3rd edn). Los Angeles, CA: SAGE Publications.

Crandall, B., Klein, G., & Hoffman, R. R. (2006). *Working minds: A practitioner's guide to cognitive task analysis*. Cambridge, MA: The MIT Press.

Daipha, P. (2010). Visual perception at work: Lessons from the world of meteorology. *Poetics 38*, 150–164.

Daipha, P. (2015a). From bricolage to collage: The making of decisions at a weather forecast office. *Sociological Forum 30*, 787–808.

Daipha, P. (2015b). *Masters of uncertainty: Weather forecasters and the quest for ground truth*. Chicago, IL: University of Chicago Press.

Davenport, C. (2014, January 23). Industry awakens to threat of climate change. *The New York Times*. Retrieved from https://www.nytimes.com/2014/01/24/science/earth/threat-to-bottom-line-spurs-action-on-climate.html

Davis, D. A., Barnes, B. E., & Fox, R. D. (Eds.). (2003). *The continuing professional development of physicians: From research to practice*. Chicago, IL: American Medical Association.

Dreyfus, H. (1972). *What computers can't do: A critique of artificial reason*. New York, NY: Harper and Row.

Dreyfus, H. L., & Dreyfus, S. E. (1986). *Mind over machine: The power of human intuitive expertise in the era of the computer*. New York, NY: Free Press.

Dreyfus, S. E. (2004). The five-stage model of adult skill acquisition. *Bulletin of Science, Technology and Society 24*, 177–181.

Dunn, W. R., & Hamilton, D. D. (1986). The critical incident technique—a brief guide. *Medical Teacher, 8*, 207–215.

Fine, G. A. (2007). *Authors of the storm: Meteorologists and the culture of prediction*. Chicago, IL: University of Chicago Press.

Fox, R. D., Mazmanian, P. E., & Putnam, R. W. (Eds.). (1989). *Changing and learning in the lives of physicians*. New York, NY: Praeger Publishers.

Gieryn, T. F. (1999). *Cultural boundaries of science: Credibility on the line*. Chicago, IL: University of Chicago Press.

Hahn, B. B., Rall, E., & Klinger, D. W. (2003). Cognitive task analysis of the warning forecaster task. Report No. RA1330-02-SE-0280. Retrieved from http://citeseerx.ist.psu.edu/viewdoc/download?doi=10.1.1.536.9648&rep=rep1&type=pdf

Heinselman, P., LaDue, D., Kingfield, D. M, & Hoffman, R. R. (2015). Tornado warning decisions using phased array radar data. *Weather and Forecasting 30*, 57–78.

Henderson, J. J. (2016). *To err on the side of caution: Ethical dimensions of the National Weather Service warning process*. Doctoral dissertation, Virginia Polytechnic Institute and State University, Blacksburg, VA.

Hitchens, N. M., & Brooks, H. E. (2014). Evaluation of the Storm Prediction Center's convective outlooks from day 3 through day 1. *Weather and Forecasting 29*, 1134–1142.

Hoffman, R. R., Coffey, J. W., Ford, K. M., & Novak, J. D. (2006). A method for eliciting, preserving, and sharing the knowledge of forecasters. *Weather and Forecasting 21*, 416–428.

Hoffman, R. R., & Hanes, L. F. (2003). The boiled frog problem. *IEEE Intelligent Systems 4*, 68–71.

Hoffman, R. R., LaDue, D., Mogil, H. M., Roebber, P., & Trafton, J. G. (2017). *Minding the weather: How expert forecasters think*. Cambridge, MA: The MIT Press.

Kahneman, D., & Klein, G. (2009). Conditions for intuitive expertise: A failure to disagree. *American Psychologist 64*, 515–526.

Karstens, C. D., Stumpf, G., Ling, C., Hua, L., Kingfield, D. M., Smith, T. M., ... Rothfusz, L. P. (2015). Evaluation of a probabilistic forecasting methodology for severe convective weather in the 2014 Hazardous Weather Testbed. *Weather and Forecasting 30*, 1551–1570.

Klein, G. (1998). *Sources of power: How people make decisions*. Cambridge, MA: MIT Press.

Klein, G., & Hoffman, R. R. (2008). Macrocognition, mental models, and cognitive task analysis methodology. In J. M. Schraagen, L. G. Militello, T. Ormerod, & R. Lipshitz (Eds.), *Naturalistic decision making and macrocognition* (pp. 57–80). Aldershot, UK: Ashgate.

LaDue, D., Karstens, C. D., Correia Jr., J., Ling, C., Hoffman, R. R., & Gerard, A. (2017, January). *Designing research to co-create a new paradigm for a continuous flow of information during severe weather*. Paper presented at the Seventh Conference on Transition of Research to Operations, Seattle, WA. Abstract retrieved from https://ams.confex.com/ams/97Annual/webprogram/Paper312903.html

LaDue, D. S. (2011). *How meteorologists learn to forecast the weather: Social dimensions of complex learning*. Doctoral dissertation, University of Oklahoma. Retrieved from http://proquest.com

Latour, B. (1987). *Science in action: How to follow scientists and engineers through society*. Cambridge, MA: Harvard University Press.

Lincoln, Y. S., & Guba, E. G. (1985). *Naturalistic inquiry*. Newbury Park, CA: SAGE Publications.

Mersereau, D. (2015, March 25). Do television meteorologists have a responsibility to prevent confusion? *The Vane*. Retrieved from http://thevane.gawker.com/do-television-meteorologists-have-a-responsibility-to-p-1693633861

Militello, L. G., & Hutton, R. J. B. (1998). Applied cognitive task analysis (ACTA): A practitioner's toolkit for understanding cognitive task demands. *Ergonomics 41*, 1618–1641.

Morss, R. E., Demuth, J. D., Bostrom, A., Lazo, J. K., & Lazrus, H. (2015). Flash flood risks and warning decisions: A mental models study of forecasters, public officials, and media broadcasters in Boulder, Colorado. *Risk Analysis 35*, 2009–2028.

National Centers for Environmental Information. (2018). *U.S. Billion Dollar weather and climate disasters*. Retrieved August 14, 2018, from https://www.ncdc.noaa.gov/billions/

National Research Council. (2010). *When Weather Matters: Science and Service to Meet Critical Societal Needs*. Washington, DC: The National Academies Press.

National Weather Service (2016). *Weather fatality, injury and damage statistics*. Retrieved July 9, 2017, from http://www.weather.gov/os/hazstats.shtml

Novak, D. R., Bright, D. R., & Brennan, M. J. (2008). Operational forecaster uncertainty needs and future roles. *Weather and Forecasting* 23, 1069–1084.

Pliske, R., Klinger, D., Hutton, R., Crandall, B., Knight, B., & Klein, G. (1997). *Understanding skilled weather forecasting: Implications for training and the design of forecasting tools*. Technical Report 1L/HR-CR-1997-0003, Armstrong Laboratory: Brooks AFB, TX. Fairborn, OH: Klein Associates.

Pliske, R. M., Crandall, B., & Klein, G. (2004). Competence in weather forecasting. In J. Shanteau, P. Johnson, & K. Smith (Eds.), *Psychological explorations of competent decision making* (pp. 40–68). New York: Cambridge University Press.

Pliske, R. M., & Klein, G. (2003). The naturalistic decision making perspective. In S. L. Schneider & J. Shanteau (Eds.), *Cambridge series on judgment and decision making. Emerging perspectives in decision making* (pp. 559–585). New York, NY: Cambridge University Press.

Rothfusz, L. P., Karstens, C. D., & Hilderbrand, D. (2014). Forecasting a continuum of environmental threats: exploring next-generation forecasting of high impact weather. *Eos* 95(9 September), 325–326.

Salas, E., & Klein, G. (2001). *Linking expertise and naturalistic decision making*. New York, NY: Psychology Press.

Samenow, J. (2015, 27 January). Why the snow forecast for New York City was so bad, and what should be done. *The Washington Post*. Retrieved from https://www.washingtonpost.com/news/capital-weather-gang/wp/2015/01/27/why-the-snow-forecast-for-new-york-city-was-so-bad-and-what-should-be-done/?utm_term=.042f6a026218

Schön, D. A. (1987). *Educating the reflective practitioner*. San Francisco, CA: Jossey-Bass.

Strauss, A. L., & Corbin, J. (1998). *Basics of qualitative research: Techniques and procedures for developing grounded theory*, 2nd edn. Thousand Oaks, CA: SAGE Publications.

Van Merriënboer, J., & Kirschner, P. A. (2007). *Ten steps to complex learning*. New York, NY: Routledge.

SECTION IV

DEVELOPING, ACCELERATING, AND PRESERVING EXPERTISE

SECTION EDITOR: PAUL WARD

CHAPTER 39

EXPERTISE FOR THE FUTURE

A New Challenge for Education

LAUREN B. RESNICK, JENNIFER LIN RUSSELL, AND FAITH SCHANTZ

INTRODUCTION

THIS chapter calls for a revived and reconsidered view of how individuals and organizations can acquire the intellectual expertise that will be needed to prosper in the twenty-first century, and of the changes in educational thought and practice that will be necessary to create such changes on a large scale. The social and intellectual skills needed to prosper in an information-rich environment include far more than mastery of particular bits of information and standardized procedures. They include the ability to reason about complex concepts, to explain these concepts to others, and to accept challenges to one's ideas as a normal and fundamental aspect of learning. By *reasoning*, we mean the ability to process, interpret, and evaluate information; to formulate hypotheses about how things work; and to investigate the ideas that lie behind the sometimes incomplete (and even contradictory) claims of peers, teachers, and authors. By *expertise* in reasoning, we mean the ability to draw fluidly and flexibly on the information at hand and on the complex set of skills and attitudes (including the willingness to change one's mind) that comprise reasoning.

Here, we will present current evidence on how these abilities can be created in children from a young age. We will also consider how teachers, and the organizations within which they function, can build the expertise necessary to support such learning in schools. We elaborate a concept of educational expertise that links the knowledge and skills of the different communities involved in education. Unless all of these "layers" of expertise are attended to, efforts to create, evaluate, and sustain new capacities in the future are unlikely to succeed.

We begin with a rather large claim: Virtually everyone can become more intelligent. The mind is capable of growing. Almost everyone can become an expert user of his or her brain. We base this claim on evidence, from a range of countries and educational institutions, that students who engaged in argumentation-based discussions with their classmates showed learning gains that went well beyond the topics they discussed (Resnick, Asterhan, & Clarke, 2015). Engaging in forms of classroom talk that involved formulating and supporting or challenging arguments appeared to develop their general abilities to reason (which can be seen as a process of internal argumentation). As we will show, this still small but growing body of research has upended traditional notions of how people learn, and how much they can learn. It has implications for how schools might be organized, what teachers should be expected to do, and what the very purpose of schooling should be (Resnick & Schantz, 2015). Because this argumentation-based form of teaching and learning has mainly been reserved for elite students, the evidence also raises ethical questions about opportunities to learn. Schools are potentially privileged settings for learning to reason, but many students never experience the kind of discussion and debate with peers that appears to develop internal processes of reasoning. In this chapter, which we write from an American perspective (although deeply influenced by an international body of research), we consider what must change in classrooms, schools, and school systems to make opportunities to create *expertise in reasoning* available to all.

In the first section of this chapter, we look at how expertise in reasoning can be systematically developed. To do this, we contrast two forms of teaching and learning: *recitation*, in which students give short answers to the teacher's questions, and *dialogue*, in which students argue and debate ideas with their classmates. We illustrate the distinction between recitation and dialogue by examining two early efforts by teachers and scholars to shift from recitation to dialogic teaching.

In the second section, we consider how *teachers* can develop their own expertise in dialogic teaching. Here, we look at how the surprising success of the early experiments led some researchers to design new forms of professional learning for teachers. We also discuss some of the challenges to making this form of teaching broadly available to students.

In the third section, we investigate how systems can build *organizational* expertise to support dialogic teaching and learning. Here, we examine the social and institutional structures within which teachers work. We present two promising examples of networks that share expertise within and among organizations.

Finally, we examine the issue of future research in the field. We call for shifting the focus from conducting tightly designed studies of individual classrooms to engaging educational systems rather than educators, and investing all parties in practice improvement.

Forms of Teaching and Learning

Most would agree that the tremendous resources of technology, and the vast social changes of the past century, have brought about the need for different skills for navigating our lives as citizens, workers, and social beings, compared to the past. Now, each individual must be able to solve non-routine problems, judge the worth of competing sources of information, engage in teamwork, and exercise personal judgment, among other skills. In broad terms, we need expertise in reasoning.

Following the work of Soviet psychologist Lev Vygotsky (1978), many learning scientists now believe that reasoning abilities are developed not by the lone brain digesting information, but through social interaction. In a related view, developed by the American philosopher and social psychologist George Herbert Mead, thought is "the conversation of the generalized other with the self" (1922/1964, p. 246; see also Mead, 1934). Thinking serves as a rehearsal for communication; therefore, thought typically occurs in relation to others. When skilled thinkers think through a problem, for example, they anticipate how others might challenge or accept their conclusions in a discussion. The idea of learning as a social act has implications that reach beyond formal schooling. Scholars now study reasoning-based talk patterns in everyday situations (Resnick, Säljö, Pontecorvo, & Burge, 1997), from family dinner table conversations (Pontecorvo & Fasulo, 1997) to online gaming forums (Gee, 2015).

Studies suggest that most people possess a certain amount of informal reasoning skill, but that the skills of sophisticated argument, e.g., making a counterargument or using an analogy, must be learned through *deliberate practice* (Kuhn, 1991; van Gelder, Bissett, & Cumming, 2004). For K–12 students, deliberate practice would consist of ongoing participation in argumentative talk, guided by a teacher and by norms for participation that promote dialogue. Please note: We do *not* mean that students use a formal language of explanation or reasoning. To the contrary, we are committed to a process that assumes participants are thoughtful listeners and speakers, but that they express their ideas in everyday language.

Schools *could* offer virtually all students plenty of practice in arguing and debating their ideas in the social space of the classroom. However, most schools and classrooms still are structured to provide students with practice in memorizing and recalling information. The difference, to a large extent, is in how teachers use the opportunity of talk, which is the main activity in most classrooms (Mehan & Cazden, 2015; Schwarz & Baker, 2016). Talk can be the mechanism through which students learn, or the mechanism through which learning is shut down.

Let us first look at the kind of daily learning opportunities most students are offered. We will use elementary school (primary or lower school in the UK) mathematics as an example because mathematics, a highly structured discipline, could be taught not only for its own sake, but also as a vehicle for learning to reason. The *inherited* curriculum in elementary school mathematics, however, is largely aimed at building computational expertise—the ability to quickly and accurately solve written arithmetic problems. Traditionally, the teacher writes a few sample problems on the board and invites students to provide answers. The teacher either approves or moves on to ask another student, searching at each step for a pupil who can provide the right answer and perhaps an explanation for how it was reached. This is an ancient form of group teaching, similar to religious catechism. It assumes that voicing the correct answer means the concept is understood.

In education research, this form of *recitation* (in any subject matter) has come to be known as IRE: initiation (question posed by the teacher), response (by the student), evaluation (again by the teacher). The evaluation move may consist of the teacher commenting, "Correct," or "Right," just nodding and moving on to the next problem, or moving to another student, implicitly judging the first student's answer as wrong. The teacher may circulate to help and evaluate individual students' work, but the bulk of teaching expertise lies in managing the whole class lesson—keeping children's attention, evoking correct answers, correcting answers that are not quite correct, and keeping up the pace so that the expected curriculum can be covered in the time allowed (which never seems to be quite enough for some children and is too much for others). The teacher does not expect to be challenged. He or she maintains control over what is considered to be knowledge, and what is considered to be worth knowing. In effect, the teacher does the thinking for the students. This form of teaching is still standard practice in most schools. It is based on a theory of learning that considers knowledge to be made up of *bits* of information, which are best learned through direct instruction and targeted practice.

Consider now a different form of learning, often called *dialogic*. In dialogic learning, students discuss ideas. They are expected to take and defend positions, and challenge other students' (and even teachers') propositions. Knowledge is assumed to be co-constructed. Dialogic learning is and has been mainly reserved for socially and economically privileged children.

However, evidence shows that dialogic teaching can produce learning gains for a variety of students (Resnick, Asterhan, & Clarke, 2015). We turn now to two examples of dialogic teaching that took place in average and low-performing schools. Both of these interventions involved teachers who were already experts in teaching using standard methods, but who felt those methods were failing their students. In both cases, the teachers collaborated with scholars to invent a new form of teaching.

Two Examples of Dialogic Teaching

One of the first such experiments in argument- and discussion-based teaching was introduced by Vic Bill, who was the math specialist in an elementary school serving predominantly low-income, African American families. Seeking new instructional techniques that she hoped would be more successful with her students, Bill spent a summer at our laboratory at the University of Pittsburgh. There, she worked with a group of international scholars, including visitors from Latin America who had discovered extraordinary arithmetic reasoning skills among *street children* in Brazil (Carraher, Carraher, & Schliemann, 1985). Bill thought that if unschooled children in Brazil could handle complex arithmetic problems, her students could also.

Gradually, Bill developed *story* problems with the potential to draw her students into a process of shared mathematical reasoning. The lessons began with whole-class discussion of a problem, followed by team-work and more discussion as each team reported their findings. Part of almost every math class was devoted to this type of lesson. She used the rest of the period for traditional activities such as timed drills and worksheets.

As one example of a reasoning-based lesson, Bill brought in a tray of cupcakes, and asked her second graders to consider whether they were arranged in three rows of seven or seven rows of three (Bill, Leer, Reams, & Resnick, 1992). Instead of asking them to find the total, she wanted them to figure out if they had enough for everyone in the class. Students jumped out of their seats to solve the problem, running back and forth to view the tray from all sides. They were working with the commutative property of multiplication (3 rows of 7 = 7 rows of 3), the relationship between repeated addition and multiplication, and the distributive property—all key mathematical concepts (although students used their own words to talk about them). During class discussions, they were being socialized into a kind of mathematical thinking that goes beyond a simple route to a solution.

Researchers who visited Bill's classroom saw dramatic change in the first semester (Resnick, Bill, Lesgold, & Leer, 1991). At the beginning of the school year, only about one-third of first graders could count to 100 or add small numbers. By December, most could add and subtract multidigit numbers, and at least half were using invented procedures that showed they understood the underlying concepts. When her students took a nationally normed test at the beginning of the second year, they performed near the top of the scale in mathematics. Their scores in reading rose also, though Bill did not teach reading. The researchers concluded that Bill had invented a new form of classroom discourse, and thus a new way of teaching arithmetic.[1]

Our second example, Project Challenge, involved many more students over a greater number of years. Project Challenge showed that teaching arithmetic as a form of reasoned conversation could have a wide impact on students' mathematics learning and more general school achievement. The intervention was designed and led by a highly skilled and dedicated teacher, Nancy Anderson, in partnership with scholars Cathy O'Connor and Suzanne Chapin (O'Connor, Michaels, & Chapin, 2015; Chapin & O'Connor, 2007). It took place in one of the lowest performing school districts in Massachusetts, among children from low-income families, many of whom spoke English as a second language.

---

[1] For a detailed description and analysis of the lesson, see "From Cupcakes to Equations: The Structure of Discourse in a Primary Mathematics Classroom" (Bill, Leer, Reams, & Resnick, 1992). For more information about students' learning gains, see "Thinking in Arithmetic Class" (Resnick, Bill, Lesgold, & Leer, 1991).

Project Challenge provided students with complex problems, projects, and arithmetic learning through games. Sociolinguist O'Connor and her colleague Sarah Michaels added training in the kind of argumentation that encourages thinking and reasoning (O'Connor, Michaels, & Chapin, 2015). During the training, the teachers learned to focus their lessons on mathematical ideas. They were provided with *talk moves* (discussed in more detail in the section "Developing Teachers' Expertise in Dialogic Teaching" and in Table 39.1) to help them guide discussions toward students' explanations of their own and others' reasoning. They were encouraged to practice *wait time*—waiting in silence while a student struggled to express an idea instead of quickly moving on. They learned to use wrong answers and misconceptions as opportunities to explore a concept in greater depth. And they developed norms for *respectful and equitable* participation, framed as students' rights (e.g., "You have the right to ask questions") and obligations (e.g., "You are obligated to agree or disagree with the speaker's comments and explain why") (Chapin & O'Connor, 2007).

In interviews, Project Challenge teachers described the changes they saw in their students. Students became better at listening to each other and building on one another's ideas, teachers said, and they were willing to engage in extended discussions of mathematical ideas. These new skills were not limited to a few students—they extended across most of the class (O'Connor, Michaels, & Chapin, 2015).

When Project Challenge students took the state test in the spring of the first year, more than half scored in the Advanced or Proficient score ranges (the top two quartiles), compared with only 38% in Massachusetts overall. After three years in the program, 82% of students scored in the Advanced or Proficient ranges, compared to only 40% for Massachusetts sixth-graders overall. In an additional study (also reported in O'Connor, Michaels, & Chapin, 2015), the researchers compared English Language Arts (ELA) scores on the state test for a smaller group of Project Challenge students with those of a control group. They found that Project Challenge students' ELA scores were also much higher than those of the control group, although the intervention took place only in math classes. Effect sizes for both math and ELA were over 1.1 using Cohen's *d*.

Expanding Research and Evidence on Dialogic Learning

The studies described in the previous section, "Two Examples of Dialogic Teaching," formed part of an early and small, but intriguing, body of evidence that thinking abilities could be taught systematically within the framework of traditional academic subject matters. Resnick and colleagues began to search for wider evidence of such teaching and its impact. We used a *recursive nomination* method in which we wrote to individual scholars, asked them to send us reports (published or not), and to suggest other scholars or working educators who might have findings relevant to our questions. By 2011 we had collected enough reports from a range of countries to warrant a rather startling claim: discussion/reasoning-based teaching often led to better initial learning and retention of the taught subject matter, *transfer* to other subject matters (one study even showed effects on a well-known non-verbal test of intelligence), and retention of learning gains for three years or more. For example, students in England who talked their way through puzzling science problems scored higher than their peers on the General Certificate of Secondary Education exams not only in science, but also in math and English (Adey & Shayer, 1990; 2015). And Scottish students scored higher than a control group on tests of both verbal and quantitative reasoning after talking about philosophical dilemmas in class (Topping & Trickey, 2015; Trickey & Topping, 2004). This collection of evidence became the basis for a conference and subsequent edited volume, *Socializing Intelligence through Academic Talk and Dialogue* (Resnick, Asterhan, & Clarke, 2015), published by the American Educational Research Association.

Structured Dialogic Learning in the Classroom

Based on this body of studies, we can describe the features of this approach to learning. It centers on a problem that requires collaboration. In whole-class discussions or in small groups, students think out loud: noticing something about the problem, questioning a surprising finding, or articulating, explaining, and reflecting on their own reasoning. The teacher works to elicit a range of ideas, which may be only partially formed in students' minds. To push a student to think further, the teacher might say, "I'm not sure I understand. Did you mean ... ?" Ideas, including the teacher's ideas, are treated as objects to be considered and shaped up in the public space of the discussion.

With teacher guidance, other students take up their classmates' statements. A teacher might ask, "Who agrees with Maria? Who disagrees?" or "Who can add to what Jon said?" Students learn to build on, clarify, or challenge a claim (including a teacher's claim); pose questions; reason about a proposed solution; and/or offer a counter claim or an alternate explanation. The reward for persisting is a *teacher-led* but *student-owned* process of shared reasoning that ultimately results in a more fully developed conclusion, solution, or explanation than any one student is likely to create on his or her own.

We came to call this form of discussion *Accountable Talk* (Michaels, O'Connor, & Resnick, 2008; Resnick, Michaels, & O'Connor, 2010).[2] Accountable Talk is so named because its features fall under three dimensions: accountability to *reasoning* (providing a rational justification for a claim), accountability to *knowledge* (getting the facts right even if it is a struggle to find the right wording), and accountability to the *learning community* (respecting the ideas and feelings of classmates).

---

[2] Accountable Talk is a registered trademark of the University of Pittsburgh.

Evidence continues to grow that dialogic learning has powerful social and cognitive benefits for students (e.g., Alexander, 2017; Asterhan & Schwarz, 2016; Education Endowment Foundation, 2015; Sun, Anderson, Perry, & Lin, 2017). More students participate when the goal is to build an argument rather than simply to state the right answer, because their emergent and half-formed ideas and statements are accepted as valid contributions. When they share their ideas, students are relieved of the obligation to speak formally, or to sound like the teacher, because the idea is seen as more important than its form of expression. Over time, students come to see discussion not as the occasion for displaying knowledge, but as a mechanism for creating it. Ultimately, students become sophisticated, flexible, and self-directed arguers. Their expertise in argumentation is an outward manifestation of an inner process of learning how to reason.

The evidence also suggests that students can begin to debate and argue from the first day of school, or even within their families before they enter school, without waiting until they have first learned the *basic facts* within a domain of knowledge (Kuhn & Udell, 2003, p. 1246). This does not mean that children should make any irrelevant statement that comes to mind. If discussion-based learning is entirely reserved for those who have already mastered the facts, however, then some students will never experience it. Students come into the classroom with varying levels of knowledge and skills, but virtually all students can say what they notice or wonder about a problem or a phenomenon, pose questions to one another, and make connections among ideas.

We have noted earlier the differences between recitation and dialogic learning. We also want to draw a distinction between dialogic learning and *discovery* learning. Scholars dating back at least to John Dewey (1916) have attempted to design ways for teachers to turn over the cognitive work to students through various forms of discovery learning. This form of learning often involves setting an authentic problem, or asking children to identify such a problem, and letting them figure out how to solve it. Broad evidence suggests that this *hands-off* approach works for some students (mostly those from highly educated families) but fails the large majority. By contrast, in dialogic learning, teachers set up the problems and engage students in dialogue about them (as we have shown), with the explicit goal of teaching students expertise in reasoning. And as we have presented here, dialogic learning shows promise for a wide variety of students.

Developing Teachers' Expertise in Dialogic Teaching

It should be clear from the descriptions we have given that teachers need a different kind of expertise to initiate and manage productive classroom discussions, compared to the skills of managing recitation. Teachers must be able to pose questions and set problems capable of evoking children's conceptions of the subject at hand. They must learn to listen to students' expressions of their ideas, no matter how ill-formed, and guide students toward shareable formulations of their thinking. And they must push discussions toward reasoning—eliciting a range of viewpoints, highlighting differences, and avoiding shallow resolutions. These skills must be backed by deep knowledge of the subject matter.[3]

To a large extent, this represents a *new* form of teaching expertise. The challenge, then, is how to develop it in many more teachers and for a wide range of students. To help teachers, researchers and working educators at the Institute for Learning (IFL) at the University of Pittsburgh analyzed the talk and broke it down into components that we called *talk moves*, which are illustrated in Table 39.1.

Table 39.1 Talk moves

Type of move	Teacher's utterance
Press for reasoning	"Why do you think that?" "What's your evidence?"
Add on	"Can anyone add to what Lily just said?" "Who can take that suggestion and push it a little further?"
Agree/disagree	"Who agrees with Jamal?" "Who disagrees?" "Are you saying the same thing as Jamal?"
Say more	"Can you say more about that?" "Can you give us an example?"
Revoice	"Let me see if I have your thinking right. Are you saying . . . ?"
Explain other	"Who can explain Lee's answer?" "Why do you think he said that?"
Restate	"Who can put in their own words what Samantha just said?"
Challenge	"Is this always true?" "Can you think of an example that wouldn't work?"

Data from Michaels, S., O'Connor, C., & Resnick, L. B. (2008), Deliberative discourse idealized and realized: Accountable talk in the classroom and in civic life, *Studies in Philosophy and Education* 27 (4), 283–297; and Resnick, L. B., Michaels, S., & O'Connor, C. (2010), How (well structured) talk builds the mind, in D. D. Press & R. J. Sternberg (Eds), *Innovations in educational psychology: Perspectives on learning, teaching and human development* (pp. 163–194).

---

[3] Of course, teachers also need a range of other skills, such as knowing how to set up norms for participation that go beyond "Sit silently until you're called on." They must be able to help students internalize the norms—e.g., students should come to believe they not only have a right to contribute their ideas and solutions, but also that they have a responsibility to do so. Teachers must be able to evaluate tasks for their potential for rich classroom discussion and problem solving. And they must be able to judge which formats for discussion—whole group, small group, pairs, or computer-based—are appropriate for any particular segment of a lesson.

In the (anecdotal) experience of IFL Fellows who worked with a range of schools and districts, teachers used these moves fairly mechanically at first. Even so, the moves worked to evoke a conversation in which students could expand their comments, elaborate on their ideas so that the teacher and other students understood them, and reason about their conclusions. No matter how the teachers used them, the moves changed the social norms of the classroom. Asking for students' ideas acknowledged that students *have* ideas, and that their ideas could be valuable contributions to the discussion. Students who were used to being told to follow a procedure without attention to why it worked, or answering a factual question about a story such as "Who is the main character?," now were being asked why their solution worked or did not work, or to use evidence from the text to defend their view of a character's motivation. Teachers noted students' surprisingly positive response to being asked to use their minds, and often commented on how intelligent their students turned out to be.

The moves also serve as an objective tool for analyzing a classroom discussion. Teachers, as well as researchers, can use the moves to track the development of their own discussion expertise.

Resistance to Dialogic Teaching

We now have evidence that when it is actually used, dialogic teaching can work to improve reasoning and academic performance of individual learners (Resnick, Asterhan, & Clarke, 2015). We know something about what kind of expertise teachers need, and we have tools (e.g., talk moves) they can use to shape new forms of discussion. But there is still little knowledge about how to spread this form of teaching to many more classrooms, schools, and districts. For one thing, we have not yet developed a systematic approach to training teachers. Rather, each investigator has developed his or her own training techniques in order to conduct their studies. The IFL offers districts and schools Accountable Talk training as part of its instructional improvement program. However, there appears to be no systematic evaluation yet of "what works" (and at what cost) in such training.

Also, when teachers and schools do agree to participate in these efforts, often they are unable to sustain dialogic teaching beyond the first year or even the first semester of the school year. These unsuccessful attempts have not been systematically investigated, but we can propose several explanations, both those that relate to individual teachers and those that relate to systems.

Perhaps most importantly, teachers may lack the necessary knowledge and skills to make the fundamental changes in teaching that are required. Few experienced this form of teaching during their own schooling, and most have not been prepared by their training and certification programs. They may have some idea of the benefits for students of talk and argumentation, but few have seen dialogic teaching in practice, much less had the chance to study with an expert practitioner during their teacher preparation programs. Compared to feeding students questions that have set answers, orchestrating productive discussions is much more difficult; learning how to do it requires extended practice and ongoing reflection. Teachers may not see immediate improvement in themselves, and they may not immediately see improvement in student outcomes. Research shows that the benefits of dialogic teaching are often delayed (e.g., Asterhan & Schwarz, 2007; Crowell & Kuhn, 2014; Howe, McWilliam, & Cross, 2005; Kapur, 2011, 2012; Schwartz & Martin, 2004). Dialogic teaching also requires deeper understanding of the subject matter to be taught than is now the norm. Especially in math and science, some teachers lack the necessary content knowledge to guide their students through the kind of discussions that result when students are encouraged to wonder aloud and share their observations.

To persist in challenging work that may only show results in the distant future, most individuals would need to be supported by a strong belief in its efficacy. However, teachers, deep down, may question whether their students are up to the task. Many teachers (as well as parents and the public at large) believe that differences in achievement can be accounted for by how much intelligence an individual has inherited, rather than by the kinds of learning they experience. In this view, intelligence is a fixed quantity, and there is not as much of it to go around as we might wish. If only a few people are thought to be able to reason at high levels, for the majority of students, dialogic teaching would be ineffective at best. In addition, many teachers (and parents, and the general public) believe that children must learn facts before they can engage in reasoning-based discussions. This view also reserves dialogic teaching for a select few.

At the system level, resistance to dialogic teaching is amplified. Without a serious commitment to new forms of education expertise, schools, school districts, and even state departments of education will (most often unintentionally) continue to reinvent current forms of teaching rather than explore and support more ambitious forms. Even in schools and districts that have agreed to participate in these efforts, teachers may find that their principals (head teachers in the UK) do not understand—and may even distrust—dialogic teaching, because they assume teaching is a matter of transmitting established information rather than engaging students in reasoning.

A further difficulty is that school districts usually allow teachers to opt in or out of professional development efforts, in part because training in the new methods involves time commitments that go beyond teachers' contractual responsibilities. This constraint also narrows the kinds of research that can be conducted. Virtually all of the research on reasoning-based teaching has been based on working relationships between scholars and teachers who volunteer (with their supervisors' approval) to participate. Typically, even the self-selected group of volunteers dwindles over time, in part because of accountability pressures. In the USA, teachers are now required to spend weeks or months preparing their students to answer short, factual questions on tests. For example, in a study of a coaching program that supported dialogic teaching in a large district, Matsumura and Garnier (2015) note that coaches described teachers as "less interested in working with them after Christmas, when the focus of their instruction shifted to 'test prep'" (p. 423).

There are also organizational structures within current school systems that interfere with the implementation of new instructional practices such as dialogic teaching. In the next section, we look at how systems resist instructional change, and consider how those systems themselves might be changed to support this form of teaching.

Building Organizational Expertise for Dialogic Instruction

In order to realize the vision of complex, dialogic instruction we have outlined, the educational field needs to devote serious attention to how organizations and systems function. This includes designing the social and organizational structures that help teachers learn to enact and sustain this type of instruction. By organizational expertise we mean an organization's collective knowledge and system-wide structures that would enable the creation, development, and sustainment of dialogic teaching expertise.

The traditional form of organizing public school systems and most large public institutions has been patterned on Weber's theory of bureaucracy, which posits that expertise can be embedded in hierarchical structures. The rise of bureaucratic administrative structures marked the transition from traditional authority (e.g., feudalism) to rational-legal authority structures (Scott & Davis, 2007; Weber, 1946). In a bureaucratic system, there is a fixed division of labor among positions or offices, and personnel are selected on the basis of technical qualifications. This was an important shift from traditional administrative forms where leaders could demand that subordinates perform any tasks and personnel were selected from those personally dependent on the leader. These and other elements of a bureaucratic system provided a more stable and predictable administrative structure for superiors and subordinates, and enabled subordinates to exercise greater independence and discretion. This transition to modern administrative structures enabled the building of the modern state and its public systems such as organized schooling.

However, the educational community has increasingly recognized the limits of Weberian bureaucracy when organizing for new visions of teaching and learning (Ogawa & Russell, 2010; Resnick & Spillane, 2006). As the purposes and nature of the work have evolved in education and other modern systems, critiques of bureaucratic structures have arisen. Particularly relevant to our argument are discussions of expertise and authority. As Scott and Davis note, "Weber argues that authority is based on technical competence" and control exercised on the basis of knowledge. However, organizational theorists have argued that there is a gap between technical expertise and authority in modern organizations. Authority is centralized but complex tasks require specialized knowledge that is built through prolonged training, which central leaders cannot have mastered given the diversity of expertise required (Thompson, 1967).

Furthermore, our current schooling structures were designed in the nineteenth century, when the purpose of education did not include pursuit of reasoning, argumentation, and conceptual understanding for all students (Ogawa, 2015). Centralized, hierarchical structures are well suited to tasks that are routine and that can be organized around compliance with standard operating procedures. For example, the division of labor in modern school systems separates curriculum selection and development from teaching, standardizes the monitoring of instruction through frequent testing, and creates demands for routine tasks that are not well aligned with instructional improvement such as lesson-planning practices that bring teachers together but ask for superficial engagement in the process.

We have mentioned contractual constraints that often do not allow schools to require teachers to participate in interventions. These constraints also affect how teachers' time can be allocated during the school day. Having their role circumscribed to providing instruction to students means teachers spend almost none of their contracted time in planning, reflection, or collegial interaction. Relatively low wages for professional work, coupled with (in many areas of the United States) unionized teaching contracts, reinforce the idea that teachers should not be working outside of their official work hours when they are directly interacting with students.

Yet, in order to make the transition to ambitious instruction, teachers will need to learn new practices. School leaders will need to learn how to incentivize and support this transition, and district and state leaders will need to learn to monitor instructional change and deploy resources to support the new practices. Current bureaucratic structures in schools and school systems are not well suited to managing the complex, adaptive systems we need to foster in educational organizations to support substantial learning at all levels of the system.

The educational organizations we need are designed to support systemic learning and ongoing improvement. Education research in the past three decades has begun to attend directly to organizational contexts and systemic structures, drawing on theory from organizational sociology, and identifying key features that contribute to instructional improvement. While not specifically focused on building the type of instructional expertise we describe in this chapter, this more organizationally focused research provides some guidance for thinking about redesigning education systems and building the kind of organizational expertise necessary to develop and sustain capacity for dialogic, reasoning-focused instruction at scale. In the following sections we explore some key insights from this growing body of organizational research.

Opportunities for Learning through Social Interaction

School systems have typically approached the promotion of new instructional practices through investment in teacher training (Birman, Desimone, Porter, & Garet, 2000;

Sandholtz, 2002). Traditional approaches to professional development tend to leverage expertise found outside of school systems by contracting with professional development providers, who typically deliver short-term workshops to large groups of teachers. For example, the New York City Public Schools spent nearly 100 million dollars from May 2011 to April 2012 on outside consultants who provide professional development (Fertig & Garland, 2012). On face value, this seems like a rational approach. However, research on teacher professional development suggests that changing instructional practice to support students reasoning and arguing about complex concepts requires more intensive support (Garet et al., 2001; Kennedy, 2016; Thompson & Zeuli, 1999).

Acquisition of the type of instructional expertise we describe in this chapter likely requires ongoing interactions with and guidance from more expert practitioners (Greeno, Collins, & Resnick, 1996; National Research Council, 2000; Putnam & Borko, 2000). One way to provide this is through instructional coaching. In our vision, coaches are educators with expertise in utilizing dialogic instruction and knowledge of how teachers learn. Skilled coaches use a variety of job-embedded approaches to support teacher learning and development. They work with teachers to plan for instruction, anticipating the likely misconceptions students will face when engaging in complex reasoning tasks, and planning questions to move students toward conceptual goals. Coaches can observe a teacher's instruction, and provide feedback and suggestions during or after a lesson. They can also model dialogic instruction in a teacher's classroom, so the teacher can not only see the practices in action, but also discover that their own students are capable of engaging in complex reasoning. A growing set of research studies suggests that coaches can support teachers in making significant shifts in their teaching practice (Campbell & Malkus, 2011; Kraft, Blazar & Hogan, 2016; Matsumura, Garnier, & Spybrook, 2012).

Another way to provide more intensive support for taking up dialogic instruction is to leverage the interactions that occur naturally among educators in schools. There is accumulating evidence that informal interactions among educators are critical to teacher learning and development (Bryk, Camburn, & Louis, 1999; Bryk, Sebring, Allensworth, Easton, & Luppescu, 2010; McLaughlin & Talbert, 2006). Teachers' informal collegial interactions are associated with the spread of new practices, transfer of complex knowledge, and reform uptake (Frank, Zhao, & Borman, 2004; Hansen, 1999; Obstfeld, 2005; Penuel et al., 2009; Reagans & McEvily, 2003). However, as we have noted, time for collegial interactions in schools is quite limited, often to one hour per week.

Attending to the social resources that teachers can access in school-based professional communities, such as expert coaching and ongoing interactions with other teachers, shifts the focus on teaching development to internal expertise in schools and the tacit, practical knowledge that educators develop about teaching and learning. This shift in perspective on professional learning attempts to make visible and support the development of both human and social capital in schools and systems. However, current ways of organizing schools do not create the conditions for this type of professional learning to occur regularly.

District Central Offices as Learning Organizations

In addition to more robust learning opportunities to build teacher expertise, we must consider how schools and school districts are organized to support the kind of professional transformation we envision. Individual teachers cannot carry the burden of enacting and sustaining ambitious teaching practices when systemic norms, routines, and practices reflect inconsistent priorities. Too often, American teachers face pressure to comply with mandates such as content pacing calendars that promote a focus on content coverage versus mastery. As we have noted, they also face pressure to focus considerable instructional time on preparing students for standardized tests that emphasize recall, procedural algorithms, and formulaic responses rather than thinking and reasoning. These pressures are endemic to our bureaucratically organized educational systems where authority and expertise are disconnected, and control is exercised through accountability for compliance with standardized procedures. Consequently, it is worth considering how we might redesign school districts to support a transition to dialogic instruction.

School districts are an important organizational unit in education because their leaders can design initiatives to implement a particular vision of teaching and learning, create structures and processes to support implementation, and provide management and oversight for change efforts (Rorrer, Skrla, & Scheurich, 2008; Stein & Coburn, 2008). Ideally, as central office administrators participate in school assistance relationships, they learn how to help schools take up new complex practices, and ultimately embed what they learn in central office policies and informal practices (Honig, 2008).

One way to conceptualize how organizations build expertise for better supporting ambitious instruction is the concept of organizational routines. In the sociological sense, routines are the repetitive and recognizable patterns of interdependent activity that characterize the work of professionals in organizations (Feldman & Pentland, 2003). It is important to note that this notion of organizational routines is not necessarily synonymous with static procedures that run counter to the notion of adaptation to complexity. While some routines are more procedural, other organizational routines facilitate collaborative work, which can in turn be a vehicle for learning and innovation. Routines play a key role in educational organizations by structuring educator practice (Sherer & Spillane, 2011). For example, Stelitano and colleagues' (2017) ethnographic research in high schools identified routines that determined how educators organized their work in order to try to meet the individualized needs of diverse learners. Many existing routines in high schools reinforced a notion that special educators are responsible for the learning of special education students, despite these students primarily attending general education courses. Consequently, teachers focused on reactive, triage-oriented strategies, like helping students make up assignments and providing extensive assistance during tests.

In contrast, there are accounts of learning-focused routines in literature and practice. A prominent example is *The Learning Walk*,[4] which was developed by the IFL and

---

[4] The Learning Walk is a registered trademark of the University of Pittsburgh.

utilized in New York City's District 2 under the leadership of Tony Alvarado and Elaine Fink. In this routine, teams of administrators and teacher leaders observed teaching in a school to identify priorities for instructional improvement (Stein & D'Amico, 2002). Critical features of the design reinforced collective responsibility for continuous instructional improvement, rather than focusing only on what teachers did or did not do. For example, teachers and administrators knew in advance when the Learning Walk would take place and what its focus would be, observers used protocols that encouraged them to make objective observations, and the stated purpose of the Learning Walk was to identify systemic supports for instruction, rather than to evaluate the skill of individual teachers.

Finally, in order to become learning organizations, schools and school systems will need to embrace routines that drive continuous improvement. In education, there is rising interest in the kind of continuous improvement work embraced in sectors such as healthcare that includes identification of high leverage problems of practice and engaging in short cycle inquiry routines that enable organizations to learn how to improve core processes (Bryk, Gomez, Grunow, & LeMahieu, 2015; Hannan, Russell, Takahashi, & Park, 2015; Lewis, 2015; Tichnor-Wagner, Wachen, Cannata, & Cohen-Vogel, 2017). In a continuous improvement model such as lesson study, for example, teachers engage in deep, collaborative work to improve lessons and core processes related to instruction (Lewis, 2015). Teachers are the agents of their own improvement.

While we know something about how to better organize systems for learning, some fixed structural constraints still challenge this vision. When the role of teachers is conceptualized strictly as classroom teaching, and teachers are compensated solely for their time with students, it is hard to imagine that even the best routines could enable the type of transformation of teachers' practice required to realize the vision of instruction we present. In places such as Japan, where teachers engage in robust routines for instructional transformation such as lesson study, teachers have considerably more time during their workday to engage in planning and reflection with colleagues (Lewis, 2002; Lewis, Perry, & Hurd, 2009).

In sum, we argue that in order to build the expertise necessary for teachers to orchestrate ambitious, dialogic instructional practices, we need to redesign educational organizations and systems. A critical take-away is that we need to see expertise in non-hierarchical ways, embedding different types of expertise in positions at the classroom, school, and district levels, and enacting routines that support learning at all levels of the system.

Interorganizational Mechanisms for Social Learning

In addition to school- and district-level structures to support reasoning-based learning, the field would also benefit from structures that enable schools and systems to learn from one another. This need is reflected in the field's interest in interorganizational networks that enable the social architecture for creating, evaluating, and sustaining new systemic capacities, and for accessing external expertise. We briefly outline two examples of continuous improvement networks in education.

One way to develop organizational expertise to support high-quality teaching and learning is to promote intentional, improvement-focused collaborations between researchers and practitioners, with the goal of generating new organizational and practical expertise. The education field is excited about the concept of research–practice partnerships in which educators become true partners in determining the practical problems that researchers address, as well as designing and testing instructional methods that promote student reasoning and argumentation (Coburn & Penuel, 2016; Penuel, Coburn, & Gallagher, 2013).

The Tennessee Math Coaching Project is an example of such a partnership (Russell et al., 2017). In this project, researchers at the University of Pittsburgh, professional development providers from the IFL, and leaders from the Tennessee Department of Education worked together to develop a model for mathematics professional development that can be a resource for districts throughout the state. The model focuses on the role of mathematics instructional coaches. Coaches leverage strong connections and substantive interactions with teachers to support their enactment of mathematics teaching practices that develop student reasoning and conceptual understanding. In addition to identifying the practices that coaches can utilize to support the shift to dialogic, reasoning-focused math instruction, the project also tested and refined tools and routines that build the organizational expertise needed to monitor and sustain instructional improvement at scale. For example, coaches identified scheduling routines that allowed them to preserve time for intensive work with teachers amidst competing demands from organizational leaders. Additionally, the team is developing monitoring tools and routines that enable district and state leaders to track the uptake of ambitious instructional practices and monitor and learn about quality coaching practice. This kind of systemic learning is supported by rapid analytic cycles that enable continuous improvement (Russell et al., 2015).

The second example is a model for interorganizational learning that aims to build a social architecture for addressing high-leverage problems of educational practice: the networked improvement community (Bryk, Gomez, Grunow, & LeMahieu, 2015; Russell et al., 2017). Networked improvement communities (NICs) provide a social, organizational, and technical structure for learning to occur among schools and districts. By bringing together different systemic actors with the expertise and authority to make changes in educational systems, NICs reshape the systemic structure of knowledge generation in education. For example, the Better Math Teaching Network is a networked improvement community aiming to promote student-centered math teaching and learning practices in Algebra I classrooms throughout New England. Teachers use systemic inquiry routines to test specific strategies for enacting dialogic instruction. They work in collaboration with school, district, and state leaders who are trying to better understand and enact systemic supports for this type of instruction.

Additionally, teachers interact with math education researchers and professional development experts in order to build their expertise. In this way, NICs bring together a diverse colleagueship of expertise in order to form an interorganizational learning community.

Next Steps

In this chapter, we have described how reasoning abilities can be developed in students through certain forms of classroom dialogue. We have considered the kind of expertise teachers need in order to teach dialogically, and we have examined some of the ways teachers have been supported in that endeavor. We have discussed a range of challenges to dialogic teaching, and presented promising examples of how systems can respond. We turn now to the forms of research that will be needed if we are to make dialogic teaching and learning more broadly available.

We believe that researchers and scholars can play a role in developing the individual and systemic expertise to enable all citizens to prosper in the twenty-first century, but as a field we must rethink the relationship between research and practice. Making the types of large-scale changes to educational systems that we describe in this chapter will not happen through accumulation of knowledge from a series of tightly designed experimental studies that is then disseminated to practitioners. Rather, we must directly address the implementation problems we face when intervening solely at the classroom level. Realizing this vision will require addressing two major issues. First, we must reshape the process whereby researchers engage with practitioners, shifting the focus to joint investment in practice improvement. Second, we need to engage educational systems rather than educators in order to promote continuous systemic improvement. We discuss each idea in turn below.

A number of emergent models for research–practice partnership in education aim to bring both sides into more productive, collaborative relationships. For example, the design-based implementation research (DBIR) approach has identified principles that help design-based researchers overcome the limitations faced when trying to move successful interventions, such as dialogic instruction, beyond individual classrooms (Fishman, Penuel, Allen, & Cheng, 2013; Penuel, Fishman, Cheng, & Sabelli, 2011). This approach emphasizes four core principles: (1) research teams that include researchers and practitioners form around persistent problems of practice from multiple stakeholders' perspectives; (2) to improve practice, teams commit to iterative, collaborative design; (3) to promote quality in the research and development process, teams develop theory related to both classroom learning and implementation through systematic inquiry; and (4) DBIR is concerned with developing capacity for sustaining change in systems. Drawing on the DBIR approach, we imagine researchers partnering with practitioners to develop, test, and implement the types of instructional practices and systemic supports that promote student expertise in reasoning.

Additionally, in order to promote this vision, we have argued that educational systems must be redesigned to enable teachers to learn to enact the pedagogical practices that promote student expertise in reasoning. In prior work Resnick has argued that in order to organize educational systems for continuous systemic improvement, we should strive to build *nested learning communities*. This idea, developed with engineer Mary Besterfield-Sacre and others, conceptualizes a school system as a series of nested layers, each of which shapes and is shaped by the other layers (Resnick, 2010). This line of work included the creation of a hypothetical flow model of processes that work throughout the system to enable or constrain the goal of producing student learning (Resnick, Besterfield-Sacre, Mehalik, Sherer, & Halverson, 2007). We return to nested learning communities in the concluding paragraph.

One example of an initiative that takes up these two ideas—the need for collaboration between research and practice and the engagement of nested systemic levels in continuous improvement and learning—is the Tennessee Early Literacy Network (TELN). TELN was launched by leaders in the Tennessee Department of Education who were thinking about new ways that states could support school districts to promote more ambitious instruction and student learning. The State Department is working with regional support offices, school districts, and school teams to increase the proportion of students who are proficient in literacy by grade three. Additionally, the network engages researchers within the State Department and beyond to create opportunities for practitioners to identify and implement ways to improve student literacy, and share the accumulated practical knowledge with the field. The structure of the network puts researchers and practitioners in close collaboration and builds their mutual commitment to achieving practical improvement. Additionally, the network is designed to engage the multiple systemic levels that influence student learning including classroom instruction, but also extending to the additional supports for struggling students provided by schools, and coherent district- and state-level instructional guidance. While TELN is not directly organized to address the aim of building expertise in reasoning, it provides a potential model for the way we might engage researchers and practitioners in systemic improvement toward this aim.

Conclusion

We have argued that current theories of learning and mental development suggest that deep learning is inherently a social process, dependent on reasoning, on sharing of ideas, and on argumentation. We have described how specific forms of reasoning can be developed in school classrooms. We have shown ways in which the institutional structures of education systems can support—but also hinder—expert teaching and learning. We have offered promising examples of networked systems and called for more fruitful and fluid partnerships among researchers and practitioners.

What will it take to bring reasoning-based teaching and learning to more classrooms? To commit to providing all students with the kinds of learning opportunities we have described, we believe both researchers and practitioners must engage in a sustained effort to analyze, understand, and learn from each layer of the *nest*. We must evaluate how each aspect of the system supports or constrains the goal of developing students' expertise in reasoning. We must determine how to modify the system to better support the goal, and build the will to make needed changes. The reward for such an effort will be more students who use their minds with confidence, and ultimately more adults prepared for the twenty-first century's civic, economic, and social demands.

References

Adey, P., & Shayer, M. (1990). Accelerating the development of formal thinking in middle and high school students. *Journal of Research in Science Teaching* 27(3), 267–285.

Adey, P., & Shayer, M. (2015). The effects of Cognitive Acceleration. In L. B. Resnick, C. S. C. Asterhan, & S. N. Clarke (Eds), *Socializing intelligence through academic talk and dialogue* (pp. 127–140). Washington, DC: American Educational Research Association.

Alexander, R. (2017). Developing dialogue: process, trial, outcomes. Paper for 17th Biennial EARLI Conference, Tampere, Finland, Symposium H4, 31 August 2017. Retrieved from http://www.robinalexander.org.uk.

Asterhan, C. S., & Schwarz, B. B. (2007). The effects of monological and dialogical argumentation on concept learning in evolutionary theory. *Journal of Educational Psychology* 99(3), 626.

Asterhan, C. S., & Schwarz, B. B. (2016). Argumentation for learning: Well-trodden paths and unexplored territories. *Educational Psychologist* 51(2), 164–187.

Bill, V. L., Leer, M. N., Reams, L. E., & Resnick, L. B. (1992). From cupcakes to equations: The structure of discourse in a primary mathematics classroom. *Verbum* 1, 2, 63–85.

Birman, B. F., Desimone, L., Porter, A. C., & Garet, M. S. (2000). Designing professional development that works. *Educational Leadership* 57(8), 28–33.

Bryk, A., Camburn, E., & Louis, K. S. (1999). Professional community in Chicago elementary schools: Facilitating factors and organizational consequences. *Educational Administration Quarterly* 35(5), 751–781.

Bryk, A. S., Gomez, L. M., Grunow, A., & LeMahieu, P. G. (2015). *Learning to improve: How America's schools can get better at getting better*. Cambridge, MA: Harvard Education Press.

Bryk, A. S., Sebring, P. B., Allensworth, E., Easton, J. Q., & Luppescu, S. (2010). *Organizing schools for improvement: Lessons from Chicago*. Chicago: University of Chicago Press.

Campbell, P. F., & Malkus, N. N. (2011). The impact of elementary mathematics coaches on student achievement. *Elementary School Journal* 111(3), 430–454.

Carraher, T. N., Carraher, D. W., & Schliemann, A. D. (1985). Mathematics in the streets and in schools. *British Journal of Developmental Psychology* 3(1), 21–29.

Chapin, S., & O'Connor, C. (2007). Academically productive talk: Supporting student learning in mathematics. In W. G. Martin, M. Struchens, & P. Elliot (Eds), *The learning of mathematics* (pp. 113–139). Reston, VA: National Council of Teachers of Mathematics.

Coburn, C. E., & Penuel, W. R. (2016). Research–practice partnerships in education: Outcomes, dynamics, and open questions. *Educational Researcher* 45(1), 48–54.

Crowell, A., & Kuhn, D. (2014). Developing dialogic argumentation skills: A 3-year intervention study. *Journal of Cognition and Development* 15(2), 363–381.

Dewey, J. (1916). *Democracy and education.* New York: The Free Press.

Education Endowment Foundation (2015). *Philosophy for Children: Evaluation report and executive summary.* London, UK: Education Endowment Foundation.

Feldman, M. S., & Pentland, B. T. (2003). Reconceptualizing organizational routines as a source of flexibility and change. *Administrative Science Quarterly* 48(1), 94–118.

Fertig, B., & Garland, S. (2012). Millions spent on improving teachers, but little done to make sure it's working. Hechinger Report. Retrieved from http://hechingerreport.org/millions-spent-on-improving-teachers-but-little-done-to-make-sure-its-working/.

Fishman, B. J., Penuel, W. R., Allen, A. R., Cheng, B. H., & Sabelli, N. (2013). Design-based implementation research: An emerging model for transforming the relationship of research and practice. *National Society for the Study of Education* 112(2), 136–156.

Frank, K. A., Zhao, Y., & Borman, K. (2004). Social capital and the diffusion of innovations within organizations: The case of computer technology in schools. *Sociology of Education* 77(2), 148–171.

Garet, M. S., Porter, A. C., Desimone, L. M., Birman, B., & Yoon, K. S. (2001). What makes professional development effective? Analysis of a national sample of teachers. *American Educational Research Journal* 38(3), 915–945.

Gee, J. P. (2015). Accountable talk and learning in popular culture: The game/affinity paradigm. In L. B. Resnick, C. S. C. Asterhan, & S. N. Clarke (Eds), *Socializing intelligence through academic talk and dialogue* (pp. 197–204). Washington, DC: American Educational Research Association.

Greeno, J. G., Collins, A. M., & Resnick, L. B. (1996). Cognition and learning. In D. C. Berliner & R. C. Calfee (Eds), *Handbook of educational psychology* (pp. 15–46). New York: Macmillan.

Hannan, M., Russell, J. L., Takahashi, S., & Park, S. (2015). Improving feedback and support for beginning teachers: The case of the Building a Teaching Effectiveness Network. *Journal of Teacher Education* 66(5), 494–508.

Hansen, M. T. (1999). The search-transfer problem: The role of weak ties in sharing knowledge across organization subunits. *Administrative Science Quarterly* 44(1), 82–111.

Honig, M. I. (2008). District central offices as learning organizations: How sociocultural and organizational learning theories elaborate district central office administrators' participation in teaching and learning improvement efforts. *American Journal of Education* 114, 627–664.

Howe, C., McWilliam, D., & Cross, G. (2005). Chance favours only the prepared mind: Incubation and the delayed effects of peer collaboration. *British Journal of Psychology* 96(1), 67–93.

Kapur, M. (2011). A further study of productive failure in mathematical problem solving: Unpacking the design components. *Instructional Science* 39(4), 561–579.

Kapur, M. (2012). Productive failure in learning the concept of variance. *Instructional Science* 40(4), 651–672.

Kennedy, M. M. (2016). How does professional development improve teaching? *Review of Educational Research* 86(4), 945–980.

Kraft, M. A., Blazar, D., & Hogan, D. (2016). The effect of teacher coaching on instruction and achievement: A meta-analysis of the causal evidence. Brown University Working Paper. Retrieved from https://scholar.harvard.edu/files/mkraft/files/kraft_blazar_hogan_2016_teacher_coaching_meta-analysis_wp_w_appendix.pdf.

Kuhn, D. (1991). *The skills of argument.* Cambridge, UK: Cambridge University Press.

Kuhn, D., & Udell, W. (2003). The development of argument skills. *Child Development* 74(5), 1245–1260.

Lewis, C. (2002). Does lesson study have a future in the United States? *Nagoya Journal of Education and Human Development* 1, 1–23.

Lewis, C. (2015). What is improvement science? Do we need it in education? *Educational Researcher* 44(1), 54–61.

Lewis, C. C., Perry, R. R., & Hurd, J. (2009). Improving mathematics instruction through lesson study: A theoretical model and North American case. *Journal of Mathematics Teacher Education* 12(4), 285–304.

McLaughlin, M. W., & Talbert, J. E. (2006). *Building school-based teacher learning communities: Professional strategies to improve student achievement,* Vol. 45. New York: Teachers College Press.

Matsumura, L. C., & Garnier, H. E. (2015). Embedding dialogic teaching in the practice of a large school system. In L. B. Resnick, C. S. C. Asterhan, & S. N. Clarke (Eds), *Socializing intelligence through academic talk and dialogue* (pp. 415–426). Washington, DC: American Educational Research Association.

Matsumura, L. C., Garnier, H. E., & Spybrook, J. (2012). The effect of content-focused coaching on the quality of classroom text discussions. *Journal of Teacher Education* 63(3), 214–228.

Mead, G. H. (1922/1964). A behavioristic account of the significant symbol. In A. J. Reck (Ed.), *Selected writings: George Herbert Mead* (pp. 240–247). Chicago: University of Chicago Press.

Mead, G. H. (1934). *Mind, self, and society.* Chicago: University of Chicago Press.

Mehan, H., & Cazden, C. (2015). The study of classroom discourse: Early history and current developments. In L. B. Resnick, C. S. C. Asterhan, & S. N. Clarke (Eds), *Socializing intelligence through academic talk and dialogue* (pp. 13–34). Washington, DC: American Educational Research Association.

Michaels, S., O'Connor, C., & Resnick, L. B. (2008). Deliberative discourse idealized and realized: Accountable talk in the classroom and in civic life. *Studies in Philosophy and Education* 27(4), 283–297.

National Research Council. (2000). *How people learn: Brain, mind, experience, and school: Expanded edition.* Washington, DC: National Academies Press.

Obstfeld, D. (2005). Social networks, the *tertius iungens* orientation, and involvement in innovation. *Administrative Science Quarterly* 50(1), 100–130.

O'Connor, C., Michaels, S., & Chapin, S. (2015). "Scaling down" to explore the role of talk in learning: From district intervention to controlled classroom study. In L. B. Resnick, C. S. C. Asterhan, & S. N. Clarke (Eds), *Socializing intelligence through academic talk and dialogue* (pp. 111–126). Washington, DC: American Educational Research Association.

Ogawa, R. T. (2015). Change of mind: How organization theory led me to move from studying educational reform to pursuing educational design. *Journal of Educational Administration* 53(6), 794–804.

Ogawa, R., & Russell, J. L. (2010, May). Organizing schools for learning: A proposal for linking the learning sciences and organization sciences. Paper presented at American Educational Research Association Annual Meeting, Denver, CO.

Penuel, W. R., Coburn, C. E., & Gallagher, D. J. (2013). Negotiating problems of practice in research-practice design partnerships. *Yearbook of the National Society for the Study of Education* 112(2), 237–255.

Penuel, W. R., Fishman, B., Cheng, B. H., & Sabelli, N. (2011). Organizing research and development at the intersection of learning, implementation, and design. *Educational Researcher* 40, 331–337. doi:10.3102/0013189X11421826

Penuel, W. R., Riel, M., Krause, A., & Frank, K. A. (2009). Analyzing teachers' professional interactions in a school as social capital: A social network approach. *Teachers College Record* 111(1), 124–163.

Pontecorvo, C., & Fasulo, A. (1997). Learning to argue in family shared discourse. In L. B. Resnick, C. Pontecorvo, & R. Säljö (Eds), *Discourse, tools, and reasoning: Essays on situated cognition*. Berlin: Springer.

Putnam, R. T., & Borko, H. (2000). What do new views of knowledge and thinking have to say about research on teacher learning? *Educational Researcher* 29(1), 4–15.

Reagans, R., & McEvily, B. (2003). Network structure and knowledge transfer: The effects of cohesion and range. *Administrative Science Quarterly* 48(2), 240–267.

Resnick, L. B. (2010). Nested learning systems for the thinking curriculum. *Educational Researcher* 39(3), 183–197.

Resnick, L. B., Asterhan, C. S. C., & Clarke, S. N. (Eds) (2015). *Socializing intelligence through academic talk and dialogue*. Washington, DC: American Educational Research Association.

Resnick, L. B., Besterfield-Sacre, M., Mehalik, M. M., Sherer, J. Z., & Halverson, E. R. (2007). A framework for effective management of school system performance. In P. A. Moss (Ed.), *Evidence and decision making: The 106th yearbook of the National Society for the Study of Education* (Part I, pp. 155–185). Malden, MA: Blackwell.

Resnick, L. B., Bill, V. L., Lesgold, S. B., & Leer, M. N. (1991). Thinking in arithmetic class. In B. Means, C. Chelemer, & M.S. Knapp (Eds), *Teaching advanced skills to at-risk students* (pp. 27–53). San Francisco: Jossey-Bass.

Resnick, L. B., Michaels, S., & O'Connor, C. (2010). How (well structured) talk builds the mind. In D. D. Press & R. J. Sternberg (Eds), *Innovations in educational psychology: Perspectives on learning, teaching and human development* (pp. 163–194). New York: Springer.

Resnick, L. B., Säljö, R., Pontecorvo, C., & Burge, B. (Eds). (1997). *Discourse, tools, and reasoning: Essays on situated cognition*. Berlin: Springer.

Resnick, L. B., & Schantz, F. (2015). Rethinking intelligence: Schools that build the mind. *European Journal of Education* 50(3), 340–349.

Resnick, L. B., & Spillane, J. P. (2006). From individual learning to organizational designs for learning. In L. Verschaffel, F. Dochy, M. Boekaerts, & S. Vosniadou (Eds), *Instructional psychology: Past, present and future trends. Sixteen essays in honor of Erik De Corte* (pp. 259–276). Oxford: Pergamon.

Rorrer, A. K., Skrla, L., & Scheurich, J. J. (2008). Districts as institutional actors in educational reform. *Educational Administration Quarterly* 44(3), 307–357.

Russell, J. L., Bryk, A. S., Dolle, J., Gomez, L. M., LeMahieu, P., & Grunow, A. (2017). A framework for initiation of Networked Improvement Communities. *Teachers College Record* 119(7), 1–36.

Russell, J. L., Meredith, J., Childs, J., Stein, M. K., & Prine, D. W. (2015). Designing inter-organizational networks to implement education reform: An analysis of state Race to the Top applications. *Educational Evaluation and Policy Analysis* 37(1), 92–112.

Sandholtz, J. H. (2002). Inservice training or professional development: Contrasting opportunities in a school/university partnership. *Teaching and Teacher Education* 18(7), 815–830.

Schwarz, B. B., & Baker, M. J. (2016). *Dialogue, argumentation and education: History, theory and practice.* Cambridge, UK: Cambridge University Press.

Schwartz, D. L., & Martin, T. (2004). Inventing to prepare for future learning: The hidden efficiency of encouraging original student production in statistics instruction. *Cognition and Instruction* 22(2), 129–184.

Scott, W. R., & Davis, G. F. (2007). *Organizations and organizing: Rational, natural and open systems perspectives.* Upper Saddle River, NJ: Pearson Prentice Hall.

Sherer, J. Z., & Spillane, J. P. (2011). Constancy and change in work practice in schools: The role of organizational routines. *Teachers College Record* 113(3), 611–657.

Stein, M. K., & Coburn, C. E. (2008). Architectures for learning: A comparative analysis of two urban school districts. *American Journal of Education* 114(4), 583–626.

Stein, M. K., & D'Amico, L. (2002). Inquiry at the crossroads of policy and learning: A study of a district-wide literacy initiative. *Teachers College Record* 104(7), 1313–1344.

Stelitano, L., Russell, J. L., & Bray, L. E. (2017). Organizing for inclusion: Exploring the routines that shape student supports. American Educational Research Association Annual Meeting, San Antonio, TX.

Sun, J., Anderson, R. C., Perry, M., & Lin, T. J. (2017). Emergent leadership in children's cooperative problem solving groups. *Cognition and Instruction* 35(3), 212–235.

Thompson, C., & Zeuli, J. (1999). The frame and the tapestry: Standards-based reform and professional development. In L. Darling-Hammond & G. Sykes (Eds), *Teaching as the learning profession: Handbook of policy and practice* (pp. 341–375). Jossey-Bass Education Series. San Francisco: Jossey-Bass.

Thompson, J. D. (1967). *Organizations in action: Social science bases of administration.* New Brunswick: Transaction Publishers.

Tichnor-Wagner, A., Wachen, J., Cannata, M., & Cohen-Vogel, L. (2017). Continuous improvement in the public school context: Understanding how educators respond to plan–do–study–act cycles. *Journal of Educational Change* 18(4), 465–494.

Topping, K. J., & Trickey, S. (2015). The role of dialogue in Philosophy for Children. In L. B. Resnick, C. S.C. Asterhan, & S. N. Clarke (Eds), *Socializing intelligence through academic talk and dialogue* (pp. 99–110). Washington, DC: American Educational Research Association.

Trickey, S., & Topping, K. J. (2004). Philosophy for Children: A systematic review. *Research Papers in Education* 19(3), 365–380.

van Gelder, T., Bissett, M., & Cumming, G. (2004). Cultivating expertise in informal reasoning. *Canadian Journal of Experimental Psychology* 58(2), 142.

Weber, M. (1946). Bureaucracy. In H. H. Gerth & C. Wright Mills (Eds), *From Max Weber: Essays in sociology* (pp. 196–224). Oxford: Oxford University Press.

Vygotsky, L. S. (1978). *Mind in society: The development of higher psychological processes*, ed. M. Cole, V. John-Steiner, S. Scribner, & E. Souberman. Cambridge, MA: Harvard University Press.

CHAPTER 40

LEARNING WITH ZEAL

From Deliberate Practice to Deliberate Performance

PETER J. FADDE AND
MOHAMMADREZA JALAEIAN

Introduction

In a classic study of chess expertise, Herbert Simon and William Chase postulated, "We would estimate, very roughly, that a master has spent perhaps 10,000 to 50,000 hours staring at chess positions, and a Class A player 1,000 to 5,000 hours" (Simon & Chase, 1973, p. 402). Twenty years later, Ericsson, Krampe, and Tesch-Römer (1993) studied young violin players in a European conservatory and found that the best students reported logging around 7,400 hours of solitary practice by the time they reached the conservatory at around age 18. Lower-rated violin students at the conservatory had logged distinctly less solitary practice, about 5,300 hours. The authors found that differences in the amount of solitary practice was the primary factor in predicting the status of the violin students and speculated that, by the time the best students became professional violinists, they would have amassed around 10,000 hours of *deliberate* practice, which Ericsson and colleagues defined as "highly structured activity, the explicit goal of which is to improve performance" (Ericsson et al., 1993, p. 368).

Deliberate practice then entered public awareness through popular books such as *Talent is Overrated: What Really Separates World-Class Performers from Everybody Else* (Colvin, 2008); *The Talent Code: Greatness Isn't Born. It's Grown. Here's How* (Coyle, 2009); *Bounce: Mozart, Federer, Picasso, Beckham, and the Science of Success* (Syed, 2010); and *Outliers: The Story of Success* (Gladwell, 2008). What became known as the 10,000-hour rule, loosely based on expertise research, claimed that achieving expert status in almost any pursuit requires an average of 10,000 hours of deliberate practice—20 hours per week, 50 weeks a year, for 10 years.

While claims in popular literature have been over generalized (Ericsson, 2008), deliberate practice has received wide awareness and acceptance. The challenge for teachers, trainers, coaches, and instructional designers, as well as researchers, is to determine which specific activities qualify as deliberate practice. We offer seven principles that are based on Ericsson and colleagues' (1993) original description but that have evolved with application in various domains. To qualify as deliberate, practice activities should:

1) Capture important aspects of complex performance;
2) Be observable and measurable;
3) Offer timely and objective feedback;
4) Enable repeated engagement for refinement of skills;
5) Be designed, assigned, and monitored by a coach, instructor, or mentor;
6) Address specific deficiencies in performance; and
7) Require concentrated effort that is not inherently enjoyable but rather is engaged in with the goal of improving performance.

Deliberate practice is associated with the thousands of hours of solitary practice that aspiring musicians and athletes invest, but it is also relevant in other domains, especially professions such as teaching and medicine, where the role and form of deliberate practice is less clear. We believe that extending the deliberate practice framework into professions can accelerate the development of expertise, defined here as the consistent, reproducible, and measurable performance of elite performers in a particular domain (Ericsson, 2004).

Learning with Zeal: Background of Deliberate Practice

The notion that high achievement is primarily due to "practice with zeal" (Thorndike, 1912) rather than in-born talent is not new. Academically, the debate revolves around the extent to which the amount of deliberate practice that performers accrue accounts for the level of performance that they attain. While advocates of deliberate practice maintain that it is the single most important factor predicting level of performance (Ericsson et al., 1993), several researchers have challenged the degree to which deliberate practice predicts expert performance (e.g., Hambrick et al., 2014). But even critics characterize deliberate practice as "necessary if not sufficient" (Macnamara, Hambrick, & Oswald, 2014), and deliberate practice provides an extremely useful framework for accelerating the development of expertise.

While deliberate practice provides a useful framework for guiding the design of practice activities, it is not always obvious or easy to apply. Beyond knowing that deliberate practice (and a lot of it) is important to attaining high levels of performance, teachers, trainers, coaches, and instructional designers need to know what expert skills to target. They need to know how deliberate practice fits into professional education (i.e., the education of professionals) and professional development in the workplace, where the logical question is: How can *practicing* professionals make time to pursue hours of deliberate practice?

Obviously, deliberate practice includes the thousands of hours that performers typically invest in acquiring a body of declarative knowledge and mastering requisite technical skills in order to become competent in a domain. However, in this chapter we focus on things beyond knowledge and skill that tend to differentiate expert performers from competent or journeyman performers (Hoffman et al., 2014). These attributes of expertise include situational awareness, pattern recognition, mental representations, tacit knowledge, and intuitive decision making (Klein & Hoffman, 1993), which generally are considered to come from experience rather than formal education or training.

Beyond Competent to Proficient and Expert

While many models depict stages in the development of performers (see Hoffman et al., 2014 for a summary), we adopt the original Dreyfus and Dreyfus (1980) 5-Stage Model of adult learning and development. This model positions *expert* as a level below the penultimate level of *master*. *Proficient*, and even *expert*, are levels that performers can reasonably expect to reach in the course of a career spent in a profession. Table 40.1 shows the stages of development, along with the mental functions associated with the stages.

Focusing on the transition points between skill levels highlights changes in mental functions that present particularly rich opportunities for accelerated development. The transition from competent to proficient, for example, is marked by a change in *recognition* from decomposed (i.e., rule-based) to holistic (situation-based), consistent with expertise theory and research concerning experts' superior awareness of their performance context (e.g., Endsley, 2006; Lajoie, 2003). The transition from *proficient* to *expert* is marked by a change from analytical to intuitive decision making (Klein,

Table 40.1 Mental functions at skill levels

	Skill level				
Mental function	Novice	Competent	Proficient	Expert	Master
Recollection	Non-situational	Situational	Situational	Situational	Situational
Recognition	Decomposed	Decomposed	Holistic	Holistic	Holistic
Decision	Analytical	Analytical	Analytical	Intuitive	Intuitive
Awareness	Monitoring	Monitoring	Monitoring	Monitoring	Absorbed

Source: Adapted from Dreyfus & Dreyfus (1980), *A five-stage model of mental activities involved in directed skill acquisition*. Supported by the U.S. Air Force, Office of Scientific Research (AFSC) under contract F49620-C-0063 with the University of California, Berkeley.

2015) that is based on performers' maturing mental representations emanating from extended domain experience (Ericsson, 2015). While stage models such as Dreyfus and Dreyfus describe a progression to expertise that comes with experience, expertise research suggests that deliberate practice may provide ways to accelerate performers' progression to expertise (Hoffman et al., 2014).

Deliberate Practice and Accelerated Expertise

The theory of deliberate practice was originally articulated, and has been validated through correlational studies, in the domains of chess (Charness, Krampe, & Mayr, 1996), music (Ericsson et al., 1993), and sports (Baker & Young, 2014; Ward, Hodges, Williams, & Starkes, 2004). Chess, music, and sports are type 1 domains that have a history of direct competition, objective feedback on performance (Hoffman et al., 2014), and an established culture of practice in which deliberate practice-type activities have long been implemented, at least intuitively. However, transitioning from competent to proficient and then to expert potentially can be accelerated by applying deliberate practice principles to aspects of expertise, such as recognition and decision making (Williams, Fawver, & Hodges, 2017), that are typically assumed to come only with massed experience.

Type 2 domains don't have cultures of practice, direct competition, or objective feedback on performance (Hoffman et al., 2014). In these domains, including most professions, performers may not start serious career preparation, including deliberate practice, until an age at which developing musicians and athletes have already amassed thousands of hours. The acceptance and appropriate use of deliberate practice in type 2 domains, therefore, depends in part on cultural assumptions within the domain about how expertise develops.

As depicted in Figure 40.1, domains with an established *culture of practice*, such as sports and music (or other performance arts), are more likely to engage in activities that have characteristics of deliberate practice. Domains such as history, on the other hand, have a *culture of study* that emphasizes declarative knowledge. Experts in these areas sometimes generate a high level of interactional expertise in other domains (Collins, 2004), knowing a lot about the domain (e.g., military history, literary criticism) but not engaging in performance in the domain, and therefore unlikely to engage in performance-oriented deliberate practice activities. Domains with a *culture of experience*, such as teaching, invest heavily in immersive experiential learning through

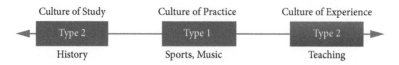

FIGURE 40.1 Continuum of cultures of expertise in professions.

internships or apprenticeships and are likely to value holistic experience over the systematic practice activities associated with deliberate practice.

Deliberate practice research has been criticized as describing social/cultural aspects (i.e., things naturally done in domains with cultures of practice) rather than revealing actual skills of expertise (Williams et al., 2017). Not surprisingly, strong correlations have been found between the amount of deliberate practice and the level of expertise in type 1 domains (e.g., chess, sports, and music) but much weaker correlations have been found between deliberate practice and expertise in type 2 domains such as education and professions (Hambrick et al., 2014; Macnamara et al., 2014). On an implementation level, teachers, trainers, coaches, and instructional designers working in type 1 domains are challenged with making better use of their familiar deliberate practice framework. In type 2 domains the challenge is to adapt the unfamiliar deliberate practice framework to fit the culture.

Deliberate Practice and Training-Based Research

The scope and purpose of this chapter do not permit a comprehensive review of expertise training research across the many and varied domains of performance. Instead, we consider training-based studies in three domains that represent different types of expertise and different cultures of expertise. We look first at the profession of teaching in public education. Teaching has a well-established formal education component (culture of study) in the form of teacher education programs at universities and colleges but also has a *culture of experience* in which holistic experience (e.g., student teaching) is highly valued (Berliner, 2001). However, research over the past decade has introduced video analysis activities that add a substantial deliberate practice character to the established experiential learning activity of reflecting on classroom teaching events. Video analysis of performance, and its degree of acceptance in teaching and teacher education, has implications for many other professions as well.

We then consider research on deliberate practice in medical and surgical education, especially as it relates to simulation-based training. Deliberate practice provides a framework that supports efficient as well as more effective simulator-based training (Causer, Barach, & Williams, 2014). With virtual reality greatly expanding use of simulated environments for training in many more domains, deliberate practice provides a counter balance to the "naive realism" (Smallman & St. John, 2005) that sometimes can supersede thoughtful design of simulation-based training.

We look next at sports, but with a focus different than the thousands of hours developing athletes typically spend on acquisition of technical skills (Baker & Young, 2014). Instead, we follow the focus of sports expertise research that, since the early 1980s (e.g., Abernethy & Russell, 1984), has investigated *perceptual–cognitive* skills in which visual perception is closely tied to cognitive processing and underlies expert athletes' exceptional ability to anticipate the actions of opponents (Williams & Ward, 2003).

Coaches and athletes often assume that perceptual–cognitive skills come from innate ability or massed experience rather than the type of deliberate practice activities typically associated with psychomotor skill acquisition. Therefore, targeted training of perceptual–cognitive skills offers an underexplored path to accelerated expertise for performers in many sports as well as in military, medical, law enforcement, and emergency response domains that also require extremely rapid decisions and actions under high stress conditions (Eccles, Ward, Janelle, Woodman, & LeScanff, 2008; Ward et al., 2008).

While there is a rich and long tradition of descriptive research on expertise and expert performance across many domains and disciplines (see Ericsson, Charness, Feltovich, & Hoffman, 2006), we focus on expertise studies—within our target domains of teaching, surgery, and sports—that are intervention-based and intended to provide prescriptive recommendations to instructional design practitioners (Williams et al, 2017).

Deliberate Practice in Teaching and Teacher Education

With the large amount of content knowledge and pedagogical skills that contribute to the complexity of teaching, it is not surprising that several teacher education theorists and researchers have been drawn to the notion of deliberate practice. However, the established culture of experience in teaching and teacher education is not readily compatible with principles of deliberate practice, as revealed by a researcher's summary comment in one of the few studies to directly connect deliberate practice and teaching. "For most of us, the word 'practice' elicits images of repeated performances aimed at refining and perfecting some skill, usually a motor skill. Teachers do not practice, they 'teach'" (Dunn & Shriner, 1999, p. 647).

Teacher expertise theorist David Berliner attempted to make deliberate practice more palatable to teachers by describing it as

> a practice where coaches help you get some aspect of performance right. It is practice where routines are honed so they need not be thought about much. This is practice where the developing expert's own thoughtfulness allows him or her to get better at what he or she wants to do. (Berliner, 2000, p. 368)

Although adapted to fit the domain's cultural view of how expertise develops, Berliner's description remains true to the essence of deliberate practice by emphasizing that: (1) a coach guides deliberate practice, (2) it targets specific aspects to improve performance through repetition and refinement, and (3) it assumes a self-motivated performer. Berliner also alludes to thoughtful reflection, a disposition that has long been associated with teacher expertise (e.g., Dewey, 1933) and may, therefore, provide an acceptable goal for deliberate practice in teaching and teacher education that favor cultures of study and experience rather than practice.

Observation of pre-service student teachers by peers, university faculty, clinical supervisors, and cooperating teachers has a long history in teacher education (Gaudin & Chaliès, 2015). When classroom teaching is video-recorded then pre-service or in-service teachers, as well as observers, are better able to reflect on teaching events. However, even when video recorded, observation and reflection activities often are not structured enough to qualify as deliberate practice.

Over the past decade, a number of research projects have implemented *video annotation* activities and technologies that greatly increase the ability of teachers and observers to provide feedback (Tripp & Rich, 2012). Video annotation involves storing video footage, either on a stand-alone computer or in the cloud, and providing mechanisms for viewers such as peers, university faculty, clinical supervisors, and the pre-service teachers themselves to mark the start and end times of specific classroom events, tag the events, add text, voice or video comments, and sort coded video clips for concentrated analysis (Rich & Tripp, 2011). In the course of tagging their own videos and reviewing the comments of others, pre-service teachers gain guidance, repeated exposure, and other aspects characteristic of deliberate practice.

The deliberate practice value of video annotation is further increased when student teachers become involved in preparing their own teaching videos for analysis (Hannafin, Recesso, Polly, & Jung, 2015). Some video analysis projects have required student teachers to identify critical incidents and edit them into video segments (Calandra, Brantley-Dias, Lee, & Fox, 2009; Fadde, Aud, & Gilbert, 2009). Of course, video analysis activities demand effort in learning to use particular technologies, which may constitute *desirable difficulty* that deepens learning (see Hoffman et al., 2014), or may simply add distraction.

Most video annotation and video analysis studies over the past decade have involved student teachers writing reflections—an established teacher education assignment—based on video-coding activities. Since many of the studies were investigating the value and feasibility of incorporating video analysis projects in teacher education programs, analysis usually has included participants' self-reported perceptions of their learning and their satisfaction with the activity (Hannafin et al., 2015). However, several studies also analyzed the quantity and quality of student teachers' reflections written after engaging in video analysis activities (Calandra, Gurvitch, & Lund, 2008). Researchers also compared the critical reflections written by student teachers who debriefed with a clinical supervisor to the critical reflections written by student teachers who edited video clips of their teaching to show critical incidents. The reflections written by student teachers that edited their teaching videos were longer and more pedagogically connected than the reflections written by student teachers who debriefed with a supervisor (Calandra et al., 2009).

Most video annotation and analyis studies have involved pre-service teachers during a student teaching internship or in-service teachers' professional development. However, research has also demonstrated the feasibility of training aspects of teacher expertise, such as classroom *noticing* (Sherin & van Es, 2005) with early-stage teacher education students (Fadde & Sullivan, 2013). Students in an introductory teacher education course were tasked with viewing short video clips of student teachers. The students then typed text into an on-screen form to note any incidents they noticed that involved classroom management or student questioning. After typing their observations, students were shown what two experienced teacher educators had noticed when viewing the same clips. Participants were told to reflect on differences between their observations and those of the experts before viewing the next video clip. Students using this form of *expert-model feedback* (Ifenthaler, 2009) increasingly noticed like the experts, both during training and on a post-training test (Fadde & Sullivan, 2013).

Video annotation and analysis activities developed by researchers provide both evidence and design ideas for technology-based guidance of teacher reflection, thereby increasing deliberate practice of reflection skills that are associated with teacher expertise. However, although video recording, editing, uploading, tagging, and replay technologies are increasingly inexpensive and easy to use, it is not clear whether and to what extent video analysis activities will be adopted for routine preparation of professional teachers (Hannafin et al., 2015). Adoption of video annotation and analysis in teacher education may be facilitated by increasing use of video-based assessment for teacher licensure and for formal evaluation of in-service teachers (Hannafin et al., 2015).

Deliberate Practice in Medical and Surgical Education

A sizable body of research in the medical area concludes that deliberate practice explains a substantial degree of variation in performance level in medicine, while acknowledging the importance of working memory and domain knowledge as other contributing factors (e.g., Kulasegaram, Grierson, & Norman, 2013). In those medical specializations where expertise would be profiled as residing primarily in skilled performance, a review of 1200 studies in anesthesiology affirmed the critical role of deliberate practice in acquiring and maintaining expertise (Hastings & Rickard, 2015).

In terms of domains of medical learning, surgery stands apart from other domains of medicine (such as internal medicine) for its emphasis on psychomotor skills, while recognizing the cognitive and affective skills that are also part of surgical performance, and therefore surgical education. Expert surgeons store knowledge with more detail and have extraordinary ability to recognize patterns, and notice deviations from typical patterns that call for heightened attention. For example, an expert surgeon senses when to slow down if something does not seem right during an operation, and anticipates the appropriate actions to take before a situation gets out of control (Alderson, 2010; Kirkman, 2013; Mahvi, 2010).

Medical students preparing to be surgeons are expected to master a body of content knowledge, then to expand their knowledge base by learning from experience in clinical placement, and also to improve their performance through deliberate practice while receiving immediate and informative feedback (Ericsson, 2004; Hashimoto et al., 2015; van de Wiel, van den Bossche, Janssen, & Jossberger, 2011). Although experience is embedded in clinical components of medical education along with medical residency, additional experience in surgery does not necessarily lead to performance improvement due to a lack of control over specific types of experience in real-world settings during medical education (Ericsson, 2004, 2015). Especially during residency, the performance goals of real-world surgical experiences may conflict with the learning goals of controlled and varied surgical experiences. For this reason, simulations and simulators long have been used in surgical education and to train working surgeons on new procedures (Causer et al., 2014).

From high-fidelity mannequins to low-fidelity part trainers, simulations and simulators provide surgeons and surgical students with opportunities to learn and practice surgical procedures in contexts in which mistakes can safely be made. Rapid growth in virtual reality (VR) and augmented reality (AR) technologies promise to deliver simulations for many more surgical procedures (Causer et al., 2014), and simulation-based medical education (SBME) combined with deliberate practice (DP) yields better results than traditional clinical education (McGaghie, Issenberg, Cohen, Barsuk, & Wayne, 2011). Causer and colleagues (2014) also recommend deliberate practice as a guiding framework for implementing simulation-based training and identified reflection, rehearsal, and trial-and-error learning as specific deliberate practice strategies that can effectively be paired with simulation. Indeed, associating simulator-based training (SBT) with learning objectives, instructional design, and clear assessment of learners makes it compliant with a long-standing emphasis in simulation theory (e.g., Salas, Rosen, Held, & Weismuller, 2009) that is highlighted when SBT is delivered in a deliberate practice framework.

Medical educators seeking guidance in designing deliberate practice activities to use with simulators can find ideas embedded in expertise research studies. For instance, an expert–novice study using a laparoscopic simulator included the representative task of clipping and cutting the cystic duct and artery during a laparoscopic procedure (Schijven & Jakimowicz, 2003). Expert and novice surgeons completed the task three times. There was no difference between the groups on the first trial, but the experts performed better and faster on subsequent trials. This research task can readily be repurposed as a training task because it satisfies the deliberate practice criteria of providing repeated opportunities to refine skills that a coach, instructor, or mentor has identified as capturing an essential characteristic of expertise. The study also provides a benchmark for surgical students by noting that experienced surgeons learning a new technique were as clumsy as a novice on their first attempt but demonstrated smooth and fast execution of the procedure within two or three trials. When a surgical student or resident is able to master a new procedure on a simulator within two or three attempts, therefore, he or she can feel successful.

Targeted Fidelity in Simulation-Based Surgical Training

Simulation environments used to acquire or test surgical skills often strive to be realistic and immersive. However, a deliberate practice framework that targets identified deficiencies in technique aligns with surgical education's traditional use of single-procedure simulators, variously called part trainers, task trainers, or box trainers, to allow surgeons or surgical students to practice psychomotor procedures and techniques, such as arthroscopic knot tying (Horeman, Akhtar, & Tuijthof, 2015). Despite the proven effectiveness of part-task simulators, however, medical educators historically have been concerned with the physical fidelity of simulators (Ward, Williams, & Hancock, 2006). Overemphasis on physical fidelity (looks and feels real) can potentially overwhelm functional fidelity (acts real) or psychological fidelity (preserves sufficient complexity to permit similar sensemaking and decision making). The principle stating that deliberate practice does not need to be inherently enjoyable (Ericsson et al., 1993) arms medical educators to resist building or buying the "wrong" simulator that has high physical fidelity but low functional or psychological fidelity (Foshay, 2006). Within a deliberate practice framework, using the minimum fidelity needed to engage the appropriate perceptual and cognitive skills not only saves money but also reduces extraneous cognitive load (van Gog, Ericsson, Rikers, & Pass, 2005) for surgeons or surgical students using simulators.

Along with cost and safety benefits, particularly when compared with using cadavers or actual patients, simulator-based education combined with deliberate practice (SMBE with DP) in surgical education provides educators and trainees with control over the learning environment and feedback mechanisms that are often not available in the real world (Ericsson, 2015). Overall, the effectiveness of incorporating deliberate practice with simulator-based training before or during surgical residency needs further research but holds promise for accelerating surgical expertise (Duvivier et al., 2011; Hashimoto et al., 2015; Kirkman, 2013).

Deliberate Practice and Expertise Training in Sports

Sports provide a core arena for deliberate practice studies, with correlational studies showing that the amount of deliberate practice accounts for differences in level of performance (Ward et al., 2004), especially at elite performance levels (Moesch, Elbe,

Hauge, & Wikman, 2011). A review by Baker and Young (2014) covering 20 years of correlational studies highlights the overall importance of deliberate practice across several sports while also noting the need to clarify what constitutes "deliberate practice" in order to guide further research as well as inform the efforts of sports coaches and administrators.

Although deliberate practice-type activities are commonly used for acquiring psychomotor skills in sports, sports expertise research reveals opportunities for deliberate practice beyond psychomotor skill acquisition that can potentially accelerate expertise. Competent performance in many sports requires strength, conditioning, nutrition, psychomotor skills, psychological skills, strategic knowledge, and perceptually-based decision making. The latter, in particular, characterizes expert quarterbacks, point guards, goalies, and batters who seem to anticipate opponents' actions and to respond in time frames that defy simple human reaction time (Broadbent, Causer, Williams, & Ford, 2014).

Sports expertise researchers have found the uncanny anticipation of many expert athletes is based on *perceptual–cognitive* skills that combine information from the senses and with rapid mental processing (Williams & Ward, 2003). For example, skilled baseball batters perceive cues in the pitcher's motion, release of a pitch, and early ball flight that allow the batter to anticipate if and when a pitch will arrive in the hitting zone (Müller & Abernethy, 2012). Thus, perceptual–cognitive skills have been a primary focus of sports expertise research since the early 1980s and a meta-analysis of 388 effect sizes generated from 42 studies in a variety of interceptive ball sports shows that experts consistently produce better response accuracy and speed than do less skilled athletes in tests of perceptual–cognitive skill (Mann, Williams, Ward, & Janelle, 2007). Yet, despite these strong findings, there has been surprisingly little application of the research to the systematic training and preparation of high-level performers (Larkin, Mesagno, Spittle, & Berry, 2015).

The expert–novice approach commonly used in sports expertise research typically compares the performance of highly skilled and less skilled athletes on a representative task. Many representative tasks involve *occlusion* in which a video display depicting an opponent serving or kicking or pitching a ball is edited to black at various points in the opponent's motion or early ball flight (temporal occlusion) or portions of the opponent's body or the ball are masked (spatial occlusion). Participants are tasked with identifying the type of serve, kick, or pitch and sometimes predicting the ball's ultimate location (e.g., forehand or backhand; ball or strike).

While a sizable body of expert–novice research using occlusion methods has affirmed perceptual–cognitive advantage, a smaller number of intervention-based studies have implemented training programs using occlusion tasks to improve perceptual–cognitive skills such as tennis serve recognition (Farrow, Chivers, Hardingham, & Sasche, 1998; Scott, Scott, & Howe, 1998) and baseball pitch recognition (Burroughs, 1984; Fadde, 2006). Larkin and colleagues (2015) identified 25 intervention-based studies that trained perceptual–cognitive skills, most of them using the temporal occlusion training method and quasi-experimental research method. Most of the training studies showed at least modest training effects based on pre-/post-test improvement on video-occlusion tests of response accuracy. Two of the studies showed transfer of gains in video-based training to improvement of in-game performance (Fadde, 2006; Gorman & Farrow, 2009).

Figure 40.2 shows a typical video-occlusion task in which a baseball batter views a video display of a pitcher throwing a pitch that is cut off (occluded) at various points at or after the release of the pitch. The batter does not simulate swinging a bat but rather identifies the type of pitch via voice recognition, key press, mouse click, or touch screen. The video display does not change in response to the choice or movement of the batter—although the system typically provides corrective feedback and a score when being used in a training mode rather than in a research or testing mode in which feedback would not be provided.

Several expertise researchers have argued that training perceptual–cognitive skills has potential application not only in sports but also in aviation, medical, law enforcement, and military contexts (Eccles et al., 2008; Roca & Williams, 2016; Ward et al., 2008). For example, surgical students may benefit from concentrated and part-task deliberate practice of the perceptual–cognitive, situation awareness, and pattern recognition skills that expertise researchers have identified as being characteristic of expert surgeons. The insight offered by a deliberate practice framework is that the targeted, part-task approach using box trainers to practice the psychomotor component of surgical procedures might also apply to part-task, perception-only practice of surgical recognition skills using video occlusion (Fadde, 2009b).

FIGURE 40.2 Video-based temporal occlusion computer program for training pitch recognition.

Photo courtesy of gameSense Sports, LLC.

Deliberate Practice in Professional Domains

Researchers who are associated with particular domains (rather than expertise studies) have conducted field-based studies to identify existing activities that have characteristics of deliberate practice in professions such as education (Bronkhorst, Meijer, Koster, & Vermunt, 2014), insurance (Sonnentag & Kleine, 2000), sign language (Schafer, 2011), engineering education (Litzinger et al., 2011), and organizational consulting (van de Wiel, Szegedi, & Weggeman, 2004). However, the practice activities described in these studies tend to be ad hoc rather than organized into distinct training programs that incorporate a deliberate practice framework (Boshuizen, 2004).

Outside the realm of academic research, a practicing and teaching psychotherapist published a book that describes a systematic program of deliberate practice to cultivate the expertise of working psychotherapists, and thereby improve their clinical effectiveness (Rousmaniere, 2017). Rousmaniere created the program for himself and his trainees to leverage more learning from their clinical cases by:

1) Observing their own work via videotape;
2) Getting expert feedback from a coach or consultant;
3) Setting small incremental learning goals just beyond our ability;
4) Repetitive behavioral rehearsal of specific skills;
5) Continuously assessing their performance via client reports and outcomes.

Rousmaniere notes, "This routine aims to help us break through a competency plateau by engaging in a never-ending gradual improvement process toward psychotherapy expertise" (p. 92). He recommends repeating the processes throughout a career, from graduate school through licensure and into middle and later career. However, recognizing that most working professionals do not have the vision or motivation to design and implement a personal program of deliberate practice, *deliberate performance* has been articulated as an adaptation of deliberate practice principles that can facilitate working professionals' progression toward levels of proficient and expert (Fadde & Klein, 2010, 2012).

DELIBERATE PERFORMANCE TO ACCELERATE EXPERTISE IN THE WORKPLACE

Integrating deliberate practice into their career work is challenging for professionals who do not work within in a culture of practice (such as sports or entertainment arts). Even highly motivated performers have limited time or inclination to pursue deliberate practice activities. Their access to performance coaches who can direct a program of deliberate practice is also limited and, in most cases, working professionals must act as self-regulated learners who assume responsibility for developing their own expertise (Zimmerman, 2006). Whether directed by a coach or by a self-regulated performer, it is difficult to manage the learning of tacit knowledge that is learned implicitly (Klein & Hoffman, 1993).

On a macro level, the progression of performers from competent to proficient and expert levels is probably most influenced by work assignments that place them in a variety of increasingly challenging contexts and facilitate learning through doing. Unfortunately, work assignments are not always under the control of developing performers (Hoffman et al., 2014) and learning goals are distinctly secondary to job performance. While it can be effective, implicit learning during job performance lacks the efficiency and focus that are characteristic of deliberate practice.

With the goal of accelerating working professionals to proficient and expert levels (see Dreyfus & Dreyfus, 1980, Table 40.1) Fadde and Klein (2010) draw on expertise research within the naturalistic decision-making paradigm (Klein, 2015) in order to incubate growth of macrocognitive skills such as *intuitive decision making* and *sensemaking* (e.g., pattern recognition, metacognition) as appropriate skills to target for accelerated expertise. Obviously, the advanced skills and knowledge of experts, as revealed through techniques such as cognitive engineering based on expert skill (CEBES, Staszewski, 2013) and a variety of cognitive task analysis (CTA) methods (e.g., Clark, Feldon, van Merriënboer, Yates, & Early, 2007; Crandall, Klein, & Hoffman, 2006), are essential components of expertise training. However, CTA-based training is usually delivered during formal professional (e.g., medical) education or requires taking working professionals offline for concentrated training programs. Deliberate performance, on the other hand, is intended for professionals to practice skills such as intuitive decision making in the context of their work, thus gaining both domain specificity and learning efficiency compared to traditional professional development activities. Deliberate performance applies principles of deliberate practice (purposeful, targeted, observable, and repeatable) while adapting them to use in the context of professional work.

Fadde and Klein (2012) propose action learning activities (ALAs) as small-scale practice activities that are embedded in routine work. For example, *estimation* involves performers predicting the outcome or cost or duration of a work project as a way to improve their situational awareness. A teacher preparing a novel lesson can estimate how long she/he expects the lesson to take and which students will "get it"—or not— and record the estimates for later review. While the teacher might naturally reflect on the effectiveness of this newly designed lesson *after* delivering the lesson, writing down predictions ahead of time leads the teacher to reflect *in* the action (Schön, 1983), a skill associated with teacher expertise (Berliner, 2001). Other ALAs include *experimentation* in which performers try different ways of doing a routine task, and thereby combat the kind of automaticity that can inhibit development of expertise (Boot & Ericsson, 2013; Feldon, 2007).

Deliberate performance has yet to be demonstrated through intervention-based research studies and is described here to demonstrate the potential for extending principles of deliberate practice beyond type 1 domains and beyond skill acquisition stages of development. On the other hand, mature deliberate practice theory and frameworks need to be applied in as many domains as possible, both to accelerate expertise and to reveal gaps that generate further theory modification and research questions. We conclude by describing several models for training expertise that apply principles of deliberate practice and performance.

Models for Designing Expertise Training Activities and Programs

Practitioners or researchers who want to design, implement, and evaluate research-based expertise training programs can choose from several models to serve as guides, including: *expertise-based training* (Fadde, 2009a), *expert performance training* (Ward, Suss, & Basevitch, 2009), and *ShadowBox* (Borders, Polander, Klein, & Wright, 2015). Each model is especially appropriate for different learning contexts.

Expertise-Based Training (XBT)

As demonstrated in a study training noticing skills in pre-service teachers (Fadde & Sullivan, 2013), focused practice of *recognition* skills can be implemented very early in development—before learners or performers have the mental library of experiences they will need to execute the full processes of classroom noticing or recognition-primed

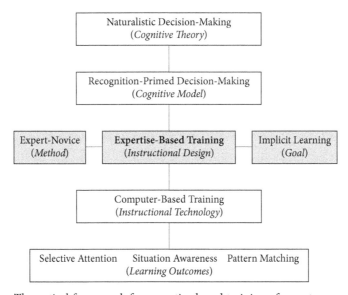

FIGURE 40.3 Theoretical framework for expertise-based training of expert perceptual skills.

decision making (RPD). As shown in Figure 40.3, XBT builds on a theoretical foundation in naturalistic decision making (NDM) along with the RPD model to train recognition skills using drill-and-practice instructional methods that work well with computer-based training (CBT) technology.

XBT presents interactive training using still or video images and representative tasks similar to those used in expert–novice studies (Chi, 2006): Recall, detection, categorization, and prediction. For example, libraries of authentic mammograms can provide stimulus materials for practice tasks such as *detecting* anomalies in mammograms, *categorizing* types of lesions detected, and *predicting* outcomes. Since the outcomes associated with archived mammograms are known (e.g., if a biopsy was ordered, the resulting diagnosis, and the outcome for the patient) then trainees can receive immediate feedback on their diagnoses. Through repetition with immediate feedback trainees or working radiologists can sharpen their diagnostic skills (Ericsson, 2015).

While case libraries of mammograms provide learning materials with known answers, XBT also uses authentic images or video recordings (e.g., dashboard video recordings from highway patrol cruisers) that don't capture everything in the original event but do provide stimulus material that expert can react to. The experts' observations, then, become the "correct" answer using the *expert model feedback* method (Ifenthaler, 2009).

Video recordings of workplace performance, such as classroom teaching, offer a rich source of XBT materials but also come with many ethical, legal, and logistic concerns (Rich & Tripp, 2011). When authentic workplace video is not available, video recordings of simulator use by students or instructors can provide source material for XBT (Razer, McIntyre, & Fadde, 2019). XBT is most appropriate for designing self-paced computer-based training (CBT) that can be accessed online and on demand (Fadde, 2013), making it ideal for busy performers in internship or work stages of their careers.

Expert Performance-Based Training (ExPerT)

"Ultimately, if the expert performance approach has validity, it should be demonstrable through the development of skill-sensitive training . . . to high levels of performance more quickly" (Charness & Tuffiash, 2008, p. 427). *Expert performance-based training* extends the expert performance approach (Ericsson & Smith, 1991) into a proscriptive instructional design model (Ward et al., 2009). As depicted in Figure 40.4, ExPerT provides a framework for identifying relevant skills, devising training activities that address these skills, and suggesting iterative rounds of evaluation and training. ExPerT provides a comprehensive framework for program development, implementation, and evaluation.

In a study that involved training peer academic coaches, Blair (2015) compared two versions of a training program to accelerate the development of undergraduate peer academic coaches—who need to progress from novice to competent to proficient and

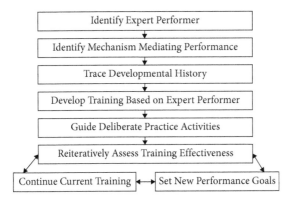

FIGURE 40.4 Conceptual framework of the expert performance-based training (ExPerT) method.

Reproduced from Kevin R. Harris, David W. Eccles, Paul Ward, and James Whyte, A theoretical framework for simulation in nursing: Answering Schiavenato's call, *Journal of Nursing Education* 52(1), pp. 6–16 doi.org/10.3928/01484834-20121107-02, © 2013, Helio. Reprinted with permission from SLACK Incorporated.

even to expert within three years on the job. The study implemented and compared two versions of the training program, one using the ExPerT framework and one using XBT activities. The ExPerT version involved "live" simulation with experienced peer tutors role-playing as student clients while the XBT version involved trainees identifying suboptimal behaviors in video recordings of simulated coaching sessions with experienced peer counselors playing the roles of both counselor and client.

Both XBT and ExPerT conditions were more effective than a control condition that consisted of the traditional curriculum. Effectiveness of the training was judged using conversation analysis of trainees' "final exam" consisting of a role-play simulation with an experienced peer academic coach acting as a client and the trainee as the counselor. The ExPerT condition was associated with the largest learning effects. Both methods have a place in an instructional designer's expertise training toolkit, with ExPerT providing greater learning effects, but requiring greater investment in instructional time than the relatively low-overhead XBT method.

ShadowBox

The *ShadowBox* method helps performers align their situational awareness and decision making with that of experts (Borders et al., 2015). As does XBT, *ShadowBox* uses expert model feedback. Learners read a scenario, presented on paper or on a computer, which pauses at various decision points. Learners are presented with a list of observations or decisions and tasked with prioritizing them. They are then shown the priorities, along with explanatory comments, made by a panel of experts who had completed the same scenario.

ShadowBox has been used to train nurses to perceive patient needs, and to train marines and army soldiers in *Good Stranger* interactions with civilians in conflict areas.

Marines completing a *Good Stranger* program aligned with experts 28 percent more than a comparison group. However, Klein (2015) points out that learners don't need to match the experts but rather to consider the experts' responses and to increasingly notice what experts notice.

Future Directions for Expertise Training and Research

The theory of deliberate practice developed in type 1 domains such as chess, music, and sports that feature direct competition, clear measures of performance, and cultures of practice. However, deliberate practice can have a much wider and arguably more important impact in type 2 domains—including professions such as education, law, medicine, and business. Practitioners and researchers should seek opportunities to add, and analyze, deliberate practice activities and programs during preparatory professional education and career professional development phases that potentially can accelerate the progression of professionals from competent to proficient and expert.

Researchers should embrace the challenge of Williams, Fawver, and Hodges (2017) to conduct intervention-based studies that provide proscriptive findings to practitioners. Ideally, studies should isolate particular aspects of instructional design or delivery and compare an optimized design with a lower fidelity design (e.g., Blair, 2015; Fadde & Sullivan, 2013) in order to build efficient as well as effective approach to expertise training that aligns with principals of deliberate practice.

When possible, researchers should conduct longitudinal studies of authentic expertise training programs that implement strategies and methods that have been validated in short-term controlled experiments. For example, Fadde (2016) designed and evaluated a multi-year program to train pitch recognition that was implemented by a competing college baseball team. The program included temporal occlusion drills on a computer and also in situ drills that targeted similar perceptual–cognitive skills. While the cooperating team showed statistically significant improvements in team batting statistics, compared to mean statistics of other teams in the same athletic conference, the training effects are not generalizable. However, the case study provides coaches with context-sensitive ideas for training perceptual–cognitive skills.

In relation to simulator-based training, the deliberate practice framework supports part-task approaches, including perception-only training (Fadde, 2009b), which emphasizes training efficiency in a way that simulation research typically does not include (Hubal & Parsons, 2017). In summary, deliberate practice and deliberate performance are appropriate not only for skill acquisition during performers' early developmental stages but also for accelerating the development of mental functions, such as patter recognition, reflection, and intuitive decision making, that are associated with expertise and expert performance.

References

Abernethy, B., & Russell, D. G. (1984). Advance cue utilisation by skilled cricket batsmen. *Australian Journal of Science and Medicine in Sport* 16(2), 2–10.

Alderson, D. (2010). Developing expertise in surgery. *Medical Teacher* 32(10), 830–836.

Baker, J., & Young, B. (2014). 20 years later: Deliberate practice and the development of expertise in sport. *International Review of Sport and Exercise Psychology* 7(1), 135–157.

Berliner, D. C. (2000). A personal response to those who bash teacher education. *Journal of Teacher Education* 51(5), 358–371.

Berliner, D. C. (2001). Learning about and learning from expert teachers. *International Journal of Educational Research* 35(5), 463–482.

Blair, L. (2015). Implementing expertise-based training methods to accelerate the development of peer academic coaches (unpublished doctoral dissertation). Southern Illinois University, Carbondale, IL.

Borders, J., Polander, N., Klein, G., & Wright, C. (2015). ShadowBox: Flexible training to impart the expert mindset. *Proceedings of the 6th International Conference on Applied Human Factors and Ergonomics*, Procedia Manufacturing, 3, 1574–1579.

Boot, W. R., & Ericsson, K. A. (2013). Expertise. In J. D. Lee & A. Kirlik (Eds), *The Oxford handbook of cognitive engineering* (pp. 143–158). New York: Oxford University Press.

Broadbent, D. P., Causer, J., Williams, A. M., & Ford, P. R. (2014). Perceptual-cognitive skill training and its transfer to expert performance in the field: Future research directions, *European Journal of Sport Science* 15(4), 322–331.

Bronkhorst, L. H., Meijer, P. C., Koster, B., & Vermunt, J. D. (2014). Deliberate practice in teacher education. *European Journal of Teacher Education* 37(1), 18–34.

Boshuizen, H. P. A. (2004). Does practice make perfect? In H. P. A. Boshuizen, R. Bromme, & H. Gruber (Eds), *Professional learning: Gaps and transitions on the way from novice to expert*. New York: Kluwer Academic Press.

Burroughs, W. A. (1984). Visual simulation training of baseball batters. *International Journal of Sport Psychology* 15(2), 117–126.

Calandra, B., Brantley-Dias, L., Lee, J. K., & Fox, D. L. (2009). Using video editing to cultivate novice teachers' practice. *Journal of Research on Technology in Education* 42(1), 73–94.

Calandra, B., Gurvitch, R., & Lund, J. (2008). An exploratory study of digital video editing as a tool for teacher preparation. *Journal of Technology and Teacher Education* 16(2), 137–153.

Causer, J., Barach, P., & Williams, A. M. (2014). Expertise in medicine: Using the expert performance approach to improve simulation training. *Medical Education* 48(2), 115–123.

Clark, R. E., Feldon, D., van Merriënboer, J. J. C., Yates, K., & Early, S. (2007). Cognitive task analysis. In J. M. Spector, M. D., Merrill, J. J. G. van Merriënboer, & M. P. Driscoll (Eds), *Handbook of research on educational communications and technology* (pp. 577–593). Mahwah, NJ: Lawrence Erlbaum.

Charness, N., & Tuffiash, M. (2008). The role of expertise research and human factors in capturing, explaining, and producing superior performance. *Human Factors* 50(3), 427–432.

Charness, N., R. Krampe, R. Th., & Mayr, U. (1996). The role of practice and coaching in entrepreneurial skill domains: An international comparison of life-span chess skill acquisition. In K. A. Ericsson (Ed.), *The road to excellence: The acquisition of expert performance in the arts and sciences, sports, and games* (pp. 51–80). Mahawah, NJ: Lawrence Erlbaum.

Chi, M. T. H. (2006). Laboratory methods for assessing experts' and novices' knowledge. In K. A. Ericsson, N. Charness, P. J. Feltovich, & R. R. Hoffman (Eds), *The Cambridge handbook of expertise and expert performance* (pp. 167–184). New York: Cambridge University Press.

Collins, H. (2004). Interactional expertise as a third kind of knowledge. *Phenomenology and the Cognitive Sciences* 3(2), 125–143.

Colvin, G. (2008). *Talent is overrated: What really separates world-class performers from everybody else*. New York: Portfolio.

Coyle, D. (2009). *The talent code: Greatness isn't born. It's grown. Here's how*. New York: Bantam Books, 2009.

Crandall, B., Klein, G., & Hoffman, R. R. (2006). *Working minds: A practitioner's guide to cognitive task analysis*. Cambridge, MA: MIT Press.

Dewey, J. (1933). *How we think: A restatement of the relation of reflective thinking to the educative process*. Boston: D. C. Heath.

Dreyfus, H., & Dreyfus, S. (1980). *A five-stage model of mental activities involved in directed skill acquisition*. Supported by the U.S. Air Force, Office of Scientific research (AFSC) under Contract F49620-C-0063 with the University of California, Berkeley.

Dunn, T. G., & Shriner, C. (1999). Deliberate practice in teaching: What teachers do for self-improvement. *Teaching and Teacher Education* 15, 631–651.

Duvivier, R. J., van Dalen, J., Muijtjens, A. M., Moulaert, V. R. M. P., van der Vleuten Cees, P. M., & Scherpbier Albert, J. J. A. (2011). The role of deliberate practice in the acquisition of clinical skills. *BMC Medical Education* 11(1), 101.

Eccles, D. W., Ward, P., Janelle, C. M., Woodman, T., & LeScanff, C. (2008). Shared interests in solving common problems: How sport psychology might inform human factors and ergonomics. *Human Factors and Ergonomics Society Annual Meeting Proceedings* 52(11), 743–747.

Endsley, M. R. (2006). Expertise and situation awareness. In K. A. Ericsson, N. Charness, P. J. Feltovich, & R. R. Hoffman (Eds), *The Cambridge handbook of expertise and expert performance* (pp. 633–651). New York: Cambridge University Press.

Ericsson, K. A. (2004). Deliberate practice and the acquisition and maintenance of expert performance in medicine and related domains. *Academic Medicine* 79(10), S70-S81.

Ericsson, K. A. (2008). Deliberate practice and acquisition of expert performance: A general overview. *Academic Emergency Medicine* 15, 988–994.

Ericsson, K. A. (2015). Acquisition and maintenance of medical expertise: A perspective from the expert-performance approach with deliberate practice. *Academic Medicine* 90(11), 1471.

Ericsson, K. A., Charness, N., Feltovich, P. J., & Hoffman, R. R. (Eds). (2006). *The Cambridge handbook of expertise and expert performance*. New York: Cambridge University Press.

Ericsson, K. A., Krampe, R. T., & Tesch-Römer, C. (1993). The role of deliberate practice in the acquisition of expert performance. *Psychological Review* 100(3), 363–406.

Ericsson, K. A., & Smith, J. (1991). Prospects and limits in the empirical study of expertise: An introduction. In K. A. Ericsson & J. Smith (Eds), *Toward a general theory of expertise: Prospects and limits* (pp. 1–38). New York: Cambridge University Press.

Fadde, P. J. (2006). Interactive video training of perceptual decision-making in the sport of baseball. *Technology, Instruction, Cognition & Learning* 4(3/4), 265–285.

Fadde, P. J. (2009a). Expertise-based training: Getting more learners over the bar in less time. *Technology, Instruction, Cognition & Learning* 7(2), 171–197.

Fadde, P. J. (2009b). Instructional design for advanced learners: Training recognition skills to hasten expertise. *Educational Technology Research & Development 57*(3), 359–376.

Fadde, P. J. (2013). Accelerating the acquisition of intuitive decision-making through expertise-based training (XBT). *Proceedings of the Interservice/Industry Training, Simulation, and Education Conference*.

Fadde, P. J. (2016). Instructional design for accelerated macrocognitive expertise in the baseball workplace. *Frontiers in Psychology 7, 292*, 1–16.

Fadde, P. J., Aud, S., & Gilbert, S. (2009). Incorporating a video editing activity in a reflective teaching course for pre-service teachers. *Action in Teacher Education 31*(1), 75–86.

Fadde, P. J., & Klein, G. (2012). Accelerating expertise using action learning activities. *Cognitive Technology 17*(1), 11–18.

Fadde, P. J., & Klein, G. A. (2010). Deliberate performance: Accelerating expertise in natural settings. *Performance Improvement 49*(9), 5–14.

Fadde, P. J., & Sullivan, P. (2013). Using interactive video to develop pre-service teachers' classroom awareness. *Contemporary Issues in Technology and Teacher Education 13*(2).

Farrow, D., Chivers, P., Hardingham, C., & Sasche, S. (1998). The effects of video-based perceptual training on the tennis return of serve. *International Journal of Sports Psychology 23*, 231–242.

Feldon, D. F. (2007). Cognitive load and classroom teaching: The double-edged sword of automaticity. *Educational Psychologist 42*(3), 123–127.

Foshay, W. R. (2006). Building the wrong simulation: Matching instructional intent in teaching problem solving to simulation architecture. *Technology, Instruction, Cognition and Learning 3*, 63–72.

Gaudin, C., & Chaliès, S. (2015). Video viewing in teacher education and professional development: A literature review. *Educational Research Review 16*, 41–67.

Gladwell, M. (2008). *Outliers: The story of success*. New York: Little, Brown.

Gorman, A., & Farrow, D. (2009). Perceptual training using explicit and implicit instructional techniques: Does it benefit skilled performers? *International Journal of Sports Science & Coaching 4*, 193–208.

Hambrick, D. Z., Oswald, F. L., Altmann, E. M., Meinz, E. J., Gobet, F., & Campitelli, G. (2014). Deliberate practice: Is that all it takes to become an expert? *Intelligence 45*, 34–45.

Hannafin, M. J., Recesso, A., Polly, D., & Jung, J. W. (2015). Video analysis and teacher assessment: Research, practice, and implications. In B. Calandra & P. J. Rich (Eds), *Digital video for teacher education: Research and Practice* (pp. 164–180). New York: Routledge.

Hashimoto, D. A., Sirimanna, P., Gomez, E. D., Beyer-Berjot, L., Ericsson, K. A., Williams, N. N., . . . Aggarwal, R. (2015). Deliberate practice enhances quality of laparoscopic surgical performance in a randomized controlled trial: From arrested development to expert performance. *Surgical Endoscopy and other Interventional Techniques 29*(11), 3154–3162.

Hastings, R. H., & Rickard, T. C. (2015). Deliberate practice for achieving and maintaining expertise in anesthesiology. *Anesthesia & Analgesia 120*(2), 449–459.

Hoffman, R. R., Ward, P., Feltovich, P. J., DiBello, L., Fiore, S. M., & Andrews, D. H. (2014). *Accelerated expertise: Training for high proficiency in a complex world*. New York: Psychology Press.

Horeman, T., Akhtar, K., & Tuijthof, J. M. (2015). Physical simulators. In M. Karahan, G. M. M. J. Kerkhoffs, P. Randelli, & G. J. M. Tuijthof (Eds), *Effective training of arthroscopic skills* (pp. 57–69). Luxembourg: Esska/Springer.

Hubal, R., & Parsons, T. (2017). Synthetic environments for skills training and practice. In L. Lin & J. M. Spector (Eds), *The sciences of learning and instructional design: Constructive articulation between communities* (pp. 1–22). New York: Taylor & Francis/Routledge.

Ifenthaler, D. (2009). Model-based feedback for improving expertise and expert performance. *Technology, Instruction, Cognition, & Learning* 7(2), 83–101.

Kirkman, M. A. (2013). Deliberate practice, domain-specific expertise, and implications for surgical education in current climes. *Journal of Surgical Education* 70(3), 309–317.

Klein, G. (2015). A naturalistic decision-making perspective on studying intuitive decision making. *Journal of Applied Research in Memory and Cognition* 4(3), 164–168.

Klein, G. A., & Hoffman, R. (1993). Seeing the invisible: Perceptual/cognitive aspects of expertise. In M. Rabinowitz (Ed.), *Cognitive science foundations of instruction* (pp. 203–226). Mahwah, NJ: Lawrence Erlbaum.

Kulasegaram, K. M., Grierson, L. E. M., & Norman, G. R. (2013). The roles of deliberate practice and innate ability in developing expertise: Evidence and implications. *Medical Education* 47(10), 979–989.

Lajoie, S. P. (2003). Transitions and trajectories for studies of expertise. *Educational Researcher* 32(8), 21–25.

Larkin, P., Mesagno, C., Spittle, M., & Berry, J. (2015). An evaluation of video-based training programs for perceptual-cognitive skill development: A systematic review of current sport-based knowledge. *International Journal of Sport Psychology* 46, 555–586.

Litzinger, T. A., Lattuca, L. R., Hadgraft, R. G., Newstetter, W. C., Alley, M., Atman, C., ... Yasuhara, K. (2011). Engineering education and the development of expertise. *Journal of Engineering Education* 100(1), 123–150.

McGaghie, W. C., Issenberg, S. B., Cohen, E. R., Barsuk, J. H., & Wayne, D. B. (2011). Does simulation-based medical education with deliberate practice yield better results than traditional clinical education? A meta-analytic comparative review of the evidence. *Academic Medicine* 86(6), 706–711.

Macnamara, B. N., Hambrick, D. Z., & Oswald, F. L. (2014). Deliberate practice and performance in music, games, sports, education, and professions: A meta-analysis. *Psychological Science* 25(8), 1608–1618.

Mahvi, D. (2010). Zen and the art of surgery: How to make Johnny a surgeon. *Journal of Gastrointestinal Surgery* 14(10), 1477–1482.

Mann, D. T. Y., Williams, A. M., Ward, P., & Janelle, C. M. (2007). Perceptual cognitive expertise in sport: A meta-analysis. *Journal of Sport & Exercise Psychology* 29, 457–478.

Moesch, K., Elbe, A. M., Hauge, M. L., & Wikman, J. M. (2011). Late specialization: The key to success in centimeters, grams, or seconds (cgs) sports. *Scandinavian Journal of Medicine and Science in Sports* 21, e282–e290.

Müller, S., & Abernethy, B. (2012). Expert anticipatory skill in striking sports: A review and a model. *Research Quarterly for Exercise and Sport* 83(2), 175–187.

Razer, A., McIntyre, C., & Fadde, P. (2019). Accelerating the noticing skills of nursing and medical students using video simulation. *MedEdPublish* 8(2). doi:10.15694/mep.2019.000078.1

Rich, P. J., & Tripp, T. (2011). Ten essential questions educators should ask when using video annotation tools. *Tech Trends* 55(6), 16–24.

Roca, A., & Williams, A. M. (2016). Expertise and the interaction between different perceptual-cognitive skills: Implications for testing and training. *Frontiers in Psychology* 7, 792.

Rousmaniere, T. (2017). *Deliberate practice for psychotherapists: A guide to improving clinical effectiveness*. New York: Routledge.
Salas, E., Rosen, M. A., Held, J. D., & Weismuller, J. J. (2009). Performance measurement in simulation-based training: A review and best practices. *Simulation & Gaming 40*(3), 328–376.
Schafer, T. (2011). Developing expertise through a deliberate practice project. *International Journal of Interpreter Education 3*, 17–27.
Schijven, M. P., & Jakimowicz, J. J. (2003). Construct validity: Experts and novices performing on the Xitact LS500 laproscopy simulator. *Surgical Endoscopy 17*(5), 803–810.
Schön, D. A. (1983). *The reflective practitioner: How professionals think in action*: New York: Basic Books.
Scott, D., Scott, L. M., & Howe, B. L. (1998). Training anticipation for intermediate tennis players. *Behavior Modification 22*, 243–261.
Sherin, M. G., & van Es, E. A. (2005). Using video to support teachers' ability to notice classroom interactions. *Journal of Technology and Teacher Education 13*(3), 475–491.
Simon, H. A., & Chase, W. G. (1973). Skill in chess. *American Scientist 61*, 394–403.
Smallman, H. S., & St. John, M. (2005). Naive realism: Misplaced faith in realistic displays. *Ergonomics in Design 13*(3), 6–13.
Sonnentag, S., & Kleine, B. M. (2000). Deliberate practice at work: A study with insurance agents. *Journal of Occupational & Organizational Psychology 73*, 87–102.
Staszewski, J. J. (2013). Cognitive engineering based on expert skill. In J. J. Staszewski (Ed.), *Expertise and skills acquisition* (pp. 29–57). New York: Psychology Press.
Syed, M. (2010). *Bounce: Mozart, Federer, Picasso, Beckham, and the science of success*. New York: Harper.
Thorndike, E. L. (1912). *Education: A first book*. New York: Macmillan.
Tripp, T., & Rich, P. (2012). Using video to analyze one's own teaching. *British Journal of Educational Technology 43*(4), 678–704.
van de Wiel, M. W. J., van den Bossche, P., Janssen, S., & Jossberger, H. (2011). Exploring deliberate practice in medicine: How do physicians learn in the workplace? *Advances in Health Sciences Education 16*(1), 81–95.
van de Wiel, M. W. J., Szegedi, K. H. P., & Weggeman, M. C. D. P. (2004). Professional learning: Deliberate attempts at developing expertise. In H. P. A. Boshuizen, R. Bromme, & H. Gruber (Eds), *Professional learning: Gaps and transitions on the way from novice to expert* (pp. 181–208). New York: Kluwer Academic.
van Gog, T., Ericsson, K. A., Rikers, R. M. J. P., & Paas, F. (2005). Instructional design for advanced learners: Establishing connections between the theoretical frameworks of cognitive load and deliberate practice, *Educational Technology Research & Development 53*(3), 73–81.
Ward, P., Farrow, D., Harris, K. R., Williams, A. M., Eccles, D. W., & Ericsson, K. A. (2008). Training perceptual-cognitive skills: Can sport psychology research inform military decision training? *Military Psychology 20*, S71–S102.
Ward, P., Hodges, N. J., Williams, A. M., & Starkes, J. L. (2004). Deliberate practice and expert performance: Defining the path to excellence. In A. M. Williams & N. J. Hodges (Eds), *Skill acquisition in sport: Research, theory and practice* (pp. 232–258). London: Routledge.
Ward, P., Suss, J., & Basevitch, I. (2009). Expertise and expert performance-based training (ExPerT) in complex domains. *Technology, Instruction, Cognition, & Learning 7*(2), 121–145.

Ward, P., Williams, A., & Hancock, P. A. (2006). Simulation for performance and training. In K. A. Ericsson, N. Charness, P. J. Feltovich, & R. R. Hoffman (Eds), *The Cambridge handbook of expertise and expert performanc* (pp. 705–722). New York: Cambridge University Press.

Williams, A. M., Fawver, B., & Hodges, N. (2017). Using the "expert performance approach" as a framework for improving understanding of expert learning. *Frontline Learning Research* 5(3), 64–69.

Williams, A. M., & Ward, P. (2003). Perceptual expertise: Development in sport. In J. L. Starkes & K. A. Ericsson (Eds), *Expert performance in sports: Advances in research on sport expertise* (pp. 220–250). Champaign, IL: Human Kinetics.

Zimmerman, B. J. (2006). Development and adaptation of expertise: The role of self-regulatory processes and beliefs. In K. A. Ericsson, N. Charness, P. J. Feltovich, & R. R. Hoffman (Eds), *The Cambridge handbook of expertise and expert performance* (pp. 705–722). New York: Cambridge University Press.

CHAPTER 41

COGNITIVE FLEXIBILITY THEORY AND THE ACCELERATED DEVELOPMENT OF ADAPTIVE READINESS AND ADAPTIVE RESPONSE TO NOVELTY

RAND J. SPIRO, PAUL J. FELTOVICH,
ARIC GAUNT, YING HU, HANNAH KLAUTKE,
CUI CHENG, IAN CLEMENTE, SEAN LEAHY,
AND PAUL WARD

Introduction

THIS chapter is about cognitive flexibility theory (CFT) as an approach to: (1) mastering complexity and combating its oversimplification; (2) fostering *readiness* to be adaptive in new cases; and (3) developing skill in *adaptively responding to novelty* by assembling aspects of prior knowledge and experience to build a *schema of the moment* to fit the needs of the situation at hand. Various forms of complexity and irregular relationship across cases that are nominally (but not effectively) of the same type present special cognitive challenges. We will illustrate CFT research on the cognitive shortcomings that inhibit performance in these complex and ill-structured aspects of domains of knowledge application and practice, and the CFT-based approach to learning and instruction addressing

Learning & Accelerating Expertise in Complex Environments: Lessons from Cognitive Flexibility / Transformation Theory by Rand Spiro, Paul J. Feltovich, Aric Gaunt, Ying Hu, Hannah Klautke, Cui Cheng, Ian Clemente, Sean Leahy, and Paul Ward

The author (Paul Ward's) affiliation with The MITRE Corporation is provided for identification purposes only, and is not intended to convey or imply MITRE's concurrence with, or support for, the positions, opinions or viewpoints expressed by the author.

them the CFT-based approach. A repertoire of theory-based modes of *deliberate practice* (Ericsson, Krampe, & Tesch-Romer, 1993) that differ in substantial ways from the kinds of practice usually stressed in the expertise literature for more routinized skills and situations is presented. At the heart of the chapter is a focus on the structures and operations of CFT that *accelerate*, directly and indirectly—and with support from the affordances of digital media for promoting context-sensitive rearrangement of knowledge and case experience—the development of adaptive skill. The work summarized here has been conducted over a span of decades and continues to grow to the present day (from Spiro, Vispoel, Schmitz, Samarapungavan, & Boerger, 1987; and Spiro, Coulson, Feltovich, & Anderson, 1988; to Spiro, Feltovich, Jacobson, & Coulson, 1992a; to Feltovich, Spiro, & Coulson, 1997, and Feltovich, Coulson, & Spiro, 2001; to Spiro, Collins, Thota, & Feltovich, 2003; Spiro, Collins, & Ramchandran, 2006, 2007; Spiro, Klautke, Cheng, & Gaunt, 2017; among others).

CFT began with two concerns: problems in complex understanding and difficulties in applying knowledge and experience to new, real-world situations. An initial motivating impulse was a dissatisfaction with *schema theories* (we will not review that extensive literature here; cf. Anderson, Spiro, & Anderson, 1978; Collins, Brown, & Larkin, 1980; Minsky, 1975; Rumelhart, 1980; Rumelhart & Ortony, 1977; Schank & Abelson, 1977; Spiro, 1977, 1980a, b). Schema theories appeared to be gravely limited by their reliance on prepackaged knowledge structures retrieved from long-term memory and instantiated by features of some present context. A need was sensed for a complementary theory of knowledge representation and use for more ill-structured domains, along with sufficiently flexible technology for case-based learning to develop such knowledge structures, and the ability to utilize them in building adaptive *schemas of the moment* for novel situations (Feltovich et al., 1997; Spiro et al., 1988; Spiro, Feltovich, Jacobson, & Coulson, 1992a, b; Spiro & Jehng, 1990; Spiro & Myers, 1984; Spiro et al., 1987). A second strand of CFT originated interactively with the first strand that addressed concerns about oversimplification of conceptual and case complexity. Considerable research has provided accounts of patterns of such oversimplification and remedies for those problems (Feltovich, Spiro, & Coulson, 1989, 1993; Feltovich et al., 2001; Spiro et al., 1988; Spiro, Feltovich, Coulson, & Anderson, 1989). All of these aspects of CFT will be overviewed in this chapter, with an emphasis on the development of adaptive skill (see Ward, Gore, Hutton, Conway, & Hoffman, 2018). Novel kinds of deliberate practice for adaptive flexibility and accelerative operations in case-based computer learning environments will receive special attention.

Circumscription of the Topic

Terms like *expertise* and *adaptive skill* are used in many different ways. In this section we delineate what CFT addresses and what it does not. One thing that will become clear is that *all* aspects of CFT address adaptive skill and its development.

A Broader View Than Just Expertise

CFT addresses more than expertise—or even the narrower designation, adaptive expertise (Hatano & Inagaki, 1986)—because the theory applies to adaptive response to novelty in two ways that are not always central to the expertise literature. First, it addresses conceptual understanding that does not necessarily involve decision-making. For example, many of the same CFT principles that apply to naturalistic decision-making apply similarly in learning for situational application of *concepts*. An example might be when biology concepts such as *adaptation* or *ecosystem stability* have to be used in the explanation of an ecology case. Second, CFT addresses sub-expert proficiency levels that require adaptive skill in knowledge/experience application appropriate to a given level of education or training. Principles similar to those used by CFT at the expert level to foster adaptive skill are used, except with lower expectations for the depth and range of adaptiveness that will be required. In other words, although the problem difficulty, background presupposed, and criteria for post-introductory adaptive performance will be different at different ages and ability levels (up through expertise), the principles—appropriately adjusted for those different contexts—will be the same or at least quite similar in their overall form. To the extent that dealing with the non-routine entails a shift from retrieving *intact* structures to assembling from aspects of prior case and knowledge structures to fit the context-sensitive needs of the situation at hand, we call this kind of *high proficiency* behavior *adaptive. The focus of CFT is adaptiveness, when it is required, rather than expertise.* CFT is a theory of learning and instruction for adaptive application of knowledge and experience in any non-routine aspect of a domain (or, for domains that are largely non-routine in the new cases one encounters, as a *primary cognitive worldview* for that domain, with switching to a more reductive worldview when regularities are detected, rather than the other way around; see Spiro, Feltovich, & Coulson, 1996).

In fact, because of the common CFT finding that *early simplifications impede the later attainment of complexity* needed for adaptive understanding and knowledge application (e.g., Feltovich et al., 1989; Spiro et al., 1989), it is important to begin as early as possible—far before expert levels of proficiency—with the use of CFT principles in any complex domain. This not only heads off the development of specific misconceptions and overly rigid knowledge representations that inhibit situation-adaptive knowledge application, but also avoids the establishment of overly simplistic and counter-adaptive worldviews (epistemic stances, mindsets, prefigurative schemes; Feltovich et al., 1989; Spiro et al., 1996) that then have to be undone and replaced by appropriately complex and adaptive mindset features. This is a much more difficult task because of how entrenched epistemic *prefigurations* become (they are lenses you look through *at* something else that you are thinking about, that you think in terms of, and thus are often not even aware that you are using; Bransford, Nitsch, & Franks, 1977; Feltovich et al., 1989; Polanyi, 1958; Spiro et al., 1996).

CFT builds in various features so that early introduction of complexity and features of complex worldviews is not overwhelming and is cognitively manageable. For example, a *New Incrementalist* sequencing logic is used that starts from compact bits of real-world complexity (e.g., a short scene from a larger case) so that it is cognitively manageable while still having complex worldview features, except in smaller instructional *doses* (e.g., multiple perspectives need to be represented, rather than a single "best" one, but with a small multiple; Spiro & Jehng, 1990). Early simplifications impede the later development of complex understanding. By starting with real but manageably sized parts of cases, tractability is possible along with an adaptive rather than reductive mindset.

An Emphasis on Adaptation to Ordinary Novelty

Another defining feature of CFT is that it is primarily intended to prepare people for *ordinary* novelty. An expert's large repertoire of stored *chunks* familiar from novice–expert studies (e.g., Chase & Simon, 1972), what might be called *infrequent but routinized expertise*, is not part of the main focus of CFT. This low frequency is one of several reasons that it takes so long to become an expert (Ericsson et al., 1993)—one needs to see a lot of such cases. With enough time and practice one can form templates or chunks for even more obscure recurring patterns. The problem of how to provide direct instruction for routine but low-frequency cases is merely one of time and effort in study, involving record-keeping and identification of the low-frequency patterns. *If* they can be identified, then, in principle, they can be directly taught in a relatively short amount of time with traditional instructional means that have been proven effective for learning routinized knowledge (e.g., Kirschner, Sweller, & Clark, 2006). We do not consider this kind of expertise to involve much if any, adaptive skill. And, as we shall see, when direct instructional approaches of the essentialized and proceduralized sort Kirschner et al. definitively require on the very first page of their 2006 article are applied in areas that require adaptive response to novelty, they hurt rather than help (Spiro & DeSchryver, 2009).

However, there is another reason one needs to see a lot of cases: When knowledge touches the unconstrained real world, with many factors coming together in nonpredetermined ways, novelty emerges with enough frequency (more in some domains than others) that an *adaptive* response is required. Such novel contexts, which occur with great frequency in the unconstrained real world[1] (especially as it has been increasing in its complexity, some would argue at an accelerating rate), and how to deal with them, are the province of CFT. As we will see later in this chapter, *CFT takes a two-pronged approach to resolving the seemingly paradoxical problem of how to use experience to prepare people to deal with novel situations for which their past experience has not already prepared them.* We will see as well how CFT uses case-based learning environments that compact, elaborate, and interconnect cases while simultaneously

---

[1] Novel contexts occur in constrained worlds like chess, of course, but far more often in the unconstrained real world.

providing cognitive support for learning and later application to provide that experience in a *greatly* accelerated time span.

The emphasis in this chapter is on the accelerated[2] development of adaptive skill. As such it falls in line with and builds upon a major recent book on expertise acceleration by Hoffman et al., (2014) and a landmark article by Ward et al. (2018), who presented a framework for the development of adaptive skill and argued persuasively for such a skill being the "*condition sine qua non*" of expertise. (Henceforth, we refer to this work as the "adaptive skill framework.")

Tabling Wicked Problems

The concern with *ordinary* novelty means CFT has not usually dealt with *wicked problems* (Rittel, 1972). We believe CFT has great promise for approaching such problems, and we are beginning to apply CFT in areas of grand social challenge; however, we have not done enough work to make claims for CFT in such contexts. We believe wicked problems should be addressed but not *emphasized*. They would be addressed in the same way as novel problems that are not wicked in CFT, but these are much more infrequent, success rates will be lower than non-wicked, pervasive, adaptive problems require, and they give the impression when emphasized that adaptive flexibility is much harder than it is for frequently encountered ordinary novelty, and thus undermines confidence needlessly.

A Complement to Other Approaches

While CFT principles are important to substantial aspects of needed processing in the development of adaptive skill, those principles and their application in situated vicarious learning are not sufficient in themselves and require supplementation by some actual experience as well as by principles from other theories.

An especially pertinent example of complementary work is that of Gary Klein, whose theoretical stance has many affinities with CFT. Klein and his colleagues' viewpoint, for example in the ideas of *flexecution*, macrocognition, and cognitive transformation theory (summarized in Klein, 2011), is either overlapping with CFT or naturally complementary, as Ward et al. (2018) and Hoffman et al. (2014) have illustrated at length and in depth. CFT and Klein's theory are the two most centrally drawn upon theoretical sources for the integrative frameworks in both Ward et al. (2018) and Hoffman et al. (2014). Klein's work has points of extensive consonance with CFT, as well as substantial non-overlap (though not inconsistency). Discussion of similarities and differences (for example, we have a slightly different view of the role and importance of *unlearning*) is a matter for another time.

In other words, some real experience (though not nearly 10 years' worth) and other approaches complementary to CFT are needed. We consider the application of CFT's

---

[2] It is worth noting that by "acceleration" we mean *both* the results of deliberate structuring of learning and practice, rather than leaving experience acquisition to serendipitous exposure and feedback over time, and expressly accelerative *moves*.

Cognitive Flexibility Theory and a Solution to the *Paradox of Novelty*

tenets to be necessary but not sufficient for the development of adaptive skill. Ward et al. (2018) and Hoffman et al. (2014) map many of these necessary complements to CFT, and for reasons of space they will not be discussed here.

The focus on adaptive skill under conditions of novelty in CFT derives from the acknowledgment of ill-structured aspects of knowledge and application domains. Deviations from routine are of sufficient frequency and importance to keep general prescriptions for understanding and action from accounting for enough of the variability in how concepts and experience have to be used in new situations. The resulting importance of dealing with *novelty* in newly encountered cases presents special cognitive challenges. In this section we begin to address those challenges, starting with the problem of novelty itself.

The Novelty Paradox: Acquiring Experience for Dealing with What Is Novel (and Thus Different from Experience)

We now face a paradox: How is learning possible under conditions of novelty? This is a paradox going back at least as far as Plato's *Meno*: How can we come to know what we do not already know? Since we are talking about *ordinary* novelty, commonly encountered in our increasingly complex and rapidly changing world, as well as always found in non-routine cases in professional domains, and frequently found in real-world naturalistic decision-making, this is a milder form of the paradox than that of more radical creative discovery. But it is a paradox nonetheless. If prior case experience does not generalize to new cases (by definition of ill-structuredness), then how can one become prepared to deal with new cases? In other words, how is acquiring adaptive proficiency *possible*?

CFT has a ready answer to the paradox of learning for novelty that is *twofold*, as described in the following two sections.

Reuse of Old Ingredients in Novel Assemblies

Cases and concepts are complex. Their parts and aspects can become the basis for old case–concept combinations to be reused in new contexts. Adaptive responses are novel; their ingredients are often not.

In these new contexts concepts can *not* retain fixed meanings and still retain their adaptive flexibility. Family resemblances, yes; rigid, predefined meanings, no. Hence learning in CFT environments places a central emphasis on the deliberate practice of open and context-dependent meaning-making so that concepts are ready to be adapted to new contexts and have greater readiness to participate more adaptively in larger-scale concept combination in situational sense-making (Ward et al., 2018). Case-based hypertext learning environments, multiply coded (as we will see later in this chapter), are ideal for demonstrating *conceptual variability*. Simply click on a key concept and various contexts in which it occurs is quickly demonstrated. And, when learning becomes more complex, they are ideal for similarly demonstrating *concept-combination variability*.[3] The result is a threefold benefit: The process of tailoring concepts to contexts is *deliberately practiced* and contextual dependencies are learned, increasing *readiness* to adapt concepts to future contexts of application. The general conceptual-variability (rather than fixed meanings) feature of what we will call, later, an adaptive worldview is learned and reinforced; and the readiness of concepts to be *combined* with other concepts in novel situation-adaptive assemblages (constructed *schemas of the moment*) is enhanced. Again, the latter process is the central act of conceptual cognitive flexibility. Another benefit of this and other CFT *moves* related to concepts (like those discussed in the following and later in the paper) is that they accelerate *conceptual* experience by countering common tendencies toward oversimplification that have been found in a large amount of CFT research (see the next major section, dealing with patterns of oversimplification of conceptual complexity). In the expertise field, *acceleration* more often refers to that aspect of expertise related to acquiring case experience, rather than experience necessary for the development of applicable conceptual understanding as such. However, in ill-structured domains, concepts are Wittgensteinian in the sense that he said meaning is *use* (1953).

Also, some CFT systems *selectively highlight* concepts active for a part/scene of a case (i.e., not *all* of a case)—the relevant concepts and their combinations *change* during the course of a case. Furthermore, and as we will see, CFT works with assembly from fragments rather than larger wholes since that allows for more adaptive flexibility (see Spiro & Jehng, 1990, for much more on both of these points). A set of key concepts for a domain are displayed, and then, as stages of a case are displayed, only those concepts pertinent at the time are highlighted, thus displaying the configuration of concepts that go together for *that* case context, with those contextual features, as a case unfolds. This kind of *dynamic context-sensitive selective highlighting* was first employed in the early *Cardio-World Explorer* (see Spiro et al., 1988). The benefits here are *deliberate practice* in tailoring of concepts to *each other* (a process in turn supported by the conceptual-variability searches just described) *in the context of overarching case constraints*; *deliberate practice* in tailoring concept *configurations* to dynamically unfolding parts of cases; and *deliberate*

---

[3] It is important to note that *all* learning with CFT-based systems occurs in the context of some form of *active learner involvement*, whether in real-world problem solving or some other way. Passive/receptive learning is rarely if ever used. Since this is a constant, we will not keep repeating it.

practice of these general features of the adaptive worldview. All of this contributes a supportive ground for using similar situation-adaptive processes in a *new* case context.

Of course, CFT instruction also includes various forms of kaleidoscopic assembly work to deliberately practice the building of *schemas of the moment* for new cases/situations/problems.

Another implication of this part of the solution to dealing with ordinary novelty—the building of new blends from old ingredients—is that learners working with CFT get lots of deliberate practice in situated assembly of *fragments*. This is not to say that CFT promotes fragmented thought. Fragments are a problem only if they are not coherently combined with other fragments in a context of use. CFT prepares learners to use fragments in many kinds of situational assemblages (Spiro & Jehng, 1990). Fragments also have the advantage of being more cognitively manageable, a very valuable side benefit of dealing with complexity without it becoming cognitively overloading.

Deployment of Common Structural Meta-features of Novel Complex Cases Organized in an Adaptive Worldview

The second aspect of CFT's solution to the novelty paradox of how to prepare for new cases that experience cannot directly prepare you for—used in combination with the re-assembly solution just discussed—is a simple one: Even though situations are new and vary a lot from one to the next (despite often being grouped together and having the same name; this is a definition of ill-structuredness), *there is a common set of meta-features of an adaptive mindset that are found across all situations in complex, ill-structured domains.* (It is what *makes them* complex and ill-structured.) Specific content schemas do not have general applicability; the general *ways* that knowledge and experience are used adaptively *do*. One can already see a hint of this in the subsection on rearranging ingredients to form new situation-adaptive assemblages, "Reuse of Old Ingredients in Novel Assemblies," where we have already mentioned or implied such features as rearrangement (rather than retrieval from long-term memory of intact, precompiled/prepackaged structures); nonlinearity of processing in support of coming up with new combinations; an emphasis on context-dependency; and so on. This is just a sample list; we will make the sample larger a little later in this chapter when we return to the adaptive worldview.

In the past we have called the adaptive worldview the expansive worldview (contrasted with a reductive one). We now prefer to emphasize its adaptive function rather than its expansive description. Either term is helpful though. It is also worth noting that although we call it a *worldview*, it is actually a combination of *ontological* features—how the world, or an aspect of the world, *is*—and *epistemological* features—how one should *learn* about and understand such a world (Spiro et al., 1996). As mentioned earlier in this chapter, we have sometimes referred to these as *pre*-figurative schemes (meta-mindsets, if you will), schemas that prefigure the *form* of content schemas (Feltovich et al., 1989; Pepper, 1942; White, 1973). So novelty is helped to be processed by ensuring that the structuring of learning and of knowledge application are guided by the features of an adaptive worldview and not a reductive worldview that cannot prepare for novelty if a domain aspect is non-reducible. The features of the adaptive worldview *are* common across cases making cases *not* novel at this level of *meta-features*. This ensures that learning and deliberate practice occur in the right overarching mental context, one suited to the development of adaptive skill. There is a virtuous circle of new cases/experiences being learned, analyzed, and represented in this more adaptiveness-ready way while old cases are operated upon in new contexts in an open and recombinable rather than rigidly intact way. In other words, by applying the adaptive worldview as learning happens, the adaptive *readiness* of resultant learning is increased.

To understand how this adaptive worldview is derived and then learned and deployed, we must first detour to a discussion of CFT research on oversimplification of complex concepts and, relatedly, psychological *shields* against correcting oversimplifications even when better information is available (Feltovich et al., 2001). In that next section it will be important to take particular note of how many of these common meta-features of the adaptive worldview are the *opposite* of those usually valued in learning and instruction, what we have called the reductive worldview (Spiro et al., 1996). That is because so much learning and instruction, research and practice, has been predicated on an assumption of domain well-structuredness and reductive potential. However, what helps for one hurts for the other, and vice versa. So that assumption, where it is not appropriately applied (that is, in all the kinds of domains and aspects of domains discussed in this chapter, that require larger amounts of adaptive processing), is harmful to an egregious degree. Why has the adaptive worldview been so relatively neglected? We argue that this is because so much of the study of learning has been conducted in well-structured domains (Spiro & DeSchryver, 2009), which also happen to be the easiest ones to do research in, as well as to learn, teach, and test in (Feltovich et al., 1989; Spiro et al., 1987).

Cognitive Flexibility Theory, Patterns of Oversimplification of Conceptual Complexity, Shields from Dealing with Complexity, and the Importance of Epistemic Worldviews in Adaptive Skill

A crucial aspect of the development of adaptive skill (and expertise generally) is the mastery of domain-central complex concepts and the avoidance of conceptual oversimplification (Hoffman et al., 2014; Klein, 2011. In the Ward et al. (2018) *adaptive skill framework*, this idea is emphasized in their Principle 5, *complexity preservation*, including the crucial tenet "do not oversimplify."

In this section we highlight various forms of CFT research on *conceptual* complexity oversimplification. (Later in this section and especially the one that follows, we will talk more about *case* complexity.) There is considerable theoretical and empirical detail behind the work in this section, and it has been presented in detail in numerous publications and discussed widely. We will present just the briefest of overviews of a subset of the work drawn from the larger corpus of studies.

Oversimplification of Complex Concepts: Types of Maladaptive Reductive Tendency and Patterns of Their Occurrence

We have conducted numerous studies demonstrating the ways that advanced learners acquire and apply oversimplified conceptual understanding. For example, in one study we examined in intensive detail the role of instructional analogy in *impeding* the development of complex conceptual understanding of physiological phenomena when learners are tested much later in instruction, when more advanced understandings than those of introductory learning have been presented and are required to be understood (Spiro et al., 1989). For just one example of analogy, in the context of the physiology of force production by muscle fibers (the common use of rowing crews as an analogy to represent the structure of muscle cells and the ratcheting and pulling process of sarcomere arms, etc.), we found evidence of *nine distinctive kinds of analogy-induced misconception* at more advanced stages of instruction. Any analogy is partly correct and is partly misleading and/or missing aspects of correct understanding. The misleading and missing aspects are predictive of future oversimplification of complexity.

This is a pattern we have found in a number of studies: *early simplifications (sometimes for modest introductory learning goals) interfere with the later attainment of important kinds of complexity*. The early instructional simplifications are *seductive reductions* and exert an outsized cognitive influence on limiting and narrowing the kinds of knowledge structures that are eventually developed.

We have identified a large number of types of oversimplification. The following is just a small illustrative subset (see the papers for descriptions beyond the shorthand designations provided here and for methods used to uncover them, as well as how they are evinced in concrete examples of understanding situations; e.g., Feltovich et al., 1989, 1993, 2001): separation from context; attribution of complex intrinsic causality to simple causation by an external agent; reduction to analogy; treating the multiplicative as additive; relying on surface features, and many more.

Why do these oversimplifications occur? There are many reasons, including a preference for cognitive ease and security in learners, teachers, textbook writers, and test constructors, referred to a long time ago, in an impolitic moment as an educational "conspiracy of convenience" (Spiro et al., 1987). We have even observed canons of scientific research being misapplied as mistaken justifications for oversimplified understandings (Feltovich et al., 1989). All are examples of the inappropriate application of a reductive worldview, which we will discuss later in this section.

A further problem with the oversimplified conceptual understandings caused by these reductive tendencies is that they combine and influence each other in larger contexts of understanding. The various causes of oversimplification and patterns of their interaction and mutual bolstering were detailed in an analysis of a common misconception of the nature of congestive heart failure held by medical students and many doctors (Feltovich et al., 1989). Different individual reductive tendencies produce local misconceptions that compound to form a climate of oversimplification that results in the larger scale misconception about heart failure that is made up of interlocking misconception combinations.

A particularly insidious effect of the reductive tendencies and the reductive worldview that encompasses them is that the same or very similar tendencies to those identified above (that lead to the *formation* of oversimplified understandings) are used by learners as *knowledge shields* to protect themselves from having to do the hard work of developing more appropriately complex understandings (Feltovich et al., 2001). As a result, learners are often unaware that their understanding falls short of the mark. Roughly, shields correspond to the reductive tendencies, and instead of *producing* conceptual error, they are used to *reinforce* stability in conceptual error and prevent improved understanding from occurring by *fending off* conceptual change. Some of these knowledge shields are similar to the reductive tendencies (e.g., *extirpation from context* or *reduction to analogy*), while others are convenient rationalizations for exploring no further and accepting a current, oversimplified understanding (e.g., *argument from authority*; for example, "That is what Professor X said, so it must be fully correct").

How should these reductive tendencies and negative cognitive shields that guard against having to deal with more cognitively challenging learning (or teaching, for that matter) be counteracted? One answer is to find ways to direct learners away from the reductive direction to a more expansive one. So, returning to the analogy example, should analogies *not* be used, despite the known effectiveness for learning of relating something new to something familiar? No. The answer to how to get the benefits of analogy but avoid the seductively reductive effects of overreliance on analogy is to use *multiple* analogies. For example, in Spiro et al. (1989) we suggest an alignment of what works and does not work for four different analogies useful for understanding the physiology of force production. Information is provided about an analogy's strengths and weaknesses, which provides scaffolding for the context for use for each analogy. For example, in a context in which limits on the lengthening of muscle fibers are an operative concern, the rowing crew analogy would not help, but it would point to a toy finger-cuffs analogy in which the collective limits of fiber lengthening result from constriction of an elastin coating on the myofibril bundle, with greater constriction the greater the attempts to stretch longer. This approach provides pointers from the weaknesses of one analogy to the strengths of another to aid *multiple analogy assembly/alignment in context*. It is *adaptive* to have available, and to be able to *deliberately practice*, multiple possible representations of a concept as a function of different contextual needs; and the same goes for counter-reductive antidotes for other kinds of oversimplification. This kind of approach also produces an incremental change in worldview. For example, the more one sees examples in which a single mode of representation is inadequate and coordinated multiple conceptual representations work better (and can be cognitively supported), the harder it is to maintain a worldview value for encompassing single representations rather than coming to believe in the cognitive value of multiple representations for complex concepts.

The best remedy, of course, is a more general one, and that is to promote change in worldviews, from a reductive to an adaptive worldview. This, in effect, turns negative knowledge shields into positive "knowledge *swords*" that seek out and penetrate oversimplifications, thus promoting fuller understanding and improved skill in adaptive knowledge application. We now turn to the adaptive worldview and how it is promulgated in CFT.

The Adaptive Worldview

We began talking about the adaptive worldview earlier, as part of the resolution of the novelty paradox. One way to prepare for the new is to be always thinking in terms of the general forms in which the new comes. Together, the specific reductive tendencies coalesce as a reductive worldview with *cognitive values* antithetical to the features of processing conducive to the development of adaptive skill. The opposite of each such tendency is a part of the adaptive worldview. Examples of cognitive values of the adaptive worldview (cf. Spiro et al., 1996), which very much overlap with the adaptive skill framework of Ward et al. (2018), include the following: pay attention to cases in the variegated richness while de-emphasizing the primacy of concepts (which serve a needed subsidiary function to cases in ill-structured domains); use multiple rather than single conceptual relations (as in schemas, prototypes, analogies, perspectives, etc.); treat cases as wholes with emergent properties so they are greater than the sum of their parts; increase the attunement to difference and decrease the bias toward seeing similarity; expect unpredictability, irregularity, contingency, indeterminateness; expect to return to earlier cases in new contexts to bring out facets that were hidden in the earlier context—*nonlinear* revisiting is not repeating; embrace flexibility and openness of knowledge representation over rigidity; stress context dependency over context independence; avoid rigidity in understanding, remaining *open* instead, with an appreciation for the sometimes limitless range of uses of knowledge in new combinations, for new purposes, in new situations; rely on situation-adaptive assembly of prior knowledge and experience rather than retrieval of intact knowledge structures and procedures from long-term memory; and so on (for more on the ways CFT *prefigures* understanding and readiness to adapt knowledge to new situations, see Feltovich et al., 1989; Mishra, Spiro, & Feltovich, 1996; and Spiro et al., 1996).

In one sample study (Jacobson & Spiro, 1995), these various features were combined under an omnibus experimental condition and compared in a true experiment with tight controls for content and study time while using a traditional book-like organization of concept-based chapters with cases nested under them. There was not much difference in memory for the materials presented, but *transfer* to a new situation was significantly greater for the material organized by the features of the adaptive worldview.

Direct and Indirect Teaching of the Adaptive Worldview

In the next section of this chapter, we will discuss operations in CFT and in computer-based CFT learning environments that develop specific aspects of adaptive readiness and skill in adaptive performance in context. Cumulatively, those specific moves *indirectly* help to form the adaptive worldview in learners.

However, direct instruction in worldview can also be helpful. The common metafeatures of the adaptive worldview can be taught *directly and generally* in *four* ways.

One way is to provide a non-technical, jargon-free introduction/overview of the CFT mindset and its underlying adaptive worldview, both their features and rationale (emphasizing why learning in an upcoming context should have that mindset activated as a cognitive attunement). This is common practice in learning with CFT. A study by Jonassen, Ambruso, and Olesen (1992) showed that improved transfer to new situations was induced in a controlled experiment when a CFT hypertext system for transfusion medicine was taught under an experimental condition that was first provided with such an overview that *taught* CFT as a way of thinking before the learning stage began.

In a second way, on any occasion that a general adaptive worldview feature comes up that is of central performance, repeat it in that context, as a kind of *mantra*. For example, given the prevalence of *oversimplification*, processing a concept or case in a reductive manner or jumping to an oversimplified solution, CFT instruction frequently invokes mantra-like reminders like "it's not that simple," or "it depends." When you hear these mindset mantras often enough, with a clear and present instance of oversimplification as the context and a demonstration of ways to think that would not be that simple (and how they contrast with the learner's reductive tendency), that mindset feature is quickly learned and adopted.

In a third way, we use a *four-stage model* for worldview change (Spiro et al., 2007): First, demonstrate to a learner that he or she *has* a reductive worldview by creating situations that make it salient. Often, we are not aware that we have and pervasively use these epistemic stances. Second, show how that worldview is maladaptive. Third, introduce the adaptive worldview as a substitute. Fourth, demonstrate the latter's operation and provide support for mastering it.

Finally, in a fourth way, mindsets are learned *incidentally* by the processes of CFT learning. For example, as we will see later, in a CFT learning environment one of the first moves is to do a conceptual variability search. Once a learner has seen several real world cases that use a key concept in ways that clearly have a family resemblance yet clearly have differences in patterns of use in different contexts, they quickly realize that a reductive definition or set of necessary and sufficient conditions for using the concept will not work and that a mindset for conceptual variability within families but across contexts is the cognitive attunement they need to have in these ill-structured aspects of domains.

A Question for the Future: Do We Need to Change the Dominant Mindset for Thought?

The work of CFT has always prescribed determining as best as one can (and changing if circumstances change) whether aspects of domains are primarily well- or ill-structured, because an opposite mindset and process of learning and instruction are required for the two kinds of domains, with what helps one hurting the other. A question we have never addressed until recently is *which of the mindsets should be the primary/first cognitive attunement?* Of course, whichever is the initial bias provides for a search for cues that the other is required, at least in part. However, with the world becoming increasingly complex and changing at an accelerating rate (see, for example, McChrystal, 2015), the prevalence of ill-structuredness has become so great that perhaps it is time to question the centuries-old Enlightenment assumption that regularities are pre-eminent and should instead be sought for but not expected. This is important because if orderly and predictable structure is the first expectation, then the schemas that are selected produce a screening effect for what does not fit those schemas, leading to confirmation bias. (This is what frequently happens with doctors encountering non-routine cases; Feltovich et al., 1997.) In that case, ill-structured aspects will not be noticed. However, instead starting with the adaptive mindset will naturally capture order and regularity since nothing is initially screened out (it is *expansive*, not reductive).

SOME ILLUSTRATIVE FEATURES OF LEARNING AND INSTRUCTION IN CFT-BASED ENVIRONMENTS: DELIBERATE PRACTICE AND ACCELERATION IN THE FOSTERING OF ADAPTIVE SKILL

How is learning and instruction based in CFT done in case-based hypertext learning environments? In this section we only have space for a brief overview of some key illustrative aspects of structure and process, and how they pertain to the development of adaptive skill (for more detailed treatments, see these papers, and other CFT papers that they cite: Spiro & Jehng, 1990; Spiro et al., 1987, 1988, 1992a, 1992b, 2003, 2006, 2007, 2015, 2017).

We will begin with some information about how CFT-based computer learning environments are designed, and how their features dovetail with the needs for the development of adaptive skill. We then talk about the crucial process of concept and case selection to maximize adaptive potential. At that point, reasons of space will require that we shift to a mere listing, with brief description, of kinds of deliberate practice in CFT learning and instruction (beyond those already discussed earlier), using those learning environments, that are tailored to the fostering of adaptive *readiness* and promoting more successful adaptive *performance* in new situations. (We will point to intersecting principles of the Ward et al. (2018) framework along the way.)

Notes on the Role of Flexible Media: CFT Hypertext Learning Environments

CFT learning environments have always been highly context-sensitive, nonlinear and rearrangeable, case-based *hypertext* systems: The flexible medium has been used to promote a *cognitive message* of flexibility, paraphrasing McLuhan's (1964) famous "the medium is the message." As McLuhan and others have shown, our underlying way of thinking is influenced by the properties of the media we interact with, not just, and even beyond, the content the media are carrying. We use the affordances of the medium to allow connections across different parts of cases along different conceptual dimensions to emerge in different configurations in different contexts for different purposes. This leads both to more adaptively flexible learning and to *deliberate practice* in that *way* of thinking, so that it can be deployed to deal with novelty in the manner described earlier in the discussion of the novelty paradox.

Such systems have been used for CFT-based instruction since the first one developed in the late 1980s to teach complex and adaptive multi-thematic learning of literary interpretation, using early random access (CAV) videodisc for the film *Citizen Kane*—KANE (knowledge acquisition in nonlinear environments; Spiro & Jehng, 1990). They have been called cognitive flexibility hypertext systems (*CFHs*), and with later variants that had an increased emphasis on experience and expertise *acceleration*, experience consolidation systems (*ECSs*; Spiro, 1988) or experience acceleration support environments (*EASE* systems; e.g., Palincsar et al., 2007). Though they have evolved over the years, they have also had essentially the same principled design structure. This allowed *shell programs* to be written for quick generation of a CFH as content for a domain is filled in. Again, CFT is a domain-independent theory of learning and instruction in complex domains to uncover the unique, domain-specific contours of any particular domain from the relationship between its core concepts and emblematic cases.

Unique Features in the Design of CFHs, and Their Function in the Acquisition of Adaptive Skill

There are two basic kinds of hypertext (essentially multiple texts, images, and other media explored in a nonlinear manner based on some underlying programming structure). The most familiar is the kind we see on the World Wide Web. It is *link-based*. CFHs have always taken a different path to hypertext. Our goal is that the learner should be able to actively construct an almost limitless number of text (or image or video) combinations for an innumerably large number of purposes, contexts of application, and so on, without being limited to what CFT theorists and learning environment developers could anticipate for possible uses or structures of instantiation. Thus, we have never relied on pre-programmed links (other than some occasional ones on an opportunistic, ad hoc basis, off the main programming line, when a particularly useful connection is one we especially want learners to see; these are easy to add as they come up). Instead, cases are coded in a multidimensional space of key concepts and context features that allow learner queries to *generate* alternative suggested connections across cases. These multiple tags and tag combinations (we cannot be certain, but we think this was the first use of tags of the sort so common today, although we called them *themes*). In that way learners choose for themselves, in active learning and problem-solving contexts, from a vastly large range of possibilities that the CFH narrows and makes cognitively manageable for them (many that we on the psychologist/domain expert design team never had to envision).

Compared to hypertext that is reliant on precompiled rather than CFH's more open, *intrinsically generative* pre-structuring, the CFH learning environment is cognitively engineered—prefigured even at this most basic programming level—to promote adaptive response to novel cases in the manner of the earlier discussed novelty paradox. The system is built not only to show varying kinds of deliberate practice of conceptual variability, context dependency, and ecological concept-combination-in-case-context, of the kind discussed earlier (promoting what we call *adaptive readiness*), but to provide an arena for combinatorial idea play in the deliberate practice of situation-adaptive assembly of candidate schemas of the moment. This occurs by allowing rapid tryout and selection of combinations of fragments of prior cases aligned along varying conceptual dimensions (promoting *adaptive performance*). This is a kind of evolutionary epistemology of variation and selection (Campbell, 1974) that always happened among elite thinkers but not for *all* learners (since the technology was not available to support this kind of learning), who all now have to learn how to respond to novelty encountered in an everyday way, in situations for which they were never explicitly prepared.

We will not talk about weighting algorithms that can be used for ordering the sequence of cases that come up when a given set of conceptual- or case-feature search criteria have multiple hits (used in some but not all CFHs). Part of the point is if such more advanced programming is not cost-effective, the sequence really doesn't especially matter—you will see many cases with common joint properties, and each teaches a valuable lesson. If you have more time, then look at more—additional case experience is always helpful for learning, though not always *efficient*. And you are still way, way ahead compared to naturally encountering cases in *noisy* real-world environments separated sometimes by years and never organized in any systematic way. Noticing connections and situation-sensitive assembly possibilities in real-world learning over years of experience is wildly hit-or-miss. In CFHs, that process is supported and made cognitively manageable in a radically shortened amount of time, and in that sense *all* CFT learning environments *accelerate experience and the development of adaptive skill* (again, compared to natural learning); of course, CFHs also have specially designed accelerative moves built into their intrinsic structure and the basic processes that operate on that structure as well as added on to that structure (as we will see illustrated in the section, "Illustrative Operations in CFT-Based Learning Environments").

Another important feature of CFH design that fosters the development of adaptive skill is that the *scene or part* of a case is the coding unit (usually *temporally* determined for cases that unfold dynamically over time; note there is no correct way to break the case into *mini-cases*—lots of different ways of dividing at *convenient* seams work roughly as well), *not* the case as a whole (Spiro & Jehng, 1990). This is important because cases tend not to be homogeneous over time—an important feature of ill-structuredness that CFH users quickly learn. Hence, complexity is not needlessly reduced by coding for *common denominator* features that run through the whole case, and parts of cases that are relevant to a potential schema of the moment assemblage for a new case are not missed because the pertinent features are found only in that part of the case and not in the case as a whole. Even if the pertinent part of a past case is part of the coding of a whole case, it would be highly inefficient to have to call up the whole case and then search for and choose the pertinent small part of the larger case. Getting to the pertinent part directly speeds up the process (acceleration that is intrinsic to the setup of the system), makes for more cognitively manageable processing of complexity (bite-size pieces of knowledge that nevertheless have the properties of the adaptive worldview, just *smaller* multiples of representation perspective, *fewer* multiplicative connections, and so on), and permits a much larger number of experiences to be processed in a much smaller amount of time (another accelerative feature; see below about the exponential increase in potential case combinations with each added case). This allows greater fidelity to the case at hand in assembling a schema of the moment because you are "drawing" the schema from smaller *pixel* sizes and thus can have a smoother rendering of a fit to the needs of a novel case than if one is assembling from big block precedent cases (among other benefits of coding off of mini-cases rather than the larger cases from which they are extracted.). Of course, the learners first become very familiar with the whole case before they work with these case fragments so the latter come with their fuller case context, which is also available at a click if they have forgotten it. (See Spiro & Jehng, 1990, for a long list of other virtues of structuring in small segments). A case conceptually coded monolithically is an oversimplified case.

Concepts, and concept combinations, found in one part of the case will not be salient in another part of the case.

A practical benefit of coding off of a case's features (rather than pre-coding links) is that it makes it easy to add cases indefinitely as more and more are encountered, because each new case does not have to have links added to large numbers of other, earlier coded cases. Instead each case is simply coded for its conceptual and case features; relevant additional codings are easy to add. At the same time one can go quite far in deliberate practice for adaptive readiness and performance with just a *beginning* CFH because, as we will see, we work from dense *crossroads cases* and multiple *candidate schemas* of the whole domain. If each case can be broken into several mini-cases, each of which has multiple concept and case features, the number of potential case-to-case connections increases exponentially with each case added. (Customized codings can and should be able to be added by learners and mentors to the wide-scope codings provided by the environment designers, but space does not permit discussion of this important feature.)

Illustrative Operations in CFT-Based Learning Environments

The same set of CFT principles has been used to develop learning environments for domains as disparate as the following: cardiovascular medicine; the teaching of reading comprehension strategies; literary interpretation; military strategy; natural selection in biology; twentieth-century social and cultural history (history of modernity, more broadly); controlled experimentation and measurement in high school physics; among others. How can a general theory work in so many widely different, domain-dependent areas? Again, CFT is a domain-independent theory of how to *crisscross* cases to uncover the unique contours of their domain.

We begin at the beginning, with the cases and concepts that are selected in CFT systems for initial case–concept coupling (Ward et al., 2018, Principle #2) to maximize adaptive skill development. Cases are central because in ill-structured domains you cannot generalize away from them (that is what makes the domains ill-structured). So instead of having cases nested as examples under concepts, as one would do in well-structured domains where cases/examples can be said to illustrate a concept, concepts weave through cases as cases naturally occur. When one cannot learn general principles that apply across cases, learning must be case-based.

CFT systems start from sets of 10–20 *crossroads cases*, called that because they are densely packed with many of the conceptual features crucial to the domain and that could even be considered *emblematic* of the domain. Crossroads cases are a good investment. For example:

- They allow many lessons to be taught from one source—one kind of *deliberate practice* in CFT learning is to unpack a crossroads case in as many ways as possible, producing deep learning for that case and also teaching the adaptive

worldview lesson that cases are complex and a case is not just a case of some one thing, but has many possible "title"s—cases are cases of *many* things.
- As crossroads cases are reused in new contexts, they quickly become overlearned. At that point, when showing just a small distinctive part of a case (we call these *epitomés*), the rest of the case is evoked, cognitively brought "along for the ride" with the epitomé. Once epitome mode is reached, many case combinations and contrasts can be made in a much shorter amount of time. This can be done *both*: (1) for the kind of combinatorial idea play we have discussed, *deliberate practice* for the adaptive *performance* skill of assembling a schema of the moment for the situation at hand; and (2) for multiple case contrasts that build complex understanding more quickly. CFHs have design formats to facilitate these processes. For example, epitomés format allows learners to quickly do crisscross comparisons of several cases with single clicks on each case, allowing dozens of comparisons in the time that the full experience of a new case would only allow a few, while also allowing the easy swapping in and out of cases, an important *accelerative* feature for the *deliberate practice* of combinatorial idea play. Quadrant mode allows more focused comparisons of multiple cases—how is Case A like Case B but not like Case C, and so on—so the subtleties and nuances of expert understanding are inculcated, and the ill-structured nature of the domain is illustrated (using kinds of deliberate practice tasks like "Find surprising *differences* between cases that appear similar, and find surprising *similarities* between cases that appear different on the surface").
- Because of the crossroads feature of these cases, each one can be used to reach many other cases, accelerating the building of networks of connection. This kind of connecting in CFT is deliberately practiced with *hub and spokes* type tasks. Each crossroads case is treated as a hub, and the learner has to find as many other crossroads cases it can be related to (spokes), on as many conceptual dimensions (that is, from as many conceptual perspectives) as possible. Then the hub is shifted to another crossroads case and repeated in turn with more and more crossroads cases being added. A variant is to see how what is most important about a past case changes in the context of other new cases—a mantra of CFT is that "revisiting is not repeating."

On the conceptual side, CFT systems begin with four to eight of what we call *candidate schemas of the whole*, higher-order concepts (like traditional schemas) that could reasonably be considered by experts to be the most dominant, superordinate schema for the domain. (The reason four to eight are used is so that once learned they can be kept in mind simultaneously as active *screeners*.) Of course, the complex and ill-structured domains of the adaptive worldview are heterarchies, not hierarchies, and thus no single schema is best (that is another feature that makes it an ill-structured domain). Learners in CFT systems come to realize this by *deliberate practice* of tasks that require them to look at a set of new cases and find ways that different candidate schemas work better for some than for other cases, and for some cases, combinations of candidate schemas *make the best sense* of the case. They quickly learn that if there are a number of great schemas that can be best applied in different contexts, no one single schema is the best, thus disarming that very dangerous reductive trend of thought that there is a single best schema. And, of course, adaptive potential is greatly increased by having many powerful schemas to choose from and/or combine to most ideally fit the needs of a new case.

It is worth noting that there is no *best* set. Any reasonably chosen set of crossroads cases and candidate schemas of the whole will do the job. What is important is that both within the case set and within the concept set, there is some overlap (to facilitate connecting and combining) but not too much overlap (producing inefficient redundancy), while at the same time not leaving gaps. Those are some of the criteria used for case-selection and for schema/concept-selection.

Another important kind of *deliberate practice* that supports adaptive performance with new cases is situation-adaptive assembly of knowledge and experience. Using the CFH system as support, the learner plays with different combinations of past case fragments, from varying conceptual perspectives, to work toward approximations of schema of the moment assemblies that maximize relatedness of new cases to past success contexts and minimize relatedness to past failure contexts. This approach to preparing for novelty is a very old one, albeit one that never had this kind of extensive formal support. A long-favored theory of the creation of novel ideas is *combinatorial idea play*. Einstein, asked about his creative process, said that it involved a "play" of ideas (1952). Similarly, Poincaré described his process in discovering Fuchsian equations as involving a "dance" of ideas he suspected were relevant but did not know how to put together (1952). Learners in CFT environments use just this kind of approach, except to deal with ordinary novelty and not to invent relativity theory.

There is much more that can be said about kinds of deliberate practice and acceleration in CFT, along the lines presented in this chapter as well as in areas we did not have space to cover (such as the CFT approach to mentoring, feedback, and active learning contexts). Of particular interest but finding no space here is the use of embodied, multisensory metaphorical representations to make additional complexity cognitively manageable (Spiro et al., 2003, 2007) and, relatedly, research on how affect plays a role in joining parts of cases that co-occur in a new case but have no a priori basis of connection (Bartlett, 1932; Spiro, Crismore, & Turner, 1982; for more on the role of experiential tone in understanding and memory, see Spiro, 1980a, b). However, the present purpose was to give a brief overview of the extensive work on CFT's approach to the development of adaptive skill, and the illustrative ideas presented here go a fair way toward providing a sense of CFT's concerns and how CFT-based learning happens. Naturally, much more detail will be found in past and forthcoming papers.

Change is Difficult with Existing Approaches to Testing: New Tests for Adaptive Readiness and Adaptive Performance

Finally, an obstacle to this kind of learning that people will point to is assessment. Obviously, if teacher and trainer evaluation and accountability depend on student or trainee performance on short answer tests that depend mostly on retrieval of information from memory, instructional progress in these new directions will be retarded. However, frameworks for new kinds of tests of complex understanding and transfer exist (e.g., Feltovich et al., 1993; also see discussions of "preparation for future learning" in Bransford & Schwartz, 1995 and Spiro et al., 1987). And new, performance-oriented measures are being developed all the time (e.g., Hoffman et al., 2014).

We were recently asked to write an assessment handbook chapter on CFT and the assessment of *twenty-first-century skills* (Dede, 2010), with an emphasis on adaptive skill. In that chapter, by Spiro et al. (2017), we showed examples of how the kinds of learning operations and deliberate practice tasks described in this chapter, both for adaptive readiness and for adaptive performance, have test forms analogous to the kinds of deliberate practice. Similar kinds of examples of new kinds of tests could have been provided for many other *moves* described in this chapter.

SOCIETAL IMPLICATIONS OF *COGNITIVE FLEXIBILITY THEORY* AND ITS APPROACH TO DELIBERATELY AND RAPIDLY DEVELOPING ADAPTIVE SKILL: PREPARATION FOR TWENTY-FIRST-CENTURY JOBS AND ADDRESSING GRAND SOCIAL CHALLENGES

It has become common to observe that the world has been changing at an accelerating rate, with many old fixities and assumptions no longer holding. Furthermore, increases in complexity, of life and work, have been commonly observed. (See McChrystal, 2015, for an excellent summary of these trends.)

These observations have been accompanied by the acknowledgment that learning and training will likely have to occur in qualitatively different ways to develop the ability to deal adaptively with the resultant increase in novelty that must be routinely confronted. Commonly discussed *twenty-first-century skills* for education, often at odds with predominant skill targets that have been the focus of education based on mainstream research in cognitive and educational psychology (Kirschner et al., 2006; Spiro & DeSchryver, 2009), have stressed the ability to deal adaptively with novelty (see the excellent summary of twenty-first-century skill rubrics in Dede, 2010). Reinforcing this trend in education are recent economic forecasts that have stressed the importance of adaptive skill. Good jobs will require the ability to deal with novelty adaptively as soon as five years from now, according to a recent World Economic Forum report (2016). This trend in the necessity of adaptive skill for jobs is compounded by the fast-growing successes of artificial intelligence, which will mean that more well-structured, algorithmic tasks will increasingly be done by smart machines instead of people who have made good livings doing such tasks up until recent times (see Cowen's *Average Is Over*, 2013, and Brynjolfsson and McAfee's *The Second Machine Age*, 2014, for excellent recent treatments of this and related issues and trends in employment). And, of course, the development of better ways for our educational and training systems to produce such adaptive skill will have aggregate effects on national economies' success in global markets that increasingly require almost constant innovativeness.

Furthermore, we face *grand social challenges* that pose great risks to our future. Climate change and the growing worldwide fresh water crisis are just a couple examples. These novel and complex problems will require an ability to deal with complexity and novelty to educate successful contributors to solutions, as well as to grow an informed citizenry prepared to vote on possible responses to these challenges with an awareness of the complexity of the issues and a readiness for adaptive response.

In a somewhat more prosaic realm, though still a grand social challenge, another area where educational shortcomings have had negative societal effects has been the way performance by professionals has always suffered from the inability to train for adaptation to novelty. Whether it is doctors who are excellent at solving a wide range of routine medical problems after years of experience but have great difficulty at that same advanced stage of experience with non-routine problems (Feltovich et al., 1997), experienced lawyers who have difficulty because of the variability of legal concepts across cases (Feltovich, Spiro, Coulson, & Myers-Kelson, 1995), military strategists who have to employ strategies of indirect approach and relational maneuver that were once thought to be in the province of the art or genius of war and thus not learnable (Spiro, 1988), or teachers who constantly face non-routine cases with multiple agendas that have to be juggled in novel ways as situations dynamically change over time (Palincsar et al., 2007), we should not have to continue to wait ten years for adaptive professional skill to develop.

The work described in this chapter, together with the related work captured in Ward et al. (2018) and Hoffman et al. (2014), supported by the affordances of digital media that are so perfectly suited for the necessary features of learning for adaptive skill (as we have seen illustrated in this chapter), offers immediate promise for the kind of shift in emphasis in education and training that is so urgently needed for the changing times that are already upon us.

Acknowledgments

The various research projects of Spiro and Feltovich cited in this chapter were supported at various times, for different of the projects, by the National Science Foundation, US Department of Education (Office of Educational Research and Improvement, Institute for Educational Sciences, and Dept. of Postsecondary Education), the National Institutes of Mental Health, the Interagency Educational Research Initiative (IERI), the Basic Research Office of the Army Research Institute in the US Department of Defense, the Office of Naval Research,

Defense University Research Instrumentation Program (DURIP), the Navy Personnel Research & Development Center of the Department of Defense, the Josiah Macy Foundation, the Spencer Foundation, and the IBM Watson Research Center. We wish to express our gratitude and, of course, to indicate that the views in this chapter are our own and not those of any funding agency. Especially helpful feedback and suggestions for this chapter were provided by Professor Marjorie Siegel, and are much appreciated. Dr. Richard L. Coulson collaborated *centrally* on all the papers for which he is a co-author; Spiro and Feltovich owe an endless debt of gratitude to Dr. Coulson for his innumerable and invaluable contributions to much of this work.

References

Anderson, R. C., Spiro, R. J., & Anderson, M. C. (1978). Schemata as scaffolding for the representation of information in discourse. *American Educational Research Journal* 15, 433–440.

Bartlett, F. C. (1932). *Remembering*. Cambridge, UK: Cambridge University Press.

Bransford, J. D., Nitsch, K. E., & Franks, J. J. (1977). Schooling and the facilitation of knowing. In R. C. Anderson, R. J. Spiro, & W. E. Montague (Eds), *Schooling and the acquisition of knowledge*. Hillsdale, NJ: Lawrence Erlbaum.

Bransford, J. D., & Schwartz, D. L. (1995). Efficiency and innovation in transfer. In J. Mestre (Ed.), *Transfer of learning from a modern multidisciplinary perspective* (pp. 1–51). Charlotte, NC: Information Age Publishing.

Brynjolfsson, E., & McAfee, A. (2014). *The second machine age: Work, progress, and prosperity in a time of brilliant technologies*. New York: W. W. Norton.

Campbell, D. T. (1974). Evolutionary Epistemology. In P. A. Schilpp (Ed.), *The philosophy of Karl R. Popper* (pp. 412–463). LaSalle, IL: Open Court.

Chase, W. G., & Simon, H. A. (1972). Perception in chess. *Cognitive Psychology* 4, 55–81.

Collins, A. M., Brown, J. S., & Larkin, K. M. (1980). Inference in text understanding. In R. J. Spiro, B. C. Bruce, & W. F. Brewer (Eds), *Theoretical issues in reading comprehension*. Hillsdale, NJ: Lawrence Erlbaum.

Cowen, T. (2013). *Average is over: Powering America beyond the age of the great stagnation*. New York: Penguin.

Dede, C. (2010). Comparing frameworks for 21st century skills. In J. Bannanca & R. Brandt (Eds), *21st century skills: Rethinking how students learn* (pp. 51–75). Bloomington, IN: Solution Tree Press.

Ericsson, K. A., Krampe, R. T., & Tesch-Romer, C. (1993). The role of deliberate practice in the acquisition of expert performance. *Psychological Review* 100, 363–406.

Einstein, A. (1952). Letter to Jacques Hadamard. In B. Ghiselin (Ed.), *The creative process* (pp. 43–44). New York: New American Library.

Feltovich, P. J., Coulson, R. L., & Spiro, R. J. (2001). Learners' understanding of important and difficult concepts: A challenge to smart machines in education. In P. J. Feltovich & K. Forbus (Eds), *Smart machines in education*. Cambridge, MA: MIT Press.

Feltovich, P. J., Spiro, R. J., & Coulson, R. L. (1989). The nature of conceptual understanding in biomedicine: The deep structure of complex ideas and the development of misconceptions. In D. Evans & V. Patel (Eds), *The cognitive sciences in medicine* (pp. 113–172). Cambridge, MA: MIT Press.

Feltovich, P. J., Spiro, R. J., & Coulson, R. L. (1993). Learning, teaching and testing for complex conceptual understanding. In N. Frederiksen, R. Mislevy, & I. Bejar (Eds), *Test theory for a new generation of tests* (pp. 181–217). Hilldale, NJ: Lawrence Erlbaum.

Feltovich, P. J., Spiro, R. J., & Coulson, R. L. (1997). Issues of expert flexibility in contexts characterized by complexity and change. In P. J. Feltovich, K. M. Ford, & R.R. Hoffman (Eds), *Expertise in context: Human and machine*. Cambridge, MA: MIT Press.

Feltovich, P. J., Spiro, R. J., Coulson, R. L., & Myers-Kelson, A. (1995). The reductive bias and the crisis of text (in the law). *Journal of Contemporary Legal Issues* 6(1), 187–212.

Hatano, G., & Inagaki, K. (1986). Two courses of expertise. In H. Stevenson, H. Azuma, and K. Hakuta (Eds), *Child development and education in Japan*. New York: W. H. Freeman.

Hoffman, R. R., Ward, P., Feltovich, P. J., DiBello, L., Fiore, S. M., & Andrews, D. (2014). *Accelerated expertise: Training for high proficiency in a complex world*. New York: Psychology Press.

Jacobson, M. J., & Spiro, R. J. (1995). Hypertext learning environments, cognitive flexibility, and the transfer of complex knowledge: An empirical investigation. *Journal of Educational Computing Research* 12, 301–333.

Jonassen, D., Ambruso, D., & Olesen, J. (1992). Designing hypertext on transfusion medicine using cognitive flexibility theory. *Journal of Educational Multimedia and Hypermedia* 1, 309–322.

Kirschner, P. A., Sweller, J., & Clark, R. E. (2006). Why minimal guidance during instruction does not work: An analysis of the failure of constructivist, discovery, problem-based, experiential, and inquiry-based teaching. *Educational Psychologist* 41(2), 75–86.

Klein, G. (2011). *Streetlights and shadows: Searching for the keys to adaptive decision-making.* Cambridge, MA: MIT Press.

McChrystal, S. (2015). *Team of teams: New rules of engagement for a complex world.* New York: Portfolio.

McLuhan, M. (1964). *Understanding media: The extension of man.* New York: McGraw Hill.

Minsky, M. (1975). Frame system theory. *Proceedings of the 1975 workshop on Theoretical issues in natural language processing—TINLAP '75* (pp. 104–116).

Mishra, P., Spiro, R. J., & Feltovich, P. J. (1996). Technology, representation, and cognition: The prefiguring of knowledge in Cognitive Flexibility Hypertexts. In H. van Oostendorp & A. de Mul (Eds), *Cognitive aspects of electronic text processing* (pp. 287–305). Norwood, NJ: Ablex.

Palincsar, A. P., Spiro, R. J., Kucan, L., Magnusson, S. J., Collins, B. P., Hapgood, S.,... DeFrance, N. (2007). Designing a hypertext environment to support comprehension instruction. In D. McNamara (Ed.), *Reading comprehension strategies: Theory, interventions, and technologies.* (pp. 441–462). Mahwah, NJ: Lawrence Erlbaum.

Pepper, S. (1942). *World hypotheses.* Berkeley, CA: University of California Press.

Poincaré, H. (1952). Mathematical creation. In B. Ghiselin (Ed.), *The creative process* (pp. 33–42). New York: New American Library.

Polanyi, M. (1958). *Personal knowledge.* Chicago: University of Chicago Press.

Rittel, H. (1972). On the planning crisis: Systems analysis of the first and second generations. *Bedriftskonomen 8*.

Rumelhart, D. E. (1980). Schemata: The building blocks of cognition. In R. J. Spiro, B. C. Bruce, & W. F. Brewer (Eds), *Theoretical issues in reading comprehension* (pp. 33–58). Hillsdale, NJ: Lawrence Erlbaum.

Rumelhart, D. E., & Ortony, A. (1977). The representation of knowledge in memory. In R. C. Anderson, R. J. Spiro, & W. E. Montague (Eds), *Schooling and the acquisition of knowledge*. (pp. 99–136). Hillsdale, NJ: Lawrence Erlbaum.

Schank, R. C., & Abelson, R. P. (1977). *Scripts, plans, goals, and understanding*. Hillsdale, NJ: Lawrence Erlbaum.

Spiro, R. J. (1977). Remembering information from text: The "State of Schema" approach. In R. C. Anderson, R. J. Spiro, & W. E. Montague (Eds), *Schooling and the acquisition of knowledge* (pp. 137–166). Hillsdale, NJ: Lawrence Erlbaum.

Spiro, R. J. (1980a). Accommodative reconstruction in prose recall. *Journal of Verbal Learning and Verbal Behavior 19*, 84–95.

Spiro, R. J. (1980b). Constructive processes in prose comprehension and recall. In R. J. Spiro, B. C. Bruce, & W. F. Brewer (Eds), *Theoretical issues in reading comprehension* (pp. 245–278). Hillsdale, NJ: Lawrence Erlbaum.

Spiro, R. J. (1982a). Long-term comprehension: Schema-based versus experiential and evaluative understanding. *Poetics: The International Review for the Theory of Literature 11*, 77–86.

Spiro, R. J. (1982b). Subjectivite et memoire. [Subjectivity and memory.] *Bulletin de psychologie* (Special issue on "Language comprehension") *35*, 553–556. [Reprinted in W. Kintsch & J. Le Ny (Eds) (1983). *Language comprehension*. The Hague: North-Holland.

Spiro, R. J. (1988). Experience-consolidation systems: Computer-based training in ill-structured domains. In R. Seidell & P. D. Weddle (Eds), *Computer-based instruction in military environments*. New York: Plenum. [Also published as the Proceedings of the Symposium on Computer-Based Instruction in Military Environments, Brussels: NATO, 1986.]

Spiro, R. J., Collins, B. P., & Ramchandran, A. R. (2006). Modes of openness and flexibility in "Cognitive Flexibility Hypertext" learning environments. In B. Khan (Ed.), *Flexible learning in an information society* (pp. 18–25). Hershey, PA: Information Science Publishing.

Spiro, R. J., Collins, B. P., & Ramchandran, A. R. (2007). Reflections on a post-Gutenberg epistemology for video use in ill-structured domains: Fostering complex learning and cognitive flexibility. In Goldman, R., Pea, R. D., Barron, B. & Derry, S. (Eds), *Video research in the learning sciences* (pp. 93–100). Mahwah, NJ: Lawrence Erlbaum.

Spiro, R. J., Collins, B. P. Thota, J. J., & Feltovich, P. J. (2003). Cognitive flexibility theory: Hypermedia for complex learning, adaptive knowledge application, and experience acceleration. *Educational Technology 44*(5), 5–10. [Reprinted in A. Kovalchick & K. Dawson (Eds), (2005), *Education and technology: An encyclopedia* (pp. 108–117). Santa Barbara, CA: ABC: CLIO.]

Spiro, R. J., Coulson, R. L., Feltovich, P. J., & Anderson, D. (1988). Cognitive flexibility theory: Advanced knowledge acquisition in ill-structured domains. *Proceedings of the Tenth Annual Conference of the Cognitive Science Society*. Hillsdale, NJ: Erlbaum.

Spiro, R. J., Crismore, A., & Turner, T. J. (1982). On the role of pervasive experiential coloration in memory. *Text 2*, 253–262.

Spiro, R. J., & DeSchryver (2009). Constructivism: When it's the wrong idea and when it's the *only* idea. In S. Tobias & T. Duffy (Eds), *Constructivist instruction: Success or failure* (pp. 106–123). Mahwaw, NJ: Taylor & Francis.

Spiro, R. J., Feltovich, P. J., & Coulson, R. L. (1996). Two epistemic world-views: Prefigurative schemas and learning in complex domains. *Applied Cognitive Psychology 10*, 52–61.

Spiro, R. J., Feltovich, P. J., Coulson, R. L., & Anderson, D. (1989). Multiple analogies for complex concepts: Antidotes for analogy-induced misconception in advanced knowledge acquisition. In S. Vosniadou & A. Ortony (Eds), *Similarity and analogical reasoning* (pp. 498–531). Cambridge, MA: Cambridge University Press.

Spiro, R. J., Feltovich, P. J., Jacobson, M. J., & Coulson, R. L. (1992a). Cognitive flexibility, constructivism, and hypertext: Random access instruction for advanced knowledge acquisition in ill-structured domains. In T. Duffy & D. Jonassen (Eds.), *Constructivism and the technology of instruction* (pp. 57–75). Hillsdale, NJ: Lawrence Erlbaum. [Reprinted from a special issue of the journal *Educational Technology* on *Constructivism*, 1991.]

Spiro, R. J., Feltovich, P. J., Jacobson, M. J., & Coulson, R. L. (1992b). Knowledge representation, content specification, and the development of skill in situation-specific knowledge assembly: Some constructivist issues as they relate to cognitive flexibility theory and hypertext. In T. Duffy & D. Jonassen (Eds), *Constructivism and the technology of instruction* (pp. 121–128). Hillsdale, NJ: Lawrence Erlbaum. [Reprinted from a special issue of the journal *Educational Technology* on *Constructivism*.]

Spiro, R. J., & Jehng, J. C. (1990). Cognitive flexibility and hypertext: Theory and technology for the nonlinear and multidimensional traversal of complex subject matter. In D. Nix & R. J. Spiro (Eds), *Cognition, education, and multimedia: Explorations in high technology* (pp. 163–205). Hillsdale, NJ: Lawrence Erlbaum.

Spiro, R. J., Klautke, H. A., Cheng, C., & Gaunt, A. (2017). Cognitive flexibility theory and the assessment of 21st-century skills. In C. Secolsky and D. B. Denison (Eds), *Handbook on measurement, assessment, and evaluation in higher education* (pp. 631-637). NY: Routledge.

Spiro, R. J., Klautke, H., & Johnson, A. (2015). All bets are off: How certain kinds of reading to learn on the Web are totally different from what we learned from research on traditional text comprehension and learning from text. In Spiro, R. J., DeSchryver, M., Morsink, P., Schira-Hagerman, M., & Thompson, P. (Eds), *Reading at a crossroads? Disjunctures and continuities in our conceptions and practices of reading in the 21st century* (pp. 45-50). New York: Routledge.

Spiro, R. J., & Myers, A. (1984). Individual differences and underlying cognitive processes in reading. In P. D. Pearson (Ed.), *Handbook of research in reading*. New York: Longman.

Spiro, R. J., Vispoel, W. L., Schmitz, J., Samarapungavan, A., & Boerger, A. (1987). Knowledge acquisition for application: Cognitive flexibility and transfer in complex content domains. In B. C. Britton & S. Glynn (Eds), *Executive control processes*. Hillsdale, NJ: Lawrence Erlbaum.

Ward, P., Gore, J., Hutton, R., Conway, G. E., & Hoffman, R. R. (2018). Adaptive skill as the *conditio sine qua non* of expertise. *Journal of Applied Research in Memory and Cognition* 7(1), 35–50.

White, H. (1973). *Metahistory*. Baltimore, MD: Johns Hopkins University Press.

Wittgenstein, L. (1953). *Philosophical investigations*. New York: Macmillan.

World Economic Forum. (2016). *The future of jobs: Employment, skills, and workforce strategy for the Fourth Industrial Revolution*. Available at https://www.weforum.org/reports/the-future-of-jobs-report-2018

CHAPTER 42

COGNITION AND EXPERT-LEVEL PROFICIENCY IN INTELLIGENCE ANALYSIS

DAVID T. †MOORE AND ROBERT R. HOFFMAN

Introduction

The value of expertise in intelligence analysis is an important issue given the ongoing demographic changes in the US Intelligence Community (IC). Senior experts of the *Boomer* generation (Americans born between 1943 and 1960) continue to retire as now do their younger *Gen-X* colleagues (born in the mid-1960s to the late 1970s). Members of both generations have served 40-plus years in the IC and are moving on. They are being replaced by individuals who either are still at the apprentice level or perhaps have achieved a junior journeyman level of proficiency. This demographic fact, along with the manifest need for the various intelligence agencies to align, collaborate, and develop common work methods and systems, has led to distinct concern with issues of workforce, training, mentoring, career development, and analyst performance. "The IC's number one priority should be having top people" (Lahneman, 2006, p. iv). During this ongoing workforce transition, anecdotes about how junior intelligence analysts come to a situation and see in a glimpse what has stumped the fixated experts lead to questions about whether business models need to presume expertise at all. On the other hand, there is extensive research about how individuals make wise (and valid) decisions in domains where expertise is acquirable (Kahneman & Klein, 2009). Such contrasts lead to questions about hiring new personnel. Where does it leave the community if in the near future apprentices or journeymen will commonly be sitting in the chair once held by a senior intelligence professional?

Clearly, it has been useful for the IC to consider the question of what constitutes expertise, and gradations of expertise in intelligence analysis as well as how best to move individuals from lower levels of proficiency to higher levels. In this chapter we characterize expertise in intelligence analysis, and discuss how the IC can subsequently facilitate development of its workforce. We rely on studies and models of reasoning in the paradigm of *naturalistic decision making* that have revealed what happens in rapid, high-impact decision making in a shifting environment. We ask these questions: What constitutes high proficiency in intelligence analysis? How can we measure the performance of intelligence analysts? How do expert intelligence analysts reason? How can the achievement of high proficiency be rapidized?

Who Are the Experts?

Intelligence analysis trainees (apprentices and junior journeymen) typically exhibit tremendous drive and dedication to their work. They are not completely unlettered about the kinds of issues of which they must make sense. Further, they are hired on the basis of talents, knowledge, and skills believed to be relevant to that work (Moore & Krizan, 2003). However, intrinsic motivation and basic skills such as writing and critical thinking are necessary but not sufficient for the high levels of proficiency the IC requires if it is to safeguard the nation, its people, and its interests.

We mention apprentices and journeyman because it is crucial to keep in mind that humanity is not neatly bifurcated into just the novice and expert categories. In understanding expertise and its development one must consider the full proficiency spectrum: novices, initiates, apprentices, journeymen, experts, and masters (Hoffman, 1996). Thus, we do not frame our topic in terms of the novice versus expert contrast. The US Intelligence Community Directives 601 and 652 (Office of the Director of National Intelligence, 2008, 2009) specify four levels of proficiency: basic, full performance, advanced, and expert. Based on the descriptions in Directive 652 (pp. 6-7), and comparing it to a traditional proficiency scale, the first two of these correspond quite clearly to apprentice (i.e., learning and applying fundamental skills) and journeyman (i.e., independent performance of a full range of duties). The *advanced* level seems to map to what one might call senior journeyman or junior expert ("range of complex assignments and non-routine situations that require extensive knowledge and experience"). The expert level involves "an extraordinary degree of specialized knowledge to perform complex and ambiguous assignments that normally require integration and synthesis of a number of unrelated disciplines and separate concepts." In a traditional proficiency scale this might be mapped onto either the senior expert or even the master category. (For a detailed discussion of traditional proficiency scaling, see Hoffman et al., 2014).

Within the IC there are also certification procedures that reflect levels of proficiency. For example, "Level-1 All-source Analysts" is a credential granted by the US Defense Intelligence Agency to "individuals who have demonstrated a level of competence, through their work experience and analytic knowledge, consistent with the baseline of the profession. The analytic knowledge represents those facts, concepts, principles, and tools that are appropriate for a generalist." (See for example, US Army, Credentialing Opportunities On-Line, https://www.cool.army.mil/search/CERT_CDASA-I6918.htm). This credential would perhaps map to the journeyman proficiency level. Level-1 All-Source, and additional specialist credentials, reflect the desire within the IC to professionalize intelligence analysis.

Expert analysts, often called "senior analysts," tend to be individuals who have had decades of experience, and experiences in diverse situations and subdomain contexts. This is a general finding in expertise studies (see Pliske et al., 2004). In region-specific analysis, for example, the senior analyst who works at a *regional desk* will likely be one who has spent literally an entire professional career immersed in the history, culture, languages, traditions, and policies of the region in which they specialize. Across the various subdomains and specializations in intelligence analysis, the more proficient and highly motivated analysts are always immersing themselves in the literature of case studies, historical accounts, and retrospections. These analysts are typically voracious and intensely curious.

The IC has spent considerable time developing both models of analyst competencies (Moore, 2004; Moore & Krizan, 2003; Moore, Krizan, & Moore, 2005) and an Analytic Resources Catalog (ARC) promulgated by the US Office of the Director of National Intelligence (Lowenthal, 2007), as well as several agency-specific variants. While self-reporting by intelligence professionals of their capabilities and their levels of proficiency is certainly valuable, there has been little standardization of the competency self-ratings instruments. One person with a particular level of competency might declare themselves an *expert* while another person (with perhaps greater modesty) might declare themselves a *journeyperson*. Two people with similar skill sets might characterize those skill sets quite differently. Since a primary purpose of self-report inventories is to serve a staffing function, a lack of independent empirical validation impairs usefulness and has likely led to general discontinuance.

To be sure, mere extent or duration of experience is no guarantee of expertise, but it is a meaningful indicator. We will have more to say about this as we proceed.

As for all professional domains, characteristic of intelligence analysts is that they usually have expertise in subdomains (Hoffman, 1996).

SUBDOMAIN SPECIFICITY OF EXPERTISE

Assuming that we can refer to intelligence analysis broadly as a *domain*, expert analysts are highly proficient in a subdomain or across several subdomains, but not necessarily proficient in others. Schemes for categorizing general analytic skills within the IC (e.g., Tett et al., 2007) do not include *knowledge* precisely because the knowledge requirements are determined by the analytic specialization. For instance, specific mission expertise is considered as a separate category from analytic core competencies (see Moore & Krizan, 2003; Sticha & Buede, 2009).

Many of the main scholarly books on intelligence analysis processes and cognition (e.g., Reese, 2007) refer to intelligence analysis very broadly and do not go into any depth in discussing distinct analytic roles and responsibilities. In general, the treatments of analysis tacitly reference what is called *all-source* or *open source* analysis but this designation refers only to the sorts of evidence to which a given analyst might have access, not their particular roles and responsibilities. Intelligence source disciplines—specialization in signals intelligence, human intelligence, geophysical intelligence, measurement and signature intelligence, and open source intelligence—are considered a competency of tradecraft more than a core cognitive competency. Indeed, the source specializations represent distinct roles within the IC. An expert at interpreting aerial photographs for the planning of military actions would have vastly different cognitive competencies, tools, and work methods from, say, an all-source analyst who is skimming the Web for evidence of emerging civil unrest somewhere.

Consequently, what counts as an expert in one analytical role may differ from what counts as an expert in some other role. In addition to data or sources (e.g., all-source analysis, imagery analysis) roles can be specific to regions (e.g., the Middle East), or to primary tasks that are referred to as *analytic disciplines* (e.g., influence operations, counterintelligence, counter-deception, counter-terrorism). Specializations can be with respect to academic disciplines (e.g., biology, engineering, language), collection systems and operations, information processing, analytic methods (e.g., modeling and simulation, decision-aiding techniques, software systems), customer requirements (e.g., liaising with law enforcement), and policy (Sticha and Buede, 2009).

The combinatorics of disciplines, tradecraft skills, and knowledge mean that there are literally dozens of distinctive roles in intelligence analysis, so one should always be cautious about generalizing by tacitly referencing to "all-source" analysis. Proficiency means that the analyst can successfully and accurately construct judgments about relevant situations *in those subdomains and for those information sources* in which they possess expertise or proficiency of knowledge and methodological (or tradecraft) skill.

As is true for any macrocognitive work system, the need for ever-finer grains of specialization forces recognition of the need for generalists (for more information, see Klein et al., 2003). This has long been understood in the IC: The 2004 amendment to the US National Security Act of 1947 states that the Director of National Intelligence "shall prescribe mechanisms to facilitate the rotation of personnel of the intelligence community through various elements of the intelligence community in the course of their careers in order to facilitate the widest possible understanding by such personnel of the variety of intelligence requirements, methods, users, and capabilities" (US Congress, 2004, Section102A (l) (3) (A), p. 19; see also Office of the Director of National Intelligence, 2009).

There is a presumption that the intelligence subdomains exhibit sufficient stability so as to permit the acquisition of expertise. We know that this is not the case in all domains of expertise (Kahneman & Klein, 2009; Moore & Hoffman, 2011; Shanteau & Phelps, 1977). One would therefore expect that there are subdomains in intelligence analysis where it is relatively difficult to achieve expertise. These are likely to be task specific. For example, the forecast of medium- to long-term cultural shifts and trends (so-called anticipatory intelligence) entails very little opportunity for near-term feedback. The non-trivial nature of characterizing expertise in intelligence analysis extends to the problem of performance measurement. One criterion for expertise is superior performance.

How Can We Measure the Performance of Intelligence Analysts?

Claims are often made that human performance suffers from numerous cognitive biases, and that this holds for experts, including experts in intelligence analysts (cf. Heuer, 1999). Such claims need to be backed up by convincing evidence that the individuals who are referenced are: (1) empirically demonstrated to show biases, and (2) empirically demonstrated to qualify as experts, operationally defined in terms of performance measures. These two gaps in publicly available knowledge concerning analyst performance are made salient by the emphasis placed on the procurement of new software decision support systems that are intended to assist intelligence analysts by helping them overcome biases (see Moon & Hoffman, 2005). We will have more to say about this later in this chapter.

It might be expected that junior analysts who lack knowledge might show cultural bias, but it would be debatable whether the mere lack of knowledge qualifies as a *cognitive* bias. We do, however, hope to see the psychological foundations of intelligence grounded in a positive view of expertise rather than in a limited, outdated, and negativist view that overapplies the notion of cognitive bias. The biases viewpoint has engendered a paradigm that measures performance in terms of biases. It is demoralizing to analysts. Analysts are routinely blamed for errors, especially in the popular press, stimulated when politicians play the blame game.

Performance Measurement and Evaluation

A useful performance review must avoid the blame game. At the same time, "feedback that challenges our mental models can be the most valuable because it gives us a chance to reflect and even improve" (Klein, 2009, p. 175).

A US Intelligence Community Directive specifies that the IC shall use validated performance objectives and elements, employ a process for measuring performance, and a process for rating, rewarding, and providing performance feedback (see Directive 651, Office of the Director of National Intelligence, 2012). The directive goes on to say, in essence, that performance is evaluated annually by having managers use a rating system that references the core elements of analytical activity (e.g., critical thinking, communication skill). A five-point scale is used, ranging from *outstanding* to *unacceptable*. Each department and agency must adhere to this directive, but the directive states that each agency will independently develop its own performance management system.

There may be no integrated empirical base of systematically collected and scientifically robust, multi-measure evidence that informs us about the performance of analysts at various grades or levels of proficiency (see Hoffman, 2005). The circularity is that performance measurement is held to be important, but only a single measurement approach is presented, and its method is underspecific.

Making What's Important Measurable

Should the prime measure that decides for us whether a given individual is an expert be some single measure of performance at some single, stable task, as opposed to other criteria such as career experience or sociometric analysis (e.g., who talks to whom)? Analysts engage in many different kinds of tasks, and so a performance-based criterion would not work well if it tapped only performance on some single task, however representative that task might be. Furthermore, it is not entirely clear what good performance would mean. An analyst might engage in a very systematic, thorough, and informed analytical process for a given case or tasking, present meaningful and actionable conclusions, predictions, and recommendations to the policy maker, and subsequently learn that the primary recommendations were wrong. Intelligence analysis involves understanding and anticipating individual and aggregate human activity. This is understood as *complex indeterminate causation* (Moore & Hoffman, 2011). In indeterminate causal reasoning, especially about longer-term trends, it is difficult to impossible to evaluate predictions on the basis of outcome. For instance, an intelligence analyst's forecasts may be about long-term instability in some region of the world. It is impractical to wait to determine the quality of the analytical work on the basis of outcome, and thereby provide corrective feedback to the analyst, a necessary thing to promote the achievement of expertise.

Retrospectively, one can ask whether an analyst's forecasts were accurate; that the events anticipated actually occurred. What if they did but the reasons were invalid or the method used to obtain the judgment was weak? Perhaps changes in posture occurring as a result of a report tipped an adversary that we are on to them. Perhaps they simply changed their minds. True adversarial motivations are difficult (if not impossible) to determine uniquely. So, can proficiency be scaled in terms of work processes and quality and not just final judgment or "hit rate"?

Retrospective evaluations ask such questions as: How well did you capture the causal factors that were operating? How many causal factors did you postulate that turned out not to be operating? It is possible to ask these kinds of questions prospectively. An analysis or assessment might be of high quality even though a specific prediction may not have come to pass. It is possible to have a means of judging the quality of analytical work before the outcome has occurred, and whether or not the forecast outcome occurs. For consideration, we present a *Forecaster's Scoresheet* in Table 42.1. This references how the analyst progressed from data to causal stories. This concept is merely suggestive, and the specific scoring scheme is somewhat arbitrary, but we feel that it or something like it is entirely workable. This is based on the work of Klein et al. (2011; see also Moore & Hoffman, 2011) on how people reason about complex indeterminate causation. It distinguishes causes from enabling conditions (the oxygen in a room does not cause the fire). It distinguishes specific causes from abstractions, and it refers to a number of different kinds of story structures (chains, spirals, etc.).

Though an empirical base for performance evaluation is lacking, we can nonetheless ask how one might characterize expertise in intelligence analysis.

Table 42.1 A "Forecaster's Scoresheet"

Identifying Causes	
How many causal factors did you identify?	One point each.
Taken individually, is each identified cause just an enabling condition?	Take away one point if the answer is yes.
Is the identified factor really a covariant?	Take away one point if the answer is yes.
Does the list include multiple themes (lists, clockworks, reversible events, causalchains, spirals, etc.)	Add one point for each distinct type of theme.
If the identified cause is an abstraction, does it specify the superficial differences and substantive similarities?	Add one point if the answer is yes.
Synthesis to Story	
Is the story a simple list or sequence of events?	Take away a point if the answer is yes.
Does the story include counter-causes?	Score a point if the answer is yes.
Does the story include multiple interacting causes and multiple kinds of factors? That is, is the story more like an onion in which one peels back layers of causation or like a clockwork of multiple interacting causes?	Score a point if the answer is yes.
If the final determination is of a single primary cause, is it a different single cause as was initially identified in Identifying Causes Step?	Score a point if the answer is yes.
In what ways was the final synthesis actionable?	Score one point for each possible action.

How Do Expert Intelligence Analysts Reason?

As in many other domains, expertise in intelligence analysis hinges on the capacity to develop intuition, perceptual skill, and procedural skill. It also depends on an ability to be flexible and adaptive.

Intuition

Expert analysts are expected to have good intuitions, intuitions that can be justified. Expert judgment is often deliberative but is also often intuitive, and rapid intuition-formation occurs during periods of deliberative reasoning. Klein (2009) defined intuition as experience-based judgments that are arrived at without conscious or prolonged deliberation. Intuition is thus associated with tacit knowledge. It includes the ability to rapidly recognize patterns and events, and rapidly recognize anomalies. Thus, intuition is the rapid, direct, or immediate translation of experiences into actions (Klein, 2003).

As is well documented in the literature of expertise studies, while experts rely on intuition, less proficient individuals are more likely to rely on rules and explicit procedures. This process of generating tacit knowledge from experience and internalizing or automating rules and procedures is one of the key phenomena in the development of expertise, and is noted in one form or another in nearly all models of the development of expert reasoning (see Hoffman & Militello, 2008; Hoffman et al., 2014).

We should point out that we do not advocate reliance on either intuition or analysis exclusively. Both are necessary parts of sound decision- and sensemaking. Systematic analysis based on rules and procedures can lead us to challenge and question our assumptions and intuitions. On the other hand, because analysis takes time, in tactical or rapidly changing situations where speed is critical, intuition becomes essential. Given more time to examine a complex situation we gain a great deal from a detailed systematic analysis (and the intuitive considerations it engenders).

Perception and Recognition

Expert analysts are expected to be able to recognize patterns and pattern violations. "Novices see only what is there, experts can see what is not there. Through experience, a person gains the ability to visualize how a situation developed and to imagine how it is going to turn out" (Klein & Hoffman, 1992, p. 203). While both novices and experts can exhibit situational understanding, experts are better able to perceive when data or cues are missing in a situation. The capacity to *see what isn't there* is crucial in intelligence analysis. With the ongoing exodus of expertise from across the IC the loss of a capability to see what isn't there, to make fine-grained perceptual discriminations in complex and ambiguous situations, not only impedes the IC's conduct of its various national-security missions, potentially endangering the nation and its citizens, but it also seriously limits the ability of the IC to look to the fringes of its world for anticipatory indicators of arising issues and threats guaranteeing unpleasant surprises.

Experts detect cues and patterns and this directly informs them about what to do. In other words, recognition is linked directly to an action plan, without intermediate deliberation. Klein (2003) has called this recognition-primed decision making (RPD). In research on numerous domains of expertise, he and his colleagues found that "for about 90 percent of the difficult decisions (and probably many more of the routine ones), the strategy [decision makers] use is recognition-primed decision making" (Klein, 2003, p. 19).

RPD leverages the expert's rich knowledge of past cases, cues, and patterns. The action script is sometimes tested by *mental projection to the future*. This too leverages the expert's rich knowledge and "imaginal and conceptual understanding of functional relations and physical principles that relate concepts" (Hoffman, 1996, p. 89). Klein's model provides insights into why experts may disagree, even when faced with the same situation. The RPD model allows for the differences between experts, of course. Experts will have had differing experiences and resulting differences in knowledge, while they nonetheless have shared knowledge. Thus, two experts will view the same situation and apprehend different cues leading to differing patterns and action scripts. Or they may arrive at the same action script but for different reasons.

The phenomenon in which experts perceive what the novices cannot occurs in intelligence analysis. For example, novice and journeymen intelligence professionals may not detect an anomaly that is apparent to an expert, and even if they do detect it they may fail to see that it is consistent with a case of adversarial deception—something the expert sees clearly. Klein and Hoffman (1992) also noted that expert knowledge is represented in stories of historical cases. Thus, the apparent deception might be perceived and related by the expert as a modern-day example of the deception by double agent Juan Pujol, alias GARBO, who deceived the Germans during World War II through the use of a fictitious network of agents (Pujol & West, 1986). The reasons for the modern deception are explained within a story of this past incident. The trick for the expert analyst is also to know when a pattern *does not* apply, when a deception is *not* a modern-day example of an historical event.

We do not suggest that the fresh viewpoints of novices cannot be useful. In a red-teaming exercise, a novice may provide novel and useful viewpoints precisely because she lacks the frames of reference of the expert. Red teaming is a technique used in intelligence work (and other fields) to better understand the intentions of adversaries. In red-teaming exercises, intelligence analysts deliberately adopt positions other than their own in order to explore the decision space of a potential adversary. The purpose of such exercises is to go beyond the current sets of considerations about a problem in order to identify (as yet) unknown vulnerabilities, undetected threats, and unconsidered solutions.

One of us (Moore) has found novices helpful in applying reasonableness tests to red-teaming exercise results. Novices can also be helpful in making sure ideas are clearly expressed. Lacking the necessary cues and patterns means they cannot leap (as might an expert) from one idea to the next, filling in what is missing through connective sensemaking.

Procedural Skill

Expert analysts are expected to be facile at the use of a great many specific procedures. In the IC, the common methodology is for the professional to employ certain systematic methods, sometimes called *structured analysis* or *structured thinking* (e.g., Heuer, 1999). A variety of methods have been highly proceduralized, such as the rostering of alternative hypotheses followed by an evaluation of their plausibility, or the judgment of the quality of evidence, or red teaming (for a compilation of methods, see Heuer & Pherson, 2014). Analysts are expected to be facile at a variety of structured or systematic methods.

Analysts are also expected to have good metacognitive skills, especially the ability to recognize and counter-argue their own assumptions (see Klein, 2011; Moore, 2007; Pherson & Pherson, 2016, for an alternative view). Yet, evidence suggests that too much of a focus on systematic analytical procedures, for the sake of justifying the conclusions of the analysis, does not enhance judgments, and may detract from them (see Klein, 2011).

Further, expert analysts are also expected to be flexible and adaptive.

Cognitive Flexibility vs Rigidity

Cognitive Flexibility Theory (Spiro et al., 1988, 1992) derived from studies of experts versus trainees in the domain of medicine. The theory asserts that high proficiency in a domain involves the ability to flexibly apply knowledge and overcome simplifying or *reductive* explanations. Advanced learning is promoted by emphasizing the interconnectedness of multiple cases and concepts along multiple dimensions, and using multiple, highly organized representations. Experts can understand cases or situations in terms of dynamic factors or features, whereas trainees are more likely to think in terms of statics. Experts see factors as interactive rather than separable; they see variables as continuous versus discrete; they see causation in terms of multiple rather than single factors; they see causes as parallel rather than sequential; they understand situations in terms of concepts and relations rather than superficial or surface features; they see cases in context and do not attempt to reason complexity away. All of these characterize expert intelligence analysts.

Trainees, in contrast, tend to regularize that which is irregular, which leads to failure to transfer knowledge to new cases. Trainees tend to decontextualize concepts, which also leads to failure to transfer knowledge to new cases. Over-reliance on generalizations or abstractions, removed from the specific instances, also leads to failure to transfer knowledge to new cases. Conceptual complexity and case-to-case irregularity pose problems for the trainee. Misconceptions compound into networks of misconceptions. Misconceptions of fundamental concepts can cohere in systematic ways, making each misconception easier to believe and harder to change. The opposite of cognitive flexibility is cognitive rigidity: Holding onto an initial explanation even in the face of contrary evidence (DeKeyser & Woods, 1993). Trainees apply what are called *knowledge shields* (Chinn & Brewer, 1993; Spiro et al., 1988, 1992). They will find some reason to reject the contradictory data, reasons why the data do not apply. They will rationalize putting off consideration of the data until some time in the future; they will reinterpret the contradictory data to make them less problematic; and/or they will make a small change in their hypotheses that seems to handle the data without having to reconceptualize anything. Trainees seem to be wedded to procedures and rationalize or defend their past decisions (Groen & Patel, 1988). All of these characterize trainee intelligence analysts.

However, there are circumstances in which some cognitive rigidity is advisable; it can help people mature their ideas (Chinn & Brewer, 1993). The key is to know when to hold onto an idea and when to let it go, and that also distinguishes experts from trainees. This requires expertise about the domain and a capacity to apprehend alternative explanations of the phenomena being considered. In other words, the expert possesses strong ideas but holds them weakly.

Knowledge

As is true of experts generally, expert analysts are expected to demonstrate an articulated, conceptual, and principled understanding. Expert analysts are expected to know a great deal, as we have already mentioned. This can span knowledge of history, culture, adversary strategies, politics, languages, and much, much more. Many tactical problems in intelligence and military affairs are intuitively noticed and made sense of. The patterns and levels of military communications are typical in garrison and exercise situations. Experienced intelligence professionals can scan these cues and gauge the situation. When they encounter a deviance they understand it may indicate something novel is going on.

We note also that mental models and tacit knowledge require periodic review and updating. In the realms of intelligence and military affairs, inaccurate mental models likely lead to inaccurate judgments and thus challenge and review is a part of maintaining expertise. One phenomenon that arises when this does not occur is that people become more vulnerable to deception. If an adversary knows the mental models the other side is using then they can take steps to reinforce them and mislead. This was used to great success by the Allies in World War II and by Egypt and Syria engaged in such activities prior to their 6 October 1973 invasion of Israel (see Whaley, 2016).

It is difficult to discuss knowledge without bringing in the issue of *types* of knowledge, and in particular what is meant by *tacit* knowledge (for a detailed discussion, see Hoffman & Militello, 2008). Thus, Klein (2009) uses an iceberg as a metaphor for understanding expertise and represents tacit knowledge (and routines) as being partially submerged parts of the iceberg. The following kinds of phenomena can be cited in this regard:

- As people achieve expertise, the cues they rely upon become more refined and subtle, and a pattern might be discerned without any conscious acknowledgement of the individual cues on which the pattern has formed.
- When asked about a case they might offer a judgment and when asked how that judgment came about, they might meditate for a moment and say something like, "Well now that you ask about it . . ." In other words, they knew what they knew, and could talk about it, but at the time they were reasoning in-the-moment: the knowledge was not *declarative* in the sense of being conscious and deliberative.
- While "[some] procedures can be carried out directly, as when we follow the steps in a checklist, others depend on tacit knowledge to adapt a given procedure to fit the circumstances" (Klein, 2009, p. 35) and that process of adapting a highly proceduralized or routinized method might proceed without conscious deliberation about the details.

We have witnessed all of these phenomena in the reasoning of senior intelligence analysts. They come to sound judgments by using both the critical thinking and structured methods they have learned so well that they are internalized to the degree that the expert can select, without deliberation, which approaches are most appropriate to the situation at hand.

Individual Differences and Styles

Presumably, formal job analyses that are used for selection and training of intelligence analysts are conducted by the various intelligence agencies, and we must also assume those analyses are classified (see Tett et al., 2007, p. 106). Clearly, however, there is an understanding that individual differences in personality and motivation are related to performance. Psychological theories of personality have been used in the analysis of intelligence analysts, and psychometric instruments have been used in selection and performance evaluation. In addition, recent research on expertise has revealed patterns of cognition that are related to proficiency level. We discuss each of these approaches.

Psychological Models and Psychometric Evaluations of Personality

Currently, the process of job selection for intelligence analysts relies on interviews, resumes, and recommendations, of course, but historically it has also included the use of standardized psychological tests, or personality profiles, such as the Minnesota Multiphasic Personality Inventory. In addition, analysts are periodically evaluated for psychological stability using such instruments as the Meyers–Briggs Scale. Though it would be prudent to out-select individuals who are, say, very high on anti-social dimensions, the psychometric instruments are used primarily because of historical inertia as opposed to genuine discriminative power for the analysis domain. For instance, results on the Meyers–Briggs are known to vary depending on whether the test-taker is at work or at home.

The *typology* distinction—where the individual is either one personality type or another—isn't widely used anymore in personality psychology though it is used in some leadership programs. The field of psychometrics, or psychological measurement, is in continuous flux with regard to schemes and concepts of personality types versus personality traits, how these might be factored, correlated, and linked to behavioral tendencies, and how their manifestations can change depending on situational variables (Crowne, 2009; Funder & Ozer, 1983). In personality psychometrics, perhaps the most widely cited trait model is called the Big Five, which refers to these general traits: openness, conscientiousness, extraversion, agreeableness, neuroticism. Traits are not regarded as mutually exclusive categories. Tett et al. (2007) rely heavily on the Big Five in their description of analyst competencies, noting that there are correlations between critical thinking tests and some of the personality traits, but underplaying the fact that the correlations are generally quite small. The complexities of the matter leave one hard-pressed to come to a resolution. For example, Tett et al. say, "hiring high Openness applicants to meet demands for Critical and Creative Thinking (and Influencing and Directing Others) carries the risk of reduced Trustworthiness and rule following. Similar challenges arise with other traits" (p. 130).

The episodic use of existing standardized instruments within the IC involves imposing onto the domain of intelligence analysis a set of psychological concepts and categories that originated from *outside* that domain. The clearest example is the Meyers–Briggs categories of personality types, which derive exclusively from the theories of Jungian psychology. These categories (e.g., introversion, extraversion) are manifestly applicable to any domain, including analysis. However, because their adoption and use are often reactive—due to their somewhat faddish use outside the IC, or as *the fix* to an intelligence failure—their use and applicability may not be fully understood. Also, because government bureaucracies tend to move slowly, they linger long after they've been discredited or at least passed over for something else. What seemed like a good idea at the time, may not in fact be so benign.

Individual Differences Revealed by Research on Expertise

Given that entry into analytical professions involves considerable self-selection followed by selection by the employing organization, it is not surprising that domain-specific *reasoning styles* would emerge. It is also not surprising that classifications based on psychological categorizations would be of limited use in capturing the *cognitive* essentials of analyst reasoning.

Although numerous personality *types* and *traits* theorized to reflect individual differences in tendencies to approach or avoid new knowledge have been identified (e.g., Need for Cognition, Need for Cognitive Closure), and numerous distinct *learning styles* have been discussed in the educational psychology literature (e.g., Cassidy, 2004; Sternberg et al., 2008), there has been little consideration of individual differences in distinctive methods or styles of problem solving. A consideration of such styles suggests a more nuanced assessment of differential reactions to knowledge in that it goes beyond merely approaching or avoiding information, but explicitly addresses the knowledge-seeker's specific motivations and methods of working with and applying new information.

In a report prepared by Booz Allen Hamilton (2006; see also Sticha & Buede, 2009), the following *core competencies* for analysis were listed: analytic rigor, interpersonal skills, and written communication skills. In each of these main categories were a number of more specific skills or competencies. With input from a number of analysts, Clark (2004) and Johnston (2005) listed dozens of specific intelligence analyst competencies. Examples were *connect the dots* and *marshal internal resources*. Tett et al. (2007) integrated these lists, resulting in a set of 20 *job components*. Using terminology that was more academic than the analysts themselves had used, the components were clustered as information management skills (e.g., critical thinking, planning, problem awareness), interpersonal relations (e.g., trustworthiness, teamwork, oral communication), and other (i.e., cultural awareness, stress management).

In a classic study of expertise in weather forecasting, Pliske et al. (1997, 2004) discussed a variety of distinct reasoning styles. We believe that these carry over to other domains, including intelligence analysis. Earlier we referred to the fact that senior analysts are voracious and curious. We see here an approach to proficiency scaling that focuses on reasoning styles more than measures of raw performance. Each of these styles are presented in the following.

> **Scientists:** These individuals are likely to achieve genuine expertise. They can relate their previous tough case experiences in vivid detail. They develop rich mental models of situations and possess vast knowledge, which they voraciously expand. They possess "highly developed intuitions" (Pliske et al., 2004, p. 47). They possess an extensive knowledge of procedures and rules that they use to construct their mental models of situations. Their "analytic reasoning style [is] characterized by [the] use of critical thinking and reasoning skills." They typically have experience in a variety of analytic roles (or subdomains), and this seems to have contributed substantially to their development of expertise. This perhaps characterizes (among

other roles) the community's methodologists and tradecraft specialists—of which one of us (Moore) is one. They know intuitively which rules and procedures to apply to an issue and especially how to flex the rules and procedures to make them work better in a given situation. They also know how to guide junior colleagues through the use of systematic methods.

Proceduralists: These individuals approach their work as a rule-based, procedural task. They are less likely to achieve genuine expertise but can achieve at the journeyman level. They are less likely to think in *systems* terms.

Mechanics: hese individuals complete a relatively fixed set of procedures and a limited set of information sources in order to prepare *reasonable* analytical products. Their analytical process tends to always be the same. They are less motivated to improve their forecasting skill.

Disengaged: To these individuals, analysis is just a job. They possess a very limited knowledge of procedures or methods, and apply them only specifically to their current assignment and types of tasks. They are the least proficient.

All of these types can be seen in the IC's workforce. On the one hand, there are the intelligence analysts who would qualify as scientists, and on the other hand, there are disengaged, disenfranchised, or demotivated individuals, some of whom may perform at the expert level. Fortunately, this latter group is relatively small and, as has been noted, these demotivational categories do not describe most IC novices, who, as we have noted, are typically enthused and motivated, who more likely initially fall into the Proceduralist category.

While the categories developed by Pliske et al. in their study of weather forecasters can be carried over to the intelligence analysis domain, we have also noted other individual differences in style, or perhaps even personality. Based on over two decades of projects that involved interviewing intelligence analysts (spanning a range of proficiency and experience), we have determined that existing psychometric personality measures are insufficiently descriptive of the intelligence domain. We have discerned a number of distinctive reasoning styles that warrant further evaluation and quantification. These reasoning styles are described in Table 42.2. These categories seem orthogonal to the Pliske et al. categories, but this is itself an interesting question for empirical examination.

In the course of knowledge elicitation interviews with intelligence analysts, Hoffman (2003) found what appear to be individual differences in the extent to which analysts demonstrate preferences for various analytical methods. For instance, individuals who display a *Blobber* disposition tended to not like structured analytical methods, and could be quite vocal about that. Individuals with high *Weaver* tendencies reported having difficulty getting to closure. Analysts who seem to be *Inforvores* will not seek closure until all of the minutiae are satisfied.

The scheme in Table 42.2 has been presented to individuals in the IC (analysts, analyst-managers, and analyst-trainers), who have consistently resonated to it. Indeed, the

Table 42.2 Emergent patterns in analytical reasoning style

Definition	Attitude to Structured Analysis	Example
Infovores Strong motivation to gather and remember facts. Thrive on collaboration and debate.	Neutral-to-negative attitude about structured analytical approaches.	Long argument about the name of the island on which Bobby Kennedy got married.
Blobbers Like to peruse information, absorb it, review it, and let it percolate. Themes and solutions emerge, like "insights." Crave input but then work alone.	They do not like structured analytical approaches.	Strong negative reactions when Richards Heuer and the method of "Analysis of Competing Hypotheses" were mentioned.
Weavers Strong motivation to specify patterns and link things together, but prefer to follow their own path. Hard to achieve closure. Prone to reflection and the questioning of reasoning.	Neutral about structured analytical approaches. They do their own weaving, do not feel they need any help of any sort.	Mentally constructed a concept map in his mind's eye then asked for nodes to be placed at certain places, then filled in all the nodes and then linked them.
Deep Divers Strong motivation to find some hidden truth. Stay at a general level, when they think they have some insight, they delve into details. Tend to work solo.	Somewhat positive about structured methods but only when they have achieved some level of insight or understanding.	Conducted very broad Web search, looking only briefly at individual site contents. Only after hours of "skimming" did he cue on a feature of the case and begin downloading documents specific to that aspect.

most typical reaction has been that the scheme, although simple, captured observations that analysts themselves had made. The degree to which these reasoning styles are correlated or independent of one another, as well as how they might interact with specific situations, remains unknown and is fertile ground for further research.

The discovery of these distinctive styles has clear implications for education, training, and practice. How might a valid and reliable test of reasoning style be used in selection? How might a measure of reasoning style facilitate training and the *grooming* of future analysts, to create the analysts the IC wants? Is there a *best fit* of style to analytical role (e.g., are Weavers better suited for regional/historical analysis roles)?

With regard to analytical practice, is there a *best fit* of style to the type of analytical problem (e.g., are Blobbers best suited for work on long-range questions having higher intrinsic uncertainty)? Is there a *best fit* of style to analytical method (e.g., would we want to force an Infovore to adopt the process of a Weaver)? We expect that these styles address individual differences in *whether* a person approaches or seeks new knowledge, *when* a person approaches or seeks new knowledge, *how* a person approaches new knowledge or information, and *what* they do with it. Thus, we see exploration and measurement of these styles as a potentially important bridge between our current understanding of the assessment of personality and mental ability and the complex task demands unique to professional intelligence analysts.

Training Challenges

There are a number of training challenges and these naturally relate to the matter of the development of expertise in intelligence analysis. What works and works well in the training of analysts? What might work even better; that is, how can the achievement of high proficiency be rapidized?

Training and curriculum development are a major focus of effort within the IC (see for instance the *e-Connection Newsletter* of the Joint Military Intelligence Training Center (jtmoc@JMITC.mil)). Training programs have been aimed at helping analysts recognize and avoid, or mitigate cognitive biases such as overconfidence (see Chang and Tetlock, 2016). Courses on cognitive bias, and on structured analysis methods intended to mitigate bias, are a standard element to academic curricula in the field of intelligence analysis. However, there is also a recognition that there is much more to the psychology of intelligence analysis than the matter of cognitive bias. Thus, a number of academic institutions have inaugurated new curricula, workshops, and summer sessions that emphasize practice at argumentation and critical thinking using realistically complex scenarios (see, for instance, Tecuci et al., 2010). Some academies have been able to institute curricula that are designated as "Centers for Academic Excellence" and funded by grants from the Defense Intelligence Agency (see http://www.dia.mil/Training/IC-Centers-for-Academic-Excellence/).

Training within the IC is largely designed and implemented in-house, due to considerations of secrecy, so less is known outside the IC about such training, though it is apparent that the training does include course work on critical thinking and argumentation. Training that includes practice with feedback contribute to the development of proficiency in analysis, just as they do in other domains. The IC realizes this and offers a wide array of beginning and advanced courses in critical thinking, structured analytic techniques, and the mechanics of collaboration, facilitation, and intelligence creation. Most of these classes range in length from one to five days and involve a blend of lectures, discussions, and exercises. Several IC members have gone so far as to integrate analytical games into their curricula. Often a scenario-based capstone exercise occurs on the last day of the class that *tests* what the students learned. At the completion of the course the IC students receive credit for having completed the class and return to their offices and their normal duties. Alas, there is little reinforcement, particularly of the critical thinking or structured analytic techniques.

We would not assume that training, even basic training, could or should be completely generic in the sense that all analysts-in-training for all IC organizations would get the same basic training. Differences between organizations and their missions reach to the very rudiments of required knowledge and skills. In his ethnographic analysis of the IC (in particular, the culture of the CIA), Johnston (2005) argued for cross-organization training, saying: "the intelligence agencies that do provide basic and advanced training do so independently of other intelligence organizations. A number of intelligence agencies do not provide basic analytic training at all" (p. 28).

We know from studies in various domains, such as weather forecasting, that basic capacity, not just high proficiency, is achieved only if there is significant on-the-job training and mentoring (Hoffman et al., 2017; Pliske et al., 1997, 2004). The large advantage of one-on-one instruction and mentoring over group (traditional classroom) instruction has been amply documented (Bloom, 1984). To this end, mentoring programs are supported in the IC, as Johnston (2005) noted, saying that intelligence agencies rely "on on-the-job experiences and informal mentoring" (Johnston, 2005, p. 28).

From observations made by one of us (Moore) over a 30-year career, we know that individuals become skilled at intelligence analysis by making judgments and having those judgments critiqued by peers and betters. Thus, senior apprentices and junior journeymen improve their capacity to make valid and valuable judgments by having those judgments critiqued. Notwithstanding, there is room for enhancement in both mentoring and feedback in analyst training.

A number of methods for training decision making expertise have been proven effective (see Hoffman et al., 2014; Klein, 2003). The most effective feedback is that which focuses on the processes employed in making a decision, such as "how we made the decisions or how we could have spotted patterns more quickly" (Klein, 2003, pp. 48–50). This is similar to the "lessons learned" strategies periodically employed in the IC, although that focuses mostly on intelligence errors or failures (i.e., poor decisions).

Future Directions

The US Department of Defense always has programs aimed at developing new computational technologies to assist in analytical judgment and decision making. Indeed, this reliance on technology, sometimes characterized as *technophilic hubris* (Macrakis, 2010), has its roots in the earliest days of the intelligence communities born during World War II (Jeffreys-Jones & Stafford, 2000; see also Erskine & Smith, 2011). Consistently, the challenge has been to make the right technology and not just more technology. While there is success at the latter, there is less success at the former.

Reviews have consistently shown that software decision support systems and new interfaces typically do not actually match the needs of end users. This is due, in large part, to the reliance on what is called designer-centered design. Program managers, certainly smart and well-intentioned, feel that they can fill the shoes of the workers and envision the tools and functionalities that the workers need. In response to program announcements, technologists promise computers that will work miracles, but the net result is a huge waste of taxpayer money (see Moon & Hoffman, 2005).

To examine the actual uses of technology in expert judgments, we must first ask whether the use of technology facilitates expert judgment. It seems that it often does not; it can interfere with the exercise of expertise and impede the development of expertise (Klein, 2003). If the software does not match the analyst's methods and preferences it can interfere, or at worst, impose a process on the analyst. In other words, it is a process control system rather than a decision aid. It also can slow the rate of learning, so that it takes much longer for people to build up their intuitions and expertise by working long and hard on hard problems. And further, it can teach dysfunctional skills that will actively interfere with the people's ability to achieve expertise in the future.

Technology that integrates or *fuses* data (magically turning it into *knowledge*) has clearly had a negative impact. Experts need to be able to see the indicators that something is *wrong*. Data filtering and fusing limit the analyst's ability to *drill down*. Furthermore, the filters and fusions are surrogates for the software designer's view of the data. Experts like to form their own understanding. Technology is great at producing models that summarize vast volumes of high-velocity data in recognition that too much data easily lead to sensory overload. By predefining "the data as essential or non-essential" one can let "the computer compile the data" (Klein, 2003, p. 255). However, "the appropriate level of detail depends on what you are searching for" (Klein, 2003, p. 255). Thus, a method for a priori distinguishing the essential from non-essential remains elusive. Finally, technology informs best about what is there. It is pertinent to remember Klein and Hoffman's observation that experts see what is not there. Therefore, a model that is based on the fusion of massive data is misfocused at best, and worse: trust in false positives about adversarial intentions can lead to significant intelligence failures.

Setting aside the fusion concept, the IC's obsession and massive financial investment in obtaining more and more data instead of the right data is unlikely to resolve. A result is that insufficient attention is paid to the integration of cognitive systems engineering notions and methods and the human-centered computing philosophy into the processes for procuring new technologies to aid the intelligence analyst.

A long-standing belief in the IC is that that if you collect more data you will get the right data (see Kent, 1947). We feel this—in some regards—is not too different from playing the lottery. True, unless you play, you absolutely cannot win. But most of us cannot afford to buy enough tickets to significantly increase our odds of winning. And, even if we won, we might have to share our winnings with others. In other words, we might spend more than we receive. To win the lottery we need the right ticket. In intelligence we also need the right *ticket* and unlike the lottery where everything is based on chance, finding the right data, or determining the importance of missing data, requires expertise. True, there is a certain baseline of information we must possess.

But we need more than the data; we need experts. As we have discussed, experts intuitively think about what it is they need to know and where they are likely to find it. This focuses their initial search and the results lead to refinements.

Conclusions

We conceptualize the different reasoning styles to be relatively stable and distinctive approaches to critical thinking. Accordingly, we would expect some cross-correlation with other constructs that are relevant to approaching or avoiding intellectual activity, especially *Need for Cognition, Need for Cognitive Closure*, and *Creativity*. Furthermore, the styles may not be mutually exclusive type categories. Fundamentally, though, there has to be independent validity and reliability analysis.

Experts are our best asset, especially if, as part of their expertise they're aware of their own limitations. In essence, given a continuously morphing set of challenges and issues, expert intelligence analysts constantly need to be redeveloping their expertise in order to keep pace or even pull ahead of the changes. This includes the workforce of intelligence professionals as well as the women and men who guide them. Therefore, the techniques of how to gain and maintain expertise become applicable to all levels of the IC's professional workforce.

Given the cyclical nature of its hiring patterns, the IC has experienced expert-losing periods in the past. Looking at the present situation from a broader perspective reveals that given time the IC can maintain and even reestablish the high levels of expert proficiency it needs. However, in the current environment where situational expertise rapidly changes a speedier means of developing experts is necessary. This begins, as we have outlined here, with an understanding of what constitutes expertise and a critical examination into how it can be developed. Necessary next steps include developing the specific means to more rapidly develop the experts we need and create an environment and opportunities where their expertise can constantly be tested, enhanced, and, as needed, transformed in order to keep pace with changes in the larger world. Calibrated case-based education and training repeated on a periodic basis provides expert (instructor)-led practice. Peer-reviewed judgments that look at *lessons learned* offer another means to acquire and calibrate the tacit knowledge that is so essential to the expert.

We believe the IC can position itself as a true learning organization relevantly poised to take on the tough issues it confronts today and will face tomorrow. We can recognize the value of experts and corporately encourage and reward its acquisition. A business model that also supports a wise use of technology to facilitate foraging for the right data will aid the work of experts. We suggest that in times of fiscal restraint such efforts will be more cost effective than the constant efforts to catch up with ever-increasing vast volumes of volatile data.

Disclaimer

The opinions expressed in this chapter are those of the authors and do not necessarily represent those of NSA/CSS or the US Department of Defense.

Acknowledgement and Dedication

The authors would like to thank Jordan Litman, PhD of the Institute for Human and Machine Cognition for his comments on the psychometric concepts we present in this chapter.

The second author would like to dedicate this chapter to David T. Moore. As a mentor and a thought leader, he will be dearly missed.

References

Bloom, B. S. (1984). The 2-sigma problem: The search for methods of group instruction as effective as one-on-one tutoring. *Educational Researcher* 13, 4–16.

Booz Allen Hamilton (2006). *Intelligence community analytic competency framework development and verification: Final report*. Washington, DC: Booz Allen Hamilton.

Cassidy, S. (2004). *Learning styles: An overview of theories, models and measures*. London: Routledge.

Chang, W., & Tetlock, P. E. (2016). Rethinking the training of intelligence analysts. *Intelligence and National Security* 31(6), 903–920.

Chinn, C., & Brewer, W. (1993). The role of anomalous data in knowledge acquisition: A theoretical framework and implications for science instruction. *Review of Educational Research* 63, 1–49.

Clark, R. M. (2004). *Intelligence analysis: A target-centric approach*. Washington, DC: CQ Press.

Crowne, D. P. (2009). *Personality theory*, 2nd edn. Oxford: Oxford University Press.

De Keyser, D., & Woods, D. (1993). Fixation errors: failures to revise situation assessment in dynamic and risky systems. In A. Colombo and A. Saiz de Bustamante (Eds), *Advanced systems in reliability modeling* (pp. 231–251). Norwell, MA: Kluwer.

Erskine, R., & Smith, M. (Eds) (2011). *The Bletchley Park codebreakers*. London: Biteback Publishing.

Funder, D. C., & Ozer, D. J. (1983). Behavior as a function of the situation. *Journal of Personality and Social Psychology* 44, 107–112.

Groen, G., & Patel, V. (1988). The relationship between comprehension and reasoning in medical expertise. In M. Chi, R. Glaser, and M. J. Farr (Eds), *The nature of expertise* (pp. 287–310). Mahwah, NJ: Erlbaum.

Groen, G., & Patel, V. (1991). The general and specific nature of medical expertise: A critical look. In K. A. Ericsson, and J. Smith (Eds), *Toward a general theory of expertise: Prospects and limits* (pp. 93–125). Cambridge, UK: Cambridge University Press.

Heuer, R. J. (1999). *Psychology of intelligence analysis*. Washington, DC: Center for the Study of Intelligence.

Heuer, R., & Pherson, R. (2014). *Structured analytic techniques for intelligence analysis*, 2nd edn. Washington, DC: CQ Press.

Hoffman, R. (1996). How can expertise be defined? Implications of research from cognitive psychology. In R. Williams, W. Faulkner, and J. Fleck, *Exploring expertise* (pp. 81–100). New York: Springer.

Hoffman, R. R. (2003). Use of concept mapping and the critical decision method to support human-centered computing for the intelligence community. Report to the Palo Alto Research Center (PARC) on the Project "Theory and Design of Adaptable Human Information Interaction Systems for Intelligence Work," Novel Intelligence From Massive Data (NIMD) Program, Advanced Research and Development Activity (ARDA), Department of Defense, Washington, DC.

Hoffman, R. R. (2005). *The Psychology of Intelligence Analysis* revisited: An update from developments in cognitive science post-1980. Report, Institute for Human and Machine Cognition, Pensacola, FL. [Download from http://tarf.ihmc.us/rid=1K8YFYHFB-1CFHMNR-2RJW/Psychol%20of%20Intelligence%20Analysis%20Revisited-May2005.pdf]

Hoffman, R. R., Klein, G., & Miller, J. E. (2011). Naturalistic investigations and models of reasoning about complex indeterminate causation. *Information and Knowledge Systems Management 10*, 397–425.

Hoffman, R. R., LaDue, D., Mogil, H. M., Roebber, P., & Trafton, J. G. (2017). *Minding the weather: How expert forecasters think*. Cambridge, MA: MIT Press.

Hoffman, R. R., & Militello, L. G. (2008). *Perspectives on cognitive task analysis: Historical origins and modern communities of practice*. Boca Raton, FL: CRC Press/Taylor and Francis.

Hoffman, R. R., Ward, P., DiBello, L., Feltovich, P. J., Fiore, S. M., & Andrews, D. (2014). *Accelerated expertise: Training for high proficiency in a complex world*. Boca Raton, FL: Taylor and Francis/CRC Press.

Jeffreys-Jones, R., & Stafford, D. (Eds) (2000). *American–British–Canadian intelligence relations, 1939–2000*. London: Routledge.

Johnston, R. (2005). *Analytic culture in the US intelligence community: An Ethnographic Study*. Washington, DC: Center for the Study of Intelligence.

Kahneman, D., & Klein, G. (2009). Conditions for intuitive expertise: A failure to disagree. *American Psychologist 64* (6), 515–526.

Kent, S. (1947). *Strategic intelligence for American world policy*. Princeton, NJ: Princeton University Press.

Klein, G. (1998). *Sources of power: How people make decisions*. Cambridge, MA: MIT Press.

Klein, G. (2003). *Intuition and work*. New York: Doubleday.

Klein, G. (2009). *Streetlights and shadows: Searching for the keys to adaptive decision making*. Cambridge, MA: MIT Press.

Klein, G. (2011). Critical thoughts about critical thinking. *Theoretical Issues in Ergonomics Science 12*(3), 210–224.

Klein, G. (2014). *Seeing what others don't: The remarkable ways we gain insights*. London: Nicholas Brealey Publishing.

Klein, G., & Hoffman, R. (1992). Seeing the invisible: Perceptual cognitive aspects of expertise. In M. Rabinowitz (Ed.), *Cognitive science foundations of instruction* (pp. 203–226). Mahwah, NJ: Erlbaum.

Klein, G., & Rothman, J. (2008). Staying on course when your destination keeps changing. *Conference Board Review* (November–December), 24–27.

Klein, G., Ross, K. G., Moon, B. M., Klein, D. E., Hoffman, R. R., & Hollnagel, E. (2003). Macrocognition. *IEEE Intelligent Systems* 18 (3), 81–85.

Lahneman, W. J. (2006). The future of intelligence analysis. College Park, MD: School of Public Policy, University of Maryland. [Downloaded from http://commons.erau.edu]

Lendl, C. (2012). *Bletchley Park: British cryptanalysis during World War II*. Online publication:http://www.bletchleypark.at/, accessed 1 February 2017.

Lowenthal, M. (2007). Foreword. In D. T. Moore, *Critical thinking and intelligence analysis* (pp. ix–xi). Washington, DC: National Defense Intelligence College.

Macrakis, K. (2010). Technophilic hubris and espionage during the cold war. *Isis 101*, 378–385.

Moon, B., & Hoffman, R. (2005). How might "transformational" technologies and concepts be barriers to sensemaking in intelligence analysis? Presentation at the Seventh International Conference on Naturalistic Decision Making, Amsterdam, The Netherlands.

Moore, D. T. (2004). Species of competencies for intelligence analysis. *American Intelligence Journal* 25, 29–43.

Moore, D. T. (2007). *Critical thinking and intelligence analysis*, rev. edn. Washington, DC: NIU Press.

Moore, D. T. (2011). *Sensemaking: A structure for an intelligence revolution*. Washington, DC: NIU Press.

Moore, D. T., and Hoffman, R. (2011). A practice of understanding. In D. T. Moore, *Sensemaking: A structure for an intelligence revolution* (pp. 69–94). Washington, DC: NIU Press.

Moore, D. T., and Krizan, L. (2003). Core competencies for intelligence analysis at the National Security Agency. In R. Swenson (Ed.), *Bringing intelligence about: Practitioners reflect on best practices* (pp. 95–131). Washington, DC: Joint Military Intelligence College.

Moore, D. T., Krizan, L., and Moore, E. (2005). Evaluating intelligence: A competency-based approach. *International Journal of Intelligence and Counterintelligence* 18 (2), 204–220.

Office of the Director of National Intelligence (2008). Intelligence Community Directive number 652: Occupational structure for the intelligence community civilian workforce. Washington, DC: Office of the Director of National Intelligence.

Office of the Director of National Intelligence (2009). Intelligence Community Directive number 601: Human capital joint intelligence community duty assignments. Washington, DC: Office of the Director of National Intelligence.

Office of the Director of National Intelligence (2012). Intelligence Community Directive number 651: Performance management system requirements for the intelligence community civilian workforce. Washington, DC: Office of the Director of National Intelligence.

Phelps, R. H., & Shanteau, J. (1978). Livestock judges: How much information can an expert use? *Organizational Behavior and Human Performance* 21, 213–222.

Pherson, R. (2015). *Handbook of analytic tools and techniques*. Reston, VA: Pherson Associates.

Pherson, K., & Pherson, R. (2016). *Critical thinking for strategic intelligence*, 2nd edn. Washington, DC: CQ Press.

Pirolli, P. (2007). *Information foraging theory: Adaptive interaction with information*. Oxford: Oxford University Press.

Pirolli, P., & Card, S. (1999). Information foraging. *Psychological Review* 106, 643–675.

Pirolli, P., & Card, S. (2005). The sensemaking process and leverage points for analyst technology as identified through cognitive task analysis. *2005 International Conference on Intelligence Analysis*, McLean, VA, 2–6 May, 2005.

Pliske, R., Crandall, B., & Klein, G. (2004). Competence in weather forecasting. In K. Smith, J. Shanteau, and P. Johnson (Eds), *Psychological investigations of competence in decision making* (pp. 40–70). Cambridge, UK: Cambridge University Press.

Pliske, R., Klinger, D., Hutton, R., Crandall, B., Knight, B., & Klein, G. (1997). *Understanding skilled weather forecasting: Implications for training and the design of forecasting tools.* Fairborn, OH: Klein Associates.

Pujol, J., & West, N. (1986). *Operation GARBO: The personal story of the most successful double agent of World War II.* New York: Random House.

Rees, R. L. (Ed.) (2007). *A handbook of the psychology of intelligence analysis.* Washington, DC: Central Intelligence Agency.

Shanteau, J., & Phelps, R. H. (1977). Judgment and swine: Approaches in applied judgment analysis. In M. F. Kaplan and S. Schwartz (Eds), *Human judgment and decision pocesses in applied settings* (pp. 255–272). New York: Academic Press.

Slovic, P. (1973). Behavioral problems of adhering to a decision policy. Paper presented at the Institute for Quantitative Research in Finance. Napa, CA.

Spiro, R. J., Coulson, R. L., Feltovich, P. J., & Anderson, D. (1988). Cognitive flexibility theory: Advanced knowledge acquisition in ill-structured domains. In V. Patel (ed.), *Proceedings of the 10th annual conference of the Cognitive Science Society.* Hillsdale, NJ: Erlbaum.

Spiro, R. J., Feltovich, P. J., Jacobson, M. J., & Coulson, R. L. (1992). Cognitive flexibility, constructivism and hypertext: Random access instruction for advanced knowledge acquisition in ill-structured domains. In T. Duffy and D. Jonassen (Eds), *Constructivism and the technology of instruction.* Hillsdale, NJ: Erlbaum.

Sternberg, R. J., Grigorenko, E. L., & Zhang, L. (2008). Styles of learning and thinking matter in instruction and assessment. *Perspectives on Psychological Science* 3, 486–518.

Sticha, P. J., & Buede, D. M. (2009). Cognitive factors, part II: What analysts need to possess and how they deal with uncertainty. In R. L. Rees, (Ed.), *A handbook of the psychology of intelligence analysis* (pp. 53–79). Washington, DC: Central Intelligence Agency.

Tecuci, G., Boicu, M., Marcu, D., Schum, D., & Hamilton, B. (2010). TIACRITIS System and textbook: Learning intelligence analysis through practice. In *Proceedings of the fifth international conference on semantic technologies for intelligence, defense, and security* (STIDS 2010). Sponsored by the CEUR Workshops, Technical University of Aachen. [http://ceur-ws.org/Vol-713/].

Tett, R. P., Hopper, J. E., Landis, B. D., & Swaim, B. C. (2007). Personality, motivations, and attitudes. In R. L. Rees (Ed.), *A handbook of the psychology of intelligence analysis* (pp. 106–144). Washington, DC: Central Intelligence Agency.

United States Congress, Senate. *Intelligence Reform and Terrorism Prevention Act of 2004.* Public Law 108-458-Dec 17, 2004. 108th Congress. 1st Session, S. 2845. Washington, DC: GPO, 2004. Available at: https://legcounsel.house.gov/Comps/National%20Security%20Act%20Of%201947.pdf. Accessed 11 August 2017.

Whaley, B. (2016). *Practise to deceive: Learning curves of military deception planners.* Annapolis, MD: Naval Institute Press.

CHAPTER 43

TEAM REFLECTION

A Catalyst of Team Development and the Attainment of Expertise

KAI-PHILIP OTTE, KRISTIN KNIPFER, AND MICHAÉLA SCHIPPERS

Introduction

Teamwork is prevalent in many contexts (Mathieu, Hollenbeck, van Knippenberg, & Ilgen, 2017), and effective teams are indispensable for sustaining organizational competitiveness and long-term success (Decuyper, Dochy, & Van den Bossche, 2010). An important determinant of team effectiveness is *team expertise*, that is, the ability to effectively leverage upon the knowledge and expertise of all team members (Barton & Bunderson, 2014; Hackman, 2002). Previous research points to the significance of team expertise for high team performance (e.g., Edmondson, 2002; Kozlowski & Chao, 2012; Salas, Rosen, Burke, Goodwin, & Fiore, 2006). In this chapter, we discuss how the development of expertise in teams can be facilitated or accelerated. In particular, we argue that *team reflection* is an important driver for the development and attainment of expertise in teams. In other words, we propose team reflection as a catalyst for becoming *an expert team* (Ward & Eccles, 2006; West, 1996).

Team reflection is generally conceived as collectively looking back on lived experiences (for current reviews see Konradt, Otte, Schippers, & Steenfatt, 2016; Schippers, Edmondson, & West, 2017; Widmer, Schippers, & West, 2009). It refers to a team's engagement in reviewing and evaluating their strategies, teamwork processes, team objectives, lived experience, and shared assumptions (Savelsbergh, van der Heijden, & Poell, 2009). Research has demonstrated that team reflection is positively related to team performance (e.g., Konradt & Eckardt, 2016; Villado & Arthur, 2013) and team innovation (e.g., Schippers, West, & Dawson, 2015), because it helps to develop a shared understanding of the current task (Konradt, Schippers, Garbers, & Steenfatt, 2015) and supports the adaptation to changes in the environment (Schippers, Den Hartog, & Koopman, 2007).

Among others, these findings have been attributed to a more effective sharing of unique expertise among team members (e.g., De Dreu, 2007; Konradt et al., 2015) and a better integration and use of the individual expertise of members of the team (e.g., Hoegl & Parboteeah, 2006; for a review see van Knippenberg & Schippers, 2007). Based on these findings we theorize that *team reflection can foster the development and attainment of team expertise*. This is also in line with findings on individual expertise and the role of reflection on first-hand experience in expertise development (e.g., Gray, 2007; Short & Rinehart, 1993). Yet, team reflection has rarely been subject to investigations of team expertise and vice versa, pointing to a considerable gap in the literature.

In the following, we discuss several limitations with regard to our understanding of team reflection and its role for the development of team expertise: In general, the team reflection literature is surprisingly fragmented into more or less independent fields. The number of studies within a particular field is rather small and important findings from different fields were rarely integrated. This is why we still lack a comprehensive understanding of team reflection and its effects on team processes and outcomes. Specifically, research mainly focused on establishing the basal link between team reflection and proximal outcomes such as team performance after a reflection session, thereby neglecting the role of team expertise as a potential mediator.

Moreover, team reflection has been rarely investigated over longer periods of time, which naturally limits the insights to be gained on the role of team reflection in the longitudinal development of team expertise. Furthermore, team reflection research has mainly concentrated on ad-hoc teams, that is, teams who work together for the duration of a specific task (for exceptions see Schippers, Den Hartog, Koopman, & van Knippenberg, 2008; Schippers et al., 2015). In ad-hoc teams, members rarely have high levels of unique expertise (i.e., they are novice teams) or the opportunity to develop expertise over time. However, team expertise is most likely distributed among individuals in teams, and the unique expertise of members of the team must be used and integrated to contribute to team-level expertise (e.g., Barton & Bunderson, 2014; Hollenbeck, Ilgen, Sego, Hedlund, Major, & Phillips, 1995; Salas et al., 2006). Still, the team reflection literature has hardly integrated multilevel issues, limiting our understanding of how individual expertise contributes to team expertise (and vice versa). In the next section, we will first define reflection, then make the connection with team self-regulation. Next, we discuss the research on guided team reflection, and then discuss catalysts of team reflection. Finally, we present a critical review of previous research and discuss avenues for future research.

Definition(s) of Team Reflection

Team reflection has been conceptualized in many ways, including the retrospective contemplation of the self (West, 2000), the systematic analysis of learners' behavior and its contribution to performance outcomes (Ellis & Davidi, 2005), the joint evaluation of team behaviors (Konradt et al., 2016), the discussion of team processes (Schippers, Edmondson, & West, 2014; West, 2000), and the process of extracting knowledge from work-related experience (Quinn & Bunderson, 2016). Although these definitions highlight different aspects (retrospective contemplation, analysis, evaluation, discussion, and knowledge creation) and refer to agents of learning at different levels (individual, group, or organizational level), team reflection is commonly understood as a team process that enables teams to continuously improve team effectiveness and become an expert team (Schippers et al., 2017). In other words, team reflection enables teams to share and integrate the unique individual expertise of its members.

Despite the somewhat disparate nature of the team reflection concept mirrored in different literature streams, there are fundamental similarities that are important to understand how team reflection contributes to the development and attainment of team expertise: First, team reflection typically occurs during *transition phases*, in which teams "focus primarily on evaluation and/or planning activities to guide their accomplishment of a team goal or objective" (Marks, Mathieu, & Zaccaro, 2001, 360). Thus, team reflection typically occurs when a performance episode has ended and in an effort to prepare for future action, for example in project review meetings or retreats (e.g., Marks et al., 2001). Second, previous conceptualizations have adopted an information-processing perspective (Hinsz, Tindale, & Vollrath, 1997; Schippers et al., 2014) to describe core processes involved in team reflection, namely information seeking and information evaluation (Otte, Konradt, Garbers, & Schippers, 2017). Third, team reflection is considered as an instrumental and purposeful process that allows for a valid assessment of the status quo and the development of new action to further enhance team effectiveness (Schippers et al., 2007; West, 2000). From these previous conceptualizations, we extracted the following working definition of team reflection for this chapter (cf. Konradt et al., 2016; Savelsbergh et al., 2009; Schippers et al., 2014):

> *Team reflection is the collective evaluation of prior team activities and how they have contributed to the current status of the team.*

Previous research has highlighted that team reflection can happen rather spontaneously and autonomously as a form of self-regulation (e.g., Carver & Scheier, 1998)—sometimes even *in action* (e.g., Schippers, 2003; Schmutz & Eppich, 2017)—whereas others have looked at team reflection as a formalized part of teamwork in forms of debriefing sessions (e.g., Ellis & Davidi, 2005). Hence, in the following, we chose to review the team reflection literature for two categories separately: *self-regulated* and *guided* team reflection.

Self-Regulated Team Reflection

Self-regulation is often viewed as the dominant model of learning (Kozlowski & Chao, 2012) and can be described as a dynamic process by which people manage demands and resources in order to reach desired outcomes (Neal, Ballard, & Vancouver, 2017). This stream of research is generally based on assumptions of control theory and negative feedback loops, where behavior is aimed at the reduction of discrepancies between a current and a desired state or, put differently, adjustments that are needed to stay on track (Carver & Scheier, 2011; DeShon, Kozlowski, Schmidt, Milner, & Wiechmann, 2004). Hence, teams are assumed to reflect spontaneously whenever it is appropriate or necessary in order to reduce existing discrepancies in progressing towards team goals in an effort to improve team effectiveness.

In this stream of literature, the notion of *team reflexivity* has gained much attention since being introduced by West (1996; 2000). Team reflexivity refers to "the extent to which team members collectively reflect upon the team's objectives, strategies, and processes, as well as their wider organizations and environments, and adapt them accordingly" (West, 2000, 3). It is important to mention that, while some exceptions also included the roles of team planning (e.g., Gabelica, Van den Bossche, De Maeyer, Segers, & Gijselaers, 2014; Gevers, van Eerde, & Rutte, 2001) and team implementation (e.g., Konradt et al., 2015), the majority of the team reflexivity research focused on the reflection component. Hence, although we will refer to team reflexivity when it is appropriate in the following to accurately depict the evolution of the concept, findings in the empirical research mainly examine and discuss the relationship of team reflection and outcomes (such as performance or expertise).

Following up on West's initial research, Schippers et al. (2007) advanced the team reflexivity concept and introduced two types of reflection, namely *evaluation/learning* and *discussing* processes. More recently, they proposed that the *discussing* component may be more relevant in the early stages of team development, that is, when a team's expertise is still low. The *evaluating* component may be more relevant in later stages when the team has achieved some maturity in team expertise (Schippers et al., 2017).

Konradt et al. (2016) understood team reflexivity as a core element in team self-improvement that integrates the retrospective evaluation of preceding teamwork and the development of new teamwork strategies. Based on self- and team-regulatory assumptions (Carver & Scheier, 1998; DeShon et al., 2004) and the notion of team-performance episodes (Marks et al., 2001), they argued that team reflexive processes are fundamental in the reduction of discrepancies between the current status and a desired state. They further suggested that the detection of current discrepancies in what a team has achieved and what it intended to achieve will fuel a team's reflexivity efforts, and that these efforts will be terminated as soon as the discrepancy is resolved. Konradt and Eckardt (2016) provided first evidence for this assumption. They found that an overall increase of performance of 97 teams working on a business simulation task was related to a decrease in reflection activities, indicating that teams diminish their reflection when their performance, and thereby their expertise, increases. Finally, recent research proposed to conceive team reflexivity as an umbrella framework that integrates several related but still distinct reflexive processes, namely information seeking, information evaluation, planning, and implementation (Otte et al., 2017).

A growing number of studies has linked self-regulated team reflection to team performance, particularly in contexts where the environment changes frequently (De Dreu, 2002; Hoegl & Parboteeah, 2006; Schippers et al., 2015; for a recent review see Schippers et al., 2017). These contexts are characterized by high uncertainty, rapidly changing circumstances, complex decisions to be taken, low predictability of outcomes, and a constant need for innovation, making teams very susceptible to failure if they are not able to adapt their goals and processes quickly. In line with that, positive relationships between team reflection and team learning (Quinn & Bunderson, 2016; Schippers et al., 2013), team performance (Konradt & Eckardt, 2016; Schippers, Homan, & van Knippenberg, 2013), and team innovation (Schippers et al., 2015) have been found in previous research (Konradt et al., 2016; Schippers et al., 2017).

Although there is large agreement in the literature that team reflection can increase team effectiveness, some exceptions point to the significance of boundary conditions that can leverage the impact of team reflexivity on team effectiveness (Moreland & McMinn, 2010; see also Schippers et al., 2014; Schippers et al., 2015; Schippers et al., 2017). Some studies have reported a neutral (De Dreu, 2002, 2007) or even a negative relationship (Wiedow & Konradt, 2011) of team reflection and performance. Schippers et al. (2013) examined 73 novice student teams in a longitudinal study who worked on a set of consecutive tasks (research proposal and bachelor thesis) over the course of a semester. They found a positive effect of team reflection on team performance as well as team learning—but only for those teams whose initial performance was low. In contrast, teams that were performing well initially did not benefit from team reflection. These findings indicate that the benefits of team reflection are greater in novice than in expert teams.

Furthermore, Otte et al. (2017) noted that team reflection is typically assessed by asking team members *how often* or *to what extent* they reflected, putting a focus on the frequency of reflection (i.e., how much time teams spend on reflection or how often they reflected while working on a task). This quantitative operationalization, however, neglects the depth and detail in team reflexive processes. It is plausible that this suboptimal and incomplete conceptualization of team reflexivity has contributed to the previous heterogeneous results concerning the effects of team reflexivity on team performance. In line with that reasoning, Gurtner, Tschan, Semmer, and Nägele (2007) found that teams often focus on strategies that are far too general and, as a consequence, do not benefit from team reflection. Thus, only asking teams how often they reflected might result in teams indicating equal amounts of reflection, when they actually differ on depth of reflection. Based on the work from West (2000) and Schippers et al. (2007), Otte et al. (2017) developed and validated a measure that captures both the quantity *and* quality of team reflection, allowing a more precise assessment of the reflection process (e.g., "we made very frequent evaluations of the quality of our work" vs. "we made very detailed evaluations of the quality of our work," p. 304).

To sum up, research on self-regulated team reflection concepts has provided considerable support for the positive effects of team reflection on outcomes. Yet, boundary conditions (e.g., the team context, the level of expertise) and different forms of reflection (amount and depth) must be taken into account in order to fully comprehend the effects and outcomes of self-regulatory team reflection. From our review, we conclude that researchers investigating self-regulated team reflection should (a) describe the context of their research in terms of work demands, (b) include potential moderators of the effects of team reflection on team outcomes such as maturity of the team and prior performance (e.g., Schippers et al., 2013), and (c) use fine-grained measures of team reflection that tackle not only the quantitative but also the qualitative aspects of team reflexivity.

Guided Team Reflection

In the previous section, "Self-Regulated Team Reflection," we considered team reflection as a self-regulatory team process. However, teams are inclined to prioritize action over reflection (e.g., Druskat & Kayes, 2000) or may face difficulties in making the most out of team reflection in terms of concrete lesson learned. In line with this, Hackman (1998) argued that "once a team has been formed and given its task, managers sometimes assume their work is done. A strict hands-off stance, however, can limit a team's effectiveness when members are not already skilled and experienced in teamwork" (p. 254), indicating that teams may need assistance or guidance in team reflection; otherwise, they will suffer from process losses (Eddy, Tannenbaum, & Mathieu, 2013; Hackman, 1998; Hackman & Wageman, 2005). For example, teams may be unwilling to invest cognitive resources into team reflection (e.g., Schippers et al., 2013) or feel uncomfortable in questioning previous strategies or routines (e.g., Edmondson, 1999), which may result in interpersonal conflict and even blaming (Gabelica et al., 2014).

Accordingly, some authors argue in favor of structured reflection sessions. We refer to interventions that are purposefully implemented with an effort to stimulate and guide team reflection as *formal team reflection* interventions. Formal team reflection interventions include *debriefing techniques* (Tannenbaum & Cerasoli, 2013), *guided team self-correction* (e.g., Smith-Jentsch, Cannon-Bowers, Tannenbaum, & Salas, 2008), *after-event reviews* (e.g., Ellis & Davidi, 2005), and *guided team reflection* (Gabelica et al., 2014; Gurtner et al., 2007; Konradt et al., 2015). Typically, a facilitator guides the team through a series of questions in order to aid the reflection process following a clear structure and using concrete guidelines. Moreover, Schippers (2003), and more recently Schmutz and Eppich (2017), argued that such team reflection can also be triggered while the team is working on a task by brief transition phases, which allow the team to reflect upon its previous performance in order to make subtle adjustments to the current activities. Such short bursts of reflection are also referred to as *stop and think interruptions* (Okhuysen, 2001).

The general structure of guided team reflection can be understood as *collaborative debriefing* that takes place after the team has been working on a task (Tannenbaum &

Cerasoli, 2013). Debriefing techniques have a long tradition in the USA military, but they have also become popular in other fields such as police teams (Bechky & Okhuysen, 2011), firefighting crews (Allen, Baran, & Scott, 2010) and hospital teams (Vashdi, Bamberger, & Erez, 2013). These techniques are assumed to positively impact team performance and learning by giving teams and their members a systematic tool at hand that facilitates learning from experience in order to improve their performance (Tannenbaum & Cerasoli, 2013). Examples for guiding questions can be found for instance in Gurtner et al. (2007), who asked teams that worked in a military simulation how they asked or passed information, or how the team was organized. Similarly, in a complex decision-making task, Konradt et al. (2015) instructed teams to review their performance and reflect on the knowledge each team member contributed to the task.

Guided reflection sessions can help overcome the challenges associated with team reflection described by Smith-Jentsch et al. (2008): Strong guidance supports teams to ask the right questions, leading to the deduction of rules that may facilitate strategy implementation as well as transfer to other situations or tasks later on. Guided team reflection was further shown to enhance the development of shared mental models, thereby enabling team members to communicate and coordinate effectively (e.g., Gurtner et al., 2007). As guided reflection sessions often include structured feedback about work strategies and recent performance, they allow for a more valid assessment of the status quo before providing the background of the targeted outcomes (Anseel, Lievens, & Schoellaert, 2009). In their meta-analysis, Tannenbaum and Cerasoli (2013) found an overall positive effect of debriefing techniques on team performance and concluded that organizations can improve individual and team performance by 20 to 25% by using debriefs.

Yet, most research on formal reflection interventions are targeted at individuals and not teams (e.g., Allen et al., 2010; Boet, Bould, Bruppacher, Desjardins, Chandra, & Naik, 2011; DeRue, Nahrgang, Hollenbeck, & Workman, 2012), thereby limiting the insights on the underlying mechanisms that may explain the effects of debriefings on team performance. As an exception, Villado and Arthur (2013) found positive effects of debriefings—among others—on the extent and openness of communication within the team. Similarly, Weiss, Kolbe, Grote, Spahn, and Grande (2016) provided evidence for the value of after-event reviews for promoting vocal behavior (i.e., speaking up with suggestions and concerns). Gurtner et al. (2007), Konradt et al. (2015), and Tesler, Mohammed, Hamilton, Mancuso, and McNeese (2017) have further highlighted the similarity of mental models as mediators of the link between guided team reflection and team performance (see for a review Mohammed, Hamilton, Sánchez-Manzanares, & Rico, 2017). These promising findings suggest that sharing of individual knowledge and its integration into shared team expertise may be a crucial pathway through which team reflection leverages team performance. Therefore, we encourage team researchers to include individual and team expertise as potential mediators in their future investigations of team reflection and its effects.

From the review of literature, we conclude that guided team reflection holds the potential to foster team effectiveness and the development of expertise (e.g., Okhuysen, 2001; Villado & Arthur, 2013; Waller, Zellmer-Bruhn, & Giambatista, 2002). Yet, as a boundary condition, Okhuysen (2001) found that only teams whose members did not know each other before starting on the team task improved in their performance, whereas teams whose members knew each other beforehand showed significant decreases in performance. They attributed this findings to the fact that the intervention led to a shift of focus from task work to teamwork (see also Fiore & Wiltshire, 2016; Mathieu, Heffner, Goodwin, Salas, & Cannon-Bowers, 2000). As a result, teams whose members were familiar with each other focused on social interactions that might not have contributed directly to task accomplishment.

Additionally, we know very little about whether formal team reflection sessions have effects beyond the immediate task at hand. Arsenault (2011), for example, provided evidence that guided reflection sessions improved performance on routine tasks but this benefit did not transfer to tasks that put novel performance demands on the team.

Finally, studies comparing and contrasting self-regulatory and guided team reflection are scarce. To our knowledge, only Eddy et al. (2013) compared a semi-structured debriefing session using predefined questions to induce reflection with a fully structured reflection session using an advanced online debriefing tool. They studied 35 student teams working on a series of business cases over the course of a semester. As hypothesized, the authors found that teams using the online debriefing tool engaged in more transition, action, and interpersonal processes (cf. Marks et al., 2001). Yet, this effect was fully mediated by team processes, as neither the bi-serial correlation nor the hierarchical linear regression revealed a positive direct effect of the semi- or fully structured debriefings on performance.

Future team research should further explore the differential effects of self-regulatory and guided team reflection on processes and outcomes. As briefing and debriefing could be seen as aiding team reflection, in the next section, "Catalysts of Team Reflection," we discuss factors such as the availability of feedback, psychological safety, and empowering leadership (see also Schippers et al., 2014).

Catalysts of Team Reflection

Given the significance of both self-regulated and guided team reflection for team functioning and expertise, we should understand the factors that promote or hinder team reflection in organizations. We consider the following factors to be important contextual factors that can increase team reflection to occur, namely availability of feedback, psychological safety, and empowering leadership.

Feedback Availability

In line with the self-regulatory theory that proposes that teams work towards reducing discrepancies between current and desired outcomes, the *availability of feedback* has a high impact on the degree to which individuals and teams learn from experience (DeRue & Wellman, 2009). This is particularly relevant for teams in early stages of their development (Kruger & Dunning, 1999). Feedback can be defined as "information about the actual performance or actions of a system used to control the future actions of a system" (Nadler, 1979, p. 310). Feedback provides a foundation for subsequent evaluation of performance, and permits the identification of targets for improvement (Anseel, Beatty, Shen, Lievens, & Schoellaert, 2015; Atwater & Waldman, 1998). Learning research suggests that feedback can help novices to direct their attention towards those cues that will help them identify the causes and consequences in terms of their previous performance, thereby helping them develop ideas on how to further improve (e.g., Cognitive Transformation Theory, see Klein & Baxter, 2006). Moreover, it helps to identify blind spots and misconceptions about one's own behavior by obtaining information from external sources (e.g., self-serving biases, Marsick & Watkins, 2001; Miller & Ross, 1975). Accordingly, several studies highlighted that receiving feedback is an important prerequisite for effective reflection and has the potential to increase performance (Anseel et al., 2009; Gabelica et al., 2014; Konradt et al., 2015).

Psychological Safety

Research in the field of education and learning has highlighted the significance of a safe-to-fail environment in the development of expertise, which can foster critical reflection and active experimentation (e.g., Lee, Edmondson, Thomke, & Worline, 2004). Yet, a major barrier to team reflection could be that the members of the team may feel not comfortable to voice concerns, disclose conflicts, or address failure (Tucker & Edmondson, 2003). For example, organizational norms for reporting medication errors in hospitals were shown to leverage the potential of learning from errors to a large extent (Edmondson, 1996). Accordingly, Edmondson (1999) has pointed to the relevance of *psychological safety*, that is the perception that seeking feedback, asking for help, talking about failure, and experimenting with new ideas is accepted and valued by others in the team. Psychological safety is associated with desirable consequences including an increase in employees' engagement (Nembhard & Edmondson, 2006), learning from mistakes (Edmondson, 1996), and team innovation (West & Anderson, 1996). Similarly, Mezirow (2000), as well as Fook, White, and Gardner (2006), highlighted that critical acceptance was a crucial precondition for reflection, referring to an organizational climate in which different perspectives are valued and critical reflection of commonly held assumptions is encouraged.

Senge's (1990) notion of *learning organization* is an idealized version of an organization, in which individuals and teams leverage their individual perspectives and expertise in an effort to optimize (joint) work practices, for example, in the form of communities of practice (Vince, 2002).

Empowering Leadership

Prior research has also highlighted the role of leadership in shaping a culture of reflection (e.g., Edmondson, 2003; Vera & Crossan, 2004) as leaders are particularly effective in shaping team processes and outputs (Hackman & Wageman, 2005). Leaders who succeed in creating a safe climate for their team, who provide the opportunity to get involved in decision making and change processes, and who communicate their confidence in the team's capacity to improve can effectively increase team reflection (Edmondson, 1999; Hirst, Mann, Bain, Pirola-Merlo, & Richver, 2004). Specifically, transformational and charismatic leadership were shown to motivate their teams to constantly challenge the status quo while creating feelings of control. For example, in their study conducted among 37 teams from nine organizations, Schippers et al. (2008) provided evidence that transformational leadership was related to team reflexivity and that this relation was fully mediated by the adoption of a shared vision by the team. Moreover, Knipfer, Schreiner, Schmid, and Peus (2018) tested a multilevel mediation model based on data from 196 members of 58 nascent entrepreneurial teams in an early stage of the venture creation program (i.e., the pre-founding phase). They found that charismatic team leadership was predictive of team performance; the positive effect of charismatic leadership on team performance was fully mediated by team reflection. This study points to the significance of leadership in initiating and facilitating team reflection.

While research has made some progress in understanding the factors that facilitate or hinder team reflection, we see great potential for future research on the larger organizational context of teams and the question in which specific circumstances teams are more inclined to show reflective behaviors on their own. In the next section we discuss avenues for future research, as well as issues related to levels of analysis in this kind of research. Finally, we also discuss the time element in this kind of research.

Critical Review of Previous Research and Avenues for Future Research

Although previous research has made progress in uncovering the relationship between team reflection and improved performance, several limitations are evident in the literature that should be addressed by future research to better understand the effects of team reflection on the development of team expertise. In the following, we will therefore elaborate on two important factors that have been rarely addressed in previous research, namely the level of analysis, and short- and long-term effects of team reflection.

Level of Analysis: Individual and Team Reflection and Multilevel Outcomes

Salas and collegues (2006) referred to expert teams as teams that combine individual expertise in such a way that the team creates a synergy greater than its parts. Similarly, team reflection models assume that conversations and discussions within the team provide additional value and insights that would not have *emerged* when individuals reflected alone. Emergence thereby refers to a phenomenon that originates from the individual level, but is amplified by within-team interactions and manifests as a higher-level phenomenon (Kozlowski, 2015). To date, it remains unclear (a) whether team reflection can be conceived as such an emergent phenomenon and (b) whether team reflection is superior (or even inferior) to individual reflection.

These questions are important for the following reasons. From a researcher's perspective, emergence is an interesting topic in its own right (Fulmer & Ostroff, 2016). The fundamental reason for employing teams in organizations and team research is always based on the assumption that teams are more effective than individuals working alone. It has been proposed that team reflection produces results beyond individual reflection (West, 2000). In methodological terms, however, the literature has mainly conceived team reflection as a composition process (Kozlowski & Klein, 2000), assuming that team reflection is basically the same as the average reported team reflection by individuals. Thus, research has to a large extent relied on team-level constructs in the assessment of team reflection. An implicit assumption is that each member of the team can validly assess the level of team reflection and that they largely agree in their assessment. More complex compilation models, however, would assume that the individual contributions of all team members would interact and compile on a higher level (Kozlowski & Chao, 2012). For team reflection, this assumption implies that reflecting in teams would create a quality in team reflection beyond individual-level reflection. For example, utterances from one team member would result in additional insights for another team member that would not have occurred when this member reflected alone. As a result, team reflection sessions may support the development of expertise in teams, because individual members can develop an understanding that is not only based on their own perceptions, but also on those of the other team members. As individuals share their preliminary insights with other team members, and team reflexive conversations, in turn, shape individual reflection processes (Knipfer, Kump, Wessel, & Cress, 2013), we need a comprehensive understanding of how individual reflection affects team reflection and how both are related to the development of team expertise.

Research that has examined team reflection from a multilevel perspective is very rare. Some studies have compared the effects of individual and team reflection on performance. To begin with, Daudelin (1996) found that individuals who reflected alone or with the guidance of a coach learned significantly more than a control group who did not reflect. However, teams that reflected jointly in a collective discussion did not learn much as the control group. Daudelin (1996) analyzed videotapes of the reflection sessions and found that the team reflection sessions were dominated by the search for similar experiences instead of unique experiences from which others could learn (i.e., preference for shared information, cf. Stasser & Titus, 1985). Daudelin reported that to ensure all team members had the chance to talk, the teams were not able to develop appropriate depth in their reflection (cf. Otte et al., 2017; Schippers et al., 2014).

Similarly, Gurtner et al. (2007) investigated 49 ad-hoc teams of psychology students working on a military simulation; those teams whose members reflected individually showed significantly better performance than the control group, who did not reflect at all. Teams that reflected jointly did not show this performance increase. Gurtner et al. (2007) analyzed the reflection sessions in greater detail and found that teams (compared to individuals) often focused on general strategies rather than reaching substantial depth in their discussions.

Finally, Bolinger and Stanton (2014) examined the amount of learning that occurred when reflecting alone versus reflecting in teams. Bolinger and Stanton therefore examined 150 students (i.e., novices) who participated in two consecutive decision-making tasks that were designed in such a way that deliberate reflection was likely to positively influence learning. In the first task, the participants were required to make point estimates of general knowledge questions (e.g., average male and female weights). In the second task, the students participated in a classical survival task where they had to prioritize various items in an extreme environment. Interestingly, Bolinger and Stanton found that individuals who reflected alone reported significantly higher levels of learning from the first task than participants who reflected in teams.

To summarize, the evidence surprisingly points towards inferiority of team reflection compared to individual reflection, rather than vice versa. However, this does not indicate that team reflection is not worth pursuing in the future, as various other studies demonstrated positive effects of team reflection on outcomes such as performance (e.g., Konradt & Eckardt, 2016; Otte et al., 2017; Schippers et al., 2013). Moreover, since the above results are largely based on laboratory research with teams, who may not be motivated or able to develop the appropriate depth in reflection, it remains to be seen whether teams in organizational contexts exhibit the same difficulties. For instance, team members that have an intrinsic motivation to further knowledge and expertise might be more inclined to elaborate on a particular topic in a team reflection session. In contrast, teams lacking this motivation might fall prey to various process losses (Eddy et al., 2013; Hackman, 1998), which may partly explain the inferiority of team reflection compared to individual reflection. For instance, brainstorming research has demonstrated the superiority of nominal groups (i.e., groups whose members worked individually and whose contributions are pooled afterwards) compared to actual teams (Diehl & Stroebe, 1987); in fact, production blocking appears to be the main factor for these results as listening to others may result in a *block* of original ideas. Similar effects likely account for suboptimal team reflection sessions.

Other reasons may include a desire for consensus (Nemeth & Nemeth-Brown, 2003), which holds the risk of group-think phenomena (Janis, 1989). For example, Daudelin (1996) observed that participants in the team reflection condition searched for experiences they shared rather than those they did not share. This desire for consensus and focus on shared information may result in a confirmation bias, thereby leading to reinforcement of existing knowledge structures rather than the detection of potentials for improvement, an effect frequently observed in research using hidden-profile methods (Stasser & Titus, 1985). Moreover, recent research has pointed to the risk of rumination, that is repetitive, excessive, and prolonged conversations about teamwork problems or critical events. Rumination is characterized by selective information processing and biases in the sense-making process, a narrow attentional scope, and a low extent of critical inquiry (Kump & Knipfer, 2016). This likely results in an incomplete understanding of an experience, and any implications drawn from this experience will suffer from biases in the reflection process (cf. Schippers et al., 2014).

Short- and Long-Term Effects of Team Reflection

In this chapter, we considered team expertise as an outcome of the team reflection process. The development of expertise in teams is a highly dynamic process in that it is created, modified, and discarded based on sharing and integration of team members' individual expertise. It thus can be conceived as central element of team development itself (cf. Kozlowski, Gully, Nason, & Smith, 1999). Surprisingly, previous notions of team reflection did not (explicitly) account for these developmental dynamics, which might lead to the conclusion that the relationship between team reflection and team expertise is rather static and invariable. However, research by Schippers et al. (2013) has shown that team reflection may only be beneficial in those teams whose initial team performance is low, indicating that the effects of team reflection on the development of team expertise may be dependent on the level of expertise within a team. Similarly, Gabelica et al. (2014) found that initial team performance shapes team reflection when performance (or knowledge/skill) is subsequently measured, or teams engage in subsequent reflection. More importantly, Konradt and Eckardt (2016) found that performance increased while team reflection decreased. In other words, as a team's expertise increases and it becomes better at what it is doing, reflection efforts may be reduced as a natural consequence. Oertel and Antoni (2015) examined the effects of team reflection on the development of expertise more directly. In their study, they investigated the development of knowledge in novice student teams, who jointly designed and conducted a study over the course of a semester. Interestingly, they did not find supporting evidence for the hypothesis that team reflection at the beginning fostered development of knowledge at the midpoint of the project. However, reflection after the midpoint of the project was indeed beneficial for the development of knowledge towards the end of the project.

We argue that the differential effects of team reflection on team expertise can be attributed to two reasons: first, Oertel and Antoni used transactive memory systems as a proxy for the development of expertise. Transactive memory systems are defined as meta-knowledge about who knows what in a team (Oertel & Antoni, 2015; Kozlowski, 2015). Hence, these systems can be conceived as meta-expertise on the distribution of expertise within a team. Conclusively, teams may require at least some experience with a task until they develop this form of meta-expertise. Second, in contrast to the aforementioned studies (Gabelica et al., 2014; Konradt et al., 2016; Schippers et al., 2013), the setting in the study by Oertel and Antoni did not provide the teams with performance feedback while they were working on the project. Therefore, the study might be better understood as a more detailed examination of the first performance interval in the other studies (i.e., until the teams received feedback on their performance for the first time).

To summarize, outcomes of previous reflection phases may not only shape later reflection phases but also the development of expertise. We therefore suggest that future studies should incorporate multiple points of measurement to uncover the effects of team reflection on the development of team expertise.

Conclusion

In this chapter, we reviewed the literature on team reflection to develop our argument that team reflection may be an important catalyst for the development of team expertise. Through the joint discussion and evaluation of previous team behavior, team reflection allows for the extraction of meaningful knowledge from past experiences to guide future behavior. At the same time, team reflection accelerates the development of team expertise as it involves the sharing and integration of each individual's unique expertise for the benefit of the whole team. Based on our critical literature review, we see a need to expand the conceptualization of team reflection, provided at the beginning of this chapter, to account for the effects of each team member's individual expertise onto the team reflection process. We therefore encourage research to more explicitly address the relationships of team reflection and team expertise in the future.

References

Allen, J. A., Baran, B. E., & Scott, C. W. (2010). After-action reviews: A venue for the promotion of safety climate. *Accident Analysis & Prevention 42*, 750–757. doi: 10.1016/j.aap.2009.11.004

Anseel, F., Beatty, A. S., Shen, W., Lievens, F., & Sackett, P. R. (2015). How are we doing after 30 years? A meta-analytic review of the antecedents and outcomes of feedback-seeking behavior. *Journal of Management 41*, 318–348. doi:10.1177/0149206313484521

Anseel, F., Lievens, F., & Schollaert, E. (2009). Reflection as a strategy to enhance task performance after feedback. *Organizational Behavior and Human Decision Processes 110*, 23–35. doi:10.1016/j.obhdp.2009.05.003

Arsenault, M. L. (2011). *Guided reflexivity and the importance of strategy change to adaptive team performance* (Dissertation). University of Oklahoma.

Atwater, L., & Waldman, D. (1998). 360-degree feedback and leadership development. *The Leadership Quarterly 9*, 423–426. doi:10.1111/1468-2389.00070

Barton, M. A., & Bunderson, J. S. (2014). Assessing member expertise in groups: An expertise-dependence perspective. *Organizational Psychology Review 4*, 228–257. doi:10.1177/2041386613508975

Bechky, B. A., & Okhuysen, G. A. (2011). Expecting the unexpected? How SWAT officers and film crews handle surprises. *Academy of Management Journal 54*, 239–261. doi:10.5465/AMJ.2011.60263060

Boet, S., Bould, M. D., Bruppacher, H. R., Desjardins, F., Chandra, D. B., & Naik, V. N. (2011). Looking in the mirror: Self-debriefing versus instructor debriefing for simulated crises. *Critical Care Medicine 39*, 1377–1381. doi:10.1097/CCM.0b013e31820eb8be

Bolinger, A. R., & Stanton, J. V. (2014). The gap between perceived and actual learning from group reflection. *Small Group Research 45*, 539–567. doi:10.1177/1046496414538322

Carver, C. S., & Scheier, M. F. (1998). *On the self-regulation of behavior*. New York: Cambridge University Press.

Carver, C. S., & Scheier, M. F. (2011). Action, affect, multitasking, and layers of control. In J. P. Forgas, R. F. Baumeister, & D. M. Tice (Eds.), *Psychology of self-regulation: Cognitive, affective, and motivational processes* (pp. 109–126). New York, NY: Taylor & Francis.

Daudelin, M. W. (1996). Learning from experience through reflection. *Organizational Dynamics 24*, 36–48. doi:10.1016/S0090-2616(96)90004-2

Decuyper, S., Dochy, F., & Van den Bossche, P. (2010). Grasping the dynamic complexity of team learning: An integrative model for effective team learning in organizations. *Educational Research Review 5*, 111–133. doi:10.1016/j.edurev.2010.02.002

De Dreu, C. K. W. (2002). Team innovation and effectiveness: The importance of minority dissent and reflexivity. *European Journal of Work and Organizational Psychology 11*, 285–298. doi:10.1080/13594320244000175

De Dreu, C. K. W. (2007). Cooperative outcome interdependence, task reflexivity, and team effectiveness: a motivated information processing perspective. *Journal of Applied Psychology 92*, 628–638. doi:10.1037/0021-9010.92.3.628

DeRue, D. S., Nahrgang, J. D., Hollenbeck, J. R., & Workman, K. (2012). A quasi-experimental study of after-event reviews and leadership development. *Journal of Applied Psychology 97*, 997–1015. doi:10.1037/a0028244

DeRue, D. S., & Wellman, N. (2009). Developing leaders via experience: The role of developmental challenge, learning orientation, and feedback availability. *Journal of Applied Psychology 94*, 859–875. doi:10.1037/a0015317

DeShon, R. P., Kozlowski, S. W., Schmidt, A. M., Milner, K. R., & Wiechmann, D. (2004). A multiple-goal, multilevel model of feedback effects on the regulation of individual and team performance. *Journal of Applied Psychology 89*, 1035–1055. doi:10.1037/0021-9010.89.6.1035

Diehl, M., & Stroebe, W. (1987). Productivity loss in brainstorming groups: Toward the solution of a riddle. *Journal of Personality and Social Psychology 53*(3), 497–509.

Druskat, V. U., & Kayes, D. C. (2000). Learning versus performance in short-term project teams. *Small Groups Research 31*, 328–353. doi:10.1177/104649640003100304

Eddy, E. R., Tannenbaum, S. I., & Mathieu, J. E. (2013). Helping teams to help themselves: Comparing two team-led debriefing methods. *Personnel Psychology 66*, 975–1008. doi:10.1111/peps.12041

Edmondson, A. C. (1996). Learning from mistakes is easier said than done: Group and organizational influences on the detection and correction of human error. *Journal of Applied Behavioral Science 32*, 5–28. doi:10.1177/0021886396321001

Edmondson, A. C. (1999). Psychological safety and learning behavior in work teams. *Administrative Science Quarterly 44*, 350–383. doi:10.2307/2666999

Edmondson, A. C. (2002). The local and variegated nature of learning in organizations: A group level perspective. *Organization Science 13*, 128–146.

Edmondson, A. C. (2003). Speaking up in the operating room: How team leaders promote learning in interdisciplinary action teams. *Journal of Management Studies 40*, 1419–1452. doi:10.1111/1467-6486.00386

Ellis, S., & Davidi, I. (2005). After-event reviews: Drawing lessons from successful and failed experience. *Journal of Applied Psychology 90*, 857–871. doi:10.1037/0021-9010.90.5.857

Fiore, S. M., & Wiltshire, T. J. (2016). Technology as teammate: examining the role of external cognition in support of team cognitive processes. *Frontiers in Psychology 7*, 1531. doi:10.3389/fpsyg.2016.01531

Fook, J., White, S., & Gardner, F. (2006). *Critical reflection in health and social care*. Berkshire (UK): Open University Press.

Fulmer, C. A., & Ostroff, C. (2016). Convergence and emergence in organizations: An integrative framework and review. *Journal of Organizational Behavior 37*, 122–145. doi:10.1002/job.1987

Gabelica, C., Van den Bossche, P., De Maeyer, S., Segers, M., & Gijselaers, W. (2014). The effect of team feedback and guided reflexivity on team performance change. *Learning and Instruction 34*, 86–96. doi:10.1016/j.learninstruc.2014.09.001

Gevers, J. M., van Eerde, W., & Rutte, C. G. (2001). Time pressure, potency, and progress in project groups. *European Journal of Work and Organizational Psychology 10*, 205–221. doi:10.1080/13594320143000636

Gray, D. E. (2007). Facilitating management learning: Developing critical reflection through reflective tools. *Management Learning 38*, 495–517. doi:10.1177/1350507607083204

Gurtner, A., Tschan, F., Semmer, N. K., & Nägele, C. (2007). Getting groups to develop good strategies: Effects of reflexivity interventions on team process, team performance, and shared mental models. *Organizational Behavior and Human Decision Processes 102*, 127–142. doi:10.1016/j.obhdp.2006.05.002

Hackman, J. R. (1998). Why teams don't work. *Leader to Leader 1998*, 24–31. doi:10.1002/ltl.40619980709

Hackman, J. R. (2002). *Leading teams: Setting the stage for great performances*. Harvard: Harvard Business Press.

Hackman, J. R., & Wageman, R. (2005). A theory of team coaching. *Academy of Management Review 30*, 269–287. doi:10.5465/AMR.2005.16387885

Hinsz, V. B., Tindale, R. S., & Vollrath, D. A. (1997). The emerging conceptualization of groups as information processors. *Psychological Bulletin 121*, 43–64. doi:10.1037/0033-2909.121.1.43

Hirst, G., Mann, L., Bain, P., Pirola-Merlo, A., & Richver, A. (2004). Learning to lead: The development and testing of a model of leadership learning. *The Leadership Quarterly 15*, 311–327. doi:10.1016/j.leaqua.2004.02.011

Hoegl, M., & Parboteeah, K. P. (2006). Team reflexivity in innovative projects. *R&D Management 36*, 113–125. doi:10.1111/j.1467-9310.2006.00420.x

Hollenbeck, J. R., Ilgen, D. R., Sego, D. J., Hedlund, J., Major, D. A., & Phillips, J. (1995). Multilevel theory of team decision making: Decision performance in teams incorporating distributed expertise. *Journal of Applied Psychology 80*, 292–316. doi:10.1037/0021-9010.80.2.292

Janis, I. L. (1989). *Crucial decisions: Leadership in policymaking and crisis management.* New York: The Free Press.

Klein, G., & Baxter, H. C. (2006, December). Cognitive transformation theory: Contrasting cognitive and behavioral learning. In *Interservice/Industry Training Systems and Education Conference*, Orlando, Florida.

Knipfer, K., Kump, B., Wessel, D., & Cress, U. (2013). Reflection as a catalyst for organisational learning. *Studies in Continuing Education 35*, 30–48. doi:10.1080/0158037X.2012.683780

Knipfer, K., Schreiner, E., Schmid, E., & Peus, C. (2018). The performance of pre-founding entrepreneurial teams: The importance of learning and leadership. *Applied Psychology 67*, 401–427. Advance online publication. doi:10.1111/apps.12126

Konradt, U., & Eckardt, G. (2016). Short-term and long-term relationships between reflection and performance in teams: Evidence from a four-wave longitudinal study. *European Journal of Work and Organizational Psychology 25*(6), 804–818. doi:10.1080/1359432X.2016.1160058

Konradt, U., Otte, K.-P., Schippers, M. C., & Steenfatt, C. (2016). Reflexivity in teams— A review and new perspectives. *Journal of Psychology: Interdisciplinary and Applied 150*, 153–174. doi:10.1080/00223980.2015.1050977

Konradt, U., Schippers, M. C., Garbers, Y., & Steenfatt, C. (2015). Effects of guided reflexivity and team feedback on team performance improvement: The role of team regulatory processes and cognitive emergent states. *European Journal of Work and Organizational Psychology 24*, 777–795. doi:10.1080/1359432X.2015.1005608

Kozlowski, S. W. (2015). Advancing research on team process dynamics: Theoretical, methodological, and measurement considerations. *Organizational Psychology Review 5*, 270–299. doi:10.1177/2041386614533586

Kozlowski, S. W., Gully, S. M., Nason, E. R., & Smith, E. M. (1999). Developing adaptive teams: A theory of compilation and performance across levels and time. In D. R. Ilgen & E. D. Pulakos (Eds.), *The changing nature of work performance: Implications for staffing, personnel actions, and development* (pp. 240–292). San Francisco: Jossey-Bass. doi:10.1002/hrdq.9

Kozlowski, S. W. J., & Chao, G. T. (2012). Macrocognition, team learning, and team knowledge: origins, emergence, and measurement. In E. Salas, S. M. Fiore, & M. P. Letsky (Eds.), *Theories of team cognition. Cross-disciplinary perspectives* (pp. 19–48). New York, NY: Taylor and Francis.

Kozlowski, S. W. J., & Klein, K. J. (2000). *Multilevel theory, research, and methods in organizations: Foundations, extensions, and new directions.* San Francisco: Jossey-Bass.

Kruger, J., & Dunning, D. (1999). Unskilled and unaware of it: how difficulties in recognizing one's own incompetence lead to inflated self-assessments. *Journal of Personality and Social Psychology 77*, 1121–1134. doi:10.1037/0022-3514.77.6.1121

Kump, B., & Knipfer, K. (2016). *Collective rumination at work: When "talking about the problem" hinders organizational learning and innovation*. Paper presented at Organizational Learning, Knowledge and Capabilities (OLKC), St. Andrews, UK.

Lee, F., Edmondson, A. C., Thomke, S., & Worline, M. (2004). The mixed effects of inconsistency on experimentation in organizations. *Organization Science* 15, 310–326. doi:10.1287/orsc.1040.0076

Marks, M. A., Mathieu, J. E., & Zaccaro, S. J. (2001). A temporally based framework and taxonomy of team processes. *Academy of Management Review* 26, 356–376. doi:10.2307/259182

Marsick, V. J., & Watkins, K. E. (2001). Informal and incidental learning. *New Directions for Adult and Continuing Education* 2001, 25–34. doi:10.1002/ace.5

Mathieu, J. E., Heffner, T. S., Goodwin, G. F., Salas, E., & Cannon-Bowers, J. A. (2000). The influence of shared mental models on team process and performance. *Journal of Applied Psychology* 85, 273–283. doi:10.1037/0021-9010.85.2.273

Mathieu, J. E., Hollenbeck, J. R., van Knippenberg, D., & Ilgen, D. R. (2017). A century of work teams in the *Journal of Applied Psychology*. *Journal of Applied Psychology* 102, 452–467. doi:10.1037/apl0000128

Mezirow, J. (2000). *Learning as transformation: Critical perspectives on a theory in progress*. San Francisco: Jossey-Bass.

Miller, D. T., & Ross, M. (1975). Self-serving biases in the attribution of causality: Fact or fiction. *Psychological Bulletin* 82(2), 213–225.

Mohammed, S., Hamilton, K., Sánchez-Manzanares, M., & Rico, R. (2017). Team cognition. In E. Salas, R. Rico, & J. Passmore (Eds.), *The Wiley Blackwell handbook of the psychology of team working and collaborative processes* (pp. 369–392). Chichester, UK: Wiley. doi:10.1002/9781118909997.ch16

Moreland, R. L., & McMinn, J. G. (2010). Group reflexivity and performance. *Advances in Group Processes* 27, 63–95. doi:10.1108/s0882-6145(2010)0000027006

Nadler, D. A. (1979). The effects of feedback on task group behavior: A review of the experimental research. *Organizational Behavior and Human Performance* 23(3), 309–338.

Neal, A., Ballard, T., & Vancouver, J. B. (2017). Dynamic self-regulation and multiple-goal pursuit. *Annual Review of Organizational Psychology and Organizational Behavior* 4, 401–423. doi:10.1146/annurev-orgpsych-032516-113156

Nembhard, I. M., & Edmondson, A. C. (2006). Making it safe: The effects of leader inclusiveness and professional status on psychological safety and improvement efforts in health care teams. *Journal of Organizational Behavior* 27, 941–966. doi:10.1002/job.413

Nemeth, C., & Nemeth-Brown, B. (2003). Better than individuals. In P. B. Paulus & B. A. Nijstad (Eds.), *Group creativity: Innovation through collaboration* (pp. 63–84). New York: Oxford University Press.

Oertel, R., & Antoni, C. H. (2015). Phase-specific relationships between team learning processes and transactive memory development. *European Journal of Work and Organizational Psychology* 24, 726–741. doi:10.1080/1359432X.2014.1000872

Okhuysen, G. A. (2001). Structuring change: Familiarity and formal interventions in problem-solving groups. *Academy of Management Journal* 44, 794–808. doi:10.2307/3069416

Otte, K. P., Konradt, U., Garbers, Y., & Schippers, M. C. (2017). Development and validation of the REMINT: A reflection measure for individuals and teams. *European Journal of Work and Organizational Psychology* 26, 299–313. doi:10.1080/1359432X.2016.1261826

Quinn, R. W., & Bunderson, J. S. (2016). Could we huddle on this project? Participant learning in newsroom conversations. *Journal of Management* 42, 386-418. doi:10.1177/0149206313484517

Salas, E., Rosen, M. A., Burke, C. S., Goodwin, G. F., & Fiore, S. M. (2006). The making of a dream team: When expert teams do best. In K. A. Ericsson, N. Charness, P. J. Feltovich, & R. R. Hoffman (Eds.), *The Cambridge handbook of expertise and expert performance* (pp. 439-453). New York: Cambridge University Press. doi:10.1017/CBO9780511816796.025

Savelsbergh, C., Van Der Heijden, B. I. J. M., & Poell, R. F. (2009). The development and empirical validation of a multi-dimensional measurement instrument for team learning behaviors. *Small Group Research* 40, 578-607. doi:10.1177/1046496409340055

Schippers, M. C. (2003). *Reflexivity in teams* (Dissertation). Vrije Universiteit, Amsterdam.

Schippers, M. C., Den Hartog, D. N., & Koopman, P. L. (2007). Reflexivity in teams: A measure and correlates. *Applied Psychology* 56, 189-211. doi:10.1111/j.1464-0597.2006.00250.x

Schippers, M. C., Den Hartog, D. N., Koopman, P. L., & van Knippenberg, D. (2008). The role of transformational leadership in enhancing team reflexivity. *Human Relations* 61, 1593-1616. doi:10.1177/0018726708096639

Schippers, M. C., Edmondson, A. C., & West, M. A. (2014). Team reflexivity as an antidote to team information-processing failures. *Small Group Research* 56, 731-769. doi:doi:10.1177/1046496414553473

Schippers, M. C., Edmondson, A. C., & West, M. A. (2017). Team reflexivity. In J. M. Levine, & L. Argote (Eds.), *Handbook of Group and Organizational Learning*. Oxford: Oxford University Press.

Schippers, M. C., Homan, A. C., & van Knippenberg, D. (2013). To reflect or not to reflect: Prior team performance as a boundary condition of the effects of reflexivity on learning and final team performance. *Journal of Organizational Behavior* 34, 6-23. doi:10.1002/job.1784

Schippers, M. C., West, M. A., & Dawson, J. F. (2015). Team reflexivity and innovation: The moderating role of team context. *Journal of Management* 41, 769-788. doi:10.1177/0149206312441210

Schmutz, J. B., & Eppich, W. J. (2017). Promoting learning and patient care through shared reflection: a conceptual framework for team reflexivity in health care. *Academic Medicine* 92, 1555-1563. doi:10.1097/ACM.0000000000001688

Senge, P. (1990). *The fifth discipline: The art and science of the learning organization*. New York: Currency Doubleday.

Short, P. M., & Rinehart, J. S. (1993). Reflection as a means of developing expertise. *Educational Administration Quarterly* 29(4), 501-521.

Smith-Jentsch, K. A., Cannon-Bowers, J. A., Tannenbaum, S. I., & Salas, E. (2008). Guided team self-correction: Impacts on team mental models, processes, and effectiveness. *Small Group Research* 39, 303-327. doi:10.1177/1046496408317794

Stasser, G., & Titus, W. (1985). Pooling of unshared information in group decision making: Biased information sampling during discussion. *Journal of Personality and Social Psychology* 48(6), 1467-1478.

Tannenbaum, S. I., & Cerasoli, C. P. (2013). Do team and individual debriefs enhance performance? A meta-analysis. *Human Factors* 55, 231-245. doi:10.1177/0018720812448394

Tesler, R., Mohammed, S., Hamilton, K., Mancuso, V., & McNeese, M. (2017). Mirror, mirror: Guided storytelling and team reflexivity's influence on team mental models. *Small Group Research* 49, 267-305. doi:10.1177/1046496417722025

Tucker, A. L., & Edmondson, A. C. (2003). Why hospitals don't learn from failures: Organizational and psychological dynamics that inhibit system change. *California Management Review 45*, 55–72. doi:10.2307/41166165

van Knippenberg, D., & Schippers, M.C. (2007). Work Group Diversity. *Annual Review of Psychology 58*, 515–541. doi:10.1146/annurev.psych.58.110405.085546

Vashdi, D. R., Bamberger, P. A., & Erez, M. (2013). Can surgical teams ever learn? The role of coordination, complexity, and transitivity in action team learning. *Academy of Management Journal 56*, 945–971. doi:10.5465/amj.2010.0501

Vera, D., & Crossan, M. (2004). Strategic leadership and organizational learning. *Academy of Management Review 29*, 222–240. doi:10.5465/AMR.2004.12736080

Villado, A. J., & Arthur Jr, W. (2013). The comparative effect of subjective and objective after-action reviews on team performance on a complex task. *Journal of Applied Psychology 98*, 514–528. doi:10.1037/a0031510

Vince, R. (2002). Organizing reflection. *Management Learning 33*, 63–78. doi:10.1177/1350507602331003

Waller, M. J., Zellmer-Bruhn, M. E., & Giambatista, R. C. (2002). Watching the clock: Group pacing behavior under dynamic deadlines. *Academy of Management Journal 45*, 1046–1055. doi:10.2307/3069329

Ward, P., & Eccles, D. W. (2006) A commentary on "team cognition and expert teams: Emerging insights into performance for exceptional teams." *International Journal of Sport and Exercise Psychology 4*, 463–483. doi:10.1080/1612197X.2006.9671808

Weiss, M., Kolbe, M., Grote, G., Spahn, D. R., & Grande, B. (2016). Why didn't you say something? Using after-event reviews to affect voice behavior and hierarchy beliefs in multi-professional action teams. *European Journal of Work and Organization Psychology 26*, 66–80. doi:10.1080/1359432X.2016.1208652

West, M. A. (1996). Reflexivity and work group effectiveness: A conceptual integration. In M. A. West (Ed.), *Handbook of work group psychology* (pp. 555–579). Chichester, UK: Wiley.

West, M. A. (2000). Reflexivity, revolution, and innovation in work teams. In M. M. Beyerlein, D. A. Johnson, & S. T. Beyerlein (Eds.), *Product development teams: Advances in interdisciplinary studies of work teams* (Vol. 5, pp. 1–29). Stamford, CT: JAI Press.

West, M. A., & Anderson, N. R. (1996). Innovation in top management teams. *Journal of Applied Psychology 81*, 680–693. doi:10.1037/0021-9010.81.6.680

Widmer, P. S., Schippers, M. C., & West, M. A. (2009). Recent developments in reflexivity research: A review. *Psychology of Everyday Activity 2*(2), 2–11.

Wiedow, A., & Konradt, U. (2011). Two-dimensional structure of team process improvement: Team reflection and team adaptation. *Small Group Research 42*, 32–54. doi:10.1177/1046496410377358

CHAPTER 44

LEARNING AT THE EDGE

The Role of Mentors, Coaches, and Their Surrogates in Developing Expertise

ERICH PETUSHEK, GÜLER ARSAL,
PAUL WARD, MARK UPTON, JAMES WHYTE IV,
AND ROBERT R. HOFFMAN

INTRODUCTION

STUDIES of expertise have shown that the majority of experts practice frequently under the direction of experienced teachers, instructors, trainers, coaches, or mentors who support the acquisition and development of their expertise (e.g., Ericsson, 1996; Hoffman & Ward, 2015; Hoffman et al., 2014). Nowadays, it is widely acknowledged that no matter how skilled individuals are, only a few can sustain or repeat their expert-level achievements without this kind of support.

Vygotsky (1978) popularized the idea that, irrespective of where you are on the proficiency continuum, learning needs to occur in the *zone of proximal development*—a window of opportunity between the point of one's current level of competence (i.e., what can be done without additional support) and a point just out of reach (i.e., beyond the boundary of current competence). *Learning at the edge*, by definition, requires external input—scaffolding—to support learning to reach the next milestone. Scaffolding methods must be constantly re-tailored to meet the learner's current needs, which are a moving target as individuals continue to learn and improve.

The author (Paul Ward's) affiliation with The MITRE Corporation is provided for identification purposes only, and is not intended to convey or imply MITRE's concurrence with, or support for, the positions, opinions or viewpoints expressed by the author.

Chi, Siler, Jeong, Yamauchi, and Hausmann (2001) described scaffolding as

> a kind of *guided prompting* that pushes the student a little further along the same line of thinking, rather than telling the student some new information, giving direct feedback on a student's response, or raising a new question or a new issue that is unrelated to the student's reasoning.... The important point to note is that scaffolding involves *cooperative execution* or *coordination* by the tutor and the student (or the adult and child) in a way that allows the student to take an increasingly larger burden in performing the skill. (p. 490)

Mentoring is often viewed as a quintessential scaffolding mechanism for accelerating learning, aiding professional development, and for personal growth. But what kind of scaffolding do mentors offer to the aspiring expert learning at the edge? What makes for an effective or expert mentor? How can we measure the benefit of providing such support mechanisms?

To some extent, how these questions are answered depends on the different types of mentor–mentee relationships that exist in varied educational, occupational, and social settings. When formalized, such relationships are often dictated by doctrine or tradition, rather than effective evidence-based methods. To confuse matters further, other terms, such as coaching, tutoring, and preceptorship, are also used (and sometimes interchangeably) to refer to overlapping support mechanisms. For instance, *coaching* is often used to refer to a wide range of learning and professional skill development services across many training and development contexts, whereas *preceptorship* generally refers to a practical means of support during occupational transitions and is frequently used within healthcare professions.

Our goal in this chapter is to unpack this complex interaction of support relationships, roles, and functions across a range of domains with the hope that doing so will provide some answers to the aforementioned questions. We focus primarily on the mentor–mentee relationship. However, given the similarity across roles, we also examine the coach–trainee relationship (and preceptor–trainee relationship, where applicable), especially where there are implications for developing expertise. Our goal is to increase our understanding of what makes mentoring, broadly defined, effective and to provide an evidence-based approach for how mentoring can be used to develop, accelerate, and preserve expertise.

This chapter is presented in four parts. First, we provide a detailed description of the different components and functions of mentoring, as well as related concepts, including coaching, and preceptorship. Next, we review the meta-analytic evidence supporting the effectiveness of developmental support roles on a range of outcome measures (e.g., job performance), and we identify potential qualities and functions of effective mentoring/coaching. Then, we review the empirical evidence supporting the development of expertise in mentoring-type roles. Finally, we offer some concluding remarks in which we summarize some of the major issues and make suggestions for how to advance this area of research to prepare mentees to skilfully adapt to complexity.

Disentangling the Developmental Functions and Roles Associated with Mentoring, Coaching, and Preceptorship

What is an expert mentor? How do we measure it? Before we can answer these questions, given the substantive variation in definitions of mentoring and related roles across disciplines, we must first address a more fundamental question: What is mentoring?

Mentoring and Coaching: Are they Different?

Ragins and Kram (2007) summarized traditional definitions of mentoring as "a relationship between an older, more experienced mentor and a younger, less experienced protégé for the purpose of helping and developing the protégé's career" (p. 5), whereas in healthcare, Hayes (1998) defined mentoring as "a voluntary, committed, dynamic, extended, intense, and supportive relationship characterized by trust, friendship, and mutuality between an experienced, respected person . . . and a student" (p. 525).

In other domains, coaching is the preferred term over mentoring (e.g., in sport) or is identified as a key component of mentoring (e.g., in business). For instance, Côté and Gilbert (2009) define athletic coaching as "the consistent application of integrated professional, interpersonal, and intrapersonal knowledge to improve athletes' competence, confidence, connection and character" (p. 316). In contrast, Kilburg (1996) defined coaching in business as

> a helping relationship formed between a client who has managerial authority and responsibility in an organization and a consultant who uses a wide variety of behavioural techniques and methods to help the client achieve a mutually identified set of goals to improve his or her professional performance and personal satisfaction and, consequently, to improve the effectiveness of the client's organization within a formally defined coaching agreement. (p. 142)

To disentangle the myriad ways in which terms have been used in psychology and business/management, D'Abate, Eddy, and Tannenbaum (2003) synthesized the different types of functions engaged in by those in a developmental support role. Figure 44.1 provides a summary of the most frequently mentioned developmental *roles* (including types of mentoring and coaching) and highlights how these vary with respect to the six most dominant developmental functions: *providing practical application, teaching, aiding, counseling, supporting,* and *advocating*. While many terms are used to describe the range of developmental roles in these disciplines (including different types of mentoring), most people in these roles enagage in teaching, aiding, counseling, and supporting functions (see footnote in Figure 44.1 for definition of functions). However, although mentors have traditionally engaged in more advocacy, tutors, coaches, and those supervising apprentices have usually engaged in more teaching.

In an integrative review of mentoring (cf. coaching) research in business and education, Ghosh (2013) dissected the key developmental functions of mentoring. These data are summarized in Figure 44.2. In both disciplines, coaching was almost always cited as one, if not the main or key mentoring function. In a related review of coaching-only research (including executive-, business-, and life-coaching), Hamlin, Ellinger, and Beattie (2009) indicated that helping, through facilitation or intervention, with the purpose of improving performance and aiding personal growth were commonly identified components of coaching.

In Ghosh's review, over 80% of business articles and over 60% of education articles also indicated that role modeling and counseling were key functions for mentors. However, in business, 90% of articles suggest that making a friend (e.g., accepting, relating, trusting, being available) was an integral component of mentoring, whereas

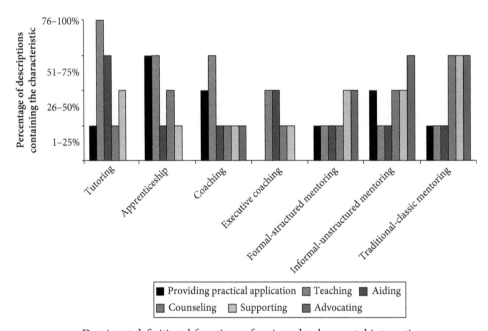

FIGURE 44.1 Dominant definitional functions of various developmental interactions.

Providing practical application: The provision of experience or practice with hands-on projects or challenging work for the learner.

Teaching: The instruction or teaching of the learner to build expertise, skills, or knowledge

Aiding: The provision of aid or help to the learner.

Counseling: The provision of counseling, advice, or guidance to the learner.

Supporting: The social, emotional, or personal (i.e., psychosocial) support of the learner.

Advocating: The sponsorship of the learner to advance in the organization or field.

Data from C. P. D'Abate, E. R. Eddy, and S. I. Tannenbaum (2003), What's in a name? A literature-based approach to understanding mentoring, coaching, and other constructs that describe developmental interactions. *Human Resource Development Review* 2(4), pp. 360–384, Tables 2–4, doi:10.1177/1534484303255033.

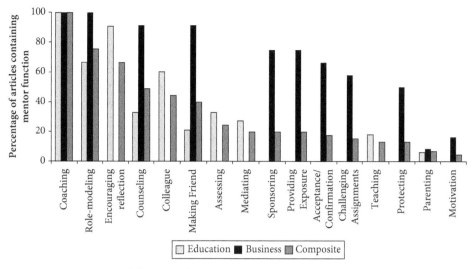

FIGURE 44.2 Dominant definitional functions of mentoring.
Data from R. Ghosh, Mentors providing challenge and support: Integrating concepts from teacher mentoring in education and organizational mentoring in business. *Human Resource Development Review* 12(2), 144–176, Tables 3 and 4, doi:10.1177/1534484312465608, 2013.

over 90% of education articles indicate that encouraging reflection was a key mentoring function (e.g., posing a problem that they will explain, explore, and challenge). The data on education mentors are consistent with a recent critical review of theoretical frameworks focused on developmental, learning, and social aspects of mentoring (Dominguez & Hager, 2013). Collectively, this research suggests that central to individual and organizational success is the mentor's role in helping develop their mentee's professional identity; acting as a role model for their mentees; transmitting knowledge and guidance; and engaging the mentee in dialogue, self-reflection, and goal setting to facilitate progress. Similar observations have been made in weather forecasting (see Hoffman, LaDue, Mogil, Roebber, & Trafton, 2017; LaDue, 2011).

We can draw a handful of conclusions about what mentoring is based on the obeserved diversity of definitions and functions. Although the role of mentor is multifaceted, coaching and role modeling appear to be the two primary behaviors in which mentors engage, irrespective of discipline. To a slightly lesser extent, helping manage intrapersonal relationships (e.g., counseling) is also a core element of mentoring. The prevalence of other key functions varies across disciplines. For instance, managing interpersonal relationships is an important aspect of mentoring in business, and self-reflection is a key component in education.

We have previously argued that when the purpose of mentoring is to specifically improve performance and promote acquisition of effective skills and strategies, *what* mentors do and *how* they do it are equally important (Hoffman & Ward, 2015). Put simply, we asserted that in the first exposure, the apprentice watches while the mentor does the job. The second time, the apprentice assists while the mentor does the job (e.g., "get me the 5-inch adjustable spanner"). The third time, the apprentice does the job and the mentor provides corrective guidance. The fourth time, the apprentice does it on his own.

Preceptorship in Nursing

Based on Hoffman and Ward's (2015) summary of how mentors mentor, and the previous distinctions between mentoring and coaching functions, one could argue that a simple way to differentiate between mentors and coaches is to say that mentors *do* (i.e., they participate in the work) and coaches *teach* (i.e., they provide instruction on how to do the work). But it is never that simple. To add to the confusion, the term *preceptorship* has been used interchangeably with mentoring (Billay & Myrick, 2008), especially in healthcare professions such as nursing, where precepting often occurs *on the job*.

Yonge, Billay, Myrick, and Luhanga (2007) argued that the difference between mentoring and precepting is in the type of relationship structure. The preceptor–trainee relationship is a formal relationship that takes place in task-oriented programs that are planned and monitored as a means to overcome the limited experience of apprentice nurses in a particular clinical area. For instance, new nurses are assigned an experienced preceptor with whom they work in a one-to-one relationship for a specified period of time, with the goal of achieving specified competencies to prepare themselves to enter independent practice (Rosenfeld, Glassman, & Capobianco, 2015; Ward & McComb, 2017). In contrast, the mentor–mentee relationship occurs in a closer and more personal format including longer-term reciprocal commitments. In this context, preceptorship can be viewed as a means to bridge the gap between education and practice (e.g., Coates & Gormley, 1997)—to facilitate the socialization of the trainee into the work setting. However, this role has also been defined by traditional mentoring functions, such as coaching, teaching, role modeling, supervising, guiding, evaluating, and being a clinical and knowledge expert (e.g., Phillips, 2006).

The ever-increasing complexities of healthcare have resulted in a need for better training and preparation of entry-level nurses. This has been driven by the incremental increase in acuity of hospitalized patients, that is, limited resources resulting in patients with progressively more complex and care-intensive problems. In response to the increasing demands and decresing resources, US hospitals have supplemented the long-standing tradition of preceptorships by instituting even more structured nurse residency programs with a built-in component of mentoring, albeit under a preceptorship framework (Ke, Kuo, & Hung, 2017; Ward & McComb, 2017)—blurring the distinction between the two developmental roles further. While one might reason that these programs should ideally focus on skill acquisition, and hence, progression towards expertise, most have focused heavily on nurse retention, reflecting the social role of the mentor (Fielder, Read, Lane, Hicks, & Jegier, 2014; Rosenfeld et al., 2015; Rosenfeld & Glassman, 2016).

The desired outcomes from instituting such programs have shifted from increases in nurse proficiency toward retention. Efforts to identify specific components of the precepting process that positively influence changes in performance in nursing have generally been unsuccessful (Blanzola, Lindeman & King, 2004; Rosenfeld et al., 2015). Likewise, the majority of studies that have examined preceptorship effectiveness have usually done so in the absence of a reasonable control or placebo group (see Ke et al., 2017), making it difficult to tease out the actual effects of current preceptorship practices.

Despite the difficulties reported in evaluating preceptorshiop effectiveness, and their evolving *raision d'être*, trainees have typically rated peer support and mentoring as the most vital elements of these programs (Fielder et al., 2014; Zhang et al., 2017). This raises the questions, how skilled are the mentors and preceptors, and on what basis are they selected into these roles? One of the most common selection procedures is peer nomination (e.g., of a senior nurse by their manager) based on perceived competence as a supervisor (e.g., Craven & Boyles, 1996; Loiseau, Kitchen, & Edgar, 2003; Messmer, Jones, & Taylor, 2004) and/or length of time spent in a clinical unit (Brunt & Kopp, 2007; Morris et al, 2007; Ward & McComb, 2017). This is the case, even though experience, for instance, may not be a good predictor of expertise (for a discussion, see Ericsson, Whyte, & Ward, 2007; Harris, Eccles, Ward & Whyte, 2013).

Although there has been a recent trend towards using a more structured approach to preceptor training (Lee, Lin, Tseng, Tsai, & Lee-Hsieh, 2017), there is considerable variation in such training. Some preceptors are considered sufficiently experienced in their pratice (as nurse, but not as preceptor) to render preceptor training unnecessary! Hence, they may lack the skills necessary to support a trainee's learning and performance and/or their own development. Where training exists, the empirical basis—necessary to reliably bring about a high level of proficiency in precepting—is often lacking (Lee et al., 2017). The crux of the problem in nursing and other applied domains is the lack of provision of empirically based training and/or lack of veracity in ascertaining the proficiency of subjectively selected preceptors, which may result in preceptors without the specific skills necessary to excel at this role.

In the next section, we review the various psychometric scales that have been used to measure mentoring and coaching, and review meta-analytic evidence supporting the effectiveness of developmental support roles on a range of outcome measures (e.g., job performance). We exclude from this section a discussion of preceptorship because no meta-analyses have been conducted, reflecting the minimal amount of empirical research in this area. Based on this evidence we identify some of the potential qualities and functions of effective mentoring and coaching.

Measuring Mentoring and Coaching and Their Effectiveness

Despite the differences in mentoring functions, and the overlap with coaching and preceptorship, an overall definition common to each role would be that these are processes aimed to develop professional knowledge and skills as well as personal characteristics. What exactly is being developed, the specific ways in which development occurs, and the effectiveness of functions may contribute to the apparent ambiguity between roles.

Using Psychometric and Other Scales to Capture the Key Components of Mentoring

Numerous scales have been developed to characterize the quality and functions of mentoring and coaching. For mentoring, eight frequently used scales have been developed (Brodeur, Larose, Tarabulsy, Feng, & Forget-Dubois, 2015; Castro, Scandura, & Williams, 2004; Fleming et al., 2013; Noe, 1988; Ragins, 2011; Ragins & McFarlin, 1990; Rose, 2003; Scandura & Ragins, 1993), resulting in twenty-two factors (see Table 44.3), with considerable qualitative overlap between many scales/items. For coaching, seven frequently used scales have been developed in business (Baron & Morin, 2009; Ellinger, Ellinger, & Keller, 2003; Gregory & Levy, 2010; Heslin, Vandewalle, & Latham, 2006; Konczak, Stelly, & Trusty, 2000; McLean, Yang, Kuo, Tolbert, & Larkin, 2005; Park, McLean, & Yang, 2008), with fourteen non-independent factors arising from various factor analytic techniques (see Table 44.4). In contrast, for coaching, six commonly used scales have been developed in *sport* (Feltz, Chase, Moritz, & Sullivan, 1999; Kavussanu, Boardley, Jutkiewicz, Vincent, & Ring, 2008; Myers, Wolfe, Feltz, & Penfield, 2006; Myers, Feltz, Chase, Reckase, & Hancock, 2008; Myers, Chase, Beauchamp, & Jackson, 2010; Zhang, Jensen, & Mann, 1997), resulting in eleven non-independent factors (see Table 44.4).

As can be seen in Tables 44.3 and 44.4, numerous factors have been proposed to describe both coaching and mentoring with effective mentor/coach qualities ranging from supporting various personal and professional development, to role modeling or fostering independence. Although it is unlikely that all of the factors described in Tables 44.3 & 44.4 would be unique if a large, cumulative factor analysis were conducted on all the items, the tables demonstrate the diverse set of skills and functions that a mentor might possess.

Four unidimensional scales have also been developed—two for mentoring (see Table 44.5; Berk, Berg, Mortimer, Walton-Moss, & Yeo, 2005; Lee, Dennis, & Campbell, 2007) and two for coaching (see Table 44.6; Ellinger et al., 2003; Smither, London, Flautt, Vargas, & Kucine, 2003)—some of which used more descriptive rather than factor analytic-type methods. These eight- to twelve-item scales (some of which have not been statistically evaluated) appear to represent a more holistic characterization of the diverse activities in which mentors and coaches engage. For example, each of these scales covers four primary areas/qualities including the following:

(a) *skill development using various pedagogical strategies*;
(b) *exemplifies a knowledgeable, caring, and professional* (e.g., approachable and accessible) *individual*;
(c) *motivates through encouragement and praise*; and
(d) *provides external resources and fosters networking opportunities*.

Taken together, all of these scales identify the types of behaviors that characterize coaches and mentors in the literature and the behaviors, one could argue, that should be employed.

Table 44.1 Relationship between Mentoring (and Functions) and Professional Development Outcomes

Author	Setting (population)	Number of studies	Time frame of review	Statistics	General outcomes	Job performance	Career prospects/ promotion	Compensation/ salary	Situation satisfaction	Turnover intent
Dickson et al. (2014)	Academic or workplace	61	1988–2010		Psychosocial support					
				N	7500					
				ρ (corrected)	0.26*					
				95% CI	[0.20, 0.30]					
					Career development support					
				N	8968					
				ρ (corrected)	0.29*					
				95% CI	[0.23, 0.32]					
Eby et al. (2013)	Youth, academic, and workplace	173	1985–2010			Psychosocial support				
				N		3267	3482	5256	5632	3827
				ρ (corrected)		0.24*	0.19*	0.03	0.26*	-0.10*
				95% CI		[0.21, 0.27]	[0.15, 0.22]	[0.0, 0.06]	[0.24, 0.29]	[-0.13, -0.07]
						Career development support				
				N		2943	5095	6760	7,627	2833
				ρ (corrected)		0.33*	0.23*	0.10*	0.36*	-0.24*
				95% CI		[0.30, 0.36]	[0.21, 0.26]	[0.07, 0.12]	[0.34, 0.38]	[-0.28, -0.21]
Kammeyer-Mueller and Judge (2008)	Academic and workplace	120	1887–2007			Psychosocial support				
				N		1203				
				ρ (corrected)		0.21*	0.03	0.05	0.28	
				95% CI		[0.11, 0.30]				

(continued)

Table 44.1 Continued

Author	Setting (population)	Number of studies	Time frame of review	Statistics	General outcomes	Job performance	Career prospects/ promotion	Compensation/ salary	Situation satisfaction	Turnover intent
					Career development support					
				N		2363				
				ρ (corrected)		0.17*	0.08	0.12	0.27	
				95% CI		[0.12, 0.28]				
					Mentored/not mentored					
				N		2432				
				ρ (corrected)		0.22*	0.12	0.06	0.16	
				95% CI		[0.14, 0.31]				
Eby et al. (2008)	Youth, academic, and workplace	116	1985–2006		*Mentored/not mentored*					
				N		10250	5833		16694	4423
				r_c		0.11*	0.09*		0.17*	-0.07*
				95% CI		[0.05, 0.17]	[0.01, 0.18]		[0.14, 0.21]	[-0.13, -0.02]

Table 44.2 Effects of Coaching on Professional Development Outcomes

Author	Setting (population)	Number of studies	Time frame of review	Statistics	Overall effectiveness	Performance/ skills	Goal attainment	Work-related attitude change
Sonesh et al. (2015)	Leadership, business, or executive coaching	24	2000–2014	N Hedges g 90% CI	3756 0.10* [0.10, 0.11]	2350 0.19* [0.15, 0.23]	216 0.21* [0.10, 0.31]	524 0.18* [0.15, 0.21]
Jones et al. (2015)	Employed working adults in an organizational setting	17	1997–2012	N Corrected population $d\ (\delta)$ 90% CI	2267 0.36* [0.16, 0.50]	1784 0.28* [0.07, 0.44]		
Theeboom et al. (2014)	Professionally trained external coaches or trained peers	18	1993–2012	N Hedges g 95% CI	2090 0.66* [0.39, 0.93]	2007 0.60* [0.04, 1.16]	789 0.74* [0.42, 1.06]	507 0.54* [0.34, 0.73]

Table 44.3 Factors and Sample Items from Mentor Function/Quality Scales

Factor	Sample items	References
Career support	Assigns me tasks that push me into developing new skills. Mentor helped me meet new colleagues	Noe (1988); Ragins & McFarlin (1990); Scandura & Ragins (1993); Castro et al. (1993)
Psychosocial support	My mentor was supportive and encouraging I share personal problems with mentor	Noe (1988); Ragins & McFarlin (1990); Scandura & Ragins (1993); Castro et al. (1993)
Role modeling	I try to model my behavior after mentor I respect mentor's ability to teach others	Scandura & Ragins (1993); Castro et al. (1993)
Integrity	Value me as a person Prefer to cooperate with others than compete with them	Rose (2003)
Guidance	Provide information to help me understand the subject matter I am researching Give me specific assignments related to my research problem	Rose (2003)
Relationship	Relate to me as if he/she is a responsible, admirable older sibling Have coffee or lunch with me on occasion	Rose (2003)
Structure	My mentor gives me useful advice when I tell him my needs, my worries, and my difficulties My mentor and I agree about the things I will need to do to help improve my situation	Brodeur et al. (2015)
Engagement	My mentor understands my needs, my worries, and my problems My mentor listens attentively to the needs, worries, and achievements I share with him	Brodeur et al. (2015)
Autonomy support	When meeting, my mentor talks more than I do (scores must be recoded) Often, my mentor takes decisions for me (scores must be recoded)	Brodeur et al. (2015)
Competency support	My mentor congratulates me when I do something right My mentor values me even after I experience failures	Brodeur et al. (2015)
Maintaining effective communication	Provides constructive feedback Active listener	Fleming et al. (2013)
Aligning expectations	Works with me to set clear expectations of the mentoring relationship Helps me develop strategies to meet research goals	Fleming et al. (2013)
Assessing understanding	Accurately estimates my ability to conduct research Employs strategies to enhance my understanding of the research	Fleming et al. (2013)

Fostering independence	Negotiates a path to professional independence with me Builds my confidence	Fleming et al. (2013)
Addressing equity and inclusion	Taking into account the biases and prejudices s/he brings to my mentor/mentee relationship Working effectively with mentees whose personal background is different from his/her own (age, race, gender, class, region, culture, religion, family composition etc.)	Fleming et al. (2013)
Promoting professional development	Helps me acquire resources (e.g. grants) Helps me balance work with my personal life	Fleming et al. (2013)
Personal learning and growth	My partner is helping me learn and grow as a person. My partner helps me learn about my personal strengths and weaknesses.	Ragins (2011)
Inspiration	My partner has inspired or been a source of inspiration for me. My partner gives me a fresh perspective that helps me think "outside the box."	Ragins (2011)
Self-affirmation	My partner is helping me become the person I aspire to be. My partner seems to bring out the best in me.	Ragins (2011)
Reliance on communal norms	In our relationship, we help each other without expecting repayment. We never keep score of who gives and who gets in our relationship.	Ragins (2011)
Shared influence and respect	My partner and I respect and influence each other. There is mutual respect and influence in our relationship.	Ragins (2011)
Trust and commitment	Our relationship is founded on mutual trust and commitment. My partner and I trust each other, and we are committed to the relationship.	Ragins (2011)

To make some sense of this literature, we conducted a thematic analysis, which identified that skill development (i.e., developing professional or occupational skill) was the characteristic assessed most frequently in all the coaching/mentoring scales—both descriptive and factor analytic. The scale evidence suggests that effective coaches and mentors report using: scaffolding, feedback, modeling, practice, and explanation to enhance learning. In addition, effective coaches/mentors report attempts to enhance motivation and confidence, build professional networks, foster independence, and develop rapport/meaningful relationships.

Table 44.4 Factors and Sample Items from Coach Function/Quality Scales

Factor	Sample items	References
Business		
Facilitate learning/ development	I use analogies, scenarios, and examples to help my employees learn In order to improve my performance, my manager serves as a role model	Baron & Morin (2009); Ellinger et al. (2003); Gregory & Levy (2010); Heslin et al. (2006); Konczak et al. (2000); Park et al. (2008)
Open/effective communication	When asked to share my feelings, I am comfortable My supervisor is easy to talk to	Baron & Morin (2009); Gregory & Levy (2010); McLean et al. (2005); Park et al. (2008)
Genuineness of the relationship/ relational	My supervisor and I have mutual respect for one another Show confidence in the person I coach	Baron & Morin (2009); Gregory & Levy (2010)
Team approach	When thinking of ways to achieve my objectives, I seek input from others My manager would rather work with others to complete tasks	McLean et al. (2005); Park et al. (2008)
Value people	In discussions with others, I focus on the needs of the individuals In daily work, my manager considers people's needs outside the workplace	McLean et al. (2005); Park et al. (2008)
Accept ambiguity	My manager views differences of opinion as constructive When seeking solutions to problems, I like to try new solutions	McLean et al. (2005); Park et al. (2008)
Accountability	My manager holds me accountable for the work I am assigned I am help accountable for performance and results	Konczak et al. (2000)
Encouragement of self-directed decisions	My manager tries to help me arrive at my own solutions when problems arise, rather than telling me what he/she would do My manager relies on me to make my own decisions about issues that affect how work gets done	Konczak et al. (2000)
Information-sharing	My manager shares information that I need to ensure high-quality results My manager provides me with the information I need to meet customers' needs	Konczak et al. (2000)
Delegation of authority	My manager gives me the authority I need to make decisions that improve work processes and procedures My manager delegates authority to me that is equal to the level of responsibility that I am assigned	Konczak et al. (2000)

Coaching for innovative performance	My manager is willing to risk mistakes on my part if, over the long term, I will learn and develop as a result of the experience I am encouraged to try out new ideas even if there is a chance they may not succeed	Konczak et al. (2000)
Comfort of the relationship	I am content to discuss my concerns or troubles with my supervisor I feel safe being open and honest with my supervisor	Gregory & Levy (2010)
Guidance	Provide guidance regarding performance expectations Provide constructive feedback regarding areas for improvement	Heslin et al. (2006)
Inspiration	Express confidence that you can develop and improve Support you in taking on new challenges	Heslin et al. (2006)
Sport		
Motivation	Motivate your athletes Build team cohesion	Feltz et al. (1999); Kavussanu et al. (2008); Myers et al. (2006); Myers et al. (2010)
Game strategy	Adapt to different game/meet situations Recognize opposing team's weakness during competition	Feltz et al. (1999); Kavussanu et al. (2008); Myers et al. (2006); Myers et al. (2010)
Technique	Teach athletes the complex technical skills of your sport during practice Detect subtle technique errors by players during practices	Feltz et al. (1999); Kavussanu et al. (2008); Myers et al. (2006); Myers et al. (2010)
Character building	Instil an attitude of good moral character Instil an attitude of fair play among your athletes	Feltz et al. (1999); Kavussanu et al. (2008); Myers et al. (2006); Myers et al. (2010)
Physical conditioning	Implement an appropriate endurance program for your athletes during the season Accurately assess your athletes' physical conditioning	Myers et al. (2008); Myers et al. (2010)
Democratic behavior	Let the athletes share in decision making and policy formulation Put the suggestions made by the team members into operation	Zhang et al. (1997)
Positive feedback behavior	Congratulate an athlete after a good play Give credit when it is due	Zhang et al. (1997)
Situational consideration behavior	Use alternative methods when the efforts of the athletes are not working well in practice or competition Adapt coaching style to suit the situation	Zhang et al. (1997)

(continued)

Table 44.4 Continued

Factor	Sample items	References
Social support behavior	Stay interested in the personal well-being of the athletes Encourage close and informal relationship with the athletes	Zhang et al. (1997)
Teaching and instruction behavior	Use a variety of drills for a practice Supervise athletes' drills closely	Zhang et al. (1997)
Autocratic behavior	Presents ideas forcefully Disregard athletes' fears and dissatisfactions	Zhang et al. (1997)

One drawback of this literature is that there are few empirical examples of observing effective mentoring or coaching and recording what they do—all of the scale-based research is based on self-report. Presumably, the items in these scales are based on observations of effective coaching; however, the evidence on which these scales were developed isn't widely reported. Another drawback is that there are frequently reported differences of the same activity by mentor and mentee. For instance, Hagen and Peterson (2015) and others (e.g., Kavussanu et al., 2008) have found substantial differences in coach versus trainee (or athlete) reports of functions/qualities. Specifically, and perhaps unsurprisingly, coaches' ratings of their own coaching efficacy were significantly and considerably higher than that of their athletes' rating of their effectiveness on all dimensions (Cohen's d = 1.16). Although this may be expected, it does raise the question of how best to assess reliability in mentoring.

To advance this area of research, future studies should aim to reduce the overwhelming number of factors and scales that have attempted to measure one super construct—coaching/mentoring effectiveness. This could be achieved through scale optimization methods. The existing scales also need to be evaluated in terms of if and/or how the associated factors and/or qualities are related to any of the aforementioned goals of mentoring. Longitudinal studies with structured mentoring interventions and competency assessments on representative tasks could be conducted to evaluate the predictive validity of these scales in terms of mentee performance, learning, and development. In addition, detailed cognitive task analyses of mentoring activities are needed to investigate the specific strategies used by mentors that are most effective and contextually sensitive, and as a basis for the disentangling the difference in mentor and trainee reporting.

Finally, two important questions remain unanswered. First, how do we verify or validate the self-report data collected using such scales? Self-report data are invaluable for gaining an insight in to the perceptions, beliefs, thoughts, actions, skills, and strategies of effective mentors. However, there are considerable differences amongst self-report methods in terms of the means and effort to check the veracity of the data collected. For instance, verbal report and introspective-type methods are a collaborative and co-constructive process and, when done properly, have built-in procedures for eliciting, scaffolding, and verifying the validity of data (see Chapter 17 on "A Historical Perspective on Introspection", this volume). However, self-report questionnaires, such as those used in the mentoring and coaching literature have few, if any, such built-in procedures to ensure quality and veracity of the reports. Instead, they rely on post hoc statistical analysis of quantifications of subjective data to assess reliability and validity. The second important question: how do we measure mentor/coach effectiveness if not through self-report questionnaires? To answer both questions, ideally, we would use multiple measures of effective mentoring, which include measuring mentor/coach characteristics, strategies and activities (especially those we *think* bring about effective learning), the extent of learning and development of the mentee, and the effectiveness of various coaching/mentoring interventions/tactics. This evaluation technique has been successfully utilized in research on tutoring and in some areas of training (as described in the section 'Surrogates for Mentors' later in this chapter), where evidence has been documented by elucidating the truly effective (measured by relatively permanent changes in skill of the mentee/trainee) individual characteristics and activities of those responsible (e.g., mentor, coach, preceptor, teacher, tutor, etc.) for the developmental of others. To provide some context for the next section, in which we review some of the alternative ways in which mentoring-type functions have been assessed, we first summarize the meta-analytic evidence that provide an assessment of the effectiveness of mentoring and coaching activities.

The Relationship between Mentoring and Performance Outcomes

Several meta-analyses have investigated the effects of mentoring-type roles and functions on a range of professional development outcomes (e.g., job performance). These meta-analyses have focused specifically on mentoring (Eby, Allen, Evans, Ng, & DuBois, 2008; Eby et al., 2013; Dickson et al., 2014; Kammeyer-Mueller & Judge, 2008), coaching (Jones, Woods, & Guillaume, 2015; Sonesh et al., 2015; Theeboom, Beersma, & van Vianen, 2014), and more recently on intelligent tutoring as a potential surrogate for mentoring (e.g., Kulik & Fletcher, 2016).

A summary of the relationship between mentoring and professional development outcomes across four meta-analyses are displayed in Table 44.1. Two meta-analyses have examined the effect of mentoring (i.e., the difference between mentored vs

Table 44.5 Non-factor Analytic Mentoring Scales

Reference	Berk et al. (2005); Mentorship Effectiveness Scale	Lee et al. (2007); nature's guide for mentor self-assessment: How good a mentor are you?
Rating scale	Items (6-point Likert scale (strongly disagree (1) to strongly agree (5)))	Open responses consist of giving an example and what could be done better for each activity
Items	My mentor was supportive and encouraging	Appreciating individual differences: Give an example of an incident that illustrates your acknowledgement of individual difference
	My mentor demonstrated content expertise in my area of need	Availability: Give an example of the strategy you use to be available to your students/staff
	My mentor provided constructive and useful critiques of my work	Self-direction: Guided independence and scientific creativity (10-point scale): Micromanagement (1): Sink or swim (10)
	My mentor motivated me to improve my work product	Questioning: Describe how you last used active questioning to lead a mentee towards a solution
	My mentor answered my questions satisfactorily (e.g., timely response, clear, comprehensive).	Celebration: When did you last celebrate a student/staff member's achievement? How did you celebrate?
	My mentor suggested appropriate resources (e.g., experts, electronic contacts, source materials)	Building a scientific community: Describe a deliberate strategy you use to build a scientific community in your group
	My mentor challenged me to extend my abilities (e.g., risk taking, try a new professional activity, draft a section of an article)	Building a social community: Describe a deliberate strategy you use to build your group as a social community
	My mentor was helpful in providing direction and guidance on professional issues (e.g., networking)	Skill development: Describe steps you take to develop the critical, writing, and presentation skills of your students/staff
	My mentor acknowledged my contributions appropriately (e.g., committee contributions, awards)	Networking: Describe one example of how you have introduced each of your students/staff into the scientific network of your research area
	My mentor was accessible	Mentor for life: How many of your past students/staff are you in contact with?
	My mentor demonstrated professional integrity	
	My mentor was approachable	

non-mentored individuals) on job performance (Eby et al., 2008; Kammeyer-Mueller & Judge, 2008). In both studies, the effect of mentoring was small ($r_c = .11$ for Eby et al., 2008; $\rho_c = .22$ for Kammeyer-Mueller & Judge, 2008). One should note that these analyses did not examine the effects of different types of mentoring functions or activities (e.g., by using mentor function type as a moderator variable) and so do not help identify effective mentoring practice. Nor did they report the proficiency level of the mentee or the level of expertise of the mentor. It is possible that these effects are explained by limited performance measures, the use of ineffective mentoring practices, and the inclusion of good and poor mentoring, and/or are influenced by the skill level of both the mentee and mentor. Indeed, the Q statistic in these analyses indicate moderate heterogeneity warranting the investigation of potential moderators such as these. One should note that these data are derived from the formal (2-factor) mentoring scales that measured self-reported perceptions of mentoring support. Likewise, a self-reported estimate of one's own performance was used as an estimate of actual job performance. Whether the observed overall effects reflect the size of effects that one might obtain using alternative assessments of mentoring and performance on, for instance, a job-sample test (e.g., Wigdor & Green, 1991) remains to be explored.

In addition to job performance, Underhill (2006) conducted a meta-analysis (including fourteen studies) to examine the effect of mentoring in corporate settings on career outcomes. Individuals who received mentoring had a slight advantage in their careers over non-mentored individuals ($ES = .24$). Similar meta-analytic effects, albeit smaller in size, have been observed when examining the effects of mentoring on career prospects/promotion ($r_c = .09$) and satisfaction (for instance, with one's job or current situation) ($r_c = .17$) (Eby et al., 2008) but have not always reached statistical significance (Kammeyer-Mueller & Judge, 2008).

Three meta-analyses explored the relationship between specific self-reported mentoring functions, such as career development support (e.g., assigning tasks that push one to develop new skills, meet new colleagues) and psychosocial support (e.g., supporting and encouraging, sharing personal problems), and various professional outcomes, including job performance and career prospects/promotion (Dickson et al., 2014; Eby et al., 2013; Kammeyer-Mueller & Judge, 2008). Overall, mentoring functions that were focused on career development support (as opposed psychosocial support) were better predictors of job performance, promotion, compensation, satisfaction, and job turnover (aggregated across all outcomes: ρ (corrected) .22 for career development support vs .17 psychosocial support). These effects were slightly larger and smaller, respectively, in the largest meta-analysis by Eby et al. (e.g., .25 vs .16), indicative of a small to medium relationship of these functions with professional outcomes.

A summary of the effects from three additional meta-analyses on coaching are displayed in Table 44.2. The results indicate that coaching can significantly improve professional outcomes (including overall effectiveness, performance, goal attainment, and work-related attitude change), with average effects ranging from small to

Table 44.6 One-Factor Coaching Behavior and Effectiveness Scales

Reference	Ellinger coaching behavior scale (Ellinger et al., 2003)	Coaching effectiveness (Smither et al., 2003)
Rating scale	Items (7-point Likert scale (almost never (1) to almost always (7)))	Items (5-point Likert scale (very ineffective (1) to very effective (5)))
Items	I use analogies, scenarios, and examples to help my employees learn	Helping you interpret your feedback results by asking questions to uncover reason for the feedback
	I encourage my employees to broaden their perspectives by helping them to see the big picture	Helping you link your feedback to your business plan/situation
	I provide constructive feedback to my employees	Offering you useful suggestions, advice, or insights to set goals for development
	I solicit feedback from my employees to ensure that my interactions are helpful to them	Helping you identify ways to share feedback with your raters and to solicit ideas for improvement
	I provide my employees with resources to they can perform their jobs more effectively	Encouraging you to coach and give feedback to others
	To help my employees think through the issues, I ask questions, rather than provide solutions	Contributing to your job performance and career development
	I set expectations with my employees and communicate the importance of those expectations to the broader goals of the organization	
	To help them see different perspectives, I role-play with my employee	

moderate/large (i.e., Hedges g = 0.10 to 0.74). The average effect size of coaching on job performance was .36, a moderate effect. Comparatively, these effects suggest that coaching, rather than mentoring per se, may have a more considerable effect on professional outcomes related to performance improvement.

Overall, it appears that functions such as career development and psychosocial support can have some effect on various outcomes related to job performance. Unfortunately, there is little consideration of the interdependence between many coaching and mentoring functions (e.g., examining the extent to which coaching moderates the relationship between mentoring and performance outcomes) and/or any other potentially important moderators of this relationship. Regardless, this information can at least partly inform our understanding of the extent to which investment in mentor–mentee and coach–trainee (and preceptor–trainee) relationships might benefit the development of expertise. Investigating the source of variability in meta-analytic effects on performance improvement would result in better understanding of the mentoring strategies that lead to a more-or less-successful developmental relationship.

It is important to note that since the primary research focus has been on investigating common (i.e., average) mentoring experiences, little regard has been given to extraordinary mentoring experiences that might constitute more effective practice of mentors we might consider expert in their own right. As a result, most of the research reviewed thus far may, in fact, tell us little about the nature of high-quality mentoring relationships and the associated mentoring activities (Chandler & Ellis, 2011). In the next section, we explore some of the actual activities of effective mentors/coaches, and draw on research from expert tutoring for comparison.

Empirical Evidence of Effective and Expert Mentors, and Their Effect on Developing Experts

The mentoring literature provides many descriptions of mentoring and mentoring functions, and some evidence of the effectiveness of these with regard to particular job/career outcomes. However, few studies have observed the activities of expert mentors per se, and even fewer have assessed whether training based on developing these features would aid in the development of skill, and expertise in particular (e.g., Collins, Brown, & Holum,1991; Pfund et al., 2014). Hence, one is left to answer the questions: What are effective and expert mentors and coaches doing and how are they doing it? Based on years of studying experts (many of whom have also been mentors), and on the literature on learning to become an expert (e.g., Feltovich, Coulson, Spiro, & Dawson-Saunders, 1992; Klein & Baxter, 2009), Hoffman and Ward (2015) highlighted some of the key characteristics of good mentors (see Table 44.7). To summarize, these authors suggest that effective mentors should understand the learner's task-specific cognitive status (level of understanding, mental model, why struggling, etc.), create representative practice situations that consider the various challenges and constraints, as well as scaffold, model, elicit understanding, and instruct the learner when necessary. Additionally, throughout this process the coach/mentor should try to create a self-regulated individual who is aware of when they need external support and able to teach themselves. Finally, the mentor should develop a personally meaningful and trusting relationship to enhance motivation and confidence as well as develop a larger support/development network for the mentee to support further growth. This combination of activities will not only foster professional development but will also create a long-lasting relationship for sustainable professional and personal development. Next, we describe some of the early research that has observed some of these types of mentoring practices in action.

Table 44.7 Characteristics of Good Mentors

Characteristics
Establish a relationship and an environment that fosters learning
Are intrinsically motivated to share their expertise, and appreciate the fact that others do not know what they know or possess the same skills
Have various strategies for conveying knowledge (for example, demonstrating the task while thinking out loud, interrogating the apprentice for the rationale behind actions, and asking the apprentice to instruct the mentor)
Use both behavioral and cognitive modeling approaches, illustrating appropriate actions and decisions in the context of the actual operational environment, and, when warranted, guiding the apprentice's own performance
Provide opportunities for the apprentice to learn in diverse settings that capture a range of contextual constraints, so that the apprentice learns to apply his or her skills in different contexts and adapt them to different circumstances
Form rich mental models of the apprentice's knowledge and skill
Provide junior apprentices with particular help in forming their initial mental models of cases, for example, by promoting information seeking strategies
Diagnose why an apprentice is struggling and tailor instruction to fit the situation
Anticipate when the apprentice will form a reductive mental model
Recognize the kinds of practice experiences that will force the apprentice to go beyond the current reductive models
Scaffold the apprentice to recognize a flawed mental model by making a richer mental model
Emphasize the interconnectedness of causes and concepts along multiple conceptual dimensions
Scaffold the apprentice to engage in sensemaking and thereby generate his or her own feedback, thus teaching the apprentice to be able to teach him or herself
Create opportunities or scenarios that encourage the apprentice to rethink concepts and mental models, by triggering a cognitive conflict and introducing a baffling event or anomaly
Provide support for managing increased uncertainty and cognitive load when the apprentice is struggling with complexity, and for dealing with the stress and frustration associated with exploring the boundaries of the mental models and transitioning between misconceived and well-conceived models
Provide instruction and practice in handling ill-structured problems with unclear or evolving goals

Data from R. R. Hoffman and P. Ward, Mentoring: A leverage point for intelligent systems? *IEEE Intelligent Systems* 30(5), 78–84, doi:10.1109/MIS.2015.86, 2015.

Masters and Apprentices: Legitimate Peripheral Participation

Early work by Lave and Wenger (1991) described various ethnographic investigations of developmental interactions and relationships between masters and their apprentices in domains such as midwifery, tailoring, and meat cutting. This work described how learners participated in communities of practitioners, and acquired knowledge and skills from experts in the context of everyday activities. Rather than receiving formal schooling (or a complete education or formal training), techniques were acquired by observing their masters engage in work, or from their own inventions in daily activities and practices under real working conditions. The effectiveness of such practices has been documented in a meta-analysis by Underhill (2006), which showed that informal mentoring (e.g., an ad hoc, emergent relationship not governed by an organization) had a greater effect on career outcomes ($ES = 0.26$) than formal mentoring programs (i.e., those established and managed by an organization, where mentors are paired with protégés based on organizational goals) ($ES = 0.06$).

Lave and Wenger (1991) showed that learners were fully immersed and actively engaged in doing the activity they were trying to learn alongside their mentor, rather than engaged in a more traditional learning environment. To develop expertise, the goal was to get novices to move toward full participation in the sociocultural practices of communities—a framework the authors called *legitimate peripheral participation*. Summarizing this mentor approach to learning in context alongside the mentor, Lave and Wenger stated that "In our view, learning is not merely situated in practice—as if it were some independently reifiable process that just happened to be located somewhere; learning is an integral part of generative social practice" (p. 35).

There are four notable characteristics of this *situated learning* perspective: First, there was little observable teaching provided by the mentor. Second, the community of practice made explicit strong goals and provided, early on in engagement, a holistic view of the deliverables/outcomes such as the masters' finished product as well as observations and communications with more advanced apprentices. Third, the curriculum was not predetermined but was emergent, and unfolded in opportunities for engagement in practice that increased in complexity and scope. Fourth, the work became more and more transparent to the mentee, permitting tacit knowledge to be actively experienced and acquired, and the level of access to all aspects of work increased—both of which were considered the true and legitimate opportunities for practice.

Lave and Wenger's (1991) work suggests that rather than provide traditional instruction (or supplementary mentoring support functions), mentors—whether in the form of a master or other members of the community of practice—provide more experiential access to and actively engage apprentices in the full gamut of work opportunistically. They argue that this should be done in a way that permits them to peel away the multiple layers of complex learning domains in an individualized manner, frequently through modeling best practice and scaffolding their engagement.

Current research suggests that practices similar to those described by Lave and Wenger are used today in sociotechnical domains, such as weather forecasting. Both mentoring and collaborative learning have been shown to be important resources for meteorologists to learn forecasting *on the job* (Hoffman et al., 2017; LaDue, 2011). After conducting a series of interviews with forecasters at various stages in their careers, LaDue (2011) showed that forecasters were regularly involved in mentoring each other. Senior forecasters helped younger forecasters and passed on their practical knowledge, but also learned from them. Being exposed to the senior forecaster's thinking helped forecasters at an early career stage to figure out things faster. This reciprocal and interactive nature of mentoring has recently been demonstrated in elite level coaches of Olympic athletes.

Elite Coaches and Olympic Athletes

To investigate the activities of elite-level (e.g., Olympic and international) athletic coaches of a variety of sports, Petushek and Ward (2016) used a series of cognitive task analysis procedures to better understand the specific contexts, situations, strategies, and cognitive processes used. In the first instance, these authors employed a modified task-diagram interview (i.e., Step 1 of the Applied Cognitive Task Analysis (ACTA) procedure; Militello & Hutton, 1998) developed in conjunction with subject-matter experts, to better understand the challenging and priority tasks/goals in which expert coaches engaged. This method was also used to inform the design of a subsequent Critical-Decision Method (CDM) interview, including the selection of relevant probe questions (for more info on ACTA and CDM see Chapter 19 on 'Incident-Based Methods of Studying Expertise', this volume).

The data elicited from initial interviews revealed four key coaching activities thought to be responsible for success. These were considered candidates for further exploration and included activities focused on: (a) *developing technical and performance skills*; (b) *enhancing interpersonal relationships*; (c) *developing personal health and well-being of athletes*; and (d) *facilitating adaptation to constraints*, such as the ability to adapt to the influence of interpersonal relationships on the development of technical skills. These initial results supported the notion that coaches operate in a complex environment where they adapt their *preferred* performance development program based on the dynamics of the current situation, including athlete behaviors and characteristics.

In the follow-on study, the CDM was used to gather empirical data about the strategies, skills, and processes used to support engagement in the aforementioned activities. Three themes emerged from the five most-utilized strategies (i.e., by >50% of elite coaches) across all types of athletic events: (a) *replanning/adaptation*; (b) *goal management*; and (c) *athlete management*. Invariably, coaches were presented with situations where they had to adapt to unforeseen events or information and routinely revise their plan for implementation. While each instance of adaptation was reported as unexpected, coaches were aware that there was always going to be something to which they had to adapt. Goal management reflected the constant need to re-evaluate and reprioritize goals, often on the fly, as well as form new plans to ensure success and skill development. This also meant coaches had to reflect on their own goals, their relative importance, and their ability to achieve them given the constraint dynamics, and to consider re-prioritization. Athlete management included empathizing with the athlete, encouraging athlete responsibility, and promoting self-regulation in developing personal health and well-being. As with goal management, this theme had a metacognitive focus since it required coaches to become effective at perspective taking. This promoted the development of empathy and reflection skills, and a core feature of the goal management theme was to scaffold the athlete's own learning about how to self-regulate their learning and coaching activities.

Given the considerable focus on adaptation and goal reprioritization, Petushek and Ward (2016) developed a template for training focused on spotting anomalies (i.e., the situational need for adaptation, and the specific constraints that likely require adaptation). Previous research indicates that when such anomalies are explained away as random events, rather than exploited and leveraged, learning opportunities are missed. When these are missed, the boundaries of current knowledge where current strategies break down are left undiagnosed and subsequent attempts to adapt when learning at the edge are less fruitful (see Klein, 2013; Hoffman et al., 2014; Ward, Gore, Hutton, Conway, & Hoffman, 2018).

Previous research suggests that case-, problem-, or scenario-based learning approaches (e.g., Jones & Turner, 2006; Morgan, Jones, Gilbourne, & Llewellyn, 2013) may provide opportunities to optimize learning if various cognitive learning principles are applied (e.g., Collins et al., 1991; Dennen, 2004; Hoffman et al., 2014). To be effective, the trainee must be exposed to a variety of tough or rare cases with external input to encourage the trainee to articulate their thought processes, reflect, and explore. In addition, the external agent (often more skilled) may model, scaffold, and provide feedback to the trainee when needed. Accordingly, for a selction of critical events identified through interviews with elite coaches, Petushek and Ward (2016) created a short descriptive case-based vignette where trainee coaches could choose a solution and provide their rationale for the strategy employed. Trainees could then compare their answers to the actual response and rationale of the elite coach (on whom the vignette was developed), and examine their context-specific sensemaking and decision-making timeline to gain a greater understanding of how, where, and why they differed from the expert.

In addition to assessing the effectiveness of the case-based vignette training exercises, an area for future research/training opportunities could revolve around the common aforementioned strategies utilized by coaches across the various contexts. Increasing our knowledge and understanding of expert reasoning strategies and associated situational complexities would be a useful basis for developing any mentoring program. Future research is also needed to understand the conditions that bring about skilled adaptation to complexity, such as reducing cognitive rigidity, preserving complexity, and improving resilience and robustness (see Hoffman et al., 2014; Ward et al., 2018).

Both Lave and Wenger (1991) and Petushek and Ward (2016) focused on understanding learning in either traditional coaching or apprenticeship systems, specifically in domains that were more focused on physical skills (e.g., tailoring, butchering, sport)—primarily because these were the domains where coaching or the apprentice system was used. The question remains whether these methods would be equally applicable to other more cognitively oriented tasks (e.g., medical diagnosis, avionics troubleshooting, computer programming, STEM learning). Fortunately, Lave's ideas have subsequently been extended to optimize learning environments in more cognitive domains, and especially instructional and academic tasks (Collins et al., 1991).

Cognitive Apprenticeship in the Classroom

Collins et al. (1991) argued that Lave and Wenger's research could be applied to reading, writing, and mathematics skill development (cited as Lave, in preparation, in Collins et al., 1991). They relied heavily on Lave and Wenger's descriptions for many of their insights into the nature of apprenticeship, and extended this approach to education, which they called *cognitive apprenticeship*.

In this framework, the teacher (or master) initially decides the problem-solving tasks, how to model the solution, how to scaffold, and when to fade assistance. Student articulation and reflection are used to integrate and construct new knowledge, and exploration is used to foster autonomy and independence. Like situated learning, cognitive apprenticeship involves learning global before local skills, and increasing difficulty and diversity over time. Collins et al. (1991) discussed the importance of making task processes visible, situating abstract tasks into relevant contexts, and varying the diversity of problems but articulating the common, generalizable aspects. Making thinking processes visible, by using strategies like think aloud or introspective-type methods, allows students to gain insights into the solution strategies which are often invisible and allows the master to gain insights into any flawed or error-ridden, problem-solving strategies used by the students (see Chapter 17 on "A Historical Perspective on Introspection", this volume). A similar apprentice-based approach has also been used to train researchers (i.e., cognitive engineers and cognitive task analysts) to become proficient in those domains in which they are to conduct subsequent cognitive task analyses (Militello & Quill, 2007).

Cognitive apprenticeship, as Collins and colleagues defined it, has many similarities to tutoring (often called coached problem solving). This concept provides another avenue to explore how human tutors (who are often domain experts)—or computer-based models of expertise, known as intelligent tutors—can help accelerate the learning of less-skilled individuals.

Surrogates for Mentors: Tutors, Intelligent Tutors, and Training Based on Analyses of Expertise

VanLehn (2011) and most recently Kulik and Fletcher (2016) evaluated the effectiveness of human and intelligent tutoring systems on learner performance gains on math or physics tasks (as compared to standard or classroom teaching). Compared to non-tutoring, both human tutors ($d = 0.79$) and step-based intelligent tutoring systems ($d = 0.76$) were shown to result in similar degrees of performance improvement. Compared to conventional instruction, intelligent tutoring systems raised test scores by slightly more than half a standard deviation. Of particular note is the greater effect of human tutors and intelligent tutoring systems compared to more traditional mentoring and coaching, presumably because tutoring is often focused specifically on learning and expertise development, whereas mentoring and/or coaching is frequently focused on both career development and/or psychosocial support.

In addition, intelligent tutoring systems seem to perform very similarly to human tutors (on which they are often modeled). The study of highly effective tutors was initiated to improve/understand effective learning processes as well as to design intelligent tutoring systems, primarily in relatively non-complex domains such as learning mathematics or physics problems (Anderson, Corbett, Koedinger, & Pelletier, 1995; Lepper & Woolverton, 2002; VanLehn, 2011). Observational studies of more and less effective tutors have documented features similar to that of learning through legitimate peripheral participation, but at a more fine-grain level of analysis (e.g., Anderson et al., 1995; Chi et al., 2001; Graesser, Person, & Magliano, 1995; Lepper & Woolverton, 2002; Merrill, Reiser, Ranney, & Trafton, 1992). For example, by analyzing tutor–tutee dialogs, tutors have been shown to guide the *learning by doing* process that is characteristic of legitimate peripheral participation or the cognitive apprenticeship approach. This allows students to engage in problem solving as much as possible, while providing just enough *help* to reduce the likelihood of frustration or confusion, and to support *learning at the edge*.

Similar to how one might think of as a mentor, expert tutors are more likely to question students for information as opposed to explaining to or informing them directly (Glass, Kim, Evens, Michael, & Rovick, 1999). Lepper, Aspinwall, Mumme, and Chabay (1990) found that tutors with extensive experience tended to draw students' attention to an error and allow them to better understand why the error occurred in addition to how to resolve it, as opposed to merely giving corrective feedback. More specifically, these individuals used scaffolding and questioning as opposed to direct instruction in a highly interactive manner.

Many of these approaches have been employed in school learning contexts and so have been more concerned with developing proficiency than expertise per se. For example, Chi et al. (2001) showed that eliciting students' constructive responses (e.g., through scaffolding and prompting) was important for eighth-grade students learning concepts related to the human circulatory system. The various tutor–tutee interactions, such as prompting the student for their thoughts, meaning, rationale, explanation, and elaboration, resulted in student self-explanations and reflection. These tutor-prompted functions facilitated integration and construction of knowledge, which led to better learning outcomes (Chi et al., 2001). Other common tutoring techniques such as detailed diagnostic assessment, individualized task selection, broader domain knowledge, and motivation play an important role in learning, but it is likely the adaptive feedback and scaffolding that make human tutors more effective than intelligent tutoring systems in these domains (e.g., see VanLehn, 2011).

One of the primary reasons tutors have been effective is the nature of the domains to which they have been applied (e.g., mathematics, physics, physiology). Do these same principles apply to domains where there is no ideal solution or where problems are ill-defined and more complex (e.g., business, defence, intelligence analysis, healthcare)? One such system that directly addressed this issue is named Sherlock—an intelligent tutoring system designed for training troubleshooting of avionics test equipment for the F-15 (see Polson & Richardson, 1988). It was designed to overcome some of the limitations of legacy training systems that too often focused on simpler methods and tasks, and that trained formal principles, including analysis of individual components. Instead, Sherlock's production rules were generated based on a cognitive task analysis of highly experienced technician's reasoning while performing direct manipulations of the system, and while assessing system-level functionalities and testing procedures. In a high-fidelity mockup of the actual technician's avionics test, Sherlock presented trainees with progressively more difficult troubleshooting scenarios in which they had to isolate faults. Trainees were provided with coaching in the form of explanations of an expert's strategic reasoning as they performed the task.

In an evaluation of training effectiveness, Gott (1995) used Sherlock to train senior apprentices (i.e., with three years of experience). The trainees received twenty-five hours of training and in a subsequent test, outperformed a group of expert avionics technicians with ten-plus years of experience (including with around four years of on-the-job training)! The large effects were attributed to the *situated* nature of the training, the ecological representativeness of the scenarios, and the interactive and hands-on approach to learning.

Building on this and similar research Klein and Borders (2016) developed a similar approach to create an effective training system, called ShadowBox, with the goal of improving decision-making skills in complex tasks (e.g., warfighting–civilian encounters). ShadowBox is a scenario-based training system in which expert feedback is embedded. The scenarios are based on incidents elicited from experts through cognitive task analysis methods such as critical incident-based techniques. The training involves having trainees progress through each scenario and, at particular decision points, interacting by responding accordingly, and ranking alternative options and their rationale for responding. Participant responses are then contrasted with expert perceptual–cognitive strategies, skills, and knowledge and their associated rationale. Particpants are then given time to reflect and contrast their responses and rationale to that of experts. Unlike most mentoring/coaching programs and/or tutoring methods, a mentor, coach, or tutor is not physically present in this learning environment; however, mentoring-type input is captured in the form of the expert feedback, which embodies the expert's cognitive model of their decision- and sensemaking processes in that scenario. Rather than just actively engage the student in receiving expert feedback, ShadowBox helps scaffold the gap between expert and trainee understanding and resulting decisions. The interactive nature of ShadowBox encourages trainees to modify their cognitive model by seeing *the world through the eyes of the experts*, by engaging them in prepared activities such as comparing their rationale statements with experts. It is also constructive, since it encourages trainees to describe the differences in their rationale (in comparison to the expert) and explicate lessons learned.

ShadowBox training for soldiers—to improve the social–cognitive skills needed to manage civilian encounters (without hostility)—was shown to improve soldier performance by approximately 25%, compared to a control, in two evaluations (Klein & Borders, 2016). ShadowBox is a nice example of how a simple system can be created to enhance learning by focusing on developing important decision-making processes. It facilitates dialogic reasoning and deep self-explanations, helps make thought processes explicit, and models and helps learners construct better cognitive models via expert feedback and learning activities that act as a surrogate for a face-to-face mentor. Unlike research on tutoring and training systems based on expertise, examples of research employing mentoring interventions are surprisingly rare.

Training by Actual Mentors

A handful of studies show that mentors and coaches can support teachers to develop more effective teaching practices that, in themselves, promote student dialogic reasoning (see Campbell & Malkus, 2011; Kraft, Blazar, & Hogan, 2018; Matsumura, Garnier, & Spybrook, 2012). Resnick et al. (Chapter 39, this volume) describe the types of coaching and mentoring activities used to develop better teachers as involving: (a) *observation of a teacher's instruction*; (b) *providing of feedback and suggestions about teaching practice*; (c) *modeling of dialogic instruction for teachers in their classroom*—to allow teachers to observe good practice in action and to facilitate realization that their own students can engage in complex reasoning; (d) *anticipation of student misconceptions of complex tasks*, and (e) *use of these strategies to shift student goals away from rote learning and toward conceptual understanding*.

Coaching designed to help teachers model and teach dialogic reasoning has been shown to improve the effectiveness of subsequent teaching practice by 0.58 standard deviations (Kraft et al., 2018). The effects of this kind of coaching on teacher practice, in particular, are larger than those observed based on veteran teachers teaching novice teachers (0.2–0.4 *SDs*; Hill, Blazar, & Lynch, 2015). Importantly, the effects of the kind of coaching employed by Kraft et al. (2018) on the associated student's achievement (i.e., 0.15 *SDs*) is equivalent to the improvements in student achievement that normally occur in the first 5–10 years of teachers' careers (0.05–0.15 *SDs*) (Atteberry, Loeb, & Wyckoff, 2015; Papay & Kraft, 2015).

One research study examined the effect of a specific mentor training program using random assignments to treatment or control group. Specifically, Pfund et al. (2014) examined the effect of a competency-based mentor training on improving the mentoring skills of clinical researchers. Long-term mentee effects were not assessed. This curriculum, taught over eight hours, included topics such as effective communication, aligning expectations, assessing understanding, addressing equality, fostering independence, and promoting professional development. Specifically, parts of the curriculum outlined how mentors should scaffold learning through breaking tasks into smaller components, using think-aloud reports for verbalizing mental processes, modeling, and questioning/cueing. The curriculum also focused on teaching mentors how to assess students' thinking processes and understanding. Short-term improvements in both mentor- and mentee-rated outcomes (utilizing the Mentor Competency Assessment, MCA; Fleming et al., 2013) were greater in the intervention compared to control group (+0.40 vs +0.18 on the 1–7 MCA scale). Further research should investigate whether this training is effective for mentees across the proficiency continuum as well as longer-term effects on measureable representative tasks.

Concluding Remarks: Advancing Theory and Future Research

Mentors have many goals and functions. However, more comprehensive research is needed that permits us to elucidate the richness of these functions and their interrelatedness. For example, very limited research has studied which functions contribute most to effective mentoring, and whether some are more important than others for skill development. Arguably, the first step in future research is for researchers to agree on what is mentoring and what it is supposed to accomplish. Many terms have been used to refer to individuals who provide feedback, direction, and support to juniors in their field. Understanding the similarities and differences in mentoring functions across domains may facilitate a critical discussion on how these mechanisms may aid the achievement of high proficiency more generally.

It will be essential to integrate theories from expertise, training, and developmental interactions to create efficient regimens for improving learning and performance. Likewise, better measures of mentoring are needed, as well as of outcomes of mentoring associated with effective performance (cf. psychosocial support). Although advances have been made in the development of psychometrically validated scales of mentoring, much less research has examined empirically the effectiveness of actual mentoring on useful indicators of expertise (e.g., skill development), rather than those that are more ancilliary (e.g., learner satisfaction) and that do little to inform how to design an effective mentoring program. Future mentoring research should focus on developing research methods that permit more verifiable, veridical, and veracious assessment of performance and learning to better identify the key components of mentoring and contextual constraints associated with performance improvement or expertise development. Without well-designed research, including component analysis, it will be difficult to efficiently deploy available resources (e.g., a week-long mentor training course may be no more effective than short two-hour workshop training essential components) and to move the field of mentoring forward.

A prime example of this need is in the use of preceptorship model in healthcare settings. Many healthcare professionals, for instance nurses and physicians, engage in such programs but there is a wide variety of ways in which these programs are implemented and an equally wide variety of outcomes in terms of skills and competencies acquired, making it difficult to discern their true effectiveness. It is plausible, for instance, that successes are equally associated with individual differences in motivation and deliberate efforts to improve, rather than the preceptorship experience itself. Until these areas are explored systematically, the real value of each program (i.e., in terms of the extent to which it caused an improvement) will remain relatively unknown. Many professions, including the medical professions, continue to engage with mentoring in this manner as a matter of tradition rather than demonstrated effectiveness.

In this chapter we focused on the key components of effective mentoring, and the role of mentoring in developing trainees' performance on both simple tasks (e.g., classroom learning) and more complex domains of expertise (e.g., sport, avionics). In a rapidly changing and increasingly complex world, in addition to having effective mentors and understanding their impact on the development of mentee expertise, it is equally vital to understand how best to train mentors to a high level of proficiency—such that new expert mentors can be developed. Unfortunately, even less research has focused on training mentors and/or the characteristics that these *master mentors* might possess. We propose that Hoffman and Ward's (2015) summary of effective mentoring, including the types of interactive dialogic reasoning described by Resnick et al. (Chapter 39, this volume), and the types of situated methods and expert reasoning feedback employed in products like Sherlock and ShadowBox, may be equally applicable here. Future studies that delve more deeply into measuring the effectiveness of these mentoring activities—both in terms of developing expert performers and in developing expert mentors—will advance the science of mentoring/coaching effectiveness. Likewise, those that explore how this kind of mentoring can be embedded within intelligent tutoring systems to create *intelligent mentoring systems* (see Hoffman & Ward, 2015) may pave the way for an increased role of technology-supported skill mentoring in tomorrow's sociotechnical systems.

REFERENCES

Anderson, J. R., Corbett, A. T., Koedinger, K. R., & Pelletier, R. (1995). Cognitive tutors: lessons learned. *Journal of the Learning Sciences* 4(2), 167–207. doi:10.1207/s15327809jls0402_2

Atteberry, A., Loeb, S., & Wyckoff, J. (2015). Do first impressions matter? Predicting early career teacher effectiveness. *AERA Open* 1(4), 1–23. doi:10.1177/2332858415607834.

Baron, L., & Morin, L. (2009). The coach–coachee relationship in executive coaching: A field study. *Human Resource Development Quarterly* 20, 85–106. doi:10.1002/hrdq.20009

Berk, R. A., Berg, J., Mortimer, R., Walton-Moss, B., & Yeo, T. P. (2005). Measuring the effectiveness of faculty mentoring relationships. *Academic Medicine* 80(1), 66–71.

Billay, D., & Myrick, F. (2008). Preceptorship: An integrative review of the literature. *Nurse Education in Practice* 8, 258–266. doi:10.1016/j.nepr.2007.09.005

Blanzola, C., Lindeman, R., & King, M. L. (2004). Nurse internship pathway to clinical comfort, confidence, and competence. *Journal of Nurses in Staff Development* 20(1), 27–37.

Brodeur, P., Larose, S., Tarabulsy, G., Feng, B., & Forget-Dubois, N. (2015). Development and construct validation of the Mentor Behavior Scale. *Mentoring & Tutoring: Partnership in Learning* 23(1), 54–75. doi:10.1080/13611267.2015.1011037

Brunt, B., & Kopp, D. (2007). Impact of preceptor and orientee learning styles on satisfaction: A pilot study. *Journal for Nurses in Staff Development* 23(1), 36–44.

Campbell, P. F., & Malkus, N. N. (2011). The impact of elementary mathematics coaches on student achievement. *Elementary School Journal* 111(3), 430–454.

Castro, S. L., Scandura, T. A., & Williams, E. A., (2004). Validity of Scandura and Ragins' (1993) Multidimensional Mentoring Measure: An evaluation and refinement. *Management Faculty Articles and Papers* 7. Retrieved from: http://scholarlyrepository.miami.edu/management_articles/7

Chandler, D. E., & Ellis, R. (2011). Diversity and mentoring in the workplace: A conversation with Belle Rose Ragins. *Mentoring & Tutoring: Partnership in Learning* 19(4), 483–500.

Chi, M. T., Siler, S., Jeong, H., Yamauchi, T., & Hausmann, R. G. (2001). Learning from human tutoring. *Cognitive Science* 25, 471–533. doi:10.1207/s15516709cog2504_1

Coates, V. E., & Gormley, E. (1997). Learning the practice of nursing: Views about preceptorship. *Nurse Education Today* 17, 91–98. doi:10.1016/S0260-6917(97)80024-X

Collins, A., Brown, J. S., & Holum, A. (1991). Cognitive apprenticeship: Making thinking visible. *American Educator* 15, 6–11.

Côté, J. & Gilbert, W. (2009). An integrative definition of coaching effectiveness and expertise. *International Journal of Sport Science* 4, 307–323. doi:10.1260/174795409789623892

Craven, H., & Broyles, J. (1996). Professional development through preceptorship. *Journal of Nursing Staff Development* 12(6), 294–299.

D'Abate, C. P., Eddy, E. R., & Tannenbaum, S. I. (2003). What's in a name? A literature-based approach to understanding mentoring, coaching, and other constructs that describe developmental interactions. *Human Resource Development Review* 2, 360–384. doi:10.1177/1534484303255033

Dennen, V. P. (2004). Cognitive apprenticeship in educational practice: Research on scaffolding, modeling, mentoring, and coaching as instructional strategies. In D. H. Jonassen (Ed.), *Handbook of research on educational communications and technology*, 2nd edn (pp. 813–828). Mahwah, NJ: Lawrence Erlbaum Associates.

Dickson, J., Kirkpatrick-Husk, K., Kendall, D., Longabaugh, J., Patel, A., & Scielzo, S. (2014). Untangling protégé self-reports of mentoring functions: Further meta-analytic understanding. *Journal of Career Development* 41, 263–281. doi:10.1177/0894845313498302

Dominguez, N., & Hager, M. (2013). Mentoring frameworks: Synthesis and critique. *International Journal of Mentoring and Coaching in Education* 2, 171–188. doi:10.1108/IJMCE-03-2013-0014

Eby, L. T., Allen, T. D., Evans, S. C., Ng, T., & DuBois, D. L. (2008). Does mentoring matter? A multidisciplinary meta-analysis comparing mentored and non-mentored individuals. *Journal of Vocational Behavior* 72, 254–267. doi:10.1016/j.jvb.2007.04.005

Eby, L. T. D. T., Allen, T. D., Hoffman, B. J., Baranik, L. E., Sauer, J. B., Baldwin, S., ... & Evans, S. C. (2013). An interdisciplinary meta-analysis of the potential antecedents, correlates, and consequences of protégé perceptions of mentoring. *Psychological Bulletin* 139, 441–476. doi: 10.1037/a0029279

Ellinger, A. D., Ellinger, A. E., & Keller, S. B. (2003). Supervisory coaching behavior, employee satisfaction, and warehouse employee performance: A dyadic perspective in the distribution industry. *Human Resource Development Quarterly 14*, 435–458. doi:10.1002/hrdq.1078

Ericsson, K. A. (1996). *The road to excellence: The acquisition of expert performance in the arts and sciences, sports, and games.* Mahwah, NJ: Erlbaum.

Ericsson, K. A., Whyte, J., & Ward, P. (2007). Expert performance in nursing: Reviewing research on expertise within the framework of the expert-performance approach. *Advances in Nursing Science 30*(1), E58–E71.

Feltovich, P. J., Coulson, R. L., Spiro, R. J., & Dawson-Saunders, B. K. (1992). Knowledge application and transfer for complex tasks in ill-structured domains: Implications for instruction and testing in biomedicine. In D. Evans & V. L. Patel (Eds), *Advanced models of cognition for medical training and practice* (pp. 213–244). Berlin: Springer-Verlag.

Feltz, D. L., Chase, M. A., Moritz, S. E., & Sullivan, P. J. (1999). A conceptual model of coaching efficacy: Preliminary investigation and instrument development. *Journal of Educational Psychology 91*, 765–776. doi:10.1037/0022-0663.91.4.76

Fielder, R., Read, E., Lane, K., Hicks, F., & Jegier, B. (2014). Long-term outcomes of a post-baccalaureate nurse residency program. *JONA 44*, 417–422. doi:10.1097/NNA.0000000000000092

Fleming, M., House, M. S., Shewakramani, M. V., Yu, L., Garbutt, J., McGee, R., . . . & Rubio, D. M. (2013). The mentoring competency assessment: validation of a new instrument to evaluate skills of research mentors. *Academic Medicine: Journal of the Association of American Medical Colleges 88*(7), 1002–1008. doi:10.1097/ACM.0b013e318295e298

Ghosh, R. (2013). Mentors providing challenge and support: Integrating concepts from teacher mentoring in education and organizational mentoring in business. *Human Resource Development Review 12*, 144–176. doi:10.1177/1534484312465608

Glass, M., Kim, J. H., Evens, M. W., Michael, J. A., & Rovick, A. A. (1999). Novice vs. expert tutors: A comparison of style. *Proceedings of the 10th Midwest Artificial Intelligence and Cognitive Science Conference, Bloomington, IN* (pp. 43–49).

Gott, S. P. (1995). Cognitive technology extends the work environment and accelerates learning in complex jobs. Technical paper. Human Resources Directorate, Manpower and Personnel Research Division, U. S. Air Force Armstrong Laboratory, Brooks AFB, TX.

Graesser, A. C., Person, N., & Magliano, J. (1995). Collaborative dialog patterns in naturalistic one-on-one tutoring. *Applied Cognitive Psychology 9*(6), 495–522. doi:10.1002/acp.2350090604

Gregory, J. B., & Levy, P. E. (2010). Employee coaching relationships: Enhancing construct clarity and measurement. *Coaching: An International Journal of Theory, Research and Practice 3*, 109–123. doi:10.1080/17521882.2010.502901

Hagen, M. S., & Peterson, S. L. (2015). Measuring coaching: Behavioral and skill-based managerial coaching scales. *Journal of Management Development 34*, 114–133. doi:10.1108/JMD-01-2013-0001

Hamlin, R. G., Ellinger, A. D., & Beattie, R. S. (2009). Toward a profession of coaching? A definitional examination of "coaching," "organization development," and "human resource development." *International Journal of Evidence Based Coaching and Mentoring 7*(1), 13–38.

Harris, K. R., Eccles, D. W., Ward, P., & Whyte, J. (2013). A theoretical framework for simulation in nursing: Answering Schiavenato's (2009) call. *Journal of Nursing Education 52*(1), 6–16. DOI: 10.3928/01484834-20121107-02

Hayes, E. F. (1998). Mentoring and nurse practitioner student self-efficacy. *Western Journal of Nursing Research 20*(5), 521–535. doi: 10.1177/019394599802000502

Heslin, P. A., Vandewalle, D., & Latham, G. P. (2006). Keen to help? Managers' implicit person theories and their subsequent employee coaching. *Personnel Psychology* 59, 871–902. doi:10.1111/j.1744-6570.2006.00057.x

Hill, H. C., Blazar, D., & Lynch, K. (2015). Resources for teaching: Examining personal and institutional predictors of high-quality instruction. *AERA Open* 1(4), 1–23.

Hoffman, R. R., LaDue, D., Mogil, H. M., Roebber, P., & Trafton, J.G. (2017). *Minding the weather: How expert forecasters think.* Cambridge, MA: MIT Press.

Hoffman, R. R., & Ward, P. (2015). Mentoring: A leverage point for intelligent systems? *IEEE Intelligent Systems* 30(5), 78–84. doi:10.1109/MIS.2015.86

Hoffman, R. R., Ward, P., Feltovich, P. J., DiBello, L., Fiore, S. M., & Andrews, D. H. (2014). *Accelerated expertise: Training for high proficiency in a complex world.* New York: Psychology Press.

Jones, R. L., & Turner, P. (2006). Teaching coaches to coach holistically: Can problem-based learning (PBL) help? *Physical Education and Sport Pedagogy* 11(2), 181–202. doi:10.1080/17408980600708429

Jones, R. J., Woods, S. A., & Guillaume, Y. R. (2015). The effectiveness of workplace coaching: A meta-analysis of learning and performance outcomes from coaching. *Journal of Occupational and Organizational Psychology* 89, 249–277. doi:10.1111/joop.12119

Kammeyer-Mueller, J. D., & Judge, T. A. (2008). A quantitative review of mentoring research: Test of a model. *Journal of Vocational Behavior* 72, 269–283. doi:10.1016/j.jvb.2007.09.006

Kavussanu, M., Boardley, I. D., Jutkiewicz, N., Vincent, S., & Ring, C. (2008). Coaching efficacy and coaching effectiveness: Examining their predictors and comparing coaches' and athletes' reports. *The Sport Psychologist* 22, 383–404. doi:10.1123/tsp.22.4.383

Ke, Y., Kuo, C., & Hung, C. (2017). The effects of a nursing preceptorship on new nurses' competence, professional socialization, job satisfaction and retention. *Journal of Advanced Nursing* 73, 2296–2305. doi:10.1111/jan.13317

Kilburg, R. R. (1996). Toward a conceptual understanding and definition of executive coaching. *Consulting Psychology Journal: Practice and Research* 48(2), 134–144. doi:10.1037/1061-4087.48.2.134

Klein, G. (2013). *Seeing what others don't: The remarkable ways we gain insights.* New York: Public Affairs.

Klein, G., & Baxter, H. C. (2009). Cognitive transformation theory: Contrasting cognitive and behavioral learning. In J. V. Cohn, D. Schmorrow, & D. Nicholson (Eds), *The PSI handbook of virtual environments for training and education: Developments for the military and beyond. Vol 1. Learning, requirements, and metrics* (pp. 50–64). Westport, CT: Praeger Security International.

Klein, G., & Borders, J. (2016). The ShadowBox approach to cognitive skills training: An empirical evaluation. *Journal of Cognitive Engineering and Decision Making* 10(3), 268–280. doi:10.1177/1555343416636515

Konczak, L. J., Stelly, D. J., & Trusty, M. L. (2000). Defining and measuring empowering leader behaviors: Development of an upward feedback instrument. *Educational and Psychological Measurement* 60, 301–313. doi:10.1177/00131640021970420

Kraft, M. A., Blazar, D., & Hogan, D. (2018). The effect of teacher coaching on instruction and achievement: A meta-analysis of the causal evidence. *Review of Educational Research* 88(4), 547–588. doi.org/10.3102/0034654318759268

Kulik, J. A., & Fletcher, J. D. (2016). Effectiveness of intelligent tutoring systems: A meta-analytic review. *Review of Educational Research* 86(1), 42–78. doi: 10.3102/0034654315581420

LaDue, D. S. (2011). *How meteorologists learn to forecast the weather: Social dimensions of complex learning* (Doctoral dissertation). Available from ProQuest Dissertations and Theses database (UMI No. 3482746).

Lave, J., & Wenger, E. (1991). *Situated learning: Legitimate peripheral participation*. New York: Cambridge University Press.

Lee, A., Dennis, C., & Campbell, P. (2007). Nature's guide for mentors. *Nature* 447(7146), 791–797. doi:10.1038/447791a

Lee, Y., Lin, H., Tseng, H., Tsai, Y., & Lee-Hsieh, J. (2017). Using training needs assessment to develop a nurse preceptor centered training program. *Journal of Nursing Education 48*, 220–229.

Lepper, M. R., Aspinwall, L. G., Mumme, D. L., & Chabay, R. W. (1990). Self-perception and social-perception processes in tutoring: Subtle social control strategies of expert tutors. In J. M. Olson & M. P. Zanna (Eds), *Self-inference processes: The Ontario Symposium*, Volume 6 (pp. 217–237). Hillsdale, NJ: Erlbaum.

Lepper, M. R., & Woolverton, M. (2002). The wisdom of practice: Lessons learned from the study of highly effective tutors. In J. M. Aronson (Ed.), *Improving academic achievement: Impact of psychological factors on education* (pp. 135–158). San Diego, CA: Academic Press.

Loiseau, D., Kitchen, K., & Edgar, L. (2003). A comprehensive ED orientation for new graduates in the emergency department: The 4-year experience of one Canadian teaching hospital. *Journal of Emergency Nursing 29*(6), 522–594.

McLean, G. N., Yang, B., Kuo, M. H. C., Tolbert, A. S., & Larkin, C. (2005). Development and initial validation of an instrument measuring managerial coaching skill. *Human Resource Development Quarterly 16*, 157–178. doi:10.1002/hrdq.1131

Matsumura, L. C., Garnier, H. E., & Spybrook, J. (2012). The effect of content-focused coaching on the quality of classroom text discussions. *Journal of Teacher Education 63*(3), 214–228. doi:10.1177/0022487111434985

Merrill, D. C., Reiser, B. J., Ranney, M., & Trafton, J. G. (1992). Effective tutoring techniques: A comparison of human tutors and intelligent tutoring systems. *Journal of the Learning Sciences 2*(3), 277–305. doi:10.1207/s15327809jls0203_2

Messmer, P., Jones, S., & Taylor, B. (2004). Enhancing knowledge and self-confidence of novice nurses: The "Shadow-A-Nurse" ICU program. *Nursing Education Perspectives, 25*(3), 131–136.

Militello, L. G., & Hutton, R. J. (1998). Applied cognitive task analysis (ACTA): A practitioner's toolkit for understanding cognitive task demands. *Ergonomics 41*(11), 1618–1641. doi:10.1080/001401398186108

Militello, L. G., & Quill, L. (2007). Expert apprentice strategies. In R. R. Hoffman (Ed.), *Expertise out of context* (pp. 159–176). New York: LEA.

Morgan, K., Jones, R. L., Gilbourne, D., & Llewellyn, D. (2013). Changing the face of coach education: Using ethno-drama to depict lived realities. *Physical Education and Sport Pedagogy 18*(5), 520–533. doi:10.1080/17408989.2012.690863

Morris, L., Pfeifer, P., Catalano, R., Fortney, R., Hilton, E., McLaughlin, J., & Goldstein, L. (2007). Designing a comprehensive model for critical care orientation. *Critical Care Nurse 27*(6), 37–48.

Myers, N. D., Chase, M. A., Beauchamp, M. R., & Jackson, B. (2010). Athletes' perceptions of coaching competency scale II-High school teams. *Educational and Psychological Measurement 70*(3), 477–494. doi:10.1177/0013164409344520

Myers, N. D., Feltz, D. L., Chase, M. A., Reckase, M. D., & Hancock, G. R. (2008). The Coaching Efficacy Scale II—High School Teams. *Educational and Psychological Measurement 68*(6), 1059–1076. doi:10.1177/0013164408318773

Myers, N. D., Wolfe, E. W., Feltz, D. L., & Penfield, R. D. (2006). Identifying differential item functioning of rating scale items with the Rasch model: An introduction and an application. *Measurement in Physical Education and Exercise Science* 10(4), 215–240. doi:10.1207/s15327841mpee1004_1

Noe, R. A. (1988). An investigation of the determinants of successful assigned mentoring relationships. *Personnel Psychology* 41(3), 457–479. doi:10.1111/j.1744-6570.1988.tb00638.x

Papay, J. P., & Kraft, M. A. (2015). Productivity returns to experience in the teacher labor market: Methodological challenges and new evidence on long-term career improvement. *Journal of Public Economics* 130, 105–119.

Park, S., McLean, G. N., & Yang, B. (2008, February). Revision and validation of an instrument measuring managerial coaching skills in organizations. Paper presented at the Academy of Human Resource Development Conference, Panama City, FL (ERIC Document Reproduction Service. ED501617).

Petushek, E. J., & Ward, P. (2016). Accelerating coaching expertise. Technical report. London: English Institute of Sport.

Pfund, C., House, S. C., Asquith, P., Fleming, M. F., Buhr, K. A., Burnham, E. L., ... & Shapiro, E. D. (2014). Training mentors of clinical and translational research scholars: a randomized controlled trial. *Academic Medicine: Journal of the Association of American Medical Colleges* 89, 774–782. doi:10.1097/ACM.0000000000000218

Phillips, J. M. (2006). Preparing preceptors through online education. *Journal for Nurses in Professional Development* 22(3), 150–156.

Polson, M., & Richardson, J. (1988). *Foundations of intelligent tutoring systems.* Mahwah, NJ: Lawrence Erlbaum Associates.

Ragins, B. R. (2011). Relational mentoring: A positive approach to mentoring at work. In K. S. Cameron & G. M. Spreitzer (Eds), *The Oxford handbook of positive organizational scholarship* (pp. 519–536). Oxford: Oxford University Press.

Ragins, B. R., & Kram, K. E. (2007). The roots and meaning of mentoring. In B. R. Ragins, & K. E. Kram (Eds), *The handbook of mentoring at work: Theory, research, and practice* (pp. 3–15). Thousand Oaks, CA: Sage.

Ragins, B. R., & McFarlin, D. B. (1990). Perceptions of mentor roles in cross-gender mentoring relationships. *Journal of Vocational Behavior* 37(3), 321–339. doi: 10.1016/0001-8791(90)90048-7

Rose, G. L. (2003). Enhancement of mentor selection using the ideal mentor scale. *Research in Higher Education* 44(4), 473–494. doi:10.1023/A:1024289000849

Rosenfeld, P., & Glassman, K. (2016). The long-term effect of a nurse residency program, 2005–2012. Analysis of former nurse residents. *JONA* 46, 336–344.

Rosenfeld, P., Glassman, K., & Capobianco, E. (2015). Evaluating the short and long-term outcomes of a post-BSN residency program. *JONA* 6, 331–338.

Scandura, T. A., & Ragins, B. R. (1993). The effects of sex and gender role orientation on mentorship in male-dominated occupations. *Journal of Vocational Behavior* 43(3), 251–265. doi:10.1006/jvbe.1993.1046

Smither, J. W., London, M., Flautt, R., Vargas, Y., & Kucine, I. (2003). Can working with an executive coach improve multisource feedback ratings over time? A quasi-experimental field study. *Personnel Psychology* 56(1), 23–44.

Sonesh, S. C., Coultas, C. W., Lacerenza, C. N., Marlow, S. L., Benishek, L. E., & Salas, E. (2015). The power of coaching: A meta-analytic investigation. *Coaching: An International Journal of Theory, Research and Practice* 8(2), 73–95. doi:10.1080/17521882.2015.1071418

Theeboom, T., Beersma, B., & van Vianen, A. E. (2014). Does coaching work? A meta-analysis on the effects of coaching on individual level outcomes in an organizational context. *Journal of Positive Psychology 9*(1), 1–18. doi:10.1080/17439760.2013.837499

Underhill, C. M. (2006). The effectiveness of mentoring programs in corporate settings: A meta-analytical review of the literature. *Journal of Vocational Behavior 68*, 292–307. doi:10.1016/j.jvb.2005.05.003

VanLehn, K. (2011). The relative effectiveness of human tutoring, intelligent tutoring systems, and other tutoring systems. *Educational Psychologist 46*(4), 197–221. doi:10.1080/00461520.2011.611369

Vygotsky, L. (1978). Interaction between learning and development (M. Lopez-Morillas, Trans.). In M. Cole, V. John-Steiner, S. Scribner, & E. Souberman (Eds), *Mind in society: The development of higher psychological processes* (pp. 79–91). Cambridge, MA: Harvard University Press.

Ward, A., & McComb, S. (2017). Precepting: A literature review. *Journal of Professional Nursing 33*, 314–325. doi:10.1016/j.profnurs.2017.07.007

Ward, P., Gore, J., Hutton, R., Conway, G., & Hoffman, R. (2018). Adaptive skill as the *conditio sine qua non* of expertise. *Journal of Applied Research in Memory and Cognition 7*(1), 35–50.

Wigdor, A. K., & Green, B. F. (1991). *Performance assessment for the workplace*, Vols I & II. Washington, DC: National Academy Press.

Yonge, O., Billay, D., Myrick, F., & Luhanga, F. (2007). Preceptorship and mentorship: Not merely a matter of semantics. *International Journal of Nursing Education Scholarship 4*, 1–13. doi:10.2202/1548-923X.1384

Zhang, C., Fan, H., Xia, J., Guo, H., Jiang, X, & Yan, Y. (2017). The effects of reflective training on the disposition of critical thinking for nursing students in China: A controlled trial. *Asian Nursing Research 11*, 194–200.

Zhang, J., Jensen, B. E., & Mann, B. L. (1997). Modification and revision of the Leadership Scale for Sports. *Journal of Sports Behavior 20*, 105–122.

CHAPTER 45

ACQUIRING AND MAINTAINING EXPERTISE IN AGING POPULATIONS

DAN MORROW AND RENATO F. L. AZEVEDO

Introduction

THE study of expertise, often defined as superior levels of performance on representative tasks or more broadly as improved performance with increasing task-related experience, is an investigation of change, typically growth in skills, knowledge, interests, and other changes that accompany improved performance. The study of aging also focuses on change, although patterns are often more complex with both growth and decline in abilities and performance. Research across the lifespan can present a somewhat pessimistic view of aging, with age-graded declines on tasks requiring fluid mental abilities such as speeded processing, working memory, and reasoning, while research rooted in meaningful, familiar situations at work and home is more encouraging. Theories of lifespan development have long grappled with this dynamic between gain and loss and how it plays out in terms of daily function and accomplishments in later adulthood. Research in expertise helps shed light on when and how experience-based gains offset losses due to biological and other changes to promote successful aging. At the same time, research on expertise across the lifespan provides insights into the development of expertise in the broader context of cognitive change. In this chapter, we first consider how experts excel on domain-relevant tasks despite cognitive limitations, and how these expertise-related advantages develop, which suggest ways in which adults offset age-related cognitive constraints to maintain performance in later years. Because costs as well as benefits are associated with expertise, we also consider whether these costs change with age.

Expertise Development

Several theories consider conceptions of expertise and how expertise develops, as well as the mechanisms underlying development that promote cognitive efficiency (see Section I of this handbook). Dreyfus and Dreyfus (1986) proposed a five-stage model of skill acquisition (novice, advanced beginner, competent, proficient, and expert), in which novices are characterized by rigid adherence to context-independent rules and procedures. As expertise develops, memory becomes more situational, recognition more holistic, and decision-making shifts from analytical to intuitive (see also Gobet & Chassy, 2009; Wilhems, Corbin, & Reyna, 2014). Finally, they characterize the mastery of expertise when individuals incorporate alternative approaches to problem-solving, perspectives, and decision options.

Hoffman et al. (2014) argue that the perceptual and knowledge advantages of experts enable adaptation, not only due to automaticity and superior performance of procedural skills, but by understanding the meaning of the skills in relation to goals and tasks (see also Feltovich, Spiro, & Coulson, 1997). Therefore, experts can anticipate and prepare for future decisions and actions or behaviors (e.g., visual, perceptual, and cognitive skills of athletes as discriminators of superior performance in sports). Novices must learn to handle uncertainty and to adopt strategies that deviate from standard procedures and methods, and therefore are pressed to learn more and improve task-related competence (journeyman), to a point in which self-regulatory and metacognitive strategies develop, which support adaptation and improvisation. Therefore, novices progressively develop superior awareness of their own knowledge base and skills, which allows them to focus their learning on what they do not know and therefore need to master. This development is also seen in changes in knowledge and understanding: novices have a limited and fragmented knowledge base and therefore are vulnerable to misconceptions and unable to build a *big picture* of representative tasks, while experts go beyond the information learned during training in order to articulate inferences and integrate new information *on the fly*. As a consequence, high-level experts are flexible and have superior metacognitive strategies, adopting more analytical or more intuitive strategies as the situation dictates or when facing new situations.

Experts' superior performance reflects *opportunistic* reasoning that incorporates new information into their diagnostic decisions as they accumulate more organized knowledge (Chi, Feltovich, & Glaser, 1981; Lesgold et al., 1988). This knowledge development reflects increasing levels of perceptual processing and attunement (Lesgold et al., 1988). Similarly, Simon (Chase & Simon, 1973; Simon & Gobet, 2000), Klein (recognition-primed decision model; Klein, 1998; Ross, Shafer, & Klein, 2006), and others have explained superior expert intuition as reflecting recognition of patterns of diagnostic cues. Hence, the more goal-relevant constraints available in the situation/tasks, the greater the expertise advantage in performance (Vicente & Wang, 1998) (see also previous chapters on ecological perspectives and methods).

Overall, despite differences among theories, development of expertise is characterized not only by more knowledge, but by changes in knowledge organization that facilitate access to knowledge when needed. These theories address the interplay between relatively fixed cognitive capacities and the cognitive efficiency derived from experience, skills, and knowledge. This tension related to how experts adapt to biological limits becomes increasingly important in later adulthood as cognitive constraints increase with declining cognitive resources, even as the experience underlying expert performance increases. In this chapter we consider to what extent efficiencies associated with expertise enable older experts to mitigate declining resources in order to maintain high levels of performance. This in turn requires considering whether the processes underlying expert maintenance of superior performance changes with age, and whether any expertise costs as well as benefits may reduce mitigation.

Facets of expertise and implications for performance

Benefits of expertise

According to Nokes, Schunn, and Chi (2010), theories of expert problem solving identify the following processes: problem categorization, constructing a problem representation, search strategies, retrieval strategies, evaluating problem-solving progress and solutions, and storage of solutions. These processes can interact (e.g., problem categorization allows more integrated mental representations and more efficient retrieval). Expertise enables efficiency in these processes that confers expertise-related advantages in problem solving. Skilled-memory theory (Chase & Ericsson, 1982) focuses on gains from experts' use of cues to encode and retrieve chunks to accomplish task goals. Compared to novices, experts often use more diagnostic cues that are relevant to the problem, supporting rapid situation assessment and problem solving (Brunswik, 1956; Klein, 1998, Shanteau, 1992) and reducing interference at retrieval (Chi, 2006a; 2006b).

Expert knowledge organization also supports efficient search. Experts typically represent problems and work from given problem variables (forward thinking), while novices start with the given goal (e.g., find an unknown variable) to generate solutions (backward thinking; Chi et al, 1981; Simon & Simon, 1978; see Patel, Arocha, & Kaufman, 1994 for similar findings in medical reasoning). Experts also learn domain-relevant information more efficiently by integrating new information with their knowledge (Ericsson & Kintsch, 1995). They read more quickly and better remember domain-relevant text in aviation (Morrow, Leirer, & Altieri, 1992), baseball (Spilich, Vesonder, Chiesi, & Voss, 1979), and other domains. Finally, experts have superior meta-cognitive skills (e.g., monitoring), investing more time evaluating solutions to make sure they satisfy task constraints (Groen & Patel, 1988) and identifying and correcting errors (Nokes et al., 2010; Wineburg, 1998).

Costs of expertise

Experts do not always outperform novices. Indeed, they sometimes perform worse, and these costs can reflect the same mechanisms that usually support superior performance. Perhaps the best documented limitation is that expertise benefits are highly circumscribed, with superior performance restricted to the domain of acquisition. The training literature abounds with studies in which skills do not extend beyond the training conditions to novel situations (far transfer), or often even slightly different ones (near transfer). Moreover, expertise can be highly specialized. For example, physics professors are unable to answer all questions on an introductory physics exam (Reif & Allen, 1992). Experts may also be handicapped because interpreting problems at a deep level can interfere with more general problem-solving strategies. Baseball experts may perform worse on tasks requiring divergent thinking when constrained by domain knowledge (Wiley, 1998). Experts may also more often falsely recognize new information as studied when it is consistent with prior knowledge (Arkes & Freedman, 1984).

Experts' reliance on readily available responses or strategies attuned to specific situations may limit performance in other ways. Highly experienced accountants in one study were more likely to rely on previous knowledge than on newly acquired knowledge about a similar case, thus failing to update their mental representations of new problems (Marchant, Robinson, Anderson, & Schadewald, 1991). Over-reliance on heuristic strategies may focus experts on the most typical rather than all available cues (cognitive tunneling; Lewandowsky & Kirsner, 2000). Compounding this cost, experts may be overly confident (Meyer, Payne, Meeks, Rao, & Singh, 2013). To sum up, while experts often excel on domain-relevant tasks by adapting to domain constraints, over-reliance on knowledge and strategies may impair performance.

EXPERTISE AND AGING

The overview of benefits and costs associated with expertise raises two issues about how expertise influences performance as we age. First, do high-level experts retain superior levels of performance with age? This issue is often addressed in fairly narrow domains such as games (chess, bridge, Go), professional sports, and music. A second, broader issue is whether benefits or costs related to domain-general as well as domain-specific knowledge change with age. While many adults never reach the heights of superior performance demonstrated by the most accomplished experts, they do become relatively expert in particular domains with increasing experience. This second issue is central to lifespan theory: to what extent does knowledge and skill associated with experience offset age-related declines in abilities and function.

Gains and Losses across the Lifespan

According to the selection, optimization, and compensation (SOC) framework (Baltes & Baltes, 1990; Riediger, Li, & Lindenberger, 2006), change in function across the lifespan is multidirectional and multidimensional: some cognitive abilities or resources increase (e.g., knowledge) while others decrease (e.g., fluid mental ability such as working memory capacity). The extent of change varies by domain, with broad declines in sensorimotor function (e.g., visual acuity, high frequency hearing, motor control) occurring alongside the more selective changes in cognition mentioned earlier. Social and emotional functions also become increasingly important across the lifespan, with older adults more likely to notice and remember emotionally relevant information, for example (Charles & Carstensen, 2010). To address how age-related gains can offset age-related loss, the SOC framework posits that older adults engage in three broad types of strategies. First, optimization strategies involve maintaining function through intensive, effortful practice. Second, selection strategies involve maintaining some functions such as cognitive abilities important for one's job (or pursuing some goals) more than others. Finally, compensation strategies involve developing new ways to accomplish task goals and maintain proficiency. In this chapter, we address whether, or under what conditions, these strategies are supported by expertise, given that older experts amass large amounts of knowledge and develop skill repertoires with increasing experience. Older experts may engage in high levels of practice that produce efficient skills (skill maintenance as a form of optimization), selectively allocate effort to more important information or tasks (expertise-based selection), and/or develop compensation strategies with experience (e.g., rely on external aids or collaboration; Bosman & Charness, 1996; Luong, Charles, & Fingerman, 2011; Salthouse, 1990).

Age-related loss

Aging is accompanied by a variety of declines in sensorimotor function. Static and dynamic visual acuity, range of accommodation, contrast sensitivity, dark adaptation, color sensitivity, and other visual functions decline (Schieber, 2006). Age-associated loss in auditory function includes declines in pure tone sensitivity and sound localization, as well as increased sensitivity to loudness (Gordan-Salant, 2005). Other sensorimotor changes include slower manual response time and declines in abilities related to continuous movement, coordination, and movement precision (Fisk, Rogers, Charness, Czaja, & Sharit, 2009).

These sensory changes can impair cognitive function by degrading the quality of perceptual input to higher-order cognitive processes (sensory deprivation) or by requiring allocation of cognitive resources to compensate for sensory loss at the expense of higher-order processing (Tun, McCoy, & Wingfield, 2009). They also impair the ability to accomplish everyday tasks at work or home that require reading and computation (Charness & Dijkstra, 1999) as well as communication, especially when speech is garbled or embedded in noise (Humes, 2007).

Aging is also accompanied by changes in cognitive abilities or resources, including gradual decline in fluid mental ability, which enables efficient information processing in novel contexts (Chen, Hertzog, & Park, 2017; Hartshorne & Germine, 2015; Horn & Cattell, 1967). Contributing to this decline is age-related slowing in mental processes that impair ability to simultaneously execute multiple processes, diminished working memory capacity, and declines in executive control processes that maintain goals, monitor performance, and coordinate processes (Kramer & Madden, 2008). These declines help explain age-related difficulty with binding concepts in memory that impair episodic recall (Kilb & Naveh-Benjamin, 2007) and difficulty with learning and decision making essential to everyday function (Wolf et al., 2012). Everyday function is also impaired by age-related differences in literacy and numeracy, the ability to make sense of numbers and reason quantitatively (e.g., Peters, 2012), although these findings may reflect cohort differences in education as well as developmental declines.

Age-related gain

Aging also involves gains in (or preservation of) mental resources that can offset declines and help explain how older adults routinely perform at superior levels. General knowledge about language and the world tends to increase at least until the early 70s (verbal or crystallized ability; Hartshorne & Germine, 2015; Horn & Cattell, 1967). These gains reflect in part accumulating literacy experience (Payne et al., 2012; Stanovich, West, & Harrison, 1995). Literacy also involves understanding language in terms of perceptual-motor processes grounded in experience. According to comprehension theories (e.g., Kintsch, 1998), people understand stories or other texts by representing the situations conveyed by the text (situation model) as well as the text's verbatim linguistic features and semantic content. The situation model is created by integrating content with knowledge about this content, which can involve mentally simulating the described situation (Zwaan, 1999). Older are as likely as younger adults to mentally simulate described situations, and sometimes more so, presumably because of their literacy experience (Radvansky & Dijkstra, 2007).

Domain-specific as well as -general knowledge increases with age (for reviews, see Beier & Ackerman 2005; Umanath & Marsh, 2014). Increased knowledge is associated with amount of domain-related experience in aviation (Morrow et al., 2009), music (Meinz, 2000), health (Beier & Ackerman, 2005), accounting (Castel, 2005), finance (Colonia-Willner, 1998, Li, Baldassi, Johnson, & Weber, 2013), politics (Trepte & Schmitt, 2017), and other domains. Age-related knowledge gains are demonstrated in several ways, including minimal age differences in semantic priming (Laver & Burke, 1993) and similar gains in recall of domain-relevant versus domain-general text (Umanath & Marsh, 2014).

Knowledge about one's own cognitive strengths and weaknesses (metacognition) may also be maintained or improve with age. Some metacognitive processes that guide and control other cognitive processes central to learning are largely intact among older adults. These include knowledge about memory function and monitoring the effectiveness of learning processes, as measured by judgments of learning. The ability to control learning processes based on monitoring, however, may be impaired with age (Castel, McGillivray, & Friedman, 2012; Umanath & Marsh, 2014). While self-regulatory processes are essential to effective function across the lifespan, the broad goals that organize these processes may change with age, with emotional and social goals becoming more important (Charles & Carstensen, 2010). Older adults are more likely to notice and remember emotionally charged (especially positive) information. Affective processes may occur with little effort and become more important across the lifespan because they are preserved relative to declining fluid mental ability that underpins deliberative processes involved in decision making (Peters, Hess, Västfjäll, & Auman, 2007). Age differences are more likely for decision tasks when older adults have little knowledge or motivation (e.g., they seek less information about options in a choice task; Finucane, Mertz, Slovic, Scholze, & Schmidt 2005), than for tasks with emotionally significant information, perhaps because older adults integrate information and decision options in terms of affective response (Mikels et al., 2010).

Expertise and Age Differences in Performance

Before considering to what extent expertise-related costs and benefits change with age, we mention several methodological issues involved in comparing adults varying in age and experience. First, a challenge to reviewing this literature is variability in how expertise is operationalized. While expertise is sometimes equated with membership in groups (e.g., pilot, accountant, physician) or with sheer amount of experience in a domain, expert status is more often validated by objective indices such as professional ratings (e.g., commercial pilot license, Accountant or Medical Board exams) or high levels of performance on tasks that are representative of the domain (Ericsson & Lehman, 1996). Second, the definition and recruitment of expert groups can include intermediate levels of expertise (e.g., journeyman) and might not reflect appropriate comparisons. Third, studies vary in their design. Some compare extreme groups to test Age × Expertise interactions (whether age differences in performance are smaller for more expert groups); others use samples in which age and experience vary continuously. Investigating relationships between age and expertise as continuous variables may provide a finer-grained picture of trade-offs between these variables. However, small samples, often reflecting challenges of recruiting from special populations, limit the ability of either design to detect age–expertise relationships. Fourth, comparing groups differing in age or expertise (cross-sectional designs) poses interpretive challenges related to potential confounding variables that can be partly addressed by attempting to equate groups on these variables. While longitudinal studies (e.g., comparing age of peak performance for experts at different levels of proficiency; Roring & Charness, 2007) can disentangle cohort and developmental changes associated with age, such studies are rare. In one example, Taylor, Kennedy, Noda, &

Yesavage (2007) investigated age and expertise effects on flight simulator performance over three years for a sample of pilots age 40–69 years at study entry. The results provided insights into how age differences in performance at different levels of expertise changed over time. A final methodological challenge in this literature lies in the fact that experience tends to increase with age. This *confound of nature* can be partially addressed by equating experience levels of younger and older groups (e.g., older and younger pilots matched on average total flight hours), although this strategy may produce unrepresentative groups (older pilots who have fewer flight hours than typical for their age group). Alternatively, studies may include samples with sufficient variance in experience and age to unravel effects of the two variables, although obtaining adequate sample size for these studies can be challenging. To sum up, understanding how relationships between expertise and age influence performance challenges researchers to clearly characterize dimensions of expertise and to tease apart the effects of age from other factors (e.g., education) that may influence performance.

We now consider whether preserved knowledge and skill offsets age-related declines in sensory and cognitive function to support performance. The interplay between the costs and benefits of knowledge use among older adults relates to two broad issues. First, how do older adults accumulate knowledge despite age declines in the cognitive resources needed for learning? For example, older adults who have had chronic illness for a longer time tend to know more about their illness despite declining processing capacity (Chin et al., 2009). Most generally, there is an age-related preference for learning about familiar topics, perhaps because learning is easier when grounded in prior knowledge (Badham & Maylor, 2015). Learning becomes more efficient as older adults increasingly rely on knowledge rather than age-vulnerable cognitive resources such as processing capacity, at least when learning does not heavily tax processing capacity (Beier & Ackerman, 2005). This spiral of increasingly efficient knowledge acquisition (Ackerman, 2007) may extend the upward spiral of reading frequency, proficiency, and comprehension gains earlier in the lifespan (Mol & Bus, 2011). The preference for learning about familiar topics is consistent with a self-regulatory view of expertise, in which learning is driven by increasingly specific interests and goals (Ackerman, 2007), as well as ecological views, where experts become attuned to domain constraints (Vicente & Wang, 1998). Both frameworks emphasize the adaptive nature of expertise. In addition, the focus of the self-regulatory view aligns with the self-regulatory mechanisms underlying development in the SOC framework. A compelling example of how learning increases cognitive efficiency, at least for perceptual-motor tasks, is the well-documented relationship between skill acquisition and automaticity. A hallmark of skilled behavior is that performance requires minimal attentional resources with high levels of practice (Logan, 1997), with neural areas subserving performance increasingly integrated and with decreased recruitment of activation (Bernardi et al., 2013). Older adults may need more practice, but can reach the same level of efficiency as younger adults (Touron, Hoyer, & Cerella, 2001). While automaticity is likely not the defining feature of expertise broadly considered, it is an important component because it frees up cognitive resources for more strategic aspects of expertise.

The second broad question relates to how knowledge and skill are deployed to mitigate age differences in performance. To address this question, we review studies in two broad areas of expertise research because mechanisms related to how expertise influences performance can differ by domain (Hoffman, Shadbolt, Burton, & Klein, 1995). These areas vary in the specificity of constraints on performance, as well as the range of abilities contributing to performance. The first area involves highly structured tasks or activities such as games, sports, and music, typically with clearly defined goals and well-specified rules that define behaviors to achieve these goals. Performance often depends heavily on perceptual and attentional skills adapted to these rule-based constraints and reflect high levels of practice in competitive environments to reach superior levels of performance. Within this broad area we distinguish two subareas that differ in the prominence of physical and perceptual-motor versus more cognitive abilities in shaping performance. The second broad area involves domains in which performance depends on complex skills such as planning and task management deployed to accomplish diverse goals. These domains often relate to work or to activities that underpin everyday competence (e.g., managing finances) that are *substantively complex* (Schooler, Mulatu, & Oates, 1999; 2004). Goals in these domains tend to be diverse and often ill-defined, requiring greater variety of plans to accommodate more contingencies to be accomplished.

For each area, we review evidence for expertise-based mitigation of age-related differences in performance, and whether such mitigation reflects the use of optimization, selection, or compensation strategies. More open-ended, complex domains may afford a greater repertoire of strategies that mitigate age-related declines in cognitive abilities (Rybash, Hoyer, & Roodin, 1986). The likelihood of mitigation may also depend on the extent to which tasks require age-vulnerable abilities and processes (physical strength, speeded response, managing uncertainty), as well as the cognitive costs associated with using knowledge to accomplish task goals (see also Morrow & Rogers, 2008). For example, while older adults tend to maintain and readily access knowledge about language, as shown by age invariance in semantic priming (e.g., Laver & Burke, 1993), the effort involved in using this knowledge to accomplish task goals can vary, influencing the likelihood of mitigation to the extent the cognitive resources required for knowledge use are age-limited and reduce performance. This pattern is consistent with dual process theories of cognitive aging: Age differences are more likely to the extent performance depends on manipulating as well as activating knowledge because of age declines in controlled versus automatic processing, which can reduce the effectiveness of SOC strategies (Craik, 2000).

Structured (highly rule-governed) domains requiring more perceptual-motor abilities

Performance in sports and related domains heavily depend on perceptual-motor processes and coordination of attention, often under time pressure. Performance frequently builds on highly practiced task components, so that skill acquisition is a core component of expertise. Experts in these domains can perform at high levels with minimal effort through high levels of practice that automatize perceptual-motor processes. In terms of lifespan theory, expertise at any age reflects optimization strategies based on extensive practice. Older experts continue to benefit from accumulating high levels of practice that reduce dependence on attentional resources (Kramer & Madden, 2008). However, age-related changes in perceptual-motor and cognitive abilities (especially speed of processing) may limit attainment or maintenance of high levels of performance among older experts in sports or other domains requiring physical strength and speeded response. A review of peak performance in sports by age found earlier peaks for more physically demanding sports requiring fast responses such as swimming and sprinting compared to slower-paced, less physically demanding sports such as golf (mid-twenties vs early thirties, Schultz & Curnow, 1988). High levels of experience in sports with lower physical demands (golf, tennis) can reduce age-related differences on some perceptual-motor tasks. For example, significant age declines did not occur for tennis players but did for nonplayers for coincident-timing tasks in one study, especially when tracking accelerating targets (Lobjois, Benguigui, & Bertsch, 2005). The authors argued that this pattern was due to decreased visual-motor delay in older players versus nonplayers, which facilitated motor planning (also see Lobjois, Benguigui, Bertsch, & Broderick, 2008). There were similar delays for younger and older expert players, suggesting high levels of practice maintained perceptual-motor skills (an optimization strategy). However, performance on domain-general tasks was not investigated, so benefits among older experts may reflect general factors such as exercise effects on cognition (e.g., Colcombe et al., 2004). An important related question is whether mitigation of age declines in specific domains such as sports has broader benefits for older adults, such that domain-specific benefits transfer to other domains. There is little evidence for this, although there is some evidence that expert benefits for some sports transfer to similar sports (baseball and cricket; Moore & Müller, 2014) and that elite athletes can outperform novices on domain-general tasks (e.g., measuring sustained attention) as well as on domain- specific tasks (e.g., Alves et al., 2013). However, these effects are small and selective, and may reflect initial selection rather than experience-based effects.

Structured (rule-governed) domains requiring more cognitive abilities

Older experts in highly rule-governed domains with more prominent cognitive skill components (music, games such as chess, bridge, Go) appear to engage in a broader range of strategies to maintain performance, with expertise more likely to mitigate age differences under certain task conditions. Master pianists do not exhibit age-related slowing on many keyboard exercises, but do on standard motor tasks (Krampe & Ericsson, 1996), consistent with the expertise literature showing expertise benefits for domain-specific but not general tasks (e.g., Clancy & Hoyer, 1994). Older amateur pianists in one study simplified temporal sequencing in complex polyrhythm tapping tasks to maintain global rate of performance, but older master pianists relied on parallel timing mechanisms (same vs sequence of hands) to accommodate to demands of changing to very rapid tempos (Krampe, Engbert, & Kliegl, 2002). Moreover, there was little evidence for age differences in these strategies among the experts. These findings provide more evidence that perceptual-motor strategies adapted to specific task conditions are maintained with age among high-level experts, highlighting the importance of optimization strategies and consistent with skill maintenance explanations.

Chess experts also depend heavily on perceptual and attentional skills. They focus on relational properties of the most relevant chess pieces, with more information extracted per fixation (Charness, Reingold, Pomplun, & Stampe, 2001), and detect more distant moves more quickly than less skilled players (Jastrzembski, Charness, & Vasyukova, 2006). Recognition of chess moves reflects rapid retrieval of knowledge of configurations from long-term memory, helping bypass working memory limitations (Ericsson & Kintsch, 1995). Player age is often unrelated to effectiveness of search for chess moves among very expert players (Charness, 1981; but see Jastrzembski et al., 2006). More generally, high levels of expertise in chess, music, and related domains depend on deliberate practice, where experts invest much effort on individualized training activities designed to improve specific aspects of performance through repetition and successive refinement (Ericsson & Lehmann, 1996). Older experts appear to maintain superior performance through deliberate practice (Krampe & Ericsson, 1996). This finding suggests older experts in chess and related domains engage in optimization strategies to maintain proficiency, and perhaps develop compensatory strategies that reduce the cognitive demands of play (e.g., searching more moves ahead of current play to compensate for slower processes associated with selecting moves; Charness, 1981). For these reasons, as well as because speeded response is less important, the age of peak performance in chess appears to be later than in many sports domains (Roring & Charness, 2007). Roring and Charness (2007) also found that chess players with higher levels of skill at the midpoint of their career experienced slower declines in performance with age, perhaps because deliberate practice promotes optimization or compensation strategies.

As we have argued, the likelihood that expertise mitigates age differences depends in part on the cognitive costs associated with using knowledge to accomplish tasks. Evidence is provided by comparing mitigation effects for tasks of varying domain-relevance. Because domain-relevant tasks are organized in terms of domain constraints (Vicente & Wang, 1998), they are representative of the conditions under which experts typically perform. Experts readily draw on their experience-based skill repertoire to perform such tasks, with little need for *overhead* operations to capitalize on their knowledge, such as developing new representations or associating new responses to stimuli. For such tasks, experts rely on perceptual-motor skills that are automatized through high levels of deliberate practice (Bosman & Charness, 1996). Less domain-relevant tasks may penalize older experts because more cognitive resources are required to form new associations between aspects of the task and domain knowledge, or to map new responses to stimuli. In contrast to the previously reviewed studies that investigated actual musical performance, tasks requiring recall of music material often do not reveal mitigation (Halpern, Barlett, & Dowling, 1995; Meinz, 2000; Meinz &

Salthouse, 1998). Similarly, chess expertise is more likely to mitigate age differences for movement search tasks (a component of playing) than for recall tasks (Charness, 1981; but see Jastrzembski, et al., 2006). Expertise in the game of Go reduces age differences on reasoning or recognition but not on recall tasks (Masunaga & Horn, 2001).

Older experts may also come to rely on compensatory strategies when performing highly structured domain-relevant tasks. Such strategies include anticipating task changes (Lobjois et al., 2008) or drawing inferences from information based on abstract problem representations (Ericsson & Lehman, 1996). Further evidence comes from investigation of skilled performance in typing. In Salthouse's (1984) classic study, highly experienced typists' performance on transcription typing tasks was unrelated to age, while they showed typical age-related slowing on standard manual reaction time measures, suggesting older experts were not selected for domain-general abilities. Information processing limits typical of serial reaction time tasks appeared to be offset because the experienced typists could *look ahead* while executing key strokes. Older typists appeared to rely more than younger typists of similar skill level on the preview strategy to offset slower motor processes as a compensatory strategy because they showed more benefit from preview for typing speed (also see Bosman, 1993).

Complex (open-ended) domains requiring broad range of knowledge and skills

Domains related to work and *enhanced activities of daily living* that underlie daily competence (e.g., health and financial management; Rogers, Meyer, Walker, & Fisk, 1998) offer many opportunities for older adults to draw on knowledge and skill derived from experience in order to mitigate age difference in performance, for at least two reasons. First, high levels of performance in these domains often depend more on sustained levels of effort over long periods of time than on the consistent, peak performance often required to excel in sports and other competitive games (see Beier & Ackerman, 2005, for a related distinction between typical and maximal performance). Second, performance depends heavily on planning, problem solving, task management, and other complex abilities that enable diverse optimization, selection, and compensation strategies.

Comprehension, which underpins many cognitive abilities required for proficient performance in many domains, is a pervasive example of a complex everyday skill that depends on the interplay of knowledge and processing capacity. We begin this section by considering the role of knowledge in comprehension and how the cost of using knowledge influences strategies related to mitigation. Comprehension involves processes such as recognizing words, activating the associated concepts, integrating these concepts into propositions that represent ideas given by the text, and integrating propositions with knowledge to create a situation model (Kintsch, 1998). Age differences in comprehension are minimal when these processes are highly practiced, with minimal demands on processing capacity. For example, word recognition and meaning activation is largely preserved with age (Laver & Burke, 1993), presumably reflecting years of reading experience (Logan, 1997; Stanovich et al., 1995). Moreover, more elaborate word knowledge among highly literate older adults facilitates recognition of longer and less frequent words (Chin et al., 2015). These largely automatic processes help explain age-related preservation of everyday language use (Stine-Morrow & Miller, 2009) and are consistent with optimization and skill maintenance strategies. Over time, older adults' accumulating reading experience increases knowledge of health (Beier & Ackerman, 2005), politics (Trepte & Schmitt, 2017), and other domains. This growth in turn facilitates further learning and knowledge use (e.g., to make decisions), presumably by decreasing the effort involved in comprehension and offsetting processing capacity constraints (Chin et al., 2017; Li et al., 2013; Payne, Gao, Noh, Anderson, & Stine-Morrow, 2012).

Age differences in comprehension become more likely as cognitive resource demands of comprehension increase, including demands of knowledge use (e.g., integrating activated knowledge in working memory with text information). For example, older adults' memory for text about familiar topics (e.g., schema for ordering in a restaurant) improves at least as much as younger adults do because they readily leverage their knowledge (for review: Hess, 1990; Umanath & Marsh, 2014). Thus, age differences are minimal when inferences that relate text information and knowledge require few resources, such as filling in a default role in a schema (e.g., "pound a nail" → hammer; Stine-Morrow & Miller, 2009). As more effort is needed to integrate prior knowledge with text, older adults may adopt compensatory strategies to maintain knowledge-based benefit. For example, when more effortful inferences are required, older adults can devote more effort than younger adults do in order to draw knowledge-based inferences (Chin et al., 2015; Miller, Stine-Morrow, Krikorian, & Conroy, 2004). However, knowledge is less likely to mitigate age differences when the relationship between text and knowledge is less apparent. Thus, while older adults are as likely as younger adults to remember schema-irrelevant information in stories (e.g., wearing a ring at a restaurant), they are less likely to remember schema-inconsistent information (e.g., the waiter rather than customer orders a meal). In the latter case, older adults are more likely to incorrectly remember the information as consistent with the schema (Hess, 1990). Similar results have been found for numeric information: age differences are greater for remembering unrealistic compared to realistic prices for grocery items (Castel, 2005). Presumably, more processing capacity is required to reconcile text information with knowledge (a *strenuous inference*; Miller et al., 2004) in these cases. Similarly, age differences are not mitigated by domain knowledge when tasks focus on recalling but not using information to accomplish domain goals. For such tasks, more processing capacity (e.g., working memory) improves performance to the same extent and accounts for same amount of age-related variance in performance at all levels of expertise (Hambrick & Engle, 2002; also see Arbuckle, Vanderleck, Harsany, & Lapidus, 1990; Morrow et al. 1992).

More generally, older adults sometimes over-rely on schemas to remember text, leading to errors such as incorrectly thinking schema-consistent information was part of the studied information, perhaps in an (unsuccessful) attempt to compensate for processing demands of comprehension (for a review see Umanath & Marsh, 2014). In other words, knowledge-related costs as well as benefits for memory can increase with age (Castel et al., 2012).

Beyond the text-processing literature, many complex domains involving communication, situation awareness, and decision making highlight the role of intensive practice as a form of optimization, as well as the use of selection and compensation strategies among older experts. For example, age differences in flight simulator performance among expert pilots are minimal for routine procedural skills such as maintaining level flight (e.g., Taylor et al., 2007). Age differences in driving are less likely to occur for highly routinized aspects of driving such as lane control (under low demand conditions such as daylight driving on straight roads with light traffic), which may help explain why accident rates are higher for younger inexperienced drivers than for older drivers even into their 70s (Tefft, 2017).

Expertise can also mitigate age differences in allocating attention to multiple tasks involving perceptual-motor skills in flying or driving. Age differences are smaller for pilots than for nonpilots when performing concurrent tasks such as manual tracking and short-term memory search, presumably because time-sharing mechanisms (e.g., flexibly allocating attention in response to shifting task priority) are supported by experience performing concurrent piloting tasks (Tsang & Shaner, 1998). However, age differences emerge for more complex cognitive components of flying, driving, and other complex tasks related to communication and decision making. Taylor et al. (2007) found that age differences in flight simulator performance (regardless of expertise level) were more apparent for demanding air traffic control (ATC) communication tasks than for routine maneuvers involving perceptual-motor skills, presumably because the communication tasks taxed working memory. Similarly, age differences were comparable for pilots and nonpilots on a readback task (pilots read back ATC instructions during flight to ensure mutual understanding between pilots and ATC), and measures of working memory accounted for much of the age-related variance for both groups (Morrow, Leirer, Altieri, & Fitzsimmons, 1994; Taylor, O'Hara, Mumenthaler, Rosen, & Yesavage, 2005). However, expertise may mitigate age differences when experts can rely on compensation strategies developed on the job (see discussion later in this section).

Open-ended domains enable selection as well as optimization strategies rooted in experience with complex daily tasks. Experts excel at identifying relevant cues that define specific task situations and responding to the situations based on these cues (Loveday, Wiggins, & Searle, 2014; Wiggins & O'Hare, 1995). Cue recognition in typical situations supports situation awareness and rapid, efficient decision making among firefighters, pilots, and other experts (Klein, 1998; Schriver, Morrow, Wickens, & Talleur, 2008). Older experts may maintain this ability to focus on and respond to relevant cues based on their accumulating domain experience. For example, older and younger airline and novice pilots in one study read flight scenarios at their own pace and then identified problems and how they would respond if they were pilot in command in the scenario (Morrow et al., 2009). The older airline pilots were as likely as their younger counterparts to identify appropriate solutions, while age differences occurred among the novice pilots. Younger and older experts were also more likely than novices to slow down when reading problem-relevant information in the scenarios, consistent with evidence that experts focus on relevant cues. More generally, expertise may more likely mitigate age differences in cue recognition than in decision making and response based on the cues, perhaps because the latter involves more processes than cue recognition such as generating and evaluating options, which require more cognitive resources (Gobet & Simon, 1996). For example, chess expertise is more likely to mitigate age effects on recognizing chess configurations (Charness, 1981) than for looking ahead/planning moves (Jastrzembski et al., 2006).

More generally, older adults are also adept at allocating limited cognitive resources to the information most relevant to their goals. According to the value-directed remembering model, older adults are able to prioritize processing of the information identified as most valuable (Castel et al., 2012). This ability may be supported by domain knowledge. For example, age differences in memory for health information that vary in value (e.g., food allergens and benefits) are minimal for high value information (e.g., Middlebrooks, McGillivary, Murayama, & Castel, 2016). Older adults also focus on and remember more important information in complex text, in part because extensive language experience sensitizes them to signals such as speech prosody and signal words (for review see Stine-Morrow & Miller, 2009). Like other adaptive strategies, selection hinges on metacognitive abilities such as awareness of cognitive limits and the need to identify information most relevant to current goals (Castel et al., 2012).

In addition to optimization and selection strategies, older adults learn to compensate for age-related loss by drawing on experience to modify existing strategies or to develop new ways to more efficiently accomplish tasks. A pervasive example is using glasses and hearing aids to enhance sensory input and offset age-related sensory declines. While audibility is the most important determinant of individual differences in comprehension for unaided listening, it is less important than cognitive factors for aided speech (Humes, 2007).

Older experts are especially likely to adopt compensatory strategies to offset sensory and cognitive declines for domain-relevant tasks. Age-related declines among medical technicians were reduced on a domain-relevant visual search task (search for bacterial targets in a stain) but not on a domain-general visual search task (Clancy & Hoyer, 1994). This mitigation was more likely when target identification was supported by congruent contextual cues (Hoyer & Ingolfsdottir, 2003), providing further evidence that expertise-based mitigation depends on the domain-relevance of the task. Similarly, expertise mitigation among ATC specialists in one study was most likely to occur for more realistic tasks requiring complex task management skills such as issuing instructions to aircraft, even though these experts showed the same age-related declines in performance as nonexpert participants on measures of the cognitive abilities required for the ATC tasks (Nunes & Kramer, 2009). Older ATC experts achieved similar operational error rates as younger experts, but issued fewer commands to aircraft, suggesting more efficient control as a potential compensatory strategy.

Financial experts also sustain high levels of competence through compensation strategies. In a study of bank managers varying in age and expertise (measured by outcomes such as salary and promotions, as well as supervisor ratings), age was unrelated to performance on a measure of management knowledge for the most expert participants, while age-related declines occurred on a fluid mental ability measure regardless of expertise level. Moreover, management knowledge predicted successful management outcomes for the experts (Colonia-Willner, 1998). More generally, older adults are more likely than younger adults to engage in a variety of compensation (as well as optimization) strategies at work, and strategy use is highly associated with older workers' performance (Abraham & Hansson, 1995; Yeung & Fung, 2009). However, very old workers (or older workers confronted by highly challenging tasks) may be less likely to actively manipulate their environment to reduce task demands (e.g., developing reminding strategies or actively collaborating with co-workers) and instead shift to strategies such as taking on fewer responsibilities (Yeung & Fung, 2009).

As with sport and game domains, mitigation is less likely for less domain-relevant work-related tasks—it does not occur when performance in work domains that require spatial ability (e.g., architecture, graphic design) is measured by standard spatial ability tests (visualization, mental rotation) rather than authentic work tasks (Lindenberger, Kleigl, & Baltes, 1992; Salthouse, 1991; Salthouse, Babcock Skovronek, Mitchell, & Palmon, 1990).

Knowledge-based compensation strategies support older adults' performance at home as well as at work. General knowledge/verbal ability predicts everyday task performance across age groups, with some evidence that knowledge explains more variance in older than younger adult performance, suggesting an age-related shift toward relying on knowledge to perform tasks as a compensatory strategy (Chen et al., 2017). However, age differences in this study were not reduced among older adults who used knowledge more (no age × general knowledge interaction). An example of compensation in everyday decision making is older adults' use of heuristic decision-making strategies that reduce processing capacity demand. Such strategies may be more likely when supported by domain knowledge (Meyer, Talbot, & Ranalli, 2007) and emotional information (Mikels et al., 2010). Emotion-based decisions may be most effective when supported by knowledge that supports gist comprehension of choice situations that capture important evaluative relations (Reyna, 2011; Wilhelms, Corbin, & Reyna, 2014). Older adults may even outperform younger adults on some decision tasks by leveraging their knowledge of social and emotional relationships (*wisdom*-based strategies; Baltes, Staudinger, & Lindenberg, 1999; Grossman et al., 2010).

A pervasive form of compensation is use of external context. Everyday performance often emerges from *distributed cognition systems*, in which effortful mental processes (e.g., search, computation) are offloaded to external representations (artifacts such as notes; procedures such as counting on fingers; Hutchins, 1995). Technology provides ubiquitous external support (e.g., spell check, GPS navigation aids). Consistent with the concept of environmental support (Morrow & Rogers, 2008), older adults offload task demands on limited cognitive resources onto external representations. Here, we consider how expertise can enhance environmental support. With experience, older adults learn to craft their environment to offload cognitive task demands. Experienced pilots take notes or set reminders on displays to reduce the working memory demands of responding to ATC instructions during flight. Age differences in readback accuracy were eliminated when pilots took notes while listening to ATC instructions (Morrow et al., 2003; Morrow & Schriver, 2007). Age differences were not eliminated either for non-pilot participants, who were less proficient in note-taking, nor for pilots who were not given the opportunity to take notes (Morrow et al., 2003; also see Taylor et al., 2005). More generally, external aids reduce age differences in remembering to perform important tasks at work and home (that is, prospective memory). Age differences are often reduced for everyday prospective memory tasks (e.g., taking medication) compared to similar tasks performed in the lab, perhaps because of greater opportunity and motivation to use external reminders in daily life (Henry, MacLeod, Phillips, & Crawford, 2004).

Finally, older experts draw on collaborative strategies to compensate for age-related limitations. Complex work is increasingly done by teams due in part to technologies that support distributed work (Morrow & Fischer, 2013). Even though age-related declines in individual resources sometimes undermine older adults' ability to collaborate (for example, hearing declines impair communication; Gordon-Salant, 2005), social interaction becomes more important across the lifespan because of shifts toward social and emotional goals (Charles & Carstensen, 2010). Collaboration is a powerful way to compensate for age-related declines in individual cognitive resources because older adults draw on social resources. Because of strategies such as partners cueing each other's retrieval of information or monitoring each other's performance, age differences may be reduced for group compared to individual performance on memory and problem-solving tasks (Gould, Kurzman, & Dixon, 1994). At a minimum, older and younger adults can similarly benefit from collaboration (Henkel & Rajaram, 2011). Older experts may especially benefit from structured procedures that support teamwork in aviation, health care, and other work domains. This is suggested by the finding that young expert pilots benefit from collaboration on memory and decision-making tasks more than novices or nonexperts do, in part because they take advantage of strategies such as cross-cueing retrieval and developing a shared mental model (Meade, Nokes, & Morrow, 2009). Older adults may be just as, or more, likely than younger adults to leverage domain expertise to promote collaborative benefits (Henkel & Rajaram, 2011).

Conclusions

Expertise, defined as superior levels of performance on representative tasks or more broadly as improved performance with increasing task-related experience and knowledge, confers a range of benefits such as accurate comprehension, decision making, and problem solving performance. At the same time, expertise may be accompanied by costs such as decreased flexibility in responding to variable task conditions, sometimes reflecting over-reliance on knowledge, familiar responses, and strategies tuned to specific situations. Because expertise benefits tend to reflect increased efficiency and reduced processing capacity constraints on performance, these benefits may increase with age to the extent that expert strategies offset age-related resource declines. We reviewed the literature on expertise and age from the perspective of lifespan theory, which focuses on the interplay of age-related gain and loss and older adults' strategies for maintaining function by managing loss, often by leveraging knowledge and other experience-related resources.

We reviewed evidence for knowledge-based mitigation in two broad domains: (1) structured (rule-governed) domains in which performance is highly adapted to perceptual-motor constraints (e.g., sports) as well as structured domains in which cognitive components often play a larger role (games, music); (2) more open-ended domains related to work and daily competence that involve complex cognitive abilities such as decision making and problem solving. The latter domains may afford a broader range of knowledge-based strategies that support mitigation of age differences. While the literature is decidedly mixed as to whether or in what circumstances expertise mitigates age differences in performance, there was some evidence that mitigation occurs for a variety of tasks in which older adult performance depends heavily on highly practiced perceptual-motor skills. Mitigation is less likely for tasks that require more age-vulnerable abilities (e.g., physical strength, speeded response, working memory capacity), and/or when the cost of knowledge use is high (e.g., drawing extensive inferences to integrate new information with prior knowledge). Under these latter conditions, costs of expertise (e.g., over-reliance on schemas and familiar responses) tend to outweigh benefits of knowledge. Mitigation may be most likely for highly domain-relevant tasks that engage maintained skills reinforced by optimization strategies; that support selection strategies (focusing on the subset of information and responses most connected to successful performance); and that underpin compensation strategies.

If maintained knowledge is a key to successful performance among older adults, then a potential drawback of knowledge-based mitigation lies in the fact that knowledge is often highly dynamic, especially in technology-driven domains (e.g., medicine, aviation, finance). Older experts may be challenged by the need to continually update their knowledge to the extent they rely on knowledge-based strategies to offset declines in other cognitive resources. Inhibiting outmoded knowledge may be impaired by declines in inhibition efficiency (Kramer & Madden, 2008), and learning new concepts or skills may be challenging to the extent they differ from or even contradict prior knowledge because older experts are unable to rely on knowledge-based efficient learning strategies. Examples include age-related differences in accepting new technology in aviation (e.g., flight management systems on commercial aircraft) and medicine (e.g., electronic health records), where evolving technology can differ substantially from earlier systems. However, there is also evidence that older adults can learn new skills and concepts to use new systems. They learn new technology applications despite typical age declines in processing capacity, and prior experience with technology is a robust predictor of learning new computer-based skills (e.g., Czaja et al., 2013). Moreover, technology itself facilitates knowledge acquisition by supporting rapid access to updated knowledge (e.g., computer-aided decision support in medicine).

Our review also suggests mitigation depends on the extent to which knowledge use is effortful. Therefore, mitigation may be more likely, or extend to a broader range of contexts, when this effort is reduced. Tailoring training strategies to older adult abilities and interests (e.g., more opportunity to practice skills) may allow older adults to incorporate newly automated skills into their expert repertoire and therefore broaden the range of conditions for optimization strategies. Selection-based strategies are enhanced by designing displays and instructions to minimize distracting or irrelevant information and perceptually enhance the most relevant cues. Finally, leveraging the environment to promote compensation strategies is a promising approach to support expertise-based mitigation, given the ubiquity of technology for offloading task demands as well as advances in using technology to capitalize on older adults' prior experience and to reduce cognitive load associated with training. Smart phone technology provides promising examples. Phone-based applications for physician treatment decisions could combine rapid access to updated illness and treatment information (reducing effort involved in activating and updating task-relevant knowledge) with a flexible *decision-making* workspace in which their expert-based gist understanding of most relevant treatment options can be reviewed in light of this updated information. On the patient side, apps could provide ready access to computer agent-based systems that deliver new illness and treatment information that builds on patients' current understanding of their illness. As with face-to-face communication, the agent would use nonverbal (e.g., tone of voice, facial expressions) as well as verbal cues to convey affective and cognitive meaning, thus meshing with the socioemotional goals that gain importance across the lifespan. Because smartphone technology provides detailed information about user interaction, research would not only be able to evaluate tool-based outcomes (relative to control conditions), but investigate how the tools engage participants varying in age and expertise, and thus shed light on strategies that underlie expertise-based mitigation of age differences.

Acknowledgments

Preparation of this chapter was supported by Agency for Healthcare Research and Quality (AHRQ) Grant R21HS022948, and the Jump Applied Research for Community Health through Engineering and Simulation (ARCHES) program, UIUC/OSF Hospital, Peoria IL. Any opinions, findings, and conclusions or recommendations expressed in this publication are those of the authors and do not necessarily reflect the views of the funding agencies.

References

Abraham, J. D., & Hansson, R. O. (1995). Successful aging at work: An applied study of selection, optimization, and compensation through impression management. *Journals of Gerontology Series B: Psychological Sciences and Social Sciences* 50(2), P94–P103.

Ackerman, P. L. (2007). New developments in understanding skilled performance. *Current Directions in Psychological Science* 16(5), 235–239.

Alves, H., Voss, M., Boot, W. R., Deslandes, A., Cossich, V., Inacio Salles, J., & Kramer, A. F. (2013). Perceptual-cognitive expertise in elite volleyball players. *Frontiers in Psychology* 4, 36.

Arbuckle, T. Y., Vanderleck, V. F., Harsany, M., & Lapidus, S. (1990). Adult age differences in memory in relation to availability and accessibility of knowledge-based schemas. *Journal of Experimental Psychology: Learning, Memory, and Cognition* 16(2), 305–315.

Arkes, H. R., & Freedman, M. R. (1984). A demonstration of the costs and benefits of expertise in recognition memory. *Memory and Cognition* 12, 84–89.

Badham, S. P., & Maylor, E. A. (2015). What you know can influence what you are going to know (especially for older adults). *Psychonomic Bulletin & Review* 22(1), 141–146.

Baltes, P. B., & Baltes, M. M. (1990). Psychological perspectives on successful aging: The model of selective optimization with compensation. *Successful aging: Perspectives from the Behavioral Sciences* 1, 1–34.

Baltes, P. B., Staudinger, U. M., & Lindenberger, U. (1999). Lifespan psychology: Theory and application to intellectual functioning. *Annual Review of Psychology* 50(1), 471–507.

Beier, M. E., & Ackerman, P. L. (2005). Age, ability and the role of prior knowledge on the acquisition of new domain knowledge. *Psychology & Aging* 20, 341–355.

Bernardi, G., Ricciardi, E., Sani, L., Gaglianese, A., Papasogli, A., Ceccarelli, R., . . . Pietrini, P. (2013). How skill expertise shapes the brain functional architecture: An fMRI study of visuo-spatial and motor processing in professional racing-car and naïve drivers. *PLoS ONE* 8(10): e77764.

Bosman, E. A. (1993). Age-related differences in motoric aspects of transcription typing skill. *Psychology and Aging* 8, 87–102.

Bosman, E. A., & Charness, N. (1996). Age-related differences in skilled performance and skill acquisition. In F. Blanchard-Fields & T. M. Hess (Eds.), *Perspectives on cognitive change in adulthood and aging* (pp. 428–453). New York: McGraw-Hill.

Brunswik, E. (1956). *Perception and the representative design of psychological experiments*, 2nd ed. Berkeley: University of California Press.

Castel, A. D. (2005). Memory for grocery prices in younger and older adults: the role of schematic support. *Psychology and Aging* 20(4), 718–721.

Castel, A. D., McGillivray, S., & Friedman, M. C. (2012). Metamemory and memory efficiency in older adults: Learning about the benefits of priority processing and value-directed remembering. In M. Naveh-Benjamin & N. Ohta (Eds.), *Memory and aging: Current issues and future directions* (pp. 245–270). New York: Psychology Press.

Charles, S. T., & Carstensen, L. L. (2010). Social and emotional aging. *Annual Review of Psychology* 61, 383–409.

Charness, N. (1981). Visual short-term memory and aging in chess players. *Journal of Gerontology* 36, 615–619.

Charness, N., & Dijkstra, K. (1999). Age, luminance, and print legibility in homes, offices, and public places. *Human Factors* 41(2), 173–193.

Charness, N., Reingold, E. M., Pomplun, M., & Stampe, D. M. (2001). The perceptual aspect of skilled performance in chess: Evidence from eye movements. *Memory and Cognition 29*, 1146–1152.

Chase, W. G., & Ericsson, K. A. (1982). Skill and working memory. *Psychology of Learning and Motivation 16*, 1–58.

Chase, W. G., & Simon, H. A. (1973). Perception in chess. *Cognitive Psychology, 4*(1), 55–81

Chen, X., Hertzog, C., & Park, D. C. (2017). Cognitive predictors of everyday problem solving across the lifespan. *Gerontology 63*(4), 372–384.

Chi, M. T. H. (2006a). Two approaches to the study of experts' characteristics. In K. A. Ericsson, N. Charness, P. Feltovich, & R. Hoffman (Eds.), *The Cambridge handbook of expertise and expert performance* (pp. 121–30), New York: Cambridge University Press.

Chi, M. T. H. (2006b). Methods to assess the representations of experts' and novices' knowledge. In K. A. Ericsson, N. Charness, P. Feltovich, & R. Hoffman (Eds.), *The Cambridge handbook of expertise and expert performance* (pp. 167–184). New York: Cambridge University Press.

Chi, M. T. H., Feltovich, P., & Glaser, R. (1981). Categorization and representation of physics problems by experts and novices. *Cognitive Science 5*, 121–152.

Chin, J., D'Andrea, L., Morrow, D., Stine-Morrow, E. A., Conner-Garcia, T., Graumlich, J., & Murray, M. (2009). Cognition and illness experience are associated with illness knowledge among older adults with hypertension. In *Proceedings of the Human Factors and Ergonomics Society Annual Meeting* (Vol. 53, No. 2, pp. 116–120). Santa Monica: Human Factors & Ergonomics Society.

Chin, J., Madison, A., Stine-Morrow, E. A. L., Gao, X., Graumlich, J. F., Murray, M. D., ... Morrow, D. G. (2017). Cognition and health literacy in older adults' recall of self-care information. *The Gerontologist 57*, 261–268.

Chin, J., Payne, B., Gao, X., Conner-Garcia, T. Graumlich, J., Murray, M. D., ... Stine-Morrow, E. A. L. (2015). Memory and comprehension for health information among older adults: Distinguishing the effects of domain-general and domain-specific knowledge. *Memory 23*, 577–589.

Clancy, S. M., & Hoyer, W. J. (1994). Age and skill in visual search. *Developmental Psychology 30*, 545–552.

Colcombe, S. J., Kramer, A. F., Erickson, K. I., Scalf, P., McAuley, E., Cohen, N. J., ... Elavsky, S. (2004). Cardiovascular fitness, cortical plasticity, and aging. *Proceedings of the National Academy of Sciences USA 101*(9), 3316–3321.

Colonia-Willner, R. (1998). Practical intelligence at work: Relationship between aging and cognitive efficiency among managers in a bank environment. *Psychology and Aging 13*, 45–57.

Craik, F. I. (2000). Age-related changes in human memory. *Cognitive Aging: A Primer 5*, 75–92.

Czaja, S. J., Sharit, J., Lee, C. C., Nair, S. N., Hernández, M. A., Arana, N., & Fu, S. H. (2013). Factors influencing use of an e-health website in a community sample of older adults. *Journal of the American Medical Informatics Association 20*(2), 277–284.

Dreyfus, H. L., & Dreyfus, S. E. (1986). *Mind over machine: The power of human intuition and expertise in the era of the computer*. New York: The Free Press.

Ericsson, K. A., & Kintsch, W. (1995). Long-term working memory. *Psychological Review, 102* 211–245.

Ericsson, K. A., & Lehmann, A.C. (1996). Expert and exceptional performance: Evidence of maximal adaptation to task constraints. *Annual Review of Psychology 47*, 273-305.

Feltovich, P. J., Spiro, R. J., & Coulson, R. L. (1997). Issues of expert flexibility in contexts characterized by complexity and change. In P. J. Feltovich, K. M. Ford, & R. R. Hoffman (Eds), *Expertise in context: Human and machine* (pp. 125-146). Menlo Park, CA, US: American Association for Artificial Intelligence; Cambridge, MA: MIT Press.

Finucane, M. L., Mertz, C. K., Slovic, P., & Schmidt, E. S. (2005). Task complexity and older adults' decision-making competence. *Psychology and Aging 20*(1), 71-84.

Fisk, A. D., Rogers, W. A., Charness, N., Czaja, S. J., & Sharit, J. (2009). *Designing for older adults: Principles and creative human factors approaches*. Boca Raton, FL: CRC Press.

Gobet, F., & Chassy, P. (2009). Expertise and intuition: A tale of three theories. *Minds & Machines 19*, 151-180.

Gobet, F., & Simon, H. A. (1996). Templates in chess memory: a mechanism for recalling several boards. *Cognitive Psychology 31*(1), 1-40.

Gordon-Salant, S. (2005). Hearing loss and aging: new research findings and clinical implications. *Journal of Rehabilitation Research and Development 42*(4), 9-24.

Gould, O., Kurzman, D., & Dixon, R. A. (1994). Communication during prose recall conversations by young and old dyads. *Discourse Processes 17*(1), 149-165.

Groen, G. J., & Patel, V. L. (1988). The relationship between comprehension and reasoning in medical expertise. In M. T. H. Chi, R. Glaser, & M. Farr (Eds.), *The nature of expertise* (pp. 287-310). Hillsdale, NJ: Erlbaum.

Grossmann, I., Na, J., Varnum, M. E., Park, D. C., Kitayama, S., & Nisbett, R. E. (2010). Reasoning about social conflicts improves into old age. *Proceedings of the National Academy of Sciences 107*(16), 7246-7250.

Halpern, A. R., Bartlett, J. C., & Dowling, W. J. (1995). Aging and experience in the recognition of musical transpositions. *Psychology and Aging 10*, 325-342.

Hambrick, D. Z., & Engle, R. W. (2002). Effects of domain knowledge, working memory capacity, and age on cognitive performance: An investigation of the knowledge-is-power hypothesis. *Cognitive Psychology 44*, 339-387.

Hartshorne, J. K., & Germine, L. T. (2015). When does cognitive functioning peak? The asynchronous rise and fall of different cognitive abilities across the life span. *Psychological Science 26*(4), 433-443.

Henkel, L. A., & Rajaram, S. (2011). Collaborative remembering in older adults: Age-invariant outcomes in the context of episodic recall deficits. *Psychology & Aging 26*(3), 532-545.

Henry, J. D., MacLeod, M. S., Phillips, L. H., & Crawford, J. R. (2004). A meta-analytic review of prospective memory and aging. *Psychology and Aging 19*(1), 27-39.

Hess, T. M. (1990). Aging and schematic influences on memory. In T. M. Hess (Ed.), *Advances in Psychology*, Vol. 71, *Aging and cognition: Knowledge, organization and utilization* (pp. 93-160). Oxford: North-Holland.

Hoffman, R. R., Shadbolt, N. R., Burton, A. M., & Klein, G. (1995). Eliciting knowledge from experts: A methodological analysis. *Organizational Behavior and Human Decision Processes 62*, 129-158.

Hoffman, R. R., Ward, P., Feltovich, P. J., DiBello, L., Fiore, S. M., & Andrews, D. (2014). *Accelerated expertise: Training for high proficiency in a complex world*. New York: Psychology Press.

Horn, J. L., & Cattell, R. B. (1967). Age differences in fluid and crystallized intelligence. *Acta Psychologica 26*, 107-129.

Hoyer, W. J., & Ingolfsdottir, D. (2003). Age, skill, and contextual cuing in target detection. *Psychology and Aging 18*, 210–218.

Humes, L. E. (2007). The contributions of audibility and cognitive factors to the benefit provided by amplified speech to older adults. *Journal of the American Academy of Audiology 18*, 609–623.

Hutchins, E. (1995). How a cockpit remembers its speed. *Cognitive Science 19*, 265–288.

Jastrzembski, T., Charness, N., & Vasyukova, C. (2006). Expertise and age effects on knowledge activation in chess. *Psychology and Aging 21*, 401–405.

Kilb, A., & Naveh-Benjamin, M. (2007). Paying attention to binding: Further studies assessing the role of reduced attentional resources in the associative deficit of older adults. *Memory & Cognition 35*(5), 1162–1174.

Kintsch, W. (1998). *Comprehension: A paradigm for cognition*. New York: Cambridge University Press.

Klein, G. (1998). *Sources of power: How people make decisions*. Cambridge, MA: MIT Press.

Kramer, A. F., & Madden, D. (2008). Attention. In F. I. M. Craik & T. A. Salthouse (Eds.), *The handbook of aging and cognition*, 3rd ed. (pp. 189–250). New York: Psychology Press.

Krampe, R. T., & Charness, N. (2006). Aging and expertise. In K. A. Ericsson, N. Charness, P. Feltovich, & R. Hoffman (Eds.), *Cambridge Handbook of Expertise and Expert Performance* (pp. 723–742). Cambridge: Cambridge University Press.

Krampe, R. T., Engbert, R., & Kliegl, R. (2002). The effects of expertise and age on rhythm production: Adaptations to timing and sequencing constraints. *Brain & Cognition 48*, 179–194.

Krampe, R. & Ericsson, K. A. (1996). Maintaining excellence: Deliberate practice and elite performance in younger and older pianists. *Journal of Experimental Psychology: General 125*, 331–359.

Laver, G. D., & Burke, D. M. (1993). Why do semantic priming effects increase in old age? A meta-analysis. *Psychology and Aging 8*(1), 34–43.

Lesgold, A. M., Rubinson, H., Feltovich, P., Glaser, R., Klopfer, D., & Wang, Y. (1988). Expertise in a complex skill: Diagnosing X-ray pictures. In M. T. H. Chi, R. Glaser, & M. J. Farr (Eds.), *The nature of expertise* (pp. 311–342). Hillsdale, NJ: Lawrence Erlbaum.

Lewandowsky, S., & Kirsner, K. (2000). Knowledge partitioning: context-dependent use of expertise. *Memory & Cognition 28*, 295–305.

Li, Y., Baldassi, M., Johnson, E. J., & Weber, E. U. (2013). Complementary cognitive capabilities, economic decision making, and aging. *Psychology & Aging 28*, 595–613.

Lindenberger, U., Kliegl, R., & Baltes, P. B. (1992). Professional expertise does not eliminate age differences in imagery-based memory performance during adulthood. *Psychology and Aging 7*, 585–593.

Lobjois, R., Benguigui, N., & Bertsch, J. (2005). Aging and tennis playing in a coincidence-timing task with an accelerating object: The role of visuomotor delay. *Research Quarterly for Exercise and Sport 76*, 398–406.

Lobjois, R., Benguigui, N., Bertsch, J., & Broderick, M. P. (2008). Collision avoidance behavior as a function of aging and tennis playing. *Experimental Brain Research 184*(4), 457–468.

Logan, G. D. (1997). Automaticity and reading: Perspectives from the instance theory of automatization. *Reading & Writing Quarterly: Overcoming Learning Difficulties 13*(2), 123–146.

Loveday, T., Wiggins, M. W., & Searle, B. J. (2014). Cue utilization and broad indicators of workplace expertise. *Journal of Cognitive Engineering and Decision Making 8*(1), 98–113.

Luong, G., Charles, S. T., & Fingerman, K. (2011). Better with age: Social relationships across adulthood. *Journal of Social and Personal Relationships 28*, 9–23.

Marchant, G., Robinson, J. P., Anderson, U., & Shadewald, M. (1991). Analogical transfer and expertise in legal reasoning. *Organizational Behavior and Human Decision Processes 48*, 272–290.

Masunaga, H., & Horn, J. (2001). Expertise and age-related changes in components of intelligence. *Psychology and Aging 16*, 293–311.

Meade, M. L., Nokes, T. J., & Morrow, D. G. (2009). Expertise promotes facilitation on a collaborative memory task. *Memory 17*(1), 39–48.

Meinz, E. J. (2000). Experience-based attenuation of age-related differences in music cognition tasks. *Psychology & Aging 15*(2), 297–312.

Meinz, E. J., & Salthouse, T. A. (1998). The effects of age and experience on memory for visually presented music. *Journal of Gerontology: Psychological Sciences 53*, 60–69.

Meyer, A. N. D., Payne, V. L., Meeks, D. W., Rao, R, & Singh, H. (2013). Physicians' diagnostic accuracy, confidence, and resource requests: A vignette study. *The Journal of the American Medical Association (JAMA) Internal Medicine 173*(21): 1952–1958.

Meyer, B. J., Talbot, A. P., & Ranalli, C. (2007). Why older adults make more immediate treatment decisions about cancer than younger adults. *Psychology & Aging 22*(3), 505–524.

Middlebrooks, C. D., McGillivray, S., Murayama, K., & Castel, A. D. (2016). Memory for allergies and health foods: How younger and older adults strategically remember critical health information. *Journals of Gerontology Series B: Psychological Sciences and Social Sciences 71*(3), 389–399.

Mikels, J. A., Löckenhoff, C. E., Maglio, S. J., Carstensen, L. L., Goldstein, M. K., & Garber, A. (2010). Following your heart or your head: Focusing on emotions versus information differentially influences the decisions of younger and older adults. *Journal of Experimental Psychology:Applied 16*, 87–95.

Miller, L. M. S., Stine-Morrow, E. A. L., Kirkorian, H., & Conroy, M. (2004). Age differences in knowledge-driven reading. *Journal of Educational Psychology 96*, 811–821.

Mol, S. E., & Bus, A. G. (2011). To read or not to read: a meta-analysis of print exposure from infancy to early adulthood. *Psychological bulletin 137*(2), 267–296.

Moore, C. G., & Müller, S. (2014). Transfer of expert visual anticipation to a similar domain. *Quarterly Journal of Experimental Psychology 67*, 186–196.

Morrow, D. G., & Fischer, U. M. (2013). Communication in socio-technical systems. In John D. Lee and Alex Kirlik (Eds.), *Oxford handbook of cognitive engineering* (pp. 178–199). New York: Oxford University Press.

Morrow, D. G., Leirer, V. O., & Altieri, P. A. (1992). Aging, expertise, and narrative processing. *Psychology and Aging 7*, 376–388.

Morrow, D. G., Leirer, V., Altieri, P., & Fitzsimmons, C., (1994). When expertise reduces age differences in performance. *Psychology and Aging 9*, 134–148.

Morrow, D. G., Miller, L. M. S., Ridolfo, H. E., Magnor, C., Fischer, U. M., Kokayeff, N. K., & Stine-Morrow, E. A. (2009). Expertise and age differences in pilot decision making. *Aging, Neuropsychology, and Cognition 16*(1), 33–55.

Morrow, D. G., Ridolfo, H. E., Menard, W. E., Sanborn, A., Stine-Morrow, E. A., Magnor, C., . . . & Bryant, D. (2003). Environmental support promotes expertise-based mitigation of age differences on pilot communication tasks. *Psychology & Aging 18*(2), 268–284.

Morrow, D. G., & Rogers, W. A. (2008). Environmental support: An integrative framework. *Human Factors 50*(4), 589–613.

Morrow, D. G., & Schriver, A. T. (2007). External support for pilot communication: Implications for age-related design. *International Journal for Cognitive Technology* 12(1), 21–30.

Nokes, T., Schunn, C., & Chi, M. T. H. (2010). Problem solving and human expertise. In P. Peterson, E. Baker, & B. McGaw (Eds.), *International encyclopedia of education*, 3rd ed. (pp. 265–272). Oxford: Elsevier.

Nunes, A., & Kramer, A. F. (2009). Experience-based mitigation of age-related performance declines: Evidence from air traffic control. *Journal of Experimental Psychology: Applied* 15(1), 12–24.

Patel, V. L., Arocha, J. F., & Kaufman, D. R. (1994). Diagnostic reasoning and medical expertise. *Psychology of Learning and Motivation: Advances in Research and Theory* 31, 187–252.

Patel, V. L., & Groen, G. J. (1991). The general and specific nature of medical expertise: A critical look. In K. A. Ericsson & J. Smith (Eds.), *Toward a general theory of expertise: Prospects and limits* (pp. 93–125). New York: Cambridge University Press.

Payne, B. R., Gao, X., Noh, S. R., Anderson, C. J., & Stine-Morrow, E. A. (2012). The effects of print exposure on sentence processing and memory in older adults: Evidence for efficiency and reserve. *Aging, Neuropsychology, and Cognition* 19(1–2), 122–149.

Peters, E. (2012). Beyond comprehension: The role of numeracy in judgments and decisions. *Current Directions in Psychological Science* 21(1), 31–35.

Peters, E., Hess, T. M., Västfjäll, D., & Auman, C. (2007). Adult age differences in dual information processes: Implications for the role of affective and deliberative processes in older adults' decision making. *Perspectives on Psychological Science* 2(1), 1–23.

Radvansky, G. A., & Dijkstra, K. (2007). Aging and situation model processing. *Psychonomic Bulletin & Review* 14(6), 1027–1042.

Reif, F., & Allen, S. (1992). Cognition for interpreting scientific concepts: A study of acceleration. *Cognition and Instruction* 9, 1–44.

Reyna, V. (2011). Across the life span. *Communicating risks and benefits: An evidence-based user's guide* (pp. 111–120). Silver Springs, MD: Federal Drug Administration.

Riediger, M., Li, S. C., & Lindenberger, U. (2006). Selection, optimization, and compensation as developmental mechanisms of adaptive resource allocation: Review and preview. *Handbook of the Psychology of Aging* 6, 289–313.

Rogers, W. A., Meyer, B., Walker, N., & Fisk, A. D. (1998). Functional limitations to daily living tasks in the aged: A focus group analysis. *Human Factors* 40(1), 111–125.

Roring, R. W., & Charness, N. (2007). A multilevel model analysis of expertise in chess across the life span. *Psychology and Aging* 22(2), 291–299.

Ross, K., Shafer, J. L., & Klein, G. (2006). Professional judgments and "naturalistic decision making." In K. A. Ericsson, N. Charness, P. J. Feltovich, & R. R. Hoffman (Eds.), *Cambridge handbook of expertise and expert performance: Its development, organization and content* (pp. 403–419), Cambridge: Cambridge University Press.

Rybash, J. M., Hoyer, W. J., & Roodin, P. A. (1986). *Adult cognition and aging: developmental changes in processing, knowing and thinking*. London: Pergamon Press.

Salthouse, T. A. (1984). Effects of age and skill in typing. *Journal of Experimental Psychology* 113, 345–371.

Salthouse, T. A. (1990). Influence of experience on age difference in cognitive functioning. *Human Factors* 32, 551–569.

Salthouse, T. A. (1991). *Theoretical perspectives on cognitive aging*. Hillsdale, NJ: Erlbaum.

Salthouse, T. A., Babcock, R., Skovronek, E., Mitchell, D., & Palmon, R. (1990). Age and experience effects in spatial visualization. *Developmental Psychology 26*, 128–136.

Schieber, F. (2006). Vision and aging. *Handbook of the Psychology of Aging 6*, 129–161.

Schooler, C., Mulatu, M. S., & Oates, G. (1999). The continuing effects of substantively complex work on the intellectual functioning of older workers. *Psychology and Aging 14*, 483–506.

Schooler, C., Mulatu, M. S., & Oates, G. (2004). Occupational self-direction intellectual functioning, and self-directed orientation in older workers: findings and implications for individuals and society. *American Journal of Sociology 110*, 161–197.

Schriver, A. T., Morrow, D. G., Wickens, C. D., & Talleur, D. A. (2008). Expertise differences in attentional strategies related to pilot decision making. *Human Factors 50*(6), 864–878.

Schultz, R., & Curnow, C. (1988). Peak performance and age among superathletes: Track and field, swimming, baseball, tennis, and golf. *Journal of Gerontology: Psychological Sciences 43*, P113–120.

Shanteau, J. (1992). How much information does an expert use? Is it relevant? *Acta Psychologica 81*, 75–86.

Simon, D. P., & Simon, H. A. (1978). Individual differences in solving physics problems. In R. Siegler (Ed.), *Children's thinking: What develops?* (pp. 325–348). Hillsdale, NJ: Erlbaum

Simon, H. A., & Gobet, F. (2000). Expertise effects in memory recall: Comments on Vicente and Wang (1998). *Psychological Review 107*, 593–600.

Spilich, G., Vesonder, G. T., Chiesi, H. L., & Voss, J. F. (1979). Text processing of domain-related information for individuals with high and low domain knowledge. *Journal of Verbal Learning and Verbal Behavior 18*, 275–290.

Stanovich, K. E., West, R. F., & Harrison, M. R. (1995). Knowledge growth and maintenance across the life span: The role of print exposure. *Developmental Psychology 31*(5), 811–826.

Stine-Morrow, E. A., & Miller, L. (2009). Aging, self-regulation, and learning from text. *Psychology of Learning and Motivation 51*, 255–296.

Taylor, J. L., Kennedy, Q., Noda, A., & Yesavage, J. A. (2007). Pilot age and expertise predict flight simulator performance: A three-year longitudinal study. *Neurology 68*(9), 648–654.

Taylor, J. L., O'Hara, R., Mumenthaler, M. S., Rosen, A. C., & Yesavage, J. A. (2005). Cognitive ability, expertise, and age differences in following air-traffic control instructions. *Psychology and Aging 20*, 117–132.

Tefft, B. C. (2017). Rates of motor vehicle crashes, injuries and deaths in relation to driver age, United States, 2014–2015. *AAA Foundation for Traffic Safety*.

Touron, D. R., Hoyer, W. J., & Cerella, J. (2001). Cognitive skill acquisition and transfer in younger and older adults. *Psychology and Aging 16*(4), 555–563.

Trepte, S., & Schmitt, J. B. (2017). The effects of age on the interplay between news exposure, political discussion, and political knowledge. *Journal of Individual Differences 38*, 21–28.

Tsang, P. S. & Shaner, T. L. (1998). Age, attention, expertise, and time sharing performance. *Psychology and Aging 13*, 323–347.

Tun, P. A., McCoy, S., & Wingfield, A. (2009). Aging, hearing acuity, and the attentional costs of effortful listening. *Psychology and Aging 24*(3), 761–766.

Umanath, S., & Marsh, E. J. (2014). Understanding how prior knowledge influences memory in older adults. *Perspectives on Psychological Science 9*(4), 408–426.

Vicente, K. J., & Wang, J. H. (1998). An ecological theory of expertise effects in memory recall. *Psychological Review 105*, 33–57.

Wiggins, M., & O'Hare, D. (1995). Expertise in aeronautical weather-related decision making: A cross-sectional analysis of general aviation pilots. *Journal of Experimental Psychology: Applied 1*(4), 305–320.

Wiley, J. (1998). Expertise as mental set: The effects of domain knowledge in creative problem solving. *Memory & Cognition 26*, 716–730.

Wilhelms, E. A., Corbin, J. C., & Reyna, V. F. (2014). Gist memory in reasoning and decision making: age, experience, and expertise. In V. Thompson & A. Feeney (Eds.), *Reasoning as memory*. New York: Psychological Press.

Wineburg, S. (1998). Reading Abraham Lincoln: an expert/expert study in the interpretation of historical texts. *Cognitive Science 22*, 319–346.

Wolf, M. S, Curtis, L. M, Wilson, E. A. H, Revelle, W., Waite, K. R, Smith, S. G., ... Baker, D. W. (2012). Literacy, cognitive function, and health: Results of the LitCog study. *Journal of General Internal Medicine 27*(10), 1300–1307.

Yeung, D. Y., & Fung, H. H. (2009). Aging and work: how do SOC strategies contribute to job performance across adulthood? *Psychology and Aging 24*(4), 927–940.

Zwaan, R. A. (1999). Situation models: The mental leap into imagined worlds. *Current Directions in Psychological Science 8*(1), 15–18.

CHAPTER 46

SKILL DECAY

The Science and Practice of Mitigating Skill Loss and Enhancing Retention

WINFRED ARTHUR, JR. AND
ERIC ANTHONY DAY

Introduction

THE recognized importance of knowledge and skill retention in the training and skill acquisition literature is consonant with its saliency and criticality in situations where individuals and teams receive initial training on skills and knowledge that they may not be required to use or have the opportunity to perform for extended periods of time (Arthur, Day, Bennett, & Portrey, 2013). Consequently, the identification and examination of the factors that enhance post-training skill retention is potentially of vital importance and value. To this end, the present chapter seeks to accomplish two primary goals. The first is to briefly review and discuss the phenomenon of knowledge and skill retention and decay, and then discuss and illustrate how research in this realm has been conducted by providing concrete examples of "how it is done." This is accomplished by presenting the "ideal" research design and then using that as a framework, discuss the factors that influence decay and retention, limiting them to only those over which the trainer has some control. Specifically, Arthur, Bennett, Stanush, and McNelly (1998) categorized a core set of major factors that influence skill decay or retention into task-related and methodological factors. Task-related factors are inherent characteristics of the task and are typically not amenable to modification by the trainer or researcher. Examples include characteristics such as the distinction between physical versus cognitive tasks, and whether said tasks are individual- or team-based. In contrast, methodological factors can be modified in the design of the training or learning environment to enhance retention. Examples of these factors include the degree of overlearning, and training/practice conditions. Our discussion is limited to methodological factors only.

The second goal builds on the review of the factors that influence decay to develop recommendations on how they can be leveraged in efforts to enhance retention and mitigate decay. Hence, we seek to demonstrate how the associated knowledge amassed in this literature might be translated into practice, intervention, and policy. Implicit to each of these goals is how to retain expertise and/or how to structure training at various levels of proficiency such that the relevant knowledge and skills are retained. However, the preponderance of empirical retention studies train college students or other novice participants for a relatively short period of time or to some designated level of task proficiency (mastery) on relatively simple tasks (or a small sampling of representative domain tasks) and then retest participants after a period of nonuse. Such studies fall short of examining the retention of established "experts" per se. Thus, given the noted differences between experts and other levels of proficiency in real-world performance domains (Ericsson, 2015; Hoffman, 1998), it needs to be acknowledged that the skill decay literature fails to consider or examine the role of expertise. In short, the empirical literature on the intersection between expertise and skill decay is practically nonexistent. With that said, where there is reasonable conceptual and logical support for doing so, this review will nevertheless speak to the potential role played by expertise.

Research Design and Methods: The *Ideal* versus the *Prototypical*

Figure 46.1 presents an illustration of what we would consider to be the "ideal" research design to examine issues pertaining to skill acquisition and retention. Such a design would be amenable to examining all levels of proficiency (expertise) with variations in the operationalizations of acquisition (much higher level for experts compared to nonexperts) and the length of the retention (nonuse) interval (longer for experts than nonexperts) to accommodate the varying levels of proficiency. So, as an example, one might implement such a design if one sought to compare the effectiveness of collaborative practice (experimental condition) versus individual practice (control condition) in terms of acquisition, retention, and reacquisition (e.g., Arthur, Day, Bennett, McNelly, & Jordan, 1997). This example is an experimental multigroup between-subjects design where if it is a laboratory-based study, the performance task would be a complex task with reasonable levels of ecological validity. Such a design can be extended to include a post-acquisition manipulation thought to influence retention and reacquisition (e.g., Kluge & Frank, 2014 (refresher interventions: practice, skill demonstration, symbolic rehearsal, and knowledge testing); Villado et al., 2013a (observations rehearsal)) or a manipulation of the length of the retention interval (e.g., Ebbinghaus, 1964; Villado et al., 2013b). Researchers may also couple these

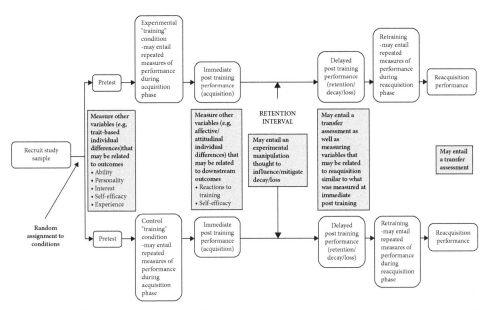

FIGURE 46.1 Illustration of the "ideal" research design to examine issues pertaining to skill acquisition and decay/retention.

experimental designs with examinations of how individual differences (e.g., ability, personality, motivation, and acquisition performance) relate to scores on tests of retention, reacquisition, and transfer after the retention interval (e.g., Day et al., 2013).

Using a design such as that illustrated in Figure 46.1, researchers can employ a mixed analysis of variance (ANOVA) to examine the interaction between the experimental manipulation and the time variable (i.e., repeated measures of performance), and compare the extent to which the experimental conditions differ with respect to acquisition, retention, and reacquisition. Alternatively, if a design involves repeated measures of performance during acquisition and reacquisition, then researchers can take advantage of more advanced statistical techniques involving discontinuous growth curve analysis (Bliese & Lang, 2016) to examine the relative degree of retention and reacquisition after the retention (nonuse) interval as well as how individual difference variables influence retention and reacquisition. Thus, advanced structural equation modeling involving latent curve analysis (Bollen & Curran, 2006; McArdle & Bell, 2000) holds considerable promise for simultaneously examining the role of experimental manipulations and multiple individual difference variables as they allow researchers to isolate the effects that variables have on retention and reacquisition independent of their effects on acquisition.

In terms of the organization of the subsequent topics in this section of the chapter, it generally follows the flow of the research design sequence, from the left side of Figure 46.1, starting with choice of the task to be trained to the assessment of reacquisition performance.

Task Complexity

In undertaking a training research study, a pivotal issue is the task on which participants are to be trained. The use of complex tasks in laboratory-based studies is important because the skill retention literature is predominantly characterized by the use of simple tasks such as word lists (Arthur et al., 1998). Consequently, the assumption that decay or retention of cognitively complex decision-making tasks is similar to that of simple tasks may be unfounded, and even more so in the context of studying expertise. Indeed, since Arthur et al.'s (1998) meta-analysis (see also Wang, Day, Kowollik, Schuelke, & Hughes, 2013), there has been a relatively small number of studies that have examined the decay of complex skills (e.g., Kluge & Frank, 2014; Stefanidis, Kordorffer, Markley, Sierra, & Scott, 2006; Woollard et al., 2006). Notwithstanding the small numbers, the general pattern of results from these studies indicate that the magnitude and rate of skill decay on complex tasks is quite different from that observed for simple tasks. For example, using a novice sample of university students and staff who were trained for 6 hours, Sauer, Hockey, and Wastell (2000) found that skill on their complex task, a computer-based simulation of the essential elements of a spacecraft's life support system, did not decay over a 32-week retention interval, indicating that the magnitude of decay for complex tasks and skills may not be the same as that observed for simple tasks. Arthur et al.'s (2010) results involving a computer-based command-and-control naval warfare simulation are consistent with this as well. Likewise, Wang et al.'s (2013) meta-analysis showed modestly better retention for more complex tasks—open- versus closed-loop tasks in particular.

In contrast to the older skill-decay research, especially that embedded in cognitive psychology, more recent research embedded in human factors and industrial/organizational (I/O) psychology is characterized by the use of more complex decision-making tasks, most of which tend to take the form of synthetic task environments and/or other forms of computer-based simulations (Salas, Cooke, & Rosen, 2008; Sitzmann, 2011; e.g., see Entin & Sefarty, 1999; Jarrett, Glaze, Schurig, & Arthur, 2017). The continued increase in the computing power of computers and other technological devices along with concomitant increases in task fidelity ensure the likely continuance of this trend. However, the absence of any attention to the study of experts is a gap that needs to be rectified.

Training/Practice Conditions: Experimental Manipulations

Concerning training and acquisition conditions, the traditional view has been that the secret to maximizing skill retention is to simply maximize acquisition. However, there is a compelling and growing body of empirical literature demonstrating that practice conditions that better facilitate performance on tests of retention and transfer do not always exhibit the best performance during acquisition or immediately at the end of acquisition and vice versa (Schmidt & Bjork, 1992). Practice conditions that enhance retention and transfer and yet inhibit skill acquisition are known as *desirable difficulties* (Bjork, 1994). Examples of desirable difficulties include spacing versus massing of practice sessions, varying the conditions of practice versus keeping the conditions similar, and limiting feedback (Schmidt & Bjork, 1992). Conceptually, desirable difficulties are those conditions that promote the use (i.e., practice) of *transfer-appropriate* processing such as (a) deeper, more effortful and elaborative rehearsal of task information, (b) metacognitive or metamemory activities, and (c) reconstruction of what was previously learned, also referred to as retrieval-based learning (Karpicke, 2012; Schmidt & Bjork, 1992). Contemporary reviews of effective learning techniques continue to point to an increasing number of studies that support the aforementioned desirable difficulties (e.g., Dunlosky, Rawson, Marsh, Nathan, & Willingham, 2013; Soderstrom & Bjork, 2015).

Operationalization of Acquisition, Retention/Decay/Loss, Reacquisition, and Transfer

It has been noted that the training literature (and indeed learning and education literatures as well) tend to treat acquisition (i.e., immediate posttraining performance or amount of skill acquired) and retention independently. However, Schmidt and Bjork (1992) point out that acquisition, retention, and transfer are indeed separate phenomena that may yield different interpretations of the extent to which learning has taken place. Therefore, to fully understand the effects of a training condition or manipulation, one must measure its effects in not only the acquisition phase but also the retention and transfer phases as well. Consequently, as illustrated in Figure 46.1, the ideal research design would be one that permits the examination of acquisition, retention/decay, reacquisition, and transfer.

Per Figure 46.1, *acquisition* would be operationalized as the difference between immediate post-training and pretest performance. It pertains to how well some knowledge domain or skill has been mastered, and is a vital prerequisite for retention (Arthur et al., 1998; Farr, 1987; cf. Schmidt & Bjork, 1992). Depending on the purpose of a given training intervention, trainees are expected to achieve a specified level of performance (i.e., criterion-based training) or to spend a specific amount of time (i.e., duration-based training) in training (Adams & Hufford, 1962; Arthur et al., 1997; Rohrer, Taylor, Pashler, Wixted, & Cepeda, 2005). Many applied researchers test performance immediately after training as an operationalization of skill acquisition (see Figure 46.1). However, the operationalization of *mastery* is quite varied with some studies training individuals to one errorless trial (e.g., Hagman, 1980a, 1980b), others to two error-free trials (e.g., Schendel & Hagman, 1982; Stefanidis et al., 2005), and yet others to three errorless trials (e.g., Goldberg, Drillings, & Dressel, 1981). For complex real-world performance domains, mastery can also be demonstrated by showing how the post-training performance of trained novices is no different from or approaches that of established experts (e.g., Moktar et al., 2016).

In summary, because the amount of acquisition is one of the strongest predictors of retention, (Farr, 1987; Hurlock & Montague, 1982; Schendel, Shields, & Katz, 1978; Soderstrom & Bjork, 2015; Wright, 1973), the operationalization of acquisition is a major methodological factor to consider in designing and conducting skill decay research. This then suggests that if the primary or major focus is on retention, then criterion-based operationalizations of acquisition may be methodologically superior to duration-based approaches because they ensure that the (training) conditions being compared are equivalent at the end of the acquisition phase. However, the use of a criterion-based operationalization requires a clear explication and reporting of how mastery or performance-to-criterion was operationalized in the specified study. Amongst others, this will permit comparisons across studies.

An issue closely associated with the amount of initial acquisition is *overlearning*, which to some, albeit to a limited extent, could serve as a proxy for expertise (cf. Ericsson, 2015). Overlearning refers to deliberate learning and practice beyond the initial point of mastery, which can be either completing the first error-free trial or reaching a set criterion performance (Ebbinghaus, 1964). In other words, overlearning is relevant only in criterion-based training. Several reasons have been proposed to explain the enhancing effect of overlearning on long-term retention. Overlearning may strengthen stimulus–response bonds, making them more resilient to interference and consequently decreasing the likelihood that the response will decay or be forgotten (Driskell, Willis, & Copper, 1992; Schendel & Hagman, 1982). Additionally, the increased repetitions and practice may provide further feedback to the trainee regarding the correctness of responses and allow for practice of performance to confirm the correctness of the response. Overlearning also enhances automaticity and subsequently reduces the amount of concentrated effort demanded of the trainee. Furthermore, it has been demonstrated that overlearning gives trainees more confidence in their performance and decreases performance-hampering factors such as stress and anxiety during retention tests (Martens, 1974).

Consistent with the preceding, Arthur et al. (1998) hypothesized a negative relationship between overlearning and skill decay, but failed to find support for this hypothesis. Because the relatively weak effect was based on only 30 studies (17% of the entire data points) with a limited range of degree of overlearning, Arthur et al. cautioned readers against drawing firm conclusions on the basis of their results. It should also be noted that a large portion of the studies included in Arthur et al.'s meta-analysis involved simple tasks. So, whereas it is easy to identify when the first error-free trial (i.e., initial mastery) occurred, the determination of initial mastery would obviously be more challenging for complex and dynamic tasks. Thus, we will have to rely on the conceptual treatise and limited published evidence observed for simple tasks to suggest that for complex tasks, decay can be reduced by overlearning (e.g., see Driskell et al., 1992).

To the extent that initial acquisition (i.e., degree of mastery or original learning) is one of the strongest predictors of retention, then overlearning or the degree of acquisition also plays an important role in the relationship between expertise and skill retention because expertise is, again, fundamentally an issue of acquisition—experts have acquired more knowledge and skill than nonexperts. This is reflected in the various definitions of expertise. So, for example, in Hoffman's (1998) description of the definition of expertise expressed in Middle Ages craft guilds terminology, a *novice* (one step higher than a *naviette*) is someone who is "new and has some, [but] minimal exposure to the domain" (p. 84). An *expert* (one step below a *master*), on the other hand, is someone "whose performance shows consummate skill and economy of effort ... [and] who has special skills or knowledge derived from extensive experience" within the domain (p. 85). Thus, experts are recognized as having acquired substantially more knowledge and skill than nonexperts (see also Schneider, 1985).

As per Figure 46.1, whereas *retention* would simply be the magnitude of the delayed post-training score (i.e., performance on the retention assessment) after the retention interval, *decay/loss* would be operationalized as the difference between immediate post-training (i.e., performance at the end of acquisition) and delayed post-training performance after the retention interval (i.e., performance at retention), with an effect size that is expected to be negative. When the focus is on the retention, experts can be expected to display higher retention scores than nonexperts. Inasmuch as original learning strongly influences retention performance, rates of decay do not differ across differing levels of original learning or degree of overlearning (e.g., Bodilly, Fernandez, Kimbrough, & Purnell, 1986; Goldberg et al., 1981; Hagman, 1980b). However, it has been argued that experts, because of their more coherent and elaborated knowledge structures, are remarkably resistant to decay effects (Schmidt & Rikers, 2007). In this vein, Coughlan, Williams, McRobert, and Ford's (2014) study of Gaelic footballers is a rare example of an intersection of the expertise and skill decay literatures. Their research showed little to no decay for experts compared to significant decay for intermediates. Thus, Coughlan et al.'s results with established experts stand in contrast with laboratory studies of acquisition and retention with naive participants and studies of overlearning which tend to show better retention performance as a function of original learning but not a reduction in decay per se. In general, there is a need for well-controlled studies of retention and decay effects with real-world experts.

As illustrated in Figure 46.1, *reacquisition* would be operationalized as the difference between the reacquisition (i.e., after retraining) performance and delayed post-training performance. Using latent curve analysis with repeated measures of performance during the acquisition and reacquisition phases, researchers can also examine the rate of reacquisition relative to the rate of acquisition. Furthermore, ideally, the same task (or test or measure) would be administered (in the same environment and under the same conditions) for the acquisition, retention, and reacquisition assessments. In contrast, for the assessment of *transfer*, a different task (or test or measure) would be administered under the same conditions or in different environments.

Retention Interval (Period of Nonuse)

As illustrated in Figure 46.1, the ideal design would also entail an extensive retention (nonuse) interval, which refers to the duration between the immediate and delayed post-tests. In *field settings*, the retention (nonuse) interval may occur in situations where individuals (and teams) receive initial training on skills and knowledge that they may not be required to use or have the opportunity to perform for extended periods of time. For instance, disaster relief and emergency first-responder teams may work for years without evacuating residents from afflicted areas, managing evacuation routes, and rescuing and treating survivors of major disasters. Likewise, military reserve personnel may receive formal training only once or twice a year with the expectation that they will need only a limited amount of refresher training to reacquire any skill that may have attenuated when they are called up for active duty. Illustrative of this, Wisher, Sabol, and Ellis's (1999) review of retention studies in the US Army compares effects for retention intervals between 1 and 13 weeks, 14 and 26 weeks, and greater than 26 weeks. However, *laboratory-based* studies are rarely characterized by such lengthy retention intervals; on the contrary, they are relatively short (less than 1 week, and often less than 1 day or even 1 hour; Arthur et al., 1998). For instance, Cepeda, Vul, Rohrer, Wixted, and Pashler (2008) noted that of the 400 reports included in Cepeda, Pashler, Vul, Wixted, and Rohrer's (2006) meta-analysis of spacing effects and retention, "only about a dozen of these looked at RIs [retention intervals] as long as 1 day, with just a handful examining RIs longer than 1 week" (p. 1095); the remainder used shorter retention intervals, thus making generalizations to real-world contexts tenuous at best.

Arthur et al. (1997) and Arthur et al. (2010) are examples of laboratory studies that used relatively long retention (nonuse) intervals, specifically, 8 weeks which is substantially longer than the 1- to 7-day mode reported in Arthur et al.'s (1998) and Wang et al.'s (2013) meta-analyses. Not surprisingly, sample sizes on tests of long-term retention tend to be smaller than the sizes in acquisition and on shorter delayed post-tests due to the attrition of participants from the study (e.g., see Villado et al., 2013b). This highlights a practical methodological trade-off. That is, to the extent that the operational (field) environment is characterized by long retention intervals, the use of long retention (nonuse) intervals are not only externally more valid, they also give the researcher the opportunity to observe the phenomenon under investigation since longer retention intervals are associated with higher levels of decay. However, they also raise threats to internal validity due to attrition from the study, which has the potential to result in non-comparable samples at the different phases of the study.

The design illustrated in Figure 46.1 uses a fixed retention interval. However, if the length of the retention (nonuse) interval is the primary variable of interest, then the design would entail a variable interval. An example of this is Arthur et al. (2015), which in a comparison of individual and team skill decay and reacquisition used retention intervals that ranged from 40 to 169 days (median = 61 days, SD = 133 days). Another important design issue arising for the retention interval is the level of assurance that participants do not have the opportunity and, indeed, do not practice the task outside the confines of the study. The extent to which this is likely to be a potential methodological threat is related to the extent that the performance task is available to the public, of which the commercially available video game *Steel Beasts Pro PE* (eSim Games, 2007) is an example. On the other hand, performance tasks which are designed for research purposes (e.g., Space Fortress (Mane & Donchin, 1989), Crisis in the Kodiak (Arthur et al., 2015)) are unlikely to engender this problem since they are unavailable to the public. The inability to control for activities during the retention interval that are likely to threaten the internal validity of the study requires that researchers record these activities and subsequently treat them as statistical controls in the data analyses.

Pertaining to its effect, the loss of knowledge or skill when not used or exercised for extended periods of time is a fairly robust phenomenon. In applied settings, this is related to infrequent opportunities to practice or perform acquired skills (Ford, Quinones, Sego, & Speer Sorra, 1992; Noe, 1986; Peters & O'Connor, 1980). So, recognizing that this effect is much weaker for complex tasks compared to simple tasks (Arthur et al., 2010; Sauer et al., 2000; Wang et al., 2013), and that various skill components may differ in their resistance to decay, in general, performance is negatively related to the retention interval with longer retention intervals, resulting in worse performance than shorter ones (Annett, 1979; Farr, 1987; Gardlin & Sitterley, 1972; Hurlock & Montague, 1982; Naylor & Briggs, 1961; Prophet, 1976). Consonant with this, Arthur et al.'s (1998) results indicated loss ranging from a d of -0.01 immediately after training to -1.40 after a retention interval of more than 365 days. In addition, the correlation between the retention interval and the corrected mean d was $-.51$. This general pattern of results was also obtained by Wang et al. (2013), who found that there was almost no decay ($d = -0.08$) when the retention interval was less than 1 day, but was as high as -0.84 and -0.71 for 14–28 days and 90–180 days, respectively. They also obtained a correlation of $-.58$ between retention interval and retention. Thus, these findings are generally consistent with the supposition that longer retention (nonuse) intervals increase the possibility of interference and forgetting during the extended delay between acquisition and retrieval (Driskell et al., 1992; Farr, 1987; Gronlund & Kimball, 2013).

In summary, both Arthur et al.'s (1998) and Wang et al.'s (2013) meta-analyses obtained a moderate to large relationship between the length of the retention (nonuse) interval and decay, albeit with minimal or no decay when the retention interval was less than 24 hours. However, over the range of intervals examined, there was no clear trend suggesting a linear relationship between decay and the length of the retention interval. It has been generally posited that most empirical retention curves tend to take on a power function (Gronlund & Kimball, 2013); however, the meta-analyses did not examine this. Nevertheless, the effects of the moderators examined are consistent with the proposition that the amount of decay is influenced by additional factors such as the operationalization of acquisition. It is also important to note that although the length of the retention interval has been cited as a powerful factor in decay, it is a factor that may operate through mechanisms other than time per se (Naylor & Briggs, 1961). In this respect, *decay* theories of forgetting per se are "intuitive but problematic" (Jonides et al., 2008) as cognitive skills do not fade simply as a result of nonuse but rather are subject to variety of interference-related processes involving events that occur during nonuse (Gronlund & Kimball, 2013).

Conditions of Retrieval

The *conditions of retrieval* are closely associated with the retention and reacquisition assessments, and can be either dissimilar or similar. Similarity of retrieval refers to the similarity between the delayed post-training performance context (i.e., retention) and the immediate post-training performance context (i.e., acquisition). The similarity between these two conditions or contexts allows the stimuli of the acquisition assessment to provide cues that enhance retrieval of information during the retention assessment. This is in accord with the tenets of identical-elements theory (Thorndike & Woodworth, 1901), which stipulates that the likelihood of information retrieval increases as the similarity between the retention and acquisition testing contexts increases. In applied contexts, this may take the form of the functional similarity of the training device (original learning) to the actual job equipment (retention test). Because similarities between these two contexts provide cues and stimuli that facilitate information retrieval, Machin (2002) suggested that in order to maximize retrieval after training, one should ensure that training incorporates tasks, procedures, and other aspects of the environment that are similar to those in the workplace. Consonant with the preceding, Arthur et al. (1998) found that decay was greater when the retrieval conditions were different from those of initial acquisition. So, in summary, in terms of long-term retention, the nature or form of a training device may be much less important than whether the trainee's performance when using the device is representative of the performance required by the task (Grimsley, 1969).

Caveats About What is "Ideal"

Although Figure 46.1 presents the *ideal* design as an experimental design, researchers may use non-experimental approaches to examine certain research questions. For example, a researcher might use a single-group (participants trained for a fixed amount of time under the same conditions), within-subjects, correlational design to examine the relationship between scores on tests of immediate (i.e., acquisition) and delayed post-training performance (i.e., retention). A researcher could use a similar design to examine the relationship between individual differences such as general mental ability, personality, or achievement motivation and retention scores. With such a design researchers would ideally statistically control for differences in acquisition performance to examine the direct effects of individual difference variables on retention independent of their indirect effects via acquisition (e.g., Day et al., 2013).

In a similar vein, if researchers were interested in examining retention as a function of real-world expertise (e.g., *masters* versus *journeymen*) using a correlational design, it would be important to include measures of other individual differences thought to influence the attainment of expertise that might also correlate with retention. For example, a variety of trait-based and more attitudinal, process-oriented variables could be measured to isolate the effects of expertise from other factors. Alternatively, one could examine a causal model of expertise and retention that specified some individual differences as antecedents of expertise and others as mediators of the expertise–retention relationship. With respect to more attitudinal and process-oriented variables, Coughlan et al.'s (2014) research with Gaelic footballers showed that experts tend to select more deliberate forms of practice—that is, effortful, less enjoyable practice that focuses on more challenging skills in greater need of improvement—than those with comparable years of experience but less proficiency. Coughlin et al. also showed that experts better retained what they practiced 6 weeks later, but they nevertheless cautioned against making causal conclusions regarding which aspects of practice selected by experts led to their better retention.

In general, characterizations of what constitutes the *ideal* research design must be balanced by logistical considerations and the specific research questions of interest. It is logistically quite challenging to conduct well-controlled experiments on complex tasks that include a variety of individual difference measures, tests of regular and transfer performance, and repeated assessments of performance during acquisition and again after a lengthy retention (nonuse) interval. Such designs also do not lend themselves to studying the decay of real-world expertise. Discussions of real-world expertise and decay among scholars and practitioners often concern highly complex performance domains where the attainment of expertise requires years of training and experience. As such, research on the decay of real-world expertise would ideally distinguish which aspects of expert knowledge and skill are more vulnerable to decay than others, explain why they are more vulnerable, and determine how best to mitigate their decay in a cost-effective way.

Mitigating Loss and Enhancing Retention

To the extent that knowledge and skills decay over extended retention (nonuse) intervals, one obvious approach to mitigating decay and enhancing retention is to manipulate methodological factors known to influence decay in an effort to mitigate it. So, consistent with the research design illustrated in Figure 46.1, we discuss a number of mitigation approaches which are presented in terms of (1) acquisition interventions,

(2) retention (nonuse) interval interventions, and (3) post-retention (nonuse) interval interventions.

Acquisition Interventions

As implied by the label, acquisition interventions are those that would be implemented *before* the commencement of the retention (nonuse) interval. Because the amount of initial skill and knowledge acquired is one of the most important determinants of decay, it would seem that maximizing the amount of initial skill and knowledge acquired would be an effective means of mitigating loss and enhancing retention. To that end, three factors that play an important role in this regard are (1) overlearning and level of expertise, (2) individual differences, and (3) manipulation of the training/ practice conditions.

Overlearning and Expertise

Overlearning refers to practice beyond the initial point of mastery, resulting in errorless performance. Accordingly, overlearning reflects high levels of acquisition, resulting in deep stimulus–response associations that are resistant to interference and decay. Furthermore, to the extent that overlearning reflects a high degree of mastery, then it is of course not divorced from a discussion of expertise since experts have acquired qualitatively higher levels of skill and knowledge than nonexperts. Indeed, there are several discernable characteristics of expertise and experts, including cognitive development, knowledge organization, and reasoning processes (Hoffman, 1998). Thus, experts are distinguishable from nonexperts in terms of the level of performance, consistency and stability of performance, coherence of mental models/knowledge structures, automatic (automaticity) versus controlled processing, and the depth and elaboration of processing. These in their totality, and not simply continued practice on tasks beyond specified criterion levels, are thought to result in not only higher levels of acquired knowledge and skill, but also better adaptability, and less susceptibility to decay (Ericsson, 2015).

Individual Differences

Although training interventions can be used to mitigate skill decay and enhance retention, identifying both cognitive and non-cognitive individual difference variables that predict initial acquisition, and retention and transfer, presents another means via which the performance decrements associated with extended periods of nonuse can be addressed. Thus, for instance, studies consistently find that higher ability individuals retain more knowledge and/or skill over periods of nonuse than lower ability individuals, with the explanation that higher ability trainees simply acquire more knowledge and skills in the same amount of time than lower ability trainees (Farr, 1987; Schendel et al., 1978; Schmidt & Hunter, 1998). For instance, Wisher (1991; see Wisher et al.,

1999) showed that aptitude and length of prior active duty were positively related to knowledge and skill retention in a sample of soldiers called to duty for Operation Desert Storm. Interestingly, aptitude and prior active duty were stronger predictors of retention than the length of the retention interval.

In summary, an individual differences approach to mitigating decay means that the selection of personnel for positions in which extended periods of downtime are common could be based on individual differences found to be predictive of skill retention and transfer. Similarly, at the end of a given training period, individuals who are more likely to need remedial or refresher training before being called into action after an extended retention (nonuse) interval can also be identified via an individual differences approach. Thus, research identifying the individual differences that are predictive of long-term skill retention and transfer is worthwhile.

Training/Practice Conditions

As previously discussed, the incorporation of desirable difficulties (Bjork, 1994) can be a viable means of enhancing retention and mitigating loss. For instance, one would recommend that practitioners incorporate a combination of spacing practice sessions and varying the conditions of practice. For complex tasks, spacing effects can be leveraged by interleaving the practice of different skills or training modules versus sequencing them in distinct blocks and requiring that learners reach a prescribed level of mastery before moving on to the next skill or module (cf. Siddaiah-Subramanya, Smith, & Lonie, 2017). And with the Holy Grail of building durable yet adaptable performance in mind, variable practice might be incorporated in each interleaved session, although it would require more practice time as speed of acquisition is also slower with variable practice (Schmidt & Bjork, 1992). Consequently, efficiently building durable yet adaptable performance is "something of a balancing act" (Healy, Kole, & Bourne, 2014). The literature on expertise reflects similar thoughts with respect to a possible trade-off between expertise and flexibility and how to maximize the acquisition of domain expertise while mitigating cognitive entrenchment and promoting adaptability (Dane, 2010), and more generally, how to enhance expert decision making (e.g., Moxley, Ericsson, Charness, & Krampe, 2012; Ward, Ericsson, & Williams, 2013). For the most part, given the paucity of retention research involving complex real-world performance domains, any specific recommendations regarding how to best combine the variety of more effective training design elements to efficiently build durable yet adaptable performance would be speculative. Examining a combination of design elements in an *ideal* factorial design is logistically quite challenging and likely cost prohibitive.

Although reviews of the training literature point to a variety of design elements that likely promote adaptable performance (e.g., Keith & Wolff, 2015; Kozlowski et al., 2001), the vast majority of the empirical studies fail to include measures of long-term retention and likewise much of this literature ignores the possibility of a trade-off between retention and generalizability (Healy et al.'s (2014) review is one notable exception). In their synthesis of recent empirical research that examined both decay and transfer effects, Arthur and Day (2013) emphasized the importance of coupling part-whole training with elaborative rehearsal. Specifically, they discussed how part-whole training (and part-training in the context of performing the whole task; Gopher, Weil, & Siegel, 1989) holds considerable promise for efficiently building durable yet adaptable performance. Arthur and Day (2013) also discussed how elaborative rehearsal should include an explicit consideration of specific components (including specific pieces of declarative knowledge) in relation to (a) one another individually, (b) their fit in a consolidated structure, and (3) other learned skills and areas of knowledge. The expectation is that coupling part-whole training with elaborative rehearsal would help individuals more deeply encode skill components in a manner that would facilitate transfer and not unduly slow skill acquisition or undermine retention. Recognizing the importance of declarative knowledge to adaptive transfer (Healy et al., 2013) and building on empirical research on beneficial *testing effects* (Carpenter, 2012; Roediger & Butler, 2011), it would also be important to repeatedly test trainees on their declarative knowledge during training versus simply emphasizing procedural knowledge and skills and providing opportunities for hands-on practice.

Regardless of the potential merits of Arthur and Day's (2013) recommendations or any other recommendations that involve combining training design elements to efficiently promote durable yet adaptive performance, the reality is that putting such recommendations into practice in real-world complex performance domains is easier said than done. They would even be more difficult to rigorously evaluate. Perhaps the most pragmatic recommendation given the state of the empirical literature is to leverage the spacing effect. For instance, Jastrzembski, Portrey, Schreiber, and Gluck (2013) in their work with warfighter readiness in military operations go as far as to provide a mathematical model designed to predict retention performance given a trainee's practice performance history and the degree to which their training sessions were spaced (e.g., 1 mission/day versus 2 missions/day).

Retention (Nonuse) Interval Interventions

Retention (nonuse) interval interventions are those that would be implemented *during* the retention interval and consist of (1) shortening the retention interval, and (2) the use of relapse prevention or refresher training.

Shorter Retention (Nonuse) Intervals

Although there is no universal decay curve per se, there is consensus in the literature that the length of the retention (nonuse) interval is negatively related to retention. Consequently, to the extent that it is practically and administratively possible to do so, ensuring that the retention (nonuse) interval is as short as possible would be an effective means of mitigating decay. This may entail providing frequent opportunities to perform the trained tasks (Ford et al., 1992). Of course, we recognize that in some domains and settings, such as emergency management and disaster teams, the frequency of operational performance, and thus the length of the retention interval, is not controllable. That being said, the adverse effects of the length of the retention (nonuse) interval can still be minimized by the use of relapse prevention or refresher training during the retention interval.

Relapse Prevention or Refresher Training

Obviously, one method of enhancing the retention of trained skills during the retention (nonuse) interval is to repeat the initial training sometime during the retention interval (tantamount to shortening the retention interval). However, although full-scale retraining may certainly enhance retention (Schendel & Hagman, 1982), employing such an approach is likely to be cost prohibitive. Thus, the interest is in alternative relapse prevention/refresher training that does not entail full-scale retraining. Unfortunately, research on alternative refresher training interventions that do not involve full-scale retraining is limited, and researchers and training professionals alike have little empirical research that provides guidance on how to best design and implement refresher training programs. That notwithstanding, the key questions of interest are (a) *what are* the alternatives, and (b) *when* should they be implemented?

Technology (i.e., e-learning) provides an opportunity for the delivery of alternative refresher training that does not entail full-scale retraining. Consequently, Internet-based training, including synthetic task environments, may be a feasible and viable alternative to full-scale refresher training. For example, Villado et al. (2013a) examined the efficacy of voluntary post-acquisition observational rehearsal during the retention interval in enhancing retention, and facilitating skill transfer and reacquisition after an extended retention interval. The observational rehearsal consisted of watching Internet-based after-action review videos of recorded task performance (Ellis, Ganzach, Castle, & Sekely, 2010). The performance task was a complex command-and-control naval warfare simulation task. There were three conditions. In the voluntary condition, trainees were given discretional control over accessing and watching the videos. In the mandatory condition, trainees were required to return to the laboratory at specified retention intervals to watch the after-action review videos. In the non-rehearsal condition, trainees were not given access to the videos. Villado et al. (2013a) found that only a small percentage (29%) of trainees in the voluntary rehearsal condition availed themselves of the opportunity to rehearse during the retention interval. Furthermore, differences in motivation (non-cognitive individual differences, e.g., self-efficacy), rather than ability, differentiated individuals who participated in voluntary rehearsal during the retention interval, from those who did not. Finally, the efficacy of the relapse prevention training was dependent on the training outcome; it improved training transfer but not retention or reacquisition.

Practitioners should weigh the benefits against the costs of refresher options (Bodilly et al., 1986). For example, Kluge and Frank (2014) examined the effectiveness of different skill-based practice refreshers and procedural knowledge-based refreshers on the retention of performance on a computer-simulated process control task.

Although the knowledge-based refreshers were less effective in mitigating skill decay than the skill-based practice refreshers, the knowledge-based refreshers led to significantly less skill decay than no refreshers. Thus, while less effective at mitigating decay than skill-based hands-on practice (whether real or simulation-based), knowledge-based refreshers may be more cost effective for real-world, complex performance domains given their lesser expense and fewer logistical challenges.

Other relapse prevention/refresher training interventions may entail the use of goal setting, self-management, and mental practice. Goal setting posits that difficulty and specificity are the two components of goals that positively influence performance via motivational mechanisms (Locke & Latham, 1990). Self-management training is a cognitive-behavioral strategy that encourages trainees to identify obstacles and strategies to overcome said obstacles, and it is related to relapse-prevention interventions that were originally advocated in the treatment of addictions (Marx, 1982). Both goal setting and self-management interventions have been found to be effective in promoting retention in training on negotiation, customer service, and supervisory skills (Gist, Stevens, & Bavetta, 1991; Hutchins & Burke, 2006; Noe, Sears, & Fullenkamp, 1990; Richman-Hirsch, 2001; Tews & Tracey, 2008). Wang et al.'s (2013) meta-analysis provides similar support for the effectiveness of relapse interventions.

Mental practice is the cognitive rehearsal of a task prior to performance, and in a meta-analysis to determine the effect of mental practice on performance, Driskell, Copper, and Moran's (1994) results indicated that mental practice has a positive effect on performance. Furthermore, the effectiveness of mental practice was moderated by the type of task, the retention interval between practice and performance, and the length or duration of the mental practice intervention. Specially, although Driskell et al. observed benefits for both cognitive and physical tasks, their results showed larger benefits for tasks involving more cognitive elements than physical. They also showed that the benefits of mental practice are substantially dissipated after 3 weeks, and thus Driskell et al. recommended implementing refresher mental practice every 1 to 2 weeks. Finally, they observed that longer mental practice sessions do not necessarily translate into better performance. They suggested approximately 20 minutes as the optimal duration of a mental practice session because longer sessions may lead to a loss in concentration. Additionally, as observed by Arthur at al. (2010), during the retention interval, individuals may be more naturally inclined to mentally rehearse or practice tasks that are more meaningful.

The second issue of interest in reference to alternative refresher training interventions that do not involve full-scale retraining is: *When* should these interventions be implemented? Unfortunately, there is very limited or any research that provides guidance on this. Nevertheless, it would seem the first step to determining this is to develop techniques for predicting the amount of knowledge and skill decay for various retention intervals. Amongst others, such models, algorithms, or equations can then be used to inform the timing and sequencing of refresher training. For example, Cepeda et al. (2008) showed that the optimal timing of a refresher study session after reaching a mastery level of memorizing trivia facts from initial study depended on the length of the retention interval. As the retention interval increases, the optimal gap between when refresher studying should be implemented after mastery training increases. However, as the retention interval increases, the ratio of the optimal gap to the length of the retention interval decreases. For instance, recall was optimized when refresher sessions were 3 days after initial study for a 7-day retention interval (i.e., 43% gap ratio), whereas recall was optimized when the refresher was 27 days after initial study for a 350-day retention interval (i.e., 8% gap ratio). Furthermore, their results showed that erring on the side of a more delayed refresher was associated with better recall then erring on the side of a more immediate refresher. Nevertheless, it is important to note that caution should be exercised when generalizing such recommendations to more real-world complex performance domains.

Post-Retention (Nonuse) Interval Interventions

Post-retention (nonuse) interval interventions are those that would be implemented *after* the retention interval, and they generally have an emphasis on the rapid reacquisition of skills. To this end, the similarity of the retrieval (delayed post-training) condition to the acquisition (immediate post-training) condition (i.e., conditions of retrieval) is one means of accomplishing this. As previously noted, the similarity between these two conditions or contexts allows the stimuli of the acquisition/learning environment to provide cues that enhance the retrieval of information from memory. Supportive transfer climates especially those involving supervisors who recognize and reward the use of the knowledge and skills targeted in training and provide opportunities to perform what was trained with feedback (that is specific instead of general) can also facilitate rapid reacquisition in the absence of full-scale retraining (Chauhan, Ghosh, Rai, & Kapoor, 2017; Rouiller & Goldstein, 1993; Svensson, Angelborg-Thanderz, Borgvall, & Castor, 2013; Tracey, Tannenbaum, & Kavanagh, 1995).

Examples of additional post-retention (nonuse) interval interventions are the use of job aids and knowledge management tools. Job aids refer to instructional material, traditionally in print or text form, located on the job to help individuals recall information presented in training and facilitate reacquisition (Wexley & Latham, 2002). Nonetheless, job aids could also be used as a substitute for initial training or to augment initial training as part of a retention plan that emphasizes the importance of knowledge and skill acquisition at the point of mobilization (i.e., the time when performance is needed; Bodilly et al., 1986). Such a "mobilization" approach recasts problems of retention and decay into a problem of building needed knowledge and skill in a just-in-time or as-needed basis. Job aids may be cost effective for relatively simple knowledge and skills in situations when it is difficult to identify and plan for a period of nonuse. Advances in instructional technologies, and synthetic task environments specifically, can be leveraged to extend conventional notions of job aids and provide just-in-time mobilization training for more complex knowledge and skills. However, the time needed to reach a required proficiency level (i.e., mastery) in relation to when actual performance is needed must be taken into account when implementing any mobilization strategy.

The broad purpose of knowledge management "is to enhance organizational performance by explicitly designing and implementing tools, processes, systems, structures, and cultures to improve the creation, sharing, and use of... knowledge that are critical for decision making" (De Long & Fahey, 2000, p. 115). Like job aids, knowledge management tools can provide just-in-time mobilization training as a substitute for more formal training programs, particularly with respect to more knowledge-based jobs. Many knowledge management tools provide individuals with access to expert knowledge that is more personalized and not easily captured through traditional job aids. For example, directories of expertise via an organization's intranet or *yellow pages* can put individuals in contact with real experts in the organization who can then provide them with needed information or resources.

Conclusion

To conclude the chapter, we summarize several conclusions that can be made about knowledge and skill decay on the basis of our review and discussion of the literature. One, decay is more a matter of interference rather than simply the forgetting of information and processes through the passage of time. Two, although decay rates are highly dependent on task and situational factors, a power function best describes decay rates such that larger decrements occur initially after nonuse with smaller decrements occurring gradually over time. Three, retention tends to be better for tasks that are more organized, meaningful, and open-looped (i.e., complex) compared to tasks that involve a discrete set of procedures or declarative knowledge—decay on complex tasks appears to be smaller than that observed for simple tasks. Four, retention is generally stronger with more practice, elaborative rehearsal, and greater mastery—expertise—of the task. Five, although related, retention and transfer are meaningfully distinct and there may even be a trade-off between the two.

With these conclusions in mind, there are a number of issues to consider in the practice of mitigating decay. Chief among the issues is weighing the benefits versus the costs of addressing mitigation via acquisition training (i.e., pre-retention interval), refresher training (i.e., during the retention interval), or mobilization training (i.e., after the retention interval or just-in-time). Certainly, a combination of acquisition, refresher, and mobilization interventions can be implemented. In many real-world complex performance domains, especially those where failure results in dire consequences, reliance on a single intervention is ill-advised, with a combination of acquisition, refresher, and mobilization strategies being a more prudent approach. Another important issue to consider is the role of adaptive performance and how to best address the possible trade-off between retention and adaptability. It is also important to consider the extent to which the performance context involves individual work or teamwork. Finally, we bemoan the lack of an empirical intersection between the study of expertise and decay in complex real-world performance domains. Clearly, intersecting these rich yet separate literatures would be of great theoretical and practical value.

References

Adams, J. A., & Hufford, L. E. (1962). Contribution of a part-task trainer to the learning and relearning of a time-shared flight maneuver. *Human Factors 4*, 159–170.

Annett, J. (1979). Memory for skill. In M. M. Gruneberg & P. E. Morris (Eds.), *Applied problems in memory* (pp. 233–247). London: Academic Press.

Arthur, W., Jr., Bennett, W., Jr., Stanush, P. L., & McNelly, T. L. (1998). Factors that influence skill decay and retention: A quantitative review and analysis. *Human Performance 11*, 57–101.

Arthur, W., Jr., & Day, E. A. (2013). "A look from aFarr (1987)": The past, present, and future of applied skill decay research. In W. Arthur, Jr., E. A. Day, W. Bennett, Jr., & A. Portrey (Eds), *Individual and team skill decay: The science and implications for practice* (pp. 405–427). New York: Taylor & Francis/Psychology Press.

Arthur, W., Jr., Day, E. A., Bennett, W., Jr., McNelly, T. L., & Jordan, J. A. (1997). Dyadic versus individual training protocols: Loss and reacquisition of a complex skill. *Journal of Applied Psychology 82*, 783–791.

Arthur, W., Jr., Day, E. A., Bennett, W., Jr., & Portrey, A. (Eds) (2013). *Individual and team skill decay: The science and implications for practice*. New York: Taylor & Francis/Psychology Press.

Arthur, W., Jr., Day, E. A., Villado, A. J., Boatman, P. R., Kowollik, V., Bennett, W., Jr., & Bhupatkar, A. (2010). The effect of distributed practice on immediate post-training, and long-term performance on a complex command-and-control task. *Human Performance 23*, 428–445.

Arthur, W., Jr., Naber, A. N., Muñoz, G. J., McDonald, J. N., Atoba, O. A., Cho, I., ... Bennett, W., Jr. (2015). *An investigation of skill decay and reacquisition of individual- and team-based skills in a synthetic training environment*. American Psychological Association Division 19 Suite presentation at the 123rd Annual Convention of the American Psychological Association, Toronto, Ontario, Canada.

Bjork, R. A. (1994). Memory and metamemory considerations in the training of human beings. In J. Metcalfe and A. Shimamura (Eds), *Metacognition: Knowing about knowing* (pp. 185–205). Cambridge, MA: MIT Press.

Bliese, P. D., & Lang, J. W. B. (2016). Understanding relative and absolute change in discontinuous growth models: Coding alternatives and implications for hypothesis testing. *Organizational Research Methods 19*, 562–592.

Bodilly, S., Fernandez, J., Kimbrough, J., & Purnell, S. (1986). *Individual ready reserve skill retention and refresher training options* (RAND Note, Contract MDA903-85-C-0030). Santa Monica, CA: RAND National Defense Research Institute, The RAND Corporation.

Bollen, K. A., & Curran, P. J. (2006). *Latent curve models: A structural equation perspective* (Vol. 467). New York: Wiley.

Carpenter, S. K. (2012). Testing enhances the transfer of learning. *Current Directions in Psychological Science 21*, 279–283.

Cepeda, N. J., Pashler, H., Vul, E., Wixted, J. T., & Rohrer, D. (2006). Distributed practice in verbal recall tasks: A review and quantitative synthesis. *Psychological Bulletin 132*, 354–380.

Cepeda, N. J., Vul, E., Rohrer, D., Wixted, J. T., & Pashler, H. (2008). Spacing effects in learning: A temporal ridgeline of optimal retention. *Psychological Science 19*, 1095–1102.

Chauhan, R., Ghosh, P., Rai, A., & Kapoor, S. (2017). Improving transfer of training with transfer design: Does supervisor support moderate the relationship? *Journal of Workplace Learning 29*, 268–285.

Coughlan, E. K., Williams, A. M., McRobert, A. P., & Ford, P. R. (2014). How experts practice: A novel test of deliberate practice theory. *Journal of Experimental Psychology: Learning, Memory, and Cognition 40*, 449–458.

Dane, E. (2010). Reconsidering the trade-off between expertise and flexibility: A cognitive entrenchment perspective. *Academy of Management Review 35*, 579–603.

Day, E. A., Arthur, W., Jr., Villado, A. J., Boatman, P. R., Kowollik, V., Bhupatkar, A., & Bennett, W., Jr. (2013). Relating individual differences in ability, personality, and motivation to the retention and transfer of skill on a complex command-and-control simulation task. In W. Arthur, Jr., E. A. Day, W. Bennett, Jr., & A. Portrey (Eds), *Individual and team skill decay: The science, and implications for practice* (pp. 282–301). New York: Taylor & Francis/Psychology Press.

De Long, D. W., & Fahey, L. (2000). Diagnosing cultural barriers to knowledge management. *Academy of Management Executive 14*, 113–127.

Driskell, J. E., Copper, C., & Moran, A. (1994). Does mental practice enhance performance? *Journal of Applied Psychology 79*, 481–492.

Driskell, J. E., Willis, R. P., & Copper, C. (1992). Effect of overlearning on retention. *Journal of Applied Psychology 77*, 615–622.

Dunlosky, J., Rawson, K. A., Marsh, E. J., Nathan, M. J., & Willingham, D. T. (2013). Improving students' learning with effective learning techniques: Promising directions from cognitive and educational psychology. *Psychological Science in the Public Interest 14*, 4–58.

Ebbinghaus, H. (1964). *Memory: A contribution to experimental psychology* (H. A. Ruger & C. E. Bussenius, Trans.). New York: Dover (Original, German work published in 1885; English translation published in 1913).

Ellis, S., Ganzach, Y., Castle, E., & Sekely, G. (2010). The effect of filmed versus personal after-event reviews on task performance: The mediating and moderating role of self-efficacy. *Journal of Applied Psychology 95*, 122–131.

Entin, E. E., & Sefarty, D. (1999). Adaptive team coordination. *Human Factors 41*, 312–325.

Ericsson, K. A., (2015). Acquisition and maintenance of medical expertise: A perspective from the expert-performance approach with deliberate practice. *Academic Medicine 90*, 1471–1486.

eSim Games. (2007). *Steel Beasts Pro PE* ver. 2.370 [Computer software]. Mountain View, CA: Author.

Farr, M. J. (1987). *The long-term retention of knowledge and skill: A cognitive and instructional perspective*. New York: Springer-Verlag.

Ford, K. K., Quinones, M. A., Sego, D. J., & Sorra, J. S. (1992). Factors affecting the opportunity to perform trained tasks on the job. *Personnel Psychology 45*, 511–527.

Gardlin, G. R., & Sitterley, T. E. (1972). *Degradation of learned skills: A review and annotated bibliography* (D180-15081-1, NASA-CR-128611). Seattle, WA: Boeing.

Gist, M. E., Stevens, C. K., & Bavetta, A. G. (1991). Effects of self-efficacy and post-training intervention on the acquisition and maintenance of complex interpersonal skills. *Personnel Psychology 44*, 837–861.

Goldberg, S. L., Drillings, M., & Dressel, J. D. (1981). *Mastery training: Effect on skill retention* (Technical Report 513). Alexandria, VA: U.S. Army Research Institute for the Behavioral and Social Sciences.

Gopher, D., Weil, M., & Siegel, D. (1989). Practice under changing priorities: An approach to the training of complex skills. *Acta Psychologica 71*, 147–177.

Grimsley, D. L. (1969). *Acquisition, retention, and retraining: group studies on using low fidelity training devices* (HumPRO Tech. Re. 69-12). Alexandria, VA: Human Resources Resource Organization.

Gronlund, S. D., & Kimball, D. R. (2013). Remembering and forgetting: From the laboratory looking out. In W. Arthur, Jr., E. A. Day, W. Bennett, Jr., & A. Portrey (Eds), *Individual and team skill decay: The science, and implications for practice* (pp. 14–52). New York: Taylor & Francis/Psychology Press.

Hagman, J. D. (1980a). *Effects of training schedule and equipment variety on retention and transfer of maintenance skill* (Research Report 1309). Alexandria, VA: U.S. Army Research Institute for the Behavioral and Social Sciences.

Hagman, J. D. (1980b). *Effects of training task repetition in retention and transfer of maintenance skill* (Research Report No. ADA1201672XSP). Alexandria, VA: Army Research Institute for Behavioral and Social Sciences.

Healy, A. F., Kole, J. A., & Bourne, L. E., Jr. (2014). Training principles to advance expertise. *Frontiers in Psychology 5*, 131. http://doi.org/10.3389/fpsyg.2014.00131

Healy, A. F., Wohldmann, E. L., Kole, J. A., Schneider, V. I., Shea, K. M., & Bourne, L. E., Jr. (2013). Training for efficient, durable, and flexible performance in the military. In W. Arthur, Jr., E. A. Day, W. Bennett, Jr., & A. Portrey (Eds.), *Individual and team skill decay: The science, and implications for practice* (pp. 176–204). New York: Taylor & Francis/Psychology Press.

Hoffman, R. R. (1998). How can expertise be defined? Implications of research from cognitive psychology. In R. Williams, W. Faulkner, & J. Fleck (Eds), *Exploring expertise* (pp. 81–100). Edinburgh: University of Edinburgh Press.

Hurlock, R. E., & Montague, W. E. (1982). *Skill retention and its implications for navy tasks: An analytical review* (NPRDC Special Report 82-21). San Diego, CA: Navy Personnel Research and Development Center.

Hutchins, H. M., & Burke, L. A. (2006). Has relapse prevention received a fair shake? A review and implications for future transfer research. *Human Resource Development Review 5*, 8–24.

Jarrett, S. M., Glaze, R. M., Schurig, I., & Arthur, W., Jr. (2017). The importance of team sex composition in team training research employing complex psychomotor tasks. *Human Factors 59*, 833–843.

Jastrzembski, T. S., Portrey, A. M., Schreiber, B. T., & Gluck, K. A. (2013). Improving military readiness: Evaluation and prediction of performance to optimize training effectiveness. In W. Arthur, Jr., E. A. Day, W. Bennett, Jr., & A. Portrey (Eds), *Individual and team skill decay: The science, and implications for practice* (pp. 153–175). New York: Taylor & Francis/Psychology Press.

Jones, M., Bourne, L. E., Jr., & Healy, A. F. (2012). A compact mathematical model for predicting the effectiveness of training. In A. F. Healy & L. E. Bourne, Jr. (Eds), *Training cognition: Optimizing efficiency, durability, and generalizability* (pp. 247–266). New York: Psychology Press.

Jonides, J., Lewis, R. L., Nee, D. E., Dustig, C. A., Berman, M. G., & Moore, K. S. (2008). The mind and brain of short-term memory. *Annual Review of Psychology 59*, 193–224.

Karpicke, J. D. (2012). Retrieval-based learning: Active retrieval promotes meaningful learning. *Current Directions in Psychological Science 21*, 157–163.

Keith, N., & Wolff, C. (2015). Encouraging active learning. In K. Kraiger, J. Passmore, N. R. dos Santos, & S. Malvezzi (Eds), *The Wiley Blackwell handbook of training, development, and performance improvement* (pp. 92–116). London: Wiley-Blackwell.

Kluge, A., & Frank, B. (2014). Counteracting skill decay: Four refresher interventions and their effect on skill and knowledge retention in a simulated process control task. *Ergonomics 57*, 175–190.

Kozlowski, S. W. J., Toney, R. J., Mullins, M. E., Weissbein, D. A., Brown, K. G., & Bell, B. S. (2001). Developing adaptability: A theory for the design of integrated-embedded training systems. In E. Salas (Ed.), *Advances in human performance and cognitive engineering research* (pp. 59–123). Amsterdam: JAI/Elsevier Science.

Locke, E. A., & Latham, G. P. (1990). *A theory of goal setting and task performance*. Englewood Cliffs, NJ: Prentice-Hall.

McArdle, J. J., & Bell, R. Q. (2000). An introduction to latent growth models for developmental data analysis. In T. D. Little, K. U. Schnabel, & J. Baumert (Eds), *Modeling longitudinal and multilevel data: Practical issues, applied approaches, and specific examples* (pp. 69–107). Mahwah, NJ: Lawrence Erlbaum Associates.

Machin, M. A. (2002). Planning, managing, and optimizing transfer of training. In Kraiger (Ed.), *Creating, implementing, and managing effective training and development: State-of-art lessons for practice* (pp. 263–301). San Francisco, CA: Jossey-Bass.

Mane, A. M., & Donchin, E. (1989). The Space Fortress game. *Acta Psychologica 71*, 17–22.

Martens, R. (1974). Arousal and motor performance. *Exercise and Sport Sciences Review 2*, 155–188.

Marx, R. D. (1982). Relapse prevention for managerial training: A model for maintenance of behavior change. *Academy of Management Review 7*, 433–441.

Moktar, J., Bradley, C. S., Maxwell, A., Wedge, J. H., Kelley, S. P., & Murnaghan, M. L. (2016). Skill acquisition and retention following simulation-based training in Pavlik harness application. *Journal of Bone & Joint Surgery 98*, 866–870.

Moxley, J. H., Ericsson, K. A., Charness, N., & Krampe, R. T. (2012). The role of intuition and deliberative thinking in experts' superior tactical decision-making. *Cognition 124*, 72–78.

Naylor, J. C., & Briggs, G. E. (1961). *Long-term retention of learned skills: A review of the literature*. ASD TR 61-390. Columbus, OH: Ohio State University, Laboratory of Aviation Psychology.

Noe, R., Sears, J., & Fullenkamp, A. (1990). Relapse training: Does it influence trainees' posttraining behavior and cognitive strategies? *Journal of Business and Psychology 4*, 319–328.

Noe, R. A. (1986). Trainee's attributes and attitudes: Neglected influences on training effectiveness: Test of a model. *Personnel Psychology 39*, 497–523.

Peters, L. H., & O'Connor, E. J. (1980). Situational constraints and work outcomes: The influence of a frequently overlooked construct. *Academy of Management Review 5*, 391–397.

Prophet, W. W. (1976). *Long-term retention of flying skills: A review of the literature* (HumRRO Final Report 76-35). Alexandria, VA: Human Resources Research Organization (ADA036077).

Richman-Hirsch, W. L. (2001). Posttraining interventions to enhance transfer: the moderating effects of work environments. *Human Resource Development Quarterly 12*, 105–120.

Roediger, H. L., III, & Butler, A. C. (2011). The critical role of retrieval practice in long-term retention. *Trends in Cognitive Sciences 15*, 20–27.

Rohrer, D., Taylor, K., Pashler, H., Wixted, J. T., & Cepeda, N. J. (2005). The effect of overlearning on long-term retention. *Applied Cognitive Psychology 19*, 361–374.

Rouiller, J. Z., & Goldstein, I. L. (1993). The relationship between organizational transfer climate and positive transfer of training. *Human Resource Development Quarterly 4*, 377–390.

Salas, E., Cooke, N. J., & Rosen, M. A. (2008). On teams, teamwork, and team performance: Discoveries and developments. *Human Performance 50*, 540–547.

Sauer, J., Hockey, G. R. J., & Wastell, D. G. (2000). Effects of training on short- and long-term skill retention in a complex multiple-task environment. *Ergonomics 43*, 2043–2064.

Schendel, J. D., & Hagman, J. D. (1982). On sustaining procedural skills over a prolonged retention interval. *Journal of Applied Psychology 67*, 605–610.

Schendel, J. D., Shields, J. L., & Katz, M. S. (1978). *Retention of motor skills: Review* (Technical Paper 313). Alexandria, VA: U.S. Army Research Institute for the Behavioral and Social Sciences.

Schmidt, F. L., & Hunter, J. E. (1998). The validity and utility of selection methods in personnel psychology: Practical and theoretical implications of 85 years of research findings. *Psychological Bulletin 124*, 262–274.

Schmidt, H. G., & Rikers, R. M. J. P. (2007). How expertise develops in medicine: Knowledge encapsulation and illness script formation. *Medical Education 41*, 1133–1139.

Schmidt, R. A., & Bjork, R. A. (1992). New conceptualizations of practice: Common principles in three paradigms suggest new concepts in training. *Psychological Science 3*, 207–217.

Schneider, W. (1985). Training high-performance skills: Fallacies and guidelines. *Human Factors 27*, 285–300.

Siddaiah-Subramanya, M., Smith, S., & Lonie, J. (2017). Mastery learning: How is it helpful? An analytical review. *Advances in Medical Education and Practice 8*, 269–275.

Sitzmann, T. (2011). A meta-analytic examination of the instructional effectiveness of computer-based simulation games. *Personnel Psychology 64*, 489–528.

Soderstrom, N. C., & Bjork, R. A. (2015). Learning versus performance an integrative review. *Perspectives on Psychological Science 10*, 176–199.

Stefanidis, D., Kordorffer, J. R., Jr., Markley, S., Sierra, R., & Scott, D. J. (2006). Proficiency maintenance: Impact of ongoing simulator training on laparoscopic skill retention. *Journal of American College of Surgeons 202*, 599–603.

Stefanidis, D., Kordorffer, J. R., Jr., Sierra, R., Touchard, C., Dunne, J. B., & Scott, D. J. (2005). Skill retention following proficiency-based laparoscopic simulator training. *Surgery 138*, 599–603.

Svensson, E., Angelborg-Thanderz, M., Borgvall, J., & Castor, M. (2013). Skill decay, reacquisition training, and transfer studies in the Swedish Air Force. In W. Arthur, Jr., E. A. Day, W. Bennett, Jr., & A. Portrey (Eds), *Individual and team skill decay: The science, and implications for practice* (pp. 258–281). New York: Taylor & Francis/Psychology Press.

Tews, M. J., & Tracey, J. B. (2008). An empirical examination of posttraining on-the-job supplements for enhancing the effectiveness of interpersonal skills training. *Personnel Psychology 61*, 375–401.

Thorndike, E. L., & Woodworth, R. S. (1901). The influence of improvement in one mental function upon the efficiency of other functions. *Psychological Review 8*, 247–261.

Tracey, J. B., Tannenbaum, I. S., & Kavanagh, J. M. (1995). Applying trained skills on the job: The importance of the work environment. *Journal of Applied Psychology 80*, 239–252.

Villado, A. J., Day, E. A., Arthur, W., Jr., Boatman, P. R., Kowollik, V., Bhupatkar, A., & Bennett, W., Jr. (2013a). Use of, reaction to, and efficacy of observational rehearsal training: Enhancing skill retention on a complex command-and-control simulation. In W. Arthur, Jr., E. A. Day, W. Bennett, Jr., & A. Portrey (Eds), *Individual and team skill decay: The science, and implications for practice* (pp. 240–257). New York: Taylor & Francis/Psychology Press.

Villado, A. J., Day, E. A., Arthur, W., Jr., Boatman, P. R., Kowollik, V., Bhupatkar, A., & Bennett, W., Jr. (2013b). Complex command-and-control simulation task performance following periods of nonuse. In W. Arthur, Jr., E. A. Day, W. Bennett, Jr., & A. Portrey (Eds.), *Individual and team skill decay: The science, and implications for practice* (pp. 53–67). New York: Taylor & Francis/Psychology Press.

Wang, X., Day, E. A., Kowollik, V., Schuelke, M. J., & Hughes, M. G. (2013). Factors influencing knowledge and skill decay after training. In W. Arthur, Jr., E. A. Day, W. Bennett, Jr., & A. Portrey (Eds.), *Individual and team skill decay: The science, and implications for practice* (pp. 68–116). New York: Taylor & Francis/Psychology Press.

Ward, P., Ericsson, K. A., & Williams, A. M. (2013). Complex perceptual-cognitive expertise in a simulated task environment. *Journal of Cognitive Engineering and Decision Making 7*, 231–254.

Wexley, K. N., & Latham, G. P. (2002). *Developing and training human resources in organizations*, 3rd edn. New York: HarperCollins.

Wisher, R. (1991). *Desert Storm mobilization: Skill decay and attitudes in the IRR* (ARI Newsletter, 7, 1–4). Alexandria, VA: U.S. Army Research Institute for the Behavioral and Social Sciences.

Wisher, R. A., Sabol, M. A., & Ellis, J. A. (1999). *Staying sharp: Retention of military knowledge and skills*. ARI Special Report No. 39. Alexandria, VA: U.S. Army Research Institute for the Behavioral Social Sciences.

Woollard, M., Whitfield, R., Newcombe, R. G., Colquhoun, M., Vetter, N., & Chamberlain, D. (2006). Optimal refresher training intervals for AED and CPR skills: A randomized controlled trial. *Resuscitation 71*, 237–247.

Wright, R. H. (1973). *Retention of flying skills and refresher training requirements: Effects of non-flying and proficiency flying* (HumRRO Technical Report 73–32). Alexandria, VA: Human Resources Research Organization: Alexandria, VA.

CHAPTER 47

EXPERTISE AND RESILIENCE

JOP HAVINGA, JOHAN BERGSTRÖM, SIDNEY DEKKER, AND ANDREW RAE

Introduction

SAFETY has a long history of reducing risk by limiting variability. In that approach, expertise was about meeting the standards set in the design phase and having the ability to keep a system working as intended. Resilience engineering, a new approach to safety, has a different view on the benefits of expertise. Resilience engineering values expertise as it can help adjust operation to varying conditions. There are still desired outcomes, but experts are not expected to work merely as planned, and can adapt to the circumstances.

In this chapter, expertise is defined as the change in how an agent approaches a problem or job, after it gains experience. An *agent* is not limited to humans, and can also include people combined with supporting technology, teams, or organizations. Experience is considered everything an agent goes through, which includes time on the job, training, and even conversations had about work. Expertise opposes designed ways of functioning, and innate capacities, like talent. Expertise is generally linked to improved performance, more sensitivity, and adjustments to conditions; things desired for resilience. However, these benefits are not integral to the definition. In fact, in this chapter we will consider the evidence for and against expertise providing such benefits. We will also consider the current state of knowledge about how expertise can be best managed to engineer resilience into different levels of an organization. Before we go into this, we first explain what resilience is, summarize its history, and explain how it is applied as an alternative approach to safety.

What Is Resilience

The notion of resilience is a multifaceted one. *Resilience* is used as a metaphor, a theory, a set of capacities or even a strategy (Norris, Stevens, Pfefferbaum, Wyche, & Pfefferbaum, 2008). The common denominator seems to be that resilience is always related to some kind of coping with stress (Kolar, 2011).

While the term resilience originated in materials science as a material's ability to regain its original shape after having been stressed, resilience as used in social and organizational science took shape in psychology, which viewed resilience as a psychological agent's ability to thrive despite adversity (Garmezy, 1971; Garmezy & Streitman, 1974). Resilience has been a topic in psychology and the health sciences since the 1950s, typically using war-traumatized children or veterans as targets of analysis (Tyhurst, 1951). The central question to answer in such resilience research is: Why do some individuals seem to thrive despite the great adversity they experience? In the 1970s resilience was introduced as a concept in ecology and defined as an ecosystem's ability to absorb and adapt to stress without going into a qualitatively different system state (Holling, 1973; Walker, Holling, Carpenter, & Kinzig, 2004). Later, research on ecological resilience expanded the boundaries of research to include research on human interaction with their environments, and with socioecological systems (Gunderson, 2010; Gunderson & Holling, 2002).

Following the 9/11 terror attacks on the USA, resilience became a target of public policy. Citizens, communities, cities, and entire nations are now supposed to be resilient in that they should develop preparedness to adapt to unforeseen natural or antagonist events (Walker & Cooper, 2011).

Looking at the relatively long history of resilience research and policy the safety science community was rather late to adopt the notion of resilience in the early 2000s. Resilience engineering was introduced in Woods' (2003) testimony on the future of the National Aeronautics and Space Administration (NASA). Woods introduced it as a school of thought dedicated to "help organizations maintain high safety despite production pressure" (p. 2).

In the book following the first resilience engineering symposium, resilience engineering was presented as a field dedicated to improve and engineer more resilience into systems. This field was borne from frustration with safety thinking based on counting errors and avoiding negative events. The new analytical path focused on how organizations anticipate and adapt to the complex and dynamic risk landscape (Hollnagel, Woods, & Leveson, 2006).

To the school of resilience engineering, resilience is a positive and desirable concept. The school takes an optimistic view "and its agenda is to develop ways to control or manage a system's adaptive capacities based on empirical evidence" (Woods & Branlat, 2011, p. 128). With this, resilience engineering stands against more negative schools of thought such as normal accident theory (Perrow, 1984). Such views suggest that complex systems cannot be tamed and there is nothing substantial that can be done to prevent accidents in these systems. Instead, with resilience engineering, scholars' attention is guided to finding things that facilitate adaptation, coping strategies, and trade-offs, things that often coincide with expertise: Finding out what makes systems function despite (production) pressures, conflicting goals, and constantly changing risks. This distinguishes resilience engineering from the traditional safety focus on negatives, such as errors or violations.

Before proceeding with any analysis of resilience, a question that needs to be answered is: Who possesses resilience? For material scientists and psychologists studying resilience, the answer would be sought in materials and humans, respectively. (Eco) system scholars, on the other hand, first need to define system boundaries, before there is *something* to study. For example, when studying resilience, do you look for it in an individual or team? Do you consider procedure handbooks or traces of equipment use part of the system? In different situations, researchers draw different system boundaries. The interactions and relations between system levels often become more fundamental than the resilience of individual levels (Bergström & Dekker, 2014).

Resilience engineering tends to focus on sociotechnical systems, which consist of people and technology that interact and (partly) self-organize. Systems can be decomposed into various levels, each of which can be a target of analysis. For instance, resilience can be analyzed in high-risk organizations as a whole, individual actors, and teams (Bergström, van Winsen, & Henriqson, 2015). At the level of frontline actors, experts are often studied in how they respond to incidents. This includes critical incident control room operators who cope with adverse events by using flexible (Grote, Weichbrodt, Günter, Zala-Mezö, & Künzle, 2009; Owen, Healey, & Benn, 2013) or adaptive strategies (Gomes, Woods, Carvalho, Huber, & Borges, 2009). At a higher level of analysis, research has focused on the resilience of teams and management (e.g. Furniss, Curzon, Blandford, & Furniss, 2016; Gomes, Borges, Huber, & Carvalho, 2014; Lundberg & Rankin, 2014; Rankin, Dahlback, & Lundberg, 2013). Resilience research can also occur at an organizational level, which was the original focus of resilience engineering.

Resilience needs to be studied at and across different levels because resilient behavior at one level can decrease the resilience at other levels, which can produce ethical dilemmas. For instance, many organizations rely constantly on the expertise of their staff to adapt beyond what a system was designed to handle, without formally recognizing or appreciating such effort. This hides potential learning opportunities, as the expert's successful adaptation can hide the stressors to others. To give one example, a Scandinavian hospital emergency ward was experiencing a high load of incoming patients. This problem is typical in Scandinavian healthcare, where staff cope by placing the incoming patients in beds located in the corridors, and giving them an alarm to sound in case of emergency. In this specific example, at one point, there were no more alarms to hand out to patients. Although nurses had never experienced this before, they were able to generate a solution: they handed out cutlery and pots to the patients so they could bang them together to alarm the staff. This might seem like a good case of resilience where experienced nurses adapted to ensure continued functioning of the system. The problem, however, is that by relying on the adaptive capacities of sharp-end experts, there is a risk that *the system* learns the wrong lesson.

In the example above, experienced nurses used their adaptive capabilities to invent a novel, low-tech alarm mechanism (i.e., banging together cutlery and pots) to deal with the stress on the system (i.e., to ensure care was delivered despite the ward being overloaded). The nurses' actions may have contributed to the preservation of system vulnerabilities (i.e., an acceptance of an overloaded emergency ward). In turn, this may increase the likelihood that the system would further drift towards the boundaries of effective operational performance, and potentially closer to a collapse of emergency care at the hospital. Stressors that are inadvertently hidden can prevent long-term investment by management in increasing the capacity of the system, which means the organization as a whole is less adaptive.

A more ethically sound way to embrace resilience is to ask whether (and how) the system provides the necessary resources to enable experts to adapt to the situations and challenges they encounter. This approach would provide support to experts who are best placed to guarantee system functioning. It reinforces the connection between experts and higher levels of an organization and shares accountability across the entire organization, rather than leaving it up to the people who happen to be closest to the action (e.g., the nurses in the earlier example).

In the next section, we examine how to manage expertise to enhance resilience at different levels of the system, including frontline workers, teams, and management. In addition, we explore the interrelationships between levels.

Frontline Workers

Sociotechnical systems are never fully prespecified or planned. This means people on the frontline will always have to adapt. There will always be a gap between how managers imagine the work in their plans and how the work is actually done (Dekker, 2005; Hollnagel, 2012). It is up to frontline workers to overcome these gaps and holes in the planned work, as they inherit these systems and form the last line of defense against accidents (Reason, 1990). To deal with the gap, humans use heuristics, work-arounds, and understanding of the system, which reflect a worker's expertise. This means any sociotechnical system relies on expertise of frontline workers (Hollnagel, 2014). In the following sections, we first discuss whether expertise creates resilience, and then discuss how training, procedures, and automation can be managed to enhance resilience through expertise.

The Effect of Individual Expertise on Resilience

In general, expertise is related to improved performance, which is desirable for any system (Farrington-Darby & Wilson, 2006). The question whether increased skill, even at the level of an expert, always leads to more adaptiveness, that is, responsiveness to changes and more varied responsiveness, is not as clear-cut. We explore this from the perspective of attention to the task and fixation of strategies or ideas of frontline workers.

In a variety of research, it has been shown that with expertise, people develop strategic attentional skills (Farrington-Darby & Wilson, 2006). At the same time, it has been said that with experience people start paying less attention to their job—partly because, with increased skill, tasks that were once fully attention demanding become less so (e.g., Fitts & Posner, 1967). Within safety research, the decrease in attention as expertise develops is often viewed negatively and considered as complacency.

Complacency as applied in safety research does not have a single definition, but often is described as a state of low suspicion and trust that things will go right, which builds during the use of a system (Moray & Inagaki, 2000). It has been established that operators in a simple signal task, like radar monitoring, decrease in vigilance after they become familiar with a system (Mackworth, 1948). In more complex simulated tasks, when people become more familiar with a system, they start to trust it more and this predicts how closely they monitor it. If a system shows faults, trust declines, and monitoring increases again (Lee & Moray, 1992; Muir & Moray, 1996). However, while monitoring frequency decreases, performance has not been found to decrease in complex tasks (Moray & Haudegond, 1998). This suggests that the decrease in monitoring in a complex task reflects a more efficient rather than less effective strategy. Trust functions as the mechanism that calibrates monitoring frequency (Moray, 2003) as expertise builds.

There are many concepts around the idea that people can get stuck on certain ideas or courses of action, because of previous experience, including the Einstellung effect (Luchins, 1942), functional fixedness (Duncker, 1945), cognitive tunnel vision (Cook & McDonald, 1988), cognitive hysteresis (Norman, 1986), and fixation errors (de Keyser & Woods, 1990). While there are differences between these concepts, one reason for the variety of concepts for similar ideas is that they originated in different types of research. The Einstellung effect, for example, was framed after experimental studies, where participants were given a task in which they had to fill water jugs. They found that if participants were first given tasks where a more complicated strategy was required, participants would stick to this approach, even when new tasks allowed simpler, more straightforward strategies. Participants that were not given the more complex tasks used simpler strategies (Luchins, 1942). Fixation errors, on the other hand, were framed around returning patterns identified in accident investigations and more naturalistic studies of work, where people stuck to one course of action or diagnosis even when they received new information that suggested the old strategy was no longer a good fit (de Keyser & Woods, 1990). Fixation errors are not necessarily caused by previous experience, but it is commonplace that skilled individuals and experts used their experience to explain away signs (Klein, Pliske, Crandall, & Woods, 2005), suggesting the expertise can enable fixation. Interestingly, it has also been found that expertise can help break fixation. In a comparison of doctors with varying levels of expertise, it was found that the most experienced doctors were better at recognizing their initial diagnosis did not fit as new evidence came in. In-depth analysis suggested that this is because the more expert doctors have richer sets of known responses and are better at recognizing when things are not *standard*, which helps them recognize the need to look for alternatives ideas (Feltovich, Spiro, & Coulson, 1997; Spiro, Coulson, Feltovich, & Anderson, 1988). This is in line with more recent research, where it is was found that considerable experience can make people rigid, but those with the highest levels of skill can benefit from their experience to make them more flexible (Bilalić, McLeod, & Gobet, 2008).

In terms of the effects of expertise on both attention and fixedness, it is unlikely that the possible downsides of expertise outweigh its benefits for resilience. While for single cases, there is the possibility for expertise to have negative effects, as a general rule, investing in expertise benefits the resilience of frontline workers.

Training

Arguably, all successful training programs increase the resilience of individuals. This happens by either giving people new ways to deal with something familiar, enabling them to deal with a higher task load, or preparing them to handle new situations. There is no question whether training can increase resilience in specified areas. In this section, we explore whether training can create expertise at being resilient more generally. Therefore, we will focus on training that facilitates dealing with unknown and unforeseen situations, applying skills more flexibly, or going beyond what was specified during the design of a system.

There are multiple types of training that prepare individuals to deal with the unknown or aim to train skills applicable to many situations, with the most popular being crew resource management (CRM) training. CRM originated in the cockpit, but has been applied to many industries. The development of CRM started in the late 1970s (Kanki, Helmreich, & Anca, 2010; Maurino & Murray, 2009). CRM is older than resilience engineering. While CRM is often promoted as a remedy for human error, resilience engineering was created to move away from the term human error. Many CRM programs aim to train individuals' skills, like decision making, situation awareness, cooperation, leadership, which are applicable in many tasks and workplaces, in both routine and emergency situations (Flin, O'Connor, & Crichton, 2008).

While CRM training is popular and has been regarded as a major success, some critical notes can be made about it. CRM has not always been defined clearly, and training programs with different goals and philosophies are grouped under the label CRM (Havinga, de Boer, Rae, & Dekker, 2017; Salas & Burke, 2001). Because of this, claims of the successfulness of CRM in general do not necessarily reflect the effect of any single training program. The evaluative studies of CRM training have repeatedly found improved attitudes, but only few studies have found changes in behavior or improved performance (Salas, Wilson, Burke, & Wightman, 2006). Also, many of the evaluation studies suffer from methodological issues (Edkins, 2002; Havinga, de Boer, et al., 2017), including no comparison groups and not analyzing questionnaire data according to their designed theoretical constructs. The uncertainty around what CRM *is* and its effects raises doubt as to whether it teaches the types of general skills some claim it does. To better understand whether CRM training produces widely applicable skills, research needs to test whether the trainees acquire these skills, whether they are applied consistently in both trained and non-trained situations, and whether the use of these skills improve performance.

As an alternative to training specified universal skills, Ward et al. (2018) proposed six training principles to make people more adaptive in the application of their skills. These principles were derived from studies of experts operating in complex environments. They include: (a) *flexibility-focused feedback* (i.e., methods to overcome cognitive rigidity and acquire knowledge flexibly); (b) *concept–case coupling* (i.e., methods that permit learners to experience the different ways in which concepts vary from situation to situation); (c) case-proficiency scaling (i.e., use of mentoring and other scaffolding methods to stretch skill); (d) *tough-case time compression* (i.e., the need to develop a bank of cases, with varying difficulty and complexity, on which to practice adapting); (e) *complexity preservation* (i.e., methods that preserve the functional complexities to be learned and avoid learning oversimplified relationships); (f) *active reflection* (i.e., methods that help learners become better calibrated in terms of what they know, and in their ability to identify competency boundaries).

However, there is limited evidence whether these practices universally lead to fewer accidents or increased performance in the typical safety-critical industries of resilience engineering. For active reflection methods, such as deliberate performance (Fadde & Klein, 2010) and collaborative cross-checking (Patterson, Woods, Cook, & Render, 2007), case studies do show that these processes change the behavior of people and long-term lessons can be drawn from this research. The intended effect of people changing their behavior based on their situation and goals in response to these types of programs and principles suggests that they create resilience. The unanswered question is whether the changes that are observable at an individual level reflect an increase in the resilience of the larger system, or a trade-off between resilience in one area for resilience in another area. With these practices the aim is on specific processes, and adapting those processes, which is also expected to lead to the development of worker expertise applicable to other situations. CRM, on the other hand, aims to instill a general knowledge base, which is expected to change behavior in multiple situations. For each of these practices, more evidence is needed to support the claim that such training works as suggested, and that it creates resilience through expertise development.

Procedures

Besides training, a major way organizations aim to influence individual is through the use of procedures. The procedures that workers have to comply with have a major impact on expertise and resilience.

Procedures and expertise have a complicated relationship. Procedures can capture strategies of experts and get other people to perform better. Scientific management was largely based on this idea (Taylor, 1911). In safety, rules and procedures are often made to comply with and seen as a barrier to stop people from making poor decisions that have been found in hindsight of accidents (Hale & Swuste, 1998). In this way, procedures can be seen as a protection against a lack of expertise. However, procedures also constrain the possible benefits from expertise. Procedures tell in a mostly context-free way what a person should do. For novices, this can be helpful, but experts are generally more sensitive to context and pick up more relevant features of a situation to adjust to than a procedure specifies (Dreyfus & Dreyfus, 1980). On top of that, experts build an understanding, or mental model, of how things work. They can use this understanding to determine what to do, even when they have never encountered a situation before. To illustrate this, Patrick and Haines (1988) compared trainees from two different training programs on their performance at fault finding in a chemical plant simulator. In one training program the trainees were taught action procedures on what to do in which situations; in the other training program the trainees received an explanation of how the technical system worked. The people trained with procedures were faster at dealing with situations matching the procedures, but the people who received the explanation of the underlying system were better at dealing with novel cases. This does not only matter for training, but also for how expertise develops on the job. Organizations enforcing compliance of procedures can be harmful to the development of an understanding of the job. To apply procedures, a worker does not need a deep understanding of the job, nor is the worker prompted to develop this deeper understanding. In compliance-focused organizations, exploring opportunities to understand the job may get workers in trouble, as experimentation means deviating from standard procedure. Recognized non-compliance from a worker can have negative consequences for the worker's career (Reason, Parker, & Lawton, 1998). By a strict focus on procedural compliance, the experiences of workers became less diverse, limiting the development of expertise.

Automation

Besides training and procedures, a worker's resilience and expertise are affected by the technology around them, especially automation. While attempts to increase expertise usually involve increasing reliance on human decision making, the rationale behind the introduction of automation is often the opposite, to keep *unreliable* humans out of the system. Like procedures, automation is specified with certain design limitations. Unlike human experts, automation does not think of different responses for new situations or decide to reorganize itself. While automation might increase the capacity of what a system can handle for stressors that are well understood, it decreases the ability of the system to handle situations that are outside its design specification.

The problem for resilience, however, goes further than the procedure-like decision-making process of automation; the automation has a larger effect on the resilience of a system. Operators that first did the job now become supervisors of the automation. Where workers before were involved in every step of the process, parts of the process can now be hidden from the supervisors. Feedback from the automation is often minimal and delayed, making it more difficult to develop effective expertise for the new job (Kahneman & Klein, 2009). On top of that, the worker's skill to manually take over will erode if not practiced somehow (Bainbridget, 1983; Sarter, Woods, & Billing, 1997). The automation can make the larger system less resilient by taking the expertise away from the human and hindering the development of expertise for the new task of supervising.

To create resilient systems that include automation, one needs to consider training design and the design of the automation. People can be trained in how to deal with automation, focusing on understanding how automation works, as opposed to just input and output knowledge. The automation needs to be designed as a team player, in that the automation is predictable and directable by signaling what it is doing. Importantly, this includes making sure the automation informs the supervising humans when it is stretched near its capacity (Bainbridget, 1983; Klein, Woods, Bradshaw, Hoffman, & Feltovich, 2004; Sarter et al., 1997). Overall, if given enough feedback and transparency, supervisors can develop expertise in handling the automation. This does not only apply for automation, but for all technology introduced to people. For resilience engineering, it is not only about how well people perform with technology, but also whether people recognize changing conditions, goals, and system boundaries, and how quickly people can learn to use technology in new ways (Hoffman, Marx, Amin, & McDermott, 2010).

Teams and Management

The literature on high-reliability organizations (HRO) provides advice on how team and management practices influence resilience, many of which involve expertise. The HRO work can be traced back to a group of researchers at the University of California, who started to study how organizations that have to deal with high risk manage to achieve safe and reliable outcomes (La Porte & Consolini, 1991). This included studies on flight deck carriers, nuclear power generation, and air traffic control, to identify what the *good ones* do (Weick, Sutcliffe, & Obstfeld, 1999). In a sense, they aimed to capture what the expert organizations do to deal with risk.

Originally the HRO work was descriptive, but later the work turned more normative and started to provide advice on how organizations should organize for reliability (Weick et al., 1999). HRO assumes that there is something different about the good organizations. Resilience engineering, alternatively, does not think success and failure have to come from different processes and instead focuses on understanding work (Haavik, Antonsen, Rosness, & Hale, 2016; Havinga, Dekker, & Rae, 2017; Le Coze, 2015).

Most of the advice of HRO aligns with the goal of engineering resilience into sociotechnical systems (Dekker & Woods, 2010). Because of that, we will discuss how managing expertise matters for resilience, along the principles of HRO that most strongly relate to expertise. These principles are deference to expertise, preoccupation with failure, and commitment to resilience. We end this section with an exploration of whether expertise at the team level can be detrimental to resilience.

Deference to Expertise

The advice to defer to expertise means giving decision authority to people who know the most about the operations and the local situation, as opposed to position in the organizational hierarchy, or to standard procedures (Weick & Sutcliffe, 2007). Rochlin, La Porte, and Roberts (1987) described how, on aircraft carriers, even the lowest ranking personnel could suspend an operation if they believed it was unsafe. This closely aligns with resilience engineering's idea of changing from original plans, adapting to circumstances, and facilitating people to do so. Not listening to experts has been linked to development of accidents (Dekker, 2014). However, deference to expertise is not always easily put into practice. For one, the phrase *deference to expertise* raises the question: Where do you find the expertise to turn to for a decision? (see Chapter 48, "Framing and Translating Expertise for Government and Business," by Conway & Gore, this volume). For example, a designer of a piece of machinery may have very detailed knowledge about the inner workings of a machine, but maintenance technicians may have a better understanding of a specific problem, or how a machine is used in practice. Even if there is no doubt which expert is best suited to the particular problem, the problem can still be beyond the expert's expertise. In that case, should that expert still be given full decision authority?

Besides who to turn to and how much decision authority should be given, there is another dilemma created for managers by deferring to expertise. Giving up control and formally recognizing people as the experts can come with undesirable consequences. Experts are recognized by organizations for their specialized technical competence, assertiveness, self-confidence, and ability to go beyond standard responses. However, these characteristics can also leave the experts insensitive to the larger organizational goals (Girard, 2005), a feeling they do not have to play by the rules (Godkin & Allcorn, 2009), and silencing diverging minority opinions (Barton & Sutcliffe, 2009). Even beyond that, recognition of expertise can lead to a sense of entitlement or narcissism, which leads to envy when their goals are threatened, exceptional belief in own successes, driving out non-conformers, managing by intimidation, and suppressing accurate testing. The localization of specialized knowledge often makes it hard for others to check on what the experts do. Especially in places of high ambiguity, *prima donnas* can assign credit to themselves, while diffusing blame away (Dekker, 2014). This is counterproductive to other principles from HRO and resilience engineering, such as being sensitive to operations and the larger picture, aligning groups to work together across levels, taking minority opinions seriously, and not taking past successes as a reason for confidence (Hollnagel et al., 2006; Weick & Sutcliffe, 2007). One could argue that the above characteristics are actually not those of true experts, but people who merely have been given, or claimed, the status of expert. However, in many situations, it is not feasible or possible, to discern between the *real* experts and those *believed to be* experts, nor are true experts immune to these effects. Deferring to expertise is a good strategy, but it should not get in the way of people working together, which we discuss next.

Commitment to Resilience

Besides managing expertise, there is the question of whether teams can develop expertise in being resilient. In HRO studies, resilience is about responding to disturbances and returning to normal operations. This is a narrower view than how resilience is used in resilience engineering, where it also includes prediction and prevention of future events. Nevertheless, the ability to recover from disturbances is an important element for both HRO and resilience engineering. By studying how emergency teams' work comes together, researchers have identified the elements that make up *emergency response* expertise, so this can be used in training (Furniss, Back, Blandford, Hildebrandt, & Broberg, 2011). We discuss the challenges emergency response teams face, and how teams meet these.

One of the most recurring issues in the domain of emergency response is the need to plan, establish goals, and distribute roles (Gomes et al., 2014; Righi, Huber, Gomes, & de Carvalho, 2016). Rather than rigidly sticking to any plan or goal, resilience is about creating the capacity to adapt. However, in emergency response teams, their plans, goals, and role distribution help to establish common ground to aid in coordination and to allow people to be proactive (Bergström, Dahlström, Dekker, & Petersen, 2011). In emergencies, it has been found that making modular plans works better, with subplans that are structured and integrated into larger plans. This adds to the ability to quickly revise and reorganize quickly (Gomes et al., 2014), which is the next point.

In crisis operations it is expected that plans change and replanning and restructuring will be required. Modular plans make it easier, but it leaves open the question: Who is updated or when is a change significant enough to update everyone (Bergström et al., 2011)? From studying teams at work, it became clear that formal centralized (re)briefings spread information quickly and help people know what others know. Wrong information that has been gathered also has the potential to spread quickly. Formal briefings create moments where wrong or implausible information can be spotted (Rankin, Dahlback, & Lundberg, 2013). Central updates, however, also take up time and impede work.

During emergency response work, adapted team structures often arise from what is necessary or possible, but might not be optimal in the long term. Teams identified that both bottom-up and top-down processes can prompt searches for more optimal structure after one pattern has stabilized (Lundberg & Rankin, 2014). Re-structuring teams often leads people to work on tasks outside their own area of expertise (Bergström et al., 2011). While it will often be impossible to cross-train everyone in all skills, it is possible to assist people in working beyond their expertise. This can include creating the expectations that role improvisation happens, teaching people how to support non-experts in their specialization, finding time during operation to prepare others working outside their expertise, and the challenges of communicating about topics outside one's expertise (Lundberg & Rankin, 2014).

At the team and management level, expertise for resilience largely lies in how teams self-organize and interact. Prior to the emergence of resilience engineering, a considerable amount of research on emergency response teams was conducted on topics like planning and communicating (e.g., CRM training; Flin, 1995). Resilience engineering has a slight shift of focus. The expertise is seen less as a level of skill or attainment that people are required to meet in order to avoid things that should not happen, such as *errors*. From a resilience engineering perspective, there are no more clear standards to meet that ensure safety. Instead, expertise is phrased more in terms of things that help people *muddle through* (see Chapter 8, "Expertise: A Holistic, Experience-Centered Perspective," by Flach & Voorhorst, this volume).

Preoccupation with Failure

HROs actively search for things that could go wrong (Weick & Sutcliffe, 2007). One way organizations do this is by assessing risk. Risk assessment ranges from estimating the rate of occurrence of historically frequent events, to judging the likelihood of rare or even unprecedented high-consequence events. For low-frequency, high-consequence events *expert judgment* is used to assess and prioritize accident scenarios (i.e., courses of action). Whilst there is strong evidence that experts have a superior ability to project situations in the short-term future, and to anticipate the effects on their own actions (Suss & Ward, 2015), and for events, like weather forecasting (McCarthy, Ball, & Purcell, 2007), there is considerably less evidence that design or operational expertise translate into superior performance on long-range forecasting of high consequence, non-repetitive, events, like industrial accidents (Rae & Alexander, 2017).

To the extent that some people make more accurate predictions than others, this appears to be expertise in translating qualitative impressions into quantitative estimates more generally, rather than a function of subject-matter knowledge or domain experience (Mellers et al., 2015). However, the value of expert risk assessments is not necessarily the accuracy of their numbers, but in how they produce these numbers.

When multiple experts decompose their estimates and discuss together what they have done, they often reach new conclusions (Rae & Alexander, 2017).

For the purpose of enhancing resilience, there is an ongoing debate about how best to identify and make use of subject-matter expertise. One school of thought suggests that superior performance on calibration problems should be the sole determinant of expertise (Cooke & Goossens, 2004). The alternative is to consider quantification as a means rather than an end, providing a focus for discussion between stakeholders with domain expertise.

Concerns about Team Expertise

As expertise is generally good for resilience, a straightforward suggestion would be to have teams develop expertise. For a team to develop expertise as a unit, one could argue that it would need to gain experience as a unit, and so benefit from becoming a stable unit. However, as with individuals, some researchers have raised the concern that too much expertise of a team, as a unified agent, can lead to decreased attentiveness and sensitivity to operations. One such idea comes from Weick and Roberts (1993). They suggest that intact teams can start to interact with each other in a *heedless* way. In this context, the term *heedless* means interacting in a way that does not take into account the larger task or organization, and that the interactions are only based on local concerns. The idea is that consistency in people can lead to thoughtless and routine responses, where each interaction is just a replication of what has been done before, instead of the product of reflection on previous actions and the current situation. These authors suggested that crew rotation and frequent introduction of new crew members on flight-deck carriers force crew members to think about why they do what they do when they explain this to the newcomers, making their interactions more mindful as opposed to routine.

However, the empirical support for the idea that intact teams become more heedless and, hence, less adaptable is mixed. This applies to both the intact team part and the heedful interrelations mediating it. For R&D teams, a curvilinear relationship between team age and team performance has been found, where intact teams perform best when they are around three to four years old, after which they start to decline (Katz, 1982). This was believed to originate from a convergence in patterns of behavior, a lack of diversity, lowering creativity, and hampering ability to come up with new solutions, showing similarity to the idea of routinization. However, this view did not hold, and it was considered more likely that successful performance relied on the team's management, specifically the combination of nurturing individual expertise and fostering social connections within the team (Allen, Katz, Grady, & Slavin, 1988).

In experiments where newly recruited three-member teams controlled uninhabited air vehicles, Gorman, Amazeen, and Cooke (2010) found that teams of non-experts that were kept intact were more corrective in their interaction patterns, while teams that replaced team members or their roles explored more solutions to encountered problems. They did not test whether this was mediated by heedful interrelating.

Grote, Kolbe, and Künzle (2010) specifically looked at heedful interrelating and analyzed the behavior of experienced cockpit crews in a simulator. They found that high-performing cockpit crews did not use more heedful interrelating, but the authors observed qualitative differences in their heedful interrelating compared to low-performing crews. Schraagen (2011) analyzed the interactions of one medical team during multiple pediatric cardiac surgeries and found that during non-routine events, the team mostly relied on explicit coordination (e.g., assigning tasks), not heedful interrelation (e.g., discussing how information relates to larger goals). In-depth qualitative analysis did suggest that heedful interrelations formed critical moments.

More research needs to be done in this area. The increased flexibility of teams who rotate team members could be explained by factors other than increased heedful interrelating, as heedful interrelating is not needed to deal with non-routine events. In addition, routines are also used to deal with non-routine situations, suggesting teams always rely on a shared background and rarely engage in heedful interrelating as a single strategy. This is supported by the idea of common ground being necessary for joint activity (Klein, Feltovich, Bradshaw, & Woods, 2005). The expertise of a team on a whole seems to allow teams to respond with less costs of coordination, while diversity of expertise between team members seems to assist finding new options. This would suggest there is no detrimental effect on teams as a whole developing expertise, but there are benefits to bringing new perspectives and types of expertise into a team.

Organizations and Systems

Woods and Branlat (2011) identified three basic patterns through which adaptive systems, including organizations, fail. These patterns are decompensation, working at cross-purposes, and getting stuck in outdated behaviors. We will discuss these three patterns, and how managing expertise can help organizations move away from these patterns.

Decompensation

A system decompensates when challenges or disturbances are stronger or faster than the system can adapt to. The system has to deal with too much in too little time. This can be a matter of ability, but also of resources it has available. From a reactive perspective, expertise can increase the ability to cope with disturbances. The pitfalls people have associated with procedural levels of expertise (e.g., journeyman, see

Hoffman, 1998), like fixation and routinization, can even be the processes that provide this extra capacity, as these processes reduce the amount of information processed. However, it is not only about having a level of proficiency, but also being able to access the necessary skills, knowledge, and resources when needed, wherever these may reside. For accessibility it is beneficial to have many connections within an organization (Carmeli, Friedman, & Tishler, 2013). More connections within an organization give agents shorter paths to expertise and the resources. However, maintaining connections can be costly in a system, especially if they are rarely used, which is why networks are rarely fully connected.

From a proactive perspective, expertise can be used in the monitoring and prediction of larger disturbances. The repository of situational representations that experts have acquired help them to quickly spot anomalies; things stand out from the norm (Feltovich et al., 1997; Klein, Pliske, et al., 2005). Expertise is also helpful for predicting how an anomaly affects a system, for considering the possible causes of failure, and with that determine whether an anomaly is relevant (Watts, Woods, & Patterson, 1996). As discussed, this deeper understanding of how a system links together makes expertise useful for risk assessments, even when it does not increase accuracy.

Working at Cross-purposes

Systems can also fail because it is working at cross-purposes. This happens when subgroups are adaptive towards their own goals, but are maladaptive towards higher-level goals. This can happen when different groups draw from a common resource pool (Ostrom, 1999) or respond to each other's work. For instance, in military operations in Iraq (Snook, 2000), teams adapted some procedures locally, including identification codes and monitoring practices, which allowed the groups to do their tasks faster. However, these changes led other groups to misidentify and misunderstand them. Hence, lower and higher goals of the system became misaligned, which led to eroding safety margins and finally resulted in a friendly fire accident.

Local adaptation is overall considered useful, necessary, and even inevitable, in sociotechnical systems. This is evident in the advice to give frontline workers the freedom to make decisions, and in deferring to expertise more broadly. The challenge on a systems level is to prevent local adaptation from becoming maladaptive at another level of the system. There are three types of solutions to this problem:

Regulations. To prevent groups at one level of the system from acting in a manner that is maladaptive to groups at another level, it is possible to constrain their actions with rules or regulations. This will limit the resilience at a lower level, but preserves the system at a higher level.

Central controller. One way to prevent maladaptation is to add a controller for higher-level goals that also oversees the lower levels. An example of this in aviation is air traffic controllers, who manage limited airport runway and air space and exert their influence on arriving and departing airplanes.

Adapting to adaptations. The third option is to make the groups adapt to the adaptations of others. This relies on setting higher-order goals and making visible how other agents behave. Ostrom (1999) found that groups tended to cooperate and form their own norms and adjust for each other's behavior, as long as there were enough face-to-face interactions within that community. Patterns that emerge from local adaptations can be just as efficient as a central organization (Hutchins, 1991; Ostrom, 1999).

All of these methods rely on a kind of expertise to manage the system. For the first, experience is translated into rules. The second, a central controller, will need an understanding of how the system behaves that it is trying to guide (Conant & Ashby, 1970), which usually is built on experience and can update with experience. In the third, the expertise is not centralized, but distributed over the agents in the system. Each approach can use experience to manage the problem of misalignment.

Getting Stuck in Outdated Behaviors

In this pattern, a system keeps doing what it has done in the past, even though the situation has changed and requires a different response now (Woods & Branlat, 2011). Cognitive rigidity (Spiro et al., 1988) and routine interactions (Weick & Roberts, 1993) represent this pattern on smaller scales. An example of this pattern is organizations that stick to a narrow interpretation in their analysis of accidents, failing to produce new learning (Cook, Woods, & Miller, 1998).

To counter this effect on an organizational level, learning needs to be encouraged. Often this is linked to the idea of instilling a learning culture. An important element to learning is that the learning opportunities are recognized and collected. Accidents and incidents, for example, provide learning opportunities; however, the opportunity can be lost if people are afraid to speak about the accident. Without a just culture that protects against blaming and shaming, organizational learning can suffer (Dekker, 2012). While this alone is not sufficient for organizations to learn from accidents, it is a first step to spread learning in an organization (Huber, Wijgerden, Witt, & Dekker, 2009).

To prevent an organization from getting stuck in outdated behaviors, it is useful if the groups in an organization are interconnected. Interconnectedness can help recognize changes in conditions and when safety margins are eroding. An outsider who does not have a preconceived notion of the process or has the resources to take a larger view of the system might spot that a stabilized task distribution is suboptimal (Lundberg & Rankin, 2014) or how work in one area affects other parts of the system (Dekker, 2011; Patterson et al., 2007).

To address the challenge of finding new ways to act, a diversity of agents can improve quality of performance more than the *quality* of single problem solvers. Individuals with similar expertise will tend to converge to similar solutions, so will contribute little to each other, while fresh perspectives add more to the group (Hong & Page, 2004). However, diversity can increase coordination cost. People from diverse backgrounds can have a harder time working together as it will be harder to establish common ground (Klein, Feltovich, et al., 2005). This can slow a system down and put it at a risk of decompensating.

Conclusion

Expertise is generally good for resilience; the hypothesized risks of expertise seem generally ungrounded, or do not outweigh the benefits of expertise. While more expertise generally leads to more resilience, how the two relate to each other does changes across different levels, which matters for how to manage expertise. At the individual level of frontline workers, resilience and expertise are almost identical. More expertise almost always increases resilience. At the level of management and teams, expertise is still beneficial, but the type of expertise that is useful, how to make experts collaborate, and political implications of recognizing experts begin to matter. At a broader systems level, it becomes harder to point to expertise, as it can take many forms. The question becomes not whether we need expertise, but in what form and how to access it.

If we look at the current picture of research on expertise and resilience, two areas have received limited attention. This includes research cutting across different levels of analysis (Bergström & Dekker, 2014), and how organizations deal with positive disturbances.

Some research makes mention of cross-scale work, like that on working at cross-purposes, but there are few empirical studies on patterns across scales. For example, do diverse groups of individuals outperform experts in risk assessment (Tetlock, 2015)? In assessment of non-repetitive, low-frequency, high-consequence events, like industrial accidents, the value of experts seems mostly in the links they make between different elements of the system and not in their numerical accuracy. If accuracy is not the goal, diverse groups of individuals with differing types of expertise are more likely to disagree on what works and which connections matter (e.g., see Shanteau, 2000), and with that more likely to build a more in-depth understanding together than that of individual experts. Further examination of the processes involved in making risk assessments and making accurate assessments may provide more answers.

Resilience engineering has largely focused on reactive behaviors (i.e., how to make systems resilient when something starts going wrong), rather than on foresight and proactive goals, which were envisioned as being central for resilience engineering in its original conception. We reiterate here that resilience engineering also encompasses creating and using foresights, leveraging available resources, and spreading insights through an organization. There has been some mention of investigating how systems can flourish (Furniss et al., 2016). To further explore such topics, resilience engineering can build on research such as divergent thinking (McCrae, 1987) and insight (Klein & Jarosz, 2011), which both seem to require expertise, but also acknowledge that more expertise is not always better for these processes (Runco, Dow, & Smith, 2006). As this might constitute a new area of research, explorative and descriptive research is most likely to help us understand the challenge of *positive* disturbances.

References

Allen, T., Katz, R., Grady, J. J., & Slavin, N. (1988). Project team aging and performance: The roles of project and functional managers. *R&D Management 18*(4), 295–308.

Bainbridget, L. (1983). Ironies of automation*. *Automatica 19*(6), 775–779.

Barton, M. a., & Sutcliffe, K. M. (2009). Overcoming dysfunctional momentum: Organizational safety as a social achievement. *Human Relations 62*(9), 1327–1356.

Bergström, J., Dahlström, N., Dekker, S. W. A., & Petersen, K. (2011). Training organisational resilience in escalating situations. In E. Hollnagel, J. Pariès, D. D. Woods, & J. Wreathall (Eds), *Resilience engineering in practice: A guidebook* (pp. 45–57). Alderstot, UK: Ashgate.

Bergström, J., & Dekker, S. W. A. (2014). Bridging the macro and the micro by considering the meso: Reflections on the fractal nature of resilience. *Ecology and Society 19*(4), 22.

Bergström, J., van Winsen, R., & Henriqson, É. (2015). On the rationale of resilience in the domain of safety: A literature review. *Reliability Engineering & System Safety, 141*, 131–141.

Bilalić, M., McLeod, P., & Gobet, F. (2008). Inflexibility of experts-Reality or myth? Quantifying the Einstellung effect in chess masters. *Cognitive Psychology 56*(2), 73–102.

Carmeli, A., Friedman, Y., & Tishler, A. (2013). Cultivating a resilient top management team: The importance of relational connections and strategic decision comprehensiveness. *Safety Science 51*(1), 148–159.

Conant, R. C., & Ashby, W. R. (1970). Every good regulators of a system must be a model of that system. *International Journal of Systems Science 1*(2), 89–97.

Cook, R. I., & McDonald, J. S. (1988). Cognitive tunnel vision in the operating room—analysis of cases using a frame model. *Anesthesiology 69*(3A), 497.

Cook, R. I., Woods, D. D., & Miller, C. (1998). *A tale of two stories: Contrasting views of patient safety*. Chicago: National Patient Safety Foundation.

Cooke, R. M., & Goossens, L. H. J. (2004). Expert judgement elicitation for risk assessments of critical infrastructures. *Journal of Risk Research 7*(6), 643–656.

de Keyser, V., & Woods, D. D. (1990). Fixation errors: Failures to revise situation assessment in dynamic and risky systems. In A. G. Colombo & A. S. de Bustamante (Eds), *Systems reliability assessment* (pp. 231–251). Dordrecht, The Netherlands: Kluwer Academic.

Dekker, S. W. A. (2005). *Ten questions about human error: A new view of human factors and system safety*. London: Lawrence Erlbaum.

Dekker, S. W. A. (2011). *Drift into failure: From hunting broken components to understanding complex systems*. Aldershot, UK: Ashgate.

Dekker, S. W. A. (2012). *Just culture: Balancing safety and accountability* (2nd edn). Aldershot, UK: Ashgate.

Dekker, S. W. A. (2014). Deferring to expertise versus the prima donna syndrome: A manager's dilemma. *Cognition, Technology & Work* 16(4), 541–548.

Dekker, S. W. A., & Woods, D. D. (2010). The high reliability organization perspective. In E. Salas & D. E. Maurino (Eds), *Human factors in aviation* (2nd edn) (pp. 123–144). Burlington, MA: Academic Press.

Dreyfus, S. E., & Dreyfus, H. L. (1980). *A five-stage model of the mental activities involved in directed skill acquisition*. Report ORC 80-2. Berkley, CA.

Duncker, K. (1945). On problem-solving (L. S. Lees, Trans.). *Psychological Monographs* 58(5), i–113.

Edkins, G. D. (2002). A review of the benefits of aviation human factors training. *Human Factors and Aerospace Safety* 31(1), 247–273.

Fadde, P. J., & Klein, G. A. (2010). Deliberate performance: Accelerating expertise in natural settings. *Performance Improvement* 49(9), 5–14.

Farrington-Darby, T., & Wilson, J. R. (2006). The nature of expertise: A review. *Applied Ergonomics* 37(1), 17–32.

Feltovich, P. J., Spiro, R. J., & Coulson, R. L. (1997). Issues of expert flexibility in contexts characterized by complexity and change. In R. R. Hoffman, K. M. Ford, & P. J. Feltovich (Eds), *Expertise in context: Human and machine* (pp. 125–146). Cambridge, MA: MIT Press.

Fitts, P. M., & Posner, M. I. (1967). *Human performance*. Oxford: Brooks/Cole.

Flin, R. H. (1995). Crew resource management for teams in the offshore oil industry. *Journal of European Industrial Training* 19(9), 23–27.

Flin, R. H., O'Connor, P., & Crichton, M. T. (2008). *Safety at the sharp end*. Aldershot, UK: Ashgate.

Furniss, D., Back, J., Blandford, A., Hildebrandt, M., & Broberg, H. (2011). A resilience markers framework for small teams. *Reliability Engineering and System Safety* 96, 2–10.

Furniss, D., Curzon, P., Blandford, A., & Furniss, D. (2016). Using FRAM beyond safety: A case study to explore how sociotechnical systems can flourish or stall. *Theoretical Issues in Ergonomics Science* 17(5–6), 507–532.

Garmezy, N. (1971). Vulnerability research and the issue of primary prevention. *American Journal or Othopsychiatry* 41(1), 101–116.

Garmezy, N., & Streitman, S. (1974). Children at risk: The search for the antecedents of schizophrenia, part I. Conceptual models and research methods. *Schizophrenia Bulletin* 1(8), 125.

Girard, N. J. (2005). Dealing with perioperative prima donnas in your OR. *AORN* 82(2), 187–189.

Godkin, L., & Allcorn, S. (2009). Institutional narcissism, arrogant organization disorder and interruptions in organizational learning. *The Learning Organization* 16(1), 40–57.

Gomes, J. O., Borges, M. R. S., Huber, G. J., & Carvalho, P. V. R. (2014). Analysis of the resilience of team performance during a nuclear emergency response exercise. *Applied Ergonomics* 45(3), 780–788.

Gomes, J. O., Woods, D. D., Carvalho, P. V. R., Huber, G. J., & Borges, M. R. S. (2009). Resilience and brittleness in the offshore helicopter transportation system: The identification of constraints and sacrifice decisions in pilots' work. *Reliability Engineering and System Safety* 94(2), 311–319.

Gorman, J. C., Amazeen, P. G., & Cooke, N. J. (2010). Team coordination dynamics. *Nonlinear Dynamics, Psychology, and Life Sciences 14*(3), 265–289.

Grote, G., Kolbe, M., & Künzle, B. (2010). Adaptive coordination and heedfulness make better cockpit crews. *Ergonomics 53*(2), 211–228.

Grote, G., Weichbrodt, J. C., Günter, H., Zala-Mezö, E., & Künzle, B. (2009). Coordination in high-risk organizations: The need for flexible routines. *Cognition, Technology and Work 11*(1), 17–27.

Gunderson, L. H. (2010). Ecological and human community resilience to natural disasters. *Ecology and Society 15*(2), 18.

Gunderson, L. H., & Holling, C. S. (Eds). (2002). *Panarchy: Understanding transformations in human and natural systems*. Washington, DC: Island Press.

Haavik, T. K., Antonsen, S., Rosness, R., & Hale, A. (2016). HRO and RE: A pragmatic perspective. *Safety Science*. Advance online publication.

Hale, A. R., & Swuste, P. (1998). Safety rules: Procedural freedom or action constraint? *Safety Science 29*, 163–177.

Havinga, J., de Boer, R. J., Rae, A., & Dekker, S. (2017). How did crew resource management take-off outside of the cockpit? A systematic review of how crew resource management training is conceptualised and evaluated for non-pilots. *Safety 3*(26), 1–20.

Havinga, J., Dekker, S., & Rae, A. (2017). Everyday work investigations for safety. *Theoretical Issues in Ergonomics Science 19*(2), 213–228.

Hoffman, R. R. (1998). How can expertise be defined? Implications of research from cognitive psychology. In R. Williams, W. Faulkner, & J. Fleck (Eds), *Exploring expertise* (pp. 81–100). Edinburgh: University of Edinburgh Press.

Hoffman, R. R., Marx, M., Amin, R., & McDermott, P. L. (2010). Measurement for evaluating the learnability and resilience of methods of cognitive work. *Theoretical Issues in Ergonomics Science 11*(6), 561–575.

Holling, C. S. (1973). Resilience and stability of ecological systems. *Annual Review of Ecology and Systematics 4*, 1–23.

Hollnagel, E. (2012). Resilience engineering and the systemic view of safety at work: Why work-as-done is not the same as work-as-imagined. In *Kongress des Gesellschaft für Arbeitswissenschaft vom 22 bis 24 Februar 2012* (pp. 19–24). Dortmund, Germany: GfA-Press.

Hollnagel, E. (2014). *Safety-I and safety-II: The past and future of safety management*. Aldershot, UK: Ashgate.

Hollnagel, E., Woods, D. D., & Leveson, N. (Eds). (2006). *Resilience engineering: Concepts and precepts*. Aldershot, UK: Ashgate.

Hong, L., & Page, S. E. (2004). Groups of diverse problem solvers can outperform groups of high-ability problem solvers. *PNAS 101*(46), 16385–16389.

Huber, S., Wijgerden, I. Van, Witt, A. De, & Dekker, S. W. A. (2009). Learning from organizational incidents: Resilience engineering for high-risk process environments. *Process Safety Progress 28*(1), 90–95.

Hutchins, E. (1991). Organizing work by adaptation. *Organization Science 2*(1), 14–39.

Kahneman, D., & Klein, G. A. (2009). Conditions for intuitive expertise: A failure to disagree. *The American Psychologist 64*(6), 515–526.

Kanki, B., Helmreich, R., & Anca, J. (Eds.). (2010). *Crew resource management*. Amsterdam: Elsevier.

Katz, R. (1982). The effects of group longevity on project communication and performance. *Administrative Science Quarterly* 27(1), 81–104.

Klein, G. A., Feltovich, P. J., Bradshaw, J. M., & Woods, D. D. (2005). Common ground and coordination in joint activity. In W. B. Rouse & K. R. Boff (Eds), *Organizational simulation* (pp. 1–42). Hoboken, NJ: Wiley.

Klein, G. A., & Jarosz, A. (2011). A naturalistic study of insight. *Journal of Cognitive Engineering and Decision Making* 5(4), 335–351.

Klein, G. A., Pliske, R., Crandall, B., & Woods, D. D. (2005). Problem detection. *Cognition, Technology and Work* 7, 14–28.

Klein, G. A., Woods, D. D., Bradshaw, J. M., Hoffman, R. R., & Feltovich, P. J. (2004). Ten challenges for making automation a "team player" in joint human-agent activity. *IEEE Intelligent Systems* 19, 91–95.

Kolar, K. (2011). Resilience: Revisiting the concept and its utility for social research. *International Journal of Mental Health Addiction* 9(4), 421–433.

La Porte, T. R., & Consolini, P. M. (1991). Working in practice but not in theory: Theoretical challenges of "high-reliability organizations." *Journal of Public Administration Research and Theory* 1(1), 19–48.

Le Coze, J.-C. (2015). Vive la diversité! High reliability organisation (HRO) and resilience engineering (RE). *Safety Science.* Advance online publication.

Lee, J. D., & Moray, N. (1992). Trust, control strategies and allocation of function in human-machine systems. *Ergonomics* 35(10), 1243–1270.

Luchins, A. S. (1942). Mechanization in problem solving: The effect of Einstellung. *Psychological Monographs* 54(6), i–95.

Lundberg, J., & Rankin, A. (2014). Resilience and vulnerability of small flexible crisis response teams: Implications for training and preparation. *Cognition, Technology & Work* 16, 143–155.

McCarthy, P., Ball, D., & Purcell, W. (2007). Project phoenix—optimizing the machine-person mix in high-impact weather forecasting. Presented at *The 22nd conference on weather analysis and forecasting/18th conference on numerical weather prediction, Park City, Utah*. American Meteorological Society, P6A.5. Available at https://ams.confex.com/ams/22WAF18NWP/techprogram/paper_122657.htm.

McCrae, R. R. (1987). Creativity, divergent thinking, and openness to experience. *Journal of Personality and Social Psychology* 52(6), 1258–1265.

Mackworth, N. H. (1948). The breakdown of vigilance during prolonged visual search. *Quartely Journal of Experimental Psychology* 1(1), 6–21.

Maurino, D. E., & Murray, P. S. (2009). Crew resource management. In J. A. Wise, D. Hopkin, D. J. Garland, & B. Kantowitz (Eds), *Handbook of aviation human factors* (2nd edn, vol. 0411, pp. 361–378). Boca Raton, FL: CRC Press.

Mellers, B., Stone, E., Murray, T., Minster, A., Rohrbaugh, N., Bishop, M., . . . Tetlock, P. (2015). Identifying and cultivating superforecasters as a method of improving probabilistic predictions. *Perspectives on Psychological Science* 10(3), 267–281.

Moray, N. (2003). Monitoring, complacency, scepticism and eutactic behaviour. *International Journal of Industrial Ergonomics* 31, 175–178.

Moray, N., & Haudegond, S. (1998). An absence of vigilance decrement in a complex dynamic task. In *Proceedings of the human factors and ergonomics society 42nd annual meeting* (pp. 234–238). Los Angeles: Sage.

Moray, N., & Inagaki, T. (2000). Attention and complacency. *Theoretical Issues in Ergonomics Science* 1(4), 354–365.

Muir, B. M., & Moray, N. (1996). Trust in automation II. Experimental studies of trust and human intervention in a process control simulation. *Ergonomics* 3, 429–460.

Norman, D. A. (1986). New views of information processing: implications for intelligent decision support systems. In E. Hollnagel, G. Mancini, & D. D. Woods (Eds), *Intelligent decision support in process environments* (pp. 123–136). Berlin: Springer.

Norris, F. H., Stevens, S. P., Pfefferbaum, B., Wyche, K. F., & Pfefferbaum, R. L. (2008). Community resilience as a metaphor, theory, set of capacities, and strategy for disaster readiness. *American Journal of Community Psychology* 41(1–2), 127–150.

Ostrom, E. (1999). Coping with tragedies of the commons. *Annual Review Political Science* 2, 493–535.

Owen, C., Healey, A. N., & Benn, J. (2013). Widening the scope of human factors safety assessment for decommissioning. *Cognition, Technology and Work* 15(1), 59–66.

Patrick, J., & Haines, B. (1988). Training and transfer of fault-finding skill. *Ergonomics* 31(2), 193–210.

Patterson, E. S., Woods, D. D., Cook, R. I., & Render, M. L. (2007). Collaborative cross-checking to enhance resilience. *Cognition, Technology and Work* 9(3), 155–162.

Perrow, C. (1984). *Normal accidents: Living with high-risk technologies.* Princeton, NJ: Princeton University Press/Basic Books.

Rae, A., & Alexander, R. (2017). Forecasts or fortune-telling: When are expert judgements of safety risk valid? *Safety Science* 99, 156–165.

Rankin, A., Dahlback, N., & Lundberg, J. (2013). A case study of factor influencing role improvisation in crisis response teams. *Cognition, Technology & Work* 15, 79–93.

Reason, J. (1990). *Human error.* Cambridge, UK: Cambridge University Press.

Reason, J., Parker, D., & Lawton, R. (1998). Organizational controls and safety: The varieties of rule-related behavior. *Journal of Occupational and Organizational Psychology* 71, 289–304.

Righi, A. W., Huber, G. J., Gomes, J. O., & de Carvalho, P. V. R. (2016). Resilience in firefighting emergency response: Standardization and resilience in complex systems. *IFAC-PapersOnLine* 49(32), 119–123.

Rochlin, G. I., La Porte, T. R., & Roberts, K. H. (1987). The self-designing high-reliability organization: Aircraft carrier flight operations at sea. *Naval War College Review* 40(4), 75–90.

Runco, M. A., Dow, G., & Smith, W. (2006). Information, experience, and divergent thinking: An empirical test. *Creativity Research Journal* 18(3), 269–277.

Salas, E., & Burke, C. S. (2001). Team training in the skies: does crew resource management (CRM) training work? *Human Factors* 43(4), 641–674.

Salas, E., Wilson, K. A., Burke, C. S., & Wightman, D. C. (2006). Does crew resource management training work? An update, an extension, and some critical needs. *Human Factors* 48(2), 392–412.

Sarter, N. B., Woods, D. D., & Billing, C. E. (1997). Automation surprises. In G. Salvendy (Ed.), *Handbook of human factors and ergonomics* (2nd ed.) (pp. 1926–1943). Hoboken, NJ: Wiley.

Schraagen, J. M. (2011). Dealing with unforeseen complexity in the OR: the role of heedful interrelating in medical teams. *Theoretical Issues in Ergonomics Science* 12(3), 256–272.

Shanteau, J. (2000). Why do experts disagree? In B. Green, R. Cressy, F. Delmar, T. Eisenberg, B. Howcroft, M. Lewis, ... R. Vivian (Eds), *Risk behaviour and risk management in business life* (pp. 186–196). Dordrecht, The Netherlands: Kluwer Academic.

Snook, S. A. (2000). *Friendly fire: The accidental shootdown of U.S. black hawks over Northern Iraq.* Princeton, NJ: Princeton University Press.

Spiro, R. J., Coulson, R. L., Feltovich, P. J., & Anderson, D. K. (1988). Cognitive flexibility theory: Advanced knowledge acquisition in ill-structured domains. Technical Report 441. Champaign, IL: University of Illinois.

Suss, J., & Ward, P. (2015). Predicting the future in perceptual-motor domains: Perceptual anticipation, option generation and expertise. In R. R. Hoffman, P. A. Hancock, M. Scerbo, & J. L. Szalma (Eds), *Cambridge handbook of applied perception research* (pp. 951–976). New York: Cambridge University Press.

Taylor, F. W. (1911). *The principles of scientific management.* New York: Harper & Brothers.

Tetlock, P. E. (2015). *Superforecasting: The art and science of prediction.* New York: Crown.

Tyhurst, J. S. (1951). Individual reactions to community disaster. *American Journal of Psychiatry 107*(10), 764–769.

Walker, B., Holling, C. S., Carpenter, S. R., & Kinzig, A. (2004). Resilience, adaptability and transformability in social—ecological systems. *Ecology and Society 9*(2), 5.

Walker, J., & Cooper, M. (2011). Genealogies of resilience: From systems ecology of the political economy of crisis adaptation. *Security Dialogue 42*(2), 143–166.

Ward, P., Gore, J., Hutton, R., Conway, G. E., & Hoffman, R. R. (2018). Adaptive skill as the conditio sine qua non of expertise. *Journal of Applied Research in Memory and Cognition 7*(1), 35–50.

Watts, J. C., Woods, D. D., & Patterson, E. S. (1996). Functionally distributed coordination during anomaly response in space shuttle mission control in Space Shuttle Mission Control. In *Human interactions with complex systems '96, Dayton, OH* (pp. 68–75).

Weick, K. E., & Roberts, K. H. (1993). Collective mind in organizations: Heedful interrelating on flight decks. *Administrative Science Quarterly 38*(3), 357–381.

Weick, K. E., & Sutcliffe, K. M. (2007). *Managing the unexpected: Resilient performance in an age of uncertainty* (2nd edn). Hoboken, NJ: Wiley.

Weick, K. E., Sutcliffe, K. M., & Obstfeld, D. (1999). Organizing for high reliability. *Research in Organizational Behavior 21*, 81–123.

Woods, D. D. (2003). Creating foresight: How resilience engineering can transform NASA's approach to risky decision making. *US Senate Testimony of the Committee on Commerce, Science and Transportation.* Washington, DC: John McCain, chair.

Woods, D. D., & Branlat, M. (2011). Basic patterns in how adaptive systems fail. In E. Hollnagel, J. Pariès, D. D. Woods, & J. Wreathall (Eds), *Resilience engineering in practice: A guidebook* (pp. 127–144). Farnham, England: Ashgate.

CHAPTER 48

FRAMING AND TRANSLATING EXPERTISE FOR GOVERNMENT

GARETH E. CONWAY AND JULIE GORE

INTRODUCTION

ONE might expect a high level of agreement on what *expertise* is, and how it influences professional workplaces. It has a dictionary definition, which usually includes skill, knowledge, occasionally professionalism, and a vast scientific literature explicating it—the latest addition is this *Handbook*. The situation, however, is not as simple as that. Many chapters in this *Handbook* reference this by providing a contextual account of definitions of expertise and discussing current rhetoric that suggests that experts are under attack; they are being questioned by politicians, by other professionals in positions of influence and power, and by members of the public (e.g., see Klein, Shneiderman, Hoffman, & Wears, Chapter 49, "The 'War' on Expertise: Five Communities that Seek to Discredit Experts," this volume). Our chapter conversely aims to shine a light of a slightly different hue on the topic.

We begin by considering the unique nature of the challenges to government, including the radical uncertainty concerning the future state of factors relevant for government, and the intent and actions of adversaries, and even who our adversaries will be. For contextual understanding, we offer brief descriptions on how the UK government works, and provide historical and current perspectives on how expertise is considered and framed within government. We then consider the difficulties one may face when trying to identify experts, and how we might seek to scrutinize the arbiters of such judgements. Concerning recent criticisms of expert judgement (see Klein et al., this volume), we explore whether it has been misrepresented and whether the arguments constitute a strawman fallacy. We then consider the increasing emphasis given to *evidence-based* practices, but balance this by reflecting on problems associated with them, especially concerning the reality of how evidence rightly interacts with issues of values in democratic government. We then propose areas for future research in order to provide a stronger basis for the consideration of expertise in support of government.

A Primer on Government

In order to provide context for the rest of the chapter, we offer a brief and necessarily simplified[1] description of how the UK government works. The UK has a type of government named a constitutional monarchy, which means that the Crown (the head of state—currently Queen Elizabeth the Second) does not make political decisions; these are instead taken by the government, and by parliament. The public elect people to represent them in parliament. The leader of the political party that is most likely to gain a majority in parliament (which could be through forming a coalition) is asked by the head of state to form a government, which then has responsibility for governing the country. This is done through government *departments*. A small number of nominated politicians (named ministers) from the government will have responsibility for setting policies for that department, usually led by a senior minister named the Secretary of State. The government of the day is supported by Her Majesty's Civil Service, whose duty it is to provide advice to ministers on particular issues. When the ministers do take decisions on what policy[2] should be adopted, the Civil Service design and implement the policy on their behalf. This brief description of the UK government is not complete until we cover accountability. Civil servants (often labeled as *officials*) are accountable to ministers,[3] who are responsible for their department. The ministers and the government in general are accountable to parliament, who as an elected body are accountable to the general public.

Challenges in Government

In order to discuss the framing and usefulness of expertise in government, we concentrate on those working in the areas of policy advice and development (known as the

---

[1] For an authoritative and comprehensive description, see Cabinet Office (2011), *The Cabinet Manual*.
[2] A working definition of policy, as used in the UK Ministry of Defence (MOD) training material, is "A choice leading to a course of action proposed or adopted by a government. A statement of intent, or a commitment to act."
[3] There are some exceptions: a civil servant always acts as the Accounting Officer for a department, accountable to parliament for the use of public money and the stewardship of public assets.

policy profession). Policy professionals specialize in providing advice to elected politicians who have a specific portfolio to lead (i.e., a department of state). By necessity we will also consider these politicians responsible for the strategic-level decisions in government.

Although there have been recent attacks on expertise, we would contend that government needs experts more than ever. Most countries are yet to recover from the worldwide financial crisis of 2008, which creates a huge problem for government in terms of budgetary decisions. One of the fundamental roles of government is to make decisions concerning how a finite amount of money is allocated to provide public services in support of the country. These resources need to be spread across the whole portfolio of public spending; from healthcare to education, income support to defense. Though government often focuses on *home* issues, by necessity it has to consider issues on a global scale. Domestic service provision has had to adjust to the large-scale migration issues associated with warfare (e.g., Syria and Iraq) and a breakdown of governments (e.g., following the Arab Spring) in many different countries. The nature of business and security threats have also changed. Where once national boundaries stood firm, states such as Russia are blurring the distinction between war and peace, and testing the international rules-based order; while globalization and the digital marketplace are also testing the concepts of national boundaries. A final, but vast challenge to government is that of Britain's secession from the European Union, which is not just a task that will require considerable effort; it is also one that will present unique and novel challenges.

The Challenge Posed by the Future in Government Policy and Planning

Government has to adapt to many challenges. However, the financial and reputational costs of *responding* to problems often far outweigh those required to *prepare* for them, and so governments cannot wait until issues emerge before they adapt. A significant part of the complexity that governments deal with is the requirement to make decisions about a future that is highly uncertain. The former governor of the Bank of England, Mervyn King, described this as radical uncertainty,[4] which he defines as "uncertainty so profound that it is impossible to represent the future in terms of a knowable and exhaustive list of outcomes" (King, 2016, p.12). Government is charged with investing money in capabilities or infrastructure that are required to last decades. The choices made in defense acquisition of platforms (e.g., submarines, ships, aircraft) need to be robust for up to 50 years; the US B52 aircraft was brought into service by the US military in 1952, and it is expected to still be in service in 2040. The complexity

---

[4] There are a variety of synonyms that describe irreducible uncertainty in addition to radical uncertainty, such as true/Knightian uncertainty, and deep uncertainty.

inherent in this choice is that government cannot provide answers to questions such as: Robust to *what* within that 50 years' timeframe? Where will wars occur in 50 years' time? Which global drivers will influence warfare? What will the security threats be? How will conflicts be fought? Who are our future adversaries?

We cannot assume that we can make inferences about any of these questions by performing a deterministic extrapolation of what we know now. Given this, we need to replace predictions and forecasts with something else. Modern *futures* approaches consider multiple possible future scenarios, against which different policy options are considered to allow a judgement as to how well they perform across the spectrum of possibilities (French, Rios, & Stewart, 2011).

The problem becomes yet more difficult if one considers what has been considered in the cognitive systems engineering literature as the *envisioned world problem* (Woods & Dekker, 2000); that is, we cannot design a system for the future based on an understanding of what is already in place, as we know that the world will change in ways we cannot predict *and* the system being designed will have an impact on whatever it comes into contact with. We believe this is also true for *policies*, given that policy interventions are *complex systems thrust amidst complex systems* (Pawson, 2006). As such, they will instigate changes to the interpretations, roles, and goals of stakeholders, that is, anyone influenced by the policy. In other words, policy interventions will cause people to *adapt* in unforeseen ways.

One obvious example of stakeholder adaptation is that of adversaries, who will also actively adapt to our intent and actions. It is possible, if not likely, that adversaries will operate under different constraints, rules, and timescales. The challenge to defense has been characterized as being like planning for a game, but where we don't know who will be on our team, who we will be playing against, or what the rules will be. However, defense policy cannot only consider the potential adaptation of adversaries; it is just as important to consider the effect of potential future policies on the positions and behaviors of coalition partners, host nation populations, and non-governmental organizations.

Of course, designing policy and plans with one eye on adaptation is nothing new, but traditional approaches have often relied on *normative* models of decision making and behavior, derived from rational decision theory. However, these approaches, such as modeling the development of relationships and confrontations via game theory, do not reflect the reality of behavior as captured by descriptive models of bounded rationality and organizational decision theory, such as the nature of uncertainty and the importance of building and maintaining trust with important stakeholders. Hence, modern approaches to dealing with the *envisioned world problem* maximize robustness by designing at the *goal* level, ensuring that the plans are flexible in order to cope with the change that will occur naturally due to the passage of time *and* as a result of system adaptation to the policy intervention[5] (e.g., Lempert, 2003). In a recent paper (Ward,

---

[5] In strategy parlance, we would describe this as a deliberate effort to ensure that the *ends* can be achieve through the flexible design of *ways* (of achieving the goals) and *means* (the resources required).

Gore, Hutton, Conway, & Hoffman, 2018), we went beyond robustness to consider skilled adaptation as part of high-level performance.[6] We observed that strategic goals are often stated in high-level or abstract terms, to allow those charged with implementation the necessary flexibility in terms of how they pursue the goals but *stay within intent*. The avoidance of over-specification (which would result in brittleness being exposed), and an acceptance of the need to manage by discovery recognizes that there will be not only emergent challenges that need to be addressed, but there will also be emergent opportunities. Hence, one might expect not only the way in which a goal is implemented to adapt over time, but also the goals themselves.

An Example of the Challenges in Government—Reasoning and Decision Making in the Iraq Conflict

The UK Ministry of Defence (MOD) has a specialist team led by Dr. Roger Hutton, a civilian director (equivalent to a military major general), which has undertaken its own assessment of the Iraq Inquiry (the *Chilcot Report*) in order to understand what the department needs to do in order to correct past problems and to improve. The team has identified a range of *themes* that are pervasive across the time period covered by the inquiry; these are set out in the Ministry of Defence publication *The Good Operation* (Ministry of Defence, 2018). The invariant across the themes is their psychological-social nature, such as poor decision making, inadequate reasoning about the future (*foresight*), and a lack of critical thinking and challenge. One issue that cuts across these themes described is "a tendency to think about what is going to happen in a linear way ... people are psychologically hard-wired to avoid changing direction" (Ministry of Defence, 2018, p. 25). This can be interpreted as an unwillingness or inability to move beyond a dominant interpretation of the situation, and to consider choices beyond the most dominant one, as they have been successful previously.

To examine each of these in more detail, the former is as a lack of effort and/or ability to generate and compare alternative explanatory frames and to challenge existing learned models, for instance when trying to anticipate adversary decisions (e.g., "they did *x* before so they are probably going to do the same again"). The latter might be interpreted as examples of recognition-primed decision making (Klein, 1993; see also Chapter 9, "Macrocognitive Models of Expertise," by Hutton, this volume), though with a failure to adequately simulate whether the choice would work in *this* situation (rather than the situations in which the choice has previously resulted in success), or the lack of ability to attend to how well the chosen action is working in the developing situation. Both of these are formalized in guidance and tools provided to policy officials, the former being supported through the use of futures methods, and the

---

[6] We defined adaptive skill as "Timely changes in understanding, plans, goals, and methods in response to either an altered situation or updated assessment of the ability to meet new demands, that permit successful efforts to achieve intent, or successful efforts to realize alternative statements of intent</i> that are not inconsistent with the initial statement but more likely to achieve beneficial results under changed circumstances" (Ward et al., 2018, p. 42; emphasis added here in italics).

latter through approaches to the *monitoring* phase of the policy cycle (see the UK Government's Magenta Book for policy evaluation: HM Treasury, 2011). As we have considered elsewhere (Ward et al., 2018) the capacity to adapt is central to expert performance under uncertainty. A particularly relevant theory for the problems described in MOD's Chilcot work is that of cognitive flexibility theory (Spiro, Feltovich, Jacobson, & Coulson, 1992), which describes the dangers of situations being characterized in an oversimplistic manner, and an associated resistance to protect and preserve misconceptions—referred to as *knowledge shields*. As knowledge shields include dismissing evidence contrary to a preferred explanation and fallacious reasoning, we believe that they underlie a lot of the issues to which *The Good Operation* refers.

The Overlap between Expertise and Values

Government decisions are rarely made through a *rational* analysis of the choices available, though such analyses may be one input. Governments are driven by their high-level goals, and these in turn are driven by *values*. Political parties differentiate themselves by the values they have. They create persuasive arguments based on these in the hope that the electorate will support them, which will in turn give them a mandate to translate the ideas into policies to develop and implement if they are elected into government.

The overlap between values and knowledge also creates a potentially unique situation concerning professional practice, with regard to who is elected into government. Many professional disciplines define what is required for someone to be recognized as a skilled practitioner, with thresholds defined by factors such as formal qualifications, levels of experience, and evidence of performing against specific competencies. However, the world of politics is often different to this. As an example, two world leaders, Donald Trump and Emmanuel Macron, had never held elected office prior to being elected as presidents of the United States and France, respectively. In the UK, it is entirely possible that the minister with responsibility for a given department has no expertise, or indeed previous experience, within that domain.

The reader should not be startled, however; ministers are not required to take decisions on their own. As previously mentioned, the government of the day is supported by Civil Service officials, who provide advice to ministers, speaking *truth unto power*, and then are charged with implementing the decisions once the minister has made up her or his mind. In the UK, the Civil Service is permanent, and is charged with having integrity, and being honest, objective, and impartial (The Civil Service Commission, 2010). In order to provide advice to ministers, the Civil Service will rely on its own experts, in addition to those outside of government, to provide research, understanding, and challenge.

Government therefore relies on expertise. But at times expertise clashes with the aforementioned values, for instance the highly publicized comments by the Conservative politician Michael Gove, on the topic of the potential exit of the United Kingdom from the European Union:

> I think the people of this country have had enough of experts with organizations from acronyms saying that... they know what is best and getting it consistently wrong because these people... are the same ones who have got consistently wrong... The organizations that many people are citing in this debate are organizations that have been wrong in the past and I think that they are wrong now. (Interview with Michael Gove; Sky News, 2016)

Comments of this kind were common in debates leading up to the UK's European Union membership referendum. Although government relies on experts, there is widespread criticism of experts, not least as experts do not have a monopoly on the factors that politicians consider, especially the aforementioned values. We next consider the varieties of expertise that government may or may not choose to utilize.

VARIETIES OF EXPERTISE

There continues to be scientific debates concerning the basis for expertise, with some researchers (e.g., Ericsson, Prietula, & Cokely, 2007) advancing a view that anyone can be an expert so long as they have sufficient motivation to engage in structured deliberate practice for a considerable period of time. However, other researchers (e.g., Hambrick, Macnamara, Campitelli, Ullén, & Mosing, 2016) contend that other factors make a significant contribution, such as genetics, personality, intelligence, and working memory capacity (for a review of this topic, see Ward, Belling, Petushek, & Ehrlinger, 2017). We believe there is merit in considering expertise from a variety of perspectives, such as considering expertise as a valid form of understanding knowledge, as a characteristic of an organization, and as a necessity for policy development.

Bohle Carbonell and Van Merrienboer (Chapter 12, "Adaptive Expertise," this volume) note the different schools of cognitive (e.g., Chi, 2011; Ericsson & Lehmann, 1996; Olsen & Rasmussen, 1989), developmental (e.g., Hatano and Inagaki, 1986; Pfeiffer, Chapter 5, "Giftedness and Talent Development in Children and Youth," this volume), and industrial (e.g., Van der Heijden, 2002) psychology, which have focused upon *individual* aspects of expertise. Hutton (this volume) describes the emergence of the study of *macrocognition* which considers the role of cognition under realistic task-environment conditions (e.g., adjusting goal trade-offs due to uncertainty and the emergence of new demands). Similarly, Naikar and Brady (Chapter 10, "Cognitive Systems Engineering: Expertise in Sociotechnical Systems," this volume) propose that expertise must be understood as part of dynamic work organizations or self-organizing systems.

Whilst these approaches have much to offer our understanding of experts and expertise there are alternative discussions that suggest that expertise is simply embodied in a *matter of doing* as well as knowing (Baber, Chapter 11, "Is Expertise All in the Mind? How Embodied, Embedded, Enacted, Extended, Situated and Distributed Theories of Cognition Account for Expert Performance," this volume); and that expertise can only be fully understood as an emergent property of experience (Flach and Voorhorse, Chapter 8, "Expertise: A Holistic, Experience-Centered Perspective," this volume). Collins (2014) describes the varieties of expertise possible as a property of social groups. These varying perspectives are important for the narrative of expertise for government.

Discussions within the public policy arena, particularly in the United States, stress that the rise of public interest in expertise presents a considerable challenge to traditional sources of expertise in policymaking that are embedded in systems of government (May, Koski, & Stramp, 2016). Complicating this further is not only the readily available knowledge that is accessible online (e.g., via Wikipedia), but also the varieties of expertise described by Collins (2014); these different varieties of *apparent* expertise may be lumped together as equally legitimate, making confusion and conflict inevitable.

Expertise and Evidence in Government

Framing Expertise and Evidence

In this section we focus on how expertise and evidence are framed, and brought to bear, on policymaking in government. Curiously, the *framing* of expert evidence supporting or refuting an issue has been relatively ignored in the mainstream theoretical and empirical literature on expertise (other than in the work of Collins). It is conceivable that this topic is too *messy* and unbounded to study with any degree of clarity, which makes generalization difficult. One way of viewing the way in which expert evidence is framed would be to (slightly mis)use Kahneman and Klein's (2009) heuristics for whether intuitive judgements in a given task domain should be trusted. These include: (i) the stability of the problem characteristics; (ii) the availability of situational cues that characterize a problem and its stability; (iii) opportunities available to the problem solver (which presumably includes the necessary goals and motivation) to study those cues; and (iv) opportunities to receive feedback on the accuracy of their judgements concerning the problem phenomenon. If the problem is sufficiently stable, cues are available, learning has been possible, and indeed successful then intuitive judgements can be trusted; if not, the advice is to rely on more of an analytic approach.

If these heuristics are applied to the challenges faced by government, then there would be a strong argument to not trust the intuitive judgements of policymakers.

This in itself might reduce interest in studying the expertise of policymakers. We believe, however, that this view, if adopted, would be misguided. Although we would agree that problems faced by governments are messy and complex, and the interaction of many factors makes each problem rare, skilled practitioners in many other domains are faced with similarly complex problems, and often achieve a successful resolution. Accordingly, the lack of research attempting to unpack the nature of policymaking expertise is unfortunate. It is important to note that valuable empirical and theoretical research focusing on the *real* nature of policymaking has been undertaken (e.g., Greenhalgh & Russell, 2005; Maybin, 2015; Stevens, 2011), but there has been little effort to explicate the expertise of those primarily responsible for forming policy. Perhaps the closest professional domain that has been the focus of expertise research is that of intelligence analysis (e.g., Baber, Attfield, Conway, Rooney, & Kodagoda, 2016; Roth et al., 2010; see also Jenkins & Pfautz, Chapter 32, "Expertise in Intelligence Analysis," and Moore & Hoffman, Chapter 42, "Cognition and Expert-Level Proficiency in Intelligence Analysis," both this volume), which plays a crucial role in supporting both the ongoing understanding of policymaker customers and the undertaking of specific assessments on their behalf.[7]

In the next section we examine where expertise sits in relation to government. Though we could provide a cross-sectional view for the current period of time, it is important to show how expertise was considered and framed through the history of the Civil Service, as it is a long-running issue, and has changed over the years.

Historical Perspectives on Expertise in Government

The first significant and most cited review of the Civil Service is the Northcote–Trevelyan report of 1854. However, this was preceded by the Report of the Macaulay Commission on the Indian Civil Service, which concerned itself with the recruitment of new employees. It recommended the selection of "men who have taken the first degree in arts at Oxford or Cambridge," and that the subjects studied should have no connection with any of the professions that they would be working in, but rather they should be subjects that "invigorate, and enrich the mind," such as knowledge of our poets, wits, and philosophers. The recommendations for the Civil Service entrance examinations were broad and included translating passages from Latin and Greek into English (and vice versa). Although used as a basis for initial selection in to the Civil Service, the emphasis on being an intellectual *generalist* only lasted up until the individual took a position on probation, which could only be passed through the demonstration of specialist knowledge and skills related to their position.

---

[7] The interested reader is guided towards Omand (2010) and Dover, Goodman, and Hillebrand (2013) for an overview of the area; and Freedman (1977) for a fascinating review of how strategic intelligence informed government on one specific topic—the Soviet nuclear threat.

The Northcote–Trevlyan report essentially created the modern Civil Service, based on values of meritocracy, rather than the nepotistic influence that had permeated the service to that date (both in terms of recruitment and progression through grades). Like the Macauley report, the Northcote–Trevlyan report also recommended that recruits be generalists. Both reports advanced the argument that *generalists* or *all-rounders* should constitute a community of middle- and upper-class men of good moral conduct and character.

The 1918 Haldane Report, undertaken with a fresh perspective following the First World War, adopted a different view, placing an emphasis on the value of specialisms and of knowledge linked to areas of departmental policy. It recognized that "adequate provision has not been made in the past for the organized acquisition of facts and information; and for the systematic application of thought, as preliminary to the settlement of policy and its subsequent administration" (p. 6). To overcome this, the report appealed to all departments to develop a specialized capacity for research and reflection before a policy is developed; and that these functions should be carried out by specialist employees. Moreover, it gave warning to ministers that they would not attain and retain public confidence unless they recognized the requirement to understand the effects on all sections of the community affected by the policy, and that public confidence would be moderated by the degree to which the advice was obtained from *within* the department. Hence, the Haldane report could be considered to be a volte-face from the previous belief in the recruitment of well-educated generalists to that of specialists who can undertake research, inquiry, and reflection to support policy and to advise ministers. It also suggested that these specialists should be within government departments, working with the generalists, to maximize public trust.

The Fulton review of 1968 contended that the Civil Service had not kept up with the demands placed on it by the modern world, and the culture of the generalist administrator, unfortunately, continued to dominate, asserting that: "the Civil Service is no place for the amateur" (p. 16). The idea of generalist administrators working alongside specialists was maintained, but the review advanced an argument that the administrators themselves needed to specialize in one of two overarching domains in which they work, namely economics/finance or social issues. The fear described in the review was a lack of some degree of specialism would be a continuation of the status quo, i.e., generalists providing advice and taking decisions on subjects that they do not fully understand. The Fulton review also provided a new direction concerning recruitment. Whereas previous reviews emphasized a good (but potentially irrelevant) degree from the Oxbridge universities, the Fulton review emphasized the recruitment of graduates who have learnt about *modern problems* (e.g., social, political, economic, scientific), which would be more likely to be found in universities formed in the twentieth century.

The Cabinet Office's (1999) *Modernising Government* review argued that there needed to be a focus on longer-term policy *outcomes* rather than implementing short-term interventions. It also placed an importance on Civil Servants collaborating with others, including outside experts, the *front-line* professions that implement policy, and the general public who are affected by policy. Perhaps the most interesting direction was to view policymaking as a continuous learning process with experimentation and evidence at its core. The review coined the phrase *what works* (used by subsequent governments), with a desire to use evidence to support a better understanding of the problem, and promoted the use of pilot schemes supporting innovation and testing whether policy options work in practice.

The Cabinet Office's (2012) Civil Service Reform Plan (CSRP) continued and advanced the dominant active themes. *Open policymaking* was the name given to the continuation of the emphasis given to the involvement of external experts and the general public throughout the policy process. Internally there was a focus on the skills that officials should be required to develop to implement effective policy, such as digitals skills, behavioral insights, and the use of data collection trials (e.g., randomized control trials; RCT) and other techniques for assessing whether observed effects can be reliably attributed to the proposed intervention.

In sum, the opinions concerning the role and type of expertise and how it has been framed has gone through considerable change over an extended period. There has been an acceptance of the need for specialist expertise, and for evidence to be used for understanding, testing of ideas, and organizational knowledge of *what works*. However, specific tensions remain, including to what degree policy officials should be specialists or generalists, and how should external expertise be assessed and used. We consider some of these issues in the following.

Current Perspectives

The Civil Service is currently structured along departmental and profession lines. There are over twenty professions, some of which operate as loose communities of interest, and some which operate in a much more formal manner, accrediting members and requiring job holders to possess accreditation for given roles. Most of the officials supporting the running of the professions do so in addition to, rather than as part of their primary job. Though the policy profession still debates the generalist versus specialist identity (with the notion of a *generalist specialist* being spoken about frequently), it has recently taken steps to codify the skills and knowledge areas required of policy officials, and is trialing a form of accreditation, whereby officials can be tested against three levels of its skills framework.

In order to provide effective advice to ministers, policy professionals have to bring three core factors together: relevant evidence, an understanding of how the policy can be implemented, and an awareness of the political landscape. To be judged to have provided effective advice to ministers, policy professionals have to bring these core factors together, and fuse them throughout the *policy cycle*. Whilst several departments have their own version, the generic version is known as the ROAMEF (rationale, objectives, appraisal, monitoring, evaluation, and feedback) cycle (HM Treasury, 2011). Ideally, the minister responsible should also be involved throughout the cycle, but their main decision will be to make a choice between competing options (in the appraisal phase), which are presented in a formal proposal.

We have already stressed the importance of having the front-line staff who deliver the policy interventions involved in the design phase (e.g., teachers for education policy and the military for defense policy). Their expertise is required to understand the realities of what will work, where, why, and how it needs to be designed and implemented. Rarely does a policy have a uniform effect, and so bringing the varying types of expertise together—from policy officials to front-line staff, members of the public to academic experts—increases the chances that the policy can be designed with these differential effects in mind (see Pawson, 2006). The politics part of the triumvirate is also nuanced, as the official has to consider what options are feasible in the current political climate, and also reflect upon the minister's own perspective and preferences.

Although we have noted the complicated role of the policy official, it is worth mentioning that perhaps the most difficult role is that of the decision maker her/himself. The ministers are the ones accountable to parliament and, of course, the general public. They have to utilize what Brown (2015) and we (Gore & Conway, 2016) have described as *hybrid judgement*; they have to consider the objective evidence against their own experience, and their judgements concerning how well parliament and the public will respond. This is not a matter of objective analysis; values have to be considered and emotional responses—personal and those of others—are important. An obvious difficulty is what to do if the evidence or advice contradicts their judgement and gut feelings about how well it will be received. In this case, they will somehow have to reconcile them and decide upon a course of action. We have taken you through a journey of how government has considered expertise within the Civil Service. So far, we have not spoken of issues concerning where expertise can be found, and how one should identify who or what is an expert. We turn to that now.

Identifying the Expert, and Who is the Arbiter?

The roles of policy officials and ministers are made more difficult because to benefit from expertise, someone has to be the arbiter of who the expert is. When experts are employed in government this can be a relatively straight-forward process; with consideration of qualifications, accreditations (e.g., chartership), tests of knowledge and skills, and experience against required competencies. However, it can be difficult for policy professionals to gauge the expertise of individuals from outside of government. In the absence of a recognized method for scrutinizing expertise, the process is far more subjective, and can involve the judgements of third parties (who are no more or less expert than the policy professionals or ministers).

There have been many debates in recent times concerning how expertise should be defined (see the many different definitions of expertise throughout this *Handbook*). Likewise, similar debates have occurred in practice about who is an expert and whether they should be believed, but we note far less consideration of who is the arbiter of the judgement about the expertise being provided. The public are exposed to myriad people described as experts via broadcast and social media, but there is little discussion of whether the media channels are trustworthy arbiters, or of what criteria they bring to bear on judgements of expertise. We believe that it is not uncommon for some of these individuals to only be self-declared experts. And yet often it is these individuals who people see via the media. Given the influence of self-proclaimed experts and of media channels who offer them a platform, we think it would be wise for government to equip itself with methods for judging the *candidate* experts. It would require more space than we can afford here to go into depth on which methods can be used, but we think the principles advanced by Hoffman et al. (2014) and Collins and Evans (2007) could be a very useful place to start.

Misrepresentation of Expert Judgement?

Accepting the premise that media channels might not be the most accurate arbiters of the bounds of an individual's professional expertise, there may also be a widespread misunderstanding about the domain- and task-specific nature of expertise. In particular, we are concerned about an erroneous view of expertise that is broader than it actually is. Hence, the seemingly increasing perception that expert judgement is untrustworthy may not reflect the actual expertise of the individual(s) in question; rather, it may reflect the misrepresentation of the generalizability of expertise; i.e., how far the expertise can be transferred or how well an expert can adapt (see Ward et al., 2018). This often takes the form of *Expert A*, recognized for their expertise in *Area A*, is questioned by the media about *Area B*. When the judgement they make about *Area B* is later exposed to be poor (relative to some objective criterion, such as a forecast compared to what actually comes to pass), conclusions abound such as *Expert A is not an expert after all*. However, we see little scrutiny of the knowledge, skills, and/or understanding required for expert performance in Areas A and/or B, and most importantly, if or how the knowledge/skill requirements for these areas overlap. Without this information and, hence, without a grasp of whether an expert *should or should not* be expected to perform well in a given area (even if they are an expert on something un/related), one could make numerous erroneous judgements that negatively impact policy (e.g., inappropriately trusting an expert who is operating outside their area of expertise, inappropriately expecting an expert in one area to be an expert in other areas, devaluing an expert's knowledge because of their poor performance on a topic that requires a different knowledge base or skill set). The problem here is not the expert, it is the *misrepresentation* of the expert, and in particular the misrepresentation of the bounds of where their professional expertise lies. It follows that there is a need for a better understanding of expertise and its limits, and a better ability to make judgements concerning expertise and its applications. We have considered current issues concerning expertise, and now we will turn our attention to evidence and its use.

Expertise vs Evidence

As described earlier, there has been an increasing emphasis on the role of evidence in policy. *Evidence-based* approaches were originally developed in the field of medicine, and represented the growing influence of clinical epidemiology and rigorous experimental methods (e.g., RCT) to supplement the judgement of clinicians. One of the main drivers for this was to reduce unnecessary variability (and sometimes, poor judgements) in clinical practice. The importance of epidemiological and mathematical methods and techniques to evidence-based medicine (EBM) is apparent when one considers the definition offered by Greenhalgh (2014, p. 1): "Evidence-based medicine is the use of mathematical estimates of the risk of benefit and harm, derived from high-quality research on population samples, to inform clinical decision-making in the diagnosis, investigation or management of individual patients." The rationale for supplementing clinical intuition is that no one clinician could possibly experience a sufficient range of cases, for all conditions, to permit them to be able to make accurate judgements and decisions across the range of possible cases they will be exposed to, which could vary along many different dimensions. EBM also provides advice on where the most appropriate evidence can be found, with possibly the best example considering the *levels of evidence* required for different decision problems, such as: How common is the problem? Is this diagnostic test accurate? And, does this intervention help? (OCEBM Levels of Evidence Working Group, 2011). However, among the sources of evidence represented, expertise plays a role, as there is recognition that in the absence of other evidence, mechanism-based reasoning (by a clinician) should be used. Experts as a source of evidence are also recognized in the field of evidence-based management (Barends, Rousseau, & Briner, 2014).

It is wise, however, to notify the reader that evidence-based approaches to decision making are not as straight forward or as linear as one might expect. EBM produces authoritative *practice guidelines* to be applied in a deductive manner in order to reduce variability in clinical judgements and decisions. However, they have been criticized for several reasons. Timmermans and Mauck (2005) identified that there is a great deal of variability with regard to adherence to the guidelines themselves, for a number of reasons, not least the fact that professional clinicians spend years qualifying to practice and honing their skills, and are then relied upon to make decisions with a great deal of autonomy. Spence (2010) identified that the deductive approach of applying guidelines has effectively produced a reductionist culture where individual problems (e.g., diseases) are treated in isolation by deferring to the guidelines. In patients with comorbidity (multiple issues/diseases, potentially interacting), for instance, the treating of the separate parts rather than the whole has caused patient deaths due to adverse drug reactions. Greenhalgh (2014) echoed this sentiment by highlighting that *evidence does not exist in a vacuum*. Rather than blindly considering epidemiological data from a population, or intervention data from randomized control trials, the clinician needs to develop a rich description of what is happening to *this* patient and how it is different to normal, what this means to them, and what is important to them in terms of outcomes.

The clinician then needs to synthesize the information, consider the nuances of variation between the factors in this case against all of the cases they have experienced before, and arrive at an initial clinical question that can be tested via further observation and testing (Montgomery, 2006). Often, clinicians will consider a number of conceptually similar issues (e.g., diseases) that could each plausibly explain the patient's story. The individual disease members of these *logical competitor sets* (Feltovich, Johnson, Moller, & Swanson; 1984) would then be used as a basis for hypotheses developed to rule in, rule out, or to narrow the number of competitors in the differential diagnosis, to eventually arrive at an accurate judgement of the responsible disease(s) rather than their conceptually similar neighbors (for an example of how this has been found in expert clinical judgement of sepsis, see Patterson et al., 2016). When viewed in this manner, we agree that evidence does not exist in a vacuum, and evidence-based approaches are rarely independent from the expertise of the practitioner interpreting them.

In comparison to evidence-based *policy* (EBP), one could argue that evidence-based medicine is relatively straightforward (though the reader may be forgiven for not believing this given the summary above). However, the treatment of EBM above provides a useful grounding in the issues one can face when attempting to develop and implement evidence-based approaches to policy. To some extent the justification for evidence-based policy is similar to that of medicine; there is often an appeal to the views that as public funds are consumed, and as real-world outcomes are influenced, policy decisions should be made based on the best available evidence. Implicit in this argument is the concept that evidence has a direct link to policy. Recall that this linkage is characterized in the interrelationship between the *three factors* of the policy profession: evidence, implementation, and politics; which should be considered throughout the policy cycle.

As mentioned earlier, there has been an increase in the emphasis of the use of evidence in policy. In addition to the various reviews of the Civil Service, influential papers have been published that set a vision for EBP and make links between it and EBM, for instance through the use of RCTs in policy (e.g., Haynes, Service, Goldacre, & Torgerson; 2012). The use of RCTs helps to deliver an understanding of what works in practice, and addresses the *two communities problem* (Caplan, 1979)—i.e., academic scientists and policymakers have different languages and cultures, and at times this may lead to problems of both supply and demand (National Research Council, 2012). Seven *What Works* centers were created in the UK to bridge this gap, effectively acting as knowledge translators and brokers between academic research and those developing or implementing government policy.

One might assume that the existence of organizations with a *foot in both camps* might address the problem. However, policy researchers such as Paul Cairney have described in great detail why there is rarely a direct and unproblematic link between evidence and policy. Cairney (2016) highlights that policymakers act under conditions of bounded rationality (following Simon, 1957). Such factors include obvious ones such as uncertainty (in a probabilistic sense) and ambiguity (different ways of interpreting a problem). However, perhaps less obvious is the way in which ambiguity interacts with persuasion and argument, as policy is as much about debates on values. Several authors have recognized the influence of external events on the consideration of evidence in policy decisions. For instance, some researchers have documented that the lack of consistency of a policymaker's attention to a given policy issue is influenced by external events that prompt or demand attention to be shifted to the issue (e.g., a disaster with many casualties that receives wide news coverage) (see the discussion of punctuated equilibrium theory, in Baumgartner & Jones, 2009). Similarly, the apparent plasticity of how evidence can be interpreted in light of effort and pressure from advocacy groups (e.g., advocacy coalition framework, see Jenkins-Smith, Nohrstedt, & Weible, 2014). These effects provide an additional challenge to (and an opportunity to develop) macrocognitive approaches to sensemaking (see Hutton, this volume), as they indicate that one's mental model or frame can be adopted, elaborated, questioned, and reframed due to political pressures and arguments around values. If this is indeed the case, then there is the opportunity to extend and assess existing macrocognitive models such as the data-frame model (Klein, Moon, & Hoffman, 2006) to account for sensemaking and associated action in the policy and political domain.

The complexity inherent in government policymaking represents a great challenge, but also a great opportunity. In particular, there is a great opportunity for interdisciplinary research to provide more accurate descriptive models of how policy is actually developed, for instance crossing the boundaries between macrocognitive theory and methods, and the policy research undertaken by people such as Cairney. If research is to provide a more useful basis for prescriptive interventions, then we believe it must capture the reality of policymaking across levels (e.g., individual, organizational) and be viewed through different disciplinary lenses, working together to consider the whole.

The Death of Expertise or Simply a *Strawman* Fallacy?

As we have flagged, there are narratives in the world-wide press that criticize expert judgement, arguably because of the erroneous judgments about expertise that we have described already. Though the Internet and global communication has provided people with more information than they have ever had before, it has also given them the *illusion* of knowledge when in fact most are skimming and confirming bias of extremely scant understanding and often ignorance. In some sections of society, ignorance, especially related to current affairs, has somehow become a virtue. And while Nichols' (2017) analysis is limited and a little pedestrian—the sentiment of his argument is worthy of acknowledgement. He notes that "To reject the advice of experts is to assert autonomy, a way for Americans to insulate their increasingly fragile egos from ever being told they're wrong about anything. It is a new Declaration of Independence: No longer do we hold these truths to be self-evident, we hold all truths to be self-evident, even the ones that are not true. All things are knowable and every opinion on any subject is as good as any other" (Nichols, 2017, p. 78). There is a danger, however, that expert advocates of expert knowledge—such as Nichols, or the UK's Brian Cox (who does much to bring science to the public, as Carl Sagan did), may themselves be dragged into the above narrative and be seen as elites who are simply self-serving academics. However, the success of science programming in the mainstream media—such as those hosted by Cox—suggests that there *is* a continued public appetite for expert knowledge.

Rod Lambert, the Deputy Director of the Australian National Centre for Public Awareness of Science (2017) provides a thoughtful review of Nichols' work and encourages readers to read the arguments very carefully. It is worth noting the recommendations made by Nichols to help bridge the expert–public gap and to encourage a greater level of thinking about claims:

- Experts should strive to be more humble;
- Readers (the public; consumers of expert knowledge) should be:
- Ecumenical—i.e., vary information sources, especially when thinking about politics;
- Less cynical—here he counsels against assuming people are intentionally lying; and
- Everyone—society, citizens, and experts—should check sources scrupulously in order to avoid confirmation bias.

These recommendations, if used in tandem with Popper's proposal that one should always try to *disprove* a claim rather than simply seek evidence to support it (Popper, 1962), are useful in helping people to consider claims generally, and also claims made about expertise.

We have heard arguments that suggest that expertise is in peril. However, we consider this to be a strawman fallacy that misrepresents experts and what it is to be an expert. Proponents of this argument criticize the fallibility of experts, but it is important to recognize that experts have rarely or never said that they always have all of the answers. This paradox was clearly reflected by Richard Feynman, Nobel prizewinner and iconic physicist, who, when asked to define science, stated that it is "a satisfactory philosophy of ignorance." Most experts know that they do not have all of the answers to complex questions; they simply have a deeper and more satisfactory domain understanding.

Conclusions

It is worth reflecting on Salas, Rosen, and DiazGranados's (2010) review of expertise-based intuition and decision making in organizations in which they propose that expertise is a construct that has value for organizations such as government. They contend that intuitive expertise plays a major role in organizations and although there is compelling evidence on how intuitive expertise works there is still much to be done to examine the conditions under which it works best and how to improve intuitive expert decision making. This is especially true in the field of government policymaking. Government has to make decisions that will commit to investments that will be in place in 50 years' time (or longer), while at the same time recognizing that it cannot predict the future, and its adversaries can seek to undermine any decisions. Government investments are not just about *what works*; they are about values too. More precisely, they are about what politicians and the electorate care about (i.e., what is *right*) and how these values change as a function of both external events and deliberate attempts to influence dominant narratives. The UK government has a deeper and more nuanced understanding of its requirement for both expertise and evidence, but challenges remain. We believe that the desire to further professionalize policymaking is commendable, and suggest that it can be supported by having theory and empirical evidence on what constitutes a policy *expert*, and how policymaking is done in reality. Having richer and more accurate descriptive models will allow for much stronger interventions, such as evidence-based training interventions, and analytic support and advice that uses more accurate assumptions concerning strategic decision making in government.

Acknowledgments

We would like to thank Paul Ward, Emilie Roth, and James (Jim) Maltby for their extremely helpful comments on previous versions of this chapter.

References

Baber, C., Attfield, S., Conway, G. E., Rooney, C., & Kodagoda, N. (2016). Collaborative sensemaking during simulated intelligence analysis exercises. *International Journal of Human-Computer Studies 86*, 94–108.

Barends, E., Rousseau, D. M., & Briner, R. B. (2014). *Evidence-based management: The basic principles*. Amsterdam: Centre for Evidence-Based Management.

Baumgartner, F., & Jones, B. (2009). *Agendas and instability in American politics*. Chicago: Chicago University Press.

Brown, R. V. (2015). Decision science as a by-product of decision-aiding: A practitioner's perspective. *Journal of Applied Research in Memory and Cognition 4*(3), 212–220.

Cabinet Office (1999). *Modernising government*. London: Her Majesty's Stationery Office.

Cabinet Office (2011). *The Cabinet manual—A guide to laws, conventions and rules on the operation of government*. London: Her Majesty's Stationery Office.

Cabinet Office (2012). *The Civil Service reform plan*. London: Her Majesty's Stationery Office.

Cabinet Office (2014). *What works? Evidence for decision makers*. Accessed at https://assets.publishing.service.gov.uk/government/uploads/system/uploads/attachment_data/file/676801/What_works_evidence_for_decision_makers_update_2018_01_12.pdf

Cairney, P. (2016). *The politics of evidence-based policy making*. London: Palgrave Macmillan.

Caplan, N. (1979). The two communities theory and knowledge utilization. *American Behavioral Scientist 22*(3), 459–470.

Chi, M. T. H. (2011). Theoretical perspectives, methodological approaches, and trends in the study of expertise. In Y. Li & G. Kaiser (Eds), *Expertise in mathematics instruction: An international perspective* (pp. 17–39). New York: Springer.

Collins, H. M. (2014). *Are we all scientific experts now?* London: Polity Press.

Collins, H. M., & Evans, R. (2007). *Rethinking expertise*. Chicago: University of Chicago Press.

Dover, R., Goodman, M., & Hillebrand, C. (Eds). (2013). *Routledge companion to intelligence studies*. London: Routledge.

Ericsson, K. A., & Lehmann, A. C. (1996). Expert and exceptional performance: Evidence of maximal adaptation to task constraints. *Annual Review of Psychology 47*(1), 273–305.

Ericsson, K. A., Prietula, M. J., & Cokely, E. T. (2007). The making of an expert. *Harvard Business Review 85*(7–8), 114–121.

Feltovich, P. J., Johnson, P., Moller, J., & Swanson, D. (1984). LCS: The role and development of medical knowledge in diagnostic expertise. In W. Clancey & E. Shortliffe (Eds), *Readings in medical artificial intelligence: The first decade* (pp. 275–319). Reading, MA: Addison Wesley.

Freedman, L. D. (1977). *US Intelligence and the Soviet Strategic Threat*. London: Macmillan.

French, S., Rios, J., & Stewart, T. J. (2011). *Decision analysis and scenario thinking for nuclear sustainability*. Working paper.

Fulton Committee (1968). *The Civil Service. Vol 1. Report of the Committee 1966–1968*. London: Her Majesty's Stationary Office.

Gore, J., & Conway, G. E. (2016). Modeling and aiding intuition in organizational decision making: A call for bridging academia and practice. *Journal of Applied Research in Memory and Cognition 5*(3), 331–334.

Greenhalgh, T. (2014). *How to read a paper: The basics of evidence-based medicine*. Chichester, UK: Wiley.

Greenhalgh, T., & Russell, J. (2005). Reframing evidence synthesis as rhetorical action in the policy making drama. *Healthcare Policy 1*(1), 31–39.

Haldane Committee (1918). *Report of the machinery of government*. London: His Majesty's Stationary Office. Retrieved from https://www.civilservant.org.uk/library/1918_Haldane_Report.pdf.

Hambrick, D. Z., Macnamara, B. M., Campitelli, G., Ullén, F., & Mosing, M. A. (2016). Beyond born versus made: A new look at expertise. *Psychology of Learning and Motivation 64*, 1–55.

Hatano, G., & Inagaki, K. (1986). Two courses of expertise. In H. Stevenson, H. Azuma, & K. Hakuta (Eds), *Child development and education in Japan* (pp. 262–272). New York: W. H. Freeman.

Haynes, L., Service, O., Goldacre, B., & Torgerson, D. (2012). *Test, learn, adapt: Developing public policy with randomised controlled trials*. London: Cabinet Office Behavioural Insights Team.

HM Treasury (2011). *The Magenta Book—Guidance for evaluation*.

Hoffman, R. R., Ward, P., DiBello, L., Feltovich, P. J., Fiore, S. M., & Andrews, D. (2014). *Accelerated expertise: Training for high proficiency in a complex world*. Boca Raton, FL: Taylor and Francis/CRC Press.

Jenkins-Smith, H., Nohrstedt, D., & Weible, C. (2014). The advocacy coalition framework: Foundations, evolution, and ongoing research' process. In P. Sabatier & C. Weible (Eds), *Theories of the policy process* (3rd edn). Chicago: Westview Press.

Kahneman, D., & Klein, G. (2009). Conditions for expertise: A failure to disagree. *American Psychologist* 64(6), 515–526.

King, M. (2016). *The end of alchemy: Money, banking, and the future of the global economy.* New York: W. W. Norton.

Klein, G., Moon, B., & Hoffman, R. R. (2006). Making sense of sensemaking 1: A macrocognitive model. *IEEE Intelligent Systems* 21(5), 88–92.

Klein, G. A. (1993). A recognition-primed decision (RPD) model of rapid decision making. In G. A. Klein, J. Orasanu, R. Calderwood, & C. E. Zsambok (Eds), *Decision making in action: Models and methods* (pp. 138–147). Norwood, NJ: Ablex.

Lempert, R. J. (2003). *Shaping the next one hundred years: New methods for quantitative, long-term policy analysis.* Rand Corporation.

Macaulay Commission (1854). *The report of the Macaulay commission on the Indian civil service.* London: Her Majesty's Stationary Office.

May, P. J., Koski, C., & Stramp, N. (2016). Issue expertise in policy making. *Journal of Public Policy* 36(2), 195–218.

Maybin, J. (2015). Policy analysis and policy know-how: A case study of civil servants in England's Department of Health. *Journal of Comparative Policy Analysis: Research and Practice* 17(3), 286–304.

Ministry of Defence (2018). *The Good Operation: A handbook for those involved in operational policy and its implementation.* Retrieved from https://www.gov.uk/government/publications/the-good-operation

Montgomery, K. (2006). *How doctors think: Clinical judgment and the practice of medicine.* Oxford: Oxford University Press.

National Research Council. (2012). Using science as evidence in public policy. Committee on the use of social science knowledge in public policy. In. K. Prewitt, T.A. Schwandt, & M. L. Straf (Eds), *Division of behavioral and social sciences and education.* Washington, DC: National Academies Press.

Nichols, T. (2017). *The death of expertise.* New York: Oxford University Press.

Northcote, S. H., & Trevelyan, C. E. (1854). *Report on the organisation of the permanent Civil Service.* London: Her Majesty's Stationary Office.

OCEBM Levels of Evidence Working Group. (2011). The Oxford 2011 Levels of Evidence. Oxford Centre for Evidence-Based Medicine. Available at http://www.cebm.net/index.aspx?o=5653.

Olsen, S. E., & Rasmussen, J. (1989). The reflective expert and the prenovice: Notes on skill-, rule-, and knowledge-based performance in the setting of instruction and training. In L. Bainbridge & S. A. Ruiz-Quintanilla (Eds), *Developing skills with information technology* (pp. 9–33). Chichester: Wiley.

Omand, D. (2010). *Securing the state (intelligence and security).* London: Hurst.

Patterson, M. D., Militello, L. G., Bunger, A., Taylor, R. G., Wheeler, D. S., Klein, G., & Geis, G. L. (2016). Leveraging the critical decision method to develop simulation-based training for early recognition of sepsis. *Journal of Cognitive Engineering and Decision Making*, 10(1), 36–56.

Pawson, R. (2006). *Evidence-based policy: A realist perspective.* London: Sage.

Popper, K. R. (1962). *Conjectures and refutations: The growth of scientific knowledge.* New York: Basic Books.

Roth, E. M., Pfautz, J. D., Mahoney, S. M., Powell, G. M., Carlson, E. C., Guarino, S. L., ..., & Potter, S. S. (2010). Framing and contextualizing information requests: Problem formulation as part of the intelligence analysis process. *Journal of Cognitive Engineering and Decision Making* 4(3), 210–239.

Salas, E., Rosen, M. A., & DiazGranados, D (2010). Expertise-based intuition and decision making in organizations. *Journal of Management* 36(4), 941–973.

Shanteau, J., Weiss, D. J., Thomas, R. P., & Pounds, J. C. (2002). Performance-based assessment of expertise: How to decide if someone is an expert or not. *European Journal of Operational Research* 136, 253–263.

Simon, H. (1957). *Administrative behavior.* London: Macmillan.

Sky News (2016). EU: In or out? Faisal Islam Interview with Michael Gove, 3 June 2016.

Spence, D. (2010). Why evidence is bad for your health. *BMJ: British Medical Journal* 341, c6368.

Spiro, R. J., Feltovich, P. J., Jacobson, M. J., & Coulson, R. L. (1992). Cognitive flexibility, constructivism, and hypertext: Random access instruction for advanced knowledge acquisition in ill-structured domains. In T. Duffy & D. Jonassen (Eds), *Constructivism and the technology of instruction* (pp. 57–75). Hillsdale, NJ: Lawrence Erlbaum.

Stevens, A. (2011). Telling policy stories: an ethnographic study of the use of evidence in policy-making in the UK. *Journal of Social Policy* 40, 237–255.

The Civil Service Commission (2010). *The Civil Service Code.* Accessed at https://www.gov.uk/government/publications/civil-service-code/the-civil-service-code

Timmermans, S., & Mauck, A. (2005). The promises and pitfalls of evidence-based medicine. *Health Affairs* 24(1), 18–28.

Van der Heijden, B. I. J. M. (2002). Prerequisites to guarantee life-long employability. *Personnel Review* 31(1), 44–61.

Ward, P., Belling, P., Petushek, P., & Ehrlinger, J. (2017). Does talent exist? A re-evaluation of the nature-nurture debate. In J. Baker, S. Cobley, J. Schorer, & N. Wattie (Eds), *Routledge handbook of talent identification and development in sport* (pp. 19–34). London: Routledge.

Ward, P., Gore, J., Hutton, R. J. B., Conway, G. E., & Hoffman, R. R. (2018). Adaptive skill as the *conditio sine qua non* of expertise. *Journal of Applied Research in Memory and Cognition* 7(1), 35–50.

Woods, D. D., & Dekker, S. (2000). Anticipating the effects of technological change: A new era of dynamics for human factors. *Theoretical Issues in Ergonomics Science* 1(3), 272–282.

SECTION V

CURRENT ISSUES AND THE FUTURE OF EXPERTISE RESEARCH

SECTION EDITOR: PAUL WARD

CHAPTER 49

THE "WAR" ON EXPERTISE

Five Communities that Seek to Discredit Experts

GARY KLEIN, BEN SHNEIDERMAN,
ROBERT R. HOFFMAN, AND ROBERT L. †WEARS

INTRODUCTION

THERE is a current movement to discredit experts. Several communities of practice are fostering mistrust of the judgments and decisions taken by experts. We call the adherents of this movement "expertise-deniers." In many ways, this feels like a war—a war for intellectual turf, scientific credibility, and even economic or political gain. This movement is having some success.

Some have noticed the downgrading of expertise. Thomas Nichols (2017), in his book *The Death of Expertise*, argues that our society is rejecting experts because of factors such as the wide availability of information on the Internet, allowing ordinary citizens to believe they know everything about any given topic. However, the changes Nichols identifies are inadvertent. They are unintended consequences of the advances of technology in our information age. Our concern in this chapter is with *intentional* behaviors—the deliberate attempts by at least five communities, each with its own interests, to diminish the credibility of experts and to devalue their contributions. The goal of this chapter is to correct these misleading challenges and to advocate for a collaboration with members of the five communities, using the challenges they raise to find ways to support and strengthen expertise.

We have spent our careers studying and advocating for individuals who have achieved the highest reaches of human potential and achievement. For example,

Klein (1998) wrote *Sources of Power* twenty years ago to describe the strengths of experts and so counter the claims that people, including experts, were prone to defective thinking. These issues continue to be relevant and, if anything, the attacks on experts have become more strident, and have been voiced by more communities. We feel our society would be far worse off were it not for experts' skills and contributions. Ours is a positive scientific psychology, not a negative one, but in this chapter we feel compelled to go on the offensive. As stated in the preceding paragraph, we hope to engage productively with each of these communities to improve the performance of experts through better training and technology. However, our primary goal in writing this chapter is to counter the misleading claims of the expertise-deniers.

Consider this example: Corporate leaders at a major petrochemical company are planning to proceduralize the decisions made by panel operators at their production plants. The rationale is that people (in general) are biased and can't be trusted. Therefore, why make any investments to build the skills of the panel operators? The company wants to rely on checklists and procedures instead. This example illustrates how organizations are devaluing their own expertise and seeking to rely on mechanical means of making inferences and decisions.

The general public seems to be of two minds on the matter. One often hears people say that weather forecasters do a lousy job. Or that economists cannot predict their way out of a paper bag. On the other hand, we have all heard people say something like "Uncle Fred has this rare cancer and so we went to the Smith Clinic where they have *the* experts." And who among us would willingly swap a randomly selected pilot for Chesley Sullenberger during an emergency landing?

When we think of experts we think of Dave Hackenberg, a Pennsylvania beekeeper wintering in Florida in 2006 who was among the first to sound the alarm about honeybee colony collapse disorder in the USA, and James Andrews, an orthopaedic surgeon renowned for his work with wounded warfighters and athletes. Our conception of experts includes Steve Jobs overseeing the design of the iPhone without conducting market research studies. Experts often take controversial positions such as Francis Oldham Kelsey, the US official in charge of the Food and Drug Administration review of the drug thalidomide, who consistently rejected efforts to gain approval for its use with pregnant women because her experience made her skeptical of its efficacy and its safety. Angela Merkel, the Chancellor of Germany since 2005, certainly qualifies as an expert political leader. Alan Mulally, who turned around both Boeing and Ford, is an organizational expert. We would also include the Apollo 13 Mission Operations Team, Thomas Paine, Glynn Lunney, Gene Kranz, Gerry Griffin, Milt Windler, and Sigurd Sjoberg, who performed heroics to bring back the crippled spacecraft. Each of these experts made a significant difference. Had they been replaced by journeymen, we would not expect the success stories that we celebrate.

Our definition of expert is simple: people who can do things that hardly anyone else can do—based on their own skills.

In this chapter we focus on five battlegrounds—the five fronts on which the attack on experts is being waged. These fronts include decision researchers seeking to develop models that can outperform experts; sociologists who see expertise as merely a social attribution; Heuristics and Biases researchers who claim that expert judgment suffers from the same biases as everyone else; practice-oriented researchers seeking to replace professional judgments with data-based prescriptions; and computer science specialists who believe that it is only a matter of time before artificial intelligence surpasses experts (Figure 49.1).

And while we are expert-advocates, eager to highlight the strengths of experts, we acknowledge that experts are not perfect and never will be. That said, it has been empirically demonstrated that experts offer distinct capabilities beyond those of less skilled decision makers. We do not go into detail about the abilities of experts because other chapters in this *Handbook* have already covered this material, but for context we do need a brief description of the strengths that experts bring.

Experts are the people to call on for handling tough situations because they often deal with novel situations and come up with continuous and disruptive improvements, they enlist the cooperation of other experts when necessary, and they take responsibility for their actions. People are identified as experts because in their field of specialty their judgments in the past have been shown to be highly accurate and reliable and these qualities continue to distinguish their performance. Experts employ more effective strategies than others, with less effort. Experts perceive patterns in data and conceive meaning in the patterns that others cannot detect. Experts form rich mental models of cases or situations to support sensemaking and anticipatory thinking. Experts' domain knowledge is extensive, detailed, and highly organized. Experts are intrinsically motivated to work hard on hard problems. They have vivid memories of their past mistakes and are driven to avoid ever making them again. They are intrinsically curious and motivated to stretch their skills, knowledge, and capabilities.

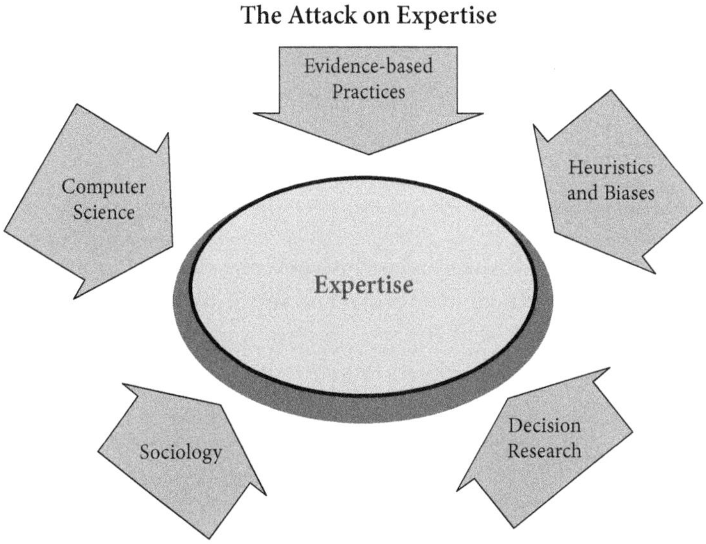

The Attack on Expertise

FIGURE 49.1 The five communities engaging in expertise-denial.

Experts are very knowledgeable about their organization and its history, culture, and operations. They are highly regarded by peers. People depend on them for mission-critical, complex technical guidance and high-stakes decision-making.

Experts are needed to construct all the mechanisms that are being suggested to replace experts—checklists, decision models, artificially intelligent actors, etc. Experts are also needed to evaluate, monitor, refine, and revise existing and improved systems, and to analyze what went wrong when the systems failed.

While our focus in this chapter is on the war being waged by scientific communities, it must be noted that the war is being waged on philosophical and political fronts as well. In philosophy, considerations of the concept of expertise, and scientific expertise in particular, have led to interesting debate on the question of why people should take the advice of scientists, technologists, or expert witnesses (e.g., Collins & Evans, 2006). This philosophical debate extends to the question of what should be the relation between authority and legitimacy. Expertise is a philosophical/political problem for a number of reasons: (a) it violates the notion of democratic equality, and (b) experts' exercise of their judgment is "out of the reach of democratic control" (Turner, 2006). Society would face the choice between rule by experts versus rule by democracy. Furthermore, if knowledge is seen as having value, then an egalitarian would insist that everyone be made expert through the dissemination of knowledge. This chapter does not examine these philosophical and political questions, and we only bring them up here to provide some context for the scientific debates about expertise.

We now discuss the war being waged by scientific communities. We describe the attacks launched by expertise-deniers and present our rebuttals of these attacks. We start with the challenge from the Decision Research community because this body of work can be traced back the furthest, to the mid-1960s. The critique made by this community echoes to this day.

The Attack from the Decision Research Community

A number of experiments in experimental psychology are used to make a strong case about human cognitive limitations broadly, and the limitations of so-called experts in particular. One of the most striking and powerful lines of research has shown that mathematical equations can outperform experts.

> During the past 30 years, researchers have categorized, experimented, and theorized about the cognitive aspects of forecasting and have sought to explain why experts are less accurate forecasters than statistical models... A number of researchers point to human biases... with few exceptions [experts] are worse at forecasting than are Bayesian probabilities based on historical, statistical models.   (Johnston, 2005, pp. 21, 65–66)

Einhorn (1972), Dawes (1971, 1979), Meehl (1954, 1965), Grove et al. (2000) and others have shown that simple formulas (linear models) can outperform experts in various kinds of judgment tasks. This research involves determining the key variables and then forming an equation in which values for each of the variables are weighted and then summed and subjected to correlational analysis. Examples are faculty ratings of graduate students versus a model based on grades and test scores, or physicians' ratings of cancer biopsy results versus statistics on survival. "Some 10 years after his book was published, Meehl (1965) was able to conclude... that there was only a single example showing clinical judgment to be superior [to linear models]" (Dawes, 1979, p. 573). To explain these results, and referencing expertise, Dawes asserted that "[P]eople—especially the experts in a field—are much better at selecting and coding information than they are at integrating it" (p. 573).

The Decision Research community wants to reduce our trust in experts because the decision researchers are convinced that their linear models should replace expert judgments. Hastie and Dawes (2001) stated that "whenever possible, human judges should be replaced by linear models" (pp. 62–63).

Rebuttal 1: Where Did the Formulas Come From?

What usually escapes mention is that the formulas were originally derived from the advice of experts about what the key variables are, the variables that experts themselves use in making their judgments. So, it is misleading to claim that formulas outperform experts as if the formulas were independently derived.[1]

Rebuttal 2: The Formulas are Brittle

The advantage of the linear models is that they are consistent, but when they fail, they fail miserably. They fail on tough cases involving departures from historical/statistical trends, but they especially fail on cases where human knowledge, reasoning, and context are crucial. For example, Meehl discussed *broken leg* cues (see also Salzinger, 2005). A linear model might do a decent job of predicting, on the basis of a host of variables, whether a given person is likely to go to the movies this weekend, but remains blind to the fact that the person in question just broke his/her leg. "If experts have privileged information that is not reflected in the statistical table, they will actually perform better than does the table" (Johnston, 2005, p. 65).

---

[1] What about a purely data-driven, inductive approach, as with some forms of machine learning? We discuss artificial intelligence in the section 'Artificial Intelligence vs People'.

Rebuttal 3: The Limitations of Controlled Experiments

Researchers typically strive to run carefully controlled experiments, but the quest for control can exclude some of the messy variables that experts must contend with (Orasanu & Connolly, 1993). These include tasks with ill-defined goals, shifting conditions, high stakes, multiple levels of team members, and ambiguity about the nature and reliability of the data. By excluding these kinds of variables, data can be collected efficiently and unambiguously, yet one of the hallmarks of expertise is to be able to handle ambiguous situations. Therefore, when we encounter carefully controlled studies that report on the weaknesses of experts, we should pay close attention to the details of the method and experimental design used to collect the data.

Rebuttal 4: Reductive Measurement

The decision research tends to reduce expertise to single measures, such as judgment hit rate, or performance on some single fixed task. Heuer (1999) provided a helpful definition of how intelligence analysts do much more than make predictions. They gather information, identify variables of interest, and identify potential consequences of courses of action. Research projects typically restrict themselves to aspects of performance that are easily measured, sometimes a single measure of performance, and thus can result in a limited and misleading portrait of the ways expertise is manifested for a given line of work.

Rebuttal 5: The Advantages Aren't Very Advantageous

Even when the linear models outperformed the experts, it is a mistake to infer that the linear models got it right and the experts failed miserably. Grove et al. (2000) found that linear models outperformed experts, which is true enough. But in the 136 studies they examined in their meta-analysis, the actuarial forecasts outperformed the experts in 63 studies, the experts outperformed the actuarial forecasts in only 8 studies, and there was no appreciable difference in the remaining 65 studies. So, another way to look at these data is that the actuarial forecasts were better than the experts less than half the time. Further, the studies examined situations that were noisy and complex, so even when they came out ahead, the actuarial forecasts were not very accurate. It was just that the experts were even worse. Thus, it is not clear that the benefits of actuarial forecasts were worth the effort.

Rebuttal 6: Insufficient Validation of Expertise

A number of studies in applied psychology do focus on expertise and not only claim to have had experts as participants, but provide some evidence that the participants were, in fact experts (see Hoffman, 1998, for a description of multiple levels of proficiency). More typical of the decision research literature, however, are studies that refer to the participants as experts (having so many years of experience) but fail to provide any convincing empirical evidence to justify this attribution—evidence that the so-called experts had actually achieved high levels of skill. Based on the information presented, they are more likely apprentices or junior journeymen. Often people will be identified as experts but will lack the necessary base of experience. To illustrate the difference between self-proclaimed experts and actual experts, consider two research projects on political expertise.

Philip Tetlock (2005) had a large number of political scientists, media pundits, and other supposed experts make forecasts about political events that would play out in the future. These "experts" showed a minimal amount of expertise—their accuracy was not much better than would be expected from a chimpanzee throwing darts. The implication is that in a complex task, expert judgment is not worth taking seriously.

Subsequently, Mellers et al. (2015) and Tetlock & Gardner (2015) made a large-scale attempt to develop forecasting skills over a two-year period, using practice, feedback and reflection, and the development of inquiry strategies. Participants were recruited "from professional societies, research centers, alumni associations, and science blogs, as well as word of mouth" (p. 4). The training and performance testing were conducted via the Internet. The researchers found that forecasting world events in human affairs was a skill that could be developed to high levels of proficiency. This research suggests that a number of studies that claim to be assessing experts, such as Tetlock (2005), are only using people who seem credible but have not engaged in the practice needed to achieve high levels of proficiency. Tetlock's (2005) initial book was titled *Expert Political Judgment*, but Tetlock and Gardner (2015) showed that the participants in the earlier research were not very expert at all. When researchers make serious efforts to develop expertise in their participants, the participants can achieve impressive results.

THE ATTACK FROM THE HEURISTICS AND BIASES COMMUNITY

The challenge from the Heuristics and Biases (HB) community dates back to the early 1970s. Led by Daniel Kahneman and Amos Tversky (Kahneman & Tversky, 1979; Tversky & Kahneman, 1974, 1981), the HB community has demonstrated that people fall prey to a wide variety of judgment biases and that even experts sometimes show these biases. The first heuristic that Kahneman and Tversky investigated was a misplaced belief in the law of small numbers (Tversky & Kahneman, 1971). They tried a set of sample problems on a group of psychologists, including two authors of statistics textbooks, and found that even these experts were vulnerable to the bias—they jumped to conclusions based on sample sizes that were too small to be valid.

Previously, Allen Newell and Herbert Simon (1972), George Polya (1957), and Karl Duncker (1945) had noted that people employ heuristics. Kahneman and Tversky's brilliant research identified a set of specific heuristics and led to the discovery of a wide range of others.

Kahneman and Tversky's findings about the judgment biases and helped create a mindset in the HB community, and diverse communities influenced by it, that experts are not to be trusted. The research agenda that HB set for psychology was to look for evidence that people aren't perfect and to expect that experts are likely to be overconfident and biased. The receipt of a Nobel Award by Kahneman, and another Nobel award in 2017 to his colleague Richard Thaler, along with the proliferation of HB research in universities, has bolstered the HB assertion that expert judgments aren't particularly accurate. The popular media echoes this assertion—Op-Ed pieces commonly explain mistakes in terms of the fallibility of experts (frequently labeled *human error*) and the biases to which they fall prey.

Tversky and Kahneman (1974) had been careful to state that "In general these heuristics are quite useful, but sometimes they lead to severe and systematic errors" (p. 1124). However, the HB field usually ignores this caveat and emphasizes the downside of heuristics.

We also see a ripple effect: Flach and Hoffman (2003) described how the positions of the HB field spread to groups of practitioners who adopt the belief that all people are necessarily and intrinsically victims of their handicapped minds. In his classic work *The Psychology of Intelligence Analysis*, Richards Heuer (1999) argued that intelligence analysts are victims of judgment biases. Heuer also referred to simple visual illusions to help make his case: not only that people (including professional intelligence analysts) can be easily fooled but that people are inescapably biased (see Hoffman, 2005a). Heuer's essays were intended as cautionary tales to trainee analysts but he leveraged the seminal HB research of the 1970s–1980s and generalized it. The stance that came to be widespread is that cognitive bias is inherent in humans, pervasive, independent of proficiency, and unavoidable. Heuer's work embedded the issue of bias in essentially all subsequent discussions of the psychology of intelligence analysis (see Johnston, 2005).

The HB community wants to reduce our trust in experts because experts are afflicted with the same biases as everyone else and are treated with a deference they do not deserve.

Rebuttal 1: HB Studies Unrepresentative Tasks

The HB paradigm typically uses participants who are not experts and gives them tasks that have little to no ecological validity. True, expert judgments are sometimes degraded by their use of heuristics but most of these effects diminish or disappear when researchers add context (e.g., Cheng, Holyoak, Nisbett, & Oliver, 1986) or have the genuine experts engage in their familiar tasks rather than artificial puzzle tasks. If researchers take experts (or anyone, for that matter) and put them in a situation in which they must perform an artificial and unfamiliar task, and remove any meaningful context, and apply a single and often inappropriate evaluation criterion, it's a mistake to conclude from the finding that they are incapable or incompetent.

Variations in the materials, instructions, procedures, or experimental design can cause bias effects to diminish or disappear (Gigerenzer, 1991; Gottlieb, Weis, & Chapman, 2007; Sanna & Schwarz, 2006). Here is one example. The Wason Selection Task (Johnson-Laird & Wason, 1970) presents the participant with four cards lying on a table. Each card shows either a number or a letter, and the participant is informed that every card has a number on one side and a letter on the other. The task is to assess the correctness of a rule such as *All cards with a vowel on one side have an even number on the other*. For this particular rule, people tend to turn over only those cards showing a vowel or an even number and this demonstrates a confirmation bias because the only way to disconfirm the hypothesis is to turn over a card showing the vowel or an odd number. However, if the four-card problem is restated in a more familiar and meaningful form, as mailing envelopes which are either sealed or unsealed on one side and either stamped or not stamped on the other, then the confirmation bias is less likely to occur (Cheng et al., 1986; Rumelhart & Norman, 1981). In other words, the bias was in the problem and task, not in the participants.

Hertwig and Gigerenzer (1999) re-examined a classic finding about the estimation of likelihoods in such problems as the *Linda Problem*: Linda is a bank teller. What is the likelihood that she is a feminist? In the classic finding people guess badly, because they do not take base rates into account (i.e., the class of bank tellers is larger than the class of feminist bank tellers). Gigerenzer found that performance on such problems depends on the format of the question and perhaps on the way terms like *probable* are understood. When the problem is restated using frequencies instead of probabilities, people answer the problems correctly (Gigerenzer, 1991). To be fair, Tversky and Kahneman (1983) had also suggested that under certain conditions people could be more accurate using frequency data than probabilities, and Gilovich and Griffin (2002) have discussed the complexity of the contrast between using frequencies and probabilities.

While some studies have shown that bias occurs in expert reasoning (e.g., Fischhoff, 1989; Holt, 1987), a number of studies have shown that bias effects are much smaller than those of the college students (e.g., studies in the domain of auditing and accountancy; see Kinney & Uecker, 1982; Olson, 1976; Shields, Solomon, & Waller, 1987). In experiments in which error-prone expert reasoning is induced, experts are more likely than non-experts to achieve a correct solution once the inadequacies of their initial problem representations have been pointed out (Johnson & Thompson, 1981).

Also needing qualification is the occurrence of bias in the example domain of intelligence analysis. In his detailed ethnographic study of intelligence analysis, Johnston (2005) concluded:

> The expert [has] the power to recognize patterns, perform tasks, and solve problems... It should come as little surprise, then, that an expert would have difficulty identifying and weighing variables in an interdisciplinary task, such as forecasting an adversary's intentions. Put differently, an expert may know his specific domain, such as economics or leadership analysis, quite thoroughly, but that may still not permit him to divine an adversary's intention, which the adversary may not himself know. (p. 65)

Rebuttal 2: HB Uses an Inappropriate Performance Criterion

The HB experiments (such as Tversky & Kahneman, 1974) rely on a single performance measure (e.g., Were the answers to the probability questions correct? Did judgments of word frequency map to their actual frequencies?) and a performance criterion that is inappropriate: "objective" truth derived from Bayesian statistics and other analytical methods. Why fault someone for not reasoning in terms of base rates when they cannot know what the base rates are? Why expect college freshmen to be facile at Bayesian analysis, which only became popular in the judgment research community in the 1980s? Why assume that reasoning is probability juggling? Laplace devised the current formulation of probability theory around 200 years ago, and most people are not fluent in applying it.

We don't need to invoke biases to explain why people are sometimes inaccurate at judgment tasks that have optimal solutions using formal analytical methods. This is just something people don't do well—but it is not the only, or much less the best indicator of rationality. If we decided that fluency in Latin was the marker for rationality, most people would come up short, but that just tells us that we are using a poor marker, asking people to converse in a language they don't know.

There are other performance criteria, such as frontier thinking (Shneiderman & Klein, 2017)—the ability to make decisions despite incomplete, incorrect, and contradictory information, when established routines no longer apply. A related criterion is speculative thinking—diagnosing what is going on under conditions of uncertainty and ambiguity, and anticipating what may happen next (Klein, 2017). Under these conditions, heuristics enable experts to engage in speculative thinking even when they only have a small sample, to use their experience to derive inferences from representative cases and instances available to them from previous events. The heuristics allow experts to make adjustments from prior data. This perspective is about *positive heuristics*, the ways that heuristics let us speculate about unfamiliar and uncharted situations.

Expert capabilities such as frontier thinking and speculative thinking are not easy to study under controlled conditions, and so they are typically neglected despite their great importance. The judgment heuristics identified by the HB community are invaluable in uncertain, ambiguous situations with missing and inconsistent information, but the HB community evaluates the heuristics against criteria that are ill-suited for the heuristics: concordance with probability theory and Bayesian statistics.

Rebuttal 3: The Confidence Game

HB advocates can still argue that decision makers must make important judgments about issues such as the likelihood of different risks, and here is where analytical methods such as probability theory and Bayesian statistics are so important. Yet experts still rely on their intuitions, which aren't well suited to such judgments. Worse yet, experts display overconfidence in their judgments. The HB research is said to present overwhelming evidence. "[C]onfirmatory behavior is a consistent finding throughout the experimental psychology and cognitive science literature" (Johnston, 2005, p. 21). The HB community claims that experts tend to be overconfident in making judgments even inside their area of competence.

However, there is mixed evidence for this claim, and the evidence stems from narrow methods for measuring confidence. Experts such as weather forecasters and firefighters are very careful to keep their judgments within their core specialty and to use experience and accurate feedback to attain reasonable levels of confidence in their judgments. Fireground commanders need to reassure their teams and to project a command presence, but that is different from their own assessment of risks. Our impression from observing fireground commanders is that they rely on their experience to keep their teams safe, withdrawing from situations that appear too dangerous, but we are not aware of any controlled studies buttressing this impression.

Expertise-deniers point to overconfidence on the part of so-called experts in areas such as predicting the movement of the stock market or predicting political events in the future—witness the pundits who gave Hillary Clinton an 80% chance of winning the 2016 election just as the polls closed. We do not consider pundits to be experts because they haven't definitively satisfied any of the criteria for expertise, including but not limited to performance on the domain's primary task (e.g., Ericsson, Charness, Feltovich, & Hoffman 2006; for a discussion of additional operational definitions of expertise, see Chapter 3 in Hoffman et al., 2014). To qualify as an expert the individual must show reliably superior performance on realistic, representative tasks. Media pundits are skilled at offering impressive explanations for events that have already happened, not at making accurate predictions. Further, the pundits and prognosticators who confidently predicted a Clinton victory were not actually using their own judgment—they were merely summarizing the results of polls and the polling data turned out to be flawed. Therefore, in the 2016 US presidential election the pundits were providing the results of careful statistical analyses—exactly the types of analyses that the HB and Big Data and Evidence-Based Performance communities advocate. The failure of the predictions illustrates the limitations of data-driven statistical analyses, not the limitations of experts.

The HB literature has pointed out eyewitness testimony as examples of overconfidence and low accuracy. "Another well-established cognitive science result concerning eyewitness testimony is that an eyewitness's degree of certainty about an identification is, at best, weakly correlated with the accuracy of the identification" (Stein, 2003, p. 296). (See also Loftus & Doyle, 1997 and Garrett, 2008). Wixted, Mickes, Dunn, Clark, & Wells, (2016) and Wixted & Wells (2017), however, found just the reverse—that high-confidence identifications were highly accurate, and low-confidence identifications were often wrong. "Confidence in an eyewitness identification from a fair lineup is a highly reliable indicator of accuracy" (p. 304). The difference was that in the past 10–20 years law enforcement personnel have developed expertise in conducting fair lineups. So now confidence levels are nicely calibrated with accuracy and the overconfidence effect has disappeared.

Therefore, we suggest that the HB community is overconfident in its assertions that experts are overconfident.

Rebuttal 4: Heuristics Are Valuable

The HB approach typically demonstrates that people rely on heuristics by using methods ans tasks that encourage participants to use heuristics even when the heuristics yield inaccurate judgments. However, this demonstration is not the same as showing that decision makers would be better off without the heuristics. Yes, in certain circumstances that researchers could design, the heuristics get in our way. But there are many other circumstances in which the heuristics are invaluable, such as for frontier thinking.

Certainly, for prediction tasks involving ample evidence, we would want decision makers to employ analytical tools as a cross-check on intuitive judgments and for the power that analytical methods can sometimes provide. But we also advocate for the reliance on expert judgment as a safeguard against the assumptions and limitations of the formal analyses.

Expert capabilities are not easy to bring under controlled conditions, and so they are typically neglected despite their potential value. Yet, with effort, expert capabilities such as those we listed earlier (e.g., perceptual skill, knowledge, sensemaking, anticipatory thinking) can be evaluated. The HB research community has failed to study the richer, broader benefits of the very heuristics they have uncovered. The HB researchers' eagerness to prove "what fools these mortals be", and their antipathy towards experts, appears to have blinded them to the importance of expertise in making real-world decisions. This antipathy opened the way for the emergence of the Naturalistic Decision Making (NDM) movement. For example, the Recognition-Primed Decision Making (RPD) model (Klein, 1998; Klein, Calderwood, & Clinton-Cirocco, 1986, 2010) depends on the heuristics of availability and representativeness, viewing these as strengths acquired through experience rather than weaknesses.

Heuer's (1999) work on the prevalence of biases in the psychology of intelligence analysis sparked over a decade of government programs aimed at developing software systems to help (that is, force) intelligence analysts to overcome the cognitive biases. In time, the software systems were found unhelpful because they served more to restrict the decision makers' process than to support their reasoning (see Moon & Hoffman, 2005).

THE ATTACK FROM SOCIOLOGY COMMUNITIES

The challenge from Sociology began in the 1970s and emerged more forcefully in the 1980s. The core of this attack is that expertise, and cognition, reside in the interaction between the individual and the community, and that it is a mistake to think of individual experts.

Sociological analysis of the consequences of occupational specialization has long considered the value of professions to society (for a review, see Evetts, 2013). Given its close association to the concept of professions, the concept of expertise was also assessed from the sociological perspective. Ethnographers and sociologists researched expertise in domains including astronomy, physics, endocrinology, and other professional domains (e.g., Collins, 1992; Latour & Woolgar, 1979; Lynch, 1993). Their resonant paradigms have been referred to as *situated cognition, distributed cognition*, the *sociology of scientific knowledge*, and *science and technology studies* (Knorr-Cetina, 1981; Lave, 1993; Suchman, 1987; Wenger, 1998) (for a review, see Hoffman & Militello, 2008, Ch. 11). Researchers have focused on the study of science (both inside and outside the laboratory), the acquisition and dissemination of scientific knowledge, and other topics (e.g., Fleck & Williams, 1996; Lynch, 1991).

The sociological attack on expertise can be thought of as an indirect attack, in that the sociological stance has often been defined as a reaction against so-called traditional cognitive psychology, which usually does consider the unit of analysis as the individual and usually does place the locus of resources for problem solving within the individual (i.e., internal mental representations). In contrast, the sociologists assert that cognition (expert or otherwise) is not a function of individual cognition or knowledge, but is socially constructed:

> If one relegates all of cognition to internal mental processes, then one is required to pack all the explanatory machinery of cognition into the individual mind as well, leading to misidentification of the boundaries of the cognitive system, and the over-attribution to the individual mind alone all of the processes that give rise to intelligent behavior.
> (Weldon, 2001, p. 76)

Jean Lave (1988) critiqued cognitive science as dividing "cognitive processes and the settings and activities of which they are part" (p. 76). To Lave and others, neither cognition nor the social world can be bracketed off as objects of study. "Learning, thinking, and knowing are relations among people engaged in activity in, with, and arising from the socially and culturally structured world" (p. 67). Olson (1994) asserted that in collaborative work "[t]he resources and constraints that affect cognitive activity have a very different profile than those associated with traditional models of individual cognition" (p. 991). Knowledge creation and other cognitive activities (e.g., remembering) are regarded as value-laden social processes (see Weldon, 2001). Proponents of the situated cognition approach offer many examples of why one should define *the cognitive system* as persons acting in coordination with a social group to conduct activities using tools and practices that have evolved within a culture.

An implication of this core claim is that expertise also needs to be seen as communal. The notion of individual experts therefore becomes suspect. Harald Mieg (2000, 2006; Mieg & Evetts, 2018) has demonstrated the limitations of an individualist view of expertise using ethnographic procedures to study experts in a number of domains. In the domain of stockbrokering, the reason that some stock brokers perform well and others do not is the disparity is accounted for by the fact that the best-performing brokers work in brokering houses that make their investment decisions according to plans, have good research resources, and have good sources of information. In other words, brokering expertise lies in the ways their organization works: "[Individual] experts are blind in that they can only try to follow some hypothesis, but lack insight in the complexity that drives the market" (Mieg, 2000, p. 122).

> A clear conclusion from all this is that the role of the expert remains socially contingent: what is judged is not so much the content of the evidence or advice, as the credibility or legitimacy of the person giving the evidence or advice; if we trust the expert, we must trust their advice.
> (Falkner, Fleck, & Williams, 1998, p. 4)

Some social activists also have questioned the concept of expertise, offering examples such as the public challenge to early explanations of the origins and nature of AIDS (Epstein, 1996) and the clash between citizen activists and "experts" in the petrochemical industry (Ottinger, 2013). They have shown how public standards can be used as a "measure of the legitimacy of experts" (Ottinger, 2010, p. 244).

Thus, we see that some sociologists regard expertise to be a misguided notion that ignores the critical resources upon which experts depend. Some segments of the Sociology community want to reduce our trust in individual experts in order to make the point that investigators need to take the broader context into account. Therefore, we consider their views part of the war on experts.

Rebuttal

One of the most valuable aspects of the sociological perspectives is to sensitize us to the importance of external resources and community relationships for the acquisition and expression of expertise. So, we respect these researchers for their contributions.

Our disappointment is with those sociologists who regard expertise as *merely* a social attribution, a matter of power and authority that goes without scrutiny (e.g., Fuller, 2006). We are also disappointed with the extreme "situated" view that discounts the importance of individual cognition, knowledge, and expertise and insists that cognition-in-practice is not isolated in the minds of individuals at all, but is distributed across individuals and their context of work (e.g., Agnew, Ford, & Hayes, 1997). Hence, all facts become social constructs and expertise exists only within a social construction (Fuller, 2006).

We disagree with this position. We can appreciate the emotional appeal of an extreme view that cognition and expertise is distributed, but we don't see how it can hold up to close scrutiny. It makes as little sense to argue that all cognition and expertise is distributed as to argue that expertise is independent of the context of the work and coordination with co-workers. There is overwhelming empirical evidence that individual knowledge and expertise are crucial for success. The fact that team dynamics and work context make a difference does not imply that individual expertise doesn't matter.

Our position is that individual expertise is impressive and is central to human accomplishments, and that experts differ markedly from journeymen and novices. If one plugs experts and non-experts into the same work settings the result will be major differences in the quality of the outputs of the groups/teams.

We agree completely that resources for cognition are in the world. We also agree that teamwork issues are an important part of naturalistic decision-making—the notion of *macrocognition* (Klein et al., 2003; Patterson & Miller, 2010) refers to such primary functions as coordinating and maintaining common ground.

Sociological research has proven valuable on such topics as apprenticeships, cognitive styles, cross-cultural cognition, the work of science teams, and other areas. But some such as Fuller (2006), respond so negatively to traditional cognitive individualist views that they brand these views as "cognitive authoritarianism" (p. 348). We believe that the pendulum has swung too far, resulting in a stance that has contributed to the attack on expertise. A moderate view is that individual cognition is an enabling condition for expertise, which just happens to be a condition that is not of particular interest in a sociological analysis.

Some advocates of the sociological-constructivist view fault cognitive-individualists for failing to consider that "knowledge" is not situated in human bodies. Had the Sociology communities said that cognition is not *just* situated in human bodies, there might be less conflict. Some of the advocates of the sociological views can perhaps be understood as not attacking expertise per se but as reminding cognitive psychologists of a broader and necessary perspective. There is no doubt that expertise, as a phenomenon, entails many interesting sociological questions (e.g., how individuals and organizations conceive of the role of expert, how organizations use experts as resources or heuristics in decision-making, and how practitioners deal with conflicts that arise between the role of the expert and the organization or bureaucracy within which they work) (see Hoffman, Ziebell, Feltovich, Moon, & Fiore, 2011). But while expertise in various domains may be more or less socially constructed and contingent, that does not mean that there is no such thing as individual expertise.

Fortunately, sociologists such as Collins (Collins & Evans, 2007, 2018) and Mieg (2000, 2006) have taken the balanced view that we advocate, describing the importance of individual expertise along with social and contextual factors that can be essential for developing and maintaining expertise. We expect that over time, this balanced sociological view will predominate.

We now describe the challenges that have emerged most recently.

The Attack from the Evidence-Based Practices Community

The Evidence-Based Practices community (e.g., Gray, 1996; Roberts & Yaeger, 2004) argues that professional fields need to evaluate their procedures, find the best scientific evidence, derive prescriptive procedures for each decision, and expect the practitioners to adhere to these procedures rather than rely on their own judgments. For example, Ghaffarzadegan, Epstein, and Martin (2013) demonstrated that obstetricians are poorly calibrated with regard to C-sections, being excessively concerned about the failure to do a needed C-section and inadequately concerned with the risks of doing an unnecessary C-section; obstetricians need to rely on the data, not their intuitions.

The takeaway message is that we should not trust expert practitioners. So-called expert practitioners aren't that good because they rely on anecdotal practices and the remedies they pursue are often out-of-date and ineffective (e.g., Haskins & Margolis, 2014). Additionally, experts get tired, distracted, or forgetful. Instead of trusting experts, we should rely on the cold, hard facts of what works and what doesn't.

The Evidence-Based Practices community is promoting data-based practices and wants to reduce our trust in experts because their intuitive judgments might conflict with the current data-based guidelines.

Rebuttal

This takeaway message seeks to replace reliance on experts with faith in scientific rigor—carefully controlled stimuli, unambiguous tasks, control groups, double-blind experimental designs, etc. However, this message ignores the limitations of scientific investigation. Although this subsection addresses evidence-based practices, our concerns run deeper, to the limitations of scientific exploration of cognition and expertise.

We use the word *faith* here purposely, because much of the enthusiasm for evidence-based practices is based on the hope that the vigorous application of "scientific rationality" will establish control over a world full of risk. This faith persists despite evidence that applications of best practices have often failed to improve outcomes, and sometimes even made matters worse. For example, Boyd et al. (2005) noted that adhering to published guidelines for care of an older woman with hypertension, osteoporosis, arthritis, diabetes, and emphysema—not at all an unusual circumstance—would require complex dietary changes and a 12-drug regimen costing about $5,000 per year while exposing the patient to over 20 potential drug–drug or drug–diet interactions. Even worse, Kavanagh and Nurok have estimated that implementation of protocols directing glucose normalization might potentially have been responsible for 26,000 additional deaths per year in ventilated ICU patients in the USA (Kavanagh & Nurok, 2016). We support doing scientific research on the efficacy of different methods and practices, but too often the scientific evidence base turns out to be shallower and more brittle than advocates of Evidence-Based Practices acknowledge. We are not arguing that scientific studies are useless. Rather, we don't think the results of such studies should be automatically trusted because the methodology was rigorous.

Klein, Woods, Klein, & Perry (2016) identified six cognitive challenges actually applying best practices:

- Characterizing the problem in the first place. Practitioners can't apply a best practice without judging what the problem is.
- Gauging confidence in the evidence. Just because data were collected and published doesn't mean that readers should trust those data; published results are often not reproducible. Too often the best practice is based on a single study. Even if several studies have been conducted they may share a flawed research design.
- Deciding what to do when the best practices conflict with professional judgments.
- Determining how to apply simple rules to complex situations. Global rules may not apply to specific cases.
- Revising plans that do not seem to be working. Expertise is needed to modify plans in progress.
- Considering remedies that are not official best practices. What should a decision maker do if there aren't any official best practices based on randomized control trials? What leeway does the decision maker have in considering courses of action that have not been validated?

Klein et al. (2016) concluded that the best way forward was to acknowledge the strengths and limitations of experts and evidence, and to blend both, balancing narrow evidence and broad experience. This approach has advocates in the United Kingdom and other parts of Europe (Barends, Rousseau, & Briner, 2014).

The best practices approach has been most seriously advocated in healthcare, where it is referred to as evidence-based medicine. However, the best practice for treating a young adult may be different than treating an elderly patient, especially if the elderly patient has additional medical conditions. Furthermore, within the group of "young adults," variation in weight/body size, anatomy, co-morbidities, etc. would modify a treatment plan. The evidence that gets turned into rules is typically collected on one medical condition at a time, but patients often are suffering from several different problems simultaneously (Boyd et al., 2005). The number of rules it would take to juggle all the permutations of multiple problems would be unmanageable.

And even when it is possible to write a rule, there must be conditional branching based on context. The best practice may be to take a medication three times a day for a month, but if the condition seems to be worsening after two weeks, should it be discontinued? Should it be modified to fit the needs of a specific patient? Best practices are based on averages across large samples; the guidelines derived from these averages may not apply to individual patients. The medical journals may advocate for one treatment but the personal physician has watched a patient's condition over time, watched the patient's reaction to and tolerance for different treatments—that is the challenge of deciding what to do when the best practices conflict with professional judgments.

Finally, we question the belief that scientific rigor will eliminate the need for people to make inferences. In 2016 three medical societies (in the USA, Canada, and Europe) published best practice guidelines on how atrial fibrillation should be managed, based on a variety of evidence. However, the three societies issued different recommendations. Out of twenty-one recommendations for emergency department management, five were completely different, six showed partial agreement (two of the three lined up), and ten were in full agreement. So, the three societies converged on fewer than half the recommendations. What should an ER physician do (Heidenreich et al., 2016; Kirchhof et al., 2016; Macle et al., 2016)?

One aspect of the evidence-based movement is to consolidate best practices into guidelines or checklists that can substitute for expertise. And while we appreciate the value of checklists (e.g., Gawande, 2010), we are also aware of their limitations. Guidelines, rules, and checklists raise the floor by preventing silly errors from being made—mistakes that even a first-year medical student would recognize as an error. But they also lower the ceiling, making it easy to shift to an unthinking, uncritical mode that misses subtle warning signs and does not serve the needs of patients. Checklists work in stable, well-defined tasks, and must be carefully crafted with a manageable number of steps. If the checklist is sequential, each step must lead to a clear outcome that serves as the trigger for the next step. However, in complex and ambiguous situations the antecedent conditions for each step are likely to be murky and will depend on the expertise of the decision maker to determine when to initiate the next step or whether to initiate it at all. Degani (2004) describes how an airplane crashed on landing because the pilots tried to systematically go through the landing checklist and encountered a situation in which the checklist sequence got tangled, with one step overlapping another rather than preceding it.

In the field of medicine, clinicians are often very enthusiastic about constructing checklists for a variety of decisions. For example, healthcare professionals have checklists such as the Ottawa ankle rule for deciding whether someone with an ankle injury needs an x-ray. But when the rules are compared to clinical judgment, the performance is roughly the same (Schriger, Elder, & Cooper, 2017; Wears & Klein, 2017). A hospital may still prefer to rely on the checklist, but it should not automatically assume that clinical judgment is inferior, and it should worry about the loss of skill that would result from making the checklist the default option.

Gawande (2010) was careful to talk about aiding experts, not replacing them; he explicitly described the importance of restricting checklists to the most important tasks that people are likely to forget. Unfortunately, expertise-deniers want to replace people, even experts, with checklists.

Checklists cannot replace experts. Furthermore, it is risky to have novices use checklists for complex tasks that depend on considerable tacit knowledge to judge when to take the next step, how to modify a step, how to decide whether the checklist is working. And reliance on checklists may impede the progression to high proficiency by forcing practitioners to adhere to existing best practices rather than working hard to enrich their knowledge and expertise and perhaps discover better practices. We recommend replacing the *best practices* community with a *better practices* approach, drawing on scientific studies along with expert advice.

THE ATTACK FROM COMPUTER SCIENCE

This attack, the most recent and currently the most active of the five battlegrounds, is being waged on three different fronts: Artificial Intelligence, Automation, and Big Data. We will consider them separately. But all of these fronts want to reduce our trust in experts because they want to replace experts with the systems they are developing, and so they want us to acknowledge that their technologies are smarter and more reliable than any expert can be.

Artificial Intelligence vs People

Artificial intelligence (AI) systems employing machine learning, search algorithms, and data-crunching capacity can be far more powerful than individual experts. AI successes have been highly visible and widely publicized, e.g., IBM's Deep Blue and Watson, and Google's AlphaGo program. Deep Blue beat Garry Kasparov, the reigning chess champion. Watson beat a panel of tested experts at the game of Jeopardy. AlphaGo trounced one of the most highly regarded Go masters. These achievements imply that AI can outperform people at any cognitively challenging task.

Weather forecasting is a good case in point (see Hoffman, LaDue, Mogil, Roebber, & Trafton, 2017). Starting in the 1980s, computer models were introduced into operational forecasting and predictions were made that computer models would soon outperform expert forecasters and would even automatically generate forecasts (e.g., McPherson, 1991). Articles in the popular and the scientific presses still present a stance of human versus machine, as when Kerr (2012) proclaimed, "All Hail the

Computer!", or when blog posts proclaim such things as "supercomputer powered models are about to make weather forecasts more accurate" (https://motherboard.vice.com; 3 Feb 2015). Articles in both the popular press and scientific outlets as well ask whether "machines are taking over" (Kerr, 2012, p. 734). "It is getting increasingly difficult for human forecasters to improve upon [computer model outputs]...[they] cannot consistently beat [computer model] precipitation forecasts for virtually all of the locations" (Baars & Mass 2005, p. 1045).

The popular myth is that if computers with more memory and faster processing speeds could be thrown at the problem, the need for humans would evaporate. This claim conveys a competition in which the expert forecasters cannot beat the computers. "[The] human's advantage over the computer may eventually be swamped by the vastly increased number crunching ability of the computer...as the computer driven models will simply get bigger and better" (Targett, 1994, p. 50).

The AI community wants to reduce our trust in experts because AI researchers are convinced, and hope to convince us, that their programs can outperform experts on any reasoning or cognitive task.

Rebuttal

The optimism of the visionaries—the AI advocates—is countered by those who look at the evidence and are not captured by the hype. According to National Weather Service statistics, "humans improve the accuracy of precipitation forecasts by about 25 percent over the computer guidance alone, and temperature forecasts by about 10 percent... these ratios have been relatively constant over time...as much progress as the computers have made, forecasters continue to add value on top of that" (Silver, 2012, p. 125). While the computer model predictions will continue to improve, and while much of the improvement will come from advances in computational systems, forecasts do not get better merely because one throws more raw computing power at the problem (for details see Hoffman et al., 2017).

Can AI outperform humans at any cognitively challenging task? Will machines crowd out human expertise? It helps to look at the less widely publicized AI failures. Medical expert systems in the 1980s were expected to replace physicians; obviously, that never happened. Similarly, expert systems for weather forecasting faded in popularity because they had to be so specific to locale and season that they were nearly useless, and the people who needed them the most (less-experienced people) were those less able to use them (Hoffman et al., 2017). AI solutions are certainly troubling when they make flawed decisions that cause harm to large numbers of individuals (O'Neil, 2016).

The widely publicized successes (chess, Go, Jeopardy) involve games that are well structured, with unambiguous referents and definitive correct answers or solutions. In contrast, most decision makers face wicked problems with unclear goals in ambiguous and dynamic situations. For example, an enemy platoon is someplace on the battlefield, or was 30 minutes ago, no telling where it is now or what its capability is. The mission may require decision makers to adapt their goals on the fly, resolve contradictory or incomplete information, and consult with others to build support for a decision.

And even the successes such as IBM's Watson may have resulted in exaggerated claims. Roger Schank (2015), one of the pioneers of artificial intelligence, stated flatly that "Watson is a fraud." He objected to IBM's claims that Watson could out-think human brains (even expert brains) to find insights within large data sets. While Watson excels at key word searches, it does not take into account the context of the passages it is searching, and as a result is insensitive to underlying messages in the material. Schank's position is that counting words is not the same as deriving insightful conclusions. AI lacks common sense, logical reasoning, and sensitivity to context. For example, even during Watson's Jeopardy triumph it missed the final question. The category was US cities, and the clue was "Its largest airport is named for a World War II hero; its second largest, for a World War II battle. The correct answer is "What is Chicago?" (O'Hare and Midway airports). Watson's answer was "Toronto," which was not only wrong, because Toronto is not a US city, but demonstrates that Watson, for all its data searching and association strengths, can't think logically.

A related problem is the "AI shuffle:" carving out a subtask that can be handled by an AI system, showing that the system outperforms humans on this subtask, and ignoring the other subtasks because they are not amenable to intelligent technology. Earlier, we discussed how the varied roles of intelligence analysts involve more than simply making predictions. Research projects and AI systems that restrict themselves to the prediction task, because accuracy is easily measured, provide a misleading impression of the expertise required to do the work.

Many of the AI attacks on experts are predicated on the assertion that the difference "between an arithmetic calculator and a human brain is not one of kind, but only one of scale, speed, degree of autonomy, and generality" (Stone et al., 2016). This assertion implies that it is only a matter of time before AI surpasses experts in any field of interest, only a matter of time before artificial intelligence will be replacing experts by being cheaper, faster, and less prone to mistakes.

Therefore, we need to make our position clear. Our thesis is that "People are *not* machines; machines are *not* people." Further, thinking is not calculation. It involves associational and pattern-matching processes as well as other processes such as causal speculation, sensemaking, problem detection, anticipation, and insight generation.

We argue for the unique human capabilities, especially the capabilities of experts. By trying to erase these distinctions, by treating thinking as reducible to calculating, the expertise-deniers have been able to cast doubt on the ability of experts.

In addition, human social skills build relationships of trust within and beyond an organization, to bring fresh ideas and deal with threats. Human relationships of trust are valuable if not essential to organizations.

AI systems will be continually improving, which will make them ever more valuable tools for people to use. It is time for system designers to stop playing games about whether humans or machines are better, and to be more serious about designing work systems in which people are in control of powerful machines that amplify human abilities (Johnson et al., 2011). These work systems are tools for improving human performance and should be designed as such. Our experience is that the actual AI developers have much greater appreciation for human expertise than the AI popularizers. The developers have first-hand experience with the frustrations of trying to design systems to perform challenging tasks.

People are uniquely capable of frontier thinking, social engagement, and taking responsibility for actions (Shneiderman & Klein, 2017). Frontier thinking involves our ability to handle wicked problems, to address unstructured tasks, to speculate in the face of ambiguity and uncertainty, to map an unfamiliar terrain. Social engagement depends on trust and coordination, the ability to develop common ground and to detect when common ground is eroding. Finally, experts take responsibility for their decisions. Intelligent systems do not—if a decision fails, there is no telling which of the designers or developers is to blame but the machine itself does not admit culpability, which is why we cannot really speak about trusting AI or automation. We can rely on these tools, but not trust them in the way we do or do not trust another person.

Automation vs People

Just as some pundits have predicted that AI will conquer the world by surpassing experts, other pundits have promulgated the myth that more automation can solve many of our problems, obviating the need for humans and experts. For example, one claim is that more automation is needed to reduce operator mental workload as modern cognitive work increases in tempo and complexity. The automation community wants to reduce trust in experts so that decision makers increasingly rely on automated systems to perform challenging tasks.

Rebuttal

The enthusiasm for technologies is often extreme (see for example, Brynjolfsson & McAfee, 2014). Failing to recognize the obvious—that *automation* is not autonomous—too many technologists succumb to the myth that automation can compensate for human limitations, the myth that automation can substitute for humans, and the myth that tasks can be cleanly allocated to either the human or the machine. These misleading beliefs have been questioned and dismissed by cognitive systems engineers for over 35 years (see Bainbridge, 1983). Nevertheless, the debunking must be periodically refreshed in the minds of researchers (see Bradshaw, Hoffman, Johnson, & Woods, 2013; Woods & Sarter, 1997). The myths persist because of the promissory note that more automation means fewer people, fewer people means fewer errors, and (especially) fewer people means reduced costs. Here is a typical example:

> Autonomy has the potential to enable U.S. forces to break out of current limitations by allowing systems to understand the environment, to make decisions, and to act more effectively and with greater independence from humans. In doing so, autonomy can augment or replace humans to enhance performance, to reduce risk to warfighters, and to decrease costs... [autonomy] presents major cost-saving opportunities in areas such as logistics, maintenance, and data analysis. (Office of Technical Intelligence, 2015, pp. iii, iv)

Nearly every funding program that calls for more automation is premised with the claim that the introduction of automation will entail a requirement for fewer expert operators at potentially lower cost to the organization. But the facts are in plain view: More automation and more advanced automation does not mean you need fewer humans; indeed, it often means you need more experts. (For detailed case studies, see Hoffman, Cullen, & Hawley 2016). Automation creates new kinds of cognitive work for the operator, often at the wrong times. Automation often requires people to do more, do it faster, or do it in more complex ways. The explosion of features, options, and modes often creates new demands, new types of errors, and new paths toward failure. Ironically, as these facts become apparent, managers seek even more automation to compensate for the problems triggered by automation (see Department of Defense, 2015, p. 8).

Further, in adversarial situations, automation may prove to be a liability if it makes the judgments and decisions predictable. The more predictable a system, the easier it is to defeat it.

Another concern is that shifting all the cognitive work into the automation may de-skill the workers and erode expertise. We know that genuine expertise is achieved after extensive hard work on difficult problems. If automation erodes expertise, then there will be a temptation to rely even more heavily on automation, further eroding expertise, in a vicious cycle (Bainbridge, 1983). One way out of this downward cycle is to treat automation as a tool for making people more effective, and to design the automation accordingly. That stance, the same that we offered with regard to artificial intelligence, is the core of human-centered computing (Hoffman, 2012).

Big Data

The claim is that Data analytics algorithms are making experts obsolete. Powerful algorithms operate on massive and varied data sets to detect trends, spot problems, and generate inferences and insights. No human, no matter how expert, can possibly sift through all of the available sensor data. And no human can hope to interpret even a fraction of these data sources. The Big Data community (also described as the Data

Analytics or Data Science community) wants to reduce the public's trust in experts so that decision makers become comfortable in outsourcing their analyses to the Big Data algorithms being developed.

Rebuttal

Big Data approaches depend on the assumptions and sophistication of the programmers. The algorithms used are often not documented or verified, much less replicated. Large numbers of predictors in even larger numbers of combinations chasing large numbers of potential hypotheses create a combinatorial explosion that makes spurious findings likely. Even if valid, not all patterns are meaningful, and statistical association does not necessarily lead to reliable prediction. At worst, data analytics can lead to *apophenia*—seeing patterns where none really exist. The quantity and novelty of data cannot trump more fundamental issues of reliability, representativeness, bias, stability, construct validity, and context.

One of the Big Data success stories was Google's FluTrends project—using data Analytics to predict flu outbreaks, but subsequently FluTrends failed so badly that it was removed from use (Lazer et al., 2014).

Algorithms can follow historical trends but may miss departures from these trends, as in the broken leg case: cues that have implications that are clear to experts but aren't part of the algorithms. Further, experts can use expectancies to spot missing events that may be highly significant. In contrast, Big Data approaches, which crunch the signals received from a variety of sources, are unaware of the *absence* of data and events, which are sometimes crucial.

Discussion

This chapter has described the attacks on expertise from five disciplines and communities of expertise-deniers. These are deliberate attacks—each community has its own reasons for wanting to discredit experts. We have offered rebuttals and counter-arguments. Now we take a step back and examine what we have learned from this exercise.

The Value of Experts

Experts are essential for the successful operations in many venues of human activity. Even though experts aren't perfect, performance and outcomes would suffer greatly by trying to replace experts with checklists, formulas, or AI. The capacity for adaptation and resilience depends on expertise.

The task of writing this chapter increased our respect for the expertise-deniers. We appreciate the challenges they are throwing out to the expertise researchers. They are forcing us—they are helping us—to think more deeply about the nature of human expertise. But we remain uncomfortable with the message the expertise-deniers give to society to devalue experts. We believe that experts are valuable and necessary, and can be made even more effective. This said, we encourage the expertise-deniers to continue their attacks, because these attacks force us to advance our understanding of expertise and develop stronger and more useful research methods.

It is tempting to question the motivation of some of the expertise-deniers, especially those who stand to gain when the government and corporations increase investment in stand-alone intelligent technologies, or enable analytical managers to wrest control from practitioners. Regardless of the motivations of the expertise-deniers we take their arguments seriously, enabling us to refine our appreciation of experts or, when appropriate, to temper that appreciation.

Part of the difficulty facing the expertise-deniers is that the features of expertise are so heavily tied to tacit knowledge and are difficult to capture and measure. It is therefore understandable that these features receive less scrutiny and less valuation. Tacit knowledge, as opposed to explicit knowledge, is very difficult to articulate and so experts usually are unable to unpack the basis for their perceptual discriminations, pattern recognition, judgments of typicality, and mental models. However, cognitive interviewing techniques (e.g., Crandall, Klein, & Hoffman, 2006; Hoffman & Militello, 2008) have been effective at eliciting these kinds of tacit knowledge.

One way to conceptualize the value of experts is to consider the consequences when experts retire from a company, as discussed by Leonard, Swap, and Barton (2014). The company loses the personal relationships built up over years that allowed a seasoned expert to pick up the phone and get a response from a customer's CEO or a timely bit of analysis from a specialist in a different field. The company suffers a reduced capacity for innovation because it no longer has experts who can appreciate how a proposed change will mesh with existing systems and practices, and how atypical events must be handled (see also Hoffman & Hanes, 2003).

Checklists are fine, but experts must interpret and adapt them. Evidence-Based Practices are fine but experts must sift through the evidence and apply the data to specific cases. Heuristics may lead to errors but they are critical for enabling experts to conduct frontier thinking. Research on artificial intelligence and related areas such as automated systems and decision support systems continues to make dramatic progress, but AI needs to be designed to support rather than replace human decision-making.

The Measurement of Expertise

Throughout the rebuttals we have expressed a theme that expert performance should never be assessed using single measures, especially just measures of correctness or hit rate. For example, in the Mellers and Tetlock research, the performance of intelligence analysts is reduced to the accuracy of their forecasts. But the job of intelligence analysts is not just to make accurate forecasts. It is also to spot anomalies that need to be monitored more carefully, to notice trends, to detect coincidences that deserve greater investigation. It involves gaining insights. Tetlock (2017) acknowledged this limitation: "forecasting tournaments and prediction markets miss the essential goals of analysis... Predictive accuracy rarely ranks high among these goals. More important are placing surprising events in context, surveying options, understanding alien points of view, and explaining why and how one thing rather than another just happened" (p. 6). When expertise-deniers restrict research to the variables they can readily measure, they do the topic and the public a disservice (see also Wears & Klein, 2017).

The Boundary Conditions of Expertise

Experts certainly aren't perfect, so critiques are useful for increasing our understanding of the boundary conditions of expertise. We believe that peaceful mediation of this particular conflict in the war—the conditions favoring expertise versus those restricting it—comes from considering differences in domains. James Shanteau, a founding father of expertise studies, distinguished between Type 1 domains that readily foster expertise (livestock judges, astronomers, test pilots, soil judges, chess masters, physicists, mathematicians, accountants, grain inspectors, photo interpreters, insurance analysts) and Type 2 domains that do not easily foster it (stockbrokers, clinical psychologists, psychiatrists, college admissions officers, court judges, personnel selectors, intelligence analysts) (Shanteau, 1992). The Type 1 domains have more predictable outcomes, and provide good feedback on judgments. They also rely more on static as opposed to dynamic stimuli. Type 1 domains are structured and tend to involve mechanical variables. In contrast, Type 2 domains require social judgments in unstructured and dynamic situations with ill-defined goals. Type 1 domains are well structured, one of the criteria identified by Kahneman and Klein (2009) as important for the development of intuitive expertise. They involve right answers, or at least very good answers, so researchers can determine whether experts are close to optimal. The Type 2 domains are not as well structured, or are so chaotic as to impede or limit the potential for the achievement of expertise (e.g., stockbrokers). They do not involve problems with single right answers; they do involve problems that lack informative feedback on judgments.

We believe that there is expertise even in Type 2 tasks, but it is less pronounced—experts at Type 2 tasks may make more errors, or different kinds of errors, than we would expect from Type 1 tasks, and their mastery will be less visible.

Researchers have trouble measuring performance on Type 2 tasks. Most of us would agree that some clinical psychotherapists are more skilled than others, and that the best might be considered experts, but what are the criteria for expert performance? Hill, Spiegel, Hoffman, Kivlighan, & Gelso (2017) claim that the top 10% of therapists could be considered experts, using criteria such as relational and technical expertise, cognitive processing, and client outcomes. None of these criteria admit to unambiguous measurement. It is a challenge to determine the best tools or measures for proficiency scaling in Type 2 domains. That said, we know of at least one clear case where individuals could be trained to high proficiency in a Type 2 domain. This is the study by Mellers et al. (2015) showing that individuals could be trained to achieve impressive accuracy in forecasting world events; the accuracy was measured using a Type 1 measure, Brier scores. However, our challenge goes further—to train and measure expertise on Type 2 tasks without relying on Type 1 measures.

Type 2 expertise might be amplified by tapping into social networks. As the sociological communities of practice remind us, experts are embedded in a social system through which they learn new methods and principles. These social systems give them resources to call on when novel problems or emergencies arise. Experts are also leaders, mentors, and trainers who pass on their skills to others, either overtly or through modeling. Human social networks amplify and improve expertise.

[Human + Machine] rather than [Human vs Machine]

One of the most important and exciting challenges regarding expertise involves design of computer systems so as to ensure effective human control, even while raising the level of automation (Bunch, Bradshaw, Hoffman, & Johnson, 2015; Johnson et al., 2014; Shneiderman, Plaisant, Cohen, Jacobs, & Elmqvist, 2016). When the design of cognitive work systems is guided by the desire to make the machines comprehensible, predictable, and controllable, they are more likely to promote better decision-making (Johnson, Bradshaw, Feltovich, & Woods, 2014; Klein, Woods, Bradshaw, Hoffman, & Feltovich, 2004).

We believe that the expertise-deniers go astray when they focus on a task that computers are well designed to do, rather than asking how people can use machines to improve the decision-making of experts. We advocate for better technology and automation, the keyword being *better*. We hope that technologists can get past the mindset of trying to build systems to replace the experts and instead seek to build useful technologies that empower experts. Effective technologies also promote learning by experts so as to improve their skills over time, enable improved social engagement when experts consult with others, and clarify the responsibilities for failures and successes.

While advocating for advanced technology, we stress that people remain responsible for their actions, even when using powerful machines. The challenge is to ensure human control while advancing the capabilities of the automation (Hoffman, Cullen, & Hawley, 2016; Shneiderman et al., 2016). After all, car drivers are in control of their vehicles, even though few drivers understand all the automotive technologies they are controlling; the designers have found ways to leave the drivers in meaningful control. Misnomers such as driverless cars will have to be changed to clarify responsibility and liability for failures, just as *unmanned aerial vehicles* are now more appropriately called *remotely piloted vehicles*. We advocate for technologies that provide experts with a better understanding of the machine's status and better control over the actions of machines.

Accelerated Expertise

We see an opportunity to formulate training methods to enable more experts to increase their proficiency. If even a few years could be shaved off the time it takes to achieve expertise, there could be a huge savings. We know that acceleration is possible (see Hoffman et al., 2014; Staszewski, 2008); for example, Barbara Mellers et al. (2015) achieved marked success in training skill at forecasting events in human affairs. We would like to see more of these kinds of efforts, finding innovative training approaches to build expertise in decision makers and to move existing experts to even higher levels.

We also see an opportunity to put a positive spin on the heuristics identified by the Heuristics and Biases community. A research program on positive heuristics would examine how commonly used heuristics benefit decision makers and permit frontier thinking.

Our Fear

One of the biggest dangers of the war on experts is its potential to degrade the decision-making and resilience of government agencies and private sector corporations. Once such organizations uncritically accept the expertise-deniers' arguments they are likely to sideline subject-matter experts in favor of statistical analysts and ever more technology. They are likely to divert funding from training programs that might produce more experts into programs intended to make decisions without requiring human intervention or responsible action. And that could lead to a downward spiral in which expertise is diminished, leading to greater reliance on the various forms of replacement, leading to further diminution of expertise. In this trajectory, systems and machines would truly gain ascendency, not through their improvements but through the permanent erosion of expertise.

Our Hope

We want to abandon the metaphor of a war, in which one side wins and the other loses. We were drawn into this metaphor because of the attacks being waged on experts and on expertise. That is why we have felt the need to defend experts, and to counter the attacks. But we are criticizing the critiques, not the communities. We don't want to return to an era when medicine was governed by anecdote, regardless of data—we think it essential to draw from evidence and from expertise. We don't want to ignore the discoveries of the Heuristics and Biases researchers—the heuristics they have discovered can have great value for fostering speculative thinking. We don't want to ignore the judgment and Decision Research community—we want to take advantage of their efforts to improve the way we handle evidence and deploy our intuitions. We don't want to turn the clock back on information technology—we want these tools to be designed to help us gain and enhance our expertise. We don't want to ignore the social aspects of work settings—we want to design work settings and team arrangements that magnify our expertise. Our hope is to encourage a balance that respects expertise while seeking various means to strengthen it.

Acknowledgements

We would like to thank Jan Maarten Schraagen and Paul Ward for their helpful comments and suggestions on several drafts of this manuscript and for their patience and encouragement. We also thank Bonnie Dorr, Hal Daume, Jonathan Lazar, Jim Hendler, Jenny Preece, and Mark Smith for their helpful suggestions.

Dedication

We would like to dedicate this chapter to our co-author, Robert Wears, who tragically died in July 2017 while this chapter was in preparation. As a researcher and a practitioner, Bob provided his many colleagues and friends with insights and inspiration.

References

Agnew, N. M., Ford, K. M., & Hayes, P. J. (1997). Expertise in context: Personally constructed, socially selected, & reality-relevant? In P. J. Feltovich, K. M. Ford, & R. R. Hoffman (Eds), *Expertise in context* (pp. 219–244). Cambridge, MA: MIT Press.

Baars, J. A., & Mass, C. F. (2005). Performance of National Weather Service Forecasts compared to operational, consensus, and weighted model output statistics. *Weather and Forecasting 20*, 1034–1047.

Bainbridge, L. (1983). Ironies of automation. *Automatica 19*, 775–779.

Barends, E., Rousseau, D. M., & Briner, R. B. (2014). *Evidence-based management: The basic principles.* Amsterdam: Center for Evidence-Based Management.

Boyd, C. M., Darer, J., Boult, C., Fried, L. P., Boult, L., & Wu, A. W. (2005). Clinical practice guidelines and quality of care for older patients with multiple comorbid diseases: implications for pay for performance. *JAMA 294*(6), 716–724.

Bradshaw, J. M., Hoffman, R. R., Johnson, M., & Woods, D. D. (2013). The seven deadly myths of "autonomous systems." *IEEE Intelligent Systems 28*(3), 54–61.

Brynjolfsson, E., & McAfee, A. (2014). *The second machine age: Work, progress, and prosperity in a time of brilliant technologies.* New York: W. W. Norton.

Bunch, L., Bradshaw, J. M., Hoffman, R. R., & Johnson, M. (2015). Principles for human-centered interaction design, Part 2: Can humans and machines think together? *IEEE Intelligent Systems* 30(3), 68–75.

Cheng, D. W., Holyoak, K. J., Nisbett, R.E., & Oliver, L. M. (1986). Pragmatic versus syntactic approaches to training deductive reasoning. *Cognitive Psychology* 18, 293–328.

Collins, H. M. (1992). *Changing order: Replication and induction in scientific practice*, 2nd edn. Sage. Chicago: University of Chicago Press.

Collins, H. M. (1996). Embedded or embodied: Hubert Dreyfus's what computers still can't do. *Artificial Intelligence 80*, 99–117.

Collins, H. M., & Evans, R. (2006). The third wave of science studies. In E. Selinger and R. P. Crease (Eds.), *The philosophy of expertise* (pp. 39–110). New York: Columbia University Press.

Collins, H. M., & Evans, R. (2007). *Rethinking expertise.* Chicago: University of Chicago Press.

Collins, H. M., & Evans, R. (2018). A sociological/philosophical perspective on expertise: The acquisition of expertise through socialization. In K. A. Ericsson, R. R. Hoffman, A. Kozbelt, and M. Williams (Eds), *The Cambridge handbook of expertise and expert performance*, 2nd edn (pp. 21–32). Cambridge, UK: Cambridge University Press.

Cook, M. B., & Smallman, H. S. (2008). The human factors of the confirmation bias in intelligence analysis: Decision support from graphical evidence landscape. *Human Factors 50*, 745–754.

Crandall, B., Klein, G., & Hoffman, R. R. (2006). *Working minds: A practitioner's guide to cognitive task analysis.* Cambridge, MA: MIT Press.

Dawes, R. (1971). A case study of graduate admissions: Application of three principles of human decision making. *American Psychologist 26*, 180–188.

Dawes, R. (1979). The robust beauty of improper linear models. *American Psychologist 34*, 571–582.

Degani, A. (2004). *Taming HAL: Designing interfaces beyond 2001.* New York: Palgrave, Macmillan.

Department of Defense (2015). *Technical Assessment: Autonomy.* Office of Technical Intelligence, Department of Defense, Washington, DC.

Duncker, K. (1945). On problem-solving. *Psychological Monographs* 58(5), i–113.

Einhorn, H. J. (1972). Expert measurement and mechanical combination. *Organizational Behavior and Human Performance 7*, 86–106.

Epstein, S. (1996). *Impure science: AIDS, activism, and the politics of knowledge.* Berkeley: University of California Press.

Ericsson, K. A., Charness, N., Feltovich, P. J., & Hoffman, R. R. (Eds) (2006). *The Cambridge handbook of expertise and expert performance.* Cambridge, UK: Cambridge University Press.

Evetts, J. (2013). Professionalism: Value and ideology. *Current Sociology 61*, 778–779.

Faulkner, W., Fleck, J., & Williams, R. (1998). Exploring expertise: Issues and perspectives. In R. Williams, W. Faulkner, & J. Fleck (Eds), *Exploring expertise* (pp. 1–27). New York: Macmillan.

Fischhoff, B. (1989). Eliciting knowledge for analytical representation. *IEEE Transactions on Systems, Man, and Cybernetics 19*, 448–461.

Fischhoff, B., Slovic, P., & Lichtenstein, S. (1978). Fault trees: Sensitivity of estimated failure probabilities to problem representation. *Journal of Experimental Psychology: Human Perception and Performance 4*, 330–344.

Flach, J., & Hoffman, R. R. (2003). The limitations of limitations. *IEEE Intelligent Systems* 18(1), 94–97.

Fleck, J., & Williams, R. (Eds) (1996). *Exploring expertise*. Edinburgh: University of Edinburgh Press.

Fuller, S. (2006). The constitutively social character of expertise. In E. Selinger & R. P. Crease (Eds), *The philosophy of expertise* (pp. 342–357). New York: Columbia University Press.

Garrett, B. (2008). Judging innocence. *Columbia Law Review* 108, 55–142.

Gawande, A. (2010). *The checklist manifesto: How to get things right*. New York: Metropolitan Books.

Ghaffarzadegan, N., Epstein, A. J., & Martin, E. G. (2013). Practice variation, bias, and experiential learning in Cesarean delivery: A data-based system dynamics approach. *Health Services Research* 48, 713–734.

Gigerenzer, G. (1991). How to make cognitive illusions disappear: Beyond heuristics and biases. *European Review of Social Psychology* 2, 83–115.

Gilovich, T., & Griffin, D. (2002). Introduction—Heuristics and biases: Then and now. In T. Gilovich, D. Griffin & D. Kahneman (Eds.). *Heuristics and biases: The psychology of intuitive judgment*. New York: Cambridge University Press.

Gottlieb D. A., Weiss, T., & Chapman, G. B. (2007). The format in which uncertainty information is presented affects decision biases. *Psychological Science* 18(3), 240–246.

Gray, J. A. M. (1996). *Evidence-based healthcare*. London: Churchill Livingstone.

Grove, W. M., Zald, D. H., Lebow, B. S., Snitz, B. E., & Nelson, C. (2000). Clinical versus mechanical prediction: A meta-analysis. *Psychological Assessment* 12, 19–30.

Haskins, R., & Margolis, G. (2014). *Show me the evidence: Obama's fight for rigor and results in social policy*. Washington, DC: Brookings Institution Press.

Hastie, R., & Dawes, R. (2001). *Rational choice in an uncertain world*. Thousand Oaks, CA: Sage Publications.

Heidenreich, P. A., Solis, P., Estes, N. A. M., Fonarow, G. C., Jurgens, C. Y., Marine, J. E., & McNamara, R. L. (2016). ACC/AHA clinical performance and quality measures for adults with atrial fibrillation or atrial flutter: A report of the American College of Cardiology/American Heart Association Task Force on Performance Measures. *Journal of the American College of Cardiology* 68(5), 525–568.

Hertwig, R., & Gigerenzer, G. (1999). The "conjunction fallacy" revisited: How intelligent inferences look like reasoning errors. *Journal of Behavioral Decision Making* 12, 27–305.

Heuer, R. J., Jr. (1999). *Psychology of intelligence analysis*. Washington, DC: Center for the Study of Intelligence, CIA.

Hill, C. E., Spiegel, S. B., Hoffman, M. A., Kivlighan, D. M., & Gelso, C. J. (2017). Therapist expertise in psychotherapy revisited. *The Counseling Psychologist* 45, 7–53.

Hoffman, R. R. (1998). How can expertise be defined? Implications of research from cognitive psychology. In R. Williams, W. Faulkner, & J. Fleck (Eds), *Exploring expertise* (pp. 81–100). New York: Macmillan.

Hoffman, R. R. (2005a). Biased about biases: The theory of the handicapped mind in the psychology of intelligence analysis. In the Panel "Designing Support for Intelligence Analysts" (p. 409) (S. Potter, Chair). In *Proceedings of the 48th Annual Meeting of the Human Factors and Ergonomics Society* (pp. 406–410). Santa Monica, CA: Human Factors and Ergonomics Society.

Hoffman, R. R. (2005b). *The Psychology of Intelligence Analysis*, revisited: An update from developments in cognitive science post-1980. Report to PARC for the Novel Intelligence from Massive Data Program of the Advanced Research and Development Activity. Palo Alto, CA: PARC. Retrieved from http://www.ihmc.us/groups/rhoffman/

Hoffman, R. R. (2012). *Collected essays on human-centered computing, 2001–2011*. New York: IEEE Computer Society Press.

Hoffman, R. R. (2017). A taxonomy of emergent trusting in the human–machine relationship. In P. Smith & R. R. Hoffman (Eds), *Cognitive systems engineering: The future for a changing world* (pp. 137–163). Boca Raton, FL: Taylor & Francis.

Hoffman, R. R., Cullen, T. M., & Hawley, J. K. (2016). Rhetoric and reality of autonomous weapons: Getting a grip on the myths and costs of automation. *Bulletin of the Atomic Scientists 72*, 247–255.

Hoffman, R. R., & Hanes, L. F. (2003). The boiled frog problem. *IEEE Intelligent Systems 18*, 68–71.

Hoffman R. R., LaDue, D., Mogil, A. M. Roebber, P. J., and Trafton, J. G. (2017). *Minding the weather: how expert forecasters think*. Cambridge, MA: MIT Press.

Hoffman, R. R., & Militello, L. G. (2008). *Perspectives on cognitive task analysis: Historical origins and modern communities of practice*. Boca Raton, FL: Psychology Press/Taylor and Francis.

Hoffman, R. R., Ward, P., DiBello, L., Feltovich, P. J., Fiore, S. M., & Andrews, D. (2014). *Accelerated expertise: Training for high proficiency in a complex world*. Boca Raton, FL: Taylor and Francis/CRC Press.

Hoffman, R. R., Ziebell, D., Feltovich, P. J., Moon, B. M., & Fiore, S. F. (2011). Franchise experts. *IEEE Intelligent Systems*, 72–77.

Holt, D. L. (1987). Auditors' base rates revisited. *Accounting, Organizations, and Society 12*, 571–578.

Johnson, J. M., Bradshaw, J. M., Feltovich, P. J., Hoffman, R. R., Jonker, C., van Riemsdijk, B., & Sierhuis, M. (2011). Beyond cooperative robotics: The central role of interdependence in coactive design. *IEEE Intelligent Systems 26*(3), 81–88.

Johnson, M., Bradshaw, J. M., Feltovich, P. J., Jonker, C. M., van Riemsdijk, B. M., & Sierhuis, M. (2014). Coactive design: Designing support for interdependence in joint activity. *Journal of Human–Robot Interaction 3*, 43–69.

Johnson, M., Bradshaw, J. M., Hoffman, R. R., Feltovich, P. J. & Woods, D. D. (2014). Seven cardinal virtues of human-machine teamwork. *IEEE Intelligent Systems 29*(6), 74–79.

Johnson, P. E., & Thompson, W. B. (1981). Strolling down the garden path: Error prone tasks in expert problem solving. In A. Drinan (Ed.), *Proceedings of the 7th International Joint Conference on Artificial Intelligence* (pp. 215–217). Los Altos, CA: Kaufman.

Johnston, R. (2005). *Analytic culture in the U.S. intelligence community: An ethnographic study*. Washington, DC: Center for the Study of Intelligence.

Johnson-Laird, P. N., & Wason, P. C. (1970). A theoretical analysis of insight into a reasoning task. *Cognitive Psychology 1*, 134–148.

Kahneman, D., & Klein, G. (2009). Conditions for expertise: A failure to disagree. *American Psychologist 64*, 515–526.

Kahneman, D., & Tversky, A. (1979). Prospect theory: An analysis of decision under risk. *Econometrica 47*, 263–291.

Kavanagh, B. P., & Nurok, M. (2016). Standardized intensive care. Protocol misalignment and impact misattribution. *American Journal of Respiratory and Critical Care Medicine 193*(1), 17–22.

Kerr, R. A. (2012). Weather forecasts slowly clearing up. *Science 338*, 734–737.

Kinney, W. R., & Uecker, W. C. (1982). Mitigating the consequences of anchoring in auditor judgment. *Accounting Review 57*, 55–69.

Kirchhof, P., Benussi, S., Kotecha, D., Ahlsson, A., Atar, D., Casadei, B.,...Hindricks, G. (2016). ESC Guidelines for the management of atrial fibrillation developed in collaboration with EACTS. *European Journal of Cardiothoracic Surgery 50*(5), 1–88.

Klein, D., Woods, D. D., Klein, G., & Perry, S. J. (2016). Can we trust best practices? Six cognitive challenges of evidence-based approaches. *Journal of Cognitive Engineering and Decision Making 10*(3), 244–254.

Klein, G. (1998). *Sources of power: How people make decisions.* Cambridge, MA: MIT Press.

Klein, G. (2017). Positive heuristics: Strategies for engaging in speculative thinking. *Psychology Today.* Retrieved from https://www.psychologytoday.com/blog/seeing-what-others-dont/201704/positive-heuristics

Klein, G., Calderwood, R., & Clinton-Cirocco, A. (1986). Rapid decision making on the fire ground. *Proceedings of the Human Factors and Ergonomics Society annual meeting 30*(6), 576–580.

Klein, G., Calderwood, R., & Clinton-Cirocco, A. (2010). Rapid decision making on the fire ground: The original study plus a postscript. *Journal of Cognitive Engineering and Decision Making 4*, 186–209.

Klein, G., Ross, K. G., Moon, B. M., Klein, D. E., Hoffman, R. R., & Hollnagel, E. (2003). Macrocognition. *IEEE Intelligent Systems 18*(3), 81–85.

Klein, G., Woods, D. D., Bradshaw, J. D., Hoffman, R. R. & Feltovich, P. J. (2004). Ten challenges for making automation a "team player" in joint human-agent activity. *IEEE Intelligent Systems 19*(6), 91–95.

Knorr-Cetina, K. D. (1981). *The manufacture of knowledge.* Oxford: Pergamon.

Knorr-Cetina, K. D. (1983). The ethnographic study of scientific work: Towards a constructivist interpretation of science. In K. D. Knorr-Cetina & M. Mulkay (Eds), *Science observed: Perspectives on the social study of science* (pp. 115–140). London: Sage Publications.

Latour, B., & Woolgar, S. (1979). *Laboratory life: The social construction of scientific facts.* Beverly Hills, CA: Sage Publications.

Lave, J. (1988). *Cognition in practice: Mind, mathematics, & culture in everyday life.* Cambridge, UK: Cambridge University Press.

Lave, J. (1993). Situating learning in communities of practice. In L. B. Resnick, J. M. Levine, & S. D. Teasley (Eds), *Perspectives on socially shared cognition* (pp. 63–82). Washington, DC: American Psychological Association.

Lazer, D., Kennedy, R., King, G., & Vespignani, A. (2014). The parable of Google flu: Traps in the big data analysis. *Science 343*, 1203–1205.

Leonard, D., Swap, W., & Barton, G. (2014). What's lost when experts retire? *Harvard Business Review.* Retrieved from https://hbr.org/2014/12/whats-lost-when-experts-retire

Loftus, E., & Doyle, J. (1997). *Eyewitness testimony: Civil and criminal,* 3rd edn. Charlottesville, VA: Lexis Law.

Lynch, M. (1991). Laboratory space and the technological complex: An investigation of topical contextures. *Science in Context 4*, 51–78.

Lynch, M. (1993). *Scientific practice and ordinary action.* Cambridge, UK: Cambridge University Press.

Macle, L., Cairns, J., Leblanc, K., Tsang, T., Skanes, A., Cox, J. L., & Mitchell, L. B. (2016). Focused update of the Canadian Cardiovascular Society guidelines for the management of atrial fibrillation. *Canadian Journal of Cardiology 32*(10), 1170–1185.

McPherson, R. D. (1991). An NMC Odyssey. *Preprints of the American Meteorological Society 9th Conference on Numerical Weather Prediction* (pp. 1–4). Boston: American Meteorological Society.

Meehl, P. E. (1954). *Clinical versus statistical prediction: A theoretical analysis and a review of the evidence*. Minneapolis, MN: University of Minnesota Press.

Meehl, P. E. (1965). Seer over sign: The first good example. *Journal of Experimental Research in Personality* 1, 27–32.

Mellers, B., Stone, E., Atasanov, P., Rohrbaugh, N., Metz, E., Ungar, L, ... & Tetlock, P. (2015). The psychology of intelligence analysis. *Journal of Experimental Psychology: Applied* 21, 1–14.

Mieg, H. A. (2000). *The social psychology of expertise*. Mahwah, NJ: Lawrence Erlbaum.

Mieg, H. A. (2006). Social and sociological factors in the development of expertise. In K. A. Ericsson, N. Charness, P. J. Feltovich, & R. R. Hoffman (Eds), *Cambridge handbook on expertise and expert performance* (pp. 743–760). New York: Cambridge University Press.

Mieg, H. A., & Evetts, J. (2018). Professionalism, science and expert roles: A social perspective. In K. A. Ericsson, R. R. Hoffman, A. Kozbelt, & A. M. Williams, *Cambridge handbook of expertise and expert performance*, 2nd edn (pp. 127–148). Cambridge: Cambridge University Press.

Moon, B., & Hoffman, R. (2005). How might "transformational" technologies and concepts be barriers to sensemaking in intelligence analysis. Presentation at the Seventh International Conference on Naturalistic Decision Making. Amsterdam, The Netherlands.

Moore, D. T., & Hoffman, R. R. (2011). Data-frame theory of sensemaking as a best model for intelligence. *American Intelligence Journal* 29, 145–158.

Newell, A., & Simon, H. A. (1972). *Human problem solving*. Englewood Cliffs, NJ: Prentice Hall.

Nichols, T. M. (2017). *The death of expertise: The campaign against established knowledge and why it matters*. New York: Oxford University Press.

Office of Technical Intelligence (2015). Technical Assessment: Autonomy. Report from the Office of the Assistant Secretary of Defense for Research and Engineering. Washington, DC: Department of Defense.

Olson, C. L. (1976). Some apparent violations of the representativeness heuristic in human judgment. *Journal of Experimental Psychology: Human Perception and Performance* 2, 599–608.

Olson, G. (1994). Collaborative problem solving as distributed cognition. In A. Ram & K. Eiselt (Eds), *Proceedings of the Sixteenth Annual Conference of the Cognitive Science Society* (p. 991). Hillsdale, NJ: Erlbaum.

O'Neil, C. (2016). *Weapons of math destruction: How big data increases inequality and threatens democracy*. New York: Crown Publishers.

Orasanu, J., & Connolly, T. (1993). The reinvention of decision making. In G. Klein, J. Orasanu, R. Calderwood, & C. Zsambok (Eds), *Decision making in action: Models and methods* (pp. 3–20). Westport, CT: Ablex Publishing.

Ottinger, G. (2010). Buckets of resistance: Standards and the effectiveness of citizen science. *Science, Technology & Human Values* 35, 244–270.

Ottinger, G. (2013). *Refining expertise: How responsible engineers subvert environmental justice challenges*. New York: New York University Press.

Ottinger, G. (2015). Is it good science? Activism, values, and communicating politically relevant science. *Journal of Science Communication* 14, C02.

Patterson, E. & Miller, J. (Eds.) (2010). *Macrocognition metrics and scenarios: Design and evaluation for real-world teams* (pp. 11–28). London: Ashgate.

Pliske, R., Crandall, B., & Klein, G. (2004). Competence in weather forecasting. In K. Smith, J. Shanteau, & P. Johnson (Eds), *Psychological investigations of competent decision making* (pp. 40–70). Cambridge, UK: Cambridge University Press.

Polya, G. (1957). *How to solve it: An aspect of mathematical method.* Princeton, NJ: Princeton University Press.

Roberts, A. R., & Yeager, K. R. (Eds) (2004). *Evidence-based practice manual: Research and outcome measures in health and human services.* New York, NY: Oxford University Press.

Rumelhart, D., & Norman, D. A. (1981). Analogical processes in learning. In J. R. Anderson (Ed.), *Cognitive skills and their acquisition* (pp. 335–359). Hillsdale, NJ: Erlbaum.

Salzinger, K. (2005). Clinical, statistical, and broken-leg predictions. *Behavior and Philosophy* 33, 91–99.

Sanna, L. J., & Schwarz, N. (2006). Metacognitive experiences and human judgment: The case of hindsight bias and its debiasing. *Current Directions in Psychological Science* 15, 172–176.

Schank, R. (2015). The fraudulent claims made by IBM about Watson and AI. Retrieved from http://www.rogerschank.com/fraudulent-claims-made-by-IBM-about-Watson-and-AI.

Schriger, D. L., Elder, J. W., & Cooper, R. J. (2017). Structured clinical decision aids are seldom compared with subjective physician judgment, and are seldom superior. *Annals of Emergergency Medicine* 70(3), 338–344.

Shanteau, J. (1992). Competence in experts: The role of task characteristics. *Organizational Behavior and Human Decision Processes* 53, 252–266.

Shields, M. D., Solomon, I., & Waller, W. S. (1987). Effects of alternative sample space representations on the accuracy of auditors' uncertainty judgments. *Accounting, Organizations, and Society* 12, 375–385.

Shneiderman, B., & Klein, G. (2017). Tools that aid expert decision making: Supporting frontier thinking, social engagement and responsibility. *Psychology Today* blog. Retrieved from https://www.psychologytoday.com/blog/seeing-what-others-dont/201703/tools-aid-expert-decision-making-rather-degrade-it

Shneiderman, B., Plaisant, C., Cohen, M., Jacobs, S., & Elmqvist, N. (2016). *Designing the user interface: Strategies for effective human-computer interaction*, 6th edn. New York: Pearson.

Silver, N. (2012). *The signal and the noise: Why so many predictions fail—but some don't.* New York: Penguin Press.

Staszewski, J. (2008). Harnessing landmine expertise. In N. J. Cooke & F. Durso (Eds), *Stories of modern technology failures and cognitive engineering successes* (pp. 9–18). Boca Raton, FL: CRC Press.

Stein, E. (2003). The admissibility of expert testimony about cognitive science research on eyewitness testimony. *Law, Probability and Risk* 2, 295–303.

Stone, P., Brooks, R., Brynjolfsson, E., Calo, R., Etzioni, O., Hager, G., ... & Teller, A. (2016). Artificial intelligence and life in 2030. *One Hundred Year Study on Artificial Intelligence: Report of the 2015–2016 Study Panel.* Stanford University, Stanford, CA. Retrieved from http://ai100.stanford.edu/2016-report. Accessed March 14, 2017.

Suchman, L. (1987). *Plans and situated actions: The problem of human-machine communication.* Cambridge, UK: Cambridge University Press.

Targett, P. S. (1994). Predicting the future of the meteorologist—A forecaster's view. *Bulletin of the Australian Meteorological and Oceanographic Society* 7, 46–52.

Tetlock, P. E. (2005). *Expert political judgment: How good is it? How can we know?* Princeton, NJ: Princeton University Press.

Tetlock, P. E. (2017). Full-inference-cycle tournaments: The quality of our questions matters as much as the accuracy of our answers. Report to Intelligence Advanced Research Projects Activity (IARPA), Office of the Director of National Intelligence.

Tetlock, P. E., & Gardner, D. (2015). *Superforecasting: The art and science of prediction.* New York: Crown.

Turner M. (2006). What is the problem with experts? In E. Selinger and R. P. Crease (Eds.), *The philosophy of expertise* (pp. 159–186). New York: Columbia University Press.

Tversky, A., & Kahneman, D. (1971). Belief in the law of small numbers. *Psychological Bulletin* 76(2), 105–110.

Tversky, A., & Kahneman, D. (1974). Judgment under uncertainty: Heuristics and biases. *Science* 185, 1124–1131.

Tversky, A., & Kahneman, D. (1981). The framing of decisions and the psychology of choice. *Science* 211, 453–458.

Tversky, A., & Kahneman, D. (1983). Extensional versus intuitive reasoning: The conjunction fallacy in probability judgment. *Psychological Review* 90, 293–315.

Wears, R. L., & Klein, G. (2017). The rush from judgment. *Annals of Emergency Medicine* 70(3), 345–347.

Weldon, M. S. (2001). Remembering as a social process. *Psychology of Learning and Motivation* 40, 67–120.

Wenger, E. (1998). *Communities of practice: Learning, meaning, & identity.* New York: Cambridge University Press.

Wixted, J. T., Mickes, L., Dunn, J. C., Clark, S. E., & Wells, W. (2016). Estimating the reliability of eyewitness identifications from police lineups. *Proceedings of the National Academy of Science* 113, 304–309.

Wixted, J. T., & Wells, G. L. (2017). The relationship between eyewitness confidence and identification accuracy: A new synthesis. *Psychological Science in the Public Interest* 18(1), 10–65

Woods, D. D. & Sarter, N. B. (1997). Automation Surprises. In G. Salvendy (Ed.), *Handbook of human factors & ergonomics*, 2nd edn. New York: Wiley.

CHAPTER 50

REFLECTIONS ON THE STUDY OF EXPERTISE AND ITS IMPLICATIONS FOR TOMORROW'S WORLD

PAUL WARD, JAN MAARTEN SCHRAAGEN,
JULIE GORE, EMILIE ROTH,
ROBERT R. HOFFMAN, AND GARY KLEIN

Introduction

THERE is no doubt that the world continues to change at an increasingly fast pace. For example, what was once considered work, such as the kind of physical activity associated with labor-intensive craft, is often now considered a recreational pastime (e.g., physical *work* outs, lifting weights, etc.). Likewise, what were once considered leisurely pursuits, such as reading, writing, analyzing problems, and trying to understand complex phenomena (which were pursued primarily by society's elite because they could afford to invest time in such pastimes), are now considered work for many of today's citizens. In brief, and generally speaking, where work was once mainly physical, it has become increasingly cognitive. What is more, advances in technology have increased the cognitive complexity of work in many domains. The pace of change in technology continues to accelerate the changing nature of this work and, frequently, the rate at which it is carried out. Hence, what worked in yesteryear may not work today, and is unlikely to work in tomorrow's world—at least for most modern work domains.

Reflections on the Study of Expertise and Implications for the Future of the Field by Paul Ward, Jan Maarten Schraagen, Julie Gore, Emilie Roth, Robert Hoffman, and Gary Klein

The author (Paul Ward's) affiliation with The MITRE Corporation is provided for identification purposes only, and is not intended to convey or imply MITRE's concurrence with, or support for, the positions, opinions or viewpoints expressed by the author.

Accordingly, what it takes to succeed in such work must continue to change. The question is must our view and definition of expertise, and how we measure it, change too?

In this chapter, we reflect on the themes that emerged from the chapters in this *Handbook*. In particular, we re-examine the definitions of expertise and the idea that expertise is, in part, about increasing one's cognitive ability to adapt to complexity (see Ward et al., 2018). Whilst we appreciate that this argument has some limitations we argue that it is worthy of further exploration. We then take a look at where we have been, as a community of communities of expertise researchers, and whether we are heading in good directions. Finally, we present some food for thought in terms of future areas of expertise studies that are required to continue to move the field forward and, ultimately, to better prepare individuals to operate effectively in tomorrow's workplace. Playing off of the penultimate chapter by Klein et al., we begin with a quick look back at the contested war on expertise and ask whether modern work domains require a new view of expertise.

Why Are Experts Not Always Revered?

Compared to some of the views of expertise presented in this *Handbook*, some researchers have argued that the value of expertise has changed markedly and that experts are becoming increasingly obsolete. One line of argument suggests that experts have simply failed to keep up with the rate of change in the nature of work, and technology's influence on it. For instance, some have devalued expertise because alternatives that are a product of technological advancements (e.g., machine-learning algorithms) can sometimes outperform expert decisions. Others have argued that experts are unnecessary, or even unwelcome in some complex domains (e.g., financial markets; voting polls; geopolitical developments) because they are not always right and are sometimes tragically wrong, which erodes public trust in expertise (e.g., see Eiser, Stafford, Henneberry, & Catney, 2009). The counter argument, however, is that experts are essential for successful operations in many (if not all) work domains—as is evidenced when an expert retires—and that many of the successes of *expert substitutes* come about because they are based, at least in part, on expert input. Hence, the apparent fallibilities of expertise may be more a question of how expertise is defined, conceptualized, and measured, and the domain in which expertise is observed, than with experts themselves (Klein et al., this volume).

A second reason why experts are revered less in the current *zeitgeist*, particularly in non-scientific circles, has to do with a politically-motivated erosion of trust in scientific knowledge. Climate change is the poster child for this polemic. Although expertise is much broader than scientific expertise, the particular example of science and its conceptualization may illustrate different views of expertise as put forth in this

Handbook. Basically, science is still viewed by many as the isolated, individualized, pursuit of truth—the discovery of objective, immutable, undisputable facts. Empirical studies of scientific practices have shown, however, that science is a much more collaborative, networked enterprise in which facts are constructed over long periods of time, requiring a lot of convincing of one's colleagues (Latour & Woolgar, 1979). The subtitle of Latour and Woolgar's book, "The construction of scientific facts," may have opened Pandora's box and has been viewed by some as providing ammunition to critics of the scientific endeavor as a whole: If scientific facts are created, then surely we should not place any more trust in them than in other facts that are likewise created (Kofman, 2018)? The ensuing misunderstandings and political implications have so many repercussions (e.g., on the debate on climate change), that Latour has recently felt obliged to mount a defense of science (Latour, 2018) that relies heavily on the important role of networks in producing and sustaining knowledge. Facts remain robust when they are supported by a common culture, much more so than by their veracity. It is precisely the networked enterprise that makes science valuable.

The various conceptualizations of expertise as discussed in this *Handbook* mirror the conceptualizations of the scientific endeavor. To some, expertise should be explained from an individualistic stance, as the result of a long process of developing and fine-tuning of mental processes and representations (Gobet, this volume), with underpinning neural mechanisms (Ullen, de Manzano, & Mosing, this volume). To others, expertise should be explained from a sociological stance (Collins & Evans, this volume), or at least from a situated, enacted, embodied (Baber, this volume), triadic (Flach & Voorhorst, this volume), or tripartite (Pfeiffer, this volume) stance, with the latter two perspectives emphasizing the intimate coupling between person, representation, and environment. To be fair, most authors, in their theoretical conceptualizations, propose some form of multifactorial perspective in which expertise is viewed as an adaptation that takes place across various levels simultaneously. Drawing a parallel with the preceding discussion on trust in scientific facts, we may advance the similar notion that trust in expertise remains robust only when expertise is supported by colleagues, peers, collaborators, parents, in short the social environment to which a person has adapted. This is a somewhat underdeveloped conceptualization of expertise that we believe is a promising venue for future research.

In the following, we will focus on expertise as adaptive skill.

Expertise as Adaptive Skill

Structured and predictable environments are more conducive to adaptation, and hence to the development of expertise, than unstructured and unpredictable environments (Kahneman & Klein, 2009; Shanteau, 1992). Second, experts always need to adapt to surprise within their domains of expertise. It is an open question what resources experts may draw upon to be able to continue to adapt to surprise. Most research so far has looked at individual resources, such as types of knowledge representations (Gobet, this volume) or problem-solving strategies (e.g., heuristics, Hoffrage, this volume). Other types of resources, such as team or organizational resources, have been studied to a far lesser extent. Therefore, based on what we know so far, the following discussion on expertise as adaptive skill relies heavily on an individualistic perspective, from a macrocognitive perspective on expertise (see also Hutton, this volume).

In many realms, expertise is defined with respect to routine mastery of a primary task or procedure (for a review of expertise definitions see introductory chapter; Ward, Schraagen, Gore, & Roth, this volume), where performance is "outstanding in terms of speed, accuracy, and automaticity of performance" (see Hatano & Inagaki, 1984, p. 31). Such definitions of routine expertise may not capture the expert's adaptive capability but they are consistent with current definitions of transfer (see Hoffman et al., 2014). Based on the way in which transfer has been measured traditionally, the research suggests that transfer of expertise is only likely to occur when engaging in similar domain-specific tasks (for a discussion of this issue, see Ward et al., 2018).

Human beings are adaptive, or otherwise they would not survive in changing environments. By definition, then, experts are also adaptive (Hoffman, 1998), but this claim goes beyond the blanket statement of adaptivity as a general human characteristic. It means that experts, by nature of their expertise, have developed specific characteristics that allow them to be more adaptive than non-experts. Experts' adaptiveness, however, does not extend to any random domain. As mentioned earlier, when discussing limits to transfer, it is well known that expertise is quite domain-specific. Setting aside the fuzzy nature of the boundaries between *domains*, the interesting empirical question is whether experts are more adaptive than non-experts in their domain of expertise. Typically, they have a good conceptual understanding that permits the development and use of a context-sensitive strategy, which allows them to readily identify, both a priori and in situ, key decision points in a specific course of action and judge when variations of an existing course of action might be in/appropriate. Having a well-indexed conceptual representation of the current situation allows the expert to immediately access opportunities to deviate from the outcome path by selecting, modifying, or generating anew, both situational interpretations and alternative courses of action *on the fly*. Hatano and Inagaki (1984, 1986) asserted that it is one's conceptual understanding of procedural skill that affords in-event adaptive thinking. We would argue that both conceptual understanding (or sensemaking capabilities) and flexible decision-making (or replanning) are two parts of an integrated dynamic system that give rise to successful adaptation in both familiar and new complex contexts within one's domain of expertise.

This characterization of expertise as skilled adaptation to complexity and novelty, at least in their domains of expertise, is consistent with the views of many of the authors of chapters in this *Handbook*. For instance, Bohle Carbonell and van Merrienboer (this volume) and Fletcher and Kowal (this volume) defined the adaptive nature of expertise, respectively, as the ability to "perform at a relatively high level in unfamiliar situations" and "successfully solve uncommon, unusually difficult, and/or strategic problems that others cannot." Naikar and Brady (this volume) drew on Rasmussen's (i.e., Rasmussen,

1986; Rasmussen, Pejtersen, & Goodstein, 1994) view of expertise and highlighted the volatility of many human–technological work environments that require adaptation, suggesting that it is about experts being "able to deal successfully with ongoing and significant instability, uncertainty, and unpredictability in their work." This point was echoed by many, including Fletcher and Kowal (this volume), who proclaimed that "unpredictability is a certainty. Dealing with surprise and the unexpected is, then, an inevitable aspect of... expertise." However, they also noted that despite *change* being "difficult to pursue, perilous to conduct, and uncertain in its success" adaptation to change is not always rewarded as a key, or even the main, component of expertise. On this point, Hatano and Inagaki (1984) suggested that in order for adaptive skill to flourish, not only must experts be given the *opportunity to explore* task variations, such exploration of system constraints must be *valued*, and the experts must have the *authority* to explore without reprisal (see also Ward et al., 2018).

These views of the expert as having a flexible and adaptive skill capacity are consistent with a view of expertise presented recently (Ward et al., 2018). Ward et al. (2018) argued that in complex work domains, adaptive skill is the essential ingredient for success—the *conditio sine qua non* of expertise. We extend this argument here by suggesting that the importance of this skill will continue to increase as the societal and human–technological challenges ahead of us proliferate and permeate every aspect of our lives. We therefore need more research to better understand the adaptive character of expertise, and what makes it more apparent and effective in experts rather than novices.

Klein (2011) suggested that in order to become a genuine expert, perhaps an expert of the future, we need to reconceptualize learning. Klein suggests that the *storehouse* metaphor of knowing more and more—that emphasizes putting knowledge *in* to memory—may need to be supplemented by the *snakeskin* metaphor—getting knowledge *out* when needed, and being prepared to shed a particular understanding or course of action for another as the context dictates. In brief, this view promotes a shift away from knowing more toward thinking dynamically, innovatively, and differently—knowing when and when not, and knowing how and why, to generate new solutions on the fly in the face of adversity and anomalies.

One method of developing expertise of this type was presented by Klein (2011), who suggested that we improve performance not just by reducing errors but by increasing insights. Klein laid out several pathways for how we generate insight, which have since been encapsulated in an integrated model of macrocognition (see Figure 50.1) (see Hoffman & Hancock, 2017; Ward et al., 2018). The integrative D/F + F model combines the Data/Frame (D/F) model of sensemaking and the Flexecution (F) model of adaptive replanning, along with the core concepts in cognitive flexibility theory (Spiro et al., 1992) and cognitive transformation theory (Klein & Baxter, 2006, 2009), all of which are based on data on how experts operate in complex and dynamic environments where adaptation is key to success. In keeping with the integrated model, Ward et al. (2018) proposed a definition of adaptive skill that, we argue here, would have to be central to any future definition of expertise, especially when those experts operate in complex work environments. According to Ward et al., adaptive skill entails:

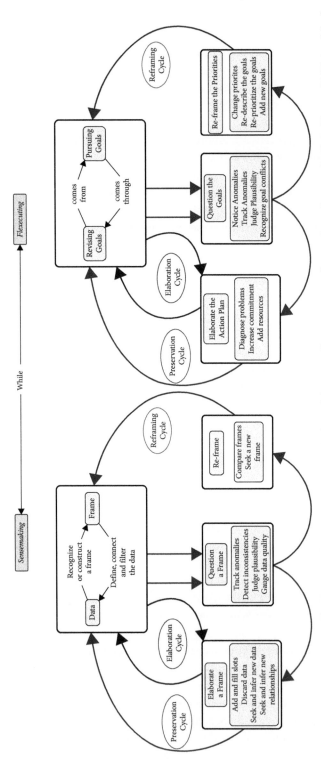

FIGURE 50.1 The integrated D/F + F model of macrocognition (adapted from Hoffman & Hancock, 2017) combining the Data/Frame (D/F) model of sensemaking and the Flexecution (F) model of adaptive replanning.

> Timely changes in understanding, plans, goals, and methods in response to either an altered situation or updated assessment of the ability to meet new demands, that permit successful efforts to achieve intent ... or successful efforts to realize alternative statements of intent that are not inconsistent with the initial statement but more likely to achieve beneficial results under changed circumstances. (Ward et al., 2018, p. 42)

This definition of adaptive skill captures the expert's requirement to update understanding on the fly in messy, complex, and dynamic environments, which can be thought of as sensemaking (i.e., build the capability to frame and reframe as appropriate, or adaptively reframe). It also captures the iterative and flexible nature of action execution, which can be thought of as a process of flexecution (i.e., build the capability to plan and replan as appropriate, or adaptively replan). Ward et al.'s (2018) definition and the integrated D/F + F model illustrate the interdependence between the sensemaking and flexecution components of adaptation, which can be thought of as managing the trade-offs within and between components. This trade-off is necessarily a highly metacognitive and regulatory process of goal evaluation relative to one's intentions and actions in the current context (e.g., see Hoffman et al., 2014). Hence, there may be fruitful opportunity to build on the D/F+F integrated model to further explore the adaptive nature of expertise.

As a means to develop adaptive skill, Ward et al. (2018) offered six training principles—based on empirical data from studies of experts in situations where there was a need to adapt—that we argue could act as an impetus for developing the kind of expertise needed for future success in the types of complex domains that are likely to be representative of tomorrow. The principles are: (a) *flexibility-focused feedback* (i.e., methods to overcome cognitive rigidity and acquire knowledge flexibly); (b) *concept–case coupling* (i.e., methods that permit learners to experience the different ways in which concepts vary from situation to situation); (c) case-proficiency scaling (i.e., use of mentoring and other scaffolding methods to stretch skill); (d) *tough-case time compression* (i.e., the need to develop a bank of cases, with varying difficulty and complexity, on which to practice adapting); (e) *complexity preservation* (i.e., methods that preserve the functional complexities to be learned and avoid learning oversimplified relationships); (f) active reflection (i.e., methods that help learners become better calibrated in terms of what they know, and in their ability to identify competency boundaries) (see also Havinga, Bergström, Dekker, & Rae, this volume; Fadde & Klein, 2010). Collectively, these principles were designed to address the need to provide practice at problems that stretch current competency and adaptive capability by promoting the opportunities outlined in Table 50.1.

Ward et al. (2016, 2018; see also Hutton, Ward, & Turner, 2018) argued that activities and practices that address the collective objectives outlined in Table 50.1, through use of the aforementioned principles, is likely to develop the requisite sensemaking and flexecution skills that are integral to any cogent definition of adaptive skill. Measuring developments in adaptive skill and ascertaining the level of expert adaptivity, however, is likely to be a far more challenging task.

Table 50.1 Collective objectives of Ward et al.'s (2016, 2018) Adaptive Skill Training Principles (adapted from Ward, Hutton, Hoffman, Gore, Anderson, & Leggatt, 2016; pp. 20–21)

Activities designed to develop adaptive skill should promote opportunities to:
- Become less cognitively rigid by overcoming the tendency to use knowledge shields;
- Learn when and under what conditions current strategies work effectively and when they do/will not;
- Experience variants of the current situation, sampled from the entire spectrum of possible ways in which the situation could manifest itself;
- Gain exposure to dynamic situations, the various ways in which those situations can change, and their effect on requirements for goal, plan, or action variation;
- Experience engaging with novel situations, contexts, tasks, and/or roles in varied ways, and evaluating the effect of those variations on the outcome and/or situational dynamics;
- Engage in dynamic sensemaking and situation assessment; and practice continual reconfiguration of understanding and dynamic reassembly of knowledge;
- Resist the temptation to be overly reductive in one's understanding of complex phenomena;
- Experience handling tough (i.e., low-frequency and/or challenging) cases of a similar/different type that require similar/different response strategies;
- Assess fit (i.e., appropriateness) between the current plan or intent and the changing nature of the current situation;
- Plan and replan a priori, replan in situ, modify responses in situ, and develop lessons learned about how plans changed and replanning occurred;
- Find boundaries of current knowledge, experience inadequacy of current knowledge at those boundaries, and provide motivation to develop new knowledge and strategies;
- Self-assess one's own ability to meet the current goal given current resources/situation, and one's ability to modify or abandon a goal, and/or generate a new one;
- Juggle/balance competing goals that emerge from changes in the situation and/or situational understanding, and to manage the resultant trade-offs between them;
- Reflect on situations that required adaptation—to reduce cognitive rigidity and to address misconceptions (rather than explain them away);
- Actively develop metacognitive and regulatory skills;
- Shift learning away from static conceptualizations of reasoning (in terms of knowledge storage and retrieval on demand) toward dynamic conceptualizations of reasoning that include predictive, active, reflective, and regulatory mental modelling.

Measuring Expertise for Future Work

In the introduction to this *Handbook*, and in many chapters contained within, several authors have discussed the challenges of measuring expertise. Naikar and Brady (this volume) suggest that by traditional standards, in terms of ability to execute an idealized sequence or set of tasks, adaptively skilled performance may be judged as *falling short*—perhaps because it deviates from the original task or results in a *good enough* outcome rather than a *reliably superior* one. However, when viewed through an adaptive lens (i.e., if assessed in terms of ability to engage in flexible action given the circumstances, which may be unfamiliar or unforeseen), performance may be considered effective or even masterful. Given the difficulties of measuring expert performance in many domains, the key question raised in the introductory chapter was: What do you do (and how should one define expertise) when working in the majority of complex domains where performance measurement is particularly challenging or impractical? We add here: What do you do when adaptation is an integral part of performance or a necessary component of expertise, as is likely the case in much future work?

Hoffman, Ford, and Coffey (2000) argued that a proficiency scale for a given domain should be based on multiple methods, and multiple types of method. Specifically Hoffman et al. argued that at least three types of method—referred to as the *three legs of a tripod* (see also Hoffman & Lintern, 2006)—that capture both the breadth (i.e., the variety of relevant experiences) and depth of experience (i.e., the length of those experiences) should be used to scale proficiency. In their research Hoffman et al. used personnel records (e.g., duty assignments), sociometry (or social interaction analysis), and career interviews to gauge skill level and were able to develop a scale that differentiated between levels (expert, journeyman, apprentice) and sub-levels (junior and senior) of expertise (see also Hoffman et al., 2014). Importantly, their method permitted experts to be differentiated from journeymen in terms of their experience and skill at being adaptive.

Hoffman (2018) extended the idea of proficiency scaling based on a three-legged tripod to that of a pentapod, to include: (i) in-depth career interviews to identify breadth and depth of education, training, and experience; (ii) professional standards or licensing to identify what it takes for individuals to become an expert at the top of their field; (iii) performance on multiple (rather than one) familiar tasks to identify competence on key problems; (iv) sociometry to identify social networks of who talks to whom about particular problems; and (v) cognitive task analyses to identify mental models of expert knowledge and strategies. Hoffman (2018) argued that one should always use at least two distinct classes of methods in order to obtain evidence that converges on a scale that is appropriate to the given domain, and that permits validation of a proficiency scale. Note that Hoffman's pentapod acknowledges the networked character of expertise and its embeddedness in social environments, as discussed in the beginning of this chapter.

Klein (2018) extended Hoffman's list to include (vi) *within-expert reliability*; (vii) *peer respect*; and (viii) *reflection,* noting that none of these measures alone are fool proof in terms of determining skill level and, like Hoffman, suggesting that multiple measures should be used to measure and scale expertise appropriately. Within-expert reliability refers to the desire for expert performance to be reliably superior, and has been a cornerstone of expert performance measurement for some time (see Ericsson & Ward, 2007).

Peer respect refers to the ability of colleagues to judge one's competence and being able to differentiate those who just talk *a good game* from those who excel in practice (cf. Collins & Evans, this volume). Peer respect should not be confused with interactional expertise since one may have mastered the language of a specialist domain, and hence may be able to interact expertly, but do so in the absence of practical competence or *contributory* expertise (see Collins & Evans, this volume).

Reflection refers to the tendency, and willingness upon request, of an expert to literally reflect on past events and critical decisions made—as a means of identifying alternative meanings and courses of actions for any given situation, particularly those that resulted in optimal and suboptimal outcomes.

Although a number of scale-based measures of adaptive skill have been developed (for a review, see Ward et al., 2016), measurement options *iv* (cognitive task analysis; Hoffman, 2018) and *viii* (reflection; Klein, 2018) above, perhaps, offer a better means of gaining insight into the adaptive nature of expertise. These options provide an opportunity to learn whether an expert noticed the need for adaptation as well as how they adapted their understanding, plans, methods, responses, or goals. Others have argued that measures of adaptation that are intimately related to measures of resilience should be incorporated into measurement of expertise, particularly Woods' (1988, 1994; for a review see, Woods, 2017) view of analyzing responses to anomalies at a systems level (see Hoffman & Hancock, 2017). According to this view, successful responding to anomalies requires a model of what is expected, making sense of deviations from those expectations without having to make additional assessment, as well as generating appropriate responses to boot. Such assessments and responses produce more expected and unexpected system responses which require further (second-order) adaptation (Woods, 2017). As such, we argue that measures that have been proposed for determining resilience capacity may also be used to measure the adaptive nature of expertise at a systems level, as discussed by Hoffman and Hancock (2017).

In future, researchers should attempt to identify which of these measures best differentiates amongst expertise levels in a particular domain, and validate the resulting proficiency scales that are developed with a view to capturing the adaptive nature of expertise within them. In the next section, we provide some additional recommendations for expertise researchers that have emerged from this *Handbook*.

Future Areas of Research Needed

In addition to a reconsideration of how we view expertise and the kinds of measures we might develop to capture adaptive skill, there are several additional themes that run through a number of chapters and that need to be addressed further in the future. The first is the discussion between structuralism and functionalism. Applied to the field of expertise, this distinction boils down to the question of whether the development of expertise can be explained by a few invariants in human cognition (e.g., the rate with which chunks can be stored in long-term memory), or whether expertise is a matter of tuning to goal-relevant constraints in the environment. The structuralist perspective is highly associated with classical views on expertise, whereas the functionalist perspective is highly associated with ecological and more adaptive views on expertise (i.e., *macrocognition*). We need not be forced to choose between the two, however. As Simon noted, both blades of the scissor, the outer as well as the inner environment, play a role in the development of expertise. People need to learn to discover what the relevant constraints are in the environment they are operating in, and then they need to somehow *learn* to recognize or perhaps *store* these constraints in memory (opinions differ as to whether mental representations are necessary constructs at all in the explanation of expertise, as witnessed by the discussion on situated or enacted/embodied cognition). Future research should take both aspects into consideration, and determine how they jointly operate in the development of expertise.

The outer and the inner environment may also be taken more broadly to include, on the one hand, the physiological environment, and on the other hand the sociological environment. Both our physiology as well as our social environment place constraints on our ability to adapt to task demands. Although both aspects were covered in chapters in this *Handbook*, future research should look into these aspects in more detail. For instance, with the advent of human enhancement techniques, we may in the not-so-distant future be able to change our physiological constraints and hence be able to learn faster. Human potential, for example in terms of giftedness, may be unleashed to a fuller extent than is currently possible. Human enhancement may be thought of as intervening in our physiology or even neurology, but it may also consist of social and motivational interventions that target our social learning environment, for instance through games or other forms of simulation or augmented reality. Furthermore, our social learning environments may be extended to include culture and subculture, schools and professional environments. To distinguish the field of expertise studies from already existing fields such as educational psychology and cultural psychology, questions would focus on how to accelerate and sustain expertise through social and motivational interventions (e.g., What is required in terms of social support to sustain deliberate practice on a regular basis over the years? How can we design environments in which scientific creativity can be increased?).

A second recurring theme throughout this *Handbook* pertains to how we should go about developing expertise and preparing individuals for success in tomorrow's workplace or, indeed, for preserving the expertise that they have at both individual and system levels. Earlier, we noted that current educational methods might inadequately prepare future experts (see Resnick, Russel, & Schantz, this volume), we highlighted a different view of learning that might lend itself to supporting the changing demands of complex work (see Klein, 2011), and proposed a set of training principles that could be used to guide the development of adaptive skill (see Ward et al., 2016, 2018). Numerous authors in this *Handbook* have proposed related ideas about what might be the best methods to develop expertise and adaptive capacity, needed not least because there will always be a gap between how management imagines work and how it is carried out (e.g., Dekker, 2005). We summarize a handful of these here and then offer some thoughts on developing expertise for the future.

Spiro et al. (this volume) argued that "learning and training will likely have to occur in qualitatively different ways to develop the ability to deal adaptively with the resultant increase in novelty that must be routinely confronted." They make several recommendations for developing twenty-first-century skills that will foster adaptivity. For instance, they recommend that we "pay attention to cases in their variegated richness while de-emphasizing the *primacy* of concepts." From their perspective, concepts are still crucial, but they must be tailored to cases rather than vice versa. In addition, they recommended that we "use multiple rather than single conceptual relations (as in schemas, prototypes, analogies, perspectives, etc.); treat cases as wholes with emergent properties so they are greater than the sum of their parts; increase the attunement to difference and decrease the bias toward seeing similarity; expect unpredictability, irregularity, indeterminateness; expect to return to earlier cases in new contexts to bring out facets that were hidden in the earlier context...; embrace flexibility and openness of knowledge representation over rigidity; stress context dependency over context independence; avoid rigidity in understanding, remaining *open* instead, with an appreciation for the sometimes limitless range of uses of knowledge in new combinations, for new purposes, in new situations; rely on situation-adaptive assembly of knowledge and experience rather than retrieval of intact knowledge structures from long-term memory; and so on" (also see Table 50.1). From our perspective, these features of cognitive flexibility theory and of adaptive skill development are likely to be the cornerstone of learning in future work environments, and the bedrock of training for any future expert.

Another recommendation may be to combine the use of acquisition, refresher, and mobilization training, such that skill relapse and maintenance training are spread throughout the pre-, post-, and after-retention intervals (Arthur & Day, this volume). However, while this approach helps mitigate skill decay, traditional overtraining methods could lead to a trade-off between retention and adaptability outcomes: greater retention, theoretically, may lead to more well-grooved behaviors that, potentially, are less conducive to being adapted. One method that may overcome this trade-off is training that emphasizes learning to spot and respond to anomalies through greater exploration and understanding of the system constraints (Havinga et al., this volume). This is consistent with our previous argument for incorporating measures of responding to anomalies (e.g., Woods, 2017) in any measure of adaptive capability.

Non-training solutions might involve, for instance, the use of regulations and centralized decision-making as means of promoting adaptability and helping avoid working at cross purposes (Havinga et al., this volume). This recommendation is not too dissimilar to current recommendations by government agencies to support policy officials' decision-making. Currently, officials are provided with relevant guidance and tools, such as using *futures* methods (i.e., considering policy options against multiple possible future scenarios). However, Conway and Gore (this volume) note that to foster adaptive decision-making such guidance should incorporate the available relevant evidence together with information critical to how this might be instantiated (e.g., as a policy, in practice) in the current context (i.e., political landscape). This should be gathered from a variety of trusted experts and iteratively fused throughout the policymaking cycle to ensure that policy is designed with policymaker values, stakeholder responses (e.g., parliament, public opinion), and differential effects in mind and, hence, can facilitate an adaptive response.

Technological solutions to developing and maintaining expertise have also been proposed, such as designing displays and instructions to minimize distracting or irrelevant information and to enhance the most relevant cues (e.g., see Morrow & Azevedo, this volume). One example, provided by Morrow and Azevedo, is a medical phone-based application that could support: (a) physician treatment decisions by providing rapid access to both up-to-date illness/treatment information *and* a flexible decision-making workspace that allows this information to be compared with an expert mental model; and (b) patient decisions by providing new illness/treatment information guided by agent-based assessment of patients' current understanding of their illness. However, Moore and Hoffman (this volume) warn against the use of technology that is more of a process control system (i.e., imposes a way of working on the expert) than a decision aid (i.e., informs and supports expert reasoning and decision-making). They highlight that such technology, especially the type that integrates and filters data as a means of providing a purported understanding to the user, rather than permitting the (developing) expert to form their own understanding, may be misguided given that the latter drives expert search.

There are several implications of this research for the design of future developmental activities. With some exceptions (for a review, see Hoffman et al., 2014), much of what we know already in terms of how experts learn is based on traditional rather than adaptive measures of expertise or on training derived from studies of non-experts (e.g., undergraduates). Hence, much research is needed to provide the empirical basis for developing experts who are sufficiently prepared to work in future complex work environments. Training that helps build an adaptive mindset (i.e., ability to generalize across cases in a content-independent manner) and adaptive readiness (i.e., skill at situation-adaptive assembly of knowledge and experience), and that does not wholly relinquish this understanding to technology, may be a good contender for developing future expertise (Spiro, personal communication).

Finally, a third recurring theme is the discussion on the generality and specificity of expertise. According to the classic view, expertise is highly domain-specific and does not transfer to novel domains. Other views have challenged this perspective, by emphasizing concepts such as general intelligence, giftedness, or the adaptive nature of expertise (e.g., Lobato, & Siebert, 2002; Ward et al., 2018). Again, expertise may be both general as well as specific, depending on one's view. What seems important here, and as we have discussed at length, is the definition of expertise one proposes. The classic view of expertise seems to reserve the concept for an end state in which there is absolute mastery of the skills and knowledge obtained. In this view, it takes at least ten years of dedicated (deliberate) practice to acquire these cognitive skills. The alternative view seems to view expertise more as a relative concept, in which experts simply possess both more knowledge and skills than non-experts. This view essentially propounds a social perspective on expertise, in which someone is considered an expert if they are considered as such by their peers. Both the classic and the adaptive view on expertise may perhaps be reconciled by taking into account the domains they typically take into consideration: the classic view has mainly studied expertise in scientific, sports, or artistic domains, that is, domains where some performance standard can easily be defined and where progress toward some gold standard can be measured. As indicated earlier, the adaptive view on expertise, in contrast, has typically focused on domains where such gold standards are not available, or where practitioners typically do not have the opportunity to spend 10 or more years in the same job—it should be noted that these domains or jobs constitute the majority of real-life situations (think of the frequent job rotations in the military).

This leads to some implications for both practical applications as well as theoretical avenues for research. Over the past fifty years, beginning in the late 1960s, we have made tremendous progress in what we know about expertise. In the first twenty years, the pendulum swung from an emphasis on domain-general characteristics to highly domain-specific characteristics. In the past thirty years, alternative conceptions of expertise have arisen, with more focus on how expertise develops under time-constrained, ill-defined, ambiguous situations with multiple, competing goals. The focus has gradually shifted from explaining expertise in terms of underlying cognitive processes to describing expertise as an adaptation to goal-relevant constraints. However, we may have thrown out the baby with the bathwater in our eagerness to focus on ecologically valid domains. Theoretical depth has given way to theoretical breadth, as an explosion of theoretical concepts occurred with each new description of expertise in a particular domain. This was actually foreseen by Langley and Simon in their concluding chapter of the 1981 book on the "acquisition of cognitive skills." They stated that as the study of expertise showed increasing domain-specificity as being key to the explanation of expertise, this would make theories of expertise barren and non-generalizable. Instead, they argued, researchers should look at the one remaining constant in the study of expertise, which was the invariance of the learning process itself. Experts in different domains may reach different end points in terms of the contents of their knowledge, what remains constant, or so hypothesized Langley and Simon, is the learning process underlying their course of expertise.

Future researchers should therefore pay more attention to the laws governing the learning process itself, as well as the various ways in which the learning process can be enhanced or accelerated. The important questions of the future are not what expertise is as an end state, but rather how we can turn relative novices into relative experts faster. We may learn some interesting novel laws on how people learn on the way.

A Potential Way Forward

What are the next steps? In this section we discuss two issues: What types of methods and approaches might help us better understand complex behavior, specifically cognitive adaptation to complexity, and the next steps we might take as a field to ensure that our science continues to mature.

In 1973, Allen Newell was challenged with trying to integrate a series of talks on varied aspects of cognition presented by some of the field's best researchers, many of whom were pioneers in the study of expertise. He lamented that it was virtually impossible to synthesize the symposium talks since the associated lines of research were disconnected from one another, and the focus was often framed by disparate polar arguments rather than some common higher goal (e.g., societal need). Despite best intentions to conduct *good science*, the tendency was for this research to be piecemeal rather than systematic, and disaggregated rather than systemic. This state of affairs was partly due to psychology's preferred method of null-hypothesis significance testing, which led to ever-more detailed hypotheses without attempts to develop overarching frameworks to integrate the results of the experiments carried out. Newell (1973) asserted that the then-current practice resulted in ideological and theoretical differences becoming less clear, and viewpoints never really being combined in a way that moves the field forward. He argued that the end product of such a process was unlikely to be a mature and cumulative science.

As a potential remedy, and with a view to developing what he would later call a "unified theory of cognition" (Newell, 1990), Newell offered three potential approaches for moving the field of cognitive science forward: (i) Develop a complete (rather than partial) processing model that provides a detailed representation of the control structure coupled with equally detailed assumptions about memories and primitive operations, such that what a participant does in completing a task, and how they do it, are fully specified (examples of such complete processing models or *architectures of cognition*, as they would later be called, are SOAR (Newell, 1990) and ACT-R (Anderson, 2007)); (ii) develop a programmatic approach (using both experimental and theoretical studies) to analyzing a single complete (rather than partial) complex task (i.e., a supertask of tasks or simply, work) such that a model of task behavior can be derived from the program; (iii) continue in the current vein of conducting disaggregated tasks but build a single processing system (rather than just multiple models) that explains them all and, as a result, their integration. Some of these recommendations have been heeded in the intervening years by researchers in the field of expertise (e.g., see Gobet, this volume; Kirlik & Byrne, this volume; Hoffrage this volume) and, arguably, the research more broadly has become more programmatic. However, the trend has frequently been toward modeling simpler tasks or, when more complex tasks have been modeled (e.g., flying a plane), researchers have focused on partial versions of the task (e.g., taxiing along a runway).

In 1993, David Woods made a call similar to Newell's but placed a greater emphasis on ways to better understand more *complex* phenomena, beyond chess, that occur outside the psychological laboratory. He was particularly interested in suggesting methods that would lend themselves to better understand cognition in natural settings that were "complex, rich, multifaceted... [rather than] ... simplified, spartan, single-factor settings" (p. 228). Understanding cognition in natural settings as a form of adaptation to complexity is of relevance to the study of expertise in light of our extensive discussion earlier on expertise as adaptive skill. Woods suggested several approaches that might be used to advance our understanding of cognitive adaptation to complexity:

(a) Establish the mapping between test and target behavioral situations such that the representativeness of the specific test situation adequately reflects the target class of situations of interest and, hence, can be generalized (see Brunswik, 1956; Harris, Foreman, & Eccles, this volume). This is not unlike Ward et al.'s (2018) principle of complexity preservation where the important complexities of target relations are maintained in any test or training;
(b) Use cognitive task analysis methods (e.g., verbal reports of thinking, action protocols, data acquisition traces) that externalize signs, which permit inferences to be made about internal processes (see Ward, Wilson, Suss, Woody, & Hoffman, this volume). These are often a precursor to building (e.g., as in verbal or protocol analyses), or are methods of validating, what Newell termed complete processing models;
(c) Retrospective analyses of critical incidents where one attempts to reconstruct the mental dynamics that occurred at the time of the event based on participants' interviews and other data (e.g., aviation black box, videos) (see Militello & Anders, this volume); and
(d) Field observations in which the observer immerses themselves in the native environment in order to obtain the point of view of domain practitioners (see Hutchins, 1995; Yardley, Mattick, & Dornan, this volume).

Woods (1993) argued that the methods he proposed—especially when studying experts—permitted complex worlds to be examined directly in a way that produces results relevant to the local problem while, simultaneously, contributing results to the generic research base on human cognition. He referred to this perspective as a complementarity assumption where one could, as Stokes (1997) later asserted, pursue both a quest for understanding and a quest for utility because these dimensions are orthogonal rather than represent ends of a single basic-applied continuum. This view promotes the use of specific behavioral contexts, which are representative of a particular class of behavior or environment, as field laboratories for examining these behaviors. Woods (2003) later referred to this approach as "staged world observation," which was more geared to verification (rather than discovery) than natural history methods (pure observation of what naturally occurs), but which shaped the conditions of observation to a lesser extent than what he referred to as "Spartan Lab" experiments (experimenter-created problem situations with low authenticity). In staged world studies, the authenticity depends on the investigators' ability to design scenarios and stage situations of interest: the scenarios need to be recognized as valid by domain practitioners (or experts in the field). Many researchers have followed this guidance and have studied experts (and expert–novice differences) in scaled-world simulations using process tracing measures (for instance, Sarter & Woods, 1995, in their studies on mode confusion in the cockpit or Schaafstal, Schraagen, & Van Berlo, 2000, in their studies on electronics troubleshooting).

Other researchers have produced cognitive models of expertise (e.g., see Burns, this volume; Hutton, this volume; Kirlik & Byrne, this volume; Matthews, Wohleber, & Lin, this volume; Ross & Phillips, this volume; Salmon, Stanton, Walker, & Read, this volume), albeit of a type qualitatively different to that proposed by Newell (1973). These models are illustrative of the maturation of the field of expertise studies (i.e., a focus on descriptive and product rather than process models) and, importantly, have played a vital role in improving our understanding of expertise in complex work settings, and in helping develop complementary theoretical advances that are more broadly applicable.

Despite this substantial contribution, there has been a tendency for some researchers to value only one type of research context or method. Paraphrasing Gigerenzer (2004), Ward, Belling, Petushek, and Ehrlinger (2017) acknowledged this tension by highlighting that "disdain has routinely been expressed by a diverse range of scientists for those 'in the other camp,' whose position, purpose, and methods have been described by those holding contrary views as having little scientific or societal value" (p. 18). Woods (1993) described such views as being *destructive,* and suggested that views that upheld one methodological strategy as having more privileged access to fundamental results than another does little to build a complete understanding of expertise. As we argued in Ward et al. (this volume), no single method or context should hold precedence if we are ever to develop a fuller understanding of expertise (see Hoffman, 2018). Despite the propensity of some to elevate one methodological approach over another (e.g., Banaji & Crowder, 1989), the complementarity between naturalism and experimentalism has long been appreciated (e.g., by Darwin) and has recently been reiterated (see Klein et al., 2003; Woods, 2003). Accordingly, we argue that both Newell's and Woods' respective recommendations provide useful guidance for how we should continue to advance the study of expertise. The complementarity assumption still holds today and is likely to prove useful guidance tomorrow.

Although some in the field of expertise studies have frequently advocated that researchers adopt a specific methodological approach or theoretical framework, few communities of practice have actually taken stock of the progress they have made towards their research being cumulative, or their community producing mature science, at least not in the reflective manner employed by Newell (1973). In a rare exception, Klein (2017) reviewed the progress of the naturalistic decision-making and macrocognition community of practice and made some recommendations for where we go next. In terms of progress, Klein stated that the communities had moved from being research-oriented to being more practice-oriented, providing services to customers in very diverse domains (e.g., Crichton, Moffat, & Crichton this volume; Moon, this volume; Roth, Naweed & Multer, this volume; Wiggins, Auton, & Taylor, this volume).

In terms of recommendations, Klein highlighted the fact that many of the models developed to this point were descriptive rather than predictive, and suggested that more effort was needed to translate these models into reliably effective interventions and tools. Second, he pointed out that much of the research had been conducted at the tactical level (e.g., examining boots-on-the-ground decision-making) rather than at the operational (e.g., command) or strategic (e.g., policy) levels, and there was little guidance to help researchers and practitioners translate between levels, or empirical research available to corroborate that what worked at one level could be successfully applied at another. Last, it had become popular for researchers from communities beyond the expertise super community to demonize experts and to devalue their contribution to society (see Klein et al., this volume). Hence, all of the expertise communities of practice needed not just to advance the field but to do so in the face of external antagonism.

Klein's reflection indicates that much progress has been made in the study of expertise. However, Klein noted many challenges still lie ahead that are relevant to most if not all expertise communities of practice: There is a need to better understand and delineate the cognitive and social processes underlying the many micro- and macrocognitive functions that have been described to date. There is a requirement to address the needs of strategic decision-makers at an organizational and policy level. There is a need for a clear public message about what the science of expertise can offer—especially in light of the war on experts—and to increase public and professional awareness and global reach of this super community. Last, there is a need to further develop instructional opportunities to better prepare scientists and practitioners to use the types of the methods discussed in this *Handbook*. These are just some of the challenges for the budding expertise researcher. Our hope is that this *Handbook* can act as a catalyst to help future researchers address some of these challenges as well as address the current issues and future directions highlighted earlier and throughout the *Handbook*.

REFERENCES

Anderson, J. R. (2007). *How can the mind occur in the physical universe?* New York: Oxford University Press.

Banaji, M. R., & Crowder, R. G. (1989). The bankruptcy of everyday memory. *American Psychologist* 44(9), 1185–1193.

Brunswik, E. (1956). *Perception and the representative design of psychological experiments* (2nd edn). Oakland, CA: University of California Press.

Dekker, S. W. A. (2005). *Ten questions about human error: A new view of human factors and system safety*. London: Lawrence Erlbaum.

Eiser, J. R., Stafford, T., Henneberry, J., & Catney, P. (2009). "Trust me, I'm a scientist (not a developer)": Perceived expertise and motives as predictors of trust in assessment of risk from contaminated land. *Risk Analysis* 29(2), 288–297.

Ericsson, K. A., & Ward, P. (2007). Capturing the naturally-occurring superior performance of experts in the laboratory: Toward a science of expert and exceptional performance. *Current Directions in Psychological Science* 16(6), 346–350.

Fadde, P. J., & Klein, G. (2010). Deliberate performance: Accelerating expertise in natural settings. *Performance Improvement* 49(9), 5–14.

Gigerenzer, G. (2004). Mindless statistics. *Journal of Socio-Economics* 33, 587–606.

Hatano, G., & Inagaki, K. (1984). Two courses of expertise. *Research and Clinical Center for Child Development Annual Report* 6, 27–36.

Hatano, G., & Inagaki, K. (1986). Two courses of expertise. In H. Stevenson, H. Azuma, & K. Hakuta (Eds.), *Child development and education in Japan* (pp. 262–272). New York: W. H. Freeman.

Hoffman, R. R. (1998). How can expertise be defined? Implications of research from cognitive psychology. In R. Williams, W. Faulkner, & J. Fleck (Eds), *Exploring expertise* (pp. 81–100). New York: Macmillan.

Hoffman, R. R. (2018). *Pentapod*. Report, Institute for Human and Machine Cognition, Pensacola, FL.

Hoffman, R. R., Ford, K. M., & Coffey, J. W. (2000). *The handbook of human-centered computing*. Report, Institute for Human and Machine Cognition, Pensacola, FL.

Hoffman, R. R. & Hancock, P. A. (2017). Measuring resilience. *Human Factors* 59(4), 564–581.

Hoffman, R. R., & Lintern, G. (2006). Eliciting and representing the knowledge of experts. In K. A., Ericsson, N. Charness, P. Feltovich, & R. Hoffman (Eds), *Cambridge handbook of expertise and expert performance* (pp. 203–222). New York: Cambridge University Press.

Hoffman, R. R., Ward, P., Feltovich, P. J., DiBello, L., Fiore, S. M., & Andrews, D. (2014). *Accelerated expertise: Training for high proficiency in a complex world*. Boca Raton, FL: Taylor and Francis/CRC Press.

Hutton, R., Ward, P., & Turner, P. (2018). *Testing the adaptive performance principles: Design of an evaluation study for Intermediate Command & Staff Course (Land) (Final Technical Report v2)*. O-DHCSTC-I309549_PT_T3_234/008. Yeovil, UK: BAE Systems.

Hutchins, E. (1995). *Cognition in the wild*. Cambridge, MA: MIT Press.

Kahneman, D., & Klein, G. (2009). Conditions for intuitive expertise: A failure to disagree. *American Psychologist* 64(6), 515–26.

Klein, G. (2011). *Streetlights & shadows: Searching for the keys to adaptive decision making* (pp. 1–337). Cambridge, MA: MIT Press.

Klein, G. (2017). *The current status of the NDM movement*. Keynote lecture presented at the 13th Bi-annual International Conference on Naturalistic Decision Making, 20-23 June, 2017. Bath, UK: University of Bath.

Klein, G. (2018). How can we identify experts: Seven criteria for deciding who is really credible. *Psychology Today* (Sep 01, 2018). Retrieved from https://www.psychologytoday.com/us/blog/seeing-what-others-dont/201809/how-can-we-identify-the-experts

Klein, G., & Baxter, H. C. (2006). Cognitive transformation theory: Contrasting cognitive and behavioral learning. In *Proceedings of the Interservice/Industry Training, Simulation, and Education Conference (I/ITSEC)* (pp. 1–9). Arlington, VA: NTSA.

Klein, G., & Baxter, H. C. (2009). Cognitive transformation theory: Contrasting cognitive and behavioral learning. In D. Schmorrow, J. Cohn, & D. Nicholson (Eds), *The PSI handbook of virtual environments for training and education: Developments for the military and beyond. Volume I: Learning, requirements and metrics* (pp. 50–65). Westport, CT: Praeger Security International.

Klein, G., Ross, K. G., Moon, B. M., Klein, D. E., Hoffman, R. R., & Hollnagel, E. (2003). Macrocognition. *IEEE Intelligent Systems* 18(3), 81–85.

Kofman, A. (2018). Bruno Latour, the post-truth philosopher, mounts a defense of science. *The New York Times Magazine*, October 25.

Langley, P., & Simon, H. A. (1981). The central role of learning in cognition. In J. R. Anderson (Ed.), *Cognitive skills and their acquisition*. Hillsdale, NJ: Lawrence Erlbaum.

Latour, B. (2018). *Down to earth: Politics in the new climatic regime*. Cambridge, UK: Polity Press.

Latour, B., & Woolgar, S. (1979). *Laboratory life: The construction of scientific facts*. Princeton, NJ: Princeton University Press.

Lobato, J., & Siebert, D. (2002). Quantitative reasoning in a reconceived view of transfer. *Journal of Mathematical Behavior* 21, 87–116.

Newell, A. (1973). You can't play 20 questions with nature and win: Projective comments on the papers of this symposium. In W. G. Chase (Ed.), *Visual information processing* (pp. 283–310). New York: Academic Press.

Newell, A. (1990). *Unified theories of cognition*. Cambridge, MA: Harvard University Press.

Rasmussen, J. (1986). *Information processing and human-machine interaction: An approach to cognitive engineering*. New York: North-Holland/Elsevier.

Rasmussen, R., Pejtersen, A. M., & Goodstein, L. P. (1994). *Cognitive systems engineering*. New York: Wiley.

Sarter, N. B., & Woods, D. D. (1995). How in the world did we ever get into that mode? Mode error and awareness in supervisory control. *Human Factors* 37(1), 5–19.

Schaafstal, A. M., Schraagen, J. M. C., & Van Berlo, M. (2000). Cognitive task analysis and innovation of training: The case of structured troubleshooting. *Human Factors* 42(1), 75–86.

Shanteau, J. (1992). Competence in experts: The role of task characteristics. *Organizational Behavior and Human Decision Processes* 53, 252–266.

Spiro, R. J., Feltovich, P. J., Jacobson, M. J., & Coulson, R. L. (1992). Cognitive flexibility, constructivism, and hypertext: Random access instruction for advanced knowledge acquisition in ill-structured domains. In T. M. Duffy & D. H. Jonassen (Eds), *Constructivism and the technology of instruction: A conversation* (pp. 57–76). Hillsdale, NJ: Lawerence Erlbaum.

Stokes, D. E. (1997). *Pasteur's quadrant: Basic science and technological innovation*. Washington, DC: Brookings Institution Press.

Ward, P., Belling, P., Petushek, P., & Ehrlinger, J. (2017). Does talent exist? A re-evaluation of the nature-nurture debate. In J. Baker, S. Cobley, J. Schorer, & N. Wattie (Eds), *Routledge handbook of talent identification and development in sport* (pp. 19–34). London: Routledge.

Ward, P., Gore, J., Hutton, R., Conway, G., & Hoffman, R. (2018). Adaptive skill as the conditio sine qua non of expertise. *Journal of Applied Research in Memory and Cognition* 7(1), 35–50.

Ward, P., Hutton, R., Hoffman, R. R., Gore, J., Anderson, T., & Leggatt, A. (2016). Developing skilled adaptive performance: A scoping study (Final Technical Report v3). In *O-DHCSTC I2 T2 077 002*. Yeovil, UK: BAE Systems.

Woods, D. D. (1988). Coping with complexity: The psychology of human behavior in complex systems. In L. P. Goodstein, H. B. Andersen, & S. E. Olsen (Eds), *Mental models, tasks and errors* (pp. 128–148). London: Taylor and Francis.

Woods, D. D. (1993). Process-tracing methods for the study of cognition outside the experimental psychological laboratory. In G. A. Klein, J. Orasanu, R. Calderwood, & C. E. Zsambok (Eds), *Decision making in action: Models and methods* (pp. 228–251). Norwood, NJ: Ablex.

Woods, D. D. (1994). Cognitive demands and activities in dynamic fault management: Abduction and disturbance management. In N. Stanton (Ed.), *Human factors of alarm design* (pp. 63–92). London: Taylor and Francis.

Woods, D. D. (2003). Discovering how distributed cognitive systems work. In E. Hollnagel (Ed.), *Handbook of cognitive task design* (pp. 37–53). Mahwah, NJ: Lawrence Erlbaum.

Woods, D. D. (2017). Reflections on the origins of cognitive systems engineering. In P. Smith and R. R. Hoffman (Eds), *Cognitive systems engineering: The future for a changing world*. Boca Raton, FL: Taylor and Francis.

Name Index

Aaron, J. 841
Abelson, R.P. 751, 952
Abernethy, B. 139, 596–9, 603, 610, 651, 931, 937
Abich, J. 496, 501, 503
Abraham, C. 504, 781
Ach, N.K. 382–3, 386, 388–9, 392, 397–8, 750
Achananuparp, P. 757
Achtzehn, S. 502
Ackerman, P.L. 6, 67, 73–4, 77, 108, 113–14, 494, 499, 650, 671, 1063, 1065, 1069–70
Adamson, M.M. 678
Adelson, B. 44, 264
Adesman, P. 41
Aguilera, C. 579
Alain, C. 596, 599–600
Allais, M. 150
Allard, F. 9, 40, 599
Alluri, V. 140
Alvarado, T. 918
Amazeen, P.G. 1121
Amunts, K. 130
Anders, S. 19, 196, 273, 358, 392, 396, 398, 475, 477, 644, 694–5
Anderson, J.R. 132, 138, 174–5, 178, 181–2, 267, 338–9, 343, 350, 493–4, 496–8, 504, 542, 745–6, 910, 952–3, 1047, 1061, 1070, 1114, 1200, 1207
Anderson, N. 907
Andrews, D. 955
Andrews, J. 1158
Antoni, C.H. 1013–14
Aristotle 1
Arora, S. 510–11
Arsal, G. 25, 299, 766, 1021
Arsenault, M.L. 1008
Arthur, W., Jr. 25, 1007, 1085, 1092–4, 1098
Asgharpour, F. 722, 728

Ashoori, M. 453–6, 462
Aspinwall, L.G. 1047
Astin, H.S. 63
Atkinson, P. 422
Atkinson, R. 799
Auton, J. 21, 1209
Avery, R. 779
Azevedo, R.F.L. 11, 25, 1205

Baber, C. 8, 13, 17, 247, 253–5, 778, 780, 865, 1139–40, 1195
Bahle, J. 37, 387–8
Bainbridge, L.C. 65, 1178–9
Baker, E.L. 801
Baker, J. 937
Baker, R.J. 64
Baker, S. 678
Ball, J.R. 842
Ballard, D.H. 244, 1004
Bangert, M. 130, 135
Barach, P. 304, 650, 931
Barnett, S.M. 263–4, 267, 270, 273, 678
Barsalou, L.W. 245, 249, 256
Barton, G. 599, 1001–2, 1118, 1181
Basak, C. 63
Bauer, D.T. 536, 540, 543, 632
Bayer, A.E. 63
Bayley, D.H. 766
Beattie, R.S. 1024
Becker, H. 418
Begg, S. 857
Behmer, L.P. 140
Bell, B.S. 264, 723
Bellenkes, A.H. 674–5
Belling, P. 56, 303, 398–9, 401, 600, 1138, 1209
Ben-Asher, N. 719, 726, 729–30
Bengtsson, S.L. 133

Benguigui, N. 603–4, 1067
Bennett, W., Jr. 183, 185, 205, 236, 598, 654, 1085–6
Berger, D.E. 61, 554
Bergström, J. 11, 26, 1111, 1119–20, 1125, 1199
Berliner, D.C. 388, 931–3, 940
Bernardi, G. 130, 1065
Berners-Lee, T. 720
Bernstein, N.A. 11, 182, 246
Besterfield-Sacre, M. 921
Bezodis, N.E. 605, 607–8
Bhaskar, R. 46
Biggs, A.T. 512, 781
Bigley, G.A. 223–4, 227–30, 236
Bijleveld, E. 509
Bilalić, M. 130, 138, 576, 1114
Bill, V. 906–7
Billay, D. 1026
Binet, A. 35–6, 61, 109–10, 120, 379, 383
Bjork, R.A. 1089–90, 1097
Blair, L. 781, 942, 944
Blekely, M.K. 676
Bloom, B. 2, 9, 585, 994
Bogdanovic, J. 225, 227–30, 236
Bohle Carbonell, K. 13, 17, 200, 272, 274, 279, 513, 768, 854–5, 1138, 1196
Bolinger, A.R. 1012
Bollwerk, N. 654, 677
Bolstad, C. 676, 681, 801
Bonaccio, S. 71
Bonaparte, N. 798, 800
Borders, M.R. 129, 442, 775, 783, 941, 943, 1048–9
Boren, T. 397
Borghini, G. 502–3, 513
Boulton, L. 775, 778, 781
Bowdle, B.F. 183, 185
Boyd, C.M. 1173–4
Braddock, J.V. 793
Brady, A. 11, 16, 200, 1138, 1196, 1200
Branlat, M. 229, 437, 1110, 1122, 1124
Bransford, J.D. 263, 271, 275, 278, 953
Bratvold, R.B. 857
Braun, A.R. 137, 584
Brentano, F. 5, 379–80, 382–3, 385–6, 401–2

Broadbent, D.P. 10, 21, 594–5, 605, 610–11, 937
Brodeur, P. 1028, 1032
Broeders, I.A.J.M. 65
Brunswik, E. 11, 151, 192, 400–401, 1060, 1208
Brunyé, T.T. 634
Bryan, W.L. 5, 36, 335–7
Bryant, D. 66
Bühler, K. 384, 386
Bull, S. 410, 416, 421
Burge, B. 905
Burge, J. 474
Burgoyne, A.P. 7, 14, 56, 58–60, 62, 64, 66, 68, 70, 72–4, 76, 78
Burns, C.M. 11–12, 19, 196, 233, 236, 452–5, 457, 459, 462, 465, 696, 744, 856, 1209
Butkovic, A. 584, 586
Butler, M. 254, 780
Byrne, M.D. 6, 8, 18, 339, 394, 556, 1207, 1209

Cacciabue, P.C. 4, 191
Cairney, P. 1146–7
Calvo, M.G. 501, 507
Calvo-Merino, B. 139, 246
Camp, L.J. 728
Campbell, D. 543
Campbell, J.M. 857
Cañal-Bruland, R. 599, 604–5, 610
Card, S. 342, 429, 740, 751
Carretta, T. 672–3, 676
Carroll, J.B. 109–10
Cassady, J.C. 504–5
Castro, S.L. 1028, 1032
Cattell, R.B. 74, 105, 109, 1063
Causse, M. 66, 673, 675, 677–8
Ceci, S.J. 61, 73, 121, 263, 267
Cepeda, N.J. 1089, 1092, 1100
Cerasoli, C.P. 1007
Champion, M. 728
Chapin, S. 907–8
Chapman, T. 647–8
Charles, M.T. 781
Charness, N. 40–1, 137, 177, 180, 792, 796, 930, 932, 1062, 1068–9, 1072, 1097, 1167

Chase, W. 4, 7, 9, 14, 35–41, 43–6, 49–51, 113–14, 137–8, 157, 177–8, 345, 576, 599, 619, 719, 752, 927, 954, 1028, 1059–60
Chatham, R.E. 793–4
Chein, J.M. 136
Chemero, A. 249, 255–6
Chen, J.L. 140
Cheng, C. 339, 920, 951–2, 1165
Chi, M.T.H. 2, 7, 41–2, 47, 49, 51, 182, 263–4, 267–9, 394, 432, 532, 752, 808, 942, 1022, 1047, 1059–60, 1138
Chipman, S. 77, 196
Chui, Y.P. 676
Cirullo, B.A. 61, 580
Clancey, W. 252, 622
Clark, A. 251
Clark, R. 792
Clark, R.M. 743
Clarke, P. 783
Clear, M. 554
Clemente, I. 951
Clinton, H. 1167
Cole, J. 778
Collins, H. 4, 8–9, 13, 15, 85–94, 96–8, 252, 263, 411, 696, 703, 759, 878, 916, 930, 952, 1041, 1045–6, 1139, 1144, 1160, 1169, 1172, 1195, 1201–2
Comte, A. 379, 383
Conway, E.M. 90
Conway, G.E. 26, 97, 156, 200, 226, 386, 611, 672, 743, 768, 854, 952, 1045, 1118, 1136, 1140, 1143, 1204
Cooke, N.J. 728–9, 1121
Copay, A.G. 781
Corbalan, G. 277
Côté, J. 1023
Coughlan, E.K. 1091, 1095
Cowan, C. 723, 745
Cowie, R. 65
Cox, D. 487
Crandall, B. 5, 10, 165, 193, 196–7, 199, 210, 378, 389, 391, 395–7, 430–2, 469, 475–7, 479, 486, 644, 775, 780, 810, 862, 883–4, 940, 1113, 1181
Crane, D. 536, 543
Creager, J.A. 63
Crewther, B.T. 512

Crichton, M.T. and L.M. 12, 23, 852–3, 857, 860–5, 1114, 1209
Cristancho, S. 421
Crocker, H. 94
Croskerry, P. 621, 627
Crothers, I. 65
Crundall, D. 774, 782
Crutcher, R.J. 393
Csikszentmihalyi, M. 578, 583–4
Curry, S. 291, 301–2

D'Abate, C.P. 1023–4
Daft, R.L. 784
Daipha, P. 878–83, 887, 889, 891–3
D'Amico, A. 728, 918
Damjanovic, L. 783
Damos, D.L. 671, 673
Dane, E. 264, 1097
Darwin, C. 401, 1209
Dattel, A.R. 676
Daudelin, M.W. 1012–13
Davidson, J. 107, 579–82, 585–6
Davis, G.F. 914
Davison, A. 301
Dawes, R. 167, 1161
Day, E.A. 6, 25, 1204
de Feijter, J. 421
de Grave, W. 421
de Groot, A. 5, 7, 36–7, 39, 49, 186, 292, 298, 345, 386, 389–91
de Manzano, Ö. 7, 15, 136–7, 1195
Deary, I.J. 64, 141, 499
Debowska, W. 134
Deffenbacher, J. 514
Degani, A. 1174
Dehais, F. 13, 22, 66, 665, 668, 673–5, 677
Dekker, S. 11, 26, 229, 851, 1111–12, 1114, 1118–19, 1124–5, 1135, 1199, 1203
Delahaij, R. 499
Delon-Martin, C. 130
d'Entremont, B. 454
Descartes, R. 244
Detterman, D. 57, 61, 580
Devold, H. 852
Dewey, J. 175, 910, 933
Dhami, M.K. 161–2, 165
Diazgranados, D. 706, 1148

DiBello, L. 23, 719, 809–10, 815, 817, 955
Diller, K.R. 265, 277
Doane, S.M. 66, 673, 675–6
Dominguez, C.O. 202, 219, 433
Dornan, T. 9, 19, 255, 409, 415, 420–1,
    435, 1208
Dosher, B.A. 345
Dover, R. 1140
Dreyfus, S.E. and H.L. 8, 13, 18, 89, 95, 245,
    251, 255–6, 313–15, 878, 882, 884, 886,
    929–30, 940, 1059, 1116
Drucker, P.F. 797–8
Dube, T. 416
Duncker, K. 5, 7, 264, 387–90, 393,
    1113, 1164
Durso, F. 676, 703, 835

Eccles, D.W. 5, 17, 294–6, 298–300, 303,
    395–6, 420, 607, 773, 775, 932, 938, 943,
    1001, 1027, 1208
Eckardt, G. 1004, 1013
Eddy, E.R 1006, 1008, 1012
Eddy, E.R. 1024
Edmondson, A.C. 1001, 1003, 1006,
    1009–10
Einhorn, H.J. 679, 696, 1161
Eisenhower, D. 799
Elbert, T. 139
Ellinger, A. 1024, 1028, 1034, 1040
Elm, W. 742, 744, 746, 748–50
Elstein, A.S. 620–4, 694
Endsley, M. 201, 219, 595, 669, 674, 676, 681,
    701, 746–7, 749, 753, 929
Engel, A. 133
Enochsson, L. 65
Enomoto, Y. 452, 457
Eppich, W.J. 1006
Epstein, A.J. 1172
Ericsson, K.A. 108, 113–14, 116, 137–8, 141,
    173, 180, 268–9, 278, 292, 303, 390–1, 393,
    395, 566, 575–7, 579–81, 583, 585, 927–8
Escalante, D. 268, 768
Eskridge, T. 473, 476
Evans, C.H. 842
Evans, R. 4, 8–9, 13, 15, 85–7, 90–4, 96–8,
    130, 156, 364, 380, 411, 878, 1037, 1144,
    1160, 1172, 1195, 1201–2

Eyre-Jackson, L. 782
Eysenck, M.W. 507

Fadde, P.J. 8–10, 24, 297, 299, 301, 304, 307,
    863, 933–4, 937–42, 944, 1115, 1199
Fairclough, S.H. 502
Faloon, S. 45, 50–1
Farrell, J.N. 67
Fatsis, S. 554
Fawver, B. 21, 594, 930, 944
Feldon, D.F. 12, 21, 273, 530–1, 535, 538–41,
    543, 940
Feltovich, J. 2–3, 7, 41–2, 182, 263, 267, 432,
    532, 619–20, 624, 719, 792, 808–9, 932,
    951–3, 955, 958–61, 963–4, 971–2, 1041,
    1059, 1114, 1117, 1122–3, 1125, 1137, 1146,
    1167, 1171, 1183
Feltz, D.L. 1028, 1035
Festa, M. 305, 645, 652
Feyer, D. 567
Feynman, R. 243, 1148
Fine, P. 569
Fink, E. 918
Fiore, S. 192, 719, 834, 845, 955, 1001, 1171
Fischer, U. 4, 23, 106, 838, 843–4, 1074
Fisher, F.T. 274
Fisher, R. 114
Fitts, P. 6, 36, 58, 671, 1113
Flach, J. 13, 16, 173–8, 181, 185, 187,
    1139, 1164
Fleishman, E.A. 67
Fleming, M. 1028, 1032–3, 1050
Fletcher, D. 506, 508–9
Fletcher, G. 858
Fletcher, J.D. 794, 802
Flin, R. 646, 859–61, 863–4, 1114, 1120
Flower, L. 43, 50
Foley-Nicpon, M. 103
Folger, J.K. 63
Folkman, S. 491, 499
Fook, J. 1009
Forbes, T. 421
Ford, J.K. 272
Ford, P.R. 602, 1091
Foreman, N.A. 5, 17, 420, 1208
Forshaw, M.J. 561–2, 569
Forster, M. 160

Fox, F.S. 268
Franco, J. 12, 21
Frank, B. 1099
Frank, J. 844
Freedman, L.D. 1140
Freeman, P. 510
Frey, P.W. 41, 137
Friedlander, K. 569
Friedman, C.P. 634
Frijda, N.H. 385, 387

Gagné, F. 77, 104, 108, 110–11, 120, 391, 574, 585
Gaillard, A.W. 499
Gaissmaier, W. 153, 166
Gallagher, A.G. 65, 106, 121, 251, 255, 511, 919
Gallagher, J. 106
Gallagher, S. 251, 255
Galton, F. 36, 105, 108
Garcia-Retamero, R. 162
Gardner, D. 1163
Gardner, F. 1009
Gardner, H. 114–15
Garnier, H.E. 913, 916, 1049
Gaunt, A. 951–2
Gauthier, I. 138, 345
Gawande, A. 1174–5
Geiss, A. 65
Gellman, L. 65
Gentner, D. 182–3, 185
Gentner, D., K.A. 187
Ghaffarzadegan, N. 1172
Ghosh, R. 1024–5, 1101
Gibbons, R.D. 64
Gibson, E. and J. 175–7, 182–3, 185–6, 192–3, 245, 248–9, 252, 347
Gibson, W. 719
Gigerenzer, G. 151–5, 157–60, 162, 166, 180, 335, 401, 651, 1165, 1209
Gilbert, W. 1023
Gilmartin, K. 40, 46
Gilovich, T. 1165
Girodo, M. 778
Gjerdingen, R. 578
Gladwell, M. 576, 585, 927
Gleick, J. 243, 873

Glencross, D. 298
Glover, K. 703
Gluck, K.A. 339, 341, 1098
Gobet, F. 6–8, 14, 37, 39, 41–2, 44, 48, 51, 137–8, 141, 173, 179, 186, 195, 346, 386, 575, 584, 675, 746, 810, 1059, 1072, 1114, 1195–6, 1207
Godwin, H.J. 306
Goff, S. 68
Goh, J. 677–8
Gonzalez, C. 296, 647, 725, 729–30, 833
Goodall, J.R. 726, 728, 730
Goodman, M. 1140
Gopher, D. 681, 1098
Gore, J. 10, 26, 138, 156, 200, 226, 386, 611, 672, 743, 768, 854, 1045, 1136, 1196, 1200, 1204
Gorman, J.C. 1121
Gorry, G.A. 622
Gough, S. 416
Gourinat, Y. 675
Gove, M. 1138
Grabner, R.H. 59
Grady, D. 567
Grahn, J.A. 140
Grande, B. 1007
Gray, W.D. 294–7, 350
Green, B.F. 69–70, 159–61, 165, 536, 540, 542–3, 627, 1039
Greenhalgh, T. 1145
Gregory, R.L. 174–7
Grenier, R.S. 270
Griffin, D. 1165
Griffin, G. 1158
Groen, G. 47, 535, 622–4
Groenier, M. 65
Grote, G. 1007, 1111, 1122
Groves, W.M. 1161–2
Guerlain, S. 632
Guggenheim, M. 225, 552–3
Gurtner, A. 1005–7, 1012
Gushin, V.I. 843

Hackenberg, D. 1158
Hackman, J.R. 1001, 1006, 1012
Hafenbrädl, S. 153, 167
Hagemann, N. 244, 605

Hagen, M.S. 1036
Hajdukiewicz, J.R. 236, 452, 455–6, 462, 744
Hallam, S. 575, 578, 582–3, 585
Halpern, D.F. 61
Halwani, G.F. 133, 135
Hambrick, D.Z. 7, 9, 14, 57, 59, 62, 64, 72–4, 77, 131, 141, 307, 443, 564, 566, 568, 576, 585–6, 928, 931, 1070, 1138
Hamlin, R.G. 1024
Hammersley, M. 414, 416, 422
Han, T.Y. 264
Hänggi, J. 130, 132, 134–5
Hanton, S. 506–8
Hardy, D.J. 672
Hardy, FletcherL. 508
Hargreaves, I.S. 554, 563
Harmon, L.R. 63
Harris, K.R. 5, 8, 10, 12, 17, 65, 196, 207, 294, 297–8, 300, 305, 307, 372, 395–6, 420, 651, 697, 773, 775, 780, 833, 943, 1027, 1208
Hart, H.B.L. 797
Harter, N. 5, 36, 335–7
Hartman, E. 506–7
Haskell, R.E. 267
Hastie, R. 1161
Hatano, G. 13, 262–5, 268–70, 272, 854, 953, 1138, 1196–7
Hausmann, R.G. 1022
Havinga, J. 11, 26, 1114, 1118, 1199, 1204
Hayes, J.R. 9, 43, 50, 397, 585, 598, 855, 1023, 1171
Henshel, D.S. 733
Hertwig, R. 151, 153–5, 158, 1165
Hesketh, B. 277, 653
Heuer, R. 741, 744, 746, 748–51, 753–4, 981, 986, 992, 1162, 1164, 1169
Hezlett, S.A. 57, 63, 74
Hill, N.M. 128, 136
Hillebrand, C. 1140
Hiller, J.H. 800
Hirtle, S.C. 44, 137
Ho, G. 779–80
Hodges, N. 9, 63, 414, 594, 599–600, 930, 944
Hofer, F. 782
Hoffman, R. 1, 3–5, 8, 13, 18, 23, 25–6, 41, 62, 74, 77, 156, 165, 190–6, 200, 203, 210–12, 219, 226, 229, 254–5, 358, 377–8, 386–7, 389, 392, 396–8, 401, 430–2, 443, 469–71, 473, 475–7, 484–6, 577, 611, 644, 650, 668, 672, 690, 692, 719, 724, 727, 733–4, 746, 753, 766, 768, 775, 780, 792, 799, 808, 810, 817, 854, 862–3, 865, 878, 883, 886–7, 889, 891, 894, 929–30, 932–3, 940, 952, 955–6, 959, 971–2, 978–9, 981–5, 988, 991, 994–5, 1021, 1025–6, 1041–3, 1045, 1051, 1059, 1066, 1086, 1091, 1096, 1117, 1123, 1132, 1136, 1140, 1144, 1147, 1163–4, 1167, 1169, 1171, 1175–6, 1178–9, 1181–4, 1193, 1196–7, 1199–202, 1205, 1208–9
Hoffrage, U. 16, 151–5, 157–8, 160, 163, 165–7, 1196, 1207
Hogarth, R.M. 157, 162, 679
Hollands, J.G. 191, 679, 745
Hollingsworth, L. 107, 112
Hollnagel, E. 4, 191, 194, 219, 229, 354, 455, 696, 1110, 1112, 1119
Horn, J. 60, 73, 109
Horsky, J. 633
Houghton, R.J. 363, 372
Huang, W. 552
Hughes, F.J. 741
Hunter, J.E. 68, 73–5, 462, 677–8
Hutchins, E. 252, 705
Hutton, R.J. 5, 10, 16, 156, 200–202, 207, 226, 323, 386, 398, 432–3, 444, 479, 611, 649, 672, 752–3, 768, 854, 883–4, 952, 1044–5, 1136, 1138, 1147, 1196, 1199–200, 1209
Hyde, K.L. 130, 133

Inagaki, K. 13, 262–5, 269–70, 854, 1138, 1196–7
Ingledew, D.K. 295, 298

Jalaeian, M. 8–10, 24, 863
James, W. 174–5, 178, 379
Jamieson, G.A. 453–4
Janelle, C. 63, 507, 597, 932, 937
Jantzen, K.J. 140
Jariwala, S. 726, 728
Jenkins, D.P. 453
Jenkins, M.P. 22, 356–7, 865, 1140, 1147
Jennings, K. 556
Jensen, R.S. 668

Jeong, S. 12, 21, 1022
Jobs, S. 1158
Johnson, C. 677
Johnson, M. 249
Johnson, R.R. 778–9
Johnston, R. 1166
Jones, C. 595
Jones, D. 674
Jones, G. 508
Jordan-Black, J.A. 65
Jordet, G. 506–7
Joung, W. 277, 653

Kahneman, D. 150–2, 156–7, 162, 174, 199–200, 262, 344–5, 621, 679, 746, 753, 797, 800, 804, 883, 977, 981, 1117, 1139, 1163–6, 1182, 1195
Kahol, K. 627, 629
Kanas, N 837, 843–4
Kannampallil, T.G. 21, 627–9, 633, 635
Kasparov, G. 1175
Katter, R.V. 745–7
Kaufman, D. 21, 63, 108–11, 115, 537, 620, 624, 626–7, 635, 1060
Kavanagh, B.P. 1173
Keegan, J. 798–9, 804
Keehner, M.M. 64
Kehrhahn, M. 270
Kelley, M. 486
Kelley, T. 719
Kellman, P.J. 252
Kelsey, F.O. 1158
Kent, A. 741, 746, 748–9, 759
Kerr, R.A. 1175–6
Kester, L. 277
Kieras, D.E. 338, 429
Kilburg, R.R. 1023
Kilgore, R.M. 453–4
Kintsch, W. 5, 7, 41, 47, 51, 66, 138, 173, 387, 394, 551, 629, 746, 751, 1060, 1063, 1068–9
Kintz, N. 843
Kirlik, A. 6, 8–9, 18, 344, 347–8, 394, 556, 596, 1207, 1209
Kirschner, P.A. 894, 954, 971
Kirsh, D. 253
Klautke, H. 951–2

Klein, G. 3, 5, 10, 26, 48, 97, 134, 156–7, 165, 181, 190–1, 193–7, 199–203, 205–11, 219–20, 313, 344, 346, 358, 378, 386, 396–7, 400–401, 430–2, 469–70, 475, 482, 484, 487, 510, 621, 642, 644, 650, 695, 706, 727, 751–2, 775, 780, 796–7, 800, 804, 808, 810, 856, 862–5, 883–4, 891, 929, 940–1, 944, 955, 959, 980–1, 983–6, 988, 994–5, 1045, 1048, 1059–60, 1066, 1071, 1113, 1117, 1122–3, 1125, 1132, 1136, 1139, 1147, 1158, 1166, 1168, 1171, 1173–4, 1181–3, 1194, 1197, 1201–3, 1209–10
Kleinbölting, H. 151
Klinger, D. 470, 773, 887
Kluge, A. 1086, 1088, 1099
Knipfer, K. 13, 25, 1010–11, 1013
Knol, J.W. 782
Kolbe, M. 1007, 1122
Koller, C.I. 782
Komlodi, A. 726, 730
Konczak, L.J. 1028, 1034–5
Konings, K.D. 855
Konradt, U. 1001–7, 1009, 1013–14
Kontogiannis, T. 277, 647
Koopmans, R. 421
Kopiez, R. 62, 141, 577, 579–80
Koslowski, B. 264, 270, 273
Kowal, D. 22, 1196–7
Kozlowski, S. 264, 272, 838, 1001, 1003–4, 1011, 1013–14, 1097
Kraft, M.A. 916, 1049
Kram, K.E. 1023
Krampe, R. 2, 49, 62, 92, 137, 268, 301, 540, 575, 577, 796, 927, 930, 1067–8, 1097
Kranz, G. 830–1, 833, 839, 841, 1158
Krauss, S. 160
Krulak, C.C. 797–8
Kubiak, D. 94
Kuhn, T.S. 88
Külpe, O. 381–2, 385
Kuncel, N.R. 63, 74
Künzle, B. 1111, 1122
Kushniruk, A.W. 631–2

La Porte, T.R. 225, 229, 1117–18
Laborde, S. 502

LaDue, D. 23, 392, 874, 878, 883, 887, 891–3, 1025, 1043, 1175
Lakoff, G.J. 173, 182, 249
Lambert, R. 1148
Landman, A. 774, 781
Lang, J. 268
Larkin, J. 42, 46, 252–3, 510, 619, 937
Latour, B. 535, 878, 1195
Lautenbach, F. 502
Lave, J. 252, 1042–3, 1045–6, 1169
Lavi, H. 505
Lawrence, G.P. 508
Lazarus, R.S. 491, 499
Leahy, S. 951
Lee, J. 62
Lee, T.Y. 60
Lefrancois, O. 675
Lehman, A.C. 36, 141, 269, 278, 339–40, 579–80, 792, 810
Lemmink, K.A. 506
Leonard, C.A. 61
Leonard, D. 1181
Lightfoot, N. 416
Liker, J.K. 61, 73
Limb, C.J. 137, 584
Lin, J. 12, 20, 499, 505, 910, 1027, 1209
Linou, N. 277
Lipps, T. 385
Loffing, F. 595, 605, 609–10
Louridas, M. 65
Loveday, T. 305, 645, 651–2, 1071
Lovell, J. 830–1
Lu, Z.L. 345
Lubinski, D. 64, 112–13
Luhanga, F. 1026
Lundberg, J. 224, 228–30, 864, 1111, 1119
Lunney, G. 831, 841, 1158
Lyons, B.D. 62

McCormack, C. 305, 645
McDaniel, M. 57, 67–8, 71, 863
McFarlin, D.B. 1032
McGeorge, P. 860
Machiavelli, N. 799
Machin, M.A. 514, 1094
McIlroy, R.C. 453–4
McKeithen, K.B. 44, 137

McKinney, E. 679
McLellan, L. 413, 415, 420
McLennan, J. 434, 643, 645
McNeese, M. 190, 1007
McNelly, T.L. 654, 1085–6
McPherson, G.E. 578, 582, 585–6
McPherson, R.D. 1175
McRobert, A. 597, 602, 607–9, 1091
McRobert, P. 602
Macron, E. 1137
Magder, S.A. 624, 626
Maglio, P. 253
Maguire, E.A. 131
Mahoney, S. 725, 728
Malhotra, V. 857
Mancuso, V. 1007
Mangio, C.A. 751
Mann, D.L. 610
Mann, F.D. 63
Manser, T. 225
Maran, A.G. 64
Marbe, K. 381–2
Marewski, J.N. 153–4, 167
Marinello, G. 162
Marshall, S.L.A. 797
Martignon, L. 153, 155, 160, 166
Martin, E.G. 1172
Martin, T. 265, 277–8
Mast, F. 165
Masunaga, H. 60, 73, 1069
Matsumura, L.C. 913, 916, 1049
Matthews, G. 12, 20, 492, 494, 496–7, 499–506, 508, 510, 512–14, 1209
Mattick, K. 9, 19, 255, 410, 415, 419–20, 435, 1208
Matton, N. 673, 675, 681
Mayer, A. 382
Mayer, R.E. 792, 803
Mayes, R.T. 257
Mbeki, T. 90, 97
Mead, G.H. 905
Meehl, P.E. 36, 164, 167, 1161
Mehr, D.R. 159–61, 165
Meinong, A. 386
Meinz, E.J. 59, 61–2, 72–3, 77, 566, 586, 1063, 1068
Mellalieu, S. 506, 508–9

Mellers, B. 1120, 1163, 1182–3
Menard, W.E. 66
Merkel, A. 1158
Merleau-Ponty, M. 244, 248
Merton, R. 86, 418, 535–6, 539, 543
Meyer, D.E. 338
Mezirow, J. 1009
Michaels, S. 907–9, 911
Michel, J.W. 62
Miedema, H.A. 65
Mieg, H. 4, 1170, 1172
Miles, T. 595
Militello, L. 3, 5, 8, 19, 165, 195–7, 207, 219, 229, 255, 273, 298, 323, 358, 392, 396, 398, 430, 432–3, 437, 439–42, 444, 469, 471, 475, 477, 479, 644, 694–5, 744, 775, 780, 810, 883–4, 984, 988, 1044, 1046, 1169, 1181, 1208
Mill, J.S. 381
Mishra, J. 9, 21, 411, 582, 963
Moffat, S. 12, 23, 863, 1209
Mohammed, S. 1007
Momtahan, K. 452, 454, 457
Montgomery, C.J. 745
Moon, B. 20, 194, 196, 470–1, 473, 475–7, 482–3, 485–7, 727, 831, 981, 995, 1147, 1169, 1171, 1209
Moore, D.T. 25, 507, 580–1, 585, 596, 757, 977–83, 986, 991, 994, 1067, 1140, 1205
Moore, E. 133
Moore, L.J. 510, 512
Moray, N. 10, 697, 1113
Morgenstern, O. 150
Morrow, D. 11, 25, 66, 73, 306, 671, 674, 680, 751, 1060, 1063, 1066, 1070–1, 1073–4, 1205
Mosier, K. 4, 23, 197, 665, 843–4
Mosing, M.A. 7, 15, 128, 130–2, 134, 136, 138, 140–2, 574, 584, 586, 1138, 1195
Moxley, J.H. 566, 796, 1097
Mueller, S.T. 21, 346, 553, 556, 560, 567
Mulally, A. 1158
Müller, G.E. 383–4, 390, 395, 397, 597
Multer, J. 22, 692, 694, 696, 698–708, 710, 1209
Mumaw, R.J. 856, 860
Mumenthaler, M.S. 66, 1071

Murdoch, J.R. 65
Murphy, C.P. 10, 21, 295, 594, 605, 652
Murray, J. 416
Myer, G.D. 65
Mylopoulos, M. 271

Nägele, C. 1005
Naikar, N. 11, 16, 185, 200, 223, 226–7, 229, 231–6, 453, 1138, 1196, 1200
Naveh-Benjamin, M. 505
Naweed, A. 22, 296, 694–700, 702–5, 707–12, 1209
Neal, A. 277, 653, 1004
Neihart, M. 103
Neil, R. 506, 508
Neisser, U. 109, 179, 186, 193, 335, 344, 349, 676
Nelson, G. 339–40
Neth, H. 166
Nettelbeck, T. 647
Neubauer, A.C. 59
Newell, A. 6–7, 37–40, 43, 50, 246–7, 335, 338, 343, 387, 390, 429, 533, 621, 650, 1164, 1207–9
Nichols, T. 1147–8, 1157
Nieuwenhuys, A. 507, 773–4, 784
Noe, R. 1028, 1032, 1093, 1100
Norman, D.A. 336–7, 346
Norman, D.O. 192
North, J. 599–601, 605, 607–9
Novak, J. 476
Novick, R. 421
Nurok, M. 1173

O'Connor, C. 863, 907–9, 911, 1093, 1114
Oertel, R. 1013–14
O'Hara, R. 66, 611, 796, 1071
Okhuysen, G.A. 1006, 1008
Olsen, S.E. 268, 272, 274, 1138
Omand, D. 1140
Omodei, M. 434, 645, 647
O'Neil, H.F. 265, 268–9, 272, 768, 801
O'Neil, H.F., Jr 801–2
Ones, D.S. 63
O'Reilly, R.C. 350
Oreskes, N. 90
Orlandi, A. 136

Orth, J. 382
Ostrom, E. 1123–4
Oswald, F.L. 7, 14, 72, 74, 141, 381, 586, 928
Ott, M. 421
Otte, K.-P. 13, 25, 1001, 1003–5, 1012
Otto, P.E. 163–4, 385
Oudejans, R. 507–8, 597, 773–4, 781, 784
Outerbridge, A.N. 68, 73, 75

Paas, F. 270, 275–7
Pachur, T. 153, 162, 166
Paine, T. 1158
Palmeri, T.J. 345
Papierno, P.B. 121
Parasuraman, R. 191, 672, 679
Parfitt, G. 508
Pashler, H. 1089, 1092
Pastor, J. 66, 673
Patel, V.L. 21, 47, 173, 619–29, 631–2, 635, 649, 987, 1060
Patton, G. 799
Pauker, S.G. 622
Paull, G. 298
Pearce, B. 453
Peirce, C.S. 174–5, 179, 623
Peluso, D.A. 10, 196, 202–3, 386, 751
Perea, M. 554, 556, 563
Perez, R.S. 268, 272, 768, 801–2
Perry, S.J. 1173
Peters, W. 841
Peterson, P. 274
Peterson, S.L. 1036
Petrov, A.A. 345
Petushek, E. 25, 56, 65–6, 398, 401, 540, 725, 766, 877, 1021, 1044–5, 1138, 1209
Peus, C. 1010
Peysakhovich, V. 675
Pfautz, J.D. 22, 747–9, 751, 753–4, 1140
Pfeiffer, S.I. 15, 103–11, 113–21, 1138, 1195
Pfund, C. 1041, 1049
Phillips, J. 8, 10, 13, 18, 196, 201–3, 205, 207, 313, 315–16, 323, 325, 386, 483, 696, 751, 783, 1002, 1026, 1074, 1209
Phillips, N.D. 166
Piaget, J. 116, 179, 265, 809
Pinard, B. 65
Pinelli, T.E. 748

Pinho, A.L. 136–7, 140
Pinkham, S.E. 783
Pirolli, P. 342, 740, 748, 751, 753
Plato 105, 956
Pliske, R. 23, 872, 883–6, 891, 893, 979, 990–1, 994, 1113, 1123
Polanyi, M. 88, 255, 953
Polgár, J. 301
Polya, G. 1164
Popper, K.R. 1148
Postlethwaite, K. 410
Proctor, J. 757
Proctor, R. 474
Proteau, L. 596, 599–600
Protzner, A.B. 554, 563
Proverbio, A.M. 136

Raab, M. 502
Raby, M. 680
Rae, A. 11, 26, 1114, 1118, 1120–1, 1199
Ragins, B.R. 1023, 1028, 1032–3
Rajivan, P. 719, 723, 729, 731, 733–4
Rall, E. 10, 196, 202–3, 386, 751, 887
Ramey, J. 397
Ramoni, M. 623–4
Rankin, A. 224, 228–30, 1111, 1119
Rasmussen, J. 7, 10–11, 175, 185, 219–20, 223, 229, 268, 369, 451, 454–5, 461, 624, 696, 743, 752–3, 800–801, 854, 1196–7
Read, G. 18, 355, 364, 368, 749, 1026, 1209
Ree, M.J. 672–3, 676
Reed, V. 413
Reeves, C.L. 71
Reingold, E.M. 177, 1068
Reisen, N. 165
Reitman, J.S. 41, 44, 50, 137
Renden, P.G. 781
Renzulli, J. 108, 110
Resnick, L.B. 8, 14, 24, 904–9, 912, 914, 916, 921, 1049, 1051, 1203
Reynolds, P.L. 175, 177–8, 180, 182
Richards, N. 555–6
Rieskamp, J. 163–4
Risucci, D. 65
Rivale, S.D. 277
Roberts, B. 63
Roberts, K.H. 223–5, 227–30, 236, 1121

Robertson, M.M. 681
Roca, A. 597, 601, 605, 607–9
Rochlin, G. 225–30, 235–6, 1118
Rodrigues, S. 507
Rohrer, D. 1089, 1092
Roring, R.W. 61, 113, 554, 1064, 1068
Rose, C. 159
Rose, G.L. 1032
Rosenbaum, D.D. 248
Rosenthal, R. 76
Ross, K.G. 8, 13, 18, 194, 209–10, 313–16, 323, 325, 483, 800, 810, 1009, 1059, 1209
Rosser, J. 65
Roth, E.M. 1, 3, 22, 26, 208, 219, 692–4, 696, 698–708, 710, 728, 743–4, 746–50, 753, 856, 860, 1140, 1193, 1196, 1209
Rousmaniere, T. 939
Rowe, J.B. 140
Rüber, T. 133, 135
Rumbold, J.L. 509
Rumelhart, D. 336–7, 346, 348, 751, 952, 1165
Runswick, O. 605–10
Russ, A.L. 434
Russell, B. 401
Russell, J. 8, 24, 74–6, 185, 596, 651, 914, 918–19, 931, 1140
Ruthsatz, J. 61, 580, 586

Sabol, M.A. 1092
Sagan, C. 1148
Sala, G. 44, 60, 584
Salas, E. 60, 197, 662, 669, 680–1, 706, 834, 837–8, 845, 861, 863, 935, 1001–2, 1006, 1008, 1011, 1088, 1114, 1148
Saleem, J.J. 436, 442
Salmon, P.M. 18, 196, 338, 355–7, 359, 362–4, 368, 372, 391–2, 478, 669, 696, 865, 1209
Salvucci, D. 338, 340–1
Sass, M. 335, 486
Satava, R.M. 65, 511
Savelsbergh, G.J.P. 773, 781, 1001, 1003
Saxberg, B.V. 248
Scandura, T.A. 1028, 1032
Scardamalia, M. 271
Schank, R. 751, 1177
Schantz, F. 8, 24, 904, 1203
Scheiner, A. 566

Schellenberg, E.G. 62
Scherpbier, A. 421
Schifferdecker, K. 413
Schinke, R. 416
Schippers, M.C. 13, 25, 1001–6, 1008, 1010, 1012–14
Schlaug, G. 130, 133, 135, 584
Schmid, E. 1010
Schmidt, F.L. 68–9, 71
Schmutz, J.B. 1006
Schneider, W. 128, 130, 136
Scholz, J. 134
Schön, D.A. 894, 940
Schön, W. 408, 410, 696
Schraagen, J.M.C. 1–2, 7, 26, 65, 77–8, 156, 190–1, 196, 213, 225, 274, 532, 810, 1122, 1193, 1196, 1209
Schreiner, E. 1010
Schriver, A.T. 306, 674, 678–9, 1071
Schum, D. 741
Schwartz, D.L. 263, 270–1, 275
Schwartz, W.B. 622
Scott, W.R. 914
Scott-Brown, K.C. 782
Seagull, F.J. 433
Segers, M. 272, 855, 1004
Selz, O. 7, 37, 383, 385–8, 390, 399
Semmer, N.K. 1005
Shadbolt, N. 389, 477, 644, 693, 810, 862, 1066
Shafer, J.L. 800, 1059
Shalin, V.L. 77, 196
Shanteau, J. 157, 315, 643, 648, 651, 679, 855, 1060, 1125, 1182, 1195
Shapiro, L. 243, 248–9
Shaunessy-Dedrick, E. 103
Sheridan, T. 10, 177
Shneiderman, B. 26, 1132, 1166, 1178, 1183
Shortz, W. 552
Sieck, W.R. 196, 201–6
Simon, H. 5, 7, 9, 14, 35–51, 109, 113–14, 137–8, 152–3, 156–7, 173, 177–9, 186, 199, 229, 254, 298, 335, 345–6, 383, 386–93, 395, 397, 429–30, 445, 533, 576, 599, 619, 621, 671, 719, 746, 752, 927, 954, 1059–60, 1072, 1146, 1164, 1203, 1206

Simonton, D. 108–9, 537, 543
Sjoberg, S. 1158
Skiles, J. 158–9
Skinner, D.B. 64
Skyles, J. 662
Sloboda, J.A. 44, 137, 574–5, 580, 584–5
Smith, J. 272
Smith-Jentsch, K.A. 834, 1006–7
Sohn, Y.W. 66, 673, 675–6
Sonnentag, S. 854
Spahn, D.R. 1007
Spearman, C. 108, 567
Spears, B. 579
Spiro, R.J. 24, 263, 267, 620, 809, 951–4, 957–63, 965, 967, 970–2, 986–7, 1041, 1059, 1114, 1124, 1137, 1197, 1204–5
Sprafka, S.A. 620, 624
Spry, K.M. 61
Stainer, M.J. 782
Staller, M. 781, 784
Stalmeijer, R.E. 272, 855
Stanley, J. 108, 112–13
Stanton, N. 18, 354–8, 362–3, 372–3, 453–4, 478, 669, 697, 863, 865, 1012, 1209
Staszewski, J. 4, 46–7, 301, 940, 1184
Steele, C.J. 133
Stefanelli, M. 623
Stelitano, L. 917
Stern, E. 59
Stern, R. 605
Sternberg, R. 107–11, 115–17, 535, 585, 625, 792, 911, 990
Stine-Morrow, E.A. 66, 1070
Stokes, A. 678
Strasser, R. 416
Stumpf, G. 379, 387
Subotnik, R. 106, 108, 111–13, 116, 121
Sulistyawati, K. 676
Sullenberger, C.B. 158–9, 662, 668, 1158
Sun Tzu 795–6
Suss, J. 3, 18, 22, 254, 297–8, 301, 306, 358, 377, 395, 398–9, 595, 600–601, 653, 668, 676, 700, 773, 777, 780, 783, 941, 1120, 1208
Swain, A. 508

Swap, W. 1181
Sweller, J. and S. 537, 543, 954

Takezawa, M. 160
Talleur, D.A. 306, 674, 1071
Tannenbaum, S.I. 105, 1006–7, 1023–4, 1101
Tatler, B. 782
Taylor, H.C. 74–6
Taylor, J.L. 66, 1071, 1074
Taylor, M. 21, 66, 114, 165, 513–14, 664, 672–3, 740, 752–3, 767, 1116, 1209
Taylor, R.G. 440
Teller, T. 66
Teresa of Ávila 379
Terman, L. 105, 107, 109, 120
Tesch-Römer, C. 2, 49, 62, 92, 137, 268, 301, 540, 575, 927
Tesler, R. 1007
Tetlock, P. 334, 993, 1125, 1163, 1182
Thaler, R. 1164
Thanasuan, K. 553, 556, 560, 567
Thompson, J.R. 745
Thomson, R. 22, 735
Thurstone, L. 109
Tiger Woods 279
Titchener, E. 173, 377, 379–85, 393, 397
Toma, M. 61, 554, 563–6
Triolet, C. 603–4
Trump, D. 1137
Tschan, F. 1005
Tschirhart, L. 45
Tuffiash, M. 61, 298–9, 301, 554, 563–6, 942
Turing, A. 93, 342, 348, 720
Turner, A. 750
Tversky, A. 150, 152, 746, 1163–5

Ullén, F. 7, 9, 15, 131, 135–7, 141–2, 584, 586, 1138, 1195
Unterrainer, J.M. 59
Upton, M. 25, 1021

Vagners, J. 192
Valentine, E. 580–1
van Dam, K. 499
van de Velde, J. 508

van der Heijden, B.I.J.M. 268–9, 1138
van Hees, S. 554, 563, 565
van Koppen, P.J. 782
van Merriënboer, J. 13, 17, 200, 270, 272–3, 275–7, 280, 768, 854, 894, 940, 1196
van Merrienboer, J.J. 653, 855
Varela, F. 250–1, 256
Veling, H. 509
Vicente, K. 11, 173, 184–5, 187, 219, 221, 223, 229, 236, 373, 451–2, 454–5, 462–3, 696, 744, 856
Vickers, J. 507, 776, 782
Villado, A.J. 1007, 1086, 1092, 1099
Vine, S.J. 502, 507, 509–10
Visscher, C. 506
von Clausewitz, C. 798, 802
von Ehrenfels, C. 379, 385
Voorhorst, F.A. 11, 13, 16, 175, 178–9, 184, 1120, 1195
Voss, M. 63, 1060
Vredeveldt, A. 782
Vul, E. 1092
Vygotsky, L. 809, 905, 1021

Waeger, D. 153
Walker, G. 18, 273, 344, 355–7, 362–3, 367, 478, 540, 669, 865, 1069, 1110, 1209
Wang, J.H. 173, 184–5, 187
Wang, X. 1088, 1092–3, 1100
Wanzel, K.R. 64
Ward, P. 1, 3, 5–11, 13, 18, 21, 24–6, 56, 63, 65, 73, 76, 156, 196, 200, 211, 226, 254–5, 273, 292–4, 297–301, 303–4, 306–7, 358, 377, 386–7, 394–6, 398–9, 401, 422, 430, 444–5, 469, 475, 507, 594–5, 597–601, 607, 611, 648–9, 653, 668, 672, 676, 679, 694–5, 700, 719, 726, 728, 746, 766, 768, 773–5, 777–8, 780, 783, 810, 854, 930, 932, 936–8, 941–3, 951–2, 955–7, 959, 962, 965, 968, 972, 1001, 1021, 1025–7, 1041–2, 1044–5, 1051, 1097, 1115, 1120, 1135–8, 1144, 1193–4, 1196–7, 1199–1203, 1205, 1208–9
Watson, J.B. 377, 383–5, 388, 393, 429, 512
Wearing, A. 434, 644–5, 647, 652
Wears, R.L. 26, 1132, 1174, 1182

Weber, M. 914, 1063
Webster, M.H. 65
Wegwarth, O. 166
Wehrens, R. 94, 97
Weick, K. 202, 229, 859, 1117–21
Weil, M. 681, 1098
Weinel, M. 90, 92–4, 97–8
Weiner, C. 243
Weiss, M. 1007
Weitzenfeld, J. 207, 209
Welsh, M. 647, 857
Wenger, E. 132, 142, 252, 1042–3, 1045–6, 1169
Wertheimer, M. 386–7
Wetter, O.E. 782
Wetzel, C.M. 511
Wetzel-Smith, S.K. 792–3
White, S. 1009
Whyte, J. IV 8, 25, 294, 299–300, 648, 679, 766, 854, 943, 1021, 1027
Wickens, C.D. 13, 22, 191, 306, 494, 498, 513, 664–6, 668–70, 672–6, 678–81, 745, 1071
Wiegmann, D. 677–8
Wiener, E.L. 663, 668
Wigdor, A.K. 69–70
Wiggins, M. 3, 21, 202, 207, 219, 305, 644–5, 650–2, 654, 668, 677, 780, 1071, 1209
Wilkinson, B.J. 751
Williamon, A. 580–1
Williams, A.M. 4, 9–10, 12, 21, 166, 297–301, 303–4, 395, 507–8, 594–605, 607–11, 644, 650, 652, 746, 773, 930–2, 936–8, 944, 1028, 1091, 1097, 1169–70
Williams, K.J. 264
Williams, R.J. 61
Wilson, J.Q. 766
Wilson, K. 3, 5, 7, 18, 131, 161, 165, 249, 254, 358, 377, 392, 430, 445, 507, 513, 538, 586, 653, 693, 696, 768, 780, 1112–14, 1208
Wilson, M.R. 510
Winch, P. 87–8
Wind, A.P. 801–2
Windler, M. 1158

Wineburg, S. 264, 273, 1060
Wisher, R. 1092, 1096
Wittgenstein, L. 87–8
Wittrock, M. 792, 803
Wixted, J.T. 1089, 1092, 1168
Wohleber, R. 12, 20, 490, 492, 494, 496, 498–500, 502, 504, 506, 508, 510, 512–14, 1209
Woike, J.K. 160, 163–4, 166
Wong, A.C. 345
Woods, D.D. 139, 177, 183, 200, 219, 221, 229, 263, 279, 338, 412, 421, 507, 513, 624, 742–4, 753, 835, 837, 987, 1037, 1110–11, 1113, 1115, 1117–18, 1122–4, 1135, 1173, 1178, 1183, 1202, 1204, 1207–9
Woody, W.D. 3, 18, 254, 358, 377, 780, 1208
Woolgar, S. 535, 1195
Woollett, K. 131
Wright, E.V. 554

Wulfeck, W.H. 792–3
Wundt, W. 379–85, 388, 397

Xiao, Y. 433

Yamauchi, T. 1022
Yankelovich, D. 402
Yardley, S. 9, 13, 19, 255, 413, 417, 420, 435–6, 694, 1208
Yesavage, J. 66, 1065, 1071
Yohannes, A. 416
Yonge, O. 1026
Young, B. 937
Young, R.M. 338

Zahavi, D. 255
Zemla, J.C. 339, 342
Zeng, Q. 632
Zhou, N. 757

Subject Index

Introductory Note
References such as '178–9' indicate (not necessarily continuous) discussion of a topic across a range of pages. Wherever possible in the case of topics with many references, these have either been divided into sub-topics or only the most significant discussions of the topic are listed. Because the entire work is about 'expertise', the use of this term (and certain others which occur constantly throughout the book) as an entry point has been minimised. Information will be found under the corresponding detailed topics.

10,000 hour rule 485, 927

abductive reasoning 205, 623
ability 12–13, 109, 119–20, 228–30, 245–8, 262–5, 582–6, 610–12, 780–3, 903–4, 984–6, 1061–4, 1199–201
  decision-making 771, 777
  groups 69, 576
  mental 6, 68, 993, 1058, 1062–4, 1094
  musical 574–6, 579–80
  psychomotor 670, 672
  spatial 676, 680, 1073
abridged introspective method 398
abridged retrospective method 399
abstraction 182, 186, 383, 385, 387, 623–4, 983, 987
  hierarchy 185, 187, 451–5, 462
  mental 182–3
  upward 436, 447
academies 300–301, 486, 766, 771, 993
accelerated expertise 767, 826, 930, 932, 940, 1184
acceleration 955, 957, 964, 967, 970, 1184
  expertise 24, 485, 955, 965
accidents 664, 668–70, 690, 693, 695, 1111–12, 1115–16, 1118, 1123–4
  nuclear 859
accommodation 177, 179–80, 1062
accountability 412, 822–3, 909, 917, 970, 1034
accountants 221, 416, 1061, 1064, 1182

accreditations 1142–3
accuracy 152–3, 272–3, 279–80, 397–8, 651–2, 676, 835, 1168, 1176–7, 1182–3
  anticipation 605, 609
  diagnostic 632, 634
  levels 603
  predictions 600, 777
  recall 434, 448
  scores 606, 609
  shooting 771, 779
ACM (applied concept mapping) 473, 475–6, 478, 481, 483–5, 487
ACPT (American Crossword Puzzle Tournament) 556, 566–8
acquired knowledge 625, 1042, 1061, 1096
  formally 625
  informally 625
acquired skills 180, 277, 1093
acquisition
  cognitive skills 18, 20, 313–14, 316, 1206
  expert performance 114, 626
  expertise 15–16, 74, 130–2, 142, 334, 343, 345, 347
  interventions 1095–6
  knowledge 271, 318–19, 878, 965, 1065, 1076
  phase 1087, 1089–90
  psychomotor skill 932, 937
  skill 4–6, 8–9, 67, 343, 496, 500, 503, 1065–6, 1085–7, 1089

ACTA (applied cognitive task analysis) 743, 1044
action scripts 985
action sequences 459, 595–6, 600
action tendencies 605, 609, 611
activity theory 415, 420–1
Actormap 369
actors 86, 90, 193, 220–3, 226–38, 369, 773–4
   interacting 220, 227, 234–5
   in sociotechnical systems 221–2, 228–9
ACT-R 155, 175, 338–44, 350, 498, 1207
adaptability 225–6, 774–5, 778, 781, 802, 854, 856, 1096–7
adaptation 135, 139, 209, 225–7, 229–30, 237, 245–6, 280, 386, 652–3, 1044–5, 1195–7, 1201–2
   cognitive 213, 1207–8
   local 1123–4
   skilled 1045, 1136, 1196
   in the workplace 223–6
adaptations, sensorimotor 139–40
adaptive capability 1112, 1196, 1199, 1204
adaptive capacities 387, 1110, 1112, 1203
adaptive control
   model 179–82
   system 16, 178–9
adaptive expertise 13, 17, 200, 262–80, 652–3, 767–8, 854–5, 859
   attitudes 274, 279
   definition 262–9
   development in informal settings 269–71
   measuring 271–4
   and related concepts 267–9
   teaching for 275–9
adaptive experts 17, 263–5, 267–9, 272–3, 279, 783–4
adaptive flexibility 952, 955, 957
adaptive performance 953, 963, 965–6, 970–1, 1098, 1102
adaptive readiness 24, 959, 963, 965–6, 968, 970–1
adaptive replanning 10, 196, 387, 1197
adaptive responses 24, 498, 953–4, 956, 966, 972
adaptive skill 952–6, 959, 962, 965–7, 970–2, 1195–7, 1199–200, 1202–3
   deliberate practice and acceleration in fostering 964–71

development 968, 1204
   expertise as 1195–9
   framework 24, 955, 962
adaptive strategies 881, 1072, 1111
adaptive systems 175, 179, 263, 915, 1122
adaptive toolbox 154–5, 165, 167
adaptive worldview 957–9, 962–3, 967, 969
adaptiveness 797, 953, 1113
adaptivity 13, 1196, 1204
ad-hoc teams 1002, 1012
administration 318, 320, 630, 722, 1141
administrators 918, 937, 1141
adolescence 111, 133, 570
adulthood 64, 1058, 1060
adults 25, 42, 60, 105, 1061, 1064
   older 1062–7, 1069–70, 1072–6
advanced beginners 245, 313–16, 319, 323, 885, 1059
adversaries 721, 724–5, 731, 982, 985, 987, 1132, 1135
adversity 1110, 1197
advocacy 722, 1024
   groups 370, 1147
affiliation 1, 35, 56, 85, 765, 770
affordances 175–6, 178, 244–53, 347–8, 742, 758
AFQT (Armed Forces Qualification Test) 58, 68–71, 74
after-action review 794, 1099
age 104–6, 112–13, 131, 133, 569, 585–6, 1060–1, 1063–9, 1073, 1075–6
   differences 42, 1064–76
   and expertise 513, 576, 1064, 1073, 1076
   of peak performance 1064, 1068
agents 16, 184–6, 253, 340, 1109, 1123–5
   external 227, 960, 1045
age-related differences 1063, 1066–7, 1075
age-related variance 1070–1
age-vulnerable abilities 1066, 1075
aggression 513, 768
aging 25, 1062–3
   and expertise 1061–74
aging populations 11, 1058–76
Air Force 674, 679, 793–4, 883–6, 929
air traffic control (ATC) 66, 341, 343, 662, 665, 669, 672, 675–6, 1071–2, 1074
aircraft 234–5, 645, 663–8, 673–4, 793–4, 799, 1072, 1134

airports 159, 342, 645, 666–7, 672
airspace 665–7, 672, 680
air-to-air combat 793–4
alarms 743, 1111, 1158
  false 726, 730
alerts 710, 726–7, 731–2
  positive 726
algorithms 112, 165, 223, 229, 252, 1179–80
all-source analysis 979–80
altitude 341, 664–5, 667–8, 670
  required 438–9
amateurs 37, 39, 46, 303, 1141
ambiguity 176–7, 458–9, 464, 1027, 1034, 1146, 1162, 1166
ambiguous situations 985, 1162, 1167, 1174, 1206
American Crossword Puzzle Tournament, see ACPT
American Psychological Association 766–7
anagrams 61, 552, 561–3, 570
analogical problem solving 264, 855
analogical reasoning 532
analogies 179, 182–3, 186, 246, 533–4, 563–4, 751–2, 960–2
analysis 265, 324, 355–9, 361–4, 372–4, 435–7, 451–4, 482–3, 743–6, 749–51, 753–6, 775–7, 988–91
  boundaries 357–8
  control tasks 452–3
  decompositional 435–7
  discourse 414–15
  functions 356–7, 362, 373–4
  integrative 437–9, 447–8
  intelligence, see intelligence, analysis
  latent curve 1087, 1091
  qualitative 324, 414, 447, 1122
  social organizational 453–4
  sociological 90, 1169, 1171
  statistical 36, 51, 69, 334–6, 532, 1167
  strategies 451–3, 458–63, 465
  task 18, 338–9, 342, 349, 394, 396–7
  video 931, 933
  work domain 451–3, 465
  worker competency 454, 461–2
analysts 25, 354–5, 358, 435–6, 452–4, 731, 740–4, 746–59, 979–83, 986, 989–93
  competencies 979, 989

expert 741–2, 744, 748–51, 753, 756, 758, 979, 984–7
  performance 742, 977, 981–2
  reasoning 25, 990
  senior 472, 979, 990
analytic(al) methods/techniques 338, 392, 399, 472, 991–3, 1166–8
analytical processes 454, 728, 740, 991
analytical skills 303, 980
anecdotal evidence 158–9
anesthesiologists 433–4
anger 503, 512–13
annotation, video 933–4
anomalies 198–9, 201, 432–3, 726, 834, 984–5, 1123, 1197–8, 1202, 1204
  potential 199, 726
anonymity 694, 769
anthropology 21, 530, 878
antibiotics 411, 419
anticipation 595, 597–606, 608–12, 697, 699–700, 777, 783
  accuracy 605, 609
  effective 604, 611
  schematic 386–7
  skilled 594–612
  in sport 605, 609, 611, 783
  superior 595, 599, 601
anticipatory behavior 598, 602, 610, 612, 699
anticipatory judgments 603–4, 609
anticipatory thinking 1159, 1168
anxiety 396, 492–3, 499–501, 503–9, 512–14, 770, 776
  detrimental effects 506–7
  high 504, 508, 703
  symptoms 508–9
  test 20, 491, 500–501, 504–6, 508, 514
applications 3–5, 18–20, 45–6, 48, 200–201, 206, 312–15, 324–5, 424–5, 487, 650–2, 955, 1023–4
applied cognitive task analysis, see ACTA
applied concept mapping, see ACM
applied contexts 346, 636, 1094
applied research 50–1, 295, 443, 494, 855, 865–6
appraisals 491–2, 494, 496, 503, 767, 1142
apprentices 485, 757, 978, 1025, 1042–3, 1163
  junior 730, 1042

apprenticeships 766, 880, 931, 1024,
        1046, 1171
    cognitive 252, 540, 1046
appropriateness 273, 434, 534, 1200
aptitudes 59, 110, 132, 330, 722, 1097
AR, *see* augmented reality
architecture 112, 163, 338–9, 344, 650, 1073
    cognitive 155, 334, 338–9, 341, 344, 350
    social 919
argumentation 908, 910, 912, 915, 919,
        921, 993
    skills 24, 539
armed and unarmed use-of-force tactics 781
Armed Services Vocational Aptitude Battery,
        see ASVAB
arousal 497, 502, 509
    sympathetic 498, 504
artifacts 251, 253, 482–3, 631–2, 635, 741
artificial intelligence 38, 1161, 1175, 1177,
        1179, 1181
assertions 197, 200, 205–6, 754, 1164,
        1168, 1177
assessment 201–2, 204–5, 271–2, 312–13,
        317–18, 328–30, 433–4, 541, 643,
        645–6, 648, 650–1, 860–1
    formal 858
    gifted 105, 120
    objective 301, 328, 643
    reacquisition 1091, 1094
    risk 356, 855, 857, 1120, 1123, 1125
    subjective 646, 652, 797
    threat 728, 780
    tools 320, 326–7, 329–30
    updated 1136, 1199
associationism 383, 386–7, 390
associations
    cue-based 651, 653–4
    semantic 554, 556, 570
astronauts 831–3, 836–9, 841–5
astronomy 401, 718, 1169
ASVAB (Armed Services Vocational Aptitude
        Battery) 59, 68
ATC, *see* air traffic control
athletes 508, 510, 595–6, 599–600, 605,
        610–11, 928, 930, 1035–6, 1044–5
    elite 508, 1067
    expert 63, 298, 598, 611, 932, 937

less-skilled 596, 599
skilled 598–9, 601–2, 612, 937
attack patterns 304, 732
attainment of expertise 2, 13, 25,
        109, 1001–3, 1095
attention
    management 191, 480, 703
    selective 178, 649, 941
attentional control 58, 73–4, 77, 497, 507
attentional focus 393, 508
attentional resources 67, 137, 493, 498, 1067
attentional skills 1066, 1068
attrition 558, 1092
attunement 175, 177–8, 181, 962, 1059, 1204
    cognitive 963–4
audit, knowledge 19, 432–3, 441, 444–5, 883
auditory regions 129–30, 133, 140
augmented reality (AR) 755, 935, 1203
Australia 358, 363–4, 369–70, 690, 692–3, 695
automated agents 221–2
automated systems 1178, 1181
automatic processing 136, 1066
automaticity 44, 671, 1059, 1065, 1090, 1096
automation 15, 17, 135–7, 296, 665, 681–2,
        710–11, 1116–17, 1178–9, 1183
autonomy 166, 223, 1145, 1147, 1177, 1179
aviation 73, 77, 305, 341, 662–82, 1060,
        1063, 1074–5
    commercial 221–2, 341, 668–9
    defining expertise 663
    domain general models 66–7
    expertise 670–81
    skills, non-technical 680–1
    tasks 22, 664–70, 672, 678
aviation professionals 22, 667–8

backward search 42, 46
balance 7, 13, 116, 227, 237, 412
ball flight 248–9, 304, 596, 603, 605–8, 937
banks 166, 664, 811, 822–3, 1134
basal ganglia 129–30
baseball 298, 303, 596, 609, 1060, 1067
basketball 507–8, 599
    shooting 508, 510
batters 298–9, 303–4, 606–8, 937–8
battlefield experience 799
Bayesian statistics 346, 1166–7

SUBJECT INDEX 1233

beginners 39, 41, 44, 60
  advanced 245, 313–16, 319, 323, 885, 1059
behavior 36, 38–40, 223–4, 226–32, 234–5,
    237–8, 329–30, 348–9, 354–7, 368–9,
    384–5, 465–6, 644–9
  change 416, 465–6
  expert 18, 354–5, 357–9, 361, 363, 367–9,
    373–4, 466, 741, 748–50
  of individuals 18, 355–6, 373–4
  outdated 1122, 1124
behavioral models 155, 161, 163
behavioral repertoires 230–2, 234–5, 237–8
behaviorism 19, 378–9, 383–5, 393
beliefs 274, 746, 754, 1164, 1174, 1178
benchmarks 378, 935
  normative 152, 237
benefits of expertise 1060–1, 1067, 1075,
    1109, 1114, 1125
BESD (binomial effect size display) 76
best practices 319, 722, 1172–5
bias, response 501, 773
biases 174–6, 746, 753–4, 757–9, 779, 981,
    1013, 1163–6, 1168–9
  cognitive 754, 857, 981, 993, 1164, 1169
  confirmation 754, 964, 1013, 1148, 1165
  judgment 1163–4
  negative 753, 758
bibliographic analysis 265–7, 279–80, 544
bibliometric analyses 537, 543
Big Data 1167, 1175, 1179–80
binomial effect size display, see BESD
biochemists 63, 88
biological motion 598–9, 603, 606, 612
black boxes 159, 597, 808, 1208
blind spots 155, 811–12, 814, 1009
blobbers 992
blood pressure 456, 502
blurring 881, 1026, 1134
boards 40, 154, 159, 177, 245, 662
Boeing 342, 645, 1158
Boggle 552–3
boom gates 358, 365, 367–8
boomer generation 889, 977
bottlenecks 481, 819
boundaries 190, 194, 1042, 1045, 1110, 1112,
    1196, 1200
  competency 1115, 1199

fuzzy 267, 280
national 1134
boundary conditions 1005, 1008, 1182
bounded rationality 16, 38, 49, 149,
    152–4, 1135
bowlers 249, 606–8
box trainers 936, 938
brain 128–30, 132, 134, 246, 249, 251, 253,
    256, 336–7
  anatomy 131, 134
  regions 129, 131–2, 135, 140–1, 343, 350
Brazil 221, 907
brevity 324, 468
bridge 40, 137, 1061, 1067
budgets 371, 619, 820, 841
building blocks 152, 154–5, 160, 194, 440
building construction 223, 228, 230
bureaucracy 914, 1171
burglars 162
burns 11–12, 19, 451–60, 462, 464–6, 744
business 20, 23, 203, 1023–5, 1028, 1031
  cases 807, 1008
  expertise 23, 804, 807–27
    future work 826–7
    research methods and findings 809–26
  experts 808–11
  models 977, 996
  plans 163–4, 1040

cabin crew 662, 669
calculation 248–9, 1177
  mental 46–7
candidate schemas 966, 968–70
capabilities 247, 487, 649–51, 742, 757–9, 793,
    796–8, 808–9, 863–4, 979–80, 1199
  adaptive 1112, 1196, 1199, 1204
  cognitive 236, 239, 800, 802, 809
  computational 38, 153
  expert 656, 792, 1166, 1168, 1177
  technological 757, 802
capacity 40–1, 175, 235–6, 238, 412–13, 493,
    498, 646–7, 651–3, 984–5, 1116–17
  adaptive 387, 1110, 1112, 1203
  cognitive 20, 744
  computational 16, 152, 167
  limited 50, 161, 409
  loss 492, 498

capacity (*cont.*)
  memory 42, 58, 61–2, 66, 73–4, 77, 676–8, 680, 745–6
  processing 491–3, 512, 1065, 1069–70, 1076
  for self-organization 236, 238
capturers 20, 474–5, 480, 486
capture-the-flag competitions 723–4
cardiac activity 502
cardiologists 456–7, 626
career development 535, 977, 1039–40, 1047
careers 323, 472, 485–6, 535, 537, 807–8, 875, 939, 942, 1029–30
caregivers 436
Carnegie Mellon University, *see* CMU
carpenters 56, 766, 1098, 1110
case experience 952, 957, 967
case-based hypertext learning environments 957, 964
case-proficiency scaling 1115, 1199
castings 817–18, 820–1, 823
catalysts 1, 13, 1001–2, 1008, 1014, 1210
categorization 154, 186, 295, 299, 728, 731
causal effects 131–2, 141–2
causal factors 664, 674, 983
causal hypotheses 392, 500
causal models 500, 1095
causal reasoning 25, 625, 982
causal relationships 385, 655–6
causation 88, 960, 983, 986
  complex indeterminate 982–3
CBT (computer-based training) 941–2
CCUs (coronary care units) 159–61, 165
CDM, *see* critical decision method
CDS (clinical decision support) 634
cells 75, 129, 372, 557, 560, 759
central nervous system activity 502
centralized coordination 220, 226–8, 237
certifications 619, 663, 670, 723, 799
Certified Information Systems Security Professional (CISSP) 723
CESC (Central Emergency Support Centre) 860
CFH design 966–7
CFT (cognitive flexibility theory) 24, 156, 951–72, 986, 1197, 1204
  and adaptive skill 959–64

learning environments 963, 965, 967–8
  and novelty paradox 956–9
  societal implications 971–2
  systems 957, 968–9
CFT learning 963, 965–6, 968, 970
chaos 796, 838, 873
charismatic leadership 1010
checklists 340, 423, 1158, 1160, 1174–5, 1180–1
  integrated 831, 841
chess 41–2, 47–9, 57, 59–62, 77, 137–8, 292–3, 297–8, 300–301, 349–50, 389–90, 930–1, 1067–8
  Chase and Simon 39–41
  de Groot 36–9
  expertise 5, 7, 57, 59–60, 132, 138, 1068–9, 1072
  grandmasters 166, 291–2, 301
  masters 345–6, 349, 810, 1182
  patterns 7, 177
  players 36, 39–40, 59–60, 138, 292, 307
  positions 41–2, 130, 292, 298, 345, 927
chest pain 159–60, 460, 629, 632
child rearing 265, 267
children 42, 103–5, 107, 109–13, 117–19, 130, 574–6, 585, 906–7, 910
  gifted 104, 106–7, 120
choking 20, 508–10
chronic stress 492, 513
chunking 7, 40, 42, 45, 178, 182
  theory 14, 35, 38–41, 45–6, 48, 51
chunks 38–43, 45–6, 48–51, 138, 178, 182, 185, 346
citizens 91–2, 780, 905, 920, 985, 1110
Civil Service 1133, 1137, 1140–3, 1146
civilians 766, 773, 779, 782–3, 943
  untrained 779, 782
classic expertise approach 6–9, 14, 35–51, 138, 195, 346
  characteristics 49–50
  chess 36–41
  community of practice 7–8
  computational model of chess memory 45–6
  future 50–1
  key theoretical work 45–7
classical expertise approach
  computer programming 43–4

games 41–2
physics 42–3
writing 43
classical music 44, 579
classification 85, 91, 160, 163, 719, 728
classmates 904, 909
classroom learning 920, 1051
classroom teaching 24, 918, 933, 942, 1046
classrooms 24, 319, 904–5, 909–12, 918–20, 922, 1046, 1049
  regular 105, 118, 326
clients 409, 411, 413, 415, 810–11, 815, 817, 892, 943, 1023
climate 872, 961
  anthropogenic 90
  change 655, 972, 1194–5
clinical decision support, *see* CDS
clinical environments 627–8, 635
clinical information 419, 635
clinical judgment 1145–6, 1161, 1174–5
clinical knowledge 620, 625
clinical practice 416, 420, 619, 630, 1145
clinical psychologists 770, 1182
clinical reasoning 273, 621–2, 624, 631
clinical supervisors 933
clinical workflow 627, 631, 634
clinicians 437, 440, 618–19, 621, 623–4, 629–35, 1145–6
clips, video 136, 303, 777, 782, 933–4
close-to-practice qualitative research methods 9, 408–25, 435, 694
  exemplar 416–17
closure 748, 991–2
  cognitive 990, 996
clues 459, 555–8, 560–1, 566, 726, 811–12
clusters 41, 265, 267, 280, 553, 565
CMU (Carnegie Mellon University) 7–8, 49, 730
COA, *see* course of action
coaches 302–3, 916, 919, 928, 931–2, 939–40, 1021–2, 1026, 1028, 1036–7, 1040–1, 1044–5, 1048–9
  academic 942–3
  elite 1044–5
coaching 322, 329–30, 1022–8, 1036–7, 1039–40, 1045, 1047–9
  activities 1037, 1045

executive 811, 1024
and mentoring 1023–5
co-citation analysis 265, 267
cockpit 156, 295, 490, 675, 705, 1114
cockpit task management 666–7, 679, 681
co-constructive processes 382, 397, 1037
coders 435–6, 439, 447
codes 432–3, 435–6, 441, 447, 874, 876
  ethical 769, 771
coding 359, 436, 439, 447–8, 967–8
  benefits 967–8
  categories 435, 876
  process 436, 439
coercion 423, 771
cognition 50–2, 173–5, 178–80, 190–3, 195, 210, 243, 247–51, 256, 338, 344, 1169–72, 1207–8
  distributed 252–3, 256, 705, 712, 1169
  embedded 250
  embodied 173, 249–50, 1203
  enactive 250–1
  expert 193, 378, 447, 809
  medical 618, 620–2
  situated 211, 251–2, 257, 344, 347–8, 1169
cognitive ability 14–15, 56–77, 671, 673, 676–7, 680, 734–5, 1062–3, 1066–7, 1069
  effects 68, 72–3
  evidence for role 58–72
  factors 57, 68, 72–4, 77, 672
  measures 57, 59, 62, 65, 72, 74–8
cognitive activities 17, 251, 253, 354, 357, 881, 886, 1170
cognitive adaptations 213, 1207–8
cognitive apprenticeship 252, 540, 1046
cognitive architectures 155, 334, 338–44, 350, 497
cognitive biases 754, 857, 981, 993, 1164, 1169
cognitive capabilities 236, 239, 800, 802, 809
cognitive challenges 432, 445–6, 951, 956, 1173
cognitive closure 990, 996
cognitive demands 209, 250, 298, 511, 651–2, 710
cognitive efficiency 1059–60, 1065
cognitive expertise 313, 504, 802

cognitive flexibility 263, 673, 776, 855, 951, 986–7
cognitive flexibility theory, *see* CFT
cognitive functions 2, 202, 255, 475, 1062, 1065
cognitive interviewing 475, 478, 482–3, 644–5, 887
cognitive load 443, 504, 619, 630, 636, 1042
  theory 607–8
cognitive mechanisms 175, 268, 300, 496, 726, 783
cognitive models 193, 195–6, 338, 340, 349, 351, 733, 735, 1048–9
cognitive overload 20, 491–2, 496–7, 633
cognitive performance 192–5, 468–9, 471–5, 477–80, 482–5, 487, 496, 499, 701, 703
cognitive processes 250–2, 355–6, 537–8, 612, 618, 620–1, 740–2, 744–7, 751–3, 775
  higher-level 609, 618
cognitive processing 47, 490, 584, 779, 932, 1183
cognitive psychology 3, 9, 14, 49, 244, 256, 268, 878–9
cognitive readiness 22, 265, 268–9, 767, 797, 801–2
cognitive resources 249, 1060, 1062, 1065–6, 1068, 1072–5
cognitive rigidity 987, 1115, 1124, 1199–200
cognitive science 265, 337, 618, 621–2, 630, 636
  radical embodied 249, 255–6
cognitive skills 502–3, 561, 598–603, 610, 612, 858, 860–1, 931–2, 936–8, 944
  acquisition 18, 20, 313–14, 316, 1206
cognitive strategies 231, 237, 272, 275, 277, 279
cognitive structures 238, 543, 618, 860
cognitive support 439, 630, 633–4, 955
cognitive systems 10–11, 14, 16, 219–21, 223, 225, 227, 229, 237, 1169–70
cognitive systems engineering (CSE) 14, 16, 200, 207, 219–39, 1135, 1138
  community of practice 10–11
cognitive task analysis (CTA) 195–6, 378, 469, 471, 477, 727–8, 733–4, 743–4, 748, 883, 887, 1048, 1201–2
cognitive tasks 276, 399, 740, 742, 745–6, 754

cognitive transformation theory 156, 955, 1009, 1197
cognitive tunneling 668, 1061
cognitive underpinnings 556, 734
cognitive work analysis, *see* CWA
cognitive workload 502, 753
cognitive-psychological perspective 498
collaboration 96–7, 455, 734, 784, 831–4, 837, 839, 841, 992–3, 1074
collaborative work 536, 839, 917–18, 1170
collage 881–2
collective evaluation 25, 1003
collective tacit knowledge 88
college students 117, 192, 561, 727, 1165
collegial interactions 915–16
collisions 159, 177, 363–4, 699
color 302, 652, 699, 709, 812
comfort 248, 483, 647, 1035
command 201, 226, 795–7, 800, 803, 831, 861, 1071–2
commanders 742, 750–1, 836
  military 197, 222, 865
commercial aviation 221–2, 341, 668–9
commitment 201, 209, 212, 256, 1033, 1118–19
common ground 487, 706–7, 839, 843, 1119, 1122, 1125, 1178
commonalities 156, 165, 183, 269, 295, 795–6
communications
  failures 364, 862
  radio 693, 702–3, 707
  skills 662, 669, 982, 990
  voice 665, 680, 743
communities of practice 1–11, 17, 197, 316, 318, 1043
  classic expertise approach 7–8
  cognitive systems engineering (CSE) 10–11
  deliberate practice (DP) 9
  Dreyfus-ian 8
  individual differences 6–7
  knowledge approach 7–8
  macrocognition 10
  naturalistic decision making (NDM) 10
  perceptual–motor expertise 9–10
  skill acquisition 5–6
  social studies of science 8–9

companies 470, 472–3, 724, 726, 807–8, 810–11, 814–15, 817–18, 822–4, 1181
compensation 1029–30, 1039, 1062, 1073
  strategies 1062, 1066, 1068–9, 1071, 1073, 1075–6
compensatory strategies 162–3, 1068–70, 1072–3
competence 8, 410, 491–5, 498–9, 799–800, 857–8, 861, 1021, 1023, 1201–2
  daily 1069, 1075
competencies 111–12, 317, 319, 451, 454, 461–3, 725, 728, 855–6, 979–80
  cognitive 733–4, 980
competency models 316–17, 330
competition(s) 506, 508, 514, 553–5, 681, 724, 1035
  capture-the-flag 723–4
  direct 930, 944
competitiveness 164, 564
complacency 1113
complementarity 191, 211, 401, 1208–9
complete processing models 1207–8
completeness 222, 382, 385, 389, 398, 693–4
complex domains 12, 19, 1066, 1071, 1194, 1199, 1201
complex environments 1, 10, 208, 210, 626, 630
complex interactions 601, 673, 1022
complex knowledge 23, 894, 916, 1101
complex performance domains 1095, 1098, 1100–1102
complex problems 193, 207, 211, 386, 531, 542
complex real-world performance domains 26, 1089, 1097, 1103
complex reasoning 916, 1049
complex situations 211, 419, 421, 424, 984, 1173
complex skills 77, 275, 278, 280, 312, 1066
complex systems 26, 220, 226, 295, 690, 693
complex tasks 295–6, 339–40, 1048–9, 1086, 1088, 1090, 1093, 1095, 1113, 1207
complexity 22–3, 26, 195–6, 222, 294–5, 410–12, 725–6, 953–4, 958–60, 967, 970–2, 1042–3, 1207–8
  functional 12, 1115, 1199
  preservation 960, 1115, 1199, 1208

compliance 371–2, 769, 855, 864, 915, 917
  procedural 856, 1116
complicated, processes 383, 852
component analyses 299, 598, 1050
component teams 832–3
composite measures 64, 72, 508
comprehension 21, 61, 618, 620, 728, 1069–74
comprehensive review 128, 313, 674, 752, 931
computation 173, 394, 1062, 1073
computational capacities 16, 152, 167
computational model of chess memory 45–6
computational modeling 18, 50, 333–5, 337, 344–6, 348–50
  research 334, 345, 348
  state of science 348–9
  techniques 335, 337
computational models 6, 18, 38–40, 333–50, 394, 556
  domain-specific 344–8
  future research directions 349–50
computer models 39, 47, 50–1, 891, 1175–6
computer programming 40, 48, 137, 1046
  classical expertise approach 43–4
computer programs 38–9, 46, 50, 263, 795
computer science 1, 37, 397, 723, 725, 1175–80
computer security 720, 723–4
computer simulations 37, 46, 155, 158, 164–5, 338
computer-based training (CBT) 941–2
computerized physician order entry (CPOE) 633
computers 93–4, 469, 720, 943–4, 1176, 1183
concept maps 473, 475–7, 482–3, 485, 992
conceptual models 211, 539, 701, 759, 889–90
conceptual understanding 263–4, 269, 272, 915, 919, 957, 960, 1196
conceptual variability 957, 964, 966
conceptual variability search 957, 964
concurrent verbal protocols 36, 358–9
confidence game 1167
confirmation 301, 756, 1025
  bias 754, 964, 1013, 1148, 1165
confirmatory evidence 630, 634
conflict resolution 778, 780

conflicts 478, 482, 742, 744, 836–7, 1135, 1139, 1171–2
conformity 174–5, 325
confusion 175, 431, 645, 669, 844, 893
connectivity, functional 15, 57, 128, 140–1
conscious deliberation 199, 988
consciousness 377, 388–90
consensus 87, 90, 154, 436, 447, 1013
consent, informed 423, 445
consistency 67, 423, 494, 754, 756, 775
  internal 165, 206, 274
constraints 152–3, 174, 176–7, 182–4, 186–7, 234, 272, 452, 552–3, 798–9, 1044–5, 1203
  contextual 400, 601, 1042, 1050
  ecological 176, 186
  external 179, 195
  functional 177, 183, 185
  goal-relevant 1059, 1202, 1206
  internal 174, 177, 179, 182
  letter 554, 556
  organizational 233, 235, 238, 400, 883
  resource 441, 750
  rule-based 555, 1066
construction 176, 348, 362–3, 648, 653, 655
  social 90, 119, 1171
constructive feedback 1032, 1035, 1040
constructivism 88, 176, 178, 184, 186, 314
constructivist approaches 86, 175–6, 178, 181–3
consultants 264, 270–1, 305, 411, 916, 939
consumers 436, 475, 740, 748, 879, 1148
content 23–4, 92–3, 317–18, 392–3, 774–5, 840, 963, 965, 1063
  knowledge 913, 932, 935
  schemas 958
  validity 317, 328
content-blind norms 152–3
context independence 962, 1204
contextual cues 694, 1072
contextual information 605–10
contingencies 438, 446, 674, 962, 1066
  local 226, 228
continuity 87, 150, 313, 656
continuous improvement 371, 918–21
continuous process operations 851–4
continuous processes 208, 642, 651, 878

continuous variables 73, 160, 1064
contractors 831, 852–3
contributory expertise 92–3, 97, 1202
control 181, 205–6, 246–7, 347, 507–8, 664, 691, 708–9, 773, 818, 935–6, 1093–4, 1183–4
  conditions 41, 531, 943, 1076, 1086
  executive 498–9, 514
  experimental 293, 643, 646–8, 655–6, 769, 774
  gaze 502, 782
  groups 60, 130, 133–4, 561–2, 782, 908–9, 1012, 1049–50
  human 712, 1183
  mission 23, 795, 830–2, 834–5, 839–45
  motor 9, 130, 134–5, 182, 248, 510
  process 19, 995, 1205
  riot 646, 783
  supervisory 347, 713, 743
  vocal 135, 578
control rooms 841, 852, 856–7, 863, 1111
control tasks 451–3, 699, 1099
  analysis 452–3
control trials, randomized 444, 1142, 1145, 1173
controlled conditions 11–12, 291–2, 296, 298, 302, 1166, 1168
controlled experiments 191, 381, 531, 963, 968, 1162
controlled processing 136, 1096
controllers 665–7, 669–70, 672, 676, 831, 833–6, 841, 845
  flight 830–5, 837, 839, 841, 843–5
  frontroom 833, 835
  mission 831, 835, 838, 842
convenience 299, 474
cooperating teams 705–8
cooperation 796, 839, 841, 1114, 1159
coordination 129–30, 133, 246, 851–2, 1062, 1119, 1122, 1170–1
  centralized 220, 226–8, 237
coordinative structures 181–3, 186
co-pilots 490, 669, 675
coping 253, 491–2, 494, 499, 511, 514
  skills 494, 497
  strategies 499, 1111
coronary care units, see CCUs

corporals 795–6, 799
correction 60–1, 68–9, 278, 560, 627–8
corrective feedback 938, 982, 1047
correctness 1090, 1165, 1181
correlational design 1094–5
correlations 59–70, 72–5, 133, 141–2, 576, 673, 676, 989, 1093
  positive 64, 673, 854
  strong 24, 931
correspondence 41, 174–5, 206, 296, 749, 756
  dyads 174, 184
cortical thickness 129–30, 141
cortisol 497, 502, 504, 510
  salivary 502, 511
costs 794–5, 817, 821, 1058, 1060–1, 1075, 1097, 1099–102
  cognitive 1066, 1068
  human 264, 277
  reduced 817, 1179
  reputational 1134
counseling 1023–5
course of action (COA) 195, 197–9, 201, 204, 314–16, 419, 796, 1196–7
court positioning 605–6, 611
co-workers 130–40, 222, 1073, 1171
CPOE (computerized physician order entry) 633
CPTs (cyber protection teams) 727
creative thinking 301, 796, 798, 989
creativity 36, 111, 114–18, 578, 797, 802
  scientific 1038, 1203
credentials 91, 535, 979
credibility 86, 880, 1157, 1170
crew and mission control 831–2, 839–40, 842, 844
crew autonomy 837, 844–5
crew communication 839–40
crew resource management, see CRM
crewmembers 832–3, 836–8, 840, 842–4
crews, train 691, 693, 695, 698, 701–8, 710–12
crew-to-ground communications 840, 843
crime 253, 722, 782, 814
crime scene examination 253–4
criminal investigators 782
crisis work 225, 228
criterion variables 75, 326
criterion-based training 1089–90

critical cues 202, 431, 437, 439–40
critical decision method (CDM) 255, 389, 391–2, 396, 431–3, 440, 443–6, 474–5, 477–8, 480–4, 486–7, 743, 1044
critical incidents 204, 439, 775, 874, 933, 1208
critical thinking skills 724, 755
CRM (crew resource management) 22, 662, 669, 680–1, 858, 1114–15
  training 1114–15, 1120
crossing environments 364, 368–9
crossing systems 18, 364, 369
  operation 362, 364, 369, 372
crossroads cases 968–70
crosswords 551, 553, 555–7, 560, 566–70
  expert solution times as function of difficulty 557–9
  players 554, 556, 565–6, 569–70
  skill 555–6
  strategy v memory skills 560–1
cryptograms 552–3
crystallized intelligence 59, 73, 109–10, 569
CSE, see cognitive systems engineering
C-sections 1172
CTA, see cognitive task analysis
cue hierarchy 151, 160, 164
cue recognition 775, 1071–2
cue utilization 437, 610, 649–52, 655–6
cue-based associations 651, 653–4
cued-retrospective interviews 19, 434–5, 447
cues 151, 156, 160–4, 166–7, 305–6, 432, 434–5, 437, 440, 650–4, 780, 985–8, 1071–2
  contextual 694, 1072
  critical 202, 431, 437, 439–40
  diagnostic 1059–60
  environmental 299, 346
  memory 46, 570
  movement-related 299–300
  musical 134
  perceptual 45, 610, 679
  postural 595, 597–8, 601, 604–6, 610–11
  proximal 342
  remote 553, 556
  visual 65, 602
cultural norms 92, 269–71, 274, 698
cultural psychology 381, 1203
culturally valued domains 104, 120–1

culture 24, 930-3, 939, 944, 987, 994, 1141, 1146
  organizational 759, 891
curricula 118, 300, 415, 993, 1043, 1050
cutlery 1111-12
cut-scores 107, 120
CWA (cognitive work analysis) 11, 19, 233, 451-66, 696, 743-4, 748
cyber analysts 722, 731
cyber attacks 720-1, 725
cyber defense 722-3
cyber domains 22, 718-35
  complexities of expertise 725-7
  current research 727
  education and training 723-5
  evolution 719-21
  future directions for research 732-5
  work roles 721-3
cyber expertise 22, 718-21, 723, 725, 727, 729, 732-5
cyber operations 22, 723, 733-4
  offensive 722, 725
cyber operators 22, 719, 723, 725-8, 731, 733-5
  defensive 725-6
  private sector 727
cyber protection teams (CPTs) 727
cyber situational awareness 727-31
cyber workforce 722-3
cybersecurity 374, 484, 719, 721, 723-4
  education 722-3
Cybersecurity Workforce Framework 721-3

dance 111, 137, 139, 246, 695, 700
dancers 132, 139
data 68-70, 200-206, 210-12, 334-6, 357-9, 394-400, 435-6, 623-4, 693-5, 747-53, 995-6, 1179-81, 1197-8
  analytics 720, 756, 1180
  saturation 323-4, 328
  sources 203, 364, 394, 756-7, 1179
data-frame theory (DFT) 16, 203, 752, 1147
data-frame model of sensemaking 196, 200-205, 207, 211
data-frame relationship 202, 204
DBD (decision-based design) 534

DBIR (design-based implementation research) 920
deadlines 750, 756
decay 25-6, 1085, 1088, 1090-7, 1101-3
  mitigation 1095, 1097-8, 1100, 1102
  rates 1091, 1102
deception 796, 985, 987
decision ladder 452-3, 457-9, 462-3, 465, 753
decision making
  expert 48, 200, 346, 440-2, 775, 778, 880, 882
  military 795-7
  recognition-based 334, 344, 346
Decision Research 1159-62, 1185
decision research community 1160-3
decision-based design (DBD) 534
decision-making skills 483, 662, 668, 774, 777
decisions
  diagnostic 632, 634, 1059
  knowledge-based 454, 463
  military 23, 197, 201, 797-8, 804
  policy 1146-7
  political 90, 1133
  recognition-primed 5, 7, 16, 48, 194-8, 985
declarative knowledge 271, 570, 929-30, 1098, 1102
decomposition, HTA goal 359-60
decompositional analysis 435-7
deduction 453, 623-4, 1007
deductive reasoning 205, 455
Deepwater Horizon 853, 861-2
defense 69, 728, 735, 994, 997, 1134-5
  policy 1135, 1143
defensive cyber operators 725-6
defensive tactics 765, 770, 781, 783
deference 859, 1118-19, 1123, 1164
definitions of expertise 3, 8, 11-14, 1091, 1194, 1196-7
delayed post-training performance 1091, 1094
deliberate practice (DP) 9, 24, 49-50, 62, 113-14, 278-9, 564-6, 575-6, 579-81, 583-7, 927-44, 957-9, 968-71
  to accelerate expertise in the workplace 939-41
  in CFT learning 965, 968
  characteristics 930, 933, 939-40
  community of practice 9

framework 49, 928, 931, 935-6,
938-9, 944
in medical and surgical education 934-6
in music 582, 586
research 4, 931
in sports 936-9
in teaching and teacher education 932-4
and training-based research 931-2
deliberation 37, 199, 315, 804, 839, 988
conscious 199, 988
deliberative thinking 8, 396, 894
delivery 294, 318, 357, 607-8, 818-19, 944
on-time 818-22
democracy 89, 96, 98, 1160
demographic information 323, 445, 730
demonstrations 142, 273, 334, 346, 543, 795
Denkpsychologie 382, 385
Denmark 451, 644
dependence 500, 542, 644, 1067
dependencies 430, 446, 893
depression 166, 512
descriptive approaches 150, 155, 643, 646,
648-9, 656
descriptive behavioral framework 855, 858
descriptive models 161, 432, 439, 441, 448,
1147, 1149
descriptive research 647, 932, 1126
descriptors 324-5, 649
design 19, 219-21, 232-5, 237-8, 302, 368,
439-40, 442, 461-6, 533-4, 1085-7,
1092, 1094-5
CFH 966-7
correlational 1094-5
experimental, *see* experimental designs
ideal 1092, 1094
implications 16, 220, 228, 232
instructional 275, 935, 941, 944
objectives 233, 238
persuasive 465-6
phases 1109, 1143
rehearsal 820
research 397, 423, 576, 879, 1086-7, 1095
designers 113, 294, 1118, 1178, 1183
instructional 928, 931
detection 198, 627-8, 720, 730-1, 1004, 1013
developers 329, 622, 817, 1178
software 323, 432, 441, 721

development of expertise 23-5, 162, 164-5,
419-20, 462, 465-6, 529-30, 584-7,
1008-9, 1013-14, 1058-60, 1115-17,
1202-3
in teams 25, 1001-3, 1011, 1013-14
developmental functions 25, 1023
developmental interactions 1024,
1042, 1050
developmental models 111, 316, 319
developmental phases 888-9
developmental progression 323-4, 330
developmental psychology 42, 50, 104
developmental stages 313-14, 320
developmental support roles 1022-3, 1027
devices 42, 167, 356, 597, 718, 720
dexterity, manual 65, 498
DF model, *see* data-frame model of
sensemaking
diabetes 270, 1173
diagnoses 198, 409, 437, 620, 623, 646,
649, 942
medical 622, 1046
required 306, 666
diagnosis loops 196, 201
diagnostic accuracy 632, 634
diagnostic cues 1059-60
diagnostic decisions 632, 634, 1059
diagnostic expertise 21, 620
diagnostic reasoning 409, 618-36
complexity and expertise 625-8
conclusions and future 635-6
expertise and deviations from standard
protocol 629-30
history 620-5
technology-mediated 620, 630-5
diagrams 233, 421, 473, 476, 483, 876
WOP 233-4, 236, 238
dialogic instruction 914-20, 1049
dialogic learning 906, 908, 910
dialogic reasoning 24, 1049
dialogic teaching 904, 906, 908, 912-14, 920
devleoping expertise in 910-14
dialogue 822, 904-5, 909-10, 1025
dichotomies 175, 184, 187, 574
difference variables 1087, 1095-6
differential effects 1008, 1014, 1143, 1205
differential psychology 36, 335

differentiated model of giftedness and talent 108, 110, 585
differentiation 176, 313, 330, 574
digit span task 39, 45, 47
digital forensics 722, 724
digital media 952, 972
digital tutoring 799
digit-symbol substitution 58, 61
direct feedback 486, 1022
direct instruction 906, 954, 963, 1047
direct interactions 92, 179
direct verbalizations 390–5, 399–400
directed probes 392, 396–400
directed thinking 386–7
directedness 397–8
directionalities 163, 508, 623
disasters 669, 812, 861, 1092, 1147
disciplinary lenses 873, 1147
disciplines 21, 424–5, 529–31, 535, 539–40, 542, 1023–5
  academic 416, 980
  operational 857
  scientific 1, 418, 424
discourse 97, 844, 878, 907
  analysis 414–15
discovery instructions, verbal 277–8
discovery learning 910
  guided 276–8
discrediting experts, see war on experts/expertise
discrimination 91–2, 96, 330, 345, 649, 651
  perceptual 445, 783, 1181
diseases 88, 160, 165, 456–7, 619, 623, 1145–6
  germ theory 88–9
dispatchers 691–3, 699, 701–3, 706–8
dissemination 627, 1160, 1169
distal perfusion 440
distillation 751, 881–2
distraction 58, 185, 504, 512, 701, 703
distractors 58, 783
distress 496, 501
distributed cognition 252–3, 256, 705, 712, 1169
distributed teams 705, 707
distribution 233, 253, 347, 537, 626, 636
  of expertise 15, 93, 731, 1014
disturbances 1119, 1122–3
  positive 1125–6

diversity 3, 20, 211, 1046, 1121, 1125
  of expertise 914, 1122
doctors 159–62, 165–6, 409–11, 415, 418–20, 437, 457, 618–19, 621–4, 629–32, 1113
  expert 409–10, 418, 420, 1113
  newly qualified 420–1
  senior 418, 421
domain constraints 187, 1061, 1065, 1068
domain experience 778, 930, 1120
domain expertise 12, 264, 272, 1097, 1121
domain experts 272, 354, 541, 772, 1046
domain general models
  aviation 66–7
  games 59–61
  music 61–2
  science 63–4
  scope and organization 56–8
  sports 62–3
  surgery/medicine 64–5
domain knowledge 273, 316, 318, 533, 535, 623, 1068, 1070, 1072–3
domain practitioners 192, 317, 475, 482, 698, 1208
domain specificity 576, 940
domain-general cognitive ability factors 56, 58
domain-general measures 15, 60
domain-general models 7, 56–78, 109, 307, 443
domain-independent theory 965, 968
domain-relevance 15, 59, 72, 141, 1068, 1072
domain-relevant tasks 57, 61, 1058, 1061, 1068, 1072, 1075
domains 20–5, 95–8, 108–12, 263–5, 297, 312–16, 338–41, 753–9, 807–10, 928–32, 968–9, 1066–70, 1182–3
  cognitive 344, 1046
  complex 12, 19, 1066, 1071, 1194, 1199, 1201
  complex performance 1095, 1098, 1100–1102
  culturally valued 104, 120–1
  dynamic 344, 808
  foundational 21, 575
  healthcare 187, 444
  ill-structured 24, 952, 957–8, 962, 968–9
  knowledge-based 618, 622

mature 327–8
medical 271–3, 304, 619–20, 625, 934, 986
musical 140, 575, 586
open-ended 1071, 1075
railroad 694, 696–7
target 319, 324, 401
well-structured 959, 968
domain-specific computational models 344–8
downdrafts 888
downtime 783, 1097
DP, see deliberate practice
Dreyfus and Dreyfus model 313–14
Dreyfus-ian community of practice 8
drilling 774, 852–3, 857, 862–3, 889, 1089
drivers 207, 209, 364–5, 367–70, 512–13, 704–5, 707, 1183
human 341
inexperienced 513, 1071
stress 500, 503, 512–13
training 369, 513
drugs 629–30, 633, 1173
dual space framework 533–4
dual-process theory 621
dual-task performance 136, 493
duties 306, 322, 503, 852, 856, 865
job-relevant 13, 792
dyads 16, 175, 178–9, 184–5, 187
dynamic domains 344, 808
dynamic environments 598–603, 610, 691, 778, 1197, 1199
dynamic interactions 190, 243, 254
dynamic processes 491, 669, 852, 1003, 1013
dynamic situations 594, 642, 654, 1177, 1182, 1200
dynamics 16, 179–80, 193, 199, 206, 248

EBM, see evidence-based medicine
EBP, see evidence-based policy
ecological approach/perspective 173, 176, 178, 182–5, 1059
ecological constraints 176, 186
ecological interface design, see EID
ecological psychology 11, 178, 596
ecological rationality 153–5, 163, 166
ecological validity 12, 22, 773, 1086, 1164
ecology 174–80, 183–6, 1110
problem 11, 16, 178, 187

economics 38, 725, 1166
economy of effort 690, 796, 1091
education 20–1, 43, 50–1, 106–7, 210, 330, 721–3, 798–9, 801–2, 873, 939–40, 971–2, 1024–6
building organizational expertise for 914–20
classroom 723, 799
cyber 723–5
cybersecurity 722–3
devleoping expertise in dialogic teaching 910–14
expertise for the future 903–22
formal 425, 726, 766, 874, 905, 929
forms of teaching and learning 905–10
future 920–1
gifted 15, 103, 105–7, 111, 115, 118–19, 121
medical 133, 408, 418, 421, 620, 935
programs 106, 113, 275, 327, 368–9, 894
teachers 9, 24, 931–4
educational organizations 915, 917–18
educational psychology 911, 971, 1203
educational research 21, 265, 530, 906, 915
educational systems 904, 919–21
educators 115, 119, 904, 916–17, 919–20, 936
medical 935–6
working 908, 911
effective mentoring 1027, 1037, 1050–1
effectiveness 24–5, 318, 936, 1022–3, 1027, 1036–7, 1039, 1045–6, 1049–51, 1099–100
efficacy 90, 632, 774, 781, 783, 1099
efficiency 269, 271, 277–8, 475, 483–4, 486–7, 634, 667, 708, 857
cognitive 1059–60, 1065
increased 607, 1075
processing 499, 507
efficient de-confliction of road and rail traffic 359, 365–8
EHRs (electronic health records) 434, 630–4
EID (ecological interface design) 462–3, 465–6
elaboration cycle 202–3, 209
elaborations, verbal 381, 390–5, 399, 445
elaborative rehearsal 26, 1089, 1098, 1102
electricity 718, 830, 851
electrocardiograms 160, 502

electronic health records, *see* EHRs
electronics troubleshooting 10, 1209
elementary schools 905–6
elicitation 18, 204, 392–4, 396, 398, 400
elite athletes 508, 1067
elite coaches 1044–5
elite performers 506, 508, 604, 928
elite players 564–5, 569
embedded cognition 250
embodied cognition 173, 249–50, 1203
embodiment 85, 338, 421, 653
emergencies 20–3, 642–56, 668–9, 850–2, 856, 859–61, 1119
  descriptive approaches to assessment of expertise 643–6
  developing, maintaining and losing expertise 652–4
  experimental approaches to assessment of expertise 646–8
  future directions 654–5
  inferential approaches to assessment of expertise 648–52
emergency conditions 642, 646, 648–50, 654–5, 859
emergency events 850–1
emergency management 222–4, 230, 643, 648–9, 652–3, 655
emergency responders 643–4, 646–7, 649–50, 654–6
emergency response 23, 207, 642–56, 861, 932, 1119
  teams 861, 1119–20
emergent properties 175, 179, 183, 185, 187, 627
eminence 15, 104–6, 111–12, 114, 120–1, 536
emotion 154, 501, 797
emotion-regulation strategies 499, 508
empirical evidence 25, 49, 190, 197, 1163, 1171
empirical findings 151, 314, 491, 539
empirical literature 1086, 1088, 1098, 1139
empirical research 23, 26, 39–40, 119, 142, 1097–9
empirical support 20, 113, 135, 197, 1121
empirical validation 210–11, 979
empiricism, radical 178–9

employees 74, 76, 690, 692–3, 699, 704, 855–6, 1040
  railroad 690, 696
employers 370, 424, 865
enactive cognition 250–1
enactivism 251, 256
encoding 41, 47, 50, 173, 498, 505
endocrinology 619, 1169
energy 456, 709, 851
  consumption 708
enforcement, law, *see* law enforcement
engagement 296–7, 470, 474, 481–2, 484–5, 794, 798, 1043
  direct 470, 474, 476
engineers 473, 530–2, 534, 698, 704, 706, 838, 841
  locomotive 690–3, 695, 697–701, 703–6, 708–11
engines 645, 665, 752, 831
  failure 158, 644, 668
enjoyment 583–4, 793
enthusiasm 1172, 1178
enthusiasts 558–9, 561, 570
environment interactions 131, 142, 400, 402
environmental conditions 158, 163, 165, 646, 652, 655
environmental constraints 13, 17, 234, 245, 247, 257
environmental cues 299, 346
environmental factors 57, 111, 142, 195, 270, 441
environmental properties 163, 166
environmental stressors 492, 496, 504
environmental support 1073–4
environments 116, 153–7, 162–4, 167, 184–6, 243–7, 249–54, 256–7, 342–3, 646–8, 653–6, 801–2, 1091–2
  complex 1, 10, 208, 210, 626, 630
  decision-making 163, 209, 882
  dynamic 598, 610, 691, 778, 1197, 1199
  emergency 643, 649, 656
  hazardous 863–4, 866
  high hazard 853, 858–60, 862–6
  inner 179, 1203
  learning 957, 964–6, 968, 1046, 1048, 1203
  natural 370, 401, 624, 846
  operating 22, 642, 742, 758, 796

operational 22, 209–10, 654, 726, 799, 865
performance 494, 499, 596
real-world 296, 648, 740, 746, 967
simulated 293, 305, 339, 342, 931, 936
simulated task 18, 292, 297, 300, 302, 304–5, 307
social 154, 411, 418, 1195, 1201
stressful 497, 511
time-constrained 612, 650
unpredictable 801, 1195
EPAM 38, 45, 47
EPAM-IV 47–8
ergonomics 18, 357, 373, 747
error
human 490, 855, 860, 1114
signals 180–1
error detection 626–8
and correction 278, 627
and recovery 625–8
error identification 356–7, 362, 373–4
error management 627–8
error monitoring 625, 635
error recovery 625–6, 628
error traps 691, 703
error-free trials 1089–90
errors 277–8, 368–9, 419–20, 624–9, 633, 635, 697, 700, 707–8, 835, 1174, 1179, 1181–2
generated 625, 628, 635
measurement 57, 72
medication 434, 1009
potential 368–9, 628
escalations 728, 865
esotericity 13, 85, 87, 92, 95, 98
ETCS (European Train Control System) 709–11
ethics 76, 769–71, 904
ethnographers 416, 1169
ethnographic interviews 874, 879, 891
ethnographies 416, 418, 879
immersive 878–9
Europe 97, 690, 692–3, 708, 710, 1173–4
European Train Control System, *see* ETCS
evaluation 356, 358, 632, 634, 654, 741, 755–7, 942, 1003, 1048–9
collective 25, 1003
comparative 346, 656

criteria 483, 1165
performance 577, 580, 983, 988
event timelines 396–7
everyday life 56, 175, 392
evidence 58–64, 72–3, 77, 113–15, 205–6, 510–11, 622–5, 732, 749–52, 908–12, 1067–9, 1142–9, 1172–4
anecdotal 158–9
confirmatory 630, 634
empirical 25, 49, 190, 197, 1163, 1171
experimental 158–9
limited 14, 511, 1115
meta-analytic 1022, 1027, 1037
evidence-based medicine (EBM) 1145–6, 1173
evidence-based policy (EBP) 1146
evidence-based practices 420, 1133, 1159, 1172–3, 1181
community 1172–5
executive coaching 811, 1024
executive control 498–9, 514
executive functions 42, 58, 66
executives 472, 809, 811–12, 814, 1024
existential threats 795, 804
expectancies 198–9, 202, 448, 475, 606, 1180
experience 22–5, 197–200, 229–32, 255–6, 314–16, 457–60, 478–82, 674–82, 732–5, 854–5, 929–33, 951–8, 1062–5
base 156, 315, 443
battlefield 799
domain 778, 930, 1120
and expertise 193, 195, 674, 679, 863
first-hand 111, 693, 1002, 1178
flight 66, 663, 668, 672, 674, 677–8
job 15, 67–71, 73, 235, 728
and knowledge 12–13, 193, 198, 200, 316, 318, 953, 1204–5
levels 433, 444, 478, 875, 1065
lived 94, 97, 254, 1001
massed 930, 932
task-related 12, 333, 647, 1058, 1074
weapons-handling 769, 773
work 274, 535, 891, 979
experience-based judgments 535, 984
experience-centered approaches 179, 183, 185–7, 535
experiential knowledge 90, 320, 347, 409, 778
experiential learning 754, 850, 863, 930–1

experimental conditions 628, 963, 1086–7
experimental control 293, 643, 646–8, 655–6, 769, 774
experimental designs 531, 533, 648–9, 1087, 1094, 1162, 1165
  cross-sectional 649, 654
experimental manipulations 393, 397, 504, 597–8, 727, 1087–8
experimental psychology 39, 115, 191, 298, 1160, 1167
experimental tasks 36, 51, 380, 388, 770
experimentalism 401, 1209
experimentation 12, 303, 380, 382–4, 388, 397–8
  controlled 381, 968
experimenters 377, 382–3, 385, 389, 392, 394–5, 397
experiments 35, 37, 39–42, 44, 382, 441, 444, 533–4, 734, 843–4
  controlled 191, 381, 531, 963, 968, 1162
ExPerT 301, 942–3
expert analysts 741–2, 744, 748–51, 753, 756, 758, 979, 984–7
expert athletes 63, 298, 598, 611, 932, 937
expert behavior 18, 354–5, 357–9, 361, 363, 367–9, 373–4, 466, 741, 748–50
expert capabilities 656, 792, 1166, 1168, 1177
expert cognition 193, 378, 447, 809
expert decision making 48, 200, 346, 440–2, 775, 778, 880, 882
expert doctors 409–10, 418, 420, 1113
expert feedback 783, 939, 1048–9
expert forecasters 873, 887, 891–2, 1176
expert groups 61, 66, 561, 564, 730, 778–9, 1064
expert intelligence analysts 743, 978, 984, 986, 996
expert judgments 421, 425, 995, 1144, 1147, 1159, 1161, 1163–4
expert knowledge 349, 386, 392, 889, 1095, 1102, 1148
expert learning 6, 156, 921, 1022
expert levels 576, 579, 800, 802, 940, 953
expert mentors 245, 1022–3, 1041, 1051
expert performance 10–11, 18–21, 113–14, 128–9, 131–4, 137–42, 297, 299–300, 302, 348–9, 575–6, 648–50, 942

expert performers 16, 18, 291–2, 295, 297, 299–301, 304, 307, 347–8
expert pilots 66, 306, 342, 420, 677, 1071
expert problem solving 37, 137–8, 264, 1060
expert reasoning 394, 454, 621, 984, 1165, 1205
expert skills 24, 454, 551, 711, 928, 940
expert status 9, 89, 927, 1064, 1119
expert strategies 459, 463, 465, 560–1, 705, 708
expert surgeons 65, 511–12, 934, 938
expert systems 4, 8, 11, 49, 622, 1176
expert teams 23, 831–2, 1001, 1003, 1005, 1011
expert thinking 10, 389, 394, 398
expert users 335, 368, 904
expertise acceleration 24, 485, 955, 965
expertise-based mitigation 1066, 1072, 1076
expertise-based training 301, 941
expertise-deniers 1157–8, 1160, 1167, 1175, 1177, 1180–4
explicit knowledge 89, 156, 165, 510, 625, 781
exploration 643–4, 1044, 1046, 1194, 1197, 1204
  long-duration missions 832, 836, 845
expressions, facial 501, 773–4, 1076
extended mind 251, 256
external agents 227, 960, 1045
external constraints 179, 195
external interventions 220, 226–8, 237, 239
external resources 1028, 1170
external stressors 491, 496–7, 500, 503, 512
external validity 397, 880
extraversion 989
eye movements 40, 177, 244, 338, 597–8, 611

FA (fractional anisotropy) 132–5
facial expressions 501, 773–4, 1076
facilitation 993, 1024
facilitators 436, 476, 821
failure 441, 504–7, 512, 664–5, 834, 987, 1009, 1117–18, 1120, 1183–4
  communications 364, 862
  engine 158, 644, 668
fallibility of experts 1148, 1164
false positives 728, 995
families 108, 112, 118, 414, 485, 694

family resemblances 957, 964
fast-and-frugal heuristics 16, 149–67, 250
　account of expertise 157–66
　conceptual framework 150–7
　program 16, 149–50, 155–8, 165, 167
fast-and-frugal trees 160–1, 163, 165–6
fatigue 512, 698–9, 701, 704, 707, 858, 862
　managing 699, 704
　mental 506, 704
feasibility 273, 756, 844, 933–4
feedback 163–4, 420, 480–1, 540, 653, 679, 816, 863–4, 886, 993–4, 1008–9, 1040, 1048–50
　cognitive 164, 276, 278
　constructive 1032, 1035, 1040
　corrective 938, 982, 1047
　direct 486, 1022
　expert 783, 939, 1048–9
　flexibility-focused 1115, 1199
　informative 935, 1182
　negative 499, 504
　objective 858, 928, 930
　performance 318, 326, 982, 1014
fidelity 335, 643, 646–7, 723, 863, 967
　physical 936
　psychological 863–5, 936
field observations 228, 393, 694, 878, 1208
fighter pilots 340, 396, 676, 681
filming 298, 303, 595–6, 599–600, 605, 773
firearms 765, 776, 781, 784
　training 772, 782
firefighters 149, 156, 224, 228, 647, 692
firefighting 5, 21–2, 204, 642–56, 804
firewalls 720, 729
first-person perspective 429, 433, 440, 446, 595, 698
fixations 177, 507, 601, 776, 782, 1113
fixedness, functional 264, 1113
flashing lights 358, 367–9
flexecution 156, 194, 208–9, 211–12, 955, 1197–9
　model of replanning 10, 16, 196, 207–11
　process 208–9, 1199
flexibility, adaptive 952, 955, 957
flexibility-focused feedback 1115, 1199
flexible decision-making workspace 1076, 1205

flight controllers 830–5, 837, 839, 841, 843–5
flight decks 226, 234–5, 662, 667, 669, 673
flight directors 831, 833, 836, 838–9, 841
flight experience 66, 663, 668, 672, 674, 677–8
flight hours 66, 306, 663, 675, 677–8, 1065
flight performance 672–3, 675
flight simulator performance 673, 1065, 1071
flight simulators 66, 293, 295–6, 342
flowcharts 208, 359
fluency 93, 1166
fluid intelligence 58–60, 62, 73, 569
fluid reasoning 60, 62
fluid thinking 552–3
flying 159, 664–6, 670–1, 677, 679, 1071
focus groups 414, 417, 421, 423, 430, 693–5
foraging 342, 996
force production 960–1
force-on-force training 784
forecaster learning 874, 879, 883
forecasters 23, 873–81, 884–7, 889, 892–3, 1043–4
　expert 873, 887, 891–2, 1176
　private sector 875, 882, 887
　senior 880, 889, 893, 1043
forecasting 872–3, 875–8, 883, 886–7, 1160, 1166
　expertise 873, 878, 883, 892
　process 884, 886
　skills 881, 991, 1163
forecasts 872–8, 884, 886, 891–3, 1175–6, 1182
forensics, digital 722, 724
formal education 425, 726, 766, 874, 905, 929
formal instruction 577, 579
formal schools 325–6
formal training 875, 1043, 1092
formally acquired knowledge 625
foundries 817–20, 823
fractional anisotropy, *see* FA
frame model of sensemaking 196, 198, 200–204, 210–11
framing 89, 174–5, 183, 741, 748, 1132–49
frontier thinking 1166, 1168, 1178
frontline workers 26, 691, 1112–14, 1123, 1125, 1143
　resilience 1112–17

frontoparietal control systems 136
frontoparietal regions 136-8
frontroom controllers 833, 835
fuel 665, 672, 692, 851, 1004
   consumption 675, 697
functional complexities 12, 1115, 1199
functional connectivity 15, 57, 128, 140-1
functional constraints 177, 183, 185
functional fixedness 264, 1113
functional MRI 136
functional neural correlates of
      expertise 135-40
functional properties 138, 140, 182-3
functional relationships 178, 295-6, 985
functional reorganization 48, 138
functional structures 178, 220, 230, 235
functionalism 173, 175, 178, 183, 1202
functions 138-9, 175-8, 180-2, 191, 193-4,
   223-6, 601-2, 609-10, 678-9, 1022-5,
   1027-8, 1050, 1061-2
   analysis 356-7, 362, 373-4
   cognitive 2, 202, 255, 475, 1062, 1065
   developmental 25, 1023
   executive 42, 58, 66
   key 1024-5
   loop 179-80
   macrocognitive 191, 193-5, 471, 1210
   mental 929, 944
   mentoring 1022, 1026-7, 1037,
      1039-41, 1050
   performance-resource 494-5
   power 1093, 1102
   sensorimotor 1062
futures methods 1136, 1204
fuzziness 267, 280, 393, 675

games 60-1, 291, 293, 300, 302, 551-5, 570,
   981, 1066-7, 1175-6
   classical expertise approach 41-2
   domain general models 59-61
   scores 605-6, 611
gaze 159, 598, 601
   control 502, 782
gender 94, 1033
generalists 411, 743, 979-80, 1141-2
generalizability 643, 880, 1097, 1144
generalizations 10, 181, 316, 391-2, 396, 499

generals 796, 798-9
generic models 142, 484, 487
genetic predisposition 585-6
genetics 21, 108, 110, 277, 575, 585-7
germ theory of disease 88-9
Germany 94, 794, 1158
Gestalt School 387
gestures, physical 384, 705
gifted children 104, 106-7, 120
gifted education 15, 103, 111, 115,
   118-19, 121
   history 105-7
gifted students 103, 105-7, 109, 112-13,
   118-21
giftedness 15, 103-21, 574, 585, 1203, 1205
   alternative ways of conceptualizing 107-17
   conceptualizion 107, 117, 120
   definition of gifted and talented 103-5
   expert performance perspective 113-14
   history of gifted education 105-7
   multiple intelligences model 114-15
   psychometric views 108-10
   talent development models 108,
      110-13, 121
   tripartite model 117-19
   WICS 115-16
goal conflicts 208, 212, 445, 1198
goal management 1044-5
goal state 276, 533, 538
goal-relevant constraints 1059, 1202, 1206
goals 42-3, 207-9, 212, 312-15, 356-62,
   390-5, 429-32, 443-6, 456-9, 1062-6,
   1117-19, 1135-6, 1198-200
   common 412, 456
   explicit 184, 580, 910, 927
   organizational 224, 1043, 1118
   project 443, 472, 477
golf 507-9, 1067
government 2, 26, 107, 121, 370, 721-2
   agencies 1184, 1204
   challenges in 1133-7
   departments 1133, 1141
   expertise and evidence in 1139-47
   framing and translating expertise
      for 1132-49
   primer on 1133
graduate school 106, 536, 939

graduate students 162, 380, 530, 539–42, 577, 772
graduate training 536, 539, 543–4
graduates 274, 387, 540, 733, 1141
grandmasters 37, 301
gray matter 129–32, 134
  anatomy 129, 131, 134
  density 129–30
  regions 130–1
  volume 130, 132
ground support 843–4
grounded theory 414–15, 436
group interviews 323, 417, 886
group structure 220, 233, 237–8
groups 41–2, 64–5, 73, 92–4, 112–13, 130–4, 277–8, 557–8, 731, 826, 907–9, 1064, 1123–5
  ability 69, 576
  advocacy 370, 1147
  control 60, 130, 133–4, 561–2, 782, 908–9, 1012, 1049–50
  expert 61, 66, 561, 564, 730, 778–9, 1064
  focus 414, 417, 421, 423, 430, 693–5
  social 15, 86–8, 92–3, 410, 416, 1139
  stakeholder 324, 411
growth 134, 262, 303, 1033, 1041, 1058
  personal 1022, 1024
guards 710, 937, 961
guidance 17–19, 355, 357, 851–2, 915–16, 1024–5, 1035, 1099–1100, 1204, 1208–10
  on collecting verbal reports of thinking 393–9
guided discovery learning 276–8
guided team reflection 1002–3, 1006–8

habits 6, 257, 416, 505, 679
hackers 719–20
handbooks 190, 193, 1194–6, 1200, 1202–3, 1210
handgun shooting 507–8, 779
handguns 776
handovers 835
hands-on job performance (HOJP) 69–71
hazard avoidance 666, 674
hazard perception 496, 512, 514
hazardous environments 863–4, 866

hazards 512, 701, 706
  potential 222, 706
HDPI (Heart Disease Predictive Instrument) 160–1
health 2, 19, 456, 465, 1063, 1069–70
health information technology, *see* HIT
healthcare 412, 419, 447, 452, 1023, 1026
  diagnostic reasoning 409, 618–36
  domains 187, 444
  environment 624, 631
  professions 473, 1022, 1026, 1051, 1174
  providers 437, 474
  teams 416–17
Heart Disease Predictive Instrument, *see* HDPI
heart failure 160, 456–7, 961
heart rate variability (HRV) 502, 507
heat maps 812, 814
helicopters 437–9, 484, 677
heritability 57, 113
heterogeneity 2, 134, 149
heuristic searches 533, 619
heuristics 16, 149, 151–5, 157–67, 180–1, 183, 453–4, 462, 746, 753, 880–1, 1163–8, 1184–5
  building 462–3
  fast-and-frugal, *see* fast-and-frugal heuristics
  judgment 167, 1166
  positive 1166, 1184
  take-the-best 151, 153, 162–4, 250
  tallying 163–4, 167
heuristics-and-biases 150–1, 157, 1159, 1163, 1166–8, 1184–5
  community 1163–9
  program 150–2, 155–6, 167
hierarchical organizations 224, 799
hierarchical structures 430, 914–15
hierarchical task analysis, *see* HTA
hierarchies 247, 356, 359, 455, 462, 577–8, 666, 668
high ability students 104, 107–8, 112, 117
high hazard environments 853, 858–60, 862–6
high hazard industries 690, 859, 863
high IQ 104–5, 107–8, 111, 118, 120; *see also* giftedness

high reliability organizations, *see* HRO
high school students 723–4
high schools 62–3, 917
higher-level cognitive processes 609, 618
high-level knowledge theories 46–7
historians 273–4, 469, 471
historical review 18, 393
HIT (health information technology) 619, 630–1, 634–6
HOJP, *see* hands-on job performance
holistic approaches 11, 173–87, 317, 612, 929, 1043
hormonal indices 502
hospital administrators 474, 631
hospitals 48, 413, 415, 625, 1009, 1112
HPA (hypothalamo-pituitary-adrenocortical) 497, 502
HRO (high reliability organizations) 229, 859, 1117–20
HRV (heart rate variability) 502, 507
HTA (hierarchical task analysis) 18, 196, 354–74, 392, 429, 451
　and expertise 357
　outputs 18, 356, 362, 373
　plan types 361
　practical guidance 357–63
　rail level crossings 355, 358–9, 361–72
　strengths and weaknesses 372–3
human cognition 17, 173, 338, 631, 1202, 1208
human control 712, 1183
human error 490, 855, 860, 1114
human expertise 16–19, 220–1, 339, 342, 759, 1176, 1178, 1181
human performance 191–2, 297, 338, 340, 630, 648
human resources 121, 300, 535–6
human tutors 795, 1046–8
human-derived knowledge 469, 477
hypothesis formation 390, 533
hypothesis generation 444, 623, 749, 757
hypothesis space 533–4
hypothetico-deductive process 623–4

ICUs, *see* intensive care units
ideal research design 1085–7, 1089, 1095
identification 202, 204–5, 317–18, 320, 535, 538, 650, 1168
　purposes 1, 35, 56, 85, 103, 128
identity, professional 875, 1025
idiosyncrasies 230, 255, 436, 650
IFL (Institute for Learning) 911–12, 917, 919
ignorance 1147–8
illnesses 88, 221, 409, 1065, 1076, 1205
ill-structured domains 24, 952, 957–8, 962, 968–9
ill-structuredness 263, 796, 798, 956, 958, 964
illusions 174–6, 1147
　cognitive 152
　visual 36, 1164
imagery 140, 509, 512
images 248, 417, 471, 719, 724, 966
imagination
　active 118
　disciplined 471–2
imitation game 15, 85, 93–4, 96–7
immersing 328, 979
immersion
　long-term 878–9
　sustained 92–3
immersive ethnographies 878–9
impairments 493, 497–500, 507
implicit knowledge 156, 625
implicit learning 510, 940–1
improvement 75, 321, 323, 913, 915, 918, 921, 1049, 1051
　continuous 371, 918–21
improvisations 136–7, 140, 223–4, 229–30, 579, 1059
incentives 769, 772
incident command 223–4, 859–60, 862
incident-based interviews 19, 430, 446
incident-based methods 5, 19, 196, 255, 273, 429–48
　analyzing data 435–9
　future 446–7
　products 439–46
　types 430–5
incidents 224, 230, 391, 396, 429–35, 437, 439–40, 443–8, 477–8, 775, 782, 851, 862
　critical 204, 439, 775, 874, 933, 1208
　high-profile 766, 861
　real-world 434, 448

inclusion 89, 294, 296, 324, 1033, 1039
  criteria 136, 348, 564
inconsistencies 134, 203, 205, 208–9, 212, 478
independence 914, 1028, 1033, 1046,
    1050, 1147
  musical 577–8
in-depth interviews 443, 747
in-depth understanding 373, 411, 1125
indeterminacy 410–11, 880
indeterminateness 962, 1204
indicators 264, 273, 320, 328–30, 529, 532
  attitude 675
  of expertise 273, 279, 537, 704, 1050
  performance 18, 312, 316–17, 319–20, 324–30
individual differences, community of
    practice 6–7
individuals 17–18, 262–4, 268–70, 272–3,
    279–80, 355–6, 543–4, 691–4, 795,
    800–801, 977–9, 989–91, 1010–12
  and teams 194, 542, 824, 858, 863–4, 866
industrial accidents 1120, 1125
industries 469, 473, 810, 851, 858, 863
  high hazard 690, 859, 863
  nuclear 852, 859
  railroad 690, 694, 711
inexperience 432–3, 490, 513, 775, 782, 1071
inferences 153, 198, 203–6, 644–6, 1070,
    1075, 1174, 1179
  correct 151–2
  logical 253, 530
inferior frontal cortex 130, 133, 139
informal mentoring 994, 1043
informal strategies 704, 707–8
informally acquired knowledge 625
information
  contextual 605–10
  evaluation 1003–4
  exchange 134, 856
  overload 632, 746, 886
  postural 603, 606–7, 609
  requirements (IRs) 236, 695, 742
  retrieval 632, 757, 1094
  structural 598, 600, 605
information technology 166, 719, 721–2, 725,
    808, 815
  health, *see* HIT

informative feedback 935, 1182
informed consent 423, 445
infovores 992
infrared beams 596–7
infrastructure 222, 371–2, 721, 725, 1134
in-game chess positions 292, 298
initiation 606, 874, 906
innate ability 113, 932
innate talent 36, 574
inner environments 179, 1203
inner loops 179–81
inner observation 379–80, 382, 393, 430
inner perception 379–81
innovations 271, 275, 278, 629, 635,
    911, 917
  technological 756, 759, 879
input 48, 51, 178, 192–3, 262, 356–7
  sensory 461, 854, 1072
in-service teachers 933–4
instability 220, 222, 229–31, 233, 1197
institutional structures 86, 904, 921
instruction 318, 377–8, 388–9, 391–2, 394–5,
    507, 585–6, 913–16, 918–19, 959–60,
    964–5
  dialogic 914, 916–17, 920, 1049
  direct 906, 954, 963, 1047
  formal 577, 579
  reasoning-focused 915, 919
instructional design 275, 935, 941, 944
instructional designers 928, 931
instructional methods 275–8, 280, 919, 942
instructional practices 544, 914–15, 918, 920
instructors 316, 318–21, 324–7, 330, 765, 771,
    783, 928
  marine 318–20
instrumental musicians 133, 135
instruments 131, 159–60, 274, 575, 577–8,
    583, 674–5, 811–12
intact knowledge structures 962, 1204
integrated models 210–11, 233, 387, 507,
    1197, 1199
integration 610, 612, 644, 649, 713, 749–50,
    1002, 1013–14
  motor 129, 133–4, 140
integrative analyses 437–9, 447–8
intellectual abilities 104, 108–9, 113

intelligenc, multiple intelligences 108, 115, 117
intelligence 6, 67-8, 105-6, 109, 115-17, 141,
    722, 756-7, 913, 981, 987, 995
  analysis 21-2, 25, 740-59, 977-85, 989-90,
    993-4, 1140, 1162, 1164, 1166
    cognition and expert-level
      proficiency 977-96
    future directions 994-6
    individual differences and styles 988-93
    performance measurement and
      evaluation 981-2
    psychology of 741, 993, 1164, 1169
    reasoning 984-8
    studies of expertise 742-59
    training challenges 993-4
    who are the experts 978-81
  analysts 742-3, 746-7, 751-2, 754, 978-9,
    981, 988-9, 991, 1162, 1182
    expert 743, 978, 984, 986, 996
    professional 993, 1164
  community 25, 742, 980, 994
  crystallized 59, 73, 109-10, 569
  fluid 58-60, 62, 73, 569
  professionals 979, 987, 996
intelligence agencies 977, 988, 994
intelligences, analysis, overview 741-2
intelligent technology 1177, 1181
intelligent tutoring systems 1046-8, 1051
intensive care units 48, 439, 624, 628
intentions 173, 179-81, 184, 364, 369, 646,
    707-8, 1166
interactional expertise 8, 15, 92-4, 96-7,
    878, 930
interactions 72-3, 142, 244, 247-50, 252-3,
    355, 363-4, 601-3, 631-2, 877, 916,
    1121-2
  between long-term memory and working
    memory 137-8
  collegial 915-16
  developmental 1024, 1042, 1050
  direct 92, 179
  dynamic 190, 243, 254
  environment 131, 142, 400, 402
  physical 17, 691, 721
  social 418, 460, 874, 878, 905, 915
interconnectedness 808, 986, 1042, 1124
interdependence 368, 832, 1040, 1199

interdependencies 832, 844, 892
interdisciplinarity 15, 21, 85, 96, 531, 543-4
interfaces 235-6, 238, 277-8, 341, 354, 356-7,
    631-3, 729
  user 442, 1183
interference 394, 398, 1090, 1093, 1096, 1102
intermediate levels 313, 453, 654, 1064
intermix training 276, 278
internal consistency 165, 206, 274
internal constraints 174, 177, 179, 182
internal logic 176, 756
internal models 176, 749, 758
internal representations 17, 181, 206, 249,
    256-7
internal validity 394, 397, 774, 1092-3
internalization 273, 384, 651-2
International Space Station, see ISS
Internet 411, 720, 1147, 1157, 1163
internships 724, 931, 942
interpersonal relationships 506, 990, 1025,
    1044, 1181
interpersonal skills 802, 990
interrelatedness 211, 1050
interrelationships 68, 744, 1112, 1122, 1146
interruptions 389, 465, 856, 1006
interventions 161, 364, 397, 503, 505-6,
    509-10, 514-15, 627, 858-9, 906-8
  effective 2, 1210
  external 220, 226-8, 237, 239
  motivational 1203
  policy 26, 1135, 1143
  for stress 509, 515
interviewees 255, 430-4, 437, 443, 445-6, 448
interviewers 255-6, 323, 416, 430-4,
    445-6, 884
interviewing 323, 418, 469, 474-5, 693, 858
  cognitive 475, 478, 482-3, 644-5, 887
interviews 255-6, 323-5, 414, 416-17, 430-5,
    443-8, 453, 483-6, 693-4, 775, 874,
    879-80, 1043-5
  career 889, 1201
  cued-retrospective 19, 434-5, 447
  ethnographic 874, 879, 891
  group 323, 417, 886
  incident-based 19, 430, 446
  in-depth 443, 747
  phenomenological 254-7

qualitative 444, 878
retrospective 430, 443, 445, 774
simulation 19, 433-5, 445-7
unstructured 430, 469
intrinsic motivation 112, 978, 1012
introspection 3, 5, 36-7, 254-5, 377-402, 430
history 379-87
methodology 19, 378
Selz to current methods of thinking
aloud 387-90
training 396, 400
types of verbal reports of thinking 390-3
Wundtian 380-1
Würzburgers 381-5, 387-90, 393
introspective methods 18, 378-9, 384, 390, 392-3, 399-402
introspective-type methods 18, 393, 397, 400, 1037, 1046
introspectors 381-3, 385, 388, 395-6
intrusion detection
cyber situational awareness 727-31
teamwork 731-2
intuition 155-7, 199, 386-7, 767, 792, 799, 984
intuitive expertise 808, 827, 1149, 1182
intuitive judgments 796, 802, 1139, 1168, 1172
intuitive-based scientists 885, 891
invariants 50, 176-7, 181-2, 1136, 1202
investigators 370, 429, 437, 440-1, 443, 445
investment 330, 344, 423, 793-4, 1149, 1158
decisions 163-4, 1170
IQ 59, 61, 63, 109, 111, 120
scores 105, 109, 111
tests 110, 113, 117, 120
irregularity 962, 1204
IRs (information requirements) 236, 695, 742
isolation 192, 211, 534, 582, 603, 648
ISS (International Space Station) 832-4, 836-7, 841-3, 845-6

job aids 1101-2
job experience 15, 67-71, 73, 235, 728
job performance 14-15, 57, 59, 67-8, 70-3, 77, 317, 940, 1037, 1039-40
domain general models 67-72
joint cognitive systems 11, 219

journeymen 887, 891, 978, 1059, 1064, 1158, 1163, 1201
judgment biases 1163-4
judgment heuristics 167, 1166
judgments 149, 152, 423-4, 603-4, 676-9, 892-3, 986, 988, 994, 1143-6, 1159-62, 1166-7, 1181-2
anticipatory 603-4, 609
clinical 1161, 1174-5
inaccurate 987, 1168
intuitive 796, 802, 1139, 1168, 1172
pilot 668, 670, 676-8
professional 1159, 1173-4
juggling 134, 208
training 131, 134
junior doctors 419-21
junior journeymen 978, 994, 1163
just-in-time 275, 1101-2
mobilization training 1101-2

key performance areas 316-20, 324-6, 328-30
knowledge
acquisition 271, 318-19, 878, 965, 1065, 1076
application 951, 953, 958, 962
associated 12, 317, 1086
audit 19, 432-3, 441, 444-5, 883
bases 418, 533, 619-20, 701-2, 834, 1059
benefits of 1065, 1075
capture
future 485-7
illustrations 472-4
praxis 474-83
as professional practice 471-2
professionalism 483-5
scope and uses 470-1
capturers 469-72, 476, 478-82, 484-7
professional 469, 471, 478, 484
clinical 620, 625
community of practice 7-8
creation 1003, 1170
declarative 271, 570, 929-30, 1098, 1102
domain 273, 316, 318, 533, 535, 623, 1068, 1070, 1072-3
elicitation 20, 320, 397, 469, 475, 693
methods/tools 5, 378, 395, 474
and experience 12-13, 193, 198, 200, 316, 318, 953, 1204-5

knowledge (cont.)
   experiential 90, 320, 347, 409, 778
   expert 349, 386, 392, 889, 1095,
       1102, 1148
   and expertise 85, 551, 832, 839, 1171, 1175
   explicit 89, 156, 165, 510, 625, 781
   holders 469–70, 475–82, 486–7
   human-derived 469, 477
   implicit 156, 625
   limited 16, 753, 991
   management 721, 757, 1101–2
   models 452–4, 477, 891, 894
   nature of 98, 274, 413
   organizational 473, 1142
   over-reliance on 1061, 1075
   prior 178, 399, 605, 609, 611, 669,
       1061, 1075
   procedural 58, 348, 504, 815, 1098
   professional 469–72, 478, 483–5, 1027
   representation 334–6, 346, 952–3, 962,
       1196, 1204
   route 700–702
   scientific 78, 86–8, 418, 538–9, 878, 1169
   shared 835, 840, 985
   shields 961–2, 987, 1137, 1200
   and skills 263, 273–4, 619, 621, 861–3,
       1069, 1085–6, 1091–2, 1101, 1205
   specialized 619, 837, 914, 978, 1118
   storage 1200
   stored 177–8, 183, 860
   strategic 570, 937
   structures 41, 46, 48, 272, 624–5, 952–3,
       960, 962
   tacit, *see* tacit knowledge
   technical 89, 419, 756
   and understanding 8, 594, 610, 612,
       861–2, 1045
   word 68, 553, 1070
knowledge-based behavior 457, 460–1, 752
knowledge-based decisions 454, 463
knowledge-based domains 618, 622
knowledge-based mitigation 1075
knowledge-based refreshers 1099–100
knowledge-based strategies 19, 622, 1075

labels 111, 113, 119, 1096
laboratory tasks 67, 295–6, 303, 650

language 48, 93, 339, 484, 551–2, 555, 692,
   695, 809, 979–80
   expertise 570
   native 85, 118
   programming 333, 339
laparoscopic procedures 293, 305, 511, 935
laparoscopic simulators 65, 305, 935
laparoscopic surgery 305, 510, 512
latencies 39, 41, 298, 742
latent curve analysis 1087, 1091
lateral/creative thinking 552
law enforcement 20, 22, 390, 765–84, 932, 938
   agencies 768–72
   challenges to conducting research 768–74
   conceptualizations of expertise 766–8
   expertise 22, 765, 767, 778
   future directions 783–4
   illustrative examples of research 774–7
   officers/personnel 294–5, 772, 1168
   task-specific expertise 777–83
laypeople 16, 149, 161–2, 164–5
LDSE (long-duration space exploration) 23,
   832, 836–8, 845–6
leaders 796, 859, 861, 914, 917, 919, 921, 1010
   organizational 631, 919
leadership 837, 839–40, 853, 856, 858, 861,
   1008, 1010
   charismatic 1010
   shared 845
learned skills 263, 1098
learners 275–9, 317, 420, 440, 484–5, 652–3,
   943–4, 958, 960–1, 963–4, 966–70,
   1024, 1041–3
learning 270–1, 318–19, 343–4, 873–8, 903–6,
   909–11, 913–22, 939–41, 953–9,
   966–8, 1063–5, 1091, 1203–4
   CFT 963, 965–6, 968, 970
   classroom 920, 1051
   curves 336, 343, 713, 815
   dialogic 906, 908, 910
   discovery, *see* discovery learning
   at the edge 1021–51
   environments 957, 964–6, 968, 1046,
       1048, 1203
   experiential 754, 850, 863, 930–1
   expert 6, 156, 921, 1022
   forecaster 874, 879, 883

guided discovery 276-8
implicit 510, 940-1
and instruction 951, 953, 959, 964-5
lifelong 262, 278, 319, 827
online 93, 504
outcomes 318, 941, 1047
perceptual 177, 181, 183, 186, 252, 345
phylogenetic 158
social 158, 918, 1203
learning organizations 917-18, 996, 1010
Learning Research and Development Centre (LRDC) 7-8
lectures 278, 318, 446, 993
legal liability 768-9
legitimacy 87, 1160, 1170
legitimate peripheral participation 1042-3, 1047
less-experienced instructors 319
less-experienced officers 395, 766, 776-7, 779, 782
less-skilled players 299, 303, 599-602, 605, 609
lethal threats 793-4, 796, 804
lethality 22, 803-4
level crossings, rail 355, 358-9, 361-4, 366, 368-9, 372
levels of expertise 9-10, 59, 231-2, 234, 443-4, 649-50, 794-5, 808-9, 885, 1070-1
leverage 13, 402, 636, 1001, 1005, 1009
lexical decision task 554, 561, 563
liability 1179, 1184
legal 768-9
lifelong learning 262, 278, 319, 827
lifespan 104, 1058, 1062, 1064-5, 1074, 1076
theory 1061, 1067, 1075
lights, flashing 358, 367-9
likelihood ratings 777
limited knowledge 16, 753, 991
linear models 1161-2
linguistic environments 342-3
linguistic socialization 93, 95
literacy 921, 1063
literary interpretation 965, 968
lived experience 94, 97, 254, 1001
local adaptations 1123-4
local contingencies 226, 228

locations 221-2, 247-8, 302-3, 697, 700, 702, 707-8, 776, 782
locomotion 176-7, 182
locomotive engineers 690-3, 695, 697-701, 703-6, 708-11
logic 153, 156, 227, 379, 552
internal 176, 756
logical norms 152, 174
logical reasoning 673, 1177
logistic regression 160-1, 163
logistics 318, 473, 771, 794-5, 798-9, 942
lone wolves 472, 815
long-duration space exploration, see LDSE
long-term memory 40-1, 43, 45-8, 131, 135, 958, 1202, 1204
and working memory 137-8
long-term practice 141, 566, 583
long-term retention 1090, 1092, 1094, 1097
long-term working memory 5, 47-8, 66, 138, 675-6, 746
loop functions 179-80
loops 180-1, 211
closed 211
inner 179-81
voice 835-6, 839
loss-exchange ratio 793-4
lotteries 150, 995
LRDC (Learning Research and Development Centre) 7-8
LTM, see long-term memory
LTWM, see long-term working memory

machine interfaces 469, 634
machines 719, 721, 1118, 1175, 1177-8, 1183-4
macroanatomical properties 15, 128, 141
macrocognition 10, 190-4, 196, 207, 210-11, 213, 484, 487
community 10, 1209
community of practice 10
definition 191-6
methodological foundations 195-6
macrocognitive functions 191, 193-5, 471, 1210
macrocognitive models 5, 16, 156, 190-213, 487, 752
macrocognitive phenomena 191, 195, 200

macrocognitive wheel 194
magnetic resonance imaging, *see* MRI
malfunctions 833, 856
malware 720
management
   emergency 222–4, 230, 643, 648–9, 652–3, 655
   error 627–8
   goal 1044–5
   knowledge 721, 757, 1101–2
   risk 356, 371–2, 514, 812
   systems 665–6, 668
   uncertainty 191, 873, 881, 1066
managers 815, 818, 823, 1027, 1034–5, 1112, 1118
manipulations 384, 500, 605, 1086, 1089, 1096
   experimental 393, 397, 504, 597–8, 727, 1087–8
manual dexterity 65, 498
manual morphometry 129–30, 132–3
maps 203, 206, 296, 298, 476, 482, 552–3, 819–20, 978–9
   concept 473, 475–7, 482–3, 485, 992
   heat 812, 814
   process 820–1
Marine Corps Training and Education Command 325–6
marine instructors 318–20
maritime surveillance aircraft 234–5
markets 808, 810, 815, 822, 1170
massed experience 930, 932
mastery 323, 326–7, 330, 1086, 1089–91, 1096–7, 1102
mastery models 18, 312–30
   application 325–7
   enhancing development and application 327–30
   foundations of approach 313–16
   potential improvements 329–30
   process of model development 320–5
   structure 316–20
mathematical models 36, 1098
mathematics 104–5, 113, 401, 504, 905, 907–9, 919, 1046, 1048
Matthew Effect 535–7, 539
maturity 319, 1004, 1006

maximal performance 493, 1069
MCC (Mission Control Center) 830, 832–5, 839, 841–2, 845
measurement 12, 109, 838, 846, 1200, 1202
   for future work 1200–1202
   sensitivity 595, 597
measurement error 57, 72
mechanics 2, 56–7, 153, 752, 991, 993
media 1, 90, 115, 370, 965–6, 1144
   digital 952, 972
   pundits 1163, 1167
   social 720, 1144
mediators 1002, 1007, 1095
medical diagnosis 622, 1046
medical education 133, 408, 418, 421, 620, 935
medical expertise 21, 418, 421, 619, 626, 635–6
medical reasoning 47, 618, 1060
medical students 65, 418, 420–1, 618–19, 621–2, 628, 630, 634, 636
medication errors 434, 1009
medications 415, 442, 444, 457, 626, 1074
medicine 88–9, 304, 306–7, 408–12, 417–21, 618–22, 631, 934, 1075–6, 1145–6
   domain general models 64–5
   evidence-based 1145–6, 1173
   representative tasks and simulated task environments 304–5
memory 36–40, 46–50, 57–8, 66, 135–8, 175–7, 379–81, 395–6, 584, 653–4, 673, 675–6, 1070–2
   aids 434, 447, 631, 705
   capacity 42, 58, 61–2, 66, 73–4, 77, 676–8, 680, 745–6
   cues 46, 570
   decay 380, 389, 393–4, 444
   encoding 45, 570
   faulty 443, 445
   lapses 701, 703–4
   search 556, 560, 1071
   short-term, *see* short-term memory
   skilled, *see* skilled memory
   skills 7, 560, 618
   systems 15, 128, 138–9, 141, 1014
   tasks 40, 138, 562, 1074

transactive 834, 1014
updating 66, 745
mental ability 6, 68, 993, 1058, 1062–4, 1094
　measures 69, 1073
mental abstractions 182–3
mental events 378, 383, 386
mental fatigue 506, 704
mental functions 929, 944
mental models 193, 206, 243–4, 268–9,
　　274–5, 277–9, 479, 644–5, 649–51,
　　653–5, 750–3, 885–7, 1041–2
　probabilistic 151, 160
　of processes 893
　repertoire 651, 653
　shared 810, 835, 840, 1007, 1074
　of situations 751, 990
　working 809, 811
mental processes 56, 74, 752, 1050, 1063, 1195
mental representations 17, 182, 243–6, 256,
　　272, 929–30
mental resources 248, 1063
mental rotation 64, 563–4, 665, 1073
mental simulation 191, 198–9, 532
mentees 1022, 1025, 1033, 1036–41, 1043, 1050
mentoring 25, 540, 875, 877, 994, 1022–8,
　　1036–7, 1039–44, 1047, 1049–51
　and coaching 1023–5
　empirical evidence 1041–50
　functions 1022, 1026–7, 1037, 1039–41, 1050
　informal 994, 1043
　programs 893, 994, 1045
mentoring relationships 1032–3, 1041
mentor-mentee relationships 1022, 1026
mentors 112, 114, 318–19, 485–6, 540, 724–5,
　　1021–8, 1032, 1036–43, 1046–50
　effective 1037, 1041, 1051
　expert 245, 1022–3, 1041, 1051
　surrogates for 1037, 1046
messiness 19, 192, 408
meta-analysis 59–60, 63, 65, 504, 506–7,
　　1037, 1039–40, 1088, 1092–3, 1100
meta-analytic evidence 1022, 1027, 1037
meta-awareness 698, 704
metacognition 268, 272, 274, 432, 802, 1063
metacognitive processes 745, 1063
metacognitive skills 884, 986
metacognitive strategies 701, 703, 1059

meta-expertises 91, 96, 1014
meta-features 24, 958–9, 963
metaphors 182–3, 186, 249, 377, 414, 1184
meteorologists 153, 873, 876, 1043
meteorology 873–4, 882, 886; see also weather
　forecasting
method variance 131, 134
methodological approaches 131, 158, 195,
　　422, 424, 1209
methodologies 37–9, 152, 154, 254, 256, 408,
　　410, 414–15, 425, 514–15
　qualitative 19, 402, 408, 410, 413–15, 418,
　　422, 424–5
　quantitative 408, 413
metrics 65, 363, 502, 655, 678, 767
　performance 656, 816
Mexico 252, 862
microcognition 191, 206
microcognitive models 190–1
microworlds 295–6, 334, 647–8
Mile Island 219, 221, 498, 853
military careers 733, 799
military commanders 197, 222, 865
military decision-making 795–7
military decisions 23, 197, 201, 797–8, 804
military expertise 22, 792–804
　categories 797–800
　as cognitive capability 800–802
　value of developing 793–5
military operations 19, 203, 294, 797,
　　801–2, 1098
military organizations 223, 225, 796, 798, 803
military personnel 22, 68–9, 412, 779, 799, 801
military simulation 1007, 1012
mindsets 953, 963–4, 1164, 1183
　adaptive 24, 958, 964, 1205
misconceptions 835, 908, 953, 987, 1009, 1059
mission control 23, 795, 830–2, 834–5,
　　839–45
Mission Control Center, see MCC
mission controllers 831, 835, 838, 842
missions 795–6, 798, 800–801, 831, 833–8,
　　840–1, 843, 845–6
　ISS 834, 836, 842–3
　LDSE 838, 845–6
mitigation 72–3, 1060, 1066–9, 1072–3,
　　1075–6

mitigation (cont.)
  expertise-based 1066, 1072, 1076
  expertise-related 72–3
  knowledge-based 1075
mnemonics 44–5, 460
mnemonists 45–8, 138
mobilization training 1102, 1204
model, talent search 108, 112–13
modelers 165, 339–40, 349
modeling 157–8, 160–1, 164–5, 343, 349–51, 455–6, 538, 1049–50
models 14–18, 107–12, 152–7, 190–3, 195–203, 205–11, 312–18, 327–30, 338–44, 346–50, 452–7, 741–5, 747–54
  behavioral 155, 161, 163
  causal 500, 1095
  cognitive 193, 195–6, 338, 340, 349, 351, 733, 735, 1048–9
  competency 316–17, 330
  complete processing 1207–8
  computational 6, 18, 38–40, 333–50, 394, 556
  conceptual 211, 539, 701, 759, 889–90
  descriptive 161, 432, 439, 441, 448, 1147, 1149
  developmental 111, 316, 319
  DMGT 110–11
  domain-general 7, 56–78, 109, 307, 443
  domain-specific computational 344–8
  Dreyfus and Dreyfus 313–14
  flexecution 10, 16, 196, 207–11
  frame 196, 198, 200–204, 210–11
  of giftedness 104, 107
  integrated 210–11, 233, 387, 507, 1197, 1199
  internal 176, 749, 758
  knowledge 452–4, 477, 891, 894
  linear 1161–2
  macrocognitive 16, 190–7, 199–201, 203, 205, 207, 209–11, 213, 487
  mastery 18, 312–13, 315–17, 319–20, 323–31
  mathematical 36, 1098
  mental, *see* mental models
  microcognitive 190–1
  multiple intelligences 114–15
  neurobiological 141–2

  normative 621, 1135
  overload 491, 514
  prescriptive 163, 165
  process 154–5, 1209
  psychological 201, 989
  quantitative 340, 697
  RPD 196–202, 210, 346, 942, 985
  stage 43, 313, 316, 884, 929–30
  statistical 161, 1160
  talent development 108, 110–13, 121
  team 23, 455, 831, 839, 844–5
  traditional 194, 1170
modularity 152, 154
mold makers 818–20
morphometry, manual 129–30, 132–3
motivation 112–13, 118, 120, 500–501, 540, 542, 583–6, 697–8, 1138–9, 1181
  intrinsic 112, 978, 1012
motor control 9, 130, 134–5, 182, 248, 510
motor coordination 662, 673
motor cortex, primary 129–30, 133
motor integration 129, 133–4, 140
motor performance 336, 507
motor skills 9, 334, 343, 498, 503, 508
movement authorities 695, 698–9, 709
movement cues 299
MRI (magnetic resonance imaging) 129, 132, 136, 246
MTS (multi-team systems) 832, 840–1, 844
multifactorial perspective 15, 128, 141, 1195
multiple actors 192, 221, 226
  collective behaviors 234, 238
  structural possibilities 228, 231–3, 236–8
multiple intelligences model 114–15
multiple stakeholders 356, 359, 920
multitasking 510–11, 757
multi-team systems (MTS) 832, 840–1, 844–5
music 12, 20–1, 73, 77, 128–30, 134, 137, 574–86, 930–1, 1066–8
  classical 44, 579
  classical expertise approach 44–5
  defining music experts 576–84
  differences between experts and novices 584–6
  domain general models 61–2
  performance 136, 140, 579

playing 57, 579
role in development of expertise theory 575–6
schools 576, 579
musical ability 574–6, 579–80
musical activities 580–1, 583
musical domain 140, 575, 586
musical education 133, 580
musical expert performance 141–2
musical expertise 9, 57, 129–31, 133, 140, 574–87
musical independence 577–8
musical notation 44, 579
musical performance 576, 579, 585, 1068
  expertise 576–8, 580
musical practice 582–4
musical skills 575, 579
musical stimuli 44, 142
musical structures 137, 140
musical training 130, 133
musicians 44, 62, 130, 132–3, 135–6,
    139–40, 255, 575–9, 581–5, 587
  aspiring 575, 928
  instrumental 133, 135
  non-classical 579
  novice 582–3
  professional 133, 412, 578

narratives 744, 749–50, 756, 1147
narrowing 176, 744, 746, 748, 751, 757
  cycle 744, 748
NASA (National Aeronautics and Space
    Administration) 342, 832–3, 841–2,
    845–6, 1110
National Aeronautics and Space
    Administration, see NASA
National Basketball Association (NBA) 291, 301
National Science Foundation 772, 810, 817,
    824–5
National Transportation Safety Board
    (NTSB) 673–4
National Weather Service, see NWS
natural environments 370, 401, 624, 846
natural language 85, 92, 95
natural selection 250, 968
naturalistic decision making (NDM) 3–4, 10,
    197–8, 200–201, 203, 624, 883,
    942, 1168
  community of practice 10, 156–7, 167

navigation 235, 341–2, 533–4, 538,
    664–5, 667
  displays 295, 675
navigational goals 665–6
navy 793–4
NBA (National Basketball Association)
    291, 301
NDM, see naturalistic decision making
negative biases 753, 758
negative feedback 499, 504
negotiations 482, 626, 1100
nephrologists 456–7
nervous system 128, 135–6, 502
network density 363, 372
network logs 726–7, 731
network security 718–35
network traffic 724, 726, 730
networked improvement communities
    (NICs) 919–20
networks 355, 363, 372, 718–22, 724–6,
    729–32, 921
  social 1183, 1201
neural activity 246, 256, 350
neural circuitry 128, 135, 139, 142
neural mechanisms 15, 128–42, 1195
  gray matter 129–32, 134
neuroanatomical correlates of expertise 15,
    129–35
neurobiological models 141–2
neuroimaging 18, 350–1
neuroscience 343, 350, 554
neuroticism 499, 989
New Zealand 695, 712
NICs (networked improvement
    communities) 919–20
Nobel Prizes 38, 531, 535–6, 543, 1164
nodes 38, 42, 45, 362–3, 372, 889, 992
  exit 160, 163
  key 371–2, 752
noise 178, 183, 185, 204–5, 491, 496, 747, 749
non-combatants 797, 803
noncompensatory strategies 163–4
non-elite performers 506, 508
non-expert performance 644, 646, 648, 753
non-experts 94, 129, 131–2, 135, 141–2,
    273–4, 305–6, 646–7, 649–51, 807–9,
    1086, 1091, 1096

non-musicians 62, 130, 133, 140, 577
non-performers 577–8
non-pilots 66, 680, 1071
nonroutine events 859–60, 865
non-technical aviation skills 680–1
non-technical skills 667, 669–71, 680–2, 857–8, 861, 864
non-time-constrained tasks 391, 395
normative models 621, 1135
norms 531, 905, 908, 911, 913, 1123–4
　content-blind 152–3
　cultural 92, 269–71, 274, 698
　logical 152, 174
　social 154, 912
notation, musical 44, 579
novelty 200, 862–3, 865, 951, 953–4, 965–6, 970–2
　ordinary 954–6, 958, 970
　paradox 956–9, 962, 965–6
novice differences 437, 439–40, 443, 551, 562, 564
novice drivers 368, 496
novice musicians 582–3
novice operators 451–2, 463
novice performance 334, 778, 884
novice pilots 66, 306, 1071–2
novice surgeons 511, 935
novices 42–4, 177–8, 270–2, 313–16, 418–21, 453–5, 457–9, 462–4, 496–8, 510–13, 560–1, 984–6, 1059–60
　and experts 42, 46, 177, 345, 453, 493
　relative 193, 719, 1206
NPPF, *see* nuclear power production facilities
nuclear industry 852, 859
nuclear power 454, 473
　nuclear power production facilities (NPPF) 850–2, 856, 859, 863–4
　plants 473, 850–1
　production 23, 851, 863
nurses 4, 8, 299, 457, 460, 510–11, 1026–7, 1111–12
　advanced practice 457, 459
NWS (National Weather Service) 872, 879–83, 886, 1176

objective assessment 301, 328, 643
objective feedback 858, 928, 930

observation
　inner 379–80, 382, 393, 430
　skills 781
observers 377, 383, 385, 794, 858, 933
occipitotemporal junction 130, 132
occlusion 306, 595–6, 605, 782, 937
　conditions 607–8
　paradigms 63, 304
　temporal 596, 611, 774, 937
offensive cyber operations 722, 725
offensive sequences 600–1
officers 203–4, 294, 299, 306, 395–6, 765–83
　less-experienced 395, 766, 776–7, 779, 782
off-nominal events 833, 835
offshore oil & gas installation operations 850, 852–4
older adults 1062–7, 1069–70, 1072–6
older experts 1060, 1062, 1067–9, 1071–2, 1074–5
omissions 389, 419, 835
one-on-one training 579, 795
online learning 93, 504
on-the-job training 671, 728, 735, 795, 886, 994
on-time delivery 818–22
on-time performance 704, 708
ontology 413, 721
OO&G installations 850–1, 853, 855, 857–9, 861, 863
openness 962, 989, 1007, 1204
operating environment 22, 642, 742, 758, 796
operating system vulnerability 720–1
operational context 334, 653–4, 697, 742, 753
operational definitions 577, 581, 670, 1167
operational environment 22, 209–10, 654, 726, 799, 865
operational expertise 726, 797–8, 1120
operational performance 648, 1099
operational personnel 642, 649, 768, 855
operationalization 116, 119–20, 729–30, 734, 1086, 1089–90, 1093
operators 356, 361, 451, 453, 455–7, 461, 465–6, 725–8, 731–4, 850–3, 855–9, 864, 1178–9
　training 277, 851
opponents 299, 303, 596, 599–605, 609, 611–12, 932, 937

optical flow fields 176-7, 181-2
optimization 153, 208, 1062, 1066, 1068-73
  strategies 1062, 1067-8, 1071, 1075-6
option generation 191, 200, 395, 774
ordinary novelty 954-6, 958, 970
organizational constraints 233, 235, 238, 400, 883
organizational contexts 873, 880, 883, 891, 1010, 1012
organizational cultures 759, 891
organizational expertise 904, 914-15, 919
organizational goals 224, 1043, 1118
organizational hierarchy 857, 1118
organizational knowledge 473, 1142
organizational level 483, 1003, 1111, 1124
organizational psychologists 57, 67, 838
organizational structures 224, 755, 759, 914
organizations 224-5, 369, 470, 482-4, 632-3, 855-6, 917-18, 1010-11, 1109-12, 1115-18, 1120-4, 1138, 1148-9
  educational 915, 917-18
  hierarchical 224, 799
  high-reliability 229, 1117
  learning 917-18, 996, 1010
orientation 104, 248, 415
  theoretical 7, 418
originality 39, 482
outdated behaviors 1122, 1124
outliers 1, 328, 576, 585, 927
overconfidence 151-2, 677-9, 754, 993, 1164, 1167-8
overgeneralization 317, 444
overlearning 1086, 1090-1, 1096
overload 12, 490-515, 753
  cognitive 20, 491-2, 496-7, 633
  information 632, 746, 886
  metaphor 490-1
  model 491, 514
overload-induced stress 490-1, 514
over-reliance on knowledge 1061, 1075
oversimplification 951-2, 957, 959-63
oversimplified relationships 1115, 1199

paint-marking ammunition 770, 776
palindromes 552-4
parents 87, 90, 333, 585, 692, 913
parietal lobes 129, 134

participants 163-5, 294-8, 323, 388-91, 394-400, 416-17, 443-6, 595-7, 729-33, 770-5, 810-12, 874-8, 1162-5
  groups of 87, 306
  less-skilled 598, 609
  multiple 293, 774
  skilled 600-601, 606
  unskilled 139, 392
participation 87, 90, 93, 95, 540, 543, 771, 905
partners 824, 894, 919, 1033
part-task approaches 938, 944
part-task practice 275-6, 278-9
passion 113, 120, 318-19, 579, 893
patient care 439, 628, 631-2, 635
patients 94, 159-61, 269-70, 411, 418-21, 442-3, 456-7, 622-3, 629-30, 632-3, 942-3, 1145, 1173-4
  data 271, 434, 623
  safety 421, 635
pattern recognition 40, 314, 432, 599-603, 610-12, 621, 623, 796-7, 799, 802
  skills 601, 938
payloads 836
pedestrians 363, 701, 1147
peer nominations 443, 1027
peers 268, 270, 318-19, 690, 693, 903-4, 933, 1201
  same-age 104, 117
penalty misses 506-7
pentagons 69, 876
perception 37, 48-9, 175-6, 182-3, 243-6, 248-9, 457, 507-8, 727-8, 777-8
  biological motion 599, 603
  hazard 496, 512, 514
  inner 379-81
  tasks 39, 41
perceptual cues 45, 610, 679
perceptual discrimination 445, 783, 1181
perceptual expertise 334, 344-6
perceptual learning 177, 181, 183, 186, 252, 345
perceptual processes 748, 757, 759
perceptual skills 199, 344, 884, 984, 1168
perceptual strategies 644, 646
perceptual-motor processes 1063, 1066
perceptual-motor skills 570, 1068, 1071
perceptual-motor tasks 1065, 1067

perceptual–motor expertise community of
    practice 9–10
performance 8–12, 16–19, 65–7, 299–305,
    312–19, 490–505, 508–12, 1004–9,
    1058–61, 1064–70, 1072–5, 1087–91,
    1093–102
  adaptive 953, 963, 965–6, 970–1, 1098, 1102
  analysts 742, 977, 981–2
  areas, key 316–20, 324–6, 328–30
  cognitive 192–5, 468–9, 471–5, 477–80,
    482–5, 487, 496, 499, 701, 703
  context 491, 502, 515, 929, 1094, 1103
  correlations 64, 66, 505
  deficits 493–4, 502, 504–5, 508, 513
  delayed post-training 1091, 1094
  deliberate 24, 863, 927, 939–40, 944, 1115
  descriptors 317, 323, 328
  dual-task 136, 493
  effects 60–3, 498, 504, 506, 509–10, 512
  of emergency responders 649, 655
  environments 494, 499, 596
  evaluation 577, 580, 983, 988
  expert 10–11, 18–21, 113–14, 128–9,
    131–4, 137–42, 297, 299–300, 302,
    348–9, 575–6, 648–50, 942
  feedback 318, 326, 982, 1014
  flight 672–3, 675
  flight simulator 673, 1065, 1071
  human 191–2, 297, 338, 340, 630, 648
  impairments 492, 500, 504, 507, 512, 514
  indicators 18, 312, 316–17, 319–20, 324–30
  of intelligence analysts 978, 981, 1182
  job 14–15, 57, 59, 67–8, 70–3, 77, 317, 940,
    1037, 1039–40
  maximal 493, 1069
  measurement 12, 981–2, 1201
  measures 63, 65–6, 73, 297, 649, 654
  metrics 656, 816
  motor 336, 507
  music 136, 140, 579
  non-expert 644, 646, 648, 753
  novice 334, 778, 884
  on-time 704, 708
  operational 648, 1099
  outcomes 193, 1003, 1037, 1040
  problem-solving 535, 1075
  reacquisition 1087

  retention 1091, 1098
  school 109, 120
  shooting 774, 781
  skilled 20–1, 298, 300, 490–3, 495, 497–9,
    501, 503, 505, 509
  skills 512, 575, 969, 1044
  sports 20, 491, 503, 506–7
  and stress 491, 503, 509
  surgical 64–5, 934
  team 192, 1001–2, 1005, 1007, 1010, 1013
  train crews 697, 710
performance-based training 301, 942–3
performance-resource functions
    (PRFs) 494–5
performers
  elite 506, 508, 604, 928
  expert 16, 18, 291–2, 295, 297, 299–301,
    304, 307, 347–8
perfusion, distal 440
periodic table of expertises 85, 91–7
persistence 112–13, 118, 120, 754, 889
personal life 2, 1033
personality 586, 988–9, 991, 993,
    1087, 1094
  characteristics 104, 119, 512, 836, 843
  types 120, 670, 989–90
personnel 224, 325, 472, 474, 643–4, 669,
    799–800, 852, 855–6, 858–9, 865,
    980, 1111
  enlisted 228, 798
  operational 642, 649, 768, 855
  selection 74, 800, 1097
persuasive design 465–6
phenomenological interviews 254–7
phenomenology 244, 254, 415
philosophers 4, 105, 1140
philosophical foundations of expertise 244–7,
    249, 255
philosophy 4, 87, 244, 698, 856, 911
photos 306, 342, 417, 666, 816
phylogenetic learning 158
physical interactions 17, 691, 721
physical models 253, 815
physical objects 248, 253, 704–5
physical strength 1066–7, 1075
physicians, see doctors
physicists 531, 1182

physics 46–8, 50–1, 248, 384, 401, 531
  classical expertise approach 42–3
physiology 379, 383, 960–1, 1048, 1203
pianists 44, 62, 133, 135–6, 139
pilot judgment 668, 670, 676–8
pilots 66, 159, 341–2, 437–9, 645, 662–72,
    674–6, 678–80, 836, 1071, 1074
  expert 66, 306, 342, 420, 677, 1071
  fighter 340, 396, 676, 681
  navy 794
  novice 66, 306, 1071–2
pioneers 4, 11, 18, 378, 387, 1177
Piper Alpha 853, 861
pitches 298–9, 303–4, 582, 584, 600–601, 664,
    669, 937–8
plasticity 345, 1147
players 37, 39–41, 60–1, 93–4, 302–3, 555–61,
    565–70, 597, 599–601, 603–5
  best 558–9, 566, 570
  elite 564–5, 569
  less-skilled 299, 303, 599–602, 605, 609
  soccer 63, 87, 111, 139, 599–601, 609
pointers 40–1, 961
poker 38, 61, 73, 77
police 162, 766–70, 773, 777–9, 781–4
  firearms training 775–6
  research 769–70
  work 766, 768, 783, 804
policy 325–6, 979–80, 1133, 1135, 1137,
    1140–7, 1204–5, 1210
  cycle 1137, 1142, 1146
  decisions 1146–7
  interventions 26, 1135, 1143
  officials 1136, 1142–3
  professionals 1134, 1142–3
  public 89, 1110, 1139
policymaking 1139–40, 1142, 1146–7, 1149
political decisions 90, 1133
politicians 333, 981, 1132, 1134,
    1138, 1149
politics 90, 1063, 1070, 1137, 1146, 1148
popularity 93, 109, 355, 1176
positive disturbances 1125–6
positive heuristics 1166, 1184
post training performance 1087
post-hoc explanations 159, 202, 628
posttraining performance 1089, 1094

post-training performance, delayed
    1091, 1094
post-truth society 15, 85, 97–8
postural cues 595, 597–8, 601, 604–6, 610–11
postural information 603, 606–7, 609
potential errors 368–9, 628
power 408, 411–12, 662, 665, 1132, 1166,
    1168, 1171
  function 1093, 1102
practitioners
  domain 192, 317, 475, 482, 698, 1208
  skilled 92, 1137, 1140
pragmatics 174, 179, 183
prayer 378–9
precedence 184, 401, 1209
preceptors 1022–3, 1026–7, 1037, 1040
precision 37, 47, 324, 654, 803
prediction accuracy 600, 777
prediction tasks 1168, 1177
predictions 47–8, 50, 119, 306, 599–600,
    675–6, 700, 810, 813, 982–3,
    1119–20, 1167
  quantitative 350, 697
  reliable 2, 1180
predictive validity 113, 151, 1036
predictor variables 65, 75, 161–2
predictors 66, 69, 71, 74, 536–7, 676
  strong 264, 675
  valid 57, 59, 67
predispositions 263–4
  genetic 585–6
prefrontal cortex 136–7, 493, 497
pregnant women 436, 1158
premature closure 744, 750, 754, 756
premotor areas 129, 139–40
preoccupation 703, 1118, 1120
preparedness 800, 851, 1110
prescriptions 415, 674, 766, 956
prescriptive model 163, 165
presentation 35, 41, 44, 292–3, 325, 329, 632
preservation 212, 469, 512, 1063, 1112, 1198
  complexity 960, 1115, 1199, 1208
pre-service teachers 933–4, 941
pressure 222, 225, 231, 418, 507–8, 510,
    514, 917
  competitive 507
PRFs, *see* performance-resource functions

prices 42, 531, 817, 819, 821–2, 1070
primary motor cortex 129–30, 133
prioritization 245, 315, 448, 666, 750, 1012
private sector 727, 731, 874, 892, 1184
  cyber operators 727
  forecasters 875, 882, 887
probabilistic mental models 151, 160
probabilistic reasoning 625–6
probabilities 98, 152, 160–1, 537–8, 599–600, 1165
probability charts 161, 165
probes 389, 391–2, 395–400, 431–3, 478–9, 482, 879, 884
  directed 392, 396–400
  undirected 391, 396–8
probing questions 20, 255, 469, 473–5, 478, 487
problem definition 421–2
problem detection 191, 1177
problem ecology 11, 16, 178, 187
problem identification 202, 533
problem representations 7, 42, 1060
problem solving 36–8, 46–7, 49, 194, 207–8, 252–3, 264, 387–8, 421, 475–6, 618–21, 1059–60
  abilities 117, 267, 557
  adaptive 220, 878
  analogical 264, 855
  expert 37, 137–8, 264, 1060
  medical 618, 622
  strategies 7, 57, 695, 1046, 1061, 1196
  tasks 253, 392, 622, 1046, 1074
problem spaces 222, 392, 533–4, 538, 621, 623
problems
  substantive 740–1, 744, 746, 749–50, 752–3
  wicked 209, 704, 955, 1176, 1178
procedural knowledge 58, 348, 504, 815, 1098
procedural skills 13, 263, 269, 568, 984, 986
proceduralization 493–4, 497, 511, 719
procedures 224–5, 262–4, 269–71, 458–9, 464, 768–9, 842–4, 850–2, 855, 862–4, 986–8, 990–1, 1115–16
  selection 77, 152, 1027
process control 19, 995, 1205
process maps 820–1
process models 154–5, 1209

processing 15, 58–9, 62–3, 66, 135–9, 347, 457, 492–3, 498–9, 672–3, 745, 962–3
  capacity 491–3, 512, 1065, 1069–70, 1076
  cognitive 47, 490, 584, 779, 932, 1183
  controlled 136, 1096
  efficiency 499, 507
  speed 58
process-tracing 113, 694
prodigies 115, 585
productive thinking 37, 185, 386, 388
productivity 226, 229, 510, 536, 690
  scholarly 111, 542
professional development 325, 330, 916, 919, 929, 1033, 1037, 1041
professional domains 612, 939, 956, 979, 1140, 1169
professional experience 470, 484, 836
professional expertise 3, 141, 268–9, 408, 410, 1144
professional identity 875, 1025
professional judgments 1159, 1173–4
professional knowledge 469–72, 478, 483–5, 1027
professional musicians 133, 412, 578
professionalism 318, 483–5, 1132
professionalization 26, 325, 418, 485, 766
professionals
  aviation 22, 667–8
  healthcare 473, 619–20, 1051, 1174
  intelligence 979, 987, 996
  policy 1134, 1142–3
proficiency 1–2, 22, 323–6, 663, 670–1, 677–8, 732–3, 977–80, 982–3, 1027, 1064–5, 1086, 1163–4
  continuum 1021, 1050
  expert-level 977, 979, 981, 983, 985, 987, 989, 991, 993, 995
  high 670, 978, 986, 993–4, 1175, 1183
  tactical 318, 767
programming, computer 40, 43, 48, 137, 1046
progression 235, 312–13, 317, 538–40, 649–50, 652, 930, 939–40
project goals 443, 472, 477
promotion 280, 771, 915, 1029–30, 1039, 1073

properties 15–16, 162–3, 176, 178, 183–5, 187, 209, 530, 532, 965
  emergent 175, 179, 183, 185, 187, 627
  environmental 163, 166
  functional 138, 140, 182–3
  macroanatomical 15, 128, 141
  ultrastructural 133, 141
protocols 94, 96, 382, 385, 389, 482, 487, 629–30
prototypes 182, 198, 962, 1204
proxies 63, 117, 130, 677, 778, 1014
proximal cues 342
psychobiological perspective 497–8
psychological fidelity 863–5, 936
psychological laws 5, 380
psychological models 201, 989
psychological safety 1008–9
psychologists 35, 40, 103, 107, 377, 379, 766, 770
  clinical 770, 1182
  cognitive 246, 1171
  experimental 334, 379
  organizational 57, 67, 838
psychology 1, 4, 173–6, 179, 378–9, 381–4, 400–401, 769–70, 1110
  cognitive 3, 9, 14, 49, 244, 256, 268, 878–9
  cultural 381, 1203
  developmental 42, 50, 104
  differential 36, 335
  ecological 11, 178, 596
  educational 911, 971, 1203
  experimental 39, 115, 191, 298, 1160, 1167
  of intelligence analysis 741, 993, 1164, 1169
  sports 9, 609
psychometric instruments 988–9
psychometric views, traditional 108, 110, 112
psychometrics 36, 108–10, 115–16, 989, 1028
psychomotor abilities 670, 672
psychomotor skills 13, 412, 932, 934, 937
psychophysiological measures 501–3, 512–13
psychosocial support 1032, 1039–40, 1047, 1050
psychosocial variables 112
psychotherapy 36, 112, 313
public policy 89, 1110, 1139
public trust 1141, 1180, 1194

pupillary response 502, 507
puzzles 390, 551–70, 735
  cognitive taxonomy 552–3

qualifications 677, 1143, 1166
  technical 855, 914
qualitative analysis 324, 414, 447, 1122
qualitative data 413, 416–17, 435–6, 439
qualitative interviews 444, 878
qualitative methodologies 19, 402, 408, 410, 413–15, 418, 422, 424–5
qualitative research
  addressing contemporary concerns 419–21
  close-to-practice methods 9, 408–25, 435, 694
  complexity argument 409–10
  conclusions, future directions and recommendations 424–5
  enriching conceptualization of what makes an expert 421–2
  establishing value of 418–19
  ethical aspects 423
  key considerations 422–4
  need for 409–12
  ontology and epistemology 413–14
  quality in 422–3
  research topics and questions 412–13
  researcher development and support 424
  social and professional argument 410–11
  special and different argument 411–12
  studies which have changed understanding of expertise 417–22
  theory and methodology 414–15
quantifications 402, 728, 991, 1037, 1121
quantitative methodologies 408, 413
quantitative models 340, 697
quantitative predictions 350, 697
questionnaires 357–8, 500, 583, 718–19, 727–8, 730
quiet eye 507, 509, 512

racial bias 779
racket contact 596, 603–6
radar 235–6, 347, 799, 889–90
radical embodied cognitive science (RECS) 249, 255–6
radical empiricism 178–9

radio communications 693, 702–3, 707
radios 665, 692, 699, 701, 706, 799
rail level crossings 355, 358–9, 361–72
rail traffic 358–9, 364–8, 700
railroad operations 690–713
   changes 692–3
   exhibited expertise contributing to safe and efficient operation 697–705
   frontline workers 691–2
   future 711–13
   impact of new technologies on expert strategies 708–11
   research and practice 693–7
   strategies for operating as cooperating teams 705–8
railroad technologies 22, 691, 708
railroad workers 22, 691, 702, 704, 708
random assignments 562, 1049, 1087
random sequences 44, 651
randomized control trials 444, 1142, 1145, 1173
ransomware 721, 723
ratings 328, 508, 567, 677, 837, 844
   likelihood 777
   supervisor 67–9, 328, 1073
rationality 150, 152, 174, 678, 1166
   bounded 16, 38, 49, 149, 152–4, 1135
   ecological 153–5, 163, 166
   technical 410
   unbounded 152–3
rationalization 385, 644, 880, 987
reacquisition 1086–7, 1089, 1091–2, 1099, 1101
   assessments 1091, 1094
   performance 1087
   phase 1087, 1091
   rapid 1101
readiness 799–800, 802, 951, 957, 963, 972
   adaptive 951, 959, 963, 965–6, 968, 970–1
   cognitive 22, 265, 268–9, 767, 797, 801–2
realism 647, 773, 781, 863
real-world complex performance domains 26, 1089, 1097–8, 1101–3
reasoning 379, 530–2, 620–6, 631–5, 883–4, 887, 891, 903–5, 908–13, 920–2, 1166, 1200
   abductive 205, 623

   analogical 532
   analysts 25, 990
   causal 25, 625, 982
   clinical 273, 621–2, 624, 631
   complex 916, 1049
   deductive 205, 455
   diagnostic, see diagnostic reasoning
   dialogic 24, 1049
   expert 394, 454, 621, 984, 1165, 1205
   expertise in 903–5, 922
   fluid 60, 62
   logical 673, 1177
   medical 47, 618, 1060
   probabilistic 625–6
   technology-mediated diagnostic 620, 630
reasoning-based teaching 913, 922
reasoning-focused instruction 915, 919
recall 40–2, 44–5, 244–5, 395–6, 399, 474–5, 478, 481–2, 680, 1101
   accuracy 434, 448
   tasks 36–7, 41, 45, 49, 1069
recognition 157–8, 198, 220–1, 314–16, 530, 535–6, 857, 929–30, 984–5, 1059, 1069–70
   social 268–9, 535, 542
recognitional heuristic approach 200
recognition-based decision making 334, 344, 346
recognition-primed decisions 5, 7, 16, 48, 194–8, 985
recollections 381, 396, 470, 705
reconstruction 627, 1089
recruitment 111, 423, 801, 1064, 1140–1
RECS, see radical embodied cognitive science
redescription 46, 360
redesign 436, 918
reductionist approach 400, 610, 648
reductive worldview 953, 959, 961–3
redundancy 369, 707, 711
refinement 278, 622, 653, 749, 928, 933
   cycle 890
reflection
   sessions 1002, 1012
      guided 1007–8
      structured 1006, 1008
   skills 934, 1045
reflexivity, team 1004–6, 1010–11

refresher training 1092, 1097–1102, 1204
rehearsal 706, 815–17, 819–25, 827, 905,
    1099–1100
  design 820
  elaborative 26, 1089, 1098, 1102
  mental 508, 511
reinforcement 654, 993, 1013
relapse prevention 1098–9
relationships 63, 65–6, 71–2, 203–4, 223–4,
    249, 362–3, 462–3, 491–2, 504–5,
    750–2, 1033–5, 1039–40
  close 665, 672
  data–frame 202, 204
  functional 178, 295–6, 985
  interpersonal 506, 990, 1025, 1044, 1181
  mentoring 1032–3, 1041
  oversimplified 1115, 1199
reliability 373, 389, 398, 415, 422, 1036–7
  within-expert 1201
reliance 606, 644, 646, 651, 654, 994,
    1172, 1175
remediation 726–7, 731
remote team members 832, 843
reorganization 139
  functional 48, 138
  representational 719
repetition 67, 257, 278, 933, 942, 1068
  mindless 582
replanning 16, 207–9, 211, 1119, 1196,
    1199–200
  adaptive 10, 196, 387, 1197
  flexecution model 10, 16, 196, 207–11
  rapid 209
replication 132–3, 384, 863, 1121
repositories 251, 392, 394, 759, 1123
representations 11–12, 42–4, 175–6,
    178–83, 186–7, 251–4, 256, 740,
    962, 1195
  knowledge 334–6, 346, 952–3, 962,
    1196, 1204
  mental 17, 182, 243–6, 256, 272, 929–30
  problem 7, 42, 1060
  schematic 595, 606
representative tasks 11, 13, 268, 291–307, 611,
    935, 937, 1058–9
  and simulated task environments 292–7
representative tests 291–307

representativeness 195, 415, 1168, 1180, 1208
reproducibility 769, 774, 778
reproduction 233, 380, 382
reproductive thinking 386
reputation 56, 441, 535–6, 1134
research
  community 26, 220, 742, 758
  descriptive 647, 932, 1126
  educational 21, 265, 530, 906, 915
  methods 15, 22, 85, 769, 772, 775
  participants 192, 413, 416–17, 423, 771
  projects 333, 423, 432, 733, 772, 1162–3
  resilience 1110–11
  teams 324, 418, 422–4, 435–6, 439, 540–1
research design 397, 423, 576, 879,
    1086–7, 1095
  ideal 1085–7, 1089, 1095
researchers 294–6, 320, 323, 396–7, 413,
    415–17, 423–4, 540–1, 1092, 1094
residents 64–5, 160, 437, 440, 628, 634
resilience 11, 26, 263, 265, 802, 1109–26,
    1180, 1184
  commitment to 1118–19
  definition 1110–12
  frontline workers 1112–17
  organizations and systems 1122–5
  research 1110–11
  system 691, 711, 713
  teams and management 1117–22
resistance 645, 912–13, 1093
  to dialogic teaching 912–13
resourcefulness 471–2
resources 103–5, 121, 208, 245, 807, 831–2,
    893, 1062–3, 1122–4, 1134–5,
    1169–71, 1196
  attentional 67, 137, 493, 498, 1067
  cognitive 249, 1060, 1062, 1065–6,
    1068, 1072–5
  external 1028, 1170
  limited 121, 1026
  social 916, 1074
  training 435, 877
response bias 501, 773
response phase 777, 862
responses 138–9, 227–8, 245–8, 324–6,
    594–6, 642–4, 650–2, 655–6, 750,
    844, 1075, 1090, 1202

responses (cont.)
  adaptive 24, 498, 953–4, 956, 966, 972
  stress 492–4, 496–7, 499–500, 502–3, 509, 511–13
responsibilities 235–6, 238, 316, 319, 844–5, 859–60, 980, 1133, 1178, 1183–4
responsiveness 112, 1113
restaurants 264, 270–1, 273, 1070
retention 24–6, 471, 583, 909, 1085–102, 1204
  intervals 505, 1086–8, 1091–4, 1097–102
  performance 1091, 1098
  scores 1091, 1094
  skill 1085, 1088, 1090–1, 1097
  tests 1087, 1090, 1094
  and transfer 26, 1089, 1096, 1102
retirement 159, 472–3, 889
retraining 435, 1087, 1091, 1099–101
retrieval 629, 632, 672, 675, 958, 962, 1093–4, 1101
  errors 369
retrospection 380–1, 385, 388, 395–6, 694, 979
retrospective interviews 430, 443, 445, 774
reverence, lack of 1194–5
review
  after-action 794, 1099
  comprehensive 128, 313, 674, 752, 931
  historical 18, 393
rewards 112, 251, 424, 535, 909, 922
rhetoric 151, 1132
riddles 552–3
rigidity 962, 986, 1204
  cognitive 987, 1115, 1124, 1199–200
rigor 5, 115, 415, 529, 1172, 1174
riot control 646, 783
risk 90, 270–1, 412, 419–21, 441–2, 512–13, 769–71, 812–13, 864, 873, 1013, 1167, 1172
  high 76, 192, 264, 850, 1117
  potential 306, 706, 863
road users 359, 361–2, 364–7, 369
roadway workers 691–2, 701–2, 705–8
robustness 164, 752, 754, 758, 826, 1135–6
rookies 395, 567, 766, 779–80
root-mean-squared deviation (RMSD) 341–2
route knowledge 700–702

routine expertise 13, 17, 262–3, 265, 271–2, 274, 279, 1196
routine operations 23, 850, 853–8, 861
routine tasks 915, 940, 1008
routinization 1121, 1123
RPDs (recognition-primed decisions) 5, 7, 10, 194–201, 346, 941–2, 985, 1168
  model 196–202, 210, 346, 942, 985
runways 662, 750, 1124, 1207

safe operations 451, 692, 701, 851
safety 419, 451, 499, 513, 642, 698, 706–8, 856–7, 863–4, 1109
  management, systems 855, 861, 864
  psychological 1008–9
  science 354, 356–7, 700, 1110
salaries 535, 1029–30, 1073
sample sizes 15, 61, 64, 69–72, 131, 443–4
samples 59–66, 68, 443–4, 541, 564, 566, 772, 1064–5
  large 62, 67, 327, 1174
satisficing 38, 48–9, 153, 199, 746, 750
saturation, data 323–4, 328
scaffolding 273, 1021–2, 1033, 1037, 1041–3, 1045–8
  methods 396, 1021, 1115, 1199
scales/scaling, proficiency 14, 25, 887, 891, 978, 1201–2
scanning 340, 344, 677, 680–2
  visual 669–70, 673–4
scenario-based training 440, 781, 1048
scheduling 445, 473, 842, 919
schemas 47, 49, 178–82, 951–2, 958, 962, 964, 967, 969–70, 1070
  candidate 966, 968–70
  slotted 386, 745
schemata 138, 179, 256–7, 335–6, 346, 751
schematic anticipation 386–7
schematic representations 595, 606
school districts 103, 121, 913, 917–18, 921
school performance 109, 120
school systems 904, 914–16, 918, 921
schooling 538, 904, 912
  formal 905, 1043
schools 105, 118–21, 249, 299, 724–5, 798, 903–6, 910, 912–13, 915–19, 1110
  elementary 905–6

formal 325-6
high 62-3, 917
music 576, 579
science, domain general models 63-4
science and technology studies, see STS
scientific expertise 63-4, 86, 1160, 1194
scientific knowledge 78, 86-8, 418, 538-9, 878, 1169
scientists 90, 92, 400-401, 530, 532-3, 535-7, 990-1, 1209-10
  intuitive-based 885, 891
SCM (standard capacity model) 20, 491, 491-514
  limitations 494-7
  and sports 506-10
  and surgery 510-12
  and vehicle driving 512-14
SCRABBLE 551, 553-6, 563-6, 569-70
script concordance test 273
scripts 206, 264-5, 272, 751, 775
search
  conceptual variability 957, 964
  heuristic 533, 619
  selective 38, 48, 780
  space 531, 538
  talent 108, 112-13
  visual 304, 306, 340, 561, 633, 674
security 2, 4, 423, 434, 719, 722
  computer 720, 723-4
  network 718, 730
  threats 1134-5
selection 75-6, 78, 176, 247, 325, 327, 966, 988, 990, 1071-2
  effects 60-3
  initial 1067, 1140
  natural 250, 968
  strategies 1062, 1072
selective attention 178, 649, 941
selective search 38, 48, 780
selectivity 163, 383, 754
self 316, 321, 861, 864, 1002, 1004
self-appraisals 496, 504, 513
self-assessments 274, 326, 541, 704
self-awareness 749, 753
self-confidence 514, 536, 1118
self-conscious awareness 509-10
self-consciousness 89, 509, 514

self-efficacy 76, 497, 540, 1087, 1099
self-explanations 394-5, 400, 1049
self-handicapping 505, 508
self-management 1100
self-monitoring 480, 542, 704
self-observation 381-2, 401
self-organization 16-17, 220, 226-8, 232-9
self-regulated team reflection 1003-6
self-regulation 267, 279, 497, 499, 542, 1003
self-report 256, 329, 430, 447, 511-12, 1036-7
self-selection 486, 536, 719, 990
semantic associations 554, 556, 570
semantic priming 1063, 1066
semantics 44, 599
semiotic systems, triadic 16, 174-5, 178-9, 183-5, 187
semiotic terms 174-5
senior experts 731, 891, 977-8
seniority 334, 649
sensemaking 185, 194, 196, 200-206, 208, 211-12, 747-8, 751-3, 1147, 1197-9
  frame model 196, 198, 200-204, 210-11
  process 181, 202, 204-5, 747-8, 755, 1048
sensitivity 13, 17, 257, 493, 654, 782
sensorimotor adaptations 139-40
sensors 222, 235-6, 254, 347, 490
sensory input 461, 854, 1072
separation 244, 667, 960
  maintaining 666-7
  standards 667
sepsis 435, 437, 440, 634, 1146
sequences
  film 599-600
  task 222, 228-9, 232, 235, 237, 358
sequential structures 130, 539
servomechanisms 178-9
ShadowBox 487, 941, 943, 1048-9, 1051
shared knowledge 835, 840, 985
shared mental models 810, 835, 840, 1007, 1074
shared situation awareness 707, 857
shared understanding 706, 711, 809, 815, 839-40, 1001
shingles 295, 485
ships 225-6, 799, 1134
shooting 299, 766, 772-3, 776, 779, 781
  accuracy 771, 779

shooting (cont.)
  basketball 508, 510
  handgun 507–8, 779
  performance 774, 781
short-term memory (STM) 38, 40–1, 45, 47, 50, 59–60, 745–6
signals 180, 183, 185, 370, 691–2, 695, 697–705, 854
  error 180–1
  stop 272, 695, 697, 699–700, 705
  wayside 697, 699
signatures of expertise 671–2, 770
simplicity 419, 501, 691
simulated scenarios 299, 395, 398, 433, 647–8
simulated task environments 292–7
simulation 272–3, 293–6, 300, 306–7, 420, 636, 653, 793–4, 846, 863–4, 935, 943
  interviews 19, 433–5, 445–7
  mental 191, 198–9, 532
  military 1007, 1012
  video-based 306, 770
simulation-based training 440, 931, 935
simulator-based training (SBT) 863–4, 931, 935–6, 944
simulators 341, 678, 697, 773, 793, 935–6
  flight 66, 293, 295–6, 342
  laparoscopic 65, 305, 935
singers 133, 135
situated cognition 211, 251–2, 257, 344, 347–8, 1169
situated expertise 419, 728
situation awareness 669, 673–4, 676, 701–2, 745–7, 856, 858, 880–1, 938, 1071
  shared 707, 857
situation-adaptive assembly of knowledge and experience 24, 970, 1204–5
situational factors 25, 192, 199, 440, 1102
situational judgment tests 326, 328–9, 767
skill acquisition 4–6, 8–9, 67, 343, 496, 500, 503, 1065–6, 1085–7, 1089
  community of practice 5–6
skill decay 6, 26, 1085–103
  mitigating loss and enhancing retention 1095–102
  research design and methods 1086–95
skill development 300, 302, 486, 539, 542, 663–4, 1028, 1050

skill levels 37, 41, 291–2, 302–3, 929, 1201
skill transfer 14, 781, 1099
skilled anticipation 594–612
skilled athletes 598–9, 601–2, 612, 937
skilled memory 36, 47–8
  theory 7, 45–7, 1060
skilled performance 20–1, 298, 300, 490–3, 495, 497–9, 501, 503, 505, 509
skills
  adaptive, see adaptive skill
  analytical 303, 980
  argumentation 24, 539
  attentional 1066, 1068
  complex 77, 275, 278, 280, 312, 1066
  coping 494, 497
  expert 24, 454, 551, 711, 928, 940
  forecasting 881, 991, 1163
  interpersonal 802, 990
  and knowledge 263, 273–4, 619, 621, 861–3, 1069, 1085–6, 1091–2, 1101, 1205
  learned 263, 1098
  memory 7, 560, 618
  motor 9, 334, 343, 498, 503, 508
  musical 575, 579
  non-technical 667, 669–71, 680–2, 857–8, 861, 864
  observation 781
  pattern recognition 601, 938
  perceptual 199, 344, 884, 984, 1168
  perceptual-motor 570, 1068, 1071
  performance 512, 575, 969, 1044
  procedural 13, 263, 269, 568, 984, 986
  psychomotor 13, 412, 932, 934, 937
  reflection 934, 1045
  social 411, 722, 778, 780, 859, 864
  special 149, 158, 766, 1091
  tactical 773, 778
  technical 577–8, 718–19, 722, 743, 755, 758, 929, 931
  tennis 293, 298
  training 506, 1184
  visualization 666, 675
slotted schemas 386, 745
smartphone technology 1076
SMEs, see subject-matter experts
soccer 9, 63, 596, 599–600, 602, 609

social agreement 86, 809
social contexts 279, 410–11, 415, 544
social environments 154, 411, 418, 1195, 1201
social groups 15, 86–8, 92–3, 410, 416, 1139
social interactions 418, 460, 874, 878, 905, 915
social learning 158, 918, 1203
social media 720, 1144
social networks 1183, 1201
social organizational analysis 453–4
social processes 515, 921, 1170, 1210
social skills 411, 722, 778, 780, 859, 864
social studies of science community of practice 8–9
social support 876–7, 893, 1203
socialization 85, 92, 96, 98, 530, 540
  linguistic 93, 95
sociological analysis 90, 1169, 1171
sociological perspectives 15, 85, 530, 1169–70
sociologists 86, 418, 1159, 1169, 1171–2
sociology 4, 21, 87, 471, 1159, 1169
  communities 1169–72
sociometry 1201
sociotechnical systems 16–18, 213, 219–39, 354–74, 1111–12, 1118, 1123
  characteristics of 220, 237
  workers in 16, 223, 232, 235, 237
software 441, 447, 471, 476, 482, 720–1, 727
  developers 323, 432, 441, 721
solitary practice 579, 581–2, 586, 927–8
SOPs, see standard operating procedure
source materials 324, 942, 1038
space crews 23, 830–2, 836, 839–45
spacecraft 830–2, 841
spaceflight operations
  collaboration between mission control and space crew 839–44
  future research 845–6
  mission control and space crew as multiteam system 832–9
  teamwork 830–46
spatial ability 676, 680, 1073
specialists 91, 97, 224–5, 230, 411, 1141–2
specializations 15, 25, 221, 224, 228, 979–80
  technical 743, 755
specialized knowledge 619, 837, 914, 978, 1118

specificity 160, 329, 439, 1066, 1100, 1205
speculative thinking 1166, 1185
speed trajectory 708–9
spinal cord 128–9, 133
sports 20–1, 298–9, 302–4, 506–8, 511–12, 514, 594–7, 599, 601–3, 605, 609–12, 930–2, 1066–7
  deliberate practice (DP) 936–9
  domain general models 62–3
  future 609–12
  high-performance 594, 604
  performance 20, 491, 503, 506–7
  psychology 9, 609
  representative tasks and simulated task environments 302–4
  and SCM (standard capacity model) 506–10
  skilled anticipation 594–612
  team 303, 599
stability 25, 542, 961, 981, 1096, 1139
staff, see personnel
stakeholders 320, 324, 482–3, 742, 750, 1135
  multiple 356, 359, 920
standard capacity model, see SCM
standard operating procedure (SOPs) 224, 228, 354, 357–8, 776, 855, 857, 861
standardization 880, 979
standardized instruments 810, 989
standardized tests 57, 112, 917, 989
standards 12, 323, 326, 370–1, 798, 801
Stanford–Binet Intelligence Scale 61, 109
state-of-the-science 2, 19, 379
statistical analyses 36, 51, 69, 334–6, 532, 1167
statistical models 161, 1160
statistical power 38, 62, 72, 846
statistical structures 152, 162
statistics 37, 60, 62, 66, 1161, 1163
  Bayesian 1166–7
status, expert 9, 89, 927, 1064, 1119
status quo 1003, 1007, 1010, 1141
STEM 12, 20–1, 64, 529–44, 1046
  distinct disciplines 530–1
  factors contributing to development of expertise in scientific research 538–42
  future 542–4
  graduate students 539–40

STEM (cont.)
  reasoning mechanisms 531–4
  recognition of expert performance 534–8
stimuli 44, 138–9, 173–4, 380, 382–4, 595–6, 598–9, 642–3, 646–7, 773, 1068, 1094
  musical 44, 142
  video 769, 773–4
  visual 138, 416
STM, see short-term memory
stop signals 272, 695, 697, 699–700, 705
storage 629, 649, 745, 1060
  knowledge 1200
stored knowledge 177–8, 183, 860
strategic expertise 798–9
strategies
  adaptive 881, 1072, 1111
  analysis 451–3, 458–63, 465
  coping 499, 1111
  expert 459, 463, 465, 560–1, 705, 708
  informal 704, 707–8
  knowledge-based 19, 622, 1075
  metacognitive 701, 703
  noncompensatory 163–4
  perceptual 644, 646
  problem-solving 7, 57, 695, 1046, 1061, 1196
  selection 1062, 1072
  teamwork 691, 711, 1004
strawman fallacy 1133, 1147–8
strength, physical 1066–7, 1075
strengths of experts 1158–9
streptokinase 626, 629
stress 20, 396, 490–515, 770, 772, 781, 852–3, 861–3, 865–6, 1110
  chronic 492, 513
  effects 494, 512, 514
  extreme 859, 861
  interventions for 509, 515
  measures 497, 500, 503, 512–14
  measuring 491, 500
  overload-induced 490–1, 514
  and performance 491, 503, 509
  standard capacity model 491–514
  subjective 499, 501, 510
stress symptoms 500, 509, 512
stressor effects 20, 496

stressors 490–4, 496–9, 501, 503, 506–7, 510, 514, 1111–12
  environmental 492, 496, 504
  external 491, 496–7, 500, 503, 512
structural information 598, 600, 605
structural possibilities 228, 231–8
structuralism 173, 175, 1202
structured activities 49, 580, 927
structured methods 988, 991–3
STS (science and technology studies) 85–6, 88–9, 97, 878–9, 1169
  three waves 86–91
student teachers 916, 933–4
students 116–21, 277–8, 318–19, 504–6, 536, 538–43, 723–4, 904–13, 915–18, 920–2, 934, 1022–3, 1046–50
  gifted 103, 105–7, 109, 112–13, 118–21
  graduate 162, 380, 530, 539–42, 577, 772
  high ability 104, 107–8, 112, 117
  high school 723–4
  medical 65, 418, 420–1, 618–19, 621–2, 628, 630, 634, 636
  surgical 935–6, 938
  undergraduate 61, 338
styles 25, 115, 335, 988, 990–3, 996
  reasoning 883–5, 893, 990–2, 996
subcomponents 16, 293, 295, 299
subdomains 979–81, 990
subject matter 415, 539, 906, 909, 911, 913
subjective measures 274, 279, 500–501
subjective stress 499, 501, 510
subject-matter expertise 196, 327, 357–8, 784, 833–4, 1121
subject-matter experts (SMEs) 339, 342, 349, 357–8, 600, 650, 655
submarines 793, 1134
sub-skills 754–5
substantive problems 740–1, 744, 746, 749–50, 752–3
subtasks 480, 635, 670, 1177
supervisor ratings 67–9, 328, 1073
supervisors 326, 329–30, 726, 731, 767, 771, 1034–5, 1117
supervisory control 347, 713, 743
support
  environmental 1073–4
  ground 843–4

psychosocial 1032, 1039–40, 1047, 1050
roles, developmental 1022–3, 1027
social 876–7, 893, 1203
workers 232, 234, 236, 238
surgeons 64–5, 225, 411, 421, 510–12, 935–6
expert 65, 511–12, 934, 938
novice 511, 935
surgery 73, 225, 503, 629–30, 932, 934–5
domain general models 64–5
laparoscopic 305, 510, 512
and SCM (standard capacity model) 510–12
surgical performance 64–5, 934
surgical students 935–6, 938
surgical teams 225, 228, 230
surrogates 197, 540, 648, 725, 995, 1021
for mentors 1037, 1046
surveillance 742, 765, 782
symmetry span 563–4
sympathetic activation 501–2
symptoms 160, 174, 508, 620, 623, 792
stress 500, 509, 512
Syria 988, 1134
system resilience 691, 711, 713
systems management 665–6, 668

table tennis 507, 510
tacit knowledge 85, 88–9, 92, 95–6, 984, 987–8, 1175, 1181
somatic-limit 88–9
specialist 92, 95
tactical proficiency 318, 767
tactical skills 773, 778
tactics 228, 727, 765, 798–9
defensive 765, 770, 781, 783
tagging 627, 933–4, 966
take-the-best heuristic 151, 153, 162–4, 250
talent 37, 40, 50–1, 108, 110–11, 574
see also giftedness
development 106–7, 111–12
models 108, 110–13, 121
search model 108, 112–13
tallying 163–4, 167
task analyses 18, 338–9, 342, 349, 394, 396–7
task completion 304, 382, 385, 389, 484
task environments, simulated 18, 292, 297, 300, 302, 304–5, 307

task sequences 222, 228–9, 232, 235, 237, 358
task trainers 936
task-related experience 12, 333, 647, 1058, 1074
tasks
cognitive 276, 399, 740, 742, 745–6, 754
complex 295–6, 339–40, 1048–9, 1086, 1088, 1090, 1093, 1095, 1113, 1207
control 451–3, 699, 1099
domain-relevant 57, 61, 1058, 1061, 1068, 1072, 1075
perceptual-motor 1065, 1067
problem-solving 253, 392, 622, 1046, 1074
recall 36–7, 41, 45, 49, 1069
representative 11, 13, 268, 291–307, 611, 935, 937, 1058–9
routine 915, 940, 1008
taxiing 339, 341–2, 1207
teachers 577–8, 580, 583, 585, 903–6, 908–21, 928, 931–3, 940, 1049
education 9, 24, 931–4
expertise 917, 932–4, 940
in-service 933–4
pre-service 933–4, 941
student 916, 933–4
training 912, 915
teaching 24, 904–6, 908, 911–17, 928, 930–3, 968, 1023–6
for adaptive expertise 275–9
dialogic 904, 906, 908, 910, 912–14, 920
reasoning-based 913, 922
team development 13, 1001, 1004, 1013
team members 225, 465, 628–9, 635, 731–2, 833–4, 840–4, 852, 1001–2, 1004–5, 1007, 1011–14
remote 832, 843
team performance 192, 1001–2, 1005, 1007, 1010, 1013
team processes 833–4, 1002–5, 1008, 1013–14
team reflection 13, 25, 864, 1001–14
catalysts 1008–10
definition 1002–8
guided 1002–3, 1006–8
previous research and avenues for future 1010–14
self-regulated 1003–6
team reflexivity 1004–6, 1010–11

team sports 303, 599
teammates 303, 473, 599, 601, 834, 837
teams 23, 25–6, 224–5, 627–8, 731, 811–12, 814–17, 821–5, 830–3, 837–41, 861, 1001–14, 1119–23
  component 832–3
  cooperating 705–8
  efficient 23, 827
  expert 23, 831–2, 1001, 1003, 1005, 1011
  healthcare 416–17
  and individuals 194, 542, 824, 858, 863–4, 866
  research 324, 418, 422–4, 435–6, 439, 540–1
  surgical 225, 228, 230
teamwork 481, 727, 856, 858–9, 861, 905, 907, 1001, 1003–4, 1006
  challenges 23, 831
  intrusion detection 731–2
  processes 706, 1001
  spaceflight operations 830–46
  strategies 691, 711, 1004
technical expertise 90, 228, 831, 834, 836, 839
technical rationality 410
technical skills 577–8, 718–19, 722, 743, 755, 758, 929, 931
technical specialization 743, 755
technicians 410, 794–5, 799
technocracy 87, 97
technological innovations 756, 759, 879
technologists 213, 1160, 1183
technology 20–2, 462–4, 481, 630, 632–3, 635–6, 711–12, 755–9, 802–3, 994–6, 1073–6, 1116–17, 1183–4
  information 166, 719, 721–2, 725, 808, 815
  intelligent 1177, 1181
  railroad 22, 691, 708
  smartphone 1076
technology-mediated diagnostic reasoning 620, 630–5
template theory 48, 138
  and CHREST 48
temporal lobe 129–31, 133, 138
tennis 139, 299–300, 595, 597, 603–4, 937
  players 251, 299–300, 502, 595–6, 605, 1067
  table 507, 510

tennis skill 293, 298
tensions 418, 420, 837, 1060, 1142, 1209
test anxiety 20, 491, 500–501, 504–6, 508, 514
tests 6–7, 61–2, 66–7, 72, 74–7, 109, 112, 271–3, 460, 863–4, 919–20, 1091–2, 1094–5
  standardized 57, 112, 917, 989
textbooks 43, 107, 315, 345, 1164
theoretical perspectives 3–4, 17, 184, 415, 417, 452
therapists 94, 1182
thinking
  aloud 383–4
  expert 10, 389, 394, 398
  fluid 552–3
  lateral/creative 552
  out of the box 400–402
  processes 1046, 1050
  speculative 1166, 1185
  verbal reports of, see verbal reports of thinking
threats 493, 496, 724, 728, 773, 776, 782–3, 881
  potential 306, 496, 1093
  security 1134–5
thunderstorms 888–9
time-in-service 874, 877
timeline summaries 437, 447
timelines 337, 340, 396–8, 431, 437, 447–8
  event 396–7
  verification 431
tone 479, 774, 1076
tornados 872, 881, 888
tough-case time compression 1115, 1199
tournaments 60, 553, 555–6, 560, 564, 566–7
trade-offs 208, 596, 1064, 1111, 1138, 1199–1200
traditional models 194, 1170
traditional psychometric views 108, 110, 112
traditions 14, 36, 87, 151, 578–9, 587
train crews 691, 693, 695, 698, 701–8, 710–12
  performance 697, 710
trainees 296–7, 304–5, 781, 783–4, 942–3, 986–7, 1036, 1040, 1045, 1048–9, 1089–90, 1099–100, 1115–16

SUBJECT INDEX   1275

trainers 294, 296, 299, 452–3, 455, 928,
      931, 935–6, 1085
  box 936, 938
  task 936
training 131–7, 292–4, 300–301, 304–7,
      771–3, 777–81, 784, 793–6,
      799–802, 855–9, 992–4, 1048–50,
      1114–16
  drivers 369, 513
  expertise 300, 936, 940, 944
  expertise-based 301, 941
  firearms 772, 782
  force-on-force 784
  formal 875, 1043, 1092
  graduate 536, 539, 543–4
  intermix 276, 278
  introspection 396, 400
  juggling 131, 134
  mobilization 1102, 1204
  musical 130, 133
  one-on-one 579, 795
  on-the-job 671, 728, 735, 795, 886, 994
  operators 277, 851
  performance-based 301, 942–3
  refresher 1092, 1097–1102, 1204
  resources 435, 877
  scenario-based 440, 781, 1048
  simulation-based 440, 931, 935
  simulator-based 863–4, 931, 935–6, 944
  skills 506, 1184
  teachers 912, 915
training-based research 931–2
trajectories 198, 342, 538, 663–4, 670, 680
transactional theory 491, 499
transactive memory 834, 1014
transcripts 324, 435–6, 439, 483, 532, 775
transfer 5, 164, 267, 277–9, 345–6, 570,
      1007–8, 1061, 1087–9, 1096–8, 1196
  of expertise 860, 1196
  and retention 26, 1089, 1096, 1102
  skills 14, 781, 1099
transition 111–12, 316, 323, 421, 914–15,
      917, 929
  phases 1003, 1006
transitioning 330, 425, 742, 784, 930, 1042
transitivity 150, 385
transparency 441, 757, 1117

trees 37, 160–1, 163, 165–6, 437–9, 653
  fast-and-frugal 160–1, 163, 165–6
trespassers 701–2, 710
triadic semiotic systems 16, 174–5, 178–9,
      183–5, 187
trial and error 36, 385, 815, 935
trials 36, 163, 303, 384–5, 606, 935–6
  error-free 1089–90
  randomized control 1142, 1145, 1173
trial-to-trial comparison 384
trivia 552–3, 556, 568, 570, 1100
troubleshooting 340, 666, 668, 795,
      1046, 1048
  electronics 10, 1209
trust 318–19, 749, 756, 758, 780, 839,
      1033, 1113, 1170, 1172–3,
      1175–8, 1194–5
  in experts 1161, 1164, 1172, 1175–6, 1178
  public 1141, 1180, 1194
trustworthiness 432–3, 441, 756, 989–90
truth 390, 414, 723, 881, 1147, 1166
tuning 176, 184, 225, 665, 1202
Turing Test 93, 342, 348
tutoring 799, 1022, 1024, 1037, 1046–7, 1049
tutors 795, 1022, 1024, 1037, 1046–8
typicality 479, 860, 1181

UAVs (unmanned aerial vehicles) 341, 1184
ultrastructure 129, 132–4
unanticipated situations/events 120, 222, 454,
      461, 701
uncertainty 20–2, 222, 229–31, 409–10,
      430–1, 445–6, 532, 599–600, 742–4,
      857, 891–2, 1134–5, 1137–8
  and weather forecasting 878–83
undergraduates 61, 156, 163, 270, 338, 421
underpinnings 292, 299–300, 642, 651, 1195
  cognitive 556, 734
  theoretical 190, 195
understanding 2–3, 204–7, 263–4, 269–71,
      416–19, 423–5, 734–5, 745–7, 752–3,
      756–9, 961–3, 1116–18, 1204–5
  conceptual 263–4, 269, 272, 915, 919, 957,
      960, 1196
  of expertise 401, 408, 411, 417, 1139,
      1144, 1209
  in-depth 373, 411, 1125

understanding (*cont.*)
  and knowledge 8, 594, 610, 612, 861–2, 1045
    shared 706, 711, 809, 815, 839–40, 1001
unintended consequences 419, 631, 635, 1157
United Kingdom 420, 423, 852, 856, 859, 861, 1133, 1137–8
United States 364, 766, 770, 772, 1137, 1139
  Army 197, 201, 742, 794, 979, 1092
  Marine Corps 201, 316
unmanned aerial vehicles (UAVs) 341, 1184
unpredictability 220, 222, 229–31, 233, 1197, 1204
unstabilized approaches 668, 675
upward abstraction 436, 447
use-of-force situations 767, 784
users 162, 329, 365, 368, 462–3, 465–6, 734
  expert 335, 368, 904
utility 18, 74, 85, 251, 491, 497

vaccination rates 87, 90
validation 325, 327–8, 342, 483, 533, 750
  empirical 210–11, 979
validity 63, 65, 67, 69, 71–2, 75–6, 109, 317, 392–3, 502–3
  ecological 12, 22, 773, 1086, 1164
  external 397, 880
  internal 394, 397, 774, 1092–3
  predictive 113, 151, 1036
values and expertise 1137–8
variables
  continuous 73, 160, 1064
  criterion 75, 326
  difference 1087, 1095–6
  predictor 65, 75, 161–2
  psychosocial 112
variance 61–2, 65–6, 73–4, 76, 141, 270, 336, 343
  age-related 1070–1
  method 131, 134
varieties of expertise 1138–9
vehicles 342, 361, 363–4, 366–7, 512–14, 836
veracity 397, 605, 656, 775, 1027, 1037
verbal discovery instructions 277–8
verbal elaborations 381, 390–5, 399, 445

verbal reports of thinking 304, 1208
  guidance on collecting 393–9
  types of 390–3
verbalizations 390, 392, 394–5, 400, 606
  concurrent 398
  direct 390–5, 399–400
  process-traced 5
verification 401, 583, 1208
vested interests 98, 772
Vicente and Wang revisited 184–5
video 246, 254, 433–4, 483, 933, 1099
  stimuli 769, 773–4
video-based simulations 306, 770
vigilance 496, 698, 703, 705, 891, 1113
vignettes 326, 328, 432–3, 437, 445, 447–8, 1045
virtual reality (VR) 636, 719, 824, 863, 931, 935
vision 111, 338, 554, 914, 916–18, 920–1
visual perception 344, 563, 778, 781–3, 932
visual scanning 669–70, 673–4
visual search 304, 306, 340, 561, 633, 674
visual stimuli 138, 416
visualization 201, 552, 732, 734, 1073
  skills 666, 675
vocabulary 48, 61, 67, 256, 562–3
vocal control 135, 578
voice 341, 578, 665, 933, 1009, 1076
  channel 839–40
  communications 665, 680, 743
  loops 835–6, 839
volatility 22, 796, 804, 1197
volleyball 139, 596, 609
VR, *see* virtual reality
vulnerabilities 494, 496, 504, 508, 720–1, 724–5, 727, 731
  operating system 720–1
  stress 494–6, 500, 512–13

war on experts/expertise 1157–85, 1210
wargames 340, 719
water 132, 662, 820, 830, 841, 851
wayside signals 697, 699
weapons 222, 293, 773–4, 782–3, 793, 803
  handling 769, 773
weather 370, 670, 672, 677, 698–9, 872–94
  bad 668, 677, 679
  technologies 883, 889

weather forecasting 20, 23, 872–94, 990, 994, 1175–6
  educational, training and mentoring issues 873–8
  modeling knowledge and reasoning of expert forecasters 887–91
  operations 878–9, 887
  reasoning styles and strategies 883–7
  and uncertainty 878–83
weavers 634, 992
wicked problems 209, 704, 955, 1176, 1178
WICS 115–16
Wikipedia 557, 1139
wildfires 222, 642–3, 646–7, 653
windows of opportunity 179, 643
wings 662, 664, 668
wireless telepathy 712
wisdom 36, 116–17, 346, 1073
WM, *see* working memory
women, pregnant 436, 1158
word fragments 556, 562
word games 21, 61, 551–70
  cognitive skills 561–9
  cognitive taxonomy 552–3
  expert ability 553–6
word knowledge 68, 553, 1070
work contexts 190, 192–3, 196, 213, 231–5, 238, 697, 703
work domain analysis 451–3, 465
work experience 274, 535, 891, 979
work roles 22, 721, 723, 725, 731–2, 734
workarounds 11, 418, 481, 1112
worker competency, analysis 454, 461–2

workers 196, 198, 219–24, 228–32, 234–8, 461–3, 818–19, 822–3, 995, 1115–17
  roadway 691–2, 701–2, 705–8
  in sociotechnical systems 16, 223, 232, 235, 237
  support 232, 234, 236, 238
workflow 470, 472, 627, 634, 741, 819
  clinical 627, 631, 634
workforce 724, 822–3, 827, 977–8, 991, 996
  cyber 722–3
working memory (WM) 491–4, 496, 498–9, 504–5, 507–8, 514, 671, 1070–1
  capacity 58, 61–2, 66, 1062, 1075, 1138
  limitations 753, 857, 1068
  and long-term memory 137–8
workloads 458, 460, 464–5, 492, 496, 830, 844, 846
  cognitive 502, 753
workplaces 221, 223, 225, 232–8, 262, 416, 418, 420, 423–5, 475–6
  real 815–16
workshops 359, 485, 712, 766–7, 784, 879
workstations 235, 729–30, 881
World War II 36, 798–9, 985, 988, 994, 1177
worldviews 379, 422, 953, 958, 962–3, 969
  expansive 958
  reductive 953, 959, 961–3
writing, classical expertise approach 43
Würzburgers 381–5, 387–90, 393

youth 103, 105, 107, 109, 111, 113–15, 117–19, 121, 1029–30